Proceedings

IEEE INTERNATIONAL CONFERENCE ON Computer Design: VLSI in Computers

Sponsored by **IEEE Computer Society and IEEE Circuits and Systems Society Technical Committees on:**

• Large Scale Systems • Computer-Aided Network Design • VLSI • Computer Architecture • Computer Elements • Computer Packaging • Mini and Micro Computers • Design Automation

ICCD'83

Rye Town Hilton, Port Chester, New York
October 31–November 3, 1983

1884 · 1984
A CENTURY OF ELECTRICAL PROGRESS

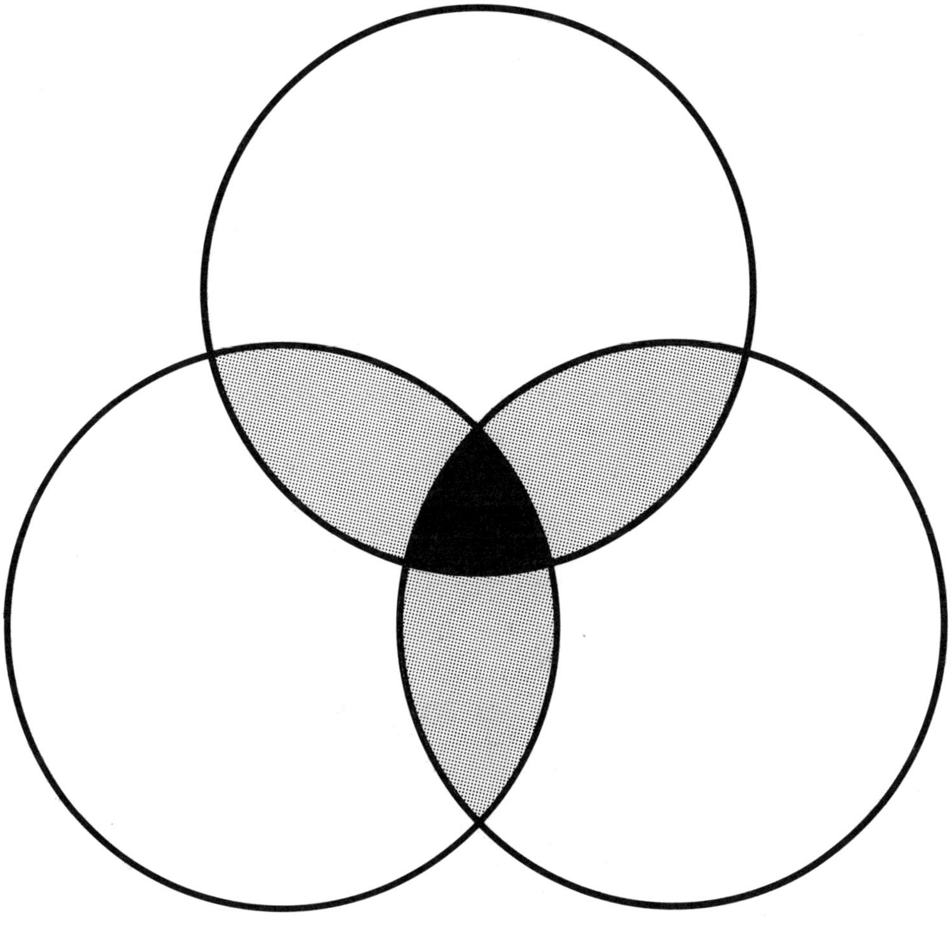

IEEE COMPUTER SOCIETY PRESS

ISBN 0-8186-0480-8
IEEE Catalog No. 83CH1935-6
Library of Congress No. 83-81680
IEEE Computer Society Order No 480

Published by IEEE Computer Society Press
1109 Spring Street
Suite 300
Silver Spring, MD 20910

ISBN 0-8186-0480-8 (paper)
ISBN 0-8186-4480-X (microfiche)
ISBN 0-8186-8480-1 (casebound)
IEEE Catalog No. 83CH1935-6
Library of Congress No. 83-81680
IEEE Computer Society Order No. 480

Order from: IEEE Computer Society
Post Office Box 80452
Worldway Postal Center
Los Angeles, CA 90080

IEEE Service Center
445 Hoes Lane
Piscataway, NJ 08854

1884–1984 A CENTURY OF ELECTRICAL PROGRESS

The Institute of Electrical and Electronics Engineers, Inc.

G. Rabbat

H. Carter

Preface

The design of computers is a discipline which has no apparent bounds. With a dichotomy of views exemplified by the desire to integrate design approaches independent of technologies on the one hand, and the divergence of technologies on the other, there is no end to the challenges confronting those actively engaged in all aspects of computer design.

This conference reflects this dichotomy. Josephson technology is becoming mature and techniques useful for other computer technologies are finding a home with this technology as well. Yet the uniqueness of Josephson technology is still very apparent. Another example of this dichotomy is the design aids for computer design. The desire is to integrate the individual tools into a cohesive usable system, but new tools continue to be developed which defy easy integration into an already existing set.

New methods for assisting the computer design process are developing with a vengence. Artificial intelligence is being sought to ease the pending crisis in the design of very large and very complex integrated circuits. Special purpose machines which more rapidly perform design and simulation tasks than the software systems that predate them are proving their usefulness.

And only recently are we seeing the fruits of the labor expended by those who have gone before us. Entire 32-bit CPUs are now being fabricated on a single chip. Coprocessor chips of large system complexity and functionality are now reality. Not only does the use of automated design aids take the credit, but a practical approach to architecture design must also take a bow.

Thus, this conference has brought together the computer design leaders of our time. Technologists, architecturalists, circuit designers, design aiders, and testers—we all share in this most exciting of endeavours. We will probably always have a dichotomy of views on the subject. And this conference has allowed us to express them.

Guy Rabbat
Harold W. Carter

General Chairman

Dr. Guy Rabbat
IBM Laboratories 70Y
East Fishkill, Route 52
Hopewell Junction, NY 12533
914-894-8126
914-297-5315

Technical Chairman

Dr. Harold W. Carter
24 Longstreet Lane
Wright-Patterson AFB, OH 45433
513-255-3576
513-253-8142

Conference Chairman

Dr. Shakir Abbas
IBM-81K
Hopewell Junction, NY 12533
914-894-3660

Secretary

Dr. Paul Horstman

Technical Enhancement

Mr. Walt Luciw
Sperry Univac
P.O. Box 500
Blue Bell, PA 19424
215-542-4011/2760

Registration

Mr. Harry Hayman
IEEE Computer Society
1109 Spring St., Suite 300
Silver Spring, MD 20910
301-589-8142

Publication

Dr. T. J. Jones
AFIT/ENG
513-255-3576
513-429-5967

Panels

Mr. V. L. Gani
Dr. Achilles Sarris

Reviewers

H. Anway
H.L. Applewhite, Ungermann
R.G. Babb II, Oregon Graduate Center
C.M. Black, Trilogy
M.A. Breuer, U.S.C.
D. Casey, Sperry
B. Courtois, IMAG
D. Davidson, Western Electric
J.P. Deschampi, MBLD
L. Egan
P.R. Gagnor, Bell Labs
G.R. Gladden, Honeywell
G. Goates, CAE
I. Gjovaay, Tektronics
W. Grant, Sperry
J.R. Gregory, Nouson
R. Guffin, HP
R.D. Hess, United Tech
H.K. Hingarh, Fairchild
R.A. Hoover, IBM
R. Iliff
T.A. Irving, Tracor
W. Kamsler, Martin Marietta
K.N. Karma, RCA
G. Kling, Sperry
K. Kobayaski, USC
A.G. Larson, Aracor
B. Malbec, IBM
A. Malik, Signetics
K. Mandl, Sperry
E.I. Muehldorf, Sperry
G.R. Nudd, Hughes
M.I. Payne, Prime
A. Peskin, Brookhaven
D. Protopapas, ITT
S. Sapiro, CAE
J. Springer, RCA
T.K. Stickel, Sperry
F. Tsui, IBM
H. Yang, Mellon Inst.
P. Yang, TI
G. Zeigler, Tektronics

Technical and Conference Committees

D. Abell—IBM
G.C. Brown—GEC
M.R. Buric—Bell Labs
R.A. Carballo—IBM
D.O. Carhoun—Mitre
I.I. Caro—Air Force
D. Casey—Sperry
D.R. Cochran—Comp. Tech
E. Cohen—Tektronics
P. Cohen—Prime
S.F. Daniels—Siemens
M. D'Abreu, Honeywell
C.M. Davis—IBM
B.I. Dervisoglu—Sperry
A. Dewey—Air Force
J.R. Distaso—TRW
G. Ditlow—IBM
D.R. Ditzel—Bell Labs
D. Edwards—Femto Tek
M. Ercegovac—Univ. CA
E.I. Ergin—TRW
P.P. Fasang—Siemens
V.L. Gani—IBM
A. Gevins—EEG Systems
S. Goto—NEC
M. Graf—IBM
H.G. Harding—IBM
W.K. Hoffman—IBM
S.C. Johnson—Bell Labs
K. Kishimoto—GTE
T.A. Larsen—IBM
R. Kung—Intel
S.C. Lee—Motorola
F.T. Leighton—MIT
D. Lynn—Los Alamos
L. Maissel—IBM
K. Mandl—Sperry
T.E. Mangir—Univ. of Cal.
J.R. Mashey—Bell Labs
B. McCormick—Los Alamos
T. Miura—Hitachi
M. Morisue—Japan
E.I. Muehldorf—TRW
A. Mukhopadyay—Univ. Cent. Fla.
M. Papa—U. of S. Cal.
H. Parks—Amdahl
P.A. Phister—Air Force
A. Ruehli—IBM
R. Russo—Stanford Univ. (Steering)
A. Sangiovanni-Vincentelli—Univ. of Cal.
A. Sarris—IBM
T. Schwarz—Control Data
C.W. Scott—ESL/TRW
M. Smith—IBM (Steering)
P.A. Subrahmanyam—U. of Utah
E. Swartzlander—TRW
F. Szorc—NCR
L. Terman—IBM
K. Torku—IBM
S. Triebwasser—IBM (Steering)
M. Varanasi—U. of S. Fla.
P. Venkatachalam—IBM
M. Wei—American Bell
K. Wu—Bell (Tutorial)
W.R. Young—Harris

KEYNOTE SPEAKERS

Dr. Kaneyuki Kurokawa, Vice-President, Fujitsu
Dr. Paul Low, Vice-President, IBM GTD
Dr. Alan Kay, Vice-President and Chief Scientist, Atari
Dr. Yuichiro Oya, General Manager, Hitachi

OPENING REMARKS

Dr. James Owens, IEEE President
Dr. Oscar Garcia, IEEE Computer Society President
Dr. Ken Laker, IEEE Circuits and
Systems Society President

PANEL ON SUPER COMPUTERS

Dr. Walter Beam, Vice-President, Sperry
Dr. Gene Amdahl, Chairman, Trilogy
Mr. Robert Schmidt, Vice-Chairman, CDC
Dr. James Daughten, Vice-President, Honeywell
Dr. L.W. Beers, Vice-President, Burroughs
Mr. Carl Conti, Vice-President, IBM
Dr. Steve Chen, Vice-President, Cray Research
Dr. John Ranelletti, Lawrence Livermore Labs

PANEL ON VLSI

Dr. Thomas Longo, Senior Vice-President, Fairchild
Dr. Walden Rhines, Sr. Vice-President, Texas Inst.
Dr. William Lattin, Vice-President, INTEL
Dr. Karl Zaininger, Vice-President, Siemens
Dr. Klaus Bowers, Vice-President, Bell Labs
Dr. Hollis Caswell, Vice-President, IBM GTD
Mr. Frank Carubba, Director, Hewlett-Packard
Mr. Jeff Kalb, Director, Digital Equipment

PROGRAM MONDAY OCTOBER 31, 1983

INVITED SPEAKERS: Dr. Levente Mate, Dr. Denis McGreivy, Dr. R. Sherbourne

9:00– AMERICAN LAW ON COMPUTERS AND SOFTWARE: *Grand Ballroom North*
Chairman: H.R. Kashani, Purdue Univ.
- Artificial Intelligence and the Law: Will Expert Systems Replace Expert Lawyers?
  Gary S. Grossman, Fried, Frank, Harris, Shriver & Kampelman, Washington, D.C.
- An American Legal Maze: Protection of Computer Programs
  James A. Coles, Barnes & Thornburg, Indianapolis, Indiana
- Your Day in Court: Liability for Computer Design
  Craig M. Walker, Alexander & Green, New York, NY
- International Information Regulation
  Robert P. Burton and Roger D. Malmrose, Brigham Young University, Provo, Utah

10:20 BREAK

11:30 WELCOMING ADDRESS: *Grand Ballroom*
"Productivity, Technology and Innovations"
Dr. Bruce Merrifield, Assistant Secretary of Commerce

11:50 LUNCH: *Grand Ballroom*

12:50 KEYNOTE ADDRESS: *Grand Ballroom*
Dr. Kaneyuki Kurokawa, Vice-President, Fujitsu Laboratories

1:30 PANEL ON SUPER COMPUTERS: *Westchester Ballroom*
Moderator: Dr. Walter Beam, Vice-President, Sperry
Dr. Eugene Amdahl, Chairman, Trilogy
Mr. Robert Schmidt, Vice-Chairman, CDC
Dr. James Daughten, Vice-President, Honeywell
Dr. L.W. Beers, Vice-President, Burroughs
Mr. Carl Conti, Vice-President, IBM
Dr. Steve Chen, Vice-President, Cray Research
Dr. John Ranelletti, Lawrence Livermore Labs

3:00 BREAK: *Westchester Ballroom Assembly*

3:30 PANEL ON VLSI: *Westchester Ballroom*
Moderator: Dr. Thomas Longo, Senior Vice-President, Fairchild
Dr. Walden Rhines, Senior Vice-President, Texas Instruments
Dr. William Lattin, Vice-President, INTEL
Dr. Karl Zaininger, Vice-President, Siemens
Dr. Klaus Bowers, Vice-President, Bell Labs
Dr. Hollis Caswell, Vice-President, IBM GTD
Mr. Frank Carubba, Director, Hewlett-Packard
Mr. Jeff Kalb, Director, Digital Equipment

5:00 BREAK: *Westchester Ballroom Assembly*

5:10 BRAIN MODELS and VLSI: *Westchester Ballroom*
- Overview of the Human Brain as a Distributed Computational Network
  Alan S. Gevins, EEG Systems Laboratory
- Silicon Brains: Whither Neuromimetic Computer Architectures
  Jacques J. Vidal, UCLA
- Semi-Autonomous Auto-Control of Input by a Large Scale Neural Assembly Through the Genesis of Images
  Walter J. Freeman, University of California, Berkeley
- A Structures for Generation and Control of Intelligent Behavior
  James S. Albus, Director, National Bureau of Standards

6:30 RECEPTION: *Grand Ballroom Assembly*

7:30 BANQUET: *Grand Ballroom*
Dr. James Owens, IEEE President
Dr. Oscar Garcia, President of Computer Society
Dr. Ken Laker, President of Circuits and Systems Society
Keynote: Dr. Alan Kay, Vice-President and Chief Scientist, Atari Inc.

Table of Contents

American Law on Computers and Software

YOUR DAY IN COURT-
LIABILITY FOR COMPUTER DESIGN

By Craig M. Walker*
Jonathan B. Whitney*

ABSTRACT This paper will discuss (1) the types of claims made in court in a computer case against a manufacturer/vendor of computer hardware or software, and (2) practical suggestions for the computer designer to use during the design phase.

Legal cases brought by disgruntled users against computer manufacturers/vendors usually involve claims under legal theories of (i) breach of contract, (ii) breach of warranty, (iii) negligence and/or (iv) misrepresentation or fraud. These legal theories and the evidence used to support them will be discussed.

As with all high-technology cases, the claimant-user of computer systems will base its case on overstated and often unrealistic assertions of what the hardware/software was promised to accomplish. The computer designer faces a dual role; first, he or she must explore and extend the capabilities of the computer system while, second, keeping in mind, and in the minds of the sales department of their company, that oral or written expectations may become guaranties. This paper will address how the computer designer can best do his job and minimize any possible confrontation in court.

* * *

In the computer field, a significant potential exists that users will make claims for damages in connection with the sale or lease of a computer system.[1] While the hardware is a tangible element, the software and performance characteristics are not always precisely established by the parties. The situation is further complicated by (a) the user's desire to have specific features or special adaptations in the system, (b) the continuing development of software programs and updates, and (c) the user's increased demands on the computer system following installation. Although the manufacturer seeks to accomodate the purchaser, at times the purchaser refuses to pay the contract price (forcing the manufacturer to sue) or commences a legal action against the manufacturer.

I. MAJOR LEGAL THEORIES

The major legal theories under which a claim may be brought against the computer designer/manufacturer are as follows:

A. Breach of Contract. The most usual cause of action against a manufacturer, whether of computers or other equipment, is breach of contract. Here the claimant user seeks to show (1) the existence of a duty arising from a specific contract, and (2) the breach of that duty, occasioned by the non-performance, total or partial, of the contractual obligation.

B. Breach of Warranty. Closely related to breach of contract is the claim of breach of express or implied warranty. Express warranties are written or oral statements or promises of performance. The law creates implied warranties. The "implied warranty of merchantability" imposes an obligation on a manufacturer that the computer goods or systems sold be fit for the ordinary purpose for which they are used.[2] An "implied warranty of fitness for a particular purpose" exists where the manufacturer-seller knew both the use for which the buyer intended the goods and that the buyer was relying on the seller's expertise to choose the right system.[3]

C. Negligence. The use of negligence, alone or as a claim of professional malpractice, has not found complete acceptance in the courts in cases involving computer sales.[4] This is due to (a) the reluctance to allow such a claim in cases involving only monetary damages and (b) by the uncertainty of the computer designer's role as a member of a defined professional community from which a specific standard of care can be determined. A number of computer cases have discussed negligence as well as malpractice claims.[5] Here the elements are (1) the existence of a duty, created by law, not contract; (2) the breach of that duty; (3) a causal connection between that breach and harm occurring; and (4) actual injury.

D. Misrepresentation or Fraud. The definitive nature of written contracts has increased the use of alternative legal theories such as fraud in an attempt by the purchaser to circumvent those contracts which are an obstacle to recovery.[6] To show fraud, the claimant must clearly establish (1) a false or misleading misrepresentation or omission made by the manufacturer which is (2) material and as to which (3) the manufacturer had the requisite

---

* Mr. Walker is a trial lawyer and member of the firm of Alexander & Green, 299 Park Avenue, New York, New York 10171. Mr. Whitney is an associate attorney of that firm.

scientor (knowledge of falsity) and that (4) the purchaser relied on the misrepresentation and (5) suffered damage as a result thereof.[7]

E. Common Elements of the Various Claims. Usually, a manufacturer may find itself defending against one or more of these legal theories applied to the same set of facts and circumstances. The common assertions made in any claim involve (1) the duty of the manufacturer, whether arising from the contract or from the law, and (2) the representations made by the manufacturer. Pre-trial discovery and the trial itself will concentrate on these points, and the computer designer will often find him or herself the focus of such discovery in connection with the designer's statements or writings concerning the research and development of the hardware or software involved. It is important for the computer designer to be aware of the effect of his actions and comments on the outcome of any action brought against the manufacturer.

II. DESCRIPTION OF TRIAL PROCEDURE

After the "pleadings" (the user's complaint and manufacturer's answer) have been filed with the court, pre-trial discovery is allowed. Pre-trial discovery consists of (i) depositions (oral examinations under oath), (ii) interrogatories (written questions) and (iii) inspection of documents. Then, at trial each side presents its case and cross-examines the opponent through witnesses, documents and physical exhibits.

The evidence such claimants will use at trial includes, *inter alia*: (1) the display of systems, including prototypes, perhaps with features still in the development stage; (2) the use of technical descriptions, sales brochures and advertisements which include features planned but which may take longer than expected; (3) oral statements made to claimant; and (4) the internal memoranda and reports generated by the computer design engineers. It is highly likely that at the trial neither the judge nor the jury will have any technical training or expertise. Accordingly, it often falls on the computer designer to explain in layman's language the computer system in dispute. Additionally, because of this lack of technical knowledge, the judge or jury will not be able to pick out relevant evidence from the many documents presented in court; thus, one cannot rely on a defense if it only comes through a jigsaw of writings and documents.

III. ENHANCEMENT OF MANUFACTURER'S CASE-BEFORE LITIGATION

Of course, the manufacturer has a variety of defenses which can be used to defeat claims of users. A manufacturer can assert that the buyer (a) was sophisticated, (b) knew what it was getting and (c) accordingly did not rely on any manufacturer statement as a warranty or representation. Following delivery and installation, it can often be shown that the buyer accepted the computer system notwithstanding any alleged non-conformity. A buyer must reject goods or revoke his acceptance thereof within a reasonable time, or give timely notice of a material or substantial breach or otherwise be barred from seeking recovery for any non-conformity or breach.[8] Additionally, a fraud claim can in some instances be defeated by the user's reaffirmance of the contract after discovery of the alleged fraud.[9]

A. The Importance of Anticipation. Nevertheless, it is better to anticipate and prevent possible litigation than to rely on the understanding of a judge or jury in a suit involving complex computer systems. In this respect the computer designer can have significant impact in defining the manufacturer's duties and in insuring the accuracy of representations made to customers.

B. Delineation of Contractual Duties. The duties and obligations of the manufacturer should be delineated in the contract between the parties as precisely as possible. It is here that many claims arise, the purchaser asserting an interpretation of the contract that was not intended by the manufacturer. The computer hardware and software should be described explicitly, and the computer designer should supply the sales department of the manufacturer/vendor with sufficient information to describe the equipment in a way that performance can be objectively determined. At the same time, of course, the description should be careful to avoid revealing any proprietary information.[10] Even with such specificity, claims may arise and, of course, the well drafted contract should include both the technical description and limitation of warranties and liability clauses.

When the contract does not specify precisely what the manufacturer must do to effect complete performance, i.e., if the equipment delivered cannot be objectively compared with the contractual description, the court in resolving a dispute must turn to other factors. Thus, it is possible that the manufacturer and the designer will be held liable on a duty, imposed by law, to act or design according to a standard prudent in the industry or in accordance with the manufacturer's promises. Of special concern is the extent of standardization of procedures, including testing in the various stages of computer design and production. The more defined these procedures are, the less likely a claim can be asserted that the designer failed to use the appropriate standard.

C. Statements and Representations to Customers. Almost all computer cases will involve a claim by the purchaser that the seller made representations about the equipment which later turned out to be inaccurate or actually erroneous.

The computer designer here is uniquely able to help prevent such claims from arising by keeping the sales department informed as to the exact progress of any particular design element in the computer system. The designers should be aware that estimates of feasibility or timing which are transmitted to the prospective purchaser may be treated in court as warranties and may be relied on to a greater extent than intended. Although statements of opinion or prediction do not constitute warranties under the law,[11] the degree between opinion and fact is not always clear and, in any event, would be determined by the uncertain process of trial by jury.

The designer also should participate in the preparation of advertisements and brochures and other forms of presentation to prospective purchasers.

D. Internal Operations. The computer engineer should be aware that his or her own notes, memoranda or internal conversations may become important to support the allegation that a representation was false when made. The scope of discovery in the courts, both on the federal and the state level, is very broad, and even a designer's telephone log may be requested and produced during litigation. Accordingly, it is important to keep even internal memoranda accurate and consistent with what is being communicated to others outside.[12]

IV. PRACTICAL SUGGESTIONS FROM A TRIAL LAWYER

Although designed for conduct during trial, the following suggestions may be helpful for the designer to keep in mind when dealing with customers.

A. Understand the Customer's Inquiry Before Responding. Misunderstandings, which eventually lead to court, often occur when a statement is understood by the customer in a context not intended by the manufacturer. Thus, as in court it is important to understand a question put forth by an attorney before attempting to answer, the same should be true in any situation which has the potential to some day involve litigation.

B. Don't Guess. Everyone has the feeling that, if asked a question, an answer should be given even if one is not known. The customer will rely on this answer. Accordingly, be careful not to guess or speculate, or, if pressed, be sure to indicate that you are guessing or expressing a mere opinion that is not a warranty or promise.

C. Be Accurate in Description. This is true of oral statements as well as written items and exhibited equipment. If a computer system is a prototype, it should be so marked. If an item in an advertisement or brochure is for future development, be sure it states that clearly.

D. Do Not Exaggerate. Customers may be impressed at the time, but they will not forget exaggerations made when they are looking for reasons not to pay for the system they continue to use. It is best to be straightforward and candid. As in court, this often impresses more than statements designed only to point out the good qualities of a system.

E. Be Careful of Admissions. Even after a customer is delivered its computer system, it is important to remember that statements may still effect possible future litigation. Do not assume you can now take the customer into confidence and, jokingly, "admit" doubts that have existed about certain features of the system. If problems existed, they should have been expressed previously; your statements of doubts or potential problems, if later expressed, can be used as admissions in court.

F. Follow Company Procedures. It is often true that the left hand does not know what the right hand is doing. It is important from a litigation standpoint to minimize this. By your following standard procedures, the sales department, and witnesses in court, will be able to accurately describe the research and design procedures. The computer designer should always be in a position to corroborate such statements to customers and to a court.

G. Use a Common Sense Approach. Finally, always use common sense when dealing with customers, or the attorney for the user in court. The customer, as with a court, wants to hear the facts and will decide for itself the connections to be made. Cleverness is not necessary and in fact is often mistaken for dishonesty. Accordingly, one should conduct oneself as common sense would dictate.

SUMMARY

This paper has discussed the major legal theories involved in litigation between a manufacturer vendor and a buyer.[13] The overall strategy to be followed in manufacturing and selling computer hardware and software is, of course, to try to accomodate the customer, satisfying as fully as possible the customer's needs and demands. Litigation, by either side, is time consuming and costly, and thus is only a last resort. If litigation occurs, however, it is best to have taken precautions to protect against claims, unjustified as they may be, which can be asserted.

The contract should adequately describe the computer system and its performance capabilities, and the designer should participate in insuring the accuracy of the description. Representations to customers should be accurate and informative and again the computer designer must apprise the other areas of the manufacturing company, especially the sales department, of the progress of development and the current status of the system. Finally, the computer designer should keep in mind that, in any litigation involving computer sales or leases, their actions, statements and writings will be focused on to prove, or disprove, allegations made by the user. Accordingly, the designer should be prepared, before litigation occurs, for a possible day in court.

FOOTNOTES

1. Citing numerous examples, the court in Earman Oil Co. v. Burroughs Corp., 625 F.2d 1291 (5th Cir. 1980) stated: "There have been an enormous number of suits in which disgruntled computer users have attempted to sort out their rights where both computer vendors and so-called lessors have been involved." Id. at 1293.

2. Uniform Commercial Code (hereinafter "U.C.C.") §2-313. The U.C.C. has been adopted in virtually all states with only slight variation.

3. U.C.C. §§2-314, 2-315. Both of these warranties can be excluded or modified by the parties by appropriate provision in the contract. U.C.C. §2-316.

4. See, e.g., Triangle Underwriters, Inc. v. Honeywell, Inc., 604 F.2d 737 (2d Cir. 1979).

5. Id. See also Chatlos Systems, Inc. v. National Cash Register Corporation, 479 F. Supp. 738 (D.N.J. 1979), aff'd, 635 F.2d 1081 (3d Cir. 1980); Chesapeake Petroleum and Supply Co. v. Burroughs Corp., 6 Computer Law Service Reporter 768, aff'd, 282 Md. 40 (1978).

6. See Walker, Computer Litigation and the Manufacturer's Defenses Against Fraud, Computer Law Journal, Vol. III, Spring 1982, No. 3, 1982. The use of a fraud theory is becoming increasingly popular due to the factual and legal difficulties of a negligence action. Moreover, in many instances a prima facie case of fraud may be made because there is a narrow distinction between sales "puffing" and fraud. Allegations of fraud, however, must be pleaded with particularity. Fed. R. Civ. P. 9(b). A fraud claim must be supported by substantial evidence and must be timely made; it cannot be a breach of contract claim couched in the legal theory of fraud. Earman Oil Co. v. Burroughs Corp., 625 F.2d 1291, 1294 (5th Cir. 1980); Investors Premium Corp. v. Burroughs Corp., 389 F. Supp. 39, 45-46 (D.S.C. 1974).

7. See W. Prosser, Law of Torts §105 (4th ed. 1977); Stone v. Farnell, 239 F.2d 750, 754 (9th Cir. 1956); Van Alen v. Dominick & Dominick, 441 F. Supp. 389 (S.D.N.Y. 1976).

8. Dunn v. Steubing, 120 N.Y. 232 (1890); U.C.C. §2-607; Restatement Second of Contracts, §§246, 247.

9. See Walker, Computer Litigation and the Manufacturer's Defenses Against Fraud, Computer/Law Journal, supra.

10. If an action is brought by or against the manufacturer, the user will often demand the production of proprietary information. A well drafted Stipulation of Confidentiality should be in place between the parties if such information is produced.

11. Royal Business Machines, Inc. v. Lorraine Corp., 633 F.2d 34 (7th Cir. 1980).

12. The computer designer's notebooks, memoranda and the like can also become crucial evidence in other types of actions such as claims for theft of trade secrets and patent and copyright infringement.

13. Other legal issues arise when relations are more complex, such as multiple manufacturer and vendor arrangements and the problem of remote diagnosis.

INTERNATIONAL INFORMATION REGULATION

Robert P. Burton
and
Roger D. Malmrose

Department of Computer Science
Brigham Young University
Provo, Utah 84602

ABSTRACT

Since 1973, twelve predominantly European countries have passed legislation restricting the use of automated and manually stored information about physical and legal persons. Nine countries are actively considering similar legislation. Thirteen member countries have signed the 1980 Council of Europe Convention for the Protection of Individuals with Regard to Automatic Processing of Personal Data. Twenty-one member countries have approved the 1980 Organisation for Economic Co-operation and Development Guidelines Governing the Protection of Privacy and Transborder Flows of Personal Data. Other international organizations, notably the Intergovernmental Bureau for Informatics, are pursuing related objectives.

Issues and terms associated with international information regulation are defined. Society's need for information is contrasted to the individual's need for privacy. The approaches taken in Europe, the United States and the Third World are compared and contrasted. Common core provisions, as well as provisions unique to certain countries or a single country, are identified and explained. International implications are discussed. The emphasis of the past decade and the shift in emphasis for the future are described.

I. INTRODUCTION

We live in an information-based society that exhibits a growing propensity for collecting and storing information. The technical problems associated with transmitting and sharing that information are disappearing. Information regulation occurs when the privacy or well-being of an individual or organization is placed in jeopardy by the potentially damaging collection, use, or dissemination of information.

II. ISSUES AND TERMS

Information regulation issues are described by several terms. A few are defined below.

"**Right to privacy:** Legal protection against unconsented intrusions by government or private parties into one's personal life.

Physical privacy: A right to privacy limited to home, procreation and family relationships recognized by the courts.

Information privacy: A concept born with the advent of computer technology (though not yet recognized by the [U.S.] Supreme Court) giving the individual the right to determine the extent to which he will share information about himself with others and giving him control over the time, place and circumstances under which he will disseminate such information, making it his own personal possession.

Confidentiality of records: Restriction of the extent to which government or others can disclose or disseminate information about another party.

Privileged communications: Legislatively created privileges against disclosure of conveyed information because of a unique relationship between the communicant and the communicatee.

Fair information practices: Practices to provide for the privacy of information giving the individual the right to be made aware of, see, copy, correct, amend, in some cases demand the elimination of, control the use and disclosure of, and limit the types of information that may be collected about himself.

Freedom of information: A concept which argues that the public has a right to gain access to information, that it has a right to know.

Transborder or transnational data flows: Transmission of data across national boundaries, usually for automated processing and storage in computer systems located on foreign soil.

Physical person: An identified or identifiable individual.

Legal person: Any corporation, public or private, or, in effect, any association.

Personal data: Any information relating to a person (data subject)."[1]

Sensitive data: Personal data concerning religious or philosophic convictions, race or skin color, political views, criminal or disciplinary records, sexual matters, or personal medical or psychological records.

"**Automated data:** Any set of data arranged for automatic processing.

Automatic processing: Any of the following operations if carried out in whole or in part by automated means; storage of data, carrying out of logical and/or arithmetic operations on the data, their alteration, erasure, retrieval or dissemination"[2]

Data haven: "an operational situation wherein data from one country are stored and processed in another country because the latter's data protection laws are lenient and permit disclosures that would be illegal in the originating country."[3]

Controller of the file: A party which maintains and is accountable for a data file.

III. SOCIETY'S NEED VS. INDIVIDUAL'S NEED

With the advent of computer technology, the ability to transfer, store, and retrieve information has greatly increased. Often, data is stored about one or more persons who have no idea that the data exists or what the data contains. This places the individual at the mercy of the data-holder -- especially if the data is incorrect or contains sensitive, personal information.

Nevertheless, society has a need for information. For example, government needs information for taxation, welfare programs, research, and other programs. An efficient analysis of information can reduce the risk an investment firm or bank must take when granting a loan or investing money. Likewise, if a company knows the credit history of a person who has applied for a credit card, the company's risk may be greatly reduced. However, credit bureaus and other information-collecting organizations have demonstrated a tendency to amass vast amounts of irrelevant and incorrect information. The information sometimes contains unnecessary data relating to a person's private life.

An individual has a recognized need for privacy. The individual claims a right to protect his personal history from being encroached upon by information-based organizations. He claims, for example, that information about his race, religious or political beliefs, health, or sexual life should not be stored and used in evaluating his ability to obtain credit, enter a university, or get a job.

In order to protect the individual from the data-holder while continuing to serve society's needs, several countries have created legislation ostensibly to protect the individual and permit the free flow of information.

IV. EUROPEAN, U.S. AND THIRD WORLD APPROACHES

The European countries, U.S., and Third World countries have taken different approaches to information regulation.

The European approach to privacy protection is, in general, to enact a sweeping 'omnibus' privacy protection act and create a new government agency empowering it to draw up rules and regulations, and giving it broad powers over data processing entities. In some countries, files containing personal data cannot be created without prior government authorization. The Europeans also tend to emphasize automatic data processing as the evil to be dealt with, and to consider sympathetically the idea that legal as well as natural persons should have protection against misuse of data. Finally, the European preference, at least nominally, is for a binding international treaty to deal with the data exchange problems between countries.[4] (emphasis added)

The Europeans have focused primarily on the privacy of information. The U.S., on the other hand, is largely concerned that the free flow of data between countries be preserved.

> The U.S. policy . . . may be summarized by the following five elements: (1) free information flow, (2) support for availability of information resources in developing countries, (3) trade in information goods and services free of tariff or non-tariff trade and policy barriers, (4) soundly planned, priced and operated telecommunication facilities, and (5) resolution of international questions of information resource use in an atmosphere of mutual respect.[5]

The U.S. is concerned that European information regulation may pose a threat to the free trade of information. Under the guise of privacy protection, European countries could create barriers to information which would protect national industries and impair foreign competition. This could put U.S.-based corporations at a severe disadvantage.

The basic U.S. approach to regulation has been a posteriori. Laws have not been created until someone has been injured and the necessity of legal protection has made itself manifest; U.S. laws have then been only sector-specific and not ´omnibus´ like European laws/legislation. The preventive European approach to regulation, on the other hand, has been largely a priori. When only a possible need for protective legislation has presented itself, laws have been created before any injuries occur.

The Third World countries do not "consider data utilization a real problem. Their vital concern [is] not protection against modern technology, but rather access to such technology."[6] For example, the Intergovernmental Bureau for Informatics (IBI) has considered the question of transborder data flows, although its concern has been primarily in the context of the Third World countries.

> The statements and declaratory principles emerging from the I.B.I. forum indicate a view different from that of European organizations. There is a pronounced emphasis on national security, sovereignty, economics and trade, and a noticeable absence of attention to fair information practices.[7]

The difference between the European, U.S., and Third World concepts of information use and regulation makes trilateral accommodation a formidable task.

V. PRIVACY-LEGISLATING COUNTRIES

In 1973, Sweden was the first country to enact national privacy laws. Before this time, however, the U.S. had enacted sector-specific laws. Since 1973, eleven other countries have enacted national privacy laws. In order, they are the United States, Germany, Norway, Canada, France, Denmark, Austria, Luxembourg, Israel, Hungary, and Iceland. In addition, nine other countries are actively considering legislation: Belgium, Finland, Italy, the Netherlands, New Zealand, Portugal, Spain, Switzerland, and the United Kingdom.

The laws/legislation of these countries contains important dissimilarities. Two organizations, the Council of Europe (CoE) and the Organisation for Economic Co-operation and Development (OECD) have focused their efforts to achieve homogeneity in laws/legislation and to facilitate the transborder flow of data among countries by providing a Convention and a set of Guidelines along which countries may structure or restructure their laws/legislation.

In 1980, the Council of Ministers of the CoE approved the "Convention for the Protection of Individuals With Regard to Automatic Processing of Personal Data." The Convention, however, is designed to be more than a simple guideline; it is to be an international contract among all countries which ratify it. The Convention is to take effect after five countries become signatories. Two countries, Sweden and France, have already become signatories. Ratification of the Convention by Norway, Iceland, Denmark, Luxembourg, and the Federal Republic of Germany is expected in the near future. Article 23 of the Convention allows non-CoE member countries to accede to the Convention should they so desire. According to William Fishman, however, it is unlikely that the U.S. will ratify the Convention.[8]

The OECD´s "Guidelines Governing the Protection of Privacy and Transborder Flow of Personal Data" are similar to the CoE Convention except that they do not contain any contractual clauses. Twenty-one of the twenty-four OECD member countries have approved the Guidelines. It is hoped that application of the Guidelines will facilitate the free flow of information between countries and perhaps obviate the need for binding conventions.

Several other organizations are examining related objectives. These organizations include the European Economic Community (EEC), the International Telecommunication Union´s Consultative Committee on Telephone and Telegraph (CCITT), the United Nation´s UNESCO, the Nordic Council, the International Federation for Information Processing (IFIC), Intergovernmental Council of Automated Data Processing (ICADP), and in particular, the Intergovernmental Bureau for Informatics (IBI).

VI. COMMON PROVISIONS AND DIFFERENCES

Information regulation throughout the world covers a wide spectrum from automated to manual data handling, from physical to legal persons, and from public to private sectors. Table 1 shows many basic aspects of privacy protection which are common to the CoE Convention and the OECD Guidelines as well as the laws/legislations of several countries. A mark indicates that the associated principle is explicitly stated or strongly implied in the country´s privacy laws/legislation.*

Although each country´s laws/legislation does not explicitly cover all the principles of the Convention and Guidelines -- in particular, the countries which have signed the Convention and Guidelines -- the laws/legislation can usually be construed implicitly to cover those principles. Furthermore, when a country ratifies the Convention, it is obligated to enact domestic legislation which accommodates the Convention unless the country claims itself exempt from certain provisions at the time of ratification. Almost every set of laws/legislation allows the government to make exceptions and changes to its laws.

Several sets of laws/legislation contain characteristics unique to an individual country or geographic region. For example, Germany allows the storage and transmission of personal data within the limits of a "contractual" or "confidential" relationship between the data-holder and the data subject. The precise definition of a contractual or confidential relationship is not specified in the German data protection law. Germany and Switzerland state that data must be "blocked" if its correctness cannot be ascertained. "Blocking" requires that data not be communicated while its correctness is in question. Austria requires that an explanatory note be attached to questionable data. An Austrian citizen can demand that personal data concerning himself be held confidential and not transmitted. Data transmission itself can occur only with written consent of the data subject unless authority is provided for by law or the data concerns statistics or financial matters. Moreover, written consent may be withdrawn at any time. Written consent for data transmission is also required in the Netherlands. Norway and Denmark state that when credit information is transmitted, the data subject cannot demand to know how the information was obtained. Norway also states that the destination of the information is confidential. At any time, the registrar in Israel, the Netherlands, Sweden, and Denmark may enter any data processing center and demand any information. Iceland states that data may be maintained until it has "lost its worth."[9] When information has "lost its worth" is not clear. Belgium devotes an entire section to aural and photographic eavesdropping. Hungary has no restraints on "open data" unless the data falls under the "office-secret" category. However, all personal information is classified as "office-secret." Nevertheless, the responsible Minister or Hungary´s Council of

*The left-hand side of Table 1 lists some aspects of privacy protection. Lines 1 to 16 list privacy principles which are found in the Convention or Guidelines. Lines 17 to 28 list the scope and details of a country´s privacy laws/legislation. The first two vertical columns show which principles are included in either or both the Convention and Guidelines. The remaining vertical columns correspond to individual countries. The two rows at the top of Table 1 indicate which countries are members of the CoE and OECD.

Table 1 - Privacy Laws and Legislation*	Convention	Guidelines	Australia	Austria	Belgium	Canada	Denmark	Finland	France	Fed. Republic of Germany	Greece	Hungary	Iceland	Israel	Italy	Japan	Liechtenstein	Luxembourg	Netherlands	New Zealand	Norway	Portugal	Spain	Sweden	Switzerland	Turkey	United Kingdom	United States
Member CoE				X	X		X		X	X	X		X		X		X	X	X		X	X	X	X	X	X		
Member OECD			X	X	X	X	X	X	X	X	X		X		X	X		X	X	X	X	X	X	X	X	X	X	X
1 Obtainment	X	X		X	X	X			X						X						X	X		X			X	
2 Collection Limitation		X				X				X								X					X					
3 Data Purpose	X			X	X				X					X	X				X		X	X		X			X	
4 Purpose Specification		X				X		X																				X
5 Relevance	X	X			X										X				X		X	X	X	X			X	
6 Accuracy	X	X		X	X	X			X				X		X			X	X		X	X		X			X	X
7 Time Limitation	X								X				X		X						X	X		X			X	
8 Use Limitation		X											X						X									
9 Sensitive Data	X				X		X	X	X				X	X	X			X	X		X	X		X				
10 Security	X	X		X	X		X		X	X					X				X		X	X		X			X	
11 Existence	X	X		X	X	X	X		X	X				X	X			X	X		X	X		X			X	
12 Access	X	X		X	X	X	X	X	X	X				X	X			X	X	X	X	X	X	X			X	X
13 Rectification	X	X		X	X	X	X		X	X				X	X			X	X		X	X	X	X			X	X
14 Remedy for Denial	X	X		X	X	X	X		X	X				X	X			X	X	X	X	X		X			X	X
15 Openness		X			X														X									
16 Controller Accountability		X		X	X		X	X	X	X			X	X				X	X		X	X		X			X	X
17 Ratified Convention									X															X				
18 Signed Guidelines				X	X		X	X	X	X	X				X	X		X	X	X	X	X	X	X	X	X	X	X
19 Automated Data Only				X	X		X											X				X	X				X	
20 Automated and Manual Data						X		X	X	X			X	X	X				X		X			X	X			X
21 Public Sector				X	X	X			X	X			X	X	X				X		X	X		X	X		X	X
22 Private Sector				X	X	X	X	X	X	X			X	X	X			X	X		X	X	X	X	X		X	X
23 Physical Person Only						X		X	X	X				X					X			X		X			X	
24 Physical and Legal Person				X	X		X						X		X			X			X		X					X
25 Burden of Proof w/Individual																		X				X						
26 Burden of Proof w/Controller of the File				X	X	X			X						X				X					X			X	X
27 Registrar				X	X	X	X		X	X		X	X	X	X			X	X		X	X	X	X	X		X	
28 Law				X		X	X		X	X		X	X	X				X			X			X				X
29 Drafted Bill					X			X							X				X	X		X	X		X		X	
30 No Law			X								X					X	X									X		

The principles of the Convention and Guidelines are listed on lines 1 to 16.

1 - Obtainment: Data should be obtained and processed fairly and lawfully.
2 - Collection Limitation: Data should be gathered with the knowledge or consent of the data subject where appropriate.
3 - Data Purpose: Data must be used for specific and legitimate purposes; they must not be used in an incompatible way.

The Guidelines also state that the purposes of data collection be specified and that the data not be used in an incompatible way. However, while the Convention does not allow for subsequent uses of data which identify an individual (see Time Limitation-7), the Guidelines provide for the subsequent use of data after the original purposes are fulfilled (see Purpose Specification-4).

4 - Purpose Specification: The purpose of the data collection should be specified no later than at the time of collection. Should a change of purpose occur, the new purpose for using the data must likewise be specified and consent obtained (see Use Limitation-8).
5 - Relevance: Data should be adequate, relevant, and not excessive.

The Guidelines, however, state only that the information must be relevant and not excessive; no reference is made to its being adequate.

6 - Accuracy: Data should be accurate and up to date.
7 - Time Limitation: Data may not identify an individual for a period of time longer than required for the data's original purpose.

In other words, data may be kept and used for statistical or other purposes provided that they do not identify a particular individual.

8 - Use Limitation: Data should not be used other than for the original purpose without the consent of the data subject or by authority of law.

The preceding principle implies that data may be stored in a form which identifies an individual longer than the original purpose requires. Therefore, the data may have subsequent uses.

9 - Sensitive Data: Data concerning racial, political or religious views, criminal history, sexual life, or health may not be automatically processed without appropriate safeguards.
10 - Security: Data must be safeguarded against accidental or unauthorized destruction, loss, access, dissemination, or alteration.
11 - Existence: An individual must be able to establish the existence and purpose of information as well as the controller of the file.
12 - Access: An individual must be able to obtain information which is at a low cost, in a readable form, and obtainable in a reasonable time interval.
13 - Rectification: An individual must be able to rectify data or erase data that are processed contrary to law.

The "processed contrary to law" clause is not stated in the Guidelines.

14 - Remedy for Denial: If an individual's request is denied, the individual must have some means of recourse.
15 - Openness: There should be a policy of openness regarding the developments, practices, and policies affecting personal data.
16 - Controller Accountability: A data controller should be accountable for complying with the Guidelines.

Lines 17 and 18 indicate which countries have signed the CoE Convention and OECD Guidelines respectively.

Lines 19 and 20 indicate laws/legislation covering automated data processing or both automatic and manual data processing.
Lines 21 and 22 indicate laws/legislation covering the public and private sectors respectively.

Lines 23 and 24 indicate whether the laws/legislation covers only physical persons or both physical and legal persons.

Lines 25 and 26 indicate whether the burden of proving that information is correct or incorrect lies with the data subject or the controller of the file.

Line 27 indicates which countries have a registrar which keeps a register of data banks, issues licenses, or grants permission for data banks.

Lines 28, 29, and 30 identify a country's laws/legislation as law, as a drafted bill, or as no legislation at all.

*In the CoE are Cyprus, Ireland, and Malta. Ireland is also a member of the OECD. Specific information is not yet available about Finland, Hungary, New Zealand, and Switzerland.

Ministers may create laws to enable the transfer of personal information.*

Each Italian citizen has the right to receive from the registrar a certificate which gives a list of all data banks likely to contain information about himself. He has the right to receive another certificate stating how, when, and why any information about himself has been transmitted. As in Belgium, an Italian data subject must be notified if data about himself has been stored for the first time. An Italian data subject must also have communicated to him any alteration of his data.

Regulations specifically concerning the transfer of data to foreign countries are included in the laws/legislation of all countries with the exception of Luxembourg, Belgium, and the U.S. The United Kingdom states that data may be transferred only to countries which have ratified the Convention or which have provided equivalent privacy protection. Portugal, Italy, and France also require equivalent protection before a data transfer. Notification to the registrar of a data transfer is required in Italy, and authorization of the registrar is required in Spain, the Netherlands, France, Norway, Sweden, Denmark, Iceland, and Austria.

The U.S. approach to privacy protection has been sector-specific. For example, the Fair Credit Billing Act and Fair Credit Reporting Act were passed in 1968 and 1970, respectively. The Family Education and Privacy Act of 1974 and the Right to Financial Privacy Act of 1978 have also addressed specific sectors. Related to these Acts is the Freedom of Information Act, passed in 1967, which allows U.S. citizens to obtain information which the government has stored. The U.S. law which is most similar to European privacy laws is the Privacy Act of 1974 which covers only the public sector, however. European legislation usually covers the private sector and the public sector.

VII. INTERNATIONAL IMPLICATIONS

Since stringent enforcement mechanisms have not yet been developed, the results of international information regulation can only be hypothesized. The total implications of information regulation for the U.S., multinational corporations, and European countries are still unknown.

> . . .many in the U.S. fear that the Europeans and Canadians are primarily motivated by a desire to protect their local industry and markets from competition. . . . The provisions in many European privacy statutes allowing governmental authorities to stop the flow of data to foreign countries are cited as a potential Trojan horse for excluding foreign competitors from the information market place. Even in Europe, there is some sense of information as an emerging front in an economic and industrial world war.[10]
>
> Broadly speaking, the international concerns focus on the major role played by the U.S. in the supply and use of information resources. There is considerable fear that existing trends, if unchecked, will lead to an unhealthy American dominance of other nations´ economic, social, and cultural lives.[11]

Multinational corporations for which international data flow is vital may experience significant effects. "It has been estimated that 70 percent of the direct and indirect costs of business involve the collection, distribution, and management of information."[12] If countries use privacy legislation as a means of preventing foreign competition with their own industries, the multinational corporations may suffer greatly. "Moreover, argues the U.S. private sector, U.S. multinationals may be crippled in their operations if they cannot freely transfer data from place to place as needed for their commercial purposes."[13]

The operating costs of these corporations will increase as they adjust to the laws of various countries. The corporations must accommodate licensing and registration, hidden costs implied in the countries´ laws, and taxes or tariffs imposed in information regulation. Multinational corporations must be prepared to deal with the risks of constrained disclosure, exposure, or forfeiture of information which may contain trade secrets, financial information, or other privileged information. As mentioned above, several countries allow that at any time, a registrar may obtain any data he wishes. Finally, a tendency of most corporations has been to store any data which they think they might need for the future. Corporations might now be able to store only information they need and can justify for the present; they may be required to destroy information after it has served its intended purpose.

However, the countries themselves may be the more significant victims of information restriction. Countries are responsible for informing the data users of the rules and regulations concerning data banks and for supporting the licensing and registration of data banks. They also have the responsibility for enforcing their laws and of adjusting to new laws created to cope with new situations.

There is an aspect of information regulation which many countries apparently have failed to consider. While attempting to protect personal privacy and perhaps regulate foreign information flow, these countries may be excluding themselves from technological growth.

> . . .certainly the availability of wideband, high speed networks may influence the design of computer systems in a fashion undreamed of only a few years ago. It would be unfortunate if this technological opportunity were restricted for reasons related to political boundaries. Furthermore, if such communications restrictions are enforced, the possibility that data processing would move to regions where highspeed [sic] computer networks were permitted, would be to the restricting nation´s loss. Analogies can be drawn to other past restrictions on technological innovation which prevented growth of a valuable industry -- delay in the introduction of railroads in certain European nations during the 19th century to protect canal and highway interests and restrictive labor practices which caused massive shifts in industrial employment.[14]

The Dark Ages came about for precisely these kinds of reasons.

VIII. EMPHASIS OF PAST AND FUTURE

In the past, privacy has presented itself as the crux of information regulation. Although the rationale behind privacy protection has been questioned, i.e. whether the motivation has been humanitarian, the protection of the individual, or economic, the protection of industry from foreign competition, many countries have implemented or have considered privacy laws.

However, it appears that the matter of privacy is being subsumed in the regulation of telecommunications built upon satellite technology and other modern communication technologies. Since these technologies are used to transmit information, telecommunications regulation also covers satellite broadcasting spill-over, the allocation of satellite orbitals and the radio spectrum, government monopoly versus free enterprise in telematic development, national sovereignty, and others. Undoubtedly, telecommunications will create the issues of the future.

*In his paper, "New Hungarian Data Law in Force," Dr. Istran Sebestin states that "The protection of privacy . . . begins by classifying all personnel information under the ´office-secret´ category, with the exception of information intended primarily for public usage (e.g., the computerized ´Who´s Who?´ database) or in the case where no link to the ´owner´ of the data can be established. . . .

"Although possible in principle, transborder data flow of personnel data nonetheless falls under the category of high security control since, as mentioned above, all personnel data is classified as ´office secret.´"

References

1Robert P. Burton, "Transnational Data Flows: International Status, Impact, and Accommodation," Data Management, 18, No. 6 (1980), 27.

2Burton, p. 27.

3Harry J. Katzan, Jr., Multinational Computer Systems: An Introduction to Transnational Data Flow and Data Regulation (New York: Van Nostrand, Reinhold, 1980), 6-7.

4William L. Fishman, "Introduction to Transborder Data Flows," Stanford Journal of International Law, 16 (Summer 1980), 11-12.

5Fishman, pp. 10-11.

6Frits W. Hondius, "Data Law in Europe," Stanford Journal of International Law, 16 (Summer 1980), 90.

7Eric J. Novotny, "Transborder Data Flows and International Law: A Framework for Policy-Oriented Inquiry," Stanford Journal of International Law, 16 (Summer 1980), 176-177.

8Private conversation between Robert P. Burton and William L. Fishman, June 1980.

9Iceland, Act No. 63/1981, respecting Systematic Recording of Personal Data, sec. III, art. 7.

10Fishman, p. 12.

11Fishman, pp. 8-9.

12Katzan, p. 94.

13Fishman, pp. 12-13.

14Ithiel De La Sola Pool and Richard Jay Solomon, "Intellectual Property and Transborder Data Flows," Stanford Journal of International Law, 16 (Summer 1980), 135.

Artifical Intelligence and the Law:
Will Expert Systems Replace Expert Lawyers?

Garry S. Grossman

Fried, Frank, Harris, Shriver & Kampelman
Washington, D.C.

The commercial availability of "expert systems" utilizing specially developed knowledge bases raises significant questions about their potential utility in the practice of law. These systems, built with the aid of recent developments in artificial intelligence research, may only prove useful in certain areas of legal practice. Counselling and interviewing are areas where expert systems are likely to effect marked changes in the practice of law. In contract, computerized legal research using a knowledge-based system is more difficult to envision. This is due to complexities presented by the multiplicity of sources of the law, and by the nuance of conflicting opinions and interpretations in the common law.

In the coming decade, use of expert systems in such diverse fields as science and medicine will grow rapidly, and attempts will continue to be made to automate the legal reasoning process. As past research efforts have demonstrated, this will not be an easy task.

CH1935-6/83/0000/0010$01.00 © 1983 IEEE

AN AMERICAN LEGAL MAZE: PROTECTION OF COMPUTER PROGRAMS

James A. Coles, Lawyer

Barnes & Thornburg
1313 Merchants Bank Building
Indianapolis, Indiana 46204

While much has been written about the insufficiency of the American legal system to provide adequate protection for computer programs, little has been said to guide the program owner in achieving protection through the legal maze of the existing methods.

The American legal system provides not one but four methods of protecting computer programs: trade secret, copyright, patent and trademark. Each method is different in the scope and nature of protection it provides, how it may be used, and the steps required to perfect and enforce the rights which it affords to the program owner. Whether the program is in the form of software or firmware, the selection of the best method involves careful consideration of the application and marketing strategy for the program. The present state of the American law dictates the need to assess individually each program and to develop a carefully guided plan for a successful approach to utilizing the four methods.

Although each method is different, they need not be mutually exclusive of each other. If applied properly, various combinations of the methods can be used successfully to protect a program. If applied improperly, a combination of the methods may result in a loss of rights to the program owner.

With carefully established procedures, the program owner can protect the underlying concepts of his program as a trade secret. For this purpose, secrecy is essential. The protective procedures must be regularly followed and closely monitored in order to preserve the secrecy. To do otherwise could result in the loss of trade secret protection. Trade secret remedies are limited to actions for misappropriation and breach of trust or confidentiality against the perpetrator of the wrongful act, and cannot prevent unauthorized copying or use of the program by innocent third parties.

Copyright protection may be used to prohibit unauthorized reproduction and distribution of copies of the program. In some instances, it may be desirable to combine trade secret and copyright protection, but care must be taken not to jeopardize the trade secret protection when the two are used in combination. In other instances it may be desirable to rely only on copyright protection.

Patent protection may be used for specific applications of programs. While patent protection is capable of providing the broadest scope of protection, it is available under the American patent system only to cover the use of the program in combination with an apparatus to control the operation of the apparatus, and only if the resultant process of controlling the apparatus with the program steps is novel and unobvious to one skilled in the art.

Trademarks are many times overlooked as another method of protecting computer programs. The selection of a good trademark and use of the mark in a marketing strategy for the program can enhance the reputation of the program owner and the demand for programs marketed by the owner.

CH1935-6/83/0000/0011$01.00

Brain Models and VLSI

OVERVIEW OF THE HUMAN BRAIN AS A DISTRIBUTED COMPUTING NETWORK

ALAN S. GEVINS

EEG SYSTEMS LABORATORY
1855 Folsom St.
San Francisco, CA 94103

Abstract

The hierarchically organized human brain is a premier example of a massively parallel, adaptive information processing and process control system (Fig 1). It consists of about 10^{12} processing elements, many of which have up to 50,000 connections. While estimates of its storage capacity and computing rate are very speculative, figures in excess of 10^{25} bits and 10^{17} bits/sec may not be excessive. Considering the disparity of scale, and because the brain's processing is not deterministic and is directed towards survival in a hostile environment, literal analogies between computers and brains are somewhat strained. However, the principles of massive parallelism, dense connectivity and self-organization of assemblies of processing elements may be relevant to the design of fifth generation VLSI computing networks. This minipaper is intended to provide a brief overview of the human brain for computer architects.

Neurons -- The Brain's Processing Elements

Different types of neurons are specialized for input, output, association and regulating nearby neurons (Fig 2). Some of the larger neurons can be compared to mainframe computers since they can integrate in complex ways the outputs of tens of thousands of other neurons. A neuron's processing can adapt to changing circumstances on short and long term time scales. Although the mechanisms have only been hypothesized, neurons may have the capacity for storing billions of bits. Neurons have both analog and digital properties. Input and processing consists of graded potentials, while output consists of fixed voltage, all-or-none pulse trains. Since neurons are so richly interconnected, the organization of neurons into larger aggregations must be considered in understanding brain functions.

Nuclei -- Special Function Processors

Assemblies of neurons of the central nervous system are organized into two major architectural types called nuclei and cortices (Fig 1). The 50-100 major nuclei each consist of tens of thousands to tens of millions of neurons. Nuclei have more or less fixed functions which are determined by their patterns of internal connections. Each of the senses and motor systems has a number of nuclei which operate in parallel and sequentially to perform stimulus registration, scaling transformations, various types of feature extraction and assembly of motor programs. For example, the successive stages of visual processing are somewhat similar to a data flow network. Visual input is transmitted in parallel and serially to many nuclei and cortical systems, including motor systems. The thalamus is a large collection of sensory and motor relay nuclei. The basal ganglia is a set of motor nuclei of the brain stem which produce impulse movements.

Cortices -- General Purpose Networks

The other type of assembly, cortex, is more recent evolutionarily. Cortices consist of thin sheets of tissue with a stereotyped local organization which is repeated over and over. Cortices receive sensory inputs from nuclei and other cortices, process them and transmit output and control signals both to sensory and motor systems. Several relatively simple types of cortex are specialized for processing smells (olfactory cortex -- see paper by W. Freeman) and for solving the differential equations needed to produce smooth movements (cerebellar cortex -- see paper by J. Albus). The cerebral neocortex is the newest brain structure. It performs the highest level processing and has the most complex internal structure. Neocortex consists of 2-3 mm thick sheets of neurons which are folded up and packed into the skull at the top of the brain. It has six main layers with neurons specialized for input, association, regulation and output arranged according to layer (Fig 2). The

CH1935-6/83/0000/0013$01.00

layers vary greatly in their patterns of input, output and feedback connections depending on whether input processing (idealized as the two "specific afferent cylinders" in layer IV of Fig 2) or output processing (shown as the large triangular cells of layers III and V on the right half of Fig 2) are performed. The actual patterns of connectivity are quite complex, but the simplified diagram of Fig 2 is quite instructive.

Neocortical Columns -- Local Network Modules

Horizontally, neocortex is organized into minicolumns of about 110 neurons (about 260 in the visual cortex). Minicolumns are about 30 microns in diameter; there are hundreds of millions of them in the human neocortex. Minicolumns in turn are organized into macrocolumns consisting of about ten thousand to one hundred thousand neurons (Fig 2). The internal structure of the hundreds of thousands of macrocolumns in the human neocortex is surprisingly similar. Because of this similarity, differentiation of processing must primarily be a "software" function. Processing in columns proceeds in parallel. For example a portion of a visual image can be processed in parallel by columns which extract lines, edges and other features. Columns are arranged with inputs from other columns at many layers ("cortico-cortical afferent" fiber of Fig 2). This allows multiple overlays of two-dimensional patterns in the horizontal plane while maintaining the topology of connections with other areas.

Communication between distant columns is via specifically labeled fiber tracts while communication between neighboring columns is more distributed. Considered as a statistical ensemble, cortical processing seems to be performed in phase with oscillations at the resonant frequency of the mass (see paper by W. Freeman). Some of the senses such as sight and smell may be modeled as sampled data systems.

Major Neocortical Divisions -- Supernetworks

The macrocolumns of the neocortex are organized into about 50 major functional divisions each consisting of some thousands to tens of thousands of macrocolumns (Fig 3). Functionally distinct divisions are also architecturally distinct. They vary greatly in their connectivity and also as to relative proportions and morphology of neuron types. While extraction of basic stimulus features such as line edges may be genetically determined and tuned by early experience, higher order cognitive functions of neocortical divisions are highly programable and adapt to the ongoing activities of the organism. Processing through the columns of a functional division is parallel and pipelined. Sensory and motor systems (and presumably higher order divisions) are subdivided to allow multiple representations of their sample space. Different aspects of the sample space are processed in parallel subdivisions and are then integrated by hierarchically higher divisions.

Multiple parallel pathways connect input and output divisions in contrast to a strictly serial hierarchy. These pathways consist of myriads of highly organized, high bandwidth fiber tracts (the brain's "white matter").

For example, about 3×10^8 fibers connect the right and left cerebral hemispheres. When considering the entire neocortex, there are so many fiber tracts that the best first approximation topological model may be a full crossbar with broadcasting. At lower hierarchical levels, it is possible to take specific interconnection pathways into account.

An Executive Supernetwork

Much of the neocortex performs spatial and temporal integration of outputs from parallel sensory, motor, "associative," internal body state, and emotional drive systems. The highest level integrations are performed by the prefrontal cortex, a massive division comprising about one-third of the entire cerebral cortex and consisting of about a dozen major subdivisions. The prefrontal cortex has connections with most cortical and subcortical systems (Fig 4). It gates outputs from other divisions and systems and can inhibit and facilitate their processing. While it performs numerous other "top-down" goal directing and control functions, the prefrontal cortex is not at the top of a strictly serial processing hierarchy since familiar behaviors can continue even when it is severely damaged. However, long term motivation, very complex processing and the acquisition of new complex behaviors may be impaired when the prefrontal cortex is damaged.

In healthy brains processing bottlenecks occur when conscious attention must be engaged to monitor and control those processes which are not overlearned (automated). Currently unknown serial processes of the prefrontal cortex may be a major limiting factor.

Differences Between Brains and Computers

Biological brains evolved hierarchically. Each successive layer or structure of the central nervous system added increased capacity to a system which was already functioning more or less adequately. The enlargement of the cerebral neocortex in mammals added greater adaptability to the largely pre-programmed (instinctual) reptile brain. This resulted in considerable redundancy of function, alternative parallel pathways between input and output, and non-deterministic processing. It also necessitated the further development of explicit communication and control protocols between hierarchical layers. The further enlargement of the neocortex, particularly prefrontal and other "association" cortices, in people added language, extensive symbol manipulation and simulation functions, and presumably the explicit concept of self-existence and the poorly understood process of intuition.

Conclusion

Since biological brains are working examples of massively parallel, densely interconnected, self-organizing computational networks, I believe that crossfertilization between the neurological and computational sciences will be as important in the 1980's and 1990's as it was in the late 1940's and 1950's when the foundation of modern computer science was created by interdisciplinary "cybernetic" scientists and engineers. The first examples of parallel VLSI processors based on neural-like principles are currently being implemented (see paper by J. Vidal).

Bibliography

The following are readable accounts of the structure and functions of the nervous system:

Albus, J. Brains, Behavior and Robotics. Byte Books: New York, 1981.

Gardner, E. Fundamentals of Neurology. Saunders Company:Philadelphia, 1963.

Kent, E. W. The Brains of Men and Machines. McGraw Hill:New Hampshire, 1981.

Shepard, G. The Synaptic Organization of the Brain. Oxford:New York, 1979.

Special Issue on the Brain. Scientific American, 241:3, 1979.

Acknowledgements

My research is concerned with imaging the neurocognitive electrical patterns of the human brain. This line of research may eventually lead to a neurocybernetic information channel which would allow people to use computers more effectively. This research was reported in Science (203:665-668, 1979; 207:1005-1008, 1980; 213:918-922, 1981; and 220:97-99, 1983), and was the subject of a popular article by OMNI 4:(1):53-59, Oct. 1981). It is supported by the Office of Naval Research, the Air Force Office of Scientific Research, the Air Force School of Aerospace Medicine and the National Science Foundation.

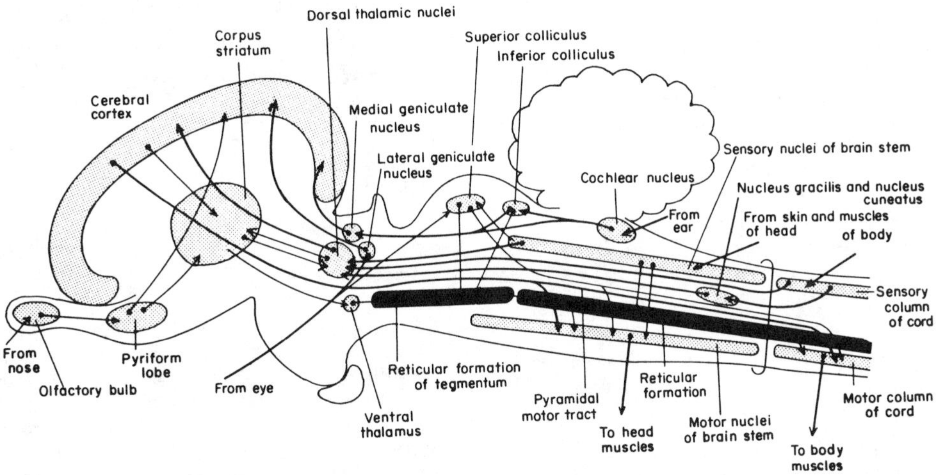

Figure 1 The central nervous system is hierarchically arranged. It consists of: a large number of special purpose processors (nuclei) which form highly parallel sensory input and motor output systems; a "nonspecific" reticular system which controls the state of arousal; the cerebral cortex (neocortex); and a number of other types of structures. The neocortex performs the highest level of sensory, motor, cognitive and executive functions.

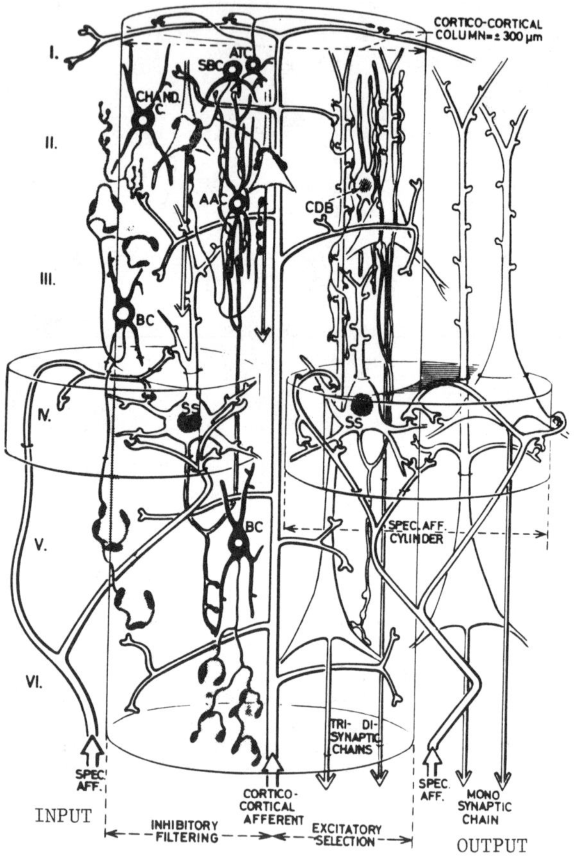

Figure 2 Various types of neocortical neurons are shown. The neocortex has a six-layered vertical structure. Sensory inputs are processed in layer 4, while motor outputs are transmitted from the large triangular shaped cells of layers 3 and 5. Horizontally, neocortex is organized into hundreds of thousands of general purpose local network modules called macrocolumns one of which is shown schematically.

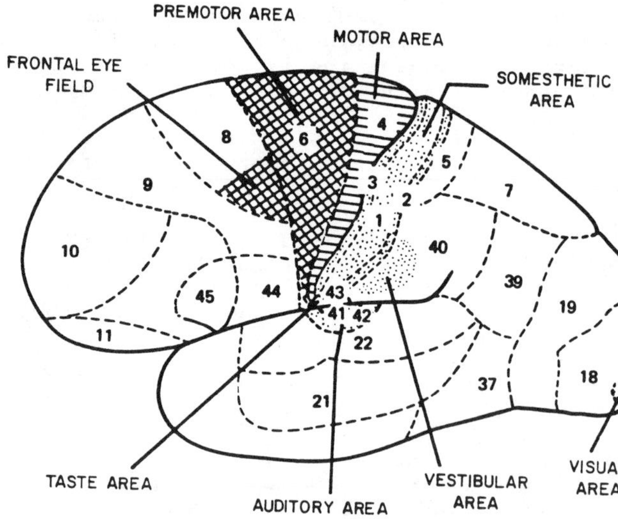

Figure 3 About 50 major structural-functional divisions of the neocortex can be distinguished based on variations in connectivity and morphology of neurons. These divisions are numbered on an outside side view of the brain, with the front of the brain at the left. As with other hierarchical levels of organization, processing in neocortical divisions is parallel and pipelined.

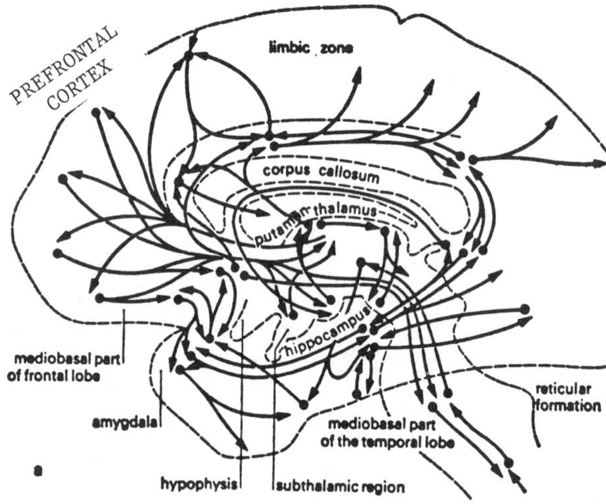

Figure 4 The prefrontal division comprises about one-third of the neocortex. It is an executive supernetwork which has input and output connections with most areas of the brain.

SILICON BRAINS: WHITHER NEUROMIMETIC COMPUTER ARCHITECTURES

Jacques J. Vidal

UCLA Computer Science Department & Brain Research Institute

ABSTRACT

This communication suggests that parallel, connectionist processor of neuromimetic inspiration are viable machine forms. Successful management of the functional and testing complexity of future VLSI systems will depend on a rigorous modularization of design coupled with the integration of local learning (hence "neuromimetic") mechanisms into the hardware. The "Universal Connection Logic Modules" are VLSI circuits developed at UCLA to experiment with the neuromimetic approach by creating dynamically reconfigurable learning networks.

1.0 RATIONALE

Conventional (so-called Von Neuman) digital machinery performs poorly in tasks that emulate the natural information processing that humans and other animals handle routinely. Effective formalisms are lacking and and in many cases, the search and retrieval of pattern information require billions of elementary operations for which serial-sequential processors are ill-adapted.

There are less recognized limitations. Ordinary computers are geared to binary arithmetics. They do not lend themselves to gradual modification of function. The positional notation achieves extreme accuracy but inherently requires multistep operations. These characteristics are ill-suited to "fuzzy" problems, which seldom require accuracy, but demand robust solutions that degrade gracefully with noise.

The obvious need for a different class of machines has spawned considerable interest in "non-Von Neumann" approaches (cellular arrays, data flow machines [5]), functional architectures all characterized by an increased use of parallelism and concurrency. These new paths can already be interpreted as converging toward "neuromimetic" computers, i.e. of machines capable of emulating brain functions at the hardware level.

It should be clear that this emulation will be more of style than of detail. VLSI devices will remain in quite a different class from that of natural automata realized in "wetware", and sets our topic apart from proposals aimed at "chemical" computers manufactured from organic molecules. The latter are still, in our view, quite remote.

Learning modules with dynamic behavior have been traditionally associated with analog rather than digital devices. Yet VLSI analog technology is not manageable on that scale, mostly because of lack of testability. Natural automata may be analog, but their built-in self-testing and repair mechanisms are completely beyond our reach. The unavoidable conclusion is that the only feasible approach is digital, with continuity broken down and approximated into small logic steps.

Retrospectively, it is ironical that the founding fathers of the information sciences, and especially von Neumann, were very much concerned with natural automata as possible blueprints for computer architectures. The technology of the time however was such that sequential machines were the only practical option.

2.0 NEUROMIMETICS

2.1 Historical Note

The groundwork for neuromimetics is rooted in the efforts of many researchers to understand how neurons and neuronal networks can perform pattern recognition and control behavior in animals. Seminal contributions can be traced to McCulloch [18], von Neumann [27], Hebb [10], and Rosenblatt. Indeed, the subject of learning machines gained notoriety around 1960 with the perceptron, an early model of parallel processing and brain pattern recognition proposed by Rosenblatt [21,19]. Perceptrons are threshold logic circuits capable of adjusting their transfer function, and hence of "learning", by changing internal analog weights.

Modeling plays a central role in developing brain theory. Many sensory or cerebellar models propose mechanisms that hold promises for computer architectures. These include for instance Lettvin et al [16], Marr [17], Arbib [4], Wilson and Cowan [25], Klopf [13], Amari [3], Albus [1], and Feldman [6].

Some authors like Kauffman [12] and Grossberg [9] have addressed the dynamic and oscillatory aspects of parallel systems with interconnected processing elements. This is a scarcely explored area that suggests a radically novel approach for the control of distributed computing structures.

2.2 Application of Neuromimetic Concepts

A hypothetical neuromimetic machine would borrow many characteristics from extant brain models: massive parallelism, a large networks of processing elements (PEs) or "cells", a regular topology,

CH1935-6/83/0000/0017$01.00 © 1983 IEEE

dense interconnections and above all a capability for self-organization and "learning" under local control.

The parallelism and the regular topology is a characteristic of contemporary cellular automata, but there are differences. In a neuromimetic perspective, the network interconnection is dynamically reconfigurable. The reconfigurability of both machine structure and module function is the key to self-organization or, in other words, of self-programming.

Another distinctive characteristic of brain-style computing is that it is "connectionist" rather that based on symbol transformations [7]. Computing results from network interconnection rather than from algorithmic processing at the nodes. The network becomes in some sense isomorphic to the structure of the problem to solve, and abstract objects in the computations are physically incarnated by separate cells or sets of cells (place coding). Furthermore, in that perspective, to represent numerical data each discrete level of a variable must be encoded by different sets of cells, where each cell covers a small range of values (coarse coding). Adjacent ranges overlap to achieve enough resolution and the ordering of the cells in the array preserves the order of the ranges (mapping) [14]. Qualitative features are also place coded and represented in ordered "feature maps" [15].

These characteristics and the fact that many objects must be represented hierarchically, at several levels of abstraction, clearly indicates that even modest problems will require an extremely large number of cells and that any hope of meaningful implementation would have to wait for VLSI.

2.3 Time Sequencing

Information processing, on any type of hardware, must unfold in time and even a massively parallel neuromimetic machine must iterate to solve problems. However there would be no central clock and operations are synchronous. Global synchronization is effected by the data driven oscillations and relaxations in the networks. This mode of control is captured in the concept of adaptive resonance [9] which has been suggested as a brain mechanism for associative pattern storage and recall.

3.0 UNIVERSAL CONNECTION LOGIC MODULES (UCLMs)

A re-introduction of circuit learning and dynamic control in computer design is not only a methodological goal for artificial intelligence, but it is also a technological necessity for VLSI. At the complexity level of future VLSI chips, it will no longer be possible to bring all control information from the outside. A measure of self organization must be available based on the broadcasting of non-specific reward signals.

Our work at UCLA started with a discussion of logically complete perceptron-related structures ("General Purpose Perceptrons") [23] and pointed out the equivalence between these and Universal Logic Modules or ULMs, a family of circuits that appeared in the theoretical computer science literature also around 1960 [26].

ULMs indeed resemble perceptrons in that they are capable of adjusting to multiple functions. They can be implemented with multiplexers in conventional rather than threshold logic. They are now reappearing in the VLSI literature because of a growing interest in modular design, as a way to manage VLSI complexity. However, when seen as simple substitutes for the conventional applications of combinational logic [24] or as building blocks for finite state automata [2], ULMs appear to be area inefficient and slow. The synthesis of specified combinational logic functions, based on Shannon decomposition theorem [22], results in cascaded structures with intermingled "data" and "control" signals, fairly difficult to program.

In our approach, modified ULMs have been designed and prototyped in NMOS to function as building blocks for connectionist structures. Accordingly, the new modules are referred to as UCLMs (Universal Connection Logic Modules). This shift of strategy results in a data flow model with physical separation of the control and data paths.

UCLMs are made up of identical dyadic "atoms"or "cells" each a complete logic function generator with two inputs, by definition capable of evaluating all 16 boolean functions of its inputs:

$$Z = c_3ab + c_2a\bar{b} + c_1\bar{a}b + c_0\bar{a}\bar{b}$$

The gate structure is the logical function block of Mead and Conway (Figure 1).

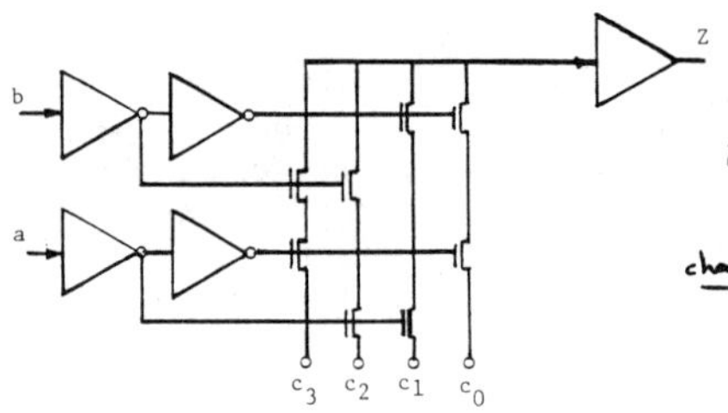

Figure 1

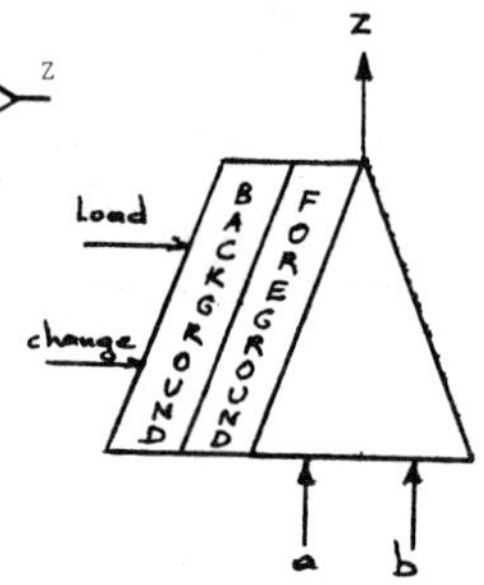

Figure 2

Each cell has a small amount of memory. and can function in a triple capacity: processing, communication (routing) and storage. A 4-bit register holds the instruction. Several designs are being evaluated: in the circuit of Figure 2. Two registers can swap control instructions. The cells are grouped into complete binary trees from which structures with arbitrary numbers of inputs can be constructed.

The simple switching functions F_9=A (left), F_3=B (right) and the constant F_0=0 (null), F_{15}=1 (one) in the set are keys to crucial capabilities: reconfiguring the structure and changing the memory content at specified nodes.

Finally, from the point of view of engineering, the observability (testability) of the tree structure is complete at the atom level. Furthermore, testing complexity increases linearly rather that exponentially with the number of components and the testing procedures are universal and independent of module function [20].

Composite atoms with more than two inputs can also be constructed [23]. Six of the former dyadic atoms, arranged into a three-layer symmetric hierarchy, yield a complete triadic atom. The new structure can compute any of the 256 possible Boolean functions of its input and thus can be treated as the elemental building block for instance to implement two-dimensional tessellations.

4.0 UCLM ARCHITECTURES

The above discussion suggests that UCLMs are suitable building blocks for a connectionist architecture with an extremely fine "granularity" and uniformly distributed memory and processing. The extreme modularity of the approach has general technological advantages: rigorously modular, repetitive structures simplify specification and facilitate normalization to a small number of "canonical" modules. Intrinsic testability which has already been mentioned may in time become the single most important feature.

UCLMs can implement combinational logic or sequential automata [2]. In a connectionist machine however, the main functions are pattern matching resulting in the recruitment of cells on the paths to corresponding targets . Messages travel on the open paths to cause activation or inhibition of activity in the set of nodes representing the information associated with the identified patterns. There are various approaches to the pattern matching problem, regular languages for instance can be recognized by pipelining modules [8].

There are also several possible designs for cutting paths inside the module's trees. The only requirement is an internal two-way communication between parent and childrens in the tree to signal the condition (load complete) of the background registers, and only two global signals ("load" and "clear") entering the module (Figure 3).

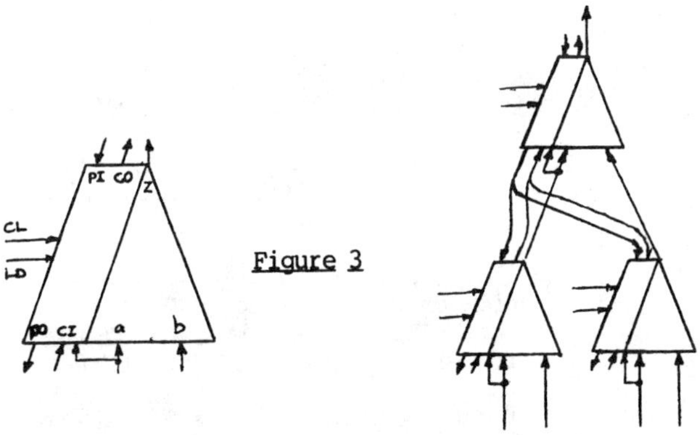

Figure 3

To make provision for local control and learning, the clusters of cells allocated to the representation of objects in the computation, must be effected by gradual changes in two directions, loosely corresponding to excitation and inhibition in neural models. This can be obtained by using the cells that lie in the path to the object as logical "filters" that can be made alternately more permissive or more restrictive so that a larger or smaller proportion of the transmitted message is received by the target cluster. [11].

Experimental data on such structures is still very sketchy and based on small scale simulations. Thus descriptions beyond this point should be considered tentative and in the nature of a "thought experiment". In our hypothetical machine, each processing array would exhibit oscillations, with attendant opportunities for resonance and relaxation. Concurrent messages would travel on multiple redundant paths in the form of propagating wavefronts. Because the data flow is free and not controlled by tokens or other mechanisms, events are time-bound and, on a global level, information become embedded in the relative synchrony between the pulse trains that travel on neighboring pathways.

Related mechanisms have been proposed to account for the consolidation of long term pattern memory and to sustain the "reverberation" needed for short-term storage. One such retrieval mechanism can be interpreted as follows: In a given state, a network array is able to store a finite number of global pattern "templates", each corresponding to one realizable oscillation mode of the array. Specific inputs from pattern matching momentarily destabilize the network and trigger global oscillations along the path associated with the input related template.

Our current plans for an experimental machine call for a "cortical" design: a hierarchical tree architecture of modules, with the trees themselves arranged in two-dimensional arrays of modules, stacked to form multiple layers and interfaced at their boundaries (with conventional processors). Neighboring trees in each array stack are interconnected both laterally, within each layer and vertically across layers.

5.0 CONCLUSIONS

A great deal of work remains to be done to establish solid theoretical and practical foundations for neuromimetic computing. Despite the prestigious attention received at the beginning by biologically motivated automata, neuromimetic concepts have not yet been recognized as potential blueprints for computer architecture or properly integrated with contemporary efforts on parallelism and data flow. To this day, only a handful of small "neuromimes" have been built, generally in computer vision experiments.

General purpose machines are still in the future. The state of the art however, suggests that medium scale neuromimetic realizations, capable of solving non trivial problems, are now feasible.

Acknowledgement

Research supported in part by a grant from the MICRO Program of the State of California with assistance from MOSIS (Multi-organization Silicon Integration Service) under DARPA and NSF.

References

[1] Albus, J.S., "A New Approach to Manipulator Control: The Cerebellar Model Articulation Controller (CMAC)," Journal of Dynamic Systems, Measurement, and Control Vol. 97(3), pp.220-227 (September 1975).

[2] Almaini, A., "Sequential Machines Implementations using Universal Logic Modules," IEEE Trans. Comp. C-27(10), pp.951-960 (October 1978).

[3] Amari, S., "Characteristics of Randomly Connected Threshold-Element Networks and Network Systems," pp. 35-47 in Proc. IEEE (Jan. 1971).

[4] Arbib, M.A., "Automata Theory and Neural Models," Proceedings of the 1974 Conference on Biologically Motivated Automata Theory, pp.13-18 (June 19-21, 1974).

[5] Dennis, J.B. and D. P. Misunas, "A Computer Architecture for Highly Parallel Signal Processing," Proc. Nat. Comp. Conf., pp.402-409, AFIPS Press (1974).

[6] Feldman, J. A., "A connectionist model of visual memory," in Parallel Models of Associative Memory, ed. J. A. Anderson, I. Earlbaum Associates, Hillsdale, NJ (1981).

[7] Feldman, J. A. and D. H. Ballard, "Connectionist models and their properties," Cognitive Science (1982).

[8] Foster, M.J. and H.T. Kung, "Recognize Regular Languages with Programmable Building Blocks," J. of Digital Syst. Vol. VI(4) (Winter 1982).

[9] Grossberg, S., "Adaptive resonance in development, perception, and cognition," in Mathematical Psychology and Psychophysiology, ed. In S. Grossberg, American Mathematical Society, Providence (1981).

[10] Hebb, D. O., in The Organization of Behavior, Wiley, NY (1949).

[11] Holland, J. H. and J. S. Reitman, "Cognitive Systems based on Adaptive Algorithms," in Pattern-Directed Inference Systems, Acad. Press (1978).

[12] Kauffman, S., "Behaviour of Randomly Constructed Genetic Nets: Binary Element Nets," pp. 18-37 in Towards a Theoretical Biology, ed. Ch. Waddington, Edinburgh Univ. Press (1970).

[13] Klopf, A. H., The Hedonistic Neuron: A Theory of Memory, Learning and Intelligence, Hemisphere, Washington, D.C (1981).

[14] Kohonen, T., Associative Memory - A System-Theoretical Approach, Springer-Verlag, Berlin (1977).

[15] Kohonen, T., "A Simple Paradigm for the Self-Organized Formation of Structured Feature Maps," in Competition and Cooperation in Neural Nets; Lecture Notes in Biomathematics, ed. S. Amari and M. A. Arbib, Springer-Verlag, Berlin (1982).

[16] Lettvin, J.Y., H.R. Maturana, W.S. McCulloch, and W.H. Pitts, "What the Frog's Eye Tells the Frog's Brain," Proc. IRE Vol. 47(11), pp.1940-1959 (Nov 1959).

[17] Marr, D., "A Theory of Cerebellar Cortex," J. Physiol. Vol. 202, pp.437-470 (1969).

[18] McCulloch, W.S. and W. Pitts, "A Logical Calculus of the Ideas Immanent in Nervous Activity," Bulletin of Mathematical Biophysics Vol. 5(4), pp.115-133 (December 1943).

[19] Minsky, M. and S. Papert, Perceptrons, an Introduction to Computational Geometry, MIT Press, Cambridge, MA (1969).

[20] Moore, D. W., "General Purpose Perceptron," UCLA Comp. Sci. Dept. Report (1983).

[21] Rosenblatt, F., Principles of Neurodynamics: Perceptrons and the Theory of Brain Mechanisms, Spartan Books, Washington D. C. (1961).

[22] Shannon, C.E., "The Synthesis of Two-terminal Switching Circuits," Bell system Tech. Jl. Vol. 28, pp.59-98 (1949).

[23] Verstraete, R., D.W. Moore, and J.J. Vidal, "The Perceptron Revisited," Submitted for Publication (1983).

[24] Voith, R.P., "ULM Implicants for Minimization of Universal Logic Module Circuits," IEEE Transactions on Computers Vol. C-26(5), pp.417-424 (May 1977).

[25] Wilson, H. R. and J D. Cowan, "Excitatory and inhibitory interactions in localized populations of model neurons," J. Biophys Vol. 12, pp.1-24 (1972).

[26] Yau, S.S. and C.K. Tang, "Universal Logic Modules and Their Applications," IEEE Transactions on Computers Vol. C-19(2), pp.141-149 (February 1970).

[27] von Neumann, J., "The general and logical theory of automata," pp. 1-32 in Cerebral Mechanisms in Behavior: The Hixon Symposium, ed. A. Jeffres, J. Wiley & Sons (1951).

SEMI-AUTONOMOUS CONTROL OF INPUT BY NERVE CELL ASSEMBLIES

THROUGH GENESIS OF IMAGES

Walter J. Freeman

Department of Physiology & Anatomy
University of California
Berkeley, CA 94720

Abstract

Analysis of the neural activity spatial patterns at the first synaptic relay in the olfactory system indicates that it performs operations of abstraction, generalization, figure completion and association on input. Synthesis of a set of coupled nonlinear integrodifferential equations to simulate the neural activity shows that the neural system operates in a distributed and interactive manner through multitudes of synaptic contacts. Input serves to drive the system from a low-level state of chaos into a limit cycle state. The oscillation serves as a carrier frequency with information conveyed in the spatial modulation of amplitude and phase. The modulation is determined by past experience embedded in the strengths of modifiable connections, by the state of the system under brain control, and by the spatial pattern of the destabilizing input. The system serves as a self-organizing adaptive spatial filter for selective control of input.

Introduction

The hallmark of neuronal VLSI is local control of input. A subsystem of the vertebrate brain that manifests this capability, that for olfaction, has been described in detail; its dynamics has been simulated with coupled nonlinear integrodifferential equations.

The essential property of the neural assembly that endows these features is distributed nonlinear feedback [2]. The parallel input in a given time frame serves to destabilize the assembly and drive it from a quasi-equilibrium or chaotic state to a limit cycle state [3]. The spatio-temporal form of the "wave packet" in the limit cycle domain is determined in part by the spatial pattern of internal connectivity that is a record of past experience (subject to updating), in part by the initial conditions set by the background "noise" or bias, and in part by the spatial pattern of input [4]. The synaptic connectivity constitutes a collection of "templates"; the wave packet constitutes an "image" of expected or wanted input. In effect it is a "search image" or a "hypothesis"; both the hypothesis and the results of the test are expressed in the spatial form of the wave packet.

Neural Structure and Activity in the Olfactory System

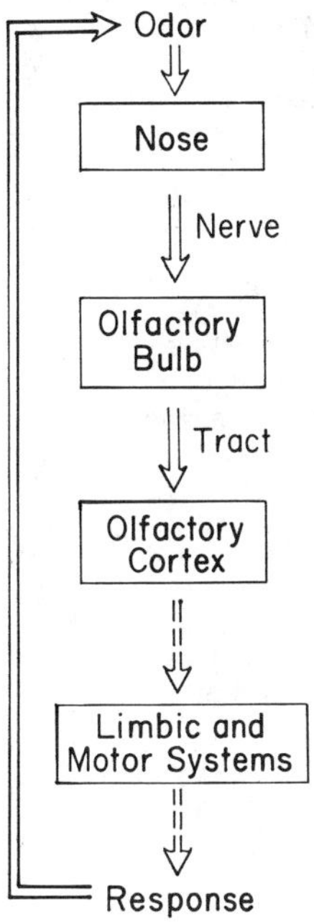

Fig.1 In main outline the olfactory system has three layers [1]. The receptors comprise a two-dimensional array in the nose where chemotransduction takes place. The olfactory bulb does initial processing of olfactory information, and the olfactory cortex does final processing before the information is admixed with input from all other sensory modalities in the forebrain. Each of the 10^8 receptors sends an output line (axon) to the bulb that synapses onto a bulbar excitatory cell with a degree of topographic order implying conformal mapping. Each axon of the roughly 10^5 excitatory cells transmits widely to cells in the cortex. Conversely each cortical neuron performs a spatial integration over a wide region of the bulb.

Transmission between each stage is simultaneous and in parallel over selected axons. The induced activity in each of the three stages is best described as a surface pattern. Each neuron in the system generates electrical potentials in two modes [1]. The input lines (dendrites) generate graded electrotonic currents that are manifested in fields of potential known as the electroencephalogram (EEG). The axons generate trains of fixed voltage impulses. Forward transmission is by excitation from odors to

CH1935-6/83/0000/0021$01.00

receptors and on serially through to targets in the brain.

Inhalation causes a monophasic wave in receptor activation and a surge of impulses on receptor axons. After a conduction delay there is a corresponding dendritic activation of bulbar cells and a surge of impulses on their axons. Concomitantly, the bulb generates a monophasic wave in the EEG. Superimposed on this wave is a patterned oscillatory burst at a frequency of 40-80 Hz that lasts on the order of 0.1 sec. Statistical analysis shows that the likelihood of impulses of single bulbar cells oscillates at the same frequency. The burst drives the cortex at the same frequency but with a phase lag by axonal conduction delay. The burst activities in the bulb and cortex comprise the wave packet with each inhalation.

The quality of a stimulus is conveyed in the location of each receptor and its axon terminals, and the intensity and timing by the impulse rate and timing. Receptors, bulbar and cortical cells respond selectively to odors. The receptors for each odor are not homogeneously distributed, so the array generates a spatial pattern of impulses (an image) to represent each odor to the bulb. In view of the mapping from receptors to bulb, a spatial pattern in bulbar activity might be expected to conform to the receptor activity pattern for each test odor.

This hypothesis was tested by placing arrays of 64 electrodes (8x8 rectangular grid with 0.5 mm spacing) surgically onto the surface of the bulb in rabbits, and after recovery by delivering odors to the awake animals while recording the 64 EEGs. After numerous experiments the surprising conclusion was reached that the bulbar activity patterns manifested in the EEG were not dependent on odor directly, but on expectation of odor established by training [5].

Neural Mechanisms of the Olfactory EEG

An explanation of these findings lies in analysis of the bulbar mechanism responsible for burst formation. The bulbar excitatory neurons (Figure 2,e) have widespread axonal branches given to each other and form a mutually excitatory assembly. This assembly has two main sets of interneurons. The outer set (not shown) is also excitatory and provides excitatory bias to the assembly.

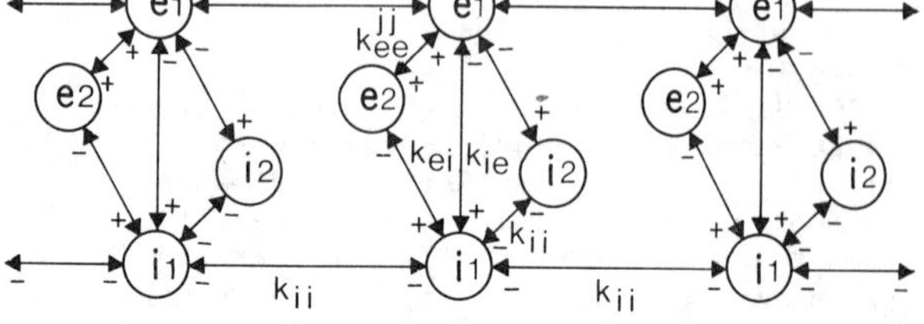

Fig. 2. The inner layers of the bulb.

The inner set receives excitatory input from the assembly, is mutually inhibitory within itself, and is reciprocally inhibitory to the assembly. The reciprocal connection constitutes negative feedback and is responsible for the oscillation in the EEG burst.

Dynamic simulation was undertaken by constructing coupled linear differential equations for an experimentally demonstrated small-signal near-linear range [1]. Open loop impulse responses were obtained under deep anesthesia with suppression of impulse and EEG activity. The data were fitted with sums of exponential basis functions that incorporated representation of afferent dispersion as well as intra-loop delays. On partitioning these elements it was feasible to express the operation of local neighborhoods of the excitatory assembly and the inhibitory set each with a second-order differential equation. The open-loop time and space constants were also evaluated experimentally.

The operation of conversion at synapses of axonal pulse activity to dendritic wave activity was represented by state-dependent gain coefficients, k_{ij}. The operation of conversion of summed dendritic currents to axonal pulse trains at trigger zones was represented by a gain coefficient K_j in piece-wise linear approximation. The solutions to the closed-loop equations with impulse input sufficed to simulate bulbar and cortical impulse responses in the analysis of closed loop dynamics.

Simulation of burst activity required analytic representation of the nonlinear operation at trigger zones. This development required explicit recognition that the module of bulbar function relating to the EEG is not the single neuron but the neural assembly. This is a macroscopic entity that forms by virtue of dense synaptic interactions among large numbers of neurons, each of which transmits to and receives from many thousands of other neurons of like kind in a region comprising a local neighborhood [1].

At this hierarchical level the state variables become pulse density and wave density. Pulse-to-wave conversion is linear and static over short time periods but is subject to step change in slope under extra-bulbar controls relating to learning behavior. Wave-to-pulse conversion is also static but is nonlinear; the operation can be represented by a sigmoid function derived in part by a statistical mechanical generalization from the Hodgkin-Huxley system. The derivative of this function times a coefficient u_o relating to the level of arousal gives the nonlinear gain. Forward gain is given by the product of the two conversion gains [2].

The critical feature of the wave-to pulse conversion is the displacement of the maximal gain to the excitatory side from rest. This implies that excitation and gain are coupled, leading to the possibility of a regenerative increase. The critical synapse relating to this feature is the mutually excitatory connection designated k_{ee}. Owing to the location of the nonlinear operation at trigger zones, any increase in mutually excitatory gain results also in increased negative feedback gain. The result is clear; sustained excitatory input to the bulb as during inhalation can destabilize it and cause it to switch from a quasi-equilibrium state to a limit cycle state, and back again upon cessation of the input [3].

Simulation of the EEG burst was done with an array of 36 simulated local neighborhoods (6x6) with a toroidal boundary condition, steady-state initial conditions corresponding to normal bias, and input corresponding to the time course of normal receptor input during inhalation [4]. When the values for k_{ee} were uniform over the set, then the amplitude pattern of burst output corresponded to the input pattern (Fig 3A). When values of k_{ee} were increased between selected neighborhood to form a spatial pattern of arbitrary shape, the amplitude pattern of output corresponded to the shape of preselected elements and not to the input (Fig 3B).

An Adaptive Spatial Filter in the Bulb

The mutually excitatory gain k_{ee} has already been identified in experimental animals as the synaptic locus of a change with learning. When animals were trained to respond behaviorally to the electrical stimulus used as a conditioned stimulus as well as to evoke impulse responses in the olfactory cortex and bulb, the responses (measured by fitting sums of damped sine waves to averaged evoked potentials) underwent a characteristic and permanent pattern of change in amplitude, frequency, phase and decay rate. This configuration was simulated with solutions of the lumped nonlinear model, only by a small increase in k_{ee}.

The pattern of change in impulse responses with habituation to the electrical stimulus differed from that for attention; moreover it was rapidly and completely reversible. It was simulated by a 40% decrease in both k_{ee} and k_{ei} together; it could be made concomitantly with the change for attentiveness and, in fact, was found in the model to be all the more effective in response suppression.

These and related findings led to the conclusion that the bulbar EEG focus of high amplitude electrical potential is the manifestation of activity of an interactive nerve cell assembly as first postulated by Hebb in 1949 [6]. It is proposed that on each trial in which an odor is presented followed by an unconditioned stimulus, those bulbar cells in the excitatory assembly that are coactive strengthen the synapses of axon collaterals interconnecting them. Over a series of several hundred odor inhalations during a training session the subset of more strongly interconnected bulbar cells forms a "template" that incorporates all covariant cells during the training session (Fig 3A). Thereafter, if input is given to any cells within the subset, the entire subset becomes active in a stereotypic spatial pattern (Fig 3B).

This mechanism can solve the problem of stimulus equivalence in olfaction; an estimated 10^6 to 10^7 receptors may be sensitive to a test odor, yet an invariant behavioral response can occur on activation of any small subset of them in arbitrary combinations. Simulation of cognitive operations of abstraction, generalization, figure completion, and association can readily be seen to follow. The template can be viewed as an adaptive spatial filter to enhance selectively the response to certain inputs, or alternatively, to selectively suppress the response by the mechanism of habituation. The EEG burst manifesting the

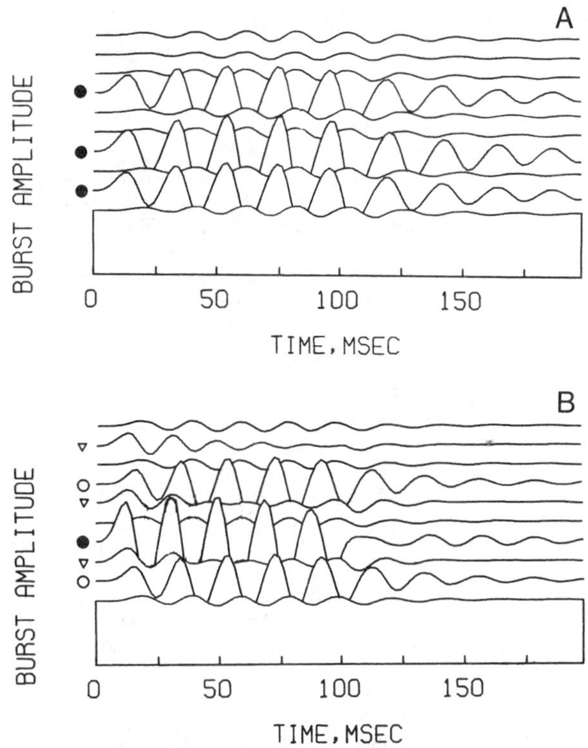

Fig.3 A. The output is shown for 10 elements of a simulated neural assembly, 3 of which receive simulated input. The coupling coefficients k_{ee} for mutual excitation are increased thereafter in proportion to the matrix of correlation coefficients of the outputs.

B. The model that now has a template corresponding to the previous input receives input to one of the three template elements. The output amplitude pattern reflects the template and not the input. From Freeman, 1979c.

spatially patterned "wave packet" can be regarded as an internal representation of that which is expected or sought, and may be linked to the "search image" which ethologists postulate as a basis for goal-directed behavior.

There is no anatomical or physiological evidence for the existence of patterned axonal input to the olfactory system from other parts of the brain ("centrifugal controls"), so that during a training session the bulb cannot be instructed on which input to select. The command must be given to form a template nonspecifically for whatever input is present at the time. Hence the template must include cells sensitive to background odors as well as whatever odor the experimenter chooses to regard as foreground. A subtemplate must thereafter be shaped by selective habituation.

EEG recordings during tests of trained animals suggest that the most common mode of operation in simple discriminations involves selective habituation of the subtemplate for the test odor and not that for the background. The animals then test for maintenance of the background odor, as shown by the occurrence of a burst on most inhalations. Arrival of a warning odor (and to a lesser extent of an unexpected or novel odor) then results in burst suppression rather than enhancement. Burst enhancement (increased spatial coherence) occurs with a test odor in more complex discriminations. The value of these modes is obvious as a broad-band receiving system that retains focussed sensitivity.

Design Considerations for VLSI

The major feature of the dynamics in operation is the destabilization by input that causes a state transition from quasi-equilibrium (actually a low-level chaotic state in the bulb) to a strong, spatially coherent limit cycle. This emergence of order from chaos gives the mechanism several desirable properties. First, it ensures the dominance of the intrinsic state of the system in the selection of input and the shaping of output, while forcing dependence on the input at the critical moment of nucleation of the wave packet. Second, the wave packet is a global event that is shaped in its larger aspects, especially its coherence, by the interrelatedness of the elements of the input and not by the elements in sum or in isolation. Third, synchronization within the wave packet is ensured by the mechanism; in relation to external assemblies the time scale becomes macroscopic and noncritical, a matter of gating input and adjusting sensitivity to bifurcation. Fourth, considering that the output undergoes spatial and temporal integration over the span of the wave packet, the system appears to be enormously tolerant of unpredictable variations of input, internal parameters, and centrifugal controls, as well as irregularities in the spatial and temporal modulation of the amplitude and frequency of the carrier wave of the burst (see also [7]).

The advantages of the transition to a limit cycle rather than to another equilibrium state are equally obvious. First, the basis for the synaptic change with attention is postulated to be correlation; simulations have shown that elements receiving input during a training trial have phase lead over the bulk of the ensemble of roughly 90°, so the oscillation may be optimal for selective synaptic enhancement. Second, the read-out is by spatial integration over coupled oscillators, so that changes in coupling strength leading to phase or frequency convergence or dispersion with "wanted" or "unwanted" input give the basis for bidirectional evaluation of the output. Third, there would appear to be a rapid quenching and return to the receptive state, so that a new sample of input can be taken.

The basic carrier frequency of 40-80 Hz depends on the passive membrane time constant of neural membranes (roughly 5 msec times 4 for two passages around the negative feedback loop). Empirically it appears that 4 to 8 cycles minimally suffice for each burst, which limits the frame rate of this sampled data system to about 10/sec; the bulb is commonly driven by respiration in rabbits at 3-7/sec. Given the constraints of biological materials nature has done well. Provided that the problems of connectivity can be solved, the operation of a heuristic solid-state device constructed to have these emergent properties but with shorter time constants might be accelerated by several orders of magnitude. Synchronization might then indeed become a problem.

References

1. Freeman, W.J. Mass Action in the Nervous System. New York, Academic Press, 1975.
2. Freeman, W.J. "Nonlinear Gain Mediating Cortical Stimulus-Response Relations." Biol. Cybern. 33:237-247, 1979a.
3. Freeman, W.J. "Nonlinear Dynamics of Paleocortex Manifested in the EEG." Biol. Cybern. 35:21-37, 1979b.
4. Freeman, W.J. "EEG Analysis Gives Model of Neuronal Template-Matching Mechanisms for Sensory Search with Olfactory Bulb. Biol. Cybern. 35: 221-234, 1979c.
5. Freeman, W.F. and Schneider, W. "Changes in Spatial Patterns of Rabbit Olfactory EEG with Conditioning to Odors." Psychophysiology 19:44-56, 1982.
6. Hebb, D.O. Organization of Behavior. New York, John Wiley, 1949.
7. d'Humières, D. and Huberman, B.A. "Dynamics of Self-Organization in Complex Adaptive Networks." To be published, 1983.
8. Supported by a grant MH06686 from NIMH.

A STRUCTURE FOR GENERATION AND CONTROL OF INTELLIGENT BEHAVIOR

James S. Albus

National Bureau of Standards
Washington, DC 20234

ABSTRACT

In this paper is presented a hierarchical architecture which has the ability to generate and control intelligent behavior. Three parallel cross-coupled hierarchies are proposed: (1) a task or goal decomposition hierarchy, (2) a sensory processing hierarchy, and (3) a world modeling hierarchy. The upper levels of these hierarchies have the ability to select goals, evaluate results, and generate plans. Intelligence is defined to be the set of computing mechanisms that enable an organism or a machine:

(1) to select good goals, and

(2) to act in a manner which tends to optimize the probability of success in achieving the selected goals.

GOAL SELECTION

The above definition of intelligence implies a mechanism for evaluating goals as "good" or "bad." In biological organisms, the goal evaluation function is provided by the limbic system. The limbic system produces rewarding signals in response to biologically "good" actions or situations, and punishing signals in response to biologically "bad." These signals steer the behavior generating system to select goals which tend to maximize the probability of the individual creature propagating its genes.

In artificial machines, good goals are those which fulfill the intent of a designer or programmer/user. It is possible for a machine, such as a robot or an Artificial Intelligence (AI) computer, to have an internal goal-selecting mechanism. If it does, it will exhibit autonomous behavior.

Most robots, however, are simply designed to accept goals as input commands from a higher level control mechanism. This is certainly the case of industrial robots today. What is required of industrial robots is consistency, speed, accuracy, and reliability. The manufacturing task is well defined at the factory floor level, and all goal selection is done at much higher corporate levels.

The higher level command input that an industrial robot receives may be as simple as a switch closure which initiates the execution of a sequence of program steps. Alternatively, the higher level command input may be as complex as a symbolic string which contains a plan (or the instructions to generate a plan) for some task of manipulation, locomotion, or perceptual interpretation.

TASK DECOMPOSITION

One fundamental element of intelligence is task (or goal) decomposition. Task decomposition is the ability to break complex tasks into sequences of simpler actions. Task decomposition implies a hierarchy of control levels which successively decompose high level tasks (or goals) into lower level actions. Finally, at the lowest level there are produced drive voltages to motors, valves, or biological muscles.

A hierarchical architecture for a factory control system is shown in Figure 1. A single chain of command from the bottom to the top of such a hierarchy is outlined by the dotted line in Figure 1. This chain of command can be further segmented, as shown in Figure 2, into three separate hierarchies: (1) a goal, (or task) decomposition hierarchy (H); (2) a feedback processing hierarchy (G); and (3) a world model hierarchy (M). This has been discussed in a number of previous papers. [1-5]

In the robot control system architecture shown in Figure 2, the bottom (or first) level of the task decomposition hierarchy is where coordinate transforms and servo computations are made.

At the second level, elemental movements such as <REACH>, <GRASP>, <MOVE TO X>, <RELEASE>, etc., are decomposed into position, force, and velocity trajectories in a convenient coordinate system.

At the third level, simple tasks such as <FETCH A>, <MATE B TO A>, <LOAD TOOL C WITH PART D>, etc., are decomposed into elemental movements.

At the fourth level, commands such as <MACHINE BATCH X>, or <TRANSPORT BATCH Y TO WORKSTATION Z>, are decomposed into simple tasks.

At the fifth level, production schedules containing commands such as <MANUFACTURE P AMOUNT OF PRODUCT Q> are decomposed into batching, routing, and batch processing commands to the fourth level.

At the sixth level, orders from customers, part designs, and process plans provide input to computing mechanisms that generate production schedules, maintain inventory, and allocate manufacturing resources such as tools, equipment, and manpower.

At the seventh and highest level, corporate policy is translated into product designs, manufacturing techniques, and marketing strategies.

FEEDBACK

A second fundamental element of intelligent behavior is feedback control. This is the ability to modify actions in response to sensed information about the task environment. Each level of the task decomposition hierarchy shown in Figure 2 is serviced by a feedback processing module which extracts from the sensory data stream and from lower level control modules the information needed for control decisions at that level. At the lowest level of the robot hierarchy, the feedback processing modules extract joint position and force data to be used for servo and coordinate transformation computations.

At the second level, touch and simple visual measurements of distance and positions of grip points are used to compute trajectory end points.

At the third level, three dimensional positions of object features such as edges, corners, holes, and surfaces are used to sequence elemental movements.

At the fourth level, recognition of objects and relationships between objects are used to sequence simple tasks.

At each successively higher level, feedback conveys information about more global, and/or more abstract, parameters of the environment.

WORLD MODELING

A third fundamental element of intelligence is the ability to store knowledge about the environment and use it to generate actions and predict what sensory experiences should be expected when those actions are carried out. In Figure 2, the world model M modules contain data and procedures that allow them to predict, based on what task is being executed, what the sensory data entering the corresponding G modules should be. This allows the G modules at each level to compare predictions with observations, and to measure both the degree of correlation and the degree of difference. A strong degree of correlation means that the proper model is being matched with the incoming sensory data. This means that the observed object or situation has been correctly recognized.

A large degree of difference between expectations generated by the model and observations derived from sensors means that either an incorrect choice of models has been made, or the model has not been correctly transformed spatially or temporally so as to generate the proper set of expected feature relationships, or that the incoming sensory data is too noisy, or is being improperly processed and filtered. In this case, the computational problem for the task decomposition module is to decide which type of error is being encountered and what is required to remedy the discrepancy.

At the lowest level, the model generates windows for filter functions that are used to screen and track regions of interest in the raw data.

At the second level, the model generates expected positions and orientations of features such as edges, corners, surfaces, and holes.

At the third level, the model generates expected positions and orientations of whole objects.

At the fourth (or Work Station) level, the model generates expected geometrical relationships between objects such as the expected layout of trays, or the stacking relationships of parts.

At the fifth (or Cell) level the model contains information about expected task times to be used in organizing parts into batches and routing batches between workstations.

At the sixth (or Shop) level the model contains information about machinability of materials, tool life, and availability of machine tools to be used in long term scheduling.

At the seventh (or Facility) level, the model contains information required to support design and process planning activities and management decisions made at that level.

PLANNNING

A fourth fundamental element of intelligence is the ability to plan action before it is initiated. Planning can be defined as conducting a search over a space of possible behavior patterns prior to action in order to find an optimum sequence of actions that is likely to achieve a goal.

Planning is similar to goal decomposition in that it generates a string of sub-goals in response to an input command. It is different in that planning takes place in anticipation of action, whereas goal decomposition occurs while action is on-going. Goal decomposition is often simply the execution of a previously defined plan.

Planning implies the ability to anticipate the potential results of a series of actions, evaluate those results, and compare that evaluation against other possible action sequences.

At each level of the hierarchy in Figure 2, a planning mechanism can use the control H module to generate a sequence of hypothesized actions; the world model M module can use these hypothetical actions to generate expected results, and the sensory G module can process and evaluate the expectations. Another string of actions can be hypothesized, and the process repeated. By this method a search over the space of possible action sequences can be conducted, and the sequence evaluated as most desirable can be chosen as the plan to be executed.

In general, the response time and cycle time requirements for real-time control of ongoing action grow longer at each successively higher level of the hierarchy. Thus, the amount of computing power needed for task decomposition decreases at successively higher levels of the hierarchy. On the other hand, there is much more need for planning at the upper levels, and the space of possible action sequences grows larger at the higher levels. Thus, the total amount of computing power needed at all levels of the hierarchy is more or less constant.

PROSPECTS FOR VLSI

The hierarchical architecture described in this paper has many requirements for special computing architectures. For example, a pipeline vision system which has several levels of world model modules, each of which generate expected images to be matched against observed images, and each of which contain feedback loops to modify the model parameters based on the difference between expectations and observations, has been proposed and is being built.[1,6] The looping activity between the G and M modules at each level constitutes a relaxation process (or in the case of a dynamically changing scene, the image equivalent of a phase lock loop). The implementation of such systems for real-time control of robots will require computing mechanisms of extraordinary power and speed. Large amounts of parallel processing will be needed and a wide variety of special purpose hardware amenable to fabrication in VLSI will be required.

ACKNOWLEDGEMENTS

The author would like to thank Marty Auman for her assistance in preparing the manuscript.

REFERENCES

1. J.S. Albus, C.E. McLean, A.J.Barbera and M.L. Fitzgerald, "Hierarchical Control for Robots in an Automated Factory." Proc. 13th/ISIR Robots 7, Vol. 2, 13:29-43, Apr. 1983.

2. J.S. Albus, A.J. Barbera and M.L. Fitzgerald, "Hierarchical Control for Sensory Interactive Robots." Proc. 11th Int'l Symposium on Industrial Robots, Tokyo, Japan, Oct. 1981.

3. J.S. Albus, A.J. Barbera, and M.L. Fitzgerald, M.Nashman, "Sensory Interactive Robots." Proc. 31st Gen. Assembly Int'l Institution for Production Engineering Research (CIRP), Toronto, Canada, Sept. 1981.

4. J.S. Albus, A.J. Barbera, and R. N. Nagel, "Theory and Practice of Hierarchical Control." Proc. 23rd IEEE Computer Soc., Int'l Conf., pp 18-39, Sept. 1981.

5. A.J. Barbera, M.L. Fitzgerald, and J.S. Albus, "Concepts for a Real-Time Sensory-Interactive Control System Architecture." Proc. 14th Southeastern Symposium on System Theory, pp 121-126, Blacksburg, VA, Apr. 1982.

6. E.W. Kent and J.S. Albus, "Servoed World Models as Interfaces Between Robot Control Systems and Sensory Data." Robotica, (in press).

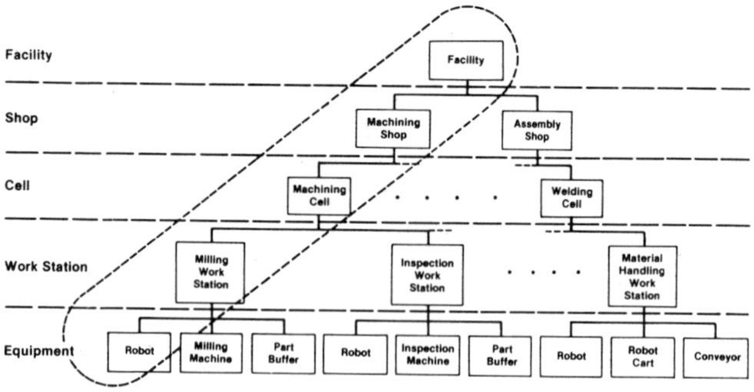

Fig. 1. A hierarchical control system architecture for an automatic factory.

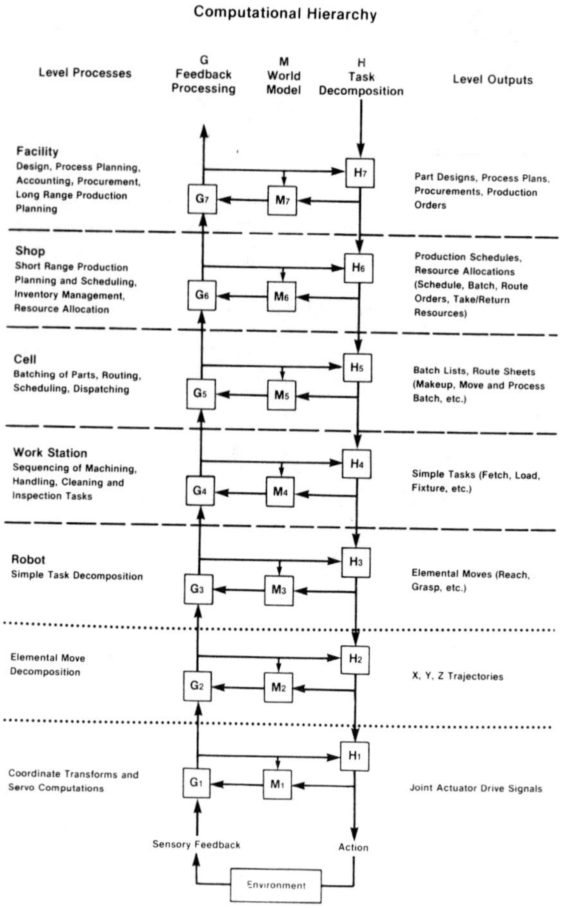

Fig. 2. The computational hierarchy for a robot corresponding to the chain of command enclosed in dotted lines in Figure 1.

Session 1-1a
VLSI Technology I

Chairman

R. Sullivan
Honeywell

IMPLICATIONS AND PROJECTIONS OF GALLIUM ARSENIDE TECHNOLOGY IN HIGH SPEED COMPUTING

R.C. Eden, B.M. Welch, F.S. Lee

GigaBit Logic, Inc.
Westlake Village, CA 91361

Abstract

The high electron mobility and the availability of the semi-insulating material of GaAs, make possible the development of GaAs IC technologies for very high speed circuit applications. Excellent switching speed (gate propagation delay time of 22.5 ps to 75 ps) have been demonstrated in MSI GaAs frequency divider circuits. LSI level, very high speed GaAs ICs, such as 1K static RAMS, 16x16 bit parallel multipliers, 32 bit ALUs etc., have also been successfully fabricated. With the high speed performance and the high level integration demonstrated to date in the GaAs IC technologies, the overall system performance of a high speed computer system should be greatly improved.

Introduction

Throughout the history of high speed electronic computers, the computational throughput or cycle time performance of the machines has fairly closely tracked the performance of the logic switching and memory devices available for this fabrication. What this means is that as the device technology changed from vacuum tubes to germanium transistors to silicon transistors to various forms of silicon integrated circuit logic, the computer designers have been able, by means of improvements in architecture, packaging and interconnection technologies, etc., to speed up the overall computer cycle times nearly apace with the device improvements. Ie., in each new generation of computers, the speeds are improved to the point where a substantial fraction of the overall machine cycle time continues to be taken up by the logic and/or memory device switching times ("semiconductor delay" in current parlance), even though simply substituting the faster devices in the previous generation machines (with no improvements in packaging etc.) would have realized only modest performance gains.

If we assume, based on extrapolation of past performance, that computer designers will (despite protestations of how difficult the task is) continue to be able to track computer performance with improved logic or memory device performance, then we see how important a new, much higher speed integrated circuit technology such as Gallium Arsenide could be to the high speed computer industry.

Because of the excellent electron dynamics in Gallium Arsenide (low electron effective mass, high electron mobility, and high electron velocities), aided by its semi-insulating substrate material, GaAs devices offer very high f_T's (current gain - bandwidth products), and GaAs digital integrated circuits show switching speeds from 3 to 10 times faster than silicon IC's with similar geometries. The economic implications of realizing these types of performance enhancements in computers would, of course, be enormous. This has motivated the Japanese mainframe computer manufacturers to generate a major thrust in GaAs integrated circuit technology development aimed at new generations of computers, with impetus in part from a large government (MITI)-funded program.

Device Approaches for GaAs IC's and Current Performance Results

In order to understand the directions and strategies being pursued for GaAs ICs in computers, it is worthwhile to review some of the principal device approaches for GaAs integrated circuits, their characteristics, and the performance results obtained to date with each. The depletion-mode Schottky - gate (metal - Semiconductor gate) field effect transistor (D-MESFET) has been pursued for GaAs IC's for about 10 years, and is clearly the leader in terms of manufacturability and commercial availability. Many circuit approaches have been used with D-MESFETs, including buffered FET logic (BFL), Schottky diode - FET logic (SDFL), capacitor-coupled FET logic (CCFL), source - coupled logic (SCL) and others. Circuits as large as 1008-gates have been fabricated with SDFL[3], and D-MESFET random-access memories using a novel 3-FET cell[4] with capacities up to 256 bits have been fabricated, with access times as short as 1ns claimed[5], with efforts toward 1K, 4K and 16K-bit versions in progress. The GaAs D-MESFETs give excellent performance, with current gain-bandwidth products for standard Lg=1μm devices of the order of f_T= 15 to 20 GHz. In minimally loaded BFL or SDFL ring oscillators, gate propagation delays as low as $\tau_d \sim 50$ ps have been obtained at Lg=1μm (at $P_D \sim 1$m

W/gate for SDFL, higher for BFL). In a standard edge triggered type-D flip-flop circuit (complement to a TTL 7474), toggle frequencies in ÷2 mode ($\bar{Q}$ connected to D_{in}) for L_g=1μm SDFL circuits of the order f_c=2.6 to 2.9 GHz have been obtained. This corresponds to τ_d=70-75ps (at $P_D \cong$3m W/gate) in this $(f_c)_{max}=1/(5\tau_d)$ divider architecture for these SDFL D-MESFET circuits, only about a factor of 1.3 to 1.4 times the minimally-loaded ring oscillator results.

The D-MESFET circuits generally require two power supplies for proper operation (for example, the SDFL circuits typically use V_{DD}=2.5V, V_{SS}=-1.5V supplies), and have some kind of voltage level shifting built into the logic gates to generate from the (positive) drain voltage of these N-channel MESFETs, voltages more negative than the pinchoff voltage, V_p, the D-MESFET devices (typically V_p=-1.0V for SDFL, for example). One approach which avoids the need for an extra power supply to the chip and level shift circuitry on chips is to use MESFETs with slightly positive (typically $V_p \cong$+0.1V) pinchoff voltages. These enhancement-mode MESFETs (E-MESFETs) are usually used in direct-coupled FET logic (DCFL) circuit approaches, but are restricted to very limited logic voltage swings (typically $\cong$0.5V) because the MESFET begins to draw excessive gate currents for V_{gs}> 0.7 or 0.8V, effectively clamping the swing. This necessitates achieving pinchoff voltage uniformities of the order of $\sigma_{Vp} \cong$ 0.025 V in order to achieve LSI complexity circuits[1] (the logic swing must exceed the standard deviation of V_p by about a factor of 20). This uniformity has proven exceedingly difficult to achieve on any consistent basis (particularly since the mean V_p for the wafer must be correct within 0.05 or 0.1V as well). However, Fujitsu has demonstrated σ_{Vp}=0.022V uniformities over the central portion of E-MESFET wafers and have demonstrated fully working 3168-gate (10,000 transistor) E-MESFET 16x16 bit parallel multiplier chips[6] and also fully working 1K bit x 1 bit GaAs E-MESFET RAMs.[7] The equivalent gate delay in the 16 x 16 multiplier was 150ps/gate, the same as obtained for the SDFL 8x8 bit D-MESFET chip, and at a lower power/gate (~0.3 m W/gate at V_{DD}=1.6V for the DCFL E-MESFET, as compared to about 1mW/gate for SDFL gates in D-MESFET multipliers).

In addition to the problems of severe uniformity requirements for E-MESFETs, manufacturability is hindered also by more complex device processing requirements than D-MESFETs. The problem with E-MESFETs (or very low pinchoff voltage, $V_p \cong$ -0.5V, D-MESFETs) is that the substantial depletion region at the surface of the GaAs between the source and gate (or gate and drain) tends to pinch the very lightly - doped channel off, leading to high (sometimes almost infinite) source resistance values which degrade the FET transconductance, g_m. One way to avoid this is to recess the gate slightly by etching into a thicker channel, but this can aggravate V_p uniformity problems. A self-aligned implant approach for these S-G and G-D gap regions using a refractory metal (eg. W-Si) has been demonstrated by Fujitsu in their work.[6,7] Some diffusion or implant straggle in this process tends to degrade the performance of these self-aligned implant devices at short gate lengths, but a number of labs are working on ways to avoid this problem. The highest performance E-MESFET ring oscillators were obtained at Cornell[8] with T-shaped gates (~0.1μm overhang) with L_g=0.55μm at top) shadowing the S-G and G-D channel implant. These circuits gave gate delays as low as τ_d=15ps @ 5.6m W/gate or τ_d=24ps @ 0.4mW/gate in minimally-loaded E-MESFET ring oscillators. Note, however, that enhancement-mode ring oscillators have very low parasitic capacitances. Typical enhancement-mode MSI circuits with normal fanouts (2-3) run with gate delays about 3 times the minimally - loaded ring oscillator delays (as compared to a degradation factor of 1.3 to 1.4 for D-MESFET circuits).

A variant of enhancement-mode GaAs FET technology that offers substantial performance improvements over the standard MESFET devices, particularly at low temperatures, is called the high electron mobility transistor, or HEMT (also referred to as TEGFET, SDHT, etc.). This device structure, usually fabricated by molecular beam epitaxy, uses an undoped GaAs chennel covered with an ~700A° thick doped $A_{L_x}Ga_{1-x}As$ layer under the metal gate which serves to furnish the electrons for the channel, and also acts as a gate insulator, so that the device acts somewhat like an N-channel MOSFET (although the gate will begin to conduct at V_{gs}> 1.2-1.5V). The principle advantage of the device is that the electron mobility in the channel is higher in the HEMT than in a MESFET because there are no doping ions in the channel to scatter carriers (electron mobilities of μ_e=8500 cm2/vs are achieved in HEMT's at room temperature, and over $\mu_e \geq$50,000 cm^2/Vs at 77°K, are achieved, as compared with typical GaAs MESFET channel mobilities of 4000-5000 cm^2/Vs at 300°K). This gives the HEMT devices a fast "turn-on" characteristic, that is, they develop nearly their full transconductance with gate logic voltage swings only a small amount above threshold[1] (a few tenths of a volt at 77°K). Further, the thin gate insulator gives these devices comparitively high transconductance values (although this also does not directly improve f_T), and the effective electron velocities achieved are excellent (~3x10^7cm/sec at 77°K; about twice that of room temperature MESFETs) so that the current gain - bandwidth products are also better. The highest switching speeds reported in minimally loaded HEMT ring oscillators to date are by Rockwell,[9] with τ_d=12.2ps obtained with L_g=1μm HEMT DCFL inverters at P_D = 1.1m W/gate. Workers at Bell Labs have fabricared D-FF's using 6 L_g=1μm HEMT DCFL NOR gates (same $(f_c)max = 1/(5\tau_d)$ circuit as used with SDFL) which operate up to f_c=3.7 GHz at room temperature (τ_d=54ps @ P_D=2.4m W/gate) and f_c=5.6 GHz at 77°K (τ_d=36ps @5.1m W/gate). This room temperature HEMT effective propagation delay result for these dividers, τ_d=54ps, is about 25% less than the D-MESFET SDFL gate delays for similar gate lengths and power dissipations in the same circuit, while the 77°K HEMT result (τ_d=36ps) is about 50% faster than the room temperature D-MESFET results. Fujitsu[11] has fabricated some even shorter gate length HEMT divider circuits, also claimed to be of an $(f_c)max = 1/5\tau_d)$ architecture, which gave f_c=5.5 GHz for L_g=0.5μm frequency dividers at room temperature (τ_d=36ps at P_D=3m W/gate), and a remarkable f_c=8.9 GHz in 77°K operation (τ_d=22.5ps at P_D=3mW/gate). When you pause to consider that this 8.9 GHz GaAs divider result is 350 times the toggles frequency of a standard 7474 TTL D flip-flop, and has lower power per gate, or is about 20 times faster than an F100131 ECL D-FF (at one tenth the power), the interest in GaAs for ultra high speed computers becomes clear.

Strategies and Approaches to the Application of Ultra High Speed GaAs IC's for Computers

The achievement of such remarkable performance levels as NOR gate delays in GaAs IC frequency dividers from τ_d=22.5ps to 75ps per gate (at relatively low power levels) is extremely exciting to the computer field, providing that these semiconductor performance increases can be translated into proportional increases in computer speed. Of course, historical precedent notwithstanding, this will not be easy (though the same was probably said of earlier "new generations" of computers). The most serious problem, of course, is the electromagnetic propagation delay along lines interconnecting various portions of the central processing unit (CPU) of the computer. In many computer architectures it is necessary that within one cycle time, in addition to the semiconductor delays, a signal must be able to traverse the full physical extent of the CPU (including cache memories). Inasmuch as mainframe computer CPU sizes have gone up from about 100,000 gates 5 - 10 years ago and are now pushing towards 1 -3 million gate complexities, there is obviously a formidable packing density and heat extraction problem as cycle times approach the few nanosecond level and CPU sizes shrink to of the order of a foot cube. There are, of course, other architectural approaches to computer design, perhaps best typified by the Cray designs, which make use of pipelining, etc., so that a signal need traverse only a small portion of the CPU in any one cycle time. While these concepts are extremely effective in a scientific computational computer such as a CRAY I, their adaptability for general purpose business computers, for example, is less clear. Whether new generations of ultra high speed computers using GaAs IC's for logic and cache memory will achieve their super performance through new architectures or through breakthroughs in high density packaging and interconnection technology, or through a combination of these (perhaps with other factors we haven't considered here, such as software), we certainly cannot predict at this point. The achievement of such performance breakthroughs is initiated by one thing - an idea. The world of business being what it is, innovators are not necessarily well rewarded for sharing such key ideas until they have had time to turn the idea into a product. Whether there are under development at this time promising solutions to this problem is a matter of conjecture. That a great deal of R & D money is being spent on the development of GaAs IC's for ultra high speed computer applications, by both Japanese (Fujitsu, Hitachi, Toshiba, OKI, NEC, NTT, etc.) and U.S. (Cray, IBM, Honeywell, Bell Labs, etc.) computer manufacturers is a matter of fact.

While we may not have access to the details, we can infer some general directions being persued by some of these computer manufacturers from our understanding of the implications of the various GaAs IC technologies and the interest the companies show in these different technologies. Cray, for example, has announced their pilot production/R&D GaAs IC facility activities. A reasonable inference from the present manufacturability status of GaAs IC technologies, as well as from announced business relationships with D-MESFET IC suppliers, that they will be using, at least initially, a D-MESFET approach, which should be quite manufacturable in the near term in at least the 100 gate - 1000 gate LSI chip complexity range. For a company capable of designing and bringing out the fastest cycle time computer in the world (the CRAY I), implemented with only 2 ECL NOR gates per package logic, much faster 10^2-10^3 gate circuits could do wonders for their products. It would appear that the computer design ideas are there, and hence their focus should be on getting the circuits manufactured quickly in order to get a product into the field as soon as possible. Capitalizing on this "now" D-MESFET GaAs IC technology could prove an extremely effective strategy for Cray.

The situation in the principal Japanese computer companies (Fujitsu, Hitachi, etc.) and IBM and Bell Labs in the U.S. appears significantly different. It appears that they see VLSI chip complexity levels in GaAs IC's as a key strategy for reducing the CPU physical size, presumably for implementing an architecture in which it is assumed necessary to have signals traverse the entire CPU in one cycle time. One infers this from the consistent commitment of research activities to more technologically difficult ("further out") GaAs IC technologies, but ones which promise the potential for achieving very low gate dissipation levels with simple circuitry, obvious prerequisites for 10,000- to 100,000-gate VLSI GaAs chips. Fujitsu, for example, has expended great efforts with excellent results in both E-MESFET [6,7] and HEMT [11] GaAs IC technologies, and have published some projections on the applications of GaAs chips for future computer generations[12] (apparantly targeted for the late 1980's). The significant technological problems to be solved before manufacturability of VLSI-complexity E-MESFET or HEMT circuits can be established precludes, of course, a nearterm product entry with this GaAs VLSI technological approach, but these compter manufacturers appear convinced that this is the best approach for their product interests.

One potentially interesting approach which could combine the circuit densities of the VLSI approach for small CPU sizes-without the technological problems of yielding 10^4-10^5 gate/chip VLSI GaAs IC's would be to develop a high resolution (~5μm lines?) multi-layer inter-chip interconnection approach. If one assumes a chip size of 3mm square for a 2000-gate LSI GaAs chip, then theoretically a 10^6-

gate CPU could occupy an area of 2.5 to 3 inches square, which would have ~0.5ns electromagnetic propagation delay from corner to diagonally opposing corner. Developing such an interconnect technology would be challenging, of course, but state of the art computer interconnect problems always are.

Summary (or Moral)

Perhaps the thesis of this paper can best be summarized as follows:

If a computer manufacturer concludes that the ultra high speed GaAs logic and subnanosecond memory ICs which are becoming available are of no value in his products because the electromagnetic propagation delays make the semiconductor delays irrelevant, then nothing happens to his products.

If a computer manufacturer accepts on principle that faster logic and RAM circuits are good for his products and makes the committment to get these GaAs IC technologies in hand, then he is only an idea away from a new generation of ultra high speed computer.

References

1. R.C. Eden, "Comparison of GaAs Device Approaches for Ultrahigh-Speed VLSI," Proceedings of the IEEE, Vol 70, No.1, Jan. 1982, pp. 5-12. See also pp. 3-4.

2. P.M. Solomon, "A Comparison of Semiconductor Devices for High Speed Logic," Proc. of the IEEE, Vol. 70, No.5, May 1982.

3. F.S. Lee, G.R. Kaelin, B.M. Welch, R. Zucca, E. Shen, P.Asbeck, C.-P. Lee, C.G. Kirkpatrick, S.I. Long, and R.C. Eden, "A High-Speed LSI GaAs 8x8 Bit Parallel Multiplier," IEEE J. Solid State Circuits, Vol.SC-17, No.4, Aug. 1982, pp. 638-647.

4 G.R. Kaelin, R.P. Vahrenkamp, F.S. Lee, S.J. Lee, B.M. Welch and R.C. Eden, "An Ultra-Low Power, Fast GaAs RAM for Space Applications," Proceedings of 1982 First Annual Pheonix Conf. on Computers and Communications, May, 1982 (IEEE Cat.82-ch1756-6), Paper D5-2, pp. 414-416.

5. S.J. Lee, R.P. Vahrenkamp, G.R. Kaelin, R. Zucca, L.D. Hou, C.P. Lee and C.G. Kirkpatrick, "An Ultra-Low Power, High Speed GaAs 256-Bit Static RAM with Depletion Mode MESFETS," 41st Annual Device Research Conference, June 20-22, 1983, Burlington, Vt, Paper IVA-4.

6. Y. Nakayama, K. Suyama, H. Shimizu, N. Yokoyama, A. Shibatumi, H. Ishikawa, "A GaAs 16x16 Bit Parallel Multiplier using Self-Alignment Technology," Technical Digest of 1983 Intl. Solid State Circuits Conf, New York, Feb. 23, 1983, Paper WAM 3.6, pp. 48-49.

7. N. Yokiyama, T. Ohnishi, H. Onodera, T. Shinoki, A. Shibatumi, H. Ishikawa, "A GaAs 1K Static RAM using Tungsten-Silicide Gate Self-Alignment Technology," op. cit. Paper WAM 3.4, pp. 44-55, p283.

8. R.A. Sadler and L.F. Eastman, "Self-Aligned Submicron Ion-Implanted GaAs MESFETs for High Speed Logic," 41st Annual Device Research Conf., June, 1983, Burlington, VT.) Paper IVA-2.

9. C.P. Lee, D.L. Miller, D.Hou and R.J. Anderson, "Ultra High Speed Integrated Circuits Using GaAs/GaAlAs High Electron Mobility Transistors" op.Cit., Paper IIA-7.

10. R.A. Kiehl, M.D. Fever, R.H. Hendel, J.C.M. Hwang, V.G. Keramidas, C.L. Allyn, and R. Dingle, "Selectively Doped Heterostructure Divide-By-Two Circuit," op. cit., Paper IVA-3.

11. K. Nishiuchi, T. Mimura, S. Kuroda, S. Hiyamizu, H. Nishi and M. Abe, "Device Characteristics of Short Channel High Electron Mobility Transistor (HEMT)," op. cit., Paper IIA-8

12. T. Misugi, K. Kurokawa and M. Mukai, "Perspective of S_i and GaAs Devices for Computer Application," High-Speed Digital Technologies Conf. at San Diego, Jan. 1981.

HMOS III TECHNOLOGY AND ITS APPLICATIONS

S. S. Liu, G. E. Atwood, H. Dun, S. L. Smith, and N. Hall

Intel Corporation
Santa Clara, California 95051

ABSTRACT

A third-generation single-poly high-performance NMOS technology has been developed. The technology features scaled MOS devices with 250 Å gate and 0.25pJ speed-power product. The technology integrates advanced processing capabilities in lithography and dry-etching to achieve a depletion-load static memory cell size of 0.98 mil^2. The technology has a major application in the shrink of existing micro-computers designed on previous HMOS technologies.

The scaling of the MOS transistors has been exercised over the past few years, first in 1977[1] and then 1979[2], to achieve dramatic improvement in device performance without the necessity of reducing the supply voltage. Continuing the effort in this direction, a third-generation high-performance NMOS technology, namely HMOS III, has been developed[3],[4]. By integrating state-of-the-art processing capabilities, the technology offers significant enhancement in both density and performance over its predecessors.

For processing simplicity and synergy, the technology utilizes a single layer of poly silicon with depletion load. The MOS device scaling was realized through the reduction of gate oxide thickness to 250 Å and channel length to 1.5μ. High-resistivity substrate of 50Ω-cm has been used to minimize the junction capacitance and body-effect. By properly adjusting the channel implant dosage, the threshold voltage of the minimum enhancement device was maintained at 0.7^V. In addition, the processing of the scaled transistors has been optimized to provide a Miller Capacitance of 0.4ff/μ, and a saturation current of 27mA for the minimum channel length device.

The performance of the HMOS III devices can be best illustrated by the results obtained from an 11-stage ring-oscillator with a fan-out of 3. As shown in Figure 1, the technology offers a typical gate delay of less than 200ps at room temperature , with a speed-power product of 0.25pJ. This is a factor of two better than the 1979 HMOS II technology. The comparison of key device parameters for the three generations of HMOS technologies is summarized in Table 1.

The technology features 1.5μ design rules, with combined usage of wafer stepper lithograhpy and dry-etching to achieve the stringent requirements on layer-to-layer registration and dimensional control. To be consistent with transistor scaling, the process is also scaled in the vertical dimension. A smaller bird's beak was formed by reducing the field isolation thickness, and a junction depth of less than 0.3μ was obtained by doping the source-drain with arsenic while maintaining low temperature processing.

A 1024X4 bit static RAM has been designed as the development vehicle for the technology. Using conservative layout, the device occupies an area of 13650mil^2 with a memory cell size of 0.98mil^2 (20μx31.75μ). A die photo of the 4K static RAM is shown in Figure 2.

One of the major applications of the technology is for the direct photographic die shrink of microprocessors fabricated on earlier technologies. Because of the excessive random logic layout involved in a typical microprocessor design, upgrading the existing products through this shrink path is highly desirable. To ensure the success of the shrink, the technology must be capable of generating a set of design rules which would accomodate the proportionate reduction of all geometries used on the current design at a fixed ratio. Table 2 listed the comparison of the key design rules between this technology and its two predecessors. A linear shrink factor of 0.8 is allowed from the HMOS II technology; whereas a dramatic reduction of 36% linear can be realized by shrinking directly from the 1977 HMOS I technology. Using the technology, several shrink designs have been realized. One example is illustrated in Figure 3, where the die size of a microprocessor designed on HMOS I was reduced from 110000mil^2 to 45000mil^2.

In summary, a third-generation high-performance NMOS technology has been developed featuring scaled MOS transistors with gate oxide thickness of 250 Å. With a speed-power product of 0.25pJ, the technology offers a significant performance improvement over its predecessors. The main application of the technology lies in the area of

CH1935-6/83/0000/0034$01.00

microcomputer shrink, where direct photographic die size reduction has been realized on both 1977 and 1979 designs.

ACKNOWLEDGEMENTS

The authors appreciate the efforts of J. Harmann, I. Hurst in assisting the process development, and R. Fulk, T. Pham, W. Richardson, B. Verzi in supporting the electrical analysis.

REFERENCES

(1) Pashley, R.D., et al., "A High Performance 4K Static RAM Fabricated With An Advanced MOS Technology," ISSCC DIGEST OF TECHNICAL PAPERS, P22-23; Feb., 1977.

(2) Jecmen, R.M., et al., "A 25ns 4K Static RAM," ISSCC DIGEST OF TECHNICAL PAPERS, P100-101; Feb., 1979.

(3) Liu, S.S., et al., "HMOS III TECHNOLOGY," ISSCC DIGEST OF TECHNICAL PAPERS, P234-235; Feb., 1982.

(4) Liu, S.S., et al., "HMOS III Technology," IEEE J. SOLID STATE CIRCUITS, P810-814; Oct., 1982.

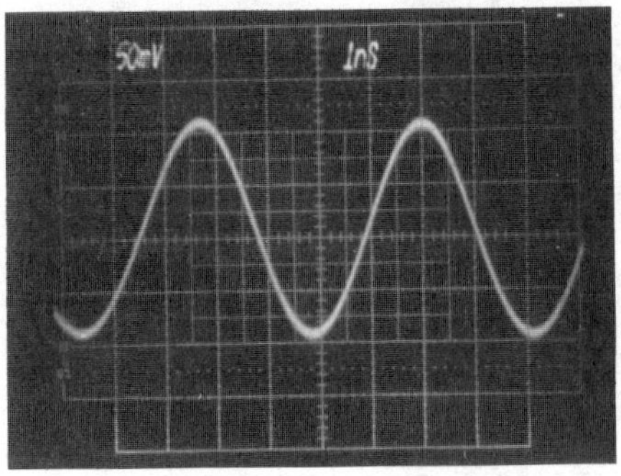

Figure 1: HMOS III RING-OSCILLATOR WAVEFORM

Table 1: Comparison of Key Device Parameters For the HMOS Technologies

	HMOS I	HMOS II	HMOS III
Gate Oxide Thickness (Å)	700	400	250
Channel Length (μ)	3.0	2.0	1.5
Idsat (mA)	8.0	14.0	27.0
Minimum Gate-Delay (ps)	1000	400	200
Speed-Power Product (pJ)	1.0	0.5	0.25

Table 2: Comparison of Key Design Rules

	HMOS I	HMOS II	HMOS III
Diffusion Pitch	8.0μ	6.4μ	5.0μ
Poly Pitch	7.0μ	5.6μ	4.0μ
Metal Pitch	11.0μ	8.0μ	6.4μ

Figure 2: Photomicrograph of 4K Static RAM.

HMOS I

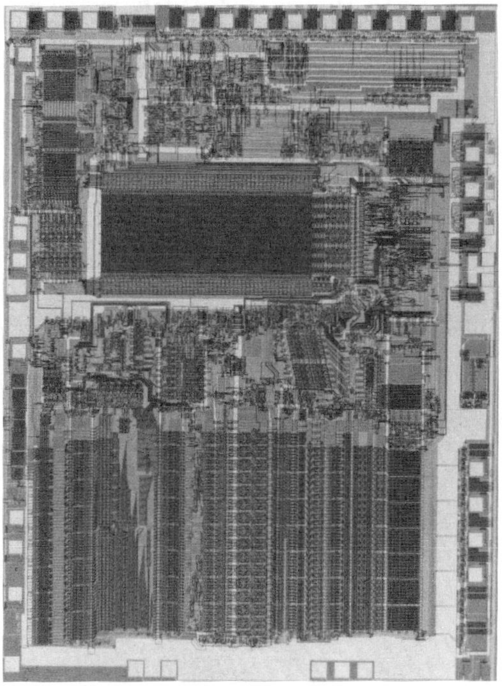

HMOS III

Figure 3: Microprocessor Die Size Comparison, HMOS I vs. HMOSIII

The QUITERON: Novel Superconducting Switching Element

Sadeg M. Faris

HYPRES, Inc. Elmsford New York

The QUasiparticle Injection Tunneling Effect device, QUITERON [1,2], is the first three terminal superconducting device possessing transistor-like characteristics as well as ultra-low swithching energy. The QUITERON action is based on non-equilibrium superconductivity in the reqime of heavy external and self injection of quasiparticles leading to gap suppression and drastic modification of the tunnel conductance.

The device consists of three superconducting electrodes (S1,S2,S3) separated by two tunnel barriers (I1,I2). The S1/I2/S3 junction, known as the acceptor, is a switch that is controlled by the S1/I1/S2 injector junction. Heavy quasiparticle injection supresses the gap of the thin film S2 to zero when a threshold power density, Pc, is exceeded. This changes the acceptor from and SIS junction to an NIS-like junction and provides the basis for the QUITERON switching action.

We present experimental confirmation of the operating principle including the demonstration of the transistor-like features such as large signal power gain, digital inversion, non-latching, high switching speed and low power dissipation. We also present observations of new and interesting physical phenomena which point to new challenges and opportunities to phycisits and technologists alike that could lead to better understanding of the physics of non-equilibrium superconductivity as well as optimized QUITERONS for practical applications .

[1] S.M. Faris, US Patent # 4,334,158

[2] S.M. Faris, S.I. Raider, W.J. Gallagher, and R.E. Drake, IEEE Tran. Mag., MAG-19, 1293 (1983).

Session 1-1b
VLSI Testing

Chairman

S. Waaben
Bell Labs

Testability Conditions for Bilateral Arrays of Combinational Cells

Anastasios Vergis and Kenneth Steiglitz

Department of Electrical Engineering and Computer Science
Princeton University, Princeton, New Jersey 08544

ABSTRACT

Two sets of conditions are derived that make one-dimensional bilateral arrays of combinational cells testable for single faulty cells. The test sequences are preset and, in the worst case, grow quadratically with the size of the array. Conditions for testability in linear time are also derived. The basic cell can operate at the bit or at the word level. An implementation of FIR filters using (systolic) one-dimensional bilateral arrays of cells, which can be considered combinational at the word level, is presented as an example.

1. Introduction

The use of iterative arrays of identical cells in current VLSI technology is becoming more frequent due to their many advantages, like ease of design, fabrication and testing. Moreover, many problems are efficiently solved with the use of "systolic arrays", which are highly iterative structures operating synchronously. An important problem associated with these structures is fault detection; that is, derivation of test input sequences to the array, such that the output sequences of the normal and any faulty array (under some fault assumptions) are different. In this paper we give some testability conditions for a special class of arrays (defined below) that improve upon the condition reported in [3]. Derivation of the test input sequences is also described.

2. Assumptions, Definitions and Notation

Figure 1a shows a bilateral array of combinational cells. The basic cell is shown in Figure 1b. At each time unit it produces left and right outputs, depending on its left, right and vertical inputs.

Let R be the set of right-moving signals, L the set of left-moving signals and Z the set of vertical cell inputs. Let $g_R: R \times Z \times L \to R$ be the right-moving signal mapping, and $g_L: R \times Z \times L \to L$ be the left-moving signal mapping.

A *fault* in a particular cell alters g_R, g_L, or both for one or more arguments (r, z, l). However, we assume that the cell remains combinational.

We assume initially that to test a cell completely, we must apply all input combinations $R \times Z \times L$ to that cell. This assumption makes testing of the cells independent of how they are realized and independent of the fault model for permanent faults [1]. We shall examine later the case when only a subset of $R \times Z \times L$ suffices to test the basic cell. We further assume that to test the array completely (for single faulty cells), we must test completely every cell in the array.

The left, vertical and right inputs of cell j at time t are denoted as $r^j(t)$, $z^j(t)$, $l^j(t)$ respectively. Hence, cell j at time $t+1$ will produce left and right outputs $l^{j-1}(t+1)$, $r^{j+1}(t+1)$.

Figure 2 shows the *space-time transformation* [2] of the array in Fig. 1a. Each row represents the array at each time unit. This makes the operation of the array easier to visualize. Note that this transformation maps the one-dimensional synchronous bilateral array into a two-dimensional asynchronous unilateral array.

If L_0 is a subset of L, define $g_R(r, z, L_0) = R_0$, where R_0 is the set $\{ g_R(r, z, l) \mid l \in L_0 \}$. If R_0 contains just one element r', we write $g_R(r, z, L_0) = r'$ (instead of $\{r'\}$). Similarly for $g_R(R_0, z, l)$, $g_R(r, Z_0, l)$, and similarly for g_L.

If $r \in R$, define $R_r = R - \{r\}$. Similarly $L_l = L - \{l\}$. An input or output labeled r/R_0, where $R_0 \subset R$, and r does *not* belong to R_0, means normal input or output r, and faulty input or output some member of R_0. If R_0 contains just one element r', we write r/r'. Sometimes, when no confusion arises, r/R_0 will also be called a *fault*. For simplicity r/R_r will sometimes be written as $r/*$.

We also define $g_R(r_1/r_2, z, l) = r'_1/r'_2$ if $g_R(r_1, z, l) = r'_1$ and $g_R(r_2, z, l) = r'_2$ *and* $r_1 \neq r_2$, $r'_1 \neq r'_2$. We define similarly $g_R(r, z, l_1/l_2) = l'_1/l'_2$, and similarly for g_L. Another definition is: $g_R(r/R_0, z, l) = r'/R'_0$ if $g_R(r, z, l) = r'$, and $g_R(R_0, z, l) \subset R'_0$. Similarly for $g_R(r, z, l/L_0)$ and for g_L. (Note that $g_R(r/R_0, z, l)$ is not uniquely defined, since R'_0 may be any superset of $g_R(R_0, z, l)$, not containing r'.) If, according to the above, the input of a cell is a fault r/R_0 and its output is r'/R'_0 we say that this cell *propagates* the fault r/R_0. Notice that according to our convention r' does not belong to R'_0, so we can distinguish between the normal and the faulty output.

In the sequel, p denotes the total number of cells in the array.

3. Some General Testability Conditions

Let V be the following set of conditions:

C1: for every $r \in R$ there exist $r' \in R$, $z \in Z$ such that $g_R(r', z, L) = r$.

C2: for every $l \in L$ there exist $l' \in L$, $z \in Z$ such that $g_L(R, z, l') = l$.

O1: for every $r_1, r_2 \in R$ with $r_1 \neq r_2$, there exist $l \in L$, $z \in Z$ such that $g_R(r_1, z, l) \neq g_R(r_2, z, l)$.

O2: for every $l_1, l_2 \in L$ with $l_1 \neq l_2$, there exist $r \in R$, $z \in Z$ such that $g_L(r, z, l_1) \neq g_L(r, z, l_2)$.

Theorem 1: Any bilateral array of combinational cells that satisfies conditions V is testable for single faulty cells.

The proof is similar in spirit to the proof of the next theorem, and it is actually simpler. So it is omited. We only mention that, under conditions V, to test the entire array completely we need $p\cdot(p+1)\cdot max(|R|,|L|)\cdot|R|\cdot|Z|\cdot|L|$ tests.

Let W be the following set of conditions (see Figs. 3 and 4):

C1: for every $r \in R$ there exist $r' \in R$, $z \in Z$ such that $g_R(r'/*, z, L) = r/*$.

C2: for every $l \in L$ there exist $l' \in L$, $z \in Z$, $r' \in R$ such that $g_L(r', z, l') = l$ and $g_R(r'/*, z, l') = r/*$.

O1: for every $r \in R$ there exist $z \in Z$ such that $g_R(r/*, z, L) = r'/*$.

O2: for every $l_1, l_2 \in L$ with $l_1 \neq l_2$, there exist $r \in R$, $z \in Z$ such that $g_L(r, z, l_1) \neq g_L(r, z, l_2)$.

Conditions W hold if 1) the basic cell just transmits unaltered the right-moving signal, 2) $g_R(R,Z,L) = L$, 3) for any two different right inputs there exist r, z that produce two different left outputs. This is a quite reasonable set of assumptions.

Theorem 2: Any bilateral array of combinational cells that satisfies conditions W is testable for single faulty cells.

Proof: Assume we want to test cell j for inputs (r_0, z_0, l_0). These test inputs will be applied at time $t=2p-j$ if the test begins at time $t=1$. Hence $r_0 = r^j(2p-j)$, $l_0 = l^j(2p-j)$, $z_0 = z^j(2p-j)$ (see Fig. 5, shaded cell).

First we must make the left input of cell j at time $2p-j$ be $r_0 = r^j(2p-j)$. Condition C1 guarantees the existence of $r^{j-1}(2p-j-1)$, $z^{j-1}(2p-j-1)$ such that $r^j(2p-j) = g_R(r^{j-1}(2p-j-1), z^{j-1}(2p-j-1), L)$; hence it suffices to apply $r^{j-1}(2p-j-1)$, $z^{j-1}(2p-j-1)$ as left and vertical inputs to cell $j-1$ (at time $2p-j-1$). Inductively, apply $r^{j-2}(2p-j-2)$, $z^{j-2}(2p-j-2)$ to cell $j-2$, etc., until we reach the leftmost cell. The tricky part is to apply right input $l_0 = l^j(2p-j)$ to cell j. Condition C2 guarantees the existence of $r^{j+1}(2p-j-1)$, $l^{j+1}(2p-j-1)$, $z^{j+1}(2p-j-1)$ such that $l^j(2p-j) = g_L(r^{j+1}(2p-j-1), z^{j+1}(2p-j-1), l^{j+1}(2p-j-1))$, and $g_R(r^{j+1}(2p-j-1)/*, z^{j+1}(2p-j-1), l^{j+1}(2p-j-1)) = r^{j+2}(2p-j)/*$. Hence it suffices to apply input $(r^{j+1}(2p-j-1), l^{j+1}(2p-j-1), z^{j+1}(2p-j-1))$ to cell j+1 (at time $2p-j-1$). Left input $r^{j+1}(2p-j-1)$ and right input $l^{j+1}(2p-j-1)$ can be applied (recursively) in the same way we applied inputs r_0 and l_0 to cell j (using C1). This solves the controllability problem.

We have not yet used the strong part of condition C2, namely the fact that $g_R(r'/*, z, l') = r/*$. The usefulness of this will become apparent in the sequel.

Assume that the normal right and left outputs of cell j (on input (r_0, z_0, l_0)) are r and l respectively; assume that we test for the error $l/\hat{l}$ in the left output. We can simultaneously test for all errors $r/*$ in the right output. Propagation of the error $l/\hat{l}$ to the leftmost output is done using O2 and C1.

We have not yet discussed the "southeast" portion of Fig. 5, that is the portion below the right-to-left diagonal that passes through the shaded cell. First we have to propagate the fault $r/*$ ($= r^{j+1}(2p-j+1)/*$) to the rightmost output. But cell j may fail to function correctly at any previous time, so for instance (see Fig. 5) cell j on input $r^j(2p-2m+j)$ (for some m in $\{j+1, j+2, \cdots, p\}$), may not output $r^{j+1}(2p-2m+j+1)$, so cell m may not output $l^{m-1}(2p-m+1)$, hence cell j may not receive l_0 as right input, and due to a fault, it may output the expected outputs l, r. Under this worst case scenario the two faults will be masked and we will get the expected observable outputs $l^0(2p)$ and $r^{p+1}(3p-2j+1)$ (assuming $j \geq (p+1)/2$). This is avoided as follows:

First, to propagate the fault $r/*$ to the rightmost output, using condition O1 we find $z^{j+1}(2p-j+1)$ such that cell $j+1$ on input $r^{j+1}(2p-j+1)/*$ outputs $r^{j+2}(2p-j+2)/*$; inductively we propagate this fault to the rightmost output ($r^p(3p-2j-1)/*$). Similarly, we propagate the fault $r^m(2p-m)/*$ for $m=j+1, j+2, \cdots, p$. Notice that the potential previous fault $r^{j+1}(2p-2m+j+1)/*$ of cell j has been "automatically" propagated to cell m as $r^m(2p-m)/*$ by the strong part of condition C2 "when" we were solving the controllability problem. So, if cell j outputs something different from $r^{j+1}(2p-2m+j+1)$, we shall detect it at the rightmost output by getting something different from the expected $r^{p+1}(3p-2m+1)$. This solves the observability problem.

The above procedure is repeated for every $\hat{l}$ in L_l; then we have tested cell j for input (r_0, z_0, l_0). This is repeated for every (r, z, l) in $R \times Z \times L$; then we have tested cell j completely. ∎

The testing time shown in Fig. 5 is $2p$. Hence to test cell j for input (r_0, z_0, l_0) we need $2p\cdot|L|)$ tests, hence to test completely cell j we need $2p\cdot|L|^2\cdot|R|\cdot|Z|$ tests, and to test completely the array we need $2p^2\cdot|L|^2\cdot|R|\cdot|Z|$ tests. Note that if some cell in the space-time transformation is "used" at time t, it is never used at time $t+1$, hence the obvious pipelining reduces the testing time to one-half of the above number of tests.

4. One-step Testability

Parthasarathy and Reddy [4] introduced the notion of one-step testability for unilateral arrays. We extend this notion to bilateral arrays as follows:

Definition: A cell in a bilateral array of combinational cells is one-step testable for input (r, z, l) if the number of time units needed to test this cell for input (r, z, l) is independent of $|R|$, $|L|$.

Definition: A cell in a bilateral array of combinational cells is one-step testable if it is one-step testable for all inputs (r, z, l) in $R \times Z \times L$.

Definition: A bilateral array of combinational cells is one-step testable if all its cells are one-step testable.

The notion of one-step testability is important for the following reason: if an array is one-step testable, the time needed to test it is greatly reduced since, if the expected output of a cell under test is, say, l, it is not necessary to apply different test inputs for each fault $l/\hat{l}$, $\hat{l} \in L-\{l\}$.

The following conditions are useful for one-step testability (see Fig. 6):

OST1: for every $r \in R$ there exist $l \in L$, $z \in Z$ such that $g_R(r/R_r, z, l) = r'/R_{r'}$.

OST2: for every $l \in L$ there exist $r \in R$, $z \in Z$ such that $g_L(r, z, l/L_l) = l'/L_{l'}$.

Let OST be conditions OST1 and OST2 together. Let V-OST (W-OST) be conditions V (W) and OST together.

If W-OST hold, instead of testing for the fault $l/\hat{l}$ for each $\hat{l}$ in L_l as in the proof of theorem 2, we can test for the fault l/L_l. Thus, to test cell j for input (r_0, z_0, l_0), we only need $2p$ time units. Therefore, if we want to test all cells for inputs a subset I of $R \times Z \times L$, we need

$2p^2 \cdot |I|$ time units. If V-OST hold, similarly as above, by testing for all output faults simultaneously (on a given input), we need $p \cdot (p+1) \cdot |I|$ tests if we want to test all cells for a subset I of $R \times Z \times L$.

5. Conditions for Testability in Linear Time

Let L be the following set of conditions (see Fig. 7):

OC1: For every $r_1, r_2 \in R$ with $r_1 \neq r_2$ and for every $l \in L$ there exists $l' \in L$ and $z \in Z$ such that $g_R(r_1, z, l') \neq g_R(r_2, z, l')$ and $g_L(r_1, z, l') = l$.

OC2: For every $l_1, l_2 \in L$ with $l_1 \neq l_2$ and for every $r \in R$ there exists $r' \in R$ and $z \in Z$ such that $g_L(r', z, l_1) \neq g_L(r', z, l_2)$ and $g_R(r', z, l_1) = r$.

Theorem 3: Condition C1 of V and L is a sufficient set of conditions for testability of the entire array in a number of time units proportional to the length of the array.

Proof: We shall give a constructive proof. To keep things simple, we shall work on an example; the generalization is straightforward. Consider the space-time transformation of an array of 6 cells as in Fig. 8. Assume that each cell must undergo 2 tests, that is it must be tested for inputs (r_i, l_i) $(i=1,2)$ for faults $r'_i/\hat{r}_i$, $l'_i/\hat{l}_i$ (i.e. r'_i, l'_i are the normal outputs and $\hat{r}_i$, $\hat{l}_i$ are the faulty outputs). In Fig. 8 cells under test are depicted as full squares; so the inputs and outputs of these cells are "given".

Let's group the cells of the space-time transformation diagonally, forming different groups on the left and right of the cells under test, as in Fig. 8. Call these groups left and right diagonal groups. Note that each left diagonal group of c cells has one left input, one right output, c right inputs and c left outputs. Its right inputs will typically be faults that have to be propagated to its left outputs. Fig. 9 shows a left diagonal group containing cells k through m, at time units t through $t+m-k$. Assume that the faults $l^{k+i}(t+i)/\hat{l}^{k+i}(t+i)$ for $i=0,1, \cdots, m-k$ are given and have to be propagated to its left outputs; also output $r^{m+1}(t+m-k+1)$ must be generated. We shall show how to find its left input $r^k(t)$ that does that. Using condition OC1 we can find input $r^m(t+m-k)$ that propagates the fault $l^m(t+m-k)/\hat{l}^m(t+m-k)$ and generates the output $r^{m+1}(t+m-k+1)$. Now $r^m(t+k-m)$ has to be generated, so inductively we can find $r^k(t)$. Hence each left diagonal group can propagate to its left outputs any faults on its right inputs and, at the same time, it can generate any right output. Entirely symmetrical things hold for each right diagonal group.

We apply the construction stated above, starting from the left diagonal group at the top of Fig. 8 (which contains just one cell); that is, we find its left input so that the fault $r'_1/\hat{r}_1$ is propagated, and output r_2 is generated. Then, for the next left diagonal group (second from the top), we know the faults that it has to propagate. If we were testing the leftmost cell for one more input, we would also know the output that has to be generated by this diagonal group; since this is not the case here, we can choose it arbitrarily. Therefore, we can apply again that construction to find its left input that does this. We proceed this way, from top to bottom. (Inductively, each diagonal group will have to propagate some faults and generate an output; as we proved above, there exists an input that does this.) We repeat the same procedure for the right diagonal groups, proceeding again from top to bottom. Obviously it does not hurt if some of the left inputs of a right diagonal group are not faults that have to be propagated, as for the first, third etc. groups in Fig. 8.

For the first $p-2$ left diagonal groups their left inputs are "inside" the array, so we have the additional problem of how to generate them. Condition C1 of V takes care of this, (as in theorem 2).

If each cell requires T tests, $(T+1) \cdot p$ is the number of inputs that have to be applied to the left boundary of the array (see Fig. 8); additionally, $p-1$ time units are required for the propagation to the right boundary of the faults in the leftmost cell. Hence the entire array can be tested in in $(T+1) \cdot p+p-1$ time units. Again, the obvious pipelining saves about half of these time units. ∎

6. Conditions for One-step Testability in Linear Time

Let us consider now one-step testable arrays. It turns out that it suffices to replace the fault r_1/r_2 (resp. l_1/l_2) in conditions L by the fault $r/*$ (resp. $l/*$). This way we obtain the following set of conditions L-OST.

OC1: For every $r \in R$ and for every $l \in L$ there exists $l' \in L$ and $z \in Z$ such that $g_R(r/*, z, l') = g_R(r, z, l')/*$ and $g_L(r, z, l') = l$.

OC2: For every $l \in L$ and for every $r \in R$ there exists $r' \in R$ and $z \in Z$ such that $g_L(r', z, l/*) = g_L(r', z, l)/*$ and $g_R(r', z, l) = r$.

Theorem 4: Condition C1 of V and L-OST is a sufficient set of conditions for one-step testability of the entire array in a number of time units proportional to the length of the array.

Proof: Analogous to the proof of theorem 3. If each cell is tested for a subset I of its inputs, the number of time units required is $(|I|+1) \cdot p+p-1$. ∎

Remark: All the above results are easily generalized for the case when g_R, g_L are not identical for every cell, that is we have g_R^i, g_L^i for the i-th cell; it suffices to replace the conditions for g_R, g_L by conditions for g_R^i, g_L^i for every i.

7. Application

Figure 10 shows the basic cell of a two-way pipeline systolic array used for FIR filtering [5], [6]. For this cell we have $|Z| = 1$ (no z-inputs), $g_R(r, l) = r$, $g_L(r, l) = l + a \cdot r$. This array can be considered as a bilateral array of combinational cells at the word level (the basic time unit is the time required to produce the outputs). It easy to see that conditions W-OST are satisfied. (Here we have the case when g_R, g_L depend on the cell.) Therefore, if a subset I of $R \times L$ suffices to test the basic cell, $2p^2 \cdot |I|$ tests suffice to test the array. Also, if we ignore overflow problems assuming for all r, $l_1 + a \cdot r \neq l_2 + a \cdot r$ for $l_1 \neq l_2$, conditions L-OST hold; hence for this case $O(p \cdot |I|)$ tests suffice to test the array. Note that the inequality above does not have to hold for all r, l_1, l_2; obviously, it suffices to hold for the signals that appear as inputs or outputs in the test described in the proof of theorem 3.

ACKNOWLEDGEMENT

This work was supported in part by NSF Grant ECS-8120037, U. S. Army Research-Durham Grant DAAG29-82-K-0095, and DARPA Contract N00014-82-K-0549.

REFERENCES

[1] M. A. Breuer and A. P. Friedman, *Diagnosis and Reliable Design of Digital Systems*, Computer Science Press, 1976.

[2] F. C. Hennie, *Iterative Arrays of Logical Circuits*, Cambridge, MA: MIT Press, 1961.

[3] F. G. Gray, and R. A. Thomson, "Fault Detection in Bilateral Arrays of Combinational Cells," *IEEE Trans. Comput.*, vol. C-27, pp. 1206-1213, 1978.

[4] R. Parthasarathy and S. M. Reddy, "A Testable Design of Iterative Logic Arrays," *IEEE Trans. Comput.*, vol. C-30, pp. 833-841, 1981.

[5] H. T. Kung, "Let's Design Algorithms for VLSI Systems," *Proc. Conf. on Very Large Scale Integration: Architecture, Design, Fabrication*, California Institute of Technology, Jan. 1979, pp. 65-90.

[6] H. T. Kung and C. E. Leiserson, "Algorithms for VLSI Processor Arrays," in *Introduction to VLSI Systems*, C. Mead and L. Conway, Addison-Wesley, 1980.

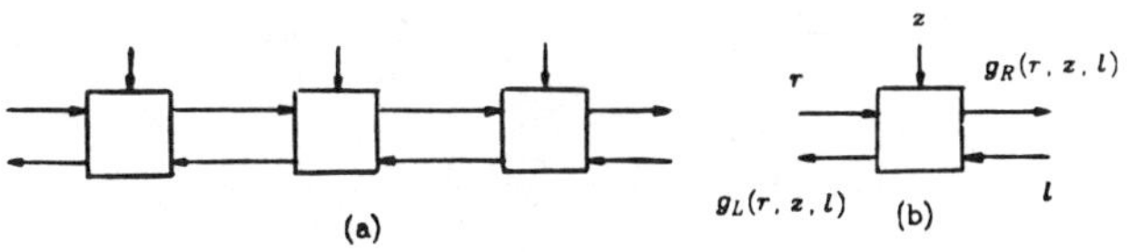

Fig. 1. (a) A synchronous bilateral array. (b) The basic cell.

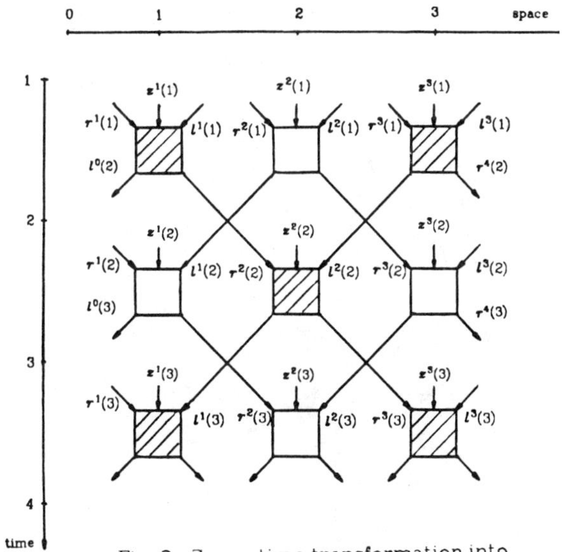

Fig. 2. Space-time transformation into asynchronous unilateral array.

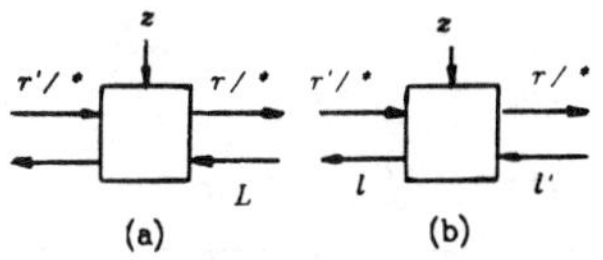

Fig. 3 (a) Condition C1 of W
(b) Condition C2 of W

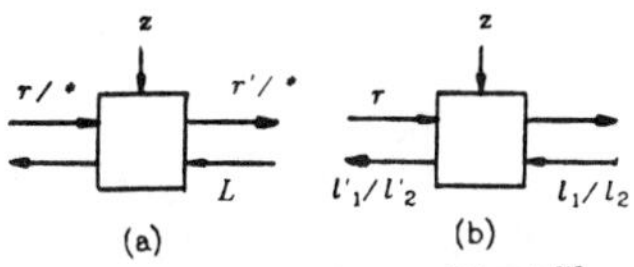

Fig. 4. (a) Condition O1 of W
(b) Condition O2 of W

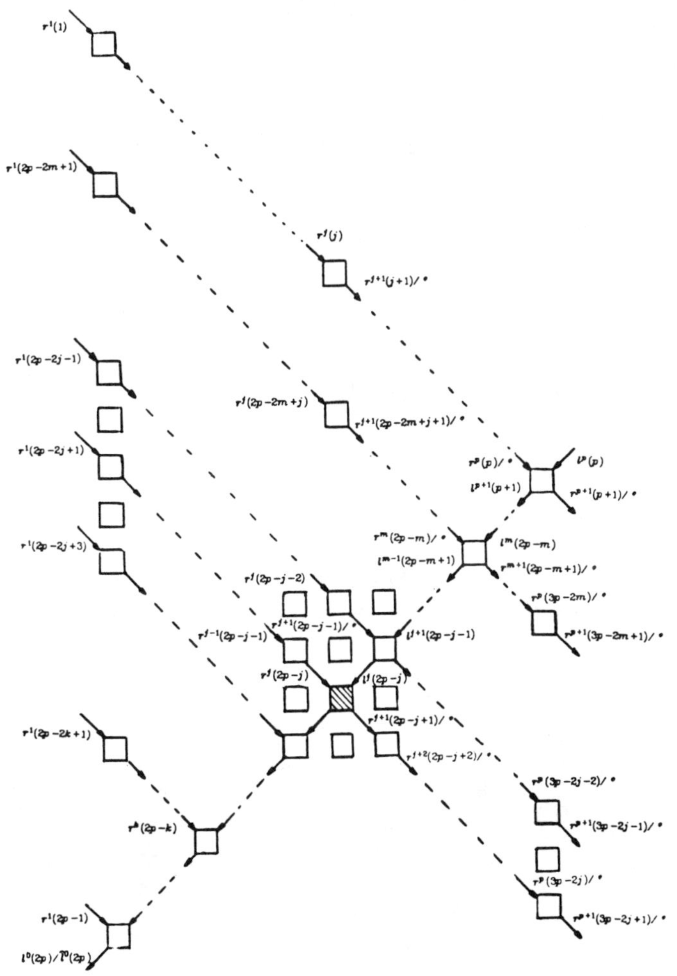

Fig. 5. The test described in thm. 2 (vertical inputs are not shown for simplicity).

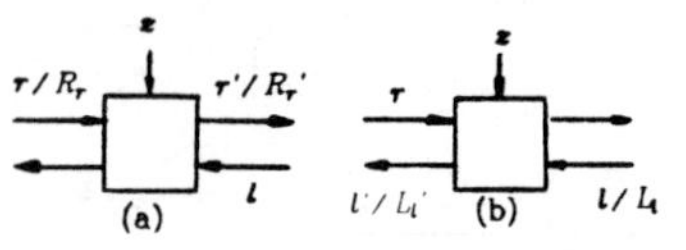

Fig. 6. (a) Condition OST1.
(b) Condition OST2.

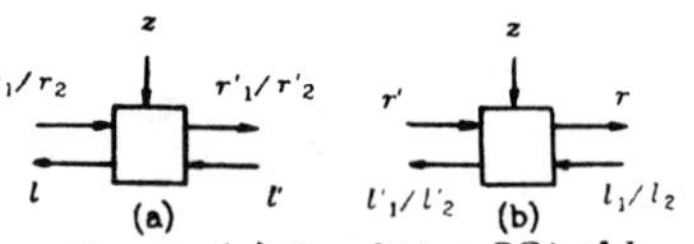

Fig. 7. (a) Condition OC1 of L.
(b) Condition OC2 of L.

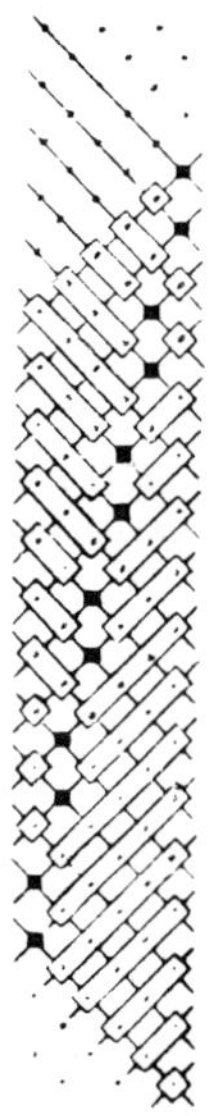

Fig. 8. Testing an array that satifies conditions L. Cells under test are depicted as full squares; the other cells are depicted as points.

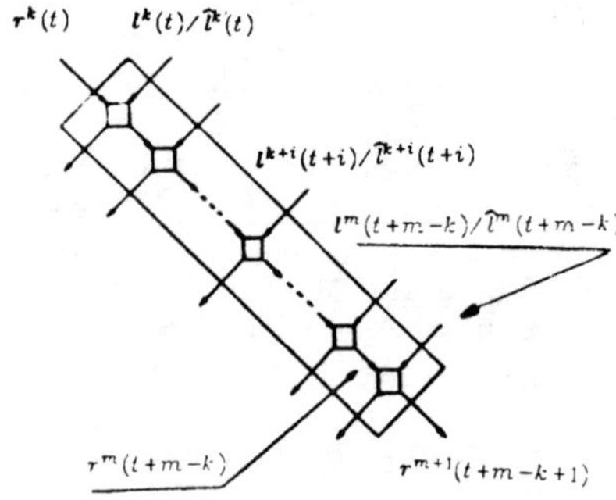

Fig. 9. A left diagonal group

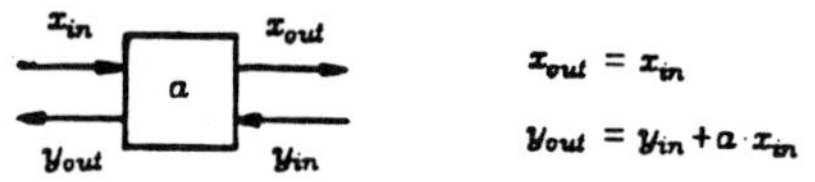

Fig. 10. The basic cell of a two-way pipeline systolic array for FIR filtering.

OVERVIEW OF MICROPROCESSOR TESTING

Alfred K. Susskind

Department of Electrical and Computer Engineering
Lehigh University
Bethlehem, PA. 18015

Abstract

A survey is given of current test-generation practices for microprocessors and of recent research and development efforts toward design for testability. Elaborations on the necessarily brief presentation can be found in the references cited.

I. Introduction

There are two major roads followed in laying out tests for microprocessors. One road--the one currently most used--is based on informal data-flow specifications and exercises normal instructions in an attempt to verify that the processor does comply with the manual of operation. The alternative road attempts to verify proper functioning of the hardware and requires that there be provisions made in the logic design to facilitate the test process. Sections II and III to follow survey each of the basic approaches and outline the current state of the art.

Difficulties in test planning come about from several sources. First of all, microprocessors are becoming remarkably complex. Not only is the gate count large (on the order of 10^4 gates are involved in currently evolving designs), but the logic structure is intricate and far more difficult to describe than in the case of other widely used chip types such as memories. Indeed, microprocessor complexity has grown to such an extent that some have held that a gate-level description has outlived its usefulness. Thus, the lack of a detailed logic description, which has previously handicapped customers in attempting to lay down test procedures, has lost much of its impact. But there still remains an "information gap". Truly adequate functional descriptions, such as detailed register transfer statements, are apparently not readily available. (There is reason to hope, however, that either this information gap will become smaller, or more help will be available for user testing.)

A second difficulty (common to all testing efforts) is based on the lack of adequately justified fault models. Most will agree that the widely used stuck-at fault model has serious shortcomings. But better alternatives are apparently not at hand, and so there is merit in attempting to formulate tests that verify correct functioning rather than confirm that stipulated faults are not present. In some circumstances, that is not excessively difficult to do. In others, it cannot be done in a pure form.

Still another concern in test planning is the complexity of the test procedure itself and the attendant cost of execution. A sophisticated modern test system can have a price tag of up to one million, and it takes strong arguments to justify the use of such powerful machinery in testing a mass-produced product with a value several orders of magnitude lower.

II. Conventional Methods for Finding Tests for Microprocessors

The microprocessors currently mass-produced and marketed have been designed without regard to testability. The user who wishes to test these designs is faced with a formidable task. One widely followed approach in planning tests is to become familiar with the capability of the microprocessor, study whatever information is available about the architecture, hypothesize, where necessary, details sufficiently to formulate a reasonably complete picture of operation, and then proceed to design tests that verify proper operation. This process is described by W.E. Sohl in [1], where the illustrative examples used are the Motorola 6800 microprocessor and the 6854 Advanced Data Link Controller. Sohl summarizes the test generation process as follows: "In order to successfully verify that all instructions are operational and that all internal data paths of the device are valid, the utmost care must be exercised in the design of the test sequence. A logical approach is to first verify that each element is totally functional, independent of all other elements. Second, that no destructive interaction exists between element to element. Third, that all output pins are capable of driving both logical 1 and 0 states. Fourth, the test sequence must assume that no logical element of the device is functional unless previously tested. Of course, a minimal level of functionality must be assumed in order to initiate testing."

In the conclusion section of [1], Sohl declared "It is no longer possible to review a device data sheet, define all possible test conditions and implement that test. Instead, the problem must first be approached with the philosophy that the device is basically untestable by the previous

CH1935-6/83/0000/0045$01.00 © 1983 IEEE

definition. Now that that fact has been recognized and understood, an effort is put forth to best define an approach that will result in a competent test being performed on the device. This process of test definition places heavier emphasis on a test engineer or engineering manager's ability to digest the function of that device in a general or specific application..."

Another example of testing by exercising the various functions is given in [2]. The illustration used there is based on the Intel 8080 microprocessor.

Some would insist that an adequate solution to the problems of testing of microprocessors can be obtained only through design for testability. But others appear to be quite content with the current state of affairs, where users take a stab at specifying a test procedure and are fully prepared to modify and augment it as experience is gained.

The literature contains solid evidence that there is an orderly way, more formal than that described in [1], for finding effective tests on the basis of no more than the information on data flow that is commonly available. Noteworthy is the pioneering study by Thatte and Abraham [3]. Their approach employs a user-derived register-transfer model that is presented in the form of a directed graph. A functional level fault model is stipulated and the detection of these faults is achieved by the execution of carefully constructed instruction streams. Procedures are given for determining the appropriate instruction sequences on the basis of the graph model of information flow. This methodology was successfully applied to an actual microprocessor, and its effectiveness was evaluated against the stuck-at fault model so widely used as a yardstick of fault coverage. The chip involved leads to about 2,200 single stuck-at faults (including undetectable ones). Of these, 96% were detected by the program sequences derived.

The major drawback in the methodology lies in the length of the test stream. The dominant component is the part devoted to verifying proper instruction decoding. One upper bound given for the length of that portion is on the order of the square of the instruction variety, which is a number that typically ranges between 50 and 200. Thus the bound lies between a few thousand and tens of thousands of instructions.

Success with a test methodology that uses as its starting point no more information than instruction manuals routinely provided by manufacturers has been reported in [4]. The test-generation method consists of algorithms and a library of test operands that allow a test-generation program to generate automatically the test program, given a user-derived register-transfer description of the microprocessor. The tester hardware has been kept moderately simple by making use of actual microprocessor instructions and compressing test responses by the use of a technique which is described in III below.

In those cases where the microprocessor is to be dedicated to a particular application, a possible test methodology is based on the actual application program. If all possibilities of control flow can be defined adequately so that all contingencies can be anticipated, an adequate test would be to run instruction and operand sequences that cover all program paths (as in program verification) and determine that along the paths data are handled properly. Ref. [5] discusses the use of graphs in formulating tests based on application programs.

Clearly one source of test sequences lies in the design verification cases simulated in the chip-design process. Those runs which involve inputs from external pads are good candidates for inclusion in a test set. This is the approach advocated in [6], where it is pointed out that the transmittal of test vectors to customers would not disclose proprietary information. But the author points out that "In the past, each user who developed a test program for a given component took a different approach...In many cases, users found that the vendor was not testing all the functions of a component...The user needs to be convinced that the vendor's chip design methodology produces a complete set of test patterns."

There have been cases where microprocessor test vectors have been purchased by users, but this practice does not appear as widespread as one might expect. Relevant here is the observation by Rice in [7]: "I believe that all the necessary functional information to establish tests is well known to the logic designer and is implicitly contained in modern simulation systems. It would seem that having to reconstruct this knowledge again later in a separate test engineering group is a gross waste of talent and elapsed development time."

III. Making Microprocessors More Testable

The ever widening stream of microprocessor-based products coming to market and gaining acceptance certainly indicates that the current state of testing practice is not posing a barrier to successful product development. Yet there is ever increasing concern over the need for greater testability. At this time, most of the response to the dissatisfaction is still coming from the captive semiconductor operations [8]. Chip merchants still appear unwilling to make provision for facilitating user tests.

Several reasons for dissatisfaction are given. First of all, in some circles quality assurance programs demand that test engineers be more knowledgeable about the effectiveness of the tests. It is indeed hard to present proof of fault coverage obtained under a good many of the test generation procedures sketched in the previous section. Second, the procedures require a substantial human effort, so they entail significant cost as well as long development cycles. With the approach to test-generation described in [1], it takes typically several months before a satisfactory test procedure is in operation. Third, the cost of testers is high.

These difficulties are not new and arose when LSI was first introduced. The successful solution then developed was a form of scan-in/out logic-design technique. The best-known variant of it goes under the acronym LSSD (level-sensitive scan design) [9]. When properly applied, LSSD achieves two major results. It transforms the problem of test generation for sequential circuits, which is difficult, to the much easier one of test generation for combinational (memory-less) circuits. Second, LSSD leads to dynamically "clean" design that reduces the need for running tests at system speed. Refinements of LSSD have been developed over the years [10], and systems for automatic test generation (ATG) have been placed in operation. Under the assumption of stuck-at faults, fault-simulation programs have been written that evaluate fault coverage. These developments took care of all difficulties except the high tester cost, to which we will refer below.

An example of a powerful automatic test-generation and evaluation system is given in [11]. This system is reported able to handle networks with almost 50,000 gates designed according to LSSD rules.

Experience with "classical" LSSD has indeed been most satisfactory. But the quality of the algorithmic test generation hinges critically on the utilization of gate-network models. While the ATG algorithms can be made to handle large networks, computer time required does, of course, become significant. Ref. [11] reports run times ranging up to 20 hours and the dependence on gate models causes repetitions of expensive computer runs whenever engineering changes are made. Moreover, while the ATG-derived test vectors are cleverly chosen, the set of vectors has no structure in the aggregate. Each vector is unrelated to the next one and can be injected into the test fixture only by reading a stored listing. This plus the needed listing of the expected (fault-free) response leads to a significant data-handling task in the tester, which in turn results in substantial cost.

There is a growing belief that "classical" LSSD has outlived its test-generation utility in dense VLSI designs. But significant mutations have been developed that promise to extend the utility of LSSD [12,13]. The simple shift registers, which are the key components of LSSD, where they are used to inject the test vectors and to observe the response of the (combinational) network under test, are replaced by more complex ones. Connections to the shift register input are made from selected tap points in the shift register via XOR gates. The resultant configuration is known as a linear feedback shift- register (LFSR). A LFSR is used as a source of test vectors by processing (shifting) many times the initial condition and using part or all of its contents as test inputs. The network response to be observed is (linearly) superposed on the shift action of a second LFSR [14]. The test-vector source register has its feedback connection so chosen that an $\underline{s}$-stage register contains all but one of the 2^s possible words as successive shift operations occur. The ordering of these words (in time) is such that little correlation exists between successive words; hence the generated word sequence is often called <u>pseudo-random</u>. The LFSR that receives response signals has its feedback similarly chosen, but adds (by means of XOR gates) the response signals. The register contents are commonly called <u>signatures</u> because for a fault-free network the contents at the end of the test are different from those of a faulty network with very high probability over the class of all possible faults [15]. To assess this probability is a very difficult task and the subject of current research. But simulation and actual field experience indicate that little information is lost through data compression by means of signature analysis.

Because the hardware for test-vector generation and response processing is so simple, tests can be run at normal operational speed and testers become much simpler, since data need not be stored, read, and compared continually. In fact, the tester becomes so simple that most of the hardware, if not all, can be implemented on the chip itself. Thus an implementation of self-test can be achieved.

Other means of data compression have been proposed, but at this time the most utilized are those based on pseudo-random number generation and signature analysis. Evidence is mounting that various degrees of self test will be incorporated into future chips, particularly for high dependability applications such as military electronics and professional equipment.

The substantial level of research activity in the area of design for testability is reflected in [16,17], workshops, and conference papers. A variety of concepts that make the testing task significantly easier have been proposed. Because of space limitations, we cannot be encyclopedic here and limit ourselves to considering design-for-testability approaches that are specifically tailored to microprocessors.

In the first method [18], a particular architectural structure is given to the control unit, which is divided into two parts and is similar to a nanoprogrammed structure. The chip is made capable of running in four modes and the address bus, which is bidirectional, serves as an input port during the three test modes. Functional (rather than stuck-at) fault models are used in the spirit of [3]. A simulation of a microprocessor laid out in accordance with the stipulated control structure and tested in the multimode manner prescribed showed "essentially 100%" fault coverage of single and multiple stuck-at faults.

A strategy for microprocessor testing is sketched in [19] that is a combination of established basic approaches. Extra observables are used to check proper control action by monitoring the actual load and gate (LG) signals that set up data paths. The data transfers are verified by algorithmically prescribed data movements. Manipulation of operands is checked by essentially exhaustive techniques (see below). Finally, the clock system and interaction with the surrounding environment are tested on the basis of classical machine-identification theory, which also provides techniques for increased testability of state machines.

The number of LG signals can be large (up to about 100 have been encountered in actual chips) and to make them pin observable, it is proposed that they be first compressed internally (e.g., by a signature register) and then multiplexed over existing pins such as the address bus, which is not needed during part of the test. Proper LG signals plus algorithmic verification of their execution assure proper control flow and register storage and bus operation.

Exhaustive data manipulation checking need not be excessively cumbersome. For one thing, one can compress the response data to be compared with the fault-free result, as shown in [20]. Alternatively, the ALU can be designed as an iterative circuit, which permits exhaustive testing of the component cells without growth in the test length as the number of cells is made arbitrarily large. (A simple example is an adder of arbitrarily long numbers that can be exhaustively tested in just eight steps.) See [21].

Bit slicing, which is based on iterative circuit techniques and eases so much the task of exhaustive data-manipulation testing, is one part of the approach to a testable microprocessor design described in [22]. The other major parts are two added functional blocks that are variants of the signature analysis register and pseudo-random number generator of [14]. The first block added compresses the output data on the data path. The second block, called control monitor, gathers the control signals emanating from the microprogrammed control section. The control monitor overcomes the great complications that arise when proper control operation must be inferred from the normally available observables, which are the data passed through and so reveal only indirectly the action of the control unit. As mentioned in Section II, this is the major problem with the test scheme in [3]. Both [19] and [22] overcome it by providing extra observables. The authors of [22] estimate that the increase in the silicon area required for the two monitors will be 4% in the case of a 16-bit microprocessor and 2.2% for a 32-bit design. The performance penalty is expected to be less than 5%.

The various approaches to design for testability--and even self-test--sketched above give clear evidence that there is on hand an adequate tool kit for achieving testable design. We may thus conclude that problems of testing need not impede further advances in the exploitation of ultra-dense circuit techniques.

References

[1] W. E. Sohl, "Microprocessors and Peripheral Device Testing", IEEE Tutorial: LSI Testing, Second Edition, 1978, pp. 47-57.
[2] J. P. Hayes and E. J. McCluskey, "Testability Considerations in Microprocessor - Based Design", Computer, XIII, 3, Mar. 1980, pp. 17-26.
[3] S. M. Thatte and J. A. Abraham, "Test Generation for Microprocessors", IEEE Trans. Comp., C-29, 6, pp. 429-441, June 1980.
[4] C. Bellon, A. Liothi, et. al., "Automatic Generation of Microprocessor Test Programs", ACM-IEEE 19th Design Automation Conf. Proc, June 1982, pp. 566-573.
[5] C. Robach and G. Saucier, "Application Oriented Microprocessor Test Method", Digest of Papers, 10th Int. Sym. Fault Tolerant Comp., 1980, pp. 121-125.
[6] W. S. Richardson, "Test Pattern Portability for Microprocessors", ACM-IEEE 19th Design Automation Conf. Proceedings, June 1982, pp. 587-593.
[7] R. Rice, "Computer-Aided Design, Testing, and Packaging", IEEE Tutorial, 1982, p. 19.
[8] R. P. Davidson, M. L. Harrison, and R. L. Wadsack, "BELLMAC-32: A Testable 32-bit Microprocessor", Digest of Papers, 1981 International Test Conference, pp. 15-20.
[9] E. B. Eichelberger and T. W. Williams, "A Logic Design Structure for LSI Testability", Proc. 14th Design Automation Conf., June 1977, pp. 462-468.
[10] S. DasGupta et. al., "A Variation of LSSD and Its Implications on Design and Test Pattern Generation in VLSI", 1982 IEEE Test Conference, pp. 63-66.
[11] P. Goel and B. C. Rosales, "PODEM-X: An Automatic Test Generation Systems for VLSI Logic Structures", Proc. 18th Design Automation Conf., 1981, pp. 260-268.
[12] P. H. Bardell and W. H. McAnney, "Self-Testing of Multichip Logic Modules", 1982 IEEE Test Conference, pp. 200-204.
[13] D. Komonytski, "LSI Self-Test Using Level Sensitive Scan Design and Signature Analysis", 1982 IEEE Test Conference, pp. 414-424.
[14] B. Konemann, J. Mucha, and G. Zweihoff, "Built-in Logic Block Observation Techniques", 1978 Semiconductor Test Conf., pp. 37-41.
[15] T. Sridhar et. al., "Analysis and Simulation of Parallel Signature Analyzers", 1982 IEEE Test Conference, pp. 656-661.
[16] Special Issue on Design for Testability, IEEE Trans. Comp., C-30, 11, Nov. 1981.
[17] T. Williams and K. Parker, "Design for Testability, a Survey", IEEE Trans. Comp., C-31 pp. 2-15, Jan. 1982.
[18] S. Nanda and S. M. Reddy, "Design of Easily Testable Microprocessors - a Case Study", 1982 IEEE Test Conf., pp. 480-483.
[19] S. K. Jain and A.K. Susskind, "Test Strategy for Microprocessors", 20th Design Automation Conference, Miami, Fla., June 1983.
[20] M. G. Karpovsky, "Error Detection in Digital Devices and Computer Programs with the Aid of Linear Recurrent Equations over Finite Commutative Groups", IEEE Trans. Computers, C-26, 3, pp. 208-218, Mar. 1977.
[21] T. Sridhar and J. P. Hayes, "A Functional Approach to Testing Bit-Sliced Microprocessors", IEEE Trans. Comp., C-30 pp. 842-854, Nov. 1981.
[22] S. M. Thatte, et. al., "An Architecture for Testable VLSI Processors", 1982 IEEE Test Conference, pp. 484-492.

VLSI TEST PROGRAM GENERATION :
A SYSTEM FOR INTELLIGENT ASSISTANCE

C. BELLON, Ch. ROBACH, G. SAUCIER

IMAG Laboratory - BP 68 - 38402 SAINT MARTIN D'HERES Cédex FRANCE

The automatic generation of test programs for VLSI circuits and systems remains an unanswered wish. The solution seems to be a computerized expert aid, integrated in a general system. This paper presents the concepts of an intelligent agent whose aims are to help the designer in appraising the diagnosability of the device or system under design, suggesting design modifications, generating test patterns for elementary modules and assembling the basic test entities into a test program for the whole part.

INTRODUCTION

The design of efficient test/diagnosis programs with good fault-coverage and fault-location is an unsolved problem for VLSI circuits and systems. Therefore, the automatic generation of test patterns and test programs for a high level of complexity remains an unanswered wish.

Fully automatic test generation (ATPG) tools have been achieved for circuits characterized by either a medium complexity or a repititive structure (random logic, memory, programmable logic array). None of the ATPG tools could be reasonably extended to cope with the actual and future integration and diversity of electronic parts. Thus, the need arises for a computerized expert aid which should be integrated in a general system. This paper presents the concepts of an intelligent agent, whose goals are to help the designer in appraising the testability / diagnosability of the device or system under design, suggesting design modifications, generating test patterns for elementary modules, assembling the basic test entities into a test program for the whole part.

The outlines of such an intelligent agent are given in the following. The intelligent agent has to compile large knowledge and expertise, related to various sources of information : level and type of description, fault models, test methods and strategies, testability measures, ...

So, the intelligent agent is based upon the cooperation of dedicated test experts. A test expert may be defined by the type of abstract model used to represent a given electronic part and the kind of information it issues to the test engineer and/or designer.

We may first define "high level" experts which consider the device or system as a whole, and which derive predictive information about the overall testability / diagnosability, the test strategy and design modifications requirements.

"Low level" experts are designed to work on more detailed descriptions or parts of the device and to give more accurate results, which will be the basis of the test procedures themselves. Those experts generate test patterns for (a part of) a circuit without any decomposition of the problem or previous estimation. Well-known and available ATPG tools or algorithms will be typically low-level experts, working on structural descriptions (D-algorithm at gate level, test patterns generation algorithms for RAMs and PLAs,...). The designer may call independently and/or successively such experts and act upon their results ; the information issued by a given expert may be used as a partial input for another one, in order to specify a result for a given sub-part, solve a peculiar problem or take into account additional specifications.

The overall structure of the SUPERCAT system, given in Figure 1, shows the basic test experts and their associated abstract models ; these models are extracted from a general data-base which contains all the levels of description and specification of the circuit (presently the data base of the French Telephone Company CASSIOPEE). This organisation is detailed in the following sections.

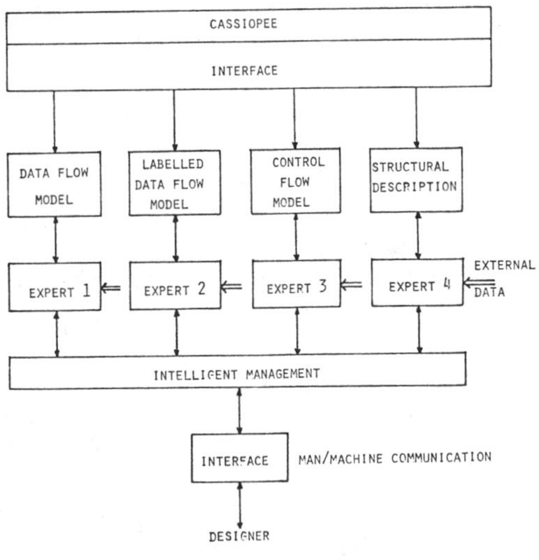

Figure 1 : SUPERCAT structure

EXPERT WORKING ON A DATA-FLOW MODEL : CATA (1)

This expert is mainly dedicated to testability evaluation and test organization for data-driven systems and hybrid circuits.

The abstract model

The abstract model of CATA may be used at the various levels of design and analysis. Concurrently with the design hierarchy, a test hierarchy is needed as design progresses and according to the test context (field maintenance, manufacturing verification). So a partition is defined in the system circuitry, of the size of a sub-system, of a board or (a part of) an integrated circuit. The partition is defined in terms of modules which characterize a minimal set of hardware that perform a basic function and which is the minimal isolable part.

The abstract model, based upon the partition, is given by a bipartite graph which is entirely defined by places, transitions and arcs :

* the places are either the modules defined by the partition, or the test data inputs (sources) or the test results outputs (traps).

* a transition between two places expressses the conditions of information transfer between those places,

* the arcs connecting places and transitions represent the information media throughout the system.

Basic transfer modes are defined in order to take into account both the distribution of the information throughout the system and test requirements.

Test expertise

The complete CATA system includes a number of computer programs and automated aids which may be categorized as follows :

Flows generation : testing an electronic part is achieved by sending specific information i.e. "test data", from the system sources which generate this information, towards the system traps which issue the test results towards the external world. This requires to define the set of information paths throughout the system, thus the set of "generalized paths" in the associated graph (called "flows").

A flow may be viewed as an elementary function or activation of the considered part and it corresponds to the intuitive notion of minimal set of hardware that can be activated completely and independently from the remainder of the system.

CATA is able to provide all the flows for a part of the system (a board, for instance) or for the whole system or device.

Testability evaluation : Testability is appraised by the number of functions or modules which are not independently fault-isolable. The testability evaluation process gives the diagnosis resolution in terms of indiscernability groups, thus allowing the designer to add test points on the concerned paths.

Test paths generation : According to the desired location level, the status of the system/device testability is examined under the implementation of a given test strategy.

The test strategy allows test paths to be determined among the set of flows and the test functions to be ordered in case of a Start-Small strategy ; the test procedure is parametered by the required location level.

A data base is built in order to relate the testability model to the physical design, and to automatically provide test documentation.

Application results

The software package CATA (designed in SCHLUMBERGER-France) has been applied with success to various circuits and equipments : hybrid circuits, boards and systems. It is implemented on a VAX 11-780 computer and the software procedures are written in PASCAL language.

Theoretical example

Consider the following abstract model where S1, S2 are inputs to the device, P1, P2 are outputs from the device and M1 through M5 are hardware constituents.

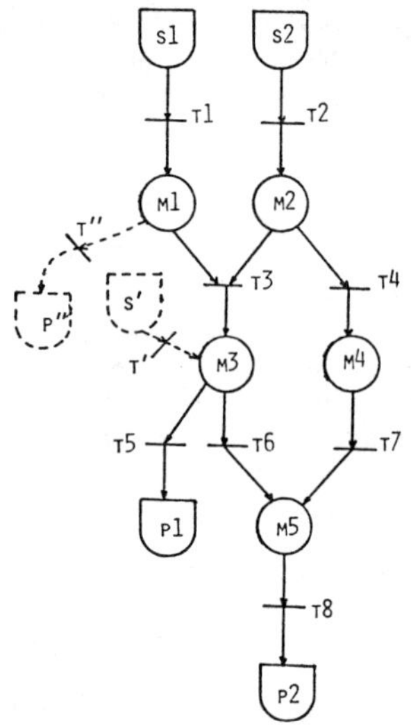

Comments

. M3 needs information from both M1 and M2 to work

. M5 receives its data from either M4 or M3

Flows generation : Three flows are determined

F1 = (S1,S2,M1,M2,M3,P1 ; t1,t2,t3,t5)

F2 = (S1,S2,M1,M2,M3,M5,P2 ; t1,t2,t3,t6,t8)

F3 = (S2,M2,M4,M5,P2 ; t2,t4,t7,t8)

Testability evaluation : CATA indicates an indiscernability group (M1,M3) and suggests to add either a control point at M3 input or an observation point at M1 output. So a new flow is created.

Control point F' = (S',M3,P1 ; t',t5)

Observation point F" = (S1,M1,P" ; t1,t")

Test strategy (Multiple clue)

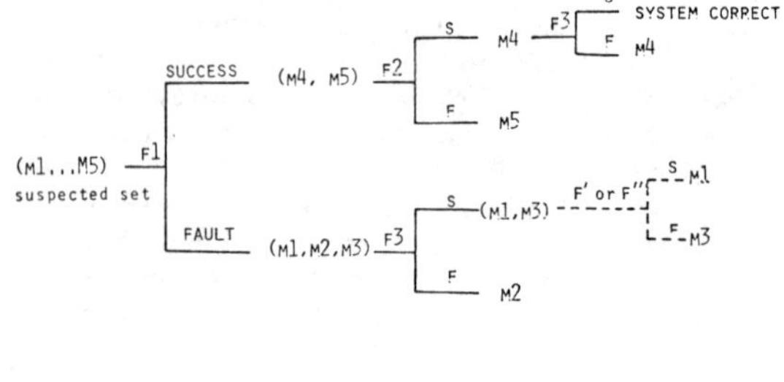

EXPERT WORKING ON A LABELLED DATA FLOW MODEL (2)

This expert aims at providing a quantitative measure (additional information) about the model described in CATA.

The abstract model

The abstract model used by this expert is the basic abstract model of CATA expert with an additional information upon the graph edges : information capacity of a line.

Information theory tools allow three parameters to be derived :

* the controllability of a module in a flow,
* the observability of a module in a flow,
* the testability of a module in a flow.

The controllability of a module in a given flow is defined as the maximum information quantity this module may receive in the flow.

The observability of a module in a given flow is defined as the maximum information that may be obtained at the unknown module output, knowing the primary inputs and outputs of the flow.

The testability of a module is empirically defined as

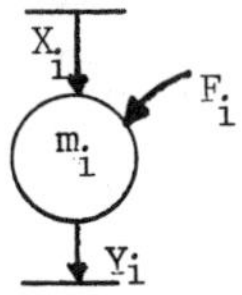

$$Ti = \frac{H(Fi) - H(Fi/Xi,Yi)}{H(Fi)}$$

where Xi, Yi and Fi are respectively the input, output and health state of the module, and H(X) is the Shannon entropy.

The testability generalizes the inverse of the minimum number of test vectors required to check this module. It may be enlarged in order to define the testability of a module in a flow, then the testability of a flow.

The controllability, observability and testability factors are computed by using a modified entropy which is an overvaluation of the Shannon entropy ; the transformation rules are the following :

* the information of a source is given by its capacity and the source variables are assumed to be independent ;
* the information on a module output is the smallest value of the input information and the capacity of this output ;
* the information on a transition output is the smallest value of the joint information of the input lines and the capacity of its output line ;
* the joint information on the set of outputs of a transition is the smallest value of the joint information on the set of inputs and the sum of the capacities of the outputs.

H'(Xj) = min(H'(Yi),CXj)
H'(Xk) = min(H'(Yi),CXk)
H'(Xj,Xk) = H'(Yi)

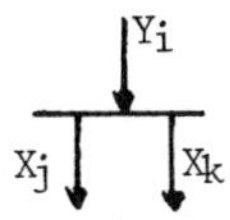

H'(Xk) = min(H'(Yi,Yj),CXk)

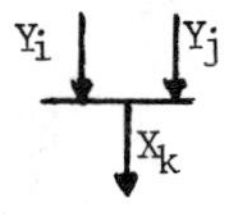

H'(Xk) = min(H'(Yi,Yj),CXk)
H'(Xl) = min(H'(Yi,Yj),CXl)
H'(Xk,Xl) = min(H'(Yi,Yj),C(Xk) + C(Xl))

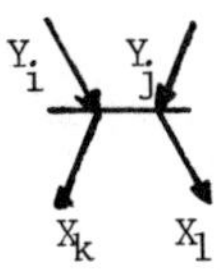

Test expertise

All the flows that are computed by the CATA expert are considered as equivalent with respect to the test completeness of a module. The introduction of information quantity on the abstract model and the controllability / observability factors that are derived, allow a discrimination between the flows. Furthermore, the testability factor allows flows that are equivalent with respect to the cost of test implementation, to be compared :

* For a given test strategy, the expert determines the set of flows which give the maximum sytem testability ;
* When several test strategies may be chosen, the expert determines the one with the maximum testability or the most economical one ;
* When the testability is assumed to be insufficient, the controllability and observability factors allow additional test points to be determined.

Theoretical example

Consider a module m which is a 4 bits adder with a 5 bits output ; the number of possible faults is 2^6. In this example, the controllability / observability / testability factors are computed for the module in a flow.

Controllabity :
H'(X1) = H'(X2) = 4
H'(X1,X2) =4
H'(Y1,Y2) = 4
$C_m^F = H'(X) = 4$

Observability :
H'(X/Xs) = 0
H'(Y/Xs) = min (6,5) = 5
$O_m^F = 5$

Testability :
$T_m^F = \frac{5}{6}$

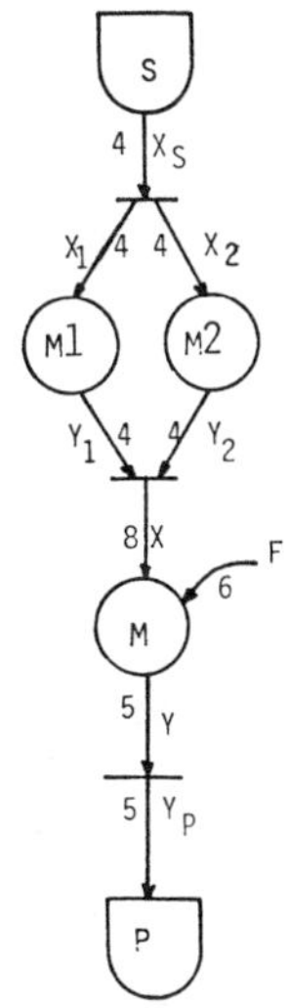

TEST EXPERT WORKING ON A CONTROL FLOW MODEL (3)

This expert is efficient for a circuit or a sub-circuit made up of a controller and a set of controlled resources.

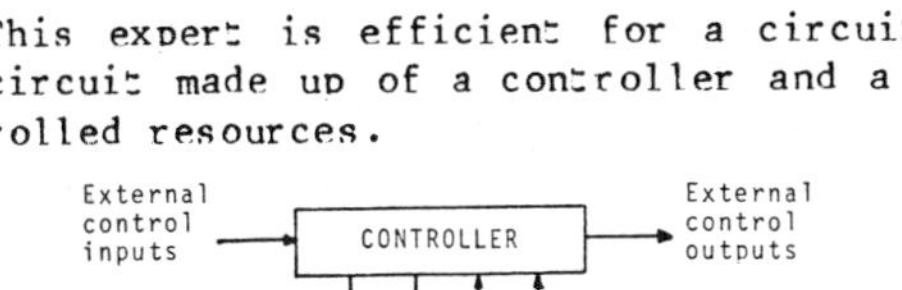

This expert aims at determining the set of test cases that can be generated under normal functioning for both the controller and the controlled resources ; its goal is to answer the following questions :

- what report values are necessary for the controller to go through a given sequence of states ?

- what input values the controlled resources may receive under this normal control ?

- are a test mode and test devices necessary ?

The abstract model

The abstract model which is suitable to this expert is an interpreted control flow graph. This graph is a bipartite graph, which presents the same features as the data flow graph, with the following interpretation :

. The places represent the controller states,

. The transitions are labelled with predicates whose variables are either external control inputs or reports from the controlled resources.

Procedures are associated with the places, which specify the transformation of the processed data when the controller is in the corresponding state. Data processing may be expressed at various levels:

- a functional level, which only requires registers and operations,

- a structural level where the inerconnections between registers and operators are specified.

For each of those levels, the duration and timing of the control signals, needed to perform the functions associated with a place or to fire a transition, may be defined along with the model.

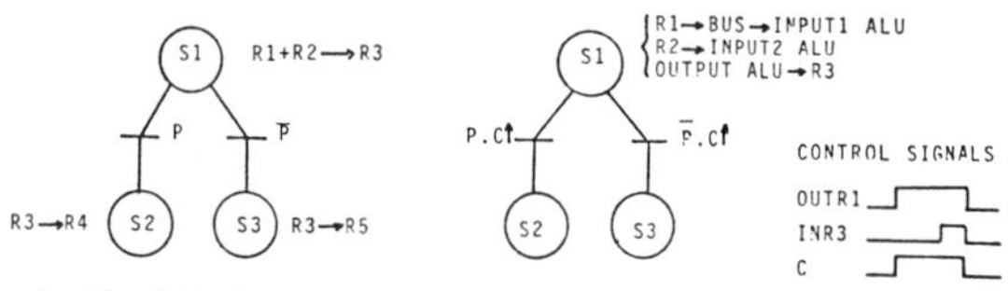

Functional level
R1--R5 are registers
P is a predicate

Structural level with timing
C↑ means the rising edge of the clock

Test expertise

Test of the controller : the test expert will be able to find the symbolic execution tree in the functional control flow graph. The path conditions will be expressed in terms of external control inputs and controlled resources reports (the translation into external data input values is left to the designer, in order to avoid heavy formal manipulations). Reductions of this symbolic execution tree have been studied, to avoid test redundancy.

The set of paths in the symbolic execution tree represent the set of sequences of states the controller may go through, under normal functioning. This helps in determining test cases when using either a structural or a functional method. In the first case (structural test method), this answers the following questions : How many of the test sequences, determined by a structural study, can be generated under normal funcioning ? Is the adjunction of a device for the generation of extra configurations necessary ? In the second case (functional test method), the symbolic execution tree allows to determine the test activations necessary to go through every state and/or every transition of the control flow graph.

The expert estimates the observability of the controller states (number of controlled resources to be used before reaching observation points), the diagnosability (distinction between faults in the controller and in the controlled resources), and suggests the introduction of test points or a scan-path to observe the controller.

When using the timed level of description, timing diagrams are established for all test paths and will constitute test files for dynamic testing.

Constraints from the controller on the test of the controlled resources : The expertise mainly concerns the determination of the input value domain of a controlled resource. The expert first determines which path of the symbolic execution tree activates a given resource, and then, which values may be sent to a controlled resource for a given path. It may therefore answer to either of the following questions, both related to the use of the low-levels experts :

- let us suppose that an efficient set of test patterns has been determined for a controlled resource, will it be possible to send these values through the controller ?

- in which input domain must work a low level expert (random generator, for instance) for the determination of test patterns ?

If the global determination of the input domain is too complex, the designer may adopt a bottom-up approach : for a given test input to a controlled resource, the expert will look for a path in the control flow graph that allows this value to be generated.

Generation of the test programs for the controlled resources : Let us assume that a low level expert has determined test patterns for every controlled resource ; the job of the expert will be to find a set of paths in the control flow graph allowing to send these patterns to each resource, and to propagate the results to observation points. It will take into account the constraints from the controller and assemble the different paths in order to obtain a test program.

Diagnosability expertise : If the normal outputs of the circuit do not provide a good diagnosis power, optimal additional observation points may be determined, according to the observation mode (optical observation through Electron Beam Microscope or additional I/O pins).

REFERENCES

(1) C. ROBACH, M. DELAUNAY, P. MALECHA, G. MICHEL, "A Computer Aided Test Analysis system : CATA", IEEE, Fault Tolerant Computing Symposium, Milano (Italy), June 1983.

(2) P. CASPI, A. MILI, Ch. ROBACH, "An information measure on nets. Application to the testability of digital systems", Information and systems, edited by B. Dubuisson, New York: Pergamon, 1978, pp 35-39.

(3) P. AMBLARD, M. CRASTES, J. RARIVOMANANA, G. SAUCIER, "CADOC : a functional specification and simulation tool", 1983 International Conference on Computer Aided Design (ICCAD), Santa Clara (USA), September 1983.

TEST GENERATION FROM MACPITTS DESIGNS

VASON P. SRINI

Computer Science Department
University of Alabama in Birmingham
Birmingham, AL 35294, 205-934-2213

ABSTRACT

Automatically generating test patterns for VLSI designs based on MacPitts design methodology is discussed. A structured testing scheme using D-propagation through the operators and constructs in MacPitts is described. A functional testing scheme based on test pattern tables for the operators is also described.

MacPitts

MacPitts [1] is a VLSI design description language that allows explicit specification of concurrent computation. Multiple data paths in the design operate concurrently to provide high throughput. A MacPitts program is a collection of processes that operate concurrently and a set of definitions. Each process has a separate data path and a set of states that are executed sequentially. All the input pins, output pins, registers, signals, parallel buses, word length of data paths, and internal variables (internal ports) are specified as a set of definitions. Constructs are provided for setting states and making state transitions. Primitives are provided for performing transformations on data. The primitives initially available are: and, or, not, nand, nor, xor, equ, 1+, 1-, +, -, =, <, <<, >>. New primitives can be added by writing layout generator routines and including them in the MacPitts compiler.

A MacPitts program is shown in Figure 1 to illustrate the constructs. The program describes a chip with 18 pins and three data paths corresponding to the three processes: add1, and1, and add2. The process add2 waits for data from add1 and and1. This is shown by the cond statement that checks the conditions on flags (one bit registers) go1 and go2. If both the flags are true, then the contents of registers t1 and t2 are added and the result is produced on the port output1.

Two observations have been made about MacPitts designs and the layouts generated from them. Most of the paths in the layout generated by the MacPitts compiler will be in the path from an input pin to an output pin. Internal signals and internal ports will also be on a path to an output pin. These observations mean that the tests for several of the faults might be single patterns instead of sequences of patterns.

STRUCTURED TESTING

The structured testing approach uses the single stuck-at fault model and the D-algorithm [2,3,4]. It is assumed that at most one line is faulty in the data path corresponding to an operator or a bit of a register, internal port, or an internal signal. Instead of using logic gates, the primitives (operators) of MacPitts are used in D-propagation [2,4]. So the first step is to develop tables for D-propagation through the operators of MacPitts and D-propagation through the constructs.

D-propagation through MacPitts constructs: MacPitts has a small number of constructs to specify the operations to be performed in a data path during a state, the state transitions, and concurrent data paths. The semantics of the constructs of MacPitts and the D-propagation through them are now described.

The setq construct ((setq <name> <formula>)) causes the data path to evaluate several operations specified in the <formula> part and stores the result in a register or outputs the result to the external environment. It also allows signals to be set. D-propagation through setq comprises D-propagation through the operators in the formula part using the D-propagation tables. The output from the formula part is then propagated through the register in the <name> part. Since no transformation is performed on the data in a register, D-propagation through the register is trivial using the identity operation. For example, consider the statement

(setq t1 (>> (+ r1 r2)))

in Figure 1. Let the contents of r1 be D 000. Let the D in the most significant bit of r1 be propagated through the + operator by using the tables for and, and xor. One way to do the D-propagation is by applying 0000 to r2. The input to the >> operator is D 000 and the output is 0 D00 which is stored in t1. This value eventually gets into another statement and thus propagating the D through the setq construct.

The cond construct allows conditional execution of other constructs. It is one of the key constructs in MacPitts and its semantics is in [1]. A cond statement comprises a list of guards. Each guard comprises a condition and a collection of formulas. Nesting of cond statements is allowed. The semantics of cond is such that the <u>first</u> <u>guard</u> whose condition is true starts the execution of the formulas following the condition. An example of a MacPitts design containing several cond statements is shown in Figure 2.

D-propagation through the cond construct is shown by considering the following form for the construct:

```
cond  x1:   z1 <- f1(y1)
      x2:   z2 <- f2(y2)
      xi:   zi <- fi(yi)
       .
       .
      xn:   zn <- fn(yn)
```

To propagate a D from yi to zi, we require xi = 1 and xj = 0 for all j, 0 < j < i. Satisfying the above condition might in general require the application of sequences of patterns. For example, in Figure 2, to propagate a D from the input port rcvdata to the register inbuf2, the following values must be present in the registers and the signals:
count <- 1
rcv <- 1, init <- 0.
Note that since there are no setq statements in the program that is of the form (setq count <formula>), it is not possible to set the value 1 to count. So count has to be set equal to 0 by using the statement (setq count 0) in the rcvtest process and then incrementing the contents of count by 1 using the statement (1+ count) in the set of states starting with the label testrcvr. The sequence of test patterns that are to be applied to the input pins for propagating a D on pin 7, and the values that are expected on the output pins are shown below. The numbers under values such as 0, 1, D are the pin numbers of the chip.

Time	D	$\underline{D}$	0	1	Don't Care	Clock/power
t	7		(8..13), 15,(18..27), (29..36),39, (41..43),44	5,6,14,16,17, 28,37,38,40		(1..4),45
t+1	7		5,(29..39), (8..14), (41..43),44	6,15,28,40	(16..27)	(1..4),45
t+2	7		5,39, (41..43),44	6,40	(8..38)	(1..4),45

The test sequences at time t is based on the following (please see Figure 2 for an explanation of the names used here):
|00000001|01|10000000000|10000000011|
where, receiver = 00000001,
increment = 01,
lower bound = 10000000000, and
upper bound = 10000000011.
The test sequence at time (t+1) is based on:
|D|00000001|xxxxxxxx|xxxx|10000000000|
where x is don't care and the value for the receiver field in the input packet is 00000001.

Since the circuits realizing the primitives may contain feedback paths, sequences of test patterns may be needed for testing. An algorithm for generating test pattern sequences is now shown. The algorithm is an extension of the D-algorithm [2-5].

Input to the algorithm is a MacPitts program. The algorithm uses D-propagation tables and calls routines to do the D-propagation for the operators and constructs discussed in the previous paragraphs. The functions SUCCESSOR, D-DRIVE, CONSISTENCY, and BACKUP are used in the algorithm. The SUCCESSOR function finds the operator or construct to which the D ($\underline{D}$) enters. The D-DRIVE function propagates the D ($\underline{D}$) through an operator or construct. The D-propagation through the cond construct requires certain conditions to be satisfied. When these conditions cannot be met, the BACKUP function is used to generate sequences of patterns so that eventually the conditions can be satisfied. The CONSISTENCY function justifies in terms of input pin values all 0, and 1 values imposed by the D-DRIVE.

The input pins of a chip are the pins that receive signals from the external environment, except the ones used for clocks, power, and ground. The following algorithm is repeated for each of the input pins. The algorithm is also repeated for each of the bits of registers, internal ports and signals that are not on sensitized paths. Each of the above mentioned sites are tested for stuck-at-0 by injecting a D, and stuck-at-1 by injecting a $\underline{D}$.

<u>Algorithm 1</u>: <u>Test pattern generation using D-propagation</u>.

1. Inject a D or $\underline{D}$ at the test site.
2. Call SUCCESSOR to find the operator or construct to which the D or $\underline{D}$ enters.
3. Call D-DRIVE to propagate the D or $\underline{D}$.
   - 3.1 If the D-DRIVE is successful then
     - 3.1.1 If the D or $\underline{D}$ propagates to an output pin then call CONSISTENCY to justify the values.
     - 3.1.2 Else go to 2.
   - 3.2 Else (D-DRIVE is not successful due to constructs such as cond) call BACKUP.
     - 3.2.1 If BACKUP fails to satisfy the conditions needed by the D-DRIVE then report "no test can be found" and end.
     - 3.2.2 Else output the sequence of test patterns and go to 2.
4. Output the test pattern and end.

FUNCTIONAL TESTING

The functional testing approach uses the generalized fault model. This approach tests a chip for more classes of faults than the previous approach. This approach uses the operators and constructs of MacPitts as functions. It assumes that at most one function in a chip has failed. Note that multiple failures could occur within the function. For each operator, there is a table containing a set of selected input data for the operator and the expected results. If the execution of the operator outputs the expected results for the selected input data, then the operator is said to be functioning correctly. The propagation of patterns of 0s and 1s through the constructs in MacPitts is done using the techniques discussed under structured testing. The strategy used in functional testing is shown below:

1. Find all paths from input pins to output pins; internal ports, registers, and internal signals to output pins. Let PATH be the routine that performs this.
2. For each path find the constructs and operators on the path. Let FIND be the routine that performs this.
3. Test each construct and operator in a path using the selected input data available in tables.

There are two problems with the above strategy. They are: controllability and observability of operators and constructs that do not have data entering directly from input pins or results coming out on output pins. There are at least two solutions. One solution is to make all the registers controllable and observable using BILBO [6] and including test in and test out pins in the design. This approach significantly increases testing time since state setting and state observation have to be done in a bit-serial manner. The second solution is to include entries in the tables so that for every pair of operators the expected results of one operator are also in the set of selected input data for the other operator. Since the data paths in a MacPitts design have the same width, the loss of bits situation that might occur in going from a wider to a narrower data path is not present. So the above requirement can be met in constructing the tables. Although the table sizes for the operators are increased, it is not significant considering the small number of operators in MacPitts. Note also that the interconnection of the operators is needed to construct test tables for the operators. For example, let function f2 appear at two places in a design. Let the output of function f1 be connected to f2 in one place. Let the output of function f3 be connected to f2 in the second place. Then the test data for f2 in the first place will be different from the one for the second place assuming that f1 and f3 are different functions.

An algorithm for generating sequences of test patterns for functional testing is now developed using the strategy outlined above and the second solution for the controllability and observability problems. Input to the algorithm is a MacPitts program. The algorithm uses tables containing sets of data for operators and constructs. The algorithm calls routines such as DRIVE, SUCCESSOR, BACKUP, and CONSISTENCY. The DRIVE routine propagates a pattern of 0s and 1s through an operator or construct using the table for the operator or the routine for the construct. The other routines are the same as those used in the structured testing using D-propagation. The following algorithm is repeated for each of the operators or constructs in the MacPitts program. These are obtained using the routine FIND.

Algorithm 2: Test pattern generation for functional testing

1. Pick an input pattern from the table corresponding to the operator or construct that is under test.
2. Call DRIVE to advance patterns of 0s and 1s through an operator or construct.
   - 2.1 If DRIVE is successful then
     - 2.1.1 If the patterns have reached output pins then call CONSISTENCY to justify the values.
     - 2.1.2 Else go to 4.
   - 2.2 Else (Drive is not successful due to constructs such as cond) call BACKUP
     - 2.2.1 If BACKUP fails to satisfy the conditions needed by the DRIVE then report "no test can be found" and end.
     - 2.2.2 Else output the sequence of test patterns and go to 4.
3. Output the test pattern. Pick the next input pattern for the operator or construct under test from the table.
   - 3.1 If no input pattern then end.
   - 3.2 Else go to 2.
4. Call SUCCESSOR to find the operator or construct to which the pattern of 0s and 1s enter next. Go to 2.

ACKNOWLEDGEMENT

I am thankful to Jay Southard of MIT Lincoln Laboratory, Lexington, MA, for compiling MacPitts programs, and explaining some of the intricacies of MacPitts.

REFERENCES

1. J. M. Siskind, J. R. Southard, and K. W. Crouch, "Generating custom high performance VLSI design from succinct algorithmic descriptions", Proc. of 1982 Conference on Advanced Research in VLSI, MIT, Cambridge, MA, Jan. 1982, pp 28-40.
2. J. P. Roth, "Diagnosis of automata failures: A calculus and a method", IBM Journal of Research and Development, Vol. 10, July 1966, pp 278-291.
3. J. P. Roth, Computer Logic, Testing, and Verification, Computer Science Press, Woodland Hills, CA, 1980.
4. W. Bouricius et al., "Algorithms for detection of faults in logic circuits", IEEE TC, Vol. C-20, No. 11, Nov. 1971, pp 1258-1264.
5. Y. H. Levendel, and P. R. Menon, "Test generation algorithms for computer hardware description languages", IEEE TC, Vol. C-31, No. 7, July 1982, pp 577-588.
6. T. W. Williams, and K. D. Parker, "Design for testability - A survey", IEEE TC, Vol C-31, No. 1, Jan. 1982, pp 2-15.
7. G. Gordon, and H. Nadig, "Hexadecimal signatures identify troublespots in microprocessor systems", Electronics, March 3, 1977, pp 89-96.


```
(program test_egl 4 ; Word length is 4.
( def 1 ground)
(def 2 phia)
(def 3 phib)
(def 4 phic)
(def input1 port input (5 6 7 8))
(def input2 port input (9 10 11 12))
(def output1 port output (13 14 15 16))
(def r1 register)
(def r2 register)
(def r3 register)
(def r4 register)                   ___       ___
(def t1 register)                  | A |     | S |
(def t2 register)               -->| D |     | H |
(def go1 flag )                 -->| D |     | I |
(def go2 flag )                    |   |---->| F |----|
(def reset signal input 17)        |   |---->| T |--- |
(def 18 power)                     |___|     |___|  | |
;                                                   | |
(process add1 0                     ___             | |     ___
  loop                             | A |            | |--->|   |
(setq r1 input1)                -->| N |            |----->| A |
(setq r2 input1)                -->| D |                   | D |
(setq t1 (>> (+ r1 r2)))           |   |------------------>| D |--->
(setq go1 t)                       |   |------------------>|   |--->
(go loop))                         |___|                   |___|
;
(process and1 0                 --->
  loop                          --->    4 bits wide path
(setq r3 input2)
(setq r4 input2)
(setq t2 (word-and r3 r4))
(setq go2 t)
(go loop))
;
(process add2 0
  loop
(cond ((and go1 go2) (setq output1 (+ t1 t2))))
(go loop)
))
```

; Figure 1 An example design for test generation purposes.

```
;  MacPitts program for asynchronous packet receiver using 4 cycle
;  timing.
(program receiver 32  ; word length is 32.
  (def 1 ground)
  (def 2 phia)
  (def 3 phib)
  (def 4 phic)
  (def init signal input 5)
  (def rcv signal input 6)
  (def rcvdata port input (7 8 9 10 11 12 13 14 15 16 17 18 19
    20 21 22 23 24 25 26 27 28 29 30 31 32 33 34 35 36 37 38))
  (def last signal input 39)
  (def ack signal output 40)
  (def rela signal output 41)
  (def output1 signal output 42)
  (def reference register)  ;The bits from left to right are used for
                            ; the following.
                            ;8 for receiver, 2 for increment,
                            ;11 for lower bound, 11 for upper bound.
  (def count register)  ; Counter for max. number of packets.
  (def maskpid register) ;Mask register.
  (def numpac register) ; Number of packets in a token.
  (def temp1 register)
  (def temp2 register)
  (def inbuf1 register) ;Input buffer for packet 1.
  (def inbuf2 register) ;Input buffer for packet 2.
  (def inbuf16 register) ;Input buffer for packet 16.
  (def completed signal output 43) ;Read for next token.
  (def reset signal input 44)
  (def 45 power)
 (process rcvtest 1
   (cond  ((and rcv init) ;Initialize the reference register.
             (setq reference rcvdata)
             (setq count 0)
             (setq ack t))
          ((and rcv (not init)) ;Data packets.
             (cond  ((= count 0) ; First packet.
; Format of the first packet:
; ---------------------------------------------------
; | type | receiver | sender | sys type | PID  |
; ---------------------------------------------------
;    1        8          8          4         11
;
                        (call testupr))
                      ((< count 16) ;Second through 16th packet.
                        (call other))
                      (last   (call finish))
             )
          )
   )
 )
other  (cond ((= count 1)  (setq inbuf2 rcvdata) (setq ack t)
                            (1+ count) (go last1))
             ((= count 2)  (setq inbuf3 rcvdata) (setq ack t)
                            (1+ count) (go last1))
             ((= count 15)  (setq inbuf16 rcvdata) (setq ack t)
                             (setq numpac count) (setq rela t)
                             (go last1)))
last1  (return)
;
;
finish  (setq ack t) (setq rela t) (setq numpac count)
        (setq count 0) (setq output1 t) (return)
;
;
; Test for PID <= upper bound register.
;
testupr  (setq maskpid 2047) ; Set the maskpid so that the least
                             ; significant 11 bits are ones.
         (setq temp1 (word-and maskpid rcvdata))
         (setq temp2 (word-and maskpid reference))
         (cond  ((< temp1 temp2) (go testlwr))
                ((= temp1 temp2) (go testlwr))
                (t (go reject1)))
;
; Test for PID >= lower bound register.
;
testlwr  (setq temp2 (3>> (8>> reference)))
         (setq temp2 (word-and maskpid temp2))
         (cond  ((< temp2 temp1 ) (go testrcvr))
                ((= temp2 temp1 ) (go testrcvr))
                (t (go reject1))    )
;
;
;Test for rcvr=receiver in reference register.
;
;
testrcvr  (setq maskpid 4278190080)
          (setq temp1 (<< rcvdata))
          (setq temp1 (word-and maskpid temp1))
          (setq temp2 (word-and maskpid reference))
          (cond  ((= temp1 temp2)
                     (setq ack t)
                     (1+ count)
                     (setq inbuf1 rcvdata))
                 (t (go reject1))
          )
          (return)
;
;
reject1   (setq rela t)
          (return)
)   ; End of program.
```

Figure 2 MacPitts program to show the use of cond construct.

Session 1-1c
VLSI Placement

Chairman

J. Rowsen
VLSI Technology

A CONSTRUCTIVE PLACEMENT ALGORITHM FOR LOGIC ARRAYS

STEVEN JAY MEYER

LSI LOGIC CORPORATION

ABSTRACT This paper describes a new constructive phase only placement algorithm for CMOS logic arrays. This algorithm is based on the idea of epitaxial type growth of placed cells from I/O pads inward. It requires neither initial partition or final iterative improvement phases but may only apply to one particular type of master slice logic array organization. Both the algorithm's efficiency and the advantages of constructive phase only placement are discussed.

Keywords and Phrases:

Physical Computer Aided Design
Layout
Automatic Placement
Logic Array Placement

1.0 Introduction

Advances in integrated circuit technology and logic design aids have made 10,000 gate logic arrays practical in 1983, but computer aided physical design must still be improved to efficiently utilize advances in silicon technology. Since most placement and routing algorithms are NP complete (probably not solvable in polynomial bounded time) in their worst cases, progress has been limited by computing time requirements alone.

This paper describes a constructive phase only placement algorithm which is part of LSI LOGIC's LDS II logic array design system. It is routinely used to place CMOS circuits with up to 6000 potentially usable gates. The algorithm is limited in the sense that it is optimized for one particular array geometry, but it requires no iterative improvement phase and uses only a few minutes of computer time.

The main scientific contribution of this paper is in showing that a constructive placement algorithm without either pre-placement circuit partitioning or iterative improvement is much more promising than currently believed and in offering reasons why this may be so.

2.0 The Basic Placement Problem

The logic array (sometimes known as gate array) placement problem involves assigning circuit elements to physical integrated circuit transistor gate locations. A circuit element is a cell which implements some logic function and ranges from a simple two-pin inverter to high-level, multi-pin macro functions such as flip-flops or multiplexers.

Following Fiduccia and Mattheyses (Fiduccia (1982)) (Schweikert (1979)), problem input is presented as a set of C cells (circuit elements) connected by a set of N nets (signals or wire lists). The point at which a net connects to a cell (internal macrocell) is called a pin and by assumption each net contains at least two pins, and each cell is in at least one net.

This formulation implies that the size of a given problem is bounded by the number of cell pins P in a given circuit and means any simple operation such as circuit input is O(P) (roughly bounded by P) except that some input schemes may require input pin sorting which is O(P log P).

Problem input also includes two data structures which are assumed to have been constructed during circuit input. One is a list for each cell giving each net that cell is in. The other is a list for each net giving each cell that net is connected to.

3.0 The CMOS Logic Array Placement Problem

The particular constructive placement algorithm described here is embedded in LSI LOGIC's LDS II automated design system (Koford (1981)) (LSI LOGIC (1983)) and is similar to the LTX system (Perky (1977)) in intent. Further, the current algorithm only applies to the 5000 and 7000 series of CMOS logic arrays (Shiraishi (1980)) (Kanada (1981)) (Werner (1982)). Arrays in the series use two layer metal for separate horizontal and vertical wiring and only one uniform cell placement area. The cell placement area is a rectangle of R cell rows each of which is G gates wide. The total internal array area A is R^*G. All external (off chip) signals must be routed through I/O pads (package pins).

The 5000 series consists of arrays ranging in size from 800 to 6000 gates (LSI LOGIC (1982)). The equal length gate rows are separated by fixed height horizontal wiring channels. For example, the 3200 gate member of the array family contains nineteen cell rows with each row containing one hundred sixty eight gates. The area between the chip I/O pad region and the internal cell region also contains usable wiring tracks both above and below the cell rows and to the right and left of the row and channel ends. These areas are somewhat larger than the wiring channels.

Each horizontal wiring channel currently contains sixteen tracks (nineteen for the 7000 series) and there are no over cell row vertical wiring channels. This causes what is known as the interchannel feed-through problem (Perky (1977)) (Kanada (1981)). It means that feed-through wires must be routed either over unoccupied gate locations or over free tracks inside cells (usually this means not occupied by a pin). The lack of vertical wiring channels implies utilizations higher than 70 to 80 percent are rarely attainable and for circuits with unusually complex wiring (many pins and large nets) utilization can be as low as fifty percent. But wire complexity and feasible utilization estimates are available to the circuit designer.

4.0 Algorithm Description

The algorithm described here is an improved version of the "epitaxial growth" constructive placement algorithm described in Soukup's 1981 overview paper (Soukup (1981)).

As described in Hanan (1972), a constructive placement algorithm selects and places cells according to some evaluation criteria which is defined only in terms of already placed cells and circuit connectivity information. The approach requires a method for selecting and placing starting seed cells, an evaluation criteria for selecting the next cell to place, and an evaluation criteria for determining where to place a cell.

The basic idea is that at any given stage of the placement process, the cell being placed is the one about which the most is known. This means the cell in the net which has the highest percentage of already placed cells. Also when placing a cell, the largest possible sample of connectivity information is used.

This algorithm is termed "epitaxial" since cells "grow" from the edges inward in a planar pattern even though isolated islands often appear early and not necessarily near the chip edges. It is an improvement over the algorithm sketched by Soukup in using a two step seed placement phase and in usually completing the placement of all cells in a net before starting on the cells in the next net.

4.1 Seed Cell Selection

The algorithm assumes I/O cell pad assignment has been specified as part of the circuit design. It then uses pads to place some cells in the internal cell region which are then used as seeds for the rest of the placement. For every net containing a pad cell, the largest non-pad cell in that net is placed as close to the pad as possible using exact wire grid distance. The pad cell containing nets are selected in arbitrary order. Nearly all cells are placed in their true minimum wire distance locations and usually from ten to fifteen percent of the circuit's gates are placed in this phase. The algorithm is normally insensitive to chip I/O pad assignment as long as pads are uniformly spaced around the chip. After completion of seed placement, pad cells are simply ignored.

4.2 Next Cell Selection

Cell selection is accomplished by selecting a net and then selecting the largest unplaced cell in that net. The selected net is the one which has the largest percentage of cells already placed. Ties are broken by choosing the largest net. Largest means containing the most cells, and if there are still ties, largest means occupying the most total gates. After a net to place is selected, the largest unplaced cell from that net is placed and the selection process is repeated. In general, placing one more cell in the already most completed net insures that the same net will be selected again, but sometimes small nets are finished before the current net is completed.

An earlier version of the algorithm placed all unplaced cells whenever a net was selected and in all but a few cases seemed to do as well. The reason behind this selection criteria is that the important idea is not how good the next cell placement is in terms of already placed cells, but in making sure the next cell has the most possible location information known about it.

4.3 Cell Placement

Once a cell is selected, the algorithm tries all unoccupied internal array locations (contiguous gate regions) in which the cell will fit. For each such location, it attempts to minimize the sum of the weighted minimum spanning tree distance for each net the given cell is connected to. In effect, net distance is computed by finding the closest already placed cell in each connecting net.

The evaluation function is measured only to gate, not wire grid, distance but otherwise simply minimizes distance weighted by user supplied weights. The weights are based on real wire grid distance but are usually modified to improve wireability. They were discovered through trial and error and seem to increase the frequency of cross channel connections and reduce the number of cell row feed through wires.

5.0 Pragmatic Considerations

The algorithm has several user selectable parameters which allow compensation for differences between array sizes and individual circuits. They also aid in producing placements which take advantage of the available automatic router's strong points. Circuits are sometimes placed more than once to improve routability.

6.0 Running Time

This algorithm's running time is bounded by the cell placement phase since the cell selection phase simply involves a minimum calculation over the as yet uncompleted nets performed once for each cell. The number of steps required to place all the cells is loosely bounded by:

$$A*p(c_i))*n_{max}$$

where A is the number of gates in the array, $p(c_i)$ is the number of pins cell i has, and n_{max} is the size of the largest net in the circuit. Since A and n_{max} are constants for a given circuit, the bound becomes $A*P*n_{max}$ where P is the total number of internal macrocell pins in the circuit. Remember that the sum for all cells C of $p(c_i)$ is P. In practice, this value is bounded by $O(P^2)$ since the number of gates is usually comparable to the number of pins and nearly all nets contain less than five or ten pins.

The running time for various placements on an Amdahl 470 V6 using IBM Pascal with moderate use of inline procedures ranges from 30-45 second for 1400 gate arrays to 8-10 minutes for 6000 gate arrays. I/O and data structure construction times are included.

7.0 Results

The algorithm works well in the sense that it produces wire completion results better than those produced by a random initial placement, net span minimization iterative improvement algorithm which is similar to those described in Hanan (1978). It also produces reasonable wire lengths and spreads wire congestion out uniformly. In the one case it was compared to a manually placed circuit, it produced better wire length but required more manual clean-up routing.

8.0 Discussion

The described algorithm seems to work well because it makes good use of the edge area of the chip while doing a reasonable job of wire length minimization. The idea of epitaxial growth from the edges inward may work because it corresponds to the way circuit designers think of their circuits. The problem of array center congestion, as described in Burstein (1982) seems to be avoided by using pad net seed cells. Finally, epitaxial growth with distance minimization seems to generate more equally good possible cell locations than iterative distance minimization schemes and, therefore, seems to work well over a large range of different circuit types.

The success of this algorithm seems surprising in light of the commonly accepted belief that the most promising placement approach involves the three-step process: circuit partition, constructive initial placement, and finally iterative improvement (Hanan (1972)), (Hanan (1978)), (Soukup (1981)). On the other hand, it is also accepted that placement is not a solved problem and work is continuing both on other placement techniques (Perky (1977)) and enhancements to iterative improvement (Kirkpatrick (1982)), (Burstein (1982)). In general, the three-step approach may be valuable, but in the case of CMOS logic arrays, there seem to be so many conflicting requirements which must be balanced, that no simple mathematical evaluation function may exist, and possibly the best evaluation function may be pad cell driven epitaxial growth itself.

9.0 References

Burstein (1982)

Burstein, M., Hong, S., and Nair, R. Spatial Distibution of Wires in Master Slice VLSI. Proceedings IEEE International Conference on Circuits and Computers, 1982, 265-260.

Cote (1980)

Cote, L., and Patel, A. The Interchange Algorithms for Circuit Placement Problems. Proceedings 17th Design Automation Conference, 1980, 528-534.

Fiduccia (1982)

Fiduccia, C., and Mattheyses, R. A Linear-Time Heuristic for Improving Network Partitions. Proceedings 18th Design Automation Conference, 1982, 175-181.

Hanan (1972)

Hanan, M., and Kurtzberg, J. Placement Techniques. In M. Breuer, Ed. Design Automation of Digital Systems, Vol. 1, Prentice Hall, 1972, 213-282.

Hanan (1978)

Hanan, M., Wolff, P., and Agule, P. A Study of Placement Techniques. Journal of Design Automation & Fault-Tolerant Computing. 2, 2 (May 1978), 28-61

Kanada (1981)

Kanada, H., et. al. Channel-Order Router--A New Routing Technique for a Masterslice LSI. Journal of Digital Systems 4, 4 (September 1981), 427-441

Kirkpatrick (1982)

Kirkpatrick, S., Gelatt, C., and Vecchi, M. Optimization by Simulated Annealing. IBM Research Report, RC 41093, April 1982.

Koford (1981)

Koford, J., and Jones, E. A Development System for Logic Arrays. Midcon 1981, Chicago.

LSI LOGIC (1981)

LSI LOGIC CORP., Silicon Gate HCMOS Macrocell Array LSI 5000 Series Data Sheet. September 1981.

LSI LOGIC (1983)

LSI LOGIC CORP., CMOS Macrocell Array Design Manual. March 1983.

Perky (1977)

Perky, G., Deutsch, D., and Schweikert, D. LTX--A Minicomputer-based System for Automated LSI Layout. Journal of Design Automation & Fault Tolerant Computing, 1, 3 (May 1977) 217-255.

Shiraishi (1980)

Shiraishi, H., and Hirose, F. Efficient Placement and Routing for Master Slice LSI. Proceedings 20th Design Automation Conference, 1980, 458-464.

Soukup (1981)

Soukup, J. Circuit Layout. Proceedings of the IEEE, 69, 10 (October 1981) 1281-1304.

Schweikert (1979)

Schweikert D., and Kernighan, B. A Proper Model for the Partitioning of Electrical Circuits. Proceedings 9th Design Automation Workshop, June 1972, 57-62.

Werner (1982)

Werner, J. Computer-Aided Design and Design Automation for ICs in Japan. VLSI Design, 3, 3 (May/June 1982) 206-214.

ORDER OF CHANNELS FOR SAFE ROUTING AND OPTIMAL COMPACTION OF ROUTING AREA

Yoji KAJITANI

Dept. of E. E. Engrg.
Tokyo Institute of Technology
Ookayama, Meguro-ku, Tokyo 152, Japan

ABSTRACT The concept of safe ordering of the channels in completing the channel routing is introduced. Following a safe order if it exists, the channels can be completed the wiring requirement one by one without any prediction of necessary width. Thus the compaction of the routing area can be made utmost each time. An efficient algorithm to find a safe order is given.

1. INTRODUCTION

The layout problem in designing integrated circuits composed of a number of rectangular blocks of various dimensions on a board is usually considered in three phases, namely, placement of blocks, global routing and channel routing. (See [1] for an overview of the layout problem.)

In the final phase of channel routing, the channels are picked up one at a time and routed. Usually, the initial placement of blocks is floating, that is, fixes only the relative position of blocks. But the completion of channels one by one reduces the allowance of the move of blocks. Furthermore, the necessary width of a channel is not known before the actual routing. Thus, at some step of the channel routing procedure, it is probable that a channel is found impossible to be completed its required wiring by the adopted channel router. Once this situation is met, the designer is forced to revise considerable part of the preceding result. This is not only because of the ill placement or the inefficient router but also the channels that have been routed before.

It has long been noticed[2] the above fact that the ordering of channels for routing affects fatally the routability, the 100% routing of nets, and moreover the whole area and density distribution of wires of channels as well. Several considerations[2,3,4] have been given to this problem and some ideas have been realized as the design rule.

The main purpose of this paper is to introduce and characterize graph theoretically the concept of safe order of the channels for a given placement with global routing. Following a safe order if it exists, the routing of each channel can be completed without any prediction of necessary width. It is a direct consequence of the definition that following a safe order, compaction of the width of each channel after the routing can be made utmost each time.

It should be remarked that part of the conclusion here is found in [2] though the formulation and reasoning are pretty different, particulaly in that they imposed the idea as the design rule and hence their model always has a safe order while this paper is to present an analysis tool for a given placement and global routing.

2. SAFE ORDERING OF CHANNELS

We begin with the definition of fundamental terms.

I. PLACEMENT OF BLOCKS

On a board, n rectangular blocks labeled with 1, 2, ...,n of various dimensions are arbitrarily layed such that no two of which do not overlap. Two blocks are said to be *adjacent* with each other if they have the sides confronting each other with no blocks between. A *placement* defines uniquely the adjacent realtion. This in turn means that any block can move arbitrarily within a range that does not alter the adjacency relation.

II. CHANNELS

Each block has signal terminals on its sides that are to be connected to others by wires which run on zones between blocks, called the channels. There may be different definitions of unit channels. The channel as indicated in Fig.1(A) which is the maximal rectangular zone between exactly two adjacent blocks is called the *element channel*. While in (B), the channel is composed of several element channels. In this paper, the channel is an element channel. If it is necessary to consider a composite channel as in (B), the condition is imposed that the corresponding element channels are to be routed simultaneously.

There remain the intersection zones that are not covered by the element channels but need to be routed (*switch boxes*[1]). Until Sec.5, they are neglected for clearness of the theory.

By definition, each channel is represented by a pair of blocks as (i,j). A channel (i,j) where block i and j are upper and lower (left and right) of the channel, respectively, is called the *X- (Y-) channel*. In the case of Fig.1(A), the set *C* of the *chnnels* is {(1,2),(1,3),(2,3),(2,4),(3,4)} where (2,3) and (3,4) are X-channels. For brevity, the placement itself is also denoted by C.

(A) Element channels (B) Composite channels
Fig.1 Definition of element channels

CH1935-6/83/0000/0062$01.00 © 1983 IEEE

III. SAFENESS

Our channel router is a general one characterized by the following three features.

A. Parallel rigidity

Suppose that an X- (Y-)channel (i,j) has been routed. Then, block i and j are forbidden to alter their relative X- (Y-)coordinates.

B. Width expandability

Suppose that (i,j) has been routed. Blocks i and j have to be at least in certain distance (the number of tracks which the router used). But they can extend the distance without dismissing the completed routing of the channel (for example, by extending the near terminal lines).

C. Safeness

If a block i can move in a direction as much as it would be permitted if no routing started (within a range of not changing the placement), then it is said that block i can move freely in that direction. An unrouted channel (i,j) is called safe if and only if one of i and j can move freely in such a direction as to expand the distance of i and j.

Consider a stage of a channel routing procedure. Let C* and C** be the sets of routed and unrouted channels, respectively. Then, the stage is represented by C*.

An ordering of unrouted channels

$$O(C^{**})=(c_{\alpha 1},c_{\alpha 2},\ldots,c_{\alpha p})$$

at stage C* is called safe if and only if $c_{\alpha k}$ for $k=1,2,\ldots,p$ is safe at stage $C^{*}\cup\{c_{\alpha 1},\ldots,c_{\alpha k-1}\}$. If there exists a safe order of all channels at the initial stage, the placement C is safe.

When it is said that a channel (i,j) (εC^{**}) is safe, it is assumed that the channels router can be applied to it and always accomplishes 100% the connection requirement in the channel. Of course, even if the channel is not safe, its routing will possibly be successful. However, the idea in this paper is that the routing should not be excuted expecting such a luckiness. Once a channel is found impossible to be completed its wiring requirement in the procedure, the designer has to pay a considerable labor for rearranging then obtained result. The safeness of a channel is the minimum to secure the routability of the channel when no prediction is made.

EXAMPLE 1 For a given placement C shown in Fig.2(A), the move of block 6 down to level y for example changes the adjacency relation of 6 and 4. Hence block 6 can shift down up to level x but no lower. Still by definition, we regard that block 6 can move " freely" . Hence channel (2,6) is safe. After its routing, 1, 3, 5, and 7 can move freely upwards, right, left, and right respectively and independently. Thus any sequence of (5,6), (1,5), (1,3), (2,7), and (6,7) is safe. The placement after their routing is shown in Fig.2(C). At this stage, blocks 1 and 5 can move left together as illustrated in (D). Thus (1,2) is safe. After its routing, (4,5) is safe. Tehy are routed as shown in (E). Now the safeness of (4,6) is understood by the combined shift of 1, 4, and 5 left as in (F).

All other channels (2,3), (3,5), (2,4), and (3,4) are not safe. The reason is: The distance of 2 and 3 cannot be expanded because of the rigidity of the routed channels (2,1) and (1,3). The distance of 3 and 5 is not expandable because the transitivity of the rigidity of routed channels (3,1), (1,2), (2,7), (7,6), and (6,5) fixes 3 and 5, vertically.

(2,4) is not expandable freely because the distance of 2 and 5 is fixed (as explained above) and the routing of (4,5) occupies a certain amount of width. (3,4) is not expandable freely by the rigidity of (4,5) and (5,1) and the occupied width of (3,1).

Remember that the explanation of unsafeness for (2,3) and (3,5) and that for (2,4) and (3,4) are a little differnet. □

3. THE SAFENESS THEOREM

For a given placement C, the block-channel graph G(C)=(V,E) is a directed graph defined as follows. The vertex set is $V=\{v(1),v(2),\ldots,v(n)\}$. Each block i corresponds to v(i). Each channel (i,j) corresponds one-to-one to an edge e(i,j) that connects v(i) and v(j). e(i,j) has the direction from v(i) to v(j) if and only if block i is on the upper (left) of the channel if it is an X- (Y-)channel.

If (i,j) is an X- (Y-)channel, e(i,j) is called the X- (Y-)edge. Thus E is partitioned into $E=X\cup Y$ with $X\cap Y=\emptyset$ where X and Y are the sets of all X-edges and Y-edges, respectively.

For a subset $D\subset C$, E[D] denotes the set of corresponding edges of G. The subgraph of G(C) whose edge set is E[D] is denoted by G(D). On the oth-

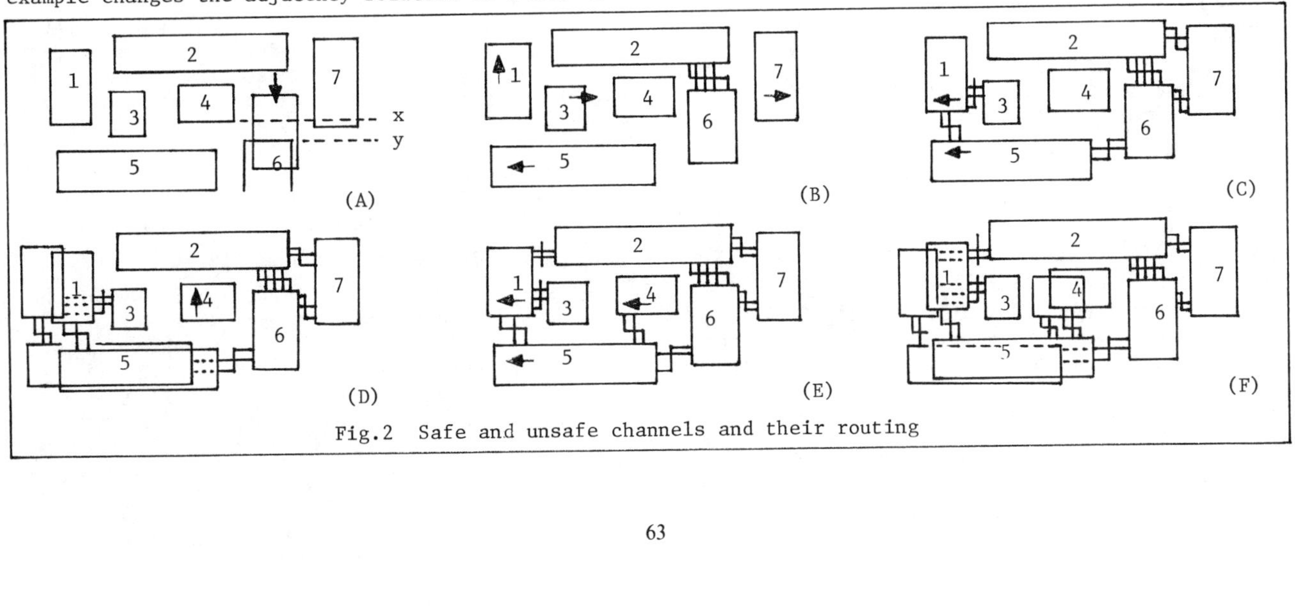

Fig.2 Safe and unsafe channels and their routing

er hand, for $F \subset E$, $C[F]$ denotes the set of channels corresponding to F.

Examples of C, C*, and G(C*) are found in Fig. 3 where double lines and single lines denote X-edges and Y-edges, respectively.

Foe $e \varepsilon E$ and $F \subset E$, e is said to be *spanned* by F if either $e \varepsilon F$ or its two end vertices are connected by a path in F. It is a well-known fact that e is spanned by F if and only if any cut containing e has at least one edge of F.

An edge set $F \subset E$ is called *X- (Y-)spanned* if every edge of F is spanned by $X \cap F$ ($Y \cap F$). When F is both X-spanned and Y-spanned, it is called *XY-spanned*. A cut (usally, cutset) $\gamma \subset E$ is called *X- (Y-)cut* if $\gamma \subset X$ ($\gamma \subset Y$).

THEOREM 1: At routing stage $C^* \subset C$, suppose that an unrouted channel $(i,j) \varepsilon C^{**}$ is an X- (Y-) channel. Then, (i,j) is safe if and only if there exists a directed X- (Y-)cut containing e(i,j) in graph $G(C^* \cup \{(i,j)\})$. □

For the proof, see [5].

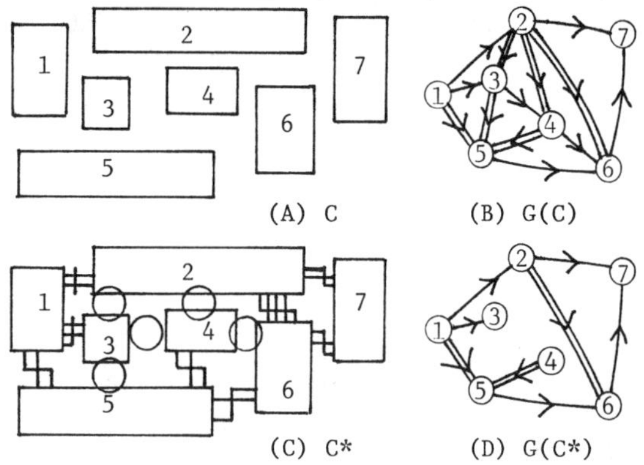

Fig.3 Placement C and stage C*

EXAMPLE 1 For stage C* shown in Fig.3(C), the theorem applies to $C^{**}=\{(4,6),(2,3),(3,5),(2,4),(3,4)\}$ as follows. Y-edge e(4,6) belongs to a directed Y-cut $\{e(4,6),e(5,6),e(1,2)\}$ of $G(C^* \cup \{(4,6)\})$. Hence (4,6) is safe. X-edges e(2,3) and e(3,5) are spanned by Y-edges of $E[C^*]$. Thus they are not safe. e(2,4) is contained in the unique X-cut $\{e(2,4),e(4,5)\}$ in $G(C^* \cup \{(2,4)\})$. But it is not directed. Hence (2,4) is not safe. Y-edge e(3,4) is contained in several Y-cuts but none of them is directed. Hence (3,4) is not safe.

Note that these conclusions are coincident with those given in Ex.1. Furthermore, the explanation for (2,3) and (3,5) was to show the nonexistence of X-cuts while that for (2,4) and (3,4) was to show the nondirectedness of single-kind-element cuts. □

4. ALGORITHM FOR A SAFE ORDER

Three corollaries are given without proof[5].

COROLLARY 1: If G(C) contains a nonempty XY-spanned subset of edges, C is not safe. □

This corollary proves that C in Fig.3(A) is not safe because G(C) contains XY-spanned subsets, for example, the one composed of edges connecting the vertices v(1), ...,v(6).

COROLLARY 2: If C is safe, there is a directed X-cut or Y-cut. □

COROLLARY 3: For a safe placement C, let

$$O(C)=(c_{\alpha 1},c_{\alpha 2},\dots,c_{\alpha m})$$

be a safe order of all channels. Let γ be any directed X- (Y-)cut. Then, an order O'(C) which is obtained from O(C) by letting the channels of $C[\gamma]$ be rearranged in the last part arbitrarily is also safe. □

From these corollaries, it can be considered an algorithm that starts by finding a directed X- or Y-cut. A conventional way to find a directed X-cut will be as follows. Find the maximal components connected by Y-edges in the underlying undirected graph. Contract each component into a vertex. In the resultant graph composed only of X-edges, take a vertex and start a walk from it along any directed path. If a vertex appears twice on the way, short all the edges between. Take another vertex and repeat. If no edge left, G(C) has no directed X-cut. If no vertex appears twice, we reach a sink vertex. The set of edges incident with it corresponds to a directed X-cut of G(C). Whole the procedure takes complexity O(mn) where m is the number of channels. Thus, an algorithm will be as follows.

Algorithm: SAFE ORDER O(C)

Output: NO: C is not safe.
YES: C is safe and the safe order is

$$O(C)=(c_{\alpha 1},c_{\alpha 2},\dots,c_{\alpha m})$$

Step 1: $G_0=G(C)$, $i=0$.

Step 2: If G_i contains no edge, go to Step 3. Otherwise, find a directed X-cut or Y-cut of G_i. If there is no such cut, output NO. If it exists, let any one be γ_i.

Step 2: Get G_{i+1} from G_i by deleting the edges of γ_i. Let $i=i+1$. Return to Step 1.

Step 3: Construct O(C) arbitrarily as far as the following condition is satisfied. In the cut sequence $(\gamma_0,\gamma_1,\dots)$ obtained, let $c_{\alpha p} \varepsilon \gamma_\delta$ and $c_{\alpha q} \varepsilon \gamma_\tau$. Then, $p > q$ if and only if $\delta < \tau$. That is, a channel comes earlier in O(C) the later it comes in the cut sequence. □

If there is no safe order for C, it occurs the idea of dropping certain channels from the list C. The channels neglected from C are to be routed at the final stage. Therefore, such channels should have enough space compared with the density.

EXAMPLE 3 The placement in Fig.3(A) is not safe because, as was shown before, it contains an XY-spanned subset. But suppose here that (2,6) is the one that can be neglected (by some reason, for example that somehow very few wires are assigned to it). In this situation, we have only to consider the subplacement $C'=C-\{(2,6)\}$. The graph G(C') contains no XY-spanned subset. In fact, if the algorithm is applied to G(C'), cut sequence $(\gamma_0,\gamma_1,\dots,\gamma_5)$ is obtained as shown in Fig.4(B). (See from the top.) The channel routing following the order determined by this sequence is shown in Fig. 4(A). (See from the bottom. At each step, the order of channels is arbitrary.) □

This kind of consideration leads to a hard problem of finding a maximum safe subplacement[5].

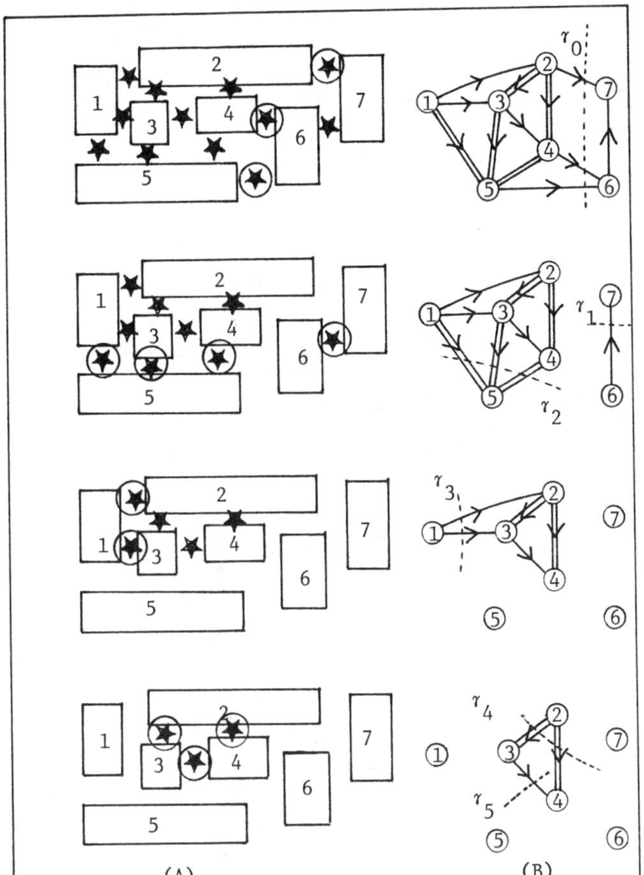

Fig.4 Routing following a safe order of $C'=C-\{(2,6)\}$

5. SWITCH BOX ROUTING

The ordering principle roughly sketched above has prospects but as well conceives many problems to be answered. They are considered to some extent in [5]. Here only the switch box routing is considered.

The typical example of a switch box is shown in Fig.5(A) which is bounded by sides of 4 element channels c_1, c_2, c_3, and c_4. If we follow strictly our frame work of theory, it is to be routed after all other channel routings. But it may cause the unroutability.

Here an idea is proposed by which the difficulty is avoidable with little change of the theory.

Suppose that the switch box in Fig.5(A) is part of a safe placement C. In a safe order O(C), let c_4 be the last of these four element channels. Then let c_1, c_2, and c_4 be as they are element channels, and c_4 be expanded to c_4' to include the remaining area as illustrated in (B). Suppose a stage when all the element channels before c_4 have been routed. In routing c_4', the width w of (3,4) indicated in (C) is expandable by definition. This means that the width of (1,2) is also expandable because blocks 1 and 2 are horizontally rigid with respect to 4 and 3, respectively. Hence, it is easy to see that c_4' is safe. Finally, it has to be proved that after the routing of c_4', the rest of C is still safe. To show this, it will suffice to prove that blocks 1 and 2 can shift upwards together vertically without dismissing the routing of c_4'. But if we note that the terminals of the equivalents on sides L and R as indicated in the figure are fixed only in the order with respect to the vertical levels, the proposition is clear.

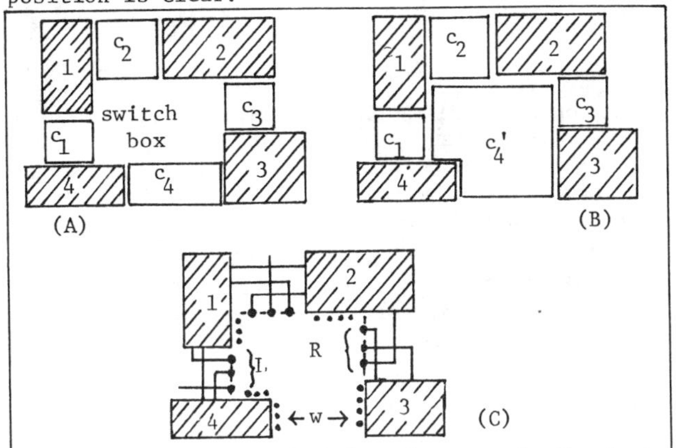

Fig.5 Routing of a switch box where c_4 is the last of c_1, c_2, c_3, and c_4 in a safe order O(C)

6. CONCLUSION

The channel ordering problem is essential in the layout since it organizes the results of subproblems of the placement of blocks, global and loose routing, and the channel routing. This paper established a theory to treat the problem by use of graphs based on the study of common features of the channel routers. It will be used as a tool to analyse a given placement with global routing for safe channel routing and for the acceptable compaction of the routing area.

ACKNOWLEDGEMENTS

The author is indebted to Dr. S. Ueno of Tokyo Institute of Technology, Dr. K. Kawakita and Dr. S. Goto of Nippon Electric Co., Ltd., and Prof. T. Ohtsuki of Waseda University, for their helpful discussions during the course of this work.

REFERENCES

[1] J. Soukup," Circuit layout," Proc. IEEE, 69, 10, pp.1281-1304, 1981.

[2] H. Kawanishi, S. Goto, T. Oyamada, and K. Kani, " A routing method of building block LSI," Rec. 7th Asilomar Conf. Circuits, Systems and Computers, 1973.

[3] K. Sato and T. Nagai," A method of specifying the relative locations between blocks in a routing program for building block LSI," Proc. Int. Symp. Circuits and Systems, pp.673-676, 1979.

[4] B. T. Preas and W. M. vanCleemput," Routing Algorithm for hierarchical IC layout," Proc. Int. Symp. Circuits and Systems, pp.482-485, 1979.

[5] Y. Kajitani," Order of channels for safe routing," IEEE Trans. CAD, Oct. 1983 (to appear)

EFFICIENT AUTOMATIC LAYOUT FOR SEMI-STRUCTURED "BAND-LOGIC"

G. SAUCIER, G. SERRERO, J.F. PAILLOTIN

IMAG Laboratory

BP 68 - 38402 SAINT MARTIN D'HERES Cédex - FRANCE

An operational automatic layout system for semi-structured logic is presented here. Such a logic is organized in "bands" made up of rectangular gates placed along the power supply lines, the connections going over the bands. The algorithms are very efficient ($O(N^2)$ or $O(N^3)$) and never fail to find a solution. Parallelization of the approach is exhibited. Applications concern several technologies.

I - SEMI-STRUCTURED BAND-LOGIC

1.1 Definition

We call semi-structured band-logic a set of interconnected gates along power supply lines. The gates are rectangles of variable length and the connections are routed along a grid within the band width. The gates can be lengthened for routing purpose. This allows layout strategies which never fail and, thus, appear to be very efficient.

The whole circuit is usually organized into cells made up of 2 or several bands (figure 1).

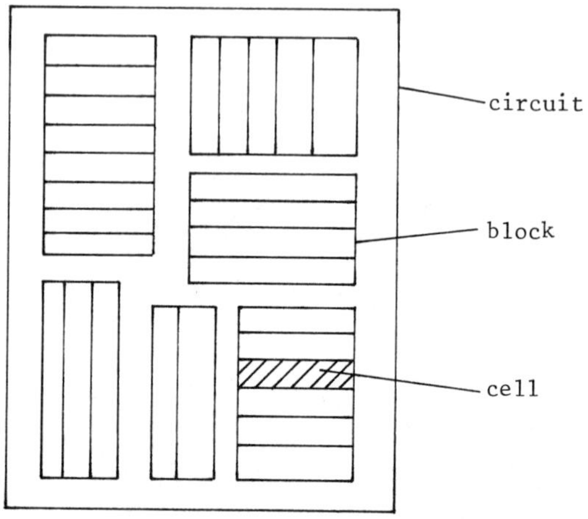

Figure 1

1.2 Examples for different technologies

This band structure has been studied for several technologies : C-MOS, N-MOS, bipolar (I2L) In CMOS technology, a commonly used structure is the Weinberger's layout (1). The gates are NOR gates in which the bottom part is made up of parallel N-transistors, the upper part of serial P-transistors. The inputs and the outputs are duplicated on both parts. The band is, here, composed of 2 parallel bands. It is sufficient to process one of them as the other one is automatically obtained by symmetry (figure 2).

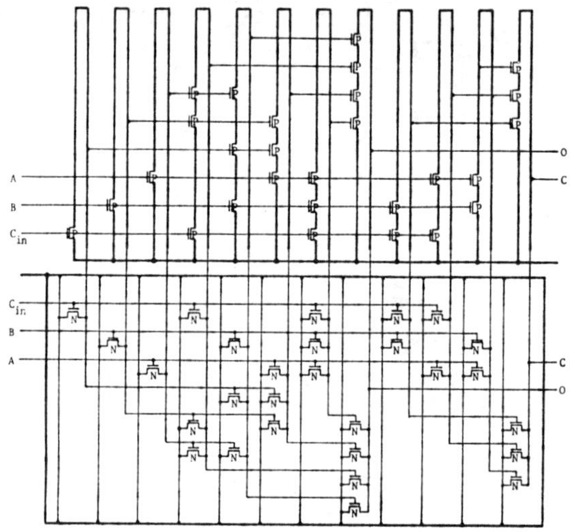

Figure 2

In bipolar technology, a band-structure is used in I2L (2). This technology enables "wired" AND's. The basic gate is an inverter with replicated outputs. Each output is electrically isolated from the others. The ground is the bulk of the circuit (see the gate figure 3).

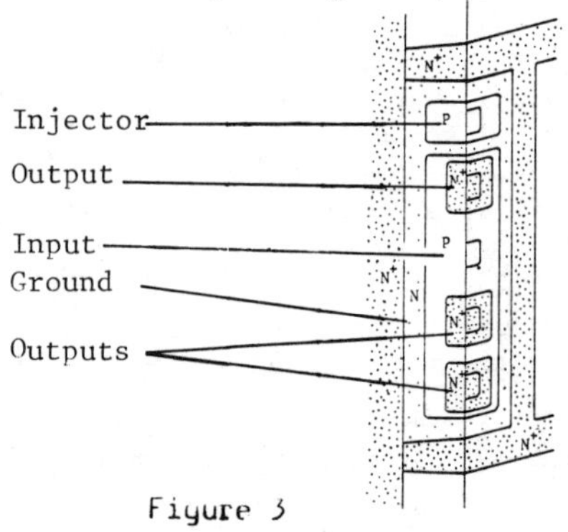

Figure 3

In N-MOS technology, the Weinberger's layout is also used. An other band-structure based on the "multidrain gate" (3) is widely used in France and has been extensively studied for automatic layout purpose. The topological structure is similar to I2L (figure 4).

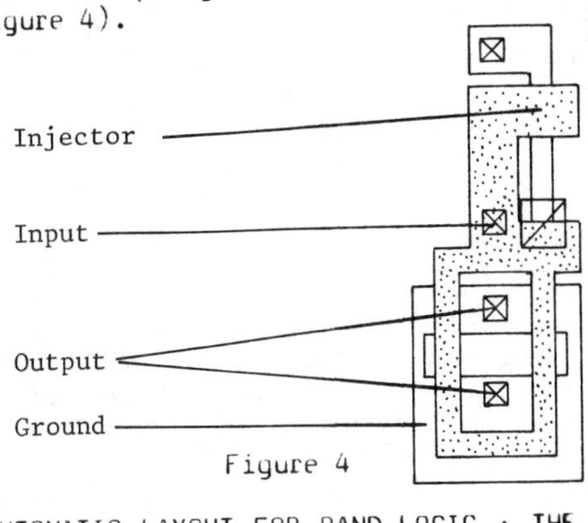

Figure 4

II - AUTOMATIC LAYOUT FOR BAND-LOGIC : THE OASIS SYSTEM

Automatic layout consists in embedding a logic network (here a network of gates) into a target structure. Two main steps will be discussed : the placement and the routing of the connections.

2.1 The placement in OASIS

The placement problem consists in assigning the gates to the places of the structured target. A gate is characterized by the cartesian co-ordinates of the place to which it has been assigned. After placement, the initial logic network becomes a labelled network. The labelling consists in associating the co-ordinates with the elements.

A cost vector is attached to a labelled network. The components of the cost vector are functions that are defined on the labelled network. They mainly concern the predictive length of the connections and the predictive width of the band. When defining the cost vector $C = \{Ci\}$, $i \in [1,k]$, ci is of higher priority than ci+1. According to that definition, a total order is defined and allows placements to be compared (4) :

C and C' being two cost vectors of R^k

$$C < C' \text{ iff } C \neq C' \text{ and } \exists j \in [1,k] \,/ \left\{ \begin{array}{l} C_i = C'_i \;\forall i < j \\ C_j <_R C'_j \end{array} \right.$$

2.1.1.Partial cost

The partial cost related to a subnetwork is a vector whose components are defined on the labelled subnetwork.

OASIS performs a constructive placement and an iterative improvement for one and two bands. The problem of constructive placement may be stated as follows : given the set E of elements to be assigned to the places P and ($E_o \times P_o$) being the set of couples of already placed elements, try the couples of (E' x P') and choose the "best" solution according to a given cost vector

$E' \in E-E_o$ $P' \in P-P_o$

2.1.2 One band-logic

In this problem, the elements will be assigned to the places from the left to the right (or from the right to the left); input and output connections of the cell are on the right and left sides, and dummy elements with fixed places are defined at the right and left. At the ith step, we look for the best element to be assigned to the ith place :

$$|P'| = 1 \text{ and } E' = E - E_o$$

The placement is then improved by pair exchanges between the elements. The complexity of such a placement is therefore $O(E^2)$. The components of the cost vector (4) to be optimized are :

- the peak function : maximal value of vertical density connections,
- the number of occurences of the peak function,
- the global length of the connections.

2.1.3. Two bands-logic

A different strategy is used for the 2 bands-logic. A partitioning is defined on the set of gates in 2 steps :

- partitioning the gates into m classes of about 10 gates ($E_1,\dots,E_m$),
- splitting each class E_i into two subclasses E_i^1 and E_i^2

The target structure is decomposed similary into m couples of macro places (P_i^1, P_i^2).

$E_i = (E_i^1, E_i^2)$ is assigned to $P_i = (P_i^1, P_i^2)$.

P_1^1	P_2^1	...	P_n^1
P_1^2	P_2^2	...	P_n^2

Partitioning of the macro places

Figure 5

In each sub-class, whose size is limited to 3-5 elements, exhaustive trials are performed. The complexity of this placement strategy is $O(E^3)$. Additional cost vector components are optimized during this placement :

- number of connections between the bands,
- number of crossovers between these connections,
- length of these connections.

The OASIS system converses with the designer to allow him :

* to put placement constraints upon the cell inputs/outputs,

* to interactively choose, at any time the cost vector among 12 proposed vectors,

* to interactively select, at any time the set of exchanges during the interative improvement.

2.2 The routing

The interconnections between the gates of each logic-band are automatically routed. They are routed as segments, and as bent connections in the three grid steps area, starting from the VDD line (no drain is allowed at less than three grid steps from VDD lines).

As the layed-out circuit is stored into the CASSIOPEE data base (5), the designer may complete the routing of the transverse connections (between the bands) and the inputs/outputs connections in an interactive way by using the graphic editor. No automatic routing routine has been developed for these connections (there are usualy few to them): manual routing comes out as being more efficient in that case and cannot fail.

2.3. Results

Cells made of 20 to 50 gates (i.e. 50 to 200 transistors) have been processed through OASIS. The average processing times necessary to their layout are estimated on a VAX 780 as follows :

- for the placement, 3 to 10 mns of CPU time, i.e. 10 to 30 mns. of the designer's waiting time.

- for the routing, 0.3 to 2 mns of CPU or 1 to 5 mns of the designer's waiting time.

The completion of the layout (transverse connections routing ...) takes about 30 minutes.

The cells are therefore layed out in less than an hour whereas many days are necessary if manually performed.

The area increase varies from 10 to 30 % when compared to an entirely manual layout.

III - PARTITIONING AND PARALLELIZATION OF THE LAYOUT PROBLEM

3.1. General problem

A large interest has concerned, in the past years, the layout parallelization problem in order to cope with futur complexity (6,7). We consider here the basic problem of the parallel placement in one band-logic.

Let us consider a partition into n classes of the set of elements (gates) of the logic network, these elements being embedded in the target structure. We call macro-placement the procedure which defines an ordering of these classes and assigns them to successive "sub-bands" of the band (figure 6).

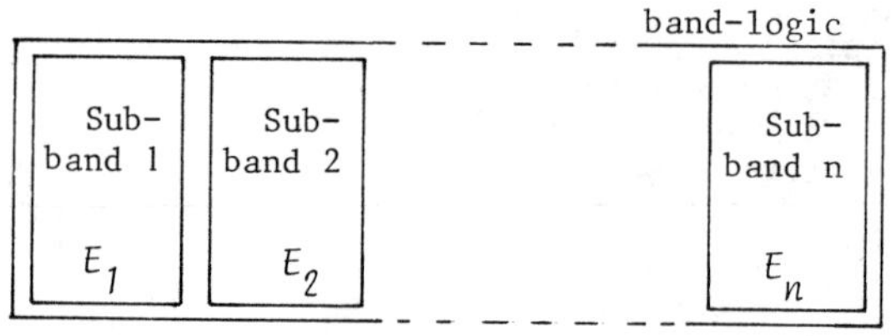

Figure 6

Let us suppose that such a partition and an ordering of the classes have been proposed ; we look, in a second step, if the parallel processing in each sub-band (placement of the gates within the sub-band) is suitable and does not lead to a worse global result.

We shall therefore discuss the 2 following problems :

* choice of the partition of E and ordering of the classes,

* discussion of the cost vector and the global efficiency.

3.2. Partition of the elements

A partition of the elements into n classes ($E_1,\ldots,E_n$) has been defined by the following constructive procedure :

- E_1 contains the elements connected to the left dummy element ;

- The (i-1) first classes being defined, the ith class contains all the elments connected at least to one element of the (i-1)th class.

Such a partition has the following interesting properties :

- the ordering of the classes is obvious,
- an element of a class is connected to elements of the same class or of adjacent classes.

This second property allows parallelization for the chosen cost vector as explained in the following part.

Remark : Extensions of this type of partition have been studied. In order to perform exhaustive placement within a class, it appears to be interesting to reduce the number of elements of a class. Therefore, different partitions have been considered by assigning to the first layer, not necessarily all the elements connected to the I/O pins, but any subset of this set.

Partitions where the number of elements of each class is bounded are therefore chosen.

3.3. Cost vector and parallelization

Let us assume that the macro-placement has

been achieved. The problem is to know if the placement may be processed in parallel within each sub-band, without loosing efficiency. The efficiency of a placement is appraised by the cost vector. There is no decrease of efficiency if the global cost can be expressed by the following expression :

$$C = f(C_{E_1}, C_{E_2}, \dots C_{E_m}) \quad (1)$$

where C_{E_i} is a restriction of the cost vector on the labelled subnetwork, defined by the element of E_i with the following properties :

- f is monotonic with respect to the variables C_{E_i} ; improvement in a sub-band will induce a stable global cost or an improvement of this global cost.

- C_{E_i} and C_{E_j} are independent (∀i,j).

In our approach, the cost vector most currently used, is defined by 3 components as follow :

C1 : the peak function maximum value which is the maximal number of connections going through a place ;
C2 : The number of occurences of the peack function maximum value ;
C3 : The global length of connections.

Let us note that the peak function maximum value is never reached "between" the sub-bands with the type of partition considered here (figure 7)

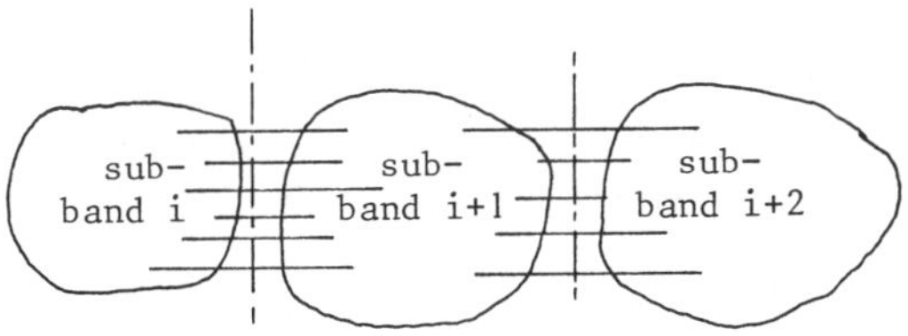

Figure 7

On the other hand the length of a connection is the sum of the lengths of this connection in each sub-band.

It is therefore obvious that the global optimization may be obtained as a result of the optimization within the sub-band and that the cost function is expressed as stated in (1).

CONCLUSION

Semi-structured logic seems an interesting compromise between strongly constrained structure such as uncommitted logic arrays and optimized full custom design. Automatic layout is easy and efficient, and the waste of area may be considered as acceptable (about 10-20 %). Moreover, it has been shown that the layout approach may easily be parallelized and this will lead in the future to very fast layout procedures.

REFERENCES

(1) A. WEINBERGER, "Large scale integration of MOS complex logic : a layout method", IEEE JSSC, SC-2, pp 182-190, Dec. 1967.

(2) E. WITTENZELLINER, "Computer aided design of large scale integration I2L Logic circuits", IEEE JSSC, SC-12, pp 199-204, April 1977.

(3) J. MAJOS, J.C. LARDY, "The multidrain MOS transistor" 4th ESSCIRC, Amsterdam, pp133-135, Sept. 1978.

(4) G. SERRERO, "Implantation symbolique automatisée de circuits intégrés", Ingeneer Doctor Thesis, INP Grenoble, March 1982.

(5) A.M. BEYLS & al. "A design Methodology based upon Symbolic layout and integrated CAD tools", 19th DAC, Las Vegas (Nevada), pp872-878, June 1982.

(6) D.J. CHYAN, M.A. BREUER, "A placement algorithm for array processors", 20th DAC, Miami Beach (Florida), June 1983.

(7) S.J. HONG, R. NAIR, " Wire routing machines. New tools for VLSI physical design", IBM Research report, RC 9374, April 1982.

Formal Semantics for a Symbolic IC Design Technique

Sanjay V.Rajopadhye and **P. A.Subrahmanyam**[1]
Department of Computer Science
University of Utah
Salt Lake City, Utah 84112

Abstract

A formal modelling scheme for studying the behavior of circuits designed using a symbolic IC design technique is developed. The formalism is of value when attempting to synthesize such circuits proceeding from higher level specifications, for verifying the correctness of such designs, and in the development of provably correct optimization strategies for these circuits.

I Introduction

We have been exploring a methodology for automated design of Very Large Scale Integrated (VLSI) circuits/systems based upon a formal (algebraic) framework and associated design aids. This methodology consists of transforming initial high level specifications of a system into a layout description for single or multiple chips that meet the criteria detailed in the specification. The following levels of system description form important stages in the transformation process.

1. Functional and performance specification
2. System architecture description
3. Chip architecture description
4. Chip Layout

In order to ensure that the transformations used in the synthesis process are *meaning–preserving* it is important to be able to assign semantics at each of these levels of system description. An appropriate formalism is also essential for verifying the correctness of local ("peephole") optimizations that are performed. Here, we focus on one aspect of this problem, namely, assigning formal semantics to a system specified at the chip architecture level.

In this paper, a technique for the *symbolic design* of integrated circuits that uses a generalized form of the design principle introduced by Programmable Logic Arrays (PLAs) is modelled using an algebraic framework. This technique, called Path Programmable Logic (PPL) design [5, 7] in conjunction with customized design techniques for standard devices such as RAMs, ROMs and multiplexors, has been found to be useful at the chip architecture level, mainly because its use elevates VLSI design to a symbolic level.

The formal basis that we use to model PPL circuits develops upon Milner's Calculus for Communicating Systems [4](CCS), wherein the behavior of a system is characterized by using "behavior expressions". Behavior Expressions (BEs) describe a system in terms of its observable input/output (I/O) actions at global I/O ports, in conjunction with its future behavior (which is another BE). Operations of *composition*, *restriction* and *relabelling* on BEs correspond to wiring up (i.e., static interconnection of) modules or subsystems, while *action* (responding to external signals) and *inaction* specify "execution time" behavior.

Limitations on space force this to be just a brief summary of the work. The interested reader is referred to [6] for a detailed report on the methodology. The rest of this paper is organized as follows. In the following section we summarize the salient features of the PPL system for circuit design. In Section III we give a brief description of CCS — its syntax and semantics along with the extensions that we propose for handling special kinds of communication. In Section IV we present our modelling scheme, principally in the first subsection, where we work out a simple example that involves only the basic PPL cells. Later on in the same section we indicate schema for other PPL cells.

II Path Programmable Logic: A tool for circuit design

Path Programmable Logic (PPL) is an extension of the PLA concept and is a form of structured logic that enables symbolic design of integrated circuits. PPLs were originally developed as Stored Logic Arrays (SLAs) by Patil [5] and consisted of a conventional PLA with the AND and OR planes "folded" into a single AND/OR plane. Segmenting PPLs with arbitrary row and column breaks allows for functional duality of rows and columns, and in particular, permits memory elements to be embedded *within* the PPL.

A PPL circuit can be thought of as a rectangular array that can contain *different* kinds of cells (unlike PLAs). Many of the cells are similar to the crosspoint devices in PLAs in that they either sense signals on the row wires and affect the column wires, or vice–versa. The important distinction is that the restriction on placement imposed by the existence of separate AND and OR planes is removed. This is made possible by permitting every location to behave either as in an AND plane or as in an OR plane, i.e. to sensitize either a row wire or a column wire. In particular, this permits localized feedback e.g., special flip–flop cells are available, and consequently both combinational and sequential circuits can be easily designed.

In a PPL circuit the logical OR function is achieved by combining data along a single column and logical AND is achieved on a single row. Columns may also represent boolean functions of state values, thereby permitting the user to have multiple levels of logic. This, along with the availability of localized feedback permits the design of independent state–machine and data–path modules. The interface between the data paths and control modules is optimized by topological alignment of control signals. A library of PPL cells is available to the user [1] so that various synchronous circuits can be designed. Work is also currently under way [2] to develop an asynchronous cell set, and to generate hybrid circuits on the same chip.

The above has perforce been a brief outline of the PPL design methodology. For a detailed description of the PPL cell set the reader is referred to the PPL User Manual [1].

III A Calculus for Communicating Systems

In this section a brief outline of CCS is presented along with an extension that permits/forces parallel communication. We find such a

[1]This research was sponsored in part by Defense Advanced Research Projects Agency, US Department of Defense, Contract No.MDA903–81–C–0414 and in part by the Office of the Naval Research under contract N00014–83–K–0317

construct extremely useful in modelling physical systems. Formal development of the theory, and proof that the power of the original CCS is not lost are presented elsewhere [6].

The static operations by which BehaviorExpressions can be formed represent the ways in which processing elements (BEs) can be strung together physically; these operations include composition (of processing elements), relabelling (renaming of I/O ports), and restriction (making certain ports externally inaccessible). The application of static operations thus yields a BE which has characterizable static properties e.g., port labels, component BEs, the way these are hooked together, and so on. The dynamic operations (actions, inactions, summation) result in an evolution of the internal configuration of the BE in response to external or internal I/O communications. Conducting experiments —applying signals to the BE— corresponds to applying functions on the types that are associated with these signals. This results in an evolution of the internal configuration of the BE, and, optionally, possible "extractor" functions being applied, i.e., external output signals being generated. Thus, the semantics of applying the dynamic functions may be represented by rewrite rules or derivations that describe the evolving configurations of processing elements.

We now define the type BehaviorExpression in greater detail.

III..1. Definition of Behavior Expressions

Behavior Identifiers denote a specific instance of a particular BE type. Each BehaviorExpression has a set of ports associated with it. Each port has a unique name and an associated type that defines the kinds of items that can be input/output at this port. All relabellings have to be consistent with the type of objects a port can handle.

BehaviorExpression are formed by (1) applying the behavior operators Composition, Relabelling, Restriction —the static operations— and Choice (or summation), Action, and Inaction (or termination) —the dynamic operations; (2) parameterizing behavior identifiers (i.e., instantiating other BehaviorExpressions from a library); (3) using conditionals, to make the future behavior depend upon the outcome of some computations.

BehaviorExpressions typically involve function symbols defined on a set of types $\{T_i\}$. That is, the type BehaviorExpression is typically *parameterized* by the set of types $\{T_i\}$. The semantics of the expressions formed by these function symbols and variables of different sorts are therefore given by the semantics of the types $\{T_i\}$.

The set of port names of a BE is also said to be its *sort set*. The informal semantics of the basic operations that are used in constructing behavior expressions are given below.

- *Composition*: Denoted by the operator |. B1 | B2 composes the agents B1 and B2 by connecting ("hooking up") complementary input and output ports. As a convention, if there is an input port named *Port1* and an output port named $\overline{Port1}$ (i.e., an overbar–ed version of the name of an input port), then the two names are called "co–names" of each other, and the two ports are implicitly connected to each other.
- *Restriction*: Denoted by B \ Set–of–PortNames. Restriction signifies that the port names in the Set–of–PortNames are made invisible to the outside world.
- *Relabeling*: B$\{l_1|l_2\}$ denotes the renaming (relabeling) of port l_1 by l_2 in the BehaviorExpression B.
- *The null action:* Denoted by the BehaviorExpression NIL, signifies termination of a process.
- *Choice Operation:* The choice operation is denoted by the operator +, and introduces nondeterminacy. The BehaviorExpression B1 + B2 offers experiments that are the conjunction of those offered by B1 and B2. On acceptance of an experiment by B1, the ensuing behavior is B1' where B1' is the future (or *mutated*) behavior of B1. This is analogous to "accepting" the invocation of an entry corresponding to B1, whence the behavior B2 is now no longer available.
- *Action:* Performing input at an input port or output at an output port i.e., performing an input/output experiment, is called an action. If, as a result of an input action, a variable *InputVariable* receives a value from an input port named *InputPort*, then this is denoted by *InputPort.InputVariable*. In the ensuing behavior, the variable *InputVariable* is bound to the actual input value.

  If, on the other hand, it is desired to output a value *OutputValue* via a port $\overline{OutputPort}$, this action is denoted by $\overline{OutputPort}$*.OutputValue*. At "execution time", i.e., when

  1. there is an input port *InputPort* associated with one processing element connected to the output port $\overline{OutputPort}$ of some other processing element, and
  2. complementary input actions *InputPort.InputVariable* and output actions $\overline{OutputPort}$*.OutputValue* are executed by two different processes

  the value *OutputValue* is bound to *InputVariable*.

  We also permit *simultaneous actions* to occur; these are indicated by tupling $<action_1, ..., action_n>$.

We will not discuss here the formal semantics of BehaviorExpressions. Semantic domains appropriate for modelling primitive kinds of BehaviorExpressions are discussed at length in [3].[2]

IV Expressing PPL circuits as Behavior Expressions

One of the primary features of a PPL circuit description is the set of topological constraints imposed on the designer. The whole circuit is, in essence, a rectangular grid with certain devices at particular crosspoints. We capture this notion by describing the row and column wires in the PPL by individual BEs that are indexed by their coordinate location. Each row BE offers/accepts "experiments" at every intersection with a column. Similarly, the BE for each column wire can interact with other BEs at every grid location corresponding to an intersection with a row. Thus, placing a cell such as a ZERO or ONE at a particular location corresponds to instantiating a BE for the cell, indexing it with the required coordinate location, and then

[2] These domains need to be extended here, since we admit simultaneous actions (represented by tuples of actions), and in addition allow the passing of process descriptions as values.

performing a restriction operation so that these ports become "invisible" (this thus prevents any other cell from being placed there). The BE for a typical row wire with a pull–up cell on it is as follows.

$$\mathrm{ROW}_i(r_i) <= \\ <\rho_{in\,i\,1}\,x_1, \rho_{in\,i\,2}\,x_2 \,..\, \rho_{in\,i\,n}\,x_n> \mathrm{ROW}_i(\mathrm{NOR}(X_1, X_2 \,..\, X_n)/r\,\} \\ + \{\bar{\rho}_{out\,i\,1}\,r_i, \bar{\rho}_{out\,i\,2}\,r_i \,..\, \bar{\rho}_{out\,i\,n}\,r_i\}\,\mathrm{ROW}_i(r_i);$$

The salient features of this BE can be described as follows. The BE represents an agent that can perform one of two actions at any instant. It either accepts input experiments at its ρin ports (*all in parallel*) and then modifies its internal state accordingly, or alternatively it accepts *any number* of output requests at its ρout ports and returns its internal state as an output value. Once it decides to accept one kind of experiment it rejects the other alternative until it has responded to the first experiment. This is intuitively reasonable, since the value on the row wire is indeterminate at the instant that a device is *in the process of* modifying it.

Since there are two wires corresponding to each logical column, we accordingly have two BEs for the column wire as shown below.

$$\mathrm{LCOL}_i(l_i) <= \\ <\alpha_{in\,i\,1}\,x_1, \alpha_{in\,i\,2}\,x_2 \,..\, \alpha_{in\,i\,n}\,x_n> \mathrm{LCOL}_i(\mathrm{NOR}(X_1, X_2 \,..\, X_n)/l\,\} \\ + \{\bar{\alpha}_{out\,i\,1}\,l_i, \bar{\alpha}_{out\,i\,2}\,l_i \,..\, \bar{\alpha}_{out\,i\,n}\,l_i\}\,\mathrm{LCOL}_i(l_i);$$

$$\mathrm{RCOL}_i(c_i) <= \\ <\beta_{in\,i\,1}\,x_1, \beta_{in\,i\,2}\,x_2 \,..\, \beta_{in\,i\,n}\,x_n> \mathrm{RCOL}_i(\mathrm{NOR}(X_1, X_2 \,..\, X_n)/c\,\} \\ + \{\bar{\beta}_{out\,i\,1}\,c_i, \bar{\beta}_{out\,i\,2}\,c_i \,..\, \bar{\beta}_{out\,i\,n}\,c_i\}\,\mathrm{RCOL}_i(c_i);$$

When a set of such BEs are instantiated we have a bare (unpopulated) grid that corresponds to the skeleton of the PPL circuit being described. Circuit design in PPL consists of placing individual cells at specific grid locations. In our model this corresponds to instantiating BEs for these cells. We now describe BEs for the different classes of cells in the PPL cell set.

A. Primitive Row Cells

As their name suggests these form the simplest kinds of cells. They occupy exactly one grid location and correspond to the simple crosspoint devices of conventional PLAs. They are made up of exactly one transistor, and depending on its configuration they sense one of the column wires and pull down the row (*viz* 0 and 1 cells) or sense the row wire and activate one of the column wires (*viz* R and S cells). The BEs that describe these cells are correspondingly simple and are as follows.

$$\mathrm{ZERO}_{i\,j} <= \alpha_{out\,i\,j}\,x < \bar{\rho}_{in\,i\,j}\,x\,,\, \bar{\alpha}_{in\,i\,j}\,0, \bar{\beta}_{in\,i\,j}\,0> \mathrm{ZERO}_{i\,j}$$

$$\mathrm{ONE}_{i\,j} <= \beta_{out\,i\,j}\,x < \bar{\rho}_{in\,i\,j}\,x\,,\, \bar{\alpha}_{in\,i\,j}\,0, \bar{\beta}_{in\,i\,j}\,0> \mathrm{ONE}_{i\,j}$$

$$\mathrm{RESET}_{i\,j} <= \rho_{out\,i\,j}\,x < \bar{\rho}_{in\,i\,j}\,0\,,\, \bar{\alpha}_{in\,i\,j}\,0, \bar{\beta}_{in\,i\,j}\,x> \mathrm{RESET}_{i\,j}$$

$$\mathrm{SET}_{i\,j} <= \rho_{out\,i\,j}\,x < \bar{\rho}_{in\,i\,j}\,0\,,\, \bar{\alpha}_{in\,i\,j}\,x, \bar{\beta}_{in\,i\,j}\,0> \mathrm{SET}_{i\,j}$$

Note that the names of the ports on the BEs for the cells are the duals of those on the row and column BEs. Thus, what is an input port for a row or a column, is an output port on the cell BE, and *vice versa*. Although this may seem odd when one looks at such a BE in isolation, it guarantees exactly the kind of communication that we desire. Invoking these BEs in parallel composition with the skeleton grid is followed by restricting the new BE so that all ports labelled $\rho_{i\,j}$, $\alpha_{i\,j}$ and $\beta_{i\,j}$ are rendered invisible.

Also note that each BE performs two "dummy" experiments (i.e., experiments where the value that is output is always 0) on the two wires that the cell is *not* intended to affect. We call these "dummy" experiments because they do not affect the row or column which accepts these values (since NOR(x, y, 0) = NOR (x, y)). The reason that these experiments are part of the BEs of primitive row cells, is because the BE for row and column wires expects an input from *every* crosspoint device, in order to be a general description of a row or column wire. When the BE for the entire circuit is formed by composition of the constituent BEs these dummy experiments are dropped.

The above ideas are clarified by a simple example. Consider the PPL segment shown in Figure IV–1 which could be part of a state machine's transition logic. Let us assume that adequate row and column pull–up devices have been placed on each row and column, and that they have been properly handled (as will be detailed later). The set of BEs describe the constituents of this circuit is also given as part of the figure.

The BE for the whole circuit is the composition of the above, followed by a restriction operation to hide all the internal ports (in this simple example *all* the ports are internal), as shown below.

CIRCUIT <= ROW1(r1) | ROW2(r2) | LCOL1(l1) | LCOL2(l2) | LCOL3(l3) | RCOL1(r1) | RCOL2(r2) | RCOL3(r3) | ONE11 | ZERO21 | ZERO12 | ZERO22 | RESET13 | SET23 $\backslash\alpha_{in\,i\,j}\ \beta_{in\,i\,j}\ \rho_{in\,i\,j}\ \alpha_{out\,i\,j}\ \beta_{out\,i\,j}, \rho_{out\,i\,j}$

We now simplify this BE and reduce it to a form that is intuitively cleaner and much more tractable. First we drop dummy experiments in the special case when the output value from the cell BE does not affect the value carried on the row or column wire. These are present because we had originally introduced dummy outputs from the cells in order to have uniform BEs for the row and column wires. We thus effectively look only at those cells that can actually affect the row or column wire, significantly alleviating the problem of combinatorial explosion. We can further simplify the special case of communication between two agents where one of the experiments never occurs (e.g., the output experiments of row and column wires at crosspoints where the cell does not sense that wire). After removing these classes of experiments the set of BEs now reduces that to the following.

$$\mathrm{ROW1(r1)} <= <\rho_{in\,1\,1}\,x1, \rho_{in\,1\,2}\,x2>\mathrm{ROW1(NOR}(x_1\quad x2)/r1) \\ + \bar{\rho}_{out\,1\,3}\,r1\ \mathrm{ROW1(r1)}$$

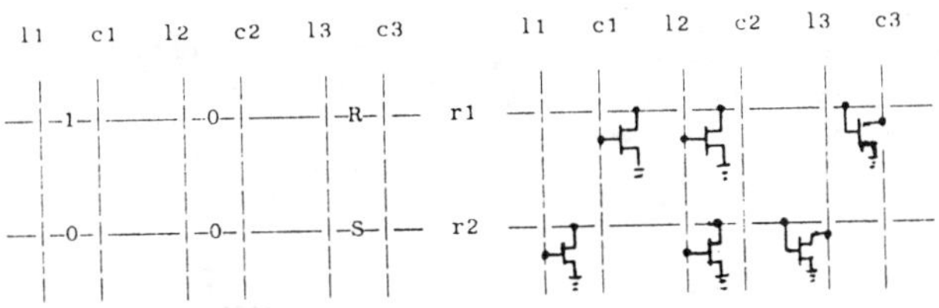

(a) Symbolic (b) Schematic

Figure IV–1: A first example — Simple PPL segment

The following set of BEs describe the constituents of this circuit.

$$\mathrm{ROW1(r1)} <= \ <\rho_{in\,1\,1}\, x_{1\,1}, \rho_{in\,1\,2}\, x_{1\,2}, \rho_{in\,1\,3}\, x_{1\,3}> \mathrm{ROW1(NOR(}x_{1\,1}, x_{1\,2}, x_{1\,3})/\mathrm{r1)}$$
$$+ \{\bar\rho_{out\,1\,1}\,\mathrm{r1}, \bar\rho_{out\,1\,2}\,\mathrm{r1}, \bar\rho_{out\,1\,3}\,\mathrm{r1}\}\ \mathrm{ROW1(r1)}$$

$$\mathrm{ROW2(r2)} <= \ <\rho_{in\,2\,1}\, x_{2\,1}, \rho_{in\,2\,2}\, x_{2\,2}, \rho_{in\,2\,3}\, x_{2\,3}> \mathrm{ROW2(NOR(}x_{2\,1}, x_{2\,2}, x_{2\,3})/\mathrm{r2)}$$
$$+ \{\bar\rho_{out\,2\,1}\,\mathrm{r2}, \bar\rho_{out\,2\,2}\,\mathrm{r2}, \bar\rho_{out\,2\,3}\,\mathrm{r2}\}\ \mathrm{ROW2(r2)}$$

$$\mathrm{LCOL1(l1)} <= \ <\alpha_{in\,1\,1}\, x_{1\,1}, \alpha_{in\,2\,1}\, x_{2\,1}> \mathrm{LCOL1(NOR(}x_{1\,1}, x_{2\,1})/\mathrm{l1)}$$
$$+ \{\bar\alpha_{out\,1\,1}\,\mathrm{l1}, \alpha_{out\,2\,1}\,\mathrm{l1}\}\mathrm{LCOL1(l1)}$$

$$\mathrm{LCOL2(l2)} <= \ <\alpha_{in\,1\,2}\, x_{1\,2}, \alpha_{in\,2\,2}\, x_{2\,2}> \mathrm{LCOL2(NOR(}x_{1\,2}, x_{2\,2})/\mathrm{l2)}$$
$$+ \{\bar\alpha_{out\,1\,2}\,\mathrm{l2}, \alpha_{out\,2\,2}\,\mathrm{l2}\}\mathrm{LCOL2(l2)}$$

$$\mathrm{LCOL3(l3)} <= \ <\alpha_{in\,1\,3}\, x_{1\,3}, \alpha_{in\,2\,3}\, x_{2\,3}> \mathrm{LCOL3(NOR(}x_{1\,3}, x_{2\,3})/\mathrm{l3)}$$
$$+ \{\bar\alpha_{out\,1\,3}\,\mathrm{l3}, \alpha_{out\,2\,3}\,\mathrm{l3}\}\mathrm{LCOL3(l3)}$$

$$\mathrm{RCOL1(c1)} <= \ <\beta_{in\,1\,1}\, x_{1\,1}, \beta_{in\,2\,1}\, x_{2\,1}> \mathrm{RCOL1(NOR(}x_{1\,1}, x_{2\,1})/\mathrm{c1)}$$
$$+ \{\bar\beta_{out\,1\,1}\,\mathrm{c1}, \beta_{out\,2\,1}\,\mathrm{c1}\}\mathrm{RCOL1(c1)}$$

$$\mathrm{RCOL2(c2)} <= \ <\beta_{in\,1\,2}\, x_{1\,2}, \beta_{in\,2\,2}\, x_{2\,2}> \mathrm{RCOL2(NOR(}x_{1\,2}, x_{2\,2})/\mathrm{c2)}$$
$$+ \{\bar\beta_{out\,1\,2}\,\mathrm{c2}, \beta_{out\,2\,2}\,\mathrm{c2}\}\mathrm{RCOL2(c2)}$$

$$\mathrm{RCOL3(c3)} <= \ <\beta_{in\,1\,3}\, x_{1\,3}, \beta_{in\,2\,3}\, x_{2\,3}> \mathrm{RCOL3(NOR(}x_{1\,3}, x_{2\,3})/\mathrm{c3)}$$
$$+ \{\bar\beta_{out\,1\,3}\,\mathrm{c3}, \beta_{out\,2\,3}\,\mathrm{c3}\}\mathrm{RCOL3(c3)}$$

$$\mathrm{ONE11} <= \ \beta_{out\,1\,1}\, x <\bar\rho_{in\,1\,1}\, x, \bar\alpha_{in\,1\,1}\, 0, \bar\beta_{in\,1\,1}\, 0> \mathrm{ONE11}$$
$$\mathrm{ZERO21} <= \ \alpha_{out\,2\,1}\, x <\rho_{in\,2\,1}\, x, \alpha_{in\,2\,1}\, 0, \beta_{in\,2\,1}\, 0> \mathrm{ZERO21}$$
$$\mathrm{ZERO12} <= \ \alpha_{out\,1\,2}\, x <\rho_{in\,1\,2}\, x, \alpha_{in\,1\,2}\, 0, \beta_{in\,1\,2}\, 0> \mathrm{ZERO12}$$
$$\mathrm{ZERO22} <= \ \alpha_{out\,2\,2}\, x <\rho_{in\,2\,2}> x, \alpha_{in\,2\,2}\, 0, \beta_{in\,2\,2}\, 0> \mathrm{ZERO22}$$
$$\mathrm{RESET13} <= \ \rho_{out\,1\,3}\, x <\rho_{in\,1\,3}\, 0, \alpha_{in\,1\,3}\, 0, \bar\beta_{in\,1\,3}\, x> \mathrm{RESET13}$$
$$\mathrm{SET23} <= \ \rho_{out\,2\,3}\, x <\rho_{in\,2\,3}\, 0, \alpha_{in\,2\,3}\, x, \bar\beta_{in\,2\,3}\, 0> \mathrm{SET23}$$

$$\mathrm{ROW2(r2)} <= \ <\rho_{in\,2\,1}\, \mathrm{x1}, \rho_{in\,2\,2}\, \mathrm{x2}> \quad \mathrm{ROW2(NOR(}x_1\ \mathrm{x2})/\mathrm{r2)}$$
$$+ \bar\rho_{out\,2\,3}\,\mathrm{r2\ ROW2(r2)}$$

$$\mathrm{LCOL1(l1)} <= \bar\alpha_{out\,2\,1}\,\mathrm{l1\ LCOL1(l1)}$$

$$\mathrm{LCOL2(l2)} <= \{\bar\alpha_{out\,1\,2}\,\mathrm{l2}, \bar\alpha_{out\,2\,2}\,\mathrm{l2}\}\ \mathrm{LCOL2(l2)}$$

$$\mathrm{LCOL3(l3)} <= \mathrm{NIL}$$

$$\mathrm{RCOL1(c1)} <= \bar\beta_{out\,1\,1}\,\mathrm{c1\ RCOL1(c1)}$$

$$\mathrm{RCOL2(c2)} <= \mathrm{NIL}$$

$$\mathrm{RCOL3(c3)} <= \ <\beta_{in\,1\,3}\,\mathrm{x1}, \beta_{in\,2\,3}\,\mathrm{x2}> \mathrm{RCOL3(NOR(x1,\ x2)/c3)}$$

$$\mathrm{ONE11} \quad <= \beta_{out\,1\,1}\, x\, \bar\rho_{in\,1\,1}\, x\ \mathrm{ONE11}$$

$$\mathrm{ZERO12} \quad <= \alpha_{out\,1\,2}\, x\, \bar\rho_{in\,1\,2}\, x\ \mathrm{ZERO12}$$

$$\mathrm{ZERO21} \quad <= \alpha_{out\,2\,1}\, x\, \bar\rho_{in\,2\,1}\, x\ \mathrm{ZERO21}$$

$$\mathrm{ZERO22} \quad <= \alpha_{out\,2\,2}\, x\, \bar\rho_{in\,2\,2}\, x\ \mathrm{ZERO22}$$

$$\mathrm{RESET13} \quad <= \rho_{out\,1\,3}\, x\, \bar\beta_{in\,1\,3}\, x\ \mathrm{RESET13}$$

$$\mathrm{SET23} \quad <= \rho_{out\,2\,3}\, x\, \bar\alpha_{in\,2\,3}\, x\ \mathrm{SET23}$$

Note that all that the primitive row cell BEs do is to accept an input from a row or a column (at a particular port) and transfer it to a column or a row (respectively) at its corresponding input. We can therefore use the Expansion Theorem of CCS (actually a property of the chaining operator developed in [4]) to further simplify these BEs. The primitive row cell BEs then do not occur at all in the resulting BE, which can be written as follows.

$$\mathrm{ROW1(r1)} \ <= \mathrm{ROW1(NOR(c1\ l2)/r1)} + \bar\rho_{out\,1\,3}\mathrm{r1\ ROW1(r1)}$$

$$\mathrm{ROW2(r2)} \ <= \mathrm{ROW2(NOR(l1\ l2)/r1)} + \bar\rho_{out\,2\,3}\mathrm{r2\ ROW2(r2)}$$

$$\mathrm{RCOL3(c3)} <= \ <\rho_{out\,1\,3}\,\mathrm{x1}, \rho_{out\,2\,3}\,\mathrm{x2}> \mathrm{RCOL3(NOR(x1\ x2)/c3)}$$

Performing the same simplification once more we get the final BE as

$$\mathrm{RCOL3(c3)} <= \mathrm{RCOL3(NOR(NOR(c1\ l2)\ NOR(l1\ l2))/c3)\ RCOL3(c3)}$$

which is what we expect intuitively! Note that in this simple example all that we were modelling was a *section* of the PPL, with no contact with the external world. When we develop the formalism for other cells we shall incorporate these features in the model that we develop.

B. Complex row cells

Complex row cells are cells that take up more than a single grid location. Such cells typically have an "origin", usually their lower left grid location, and placing them with their origins at certain locations causes some of the adjacent grid locations to become unavailable for placing other cells. We also include inverters here because although they occupy just one grid location they are functionally more complex than the earlier primitive row cells.

Typically complex row cells have placement restrictions e.g., latches and flip flops can only be placed with origins on even row and column locations. These features are expressed in the BEs that represent them.

In the interest of brevity we just present here an example for a single read enable flip–flop (REFF) cell. The BE that describes it is as follows.

$$
\begin{array}{l}
\mathrm{REFF}_{i\,j}(v) <= \\
<\rho_{out\ i+5\ j}\ \mathrm{phi1},\ \rho_{out\ i+2\ j}\ \mathrm{phi2}> \mathrm{if\ phi1} = 1 \\
\quad \mathrm{then}\ <\beta_{out\ i\,j}\ \mathrm{reset},\ \alpha_{out\ i\,j+1}\ \mathrm{set}>\ \mathrm{REFF}_{i\,j}(\mathrm{NewValue(set\ reset)}/v) \\
\quad \mathrm{else\ if\ phi2} = 1 \\
\qquad \mathrm{then}\ \rho_{out\ i\,j}\ \mathrm{enable\ if\ }\overline{\mathrm{enable}}\mathrm{\ then} \\
\qquad\qquad\qquad \{\beta_{in\ i\,j}\ \mathrm{not}(v),\ \overline{\alpha}_{in\ i\,j+1}\ v\}\ \mathrm{REFF}_{i\,j}(v)
\end{array}
$$

Here NewValue(set reset) is a function which evaluates to

```
(if (set and ˜ reset)
        then 1
        elseif (reset and ˜ set)
                then 0)
```

Thus a read–enable flip–flop first accepts the two clocks as input experiments. Depending on the phase of the clock it either accepts input experiments from the set and reset lines and updates its internal value, or accepts an input experiment from the read enable row wire. If the enable line is high it offers an output experiment on the corresponding column wires. This BE also causes the restriction of all ports in the rectangle defined by (i j) and (i + 5, j + 2).

Other complex row cells are modelled in a similar manner.

C. Other Cells

As mentioned earlier we envisage that the modelling system that we develop here will be used in an automatic simulation/synthesis kind of environment. The other cells in the PPL cell library i.e., interconnect cells, pull–up cells, bussing cells and pad cells perform an essentially static function in the system. They specify interconnections, inputs and outputs to/from the external world, and ensure that rows and columns behave as they are expected to. Since the functions that these cells perform are static, the way we handle them corresponds to type checking in conventional programming. These cells affect the *instantiation* of BEs for other cells rather than providing additional BEs for interaction

The interconnect cells enable rows and columns to be "broken" at arbitrary places, thus permitting multiple logic components. The two interconnect cells are *row break* and *column break* cells. The presence of a break cell thus causes the BE for the row or column to be "split" into two independent BEs. The effect of interconnect/breaks may be implicit. For example, in the read–enable flip–flop above, the i + 1, i + 3 and i + 4 th rows are split into two even though there is no explicit break. Such features are also handled by "type checking".

The function of pull–up cells is to ensure that rows and columns behave as NOR gates. Thus they cause the typical row/column BEs to be instantiated, while the grid location where a pull–up cell is placed is rendered invisible by restriction. Also, column pull–up cells cause an implicit break in the row.

Similarly, bussing cells (whose electrical function is to ensure that the power and ground wires are available at all grid locations) are very easily modelled by having all BE instantiation dependent on all peripheral cells being of this type (or possibly pad cells). Pad cells specify how and where the circuit communicates with the external world. These are the only cells other than primitive row cells and complex row cells that specify the instantiation of a BE. However, the ports of these BEs are not hidden from the external world.

This constitutes a complete description of the cells that can be placed in circuit. Given a PPL circuit, the appropriate BEs corresponding to the cells placed in the underlying grid are instantiated. The resulting BE when simplified yields a "higher level" description of the circuit behavior, as shown in the example earlier.

V Conclusions

We have presented a formal modelling scheme for studying the behavior of PPL circuits. This formalism is intended to be used as a basis for synthesizing such circuits automatically from higher level specifications. [6] illustrates the application of these techniques to other nontrivial examples e.g., a counter, and a state–machine. We are also exploring the use of this formalism in optimization of such circuits, particularly the use of "peephole" optimization.

VI References

1. Carter, T.et al. PPL User Manual. Internal Report, Department of Computer Science, University of Utah, 1982

2. Hayes, A.B. Self–Timed IC Designs with PPL's. Paper submitted for 1983 Cal Tech VLSI Conference

3. Milne, G.and R.Milner. "Concurrent Processes and Their Syntax." *JACM 26*, 2 (April 1979), 302–321.

4. Milner, Robin. *A Calculus of Communicating Systems*. Springer–Verlag, Berlin, Heidelberg, 1980.

5. S. S.Patil and T.A. Welch. "A Programmable Logic Approach for VLSI." *IEEE Transactions on Computers C–28*, 9 (September 1979), 594–601.

6. Rajopadhye, S.and P.A. Subrahmanyam. Formal Semantics for a Symbolic IC Design Technique: Examples and Applications. Tech. Rept. UTEC # 83–041, University of Utah, May, 1983.

7. K. F.Smith; T.M. Carter; and C.E. Hunt. "Structured Logic Design of Integrated Circuits Using the Stored Logic Array." *IEEE Transactions on Electron Devices ED–29*, 4 (April 1982), 765–776.

Session 1-1d
Custom Architecture

Chairmen

E. Ergin
TRW

C. W. Scott
TRW

STEREOSCOPIC DEPTH PERCEPTION FOR ROBOT VISION: ALGORITHMS AND ARCHITECTURES

R. J. Safranek and A. C. Kak

School of Electrical Engineering
Purdue University, West Lafayette, IN 47907

ABSTRACT

This paper will focus on the implementation of depth perception algorithms for computer vision. In automated manufacturing, depth information is vital for tasks such as path planning, and 3-D scene analysis

The presentation will begin with a survey of computer algorithms for stereoscopic depth perception. The majority of time will be spen on the Marr-Poggio paradigm of human stereo vision and its computer implementation as developed by Grimson. In addition, the stereo matching algorithm proposed by Barnard and Thompson will be examined Their approach is based on the relaxation labeling technique.

A computer architecture that is designed to efficiently implement stereo matching algorithms will be presented. This system is an MIMD array interfaced to a global memory.

1. Introduction

A lack of adequate sensory feedback remains a major limitation in the development of mechanical systems capable of intelligently interacting with their environment. Depth information is vital for both vision and proximity sensing, Depth information can serve as a feature for scene segmentation, and use to construct three dimensional moments for shape analysis.

Depth information may be extracted by several competing technologies. The most fascinating of these, and also the one that has suffered the most from program complexity and high computational cost, uses a pair of cameras for stereo perception. The computational cost of traditional stereo is precipitated by the large number of correlations required to solve the correspondence problem. This consists of selecting "candidate pixels" in one image and than locating their corresponding partners in the other. The correspondence problem is then solved by a feature comparison matching algorithm that is based on the global properties of matter. Such comparisons can again be carried out rapidly using specially designed parallel hardware. At this time, other depth extraction method such as ultrasound, infrared (and, perhaps, lasers) may appear attractive due to their simplicity. But, since computer hardware is continually becoming more sophisticated and less expensive (for the same computing power), the more computationally demanding stereo may come into vogue further down the line.

This paper will examine the use of a parallel architecture in performing stereo depth extraction. This system is designed to meet the following requirements:

1 It must have near real-time response. The use of a parallel architecture optimized to the stereo matching task, along with the use of some dedicated special purpose hardware, will allow much higher throughput than a conventional single processor system.

2 It must be economical. A range mapping system would be only part of a complete vision processor. By using readily available components, in the range mapping subsystem, total system cost may be keep to a minimum.

3 It should be fault tolerant. The failure of a portion of the parallel system should not be fatal to execution of the task.

2. Human Visual System Stereo Processing

One recently proposed stereo matching algorithm, which has generated interest, is the Marr-Poggio-Grimson (MPG) matcher[1,2]. The MPG algorithm is an attempt at simulating, at least at a computational theoretic level, the human visual system. In its complete implementation, which is not necessary for most engineering applications, it is a multichannel process. For each channel, a different operator extracts candidate pixels for stereo matching. These operators take the form of Laplacian-of-a-Gaussian (LoG) convolutional kernals. The channel with the widest kernal yields candidate pixels in a defocussed version of the stereo images. This defocussing is equivalent to representing each multicomponent object in a scene by a few candidate points. The largest channel, therefore, generates depth information with low spacial resolution. This may be used to determine the relative positions of objects in a scene.

Several Channels are utilized to reduce error in the matching process. By starting with a simplified description of the scene, which is obtained from the coarse channel, it is possible to obtain the approximate depth of complete objects in the scene. This may be accomplished with a low error probability since the coarse filter removes detail information from the input. Once there is some idea of the approximate depth of regions in the image, input with more detail may be handled. To accomplish this, the rough depth estimates are used as starting values to be modified by information contained in the next finer channel.

Candidate points for matching are now extracted. These points are the zero crossings of the filtered scene. In addition to the location of the zero crossings, their sign (whether the zero crossing goes from negative to positive, or from positive to negative), and local orientation if recorded.

These features are now matched. Matching is accomplished as follows: 1) a target zero crossing is located in one image; 2) the same location is fond in the other image; 3) a one dimensional region in the other image is searched for zero crossings. If only

CH1935-6/83/0000/0076$01.00

one zero crossing is found with the same sign as the target zero crossing, it is taken as a match. If more than one are found, the possible matches are saved for further processing. The ambiguity in the matches is resloved by using the property that, except at occluding boundaries, depth values cannot change abruptly.

Up to this point the process is similar to humans examining a scene without focussing at any particular region. It provides a rough idea of the relative locations of the major objects in the scene. Vergence is then exercised to focus on each object of interest. In a computer implementation, vergence consists of merely shifting portions of one filtered image with respect to the other. After vergence, the output of the next smaller channel is used to compute depth information on a finer scale.

Figure 1 Example stereogram

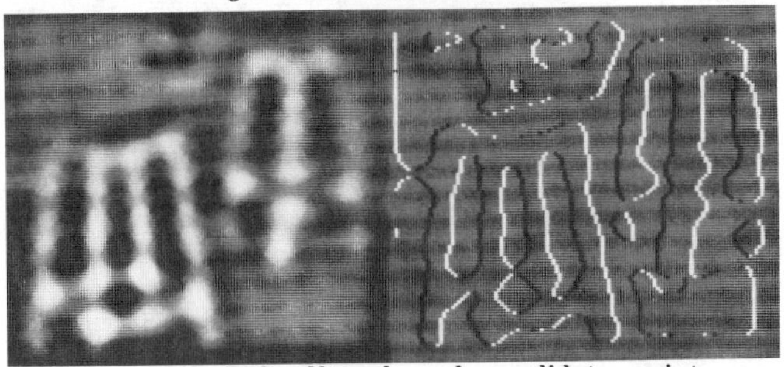

Figure 2 Left filtered and candidate point images: coarse channel

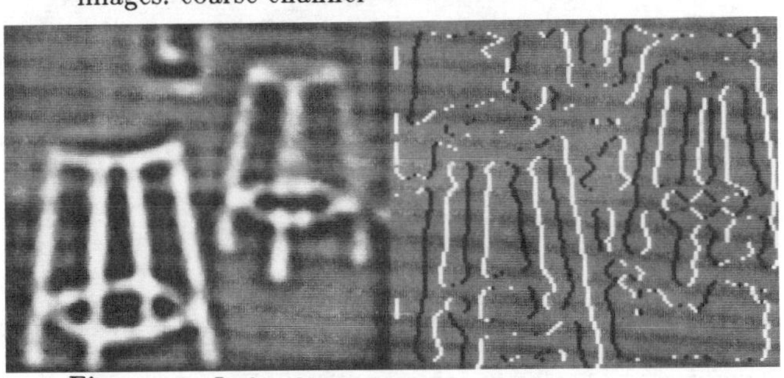

Figure 3 Left filtered and candidate point images: fine channel

To illustrate this process, an example of the algorithm will be presented. The input stereogram is shown in Figure 1. The scene consists of two stools, with the focal point of the imaging system set midway between them. These images were digitized to a resolution of 128x128 pixels. Figures 2 and 3 illustrate the output of the LOG filtering and feature extraction stages. The left image in each figure is the filtered version of the left input image. The right image shows the zero crossings in these filtered images. Figure 2 was obtained from a coarse filter

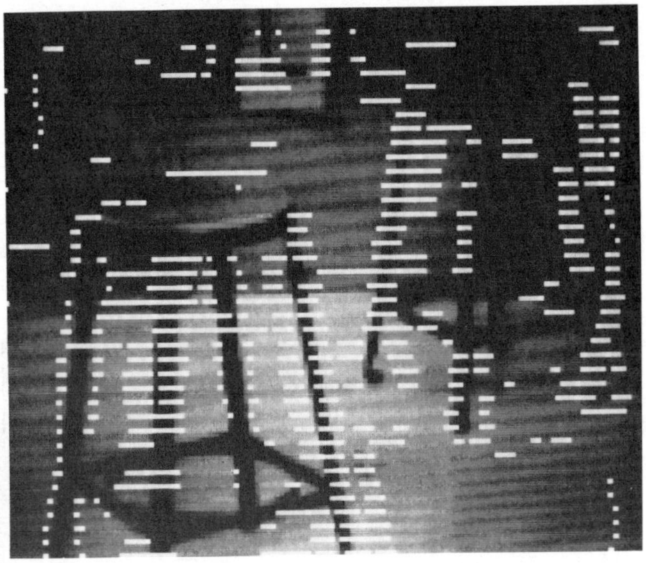

Figure 4 Disparity map (Every third line displayed)

channel, while Figure 3 was produced using a finer channel. Notice that the coarse channel provides information on the location of the entire stool, while the finer channel shows more of the features on the stools. Matching of the coarse filter data will provide rough information on the depths of the stools, while the finer channel will refine these estimates and fill in detail. The results produced by this two channel setup is displayed in Figure 4. The lines represent distance from the focal plane of the imaging system. Longer lines correspond to points that are further from the focal plane.

In the development of the MPG algorithm, a major concern was the simulation of the human visual system. Our experience has shown that stereo vision algorithms designed to simulate human stereoscopic depth perception give excellent results the a scene is digitized with a sampling rate that corresponds to the human visual system. In humans, one pixel (one receptor in the retina) corresponds to roughly 0.5 minute of arc. For a vision system with an angle of view of, let's say 20 degrees, this corresponds to digitizing the scene into a 2400x2400 matrix, which is clearly not feasible with currently available hardware. In most practical situations, the scene representation matrix should be limited to 256x256, or even 128x128 pixels [3].

Mayhew and Frisby [4] have examined some of the limitations of the MPG algorithm from the point of human visual perception. They present a strong case for using a richer set of feature primitives. In addition to the locations of the zero crossings of the filter images, Mayhew and Frisby propose that the peak and valley locations in these images be utilized. With this in mind, the proposed hardware system should be flexible enough to allow some degree of algorithm modifications without major hardware modifications.

3. A Parallel Architecture for Stereo Matching

3.1. System Overview

A system that is capable of performing the MPG matching procedure is shown in Figure 5.

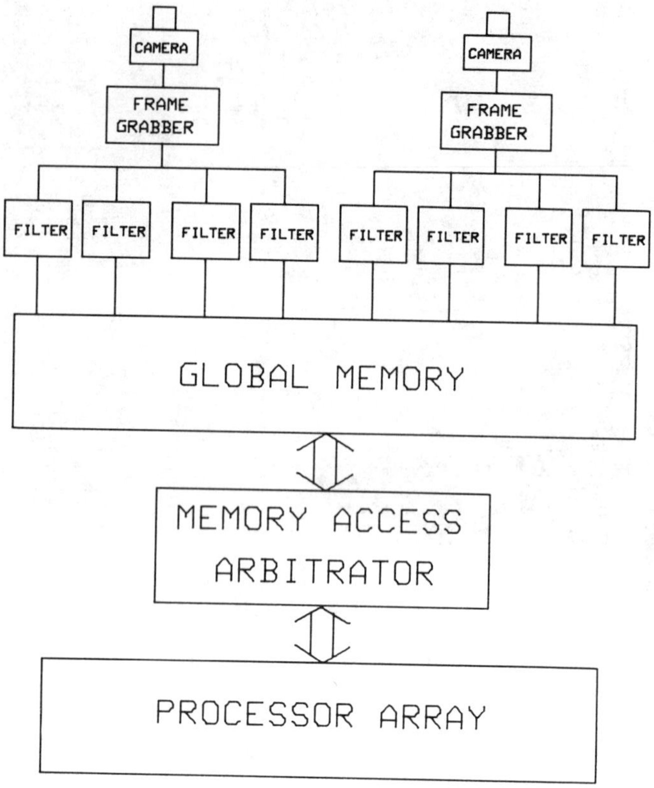

Figure 5 Architecture overview

Raw image data is captured by a pair of TV cameras and frame grabbers. This information is passed on to dedicated filtering hardware, which performs the LoG convolution. The left and right filtered images are then stored in a large **Global Memory.** Finially, the **Processor Array** performs the feature extraction and matching on the data contained in the global memory. Global memory protection and conflict avoidance is provided by the **Memory Access Arbitrator** (MAA).

The operation of this system•is illustrated by following the processing of a single pair of images through the system. Processing of the input data begins with the LoG filtering. The filtered images are stored in a global memory to allow for easy access during the following operations.

The remainder of the matching process takes place in the processor array. In the processor array, each of the processing elements operate on a sequential subset of the rows of the filtered images. First, the MAA transfers the image data from the coarsest filter into the first row processor (RP). When the transfer is complete, the RP begins processing this information. Now, the MAA loads the next RP, and begins processing. This continues until the entire image is allocated among the RPs.

When the first RP has finished processing, its local memory will contain the disparity values for the unambiguous matches in its data. In addition, the ambiguous matches will be marked. This portion of the disparity map will be moved to the global memory by the MAA. The RP will now begin to disambiguate its ambiguous matches, while the MAA unloads the disparity results from the other RPs. The disambiguation process requires a two dimensional search of the local neighborhood of the location of an ambiguous match. Therefore, to process points on the edges of the data strip in a RP, disparity data from adjoining RPs is required. This data movement will be handled by the MAA.

To obtain depth resolution at scales necessary for robotics applications, the results of 2-4 channels of input data is required. The preceeding process is now repeated for each of the finer resolution channels.

There are several reasons why this form was chosen for the architecture. First, no interconnection network is required. Interconnection networks are hardware intensive and form a large portion of the total system cost. In addition, they tend to restrict machine sizes to powers of two. Since the interprocessor communication for stereo matching algorithms is entirely with localized sets of processors, there is no need for a generalized data communication scheme.

A global memory was selected to allow other tasks to use the output of the stereoscopic imaging system. The filtered images and depth maps could be utilized by a object segmentation/recognition system. Also, a path planner for a robotic assembly system would require the depth maps. Having these reside in a global memory allows other task processors ready access to their required data.

3.2. Hardware Description

3.2.1. Laplacian-of-a-Gaussian Filter

An LSI circuit that is capable of performing the LoG convolution at video rates has been developed by Hughes Aircraft [1,p. 338]. This circuit removes an execution bottleneck from the system by allowing the matcher to receive the filtered input data at sufficient rate for near real time vision processing.

3.2.2. Memory Access Arbitrator

The MAA consists of three main portions, the global memory interface processor (MIP), the service request interface (SRI), and the bus switching circuitry (BSC). The global memory interface processor is a conventional microprocessor with a large address space. The global memory is contained in this address space.

Interface with the row processors is achieved by the SRI and BSC. The RP's and MAA communicate by a set of handshaking signals. When an RP requires service, the MIP sets up the BSC to switch a portion of the RP's memory into its memory space. This allows for data transfer utilizing fast memory to memory operations.

By using sequential processing of global memory/row processor memory transfers, it is possible to avoid the use of a large scale interconnection network. This helps improve the fault tolerance of the system by reducing system complexity. Fault tolerance of the MAA may be achieved by providing a backup global memory processor that is active only when the main cpu dies. In addition, the BSC may be built up from a small number of standard MSI logic devices, helping to keep system cost down.

3.2.3. Row Processors

Each of the row processors are essentially a conventional microprocessor There are two dedicated I/O lines for use in handshaking with the MAA. The local memory will contain both the instructions and data used by the RP.

3.2.4. System Performance

At the present time, the state of the art in microprocessor technology is rapidly changing. Currently, the top of the line commercial microprocessors have approximately 80% the performance of a DEC VAX 11/780 for non-floating point operations [5,6]. Using this figure as a rule of thumb, Table 1 gives approximate figures for the number of RPs required to obtain processing time of 2 seconds/frame for 128x128 and 256x256 images utilizing 2 input channels. These figures assume that the LoG filtering requires 0.5 seconds and include a 10% overhead to account for the time overlapping and global memory to RP memory transfers.

Image Size	Execution Time under 4.1 BSD UNIX(*)	Estimated Number of Row Processors
128 X 128	25.3	23
256 X 256	88.3	81

* UNIX is a trademark of Bell Laboratories

Table 1 Estimated system performance

4. Relaxation Based Stereo Matching

Another interesting stereo matching algorithm has been proposed by Barnard and Thompson (BT) [7]. This procedure is based on the relaxation labeling technique [8]. It has the advantage of being a single channel algorithm. That is, the depth map is produced from a single stereo image pair.

The first step in the algorithm is to extract candidate points for matching. These points are selected by using the interest operator described by Moravic [9].

For each candidate point, all possible matches are found. These are assigned initial "probabilities" based on local similarity between image regions. If the local similarity is high, a high probability is assigned to that disparity.

These estimates are then updated by examining the possible match set for neighboring matchable points. If the neighboring matchable points have possible disparities similar to the point in question, the probability of that disparty will increase, otherwise it will decrease. This updating continues until each point has only one label with high probability or until a set number of iterations has been performed.

4.1. Hardware Implementation

With minor modifications, the architecture just presented may be adapted to the BT matching procedure. As presented, this algorithm is floating point intensive. Hardware floating point capability would be needed in each of the row processors. This has the side effect of increasing the total system cost, since currently available floating point units are more expensive than the microprocessors they support. One way to remedy this situation would be to implement the algorithm utilizing fixed point arithmetic.

Our implementation of the BT algorithm runs about 10% slower than the MPG algorithm. Thus, approximately the same number of processors would be required as for the MPG version. The major drawback is that the regions used for updating the probability estimates are approximately the same as the number of image rows in each RP. This will lead to a large amount of data transfers between iterations, decreasing throughput.

5. Conclusions

Due to the variety and complexity of the real world scenes no single type of sensor can provide the information necessary for intelligent environmental interaction. Therefore, future robotic assembly systems will have a wide range of sensory input. A stereo imaging system can provide important information useful at several task levels. A stereoscopic range map can be used for path planning and collision avoidance in robotic assembly systems. In addition, this information could be an important input to an image segmentation system for use in object recognition.

For a stereo imaging to be useful in a production environment, it must be capable of producing depth maps at a rate of no less than (approximately) one every two to three seconds. Currently, there are no conventional computers that can provide this type of throughput. The special purpose architecture just presented should provide the computational power needed for proposed vision applications.

References

[1] D. Marr, *Vision,* W. H. Freeman and Company, San Francisco, CA., 1982.

[2] W. E. L. Grimson *From Images To Surfaces: A Computational Study of the Human Early Visual System,* The MIT Press, Cambridge, Massachusetts and London, England, 1981.

[3] A. C. Kak, *Implementation of the Marr-Poggio Computational Paradigm for Stereo Vision: An Engineering Perspective,* School of Elect. Engrg. Technical Report TR-EE-83-7, Feb. 1983.

[4] J. E. W. Mayhew and J. P. Frisby, "Psychophysical and computational studies towards a theory of human stereopsis," *Artificial Intelligence,* vol. 17, pp. 349-385, 1981.

[5] Amar Gupta and Hoo-min D. Toong, "An architectural comparison of 32-it microprocessors," *IEEE Micro* vol. 3, No. 1, Jan. 1983, pp. 9-22.

[6] Jim Gilbreath and Gary Gilbreath, "Eratosthenes revisited: once more through the sieve," *Byte,* vol. 8, No. 1, Jan. 1983, pp. 283-326.

[7] Stephan T. Barnard and William B. Thompson, "Disparity analysis of images," *IEEE Trans. PAMI,* vol. PAMI-2, N0. 4, July 1980, pp 333-340.

[8] A. Rosenfeld, R. A. Hummel, and S. W. Zucker, "Scene labeling by relaxation operations," *IEEE Trans. Syst., Man, Cybern.,* vol. SMC-6, June 1976.

[9] H. P. Moravic, "Towards automatic visual obstacle avoidance," *Proc. 5th Int. Joint Conf. Artificial Intell.,* Cambridge, MA, Aug. 1977, p. 584.

RESIDUE ARITHMETIC AND VLSI

Chao-Lin Chiang and Lennart Johnsson

Computer Science
California Institute of Technology
Pasadena, CA. 91125

Abstract

In the residue number system arithmetic is carried out on each digit individually. There is no carry chain. This locality is of particular interest in VLSI. An evaluation of different implementations of residue arithmetic is carried out, and the effects of reduced feature sizes estimated.

At the current state of technology the traditional table lookup method is preferable for a range that requires a maximum modulus that is represented by up to 4 bits, while an array of adders offers the best performance for 7 or more bits. A combination of adders and tables covers 5 and 6 bits the best. At 0.5 μm feature size table lookup is competitive only up to 3 bits. These conclusions are based on sample designs in nMOS.

1. INTRODUCTION

Weighted number systems such as the binary number system and the decimal number system have a carry chain. It is often limiting the performance of computer arithmetic. In the residue number system [1], a number is represented by several residue digits. The arithmetic operations addition and multiplication are performed on each digit independently. This local property allows a high degree of concurrency. However, operations such as division, magnitude comparison and overflow detection do not have the local property, [4], [6], [10].

The residue digits are obtained by evaluating a number X modulo a set of integers $m_1, m_2, .. m_p$, which are pairwise relatively prime, i.e., $\gcd(m_i, m_j) = 1$,for $i \neq j, 1 \leq i, j \leq p$. According to the Chinese Remainder Theorem [3], there exist a unique residue representation $(x_1, .., x_p)$ for any number X in the range $[0, R-1]$,

$$\text{where} \quad R = \prod_{i=1}^{p} m_i \quad \text{and} \quad x_i = X \bmod m_i$$

A number represented in residue code can be converted to binary code by the formula:

$$X = \prod_{i=1}^{p} x_j \cdot \hat{m}_j \cdot |\frac{1}{\hat{m}_j}| \quad \bmod R$$

where:

$$\hat{m}_j = \frac{R}{m_j} \quad \text{and} \quad |\frac{1}{\hat{m}_j}| = a \text{ iff } (\hat{m}_j \cdot a) \bmod R = 1$$

Let X and Y have residue codes $(x_1, .., x_p)$ and $(y_1, .., y_p)$, and be such that $X, Y, X+Y, X \cdot Y \in [0, R-1]$. Then

$$|X+Y|m_i = |x_i + y_i|m_i \quad \text{and} \quad |X \cdot Y|m_i = |x_i \cdot y_i|m_i$$

and it follows that

$$(|X+Y|_{m_1}, ..., |X+Y|_{m_p}) = (|x_1+y_1|_{m_1}, ..., |x_p+y_p|_{m_p})$$
$$(|X \cdot Y|_{m_1}, ..., |X \cdot Y|_{m_p}) = (|x_1 \cdot y_1|_{m_1}, ..., |x_p \cdot y_p|_{m_p})$$

The smaller the range of each digit is, the higher is the degree of concurrency. Minimizing the maximum modulus for a given range R maximizes the concurrency. In Table 1, M is the number of bits required for a binary encoding of a digit in the residue representation of a number. N is the number of bits required for a direct binary encoding of the same number. All moduli are of the form p^k, where p is prime. Moduli are listed in order of increasing p.

M	moduli	N	$\frac{\log N}{M}$
3	4,3,5,7	8.71	1.04
4	15,9,5,7,11,13	18.46	1.05
5	16,27,25,7,11,13,17,19,23,29,31	46.04	1.10
6	64,27,25,49,11,13, ... ,53,59,61	~90	1.08
8	256,243,125,49, ... ,239,241,251	~368	1.06

Table 1. Maximum number of bits in residue code (M) and bits in binary code (N)

It can be proved [14] that $(\log N)/M$ is asymptotically approaching 1, i.e., $M = O(\log N)$. The significance of this result is that instead of using N-bit arithmetic, $\log N$-bit binary arithmetic suffice. (Each number x_i, y_i is represented in binary code.) However, $O(N/\log N)$ such units are required instead of one N-bit unit. Also, addition and multiplication is more complex than in the binary number system in that it is performed modulo the set of integers, $m_1, .., m_p$.

2. IMPLEMENTATIONS OF RESIDUE ARITHMETIC

Addition and multiplication modulo the integers m_1, $m_2, .., m_p$ are traditionally implemented by table lookup [7], [11], [12], [13]. The table size grows exponentially with the number of bits required to represent the max-

CH1935-6/83/0000/0080$01.00 © 1983 IEEE

imum modulus. Storage can be made very dense, but a large storage contains long wires unless it is hierarchical [9]. The wire delay does not decrease with feature size, and eventually becomes a performance limiting factor. A few alternatives to the table lookup method are discussed below.

2.1 Residue Adder

Residue addition $z_i = (x_i + y_i) \bmod m_i$, where $0 \leq x_i, y_i \leq m_i - 1$, can be carried out as

$$z_i = \begin{cases} x_i + y_i & \text{if } x_i + y_i < m_i \\ x_i + y_i - m_i & \text{if } x_i + y_i \geq m_i \end{cases}$$

One M-bit adder can be used to compute $x_i + y_i$ while another computes $x_i + y_i - m_i$. The carry bit generated from the second adder indicates whether or not $x_i + y_i$ is greater than m_i. A multiplexor controlled by the carry selects the correct output, Figure 1.

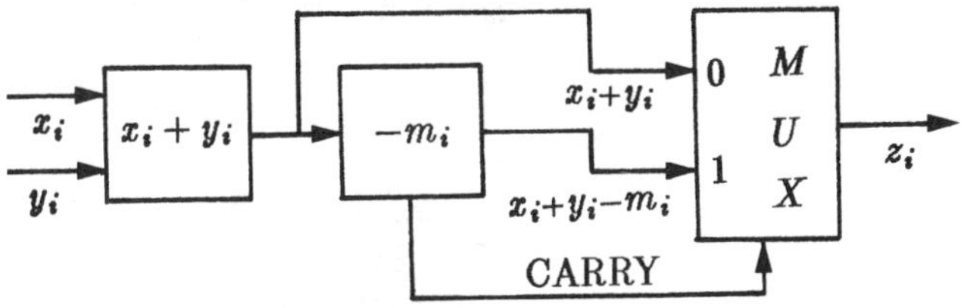

Figure 1. An implementation of residue addition

2.2 Residue Multiplier

The following three different methods to implement residue multipliers have been evaluated:

1. pure table lookup
2. combinations of tables and adders
3. arrays of adders

2.2.1 Pure Table Lookup

A table for a residue multiplier has two M-bit inputs, x_i and y_i, and produces one M-bit output, $z_i = (x_i \cdot y_i) \bmod m_i$, Figure 2. The total number of bits in the table is $2^{2M}M = 4^M M, \quad M \geq \log m_i$.

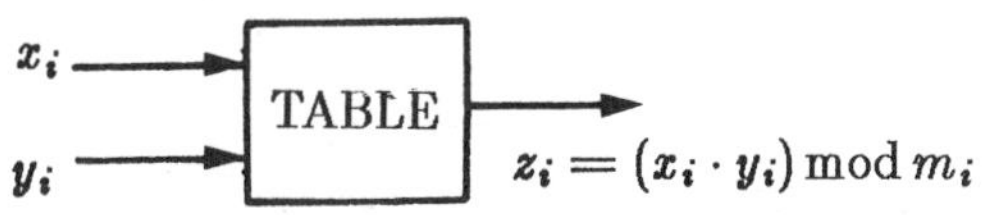

Figure 2. Table lookup residue multiplier

2.2.2.a Quarter Square Residue Multiplier [2]

Figure 3 shows another type of residue multiplier. It realizes the identity :

$$|x_i \cdot y_i|_{m_i} = \left| \left|\frac{(x_i + y_i)^2}{4}\right|_{m_i} - \left|\frac{(x_i - y_i)^2}{4}\right|_{m_i} \right|_{m_i}$$

The "modular quarter square" $|\frac{(\cdot)^2}{4}|_{m_i}$ is implemented by lookup table. Each table has $2^{M+1}M = 2M2^M$ bits. One M-bit binary adder/subtractor and one M-bit residue subtractor are needed.

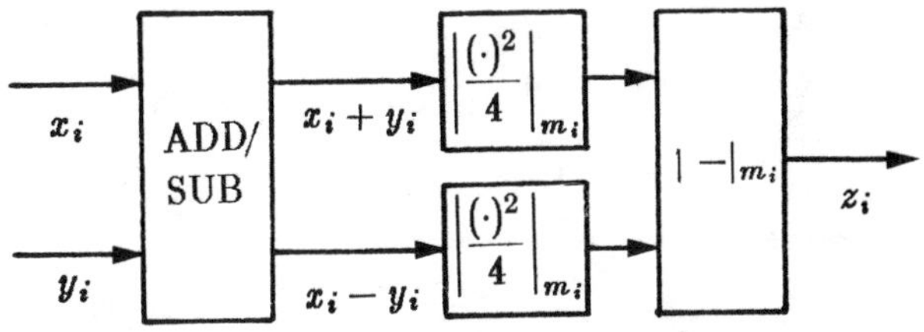

Figure 3. Quarter square residue multiplier

2.2.2.b Index Transform Residue Multiplier

One familiar method of multiplication is "log transform – addition – antilog transform" [5]. Similarly, residue multiplication can be performed by "index transform – residue addition – reverse index transform". The index transform IND_r is defined by [3]:

$$IND_r C \equiv b \bmod P \quad \text{iff } r^b \equiv C \bmod P, \ P \text{ prime}$$

An implementation based on the equation

$$|x_i \cdot y_i|_{m_i} = |r^{(IND_r x_i + IND_r y_i) \bmod (m_i - 1)}|_{m_i}$$

is shown in Figure 4. The index and reverse index transform is implemented by tables. There are two tables for the index transform and one for the reverse index transform. Each table has $M2^M$ bits.

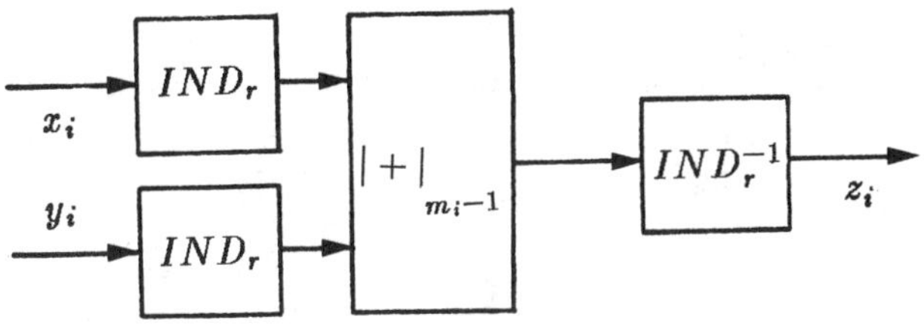

Figure 4. Index transform residue multiplier

2.2.3 Array Multiplier

A residue array multiplier can be composed of two parts: (1) a binary array multiplier and (2) a converter performing the modulo evaluation of the product. The binary multiplier has a delay of $O(M)$ and an area of $O(M^2)$. The complexity of the conversion part can be made equal to that of the multiplier by employing carry save adders [8].

Let A be the $2M$ bit product, m_i be the M bit modulus, and z_i the M bit output number, $z_i = A \bmod m_i$. To compute z_i the distributive property of the modulo operation over addition is used:

$$a_{2M-1}2^{2M-1} + \ldots + a_M 2^M + \ldots \bmod m_i =$$
$$= a_{2M-1}(2^{2M-1} \bmod m_i) + \ldots + a_{2M}(2^{2M} \bmod m_i) + \ldots$$

The M highest order bits $a_{2M-1}, a_{2M-2}, \ldots, a_{2M}$ can be used to control whether or not the M-bit constants $(2^{2M-1} \bmod m_i)$, $(2^{2M-2} \bmod m_i)$, ..., $(2^{2M} \bmod m_i)$ shall be added to the M lowest order bits, Figure 5.

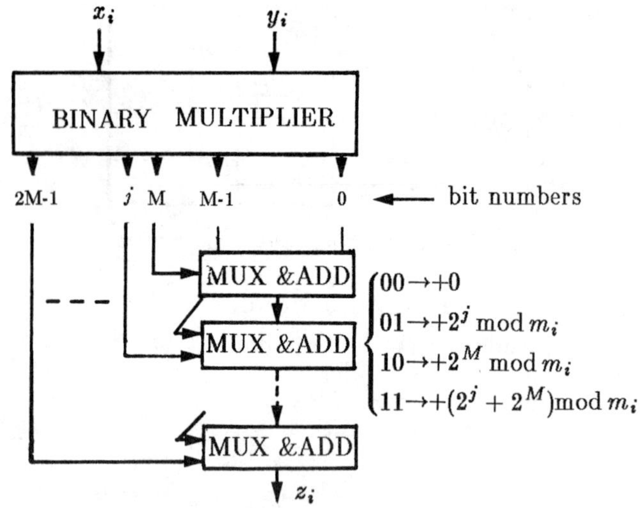

Figure 5. Residue array multiplier

The conversion part consists of M stages of M-bit adders. Each adder stage needs $O(M)$ time with the carry propagating from one bit to the next. The total delay is $O(M^2)$. However, if the carry for each bit is fed into the next stage only $O(M)$ time is needed. An additional stage is needed to convert $M+2$ bit numbers into M bit residue numbers.

2.3 Analysis of different multiplier schemes

Some features of the four residue multiplier designs discussed above are summarized in Table 2. From the table it is clear that the quarter square approach is of limited interest, and that each of the other approaches may be superior to the others depending on the value of M. To determine performance characteristics as well as more precise area requirements, and their dependence on feature sizes, some sample designs were made in nMOS.

	Table	Bin. adder	Res. adder
2.2.1	$M4^M$	0	0
2.2.2.a	$4M2^M$	M	M
2.2.2.b	$3M2^M$	0	M
2.2.3	0	$2M^2$	0

Table 2. Some features of residue multipliers

3. RESIDUE MULTIPLIER IN nMOS

3.1 Delay and area estimates

Tables 3 and 4 contain delay estimates for residue multiplier designs in $4\mu m$ and $0.5\mu m$ nMOS. The delay estimates are based on a τ-model [9], further developed to cover PLA's [15]. From the tables it is concluded that implementations based on table lookup do not scale well. The delay decreases only slightly. Measured in τ it increases significantly. The index transform approach employs much smaller tables. The delay measured in τ

Bit	Table	Ind. tran.	Array
3	198	370	900
4	400	480	1200
5	1070	700	1500
6	2400	1270	1800
7	6600	2000	2100
8	26000	5600	2400

(*unit* : τ, $1\tau \approx 0.2ns$)

Table 3. Delay in residue multipliers for $4\mu m$ nMOS

Bit	Table	Ind. tran.	Array
3	236	372	900
4	640	485	1200
5	2040	710	1500
6	7600	1400	1800
7	31000	2200	2100
8	150000	6100	2400

(*unit* : τ, $1\tau \approx 0.025ns$)

Table 4. Delay in residue multipliers for $0.5\mu m$ nMOS

increases only slightly for $M \leq 8$ bits, i.e., the delay is reduced almost to the same extent as the switching speed of the devices as the feature sizes are reduced. The areas, Table 5, expressed in terms of λ^2 [9] is to first order independent of the feature size.

From the tables we conclude that at $4\mu m$ feature size table lookup is preferable both with respect to area and delay for $M \leq 4$ bits. The delay of the index transform method is less than the other two methods for $5 \leq M \leq 7$ bits, while the array multiplier needs an area equal to or less than the other two for $M \geq 5$ bits. At $0.5\mu m$ feature size the table lookup method looses its competitive edge at $M = 3$.

Bit	Table	Ind. tran.	Array
3	45	95	115
4	147	168	207
5	524	282	297
6	1970	543	407
7	8000	1007	591
8	32000	2177	664

(*unit* : $1000\lambda^2$)

Table 5. Chip area of residue multipliers

3.2 Chip characteristics

Figure 6 shows the design of a residue array multiplier. A 4-bit version of the multiplier requires an area of $300 \times 700\lambda^2$, and an 8-bit version $540 \times 1200\lambda^2$. The designs are submitted for fabrication through the DARPA MOSIS foundary service.

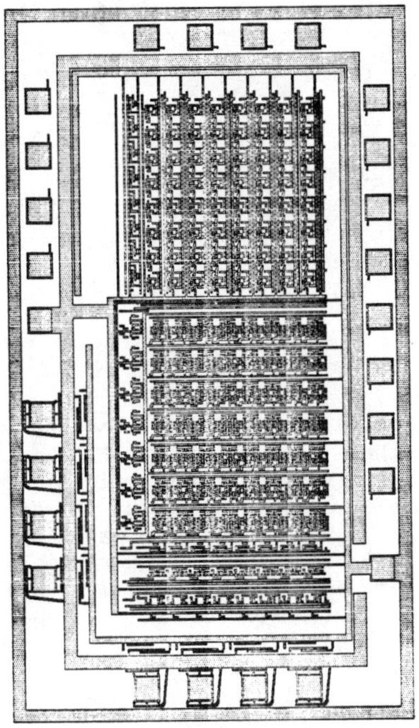

Figure 6. The metal layer of an 8-bit residue multiplier

Simulations indicates that carry and sum bit generation require 10~15 ns. The critical path in each M-bit residue multiplier goes through $3M$ adders. Therefore, the total delay is $30M$ to $45M$ ns, and hence about 240~360 ns for an 8-bit multiplier.

The performance of the design can be improved in several ways. For instance, the sample design does not have any form of pipelining. Optimizing the residue array multiplier design should make it competitive for smaller values of M.

4. CONCLUSION

The evaluation of residue multiplier designs in nMOS shows that at the current state of the technology, implementations based on table lookup are preferable both with respect to performance and area for a range that can be encoded with 18 bits in a binary number system. If a larger range is needed, index transform and array designs both have smaller area requirements and delays. The array design is preferable with respect to area, and also with respect to delay if the range requires a large number of bits for a binary encoding. However, if the design is optimized it can be operated at a clock rate that is comparable to or less than that of the index transform approach. At feature sizes of $0.5\mu m$ table lookup designs are preferable only for ranges of up to 9 bits, when binary encoded.

In comparing with computer arithmetic based on the binary number system it shall be noticed that our residue array multiplier is of the same complexity as a binary array multiplier.

Acknowledgment

This work was supported by the Defense Advanced Research Project Agency (DARPA) under contract N00014-79-C-0597 with the California Institute of Technology. Views and conclusions contained in this paper are the authors and should not be interpreted as representing the official opinion or policy of DARPA, the U.S. Government or any agency connected with them.

References

[1] H. Garner, *"The Residue Number System"*, IRE Tran. on Electronic Computers, June 1959

[2] L. Dadda, *"Some schemes for parallel multipliers"*, Alta. Freq. Vol.19 May 1965

[3] N.S. Szabo and R.I. Tanaka, *"Residue Arithmetic and its Applications to Computer Technology,"* New York, McGraw-Hill, 1967

[4] D. Banerji, J. A. Brzozowski, *"Sign Detection in Residue Number Systems"*, IEEE C-18, Apr. 1969

[5] N.G. Kingsburg and P.J. Rayner, *"Digital Filtering Using Logarithmic Arithmetic"*, Electron. Lett., vol.7, Jan., 1971

[6] E. Kinoshita, H. Kosako and Y Kojima, *"General Division in Symmetric Residue Number Systems"*, IEEE C-22 Feb. 1973

[7] W. Jenkins, *"The Use of Residue Number Systems in the Design of Finite Impulse Response Digital Filters"* IEEE, CAS-24 Apr. 1977.

[8] K. Hwang, *"Computer Arithmetic"*, Wiley 1979

[9] Mead & Conway, *"Introduction to VLSI System"*, Addison-Wesley, 1980

[10] D. Banerji, T.Y. Cheung V. Ganesan, *"A High-Speed Division Method in Residue Arithmetic"*, 5th symposium on computer arithmetic. May 1980

[11] C. H. Huang, D. Peterson, *"Implementation of a Fast Digital Processor Using the Residue Number System"*, IEEE, CAS-28 Jan. 1981

[12] M.A. Bayoumi, G.A. Jullien and M.A. Sid-Ahmed, *"VLSI implementation of Memory Intensive Residue Number System Architecture"*, Proceedings, 20th annual Allerton Conference on Communication, Control, and Computer. Oct 1982

[13] S.D. Fouse, G,R. Nudd and A.D Cumming, *"A VLSI Architecture for Pattern Recognition Using Residue Arithmetic"*, IEEE, Proceeding of the 6th International Conference on Pattern Recognition, Oct 1982

[14] F. Barsi, A. Di Cola, *"A VLSI Binary Multiplier Using Residue Number Systems"*, Proceeding ICCC, Oct 1982

[15] C. Chiang, *"Distributed RC delay line model and MOS PLA timing estimation"*, Technical Report, California Institute of Technology, in preparation.

THE POLYCYCLIC PROCESSOR

by Bijoy G. Chatterjee

ESL INCORPORATED
A Subsidiary of TRW Incorporated
Sunnyvale, California

Abstract

Horizontally microcoded processors (i.e., Array Processors), which can deliver several million floating point operations per second, have been programmed by those who could afford a very high system life cycle cost. There are over two dozen vendors who share a market of about 200 million dollars per year for the horizontal processors. The growth of this market has been seriously hampered by the difficulty of programming, lack of R&D investments, and architectural dissimilarity among the processors.

The ESL Polycyclic processor concept has been developed to correct some of the inherent problems with the present horizontal processors. ESL has developed the essential VLSI Interconnect circuit, which makes it possible for anyone to build a Polycyclic processor. It is claimed that industrial exploitation of this concept will rapidly expand this market. This paper explains how the Polycyclic processors will reduce system development cost and time while reducing the need for microprogramming specialists.

Introduction

Since the inception of computing devices, there have been rapid increases in computer speeds. However, at the same time our systems engineers have always found the latest computers too slow for the task in hand. They have invented clever algorithms to cope with the inadequacies of the latest computing devices. We don't see a change in this situation in the near future. In military signal processing, both in the strategic as well as the tactical field, the computing needs are overwhelming. The environment is becoming more complex because of increased electronic activities, low probability of intercept, and advanced ECCM strategies which reduce the threat of intercept and jamming. The need for supercomputer speeds at microcomputer prices with similar ease of use is widespread in the non-military signal processing and scientific computation market. Supercomputing capability coupled with ease of use for engineering work stations at affordable prices has a strong market potential.

Exploitation of the VLSI technology has increased the computing capability of the new devices while reducing the cost of computing devices for equivalent performance. This has rapidly increased the demand for qualified engineers who can build systems using these computers. In the case of general purpose computers, software tools like editors, operating systems, compilers, debuggers, data base management systems, etc., have increased the productivity of these engineers. In fact, personal computers have made it possible for everyone to utilize the power of a computer without special training.

Enormous profitability and the sales volume have allowed the industry to plow back large development funds to develop this solution and to achieve these impressive results. Unfortunately, such mass-produced general-purpose computers can compute no more than one million arithmetic operations per second, which is too slow for some applications.

Signal processing in particular often needs much higher performance rates than can be achieved with simple general-purpose computers. Fortunately, this need is so great that the government, as well as the industry, finds it attractive to invest in very high-speed integrated circuit processing elements which can be connected in parallel to meet these requirements. In addition to TRW LSI products, which hold the largest share of this market, startup companies find it profitable to offer innovative pipelined floating point circuits like Weitek 1032/1033[1]. Unfortunately, the differences in requirements and small market size for each type has discouraged the investors from developing a common architecture which will allow reusable software components and architecture-specific VLSI components.

The most popular solution to the need for programmable high performance processors is the so-called array processor. These horizontal processors are made up of a few very tightly coupled processing elements which are controlled by a wide, horizontal microinstruction. Unfortunately, programming such processors is extremely difficult and, hence, expensive[2]. These processors are often used as multifunction accelerators for attachment to general purpose computers. Sales of this type of processor are about to reach 200 million dollars per year. This limited market size coupled with architectural dissimilarities between products is discouraging the industry from investing adequate funds to develop HOL-programmable products.

CH1935-6/83/0000/0084$01.00 © 1983 IEEE

With the Polycyclic processor ESL suggests that the architecture of the typical array processor be slightly upgraded to (a) simplify development of efficient retargetable software, (b) take full advantage of VLSI/VHSIC, and (c) encourage reuse of HW/SW components (see Figure 1). This, in turn, will make these very high-speed processors affordable as well as programmable.

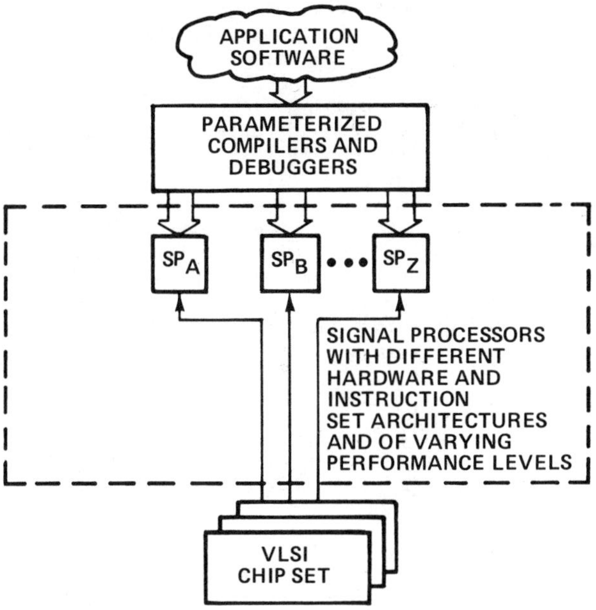

Figure 1: A Generic Architecture

Life Cycle Costs

During software development, the horizontal processor offers a great deal of hardship. These processors are extremely difficult to program because there are an enormous number of interdependent activities that must be efficiently scheduled during every clock cycle. Use of deeply pipelined processors, which are most desirable from a VLSI implementation point of view, makes scheduling even harder. An optimizing compiler is unavailable because the processor architecture makes compiler writing expensive and dissimilarities among processors negate reuse of parts of any previous compiler. Debugging of these wide-word microcodes is difficult because of the complexity of the task, the lack of software tools, and the lack of trained personnel.

The program manager is expected to strongly discourage any enhancement of the horizontal microcode during development or actual use. In spite of expensive documentation, it is very hard for a person to update the microcode written by another. It usually takes a long time and makes the verification and validation demanding. Similarly, hardware fault isolation and correction in the field is so difficult that talented engineers are called upon to perform this task.

In some cases the packaging, power, environment, or other requirements dictate development of the hardware as well as the software. Due to lack of a generic architecture, the designer freely invents new, unproven architectures. The funds available are never sufficient to develop a clean architecture for implementing the specific processor.

The following changes will significantly reduce the life cycle cost: (a) a generic architecture, (b) a higher-order language compiler, (c) a set of HOL-oriented debugging tools, and (d) hardware components to match the architecture. The positive effect of a generic architecture will be felt by the hardware designer, the compiler designer, the algorithm developer, the implementor, the service organization, and the program manager.

The Horizontally Microcoded Processors

The market for horizontally microcoded processors is increasing in spite of the inherent difficulty of programming them. There are more than two dozen different designs of horizontally microcoded processors. In addition to off-the-shelf processors (both in military as well as in commercial signal processing), custom processors are often used exclusively by the developer of that processor. Floating Point Systems holds the largest share of the market. Star Technology, on the other hand, claims to have the fastest machine. With the possible exception of FPS-164, most of the horizontal processors have been microprogrammed without the help of a HOL compiler. The processors built by different vendors are so different that software for one processor cannot be retargeted to another. The software development tools are primitive and very little debugging facility exists.

The low pin count, high device count, and faster device speeds of VLSI indicate extensive use of pipelining on the chip. Unfortunately, deeply pipelined circuits in horizontal processors make microcoding even harder. It becomes a superhuman task to keep track of many cycles of parallel activities. The programmer is then called upon to schedule these activities in such a way that all the devices are fully occupied. A compiler that can optimize this complex pipeline schedule is difficult to design. It is desirable that architectural innovation be used to simplify the task of compilation.

The requirements of high performance and low cost, along with tight size and weight contraints, indicate the need for custom, horizontal signal processors. On the other hand, software life-cycle costs can be reduced only by providing HOL compilers that generate code comparable in quality to manually generated assembly language programs. These are, apparently, conflicting objectives since custom processors imply the absence of compilers and since it is difficult to generate high quality code for horizontal processors without tools.

We suggest that the apparent contradiction be addressed by the use of a generic architecture as shown in Figure 1. A generic architecture is a set of processor architectures that may all be different from one another insofar as their performance, instruction formats, word-widths data types, physical characteristics, and cost are concerned. Each member must have sufficient commonality to make it possible to write software tools (i.e., compiler, debugger) that can be adapted to a newly designed member of the generic architecture. A generic architecture further encourages introduction of optimized VLSI circuits which will be reused on many specific processors with different capabilities.

A good example of a generic architecture is the class of register-oriented, general-purpose computers, which includes computer family architectures such as the IBM System/360-370, DEC PDP-11, and most other mainframe and minicomputer architectures. All these processors are similar in that they have general-purpose register files, serial instruction sets, etc., and as a result, they may be approached with a common strategy when building software tools or hardware components.

The Polycyclic Architecture

The compiler for any horizontal processor can be divided into a front end that translates the HOL source into a data-flow-graph and a back end that schedules (maps) this graph to a specific hardware. An optimizing compiler must exploit the available resources (processing elements, local memory, etc.) to the fullest extent. Unfortunately, as of now the compiled code for horizontal processors takes as much as 10 times longer to run than the code optimized by professional microcoders. The task of optimal scheduling will become greatly simplified if the input parameters for each resource are held in a buffer for an arbitrary length of time until the resource is ready to receive that parameter. Returning parameters to the local memory is not an effective alternative because the memory itself is one of the limiting resources. The polycyclic architecture introduces a dedicated delay element on the path between each resource output and each resource input that are connected together (see Figure 2). This, therefore, greatly simplifies the task for a compiler[3].

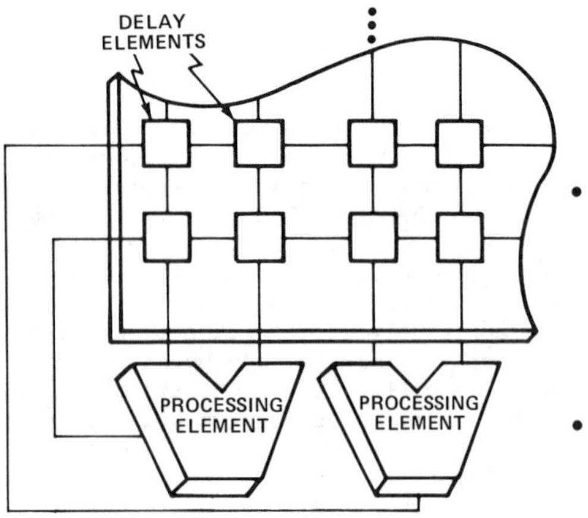

Figure 2: The Polycyclic Architecture

A polycyclic interconnect has a dedicated delay element between every resource output and every resource input that are directly connected to each other. A fragment of the interconnect of a sample polycyclic processor is shown in Figure 2. In this example, the topology of the interconnect is a complete crossbar with a delay element at each cross-point (thereby providing the polycyclic property). The interconnect has two output ports (columns) and one input port (row) for each of the two processing elements. Each cross-point has a delay element similar to that described below, which is capable of one read and one write on every cycle. A processing element can simultaneously distribute its output to any or all of the delay elements that are in the row of the interconnect corresponding to its output port. A processing element can obtain its left (right) input from any delay element in the column of the interconnect that corresponds to its left (right) input port. If a value is written into a delay element at the same time that an attempt is made to read it, the value is transmitted through the interconnect with no delay. Any positive delay may be obtained merely by leaving the datum in the delay element for a suitable length of time. Lack of this capability made the scheduling task for the previous horizontal architectures difficult[2].

The Polycyclic Compiler first identifies the critical resource on the basis of the graph to be scheduled and the target hardware. Properties of this critical resource indicate the minimum initiation interval[3]. Using this interval the graph is scheduled (mapped) to the target hardware. The delay elements are then scheduled to hold the parameters until a particular resource becomes available (see Figure 3). These delay elements in the interconnect do not appreciably increase the chip count because typical cross-bar chips are pin limited.

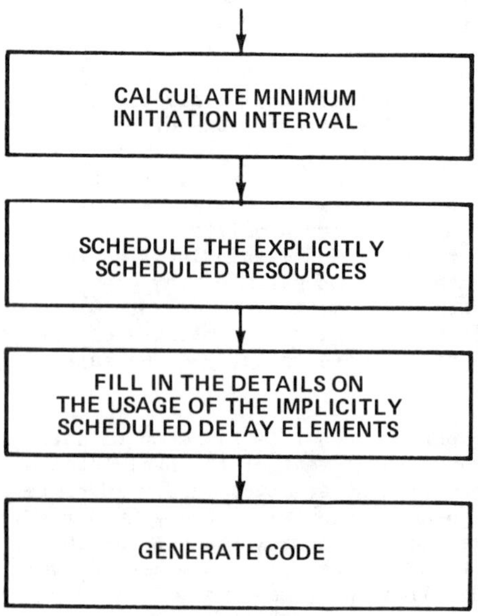

Figure 3: Scheduling the Polycyclic Processor

A previous paper[3] on this subject has discussed this simple three-step scheduling methodology. Extensive simulation has been performed with algorithms from common scientific computation libraries to verify this methodology. It has been observed that a small set of simple scheduling rules generates code that fully optimizes the use of the resources. A solution to the problem of scheduling conditional branches has also been developed.

Each delay element is a memory element which may be succinctly described as a FIFO memory with a random access read and (optional) delete capability; i.e., data are always written into the next location, but data may be read from any location and may, optionally, be deleted after being read. All the data are instantaneously compacted into the low part of the memory each time a datum is deleted[4].

The interconnect chip is a TTL compatible 2x2 crossbar with 8-bit words. (See Figure 4.) There are 16 words at each cross point. During each 100 ns cycle time, 2 reads, 4 writes, and 2 purges can be performed. There are negative clock edge triggered input and output registers and tristate output buffers. This chip can be packaged in 64-pin, dual-in-line, chip carrier or pin-grid packages. The power consumption is 650 mW at 10 MHz with 40 pF output load. These chips can be cascaded to obtain larger crossbar sizes or word widths.

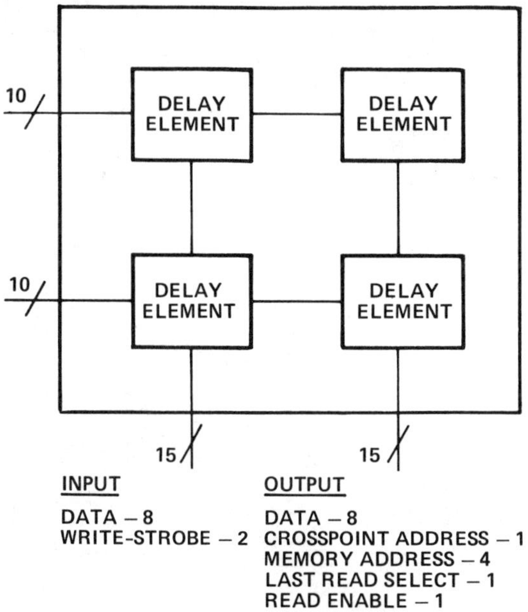

Figure 4: The Interconnect Chip

Four of the above circuits can be packaged in 132-pin hybrid packages. This produces a 4x4 crossbar with 8-bit words. These hybrid packages can then be put together on boards to build larger interconnects with more input ports, output ports, or bits per word.

Usually processing elements have a larger number of input parameters than output parameters (i.e., input A & B to obtain C as an output). To conserve on pin counts, the processing elements often take an input on each clock and produce an output on each clock. Each local memory port and each register file port is often optimized with one input and one output on each clock. The decision about the number of input and output ports for the interconnect must be made on the basis of the above and on the basis of the desired performance.

A future implementation of the interconnect circuit in 1 micrometer CMOS technology will increase the number of input and/or output ports and/or reduce the cycle time.

Conclusions

ESL has defined a generic architecture for horizontal processors, the Polycyclic architecture. ESL has also built the most essential new component, the interconnect chip. It is now possible for anyone to build a variety of Polycyclic processors to meet distinctly different performance requirements with off-the-shelf VLSI processing elements and memories together with the ESL interconnect chip. This provides an adequate business base for investors (i.e., government as well as industry) to build reusable components for the polycyclic processor. We expect some to build retargetable software and single board polycyclic processors based on this generic architecture. This, therefore, opens the door to a family of horizontal processors that can be inexpensive, have high performance, and be easy to use. The market for such a new class of product is large.

Acknowledgement

I would like to acknowledge the help of my colleagues at ESL and B.R. Rau.

References

1. B. Woo, L. Lin, and R. Owen, "ALU, multiplier chips zip through IEEE floating-point operations", Electronics, Vol. 56, No. 10, pp. 121-126, May 1983.

2. A.E. Charlesworth, "An approach to scientific array processing: the architectural design of the AP-120B/FPS-164 family", Computer, Vol. 14, No. 9, pp. 18-27, Sept. 1981.

3. B.R. Rau, C.D. Glaeser, and R.L. Picard, "Efficient code generation for horizontal architectures, compiler techniques, and architectural support", Proceedings of the 9th Annual Symposium on Computing Architecture, pp. 131-139, April 1982.

4. B.R. Rau, P.J. Kuekes, and C.D. Glaeser, "A statically scheduled VLSI interconnect for parallel processors", Proceedings of the CMU Conference on VLSI Systems and Computations, Computer Science Press, Rockville, Maryland, pp. 389-395, Oct. 1981.

DESIGN OF ALGORITHMICALLY-SPECIALIZED VLSI DEVICES

D.I. Moldovan and A. Varma

Department of Electrical Engineering-Systems
University of Southern California
Los Angeles, California 90089

ABSTRACT

In this paper a methodology is proposed for the design of algorithmically-specialized VLSI devices. The process of mapping an algorithm into hardware is divided into two main steps: 1 the mapping of algorithm into a global model describing the overall structure and operation of VLSI array and 2 the design of processing cells of the array based on the information contained in the global model. Both steps are based on the analysis and transformation of algorithm data dependences. The main focus of this paper is on the second step, the design of processing cells (the derivation of the global model is described in a previous paper). An example is given in order to illustrate the techniques proposed for the design of cell architectures.

1 INTRODUCTION

The problem of mapping computational algorithms into special-purpose VLSI systolic arrays has gained considerable attention during recent years. There were significant contributions in this area by Kung [1, 2], Guibas et al.[3], Leiserson [4] and others. However, most of the efforts were restricted to mapping specific algorithms into systolic processing arrays; heuristics played an important part in the process. Recently, Moldovan [5, 6] has suggested a general methodology for mapping algorithms into VLSI systolic arrays. The method essentially transforms the algorithm into a form more suitable for VLSI implementation. It can be shown that the architectures suggested in the above literature can all be derived in a systematic manner using this technique.

The mapping process essentially consists of two stages: (i) a mapping of the algorithm into a 'global model', and (ii) development of a 'local model' from the global model. The global model is characterized by a network geometry G, a set of functions F performed by the processing cell, and the network timing T. The formulation of the global model from the algorithm has already been discussed in [5]. This paper addresses the problem of synthesizing a local architecture from the global model.

This research has been partially supported by the Joint Services Electronics Program Contract No. F49620-81-C-0070.

2 THE MAPPING PROCESS

A systematic approach for designing special-purpose VLSI chips, is especially important due to the diversity of tasks and characteristics of VLSI design. It is almost impossible to take into account global architecture, circuit realization, and VLSI layout details all at the same time; however, these processes are interdependent. For example, implementation difficulties in the layout may demand a circuit redesign. Such wastage of effort can be minimized by being aware of the implementation limitations at an early design stage.

In figure 1 is shown the task dependency graph for the design of a special-purpose VLSI chip. The first step is the mapping of the algorithm into a VLSI global model [5]. The global model describes the VLSI arrays in terms of three parameters: the network geometry G, the set of functions F performed by the individual processing cells, and the global timing T. If the cell functions are simple enough, conventional techniques of circuit design may be adequate. If the functions are complex, the cells may be partitioned into subsystems, and designed independently. Such structural design approach is unavoidable for the development of complex VLSI structures.

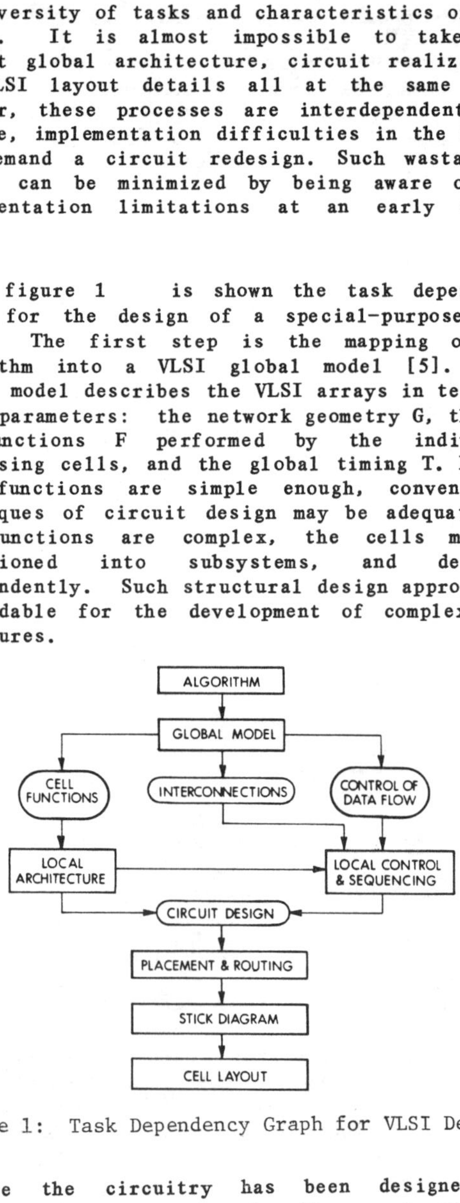

Figure 1: Task Dependency Graph for VLSI Design

Once the circuitry has been designed, an optimal placement of circuit components and routing of interconnections should be carried out.

Finally a stick diagram and the actual layout are prepared. Fairly advanced computer-aided design tools are available for transforming the circuit into a layout.

In this paper, we will address some of the issues relating to the development of the local architecture from global model, and suggest few ideas.

3 DEVELOPMENT OF THE LOCAL MODEL

In this section we suggest a general structure of a cell and explain how such structure can be obtained methodically via transformation of data dependences.

3.1 Cell Structure

The type of architecture chosen for the processing cell depends on the type and complexity of the functions to be performed by the cell. In almost all the problems addressed in [1-4] the function of the cell is simple enough to be realized by a few combinational circuits. There is very little control within the cell. However, if the cell has to compute more complex functions, for example recursions, a dedicated control unit is needed within the cell. In general, the complexity of a cell can vary from a simple combinational circuit to a full-fledged processor. In figure 2 it is shown the basic structure of a typical processing cell of moderate complexity. It consists of the following building blocks:

a. A set of registers R which hold the input data to the cell. This is characteristic of any pipeline segment.

b. A set of function units or ALUs to carry out the computations specified by the algorithm.

c. A set of register files used as delay elements to synchronize the output data from the cell.

d. A set of scratchpad registers or memory to store temporary results.

e. Interconnections between registers and ALUs for proper routing of data.

f. Control logic for sequencing the micro-operations within the cell and for implementing proper routing of data. The control logic synchronizes the operation of the cell with respect to an external clock signal or control signals and makes sure that the results are available at the right time.

Next we suggest a method for finding an optimum computational ordering by the use of local transformations.

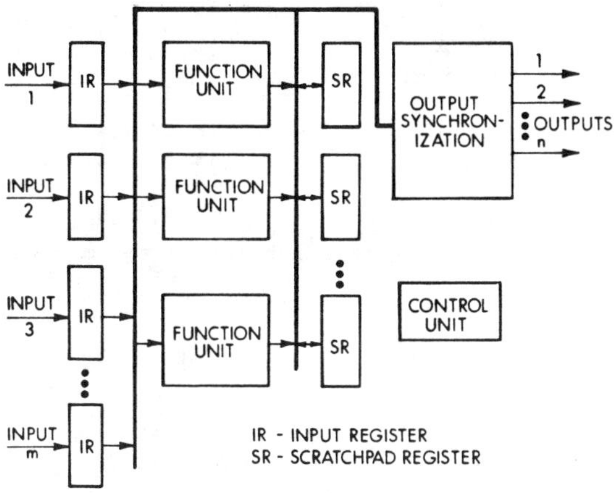

Figure 2: Elements of a Processing Cell

3.2 Local Transformations

The fundamental idea behind the application of the global transformation [5] was to express the computations in a form in which the inherent parallelism is more explicit. The same ideas, when extended to the local level, can be used to design the processing cell.

The study of dependences at the local level reveals one important fact: at the local level, the source of a data item is not significant as far as the computational ordering is concerned; what is important is the time at which data reaches the cell. This timing is entirely decided by the index variables corresponding to time in the global model. Thus we have to look only at the index variables corresponding to this partition in the global transformation matrix, for the purpose of finding an efficient computation scheme. Other index variables, which represent the spatial properties of the data can be ignored for this purpose.

Thus, at the local level, each variable can be described by means of two index variables, one representing the time at which the variable reaches the cell, and the other representing the identification of the channel along which the variable travels. A channel index of 0 can be used to denote variables generated within the same cell. Such data values are extremely important in the mapping process, for, they are available in the cell at an early point in time and hence suggest the possibility for reducing the computational burden by performing some of the operations at an early stage in time.

The idea of local transformations can best be illustrated by an example:

Example:

Consider the loop structure:

```
for i:=1 to n do
 for j:=1 to n do
   for k:=1 to n do

x(i,j,k) = x(i-1,j,k) * x(i-2,j,k) * x(i-3,j,k)
         * x(i-4,j,k) + x(i-4,j,k-1) + x(i,j-1,k)
   end.
```

The dependence matrix for this program loop is given by:

$$D = \begin{bmatrix} 1 & 2 & 3 & 4 & 4 & 0 \\ 0 & 0 & 0 & 0 & 0 & 1 \\ 0 & 0 & 0 & 0 & 1 & 0 \end{bmatrix}$$

Let us apply the global transformation:

$$T = \begin{bmatrix} 1 & 1 & 0 \\ 0 & 1 & 0 \\ 0 & 0 & 1 \end{bmatrix}$$

This results in a modified dependence matrix:

$$\Delta = T D = \begin{bmatrix} 1 & 2 & 3 & 4 & 4 & 1 \\ 0 & 0 & 0 & 0 & 0 & 1 \\ 0 & 0 & 0 & 0 & 1 & 0 \end{bmatrix}$$

At the local level, we can describe each variable with two indices, one representing time and the other representing the channel id along which the variable arrives in the cell. In terms of this representation, the computations within the cell can be written as:

$$\begin{aligned} x(0,i) &= x(0,i-1) * x(0,i-2) \\ &* x(0,i-3) * x(0,i-4) \qquad (1) \\ &+ x(1,i-4) + x(2,i-1) \end{aligned}$$

This is a recurrence relation in x. We notice that the result has to be generated in one time unit after the variable which arrives latest reaches the cell. If we postpone the entire evaluation until all the inputs are available, we need to perform 4 multiplications and 2 additions within one time unit. Even if we make use of tree height reduction, this requires 3 time stages to evaluate the entire expression, as illustrated in figure 3. An alternate idea is to pipeline the computations by applying a transformation. The expression suggests that the first four operands are available within the cell, whereas the last two have to arrive from neighbouring processing cells. Since the first four operands are local, there exists a possibility of pipelining the computations on them.

Equation 1 can be expressed as

$x^0(0,i)=1$
for j := 1 to 4 do
begin

$x^j(0,i-j) = x^{j-1}(0,i-j)$

$x^j(0,i) = x^{j-1}(0,i) * x^j(0,i-j)$

end

$x(0,i) = x^4(0,i) + x(1, i-4) + x(2,i-1)$

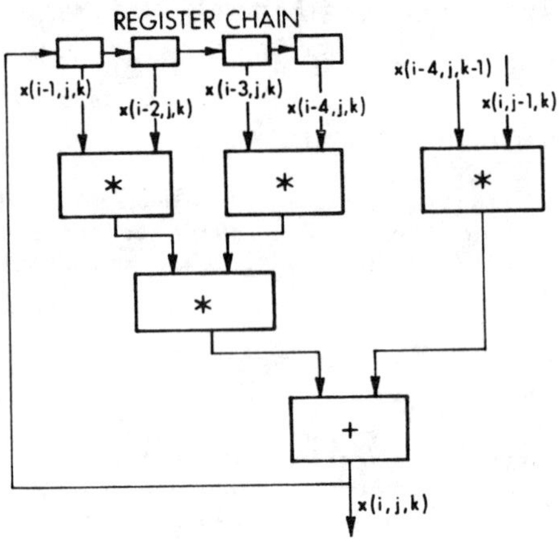

Figure 3: Computation Scheme for Example 1

Note that here the index j represents the timing within the cell and may be regarded as another dimension in time. The dependence matrix for the above loop is given by

$$D_\ell = \begin{bmatrix} 1 & 1 \\ -1 & 0 \end{bmatrix}$$

Let us use a transformation

$$T_\ell = \begin{bmatrix} 2 & 1 \\ 1 & 0 \end{bmatrix}$$

This results in a modified dependence matrix

$$\Delta_\ell = \begin{bmatrix} 1 & 2 \\ 1 & 1 \end{bmatrix}$$

This suggests the computation scheme shown in Figure 4, wherein the data generated within the cell have been pipelined. Successive values of the variable x flow through the lower data stream and the product flows in the upper data stream at twice the speed of x. The time available for each multiplication is one-half of the period of the global system clock, irrespective of the length of the recurrences. Note that this scheme results in a slight increase in hardware, but with a substantial improvement in speed and regularity of the layout. In the example, the local pipelining was restricted to only the data values generated within the cell. However, the method is not restricted to locally generated variables. Variables generated within a preceding cell and reaching at an early instant of time can be subjected to the same treatment. Thus it may be worthwhile to pass on the output from a cell to its neighbour as soon as it is generated, rather than delaying it to synchronize globally with respect to some other variables. This implies moving any inserted delays from the output of the cell to the input of the succeeding cell. This may necessitate a slight correction in the global model.

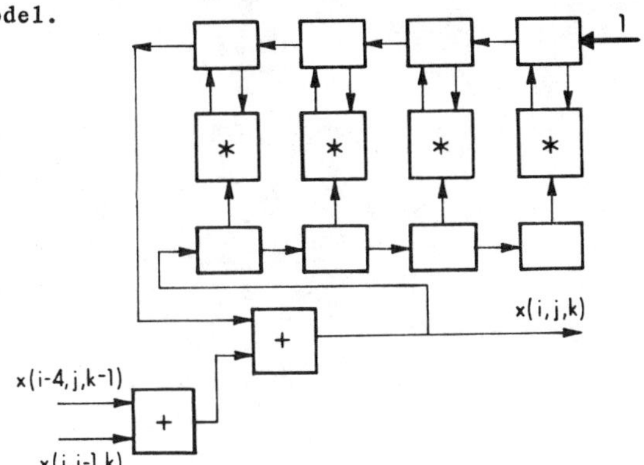

Figure 4: Pipelined Computation Scheme for Example 1

4 PROCEDURE FOR LOCAL MAPPING

If the functions of the cell are too complex and involve repetitive evaluations with fixed or variable number of steps, a complex processing cell with an "intelligent" control unit is required. However, if the functions are realizable mostly by combinational circuits, then the following steps are suggested:

1. Attempt to reduce the computations into a pipelined form by applying a local transformation. This, in general, results in a more regular cell structure.

2. Decide on the number of registers and functional units to be used. Each input channel terminating at the cell requires a register for storing the data values that arrive through the channel. The cell also needs registers for temporary storage of intermediate results, if any. The functional units should ideally be provided one for each operation, as the cost of implementing them is low, compared to the cost and difficulties associated with a complex interconnection structure.

3. Synchronize the outputs generated by the cell to the global timing - this is achieved by providing registers in the output channels to introduce delays.

4. Implement a local control scheme. The transformation in step 1 implicitly contains information about the local timing. The index "j" associated with the transformed equations represent the domain of time at the local level. This information can be used for developing the control scheme. Based on the global timing information and the local computation-sequence, the exact sequence of microoperations within the cell can be obtained.

5 CONCLUSION

In this paper, we addressed some of the issues relating to the development of the architecture of a VLSI processing cell based on the information from the global model and the algorithm. It is hoped that, together with [5], this paper will form a good starting point for automating the process of mapping computational algorithms into dedicated VLSI architectures. It may not be too far when one is able to put into a computer all the information about an algorithm and get back all the intricate design details of a VLSI chip , with an architecture tailored for the fastest execution of the algorithm. Such a package will be an important landmark in the evolution of the silicon compiler.

6 REFERENCES

[1] Kung, H.T., "Let's Design Algorithms for VLSI Systems," Proceedings of the Caltech Conference on VLSI, 1979, pp. 65-90.

[2] Kung, H.T., "The Structure of Parallel Algorithms," Advances in in Computers, Academic Press, 1980.

[3] Guibas, L.J., Kung, H.T. and Thompson, C.D., "Direct VLSI Implementation of Combinatorial Algorithms," Proceedings of the Caltech Conference on VLSI, 1979, pp. 509-525.

[4] Leiserson, C.E., "Systolic Priority Queues," Proceedings of the Caltech Conference on VLSI, 1979.

[5] Moldovan, D.I., "On the Design of Algorithms for VLSI Systems," Proceedings of the IEEE, Jan. 1983, pp 113-120.

[6] Moldovan, D.I., "On the Analysis and Synthesis of VLSI Algorithms," Trans. on Computers, Nov. 1982, pp. 1121-1125.

Session 1-2a
VLSI Technology II

Chairman

R. A. Heald
Fairchild

A SONOS/CMOS NON-VOLATILE RAM

Roger A. Haken, William E. Feger, Donald J. Coleman
and Chung S. Wang

Texas Instruments Incorporated
P. O. Box 225621/MS 944
Dallas, Texas 75265

ABSTRACT

The fabrication, design and performance of a SONOS (silicon/oxide/nitride/oxide/silicon)/CMOS direct write NVRAM is described. Low voltage and fast write speeds (13 V and 1-10 μs) have been realized by aggressive scaling of the traditional MNOS structure plus the introduction of a thin barrier oxide to create a SONOS structure ~ 130 Å thick. By using a differential cell arrangement an endurance of 10^{11} write/erase cycles is possible using a 2 μs write pulse, with a retention of 3-10 days. The technology is also applicable to low voltage E^2PROMs, since the same SONOS devices can be programmed using an 8 V, 1 ms write pulse.

1. INTRODUCTION

The realization of a direct write NVRAM would be a major breakthrough in semiconductor device technology. Currently the choice of NV memory is limited to core, plated wire, EPROM, E^2PROM, shadow and CMOS with battery backup. Each has serious drawbacks and none of them meet the specification of the ultimate NVRAM, namely < 500 ns write cycle times, low write voltage, ~ 10 V, high endurance, $>1x10^{13}$ W/E cycles, and 3 to 10 day retention. Perhaps the nearest to reaching the NVRAM goals is CMOS with battery backup, but battery maintenance is a serious drawback for most military and some commercial systems. The shadow RAM emulates the NVRAM but requires complex power down detection and data transfer sequences, and is limited in bit density because of the requirement that each static cell have its own E^2PROM backup.

In this paper an alternative technology for realizing a NVRAM is presented. The technology consists of aggressive scaling of the traditional MNOS E^2PROM multidielectric stack, plus the addition of a top barrier oxide between the silicon gate and the nitride layer to create a SONOS[1] structure ~ 130 Å thick. The approach is similar to that independently conceived by Suzuki et al.[2] except their goal was to achieve a low voltage E^2PROM structure. Although the technology presented here does not reach all the NVRAM goals discussed above it does significantly advance the state-of-the-art and is suitable for many applications. The feasibility of the technology has been demonstrated by the fabrication of both a 4K and 2K NVRAM.

2. SONOS NVRAM ELEMENT

To achieve the programming speed and voltage appropriate to a NVRAM very high fields and hence ultrathin insulating layers are required. This requirement tends to preclude floating gate technology because the transverse conductivity of the floating gate allows discharge of the stored charge through defects in the ultrathin insulator. On the other hand MNOS devices are relatively free from total charge loss at pinholes because the charge is stored in traps within the Si_3N_4 insulator.

We have therefore taken the MNOS approach to realizing a fast write, low voltage NVRAM element by aggressively scaling the nitride thickness, currently ~400 Å in E^2PROM's, to <100 Å. However, with such a thin nitride layer tunneling of charges all the way through the nitride layer can occur with a significant reduction in trapping efficiency and hence programmed window size. By introducing a thin oxide barrier layer on top of the nitride a potential barrier is created which tends to confine the carriers to the nitride layer with a corresponding improvement in capture efficiency. This structure is referred to as SONOS, see Fig. 1.

Since high programming speed is required the multidielectric stack must operate at fields close to the intrinsic breakdown value, ~ 10 MV/cm. This implies that the SONOS transistors must be fabricated with very low defect density. The approach used for meeting this requirement

CH1935-6/83/0000/0093$01.00

has been to prepare the SONOS transistors totally in-situ, at constant temperature, by using a modified Si_3N_4 LPCVD furnace that has the capability of:

- Tunnel oxide growth in a low pressure O_2 ambient.
- Deposition of LPCVD nitride for the charge storage layer by the decomposition of $SiCl_2H_2$ and NH_3.
- Deposition of SiO_2 for the barrier layer by decomposition of $SiCl_2H_2$ and N_2O.
- Deposition of the polysilicon gate electrode by the decomposition of $SiCl_2H_2$.

This procedure minimizes slice handling that can lead to particulate contamination, with the result that highly reliable SONOS transistors are fabricated.

The barrier oxide has also been grown by steam oxidation of the nitride layer. Whilst this technique does offer the attractive feature of sealing pinholes in the nitride layer it does not form such a sharp barrier to charge transport as does the thicker LPCVD oxide. Furthermore, the procedure defeats the concept of in-situ stack preparation. Experiments have demonstrated that memory retention for the steam barrier oxide is inferior to the LPCVD oxide and that the size of the programmed window is larger. This latter characteristic is probably due to the introduction of additional traps at the nitride/oxide interface, that occurs during oxidation. Associated with a larger window is a corresponding increase in the time or voltage necessary to program a saturated V_T resulting in a speed penalty for the NVRAM, see section 4. This, in conjunction with reduced retention for the stacks with a steam barrier oxide makes the total in-situ method of SONOS transistor fabrication the preferred technique.

3. NVRAM CELL OPERATION

The CMOS/SONOS NVRAM cell is shown in Fig. 2, together with operating waveforms. The cell comprises a P-channel NV device with three N-channel isolation transistors plus three control lines. Unipolar voltage programming is used to program both the positive and negative threshold states. A $+V_T$ is programmed by applying 15 V to the column line and 0 V to the bit line. When the word line is opened ~ 13 V appears on the gate of the NV transistor and 0 V on the tank. Reversing the voltages on the bit and column lines programs a $-V_T$. In either case the cell dissipates no D.C. power during writing. The cell is read by applying a read voltage to the column line and connecting a load and sense amplifier to the bit line. The read operation results in enhancement of each written state, no read disturb, and provides a static output signal.

4. NON-SATURATED STACK WRITING

A typical plot of the NV cell output voltage versus programming speed, with a 130 Å multidielectric stack, is shown in Fig. 3. The top and bottom horizontal lines show the cell output voltage when the stack is saturated with a 1 ms, 13 V programming pulse. The two other curves show how the cell output changes from the opposite saturated threshold state with cumulative 200 ns programming pulses. A single 200 ns write pulse alters the saturated threshold state by ~ 1.5 V.

Clearly the fastest memory write cycle times are achieved with non-saturating write pulses. However, because the NV stack can eventually be saturated by a large number of short nonsaturating pulses, some bits in a memory could conceivably be written to saturation. A restriction, therefore, on the minimum write pulse width is that it must be capable of writing from a saturated threshold state to a point on the curve where the sense amplifier will make a correct decision when the cell is read. In both the single ended and differential cell NVRAM design, a comparator is used to sense the cell output signal. The minimum write pulse width must therefore be capable of writing from a saturated V_T state to > mid-window since for the single ended part the comparator uses the average of two dummy NV cells programmed in opposite states as the reference. In the differential approach each cell has its own dummy cell programmed in the opposite state. For the NV stack programmed in Fig. 3 this constraint means that the minimum 13 V write pulse is ~ 4 μs. A typical cell retention plot at room temperature and 125°C after writing with a 10 μs, 13 V pulse is shown in Fig. 4.

5. ENDURANCE

Fig. 5 is a retention plot before and after the NV cell has been written with 10^9 W/E cycles. Clearly the rate of window decay before and after endurance is approximately the same, however, there is significant shift in window position. The window shift results in a serious degradation in retention for a single ended device that has a reference voltage generated by two NV cells programmed in opposite states. This results because the reference cells are not cycled as frequently as the NV cells in the memory and therefore the reference signal will not be the correct value for every bit in the memory.

The solution to this problem is a

differential cell approach so that each cell has its own dummy which undergoes exactly the same endurance as the real cell. This technique also offers the advantage that twice the output signal is available which in turn means that the same threshold window can be written with a faster write pulse. Clearly the biggest draw back is that the packing density is reduced by a factor of two. Using the differential cell approach a NVRAM with a write speed of 5 μs can be realized using 13 V with an endurance of > 10^{11} write/erase cycles and a retention of 10 days. By increasing the write voltage to ~ 16 V the write time can be decreased to ~ 500 ns. A photomicrograph of a 2K differential CMOS/SONOS NVRAM is shown in Fig. 6, this part has a read access time of 250 ns.

6. PERIODIC REFRESH

At first the terms periodic refresh and NV memory might appear inconsistent, however, a DRAM could be considered as a NV memory with a retention of a few milliseconds. Since the NVRAM considered here only requires a power down retention of between 3-10 days for a wide range of applications, consideration must be given to the power up retention. Clearly there are likely to be many applications that require continuous powered up use of a NVRAM for times that are much longer than the RAM's NV retention of a few days. To overcome this problem a periodic refresh capability has been added to both the 2K and 4K parts. This allows periodic rewriting on a row-by-row basis at a user determined interval that can vary from seconds to hours without impacting endurance. Having included a periodic refresh capability a further mode of operation of the NVRAM is possible to give maximum retention. As mentioned in section 4 the fastest operation of the NVRAM is accomplished with non-saturating write pulses. However, this mode of operation doesn't give the longest retention because the full threshold window isn't written; greatest retention is achieved with saturating write pulses. A mode of operation that achieves the fastest write times with maximum retention is to normally write the memory with pulses that alter the NV cell's threshold from saturation to just the other side of mid-window. Upon loss of power the entire memory is rewritten with saturating pulses to achieve maximum power down retention. Power up retention need only be slightly longer than the periodic refresh cycle time.

Although this mode of operation does detract from the simplicity of operation of a direct-write NVRAM, operation is now analogous to a shadow NVRAM, the part can now be used for those applications that require fast writing and long retention.

7. CMOS/SONOS PROCESS

To realize a NV memory the SONOS technology must be integrated into a baseline CMOS process. This has been accomplished using a 15 V, twin-tub, P^- epi on P+ substrate CMOS technology[3]. The process uses 10-mask levels with two-levels of poly, a regrown gate oxide and a phosphorus N-tub ~ 4 μm deep. The first poly is used exclusively for the P-channel SONOS transistors while second poly is used for regular transistors and interconnect. A process cross-section of a regular N-channel device and a P-channel NV device are shown in Fig. 7.

CONCLUSIONS

A SONOS/CMOS NVRAM technology using a multidielectric NV stack ~ 130 Å thick has been demonstrated. The NV transistor fabrication process has been developed to the point where the complete SONOS stack can be deposited as a single, automated, furnace operation. The greatest advantage of this approach is that the memory characteristic can be "dialed-in" prior to the stack deposition, from NVRAM to low voltage, VLSI compatible, E^2PROM.

The fast write, high endurance and limited retention NVRAM's produced by this technology make them suitable for applications in military systems, such as flight control, and commercially for storing system configuration data in terminals and printers, and on board with microprocessors.

REFERENCES

1. P.C.Y. Chen, "Threshold-Alterable Si-Gate MOS Devices," IEEE Electron Devices, Vol. ED-24, pp. 584-586, May 1977.

2. E. Suzuki et al, "A Low-Voltage Alterable EEPROM with MONOS Structures," IEEE Electron Devices, Vol. ED-30, pp. 122-128, Feb. 1983.

3. R. A. Haken et al, "An 18 V Double-Level Poly CMOS Technology for Nonvolatile Memory and Linear Applications," ISCC pp. 90-91, Feb. 1983.

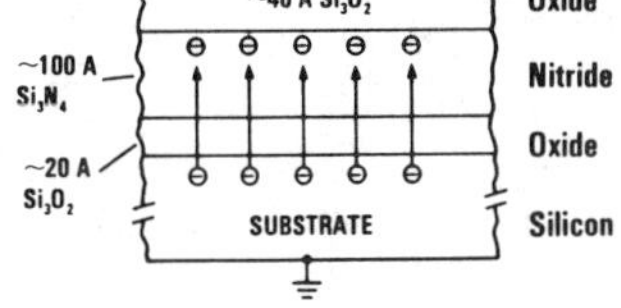

Fig. 1 The SONOS NVRAM element.

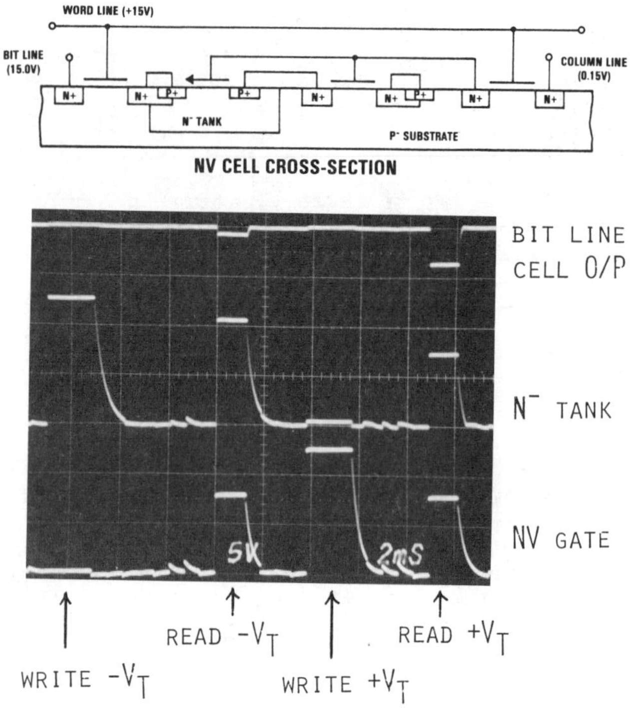

Fig. 2 NVRAM Cell Operation

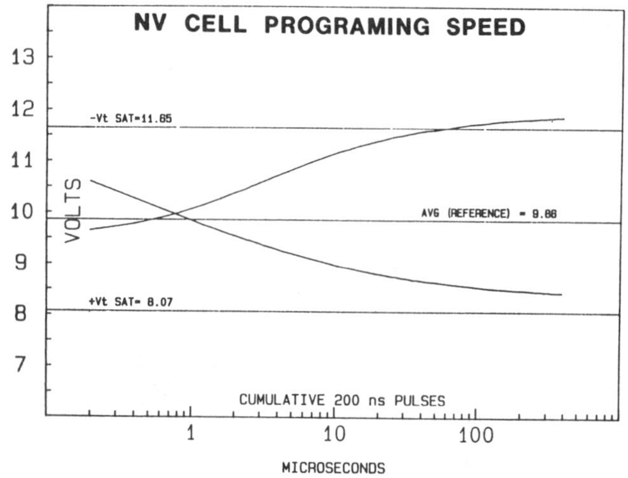

Fig. 3 NV Cell Programing Speed

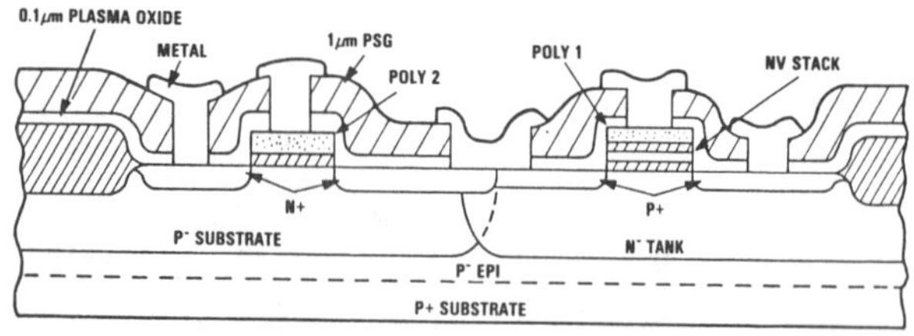

Fig. 7 Process Cross-Section of an N-Channel Device, and a NV P-Channel Device.

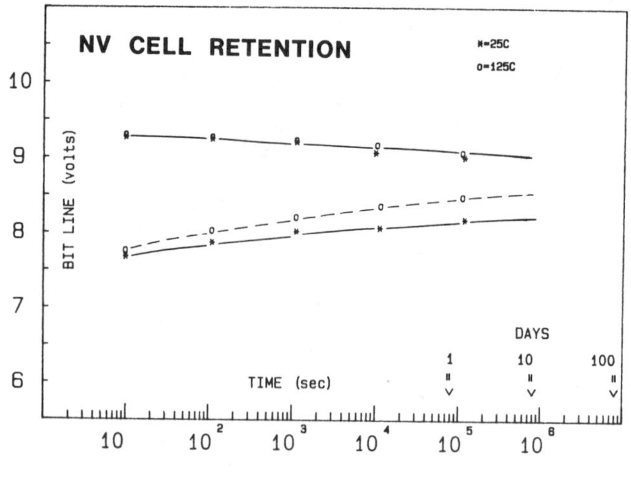

Fig. 4 NV Cell Retention

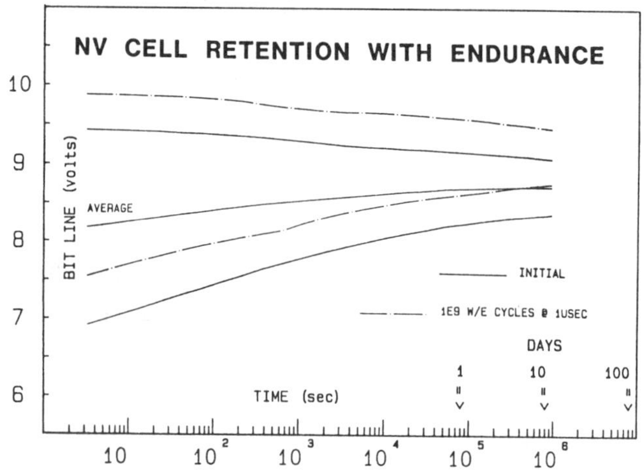

Fig. 5 NV Cell Retention with Endurance

Fig. 6 Photomicrograph of a 2K CMOS/SONOS NVRAM

Sub-picojoule bipolar LSI with a 350ps gate delay and 2000 gates

Yasunori Ohuchi*, Masanobu Sano**, Fumihiko Sato***, Hiroshi Nakashiba***,
Masahiko Nakamae***, Hiroyuki Shiraki*** and Nagao Yoshimura****

* 2nd Office Automation Division, ** Computer Engineering Division
*** 1st LSI Division, **** System LSI Development Division

NEC Corporation 10, Nisshincho 1-chome, Fuchu city, Tokyo 183, Japan

ABSTRACT

Low energy and high speed LSIs having 2000 equivalent gates have been fabricated utilizing a cell array masterslice and low energy CML technology. A gate delay of 350ps with a 0.42 mA current switch has been achieved. The design automation system has supported all the design activities and has remarkably reduced the design time for option developments.

1. Introduction

Computer and communication applications require high performance, low energy, high density and easy design LSIs. Low energy and high density LSIs have been developed with CMOS technology. However, the sub-nanosecond delay performance is still achieved only with the high speed bipolar technology.

But the standard ECL devices and gate array were not sufficentry fast and used too much power for a new large computer system.

This paper introduces a masterslice approach which has been adopted to meet the demands for system-oriented LSIs.

A sub-nanosecond delay with sub-picojoule energy has been achieved with low energy CML technology utilizing an advanced bipolar process (1).

The masterslice approach and the design automation system have created to develop high performance system-oriented LSIs offering short turn around time and a low development cost.

2. Circuit design

The basic internal circuit is a low energy CML (LCML), which has a single level CML structure with an emitter follower circuit. The LCML circuit is featured by its small logic swing of 560mv in internal area, and low power supply voltage of -3.3v. The reference voltage of -0.26v is also supplied to the chip to avoid the influence of temperature differences on the system.

A single level LCML structure requires more current sources than a 2 level series gated LCML structure (2). However, a 2 level series gated structure requires a higher power supply voltage because of its cascade current switches between ground and power supply, while a single level structure can be set at low power supply voltage of -3.3v or less.

The functionability of the single level LCML circuit have been improved with the usage of emitter dotting and collector dotting. Emitter dotting and collector dotting with a diode clamp have also improved the efficiency of transistor usage and reduced the number of logic stages.

The selective current sources of 0.42mA/0.21mA for internal current switch and 0.09mA/0.2mA/0.4mA/1.2mA for emitter follower are prepared to optimize power dissipation and performance.

The function blocks such as AND/NAND gates, AND-OR gates, Exclusive-OR's, flip-flops, etc., have been designed for option developments. Fig. 1 shows an example of 2-input Exclusive-OR block.

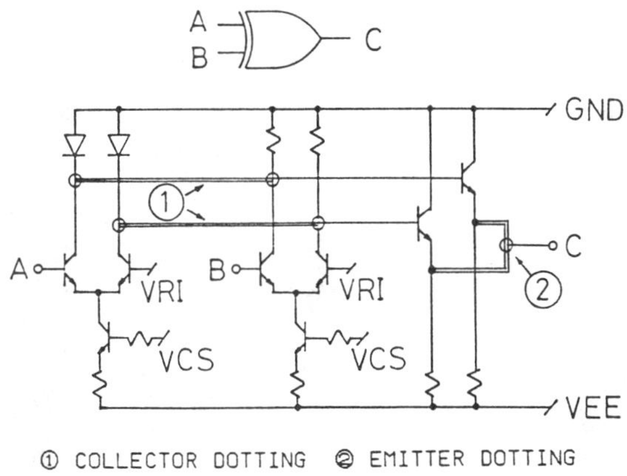

Fig. 1 An example of internal block circuit.

The output buffer circuit which is driven by a 1.2mA emitter follower is featured with a current switch of 8.47mA, logic swing of 500mv and output impedance of 59 ohms. It has a one or two-input AND/NAND function.

3. Chip layout

The masterslice consists of 832 internal cells, 56 output cells, 112 input cells, 6 current source voltage (VCS) generators and 52 internal reference voltage (VRI) generators. The internal cells are arrangd in a 26 x 32 matrix form, while the output cells are located on two opposit sides of the chip. An internal cell contains 8 small size transistors for current switch, 4 middium size transistors for emitter follower and 26 resistors. Each internal cells are equivalent to 2.5 gates in an average. The chip has 132 external pads for film carrier bonding system. Each pad has a double-bump structure to avoid the edge contact between lead and substrate without any lead formation. These pads include 8 pads for ground (GND), 8 pads for power supply (VEE) and 4 pads for reference voltage supply (VREF).

The masterslice has two metalized layers of aluminum; the first layer of aluminum has 320 channels, the second layer of aluminum covered with gold has 370 channels. The chip layout which is shown in Fig. 2 produced a chip size of 7.6 x 7.3mm.

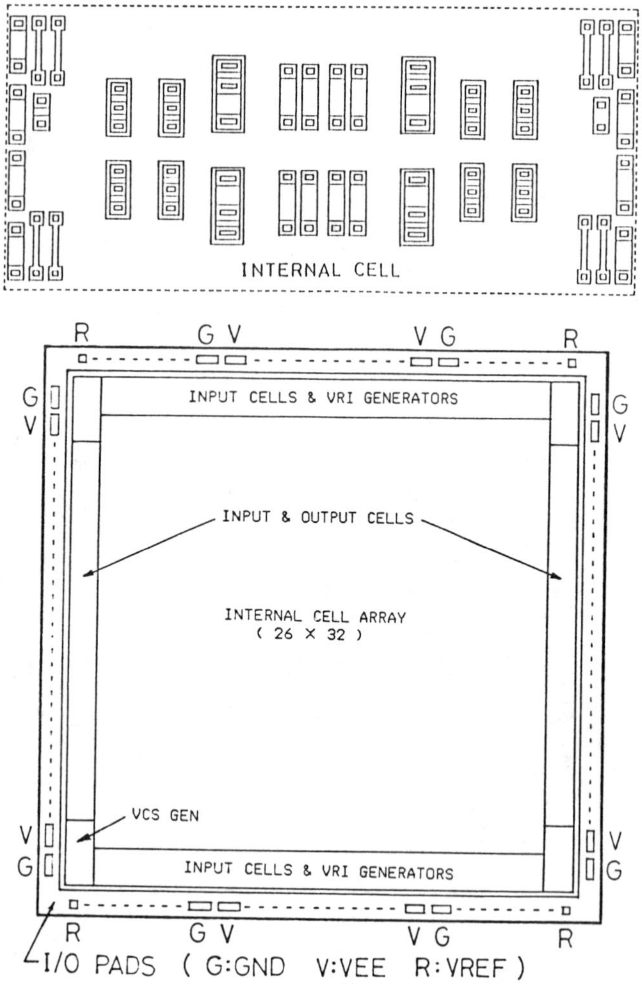

Fig. 2 The chip layout

4. Design automation

All the design activities have been supported by the DA system (3). The initial inputs to the DA system are the logical interconnection between each function blocks and the functional description of the chip using functional description language (4). The final outputs of the DA system are a pattern generation tape or an electron beam tape and the test data such as a test program and test patterns. These activities include logic simulation, block allocation, routing, delay analysis, test data generation and fault simulation. The block diagram is shown in Fig. 3.

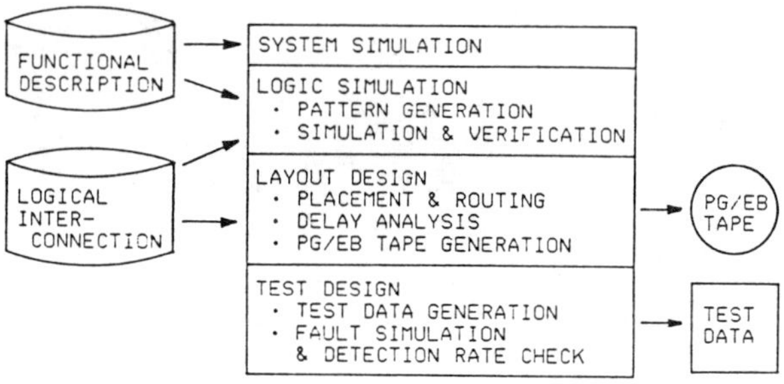

Fig. 3 The DA system

The mixed level logic simulator (5) verifies the logical operation of the two input files called the logical interconnection file and the functional description file. In this step, two input files are simulated in parallel using the antomatically generated simulation patterns, then the simulated results are compared and checked. The logical interconnection file is used for layout design of the chip and the functional description file is used for system simulation in the next design step.

The layout design includes the automatic two-dimensional placement, routing and delay analysis. The layout system generates the PG or EB tape for individual masking layers. Specific items related to the emitter dotting are processed by the layout system. These items include the selection of an emitter follower resistance, line width modification to reduce the series resistance between emitter follower sources, current density and line length check, power calculation and other circuit design rule checks.

The test patterns are generated by using the scan path technique (6) and fault detection rate is simulated. The standard device test program is automatically modified by the individual data for each option development. Finally, the test patterns and program are converted to the LSI tester language.

Using this DA system, the total turn around time of all the design activities for option development has been reduced to half when compared with the conventional system.

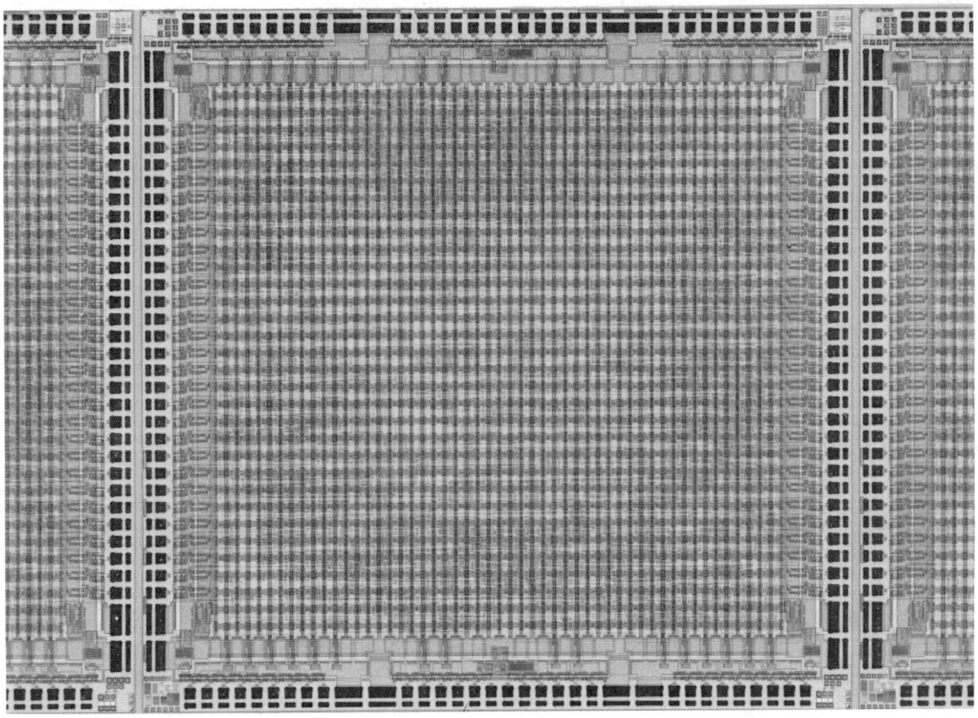

Fig. 4 A photograph of the chip

5. Application

The masterslice has been applied to the option developments such as multiplier, content addressable memory, shifter etc., for a large computer system. The representative design data for the LSIs show that they utilize 90% of internal cells and have an average equivalent of 2000 gates. These chips contain an average of 800 networks with 2.5 fan-outs, which means 2000 interconnections are routed on the chip. Total line length is about 1500mm. Typical power dissipation of the chip is 4.5w. It can be cooled by a forced air cooling system.

The basic internal gate delay of 350ps has been achieved with a 0.42mA current switch and 0.4mA emitter follower. Fig. 4 is a photograph of the chip. Fig. 5 is a photograph of the film carrier.

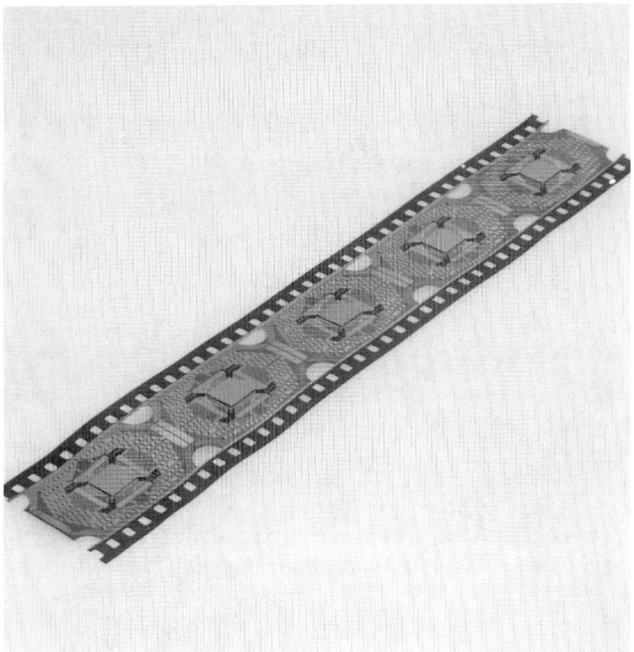

Fig. 5 The film carrier bonding

6. Conclusion

The LSIs utilizing a low energy and high performance masterslice has been developed with 2000 equivalent gates on an average and a 350ps gate delay. The DA system has supported all the design activities for option developments and has dramatically reduced the turn around time.

Acknowledgment

The authors would like to thank H. Yamamoto, T. Kubota, M. Yawata and T. Nakamura for the process development, T. Yanagawa and K. Kitano for the DA system support, and Y. Murano, K. Kimura and S. Yano for their helpful suggestions and management efforts.

References

(1) H. Yamamoto, et al, "High speed performance of a basic ECL gate with 1.25 micron design rule", VLSI symposium, pp. 38-39, Nov., 1981.

(2) S. Yano, et al, "A masterslice approach to low energy high speed LSIs", Proc. ICCC '80, pp. 203-206, 1980.

(3) H. Yoshizawa, et al, "A CAD system for gate array automated design", Proc. 4th CICC '82, pp. 260-262, 1982.

(4) S. Kato et al, "FDL: A structural behavior description language", Proc. 6th CHDL, pp. 137-152, May, 1983.

(5) T. Sasaki, et al, "MIXS : A mixed level simulator for large digital system verification", Proc. 17th Design Automation Conf., pp. 626-633, June, 1980.

(6) M. Kawai, et al, "Application of shift register approach and its effective implementation", 1980 IEEE Test Conf., Cherry Hill, pp. 22-25, Nov., 1980.

AN AUTOMATED LASER PERSONALIZATION SYSTEM FOR PLAS

by D. W. Still and R. C. Weintritt

Honeywell Large Computer Products Division
Phoenix, Arizona 85066

ABSTRACT

This paper describes a state-of-the-art automated personalization system used to define, burn, and test programmable logic arrays (PLAs). The main component of the system is a software controllable laser scriber which selects and burns the user specified PLA patterns. A description of the PLA is included before proceeding with a discussion of the personalization process. The process described allows a user to generate the necessary files needed to burn and test the PLAs from logic equations.

INTRODUCTION

In today's large-scale digital systems, there is a significant portion of the central processor consisting of repeated arrays of logic which readily lend themselves to implementation in customized LSI/VLSI circuitry. However, there is also a significant amount of complex, specialized control logic which is not well suited to LSI customized implementation, due to both the cost, and the difficulty in making design changes once a chip has been fabricated.

Until a few years ago, there were only two choices open to the logic designer implementing this complex control logic. The LSI approach, with its inherent drawbacks, could be used. Or the designer could use the space demanding SSI and MSI chips available on the market. Fuse programmable devices were introduced in the late '70s as an alternative to the above two choices. They offered both the complexity and the space saving qualities required for control logic implementation. In addition, they offered the fast turnaround capability needed during design and debug.

Fuse programmable devices contain unspecified arrays which can be programmed by the logic designer to represent any combination of logic functions desired. The programming is accomplished by breaking specified bused connections (fuses) within the array with the application of an increased amount of current. Before programming, all fuses are unbroken. After programming, the selected fuses have been broken. The result is an array with some paths opened and others left closed, personalizing the array to the desired logic function.

There are many choices of fuse programmable devices available to systems designers today. The most fundamental of the programmable devices is the Programmable Logic Array (PLA). PLAs consist of a matrix of input terms and their complements connected to the output terms through an array of AND and OR states. This allows any output to be programmed as the sums-of-products of the input.

The problem with most fuse programmable PLAs is that they are electrically slower than the gates they are replacing. In fuse programmable devices, the added features needed for personalization contribute to larger circuit delays during normal operation. Most often, a reduction of component numbers is offset by an increase of signal delay. This drawback has extensively limited the application and usage of fuse programmable devices in very fast mainframe computers.

A novel PLA has been designed and is currently being utilized by Honeywell's large computer products. Personalization of the PLA is accomplished by an automated laser system which selects and vaporizes specified fuses. Circuitry needed for personalization is not required in the PLA and as a result, the access time is significantly reduced. The personalization system allows logic designers to receive tested parts from logic equations in less than one hour, a criteria desperately needed to meet design schedules.

DEVICE DESCRIPTION

The PLA, figure 1, is a bipolar current mode logic (CML) circuit organized into 22 inputs, 40 product terms, and 12 outputs. The product and sum arrays are implemented from CML differential gates whose inputs are tied together through common emitter connections. Gates configured as described produce a logical

CH1935-6/83/0000/0101$01.00 © 1983 IEEE

AND of all the inputs (assuming negative logic conventions). By disconnecting emitters from these gates, the array outputs are personalized. A nand gate is formed by inverting each array output. Connecting the output of one nand gate (the product term) to the input of the other nand gate (the sum term), a sums-of-product function is produced. The first array has 44 inputs and 40 outputs for a total of 1760 crosspoints or links. The second array has 40 inputs and 12 outputs for a total of 480 links. 12 links are also available to allow the output buffers to invert the sum term. A total of 2252 links are present in the circuit.

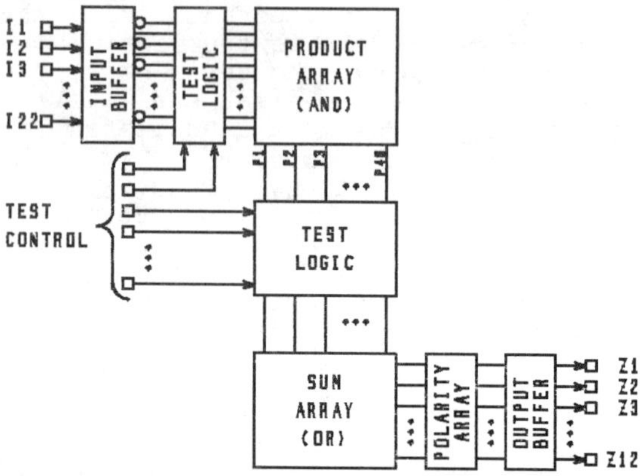

Fig. 1. PLA Block Diagram

The PLA characteristics are shown in Table 1. The speed of the circuit is due to both the circuit technology and the process technology. CML gates are current steering, non-saturating differential switches which typically have sub-nanosecond gate delays[1]. In addition, the circuit utilizes an oxide-isolation process to achieve higher performance by the reduction of parasitic capacitances.

Table 1. PLA Characteristics

Logic Type	Bipolar CML
Logic Configuration	Nand-Nand
Access Time	4.0ns at Room Temp
Die Size	3.0mm by 3.3mm
Power Dissipation	1.0 watt
Number of Links	2252
Personalization	Laser Vaporization

Circuitry has been added to the device which will assist in the testing before personalization. The circuitry allows each product and sum term to be tested individually by isolating the terms from other product and sum terms through a demultiplexor. The true and complement input terms to the product array can be disabled through a selector. One hundred percent detection of all "stuck-at" faults as well as an AC access time is achievable with the added circuitry which occupies less than three percent of the total silicon area[2].

PROCESS DESCRIPTION

The laser personalization system, figure 2, can be broken down into three distinct phases: the definition phase, the burn phase, and the verification phase. The definition phase transposes the logic design equations into burn and test files. The burn phase analyzes the information from the definition phase and commands the laser scriber to vaporize the proper links. The verification phase examines all unpersonalized devices for usable PLAs and tests all personalized parts using the information from the definition phase.

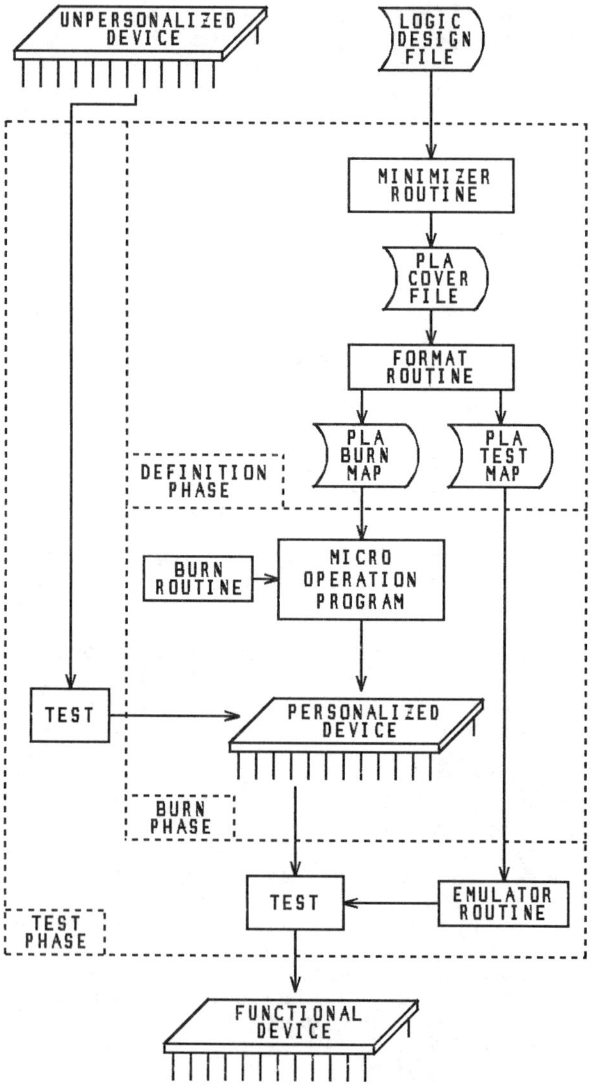

Fig. 2. PLA Personalization Process

DEFINITION PHASE

The definition phase is responsible for converting a designer's logic equations into usable data for the burning and testing process. The designer inputs a logical functional description of the device into the design automation system. The unformatted description, which includes a device identifier, is reduced into sums-of-products terms by a minimizer routine and stored in a master PLA cover file. Various design automation programs access the master PLA cover file to produce data for logic simulation, timing simulation, network calculations, and layout. In addition, two format routines access the PLA cover file to produce data for the laser burn system and the microcomputer emulator.

The laser burn system requires a data file which will convey the link connect and disconnect information to the micro-operation program. A design automation program converts the cover file into a burn file which is down-loaded to the laser system's computer. The main components of the burn file are the descriptions of the product, sum and polarity arrays. Also included in the burn file is a device identifier and various address pointers to assist the micro-operation program.

The microcomputer emulator requires a data file which contains a recognizable encoded test map of the minimized logic equations. A design automation program converts the cover file into a test file which is down-loaded to the emulator's computer.

BURN PHASE

The burn phase is responsible for disconnecting the proper links to form a useful PLA. Disconnection of the links is accomplished by a high-accuracy automated laser scribe system which vaporizes the aluminum links with a single pulse from an infrared laser. The laser system hardware and control electronics was purchased from Teradyne, Inc. The user control software for the PLA scriber was developed by Honeywell. Together the two form an integrated system for state-of-the-art personalization of PLAs.

The laser system is diagrammed in figure 3. The laser is a Q-switched Nd:YAG (Neodymium Yttrium Aluminum Garnet) solid-state laser with a wavelength of 1.06 um. The beam positioner uses optical guides to direct the laser beam to a specific point on the device. The beam attenuators reduce the laser beam's energy to a desired power level. The device positioner places the device under the beam positioning optics. Viewing optics allow magnified visual inspection of the device during setup and execution procedures. An innovative photodiode system analyzes a reflected laser beam to provide high-accuracy focusing, positioning, and alignment. The control electronics directs the operating mechanisms of the laser, beam positioner, and device positioner. The electronics are driven by a host computer which constitutes the main user interface for the generation, maintenance, and storage of operational software.

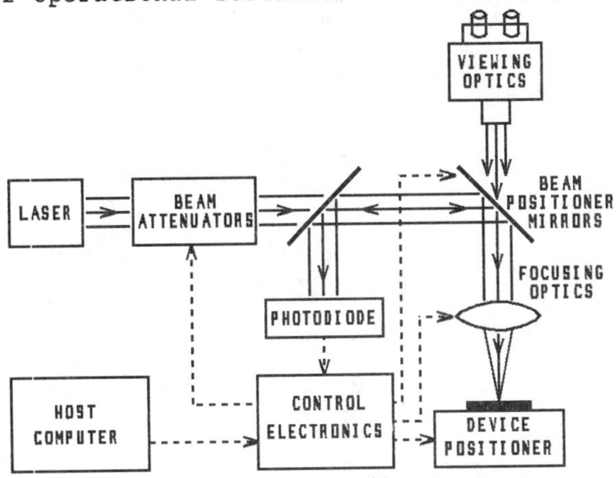

Fig. 3. Laser Scribe System

The PLA device contains various alignment targets which assist the control electronics to position itself over the product, sum, and polarity links. Once properly positioned, the aluminum links are vaporized by a single 100 nanosecond pulse from the laser (figure 4). Links with a width of 5.0 micrometers and a thickness of less than 1.0 micrometer require only 6.0 microjoules of energy to vaporize. Links are pitched in intervals as close as 15 micrometers, a spacing which has been proven experimentally and theoretically sufficient to ensure that neighboring links are not damaged. A thick oxide region under the links prevents damage to the substrate.

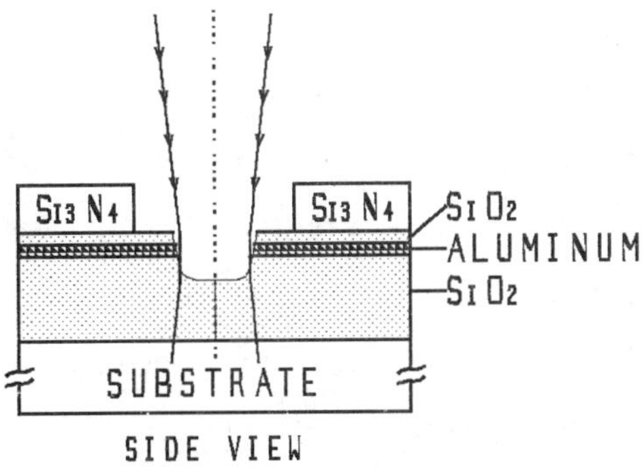

Fig. 4. Link Vaporization

The operational burn routine software analyzes the PLA burn map and commands the control electronics to position and vaporize the links. The operator places the unpersonalized device in the system and walks the device to a ground-zero (datum) location. The scribe routine is invoked and the operator becomes uninvolved until completion of the personalization. The software first commands the system controller to focus the device. A fine alignment in x, y, and theta follows until an accuracy of less than 2 micrometers is achieved. The laser power is set before proceeding to the vaporization. The software indexes through a table of link coordinates (2252 entries) and decides whether to skip the coordinate or command the laser to move to the coordinate and vaporize the link. The decision is based on the burn map code. After all link coordinates in the table have been scanned, a pattern identifier code is written on unused portions of the device. The total personalization process requires less than one minute.

VERIFICATION PHASE

Unpersonalized devices must be tested before entering the burn phase to allow greater success and faster turnaround. Devices are tested before and after packaging for both "stuck-at" faults and AC access time. Extra circuitry designed into the PLA assists the test routine. Good PLAs are delivered to the burn system for personalization.

Personalized devices must be tested to ensure that the links specified in the burn map were completely blown and that all other links remain intact. A microcomputer emulator (figure 5) is used to cycle through all possible input combinations and verify that the outputs are correct. The PLA test map produced in the definition phase contains an encoded form of the minimized logic equations. For each combination of inputs, the emulator invokes a software interpreter which reads the test map and determines the proper outputs. The emulator then compares the predicted outputs to the actual device outputs. This scheme provides for a 100 percent functional test of personalized devices.

The emulator also has the capability to compare a personalized device against a PLA which is known to be functional. In this mode, the emulator sequences the inputs and compares the outputs of the two devices, thus bypassing the test map interpreter. A 100 percent functional test is maintained while reducing the total test time to a level suitable for use in a production environment.

CONCLUSIONS

The PLA Laser Personalization System has proven effective in providing single-day turnaround of design for prototype system development. The yield through the laser system has been calculated to be greater than 85 percent. The turnaround from logic equations to functional devices is usually less than two hours. The time limiting factor is the software emulator speed. The emulator is currently being redesigned as a firmware-driven hardware emulator. This should reduce personalization time to less than one hour.

This system is currently being modified for use in production development. PLAs will be personalized on the wafer with pre-test and post-test performed by an integrated test system while the wafer remains in the laser system. If this proves as successful as the single-device system currently in place, production costs could be significantly reduced.

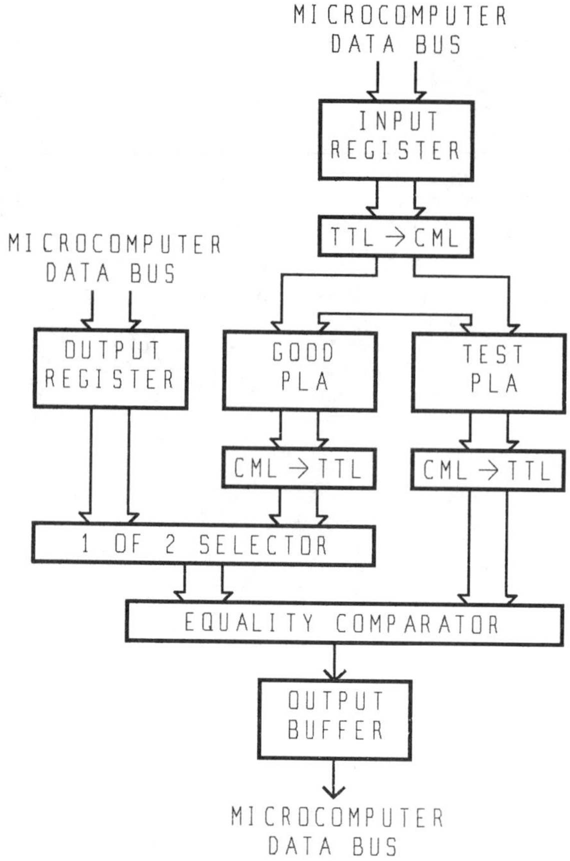

Fig. 5. PLA Microcomputer Emulator

REFERENCES

[1] D. Buhanan, "CML Scraps Emitter Follower for ECL Speed, Lower Power", ELECTRONICS, pp. 93-96, Nov. 3, 1982.

[2] D. Still, "A 4.0ns Laser Customized PLA with Pre-Program Test Capability", 1983 IEEE International Solid-State Circuits Conference Digest of Papers, pp. 154-155, Febr, 1983.

A 3.0ns Access time 1kb LCML RAM

Shigehisa Wakamatsu, Kazuo Kuno, Tadashi Ozawa
Hiroshi Kunita, Masahiko Nakamae, Hiroyuki Shiraki

NEC Corp.
Kanagawa, Japan

ABSTRACT

A 1kb Fast Bipoler static LCML (Low Level CML) RAM with address access time 3.0ns will be described.

The demand for higher speed RAM has been increasing to improve speed performance of large computer systems where devices are used in cache memories or control store.

The LCML RAM with 256 words x 4b organization has an address access time of 3.0ns, a chip select access time of 1.6ns and a write pulse width of 1.6 ns with a typical power dissipation of 1.35w. It is designed to operate in low power supply voltage, (VEE = -3.3v). As a result a stable RAM chip operation over VEE = -2.6v to VEE = -4.0v has been obtained in spite of low voltage power supply. And a high alpha particle immunity also has been obtained. The LCML RAM is gagbonded on a 52 - pin TAB (Tape Automated Bonding) tape to achieve higher packing density, smaller media delay for systems. A very small shallow transistor developed to realize the RAM with an emitter width of 1.5μm has provided a f_T value of as high as 9GHz. Oxide isolation and tripple layer metalization technologies have been employed. The line width and spacing pitch in first and second level metal layers are 5μm and 9μm, respectively. The third level metal layer is employed for power supply.

To accomplish those fine device geometries, state-of-the-art process technologies, such as advanced dry etching technology, have been utilized. Typical process condition and resultant transistor parameters are listed in the table-2. A microphotograph of the LCML RAM chip, 4.4mm x 4.4mm, is shown in Figure 1.

The memory cell is a SBD (Schottky Barrier Dode) clamped type which is suitable for the fast read/write operation. A cross section of the memory cell is illustrated in Figure 2. A minimum emitter size is 1.5μm x 3μm. The load resistor Rc is 8kΩ, formed by ion implanation. The standby current and the read current are 60μA and 0.75mA, respectively.

For wide range RAM operation an emitter follower driver serves as a fast word line driver in order to hold a sufficient cell voltage and the read level of memory cell is designed to obtain the wide tolerance of the forward voltage of the SBD (V_{FSBD}).

To assure sufficient alpha particle immunity, the following layout techniques has been employed.

The capacitances of the SBD (C_{SBD}) and the Collector-base junction (C_{cb}) reduce the switching time and enhance noise immunity against alpha particles. Therefore, these capacitance parameters are chosen to be as large as possible: i.e., C_{SBD} = 0.19pF, C_{cb} = 0.15pF. In Figure 3, output and input waveforms are shown. In Figure 4, a graph showing operating margins of the RAM is shown.

CH1935-6/83/0000/0105$01.00

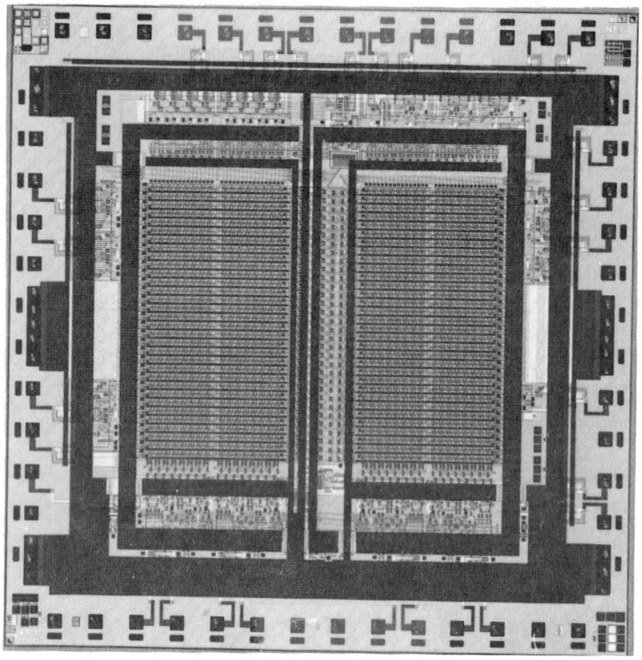

Figure 1. Microphotograph of the LCML RAM chip

Organization	256 words x 4b
Input and output level	LCML
Power dissipation	1.35w typical
Address access time	3.0ns
Chip select access time	1.6ns
Write pulse width	1.6ns

TABLE 1 - Major electrical performance

Epitaxial layer thickness		1.0u
Base depth	Xjb	0.2u
Emitter depth	Xje	0.08u
Cut off frequency	f_T	9GHz
Base resistance	rbb	250
Collector-substrate capacitance	Ccs	0.069pF
Collector-base capacitance	Ccb	0.085pF
Emitter-base capacitance	Ceb	0.036pF

TABLE 2 - Typical process condition and transistor parameter

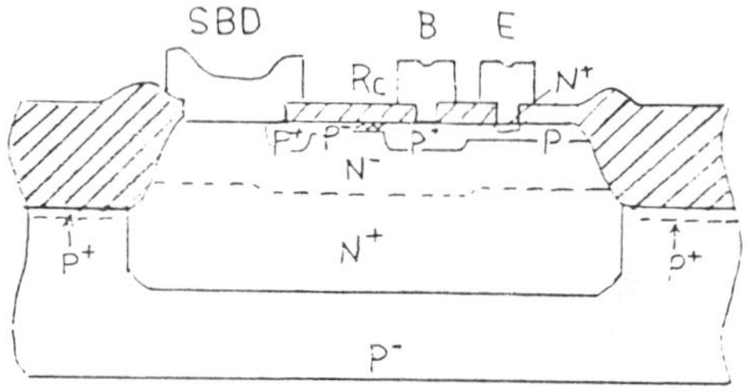

Figure 2. Cross Section of Memory Cell

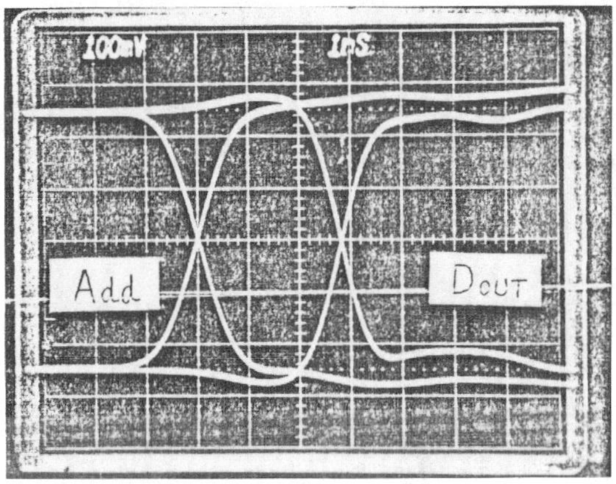

Figure 3. Output and Input Waveform.

```
VEE 0        3           4           5    [NS]
    +---//---+-----------+-----------+-----+
-4.000V      *******************************
-3.900V      *******************************
-3.800V       ******************************
-3.700V       ******************************
-3.600V       ******************************
-3.500V       ******************************
-3.400V       ******************************
-3.300V       ******************************
-3.200V        *****************************
-3.100V        *****************************
-3.000V         ****************************
-2.900V         ****************************
-2.800V          ***************************
-2.700V           **************************
-2.600V           **************************
```

Figure 4. Operating Margins

Session 1-2b
Testability

Chairman

S. Knight
Bell Labs

THE IMPACT OF RELIABILITY AND PERFORMANCE REQUIREMENTS ON VLSI DESIGN

by R. L. Barto

Martin Marietta Denver Aerospace
Denver, Colorado

Abstract

In researching the design of a spaceborne processor, it became apparent that the partitioning of the design could have as great an effect on its performance and reliability as would the basic architecture. While partitioning a processor so as to optimize the usage of integrated circuit real estate may result in a high performance processor, the implication for highly reliable processors is that they will be of either the voted redundancy or multiprocessor type. Both of these architectures have limitations making them undesirable for a space application. The architecture suggested here is internally redundant and parallel, and includes fault tolerance as a basic feature, providing rapid recovery from failures, tolerance of higher numbers of failures, and more graceful degradation. It will be partitioned into VLSI components so as to maximize its reliability and performance.

Highly Reliable High Performance Computer Requirements

Some of the the requirements for a highly reliable computer system depend on its intended environment which determines, for example, the availability of maintenance and the time permissible for recovering from a failure. This paper is based on research centered on the design of a highly reliable,high performance spacecraft computer. Here there will be no real possibility of maintenance and, in some applications, the time required to recover from a processor failure may make the difference between the success and failure of a mission.

Performance is also an issue for a spacecraft application, as it is for any other. The trend in both manned and unmanned spacecraft has been towards increasing autonomy, both from the standpoints of spacecraft guidance and control, and for experiment control and communication. The requirements for a spacecraft computer will include the following:

1. high performance: in this case, anexectution speed of one million instructions per second;
2. high reliability: the system is expected to operate reliably for several years with no maintenance;
3. extremely fast recovery from error with minimum system overhead;
4. low power consumption and compact size.

The change in performance with respect to the number of failures is also a concern. Ideally, the initial performance of the processor would be high and would not decrease as components fail. The cost and power requirements of such a processor would undoubtedly be prohibitive, and thus it is more reasonable to design a processor whose initial performance is high, but will decrease somewhat as failures occur. It would obviously be desirable to have the degredation of performance with each successive failure be as small as possible.

Fault Tolerance and High Performance Design Techniques

The techniques for designing fault tolerance into a computer system always involve the use of redundant elements and error checking. These techniques can be broken down into two broad categories. The oldest approach is N-modular redundancy (NMR) in which at least three of the N identical processors comprising the system operate with the same inputs and vote on their results, the majority opinion being the one used by the system. There is at least one system available commercially based on this technique[1]. Although an NMR system can recover quickly from a failure, it does have a number of undesirable properties:

1. the performance of an NMR system can be no more than the performance of a single processor and may be lower due to the time required for voting;
2. the reliability of an NMR system actually decreases with increasing N, beyond a certain point, due to the switching and voting circuitry being a possible source of a single point failure for the system[2];
3. the power required by an NMR system is clearly greater than 3 times the power required by a single processor, since the minimum number of voting processors is 3, and the switching and voting logic must always be powered.

Multiprocessors, the second category, were initially conceived as a means of construc-

CH1935-6/83/0000/0109$01.00

ting a system whose performance would be greater than that available from a single processor, and more recently have been considered for highly reliable systems. A system of this type is also commercially available[3]. Operationally, multiprocessors are complex and may require considerable system overhead in recovering from a failure. If a task is running on a processor which subsequently detects an internal error, the task must be restarted on some other processor at a point which will depend on the most recent time at which the failed processors state was stored.

Techniques for designing high performance into a system also involve the use of redundant elements and exploiting the natural concurrency of the computing process. Multiprocessing exploits concurrency at the software task level. Alternatively, it is possible to exploit concurrency at the instruction level, by using an internally parallel structure, and/or at the instruction operation level, by using a pipelined structure. The last two techniques have long been used in the design of main-frame processors.

Design Philosophy

Multiprocessing and NMR were rejected for this design on the basis of their deficiencies. Both of the commercially available systems mentioned are based on microprocessors,an easy to use building block, from which one may construct multiprocessing or NMR systems only. Thus the most appropriate VLSI partitioning for performance and reliability became a central issue.

The processor's design will blend fault tolerance techniques with high performance techniques so that fault tolerance will pervade its structure, as opposed to being an add-on feature. The major points of the design philosophy are:

1. the processor will be internally redundant, as opposed to externally, as are NMR and multiprocessors: whatever the functional modules of the processor are, ALU, microcode controller, etc., there will be N of each, and, for this paper, we assume that N=4;
2. there will be instruction level parallelism, allowing all of the modules to be operate concurrently;
3. the modules will be self checking;
4. the processor operation will be pipelined;

It is anticipated that this will produce a processor with the desired properties by:

1. minimizing the size of the replaceable modules, thus minimizing the change in performance with the occurrence of failures, and maximizing the number of reconfiguration modes, hence the number of failures tolerated;
2. exploiting the natural concurrency of the computing process;
3. minimizing the overhead of recovering from failures by appropriate pipelining;
4. integrating the performance and fault tolerance features so that the power and size penalties associated with NMR may be avoided.

The major topic of this paper will be the basic philosophy by which the processor is partitioned, ignoring its detailed architecture and operation. We will begin by sketching the module interconnection scheme followed by an example of the definition and partitioning of the functional modules.

Module Interconnection

The reliability and performance of the processor will in part depend on the quality of the module interconnections. The number of internal failures that the processor may tolerate is a function of the number of ways that its structure can be reconfigured, since the occurrence of a failure will require deletion of the responsible component and resumption of processing with some modified structure. The performance of the processor will depend on the time required for switch configuration and the ability of the scheduling mechanism to keep all the modules busy. This implies that the module connecting switch should provide a path for any module in one functional group to communicate with any module in its associate groups and that the operation scheduling, the error logging, and the switch configuring mechanisms be intimately related.

The switch's ability to isolate failures will be partially a function of the output driver structure and the partitioning of drivers and receivers into packages. An output driver may be driven into a high impedance mode by turning off both its source and sink transistors. Alternatively, a high impedance output may be obtained by powering down the driver, if the output has the correct structure. We will assume that each module and its associated output drivers may be powered down independently, and that the outputs have an appropriate structure. Catastrophic integrated circuit failure modes for which two or more of its inputs, or outputs, may short together or fail in a stuck-at condition are likely to exist. The drivers fanning signals out to modules in a receiving group and receivers collecting signals from modules in a sending group must therefore be partitioned so as to prevent such a failure of one integrated circuit from affecting more than one sending or receiving module.

Assume that we have broken the processor into M functional groups each comprising four identical modules, yielding F[i,j], i=0,...,M-1, j=0,...,3. A number of interconnection schemes having varying capabilities of fault isolation and reconfiguration may be used. Figure 1 shows the switching mechanism chosen for this design, comprising two 4x4 cross-point switches based on the tri-state buses labeled a, b, and c. The switch will be partitioned so that no switch elements share common packages and each element may be independently powered down. This will yield the following properties:

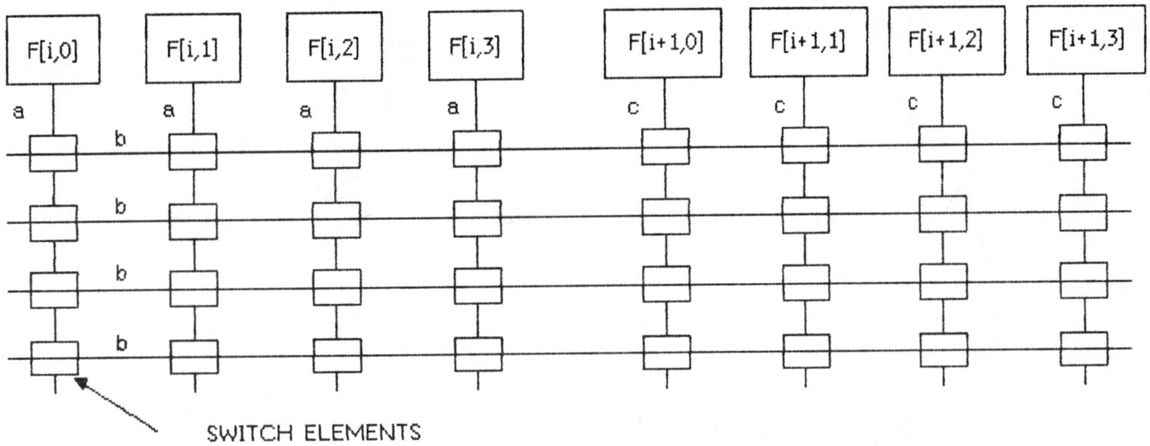

Figure 1

1. any module in group [i] may communicate with any in group [i+1];
2. a failure of any type in bus F[i,j](a) may affect module F[i,j], but no others. Failures in F[i+1,j](c) will be similarly contained;
3. a failure in bus (b) may disable that bus, reducing the switch's bandwidth but not its interconnection capability;
4. the failure of a switch element to pass signals will not disable its driving or recieving modules if the buses are intact.

The control of this switch will be integrated with the instruction scheduling mechanism. Each instruction will be given a token when fetched which will be used to schedule its execution relative to other instructions and to configure the switches allowing the modules executing it to receive the required data and commands.

Since VLSI packages with 179 pins are now in use[4], each switch element could easily be contained in a single package for any reasonable bus width. A switch element will clearly not approach VLSI complexity, given that it will mostly comprise buffers, providing room for the necessary error checking logic. Furthermore, the switch could be used as an intermediate storage point to reduce the required processing re-work in event of a failure in a receiving module, and as an error logging and operation scheduling unit.

Module Partitioning

The determination of a processors functional modules is often driven by the components that are commercially available. For example, a designer using the AMD-2901 register file (RF) and arithmetic/logic unit (ALU) combination might consider the RF/ALU to be one module. Although this may make the design easier, it may not be the most optimum partitioning from the standpoints of performance or reliability. The module partitioning philosophy developed in this research will be discussed using the RF/ALU combination as an example.

Analogous to NMR systems, a fault tolerant RF/ALU could be constructed by voting on the outputs of at least three RF/ALU modules built from the bit-slice components. The resultant design would suffer from the same performance and reliability limitations and would incur the same power and size penalties as would an NMR system. Alternatively, we may consider the ALU and the register file to be separate functional groups of four modules each, and allow the modules to operate concurrently on different instructions. This assumes the use of a mechanism for scheduling the execution of inter-dependent instructions. The one requirement to which the modules must adhere is that any RF/ALU combination must be able to accept any incoming instruction. This in turn requires that:

1. each ALU module possess the same computational capabilities;
2. each register file module hold a current copy of the registers;
3. any register file module be able to work with any ALU module.

Item (2) implies that each register file module must receive the outputs of all the ALU modules. If each register file module is placed on a single IC, then each package must have in excess of 250 signal pins, pushing the limits of experimental packaging techniques[4]. Having each module comprise more than one IC would be more workable, even though the circuitry of the intended 16 by 16 register file would easily fit onto one integrated circuit.

Some partitionings of the modules will likely tolerate more faults than others. There are two reasonable partitionings into two packages:

1. both bytes of half of the registers in each IC;
2. one byte of all of the registers in each IC.

Calculations show that option (1) requires the same number of pins per package as the single IC partitioning, while option (2) requires on the order of 170 pins per package. In either case a catastrophic IC failure would wastefully disable that entire module. Enhanced fault tolerance may be obtained by partitioning each module into four ICs, each containing one byte of half of the registers. Assuming that we may log failures of individual registers and schedule operations based on a module's partial capabilities, a catastrophic IC failure would leave a module crippled but functioning. Proper design would prevent the effects of the failure from spreading to the rest of the module. Although the four modules constituting the register file would require 16 packages, the system would be able to both tolerate more failures and degrade less with each because of its higher number of reconfiguration modes.

Similar considerations may be applied to the ALUs. Although it is obviously possible to package an entire ALU on a single IC, the effects of this on performance and reliability must be weighed. If the functions of each ALU were contained in several packages, and we could schedule ALU operations based on partial capabilities, then the system could tolerate a larger number of failures. Furthermore, higher performance might be obtained by optimizing the functional logic, if there is sufficient integrated circuit real estate available for this purpose.

Conclusion

The design and partitioning philosophy presented here includes fault tolerance as a pervading architectural feature. Its implementation would be more difficult than that of the more prevailing fault tolerant processor types, but would not suffer their limitations. Furthermore, integrated circuit real estate might not be optimally used in this design. Difficult implementation and inefficient integrated circuit usage may, however, be an acceptable price to pay for high reliability and performance.

References

1. August Systems, 2757 19th Street S.E., Salem, Oregon 97302
2. A.D. Ingle and D.P. Siewiorek, "A Reliability Model for Various Switch Designs in Hybrid Redundancy", IEEE Trans. Comput., vol. C-25, pp. 115-133, February 1976
3. Parallel Computers, 1201 Shaffer Road, P.O. Box 1187, Santa Cruz, California 95061
4. S. Orr, "Soaring I/O Counts and Larger Chips Squeeze Out Dual In-line Packages", Electronic Design, vol. 31, no. 12, pp. 135-146 June 9, 1983

Proceedings of the 1983 IEEE International Conference on Computer Design

DESIGN FOR TESTABILITY IN A VLSI CO-PROCESSOR

Milind M. Bhat, Christopher L. McAfee

Intel Corporation

Abstract:

The challenge of testing complex VLSI components is exemplified by the 82730 Text Co-Processor, which incorporates two independent, asynchronous processing units and a sizeable amount of RAM and other storage elements. Attention to testability early in the design phase has resulted in the incorporation of several features that simplify the formidable task of testing this complex peripheral.

The test philosophy aimed at providing support for different test environments such as manufacturing test, in-system diagnostics, etc., and simplifying the testing of all major functional blocks. The designed-in features include a built-in Self Test capability which allows exercising of major functional blocks without long functional sequences. Also included are provisions for forcing logical states on all output pins and a RAM Test routine. On-chip data compaction allows convenient, easy reporting of the test results.

I. INTRODUCTION

The 82730 Text Co-Processor presents a formidable testing challenge, as is the case with most complex, VLSI components and especially peripherals. Apart from the normal difficulties resulting from size, complexity and restricted observability due to pin limitations, it also poses some unique challenges of its own. Testability considerations as an integral part of the design process formed the key strategy for meeting this challenge.

The design approach aimed at implementing the testability features as an integral part of design, rather than as add-on features. Primarily the internal logic was used for this purpose, alongwith additional microcode, making full use of the existing control and data paths and on-board intelligence. A brief functional overview of the 82730, alongwith a discussion of specific testing considerations are included below to facilitate a comprehensive discussion of testability features.

II. FUNCTIONAL OVERVIEW

The 82730 Text Co-Processor is an intelligent peripheral intended for high performance alphernumeric display control. It consists of two internal processing units runnig on independent, asynchronous clocks. The Micro Controller Unit (MCU) incorporates the intelligence for fetching and pre-processing display data and executing high level commands. It provides various features that aid in text manipulation and display data management. The Display Generator (DG) performs as a high performance CRT controller and it also controls the video display format and character attributes. Two row buffer memory arrays are contained on the chip. During normal operation, the DG displays data from one row buffer, while the MCU fills the other row buffer with display data fetched from system memory. A block diagram of the 82730 is shown in Figure 1.

III. TESTABILITY CONSIDERATIONS

As a peripheral responsible for high performance CRT control, most of the control logic inthe Display Generator (DG) section is closely tied to the video timing. Thus, it would require long sequences to exercise this logic in the normal functional mode. The DG employs two independent clocks, a reference clock (RCLK) and a character clock (CCLK) in order to provide proportional character spacing. The Micro Controller Unit (MCU) runs on a third asynchronous bus clock (BCLK). This unit is microprogrammed, incorporating the intelligence for executing commands. However, the MCU and DG are independent units and most of the display control logic is isolated from the MCU.

The two row buffers, each capable of storing up to 200 words of character data and a number of display characteristics registers constitute a sizeable amount of on-chip storage. The 82730 has 68 pins and it is capable of running at 8MHz bus clock and 10MHz character clock rate, which are also factors for consideration.

IV. DESIGN OBJECTIVES

The 82730 objectives aimed well beyond the common goal of designing test features to ease the component testing burden. They aimed at providing a comprehensive set of features to facilitate testing in all environments, including board and system level testing. So the testability considerations addressed different test environments such as initial debugging (bench

CH1935-6/83/0000/0113$01.00 © 1983 IEEE

test), component test during manufacturing, board level testing and in-system testing (system self diagnostics).

In addition, the design objectives included the following goals to enhance testability:

o Provide a built-in self test capability to exercise all major functional blocks

  (Read-write test for arrays, exercise control logic, making all storage elements readable, etc.)

o Provide on-chip data compaction

o Provide electrical test support

V. OVERVIEW OF TESTABILITY FEATURES

Using the existing on chip intelligence, two additional commands, Self Test and RAM Test were incorporated. These are used in the normal operational configuration and require the presence of a host processor. Also a stand-alone self test capability is provided for component testing in the absence of host processor.

An ability to force all output pins to a high, low or high impedance state is provided to simplify parametric testing for the pins. It also allows easy deactivation of all 82730 outputs during board testing, without unplugging the part.

VI. IMPLEMENTATION

Design Strategy:

The design of testability features sought to make the maximum possible use of the existing structure, intelligence and control and data paths. The microporgrammed MCU provides the control, stimulus and communication function by devoting additional microcode for test commands.

During normal operation, the host processor communicates with the 82730 through a message block in system memory, designated as the Command Block. There is provision for switching to a new command block, using a Load CBP (Command Block Pointer) command. A similar message block, named Test Block, is used for issuing test commands. It differs from the normal command block in the parameter locations and function (Figure 2). Thus, test commands can be issued in the normal system environment, just like operational commands. Load CBP Command allows easy switching between command and test blocks.

Some of the existing control logic is reconfigured for test purposes. A natural demarcation separates the test mode and operational mode. Following chip reset, the 82730 cannot perform its display function until all the necessary parameters are properly set up by issuing a Modeset command. The interval between chip reset and completion of the Modeset command is desginated as the test mode interval. During this interval, the logic is configured for test purposes and test commands can be executed, while display commands are disabled (result in a NOP). Following the completion of the first Modeset command after reset, normal operation begins and display commands can be executed, while test commands are disabled.

Test Pattern and Data Compaction:

The built-in self test requires an on-board test pattern for testing the on-chip memory arrays and registers. An alternating checkerboard pattern is provided for that purpose. This is primarily for supporting a self contained check for go/no go testing. In addition, the RAM Test command allows more elaborate testing of the row buffer memory arrays with user defined test patterns. More sophisticated patterns like GALPAT could be used for row buffer testing.

Having provided the internal stimulus capability, a means of observing the logic behavior internally would be needed for a self contained check. Collecting data internally allows better observability and simpler user support. So the 82730 collects data onto its 16-bit internal bus (IB) and uses an on-board 16-bit register, configured as a multiple input signature register, to compact the test data. The signature register is implemented as a 16-bit polynomial counter, with parallel inputs, using an irreducible polynominal. The input to each stage of the polynomial counter is an exclusive OR of the output from the previous stage and one bit of the data input to be sampled. Thus the state of the internal bus is added without carry, each test cycle, to a pseudo-random value that represents all activity that has taken place before. Any error or combination of errors will vector the signature onto a different random sequence unrelated to the "correct" one resulting in an incorrect final value at the end of the test sequence. An example of a multiple input signature register is shown in Figure 3, using a 4-bit counter for illustration. The signature register is preset to a known state at the beginning of the test command execution. The data on the internal bus (IB) is sampled by specific control strobes during the execution of test commands. At the end of the test, the accumulated signature is reported as the Test Result. This can be used conveniently as a pass/fail indication in testing. No further data reduction is done in reporting the result to guard against untested logic in the result reporting path. For instance, if a single good/bad bit output was reported to a pin, then a false "good" indication would result if the reporting pin were stuck true. Using a parallel 16-bit path guards against this type of corruption.

Built-In Self Test:

The built-in self test is intended to exercise all major functional blocks. It primarily consists of writing the internally stored checkerboard pattern to all the storage elements

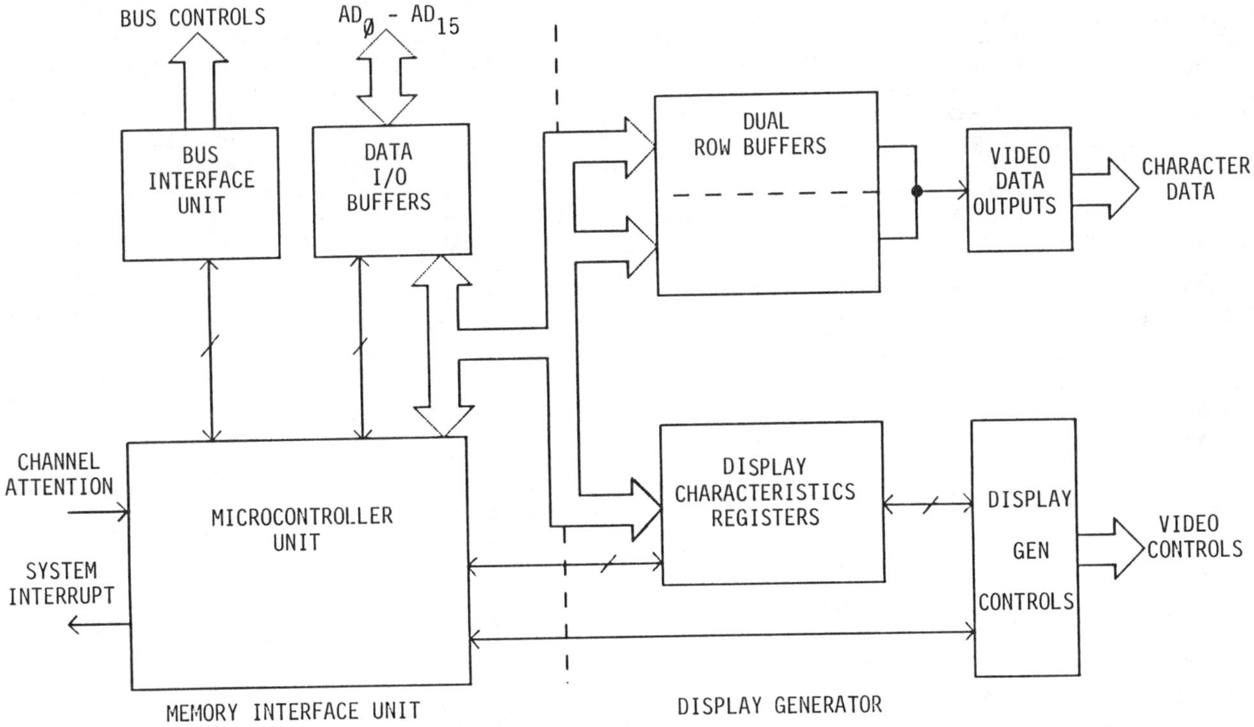

Fig. 1 82730 Text Co-Processor Block Diagram

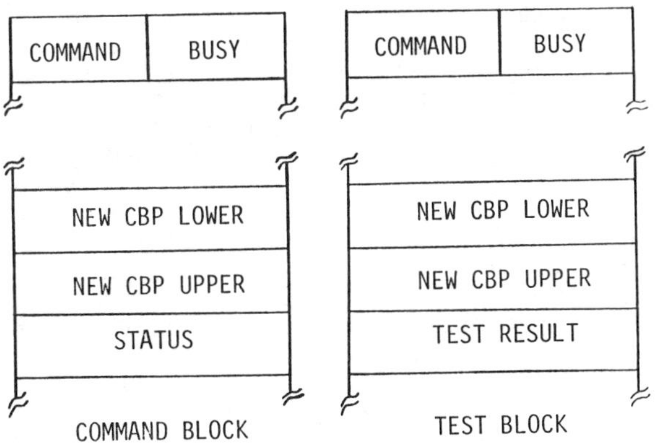

Fig. 2 Communication Blocks

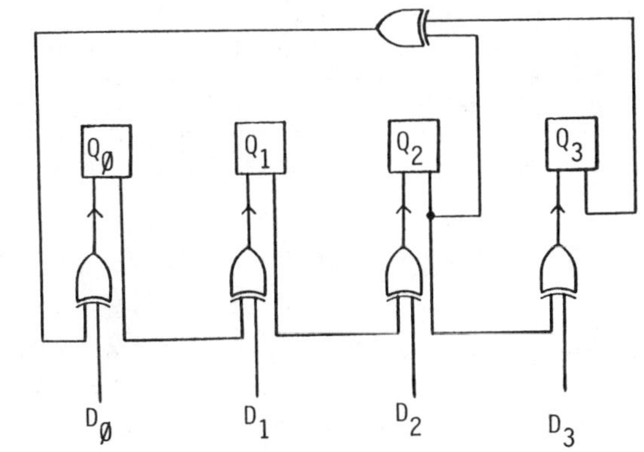

Fig. 3 Multiple Input Signature Register Example

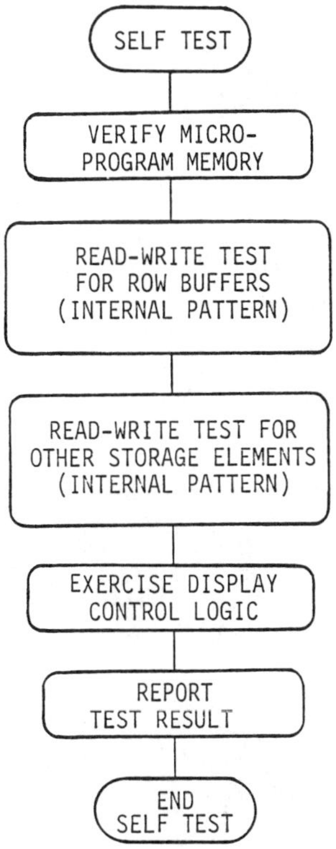

Fig. 4 Built-In Self Test

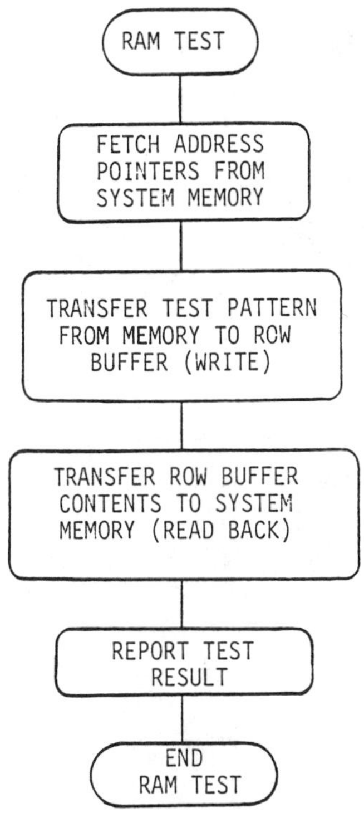

Fig. 5 RAM Test

on the chip, including the row buffers, and then reading back the contents. The test is repeated after reversing the order of alternation of the checkerboard. Most of the control and data paths are exercised in this process. In addition sections of display control logic, which are normally isolated from the MCU are also specifically exercised. Furthermore, the microprogram memory contents are also read out onto the internal bus (IB) and are incorporated in the signature generation.

Most of the self test is implemented using existing control and data paths through additional microcode. However, some special hardware provisions also had to be made. The three indenpendent clocks used by the 82730 during normal operation are tied together for test purposes. The bus clock (BCLK) controls the other two clocks (RCLK, CCLK) instead of their respective external sources during the test mode interval. The test is controlled by the MCU using the IB as the primary data path. Several storage and control elements such as registers, counters, latches in the display logic, which are not normally accessible via IB, were connected to the IB for test purposes. Also provisions were made for reading out several elements which are not readable in the normal operation.

The display logic incorporates several counters used in video timing. These counters are reconfigured to allow loading and incrementing by test control signals under microprogram control. Also wider counters are partitioned into separate 4-bit wide counters to reduce the length of the test sequence. A portion of the self test loads the counters and increments them successively, reading the counter carry and other relevant information from the logic they control.

Additional control logic was designed to read out the contents of microprogram memory sequentially. The microprogram itself initiates this readout. Upon completion, the control is passed back to the microprogram which resumes the self test execution. A flow chart of the self test command is shown in Figure 4.

Stand-Alone Self Test:

In the stand-alone mode, the built-in self test is invoked by applying an external stimulus to pins LCO-LC2 during chip reset. Upon completion of the test, a completion signal is presented via the LC3 pin, while the resulting signature is presented on pins DATO-DAT14, WDEF.

RAM Test:

The RAM test commands allows the testing ofon-chip RAM arrays (row buffers) with a user defined test pattern. Two address pointers are specified as parameters in issuing the RAM Test command. The Load Pointer points to the area in system memory where the user defined test pattern is stored. The test pattern data is fetched and written to the row buffers. The data read out from the row buffers is transferred to the area in system memory pointed to by the Store Pointer. A signature is also generated internally during the test. The resultant signature is reported as the Test Result through the Test Block. A flow chart for the RAM Test command is shown in Figure 5.

Electrical Test Support:

Pins LCO-LC2, which are output pins during normal operation, are monitored during chip reset for external stimulii. This mechanism is used for invoking the stand-alone self test as well as for forcing logic states on output pins. Dependent upon the stimulus received, appropriate controls signals are generated to force all output pins to a high, low or high impedance state. The outputs remain in this state until next chip reset.

VII. RESULTS AND APPLICATIONS

The test capabilities designed in the 82730 have been successfully used on the first silicon. The built-in self test capability was used for initial sorting and also for prescreening during subsequent tests. The RAM test capability was used in the initial debugging and will be used in testing. The output pin forcing capability will be used for parametric testing.

The cost of implementation of the test features is mainly in terms of additional microcode and some amount of control logic. The cost in terms of silicon area is 3-5%. The benefits are extensive.

VIII. CONCLUSION

Design for testability can be seen to be the key to meeting the challenge of testing complex VLSI components. Incorporation of testability features in the design objectives for the 82730 Text Co-Processor allowed the implementation of a number of vital test features at a moderate cost. The designed-in features of the 82730 address the needs of various test environments from component test through in-system diagnostics. They simplify the formidable testing task and help reduce testing cost and time for manufacturers, component customers as well as end users.

SELF - TESTING AS AN EXTENSION OF SERIAL TESTING

Fazil Osman, Member IEEE

Advanced Systems Group
Burroughs Corporation, San Diego, Ca 92128

ABSTRACT Even though some proposed approaches to self-testing VLSI circuits and systems, e.g., the BILBO approach, are the ultimate way to testing VLSI circuits, they involve too high an area overhead for practical implementation on present day state-of-the-art technology. This paper presents an approach to self-testing that extends serial testing of VLSI circuits. This technique is implementable on present day VLSI technology with an acceptable area overhead. With such a method, it should be possible to design self-testing systems and to reduce the cost of testing and manufacturing of VLSI systems today.

1. Introduction.

Mead and Conway [1], caused a revolution in the design of integrated circuits. Up to then, custom circuits were only possible for companies that could afford the $100K or more in development cost. It is currently possible to design a custom circuit for $20K or less, using the Mead and Conway approach and by using the capabilities of one of the many silicon foundries.

The major impact of Mead and Conway has been in universities where it has enabled researchers to have working prototypes of architectural concepts within a few weeks rather than having to wait for many months, by which time the designer has moved to something else. The combination of ease of design using the Mead and Conway approach, quick turn-around (QTA) and low cost of runs through silicon foundries has also attracted a lot of system design companies. One of the major differences between the researchers and commerical designers is that the latter has to get the design into production after debugging. Testing thus becomes one of their major problems as it involves cost of labor, hardware and, software, etc.

The ideal is to have a totally self-testing IC. The most promising approach is the BILBO method [2], which uses a shift register as a source of pseudo-random sequence of test patterns and another shift for signature analysis of the output.

The major advantage of this technique is the lack of test patterns. It does have disadvantages however, including high area overhead and difficulty in testing high fan-in circuits such as PLAs, a problem that can be solved with a high area overhead. It also needs software not presently available to generate the pseudo-sequences and checking the coverage.

STEST (Self-Testing Extension to Serial Testing) provides an intermediate solution and minimizes the area overhead on-chip by external control. However, it produces an overall lower cost by its use at different testing stages. Section 2 describes the operation of STEST. In section 3, the discussions focuses on the different steps in testing a VLSI chip from design to shipping of a completed product, as well as how STEST can reduce the total cost of testing.

2. Operation of STEST.

For the purposes of this discussion, it will be assumed that a MOS implementation is used with a 2-phase clock, Ø1 for clocking data into registers and Ø2 for clocking the data out. Figure 1 shows a block representation of a circuit. It consists of registers with blocks of combinational logic in between. In serial testing, it is possible to shift data in and out by serializing the registers as illustrated in Figure 2. To start a serial test, data is shifted in by enabling SHIFT. When it is at the appropriate register, SHIFT is disabled. The data ripples through, from the output of one register through the combinational logic into the next register. The data is then shifted out for analysis. Serial testing can be used in combination with parallel testing by shifting initialization data in, or by shifting data out at the appropriate time.

For STEST, an extra latch is added for each set of registers. This latch is used to indicate if the register contents are to be compared with the output of the combinational logic or whether that register is to act as a source of test patterns. Figure 3 illustrates the general set-up for a circuit. Figure 4 illustrates the operation of one set of registers when all the data has been shifted in. The control section is illustrated in Figure 5.

With the logic shown, it is possible to do the following:

- Operate in the normal serial mode. This is done by loading the compare latch with a '0' (zero). In Figure 5, this will disable ROM addressing.

- Load the output of any block of combinational logic with the expected output and setting a valid EQUAL when appropriate input is present. This can be used as a breakpoint.

- Operate in a self-test mode of the on-chip control logic and the external logic is as shown in Figure 5. When the reset signal is sent, the cycle after the ROM address counter is started if EQUAL is not valid. This is to ensure that the line is not stuck at a '1'. The ROM is 2-bits wide. Bit 1 is the data to be loaded in. Bit 2 contains '0' except at the end of each set of test patterns, when a '1' indicates an end-of-this-set. If *self-test-successful* is not reached, the ROM will hang up at the address of the end of the test patterns that failed else, it will reach the address the counter is programmed to stop.

Using the method described, it is possible to add a self-testing capability to a VLSI chip with little overhead to the chip itself. The highest overhead is an external ROM or RAM and some control logic that can all be integrated into one IC.The next section describes how this can be used to reduce the cost of manufacturing and testing VLSI chips.

3. Use of STEST.

The steps in the design and manufacture of an IC are illustrated in Figure 6. As can be seen, testing is done at least six times, from the start of a VLSI design to use of it in an actual system. In five of these steps, a dedicated tester is needed and the final power up confidence testing is usually done by software.

One additional problem, apart from complexity of the circuit, is the fact that the pin count on VLSI circuits is increasing. A great deal of custom VLSI circuits currently have 64 or 68 pins, and packages of 100-200 pins are being designed and used. At wafer testing this requires the use of an expensive probe card and a tester with sufficient I/O. It also reduces the yield, since damage to pads is higher. Board testing is also becoming more complex with the use of multilayer boards. There is also high density systems that use ceramic substrates with multilayer interconnect that can only be tested after the ICs have been assembled.

Using STEST, most of the testing described above could be done using a single board RAM based test system. The scenario would be as illustrated in Figure 7, where the same tester is used at all stages.

4. Conclusions.

The BILBO approach may be the long term solution to self-testing. However, it requires too high an overhead for present day technology and a different design approach from the popular serial testing. STEST offers a stepping stone between serial testing and self-testing. The overhead, on-chip, is minimized by having an external ROM or RAM with the test vectors. It has been shown that with STEST, the production costs of VLSI circuits could be reduced by cutting down testing costs.

REFERENCES

[1]. C. Mead and L. Conway, "Introduction to VLSI systems", Addison-Wesley, 1980.

[2]. B. Konemann elal., "Built-in Logic Block Observation Techniques", 1979 IEEE Test Conference, Cherry Hill, New Jersey.

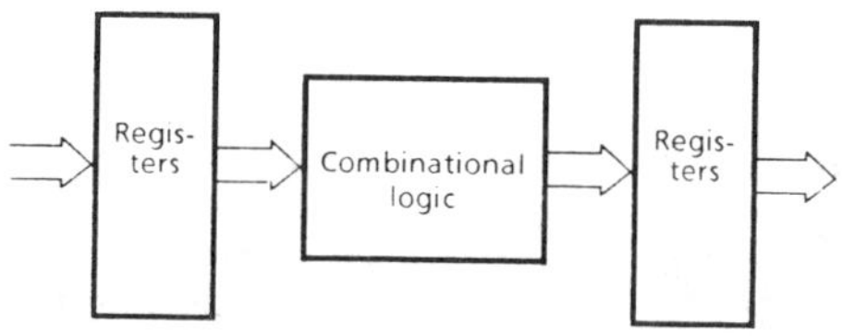

Figure 1. Block representation of a Circuit

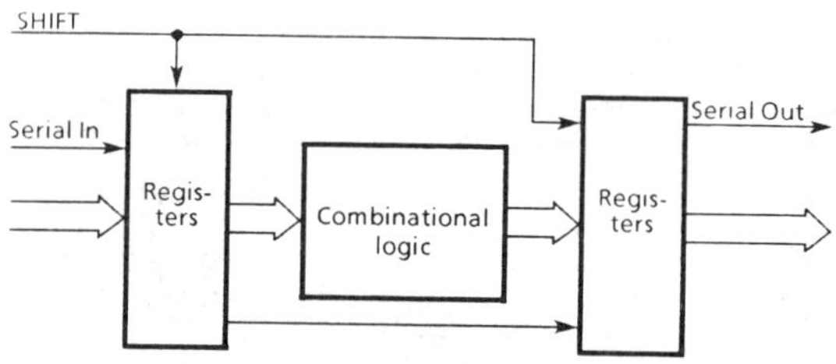

Figure 2. Adding Serial Test Line to the Circuit

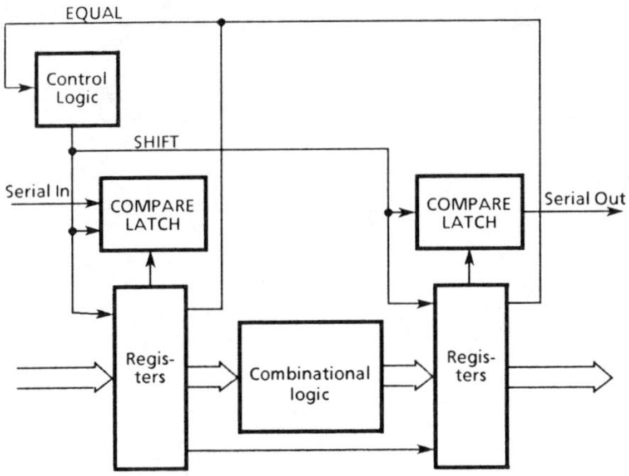

Figure 3. Block Diagram with STEST Added

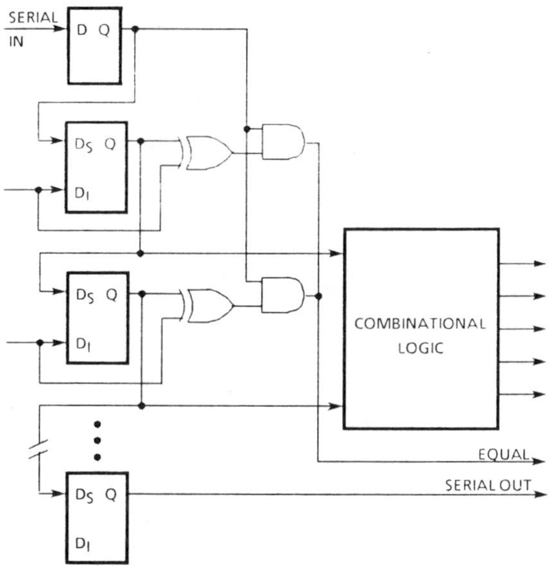

Figure 4. Details of STEST operation

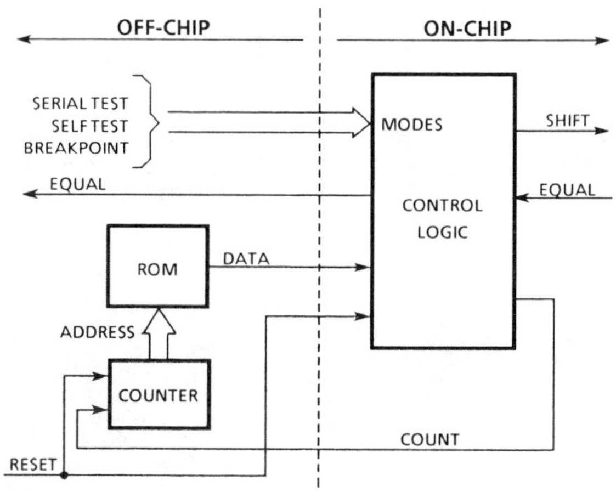

Figure 5. Control logic signals

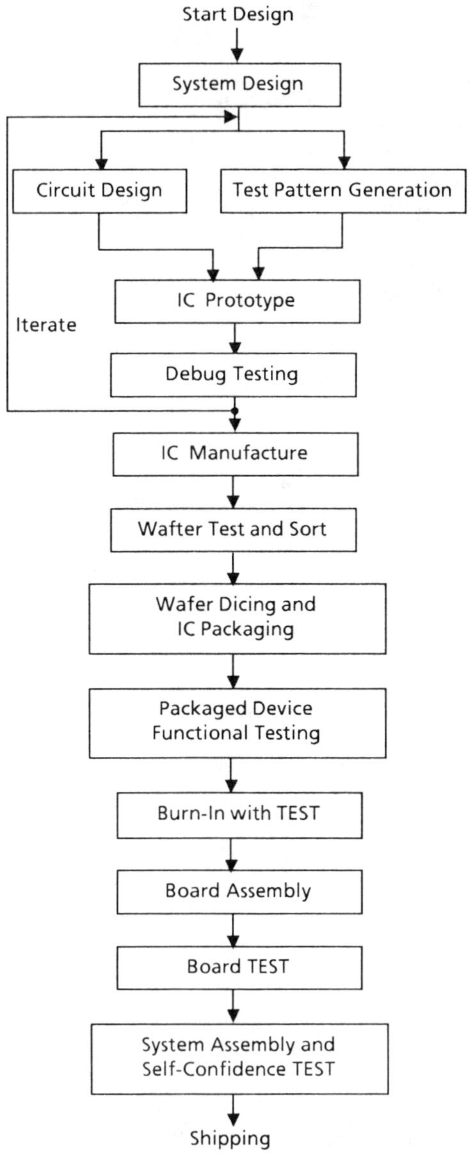

Figure 6. Typical IC Manufacture Stages

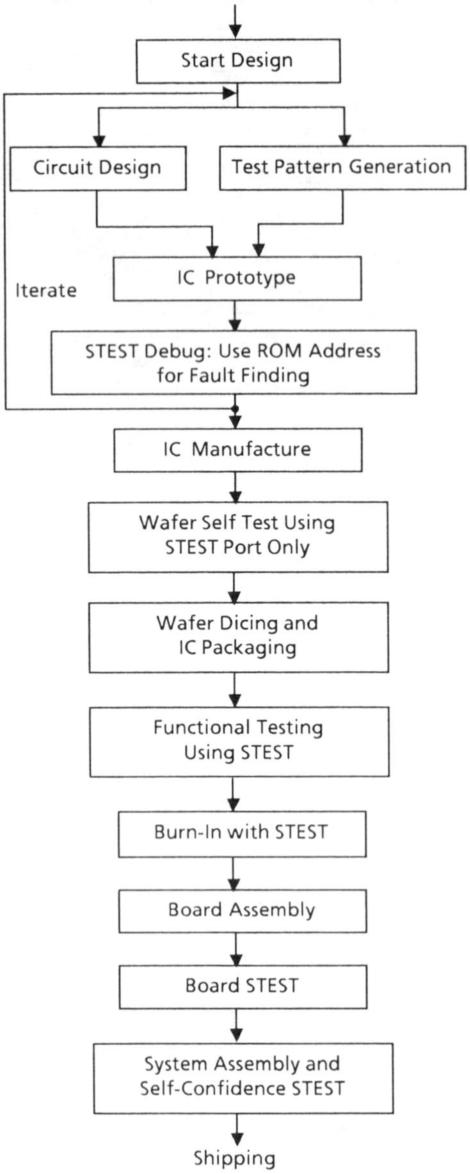

Figure 7. IC Manufacture with STEST

Session 1-2c
Automated Techniques

Chairman

W. Heller
IBM

HEURISTIC IMPROVEMENT TECHNIQUE FOR BISECTION OF VLSI NETWORKS

Mark K. GOLDBERG
Clarkson College of Technology
Potsdam, New York

Michael BURSTEIN
IBM T. J. Watson Research Center
Yorktown Heights, New York

ABSTRACT

We address the mincut partitioning problem for VLSI networks and experimentally evaluate the performance of several heuristic algorithms. For example, we observed the the quality of final partition obtained by Kernighan-Lin algorithm for graphs is heavily dependent on the ratio of the number of edges to the number of vertices: namely, the algorithm performs poorly when this ratio is less then 3 and produces nearly optimal solutions when the ratio is higher 5. We also develop a hierarchical improvement technique that can be applied to any mincut algorithm and results in an improved algorithm. We use the Kernighan-Lin-Fiduccia-Mattheyses (KLFM) technique as a base for the improvement. The resulting algorithm outperforms the KLFM precisely in the "poor performance areas" for the latter.

INTRODUCTION

The problem of partition of the set of nodes of a network into subsets arises at several stages of VLSI design and layout. The objective of such partition is usually to reduce the number of external interconnections. The importance of the problem is well recognized. We consider here a particular instance of the problem where the nodes of a network have to be distributed among two subsets in such a way that the number of dissected nets is minimal. We address it as a bisection problem in order to distinguish it from general partition. Bisection is a key feature of the several automatic gate array layout systems ([3,7,8]). The problem can be formalized in the following way. Given a network $H = (V, E)$, where V is a set of ***nodes*** and E is a set of subsets of V (called ***nets***) consisting of at least two nodes. The problem is to partition the set V into two parts $V = A \cup B$ of equal size ($|A| = |B|$) such that the number of nets having nonempty intersection with both A and B is minimal. For the sake of simplicity we assume that the total number of nodes is always even. In practice it may not be so: the nodes may be of different weight, they may not be interchangeable or free to be assigned to any subset, the desired partition may not be one that dissects the set into exactly equal parts. But the problem, as it is formulated above, is an important extraction of the practical problem and preserves the key difficulties pertinent to general partitioning.

The first nontrivial heuristic algorithm for graph bisection problem was developed by Kernighan and Lin [1]. This algorithm, when implemented straightforward results in a $O(n^{2.5})$ execution time routine. Recently Fiduccia and Mattheyses developed a modification of Kernighan and Lin method [2] but utilization of efficient data structures enabled them to come up with linear time heuristic practically without any loss in quality of partitions and applicable to general networks. This bisection technique, which further will be referred to as KLFM, is superior to several local optimization algorithms [4]. However there was still no significant performance study of this technique. As a result of numerous experiments we concluded that the KLFM performance is heavily dependent on certain network parameter, which we call ***network ratio*** and define as follows: for a network $H = (V,E)$, the ratio

$$r(H) = \frac{\sum_{e \in E} |e| - 1}{|V|}.$$

It turns out that KLFM performs relatively poor when $r(H) < 3$ but performs nearly optimally when $r(H) > 5$. Practical networks, however lie in the range of $1.8 < r(H) < 2.5$.

We also develop a mechanism that can be applied to any bisection algorithm and results in another bisection algorithm which we call a ***derivative.*** Our main result is the experimentally derived conclusion: the derivative of KLFM significantly outperforms KLFM in the $r(H) < 3$ area.

TEST MODEL

Since an analytical technique for evaluating performance of bisection algorithms does not exist, experimental methods must be employed. This is explicitly pointed out in [6] ; moreover [6] suggests that the bisection algorithms should be evaluated on specific networks for which the mincut value is known, and describes the construction of such networks. However, we argue that the suggested networks are still not very suitable for algorithm testing, since the ratio of the generated networks is very high. In order to demonstrate this denote A and B two sets with p elements, $n = 2p$ and let k be an integer. Let U_i be a random $(p + 1)$ element subset of $A \cup B$, $(i = 1, 2, ,..., k)$. Let T_i be an arbitrary tree connecting the vertices of U_i such that only one edge of it connects a node from A with a node from B. Let $KM(n,k)$ (for Krishnamurthy & Mellema) denote a graph which is the union of these $T_i - s$ on the node set $A \cup B$. It is easy to see that the mincut number of KM is exactly k. However, $r(KM) = k/2$, which easily puts us into the "good performance area" of KLFM, as it will be demonstrated in the next section (since practical values of k are larger then 10). The networks generated in [6] in our optinion are not good test cases for bisection algorithms. They practically will never occur in real logical networks. It has been observed that in practical networks the total number of nets is approximately equal to the number of active nodes. Moreover, the number of 2-terminal nets is about 45% of the total number of nets, the number of 3-terminal nets is about 15% and so on. In general [7] the following "net-size distribution" criteria is approximately satisfied. Denote m_i the number of nets connecting i nodes of the network, $M(H)$ - the vector, whose i -th component is m_i/m , where $m = \Sigma m_i$. Then

$$M(H) = \; 0 \; 0.45 \; 0.15 \; 0.12 \; 0.11 \; 0.8 \; 0.6 \; 0.3 \; .$$

We suggest that the test networks be generated according to this criteria. It was pointed out in [4] that the parameters m_i play an important role in bisectioning. The average number of dissected nets over all possible bisections (an upper bound for mincut) can be expressed in terms of these parameters.

In order to evaluate the bisection algorithm performance we need not know the exact value of the mincut. It is enough to know a good upper bound. So, we suggest the following construction. Denote $G(n,m)$ a random network on n nodes and m nets satisfying the above distribution formula. Let $B(n,m,k)$ be a network constructed as follows: take two copies of $G(n/2,(m - k)/2)$ (assuming that n and $(m - k)$ are even) and add k random nets connecting these two networks and also satisfying the size distribution. Then obviously k will be an upper bound for the mincut of $B(n,m,k)$. We will be testing the bisection algorithms in these $B(n,m,k)$ -s. The quality of the algorithm will be evaluted by the difference between the resulting number of dissected nets and k. The resulting number may easily be less then k if k is large enough. In order to model practical networks the values of n,m and k should be selected in such a way, that $m \sim 1.1n$ and $k \sim O(\sqrt{m})$.

KLFM

We have implemented a simplified version of Fiduccia-Mattheyses algorithm [2] under the assumptions that all nodes are equivalent and free to move between both buckets. The algorithm runs in linear time and our testing experimentally confirmed that fact. Using the same data structures we also implemented the original Kernighan-Lin algorithm, which theoretically may have on the average $O(n \log n)$ running time, but practicly runs almost as fast as Fiduccia-Mattheyses. It usually takes more time per pass, but requires slightly less passes. The test results on Krishnamurthy-Mellema graphs $KM(n,k)$ are presented in the Table 1. We have performed 6 runs for each setting of parameters n and k : $n = 500 \; 1000 \; 1500$ and $k = 10 \; 20 \; 30$. The optimal solution was found in each of these 6 experiments. Running times are given per pass in seconds for PL/1 implementation on IBM 370/3081.

Table 1.

	Kernighan-Lin		Fiduccia-Mattheyses	
	n. passes	time per pass	n. passes	time per pass
KM(500,10)	3	0.8	3	0.7
KM(500,20)	3	1.4	3	1.2
KM(500,30)	3	1.7	4	1.6
KM(1000,10)	4	0.9	5	0.8
KM(1000,20)	3	1.8	5	1.4
KM(1000,30)	3	2.1	3	1.9

We also tested the algorithms on several different $K(2000,50)$ -s and the optimal solution was obtained every time (with perhaps different numbers of passes) by both of them.

The results of testing on $B(n,m,k)$ -s are, however, more interesting. Table 2 contains the results of typical runs with different n,m,k setting. We have performed 5 runs with each parameter setting and the results are averaged over these 5 runs. The deviation from these average figures was also small. These results led us to the conclusion that the quality of KLFM performance increases with the growth of ratio. Figure 1 contains an illustration of this phenomenon, when on X axis we measure the ratio, and on Y the $\frac{c_r}{c_o}$, where c_r is the resulting number of dissected nets by KLFM and c_o the minimal number of them.

Table 2.

	Kernighan-Lin		Fiduccia-Mattheyses	
	avrg cut	total time	avrg cut	total time
B(500,500,50)	64	1.9	63	2.1
B(500,1000,150)	149	4.3	148	4
B(500,1500,200)	194	5.8	194	6.3
B(500,2000,250)	246	9.1	246	8.7
B(1000,1000,100)	123	5.6	126	5.4
B(1000,1600,200)	200	6.7	200	6.4
B(1000,3200,300)	298	10.1	297	9.3
B(1000,6400,400)	381	16.3	382	16.2

We conclude that $\frac{c_r}{c_o} \rightarrow 1$ when ratio increases.

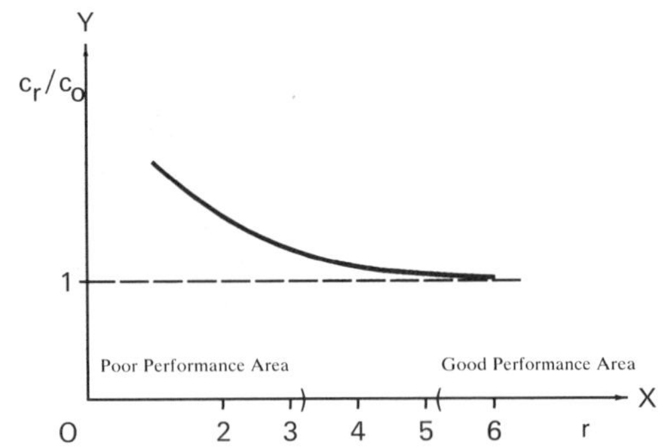

Figure 1.

DERIVATIVE

In this section we describe a mechanism for generating bisection algorithms: starting with an arbitrary algorithm we generate another one which we call a ***derivative.*** The idea of this construction was suggested in [3]. We need the following definitions. Let $H = (V,E)$ be a network. Notation: if $X \subseteq V$, then H_X denotes a subnetwork of H having X for a set of nodes and a set $\{X \cap e \mid e \in E$ and $|X \cap e| \geq 2\}$ for a set of nets.

A ***matching*** of the set V is the partition of it into $\lfloor V/2 \rfloor$ disjoint 2-element subsets. For a matching F, H^F denotes a factor-network of H in which F serves as a set of nodes and nets are the contracted nets of H (2-element nets that contracted completely are deleted).

Let Y be an abstract algorithm that can be applied to any network $H = (V,E)$ and the result of this application is the partition $V = A \cup B$ which we denote $Y(H)$. Let F be an arbitrary matching of V. Then the derivative of Y (denoted by dY) is defined as follows.

1. **Construct H^F ;**
2. **Apply Y to H^F ; let $A' \cup B'$ be the resulting partition of H^F ;**
3. **Generate $A \cup B$ - reconstructed partition of V ;**
4. **Construct H_A and H_B and apply Y to each to those networks; let $A = A_1 \cup A_2$ and $B = B_1 \cup B_2$ denote the resulting partitions: $Y(H_A)$ and $Y(H_B)$;**
5. **Select arbitrary matchings F_1, F_2, F_3, F_4 of each of the sets A_1, A_2, B_1, B_2 ; let $F_* = \bigcup_i F_i$ (matching of V);**
6. **Construct H^{F_*} and apply Y to it; let $V = A'' \cup B''$ be the resulting partition.**
7. **Announce $V = A'' \cup B''$ to be the resulting partition of the derivative algorithm dY .**

Remark: the definition of the derivative algorithm is dependent on the matching selection.

The derivation mechanism may be very useful in case when the initial algorithm Y is space consuming, since if the initial network has n nodes, the original algorithm Y is applied to $n/2$ node networks only. The derivative of Y is generally faster than Y in case when the computational complexity of Y is more than $O(n^2)$. If Y is a linear algorithm, then dY is linear as well (with a larger constant).

Our main purpose is to investigate the performance of the derivative of KLFM. Since KLFM is an improvement routine for a given partition, we specify two variations of it. KLFM1 starts with a random bisection and iteratively applies KLFM until no further improvement is made. dKLFM1 denotes a derivative of KLFM1. This derivative also assumes a starting bisection, so it may be iterated (applied to the result) as well. KLFM2 denotes only one pass of KLFM and dKLFM2 is the corresponding derivative. We applied those derivatives in iterative mode until the iterations showed no improvement. The test networks were the $B(n,m,k)$ -s with net size distributions described above. We have performed 10 runs with each of the three parameter settings:

A: $n = 512$, $m = 600$, $k = 100$;

B: $n = 1024$, $m = 1200$, $k = 200$;

C: $n = 2048$, $m = 2400$, $k = 400$;

Table 3 contains the data on execution time averaged over the 10 runs (times are given in seconds both total and per pass).

Table 3.

	KLFM		dKLFM1		dKLFM2	
	ttl	p. p.	ttl	p. p.	ttl	p. p.
A	3.5	0.7	5.2	1.3	5.1	0.4
B	8.1	1.9	10	2.9	12.3	0.8
C	20.4	4.2	23	6.1	24.5	2.2

Table 4 presents the averaged results and averaged numbers of passes.

Table 4.

	KLFM		dKLFM1		dKLFM2	
	cut	n. p.	cut	n. p.	cut	n. p.
A	96	5	84	4	83	9
B	205	4	190	3	196	12
C	418	5	360	3	361	10

The deviations from these average results are not significant. In none of these experiments KLFM resulted in better bisection then both derivatives. Note, that the ***ratio*** of networks in these experiments belongs to the "poor performance area" of KLFM. We believe that the derivatives will perform similar to KLFM when the ratio is greater than 5.

We performed another five tests on $B(512, 2500, 1000)$ -s (ratio is 12.3). Table 5 contains the averaged results for these tests (average cut and number of passes).

Table 5. Tests on B(512,2500,1000).

KLFM		dKLFM1		dKLFM2	
cut	n. p.	cut	n. p.	cut	n. p.
749	5	750	4	753	12

The following explanation for the improved performance of derivatives in the *ratio*<3 area is suggested. At the derivation step 2 we apply KLFM to H^F. The ratio of H^F is approximately twice as higher than the ratio of H, because its number of nodes is twice as less, and, at the same time, the numerators are changed insignificantly. The numerator of the ratio for H^F differs from the numerator of $r(H)$ by the number pairs of adjacent nodes of H that got contracted (two nodes of a network are adjacent if they belong to the same net). Let q denote the total number of adjacent node-pairs of H. Since total number of node-pairs is $n(n-1)/2$ the expected number of contracted pairs will be

$$\frac{|F|}{\frac{n(n-1)}{2}}q = \frac{q}{n-1}.$$

It is easy to see, that in our case $q = \sum_i m_i(m_i - 1)/2 \sim 6m$, and since $m \sim n$, on the average only 6 or 7 pairs will be contracted. This means that KLFM is applied to H^F in the "better performance area". However, H^F is a different network, and the mincut value of it may be significantly higher than the original (we might have contracted the edges of the real mincut). But we reconstruct H at step 3 and try to eleminate as many disection candidates as we can in step 4. Finally, in the step 6, we apply KLFM for H^{F*} which has almost doubled ratio as well.

ACKNOWLEDGEMENT

We are grateful to Jerome Kurtzberg, Jonathan Turner and Jacob Peysackh for helpful comments, suggestions and assistance.

REFERENCES

[1] B. W. Kernighan and S. Lin, "An Efficient Heuristic Procedure for Partitioning Graphs", Bell System Technical Journal, Vol. 49, February 1970, pp. 291-307.

[2] C. M. Fiduccia and R. M. Mattheyses, "A Linear-Time Heuristic for Improving Network Partitions", Proc. 19th Design Automation Conference, Las Vegas, June 1982, pp. 175-181.

[3] M. K. Goldberg and R. Gardner, "On the minimal cut problem", Proc. of the Silver Jubilee Conf. on Combinatorics, 1983.

[4] M. Burstein, "Analysis of a Network Partitioning Technique", Proc. ISCAS-82, Rome, May 1982, 477-480.

[5] M. A. Breuer, A. D. Friedman, A. Iosupovicz, "A Survey of the State of the Art of Design Automation", Computer, Vol. 14, No. 10, October 1981, pp. 58-75.

[6] B. Krishnamurthy and P. Mellema, "On the Evaluation of Mincut Partitioning Algorithms for VLSI Networks", Proc. ISCAS-83, Newport Beach, May 1983, pp. 12-15.

[7] M. Burstein, S. J. Hong, R. Pelavin, "Hierarchical VLSI Layout: Simultaneous Placement and Wiring of Gate Arrays", Proc. IFIP VLSI-83, Trondheim, August 1983.

[8] A. E. Dunlop, Automatic Layout of Gate Arrays, Proc. ISCAS-83, Newport Beach, May 1983, pp. 1245-1248.

Routing on one layer - an algorithm and its implementation

Jiri Soukup

Bell-Northern Research, Ottawa, K1Y 4H7, Canada

ABSTRACT:

The paper describes an implementation of a one layer router based on the algorithm originally proposed by Akers. The router is applicable to routing of printed circuit boards and of VLSI.

INTRODUCTION

The most usual style of routing connections of electronic circuits, and this applies both to the printed technology and to silicon circuits, is to use two layers: one in the horizontal direction and one in the vertical direction. Years of experience have verified that this is a good strategy for implementing complicated circuits in a small area. The connections are well packed, with a low level of mutual blocking between the nets. The strategy is similar to that used in manual routing. Various automatic routers are available to support this style of routing (see |1|).

The necessity to transfer frequently from one layer to another is inherent to the two layer routing. The transfer points are called via holes on the printed boards, and contact windows on the silicon chips. In both cases, the transfer points decrease the reliability of the circuit and the production yield. Also, via holes and the contact windows are usually larger than the width of the routing path, force the routing apart, and increase the required routing area.

Some routing systems allow vias or contact windows only at certain locations. The advantage is less blocking, the disadvantage is a more complicated topology of the routing.
The requirement to route without via holes or contact windows occurs frequently for military circuits or for circuits to be produced in large series, where the high yield is required. In general, such routing is more frequently required for analog circuits than for digital circuits.

Routing of hybrid circuits (thick film printed technology) is special. Any routing on the second or subsequent layers decreases the realiability and the yield. The requirement is to route as much as possible on the first layer, and transfer to other layers only when necessary.

Automatic single layer routing is not offered by most commercial packages. The reason is that the available algorithms are difficult to implement and require a large amount of CPU time. Also, in complicated real circuits, the user can never be sure that a given circuit can be implemented on the available number of layers.

Recently, progress has been made in so called "river routing" (see |3|). This is a rather special case of one layer routing, where the circuit is planar and all the connecting points (terminals) are on the boundary of a given routing area. This configuration results in a simple flow-like topology of the routing, hence its name.

Several serious attempts have been made to solve the general problem (see |2|), but the author is not aware of any algorithm that is used in a production environment of real large circuits.

The problem has two parts. The ordering of the connections (or nets) is essential to avoid blocking of pins. The routing itself must follow strategy entirely different from the two layer routing. The prime objective is to route in a way that would leave space for the remaining connections. The topology is most important, and the topology is the part of routing that a human can handle easily, but which is difficult for a computer. A close coordination between the two parts of the routing is necessary, but they are quite different algorithmically, and are therefore difficult to link together. The result is that the routing without vias is usually done by hand, interactively.

This paper describes an algorithm for one layer routing derived from the original comments by S. Akers (See |1|). The algorithm has been implemented as one of the routers in the Circuit Board Design System, BNR's CAD package for the design of printed circuit boards. The paper describes the algorithm, its implementation, and practical results.

The algorithm has been applied to the routing of printed circuit boards. All the terminology, and most of the illustrations throughout the paper, relate to the printed board technology. However, the same algorithm can be applied to the VLSI layout, with small adaptations in the ordering as shown at the end of the paper. In fact, the algorithm is an efficient river router.

THE MAIN IDEA

The main idea is very simple. When trying to connect a pair of points, start from one point, and run toward the board boundary, following the contour of any obstacle on the way. Once you get to the boundary, follow the boundary until you get to the point that is on the boundary, but closest to the target. At this point, change the direction again, and run directly toward the target, following contours of any obstacles. Fig. 1 illustrates the idea.

The advantage of this method is that the routing is developed first in the outer parts of the routing area, leaving many of the pins that are not yet connected available for later. The routing is packed toward the outside so well that almost all available space is fully used (see the sample problem in Fig. 5a).

There are two basic methods of packing the connections toward the boundary. Either one side of the boundary can be selected (the right side in Fig. 1), or any nearest point on the boundary (as in Fig. 2).

As the board fills with the connections, the contour search becomes more difficult (see the example in Fig. 3). Fig. 3 also illustrates that most of the connections on a complicated circuit board have many unnecessary turns and bays. The main routing phase (we will call it Phase 1) is therefore followed by the postprocessing, Phase 2. In Phase 2, all the connections are reexamined one by one, starting with the last route and ending with the first route. All necessary bends and bays are removed and the route is inserted back onto the board. Also staircase sections are replaced by 45 degree segments.

Figs. 4a and 4b show one pass of the router on a small board. Fig 4a is the result of Phase 1 with the left-side bias, Fig 4b is the result of Phase 2. Phase 2 usually clears up the routing area significantly, and more connections can be implemented by calling the router again. Note that the postprocessing shown here includes replacing of staircase sections of routing by 45 degree segments.

Fig. 5 shows routing of all nets connected to a selected device (the connector in this case).

Fig. 6 shows the result of Phase 2 for the same board as Fig. 4b, but with all-sides bias. Compare the different topology in Fig. 6 and Fig. 4b.

The ordering of the nets is critical, and a special paragraph will deal with the subject. For now, look at Fig. 1 and try to construct the connections using the same algorithm, but in a different order. The routes will block more pins and will have different shapes. Note that different ordering is required when routing the whole board, or when routing a single device (like in Fig. 5).

LIMITS AND ORIENTATION OF THE PATH

The router operates on two arrays that represent a regular grid structure covering the routing area.

S(i,j) or the signal array

= k>0 if net number k has a connection connection passing through cell (i,j),
= 0 if cell (i,j) is not available for routing,
< 0 if cell (i,j) is available for routing; the value represents the traceback code linking the path found by the router.

C(i,j) or the cost array

= 0 for the start subnet,
= 1 for cell (i,j) already examined by the router,
= 2 for all possible target subnets.
= 3 for cell (i,j) available, but not examined for this connection.

Array S is the permanent record of the routing, and the negative numbers are used when building up new connections. Array C is used only as a working array, and it must be reset after one connection is made.

When a pair of points (start and target) are given, the algorithm makes three important decisions before the actual routing starts:

1. Which of the two points will be the start, and which one will be the target.
2. Additional boundaries called "limits" are added, in order to restrict the path into a smaller area. This will allow a simple search algorithm all along the path.
3. The orientation of the path is decided (clockwise or anticlockwise). The orientation is important for the routing algorithm when following the contours.

In the current version of our router, the starting point is the one closer to the boundary. That point is more likely to be blocked off, and if it is used as the start of the search, the algorithm discovers sooner that the connection cannot be done.

The limits are constructed differently for different biases (see Fig. 7a and 7b). A pair of points on a line perpendicular to the boundary is a special case. The router makes the second attempt to connect with the different orientation of the path, if the first attempt fails (see Fig. 7c and 7d).

THE ROUTING ALGORITHM

(The algorithm will be described for a clockwise-oriented path. For an anticlockwise path, some of the angles are different.)

Any real algorithm must check for many special conditions on when to stop or turn. The following algorithm does not include all the details, but it describes the major decisions involved in the routing process:

(1) Start with the direction from the start point to the limit line.

current_point = starting_point

Set C = 0 for the start subnet,

C = 2 for all target subnets,

C = 3 for all other points on the board.

(2) Turn the direction +90 degrees.

new_point is the neighbor of the current_point in this new direction.

(3) if (new_point = starting_point and new direction = starting direction) go to (8)

if (new_point out of limits) go to (2)

if (all directions tried) go to (8)

if (new_point on the stop section of the limit line, go to (8)

(4) if (S(new_point) < 0) go to (5)

if (S(new_point) = current_net and C(new_point) = 0) go to (6)

if (S(new_point) = current_net and C(new_point) = 2) go to (7)

if (S(new_point) > 0. and S(new_point) <> current_net) go to (2)

(5) if (C(new_point) = 2) record the traceback code in S(new_point)

C(new_point) = 1

current_point = new_point

go to (2)

(6) Turn direction -90 degrees

current_point = new_point

go to (2)

(7) A connection found. Trace it using the traceback code, exit.

(8) Connection does not exist, exit.

Several points are important in this algorithm:

- Some sections of the limit line are marked as the "stop section of the limit line" - see Fig. 7 and step (3) of the algorithm. This condition is not necessary, but it accelerates the algorithm. If a connection does not exist and this condition is not present, the router would find that the connection cannot be done by arriving back to the starting point, however, that may be all around the board, and it would take a long time.

- The conditional setting of the traceback code in step (5) of the algorithm is critical. Otherwise the algorithm cannot trace back over its own trail (see the example in Fig. 3).

- In step (6) of the algorithm, the -90 degrees turn compensates for +90 degrees turn in step (2). This simply means passing through new_point.

Note that our version of the router operates on a general wide track (using a strip of cells as described in |4|), and it can also work in a selected window rather than on the whole board.

ORDERING

We still experiment with different ordering schemes. The following selection seems to produce good results:

- Select pair of points with the minimum distance to the boundary (minimum of the larger of the two distances).

- If there is more than one such pair, start with the pair that will require the longest run along the boundary.

When routing a selected device (like in Fig. 5), the ordering is different:

- Sort all pins on the selected device by X and Y coordinates.

- route the connections in the order of the pins.

The difficult part of the ordering is the decision as to what to do if a connection fails. Pins totally cut off the open routing area must be removed from the list of active pins. Fortunately, the router itself can be used to decide whether a pin is disconnected from other pins. When the route limit is defined as a ray from the given pin toward the boundary (see Fig. 9a), then the router search makes a long loop around the adjacent area, before it stops at the pin where it started. If we run a similar ray from all pins of the given net, and if we count the number of crossings with the path, we can dectect disconnected pins (the count is zero or even). The procedure must take a special step if the search runs directly through a pin, or if there is another pin on the ray from the starting pin to the boundary. The algorithm may fail if the only exit from the pin is toward the boundary, in the direction of the ray (Fig. 9b). But considering the overall topology of the routing, this case is so rare that a special algorithm is not needed.

APPLICATION TO VLSI - RIVER ROUTING

The major difference in routing the silicon chips is that usually the points to be connected are on the perimeter of a routing area. The algorithm described in this paper is directly applicable to this problem, assuming minor changes in the ordering strategy.

The Ordering Algorithm is simple:

1. Collect all terminals (pins) in a circular stack, in the anticlockwise direction.

2. Search for the point with the minimum number of hops to the next terminal of the same net (if next point in the stack, you have it, stop the search).

3. Use the all-biased one-layer router to make an anticlockwise connection.

4. Remove all pins under the new route from the stack. Replace the routed pins by a single pin, and if the net is finished, remove even that pin.

5. If any pins are left, go to (2), otherwise routing is finished.

As an example, let us follow the sample problem in Fig. 8. This is an almost planar routing problem. The router has to decide which non-planar connection will be left out.

If the terminals are stored in the stack from the left lower corner, the following table shows the changes of the stack and the development of the routes:

```
5 3 4-4 2 14 7 10 7 7 15 10 15 9 9 8 8 10 7 1 13 12 6 6 14 12 13 1 2 3 1
5 3     2 14 7 10 7-7 15 10 15 9 9 8 8 10 7 1 13 12 6 6 14 12 13 1 2 3 1
5 3     2 14 7 10 7   15 10 15 9-9 8 8 10 7 1 13 12 6 6 14 12 13 1 2 3 1
5 3     2 14 7 10 7   15 10 15     8-8 10 7 1 13 12 6 6 14 12 13 1 2 3 1
5 3     2 14 7 10 7   15 10 15         10 7 1 13 12 6-6 14 12 13 1 2 3 1
5 3     2 14 7 10 7   15 10 15         10 7 1 13 12     14 12 13 1 2 3 1
5 3     2 14 7 10 7   15 10 15         10 7 1 13 12     14 12 13 1 2 3
  3     2 14 7 10 7   15 10 15         10 7 1 13 12     14 12 13 1 2 3-
        2 14 7 10 7   15 10 15         10 7 1 13 12     14 12 13 1 2-
          14 7-10-7   15 10 15         10 7 1 13 12     14 12 13 1
          14 7        15-10-15         10 7 1 13 12     14 12 13 1
          14 7----------------------------7 1 13 12     14 12 13 1
          14
                                            1 13 12-----14-12 13 1
                                            1 13--------------13 1
                                            1........FAILS.......1
```

IMPLEMENTATION:

The program was designed and implemented in Fortran on an IBM 3033 in less than one man-month. Without comments, the code has the following number of lines: router itself: 455 lines; postprocessing: 205 lines; organization (cellmap access, path record and update): 285 lines; changes and additions to the driver: about 100 lines. One-layer router is one of the routers available on our Circuit Board Design System, and it can be called with the following options:

bias (left, right up, down, all-sides),

window to which to apply the router,

signal to route,

device to route,

path width,

signal only or supplies only,

stop after a given number of connections or failures,

display the search process (for debugging or education).

The sample problem shown in Fig. 4b took 20 CPU seconds to route. However, most of this time was spent on Phase 2 (postprocessing).

Generally, we have observed that the number of connections implemented by the router on following layers decreases exponentially as in this typical example:

layer 1: 274 connections, layer 2: 158 connections, layer 3: 108 connections, layer 4: 44 connections, layer 5: 34 connections, layer 6: 19 connections.

If 274 connections were implemented on layer 1, one would expect all routing to be finished on 3 layers. The fact that it does not happen is neither the result of the bias selection nor of the ordering, and the author has no explanation for the phenomenon. Perhaps, the first pass implements all "easy and obvious" routes, and the routes that are left block each other, and only a little progress can be made on each layer.

On the other hand, the relative emptiness of the subsequent routing layers is a great advantage when attempting a manual cleanup, or when inserting the last few routes interactively.

REFERENCES:

(1) Soukup, J. Circuit Layout, Proc. of IEEE, Vol. 69, No. 10, Oct. 1981, pp. 1281-1304.

(2) Hsu, C.P., Theory and Algorithms for Signal Routing in Integrated Circuit Layout, PhD-Thesis, Univ. of Calif. at Berkeley, Dept. EECS, 1983.

(3) Futagami, S., Shirakawa, I., Ozaki, H., An Automatic Routing System for Single-Layer Wiring Boards, IEEE Transactions Circuits and Systems, Vol., CAS-29, No.1, Jan 1982, pp.46-51.

(4) Pinter R.Y., On Routing Two-Point Nets Across a Channel, 1982 DA Conference, Las Vegas, June 1982, pp. 894-902.

(5) Soukup, J. Stockburger, U.W., Routing in Theory and Practice, First Annual Conference on Computer Graphics in CAD/CAM Systems, M.I.T., April 1979, pp. 129-146.

(6) Shirakawa I., Futagami, I., A Routing Scheme for Single Layer IEEE Trans. on CAD, Special Issue on Routing, Oct. 1983

(7) Sadowska, M., Tarng, T., Single Layer Routing for VLSI, Analysis IEEE trans. on CAD, Special Issue on Routing, Oct. 1983.

(8) Tsukiyama, S., Kuh, E.S., Shirakawa, I., An Algorithm for Single Row Routing with Prescribed Street Congestions, IEEE Trans. on Circuits and Systems, Vol. CAS-27, No. 9, Sept. 1980, pp. 765-772.

(9) Armstrong, R.A., Doreau, M.T., A Topological Representation for Single Layer Routing Problems with Variable Sized Objects, IEEE Internat. Symposium on Circuits and Systems, Rome, May 1982, pp. 771-775.

(10) Dorn, T., Mlynski, D.A., Multilayer Layout Planarization Using Boolean Algebra, IEEE Internat. Symposium on Circuits and Systems, Tokyo, July 1980, pp. 462-465.

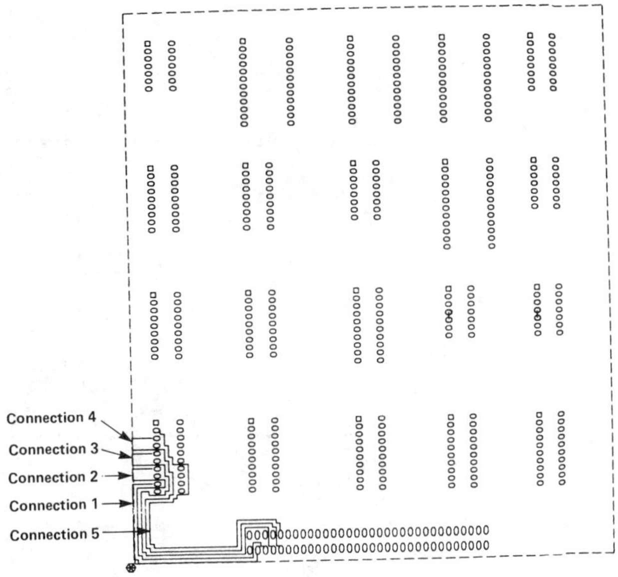

First 5 connections, when running the router with the all-sides bias.

Figure 2

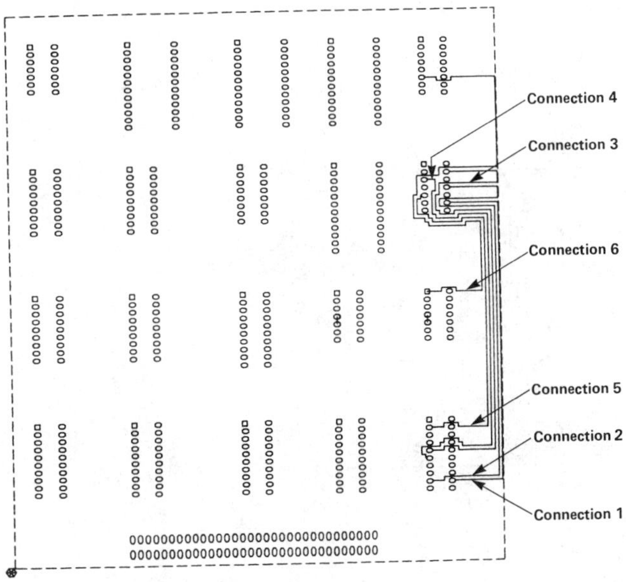

The main idea of the router is: Go toward the boundary and follow the contour. (These are the first 6 connections, when running the router with the right bias.)

Figure 1

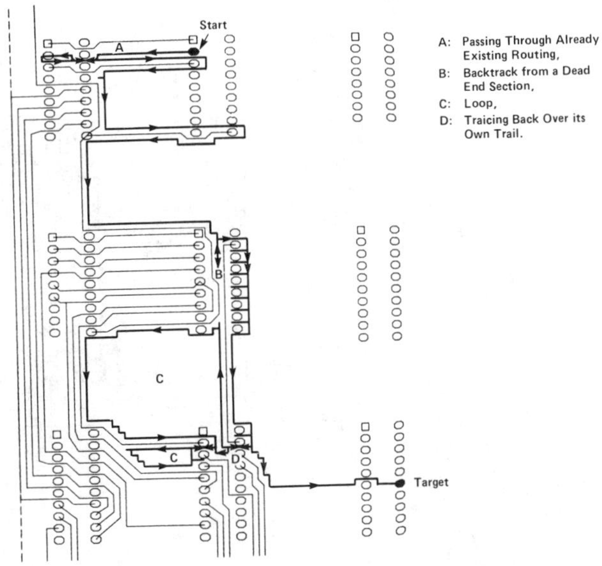

The search path for a more complex situation. Note various modes of the search.

Figure 3

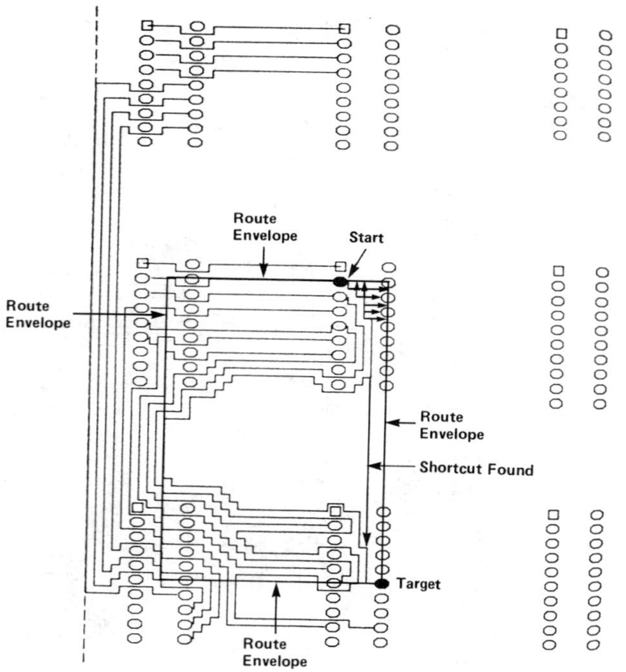

Postprocessing the first route.

Figure 4 a)

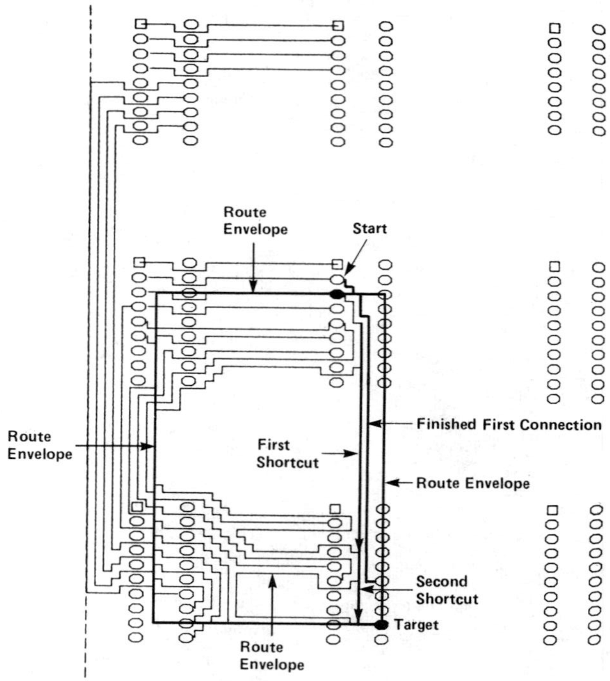

Postprocessing the second route.

Figure 4 b)

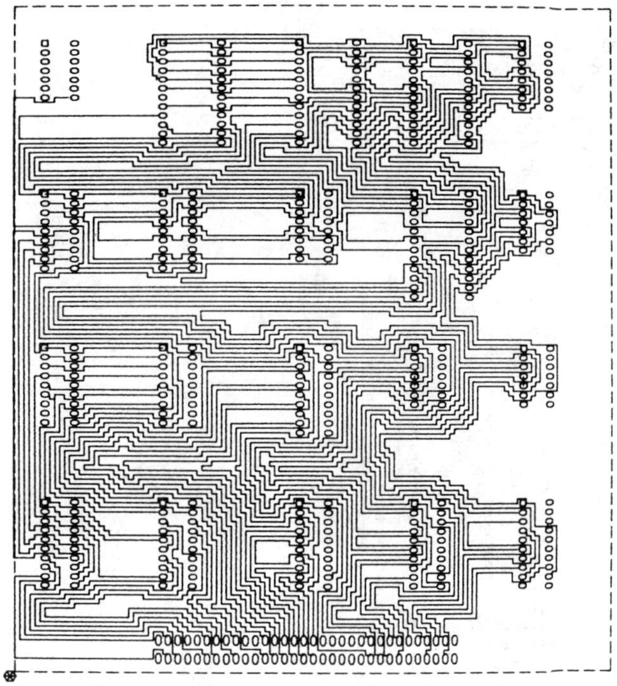

One layer routing Phase 1, left bias.

Figure 5 a)

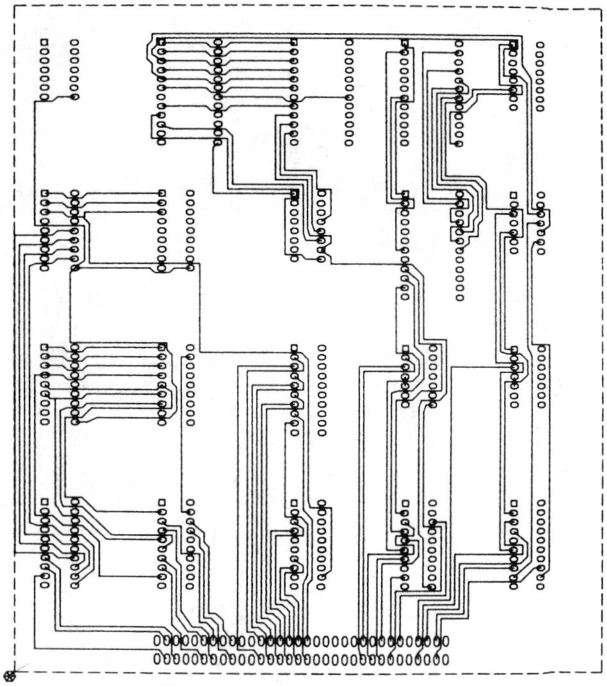

Phase 2 for the routing from Figure 5a.
(Loops and bays have been removed staircase runs replaced by 45-degree segments).

Figure 5 b)

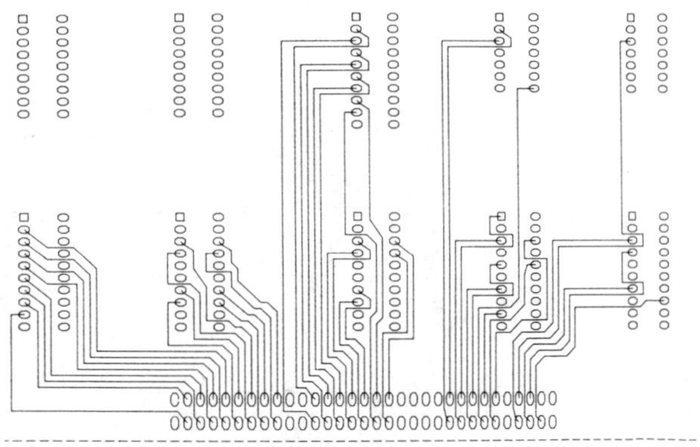

Example of routing all connections attached to a selected device (routing connector, left biased, the result of Phase 2 is shown here)

Figure 6

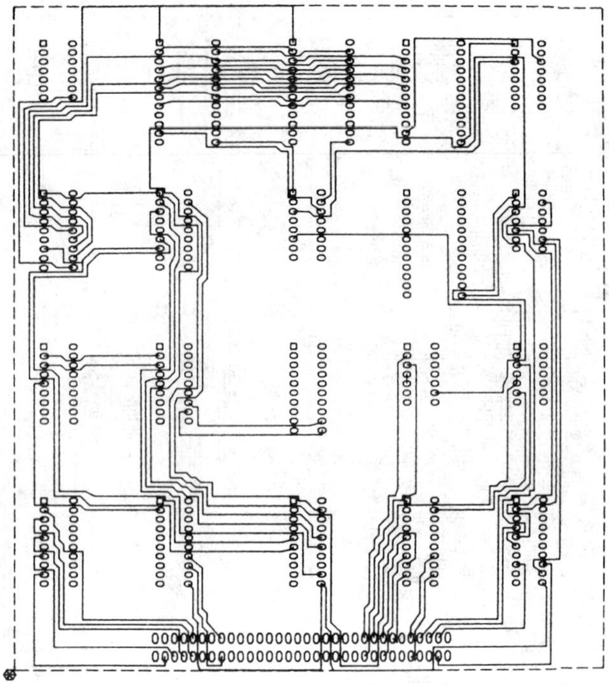

Figure 7 Routing the same board as in Figure 5 b), but using the all-sides bias. Note the different topologies.

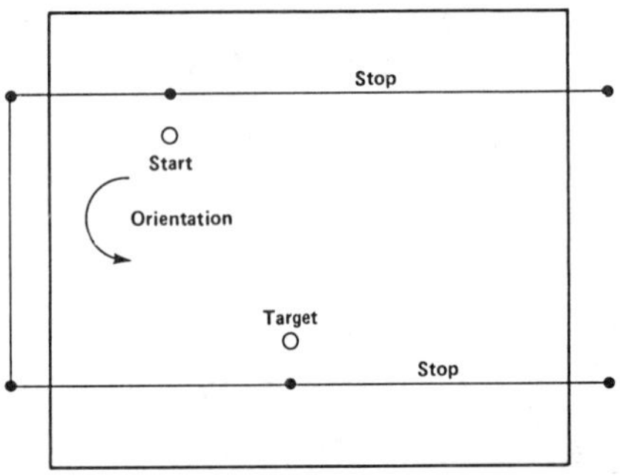

Figure 8a: Left Bias

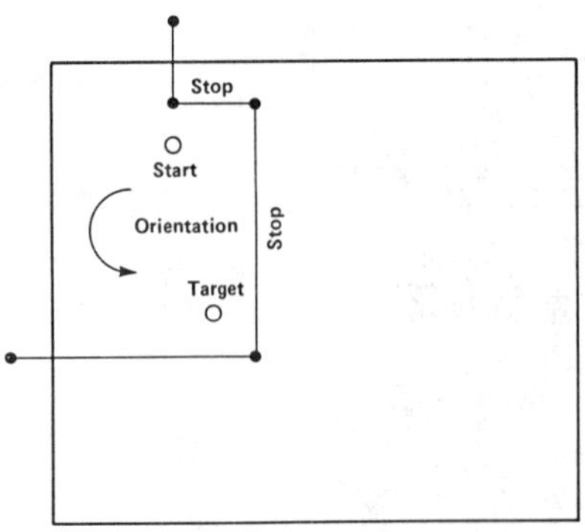

Figure 8b: All-sides Bias

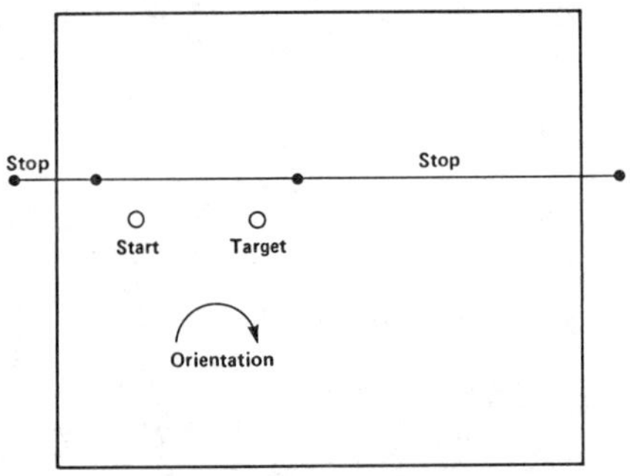

Figure 8c: Start and Target in Line, First Trial

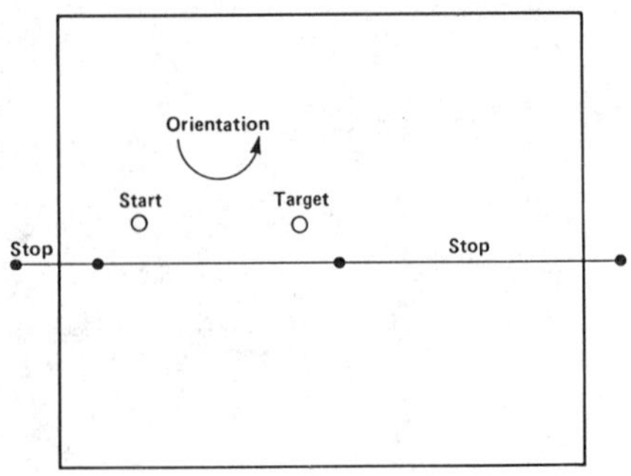

Figure 8d: Start and Target in Line, Second Trial

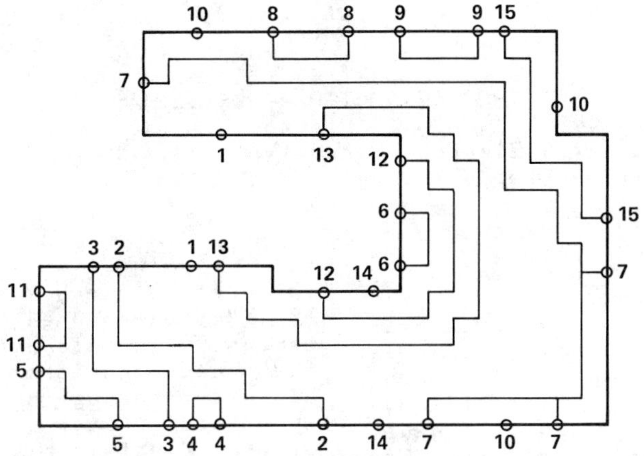

Sample problem of the VLSI-type single layer routing – Phase 1.
(Phase 2 inserts only a small change on net 7)

Figure 9

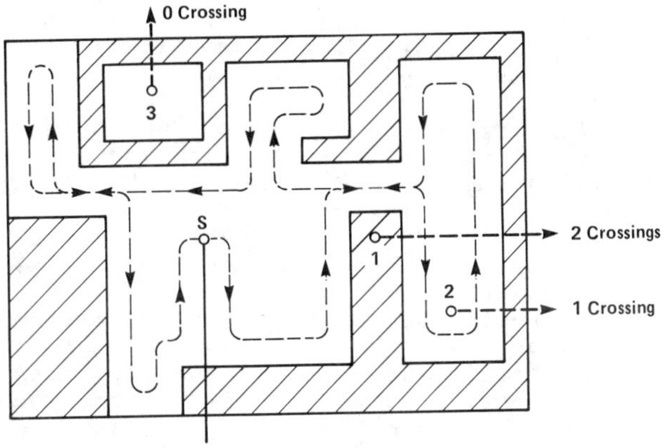

Algorithm to find disconnected pins (Pin s cannot connect to pins 1 and 3. It may, but may not, connect to pin 2.)

Figure 10a

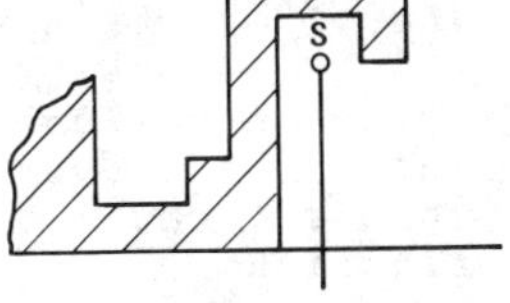

Case not covered by the algorithm

Figure 10b

THE LEE ROUTER REVISITED

David Hightower
General Electric Microelectronics Center
Research Triangle Park, NC

ABSTRACT

The Lee routing algorithm has been largely ignored for IC applications because it is slow and requires a large amount of memory; however, modern computers are fast and have more than enough memory. This paper discusses an implementation of the Lee algorithm that has proved effective on CMOS gate arrays up to 10,000 gates.

INTRODUCTION

The Lee algorithm [12] has been used extensively for automatic path routing in the electronics industry for twenty years. The algorithm requires that the routing surface be stored as a matrix which is large enough to allow routing to the finest possible resolution. Until fairly recently, the algorithm was not considered suitable for fine-resolution, large-surface routing. Its major application was in routing printed wiring boards and as a follow-on router after channel routing. For integrated circuit routing, global/channel routers [3] or line search routers [7] have dominated because of their speed and modest memory requirements.

However, the least-cost version of the Lee router, with a relatively wide range of expansion costs, can give results that are very close to the results of a channel router since the costs can control the paths used by early connections so that their effect on later paths is minimized. Further, the costs can represent the desirability of using certain structures, such as fixed poly "crossunders" on CMOS arrays, in order to control the capacitance and resistance of resulting paths.

Advances over the past few years in virtual memory computer systems have allowed a significant decrease in the speed of a Lee router by trading memory for speed. The memory requirement of the routing matrix has almost become a nonissue. Because of such advances, coupled with algorithmic improvements in the Lee router, it is time to examine its appropriateness for integrated circuit routing, in particular, for "fixed-channel" routing such as gate arrays. In this paper, the suitability of the Lee algorithm for CMOS gate arrays is discussed through the implemention of a least-cost router by the author at the General Electric Microelectronics Center (MEC).

MOORE'S ALGORITHM

Perhaps the earliest work describing a non-exhaustive path-finding algorithm with a guaranteed solution is "Shortest Path Through a Maze" by Edward Moore in 1959 [13]. The Moore algorithm operates on a graph representing a network, such as a communications network, and finds a path from a start vertex to a target vertex.

The least cost expansion algorithm for a graph starts at the START vertex and writes on each of its adjacent vertices, or neighbors, the cost of traversing the edge. Then for each of those vertices, the algorithm writes on their neighbors the current cost plus the expansion cost. If a neighbor has a cost on it already, it writes a new cost only if the new cost is less. The process continues until the cost written on any unexpanded vertex exceeds the cost written on the TARGET.

Trace-Back starts at the TARGET and looks for the adjacent vertex that must have expanded into TARGET, namely the one whose cost plus the edge cost equals the cost on TARGET. Then it finds the vertex adjacent to the new one that must have expanded into it, and so forth, until START is reached. There may be several choices for vertices that satisfy the conditions of the trace-back algorithm. Rules for describing which of several equivalent vertices to choose may be incorporated in the algorithm.

LEE'S ALGORITHM

If the vertices and edges are highly regular, the Moore algorithm is somewhat simplified since the graph is represented by a matrix, and the vertices which are adjacent to a given vertex are easily

Global routers use a concept similar to remote expansion, but global routers are acutally Moore algorithms, not gridded Lee routers so the implementation using remote expansion is trivial.

Execution Time Improvements

Speeding up the Lee algorithm is important for practical IC application. We outline below several methods of improving the execution speed.

Choosing the Next Vertex to Expand. Most of the time spent in a Lee algorithm is consumed in the expansion phase. That time is reduced by expanding the vertex having the lowest cost on it and stopping as soon as the the next vertex to be expanded is the target [1,4,8,10,14].

The trick is finding the lowest cost vertex efficiently. To solve this problem, three circular linked lists are used in a method similar to the one described by Hoel [8].

One list, FREE, links the free slots on the expansion list (EX(*),EY(*),ECOST(*)). FREE solves the problem of having to perform periodic garbage collection on the expansion list during expansion.

The second linked list, CSTLINK, links the vertices on the expansion list in order of their costs.

The third list, CSTPTR, has length equal to the maximum possible cost and serves to identify the first vertex in CSTLINK order which has a cost not less than the CSTPTR row number, i.e., if CSTPTR(K)=I, then ECOST(I)>=K.

Acceleration Schemes. The Lee algorithm can be speeded up by using "acceleration" schemes [4,11,14]. A typical scheme predicts the cost of routing from a vertex to the target vertex, adds the result to the cost on the vertex, and makes a decision about when to expand it. Most of these schemes, while reducing execution time, cannot guarantee that the least cost path is found. In CMOS applications, one cannot allow for the possibility that a path will march up a column of contacts and eliminate all of the crossunders along the path.

Another acceleration scheme expands from both points simultaneously, but this technique gives up generality in finding only one shortest path. It is better to expand from one point and have all possible least cost paths available for a clever trace-back routine.

Some researchers [1,4,8,14] have suggested storing pointers during expansion to be used during trace-back. For IC applications, using pointers is not recommended for two reasons: 1) The time spent in expansion is far greater than the time spent in trace-back; therefore, it saves time to take logic out of the expansion phase and put it in the trace-back phase. 2) If pointers are used, deciding which way to trace back when there are choices is impossible.

Dynamic Cost Schemes

It is possible to alter costs during routing to attempt to create certain types of routing. For example, after routing a wire, the expansion costs of cells along the wire can be modified to encourage or discourage wires from "hugging" it. The regions near terminals can be weighted so that they will not be used or will be used only as a last resort. This protects avenues of egress for the terminals. If several avenues are protected, the unused ones are returned to normal costs after the net is routed.

The costs for expanding through a channel can be modified as the channel becomes congested. In fact, the possibilities for experimenting with various costing schemes is limited only by the imagination emphasizing the power of the algorithm.

MEC IMPLEMENTATION

In the spring of 1981, the author began developing a Lee router for CMOS gate arrays, and, to date, it has been used on over 30 production gate arrays. It can route single-layer metal CMOS arrays using fixed contacts (the contact mask is fixed) or floating contacts (the contact mask is programmable). In the floating contact mode, only the contacts that are used in routing appear on the contact mask. The routing topology is described in a language called GAMDL (Gate Array Master Design Language).

The statements in GAMDL define a routing matrix for a portion, called a "GCELL", of the master and establish the costs for expanding from each grid point to each adjacent grid point. The costs, which may range from 1 to 63, represent a combination of the incurred capacitance for using some structure or the relative goodness of putting a wire with a given orientation in a particular cell. GCELLs are replicated with placement statements to form the completed gate array master.

The router can also route two-layer CMOS arrays with fixed or floating contacts. In the two-layer mode, the router reads via spacing values and protects vias and metal-2 from adjacent vias and metal-2 where necessary.

examined by adding or subtracting one from one of the subscripts of the matrix. We expand in this routing matrix with a "Lee" algorithm [12].

The expansion costs can be computed during expansion or stored in a matrix corresponding to the routing matrix. If they are computed, the expansion rules are programmed into the router, and therefore are severely limited in flexibility. If the expansion costs are stored in a matrix over which the user has control, a large range of cost types is possible, as is dynamic costing.

Two Layer Routing

The standard way of extending the Lee algorithm to two layers is implemented by adding another plane to the routing matrix [2, 5], doubling the matrix size.

The memory requirement of a two-layer problem can be reduced by having only one expansion layer and two bit maps to contain blockoffs and some special codes. Two bits are needed for each routing cell. If the expansion matrix uses 16 bits per cell, the bit maps add 25% more memory.

Call the two bit maps LAYER1 and LAYER2. LAYER1(Y,X) bit 0 is on (TRUE) if cell Y,X contains a metal-1 blockoff; bit 1 is on if Y,X cannot contain a via because of an adjacent via. LAYER2(Y,X) bit 0 is on (TRUE) if Y,X contains a metal-2 blockoff; bit 1 is on if Y,X cannot have a metal-2 wire because of an adjacent metal-2 wire.

Assume metal-1 is predominantly vertical and metal-2 is predominantly horizontal. The expansion algorithm is modified by adding the following logic: If EXPX,EXPY is the vertex under expansion and NX,NY is its neighbor, do not expand north or south on metal-1 if LAYER1(NY,NX) is TRUE or LAYER1(EXPY,EXPX) is TRUE. Do not expand east or west on metal-2 if LAYER2(NY,NX) is TRUE or LAYER2(EXPY,EXPX) is TRUE. Do not change layers if LAYER1(NY,NX) bit 1 is on. Do not expand on metal2 from or into a cell that has LAYER2 bit 1 on.

Trace-Back changes by adding logic to decide which layer the segments of the path go on. Short lines are put on the "wrong" layer to eliminate vias. The trace-back algorithm stores blockoff codes and via/metal-2 protection codes in LAYER1 and LAYER2.

The trace-back algorithm produces lines that can have pieces on different layers so it decides which layer the pieces go on and whether or not to put vias down. The situation in Figure 1a arises frequently: trace-back produces the short horizontal metal-2 line because of the adjacent via. The via-horizontal metal-1 blocks metal-1 in two cells instead of one. Trace-back is modified by adding a look-ahead step that produces the path in 1b: if cell EXPX,EXPY is the current cell, EXPX+1,EXPY cannot have a via, EXPX,EXPY+1 is a success, i.e., the cost on EXPX,EXPY+1 plus the south expansion cost from EXPX,EXPY+1 equals the cost on EXPX,EXPY, and a via can go at EXPX,EXPY, then take the north turn.

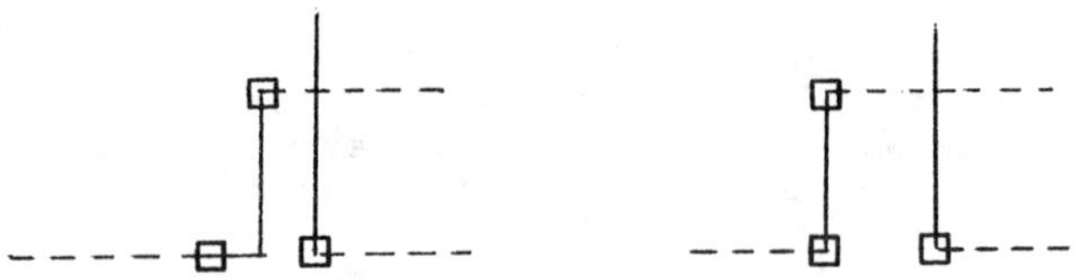

With trace-back look-ahead

1a

Via prevents adjacent via

1b

FIGURE 1

"Excursions" can be made on the "wrong" layer by modifying the expansion algorithm to permit expansion in the "wrong" direction when there is no expansion possible in the "right" direction. During expansion, the "wrong way" cost is added to the normal cost, and is subtracted during Trace-Back.

"Remote" Expansion

Expansion into a "remote", i.e., non-adjacent, cell is possible by generalizing the algorithm to allow for the possibility of expansion into more than just the adjacent cells. Remote expansion arises in a CMOS gate array because there are fixed poly structures which are the gates of the transistors, diffusion regions with contacts on either side of the buses, and, in the case of single-layer gate arrays, fixed poly "crossunders" used as the primary second layer.

The costs for each of the five ways of expansion can be stored in five fields in a matrix. The fields correspond to: remote expansion pointer, and north, east, west, south expansion costs. The pointer is an index into a list containing: the cost of expansion and deltax,deltay values to determine its immediate equipotential neighbor. Thus, if EX,EY has a nonzero remote expansion value, say M, then the translation operation NX=LIST(M,1)+EX, NY=LIST(M,2)+EY gives the neighbor with NCOST=ECOST+LIST(M,3). Note that the delta values may be negative. When expanding into a vertex with equipotential neighbors, all neighbors are added to the expansion list at once.

The router can "see" inside macro boundaries and use the existing metal, poly, and diffusion as part of the net routing. This feature permits very efficient net routing and generally smaller arrays than other systems that treat macros as black boxes.

In comparison with other commercially available routers, this router has consistently performed to a much higher completion rate on the same placement and executes in roughly 20% of the CPU time of the other routers. In a typical application, an 87% occupied array of 1300 gates takes 30 CPU minutes on the VAX 11/780. While this may seem like a long time to those familiar with channel router execution times, the extra 15 minutes or so are only of theoretical interest considering the difference in the ability to use existing structures such as diffusion and poly.

In the MEC gate array family, the single layer arrays tend to have nine vertical tracks and 2 horizontal (poly) tracks between adjacent three transistor pair cells. The two-layer arrays have four transistor pairs per cell with nine verticals and no horizontals between cells. A 10000 gate array has been routed and takes about 13MB and 3.5 hours without optimizing the router for memory useage or "tuning" the router for maximum performance on the VAX.

CONCLUSION

The results achieved with the implementation of the least-cost Lee algorithm described above indicate that the Lee router is excellent for gate arrays and 2-layer IC routing with fixed width channels. The traditional reasons for not using a Lee router for IC routing, speed and memory, seem to have vanished as a result of the progress in modern virtual memory computers.

REFERENCES

1. Dijkstra, E.W., "A Note on two Problems in Connexion With Graphs." Numerische Mathematik 1(1959), pp. 269-271.

2. Geyer, J.M., et al., "Connection Routing Algorithm for Printed Circuit Boards." IEEE Trans. on Circuit Theory, Vol. CT-18, (Jan. 1971), pp. 95-100.

3. Hashimoto, A. and Stevens, J., "Wire Routing by Optimizing Channel Assignment Within Large Apertures." Proc. 8th Design Automatiion Workshop, 1971.

4. Hart, P.E., Nilsson, N.J., Raphael, B., "A Formal Basis for the Heuristic Determination of Minimum Cost Paths." IEEE Trans. Syst. Sci. Cybern., Vol SSC-4, pp. 100-107, (July 1968).

5. Heiss, S., "A Path Connection Algorithm for Multi-layer Boards." Proc. Design Auto. Workshop, (June, 1968), pp. 6-1 - 6-14.

6. Hightower, D.W., "A Router for Single Layer CMOS Gate Arrays." Proc. 1982 International Symposium on Circuits and Systems, Rome, IEEE82CH1681-6, (May, 1982).

7. Hightower, D., "A Solution to Line Routing Problems on the Continuous Plane of Line Routing Problems." Proceedings 6th Design Automation Workshop, 1969.

8. Hoel, J.H., "Some Variations of Lee's Algorithm." IEEE Transactions on Computers, Vol. c-25, No. 1, (January, 1976).

9. Ji Guang, X. and Kozawa, T., "An Algorithm For Searching Shortest Path By Propagating Wave Fronts in Four Quadrants." ACM IEEE, 18th Design Automation Conference Proceedings, Nashville, Tennessee, (June, 1981).

10. Johnson, D.B., "A Note on Dijkstra's Shortest Path Algorithm." J. Assoc. Comput. Mach. 20, 385-388 (1973).

11. Korn, R.K., "An Efficient Variable-Cost Maze Router." ACM IEEE 19th Design Automation Conference Proceedings, Las Vegas, Nevada, IEEE 82CH1759-0, (June, 1982).

12. Lee, C., "An Algorithm for Path Connections and Its Applications." IRE Trans. on E.C. (September, 1961), pp. 346-365.

13. Moore, E., "Shortest Path Through a Maze." Proc. Int. Symp on Theory of Switching Circuits, in the Annals of the Harvard Computation lab, Harvard University Press, Cambridge, MA, vol. 30, pp. 285-292 (1959).

14. Rubin, F., "The Lee Path Connection Algorithm." IEEE Transactions on Computers, Vol. c-23, No. 9, (September, 1974).

Session 1-2d
Impact of VLSI

Chairmen

K. Ishihara
Hitachi

C. M. Davis
IBM

A HIGH-SPEED MULTI-PURPOSE VLSI DIGITAL PROCESSOR

by H. E. Tsou, A. S. Hamori and B. L. Troutman

TRW Electronic Systems Group
Redondo Beach, California

ABSTRACT

This paper describes the architecture and design of a high-speed, multi-purpose VLSI digital processor for monolithic C-MOS implementation. The architecture is optimized for high-speed arithmetic operations and for computer system applications either as a stand-alone processor or as an element in a network of processors. The internal three-stage pipeline organization enhances the throughput rate with most of the instructions having an execution time of 200 nanoseconds. The arithmetic unit consists of a full arithmetic and logic unit and a 16 x 16-bit multiplier. It has an on-chip three-port register file, three address generators and an external data memory addressability of 65K 16-bit words. The control section consists of the basic timing and control functions, a microcontroller, and a program stack with an external program memory addressability of 65K 24-bit words. The I/O consists of a 16-bit parallel I/O bus, a serial I/O channel, discrete I/O and a hardware interrupt. The latter allows the device to be interconnected to a network for array processing. The chip design is based on 2-micron bulk C-MOS with an estimated device count of 28,300 devices, power of 400 milliwatts, and chip size of 250 x 265 mils.

INTRODUCTION

The architecture is a result of two generations of high-speed processor development. In 1972-74, a multi-purpose LSI processor[1-2] was developed which utilized five functionally-partitioned LSI devices of which the TRW 16 x 16-bit multiplier[3] is one; the 4-bit sliced RALU and the address generator are the other two. This was the first high-speed processor to use the 16 x 16-bit LSI multiplier[3]. Today, the multiplier is used in many high-speed processor applications[5]. In 1975-76, MMI and AMD both produced 4-bit RALUs and TRW used the AMD-2901 in the microprogrammable signal processor[2] development. This represents the first marriage of the LSI multiplier and the LSI RALU in high-speed processor design[4]. Since then, VLSI technology has opened a new frontier for new architectures and new applications[6-10]. This paper describes a new processor architecture aimed at achieving high throughput rate, modularity, and monolithic VLSI implementation.

The architecture is optimized for high-speed arithmetic operations either as a stand-alone processor or as an element in a network of processors. The internal three-stage pipeline organization enhances the throughput rate with most of the instructions having an execution time of 200 nsec. The arithmetic unit consists of a full arithmetic and logic unit and a 16 x 16-bit multiplier. It has an on-chip three-port register file, three address generators and an external data memory addressability of 65K 16-bit words. The control section consists of the basic timing and control functions, a microcontroller, and a program stack with an external memory addressability of 65K 24-bit words. The I/O consists of a 16-bit parallel I/O bus, a serial I/O channel, discrete I/O, and a hardware interrupt. The latter allows the device to be interconnected to a network for array processing. The chip design is based on 2-micron bulk C-MOS with an estimated device count of 28,300 devices, power of 400 milliwatts, and chip size of 250 x 265 mils. The chip replaces approximately 100 IC, MSI and LSI devices and saves approximately 15 watts.

VLSI PROCESSOR ARCHITECTURE

In order to develop an optimum architecture for high-speed arithmetic processing requirements, various algorithms have been analyzed in the area of digital filtering, correlation, matrix computation, various transforms, spectral analyses, windowing functions, minimum distance vector search, and pitch extraction and smoothing. Other algorithms which have been analyzed include complex decision logic, tracing pointers, vector quantization, parameter encoding and decoding, range binning, content addressing, etc. These analyses were made in terms of data type and format, addressing and indexing, program and mode control, arithmetic operations, concurrency of processings and data flow control. Typical applications include digital signal processing, voice and modem processing, communication controllers, and microprogrammable emulators.

Results of the requirements analysis led to the development of a processor architecture for a multi-purpose VLSI digital processor which is optimized in terms of speed, efficiency, and flexibility to perform a wide variety of applications. The architecture was also driven by the goal of being able to implement the processor on a single VLSI chip. A top-level view of the architecture is illustrated in Figure 1. In this architecture, the speed required is realized by a high degree of parallelism. This parallelism is achieved by providing i) separate arithmetic units for arithmetic processing, program address generation and data address generation, ii) multiple data paths to allow simultaneous transmission of operands, results and instructions, and iii) separate program and data memories. Flexibility is achieved by means of powerful I/O and interface provisions, expansion capability for both

CH1935-6/83/0000/0141$01.00 © 1983 IEEE

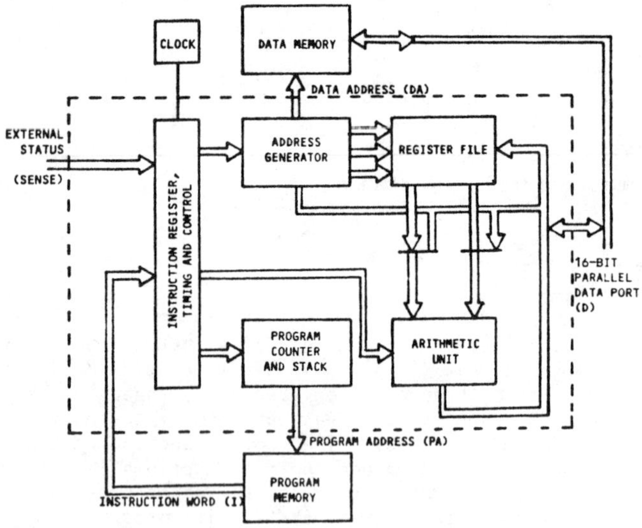

Figure 1. Multi-Purpose VLSI Processor Block Diagram

program and data memory, and a very general-purpose instruction format. Higher level parallelism is provided by means of networking. Interface provisions allow the synchronization of multiple processors each working on parts of the same problem. For processing algorithms which can be partitioned among several processors, this is a powerful feature.

The salient features of the multi-purpose VLSI digital processors architecture are:

- Stand-alone, network, or attached signal processor configuration
- On-chip programmable micro-code or macro-instruction from host processor
- Pipeline processing of array data with three address generators
  - Three base address registers/counters
  - Two index registers/counters
- On-chip register file or external data memory expandable to 65K
- On-chip program stack and external program memory expandable to 65K
- Arithmetic unit provides standard arithmetic and logic functions such as +, −, AND, OR, INV, XOR, ABS, and 16 x 16-bit multiply and accumulate
- Twelve powerful micro-instruction types, branch conditions selectable from internal and external sources
- Hardware interrupt for loosely-coupled multi-processing and real-time applications
- Flexible I/O transfers data between memory and I/O or between arithmetic unit and I/O

MEMORY UNITS

Data memory consists of an on-chip multi-port register file and an off-chip data memory. The register file is ideal for use in intensive computational loops since it has the capability of accessing two operands and storing one result simultaneously. On the other hand, external data memory may be used to store larger blocks of data and may consist of any combination of RAM, PROM, or ROM. The external data memory is accessed through the 16-bit data port and addressing is provided by a 16-bit data address port which may address either data memory or I/O ports.

Although is is desirable to have on-chip ROM or PROM program memory for stand-alone processing, in the majority of applications it is more efficient and flexible to use external program memory. The external program memory is accessed by means of a 16-bit program address from the on-chip program counter and a 24-bit instruction input from the external program RAM, PROM, or ROM. Each microinstruction is 48 bits wide; thus, two 24-bit instruction bytes are required for each instruction.

Arithmetic Unit

The arithmetic unit (AU) receives data input from the register file and from the arithmetic unit bus; the results are then output to the AU bus. As shown in Figure 2, it consists of a multiplier,

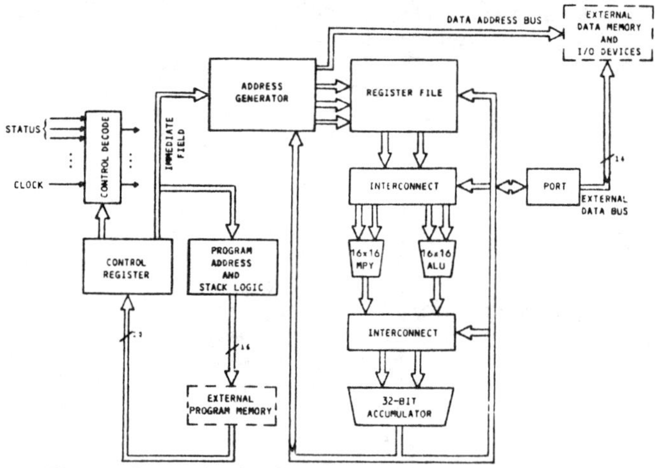

Figure 2. Block Diagram of the Arithmetic Unit

arithmetic logic unit (ALU), 32-bit adder-subtractor, register and data routing multiplexers. Its architecture and data flow have been optimized to perform the types of signal processing computations investigated in the requirements analysis.

The multiplier is 16 x 16-bit producing a 31-bit product. A round mode allows the accumulation of rounded and 16-bit truncated products. This mode can be enabled by loading the condition latch with a round bit.

The 16-bit arithmetic and logic unit provides a complete set of arithmetic and logic functions. Two 32-bit operands are input to the 32-bit adder-subtractor. Used in conjunction with two accumulator registers, the adder/subtractor facilitates sum of products and sums and differences computations which are so important in signal processing. Built into the 32-bit adder/subtractor and the two accumulator registers is a scheme to perform in-place shift/subtract division of two numbers.

Timing and Control Unit

Control of the processor resides in the instruction register, timing and control, and the program counter/program stack. Instructions are received from the program memory by the instruction register and decoded by the timing and control. Program flow is determined by the program counter based on the instruction decode and the machine status. The next program address may be determined in several ways: incrementing the program counter, branch address, call address, interrupt vector and stack. Both internal and external status may be tested. Internal status consists of machine status such as arithmetic conditions and address counter overflow. External status consists of user definable inputs which allow for sensing external events.

The processor has a single hardware interrupt. On-chip hardware features are provided to facilitate expansion of interrupt capability. The processor has a five-level push-down program stack used to provide return linkage for interrupts and call instructions and to implement program loops.

The microinstruction is 48 bits wide and horizontally coded. Figure 3 shows the typical instruction set of the processor. For high-speed digital processing application, a single-level microprogram is used rather than a higher level instruction in order to achieve maximum speed and flexibility. On-chip hardware features are provided to allow external implementation of macroinstructions.

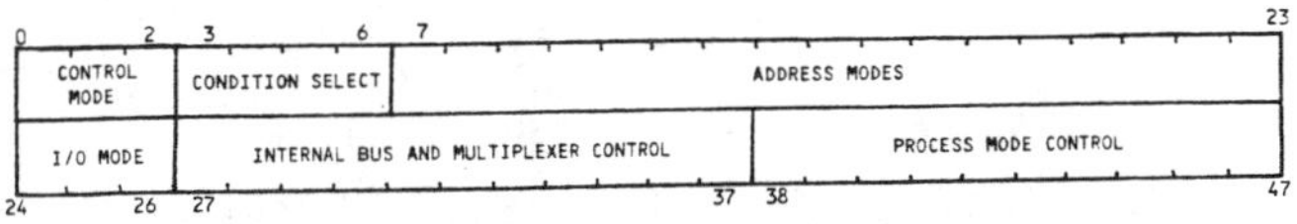

Figure 3. General Instruction Format

Interface

A flexible interface is provided for the processor. Data transfers between the processor and data memory or devices such as A/Ds, D/As, and other processors, etc., are provided by the 16-bit bidirectional data port. All data port transactions are programmed I/O. A bidirectional serial I/O channel is also included. System timing and synchronization in a multiprocessing environment are provided by program interrupt and external sense facilities.

C-MOS VLSI TECHNOLOGY

The processor architecture is currently being implemented with TRW 2-micron bulk C-MOS. The selection of this technology is based on the following considerations

- Device Density
  - C-MOS is approximately twice as dense as bipolar for equal feature size
- Functional Throughput Rate
  - Product of functional density and throughput rate: $10^{11}-10^{12}$ gates times Hertz/chip
  - Increases at a faster rate as a function of gate size than bipolar
- Power
  - One order of magnitude lower than bipolar for processing functions
  - Primarily dynamic power at the interface; very low standby power
- Yield, Cost and Multiple Source
  - Bulk C-MOS has a much higher yield than C-MOS SOS
  - Cost of bulk C-MOS is decreasing faster than C-MOS SOS and bipolar devices

TRW 2-micron C-MOS technology is summarized in Figure 4.

TECHNOLOGY	PERFORMANCE ADVANTAGES
• THIN GATE OXIDE	HIGHER SPEED, RADIATION HARDNESS
• DRY PLASMA ETCHED SILICON GATE	DIMENSIONAL CONTROL, UNIFORMITY
• SHALLOW SOURCE-DRAIN JUNCTIONS	DIMENSIONAL CONTROL, UNIFORMITY
• NON-PENETRATING CONTACT METAL	YIELD, MANUFACTURABILITY
• LOCOS FIELD OXIDE	HIGHER SPEED
• ION-IMPLANTED THRESHOLD ADJUST, P-WELL, N+, P+	BALANCED INVERTERS FOR OPTIMUM SPEED AND LOW POWER, UNIFORMITY AND YIELD
• PROJECTION PHOTO-LITHOGRAPHY	YIELD, MANUFACTURABILITY

Figure 4. TRW 2-Micron C-MOS Technology

CHIP DESIGN AND LAYOUT

The processor contains the following functional modules

- Address Generator
- Program Counter
- Register File (65 Words x 16 Bits)
- Multiplier (16 x 16 Bits)
- Arithmetic and Logic Unit (16-Bit)
- Adder/Subtractor and Dual Accumulator (32-Bit)
- Instruction Decoding, Timing and Control
- I/O Control and Interface

The technology, design and methodology, and standard cell library have been developed in another TRW VLSI device, the 32-point FFT chip[6-7]. In fact, approximately 50% (by area) of the processor chip functional modules are derived from that chip and will be used directly. These modules include the 16 x 16-bit multiplier, the accumulator, the register file, and part of the arithmetic and logic unit. A preliminary design of the processor chip has been completed and the floorplan has been layed out. The resultant CAD output is shown in Figure 5.

The left-hand side of the chip is devoted to the address outputs for two external data memories. The program addresses are generated in the program fetch unit which includes a program counter register, an incrementor, a program stack register, a stack pointer, and various routing circuitry. The timing and control block receives words from the addressed program memory every 100 nanoseconds.

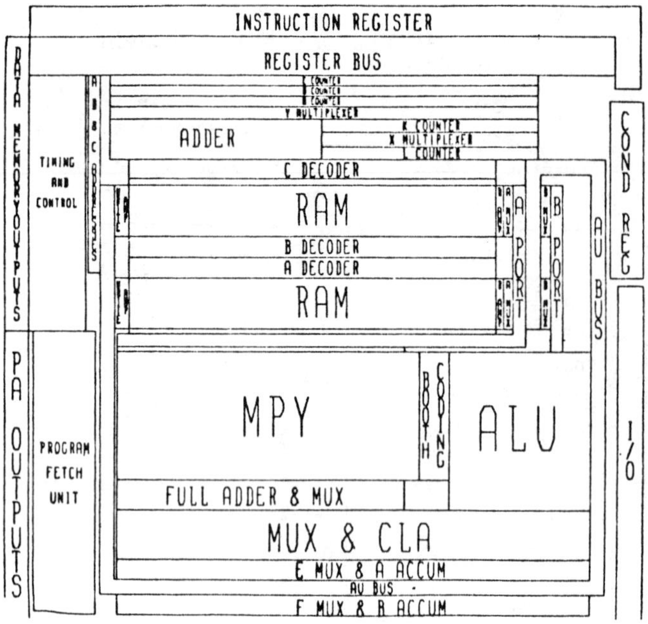

Figure 5. Processor Floorplan

instruction register collects two adjacent words before executing the instruction.

Another major block is the address generator which produces addresses to the external data memory as well as the on-chip register file which comprises part of the arithmetic unit.

The four largest areas are occupied by the register file, the multiplier, the arithmetic logic unit, and the 32-bit adder/subtractor and associated accumulators. The multiplier is a 16 x 16 Booth array which produces a 31-bit 2's complement product. There is a round option which permits either rounded or truncated products. The row of full adders at the bottom of the multiplier array permits the multiplier to directly interface with the 32-bit adder/subtractor with a minimum speed penalty given that it is selected as a source. The ALU section contains various logic functions such as AND, OR, EXCLUSIVE OR, NOT, etc. In addition, the ALU contains a completely separate 16-bit adder/subtractor.

SUMMARY OF CHIP CHARACTERISTICS

Silicon area, power dissipation, and critical speed estimates have been made for the processor chip design. They are based on data collected from previously characterized 2-micron circuits and computer-aided simulation and analysis. The chip characteristics are summarized in Figure 6.

This chip replaces approximately 100 ICs, MSIs, and LSIs and reduces power by approximately 15 watts when compared with a processor of the same architecture which is implemented with off-the-shelf devices.

TECHNOLOGY	2-MICRON BULK C-MOS
CHIP SIZE	250 x 265 MILS
OPERATING POWER	400 MILLIWATTS (5 VOLTS)
DEVICE COUNT	28,300 FETs (7100 GATES)
CLOCK RATE	20 MHz
LOGIC LEVEL	TTL COMPATIBLE
INSTRUCTION CYCLE	200 NANOSECONDS
INSTRUCTION WORD	DUAL 24-BIT WORD
DATA WORD	16-BIT
ARITHMETIC UNIT	16-BIT ALU AND 16 x 16-BIT MULTIPLIER
ON-CHIP RAM	64 16-BIT WORDS
INPUT/OUTPUT	BIDIRECTIONAL 16-BIT PARALLEL INPUT/OUTPUT, SERIAL INPUT/OUTPUT, DISCRETES, INTERRUPTS
EXTERNAL DATA MEMORY ADDRESSABILITY	65K 16-BIT WORDS
EXTERNAL PROGRAM MEMORY ADDRESSABILITY	65K 24-BIT WORDS

Figure 6. Summary of VLSI Processor Characteristics

REFERENCES

1. H. E. Tsou, "A MOS/LSI Microprocessor," a paper presented at the Microprocessor Symposium, NASA Manned Spacecraft Center (now Johnson Spacecraft Center), Houston, Texas, November 1969
2. H. E. Tsou, T. C. Berg, and S. E. Hutchins, "A High Speed LSI Processor for Voice Signal Processing," 1974 IEEE NAECON, Dayton, Ohio, May 1974
3. G. W. McIver, R. W. Miller, and T. G. O'Shaughnessy, "A Monolithic 16 x 16 Multiplier," IEEE ISSCC, Philadelphia, Pennsylvania, February 1974
4. H. E. Tsou, M. L. Kushner, and S. E. Hutchins, "A Microprogrammable Signal Processor using LSI Building Blocks," IEEE EASCON, Philadelphia, Pennsylvania, September 1976
5. H. E. Tsou, "Impact of VLSI on Voice Processing," IEEE 10th Annual Communications Theory Workshop, Orlando, Florida, April 1980
6. B. Dunbridge, B. L. Troutman, G. W. McIver, W. E. Kingsley, "2-Micron C-MOS/LSI 32-Point FFT Processor," Solid State Circuits Conference, 1982
7. J. M. Anderson, R. A. Allen, B. L. Troutman, "C-MOS/LSI 16 x 16 Multiplier/Accumulator," Solid State Circuits Conference, 1982
8. E. Greenward, et al, "A High Performance VLSI C-MOS Arithmetic Processor Chip," IEEE ICASSP 82, May 1982
9. E. R. Caudel, et al, "A Chip Set for Audio Frequency Digital Signal Processing," IEEE ICASSP 82, May 1982
10. M. Yano, et al, "An LSI Digital Signal Processor," IEEE ICASSP 82, May 1982

A CASE STUDY OF THE IMPACT OF VLSI ON COMPUTER ARCHITECTURE: THE MASSIVELY PARALLEL PROCESSOR

Paul B. Schneck

NASA/Goddard Space Flight Center
Greenbelt, MD 20771

ABSTRACT

A computer organization with a high degree of regularity can take particular advantage of the opportunities of VLSI. The Massively Parallel Processor (MPP) is a 128 x 128 array of processing elements. A custom chip, with eight processor elements, is the basis of the machine's implementation. The MPP was installed at Goddard Space Flight Center in May 1983. It was conceived and the architecture designed by NASA for use on image processing applications. The MPP's 16384 1-bit processor elements operate at a 10 Mhz rate and will perform six billion 8-bit additions per second. This paper details the architecture of the system and of the processor elements. Explicit parallel architecture, with high regularity, will enable effective use of VLSI components in high speed computers.

INTRODUCTION

The latest generation of earth-imaging satellites, typified by Landsat-4 launched July 1982, contains sensors with data rates of $8.5 \cdot 10^7$ bits per second. A single frame of Landsat-4 imagery contains over 10^8 bits. In order to process these data in a cost effective and timely fashion extremely high computational rates are required. Conventional computers contain a great deal of circuitry devoted to performing word-oriented operations (including relatively complex floating point arithmetic) which is unnecessary for satellite image processing. The Massively Parallel Processor (MPP) was designed with the following characteristics in mind:

1. image processing--one-bit wide data path; interconnected processor elements (PE) in a rectangular (subscene) array.

2. high speed--all elements of the array operate simultaneously on image data.

3. low cost--use of readily available CMOS technology at a conservative clock speed, with a custom chip containing several PEs.

Goodyear Aerospace Corporation of Akron, Ohio was selected to design and fabricate the MPP based on the NASA concept. Among the significant enhancements introduced by Goodyear were the variability of the shift register, the redundant column-pairs, and the corner-turning memory (Batcher, 1979).

The MPP system contains a 128 x 128 array, or 16384, one-bit wide processing elements which operate simultaneously. The 10 Mhz clock rate yields a peak instruction rate of $1 \cdot 6 \cdot 10^{11}$ instructions per second. The computational capability, coupled with an I/O rate of $1.2 \cdot 10^9$ bits per second allows near-real-time processing, thus decreasing operational complexity and cost. The array is managed by an Array Control Unit (ACU) and communicates through a Program and Data Management Unit (PDMU) to a host computer. The role of each subsystem is described below.

ARCHITECTURE

The architecture of the MPP array unit and its design and implementation result in a machine which can operate on 16384 operands each 100 nanosecond (ns) cycle. In order to make effective use of this processing power, the system configuration must be able to rapidly transfer data to and from the processor in an appropriate format. A corner turning memory is provided for reformatting and staging data. This section describes the architecture of the MPP array unit and of the MPP system.

The Array Unit

The MPP array unit contains 16384 PEs one-bit wide and performs instructions in lockstep at a 10 MHz rate. Few machines have been built in the single-instruction-multiple-data (SIMD) style of the MPP. Most notable is the ILLIAC-IV (Slotnick, 1968), constructed by Burroughs Corporation for the Department of Defense Advanced Research Projects Agency in the late 1960's. The ILLIAC had an 8 x 8 array of PEs with a 16 MHz clock. Each of the 64 PEs had the complexity and capability of a moderately powerful system of that time, such as the CDC 6400.

The ILLIAC was designed for numerical modelling and the solution of systems of partial differential equations and utilized a small number (64) of complex (floating point) PEs at moderate speed. The MPP is focused on image processing and employs a large number (16384) of simple one-bit wide) PEs at moderate speed.

Figure 1 depicts the organization and interconnection of the array of processor elements. Each PE can communicate with its four nearest neighbors in one cycle. This includes those PEs in the columns on the sides, which may be connected to corresponding PEs on opposite sides. The array is topologically equivalent to a cylinder. If desired the array may also be treated as a square without connections between opposite edges or even as a

CH1935-6/83/0000/0145$01.00 © 1983 IEEE

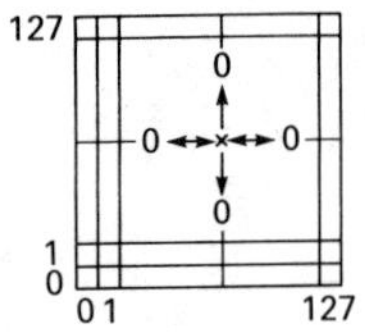

a) THE 128 × 128 ARRAY; THE PROCESSOR AT X IS CONNECTED TO ITS NEAREST NEIGHBORS HORIZONTALLY AND VERTICALLY.

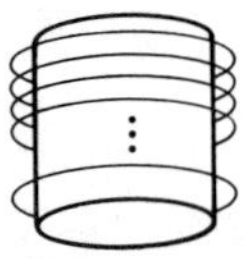

b) IN THIS CONFIGURAITON THE ARRAY ACTS AS A CYLINDER.

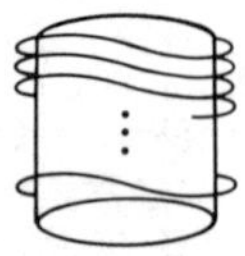

c) THE CONFIGURATION IS A 16384 ELEMENT VECTOR.

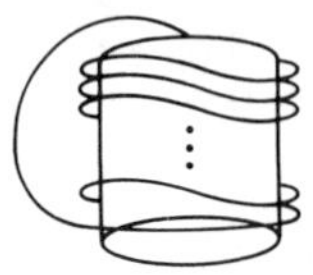

d) THIS CONFIGURATION IS A 16384 ELEMENT CYCLE.

FIGURE 1. ARRAY INTERCONNECTION MODES

1 x 16384 vector. The array performs input, output, or both input and simultaneous output by shifting an entire plane of data sideways to the right, a column at a time. Input data are accepted at the left column of the array, while output data are produced at the right column. A full column of data are transferred each 100 ns cycle, yielding an I/O rate of $1.28 \cdot 10^9$ bits/second. Computation is interrupted only when a plane of data is transferred to or from the PE I/O register.

In addition to the local connectivity already described, the PEs of each 16 x 16 subarray[1] (of which there are 64) are 'or'ed and the result transmitted to the Array Control Unit available for use in various searching algorithms.

Not all PEs are required to participate in each operation. To permit dynamic (data dependent) control of participation by each PE a <u>mask</u> register is provided within the PE. The mask register is loaded from the data bus (giving it access to all data within the PE) and allows the PE to operate on "masked operations" only when it contains a one.

Local random access of 1024 bits (with provision for increasing to 64K) is associated with each PE. Because of the machine's initial application to image processing we speak of 1024 planes of memory, considering the 1024 bits for each PE arranged with all other PE memory bits, as illustrated in Figure 2. The data are parity checked (across 8 PE memories) and a program may interrogate an error latch. Use of the parity feature is not seen as essential for many of the image processing tasks on which the MPP is to be applied. Thus, the implementation requires an ACU instruction to test the parity latch (which is not available for several cycles due to the fan-in delay as the 2048 parity bits are 'or'ed together.

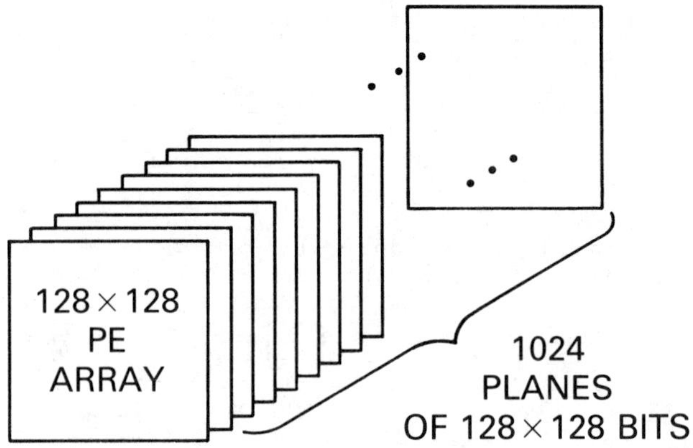

FIGURE 2. BIT-PLANE MEMORY

When an error is detected, the machine state may be saved on a file device and diagnostics loaded to determine the nature and location of the failure. Isolation to the failing PE or memory element allows maintenance personnel to replace the failed unit. To minimize the occasions when it is necessary to remove the machine from productive use for service, two additional pairs of columns of PEs are provided. The MPP is actually a 128 x 132 array of PEs.

Up to two column-pairs may be removed from use when they are found to contain a failing element(s). This reconfiguration is performed by the diagnostic software and is not visible to the user. Figure 3 illustrates this aspect of the MPP.

The ability of the system to be reconfigured to bypass failed PEs, coupled with checkpoints taken at reasonable intervals[2], permits operation to continue with minimal loss of productive time (only those computations performed since the last checkpoint are lost). Additionally, it will usually be possible to defer remedial maintenance until the next regularly scheduled maintenance period.

<u>THE PROCESSOR ELEMENT</u>

The technological challenge in development of the MPP was centered on distributing signals to each of the 16896[3] PEs. Thus, a premium was placed on minimizing the physical and electrical distance signals would be required to traverse. This resulted in placement of eight processor elements on a single CMOS custom chip. There are 2112 processor

[1]The naturally occurring set of subarrays anchored at (0,0) is used.

[2]Because of file manipulation, checkpoints are the responsibility of the user, not the system.

[3]16896 = 128 rows x 132 columns.

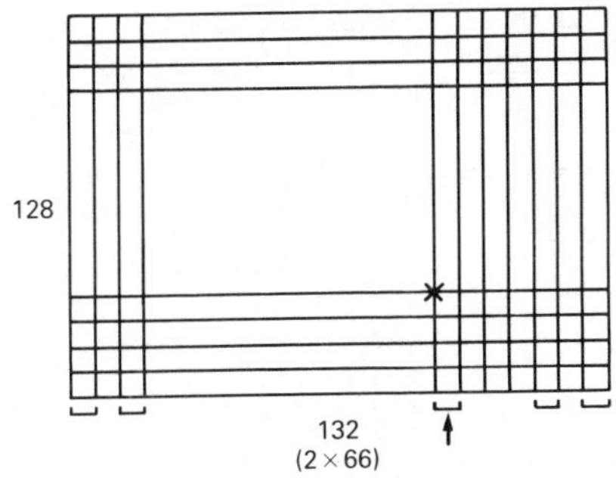

X IS A FAILING PE; THE COLUMN PAIR OF PROCESSING ELEMENTS INDICATED BY THE ARROW IS CONFIGURED OUT OF THE SYSTEM; ANOTHER COLUMN PAIR IS AVAILABLE IN CASE OF AN ADDITIONAL FAILURE. THE PROGRAMMER SEES A 128×128 MACHINE.

FIGURE 3. REDUNDANCY AND RECONFIGURATION

element chips, mounted on 88 boards. This high level of integration enabled the entire ACU to be contained within a single frame, immensely simplifying the task of distributing control signals to each PE.

The chips were produced using the HCMOS process by Solid State Scientific Corporation. Figure 4 illustrates the layout of the 10 Mhz chip.

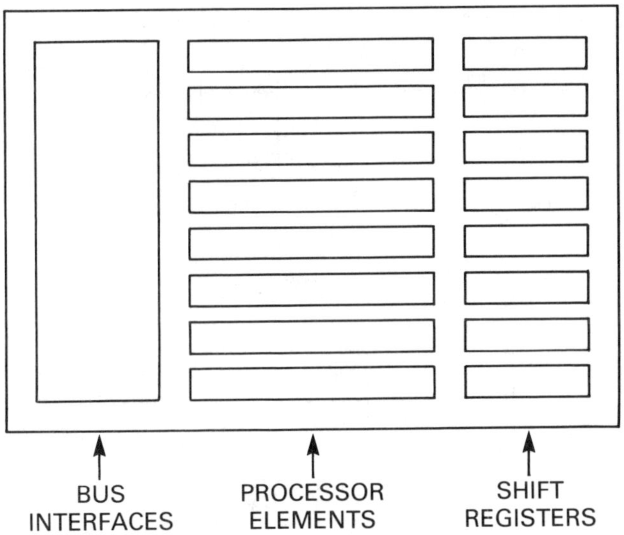

FIGURE 4. PROCESSOR ELEMENT CHIP (8PEs) LAYOUT

PE logic accounts for approximately 2/3 of the chip area while the shift registers account for the remaining 1/3. Each chip is 235 x 131 mils, with 52 pius and 8000 transistors.

Figure 5 depicts the architecture and data paths of a PE. The central component is the data bus, through which all other components communicate. The one-bit wide processing of data permits high speed execution units. The PE implementation requires approximately 350 gates.

The logic unit can compute any of the 16 functions of two binary variables. One variable is obtained from the "P" register, the other is obtained from the data bus (allowing access to all PE components), the result replaces the value in the "P" register. The "P" register may also be loaded from the data bus or any one of the four nearest neighbor PEs.

The (full) adder produces the sum and carry of two inputs (plus the "C" register, used to propogate a lower order digit carry). The inputs are the "A" register, which may be loaded from the data bus or the shift register, and the "P" register discussed above. The carry output replaces the value in the "C" register and the sum output replaces the value in the "B" register from which it can either be placed on the data bus or sent to the shift register.

The shift register provides a variable length storage element for use in iterative operations such as multiplication, division, and square root. The effective length of the path provided by the shift register may be 2, 6, 10, 14, 18, 22, 26, or 30 bits.

The mask register is used to control the participation of an individual PE on instruction execution. When masked-mode operations are issued, only those PEs participate whose mask registers contain a one-bit.

The I/O, or S, register of each PE is connected to the I/O registers of the PEs on its left and right. Whenever an I/O operation is specified, the contents of the rightmost column of I/O registers is made available at the 128-bit wide output register column. Simultaneously each I/O register transfers its datum to the right, and the leftmost column of I/O registers accepts data from the 128-bit wide input register column. I/O operations take place in parallel with computation under microprogrammed control. In 128 cycles, the entire array of 16384 data elements can be output or input or both operations may be accomplished in the same period.

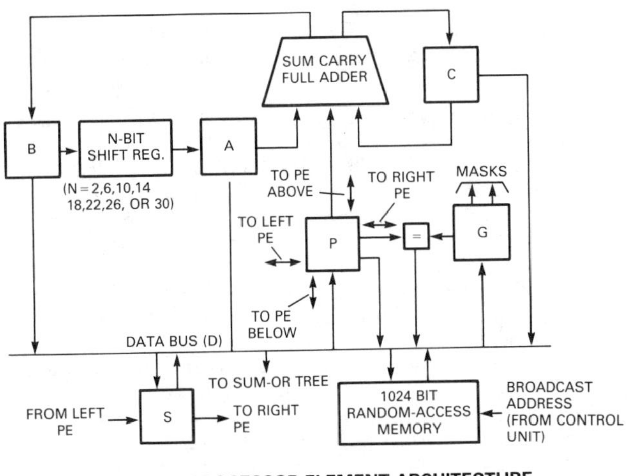

FIGURE 5. PROCESSOR ELEMENT ARCHITECTURE

The Array Control Unit

The high speed of the MPP is obtained within the array unit; the array control unit (ACU) is responsible for instruction issue to the array, and for execution of all non-parallel aspects of a program: scalar arithmetic, loop control, address calculation, and I/O. The ACU operates on a 100 Mhz clock, implemented in ECL. The three subunits of the array control unit are: the PE control unit, a microcoded "subroutine library" of array sequences; the I/O control unit, responsible for data transfer to and from the array memory and the corner-turning memory; and the main control unit (MCU), with its own execution capabilities which issues PE array and I/O requests. As noted above, the array control unit receives the status of various special conditions within the array, e.g., the parity failure latch. One of the key programming challenges for users of the MPP will concern program branching on the basis of the restricted data available to the MCU.

The Program and Data Management Unit

The program and data management unit (PDMU) may be thought of as the front-end processor for the user/host-computer. In this role the PDMU sets up jobs for execution on the MPP, transfers data to the corner-turning memory, loads programs into the ACU, and performs other job management and executive functions (e.g., debugging support).

The Corner-Turning Memory

A major use of the MPP will be the processing of data returned from imaging satellites such as Landsat-4. Such a data stream is organized according to the mechanical and electrical requirements of the imaging instrument and may contain interleaved data from different sensor elements, bands, and bit positions. The corner-turning memory provides a convenient and high speed mechanism for reformatting such data into planes of picture elements.

Three subsystems make up the corner-turning memory: a main memory unit, an input staging unit, and an output staging unit. The main memory is implemented in MOS RAM, is word organized, and interleaved for high speed. The input and output staging units have mirror image functional symmetry--the input staging unit operates on data from the input port prior to storage in the main memory, while the output staging unit operates on data between the main memory and the output port.

Figure 6 illustrates the corner-turning memory and its application in converting Landsat data from a "pixel interleaved by band" format where each picture element (pixel) contains data from four bands to bit plane format ready for processing. Because this process can be handled "on-the-fly" the MPP is capable of real time processing of satellite data streams.

PERFORMANCE

The MPP is a bit serial machine, performing most rapidly on short, integer operands, as are found in satellite imagery. Because of its parallelism it also performs well on arrays of floating point numbers. The following table indicates peak speeds when all PEs are operating on an array of data:

Operations	MPP Execution Speed (MIPS)
Addition of Arrays	
8-bit integers (9-bit sum)	6553
32-bit integers (33-bit sum)	1689
32-bit floating-point numbers	470
Multiplication of Arrays (Element-by-Element)	
8-bit integers (16-bit product)	1861
32-bit integers (64-bit product)	146
32-bit floating-point numbers	291
Multiplication of Array by Scalar	
8-bit integers (16-bit product)	2824
32-bit integers (64-bit product)	280
32-bit floating-point numbers	373
Multiplication/Addition	
8-bit integers	1450
32-bit integers	134
32-bit floating-point numbers	180

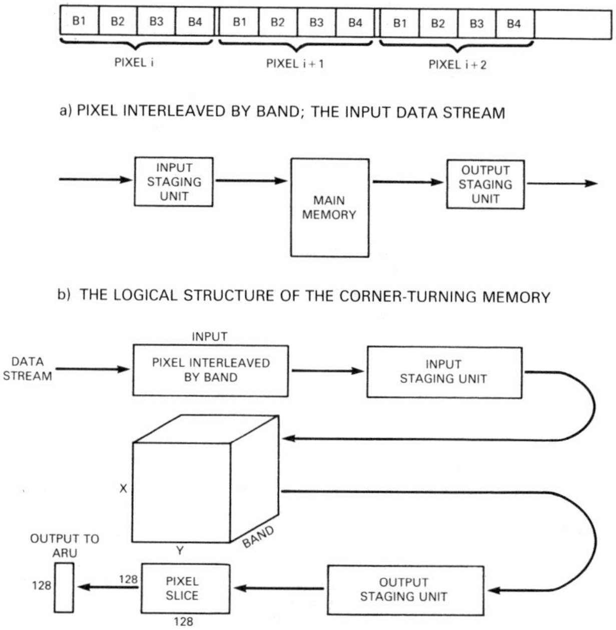

FIGURE 6 THE CORNER TURNING MEMORY

REFERENCES

Batcher, K. E., "MPP-A Massively Parallel Processor," Proc. 1979 International Conference on Parallel Processing, IEEE.

Slotnick, D. L., et al, "The ILLIAC-IV Computer," IEEE Transactions on Computers, VC-17, No. 8, August 1968, pp. 746-757.

LOCAL MEMORY IN RISCS

Robert W. Sherburne Jr., Manolis G.H. Katevenis,
David A. Patterson, and Carlo H. Séquin

Computer Science Division
Department of Electrical Engineering and Computer Sciences,
University of California, Berkeley, CA 94720

ABSTRACT

Performance of the RISC II microprocessor is investigated as a function of local memory size. A larger local memory effectively reduces data I/O traffic arising from procedure calls and returns. Datapath bandwidth, however, is also reduced due to the increased register cycle time. A conflict then exists between the desire for maximum datapath bandwidth, and the need to reduce data I/O overhead.

Since the local memory occupies a significant portion of the chip, it is important to ensure effective usage of limited silicon resources. Programming environments which include many nested procedures benefit significantly from a multiple-bank local memory scheme. On the other hand, those with few procedures may suffer from the increased register cycle time. In addition to the programming environment, the register-bank swapping strategy, bank overflow interrupt overhead, and swap I/O support affect optimal memory size.

Introduction

Traditionally, decisions in computer design have been based in two different camps: architecture and implementation. Architecture concerns the design as it is reflected in the instruction set: instruction and data word sizes, data types, and addressing modes. Implementation concerns the microarchitecture: system partitioning, placement, communication and timing; and, at a lower level, circuit design and device technology. Low levels of integration have permitted a wide semantic gap to exist between the architect and the chip designer: the limited repertoire of SSI/MSI chips offered little input for architectural influences.

Enter LSI and VLSI, and we suddenly find ourselves confronted with the prospect of integrating an entire system on a few chips. Each chip can now take on architectural personality. Thus, the architecture-implementation semantic gap is reduced. What this means is that more direct interdependencies exist between the two levels. The microarchitect must now evaluate architectural decisions in the context of limited chip resources.

Performance of a given architecture is limited by memory I/O and datapath cycle times. From the architect's point of view performance is I/O limited, and datapath delay can be neglected. Conversely, the designer considers the datapath bandwidth to be the performance limit, with memory I/O neglectable. Thus, two solutions may exist for design of a particular system. The goal of this paper is to examine these solutions in the context of the local memory scheme used in RISC II.

Ideal Performance of RISC II

The RISC II is the second in a series of 32-bit, NMOS microprocessors developed at U.C. Berkeley [1]. It runs at 500ns per cycle, using an 8Mhz clock rate, and was fully functional in its first silicon run. The RISC instruction set consists solely of single register-to-register operations [2]. This simple and regular implementation reduces control complexity, chip area, and design time while supporting pipelined execution [3]. The simple RISC instruction set is a better match to highly optimizing compilers than complex instruction sets [4]. Such optimization can also reduce dependency overhead inherent in pipelined implementations [5], allowing more effective use of available datapath bandwidth.

A drawback of such an instruction set is the high I/O bandwidth required for external memory. In order to reduce this penalty, the RISC microarchitecture includes support for subroutine call and return, one of the most time-consuming operations in typical high-level language programs [6]. Register saves and restores are reduced by employing a multiple-bank register file, or local memory. The banks are organized as stack levels, so that register save or restore need be performed only during stack overflows or underflows.

RISC II includes eight register banks, or windows, one of which is reserved for interrupt processing. At any one time there are ten registers local to the present procedure level. Additionally, there are six each high and low registers which overlap adjacent procedure levels; these are used primarily for parameter passing. Each window swap (for save or restore) involves sixteen registers: the ten locals and one set of overlaps. Ten global registers are accessible from any procedure level, forming a total of 32 addressable registers.

Table I shows the relative performance of two C programs as the number of windows on the chip varies. These results are based on earlier studies of procedure behavior and register file management overhead for RISC[7,8]. Both benchmarks, Tower and Puzzle, nest to a depth of twenty. However, Tower has significantly more procedure calls (19% versus 0.7% for Puzzle), and makes much more intensive use of the multiple windows. Performance in such a programming environment improves considerably through the use of, say, seven windows. Puzzle, on the other hand, performs well with only two windows.

WINDOWS	1	2	3	5	7	9	∞
TOWER 19% Dynamic Call & Return	7.08	3.02	2.52	1.38	1.10	1.02	1.00
PUZZLE 0.7% Dynamic Call & Return	1.17	1.02	1.00	1.00	1.00	1.00	1.00

TABLE I: Normalized RISC II Execution Time
(relative to case of infinite windows)

Typical programs have a procedure call or return every twenty instructions, so the benchmarks shown here represent extremes [9]. In consideration of limited chip resources, it may be attractive to tailor the local memory size to specific environments. This allows resources to be freed for performance improvement in other areas.

Cost of Fixed-Size Window Swaps

The cost of register window overflow is determined by two factors: the overhead of servicing an interrupt, and the data I/O bandwidth. The RISC II microprocessor incurs a penalty of about thirty instructions for the window overflow/underflow interrupt routine, in addition to the register swap cost. The single I/O bus implementation supports one I/O operation per cycle, meaning that each Load or Store takes two cycles. Although each window swap is costly, there is a sufficient number of windows on chip so that the swaps are few. Reducing the local memory size degrades performance significantly due to this high cost of swapping. With fewer windows, better swap interrupt and I/O support is important.

At compile time, the dynamic procedure profile is not known. Therefore the compiler cannot anticipate window overflows. For this reason, overflows must be detected on chip. It is the cost of handling this interrupt which accounts for thirty RISC II instructions. The compiler can, however, anticipate swaps for a single window implementation: every executed call or return would require a save or restore, respectively.

Since each swap utilizes the same protocol (sixteen adjacent registers swapped to/from the current window) better data I/O support can be easily obtained. For example, a single instruction may provide all the necessary information for the swap. Then it is not necessary to fetch individual Load or Store instructions for each of the registers. Two data words may be passed on the bus each machine cycle, instead of one data word every two cycles interleaved with instruction fetching.

Table II presents RISC II performance as a function of data I/O support and local memory size. The cost of each swap includes the thirty cycles for interrupt overhead, as well as the sixteen data word transfers. Since swap overhead for Puzzle is small, only Tower is considered here. As before, seven or more windows are desirable for high performance, regardless of the level of data I/O support. This is because the interrupt overhead is so high. An exception is the single window case, which is seen to provide best performance for implementations with fewer than five windows.

WINDOWS	1	2	3	5	7	9	∞
One-Half Data I/O Per Cycle	7.08	4.91	3.95	1.74	1.19	1.05	1.00
Single Data I/O Per Cycle	4.04	3.90	3.19	1.55	1.14	1.03	1.00
Dual Data I/O Per Cycle	2.52	3.39	2.81	1.46	1.12	1.03	1.00

TABLE II: Tower Execution Time with Varying Swap Support
(includes interrupt overhead for multiple window cases)

Improved Swapping Strategies

Thus far, we have discussed fixed-size swap overhead of RISC II. This scheme is attractive due to its simplicity, as well as its amenability to better I/O support. However, such a scheme swaps all registers in the window, whether they were used or not. A study of several C programs has determined that a mean of four registers are used per procedure in RISC [7]. Therefore, the fixed-swap scheme performs four times the number of save and restore data I/O actually necessary.

In order to keep track of which registers have been used, a "dirty bit" scheme may be implemented. During each register write, a bit is set to indicate a potential swap candidate. Swaps would vary in length, depending on the number of bits set. The increased hardware complexity necessary to support such an approach, however, is undesirable.

A single window implementation does not require this hardware. The compiler can insert code before each call which saves utilized registers. Restoring registers after a return can be done on demand. Since not all registers may need to be restored, further reduction in I/O is anticipated. Save overhead may be eliminated by performing a data memory write in parallel with all register file writes. This requires a dual-bus microarchitecture which can perform an instruction and data access each cycle, such as the MIPS microprocessor [10]. Overall swap overhead may then reduce by more than a factor of eight.

Although we have described alternative strategies for local memory management, we have not yet addressed effective use of on-chip memory area. These alternatives reduce off-chip register save space by a factor of four, but on-chip memory still attains only 25% utilization. A variable size window scheme addresses this issue by allowing dynamic allocation of window size. Several register banks can be implemented for efficient replacement strategy, and procedures may span bank boundaries. This scheme requires additional hardware, in the form of pointers for each procedure domain, and an adder to calculate the physical addresses of the registers.

Performance comparison of these schemes is presented in Table III. All cases assume single data I/O per cycle support and four registers per procedure. Interrupt overhead is included in all multiple window, as well as the variable size, implementations. With few windows, significant performance improvement is observed. Using more efficient swapping strategies, high performance may be attained with fewer windows.

WINDOWS	1	2	3	5	7	9	∞
Full-Bank Register Swaps	4.04	3.90	3.19	1.55	1.14	1.03	1.00
Partial Register Swaps	1.76	2.16	1.62	1.41	1.10	1.02	1.00
Partial Swaps with Write Save	1.38	1.58	1.31	1.21	1.05	1.01	1.00
Variable Size with Half Bank Swaps	3.91	1.60	1.25	-	-	-	1.00

TABLE III: Tower Execution Time for Various Swap Schemes
(one data I/O per cycle assumed for all cases)

Register File Delay

Up to this point, only I/O limited performance has been discussed. From the designer's point of view, attention should also be focused on datapath bandwidth. Especially for RISCs, where each execution cycle consists of a standard register-to-register operation, the datapath cycle time determines maximum system performance.

The machine cycle of the RISC II consists of a dual-port register read, followed by an ALU or shift operation; these latter operations overlap register write and bitline precharge of the previous instruction. The machine cycle is then limited directly by the register file read-write-precharge cycle time.

Increased local memory size is accompanied by proportionally greater bitline loading. To reduce the bitline discharge delay, the wordline transistors may be widened at some cost in addressing delay. Optimal wordline transistor size yields a register file cycle delay which goes as the square root of memory size [11]. Datapath bandwidth can then be traded off for reduced I/O with a larger local memory.

Table IV includes the effect of variable register cycle delay. The partial register swap schemes using a single window yield the best performance. A single window, fixed-swap implementation with dual data I/O per cycle approaches the performance of the seven window RISC II with half data I/O per cycle. Execution time for Puzzle, with little swap overhead, follows the register cycle time dependence with memory size; it executes nearly twice as fast with one window as it does with seven.

WINDOWS		1	2	3	5	7	9
Full-Bank Register Swaps with Varying Data I/O per cycle	$\frac{1}{2}$	7.08	6.00	5.57	3.01	2.38	2.35
	1	4.04	4.76	4.50	2.68	2.28	2.31
	2	2.52	4.14	3.96	2.53	2.24	2.31
Partial Register Swaps		1.76	2.64	2.29	2.44	2.20	2.28
Partial Swap with Write Save		1.38	1.93	1.85	2.09	2.10	2.26
Variable Size with Half Bank Swaps		3.91	1.95	1.76	-	-	-
Normalized RISC II Register Cycle Time (ignoring swaps)		1.00	1.22	1.41	1.73	2.00	2.24

TABLE IV: Datapath Bandwidth Limited Execution Time
(smaller local memory is faster; using Tower benchmark)

Conclusions

Chip design tradeoffs in VLSI must be made using both architectural and circuit design considerations. One measure of local memory cost, delay, yields two minimum execution time scenarios: I/O limited, and datapath bandwidth limited. Because of the limited number of pads which can be placed on a chip, memory I/O is a severe bottleneck in system performance. For this reason, a large local memory was chosen for RISC II. Presently, memory speed is increasing, making datapath bandwidth a more critical determinant of system performance. In the future, increased chip resources will support greater local memory hierarchy [12]; the I/O bottleneck may then be replaced by datapath limited performance.

References

[1] M. Katevenis, R. Sherburne, D. Patterson, C. Séquin: "The RISC II Micro-Architecture," presented at the VLSI 83 Conference, Trondheim, Norway, August 1983.

[2] D. Patterson & C. Séquin, "A VLSI RISC," IEEE Computer, vol. 15, no. 9, pp. 8-21, September 1982.

[3] J. D. Wright: "Relation of Microcode to Future Machine Design," COMPCON Digest of Papers, March 1983, pp. 104-106.

[4] G. Radin, "The 801 Minicomputer," Proc. ACM SIGARCH, pp. 39-47, March 1982.

[5] T. Gross: "Code Optimization Techniques for Pipelined Architectures," COMPCON Digest of Papers, March 1983, pp. 278-285.

[6] D. Patterson & C. Séquin, "RISC I: A Reduced Instruction Set VLSI Computer," Proc. ACM SIGARCH, pp. 443-457, May 1981.

[7] D. Halbert & P. Kessler, "Windows of Overlapping Register Frames," CS 292R Final Reports, UC Berkeley, Spring 1980.

[8] Y. Tamir & C. Séquin: "Strategies for Managing the Register File in RISC," to be published, IEEE Trans. on Computers, 1983.

[9] D. Ditzel & H. McLellan: "Register Allocation for Free: The C Machine Stack Cache," Proc. ACM SIGARCH, pp. 48-56, March 1982.

[10]J. Hennessy, N. Jouppi, S. Przybylski, C. Rowen, T. Gross: "Design of a High Performance VLSI Processor," Third Caltech Conference on Very Large Scale Integration, Computer Science Press (Rockville, 1983), pp. 33-54.

[11]R. Sherburne, "Processor Design Tradeoffs in VLSI," Ph.D. Dissertation (in preparation), UC Berkeley.

[12]D. Patterson & C. Séquin: "Design Considerations for Single-Chip Computers of the Future," IEEE Journal of Solid-State Circuits, vol. SC-15, no. 1, February 1980, pp. 44-52.

Performance Issues in VLSI Processor Design

John Hennessy, Norman Jouppi, Steven Przybylski,
Christopher Rowen, and Thomas Gross

Computer Systems Laboratory
Departments of Electrical Engineering and Computer Science
Stanford University

Abstract

MIPS is a new single chip VLSI microprocessor that achieves high performance with the use of a streamlined instruction set. The software system for MIPS helps obtain a high effective instruction rate through the pipeline, while the streamlined pipeline structure allows a high clock speed. Performance limitations come from a variety of factors including off-chip communication requirements, internal bus communication delays, and the need for the processor to correctly respond to asynchronous events.

Introduction

MIPS (Microprocessor without Interlocked Pipe Stages) is a new general purpose microprocessor architecture designed to be implemented on a single VLSI chip[1]. The main goal of the design is high performance in the execution of compiled code. The basic philosophy of MIPS is to present an instruction set that is a compiler-driven encoding of the microengine. Thus, little or no decoding is needed and the instructions can be dispatched quickly (in one stage). The processor is pipelined but provides no pipeline interlock hardware; this function must be provided by software.

The MIPS architecture presents the user with a fast, streamlined machine with a simple instruction set. This approach has been used by the IBM 801 project[2] and the RISC project at Berkeley[3]; it is directly opposed to the approach taken by architectures such as the VAX.

MIPS is designed for high performance. To allow the user to get maximum performance, the complexity of individual instructions is minimized. This allows the execution of these instructions at significantly higher speeds. To take advantage of simpler hardware and an instruction set that easily maps to the microinstruction set, additional compiler support is needed. The shifting of the complexity from the hardware to the software has major advantages[4].

The design of a high performance VLSI processor is dramatically affected by the technology. Among the most important design considerations are: the effect of pin limitations, available silicon area, and size/speed tradeoffs. Pin limitations force the careful design of a scheme for multiplexing the available pins, especially when data and instruction fetches are overlapped. Area limitations and the speed of off-chip intercommunication require choices between on- and off-chip functions. Choosing what functions are migrated off-chip must be done carefully so that the performance effects of the partitioning are minimized.

The MIPS project has been supported by the Defense Advanced Research Projects Agency under contract # MDA903-79-C-0680. Norman Jouppi is supported by a NSF Graduate Fellowship. Steven Przybylski is supported by a Natural Sciences and Engineering Research Council of Canada Scholarship. Thomas Gross is supported by an IBM Graduate Fellowship.

Systems issues

Any attempt to build a high-performance processor is fruitless if the resulting processor is unable to support real systems. Therefore, systems issues have to be given adequate attention right from the start. The key systems issues are the memory system and the handling of asynchronous events, for example internal traps and external interrupts.

The design for the MIPS memory system is straightforward and includes an instruction cache. The MIPS architecture supports asynchronous events correctly; because the MIPS instruction set is not interpreted by a microengine (with its own state), hardware support for page faults, external interrupts, and internal traps (for example, arithmetic overflow) is significantly simplified.

To handle interrupts and page faults correctly, two important properties are required. First, the architecture must ensure correct shutdown of the pipe, without executing any faulted instructions. Most present microprocessors can not perform this function correctly (e.g. Motorola 68000, Zilog Z8000, and the Intel 8086). The MIPS pipeline has been designed carefully to accomplish this goal: Instructions that can page fault do not write to any storage, either registers or memory, before the fault is detected. The occurrence of a page fault need only turn off writes generated by this and any instructions following it which are already in the pipe. Second, the processor must be able to correctly restore the pipe and continue execution as if the interrupt or fault had not occurred. The MIPS multistage program counter tracks the last three instructions and aids in correctly restarting them after a page fault or interrupt.

Architectural issues

The fastest execution of a task on a given processor would be one in which all hardware resources of the processor were used at a 100% duty cycle performing a nonredundant and algorithmically efficient encoding of the task. The MIPS architecture attempts to achieve this goal. The user instruction set is an encoding of the microengine that makes a maximum amount of the microengine available; the pipeline structure is designed to execute a pipeline stage every machine cycle and to maximally utilize the resources. The goal to sustain 100% resource utilization motivated many of the design decisions found in the architecture.

CH1935-6/83/0000/0153$01.00 © 1983 IEEE

The major pipestages and their tasks are shown in Figure 1. The MIPS five stage pipeline contains three active instructions at any time; either the odd or even pipestages are active. Figure 2 shows the flow of instructions through the pipeline.

Stage	Mnemonic	Task
Instruction Fetch	IF	Send out the PC, increment it
Instruction Decode	ID	Decode instruction
Operand Decode	OD	Compute effective address and send to memory if load or store, use ALU
Operand Store/ Execution	OS/ EX	Store: write operand/ Execution: use ALU
Operand Fetch	OF	Load: read operand

Figure 1: Major pipestages and their functions

MIPS is a load/store architecture, that is data may be operated on only when it is in a register and only load/store instructions access memory. If data operands are used repeatedly in a basic block of code, having them in registers will prevent redundant load/stores and redundant addressing calculations; this allows higher throughput since more operations directly related to the computation can be performed. For a load or store instruction, the ALU is used in the third pipestage to compute the effective address. Another ALU operation available in the fourth stage of every instruction can be used for a (possibly unrelated) computation.

The major benefit derived from the load/store architecture is simplicity of the pipeline structure. The pipeline has a fixed number of pipestages, each of the same length. Also, the absence of synchronization between stages of the pipe increases the performance of the pipeline, simplifies the hardware, and eases the handling of both interrupts and page faults. Lastly, a load/store architecture presents the compiler with a natural problem decomposition. First, get the operands, then manipulate them.

Although MIPS is a pipelined processor it does not have hardware pipeline interlocks on any synchronous processor resources. Interlocks that are required because of dependencies brought out by pipelining are *not* provided by the hardware. Instead, these interlocks must be statically provided where they are needed by a pipeline reorganizer either by rearranging instructions or by inserting no-ops. This has two major benefits:

1. A more regular and faster hardware implementation is possible since it does not have the usual complexity associated with a pipelined machine[5, 6]. The lack of such hardware is especially important for VLSI implementations, where regularity and simplicity is crucial.
2. Rearranging operations at compile time is better than delaying them at run time.

In cases where rearrangement of instructions is not possible, no-ops must be inserted into the code. This does not slow down execution compared to a similar machine with hardware interlocks, but does increase code size.

The instruction set

MIPS has one instruction size (32 bits), and all instructions execute in the same amount of time (one data memory cycle). This choice simplifies the construction of code generators for the architecture (by eliminating many nonobvious code sequences for different functions) and makes the construction of a synchronous regular pipeline much easier. Additionally, the fact that each macroinstruction is a single microinstruction of fixed length and execution time means that a minimum amount of internal state is needed in the processor.

Multiple simple (and possibly unrelated) instruction pieces are packed together into an instruction word. The basic instruction pieces are:

1. ALU pieces - these instructions are all register/register (2 and 3 operand formats). They all use less than 1/2 of an instruction word.
2. Load/store pieces - these instructions load and store memory operands. They use between 16 and 32 bits of an instruction word. When a load/store instruction is less than 32 bits, it may be packaged with an ALU instruction, which is executed during the SX stage of the pipeline.
3. Control flow pieces - these include direct jumps and compare instructions with relative jumps. All branch instructions are delayed branches of length one or two; that is, the next sequential instruction is always executed[7].
4. Other operations - this group includes procedure and interrupt linkage instruction pieces.

Software support

Software support is an important part of the MIPS architecture becuase it is responsible for making the low-level microarchitecture into a usable and comprehensible instruction set. Since the exact details of pipeline interlocks and branch delays may change between implementations, the architecture is actually defined by the input to the pipeline reorganizer.

The major novel component of the MIPS software system is the pipeline reorganizer. The input to a pipeline reorganizer is a sequence of simple MIPS instructions or instruction pieces generated without taking the pipeline interlocks and instruction packing features into account. This relieves the compiler from the task of dealing with the restrictions that are imposed by the pipeline constraints on legal code sequences. The reorganizer statically schedules the instructions to make maximum use of the pipeline while enforcing the pipeline interlocks in the code[8]. It also packs the instruction pieces to maximize use of each instruction word. Lastly, the pipeline reorganizer handles the effect of branch delays[7].

Resource utilization

Figure 2 shows the statically assigned utilization for the major resources within the MIPS pipeline structure. To achieve 100% utilization, primitive operations in the microengine (that is, instruction pieces) must be completely packed into instruction words. This is not possible for three reasons:

1. A horizontal encoding that preserved all the parallelism (that is the microcontrol word itself) would be too large. This is not a serious problem since many of the possible instructions are not useful.
2. Data dependencies can prevent full usage of the micro-

engine.

3. The effect of some delayed branches cannot be optimized.
4. The encoding of the microengine presented in the instruction set sacrifices some functional specification for immediate data. In this case the memory interface and ALU can not be used during the pipe stage for which they are dedicated.

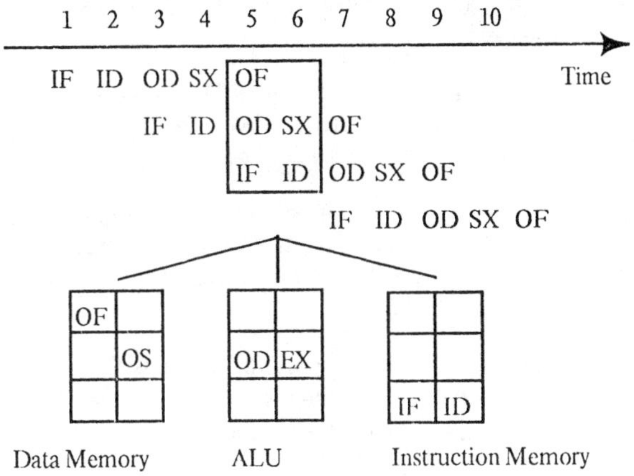

Figure 2: Resource utilization of the MIPS pipeline

Nevertheless, many instructions fully utilize the major resources of the machine. Other instructions, such as load immediate, which use few of the resources of the machine, would mandate greatly increased control complexity if overlap with surrounding instructions was attempted in an irregular fashion.

Table 1 summarizes preliminary data on resource utilization for the instruction set and reorganizer; the benchmarks are a set of Pascal programs. The table gives the percentage of wasted instructions. An instruction cycle is wasted if the instruction contains only no-ops or if the branch optimizer predicted the outcome of the branch incorrectly. Although some instructions are no-ops, other instructions contain two instruction pieces and thereby allow the software system to compensate for the no-ops.

	Static	Dynamic
No-ops due to:		
Data dependencies	0.4 %	0.2 %
Wasted branch delay	7.1 %	10.2 %
Total	7.5 %	10.4 %

Table 1: Wasted cycles

Implementation issues

Many different implementation issues are crucial to the performance of a processor. These issues affect the cycle time of the design. Of course, architectural issues determine to a large extent the feasibility of a high performance implementation.

The first MIPS implementation had a clock cycle goal of 250ns. Based on analysis performed using TV[9, 10], we expect a cycle time of 500ns for our first implementation. Several factors contributed to our failure to achieve our design goal; these are discussed in the following sections.

Micromachine organization

In each clock cycle MIPS reads two registers, performs an ALU operation on them, and writes back the result. This organization was chosen because we expected the add time to dominate the cycle. An alternative organization that delays the write back of the result until the next cycle, with bypassing as needed, could have sped up the processor cycle time by up to 30% by reducing the length of the phase 1 critical path[11]. This organization was not investigated until after the layout of the current version was almost complete. It would have required a total redesign of the processor, including a three bus instead of two bus data path and a different control structure.

Short critical paths

Short critical paths are a distinguishing feature of most high performance machines; a good example is the CRAY-1[12]. Great care was taken to make all the paths in the CRAY-1 the same length: 8 gate delays[13]. In contrast, the critical path in MIPS is 47 nodes long on phase 1 and 22 nodes long on phase 2. A direct comparison of these numbers is not appropriate since several nodes may make up a gate in the nMOS case and because the CRAY-1 takes three cycles to perform an addition. A CRAY-1 instruction can be issued every cycle, however, which mitigates the three cycle latency if reasonable amounts of local parallelism are present. Even with these caveats, MIPS has around seven times more gate delays per cycle than the CRAY-1.

The MIPS critical path could be shortened in several ways. For example, the use of a carry-bypass adder instead of a carry-lookahead for 32 bit word lengths could shorten the critical path by approximately 10%. Another way to reduce the MIPS critical path would be to try to move some control computations from phase 2 to phase 1 in parallel with ALU operations. This would cut down on the required length of phase 2. To facilitate a good partition of the computation between phase 1 and phase 2, it would be necessary to perform timing analysis on the processor at the schematics level before layout was begun.

Timing analysis at the schematics level

Timing analysis at the schematics level is not stressed within the the Mead-Conway[14] design methodology. Our experience with designing a VLSI chip, however, is that timing analysis at the schematics level is *essential* for a high performance design. Several difficult timing problems were not found until the layout was complete. To resolve these in the best possible manner, a significant redesign of the chip would be required. If we would have been able to analyze the design at the schematics level before layout, we would have saved ourselves the tedious task of relayout or the compromise of settling for reduced performance.

The best example of this involves the MIPS data path busses. The original circuit design is shown in Figure 3. To reduce the time required to transfer high values, it is important that the bus reaches a high value before the start of each clock phase. If the duty cycle of the clocks is small, the pull-ups on the bus can pull it high before the next clock phase starts. Our performance estimates at the beginning of the project relied on this property of the clocks. What we did not find out until the end of the design was that the worst critical paths in the machine occurred between functional units and were latched by the same phase they were enabled. This requires that one of the clocks be high for most of the cycle, vastly degrading bus performance. The use of TV on a schematics level representation of the chip, even if not completely accurate, would have prevented us from making this mistake.

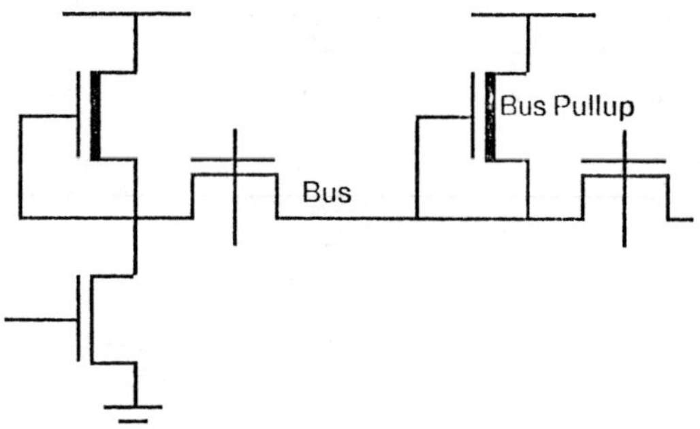

Figure 3: MIPS Bus Structure

Once we discovered the bus timing problem above, we tried using different circuits to speed up the bus transfer time. We replaced the bus pull-ups with a conventional precharged bus. Since the bus is used on both clock phases, this necessitates the addition of an extra precharge clock phase before every clock phase. This reduces the bus transfer time from 90ns to 70ns. After we submitted the design for fabrication we learned of another bus circuit which would only require 20ns for the bus transfer. This showed us the necessity of using sophisticated circuit designs in the critical paths of the chip.

Concluding remarks

The MIPS microprocessor is an example of a careful tradeoff between system support, compiler technology, and hardware implementation issues. The goal of high performance without loss of functionality lead to a simple pipeline structure that makes the correct handling of asynchronous events possible. The simple pipeline executes at high speed but needs additional compiler support to operate efficiently. Our data indicate that MIPS achieves high performance by sustaining high resource utilization and a fast processor cycle. Nevertheless, with timing analysis at the schematics level and better circuit designs, we could have improved the performance of the first implementation by a factor of two.

Acknowledgments

Many other people have contributed to the success of the MIPS project, especially John Gill, Forest Baskett, and Judson Leonard.

References

1. Hennessy, J.L., Jouppi, N., Baskett, F., and Gill, J., "MIPS: A VLSI Processor Architecture," *Proc. CMU Conference on VLSI Systems and Computations,* Computer Science Press, October 1981, pp. 337-346.

2. Radin, G., "The 801 Minicomputer," *Proc. SIGARCH/SIGPLAN Symposium on Architectural Support for Programming Languages and Operating Systems,* ACM, Palo Alto, March 1982, pp. 39 - 47.

3. Patterson, D.A. and Sequin, C.H., "A VLSI RISC," *Computer,* Vol. 15, No. 9, September 1982, pp. 8-22.

4. Hennessy, J.L., Jouppi, N., Baskett, F., Gross, T.R., and Gill, J., "Hardware/Software Tradeoffs for Increased Performance," *Proc. SIGARCH/SIGPLAN Symposium on Architectural Support for Programming Languages and Operating Systems,* ACM, Palo Alto, March 1982, pp. 2 - 11.

5. Lampson, B.W., McDaniel, G.A. and S.M. Ornstein, "An Instruction Fetch Unit for a High Performance Personal Computer," Tech. report CSL-81-1, Xerox PARC, January 1981.

6. Widdoes, L.C., "The S-1 Project: Developing high performance digital computers," *Proc. Compcon,* IEEE, San Francisco, February 1980, pp. 282 - 291.

7. Gross, T.R. and Hennessy, J.L., "Optimizing Delayed Branches," *Proceedings: The 15th Annual Microprogramming Workshop Micro 15,* IEEE, October 1982, pp. 114-120.

8. Hennessy, J.L. and Gross, T.R., "Postpass Code Optimization of Pipeline Constraints," *ACM Trans. on Programming Languages and Systems,* Vol. 5, No. 3, July 1983,.

9. Jouppi, N., "TV: An nMOS Timing Analyzer," *Proceedings 3rd CalTech Conference on VLSI,* Randal Bryant, ed., California Institute of Technology, March 1983, pp. 71-85.

10. Jouppi, N., "Timing Analysis for nNMOS VLSI," *Proc. 20th Design Automation Conference,* IEEE, June 1983, pp. 411-418.

11. Hennessy, J., Jouppi, N., Przybylski, S., Rowen, C. and Gross, T., "Design of a High Performance VLSI Processor," *Proc. of the Third Caltech Conf. on VLSI,* Calif Institute of Technology, Pasadena, Ca., March 1983, pp. 33-54.

12. Russell, R. M., "The CRAY-1 Computer System," *Comm. ACM,* Vol. 21, No. 1, January 1978, pp. 63 - 72.

13. Kolodzey, J.S., "CRAY-1 Computer Technology," *IEEE Trans. on Components, Hybrids, and Manufacturing Technology,* Vol. 4, No. 2, June 1981, pp. 181-186.

14. Mead, C. and Conway, L., *Introduction to VLSI Systems,* Addison-Wesley, Menlo Park, Ca., 1980.

Session 1-3a
Regular Arrays

Chairman

W. Luciw
Sperry Univac

DESIGN OF REGULAR ARRAYS USING CMOS IN PPL

Kent F. Smith
Department of Computer Science
University of Utah
Salt Lake City, Utah 84112

Abstract

The design of regular arrays using a methodology known as Path Programmable Logic (PPL) with a single layer metal, P well, CMOS process is discussed. PPL cell sets for both static and dynamic circuits using this process are presented. The static CMOS cells are similar to those cells previously designed for an NMOS process but the dynamic CMOS cells are a unique departure from these static designs. A simple counter circuit is used as an example to demonstrate the relationship between the present CMOS designs and previous NMOS designs.

Introduction

A methodology known as Path Programmable Logic (PPL) for the design of integrated circuits (ICs) using predefined cells for an NMOS Silicon Gate process has been developed by the VLSI group at the University of Utah[1, 2, 3]. The methodology has proven to be useful as the foundation upon which CAD tools for structured logic can be built. An example of the use of this methodology for a CAD tool is the research done by Carter[4] on the compilation of linguist state machine descriptions directly to the composite layout of an integrated circuit.

Design of an IC using this methodology is performed by placing small circuit modules (cells) which can be represented by logic symbols on a grid representing the integrated circuit. When the grid is completely populated, it is both the logical representation and the topological layout of the circuit. The circuit modules have predefined schematic and composite representations. They are custom designed to optimize performance and size for any specific integrated circuit process. The conversion from high level descriptions to the PPL cells is easily accomplished because of this tight coupling between the topology and the logic.

The PPL methodology yields circuits which are similar in structure to that found in the Programed Logic Array (PLA). In the PPL however, the AND and OR planes of the PLA are folded together into one plane. The AND conditions of the input signals are formed on the rows of the PPL and the OR conditions are formed on the columns. Complex cells other than those needed to perform the AND/OR products of a function are provided and may be inserted into the grid at arbitrary locations. These include flip flops, inverters, loads, row and column connections, pass transistors, etc. The AND/OR cells are of unit size, one row high and one column wide, but the complex cells are composed of multiples of the rows and columns.

The row and column wires that connect the cells may be interrupted at any cell boundary. In contrast to a PLA, the columns and rows in a PPL can be divided into any number of segments desired, allowing greater design flexibility. Discrete modules, which may include memory as well as combinational logic, can be placed anywhere on the grid and segmented from the other modules on the grid by row and column breaks. Special ohmic contact cells are used to directly connect rows with columns, allowing the path of a signal to be routed between different modules.

An Example

In order to clarify the problems associated with design of PPL structures using a CMOS process, a design of a simple 2 Bit Counter was carried out using a standard PLA structure, an NMOS PPL structure, a static CMOS PPL structure, and a dynamic CMOS structure. A truth table of the machine which was used for this demonstration is shown in the table below.

Truth Table For Two Bit Counter

	K	C	A_1	A_0	D_1	D_0
Hold Present Count	0	0	0	0	0	0
	0	0	0	1	0	1
	0	0	1	0	1	0
	0	0	1	1	1	1
Count	0	1	0	0	0	1
	0	1	0	1	1	0
	0	1	1	0	1	1
	0	1	1	1	0	0
Clear	1	0	0	0	0	0
	1	0	0	1	0	0
	1	0	1	0	0	0
	1	0	1	1	0	0
	1	1	0	0	0	0
	1	1	0	1	0	0
	1	1	1	0	0	0
	1	1	1	1	0	0

The K and C signals are clear and count inputs respectively to the counter. The A_1 and the A_0 are the present state of the

CH1935-6/83/0000/0158$01.00 © 1983 IEEE

two D type flip–flops in the counter. D_1 and D_0 are the next state D inputs to the counter.

PLA AND PPL DESIGNS

The physical implementation of this counter using a conven–tional PLA for the realization of a Finite State Machine (FSM) is illustrated in Figure 1. The AND and the OR planes are physi–cally separated and the memory elements are placed above the two planes. These AND and OR planes are compact structures and have a similar vertical and horizontal pitch which allows very compact layouts for the PLA.

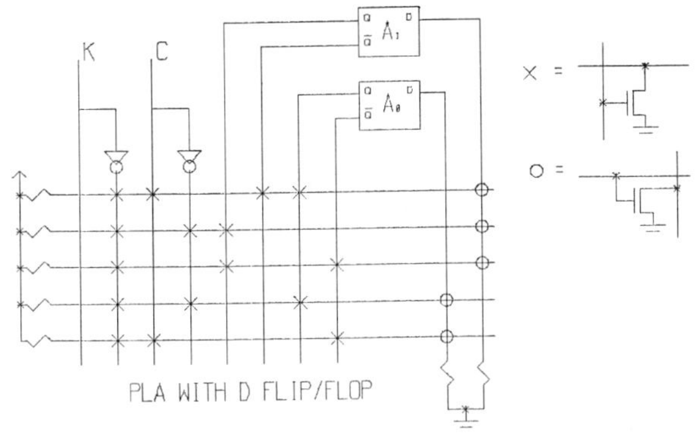

Figure 1: 2 Bit Counter Using PLA with a FSM

This same counter implemented using PPL cells is represented in Figure 2. The AND and OR planes have been folded together and symbols have replaced the transistors result–ing in a functional description of the part at the layout level. This structure is relatively compact in NMOS since each of the 1, 0, S, R, +, and RPU terms are only single transistors. All of these cells are single enhancement transistors whose sources are tied to ground except the RPU cell which is a single depletion transistor whose drain is tied to the power supply line. The 1 and 0 transistors form the AND plane and are 90 degree rota–tions of the S, R, and + transistors in the OR plane. The space occupied by these transistors in this structure is larger than that in the conventional PLA structure because both AND and OR elements must be placed in the PPL grid cell. This space dis–advantage is at least partially overcome by the added advantage of being able to break rows and columns at arbitrary locations and to place memory elements within the array.

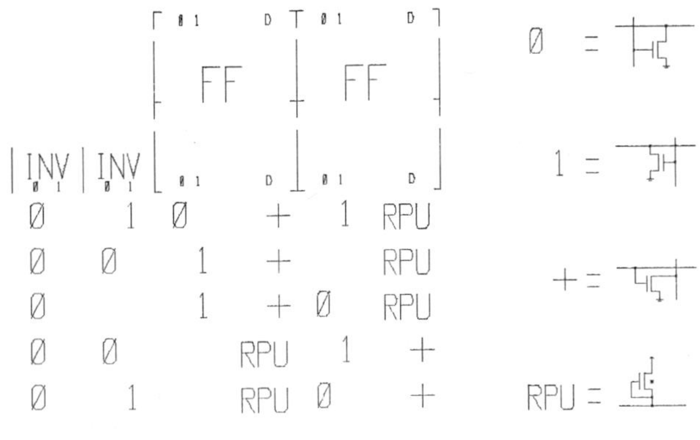

Figure 2: 2 Bit Counter Using PPL Methodology

STATIC AND DYNAMIC CMOS DESIGNS

Figure 3 represents an implementation of the same counter in static CMOS. The P type transistors are shown with circles at the gate as compared to N type transistors with no circle. There are no load resistors in this implementation but each of the 1, 0, S, R, and + terms are composed of two transistors rather than the single transistor found in the NMOS implementation. The 1 and 0 terms are 90 degree rotations of the S, R, and + terms as in the NMOS representation. These cells occupy a significantly larger space than the equivalent NMOS terms. Part of the problem with layout is the fact that the AND plane pair of tran–sistors must be rotated by 90 degrees from the OR plane pair. This coupled with the associated problem of defining the P–well result in large cells.

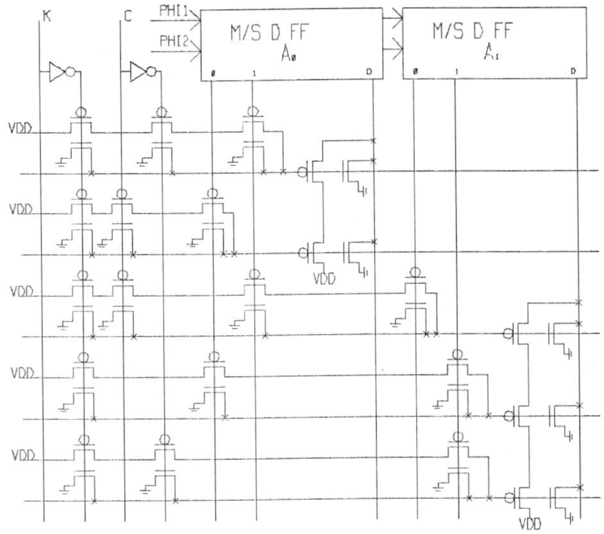

Figure 3: 2 Bit Counter Using Static CMOS

A dynamic CMOS implementation of this example is shown in Figure 4. The AND and the OR plane are folded together as before but the transistors in the array all have the same orien–tation i.e. the gates of both AND and OR transistors all run in vertical columns. The rows are all precharged when PHI1 is low and transistors Q17 to Q21 are turned on and Q22 is turned off. The rows become true when PHI1 is high and Q17 to Q21 are turned off and Q22 is turned on. The AND plane in Row 1 of the PPL program of Figure 2 is formed by transistors Q1, Q2 and Q3. The AND plane of Row 2 is formed by transistors Q4, Q5 and Q6. The OR plane is made simply by connecting the two rows of transistors, Row 1 and Row 2, in series. The output, D_0, of the AND–OR combination is taken from the top row wire. The out–put, D_1, is taken from the bottom row wire of the array. The memory elements for the array are implemented as dynamic D flip–flops which for this case are simply dynamic shift registers.

The advantages of this scheme over the static scheme are ob–vious: (1) the dynamic circuits are very low power, (2) the array transistors are all of one type, N or P and are all contained in a single P– or N– well and (3) these array transistors all have their gates driven by vertical wires and do not have to be rotated by 90 degrees for the AND and OR plane connections as was the case for the static implementations. The disadvantages are (1) only a single AND/OR combination of terms can be formed during any clock cycle and (2) the AND terms formed at any row are not accessible for use in other OR combinations as was the case in the static implementations. This may turn out to be a significant disadvantage but the space savings which are realized by this configuration may out weigh most of these disadvantages.

COMPOSITE DRAWINGS OF PPL CELLS

Figure 5(a) shows a composite drawing of a single 1 element using a four micron NMOS process. This cell contains two column signal wires with the gate of the transistor connected to the left signal wire. A 0 element would be identical in size but would have the gate of the transistor connected to the right column wire. A composite drawing of an equivalent functional static CMOS cell of a single 1 element for a four micron process is shown in Figure 5(b). Figure 5(c) is a composite drawing of the same functional cell for dynamic CMOS. Figure 5(d) is a composite drawing of three rows and three columns of these dynamic cells connected together.

The NMOS and the static CMOS cells shown in Figures 5(a) and (b) both contain a pair of column signals. These signals are usually the TRUE and NOT output signals coming from inverters or flip–flops or the signals going to the set and reset inputs of the R/S flip–flops. They are carried in pairs partly because of layout constraints (the 90 degree rotation of the transistors between the 1 and S functions of the folded AND and OR planes required the width of two column wires) and partly as an attempt to keep the TRUE and NOT signals together as a logical pair. The dynamic CMOS cell does not have this layout constraint since each transistor has its gate tied to the column wire and a 90 degree rotation of the transistor is not required. In addition, the dynamic CMOS implementation uses D–type flip–flops rather than the R/S flip–flops found in the NMOS and static CMOS cells and thus two inputs are not required.

The NMOS cell design contains metal column wires for power, ground and two signal wires and a single polysilicon row wire. The static CMOS cell contains similar power and ground column wires. It also contains a pair of row wires (which function as a single row) for the parallel and series connections to the OR plane transistors. The row wires are composed of a polysilicon wire for parallel connections between transistors and a diffusion wire for series connections . There are two pairs of metal column wires in this static CMOS cell. Each pair of wires is used for the parallel and series connections to the OR plane transistors. The dynamic CMOS implementation does not require power and ground in the array and it is necessary to run only a single metal wire for the row and a single polysilicon wire for the column. The row wires are shared between two rows.

The NMOS cell is 44 by 28 microns (1232 square microns), the static CMOS cell is 66 by 91 microns (6006 square microns), and the dynamic CMOS cell is 18 by 12 microns (216 square microns). These numbers translate into 20,292 atomic cells (a single row and a single column) in the NMOS, 4,180 cells for the static CMOS and 115,741 cells for the dynamic CMOS in an active area of 5X5 millimeters.

COMPARISONS AND CONCLUSIONS

It is important to make comparisons between circuits designed with the PPL methodology and those designed as gate arrays and as full custom circuits. It is difficult, however, to compare PPL circuits with these more conventional designs because of the very different design techniques required in each methodology.

A possible way to compare a PPL design with a gate array design would be to find the number of single cells required to implement a specific function in PPL and the number of gates required to implement the same function in a gate array. It should then be possible come up with a number of equivalent PPL cells to replace a single gate. This method has serious disadvantages because the function may not be representative of a general function but at least it would give some frame of reference to compare the two methodologies.

Consider the NMOS PPL design in Figure 2. The flip–flops occupy 8 atomic cells and the inverters occupy 1 atomic cell or a total of 18 atomic cells for the memory and inverter cells. The array occupies 30 atomic cells so the total function requires 48 atomic cells. This same function would require 18 two input NAND gates to form the D1 and D0 inputs to the flip–flops and 9 two input NAND gates to build each D flip–flop (assuming two clocks are present for a master–slave flip–flop). Using this as the criteria, there are 36 two input gates in the space of 48 atomic cells or 1.3 atomic cells per gate. The 20,292 cells in the 5X5 millimeter active area could then be thought of as being equivalent to 15,609 two input gates.

This same reasoning could be used to make predictions on the static CMOS designs for the circuit of Figure 2. If it is assumed that the ratio of PPL cells per two input logic gate remains the same for the static CMOS design as for the static NMOS design, there would be 3,215 two input gates in a 5X5 millimeter active area.

The ratio of gates to PPL cells must be recomputed for dynamic CMOS. Each memory cell (D flip–flop) and inverter has been estimated to occupy a space of 42 atomic cells for a total space of 168 cells for the memory and inverters. The array occupies 60 atomic cells. Using these estimates and adding space for the pullup and pulldown transistors, the circuit in Figure 4 would take a total space of 248 atomic cells. This gives a ratio of 6.9 atomic cells per two input gate or the equivalent of 16,801 two input gates in a 5X5 millimeter active area. This number is very dependent upon the particular function which is being implemented since memory and inverter elements occupy large areas when compared to the atomic array cell. This was not the case in the static NMOS design where the ratio of atomic cells to two input gates was 1.3.

There are a number of other factors which would have to be considered to give more validity to these predictions. For example, the interconnect between PPL modules would be have to be considered as well as the per cent utilization for the PPL cells. Most importantly, more functions would have to be considered.

Comparisons between the PPL circuits and full custom chips could be made in a similar manner. The results would obviously show that the PPL circuits are larger than equivalent custom chips. The main advantage of the PPL design methodology would then be its functional representation of the circuit and the ease with which CAD tools can be implemented to minipulate the PPL designs.

Acknowledgments

The author would like to acknowledge the suggestions made by Bill Knapp formerly with General Instrument Corp. for the initial definition of the static CMOS cells and from Tony M. Carter, a colleague at the University of Utah on the implementation of the dynamic CMOS cells. This work was sponsored in part by a contract with General Instrument Corp.

References

1. K. F. Smith; B. E. Nelson; T. M. Carter; A. B. Hayes, "Computer-Aided Design of Integrated Circuits Using Path-Programmable Logic," *IEEE Electro 83 Professional Program Session Record, New York, New York,* April 1983.

2. B.E. Nelson; K.F. Smith; and T.M. Carter, "Cost Effective VLSI Design System," *IEEE International Symposium on Circuits and Systems,* May 1983.

3. K.F. Smith; T.M. Carter; and C.E. Hunt, "Structured Logic Design of Integrated Circuits Using the Stored Logic Array," *IEEE Transactions on Electron Devices,* Vol. ED-29, No. 4, April 1982, pp. 765–776.

4. T. M. Carter, "ASSASSIN: An Assembly, Specification and Analysis System for Speed-Independent Control-Unit Design in Integrated Circuits Using PPL," Master's thesis, Department of Computer Science, University of Utah, June 1982.

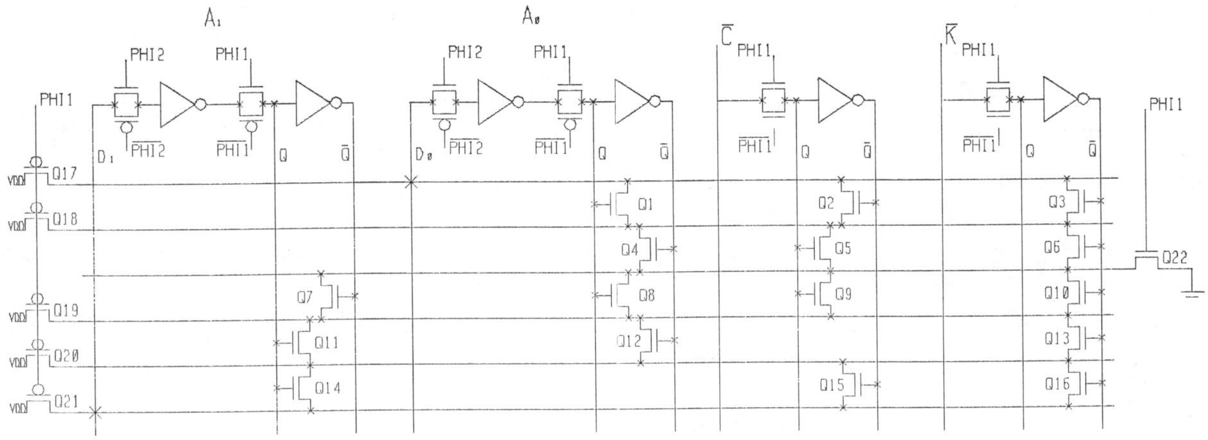

Figure 4: 2 Bit Counter Using Dynamic CMOS

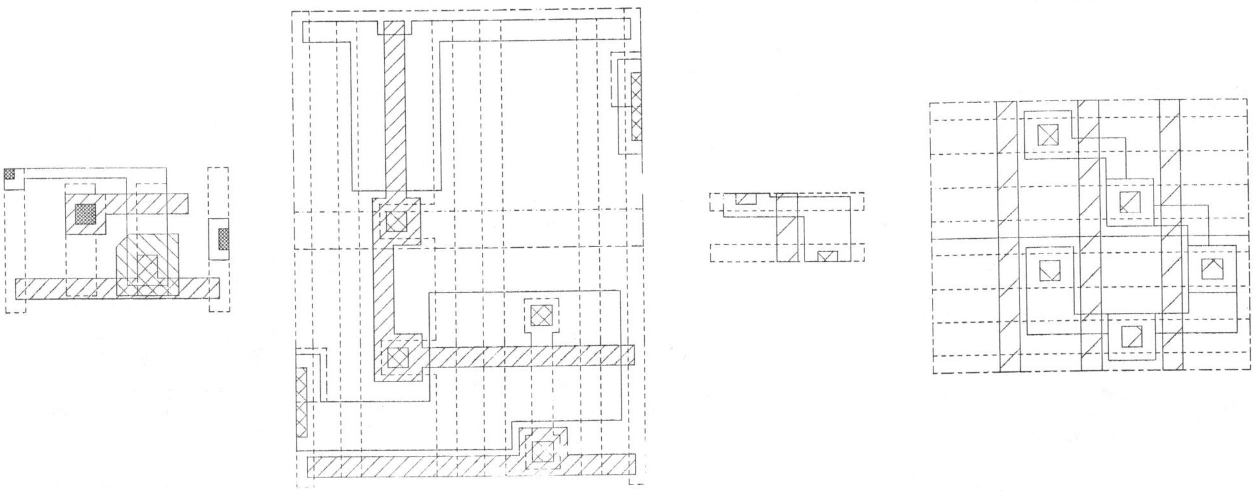

A. NMOS Cell

B. Static CMOS Cell

C. Single Dynamic CMOS Cell

D. Multiple Dynamic CMOS Cell

Figure 5: Composites of NMOS and CMOS Cells

ANALYSIS OF INTERCONNECTIONS IN TRADITIONAL ICs AND DERIVATIVE PLA-LIKE STRUCTURES

David L. Greer
Military Electronic Systems Operation
General Electric Company
P.O. Box 4840
Syracuse, NY 13221

ABSTRACT

This report provides an analysis of space required by interconnections in logic networks realized in integrated circuit form. Both traditional and derivative PLA-like architectures are considered and compared.

INTRODUCTION

Structural regularity is a necessary precondition for achieving effective automated layout of very-large scale integrated circuits. The importance of regular structures is implicit in the notion of a silicon compiler[1] and is reflected in the results of past research in the form of the bristle-block concept[2], the gate matrix layout method[3], multiple programmable logic array (PLA) building blocks[4], and the Mead-Conway design approach[5] to cite some specific examples. However, the ultimate in regular logic structures are those that combine the memory-like replication of PLAs with additional features of line segmentation (folding) and internal functions (feedback). These structures are known by various names including associative logic array[6,7], storage/logic array[8], mixed logic array[9], and folded PLA[10]. It is this group of structures that are emphasized in this paper.

LOGIC NETWORK INTERCONNECTION MODEL

Obvious physical differences exist between traditional structures formed by the discretionary interconnection of discrete gates and rigid PLA-like orthogonal structures in which gate and interconnection functions are combined. However, analysis reveals that despite these apparent differences, the distribution of interconnection lengths and the rate of increase in die area with network size is essentially the same for both.

While separate models are used to represent traditional and associative logic array (ALA) structures, they are each based on the same empirical relationship

$$P = KG^r \qquad (1)$$

which expresses the number of input and output signals, P, required by a network of G gates[11]. In traditional structures this relationship leads directly to the expression

$$L = (K-1)\,G - cKG^r \qquad (2)$$

from which L, the number of interconnections within the network can be determined. Network constants are represented by K, r, and c, the latter expressing the ratio of network inputs to total inputs plus outputs, or P.

In order to determine the distribution of interconnections by length, it is assumed that network gates are located on a uniform grid of minimum aspect ratio. If l_i/i represents the difference between the average number of interconnections per gate in an i gate and an i-1 gate subnetwork, then

$$\frac{l_i}{i} = \frac{L_i}{i} - \frac{L_{i-1}}{i-1}. \qquad (3)$$

The value of l_i is significant because it expresses the number of interconnections of greatest length in an i gate subnetwork. The number of interconnections of the length of l_i in a network of G gates is therefore

$$S_i = G\left(\frac{l_i}{i}\right). \qquad (4)$$

The total number of interconnections in the network can be obtained either through the use of Equation (2) or by summing all values of S_i from i = 1 to G.

The concatenated length of all interconnections, TL, in a G gate network will be

$$TL = \sum_{i=1}^{G} (S_i)\,(q_i) \qquad (5)$$

where q_i represents the Manhattan distance (measured on the grid) between the most distant gates in the i gate subnetwork. Assuming that the subnetwork gates are located so as to form a rectangle of minimum aspect ratio,

$$q_i = m_i + n_i \qquad (6)$$

where

$$m_i = (i-1)^{\frac{1}{2}}\Big\rfloor \text{ Truncate } \quad (7) \qquad n_i = i - m_i - 1^{\frac{1}{2}}\Big\rfloor \text{ Truncate } (8)$$

To convert relative dimensions expressed in units of grid spacing to units of physical distance requires knowledge of gate area, GA, and minimum center-to-center conductor spacing, w.

Let d represent the physical distance between gates assuming minimum allowance for interconnections. Then

$$d = (GA + IA)^{\frac{1}{2}} \qquad (9)$$

where IA represents interconnection area per gate. Recognizing that TL, the relative length of interconnenctions, can be converted to physical units by multiplying by d permits interconnection area to be expressed as

$$IA = (TL)\ (d)\ (w)/(G)\ (Z) \qquad (10)$$

where Z represents the influence of layout packing efficiency. The solution of Equations (9) and (10) will yield the values of d and IA.

Figure 1 is a log plot of the distribution of interconnection lengths representing each of four network sizes as obtained from application of the above analysis. The straight

CH1935-6/83/0000/0162$01.00 © 1983 IEEE

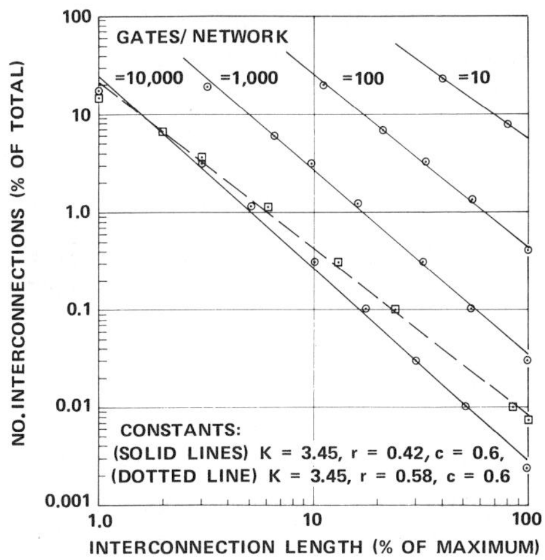

Figure 1. Distribution of Interconnections as a Function of Length

lines each represent a best fit to corresponding calculated data points, the locus of which describes a downward sloping curve. Slopes of the two lines representing 10,000 gate networks can be expressed in terms of the exponent of Equation (1) as 2.65 - (1.61) r. However, because the slopes of lines representing all the networks shown in Figure 1 are not constant, -- they range from about 1.6 to 2.0 -- the generality of this relationship is found to be only approximate at best. With this exception, results of the above analysis are generally consistent with experience and with the findings of other researchers[12,13].

The area required for the conductors used to interconnect logic gates, normalized to the area of the gates themselves in a typical IC layout, is shown in Figure 2. Two different gate areas are represented. The interconnection area requirements of typical integrated circuits can be expected to be bounded by the two curves shown in this figure.

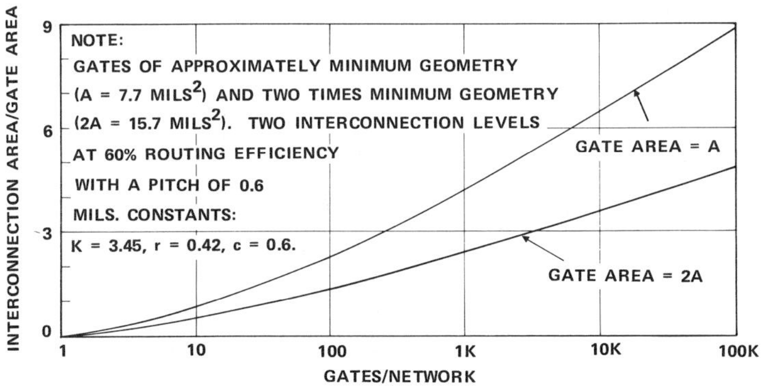

Figure 2. Interconnection Area-Gate Area Ratio vs. Number of Gates Per Network

ALA INTERCONNECTION MODEL

The above interconnection model is straightforward since it applies to tradiational networks in which gate and interconnection functions are clearly separate. However, this differentiation does not exist in ALA structures in which orthogonally disposed interconnections serve also as distributed logic gates.

In the following analysis of ALA topology, consideration will be limited initially to the partial segmentation of word lines. The results will then be generalized to include partial segmentation of bit lines. Full segmentation involves the separate application of word and bit line models.

Word Line Segmentation

A general topological representation of a word line segmented structure is shown in Figure 3. While word line segments can in theory be of any length, practical considerations limit line terminations both in number and location. Typically word line segments will be confined to designated columns separated by a number of bit lines. Each of the several clusters of bit lines so formed will be referred to as a bit line block, or simply as a block where no confusion will result. Three blocks are represented in Figure 3.

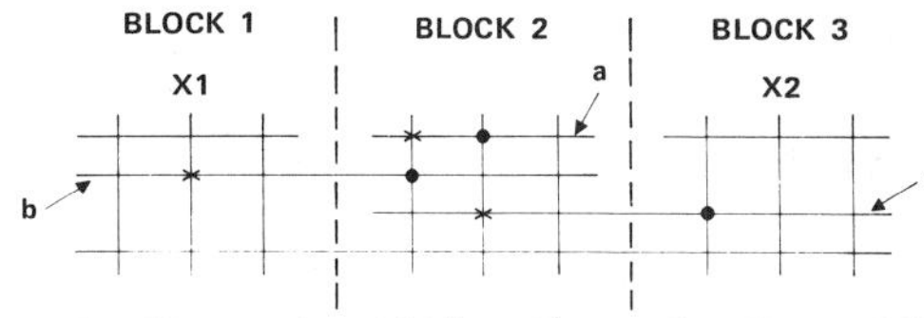

Figure 3. Illustration of Lines Extending Beyond Block Boundaries

If a matrix of N_b blocks is used to realize a network of G gates, each block will represent an average of G/N_b equivalent gates. However, only a fraction of the word lines will be entirely contained within the block's boundaries. This is illustrated in Figure 3 by the three lines identified as (a), (b), and (c). Note that line (a) is entirely contained within the second block but lines (b) and (c) are not. Line (b) extends into block one to intersect the output variable X1 while line (c) intersects the input variable X2 in block three.

Let a cell be defined as a group of one or more blocks in a partially segmented matrix as shown in Figure 4. If cell j contains J blocks, (j = 1 to N_b), it will have I inputs and O outputs where

$$I_j = (c)\left(P_j\right) \quad (11) \qquad O_j = (1-c)\left(P_j\right) \quad (12)$$

and

$$P_j = (K)\left(G_j\right)^r \quad (13) \qquad G_j = (j)\,(G)/N_b. \quad (14)$$

The constants K, r, and c remain as previously defined.

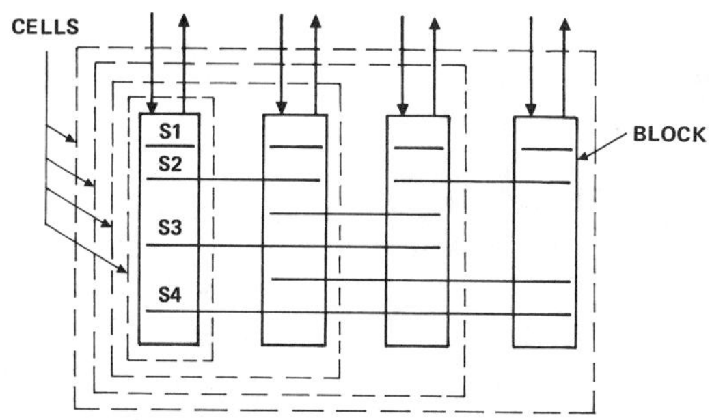

Figure 4. Matrix Cell and Block Structure

Of the total number of cell inputs and outputs, respectively, let NI_j and NO_j represent that fraction having origins or destinations external to the G gate network. Also let WI_j and WO_j represent the remaining cell inputs and outputs -- those provided by word line segments extending beyond cell boundaries.

Thus,

$$I_j = NI_j + WI_j \quad (15) \qquad O_j = NO_j + WO_j. \quad (16)$$

If one assumes that network inputs and outputs are distributed evenly between blocks, then

$$NI_j = (I)(j)/N_b \quad (17) \qquad NO_j = (O)(j)/N_b . \quad (18)$$

Figure 5 shows the correspondence between a conventional and distributed gate representation of a word-line segmented matrix. Note that a single cell input realized using conventional gates can represent two or more cell inputs with distributed gates.

If Q_W represents the fanout of each WI_j input within cell j, the fraction of AND gates receiving the first input will be Q_W/AG_j where AG_j represents the number of AND gates in cell j. Thus the fraction of remaining gates -- those not receiving this input -- will be $1-(Q_W/AG_j)$. Each of the remaining input lines will similarly go to Q_W AND gate inputs. Until the fanin of any one gate is exhausted, the location of each input will be independent of the others. For simplicity, although not strictly correct, all input placements are assumed to be independent events. Thus the fraction of gates not receiving a WI_j input will be approximately

$$F1_j \simeq \left[1-\left(Q_W/AG_j\right)\right]^{WI_j} . \quad (19)$$

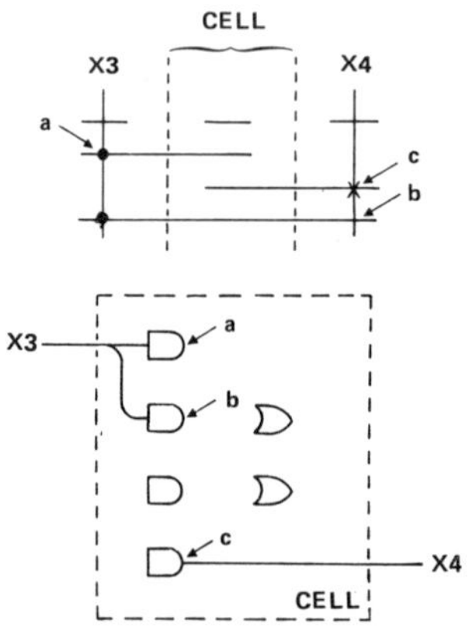

Figure 5. Word Line Segmentation

The value of Q_W is a function of AND gate fanout, the value of j, and the distribution of word line segments which, in turn, is a function of Q_W. Upper and lower bounds on Q_W are AND gate fanout and one, respectively. Typically the value of Q_W will be marginally greater than one.

The fraction of AND gates within a cell that are tied to a WO_j output will be WO_j/AG_j, or the fraction of AND gates not connected to these outputs will be

$$F2_j = 1 - WO_j/AG_j . \quad (20)$$

The probabilities of a WI_j or WO_j connection to any gate are independent. Therefore, the fraction of gates in a cell not receiving either a WI_j input or a WO_j output connection will be the product of $F1_j$ and $F2_j$. If this product is F_j the number of AND gates or word line segments contained entirely within a cell will be

$$L_j = (F_j)(AG_j) . \quad (21)$$

From values of L_j, the number of word lines segments spanning j blocks in cell j will be found to be

$$S_j = L_j - 2L_{j-1} + L_{j-2} . \quad (22)$$

The physical nature of S_j is illustrated in Figure 4.

Calculation of S_j as described results in fractional line quantities. These can be converted to integers by truncation if the fractional part, properly weighted, is then added to a longer line segment, S_i (where i > j). With reference to Figure 4, it can be shown that $(N_b - j + 1)$ word line groups exist in the matrix, each containing S_j word line segments. Therefore, if the fractional part of S_j, say SF_j, is deleted from S_j, the quantity required to be added to S_i to keep total number of network gates constant is

$$(SF_j)(N_b - j+1)/(N_b - i+1) . \quad (23)$$

Bit Line Segmentation

The analogy between word and bit line segmentation is evident from a comparison of Figures 5 and 6.

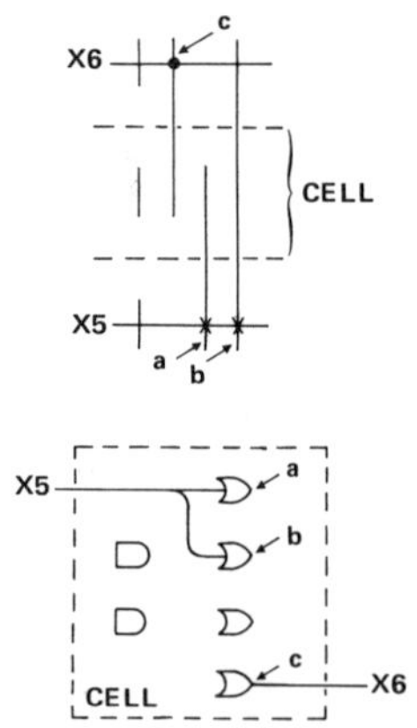

Figure 6. Bit Line Segmentation

However, the analogy is not perfect due to the fact that the external outputs, NO_j, are realized only by OR gates, a situation not paralleled in the case of word line segmentation. Thus, if bit line segmentation is to involve only internal function lines, the corresponding number of OR gates -- say FG_j -- will also be less than the network total or

$$FG_j = OG_j - NO_j \ . \qquad (24)$$

Taking into account the above adjustment in number of OR gates, and with appropriate substitution of variable names, the preceding analysis of word line segmentation can be applied also to bit line segmentation. Specifically, the substitutions required are FG_j for AG_j, N_w for N_b, BI_j for WI_j and BO_j for WO_j.

The foregoing segmentation analysis does not explicitly define the location of network input and output interconnections. Conventionally these are bit lines with access at the periphery of the matrix. However, it is inevitable that techniques will be developed to permit the location of external connections anywhere on the face of a die. When such techniques become routinely available, segmentation of input-output lines will provide the advantages not now possible.

ESTIMATION OF ALA AREA AND OPTIMUM BLOCK STRUCTURE

Without segmentation, the width of an ALA must accommodate all bit lines -- inputs, outputs, and internal functions -- while height will equal the number of word lines. Word lines correspond approximately to the number of AND gates in the network realized. While it is apparent that array segmentation will reduce one or both of these dimensions, by precisely how much remains to be determined.

A plot of the relative area vs. block product of fully segmented arrays is shown in Figure 7. Relative area (RA), is the area of the segmented structures divided by the area of an equivalent unsegmented structure. Block product, N, is the product of bit line and work line blocks.

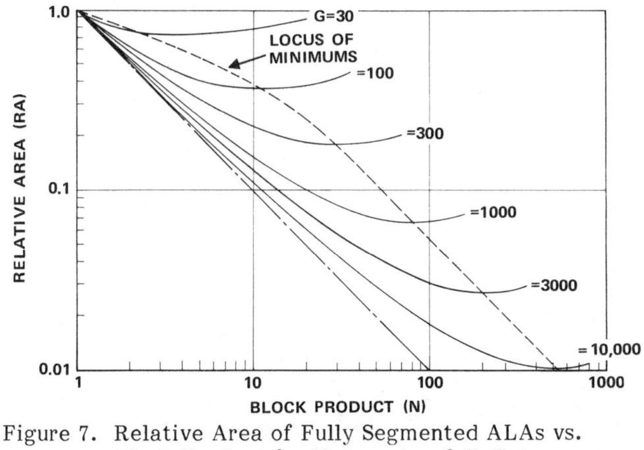

Figure 7. Relative Area of Fully Segmented ALAs vs. Block Product for Netwroks of G-Gates (Nonsegmented I/O)

With reference to Figure 7, note that each curve defines, by its minimum, the block product yielding minimum array area for networks of corresponding gate count. The locus of all points so defined is shown by means of a dotted line. The location of this line for arrays of over 300 equivalent gates can be described by the expression

$$AR = (K1)/N \ . \qquad (25)$$

In addition, from data provided in Figure 7, it can be shown that

$$N = (k2)G^p \qquad (26) \qquad AR = (K3)/G^p \qquad (27)$$

While Figure 7 represents fully segmented structures, similar plots corresponding to partially segmented arrays have the same form although the curves are displaced upward somewhat. Therefore, the above expressions, with different constant values, apply also to partially segmented arrays.

Data supporting the above discussion was obtained using the ALA interconnection model and constants having the following values: K = 3.45, r = 0.42, c = 0.6, and AG/OG = 1.941. Corresponding values of the constants p, K1, K2, and K3 appearing in Equations (25-27) were found to have the following values respectively: For word line segmentation, 0.5637, 3.70, 0.776, 4.77; for full segmentation (nonsegmented I/O) 0.8143, 5.40, 0.293, 18.4.

Results obtained using the ALA model were compared to a number of partially segmented arrays representing practical network designs. Prediction of the height of these arrays -- the maximum number of rows containing one or more word line segments -- was within 15% in all cases. The two most complex network examples involved 450 and 717 equivalent gates realized using 26 and 30 bit line blocks respectively. The height of the 450 gate network was 46 rows while that of the 717 gate network was 59 rows. The corresponding predicted heights were 51 and 58 for errors of +11 and -1.7 respectively.

CONCLUSION

Having derived the above models, it becomes a straightforward exercise to compare segmented ALA to interconnected gate structures with respect to their rates of area increase as network complexity is increased. Figure 8 is the result of such a comparison. It shows the row-column products, or line crossings, of segmented ALA structures normalized with respect to the corresponding areas of equivalent conventional interconnected gate structures, these ratios plotted against network complexity expressed in gates per network. Curves representing two gate areas are shown as was done in Figure 2.

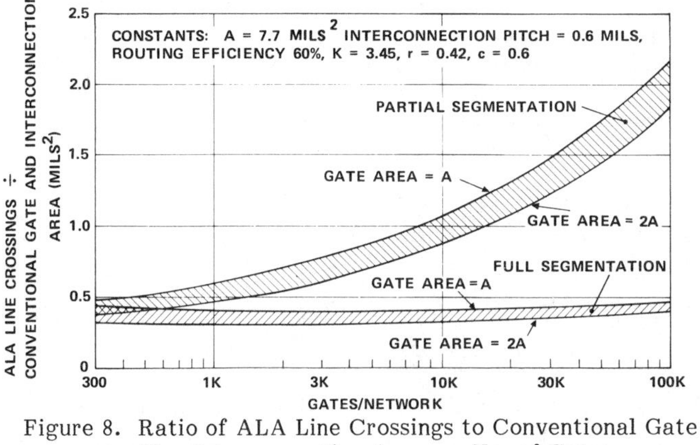

Figure 8. Ratio of ALA Line Crossings to Conventional Gate Plus Interconnection Area vs. No. of Gates

With reference to Figure 8, the unity slope characteristic of the curves representing fully segmented ALA configurations indicate that essentially no difference exists in the rate of area increase as a function of network complexity between fully segmented ALA and conventional interconnected gate structures. This is not true in the case of partially segmented ALAs which are seen to increase in size at a greater rate then do the conventional structures.

Finally it is to be noted that the reciprocals of the numbers shown on the ordinate axis of Figure 8 represent the area per line crossing (in square mils) necessary in order that corresponding ALA and interconnected gate structures have equal areas. That such upper bounds on effective line crossing areas can be achieved in practice has been demonstrated in previous work[7].

REFERENCES

1. R. Ayres, "Silicon Compilation - A Hierarchical Use of PLAs," Proc. 16th Design Automation Conf., June 1979, pp. 314-326.

2. D. Johannsen, "Bristle Blocks: A Silicon Compiler," Proc. 16th Design Automation Conf., June 1979, pp. 310-313.

3. A.D. Lopez, H.F.S. Law, "A Desnse Gate Matrix Layout Method for MOS LSI," IEEE Trans on Elect Devices, Vol 27, No. 8, August 1980, pp. 1671-1675.

4. P.W. Cook, S.E. Schuster, D.R. Freedman, J.T. Parrish, V. DiLonardo, " One Micron MOSFET PLAs," ISCC Digest of Technical Papers,February 1979, pp. 62-63.

5. C. Mead, L. Conway, "Introducttion to VLSI Systems, "Addison-Wesley Publishing Co., 1980.

6. D.L. Greer, "An Association Logic Matrix, " IEEE J. Solid-State Circuits, Vol. SC-11, No. 5, October 1976, pp. 679-91.

7. D.L. Greer, "Orthogonally Interconnected Array Technology Development," Proc. Semicustom IC Technology Symposium, Institute for Defense Analyses paper P-1593, June 1981, pp. 177-190.

8. K.F. Smith, T.M. Carter, C.F. Hunt, "Structured Logic Design of Integrated Circuits Using the Storage/ Logic Array (SLA)," IEEE J. Solid-State Circuits, Vol. SC-17, No. 2, April 1982, pp. 395-406.

9. J.L. Burrows, S. Forde, V. Vijums, "Memory Processor Study," Final Report on Air Force Contract F33615-80-C-1192, prepared by Sanders Associates, Inc., Nashua, N.H., 21 December 1981.

10. R.A. Wood, "A High Density Programmable Logic Array Chip," IEEE Trans Comput., Vol. C-28, No. 9, September 1979, pp. 602-608.

11. B.S. Landman, R.L. Russo, " On a Pin Versus Block Relationship for Partitions of Logic Graphs, "IEEE Trans Comput., Vol. C-20, No. 12, December 1971, pp. 1469-1479.

12. W.E. Donath, "Wire Length Distribution for Placements of Computer Logic, " IBM J. Res. Develop., Vol. 25, No. 3, May 1981, pp. 152-155.

13. M. Feuer, "Connectivity of Random Logic," IEEE Trans Comput., Vol. C-31, January 1982, pp. 29-33.

Storage/Logic Array (SLA): A High Density Standard Cell Design Method

Suhas Patil, Clement Leung, Bill Knapp and H. Ravindra

Patil Systems, Inc.
391 Chipeta Way, Suite C
Salt Lake City, Utah 84108
(801) 582-3040

ABSTRACT

Storage/Logic Array (SLA) combines the distributed gates of PLAs together with cells of the standard cell design method to achieve high density for digital system realization in integrated circuits. Digital circuits in Bipolar, HMOS, and CMOS technologies can be implemented with equal facility with this method. In this paper we compare and contrast this method of design with other methods of design and explain what contributes to the ability of SLAs to achieve higher density than the conventional standard cell design method.

INTRODUCTION

The Storage/Logic Array (SLA) is a form of structured logic array in which storage elements are embedded in the body of the logic array and the logical rows and columns of the array can be subdivided into independent logic segments. SLA is closely related to Programmable Logic Arrays (PLAs) and Programmable Array Logic (PALs (TM) Monolithic Memories) but are distinctly different from Gate Arrays and integrated circuits designed with the conventional Standard Cell method. There is evidence to show that with SLA one can achieve functional density in digital integrated circuit approaching that of full custom circuits. SLAs are directed towards the design of digital integrated circuits. Analog circuits are not in the domain of SLAs.

HOW DOES SLA DIFFER FROM CONVENTIONAL STANDARD CELL BASED IC

The Storage/Logic Array method of design entails both a symbolic notation (language) of design and a physical structure that can be implemented in integrated circuits. The design is expressed in a notation called an SLA program (layout) and then automatically translated into integrated circuit composites for the manufacture of the integrated circuits. Both master-slice and standard cell technique of manufacture can be used. In the master-slice approach one or two layers of interconnect are defined on a fixed array of semiconductor devices to realize the desired digital system in an integrated circuit. On the other hand in the standard cell technique, all the mask layers of the IC are defined.

The primary difference between the SLA and the conventional gate-array and standard cell is that SLA utilizes both localized and distributed gates. The distributed gates perform the function of both logic and interconnect. Unlike the gate-arrays and conventional standard cell approach there is no need to set aside wide interconnect channels in the case of SLA. The ability to rout a signal from one place to another is present but invariably the task of distribution of signal and performance of logic overlap in space.

Unlike the gate-arrays and conventional standard cell approach, in the context of SLA it is quite meaningful to talk about silicon architecture. In designing ICs with SLA it is relatively easy to arrive at a good silicon architecture for the chip. Often the layout that results is as good as, and sometimes much better than, that achieved in design of ICs using the full custom method. Higher density results primarily from use of distributed gates and from achieving a superior silicon layout.

MACRO CELLS AND MICRO CELLS

The macro cells implement complete functions such as flip-flops, input/output buffers and any other full custom piece of layout such as might be used in the conventional standard cell design. The micro cells, on the other hand, consist of a layout that is only a piece of a larger structure, such as a distributed gate, which is constructed from more than one micro cell. In SLA some of the cells have just a single transistor and there are micro cells, such as the interconnect cells, that have no transistors at all.

Larger structures such as counters, arithmetic units and other MSI level functions are built using a number of micro cells and some macro cells. As the MSI functions are implemented with a collection of cells which are all much smaller than the MSI function, and because the cells can be arranged in many ways to realize the same MSI function, the

CH1935-6/83/0000/0167$01.00 © 1983 IEEE

shape of the MSI function can be altered easily to suit the available space. Also the location of its inputs and outputs can be arranged to be optimally located from the point of view of interconnection with other parts in the design.

RELATION TO PLAs and PALs

Storage/Logic Array combines the features of PLA and standard cell design in one structure. As an array it has the structure of a folded PLA in which the AND plane and the OR plane are folded onto one another. Furthermore, the resulting columns and rows can be broken into several electrically independent segments. Each segment acts like an independent gate.

As a result of folding the input and output array into one single array the input and output wire are located near each other and one can attach standard cells like flip-flops to them without any routing of wires. These storage elements (flip-flops) can be located inside the body of the array.

PLA and PALs (Programmable Array Logic, (TM) Monolithic Memories) do not have storage elements in the body of the array. PALs use a folded array but the flip-flops are located only at the edge of the array.

Higher density of digital circuit realization is achieved by SLAs because of the storage elements, folding of the array and the ability to segment the columns and rows. The individual cells, the space at the interconnection of a column and a row, may not be as dense as in a PLA but the overall density achieved by SLA is often much superior because of a higher utilization of the array and because the storage elements are well packed standard cells.

EXAMPLES

The 74163 Synchronous 4 Bit Counter shown in Figure 1 shows an example of an SLA program, which implements the 74163 synchronous 4 bit counter. This program specifies the location of input and output signals, the placement of macrocells and microcells, and the sharing of column and row segments among their occupants. The macrocells used in this example are D flip flops. The microcells occupy single squares and are labelled "0", "1" or "+". The D flip flops also receive clock signals which synchronize their operations with those of combinational operators. All feedback loops in the SLA program pass through these clocked flip flops and the counter operates as a synchronous state machine. Distribution of the clock signals is not shown.

A design is captured in an SLA program as shown in Figure 1. The design can be exercised by conducting simulation runs using the SLA program. An SLA program can also be assembled into a set of mask specifications, with the clock distribution and power and ground distribution paths inserted either manually or automatically. For simplicity signals are propagated along columns, operators are constructed on rows, and macrocells are used as storage elements in this design. Using other types of microcells, signals can be propagated on rows as well as columns, operators can receive their inputs and distribute their outputs in both dimensions, and other types of storage elements can be made available as macrocells or constructed out of microcells.

In the counter design, signals are propagated along the columns and operators are constructed on the rows. If more than one signal is propagated along a column, or more than one operator is constructed on a row, that column or row is segmented. Each segment is enclosed within a pair of parentheses (Figure 1). An operator in a row segment is activated when its input signals satisfy the conditions specified. Consider the operator in the segment to the left on row 11. This operator receives two input signals, one distributed on the lower segment of column 4 (the output signal of a D flip-flop) and the other distributed on the lower segment of column 6. When both of these signals are asserted, this operator is activated and its output signal, distributed on column 5, is asserted. Assertion of the signal carried on column 5 sets the D flip flop at the next clock event.

As another example, consider the operator on row 2. This operator is activated if the signal carried on column 2 (the input signal "clear") is not asserted and the signal carried on column 3 (the input signal "load") is asserted. Activation of this operator causes the signals carried on the upper segments in column 6 and column 12 to be asserted.

The 8251A USART

The SLA methodology has also been applied at Patil Systems to the design of a popular USART (Universal Synchronous Asynchronous Receiver Transmitter), the 8251A. The block layout of an SLA program for this design is shown in Figure 2. Each block in Figure 2 consists of either a state machine, a functional unit such as a programmable comparator or a programmable counter, or a set of data/control latches. The design occupies about 100 rows and 75 columns and uses a set of macrocells and microcells designed for a 4 micron HMOS process. The automatically assembled chip layout is approximately 200 mil by 200 mil. Prototypes of this chip have been fabricated and have been observed to operate at 6 megabaud.

WHAT CONTRIBUTES TO HIGHER DENSITY

There are several factors that contribute to the realization of higher density of digital logic with SLA. Among these the important factors are:

i) The ability to choose just the functionality desired and to mold the physical shape of the module and determine the location of the input and

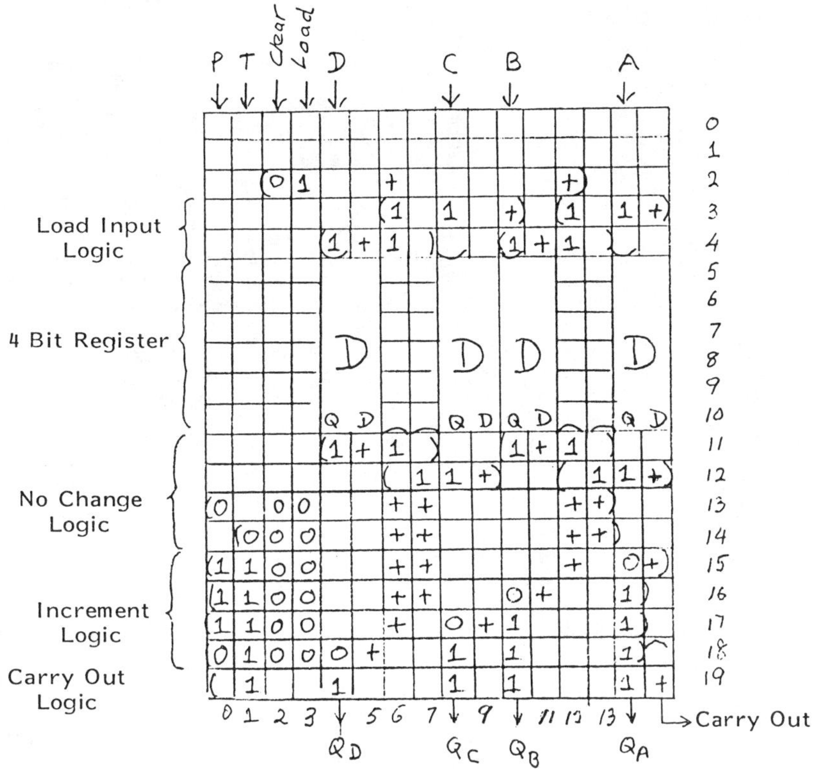

Figure 1. 4 Bit Counter 74163

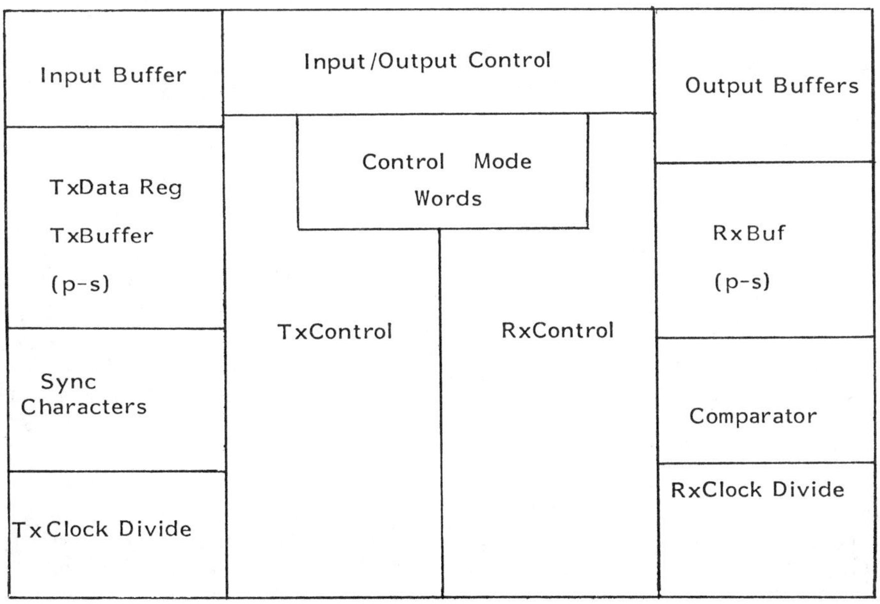

Figure 2. 8251A Design

output wires: This flexibility results in superior match between one module and another both in terms of functionally and in terms of physical interconnect. The former leads to optimization of the logic to be implemented in each module and the latter results in a reduced amount of interconnect wiring. In the conventional standard cell approach the functionality and physical shape of the MSI modules is fixed and so are the locations of the input and output wires.

ii) The use of distributed gates which perform the logic, as well as do the task of interconnect: As a result of the use of distributed gates and the flexibility to arrange the location inputs and outputs there is no need to allocate wide channel space for routing purposes. In comparison with the conventional standard cell approach, the routing in SLA is far better organized. From the point of view of the quality of routing SLA designs often have superior layout than typical full custom designs.

iii) The ability of the designer to achieve silicon architecture well matched to the design on hand: The SLA method does not force a particular silicon architecture on the designer, he is free to choose the architecture that suits the design and is able to implement it with the SLA method.

iv) The use of standard cells well matched to the discipline of design: It has been possible to design macro cells that are well matched for the SLA method of design which helps the designer in creating good layout for the design.

v) Reasonable compactness of the basic cells of the SLA: The micro cells for SLA are reasonably regular and compact. Any improvement in the compactness of these cells directly results in savings in the silicon area. Furthermore, as there are only a handful of such cells it is possible to work on them intensely and obtain excellent packing. The replication factor for the micro cells in an SLA design is generally very large.

All of the factors above together help SLA based designs achieve higher density in comparison with the conventional standard cell method of design. In some cases the density that is achieved is as good as full custom circuits.

WHAT NOT TO EXPECT FROM SLA

Storage/Logic Array Method of design is directed towards the design and implementation of digital systems. The focus here is on the design of the system and not on the design of standard cells. SLA is not appropriate for the design of specialized structures such as RAM, ROM or analog circuits. SLA can use such specialized structure as macro cells in the design of larger systems, but these specialized structures must be designed using other design methods directed towards circuit design and full custom layout. Thus one would not design a 64K RAM chip with SLA but SLA could use a 64K RAM cell in designing a system benefiting from the use of such memory.

In some of the chip designs, the SLA method of design has resulted in superior density and performance than hand-crafted designs. It would be a mistake to conclude from these examples that the SLA method will always lead to chips of higher density or higher performance than full custom designs. Given enough time the full custom circuit can always achieve better density and performance. Given a time constraint, however, the SLA method competes well against the full custom method. Under severe time constraints the only way to get a reasonably dense circuit of good performance might be by use of a structured approach to the design of integrated circuits, such as the SLA.

WHAT CAN ONE EXPECT FROM SLA

An engineer who is able to design digital circuit using SSI and MSI parts could expect to be able to design digital ICs of reasonable density and performance using the SLA method without having to become knowledgeable in semiconductor physics, electrical circuits or geometric layout techniques.

One could design chips targeted for implementation in Bipolar, HMOS or CMOS technologies using essentially the same design and conceptual framework using the SLA method of design.

Using 3 micron CMOS process rules, one could expect to design digital systems of the complexity of 15,000 gates in 250 X 250 mill square chip that might operate with a 30 megahertz clock.

One could expect the cost of obtaining a working prototype chip for a digital system designed with the SLA method to be the same as is presently required to obtain a working wire wrap board for the digital system using SSI, MSI and standard LSI parts.

PRACTICAL EXPERIENCE WITH SLA DESIGN METHOD

In the last year a total of nine chips were designed using the SLA method at Patil Systems and General Instrument Corp. The largest of these were an eight bit by eight bit multiplyer and the Universal Synchronous and Asynchronous Receive and Transmit (USART) data communication chip (8251A). These chips were designed by individuals with widely varying technical background which included a programmer who had never previously designed a digital system, a technician, an electrical engineer, and a computer scientist. All chips functionally worked the first time the silicon was produced. The availability of a simulation system that automatically extracts delays of signal propagation paths, and the designers ability to simulate the whole chip at one time were very important factors in the success of the chip designed. This simulation system is an integral part of the STRuctured IC DEsign system (STRIDE (TM) Patil Systems, Inc.) which has been designed specifically to support the SLA design method.

ACKNOWLEDGEMENTS

The work on SLA at Patil Systems, Inc. was supported by General Instrument Corporation and the investors in Patil Systems, Inc.

REFERENCES

G.B. Goates, "Storage/Logic Array Design: Reducing Theory to Practice", VLSI Design, Vol 3, No.4, July/August, 1982.

G.B. Goates, and S.S. Patil, "ABLE: A Lisp-Based Layout Modeling Language with User-Definable Procedural Models for Storage/Logic Array Design". Proc. of the 18th Annual Design Automation Conference, IEEE, 1981.

R.B.Haslam, "Design Goals of an Editor for Storage Logic Arrays", M.S. Thesis, Dept. of Computer Science, U. of Utah, June 1981.

E.Lin, "A Study of Loading Constraints of Existing Integrated Injection Logic Realizations of Storage/Logic Arrays", M.S. Thesis, Dept. of Computer Science, U. of Utah, August 1980.

Z.Navabi, C.Chiang, F.Hill, "Storage Logic Array Realization of RTL Descriptions", Proc. of the 1983 International CHDL Conference, IEEE, June 1983.

S.S.Patil, "On Testability of Digital Systems Designed with Storage/Logic Arrays", Proc. of the International Conference on Circuits and Computers, IEEE 1980.

S.S.Patil, and T.A.Welch, "A Programmable Logic Approach for VLSI", IEEE Transactions on Computers, Vol.28, No.9, September 1979.

K.F.Smith, "Design of Stored Logic Arrays in I2L", Proc. of the 1981 IEEE International Symposium on Circuits and Systems, IEEE, April, 1981.

K.F.Smith, "Design of Integrated Circuits with Structured Logic Using the Storage/Logic Array (SLA) Definition and Implementation", Ph.D. Thesis, Dept. of Electrical Engineering, U. of Utah, March 1982.

K.F.Smith, "The CMOS SLA Implementation and SLA Program Structure", in "VLSI Systems and Computations", edited by Kung, Sproull and Steele, Computer Science Press, 1981.

K.F.Smith, T.M.Carter, and C.E.Hunt, "Structured Logic Design of Integrated Circuits Using the Stored Logic Array," IEEE Transactions on Electron Devices, Vol. ED-29, No.4, April 1982.

Session 1-3b
Design Concepts

Chairmen

S. Johnson
Bell Labs

H. G. Harding
IBM

THE IMPACT OF AN ADVANCED CHDL ON VLSI DESIGN

Dieter Frantz*
Franz J. Rammig°

* Siemens AG, ZT ZTI, München FRG
° Universität Dortmund, Abt. Informatik, FRG

ABSTRACT It is shown how a multilevel broadband hardware description language supports VLSI design. It is indicated which design style is emphasized by the language. The entire system is in use at SIEMENS AG since a couple of years. So experiences with this system and some examples of designed systems are given.

1. Principles of CAP/DSDL

1.1 Basic ideas

Today there should be no doubt that in order to design a VLSI system a hierarchical method, covering a couple of abstraction levels, has to be used. From our point of view the following levels of abstraction have to be considered:

- Functional level

  At this level only the global behaviour is of interest. As far as possible a first modularisation into semiautonomous modules should be carried out. These modules are viewed at as black boxes that perform certain operations on request. In [SC1] this level is called LMA while we prefer a more general point of view as offered by concurrently cooperating Abstract Data Types.

- Algorithmic RT level

  At this level the basic module of the hardware to be designed are already present. Control and data manipulation still form an integral description as concurrent algorithm.

- Structural RT level

  That's what usually is called RT level. Control and data manipulation now are separated. The control part is implemented in any way and the algorithm itself is no longer present. In [SC1] a special case of this level is given as CRL.

- Gate level

  The basic concept of this level is the same as at the level above, only the primitive objects are simpler now. This level is not completely disjoint with CRL of [SC1].

- Switching level

  At this level the primitive objects are transistors, attenuators and lines. The transistor is viewed at not as an analog device but just as a bidirectional switch. A special case for this level is CPS in [SC1].

- Topology level

  The transistor network now has been arranged as symbolic layout without absolute scaling.

- Geometry level

  At this level the final layout for the specific technology is present.

It should be mentioned that within each level of abstraction hierarchical descriptions are allowed as well.
Traditionally the upper levels are supported very rarely by formal description tools while for the lower levels individual languages have been designed. This causes a lot of friction losses during the design process. Therefore today two main approaches are favored: Either the language family approach (mainly CONLAN [PB1]) or the broadband language approach, preferred by the authors.

Figure 1 illustrates this multilevel point of view for the design process. At the right the design activities can be identified while at the left the description tools are indicated. Per level of abstraction we distinct between generation and verifikation, forming a feedback loop. These loops are expanded to multilevel loops especially by implementation activities.

A prerequisite for a broadband language is a very small set of basic principles, valid for all levels covered by the language. In the case of CAP/DSDL (Concurrent Algorithmic Programming Language/Digital Systems Description Language) [RA1] there exists exactly one basic principle: Timed Interpreted Petri Nets [PE1, RA2] .

A Petri net is given by its static description (the Petri net graph) and its dynamic behaviour (via markings and firing rules). A Petri net graph consists of two kinds of modes: "places" to model conditions and "transitions" to model actions. The nodes are connected by directed arcs from places to transitions and from transitions to places. Each place may contain tokens. A transition is called "activated" if each of its input places contains at least one token. An activated transition may fire. If it fires it withdraws a token from each of its input places and adds one token into each of its output places. See fig. 2 as an selfexplanatory example.

CH1935-6/83/0000/0173$01.00 © 1983 IEEE

An interpreted Petri net is obtained if operations (data manipulation) are attached to the firing of transitions. They may be extended to timed Interpreted Petri Nets if in addition delays are attached as well.

In CAP/DSDL Petri nets may directly be formulated in order to describe the control structure of a digital system. On the other hand all the language constructs with control information envolved are semantically based on Timed Interpreted Petri Nets. In this context constructs for the structured concurrent programming and a monitor concept for encapsulation have to be mentioned especially [RA3] .

1.2 Refinement steps in terms of CAP/DSDL

At the top level (functional description) we have an abstract point of view of both, the data path (represented by abstract data types) and the control part (represented by a concurrent algorithm) with relative timing only. Abstract Data types are described in CAP/DSDL as "export procedures" (they export operations on a hidden internal structure.) Like any procedure in CAP/DSDL they constitute a critical section like encapsulation (like LMA moduls [SC1]). The concurrent algorithm is given either by directly formulating a Petri net or using structured concurrent constructs like segbegin ... end, conbegin ... end, case .. of ... end, while ... do ... (PASCAL like notation). If the data part is refined in such a way that the abstract data types are replaced by implementation objects (registers, combinatorial logic, busses) but the concurrent algorithm point of view is maintained, then the algorithmic RT level is reached. In CAP/DSDL registers and terminals of arbitrary length and structure may be declared and used and expressions of high complexity may be used in order to describe combinatorial logic. If desired the designer may underlay a clocking structure at this level of abstraction but this constitutes only a "second order" control structure. Clocking at this level is introduced by describing a clock generator and prefixing arbitrary statemets by constructs like at up (<clock name>) do (edge triggered) or when <clock name > do (level controlled).

In order to obtain the structural RT description the entire concurrent algorithm has to be separated into the data path with incoming control lines and outgoing information lines and a isolated control part, now only controlling these lines. Usually (but not necessarily) at this level of abstraction clocking is introduced. This is a key transformation as now the data manipulation is no longer controlled by an abstract algorithm but by events on certain data objects.

At the gate level the information about what is data manipulation and what is control is no longer present. It is the first level of abstraction where we have a pure structural description. The control part has to be transformed into any implementation of a finite state machine (as micro engine, via PLA, via random logic or by directly implementing the Petri net). In order to implement devices like registers this level may be skipped.

At this level of abstraction the rich set of logic operators including bundle operations and reductions offered by CAP/DSDL are especially valuable.
In order to obtain the switch level, all circuits have to be expanded into the interconnection of pullup, pulldown and transfer gates. This is the lowest level of abstraction offered by CAP/DSDL. Using a logic with different signal strengths constructs for a precise description like "bidirectional transfer" gate in various technologies and "charge decay time" are included [LR1] .
As a common concept for all levels (like PASCAL) CAP/DSDL distincts between constants, types of objects (that may be of arbitrary complexity, including procedures and abstract data types), and objects of such predefined or standard types.

Figure 3 shows the design style as supported by the CAP/DSDL system, while figure 4 points out the overall structure of every CAP/DSDL module (which may contain a couple of such modules).

2. The role of the CAP system within the VLSI logic design

Starting from given product requirements and target specification the designer begins modelling and simulating his design with the aid of the CAP system. The numerous refinement steps in the logic design process are supported by modelling and simulating at each abstraction level by this system. In the early design phase the models based on Petri nets, state diagrams, of functional block diagrams can be evaluated and verified by simulation. So the CAP system helps the designer not only to prove the correct behaviour of the circuit but also to optimize the design. Design debugging becomes easier, as the detection of failures takes place in the phase where the failures originate.

The timing behaviour especially at the logic gate level can be described in an accurate way: to each pin of an element individual delays can be assigned for differing rise- and fall-times and for switching from the low impedance state to the high impedance state and vice versa. The growing knowledge about the circuit during the design process results in new details, which can be added to already verified descriptions in a simple way, e.g. by superposing a clock structure to synchronize the logic control flow.

The consistency of sequential design steps can be guaranteed by a model, which is verified by simulation. The formal description of the circuit at each level not only gives the basis for simulation but also represents a unique documentation, which is necessary for the communication between the designers in a development with parallel tasks. With the aid of CAD more work can be done in parallel to get shorter development time. One example is the preparation of test data. With it the designer can already begin in the functional design phase bridging the time needed for layout, mask generation, and manufactoring first silicon by test data evalutation.

Data and results obtained by the CAP system are input to other CAD systems, which are used subsequently to CAP like fault simulation or test data generation. Input patterns for the logic and fault simulator for example can be selected from an overall system simulation at an higher abstraction level with CAP.

3. The multilevel simulation aspects

An important point for a homogenous, integrated simulation system is the fact, that at different description levels anything can be expressed with the same syntax and semantic and there is no need for multiple descriptions in different ways. To describe the logic and timing input-/output behaviour of blocks like cells at the gate level (used as pseudo-primitives by the logic and fault simulator of the integrated system) the same language as for modelling at higher abstraction levels can be used. Overall a mixed level model consisting of gate and transistor level parts and functional parts can be simulated. This contributes to more readability and efficiency, as only the parts of interest are modelled at the lower level. All other parts can be modelled in a more abstract way using already existing functional block models of the former design phase [KF1] .

The language for test data description (input patterns, internal signal changes, matching patterns) is also common to the simulators used in the integrated system and therefore also contributes to the homogenous user interface of the CAD functions used in several phases of the logic design. Simulating parts of a whole, more abstract system description at a lower level is supported by extracting parts modelled with more detail (which is a common capability) and more importantly by selecting the test data for the chosen part from the system simulation results.

Of great importance is a common command language and a common representation of the simulation results over all logic levels to get a great acceptance of the system by the user. Furthermore the system is dialog-oriented with the capability for entering subtasks as batch jobs. The system handling is easy to learn for the designer. It does not impose handling for the user, which can can be done by the CAD system itself, like file management or entering operating system commands. Instead, the system supports the designer by telling him the status of his model - for example if there are errors which need to be corrected - and by providing help functions in each situation. Another important point for acceptance is the fact that all functions from text editing over simulation to representing the results can be done within one system. There is no need to leave one program, return to the operating system and to start another program.

The simulation results can be called to the screen or printer. There are numerous commands to choose the desired signal- and time ranges and to represent the simulation results in an appropriate form.

Automatic restart capabilities and support for error handling are provided. This is important for efficient application with short turn-around-time.

4. Productive applications

The CAP system was and is still used for several LSI and VLSI developments. Examples are a peripheral board controller, a data transmission processor, and a DMA controller. These products were developed in MOS technology but some parts of the products also used bipolar gate arrays and standard TTL and ECL integrated circuits. The top down methodology, used in a nearly ideal way in the DMA controller project, was supported by the application of the CAP system in an efficient way. Each design phase, beginning with state diagrams was verified before the refinement of the design was continued. The development time of the data transmission processor could be shortened by the test data development in parallel with the layout and prototype manufactoring. Some characteristics of the circuits mentioned above are peripherical board controller with 80.000 gates (equivalents), data transmission processor with 40.000 gates, and DMA controller with 20.000 gates and four stages pipelining.

References:

[KF1] : F. Klaschka, D. Frantz: Die logische Simulation - Ein Werkzeug für den System- und Schaltungsentwurf (Logic Simulation - A tool for system and circuit design), in german in: CAD für VLSI, H.G. Schwärtzel (ed.) Springer, 1982

[LR1] : K.-D. Lewke, F.J. Rammig: Description and Simulation of MOS Devices in Register Tranfer Languages Proceedings VLSI '83, Trondheim Norway,1983

[PE1] : J.L. Peterson: Petri Nets ACM Computing Surveys, Vol. 9, 1977

[PB1] : R. Piloty, M. Barbacci , D. Borrionne, D. Dietmeyer, F. Hill, P. Skelly: CONLAN report Springer, 1983

[RA1] : F.J. Rammig: CAP/DSDL: Preliminary Language Reference Manual; Forschungsbericht Nr. 129 d. Abt. Inf. d. Univ. Dortmund, 1982

[RA2] : F.J. Rammig: An alternative method for self-timed VLSI systems; Forschungsbericht Nr. 131 d. Abt. Inf. d. Univ. Dortmund, 1982

[RA3] : F.J. Rammig: Hierarchical modular description of VLSI systems; Proceedings of IEEE Workshop on SE ∩ VLSI, Port Chester, 1982

[SC1] : M. Stefik, L. Conway: Towards the Principled Engineering of Knowledge; the AI Magazine, Summer 1982

Fig. 1

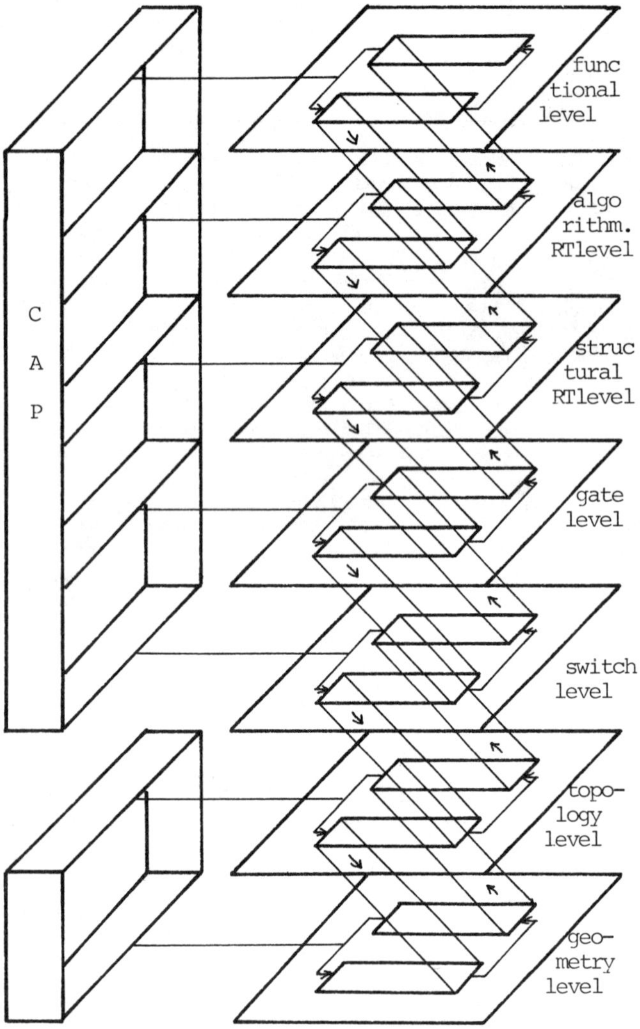

Fig. 2

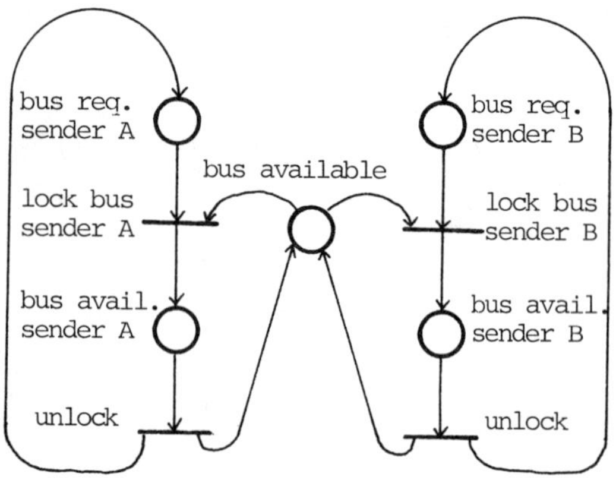

Fig.3

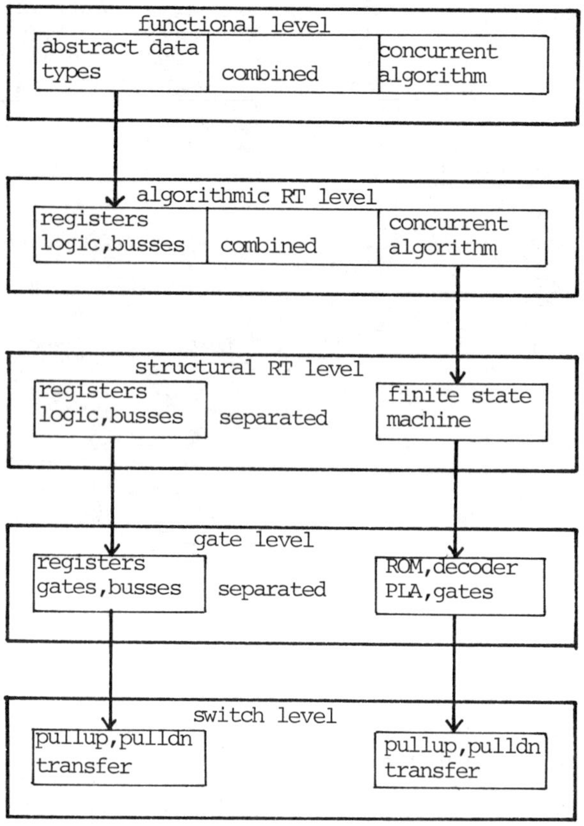

Fig. 4

interface to environment	common to all levels
constants	
object types of arbitrary complexity	
object declaration of standard types or of user defined types	
data manipulation without control part,nonprocedural (mainly used at structural RT-level and below)	level specific
algorithmic part (Petri net or structured concurrent constructs) includes data manipulation at algorithmic RT level and above, shrunk to symbolic part at structural RT-level and below	

A FAILURE RATE MODEL FOR MEMORY DEVICES

by Zoltan Papp

IBM General Technology Division, East Fishkill Facility
Hopewell Junction, New York 12533

Abstract

The use of fault-tolerant techniques (error correcting codes, on-line spare parts, etc.) increases the reliability of the memory systems. The application of these methods, however, makes the classification of the memory device failures necessary according to their effects on the reliability of the whole system. In this paper, we construct a failure rate model for the memory devices such that these "class-specific" (single cell, row, column, etc.) failure rates can be evaluated. The data for the model have to contain the following information: the failure time of the device, the cause of failure (failure mechanism) and the class of failure (piece, part, class). This type of data is available from the qualification life tests and from the incoming quality tests. The paper provides technical details of the model and the calculation methods. Examples are provided for illustration.

Introduction

The failures of memory devices can be classified in the following ways. First, according to the causes of the failures. This classification separates the failure mechanisms which are the theoretical explanations of the breakdowns. We are going to call the failures that are attributed to a failure mechanism <u>cause-specific failures.</u>

The use of fault-tolerant techniques in the memory systems (error correcting codes, on-line spares, etc.) makes another classification necessary. The breakdown of a single cell, a row, a column, or the whole chip affects the system reliability in different degrees. Consider the following example. A 16K DRAM chip has 128 rows with 128 cells in each row. Assume that a word consists of w bits, thus we have w chips linearly arranged such that the bits of a given word occupy the same relative location on each of the w chips. Assume further that we have a single error correcting mechanism. If a cell becomes disabled (e.g., a pinhole in the plate oxide of the capacitor shorts the cell), the memory system is still working correctly due to the error correcting technique. In order to obtain an uncorrectable error (UE), we need another cell to fail out of the 16,384 cells of another chip, and it has to be in alignment with the first fail. The chance of this is proportional to 1/16,384, which is rather small. If the first fail is a row fail (e.g., a row decoder fails due to an oxide fault), the chance of UE increases 128-fold since a row fail disables 128 cells. Finally, if the first fail is a chipkill (e.g., a short due to a pinhole in the gate oxide of same peripheral device), it is certain that a second fail of any kind leads to a UE. This simple example implies that in considering the failure rate model of a memory device, we need to classify the failures according to their effect on the whole memory system. In this way, we have different classes of failures and those in a given class are called <u>class-specific failures</u>.

The failure rate model discussed in this paper considers both classifications in order to estimate the failure rate of the memory devices. However, our main purpose is to obtain a method for the estimation of the class-specific failure rates.

The Failure Rate Model

Under both use and stress conditions the memory modules are exposed to several (competing) <u>risks</u>. Failure can occur in random <u>time</u> and in different <u>classes</u>. The data base obtained from the analysis of stress test failures contains the following information:

Data = (Time-to-failure, cause of failure, class of failure)

In this way, the failure of a memory device can be characterized by three random variables.

T: The time-to-failure random variable. We can assume that it is a continuous, non-negative random variable.

R: The failure mechanism, i.e., one of the risks that caused the failure of the device. It is a discrete random variable with k values, R=1,2,...,k.

C: The class of the failure. It is also a discrete random variable with m values if we are considering m failure classes. C=1,2,...,m.

The joint distribution of the triad (T,R,C) can be approached through the class and risk-specific survival, probability density, or hazard rate functions defined as follows:

$$S(t,i,j) = P(T > t, R=i, C=j) \quad (1)$$

$$f(t,i,j) = \frac{-dS(t,i,j)}{dt} \quad (2)$$

$$h(t,i,j) = \lim_{\Delta t \to o} \frac{P(t<T\leq t+\Delta t, R=i, C=j \mid T>t)}{\Delta t} = \frac{f(t,i,j)}{S(t)} \quad (3)$$

where $i=1,2,\ldots,k$; $j=1,2,\ldots,m$ and $S(t)$ is the marginal survival function where both the cause and the class of the failure are ignored.

$$S(t) = \sum_{i,j} S(t,i,j) \quad (4)$$

Similarly, we can obtain the class-specific and cause-specific marginal functions as well. For example, the cause-specific marginal hazard rate function, where the classification of the failures is ignored, is the following. (The sign "*" means that we performed the sum along that variable.)

$$h(t,i,*) = \sum_{j=1}^{m} h(t,i,j) \quad (i=1,2,\ldots,k) \quad (5)$$

Failure rate estimation can be obtained by the maximum likelihood method either under parametric or non-parametric assumptions. Observations are taken on a random sample of n devices. Let δ_1 be the censoring indicator variable. This means that when the lifetime of the 1th device is observed $\delta_1=1$, otherwise $\delta_1=0$, $(k=1,2,\ldots,n)$. Hence the observation of the 1th device has either the form $(t_1, r_1, c_1, \delta_1=1)$ or $(t_1, \delta_1=0)$, depending on whether the 1th device failed during the test or has been censored. Under the assumption of independent censoring the likelihood function has the following form.

$$L=A \prod_{1=1}^{n} \left\{ h(t_1,r_1,c_1)^{\delta_1} \prod_{i=1}^{k} \prod_{j=1}^{m} \exp\left[-\int_o^{t_1} h(u,i,j)du\right] \right\} \quad (6)$$

where A is a constant. (See: Kalbfleish and Prentice [2], p. 168).

We can rearrange the products in (6) according to the failure mechanisms (causes) of the failure.

$$L=A \prod_{i=1}^{k} \left\{ \prod_{1=1}^{n} [h(t_1,i,c_1)^{\delta_{i1}} \prod_{j=1}^{m} \exp\left(-\int_o^{t_1} h(u,i,j)du\right)] \right\}$$

where $\delta_{i1}=1$ if we observe the failure of the 1th device and it fails due to the ith failure mechanism, otherwise $\delta_{i1}=0$.

The factors of the likelihood function involving a specific failure mechanism are exactly the ones which would be obtained by regarding all failures due to other failure mechanisms censoring at the device's failure time.

Consider now the ith failure mechanism and its likelihood function.

$$L_i=B_i \prod_{1=1}^{n} [h(t_1,i,c_1)^{\delta_{i1}} \prod_{j=1}^{m} \exp\left(-\int_o^{t_1} h(u,i,j)du\right)] \quad (7)$$

We are going to put forward arguments to justify the assumption that, for a given failure mechanism i, the class-specific hazard functions are proportional to each other. This also implies that

$$h(t,i,j) = P_{ij}\, h(t,i,*)$$

where $h(t,i,*)$ is the class-specific marginal hazard rate function given in equation (5). Based on this assumption, equation (7) changes to

$$L_i=B_i \prod_{j=1}^{m} P_{ij}^{d_{ij}} \prod_{1=1}^{n} [h(t,i,*)^{\delta_{i1}} \exp\left(-\int_o^{t_1} h(u,i,*)du\right)] \quad (8)$$

where d_{ij} is the number of failures due to the ith failure mechanism in the jth class. Maximizing L_i means that we are maximizing both products in equation (8) separately. The first term

$$L_i' = \prod_{j=1}^{m} P_{ij}^{d_{ij}}$$

gives the maximum likelihood estimation of the proportions of failures in the different classes:

$$P_{ij} = \frac{d_{ij}}{\sum_{j=1}^{m} d_{ij}} \quad (9)$$

The estimation of the marginal hazard rate function $h(t,i,*)$ can be obtained from the other term of equation (8).

$$L_i'' = \prod_{1=1}^{n} h(t_1,i,*)^{\delta_{i1}} \exp\left(-\int_o^{t_1} h(u,i,*)du\right) \quad (10)$$

This multidimensional failure time pattern can be described in the "competing risk framework." In the case of k failure mechanisms, we have the random variables $T_1, T_2, \ldots, T_k$ representing the latent failure times of the devices from the different failure mechanisms. The actual failure time $T=\min(T_1,T_2,\ldots,T_k)$, and in equation (4) $S(t)=P(T_1>t, T_2>t, \ldots, T_k>t)$. The likelihood function (10) estimates the cause-specific hazard function associated with T_i if we can assume that the random variables $T_1,T_2,\ldots,T_k$ are independent. By experience, the assumption of independence for integrated devices is reasonable. See references [1] - [4] for more details on the competing risk model.

Let us continue the discussion with a given failure mechanism i. In the following, we are going to suppress the index i and the notation will reflect the classes of failures only.

The class of failures of a memory module is location dependent. This means that the failure due to a failure mechanism can cause a failure in different classes depending on the location of the breakdown. A short due to a pinhole in the gate oxide of an FET can lead to a single cell failure if it happens to a cell transistor. The same fault can lead to a column fail if it affects a transistor in a column decoder, or to a chipkill failure, if a transistor of one of the peripheral devices (e.g., date in/out latch) is damaged.

Therefore, if we list all the potential fault sites (with weights due to their frequency of use), we can obtain the proportions of potential failures in each of the m classes.

$$p_1,p_2,\ldots,p_m \text{ with } \sum_{j=1}^{m} p_j=1$$

These proportions are determined by the design of the memory chip and the manufacturing processes. They approximate the conditional probability that failure occurs in one of the classes, provided failure has occurred. The proportions should be estimated statistically from the failure data.

Let the marginal hazard rate function for the failure mechanism in question by h(t). Also $P(t < T \le t+\Delta t|T > t)$ is the probability that failure occurs in the time period $(t, t+\Delta t_{th})$ among the survivals until t. The same probability in the jth class is $P(t < T \le t+\Delta t|T > t)$. The failure event in the interval $(t,t+\Delta t)$ can be pictured as follows:

Failure in the jth class.

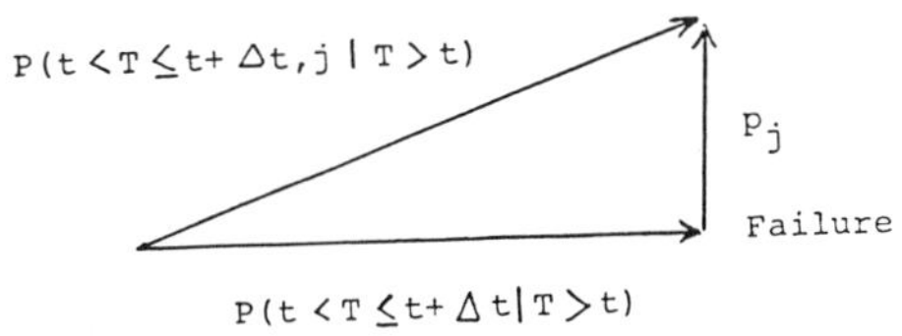

This means that our assumption implies the relation

$$P(t <T \le t+\Delta t,j\ |T >t)=p_j\ P(t <T \le t+\Delta t\ \ T >t)$$

In turn, this implies that the class-specific hazard rate function h(t,j) is the same multiple of the marginal hazard rate function h(t).

$$h(t,j) = p_j\ h(t)$$

This assumption is the basis of the "proportional hazard rate model" which we propose for the description of the failure pattern relative to the class of failures.

The consequence of this assumption is that the cause-specific marginal hazard rate functions and the proportions of failures in the classes for each failure mechanism can be estimated separately.

Failure analysis, and a maximum likelihood estimation method provide the following information which make up the <u>failure rate model</u> of the memory devices.

(A) Failure mechanism that are affecting the lifetime of the devices.

(B) The hazard rate functions of the lifetime random variables for each failure mechanism. (Cause-specific hazard rate functions.)

(C) The functional forms of the stress related acceleration factors and the necessary parameters (activation energy, etc.) for each failure mechanism.

(D) The proportion of failures in each class for each failure mechanism.

Using the above model, the failure rate calculations can be organized according to the framework presented in the following section.

Failure Rate Calculation

Let us introduce the following notations for the average failure rate of the devices over the time period (0,t).

$A_i(t,j)$ = average class specific failure rate for the ith failure mechanism in the jth class.

$A_i(t)$ = average failure rate for the ith failure mechanism.

$A(t,j)$ = average class-specific failure rate for the device in the jth class.

$A(t)$ = average total failure rate of the device.

Let also h(t) be the marginal hazard rate function of the devices, ignoring both the cause and the class of the failure. Then,

$$h(t)=\sum_{i=1}^{k} h(t,i,*)$$

where, the cause-specific hazard rate function $(h(t,i,*)$, $i=1,2,\ldots,k$ are available from parts (A) and (B) of the failure rate model. Part (D) of the model provides the proportions of failures in each class for each failure mechanism. A matrix of proportion $\mathbb{P}$ can be formed.

$$\mathbb{P} = \begin{bmatrix} P_{11} & P_{12} & \cdots & P_{1m} \\ P_{21} & P_{22} & \cdots & P_{2m} \\ \vdots & \vdots & & \vdots \\ P_{k1} & P_{k2} & \cdots & P_{km} \end{bmatrix}$$

With the use of we can obtain the class-specific hazard rate functions

$$h(t,i,j)=p_{ij}\ h(t,i,*).$$

The definition of the average failure rates and the results in Section 2 will provide the following equations.

$$A_i(t,j) = \frac{1}{t}\int_0^t h(u,i,j)du = \frac{1}{t}\int_0^t p_{ij}\ h(u,i,*)du = p_{ij}\ A_i(t)$$

$$A(t) = \frac{1}{t}\int_0^t h(u)du = \frac{1}{t}\int_0^t \sum_{i=1}^{k} h(u,i,*)du = \sum_{i=1}^{k} A_i(t) \qquad (11)$$

$$A(t,j) = \sum_{i=1}^{k} A_i(t,j) = \sum_{i=1}^{k} p_{ij}A_i(t) \qquad (12)$$

Also,

$$A(t) = \sum_{j=1}^{m} A(t,j) \qquad (13)$$

Equations (11) or (13) give the total average failure rate, and equation (12) gives the class specific failure rate of the device over the time period $(0,t)$. The use of part (C) of the model is to "translate" the parameters to the different use conditions from the stress condition.

The equations can be expressed in a short form using vector notation. The average failure rate for the failure mechanisms (i.e., marginal cause-specific failure rates) can be put into a vector of m components.

$$\mathbb{A}_{FM}(t) = [A_1(t),\ A_2(t),\ \ldots,\ A_k(t)]$$

The class-specific failure rates for the device is given by a vector of k components.

$$\mathbb{A}_C(t) = [A(t,1),\ A(t,2),\ \ldots,\ A(t,m)]$$

Then we have the following relations.

$$\mathbb{A}_C(t) = \mathbb{A}_{FM}(t)\ \times\ \mathbb{P}$$

$A(t)$ = Sum of the components of $\mathbb{A}_C(t)$

The same relations can be obtained for average failure rates over any time period (t_1, t_2) or for the hazard rate (failure rate) function.

Conclusion

The failure rate model discussed in this paper considers both the cause and the class of failures. Two basic principles are used to integrate the data into a comprehensive model. These are the competing risk and the proportional hazard assumptions. We gave reasons to justify the latter assumption in the calculation of the class-specific hazard rate functions. The consequence of these principles is the separation of the likelihood function such that the cause-specific marginal hazard rate functions and the class proportions can be estimated separately. Matrix algebra helps to calculate the failure rates from the proportions and the marginal failure rates.

Acknowledgement

The author wishes to thank Reynier W. Dam for his valuable remarks and suggestions, and to Rudolph A. Carballo and Michael J. McGowan for their support and encouragement.

References

[1] R. H. Herman and Rusi K. N. Patell, Maximum Likelihood Estimation for Multi-Risk Model, Technometrics, 13, (1971), 385-396.

[2] J. D. Kalbfleisch and R. L. Prentice, The Statistical Analysis of Failure Time Data, New York, Wiley, 1980.

[3] J. F. Lawless, Statistical Models and Method for Lifetime Data, New York, Wiley, 1982.

[4] R. L. Prentice, et al. The Analysis of Failure Times in the Presence of Competing Risks, Biometrics, 34, (1978), 541-554.

Plex: Automatically Generated Microcomputer Layouts

M. R. Buric, C. Christensen, T. G. Matheson

Bell Laboratories
Murray Hill, New Jersey 07974

ABSTRACT

A program has been developed that automatically generates VLSI layouts of microcomputers tailored to user specifications. Most major features of the generated microcomputers, such as the data-word size, the number of registers, and the instruction- and data-memory size, can be varied. The resulting microcomputers are small enough to be used as components on a custom VLSI chip. The machines have separate instruction and data space, three levels of pipe-lining, and execute an instruction every clock cycle.

1. Introduction

A program has been developed that automatically generates VLSI layouts of microcomputers. The program starts with a high-level specification of the functions that the machine is to perform and produces a complete mask-level layout. The resulting Plex microprocessors are small enough to be used as macrocell components on a custom VLSI chip.

The objective is to provide the system designer with high-level building blocks for systems on a chip. This has led us to create a design-automation methodology that is based on macrocell-layout generators. We have developed programs that produce user-defined subsystems such as RAM's, ROM's, ALU's, PLA's, counters, registers, stacks, and micro-sequencers. The most complex building block that we have developed so far is the Plex microcomputer.

Each Plex machine is a complete microcomputer containing a CPU, ROM, RAM, interrupt handling, and an input/output bus. Since the entire computer is on the same chip, the architectural constraints are different from classical microprocessors. Each Plex microcomputer is tailored to user specifications, and contains only the circuitry that is needed to execute the functions that the user specifies. Most major features can be varied, such as the data-word size, the number of registers, the instruction- and data-memory size, and the number of interrupt lines. The user specifies a Plex machine by writing the assembly or C program that it is to execute. The Plex design program evaluates the user program and calls circuit-layout programs (generators) that create the final design.

Each machine executes a fixed, reduced instruction set that is selected from a list of available instructions. The actual op-codes that each machine executes are built by the Plex design program when the machine is created. This allows the design program to minimize the number of bits used in the op-code, and thereby reduce the size of the program memory.

Plex microcomputers' architecture is hierarchically organized. This allowed us to construct the Plex-generator program from a hierarchical collection of lower-level circuit generators. The layout generators encapsulate hand-layout efficiency while providing the geometric and electrical flexibility needed to build Plex microcomputers in different configurations.

The machines all have a similar floor plan. This allows major blocks of the machine to grow and shrink to accommodate user needs. The machines have separate instruction and data space, three levels of pipe-lining, and a two-phase, non-overlapped clocking scheme. These features allowed us to partition the machine to speed instruction execution. Signals need only propagate through at most a third of the machine in a given clock cycle. This, plus the reduced instruction set, allowed us to implement a single-cycle instruction execution scheme that also increases throughput.

2. The User's View of Plex Microcomputers

The architecture of Plex processors is fixed, but all the important parameters are under user control. These are: the data width, the number of registers, the stack depth, the number of interrupt lines, the data- and program-ROM and RAM sizes, and the instruction set. All Plex processors execute a ***subset*** of a global instruction set. This subset is determined automatically by the Plex design program for each specific processor. The general architecture and the global instruction set are discussed in the next section.

From the user's point of view, a Plex microcomputer is a complete subsystem that executes a program and interacts with the rest of the chip by means of the following signals: the input data bus, the output data bus, the address bus, the input control signal, the output control signal, the interrupt lines, and the interrupt acknowledge signals (Fig. 1).

The width of input and output buses is the same as the width of the data path. Data is transferred from the processor to the output bus by means of the ***out address*** instruction. The ***address*** is placed on the address bus simultaneously, and the ***output control signal*** is pulsed to signify the transfer. External circuitry can decode the address to select the destination for the output data. The ***in address*** instruction transfers data from the input bus into the processor. The address bus may again be used to select an external device under the control of the ***input control signal.***

CH1935-6/83/0000/0181$01.00

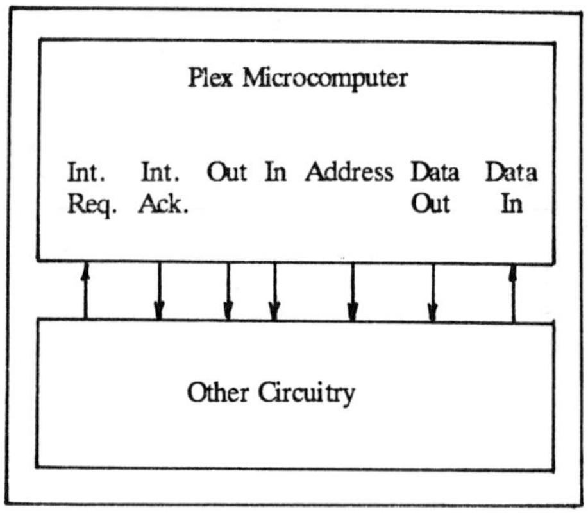

Figure 1. Plex microcomputers from the user's view.

Each interrupt line has an interrupt vector associated with it. The interrupt lines are prioritized. An interrupt is generated by an external device by a low-to-high transition on an interrupt line. As soon as the interrupt service begins, an interrupt acknowledge signal is produced and fed back to the device. The response to interrupts is fast; there are only a few instructions (such as ***jmp, jsr,*** and ***skips***) that can not be interrupted within one clock cycle.

3. Overview of Plex Architecture

A Plex microcomputer consists of three major sections: the data path, the program section, and the instruction decoder (Fig 2). Each section operates concurrently to maximize overall performance. The program memory is separate from the data memory, so instruction fetch can occur simultaneously with data-memory access and ALU operation.

The data path consists of an ALU, registers, data ROM and RAM, and an input/output bus. The data-path width is user programmable, as are the number of registers and the size of the ROM and RAM. The data width can be anywhere between 4 and 32 bits. We use dynamic RAM in the data path to minimize power consumption and layout area. The instruction decoder provides a refresh signal to the memory during all instructions that do not require memory access. The data path is controlled by the instruction decoder, and executes the following types of instructions:

```
alu_op   register1, register2
alu_op   memory, register
move     memory, register
move     register, memory
move     register, out_bus
move     in_bus, register
```

The memory address is provided by either the instruction decoder or by one of the registers. The ALU operations are all standard arithmetic, logical, and shift operations. The ALU also has a status register that consists of a carry bit, a zero bit, a minus bit, and an appended interrupt-enable bit.

The program section consists of a program counter, a stack, a program ROM, and an interrupt handler. The design parameters that are under user control here are the ROM size, the instruction size, the stack depth, and the number of interrupt lines. The program ROM is accessed every clock cycle to provide a new instruction for the instruction decoder. At the same time, the instruction decoder determines where the program address for the next instruction cycle will come from. The program address may come from the program counter, the stack, the interrupt handler, or an instruction.

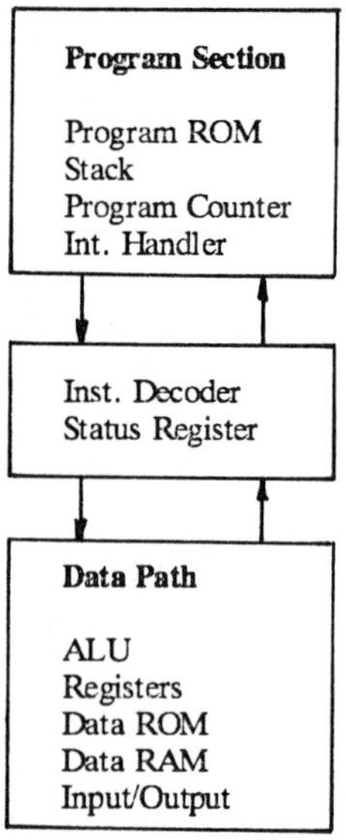

Figure 2. Major sections in Plex microcomputers.

The instruction set for a particular instance of a Plex microcomputer is drawn from a global set of available instructions, and in this sense is user defined. The instruction decoder is constructed automatically to decode only the used instructions; it is not capable of decoding the other instructions from the global set. Because of this, each Plex machine is not able to execute any Plex program, but only the one that specified it. The global set of instructions consists of three classes of instructions: data-path instructions, unconditional program-flow instructions, and conditional program-flow instructions.

The two-address data-path instructions are drawn from the following set:

```
mov   op1, op2
add   op1, op2
sub   op1, op2
cmp   op1, op2
or    op1, op2
xor   op1, op2
and   op1, op2
bit   op1, op2    (bit test)
```

In all cases except the ***mov*** instructions the second operand is one of the registers, while the first operand is either a register or a memory location. A memory address can be either absolute or register indirect. Either operand (but not both) in the ***mov*** instruction may be a memory reference. Immediate constants are handled by including them in the data-ROM space. The one-address data-path instructions are drawn from the following set:

clr reg
inc reg
dec reg
shl reg
shr reg
ror reg
rol reg

Instructions that operate on the status register are drawn from the following set:

clc (clear the carry bit)
sc (set carry bit)
eni (enable interrupt)
dsi (disable interrupt)

Interaction with the input/output bus is provided by:

out address
in address

The input/output bus consists of a separate ***in bus, out bus,*** and ***address bus.*** The address bus specifies which of the external devices is to be involved in a transaction. In both cases one of the operands is register 0, while the address can be either absolute or register indirect.

The unconditional program-flow instructions are:

jsr address
jmp address
ret
rti (return from interrupt)

The program addresses are always absolute.

The conditional program flow instructions are drawn from:

sz (skip if zero status bit is set)
snz (skip if zero status bit is clear)
sm (skip if minus status bit is set)
snm (skip if minus status bit is clear)
sc (skip if carry bit is set)
snc (skip if carry bit is clear)

Conditional branching is done with ***skip*** instructions and requires special attention because of the pipelined instruction execution. The instruction following the skip has already been fetched and is being decoded in the pipe when the branch condition associated with the skip becomes available. If the branch condition is satisfied, the instruction decoder transforms the following instruction into a ***no operation*** instruction. This ensures that the datapath will do nothing that affects the data or status registers, or the data memory.

4. User Specification of Plex Microcomputers

Since the Plex computers are used as components on a VLSI chip, and since all the important parameters are user defined, it is important to provide a natural way to specify them. A user defines a Plex microcomputer by providing ***a program*** that is to be executed by it. A two-pass compiler (the Plex design program) analyzes this program and constructs an intermediate representation of the microcomputer, which is then passed to the Plex layout generator. The compiler determines which instructions from the global set are actually used and constructs an instruction set which embodies these. The instruction size is independent of the data size, and is determined by code-minimization procedures. The instruction-decoding section is synthesized on the basis of the needed instructions and the required size of the address fields. The compiler also determines the number of registers needed in the data path. It finds the size and contents of the data ROM by analyzing the immediate constants in the program. Some Plex parameters are specified directly by the user through pseudo-ops. These are the data width, the number of interrupt lines, the stack depth, the size of the data RAM, and the address size of the input/output bus.

5. A Hierarchy of Layout Generators for Plex

Any successful silicon compilation has to offer high-performance systems in addition to complete layout design automation. We accomplish this objective in Plex microcomputers by using a hierarchy of specialized macrocell layout generators. Our generators are programs that output circuit descriptions in the language i. i is a hierarchical layout-description language developed at Bell Labs. [1] The basic primitives in i are transistors, wires, contacts and instances of other cells. Because i includes transistors, i circuit descriptions not only contain layouts, but also capture circuit diagrams.

i layout descriptions need not contain the absolute coordinates of all circuit elements, since the i parser, which is the front end of any program that manipulates i, resolves geometric constraints. If the coordinates of point X and point Y are specified, one need only say that a wire runs from X horizontally and then vertically to Y, and the i parser will determine the exact location of the rectangles that make up the wire. If the coordinates of only X are specified, one could say that a wire runs right n lambda and then up m lambda to Y, and the parser will determine the coordinates of Y. Entire blocks of circuitry can be placed this way. These features make i a natural base for writing generators. The generator programs need not calculate the exact location of every element in the layout, since the i parser can do this. This greatly facilitates building geometric flexibility into the generators.

Generators are basically C programs. C provides conditional testing and numerical computation, as well as a convenient interface to the UNIX† operating system. Flags and conditional expressions can be tested, and based on the tests, different geometric configurations can be output. Transistor sizes, power-line widths, and driver sizes can all be determined by calculations based on input parameters. Both geometric and electrical information can be woven into generators in this way. Since the generators are programs, one generator can call another. Complex generators can be produced from calls to lower-level generators and hierarchies can be built up in a natural fashion. Parameters that specify the circuitry that a generator is to produce are passed down through the C-function argument lists, and information about the circuitry that was produced by the called function is passed back up through standard data structures.

Our circuit generators are actually written in a hybrid language called ic that consists of C programs with i circuit descriptions embedded in them. The ic preprocessor recognizes lines that contain i keywords, and places these lines in standard C printf statements. Any C variables that appear in

†UNIX is a Trademark of Bell Laboratories.

the i text are automatically properly formatted in the printf statements. The output of the ic preprocessor is passed directly to the standard C compiler. When the compiled generator program is run, it outputs i text with the computed values of the C variables inserted in it.

We create generators by first laying out the lowest-level circuitry using a color-graphics design station that manipulates i layouts. Basic cells are hand packed, and a feel is developed for the different geometric configurations that are possible. The i descriptions of the layouts are taken directly from the layout editor and embedded in an ic program. Unneeded constraints that may have been inserted by the layout editor are removed, and C variables are inserted to create flexibility. Additional circuitry can be added by hand, entering i statements using a text editor. Detailed circuit simulations are done, and the empirical performance information that is derived is built into the program. When several different layout geometries are possible, all may be included in the program. The one that best fits user specifications is chosen at run time. Higher-level macrocell generators are constructed by mixing i text with calls to lower-level generators. The calling generator places the blocks of circuitry that it receives from the called functions and wires them together.

Plex microcomputers contain ten levels of hierarchy from the top level to the bottom leaves. Presently, there is one floor-plan strategy on the top level, but several new strategies are being developed. They must provide for arbitrary growth of Plex subsystems while allowing the best use of the remaining silicon area on the chip.

The present implementation of Plex generators is in 3.5 micron NMOS technology. In Figure 3 we show photomicrographs of two Plex processors in 6 mm frames. One is an 8-bit processor with 44 data-ROM and 56 data-RAM words. The program is 292 instructions long. The other is a 16-bit machine with 52 data-ROM and 96 data-RAM words. The program is 297 instructions long. Practical bounds presently are: 4- to 32-bit data-paths,1 to 8 registers, up to 512 words of data ROM and RAM, up to 2K instructions, and up to 16 interrupt lines.

6. Conclusion

We have developed a novel approach to generating microcomputer macrocells. Plex microcomputers are flexible, user-defined, and high-performance processors. The architecture is fixed in a general sense, but each instance of a Plex microcomputer is different from others. Several different designs of Plex microcomputers have been fabricated with data-paths ranging from 8 to 16 bits. All functioning devices met the clock-rate goals. Depending on the combination of specification parameters, the throughput ranged from 6 to 8 million instructions per second.

7. References

[1] Steve C. Johnson, "Hierarchical Design Validation Based on Rectangles". *Proceedings, Conference on Advanced Research in VLSI,* MIT, (January, 1982).

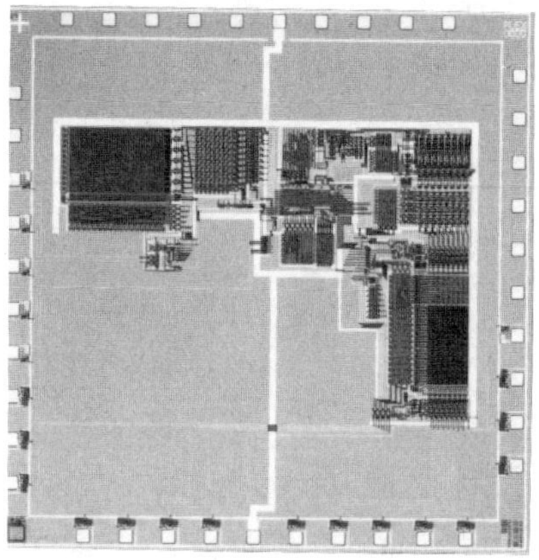

a)

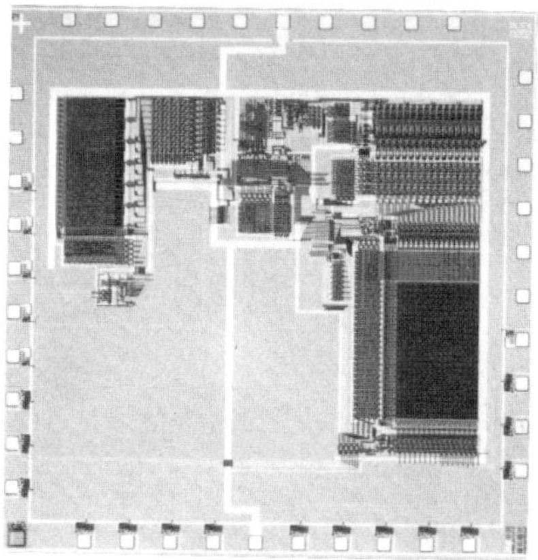

b)

Figure 3. Photomicrographs of Plex microcomputers. a) An 8-bit Plex microcomputer. b) A 16-bit Plex microcomputer.

PROCEEDINGS OF THE 1983 IEEE INTERNATIONAL CONFERENCE ON COMPUTER DESIGN

EFFECTIVE STRESS SCREENING

QUINTON O'NEAL

IBM GENERAL PRODUCTS DIVISION, SAN JOSE, CA

Abstract

Today's computer system applications require an unprecedented degree of operational reliability overall, and in the individual system components. This paper describes the use of manufacturing stress screening as a product and process design improvement tool for achieving high product reliability levels. Manufacturing stress tests applied to initial production provides vital early reliability performance and failure mechanism data. This data, used in a comprehensive failure analysis program can uncover subtle product design deficiencies as well as latent problems in assembly or test. Early data often allows corrective action implemention months before field data is available. Key elements in accomplishing this are; incisive stress screening procedures, accurate failure data collection, and disciplined analysis/corrective action techniques.

Introduction

Manufacturing Stress Screening Definition

As used here, manufacturing stress screening is defined as production testing procedures intended to improve fielded reliability and provide data for product or process enhancement. The procedures may include abnormal conditions to accelerate responses and are beyond those usually needed to assure correct product operation. Field reliability is improved by preventing marginal units from being shipped and by on-going corrective actions.

Some examples of manufacturing screening test techniques are; extended test runs, thermal cycling, burn-in and power cycling. Manufacturing stress screening is distinct from engineering design verification testing that occurs on preproduction units.

Stress Test Procedures

When establishing a stress test procedure consider the product's design, including it's semiconductor technologies, circuit interconnection methods, power supplies, and mechanical packaging. Design analysis can help predict failure mechanisms and place limits on the degree of stress applied. To find effective stress tests, design analysis will usually be augmented with experiments using a variety of stress conditions.

Complex systems may exhibit failure modes not detectable in the individual system components. An example is data transfer reliability failures in computer peripherals. One mode observed is failures whose rate decreases with time, as in Weibull distribution failures, commonly associated with semiconductors.

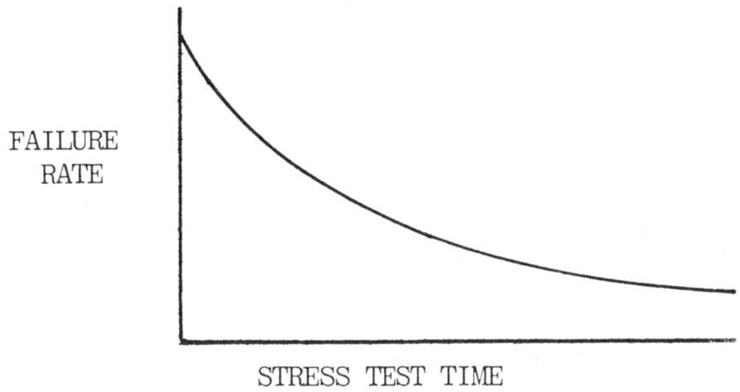

In another failure distribution mode seen, the failure rate increases with time, before decreasing. This has been found to be associated with data transfer elements and due to a variety of causes. Among the causes identified are: interaction between system components, very intermittent failures and failures requiring large data transfer volumes for detection.

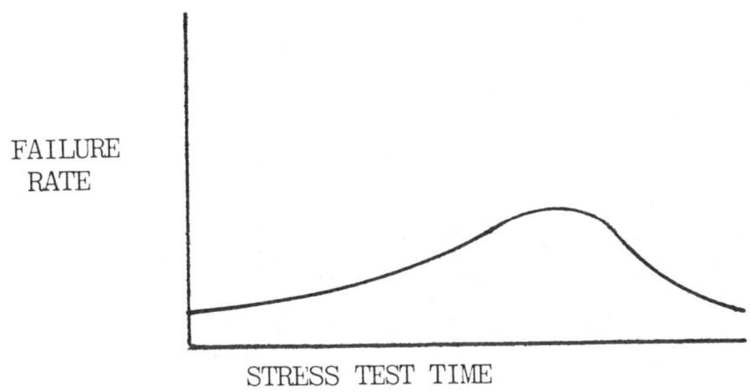

CH1935-6/83/0000/0185$01.00 © 1983 IEEE

A composite of these two major failure distribution modes yields a curve quite different from the Weibull distribution alone.

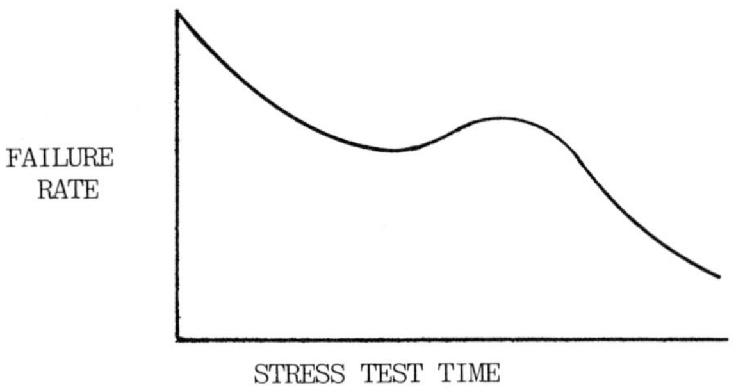

This curve may be used to set stress procedure time limits and data transfer or operational cycle guidelines.

Stress Test Data Collection

Data collection for subsequent analysis and corrective action is one of the primary reasons for performing a manufacturing stress test. Accurate, complete data collection will greatly aid failure analysis efforts.

An important element in stress test data collection is a data base with flexible sort and retrieval capability. The data base should allow searches for combinations of failure data elements, such as failing part, time to failure, and failure symptom. Other useful information to record for future analysis is the test conditions at failure, amount of testing prior to failure and an identification number.

Failure Analysis/Corrective Action

Manufacturing stress testing can be most effective only when used in conjunction with a comprehensive corrective action program. Continuous reliability improvements can be achieved by systematic elimination of defect root causes. A key step is to establish a failure analysis sequence or part flow path for failed parts. The sequence should provide for identification of the part (serialization) or tracking, verification of the defect, visual/mechanical inspection, failed component isolation and further stress tests if necessary. The failure analysis sequence should provide for appropriate data collection at each step.

- Stress Screening
- Failure Analysis Serialization
- Failure Verification
- Failure Physical Analysis

Steps that have proven beneficial in improving failure analysis discipline are:

- Verify all failures - to reduce misdiagnosis.
- After verification, trace all failures to a physical (mechanical/chemical) reason - this leads to cause identification.
- Correlate all physical failure reasons to the original failure symptom - verifies cause diagnosis.
- Track all corrective actions to implementation.

Conclusion

An effective stress screening program will significantly enhance new product field performance, yield data for future product and process design, and provide rationale for its own reduction. Along with design for reliability, failure analysis and corrective action, stress screening can be a key contributor to product reliability.

Bibliography

1. 1978 JOURNAL OF ENVIRONMENTAL SCIENCES, "Reliability of Mechanical and Electronic Products: Is a different approach necessary?" A.D.S. Carter, Vol 21, No.1

2. "An Effective Thermal Cycle Screen for Electronic Subassemblies" A.F. Bonetti, Proceedings of the National Conference on Environmental Stress Screening of Electronic Hardware 1979.

3. "Environmental Stress Screening Guidelines" 1981 The Institute of Environmental Sciences, Mount Prospect, Ill. 60056

Session 1-3c
Workstations I

Chairman

E. Rymaszewski
IBM

TOWARD A WORK STATION FOR VLSI DESIGN

B. S. Ting, B. N. Tien
Hughes Aircraft Company
Newport Beach, California

J. D. Schmidt, J. M. Youmans
Hughes Aircraft Company
Long Beach, California

INTRODUCTION

As chip size and resulting functional complexity increase with wafer fabrication technology, the use of computer aided design tools by chip designers has become mandatory. Sheer size and complexity necessitate large amounts of computational resources. Industry competition in labor costs and design/production schedules demands the engineer design more logic gates per hour. At Hughes Aircraft Company in Southern California, an LSI computer aided design system has been implemented in a hierarchical system of computing capabilities to satisfy both the high productivity needs of the LSI designers plus the CPU and disk storage intensive requirements of the CAD tools for large complex chips. This paper examines how this system provides the necessary CAD tools in an integrated manner to support the logic and circuit portions of the behavioral design and the layout aspect of the physical design of the chips.

DESIGN TASKS

The design tools are required by two classes of designers: engineers concerned with behavioral design - schematic capture, logic simulation, fault simulation, mix mode simulation, timing verification, testability analysis, test generation and circuit simulation; and engineers concerned with physical implementation - automatic/interactive placement and routing, cell library creation, and mask pattern generation. Design verification tools involve both classes of engineers, involved in mask to logic verification, electrical rule check, geometrical design rule check, and parameter extraction.

a) BEHAVIORAL DESIGN

The most time-consuming part of overall chip design is behavioral design, transforming system level specifications to detailed circuit level design. This is also the most creative part of the design process, requiring constant feedback and engineering changes. Design automation tools used in this phase perform schematic capture, logic simulation, fault simulation, timing verification, and circuit simulation.

Schematic capture is a task that requires very little computation time but needs a fair amount of data base handling and fast interaction. Medium range resolution color graphics is generally required for schematic capture.

With the timing verifier, the designer can obtain first order approximation of network delays in logic simulation speed without relying on heavy circuit simulation analysis. Circuit simulation is used primarily for component level design. The most time-consuming aspect in this design phase is fault simulation. When system requirements dictate large percentage fault coverage (typically better than 95%), extensive simulation is conducted to obtain a set of test vectors for fault grading. Using automatic test pattern generation employing the "set and scan" type of design approach, for testability, can help reduce the computation requirement.

b) PHYSICAL DESIGN

The physical design stage requires both computational resources and a high degree of interactiveness. The automatic placement task often consumes large amounts of CPU time, especially at the iterative improvement stage. The automatic routing task also has heavy needs for main memory and computation time. In addition to the CPU requirements, some physical layout tasks demand highly responsive graphical interaction either to improve layout or to complete a layout. High resolution color graphics is generally required for tasks such as wire editing and interactive placement.

c) VERIFICATION

The verification process inspects the physical design and the physical design relative to the logic design. Design rule check and electrical rule check are the two most commonly performed tasks on finished layout before mask-making. The physical design data must be compared against the original logic specification to ensure the validity of design. Post-layout timing analysis with the actual wiring interconnect should be performed to check if the design is within system specification. Each component's physical design must be verified for correctness in both function and performance analyses. The geometric design data base must be checked to meet process design rules.

Table 1 summarizes the relative demands imposed upon the computational resources of the CAD system by the various design tasks.

DESIGN TASKS	Main Memory	CPU Time	Disk Storage	Graphics Resolution/ Response
Behavioral Design				
Schematic Capture	H	M	M	M / H
Logic Simulation	M	H	M	M / M
Fault Simulation	H	VH	M	NA/NA
Testability Analysis	M	M	M	NA/NA
Test Generation	H	VH	M	NA/NA
Timing Analysis	M	M	M	NA/NA
Circuit Simulation	M	H	M	M / M
Physical Chip Design				
Automatic Placement	L	H	L	M / M
Automatic Routing	H	M	H	L / L
Interactive Placement	L	L	L	M / H
Interactive Routing Editing	H	L	H	H /VH
Cell/Macro Design	M	L	M	H /VH
Mask Pattern Generation	H	VH	VH	NA/NA
Mask Verification				
Mask to Logic Check	H	H	VH	H / M
Electrical Rule Check	H	H	VH	H / M
Design Rule Check	H	H	VH	H / M
Parameter Extraction	H	H	VH	NA/NA

LEGEND: L - Low, M - Medium, H - High,
VH - Very High, NA - Not Applicable

Table 1. Relative Computational Resources Requirements of Chip Design Tools

SYSTEM ARCHITECTURE

An analysis of the computational characteristics of the various LSI CAD tasks described previously leads to the following categories of computing.

a) PRODUCTION GRAPHICS

Examples of production graphics tasks are logic design capture and editing, and the interactive placement and routing of the physical design. Production graphics requires a significant investment for high resolution color graphics terminals, adequate CPU resources, and application software. Because of the need for an operator to maintain a continuous thought process, system response levels must be stringently maintained. Therefore, graphics data communications rates must be high and the host CPU insulated from the demands of other users and applications.

b) INTERACTIVE ENGINEERING COMPUTING

This category refers to interactive processing that the engineer must perform outside of the realm of production graphics. For interactive engineering computing, the CPU requirements are often less demanding and low cost alphanumeric terminals are ordinarily used. Examples are: circuit simulation of small circuits; logic simulation of a portion of the total design; input file editing and verification prior to large simulations; certain automatic layout functions; software development; and word processing. To maintain good interactive response levels, CPU-intensive tasks are often assigned a lower CPU priority.

c) BATCH PROCESSING

The batch processing category of work refers to previously mentioned CPU-intensive applications as logic and fault simulation, timing verification, automatic layout of VLSI circuits, design rule checking, and logic-to-layout verification. While interactive users should be insulated from the demands of these applications, the need to expeditiously perform batch processing remains.

The diverse characteristics of the three categories of LSI CAD computing suggests a hierarchical computing system architecture, as illustrated in Figure 1.

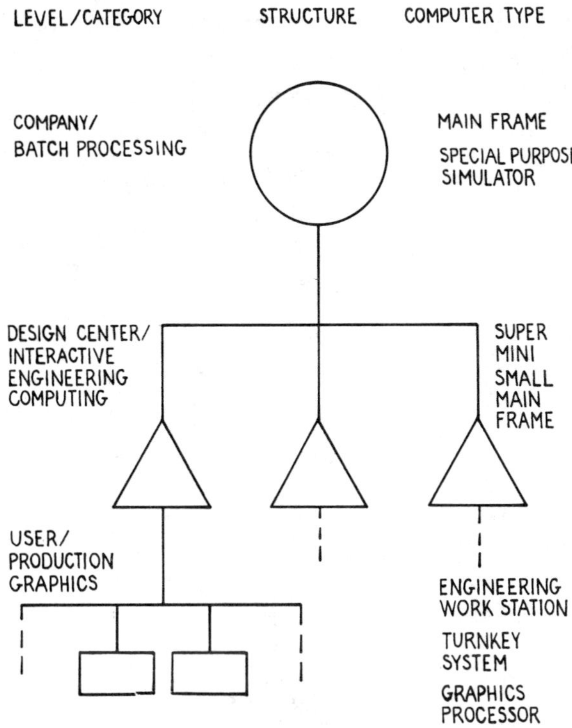

FIGURE 1. HIERARCHICAL SYSTEM ARCHITECTURE FOR LSI CAD

This multilevel architecture provides maximum opportunity for optimizing computer resource utilization while maintaining service levels essential for the LSI CAD software user community. For example, graphics processors may be sized to the need of the color graphics user without being under-utilized (as they would be for such applications as editing or word processing), or over-utilized (as they would be for large simulations). Color graphics terminals are preserved for those who need them. Users of the Interactive Engineering Computer are insulated from CPU-intensive work through "down-tuning" of CPU-intensive applications, or "down-loading" to the batch processor. Optimal utilization of the batch processor may be achieved by balancing the batch workload submitted from many design centers, gaining the "economics of scale" inherent to a large volume operation.

Effective use and data communications between computers are a critical requirement in a hierarchical system configuration. Interfaces between the user and the computer must be simple and consistent. File transfers between computers must be accomplished simply and quickly, and high speed communications links are essential. "Passthrough" capabilities, a user can access a second computer while remaining connected to a first, are desirable.

SYSTEM IMPLEMENTATION

The Hughes Aircraft Communications and Data Processing Division and Solid State Products Division have developed a hierarchical system for LSI CAD. Evolving over a period of several years, the system has been operational since June 1983, as shown in Figure 2.

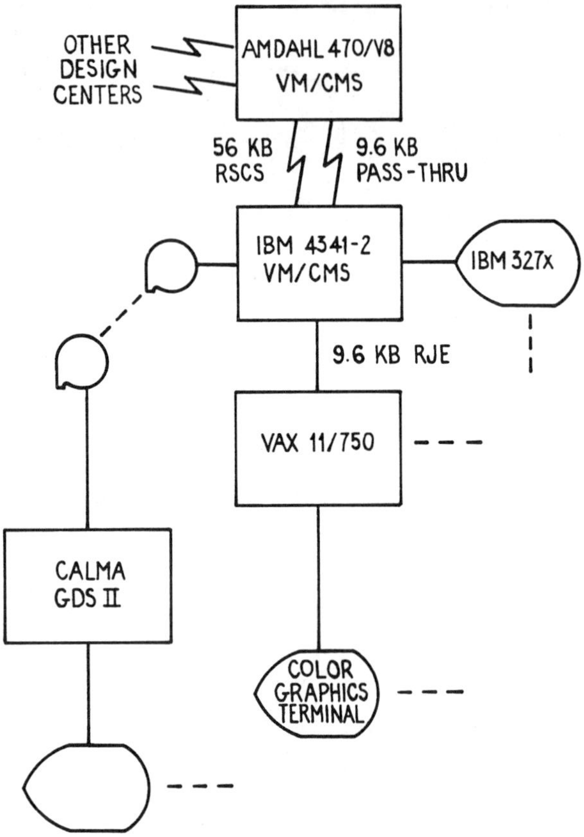

FIGURE 2. LSI CAD IMPLEMENTATION AT HUGHES AIRCRAFT

Serving as a batch processor dedicated for LSI CAD, the Amdahl 470/V8 is an IBM plug-compatible 6.25 MIP computer running the VM/CMS operating system. It is available for use by all Hughes Aircraft LSI Design Centers. It hosts numerous proprietary industry standard and internally-developed LSI behavioral and physical design application softwares. Although integrated into a single system, multiple entry flexibility is maintained for those wishing to utilize only a single application.

The IBM 4341-2, also running the VM/CMS operating system, is the hub of the network for engineers located at Newport Beach. Use of the same operating system on plug-compatible computers provides multiple advantages, including compatible communications protocols, common user interfaces, identical system and application software, and redundancy in the event one of the system fails. Use of the 4341-2 by all those in the Newport Beach Design Center affords a single location for storing design and process data, thereby facilitating data configuration management.

For production graphics, selection of the VAX 11/750 and high resolution color graphics terminals was influenced by its compatability with the selected logic design capture software, the need to use the same graphics terminals for interactive placement and wire editing tasks, displaying simulation outputs, probing of mask verificaiton results, and creation of mask design for basic component cells. CALMA GDS-II turnkey graphics system is currently used for cell mask design and pattern generation. Interface between CALMA and other computers is through tape transfer of the STRING formatted data.

Table 2 presents the CAD tools and computational resource primarily utilized to accomplish each of the LSI chip design tasks.

DESIGN TASKS	Computer System	DA Software Tools
Behavioral Design		
Schematic Capture	VAX	TEGATE(1)
Logic Simulation	AMDAHL	TEGAS,TEXSIM(1),SHIELD
Fault Simulation	AMDAHL	TEGAS(1),SHIELD(2)
Testability Analysis	AMDAHL	TEGAS,SHIELD
Test Generation	AMDAHL	COPTR/ATG(1),TESTGEN(2)
Timing Analysis	IBM/AMDAHL	DANA(2)
Circuit Simulation	IBM/AMDAHL	SPICE(1),SLICM(1)
Physical Chip Design		
Automatic Placement	IBM	GALA(2),HAL(2)
Automatic Routing	IBM	GALA,HAL
Interactive Placement	IBM/VAX	GALA,HAL
Interactive Wire Edting	VAX	GALA,HAL
Cell/Macro Design	CALMA	
Mask Pattern Generation	CALMA	
Mask Verification		
Mask to Logic Check	AMDAHL	MASKAP(1)
Electrical Rule Check	AMDAHL	MASKAP(1)
Design Rule Check	AMDAHL	MASKAP(1)
Parameter Extraction	AMDAHL	MASKAP(1)

1 - available externally, 2 - Hughes internal

Table 2 Primary Computer Systems and DA Software Tools for Chip Design Tasks

CONCLUSION

The key in total system planning is the realization that the designers will always need more, not less, computational resources. As chip complexity increases with technology, software programs supporting the design activity are requiring more computation power than hardware improvement. The implication is that the designer will need more on a larger machine due to computation requirements. However, for most front end engineering activities, designers will need for user friendly, easy access, and fast response design stations in an interactive mode. The implication is for a dedicated engineering workstation with adequate computing hardwares properly structured to meet the drastically different requirements at different stages of the design cycle. In our system configuration, we have provided various types of hardware to meet different levels of computational requirements in conjunction with supplying each designer with a user-friendly, interactive, and responsive front end engineering workstation.

We call our configuration 'workstation system' based on the concept of individual design terminals with a hierarchically networked CPUs to furnish the designers with the necessary computation and memory resources. In addition, it provides the highly responsive interactive graphics required to finish a design in a user transparent mode. The overall goal is to have a fully integrated CAD system in an user friendly environment. To the individual designer, the system is accessed through a dual screen setup consisting of one alphanumeric CRT and a high resolution color graphics terminal. To the architect/builder, the system consists of multiple levels of computer hardwares and application softwares integrated together to provide a systematic approach to assist the LSI/VLSI design task.

ACKNOWLEDGEMENT

The authors would like to express their deep appreciation to Hughes executives and management at Solid State Products Division and Communications and Data Processing Division for their enlightened support to the development of this system.

WORKSTATION DESIGN FOR VLSI LAYOUT

By Philip A. Preston and Robert J. Smith, II

VR Information Systems, Inc.
5818 Balcones Dr.
Austin, Tx 78758

ABSTRACT

In recent months, the CAD industry has increasingly focused on designer workstations. This is due in large part to rapid advances which have been made in desktop computers and graphics displays. This pace shows no signs of abating in the foreseeable future. At the same time, circuit complexity is increasing rapidly and also is expected to continue at the present pace until the end of the decade.

These two factors, coupled with ever-increasing expectations of user-friendliness, are posing interesting challenges to the designers of workstations for CAD/EE. In this paper, we examine these factors as they appear in the layout design of complex semicustom IC and PCB circuitry, and discuss a proven approach.

INTRODUCTION

During the past several years more and more computer-based services have become available to the circuit design engineer. Early tools consisted of isolated, special purpose programs which aided in various aspects of design and fabrication. These tools have progressed in breadth and sophistication, and are increasingly integrated into a fully automated CAD/EE system. An illustration of the range of design tasks which these tools address is shown in Figure 1. The processor for this integrated system is at present usually a large host mainframe. Graphics terminals are used to interact with the system and complete the designs.

Recent advances in graphics hardware, coupled with more powerful microprocessors, are making it possible to

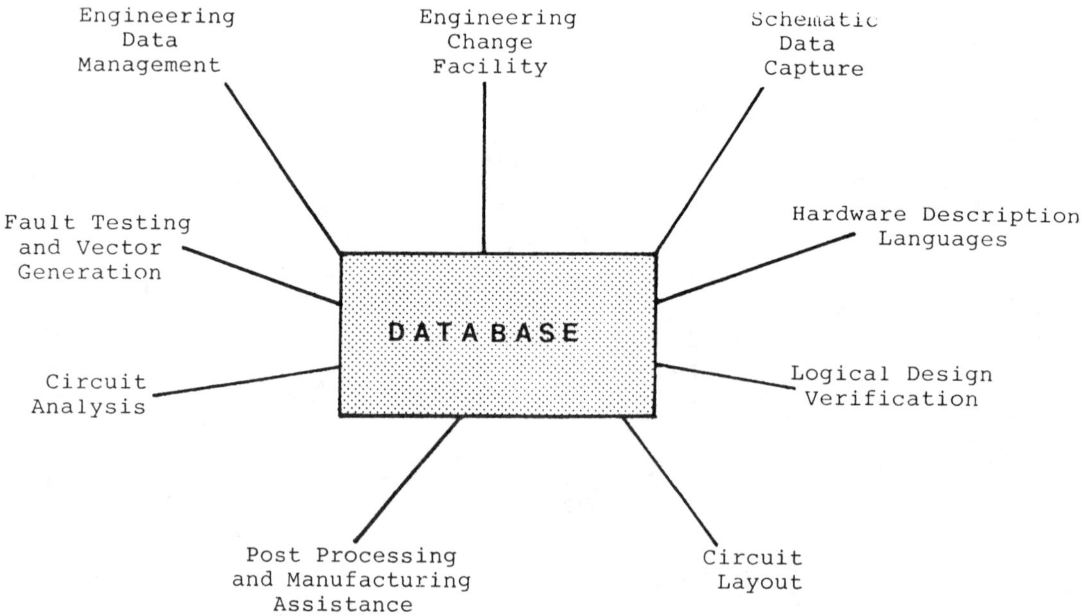

Figure I

improve this design environment through programmable workstations. These workstations allow a range of design tasks to be performed locally at the designer's desk, improving response time and increasing productivity. As processing power and local storage continue to increase, the range of design tasks which can be performed effectively on local workstations will also increase.

This increased power, however, should be seen in contrast to the demands being placed by the users on them.

First, circuit complexity for gate array and semi-custom integrated circuits is increasing rapidly. Gate counts of 40K gates have already been reported. It is anticipated that gate counts for moderate-volume IC's may double every two years at least through the end of this decade.[1] Obviously, although smaller circuits can be effectively designed on a workstation, there is also a continuing need for larger hosts for these more complex circuits.

Secondly, the desk-top design environment brings with it higher user expectations of service delivery speed and user-friendliness. CAD systems which have been quite successful in a batch or large host environment must be altered to perform adequately in conversational interactive mode or in a distributed processing environment. This has several subtle ramifications associated with program structure and data management.

In addition, the ready availability of the design station and its suite of tools will encourage the engineer to explore more design alternatives. This will inevitably lead to higher quality circuits, but places a heavy burden on the CAD tools to provide rapid response to complex design service requests.

Ideally, the new design environment would blend the increased productivity associated with workstations together with the computing power of the large mainframes, and not lose the fully integrated system environment which is available today. In this evolving environment several questions must be considered by developers and users of CAD tools alike:

- How should a CAD system be designed so that the advantages of workstations are used to facilitate the design of complex circuitry?

- What range of circuits can be effectively designed with standalone workstations?

- As circuit demands change and new graphics hardware becomes available, how does an organization protect the sizable investments already made in computer aided design hardware and software?

DISTRIBUTED CAD

In the present fully integrated design environment, engineers have access to all design tools though the host mainframe. As workstations are incorporated into this environment, several of these CAD tools will be found quite effective when operated on local workstations. Selection of specific design tools to be provided on an individual workstation is governed by hardware characteristics, the intended use by the design engineer, and software transportability.

DYNAMIC HARDWARE

It will not in the future be unusual to find more than one type of workstation in the design environment. This could be for several reasons. As an example, design engineers will have more use for schematic data capture tools than for graphical route editing. Consequently, less expensive workstations which effectively handle schematics but are insufficient for layout editing may be found in design engineering, while more sophisticated stations are used for other tasks.

In addition, as advances in workstation hardware and software continue, new products may augment pre-existing equipment, resulting in a diversity of workstation configurations in the same facility.

DYNAMIC SOFTWARE SET

Depending upon the nature of the work being performed by an engineer, some design tools will be more highly utilized than others. It will be found advantageous to move some of these tools and capabilities to the local workstation, in order to increase productivity. One problem frequently encountered by users of large host mainframes is that although the processor is more powerful, it is unavailable due to high utilization. Engineers frequently prefer to run computationally intensive jobs on a local workstation even if processing elapsed time is significantly increased on the smaller machine.

HIERARCHY

In designing large, complex circuitry, it is often essential to subdivide the problem into several subfunctions. Each of these subfunctions may at any time be at a different stage of completion by a different engineer. In a workstation environment, this could easily mean that the design is evolving locally on several workstations, and not on the host. Consequently, data integrity and synchronization is of vital importance. Compatible and easily updated database representations of evolving designs across all workstations and hosts is therefore a requirement.

TECHNOLOGY INDEPENDENCE

Computer aided design service delivery software written for the environments characterized above requires a significant investment. This can be quickly made obsolete with advancing semiconductor and packaging technology unless adequate planning is done during design of the software. One of the major challenges to developers of CAD tools is to provide in advance for evolving technology directions. To the fullest extent possible, the design tools should be independent of the groundrules of any one technology. Underlying assumptions of spacing and wire width, as an example, should be identified and handled in as general a manner as practical.

The trend which is developing will be characterized by dynamic hardware and software configurations. These configurations must be readily adaptable to the changing needs of the designer and the rapid evolution of new workstation and host hardware. Clearly, this places a large burden on the underlying CAD system and its software constituents. True technology independence will, from a business viewpoint, require generalized circuit database facilities that can be effectively distributed, as well as circuit design software that is transportable to an evolving diversity of processing and graphical environments.

PRACTICAL APPROACH

The exploitation of semicustom, high density microprocessor and memory integrated circuits is accelerating. This is producing a demand for widespread availability of comprehensive sets of circuit design tools --for most design engineers and technicians. A practical approach to providing the necessary services will, at least for larger designs, be based on distribution of services among workstations, large host computers, and perhaps special purpose CAD hardware.

Initial circuit specification will for most purposes, be workstations-based, and should allow for use of schematic- as well as language-oriented tools. These must be integrated with circuit evaluation and analysis tools that are also local to the workstation. These tools must be closely coupled to simulation facilities, which might be used on the workstation for some design work, but will need to be host resident for large circuit evaluation and test pattern grading phases of the design effort. If one is realistic, it simply is not practical to do large circuit designs in a quantity on standalone workstations. Simulation and test generation/evaluation services should, however, be conveniently accessible to the designer through the workstation: in fact, it should not (other than for administrative purposes) be evident that the computationally intensive work has been allocated to a host.

Economic factors will probably motivate the development of special purpose CAD hardware products that, at least in larger organizations, provide some of the computational services presently delivered in the most effective manner by larger host computers.

For larger circuits, it is likely that automated layout, as well as design rule checking, will also be most practical to assign to host computers accessible from workstations. The host will in many cases be found to be the only practical environment in which to maintain large historical archives of on line design data.

The interactive graphical power of the workstation will, however, be the most effective setting in which to accomplish interactive layout (placement and routing) editing. It is important to note that the designer of the future should only peripherially be aware that some of his services are being provided through usage of the host. Future CAD workstations must promote the perception that the workstation is "doing it all."

How will such products evolve? Clearly, the service delivery software must be able to reside in either host or mainframe environments, or both. Designer interfaces must provide, in so far as practical, for transparent off-loading of intensive computational service requests. Constituent software must be structured to facilitate both interfacing and transportability. Packaging, marketing and support of such products will be challenging, in the face of demands for rapidly declining design tool costs.

These products must be able to communicate with (perhaps geographically distant) reference databases that are frequently updated to reflect evolving technology, new macrocircuits, and processing changes. Mechanisms must be put in place to assure effective coordination between vendors, manufacturers, designers and users of the base circuit technologies.

PRACTICAL EXPERIENCE

Can such generalized, integrated tool sets be created and distributed to a large commercial market? The answer is clearly that they can. We have already seen that the same software can be used to provide individual classes of design services (circuit specification, simulation, layout) over a diversity of semiconductor and packaging technologies. A limited number of large organizations have "put it all together," and truly integrated full spectrum services in commercial products are on the horizon.

We have, for example, found that gate arrays and standard cell semi-custom integrated circuits can be effectively designed in this manner using a common software base, in a variety of technologies. The size of the host computer needed depends, of course, on circuit size and design volume. What does the future hold?

FUTURE DIRECTIONS

As the processing power available in workstations increases, it is likely that simulation and layout of smaller circuits can be moved to the workstation. The larger circuits will continue to require host based design aids, because feasible and necessary circuit complexity will grow at least as fast as workstation processing power. Sets of commercially viable design tools will be offered in distributable configurations to meet these needs, with selectable host-workstation configurations.

As local design of semicustom integrated circuits becomes more widespread, it will become necessary to provide our customers with a means for assuring the validity of chip-specific libraries and processing streams. This will require close cooperation between IC manufacturers and those who distribute design tools applicable to these chips. There will perhaps need to be standardized representations of models as well as design data, to facilitate economical distribution and communication of the necessary logical, physical and symbolic data.

As our market matures, distributable design tools will blur the present distinction between workstation and mainframe hosted products. It will be found that standalone workstation-based design systems will not address all design problems cost-effectively. The market will insist on distributable tools, allowing the design manager to configure the designer's environment to suit local needs.

REFERENCES

1. Moore, Gordon. "VLSI: Some Fundamental Challenges," IEEE Spectrum Magazine, April 1979, pp. 108-110.

2. Smith, Robert J. Smith, II. "Technology-Independence Circuit Layout," Proceedings of the 20th Design Automation Conference, June 1983, pp. 390-393.

The Role Of Work Stations In A Computer-Aided-Design (CAD) Environment

Dr. Douglas B. Boyle
J. Michael Connell
Patricia Rohrs

LSI LOGIC CORPORATION

Abstract

Deciding how to integrate work stations is an important factor in planning a CAD system for IC design. At LSI LOGIC Corporation the work station is used for certain tasks, while mainframe computers are used for others. The result is an efficient system for designing high-performance integrated circuits.

1. Introduction

The rapid market acceptance of CAD work stations for IC design has forced CAD system designers to consider the work station while developing strategy for new CAD systems. At LSI LOGIC Corporation, the CAD strategy has two principal goals:

- o Maintain the desirable features of LDS-II, LSI LOGIC's mainframe-based CAD system for logic arrays.
- o Provide additional capability by integrating the work station into the existing LDS-II system.

The LDS-II system's most salient characteristics are as follows:

- o It is easy for the designer to learn.
- o It permits short design time and usually results in chips that work the first time.
- o It handles large, complex logic array designs.

These CAD system features are key to LSI LOGIC Corporation's past success. Our present challenge is to capitalize on that success by combining work stations and mainframes, creating a system in which some tasks are allocated to a designer using a work station and other tasks to the IC manufacturer's mainframe system.

2. The LDS-II Environment

LSI LOGIC Corporation divides the design and manufacture of logic arrays into five major phases:

- o Network Description
- o Simulation
- o Layout
- o Manufacturing
- o Testing

These phases are divided into two categories: tasks performed by the customer and tasks performed by the manufacturer, LSI LOGIC Corporation. This task division increases the system's efficiency by allowing specialization. The interface between the task categories is well-defined, and it provides the necessary communication to ensure high-performance designs that meet the customer's requirements.

The customer's tasks--network description and simulation--result in a machine-readable description of the desired circuit function. Since test patterns are derived automatically from simulation, the test specification is also produced. The customer can also specify critical timing paths, so that the layout system can optimize performance.

- o The customer uses an alphanumeric terminal connected to the LDS-II host to:
  - describe his network with a hardware description language
  - simulate the network and obtain critical timing information.
- o The tools for performing these tasks can be learned in three days.

At LSI LOGIC, tasks performed by the manufacturer include layout, manufacturing, and testing. Because layout is closely tied to manufacturing, it is relatively unimportant to the LSI LOGIC Corporation customer but critical to the manufacturer.

CH1935-6/83/0000/0196$01.00 © 1983 IEEE

The manufacturer provides several kinds of information for the customer's tasks:

- o Logical specifications of macrocells to be used in designs
- o Performance specifications of macrocells
- o Wire lengths from the layout of specific designs

This information is automatically incorporated by LDS-II into the logic simulation, so that the customer is provided with a highly accurate computer simulation of the device's performance before committing it to silicon implementation.

The design process is complete when the manufacturer delivers tested prototypes to the customer.

3. Work Station: Approaches, Problems and Advantages

Two approaches for exploiting CAD work stations are popular in today's marketplace:

- o Create a work station environment that replicates a mainframe environment.
- o Integrate work stations and mainframe into a single environment.

Within these approaches, certain problems and advantages can be identified.

What are the problems associated with work stations? Several are worth mentioning:

- o Because work stations have relatively low processing power, many of them are required to equal the computer power of a single mainframe.
- o Because work stations are dispersed geographically, data communications is required.
- o Because work stations are owned by customers and installed at their sites, the IC manufacturer can lose control of both design methodology and proprietary information.

What are the advantages of work stations?

- o The work stations's hardware architecture provides for a graphic rather than alphanumeric method for describing and simulating networks.
- o Because the designer is thoroughly familiar with graphic representations of his circuit, he may find the work station a more desirable design tool than an alphanumeric terminal. He can display schematics and wave forms.
- o With a work station, the designer will not have to translate his schematic representation into a hardware design language. Eliminating this translation step facilitates correspondence between the designer's conception and the machine-readable description.
- o Because the station is relatively small and inexpensive, a customer may purchase it for on-site design work.
- o The graphic capability of the work station can be used for interactive layout of the IC.

4. The Role of Work Stations Within LDS-II

Having assessed both the problems and advantages of work stations, LSI LOGIC Corporation has committed itself to the second approach mentioned above--integration of work stations with mainframe computers. With such integration, LSI LOGIC can continue to divide tasks between the customer and itself. It can also provide a better tool--the work station--for the customer's performance of his specialized tasks: network description and simulation. Finally, the company can maintain control of those vital task areas in which it excels: layout, manufacturing and testing.

The work station is a highly effective customer tool for network description because a designer can enter and display schematics, without requiring significant computing power. The work station is effective for simulation because a designer can enter and display waveforms. However, because the work station's ability to simulate large designs is questionable, having a mainframe available for simulation is desirable.

LSI LOGIC Corporation's experience with logic arrays is that it can perform the layout task better than the customer. Several factors make this the case:

- o The need for customer interaction during the layout phase is minimal.
- o Because a well-designed layout system is automated rather than interactive, it requires the large computational power of a mainframe computer, rather than the interactive graphics capability of a work station.

- A layout system requires its own technical specialists, who continually optimize the placement and routing of ICs.

- During the layout phase, much proprietary information must be exchanged between layout and manufacturing specialists. This exchange is not properly the concern of a customer's logic designer.

5. Conclusion

Study of the advantages and shortcomings of work stations has defined a specific role for them within an existing mainframe-based CAD system. Work stations are used for the customer tasks--network description and simulation. Mainframe computers are used for central library maintenance, complex logic simulations, and layout. This system architecture provides a graphical user interface for the customer and allows the manufacturer to use automatic placement and routing to ensure a high-performance, error-free product.

Session 1-3d
Residue Number

Chairman

D. O. Carhoun
Mitre Corporation

VLSI DESIGN OF NUMBER-THEORETIC TRANSFORMS FOR A FAST CONVOLUTION

C.-S. Yeh, I.S. Reed, J.J. Chang

Department of Electrical Engineering, University of Southern California
Los Angeles, CA 90089-0272

T.K. Truong

Communication System Research, Jet Propulsion Laboratory
California Institute of Technology, Pasadena, CA 91103

Abstract

In this paper, a VLSI chip is designed to compute a 32-point Fermat number transform over the fourth Fermat number $F_4 = 2^{16}+1 = 65537$. Also the architecture of a generalized overlap-save algorithm is developed to realize an FIR digital filter of arbitrary length using number-theoretic transforms.

Introduction

Number-theoretic transforms (NTT's) were developed to compute cyclic convolutions [1]. The cyclic convolution of two sequences can be obtained by taking the inverse NTT of the product of the NTT's of these two sequences.

Some advantages of the NTT over the more usual FFT are: (1) For some NTT's, no multiplications are required; (2) Integer arithmetic is used in the NTT; (3) No round-off error is introduced. Some disadvantages of the NTT are: (1) There is a conflict between the long transform lengths and a large dynamic range under a reasonable word size. This conflict often limits the applicability of the NTT; (2) A modulus operation is required in the arithmetic of the transform.

The Fermat number transform (FNT) [2] and the complex number-theoretic (CNT) [3] are two important NTT's. The FNT has the advantage over most NTT's in that the transform length is a power of 2. As a consequence the FFT structure can be applied to compute the FNT.

McClelland [4] designed a hardware system to realize a 64-point 17-bit FNT that used commercially available ECL IC chips. In this paper, a VLSI chip is designed to implement a similar 32-point 17-bit FNT over F_4, where $F_4 = 2^{16}+1 = 65537$ is the fourth Fermat number. Finally the architecture of a generalized overlap-save method is developed to realize a FIR digital filter in conjunction with number-theoretic transforms.

This work was supported in part by the U.S. Air Force Office of Scientific Research, under grant number AFOSR-80-0151, and also in part by NASA, under contract number NAS 7-100.

A VLSI Design of a Fermat Number Transform

Let $F_t = 2^{2^t}+1$ be the t-th Fermat number, where $t \geq 0$. Let $\{x_n\}$ be a N-point sequence of integer numbers, where $0 \leq x_n \leq F_t-1$, $0 \leq n \leq N-1$, and N is a power of 2. The Fermat number transform $\{X_k\}$ of $\{x_n\}$ over F_t is defined as follows:

$$X_k \equiv \sum_{n=0}^{N-1} x_n \alpha^{nk} (\text{mod } F_t), \quad 0 \leq k \leq N-1 \tag{1}$$

where $0 \leq X_k \leq F_t-1$ and α is an N-th root of unity. That is, N is the least positive integer such that $\alpha^N \equiv 1 \pmod{F_t}$. The corresponding inverse FNT is the following:

$$x_n \equiv \left(\frac{1}{N}\right) \sum_{k=0}^{N-1} X_k \alpha^{-nk} \; (\text{mod } F_t), \quad 0 \leq n \leq N-1 \tag{2}$$

where $(1/N)$ is the inverse of N modulo F_t. $(1/N)$ is also an integer. In eqs. (1) and (2), exponents are assumed to be taken modulo N.

A cyclic convolution can be computed with the FNT pair in eqs. (1) and (2). The transform length N depends on F_t and the α chosen [1,2]. More details of the FNT can be found in [1,2].

Specifically in the following design, F_t and α are $F_4 = 2^{16}+1$ and 2, respectively. For such an FNT, the transform length $N = 2^{t+1} = 2^5 = 32$ [2]. The data of this FNT are integers in the range of 0 and 2^{16}. Hence 17 bits are required to represent a number.

To efficiently perform the binary arithmetic operations modulo a Fermat number, Liebowitz developed the diminished-one representation in [5]. Let A be represented by $[a_{16}, a_{15}, \ldots, a_1, a_0]$, where $0 \leq A \leq 2^{16}$ and the i-th bit a_i of A is either 0 or 1. In diminished-one representation, if A = 0, then $a_{16} = 1$ and $a_i = 0$ for $0 \leq i \leq 15$. If $A \neq 0$, then the diminished-one representation of A is the same as the normal binary representation of (A-1). For example, in diminished-one representation, 1 is represented by [00...00], 2 is represented by [00...01] and 2^{16} is represented by [01...11]. In the diminished-one representation the most significant bit a_{16} can be viewed as the zero-detection bit.

In the FNT over F_4 with $\alpha = 2$, two basic binary arithmetic operations are addition and multiplication by a power of 2, modulo F_4. In the following these two operations are described briefly.

(1) Addition: Let $S = A+B$. If $A = 0$, then $S = B$. If $B = 0$, then $S = A$. If neither A nor B equals 0, add $[a_{15},a_{14},\ldots,a_0]$ and $[b_{15},b_{14},\ldots,a_0]$. Then complement the carry and add it to the previous sum. This yields S. (2) Multiplication by a power of 2: Let $B = A\cdot 2^C$. If $A = 0$, then $B = 0$. If $A \neq 0$, left rotate $[a_{15},a_{14},\ldots,a_1,a_0]$ C bit positions, but complement the value of bit 15 when it is rotated to bit position 0, and set $b_{16} = 0$.

By virtue of eq. (1) the FNT has a mathematical structure very similar to the FFT. Also the transform length N of an FNT is a power of 2. Hence an almost exact analogue of an FFT-type structure can be used to perform a fast FNT. Fig. 1 shows an FFT-type pipeline structure [6] for computing a 32-point FNT over F_4. This pipeline structure can be viewed as a one-dimensional systolic array. The basic elements of the structure are delays, switches, and FNT butterflies. These basic elements are arranged alternately in the structure.

In Fig. 1 z^{-j} denotes a j-step delay element. SW_i in Fig. 1 is a two-input, two-output switch controlled by the control signal S_i for $1\leq i\leq 4$. For $S_i = 0$, the input wires of the SW_i are directly connected to the output wires. For $S_i = 1$, the input wires of the SW_i are cross connected to the outputs. The S_i's can be realized simply by a 4-bit binary counter and some delay elements at the outputs of the counter for the purpose of synchronization.

An FNT butterfly F in Fig. 1 performs $D=A+B(2)^C$ abd $E = A-B(2)^C$, where A and B are inputs of the FNT butterfly, D and E are outputs of the FNT butterfly and C is an exponent in Eq. (1). C is stored in the FNT butterfly. Fig. 2 shows a block diagram of an FNT butterfly. Two realizations of an FNT adder can be found in [4]. The multiplier in Fig. 2 is used to multiply a number by a power of 2. This multiplier can be realized by a modified barrel shifter circuit [7]. The negation can be realized by a set of inverters and multiplexers.

The circuit and layout design of the structure in Fig. 1 has been completed with the use of CAESER and SPICE design tools [9]. The final layout of this FNT chip is shown in Fig. 3. In the future, this FNT will be fabricated and tested. The total number of transistors in this chip is about 20,000. The area of the chip is estimated to be about 12,000,000 λ^2.

A Digital Filter Architecture Using the FNT

A practical limitation of realizing an FIR digital filter using the FNT, is the conflict between a wide dynamic range and long transform lengths. A generalized overlap-save algorithm was developed in [7] to solve this conflict associated with the NTT. This algorithm is summarized as follows.

Let $\{x_n\}$ and $\{h_m\}$ be the input-data and filter sequences of a FIR digital filter, respectively, where $0\leq n\leq N-1$ and $0\leq m\leq M-1$. The output sequence $\{y_k\}$ of the filter is the linear convolution of $\{x_n\}$ and $\{h_m\}$, where $0\leq k\leq N+M-1$ [6]. For purposes of exposition it is assumed $M = 64$ and $N \geq 32$.

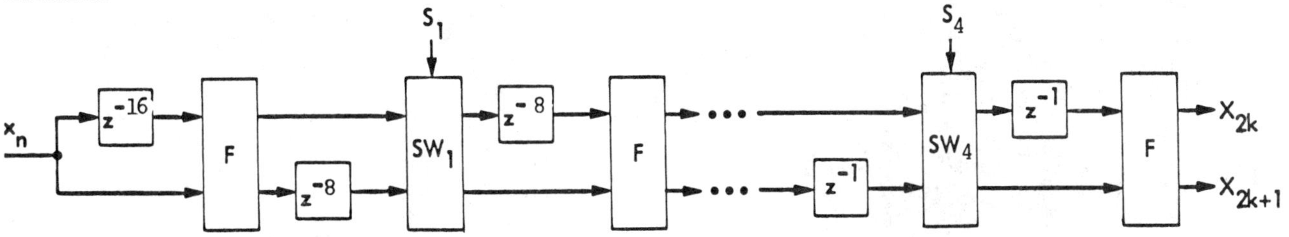

Figure 1. A pipelined structure for computing a 32-point Fermat number transforms

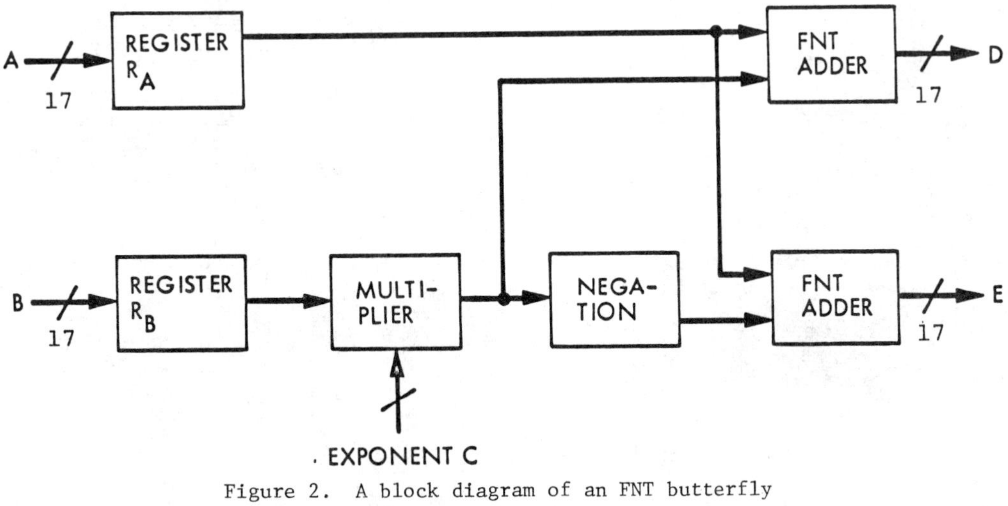

Figure 2. A block diagram of an FNT butterfly

Other cases of the generalized overlap-save method are constructed in a similar manner.

The output sequence $\{y_k\}$ can be computed by the following procedure [7]:

(1) Partition $\{h_m\}$ into 4 subfilters $\{h_m^i\}$ for $1 \le i \le 4$ with $h_m^i = h_m$ for $16(i-1) \le m \le 16i-1$ and $h_m^i = 0$, otherwise.

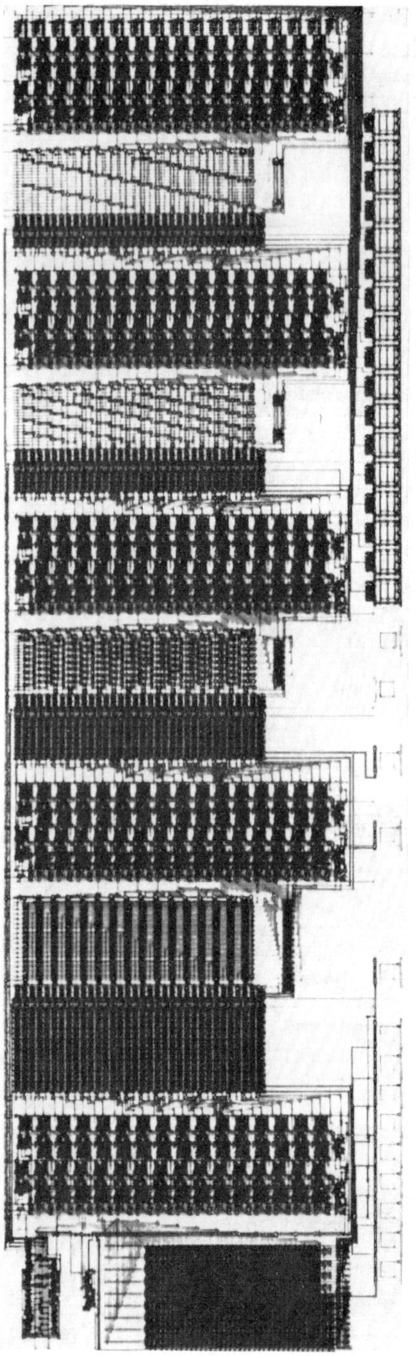

Figure 3. The layout of an FNT chip for a 32-point FNT over F_4

(2) Compute the linear convolution $\{y_k^i\}$ of $\{x_n\}$ and $\{h_m^i\}$ for $1 \le i \le 4$ using the standard overlap-save method [6] as follows:

(2.1) Section $\{x_n\}$ into 32-point subsequences $\{x_n^j\}$ for $j \ge 0$ with 16 points of $\{x_n\}$ overlapped between two consecutive subsequences. That is $x_n^j = x_n$ for $16(j-1) \le n \le 16(j+1)-1$, and $x_n^j = 0$ otherwise, where $x_n = 0$ for $n<0$.

(2.2) Compute the cyclic convolution $\{z_k^j\}$ of $\{x_n^j\}$ and $\{h_m^i\}$ using the FNT.

(2.3) $\{y_k^i\}$ is obtained by discarding the first half and saving the second half of each convolution $\{z_k^j\}$ computed in (2.2).

(3) Finally, the desired output sequence $\{y_k\}$ equals the arithmetic sum of $\{y_k^i\}$ for $1 \le i \le 4$.

In Fig. 4 is shown a parallel architecture to realize the digital filter described above using the generalized overlap-save algorithm and the FNT. The sequence $\{H_k^i\}$ in Fig. 4 is the FNT of $\{h_m^i\}$ multiplied by (1/N), modulo F_4, where $1 \le i \le 4$. In step (2.3) of the above algorithm, one of two outputs of an inverse FNT is not needed. The adders in Fig. 4 perform normal binary addition, not addition modulo F_4.

The generalized overlap-save algorithm can be used also in conjunction with the complex number-theoretic transform to realize a FIR digital filter [8].

Conclusions

A VLSI chip is designed to compute a 32-point FNT over F_4. In the future, the design will be extended to VLSI chips for computing 64-point and 128-point FNT's over F_5. A generalized overlap-save algorithm is developed to be used in conjunction with number-theoretic transforms to realize FIR digital filters. A parallel architecture is exhibited for implementing the generalized overlap-save algorithm for realizing digital filters.

References

[1] J.H. McClellan and C.M. Rader, Number Theory in Digital Signal Processing, Prentice-Hall, Englewood Cliff, New Jersey, 1979.

[2] R.C. Agarwal and C.S. Burrus, "Number Theoretical Transforms to Implement Fast Digital Convolution," Proc. IEEE, Vol. 63, No. 4, pp. 550-560, April 1975.

[3] I.S. Reed and T.K. Truong, "The Use of Finite Field to Compute Convolutions," IEEE Trans. Information Theory, Vol. IT-21, No. 2, pp. 208-213, March 1975.

[4] J.H. McClellan, "Hardware Realization of a Fermat Number Transform," IEEE Trans. Acoustic, Speech, and Signal Processing, Vol. ASSP-24, No. 3, pp. 216-225, June 1976.

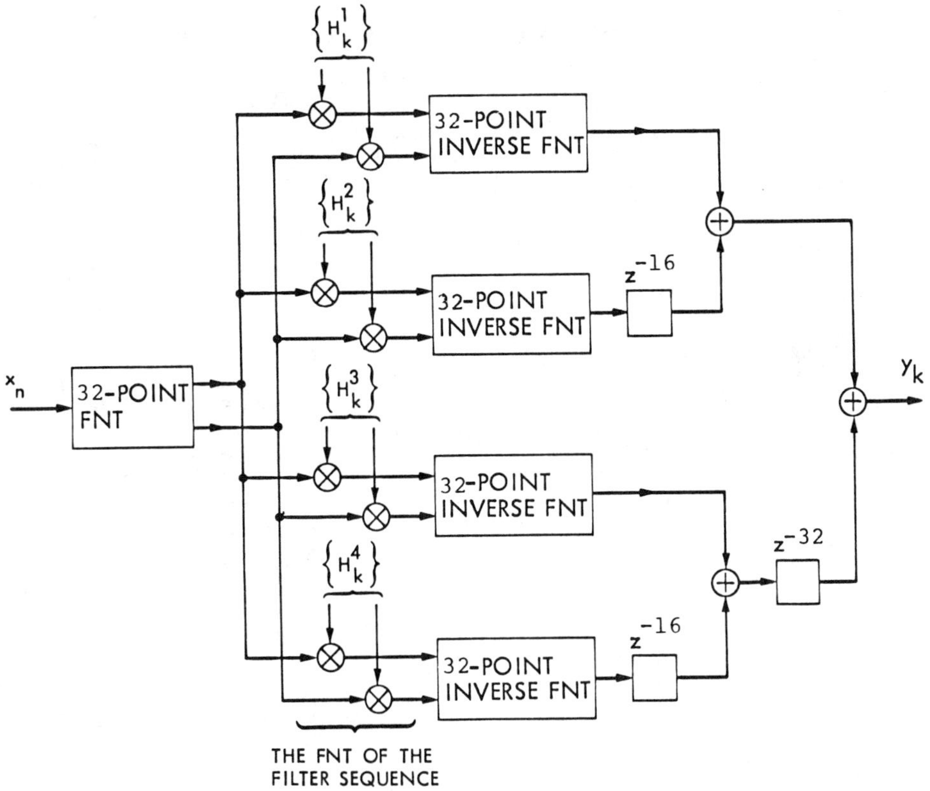

Figure 4. A realization of a digital filter with a filter sequence of 64 points using the generalized overlap-save method and the FNT technique

[5] L.M. Leibowitz, "A Simplified Binary Arithmetic for the Fermat Number Transform," IEEE Trans. Acoustic, Speech, and Signal Processing, Vol. ASSP-24, No. 5, pp. 356-359, Oct. 1976.

[6] L.R. Rabiner and B. Gold, Theory and Application of Digital Signal Processing, Prentice-Hall, Inc., Englewood Cliff, New Jersey, 1975.

[7] T.K. Truong, I.S. Reed, C.-S. Yeh and H.M. Shao, "A Parallel VLSI Architecture for a Digital Filter of Arbitrary Length Using Fermat Number Transforms," Proceedings of IEEE International Conference on Circuits and Computers (ICCC '82), Sep. 28-Oct. 1, 1982, New York, pp. 574-578.

[8] I.S. Reed, C.-S. Yeh and T.K. Truong, "A VLSI Architecture for Digital Filters Using Complex Number-Theoretic Transforms," Proceedings of 1983 International Conference on Acoustics, Speech, and Signal Processing (ICASSP 83), April 14-16, 1983, Boston, Massachusetts, pp. 923-926.

[9] L.W. Negal and D.O. Pederson, "SPICE - Simulation Program with Integrated Circuit Emphasis," Memorandum No. ERL-M382, Electronics Research Laboratory, University of California, Berkeley, April 12, 1983.

Digital Signal Processing With Residue Number Systems

Edwin L. Key

The MITRE Corporation
Bedford, MA 01730

ABSTRACT

The fundamental advantages for Residue Number System (RNS) design of digital processors, especially for the VLSI implementation of powerful linear signal processors, are discussed. A high performance digital transversal filter is used as an example to compare the complexity of RNS and conventional design methods. A detailed comparison using simple, but equitable and illustrative, rules is made.

INTRODUCTION

The purpose of this paper is to present the fundamental advantages of using residue number systems (RNS) to design digital processors, and especially to illustrate those advantages for the very large scale integrated (VLSI) electronics implementation of powerful linear signal processors. The discussion will begin with an introduction to residue number systems and their consequences for computation. Then, the main argument will be developed by comparing the complexity of a digital transversal filter designed with RNS and by conventional (radix-2) methods. A summary of the main points and some secondary issues will complete the discussion.

RESIDUE NUMBER SYSTEMS

Historical Background

Computation with residue number systems has a venerable history. It is based on the Chinese remainder theorem which originates from ancient publications concerning solutions to puzzles or parlor games, popular throughout Chinese history and typically posed in the following form[1],

> There is a certain number. When divided by three this number has remainder two; when divided by five, it has remainder three; when divided by seven, it has remainder two. What is the number?

The number in question is twenty-three. The general solution to such problems was presented as a poem in an arithmetic book (Sun Tsu Suan Ching) published circa 500 BC. Known today as the Chinese remainder theorem, it has become a fundamental theorem of modern number theory, establishing under proper conditions that a number can be represented uniquely by a set of remainders, in other words, by a residue code.

One of the earliest practical applications of the Chinese remainder theorem was for counting objects. The Han Dynasty military strategist Han Hsin is reported to have used it as an efficient method for counting the soldiers in his armies. Two thousand years later, it provides us with an efficient means of implementing digital signal processors in VLSI.

Computations with the Chinese Remainder Theorem

The Chinese remainder theorem was proved in proper generality by L. Euler in 1734. Its basic properties are easily stated. Let $M=p_1p_2\ldots p_j$ where the factors p_i are integers relatively prime in pairs. (This paper will assume prime integers.) Any integer N in the range $0 \le N \le M-1$ can be represented uniquely by a set of residues, $\{R_i\}$ where

$$N \equiv R_i \bmod p_i, \quad i = 1,2,\ldots,j \tag{1}$$

The congruence of (1) can be stated alternatively as a division algorithm, $N = Qp_i + R_i$, showing R_i as the least remainder resulting from division of N by p_i. The theorem states that N can be uniquely represented as

$$N \equiv R_1M_1 + R_2M_2 + \ldots + R_iM_i + \ldots + R_jM_j \quad \bmod M \tag{2}$$

where, by definition

$$M_i \equiv 1 \quad \bmod p_i \tag{3a}$$
$$M_i \equiv 0 \quad \bmod p_j, \quad j \neq i \tag{3b}$$

It follows from (2) and (3) that

$$M_i^2 \equiv M_i \quad \bmod M \tag{4a}$$
$$M_iM_j \equiv 0 \quad \bmod M, \quad j \neq i \tag{4b}$$

Consequently, the addition or multiplication of two integers whose sum and product lie in the range (0,M-1) can be calculated directly with the residue code by performing the computations in each residue class independently and composing the radix system result by means of (2). As an example, let M=2x3x5=30; then if N is represented

CH1935-6/83/0000/0204$01.00

by the residues $\{R_2,R_3,R_5\}$ and K by the residues $\{S_2,S_3,S_5\}$, the Chinese remainder theorem yields the compositions

$$N \equiv 15R_2 + 10R_3 + 6R_5 \quad \text{mod } 30 \tag{5a}$$
$$K \equiv 15S_2 + 10S_3 + 6S_5 \quad \text{mod } 30 \tag{5b}$$

For addition, it follows immediately that

$$N + K \equiv 15(R_2 + S_2) + 10(R_3 + S_3) + 6(R_5 + S_5) \quad \text{mod } 30 \tag{6}$$

For multiplication,

$$N \times K \equiv 15(R_2 \times S_2) + 10(R_3 \times S_3) + 6(R_5 \times S_5) \quad \text{mod } 30 \tag{7}$$

since $15^2 \equiv 15,\ 10^2 \equiv 10,\ 6^2 \equiv 6 \quad \text{mod } 30$.

RNS Functional Processing

Since the computations for both addition and multiplication can be carried out independently in the residue classes, any linear function can similarly be computed independently in the residue classes by performing the same functional operations on the members of the residue code in each class.

For applicability of a residue number system processor, the function need not be strictly linear, although it must be a homomorphism in the ring of integers mod M. The function of principal concern in this paper is linear convolution between an input sequence, x(k), and a system impulse response, h(k). For an N-point convolution, the output, y(k), is expressed as

$$y(k) = \sum_{n=0}^{N-1} h(n)x(k-n) \tag{8}$$

which can be readily implemented by a transversal filter, or tapped delay-line, structure. For RNS implementation, the computation of (8) is performed independently in the residue channels, without any propagation of carries between the residue components, as would be required in a radix-system computation. It is, of course, necessary to perform residue conversion at the input and radix-system conversion at the output in order to interface with the conventional digital environment.

For input interfacing to binary words, the binary to residue conversion can either be incorporated in the design of the analog-to-digital converter (when processing analog signals) or the operation can be organized in independent residue processing channels. Let the integer to be converted be represented as a radix-2 number of m bits.

$$a = a_{m-1}2^{m-1} + a_{m-2}2^{m-2} + \ldots + a_1 2^1 + a_0 2^0, \quad a_i \in \{0,1\} \tag{9}$$

Then, since the coefficients {0,1} are members of each residue class mod p_i, (9) can be expressed as

$$a \equiv [a_{m-1}(2^{m-1} \text{ mod } p_i) + a_{m-2}(2^{m-2} \text{ mod } p_i) + a_1 2 + a_0] \text{ mod } p_i \tag{10}$$

The computation can be implemented in a set of residue channels, each consisting of a tree of mod p_i adders driven by the logical "and" (or conjunction) of the precalculated values 2^{ℓ} mod p_i with the binary values a_{ℓ}. There is no coupling between the residue channels performing the conversion. The radix system conversion can be implemented directly from the Chinese remainder theorem, or alternatively, by a mixed-radix conversion to be discussed later.

COMPARISON OF COMPLEXITY

Rules for Comparison

The main point to be made in this paper is that there are fundamental advantages in hardware complexity and processing throughput to be gained by implementing digital signal processors with residue number systems. To support the claim of reduced hardware complexity, we will compare the numbers of logic gates used to implement a transversal filter with RNS and conventional methods. The processing throughput advantage of RNS will also become evident from this comparison.

Simple rules will be used for comparison. The logic will be restricted to "and," "or," and "not" gates. It will be assumed that all gates have two inputs and one output and that a gate can drive several other gates. Gate minimization techniques will not be employed for the purpose of comparison. While different technologies have different rules, these rules will provide an equitable basis for comparison.

RNS Gate Count Formulation

It is assumed that in both cases the input data and tap weight values are encoded as binary k-tuples. The value in each residue class requires $\lceil \log_2 p_i \rceil$ bits, where the brackets denote the least integer greater than the argument. Binary shift registers of N stages, $\lceil \log_2 p_i \rceil$ of them for each mod p_i channel, provide the delay elements for the transversal filter. It will be assumed for the adders and multipliers, however, that the mod p_i arithmetic is implemented directly from the p_i-valued truth tables describing the operations. This logic implementation of the tables, while providing an upper bound on the gate complexity, is actually quite suitable for VLSI design with programmable logic array (PLA)-like structures employing separate "and" and "or" planes, and is very practical when the prime moduli are small.

For RNS implementation, residue coders are needed at each delay tap to convert the binary-coded values to one of p_i values, while binary coders are required at the output for the inverse operation. The adders and multipliers use PLA-like arrays of logic gates as previously mentioned. Dynamic shift registers are assumed to be used for the tapped delay lines. Upper bounds on the numbers of gates required to implement these functions are listed in Table 1 for each mod p_i channel as a function of the modulus. The third column of Table 1 also lists the total number of gates used in a RNS implementation of a 128-stage transversal filter that uses the first six primes as moduli. This selection of moduli provides a dynamic range in excess of 90 db for computation of the output.

TABLE 1.

Function	Gate Complexity per Residue Channel	Number of Gates For 128-Stage RNS Filter
Residue Coder	$p_i\{\lceil \log_2 p_i \rceil -1\}$	12,672
Binary Coder	$(p_i-2)\lceil \log_2 p_i \rceil$	106
Mod p_i Adder	$2p_i^2 - p_i$	90,551
Mod p_i Multiplier	$2p_i^2 - 5p_i + 4$	73,344
Mod p_i Shift Register Stage	$2\lceil \log_2 p_i \rceil$	8,704
		185,377

Input and Output Conversion

The values listed in Table 1 pertain only to the transversal filter implementation in the residue number system and do not include the gates needed for the input and output conversions. For an equitable comparison with the conventionally designed filter, these values must also be included.

Binary-to-Residue Conversion

Implementation of the binary-to-residue conversion with trees of mod p_i adders, as previously described, provides an upper bound on gate complexity. This implementation uses a maximum number of $m(2p_i^2 - p_i)$ gates in each residue channel, where m is the number of bits of binary quantization. For a 128-stage transversal filter that uses the first six primes for the RNS, a total of 10,605 gates will support full 15-bit quantization, with proportionately fewer gates used for lesser quantization of the input.

Residue to Binary System Conversion

The radix system conversion at the output of the transversal filter will use a mixed radix system conversion as discussed by Knuth[2]. The conversion occurs in two phases; first a RNS to mixed radix conversion is performed, and then the mixed radix representation is converted to a binary word. Unlike direct implementation of the Chinese remainder theorem, the mixed radix conversion avoids modulus overflow during the computation. It is also useful for magnitude comparison, and the computation can be pipelined in the conversion hardware. The process is best illustrated with an example.

Choose the first six primes (2 through 13) as the moduli for the RNS. Then, any integer in the range $0 \le X \le M-1$ can be represented in mixed radix form as

$$X = 2310V_{13} + 210V_{11} + 30V_7 + 6V_5 + 2V_3 + V_2 \quad (11)$$

where the mixed-radix code $\{V_{13}, V_{11}, V_7, V_5, V_3, V_2\}$ is related to the residue code $\{X_{13}, X_{11}, X_7, X_5, X_3, X_2\}$ by the following transformation:

$$\begin{aligned}
V_2 &\equiv X_2 \bmod 2\\
V_3 &\equiv 2(X_3 + 2V_2) \bmod 3\\
V_5 &\equiv (X_5 + 3V_3 + 4V_2) \bmod 5\\
V_7 &\equiv 4(X_7 + V_5 + 5V_3 + 6V_2) \bmod 7\\
V_{11} &\equiv (X_{11} + 3V_7 + 5V_5 + 9V_3 + 10V_2) \bmod 11\\
V_{13} &\equiv 3(X_{13} + 11V_{11} + 9V_7 + 7V_5 + 11V_3 + 12V_2) \bmod 13
\end{aligned} \quad (12)$$

Given the residue code as input, these equations can be implemented in a pipelined structure in which each of the blocks implements a multiply-and-accumulate function modulo p_i, each row of blocks in the array representing a modulo p_i residue channel. The mixed radix-to-binary conversion of (12) requires only binary scaling and addition and is readily implemented with combinational logic. For this example, approximately 4600 gates are adequate to implement the conversion.

Aggregate Gate Complexity

Gate counts for the RNS transversal filter presented above, were formulated for very simple and suboptimum implementations of the mod p_i computational elements, selected for clarity of illustration. While they provide upper bounds on the gate complexity, combinational logic minimization techniques and algorithmic engineering tricks could provide further reduction. For example, a simple change from the mixed-radix representation (11) to the equivalent form,

$$X = U_{13} + 13U_{11} + 143U_7 + 1001U_5 + 5005U_3 + 15015U_2 \quad (13)$$

would result in a simpler hardware implementation of the radix system conversion, reducing the number of gates by one-half. Similarly, implementing the adders and multipliers with reduced combinational logic produces reductions in the gate count at each stage in the transversal filter. For example, a mod 13 adder is upper-bounded by 325 gates for the direct implementation, while a gate reduction based on the Quine-McCluskey technique produces an implementation with the NMOS transistor-equivalent of 206 gates. Similarly a mod 13 multiplier can be reduced from 277 to 185 gates. In both cases the reduced logic designs are quite suitable for VLSI implementation in programmable logic array structures. A combination of algorithmic and gate-reduction techniques can provide even bigger reductions in the residue conversion circuitry. For example, computation of the mod 13 residue of an eight-bit input sample can be reduced from the upper bound of 2680 gates to a transistor-equivalent of 160 gates. The simple model of the RNS transversal filter implementation described above provided an easily determined upper bound on gate complexity of approximately 200,000 gates, but it is apparent that a well-engineered device could be designed with fewer than 150,000 gates.

CONVENTIONAL IMPLEMENTATION

For comparison, assume a conventional binary (radix-2) implementation of the 128-stage transversal filter using ripple-carry multipliers and adders at each delay tap[3]. For comparison with the output dynamic range of the RNS transversal filter (90 db), we assume 15-bit input quantization, 15-bit multipliers and 30-bit adders. This degree of quantization is justified for comparison with the 15-bit output dynamic range of the RNS filter because allowance for computational overflow beyond 15 bits must be included at each stage of the conventional filter, while the non-overflow requirement must be met only at the output of the RNS filter. Dynamic shift registers are provided for delaying the input samples and for dynamically refreshed storage of the tap weights. Enumeration of the two-input gates required to implement such a processor yields 128 x 2280 gates for the multipliers, 127 x 266 gates for the adders and 128 x 60 equivalent gates for the registers, for a total of just under 334,000 gates.

Since ripple-carry logic is used for the adders and multipliers, in accordance with conventional practice, the conventional binary design suffers a throughput loss relative to the RNS design, the degree of loss depending on whether or not the adders are pipelined. The RNS filter, because of the completely parallel architecture, can operate at a throughput equal to the maximum clock rate determined by propagation delay. The gate complexity of a fully parallel binary filter, using parallel logic to implement the adders and multipliers in order to throughput data at the clock rate, would, of course, be too prohibitive to consider for VLSI implementation.

A comparison of the gate complexity and throughput for the RNS transversal filter and the conventional binary filter is shown in Table 2. The RNS filter demonstrates both a complexity and throughput advantage.

TABLE 2.

Design Method	Gate Count	Throughput/ Clock Rate
RNS	200K	1.0
Composite Modulus	4×10^{11}	1.0
Conventional Binary	334K	0.1 to .01

SUMMARY AND CONCLUSIONS

In this paper we have attempted to illustrate some fundamental advantages for digital signal processing with residue number systems, using a high performance transversal filter as an example for comparison. This example demonstrated that digital signal processing with RNS has great practical advantages, especially when the opportunity to implement the functions with VLSI electronics is considered. Furthermore, it is quite practical to design and implement special RNS computational elements with combinational logic in VLSI. Fundamentally, a quadratic law governing the growth of logic-gate complexity with binary word length, easily verified for ripple-carrry multipliers, establishes an absolute advantage for RNS designs, a simple manifestation that the "sum of the squares" is always less than the "square of the sum." Finally, it was demonstrated that the parallel architecture of the RNS design, by partitioning into residue-class channels of individual low complexity, makes practical a parallel logic implementation in which the filter can operate at the maximum clock rate determined by propagation delay. This throughput advantage is offset somewhat by the accumulated propagation delay in the output radix system conversion (six delays for the mixed-radix pipeline of the example) but the overall advantage is still preserved.

REFERENCES

1. Hua Loo Keng, Introduction to Number Theory, Springer-Verlag, 1982.

2. D. E. Knuth, The Art of Computer Programming; Semi-Numerical Algorithms (V2), Addison-Wesley, 1981.

3. L. Rabiner and B. Gold, Theory and Application of Digital Signal Processing, Prentice Hall, 1975.

A SYSTOLIC MULTIPLIER FOR FINITE FIELDS GF(2^m)

C.-S. Yeh, I.S. Reed

Department of Electrical Engineering, University of Southern California
Los Angeles, CA 90089-0272

T.K. Truong

Communication System Research, Jet Propulsion Laboratory
4800 Oak Grove Drive, Pasadena, CA 91109

Abstract

A systolic architecture is developed for performing the product-sum computation, AB+C, in the finite field GF(2^m) of 2^m elements, where A, B and C are arbitrary elements of GF(2^m). The multiplier is a serial-in, serial-out, one-dimensional systolic array. This multipler for GF(2^m) requires m basic cells. The average time per computation of the multiplier is m time units if a number of computations are computed consecutively. To perform an isolated computation the multipler requires 3m time units. The architecture is simple and regular and possesses the desirable properties of concurrency and modularity. As a consequence it is well suited for use in VLSI systems.

Introduction

Finite or Galois fields have many important and practical applications. Finite fields can be applied to error-correcting codes [1-3], switching theory [4] and digital signal processing [5]. For example, finite fields are used in the construction of many error-correcting codes. Reed-Solomon (RS) codes utilize the finite field GF(2^m) of 2^m elements, where m is a positive integer. The encoding and decoding algorithm of a binary RS code require algebraic operations in some field GF(2^m), rather than the usual binary arithmetic operations.

The operations of addition and multiplication in a finite field are quite different from the usual binary arithmetic operations. Because of their simplicity and practical usefulness, only the finite fields GF(2^m) are considered in this paper. Addition in GF(2^m) is bit-independent and straightforward. Thus it is easier than the usual binary addition. On the contrary, multiplication in GF(2^m) is more complex and difficult than binary integer multiplication.

Several circuits have been proposed [1-3, 6-8] to realize multiplication in GF(2^m). Unfortunately, these circuits are not suited for use in VLSI systems, due to irregular wire routing and complicated control problems as well as a nonmodular structure or lack of concurrency [9]. In this paper a pipeline architecture is designed to perform multiplication in GF(2^m).

Multiplication in GF(2^m)

A finite field must contain p^m elements, where p is a prime integer and m is a positive integer. The finite field GF(2^m) contains 2^m elements. GF(2^m) is an extension field of the ground field GF(2) of 2 elements, i.e. GF(2) = {0,1}. All arithmetic operations in GF(2^m) are performed by taking the results modulo 2.

The nonzero elements of GF(2^m) are generated by a primitive element α, where α is a root of a primitive, irreducible, polynomial $F(x) = x^m+f_{m-1}x^{m-1}+\dots+f_1x+f_0$ over GF(2). For example $F(x) = x^4+x+1$ is one such primitive, irreducible polynomial for GF(2^4).

The nonzero elements of GF(2^m) can be represented as the powers of α, i.e. GF(2^m) = $\{0, \alpha^1, \alpha^2, \dots, \alpha^{2^m-2}, \alpha^{2^m-1} = 1\}$. Since $F(\alpha) = 0$, $\alpha^m = f_{m-1}\alpha^{m-1}+\dots+f_1\alpha+f_0$. Therefore an element of GF(2^m) = $\{a_{m-1}\alpha^{m-1}+\dots+a_1\alpha+a_0 | a_i \in GF(2) \text{ for } 0\le i\le m-1\}$. In the following discussion, the polynomial representation is used to represent the finite field GF(2^m).

Let $A = a_{m-1}\alpha^{m-1}+\dots+a_1\alpha+a_0$ and $B = b_{m-1}\alpha^{m-1}+\dots+b_1\alpha+b_0$ be two elements in GF(2^m). Then $A+B = S_{m-1}\alpha^{m-1}+\dots+S_1\alpha+S_0$, where $S_i = a_i+b_i$ (mod 2) for $0\le i\le m-1$. Therefore addition in GF(2^4) is realized easily by m independent Exclusive-OR gates.

Suppose $P = p_{m-1}\alpha^{m-1}+\dots+p_1\alpha+p_0$ is the product of A and B, i.e. P = AB. P can be written as follows [1-3]:

$$\begin{aligned} P &= \sum_{k=0}^{m-1} (A\alpha^k)b_k = \sum_{k=0}^{m-1} \left(\sum_{n=0}^{m-1} a_n^{(k)}\alpha^n\right)b_k \\ &= \sum_{n=0}^{m-1} \left(\sum_{k=0}^{m-1} a_n^{(k)}b_k\right)\alpha^n \end{aligned} \tag{1}$$

where $a_n^{(k)}$ is the coefficient of α^n in $A\alpha^k$, i.e. $A\alpha^k = a_{m-1}^{(k)}\alpha^{m-1}+\dots+a_1^{(k)}\alpha+a_0^{(k)}$ for $0\le k\le m-1$. From eq. (1), one obtains $p_n = a_k^{(0)}b_0+a_n^{(1)}b_1+\dots+a_n^{(m-2)}b_{m-2}+a_n^{(m-1)}b_{m-1}$.

The computation of $A\alpha^k$ can be performed recursively on k for $0\le k\le m-1$. Intially for k = 0,

CH1935-6/83/0000/0208$01.00

$A\alpha^0 = A$, i.e., $a_n^{(0)} = a_n$ for $0 \leq n \leq m-1$. For $1 \leq k \leq m-1$,

$$
\begin{aligned}
A\alpha^k &= (A\alpha^{k-1})\alpha = \sum_{n=0}^{m-1} a_n^{(k-1)} \alpha^{n+1} \\
&= a_{m-1}^{(k-1)}\alpha^m + \sum_{n=1}^{m-1} a_{n-1}^{(k-1)}\alpha^n
\end{aligned} \tag{2}
$$

Substituting $\alpha^m = f_{m-1}\alpha^{m-1}+\ldots+f_1\alpha+f_0$ into eq. (2), yields

$$A\alpha^k = \sum_{n=1}^{m-1} (a_{n-1}^{(k-1)}+a_{m-1}^{(k-1)}f_n)\alpha^n + a_{m-1}^{(k-1)}f_0 \tag{3}$$

From eq. (3), one obtains

$$
\begin{aligned}
a_n^{(k)} &= a_{n-1}^{(k-1)}+a_{m-1}^{(k-1)}f_n \qquad \text{for } 1 \leq n \leq m-1 \\
a_0^{(k)} &= a_{m-1}^{(k-1)}f_0
\end{aligned} \tag{4}
$$

Figure 1 illustrates the step-by-step operations of a procedure for performing $P = AB+C$ in $GF(2^4)$. In Fig. 1 $a_n^{(k)}$, b_n, c_n, f_n, and p_n are the n-th bits of $A\alpha^k$, B, C, F and P, respectively, where F is the primitive, irreducible, polynomial. $p_n^{(i)}$ is the partial sum of p_n.

Systolic Multiplier for $GF(2^m)$

In this section a systolic array is developed to compute $P = AB+C$ in $GF(2^m)$. An analogous structure was proposed to multiply the usual two's complement binary number [10]. For simplicity in description the ensuing discussion is limited to the particular finite field $GF(2^4)$. In Fig. 2 this architecture is shown for $GF(2^4)$. The primitive, irreducible, polynomial is $F = f_3\alpha^3+f_2\alpha^2+f_1\alpha+f_0$. Input d_n receives the b_n of B. The n-th bits c_n, a_n and f_n of C, A and F are received serially at inputs e_0, g_0 and h_0, respectively. Two control signals, START and END, are used in the design. Inputs r_0 and t_0 receive START and END control signals, respectively.

Output e_4 serially transmits the n-th bit, p_n, of the result P out of the system. The order of the inputs and outputs are also shown in Fig. 2. The flip-flops associated with inputs t_0 and h_0 are used for the purpose of synchronization.

The circuit of cell L_i is shown in Fig. 3. The operation of flip-flops in this system is synchronized implicitly by a clock signal. In Fig. 3, when $r_i^* = 1$, $u_i = g_i^*$ at the next time unit (through switch SW). When $r_i^* = 0$, u_i sustains its value. Two principle operations of the system are the following:

$$
\begin{aligned}
e_{i+1} &\leftarrow (g_i^* d_i) \oplus e_i^* \\
g_{i+1}^* &\leftarrow (u_i h_i^*) \oplus (g_i^* t_i^*)
\end{aligned} \tag{5}
$$

where $0 \leq i \leq 3$, $\oplus$ denotes Exclusive-OR operation and the backwards arrow denotes the substitution operation.

A comparison of the procedure in Fig. 1 and the structure in Figures 2 and 3 yields the following facts: The signal u_i in L_i is equal to $a_3^{(i)}$ in $A\alpha^i$. The signal g_i^* is equal to $a_n^{(i)}$ in $A\alpha^i$ for some n. The signal e_i^* is equal to the partial sum of AB+C.

The multiplier in Fig. 2 can be generalized to the finite field $GF(2^m)$ by simply concatenating m identical cells. Extra registers and control signals are required if the b_i's are fed serially into the system in the same manner as the a_i's.

Conclusion

A systolic architecture is designed to perform the multiplication operation in the finite field $GF(2^m)$ of 2^m elements. This multiplier requires m basic cells for the finite field $GF(2^m)$. This multiplier is suited highly to VLSI implementation because of its simple and local control methodology, its regular interconnection pattern, and modular structure, and finally the complete concurrency of its micro-operations.

References

[1] P.J. MacWilliams and N.J.A. Sloane, The Theory of Error-Correcting Codes, North-Holland Publishing Company, 1978.

[2] W.W. Peterson and E.J. Weldon, Jr., Error-Correcting Codes, 2nd ed., MIT Press, 1972.

[3] E.R. Berlekamp, Algebraic Coding Theory, McGraw-Hill, New York, 1968.

[4] B. Benjauthrit and I.S. Reed, "Galois Switching Functions and Their Applications," IEEE Trans. on Computers, Vol. C-25, No. 1, pp. 78-86, Jan. 1976.

[5] I.S. Reed and T.K. Truong, "The Use of Finite Fields to Compute Convolutions," IEEE Trans. Inf. Theory, Vol. IT-21, No. 2, pp. 208-213, Mar. 1975.

[6] T.C. Bartee and D.I. Schneider, "Computation with Finite Fields," Inform. Contr., Vol. 6, pp. 79-98, Mar. 1963.

[7] B.A. Laws and C.K. Rushforth, "A Cellular-Array Multiplier for $GF(2^m)$," IEEE Trans. on Computers, Vol. C-20, pp. 1573-1578, Dec. 1971.

[8] R.G. Gallagher, Information Theory and Reliable Communication, Wiley, New York, 1968.

[9] H.T. Kung, "Why Systolic Architectures," IEEE Computers, Vol. 15, No. 1, pp. 37-46, Jan. 1982.

[10] R.F. Lyon, "Two's Complement Pipeline Multipliers," IEEE Trans. on Communications, pp. 418-425, April 1976.

*This work was supported in part by the US Air Force Office of Scientific Research, under grant number AFOSR-80-0151, and also in part by NASA, under contract number NAS 7-100.

STEP NUMBER	OPERATIONS	
1	$p_3^{(0)} = c_3$	$a_3^{(0)} = a_3$
2	$p_3^{(1)} = p_3^{(0)}+a_3^{(0)}b_0$, $p_2^{(0)} = c_2$	$a_2^{(0)} = a_2$
3	$p_2^{(1)} = p_2^{(0)}+a_2^{(0)}b_0$, $p_1^{(0)} = c_1$	$a_1^{(0)} = a_1$ $a_3^{(1)} = a_2^{(0)}+a_3^{(0)}f_3$
4	$p_3^{(2)} = p_3^{(0)}+a_3^{(1)}b_1$, $p_1^{(1)} = p_1^{(0)}+a_1^{(0)}b_0$, $p_0^{(0)} = c_0$	$a_0^{(0)} = a_0$ $a_2^{(1)} = a_1^{(0)}+a_3^{(0)}f_2$
5	$p_2^{(2)} = p_2^{(1)}+a_2^{(1)}b_1$, $p_0^{(1)} = p_0^{(0)}+a_0^{(0)}b_0$,	$a_1^{(1)} = a_0^{(0)}+a_3^{(0)}f_1$ $a_3^{(2)} = a_2^{(1)}+a_3^{(1)}f_3$
6	$p_3^{(3)} = p_3^{(2)}+a_3^{(2)}b_2$, $p_1^{(2)} = p_1^{(1)}+a_1^{(1)}b_1$,	$a_0^{(1)} = a_3^{(0)}f_0$ $a_2^{(2)} = a_1^{(1)}+a_3^{(1)}f_2$
7	$p_2^{(3)} = p_2^{(2)}+a_2^{(2)}b_2$, $p_0^{(2)} = p_0^{(1)}+a_0^{(1)}b_1$,	$a_1^{(2)} = a_0^{(1)}+a_3^{(1)}f_1$ $a_3^{(3)} = a_2^{(2)}+a_3^{(2)}f_3$
8	$p_3 = p_3^{(4)} = p_3^{(3)}+a_3^{(3)}b_3$, $p_1^{(3)} = p_1^{(2)}+a_1^{(2)}b_2$,	$a_0^{(2)} = a_3^{(1)}f_0$ $a_2^{(3)} = a_1^{(2)}+a_3^{(2)}f_2$
9	$p_2 = p_2^{(4)} = p_2^{(3)}+a_2^{(3)}b_3$, $p_0^{(3)} = p_0^{(2)}+a_0^{(2)}b_2$,	$a_1^{(3)} = a_0^{(2)}+a_3^{(2)}f_1$
10	$p_1 = p_1^{(4)} = p_1^{(3)}+a_1^{(3)}b_3$,	$a_0^{(3)} = a_3^{(2)}f_0$
11	$p_0 = p_0^{(4)} = p_0^{(3)}+a_0^{(3)}b_3$	

Figure 1. A procedure for computing P = AB+C in the finite field $GF(2^4)$, where A, B, C and P are elements of $GF(2^4)$

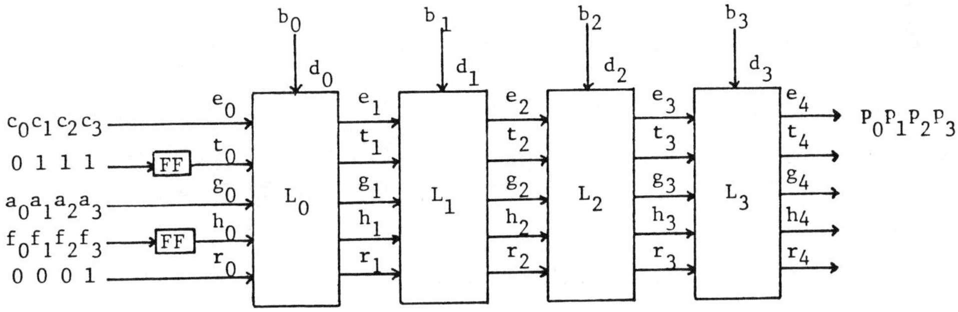

FF: Flip-Flop

Figure 2. A systolic multiplier for the finite field $GF(2^4)$

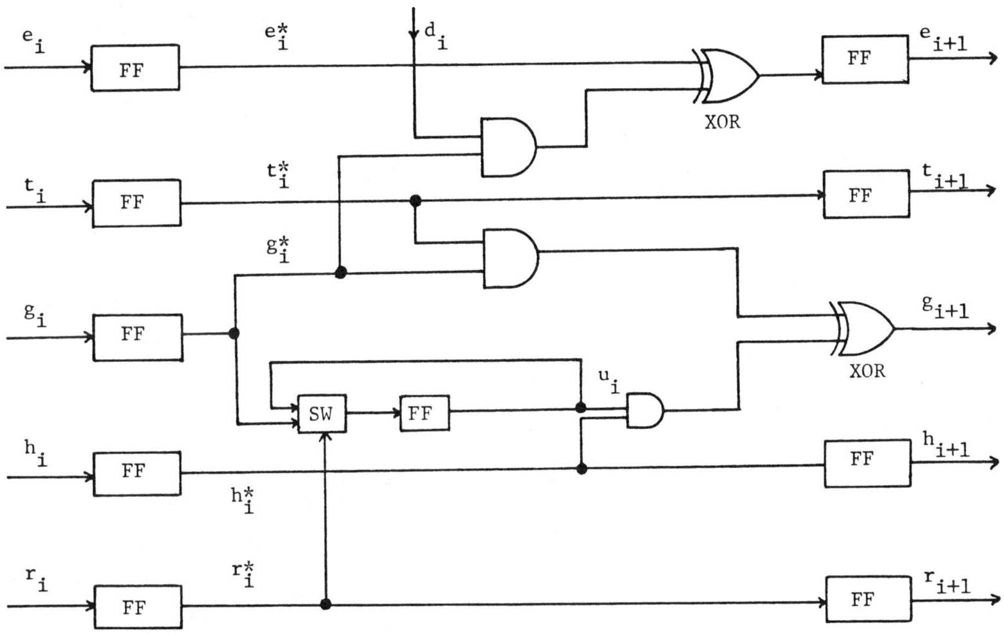

FF: Flip Flop, SW: Switch
XOR: Exclusive-OR

Figure 3. The circuit of the cell L_i used in the multiplier shown in Fig. 2

DESIGN FOR CONCURRENT TEST OF A PROGRAMMABLE GALOIS FIELD MULTIPLIER

A. L. Bequillard
D. Moulin

The MITRE Corporation
Bedford, MA 01730

ABSTRACT

The design of a $GF(2^m)$ finite-field multiplier that is programmable for m = 4,5,6,7, and 8 and is capable of concurrent error detection has been developed. The failure model used is consistent with an existing VLSI layout of the multiplier. All output errors caused by single errors in the internal logical variables of the circuit or shorts between adjacent parallel data lines are detected.

Introduction

Some signal processing operations are carried out over the Galois field $GF(2^m)$, where m is a positive integer. We are particularly interested in algebraic error-correction coding for communications systems, where fast multiplication over $GF(2^m)$ is a prime concern. Furthermore, because the coding operation is intolerant to multiplication errors, accuracy in the output of the multipliers is an equally important concern. This paper presents the design of a $GF(2^m)$ multiplier programmable for m = 4,5,6,7, and 8 that is both fast and self-checking.

An existing very-large-scale integrated (VLSI) design of a $GF(2^m)$ multiplier has been redesigned to be self-checking by incorporating time redundancy. The failures described here are those most frequently considered in the literature.

$GF(2^m)$: Definitions

The Galois field GF(2) can be represented by the set {0,1} with multiplication and addition of its elements defined by the AND and Exclusive-OR (X-OR) operations, respectively. When m is a positive integer greater than 1, $GF(2^m)$ can be thought of as the set of polynomials of degree less than m with coefficients in GF(2). Addition can be defined as polynomial addition over GF(2), and multiplication can be represented by polynomial multiplication mod $p_m(x)$, where $p_m(x)$ is of degree m and is irreducible over GF(2). [1]

The multiplicative group of $GF(2^m)$ is cyclic, generated by a primitive element α. A standard basis for $GF(2^m)$, considered as a vector space over GF(2), is $\{\alpha^{m-1},\alpha^{m-2},\ldots,\alpha^0\}$, so that any field element α^i can be represented uniquely by an m-tuple of elements of GF(2). If α is a root of $p_m(x)$, then this m-tuple coincides with the m-tuple of coefficients of the polynomial representation of α^i.

Implementation of $GF(2^m)$ Multiplication

We select α to be a root of $p_m(x)$, so that if

$$p_m(x) = 1+p_1+ \ldots +p_{m-1}x^{m-1}+x^m$$

then

$$\alpha^m = p_{m-1}\alpha^{m-1}+ \ldots +p_1\alpha+1 \qquad (1)$$

since the negative of an element of $GF(2^m)$ is the element itself.

Equation (1) can be used to find the product of two elements represented by a(x) and b(x) of $GF(2^m)$. First, form the polynomial product a(x)b(x). Then, by repeated use of equation (1) on the terms of degree m or higher, reduce $a(\alpha)b(\alpha)$ to a polynomial in α of degree less than m, whose coefficients constitute the binary vector representation of the product $a(\alpha)b(\alpha)$.

Figure 1 illustrates a design of a $GF(2^m)$ multiplier programmable for m = 4,5,6,7 and 8. [2] All field elements are represented by eight bits, the most significant 8 − m bits being zero. The pairwise-product array is an array of AND gates which accepts the eight-bit inputs a(x) and b(x) and produces all pairwise products a_ib_j. The partial-product accumulator is a bank of X-OR's that sums appropriate groups of pairwise products to yield the 15 coefficients of P(x)=a(x)b(x); some of these should be zero if m<8. The field-reduction circuit, which performs polynomial evaluation at α, reduces P(x) via equation (1) to

The work reported was supported by the Rome Air Development Center, Electronic Systems Division, AFSC, under Contract F19628-82-C-0001.

CH1935-6/83/0000/0212$01.00 © 1983 IEEE

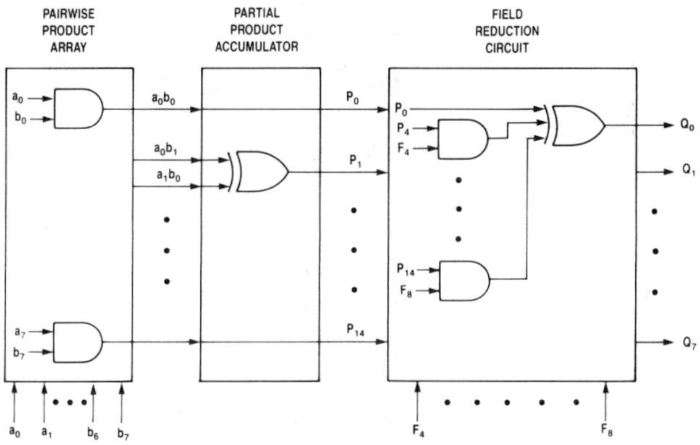

Figure 1. Nonredundant Programmable GF(2m) Multiplier

$Q(x)$, a polynomial of degree less than m that corresponds to $P(\alpha)$. Each coefficient of $Q(x)$ can be expressed as a linear combination of the coefficients of $P(x)$, where the linear combination depends on m. Thus the field-select lines $F_4,\ldots,F_8$ are needed as inputs to the field-reduction circuit, and only one can be a logic-1 at any time

Time Redundancy and the Self-Checking Multiplier

The concept of time redundancy for concurrent testing usually consists of repeating the computation in such a way that both results can be compared to detect the presence of a failure. The failure usually affects the computation in a different way each time.

In view of the vertical regularity of the multiplier circuit, the most appropriate method of time redundancy appears to be recomputing with shifted operands (RESO), a method that has previously been applied by others to a conventional binary multiplier. [3] RESO works for the $GF(2^m)$ multiplier in a manner very similar to the way it works for the binary multiplier (except that in the binary multiplier the field-reduction section is absent). Here, however, the algebraic properties of $GF(2^m)$ make the error coverage powerful and the analysis simple.

Figure 2 shows how the $GF(2^m)$ multiplier is modified to handle RESO.

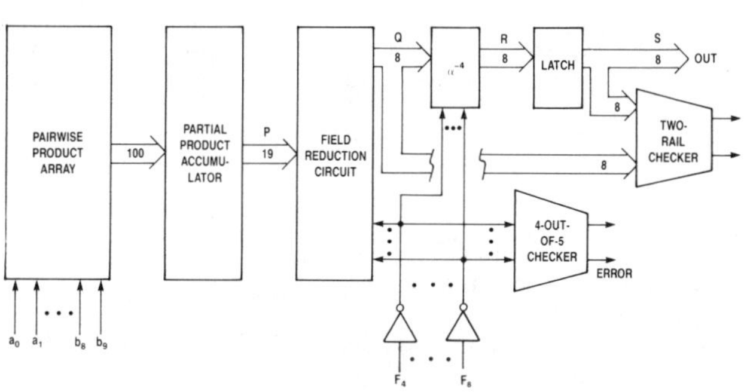

Figure 2. Self-Checking Multiplier

Note that:

(1) The pairwise-product array and the accummulator are extended to accept input polynomials, each of degree nine, via input lines called $A_0,A_1,\ldots,A_9$ and $B_0,B_1,\ldots,B_9$, to produce a product polynomial of degree 18 at lines $P_0,\ldots,P_{18}$.

(2) The field-reduction section is extended so that it can reduce the terms of degree 2m-1, 2m, 2m+1, and 2m+2 for each $GF(2^m)$. This implements a new set of linear equations.

(3) The output $Q(x)$ of the field-reduction section is then passed through a circuit that divides it by α^4 to yield a polynomial $R(x)$ of degree m-1, which represents the field element $\alpha^{-4}Q(\alpha)$. $R(x)$ is then passed to an eight-bit latch whose output is labeled $S(x)$.

(4) Finally, an eight-bit totally self-checking two-rail comparator [4] is used to detect any difference in the vectors $\{Q_i\}$ and $\{S_i\}$, and a totally self-checking 4-out-of-5 checker [5] is used to detect errors in the F_i.*

To multiply two field elements represented by $a(x)$ and $b(x)$, the coefficients of $a(x)$ are applied to lines A_2 through A_9 and B_2 through B_9 during the first computation, and lines A_0, A_1, B_0 and B_1 are set to logic-0. This is equivalent to applying $x^2a(x)$ and $x^2b(x)$, so that $P(x)=x^4a(x)b(x)$, $Q(\alpha)= {}^4P(\alpha)= {}^4a(\alpha)b(\alpha)$ and $S(\alpha)=a(\alpha)b(\alpha)$.

Next, the inputs are shifted by two bits to lines 0 through 7, and a zero is applied to lines A_8, A_9, B_8 and B_9. This corresponds to an input of $a(x)$ and $b(x)$, so that $Q(\alpha)=a(\alpha)b(\alpha)$. At this time S, which contains $a(\alpha)b(\alpha)$, is compared with Q. A mismatch signals a fault.

In the VLSI implementation of the self-checking multiplier (see figure 3), the first half of the pairwise-product array, the first half of the field-reduction circuit, and the first half of the α^{-4} multiplier are implemented as the AND

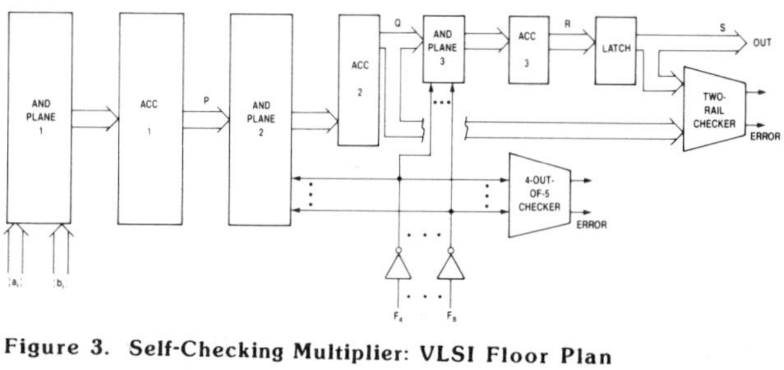

Figure 3. Self-Checking Multiplier: VLSI Floor Plan

*In reference [4] a gate-level design of a two-rail comparator that is totally self-checking for all unidirectional multiple faults is given. Reference [5] presents a PLA implementation of an m-out-of-(m+1) checker that is totally self-checking for stuck and crosspoint defect failures.

plane of a NOR-NOR programmable logic array (PLA). They are called AND planes 1,2, and 3, respectively. The partial-product accumulator, the second half of the field-reduction circuit, and the second half of the α^{-4} multiplier are banks of X-ORs. They are referred to as accumulators 1,2, and 3 (ACC 1, ACC 2, ACC 3), respectively.

In AND plane 1, the bit lines are the complements of the $\{a_i\}$ and $\{b_i\}$ and the product lines are all the pairs $\overline{\overline{a_i}+\overline{b_j}} = a_i b_j$. A section of AND plane 1 is shown in figure 4. ACC 1 is a bank of X-ORs that has as inputs the product lines of AND plane 1.

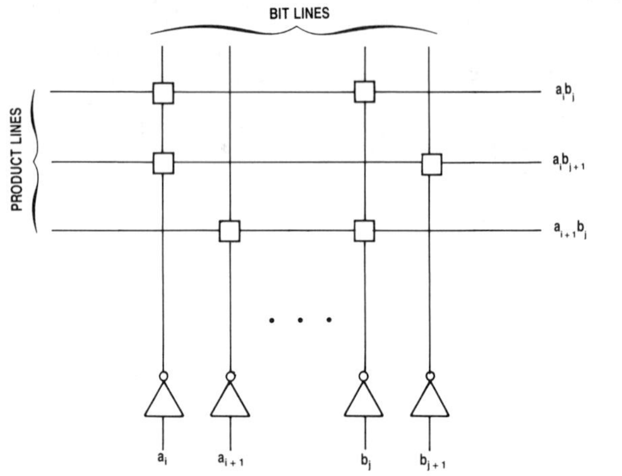

NOTE: SQUARES REPRESENT NOR GATES.

Figure 4. AND Plane 1

AND plane 2 and ACC 2 constitute the field-reduction circuit. As seen in figure 5, AND-plane 2 is similar to AND plane 1, but it has the P_i as inputs as well as the field-select lines F_i and it outputs a subset of all pairs P_iF_j, which are then added in groups by ACC 2.

The α^{-4} multiplier also implements a set of linear equations and is composed of AND plane 3 and ACC 3, which are similar to AND plane 2 and ACC 2. The inputs to AND plane 3 are the Q_i and the F_i.

Failure Model and Error Coverage

The self-checking multiplier can be partitioned into three AND planes, three accumulators, a latch, and two totally self-checking checkers. The accumulators and latch are bit-sliced circuits whose slices can be labeled H_i, so that H_i is either an X-OR cell in an accumulator or a one-bit slice of the eight-bit latch.

The failures analyzed are divided into five classes:

1) a random error in the output of any one H_i
2) a random error in a single input bit line of any AND plane
3) a random error in a single product line of any AND plane
4) a single short between adjacent bit or product lines
5) a single extra or missing crosspoint device in an AND plane

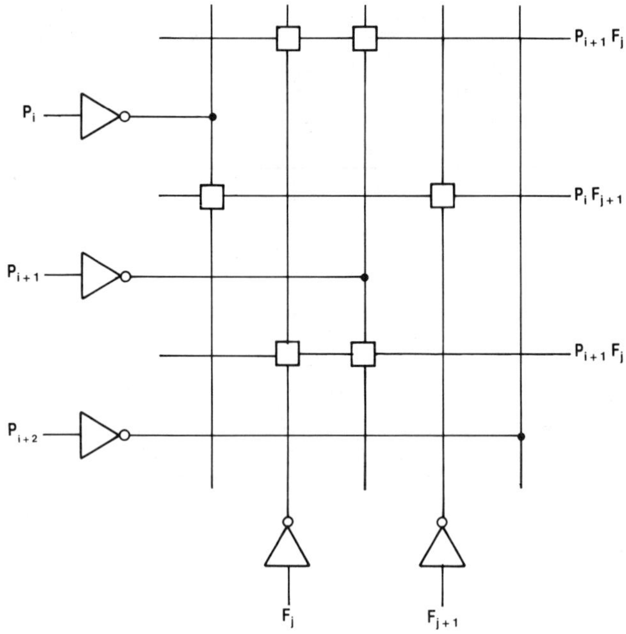

Figure 5. AND Plane 2

Note that a random error in any single line of the whole circuit falls within classes 1), 2), and 3). Similarly, all single shorts between adjacent parallel data lines belong to class 4), with the exception of shorts among the outputs S_i of the latch. But since these shorts affect only one of the two results to be compared, they are detected. Hence we need only consider the above five failure classes.

These failure classes are consistent with the failure model used in the usual treatment of PLAs in the literature. Concerning a short between two bit lines or product lines, we only assume that it causes the two lines to take on a valid (and the same) logic level. [6]

Any error caused by these failures can be detected by the self-checking multiplier. Since the 4-out-of-5 checker detects single errors and shorts in the complemented field-select lines, F_i, these do not have to be considered. We will now prove the error coverage for the five failure classes mentioned above.

Proof of 1): Let h_i denote the output of H_i and consider three cases:

a) h_i is some P_i (output of ACC 1): If h_i is incorrect at t_1 but correct at t_2, this error can be modeled as an additive polynomial with one non-zero term x^i, where $i=0,1,\ldots,18$. Thus, $P(x)$ equals $x^4a(x)b(x)+x^i$ at t_1 and $a(x)b(x)$ at t_2. The output of the multiplier then is $\alpha^{-4}P(\alpha)=a(\alpha)b(\alpha)+\alpha^{i-4}$ at t_1 and $a(\alpha)b(\alpha)$ at t_2, which differ because $\alpha^i \neq 0$ for all i; therefore the comparator will flag an error signal. If h_i is incorrect at t_2 but correct at t_1, then the results compared are $\alpha^{-4}(\alpha^4 a(\alpha)b(\alpha))$ and $a(\alpha)b(\alpha)+\alpha^i$; again the error will be detected. If h_i is in error at both times, the outputs are $\alpha^{-4}(\alpha^4 a(\alpha)b(\alpha)+\alpha^i)=a(\alpha)b(\alpha)+\alpha^{i-4}$ and $a(\alpha)b(\alpha)+\alpha^i$, which again differ because $\alpha^i \neq \alpha^{i-4}$ for every i.

b) h_i is some Q_i (output of ACC 2): In this case the error can be modeled as adding α^i to $Q(\alpha)$. The proof proceeds as before.

c) h_i is some R_i or S_i (output of either ACC 3 or latch, respectively): Because ACC 3 and the latch are used only in the first computation, an output error at t_1 will not match the correct output at t_2, so that an incorrect result is detected.

Proof of 2): A random error in a bit line of AND planes 2 and 3 is equivalent to an error in the h_i from which it comes, so by 1) it is detected. An error in one of the inputs to AND plane 1, say a_i since the case for b_i is symmetric, can be modeled as inputs of $x^2a(x)+x^i$ and $x^2b(x)$ at t_1, and $a(x)+x^i$ and $b(x)$ at t_2. These produce respective outputs of $a(\alpha)b(\alpha)+\alpha^{i-4}$ and $a(\alpha)b(\alpha)+\alpha^i$. Thus, the discrepancy is detected.

Proof of 3): A random error in a product line is equivalent to an error in the output of the H_i it feeds. Thus, by 1), it is detected.

Proof of 4): If bit lines a_i and a_{i+1} (similarly, b_i and b_{i+1}) of AND plane 1 are shorted, then at most one of them will be in error; if one is in error, the inputs can be represented by $x^2a(x)+x^i$ or $x^2a(x)+x^{i+1}$ and $x^2b(x)$ at t_1, and $a(x)+x^i$ or $a(x)+x^{i+1}$ and $b(x)$ at t_2. The outputs are

$$\alpha^{-4}(\alpha^2 a(\alpha)+\alpha^i)(\alpha^2 b(\alpha)) = a(\alpha)b(\alpha)+\alpha^{i-2}b(\alpha)$$

or

$$\alpha^{-4}(\alpha^2 a(\alpha)+a^{i+1})(\alpha^2 b(\alpha)) = a(\alpha)b(\alpha)+a^{i-1}b(\alpha)$$

at t_1, and

$$(a(\alpha)+\alpha^i)b(\alpha) = a(\alpha)b(\alpha)+\alpha^i b(\alpha)$$

or

$$(a(\alpha)+\alpha^{i+1})b(\alpha) = a(\alpha)b(\alpha)+\alpha^{i+1}b(\alpha)$$

at t_2. Both outputs are different in either one of the possible cases. A short between adjacent product lines is treated similarly, as is a short in the bit lines and product lines of AND plane 2. A short anywhere in AND plane 3 affects only one of the two computations; it would, thus, be detected.

Proof of 5): A single extra or missing device is detected if it results in an error, for it affects only one h_i.

VLSI Implementation

The nonredundant multiplier (figure 1) has been implemented in 5.0 micron n-channel metal oxide semiconductor technology. The chip is approximately 8.0 mm^2 and contains about 1,800 transistors. It was designed to perform a-synchronous multiplication and it can multiply two field elements in 100 ns (10 MHz continuous multiplication rate) while consuming an average of 140 mW.

A direct application of RESO requires two computations per input pair, effectively doubling the multiplication time. However, since the propagation times through the partial-product accumulator and through the remaining circuitry are approximately equal, the self-checking multiplier could be pipelined by inserting a 19-bit latch at P. It is estimated that each half would operate at 15 MHz, producing an effective multiplication throughput of about 7.5 MHz.

Conclusion

This paper has described the redesign of a programmable $GF(2^m)$ multiplier to detect any one of a set of internal failures without interrupting normal operation. The failure classes considered, which have been drawn directly from the VLSI layout, include all single errors at the logical nodes of the circuit and all shorts between adjacent parallel data lines.

References

[1] W. W. Peterson and E. J. Weldon, Error Correcting Codes, second ed., The MIT Press, Chap. 2 and 6, 1972.

[2] D. O. Carhoun, B. L. Johnson, S. J. Meehan, "VLSI Architectural Design for a Reed-Solomon Transform Decoder," 1981 IEEE International Symposium on Circuits and Systems, Chicago, IL; April 1981.

[3] J. H. Patel and L. Y. Fung, "Concurrent Error Detection in Multiply and Divide Arrays," IEEE Trans. Comput. Vol. C-32, No. 4, pp. 417-422, April 1983.

[4] D. A. Anderson, "Design of Self-Checking Digital Networks Using Coding Techniques," Tech. Rpt. R-527. Urbana: U. of Illinois Coordinated Science Laboratory, 1971.

[5] S. L. Wang and A. Avizienis, "The Design of Totally-Self-Checking Circuits Using Programmable Logic Arrays," Proc. 9-th Annu. Symp. Fault-Tolerant Computing, pp. 173-180, Madison, WI; June 20-22, 1979.

[6] D. L. Ostapko and S. J. Hong, "Fault Analysis and Test Generation for Programmable Logic Arrays," IEEE Trans. Comput. Vol. C-28, No. 9, pp.617-626, Sept. 1979.

Session 1-4a
Arrays Panel

Chairman

W. Luciw
Sperry Univac

Panelists

C. M. Davis, *IBM*
D. L. Greer, *General Electric*
S. S. Patil, *PSI*
K. F. Smith, *University of Utah*

Session 1-4b
Design & Packaging

Chairmen

J. K. Hassan
IBM

R. J. Petschauer
Sperry Univac

ANALYSIS OF SIGNAL PROPAGATION DELAYS AND CHIP LEVEL PERFORMANCE DUE TO ON-CHIP INTERCONNECTIONS

D. L. Carter and D. F. Guise

Bell Laboratories, Murray Hill, N.J.

ABSTRACT

We model driver gate-interconnect-receiver gate combinations in VLSI as RC ladder networks (RCLN's) with mixed continuous and lumped R's and C's. Techniques to estimate the delay time are derived and applied to the problem of on-chip delay degradation relative to the unloaded gates in MOSFET and MESFET chips with various interconnect material parameters and design rules. The effects on relative delay are shown for Al, polysilicon and silicide interconnects for Si FET's as a function of interconnect length at 1 micron and submicron rules. An effect of line resistance on performance due to placement is illustrated by showing that placing a high resistance polysilicon line near a driving gate rather than far from the gate can incur a substantial delay penalty. We use the relative delay model to compare delays in GaAs and Si FET LSI based on published ring oscillator delays, transconductances and interconnect capacitance models. Coupling this with a model for average on-chip interconnection length as a function of total number of gates, GaAs loses its speed advantage over Si at random logic densities of 10,000 gates per chip unless higher transconductance GaAs transistors are employed in the future.

I. INTRODUCTION

As design rules shrink and chips grow larger the resistance in the interconnects between devices on chip become an increasingly important factor in chip performance.[1] The interconnect lines on chips for MOSFET or MESFET technologies can be modeled as series resistance, shunt capacitance lines; the inductance can be neglected even for very fast rise time signals as long as the driving point impedances are larger than the line characteristic impedance so that no ringing or overshoot is expected.[2,3] Under such conditions on-chip interconnects can be considered RC ladder networks (RCLN's) to first order. The voltage time response of RCLN's in general is a complicated function of time. We have studied approximations to the step response of single unbranched RCLN's with arbitrary parameters[4] the results of which are presented here.

Considering a lumped parameter RCLN as shown in the inset of Fig. 1 an accurate approximation to the unit step response is

$$v(t) = 1-[s_2\exp(-s_1t)-s_1\exp(-s_2t)]/(s_2-s_1) \quad (1)$$

where

$$s_{1,2} = b_1/b_2 \mp [(b_1/2b_2)^2-1/b_1]^{1/2} \quad (2)$$

$$b_1 = \sum_{i=1}^{n}\sum_{j=i}^{n} R_iC_j \quad (3)$$

$$b_2 = \sum_{i=1}^{n-1}\sum_{j=i}^{n-1}\sum_{k=j+1}^{n}\sum_{\ell=k}^{n} R_i\,C_j\,R_k\,C_\ell \quad (4)$$

where b_1 and b_2 are the first and second order coefficients of the reciprocal of the frequency transfer function and $s_{1,2}$ are its poles when it is truncated to second order in frequency. The *b's* are also related to the Elmore[2] definitions of delay and rise times

$$T_d = b_1 \qquad (Delay\ time) \quad (5)$$

$$T_r = \left[2\pi\left(b_1^2 - 2b_2\right)\right]^{1/2} \qquad (Rise\ time) \quad (6)$$

Figure 1 shows the "exact" response obtained by ADVICE[5] simulation and the response, Eq.(1) for a 13-section RCLN with randomly chosen R's and C's as given in Table 1. Also shown in Fig.1 is the voltage at the time point $T_d, v(T_d)$ and upper and lower bounds for $v(T_d)$. We determined that $.59 \leqslant v(T_d) \leqslant .63$ if $s_{1,2}$ are real and $.53 \leqslant v(T_d) \leqslant .63$ otherwise. T_d computed by Eq. (3) then represents the time for the output voltage to reach about 60% of its final value and can thus serve as a measure of delay time.

Interconnections on IC's are mostly not single RCLN's but are in general tree circuits for which Rubinstein, Penfield, and Horowitz[6] have derived T_d. We have found that for tree circuits the bounds on $v(T_d)$ are wider than for single RCLN's, although not as wide as would be predicted by the bounding time functions found in reference 6. However, T_d still provides a useful measure to assess the resistive effects of interconnections on delay time and we use it to analyze the effects of scaling, interconnect materials, transistor technology and special placement effects on on-chip delay in the following sections.

II. ON-CHIP INTERCONNECTION PROPAGATION DELAY

We first develop a model of an MOS gate driving a uniform distributed RC network (URC) connected to a receiving gate. The RC ladder representation of the circuit is shown in the inset of Fig. 3. Modeling the driving gate with suitably chosen linear resistance R_g = 1 / (large signal transconductance of the inverter) and parasitic capacitance C_g gives an accurate measure of the delay time, T_d (62% rise criterion, average of pulling up and down) of n-MOS inverter gate-line-receiver gate combinations as simulated by ADVICE [5]. From Eq.(3) and Eq.(5) the expression for the delay is

$$T_d = R_g(C_g + \ell c + C_o) + \frac{1}{2}rc\ell^2 + r\ell C_o \quad (7)$$

CH1935-6/83/0000/0218$01.00 © 1983 IEEE

where r and c are respectively the resistance and capacitance per unit length of the uniform line of length l. C_o represents the input capacitance of the receiving gate (fanout). The five additive delay terms of Eq.(7) are as follows. The first term is the intrinsic unloaded delay of the driving gate. The second and third terms are the delays due to the line and output (fanout) capacitance, respectively. The fourth term of Eq.(7) is the intrinsic delay of the URC line and the form shown was derived by application of Eq.(3) to an n-section lumped equivalent of the uniform line, and then taking the limit $n \rightarrow \infty$. The last term of Eq.(7) is the delay formed by the product of the total resistance of the interconnect line and the fanout capacitance. Let $\tau_g = R_g C_g$ be the unloaded gate (ring oscillator) delay and $D = T_d/\tau_g$ be the relative delay. Express the line length, l, in terms of circuit pitches n and design rule λ as $l = 2\lambda n$; i.e., a circuit pitch is a line width plus line spacing, both of which are taken as λ wide. Using the following parameters and relationships:

- w = width of the driver
- C_g = $C_{gmin} w/\lambda$, where:
  - C_{gmin} = parasitic capacitance of a minimum size gate
- c = $c(\lambda, H, T)$, where:
  - λ = line width (design rule)
  - H = height of the line from the Si ground plane
  - T = thickness of the line
- ρ = line resistivity
- C_o = $C_{gmin} q$, where:
  - q = load capacitance size factor, relative to a minimum size gate
- FO = fanout number
- g_m = $1/(R_g w)$, (large signal transconductance per unit width of the driver)
- s = $R_g C_g/\lambda$ (the slope of a linear fit of the ring oscillator delay time versus design rule plot)

$$D = 1 + \frac{2}{g_m s w} c(\lambda, H, T) n + q\ FO \lambda / w + \frac{2\rho c(\lambda, H, T)}{sT} n^2 + 2\rho\ q\ FO\ g_m \left(\frac{\lambda}{T}\right) n \tag{8}$$

To show the general properties of Eq.(8), we determine the relative signal delay as a function of line length in circuit pitches for polysilicon lines for ½ micron design rules in silicon FET technology. Values for line resistivity were taken from Sinha, et al.[1] and capacitance values were obtained from Cooper.[7] The capacitance values in Table 2 were determined by rigorous solution of the Laplace equation and include the effects of capacitive fringing to the ground plane and adjacent wires, and are realistic with respect to scaling of insulator and conductor dimensions shown in Fig. 2. In this demonstration a value of 100 mS/mm is used for g_m; other parameter values are: s = 106 ps/micron,* q = 1, and the fanout FO = 2.5.

Figure 3 is a plot of Eq.(8), $D(n)$ for two extreme values of w and in addition shows graphically the various terms of the equation. The curve $D(n)$ labeled D_{MAX} is the maximum delay as a function of line length for a minimum size gate. The curve labeled D_{MIN} is the minimum delay corresponding to a driving gate of infinite width, in which case purely capacitive load terms of Eq.(8), the second and third, do *not* contribute to the delay. For small line lengths and driver widths, $D(n)$ is slowly varying and is dominated by the "SELF" and "FANOUT" delays, the first and third terms of Eq.(8), respectively. For long line lengths the intrinsic delay of the line, which varies as the square of the line length (the fourth term of Eq.(8)), becomes dominant and is always the asymptotic form for the delay. For intermediate line lengths and small driver widths the dominant terms of Eq.(8) are the second, denoted in Fig. 3 as c_{LINE} and the last term, denoted as $R_{LINE}\,C_{FANOUT}$. For the choice of parameters in this demonstration, the contributions of these two terms are nearly the same. A 0.5 mm long polysilicon line (length 500 pitches) is already in the intrinsic wire dominated delay regime for any driving gate at λ=½ micron.

FIG. 1 RCLN STEP RESPONSE

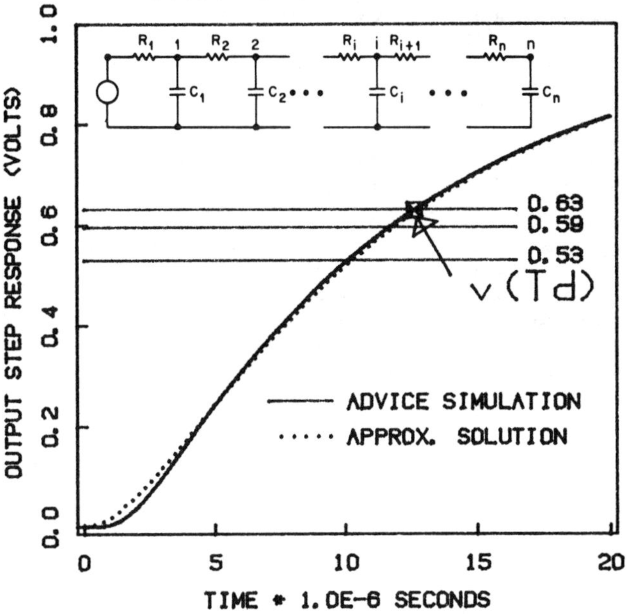

TABLE 1. VALUES OF R AND C FOR FIG. 1.

Node	Rx10^3(ohm)	C (pF)
1	30	0.1
2	36	6.6
3	32	3.7
4	51	0.8
5	74	6.4
6	48	1.1
7	26	8.9
8	20	0.3
9	2	4.4
10	3	1.8
11	98	7.9
12	18	2.3
13 (output)	52	1.9

FIG. 2 LINE DIMENSIONS

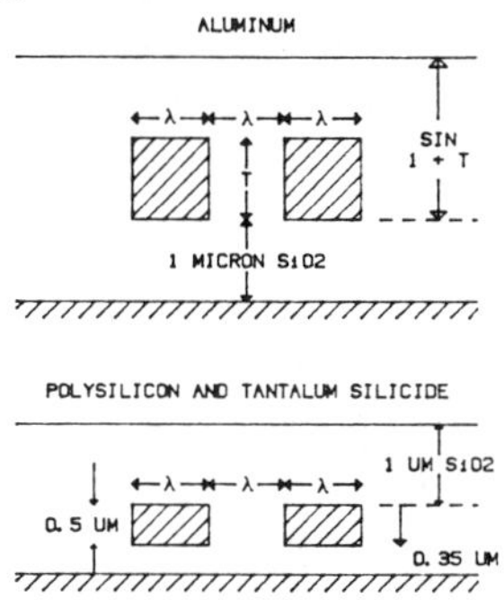

Fig.3 DELAY vs. LINE LENGTH

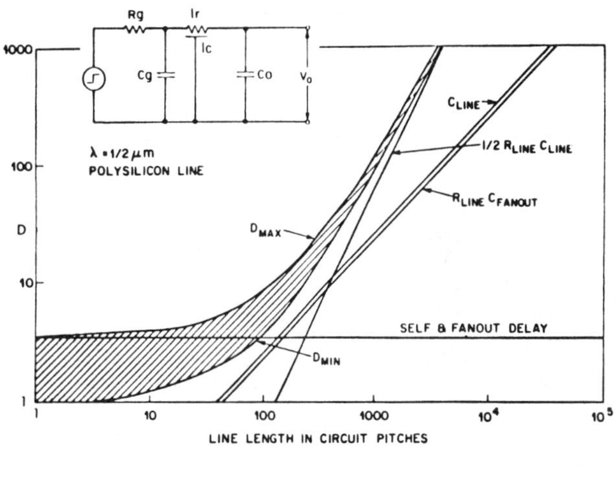

* This is our fit to various published ring oscillator delays (as a function of λ).

III. EFFECTS

To compare the performance of polysilicon, tantalum silicide, and aluminum lines, we find the driver size needed to obtain a desired relative delay as a function of line length as given by Eq.(8). For a fixed relative delay of 5, Fig. 4 shows w/λ as a function of line length in circuit pitches for polysilicon, tantalum silicide, and aluminum lines at design rules of 1 and ½ micron. Values of line resistivity, capacitance, and line thickness are summarized in Table 2. The thickness of polysilicon and tantalum silicide lines (first level interconnect) is taken as ½ micron for both 1 and ½ micron design rules. The thickness of aluminum is assumed to be 1 and 3/4 micron for 1 and ½ micron design rules, respectively, and capped with Si_3N_4 to represent a top level interconnect. Values of the other parameters are the same as those for Si FET technology given in the previous section.

Figure 4 shows that as the line length increases the relative driver size, w/λ, needed to maintain a fixed relative delay, monotonically increases to the point where $w/\lambda \rightarrow \infty$. At this point it is no longer possible to maintain the delay by increasing the driver size because the delay is determined solely by line resistance and capacitance and load capacitance. The knees of w/λ versus n curves in Fig. 4 represent the line lengths at which control of delay begins to be lost.

As the design rule changes from 1 to ½μm, this analysis predicts that the relative driver sizes to maintain the same delay will be larger where control of delay is still possible. Part of this is due to g_m being assumed constant as λ decreases and in the case of aluminum is also due to an increase in capacitance per unit length at design rules of ½μm compared to 1 μm. This is caused by the higher sidewall capacitance contribution from the thicker line and higher cap dielectric compared to poly or silicide lines. Loss of delay control occurs at slightly longer lines (in pitches) for poly and silicide lines and slightly shorter lines for aluminum comparing 1 μm and ½μm design rules. This too is primarily due to the differences in line capacitance at the two design rules. Figure 4 also shows that traditional line and load capacitance for aluminum at 1 μm and ½μm dominates delay to line lengths of about 1cm and 0.4cm respectively, after which line resistance effects begin to dominate. The need for reduced aluminum thickness and cap dielectric constant and higher g_m is manifest in Fig.4. A 10x driver driving a 1000 pitch line would have to be a 24x driver at ½μm rules to maintain the predicted 2x speed advantage (sacrificing area for speed).

As another example of the effects of line resistance we show the asymmetry in delay time that can occur due to the placement of polysilicon and aluminum lines. Consider the following two cases that differ only in the placement (ordering) of the polysilicon lines. In case 1 a long polysilicon line having total resistance R_{pl} and total capacitance C_{pl} is adjacent to the driver and a short polysilicon line with R_{ps} and C_{ps} is adjacent to a receiving gate with a long aluminum line with C_{Al} and negligible resistance connecting the polysilicon lines. In case 2 the positions of the polysilicon lines are interchanged. Case 1 has the longer delay even though the total R and C of the two circuits is identical. This is a result of resistance closer to the driver having more effect on delay. Assume that the long polysilicon line is about 100 pitches long and the short polysilicon line is about 5 pitches long. From Table 2, assuming 1 micron design rules, we have $R_{pl} =$ 4800 Ω, $C_{pl} = 0.04$ pf, $R_{ps} = 240$ Ω, and $C_{ps} = 0.002$ pf. The aluminum line will be assumed to be about 1000 pitches long, as for example a clock line extending along an entire row of standard cells, so that $C_{Al} = 0.5$ pf. The driver and receiver gates will be assumed to be one micron design rule 5x minimum e/d n-MOS and 1x e/d n-MOS inverters, respectively. Application of Eq.(3) with driver parameters $R_g = 1 / g_m w = 2000$ Ω, $C_g = 0.053$ pf ($\tau_g = 106$ ps) gives delays of 3.8 ns and 1.5 ns for case 1 and case 2, respectively, a factor of 2½ in delay for the two cases. The delay asymmetry ratio is *not* sensitive to design rule. Also, if the polysilicon were replaced by tantalum silicide, the line lengths to get the same absolute delay difference in the two cases would have to be about 10 times longer, including the aluminum segment.

TABLE 2: RESISTANCE AND CAPACITANCE DATA FOR INTERCONNECT LINES

Material	ρ [Ω-cm]	C[pf/cm],T[μ] λ=1μ	λ=½μ
Aluminum	2.5×10^{-6}	2.36, 1	3.47, 0.75
Tantalum Silicide	1.0×10^{-4}	1.96, 0.5	1.67, 0.5
Polysilicon	1.2×10^{-3}	1.96, 0.5	1.67, 0.5

FIG. 4 DRIVER WIDTH VS. N (D=5)

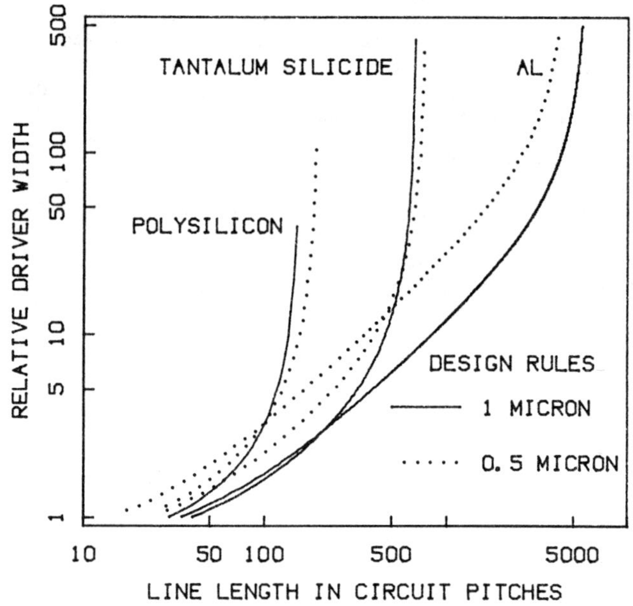

IV. COMPARISON OF GaAs AND Si TECHNOLOGIES IN THE WIRE DOMINATED CHIP

Very high speed gates, ring oscillators, multipliers, counters, all with small to moderate levels of integration, have been fabricated using GaAs MESFET technology.[8] These types of structures are quite regular: the on-chip interconnections are mostly between nearest-neighbor gates. Contrast this to future large gate arrays or random logic-like structures in which the loads on individual gates are wire dominated rather than fanout dominated. How will GaAs circuits perform compared to fine line Si MOS under these circumstances? Clearly, under the conditions of maximum performance it will be important to have a high g_m to charge the wire capacitances as fast as possible. Interestingly, the reported GaAs high speed working chips to date have transistors with g_m's that are comparable to those of fine line n-MOS, about 100mS/mm at one micrometer gate length.[9] This implies that the superior performance of these GaAs chips was due to their lower gate parasitic capacitance compared to their Si MOS counterparts.

We now estimate the delay time for both E/D GaAs and Si n-MOS circuits as a function of wire loading using the linear delay theory developed above and published experimental ring oscillator delays at 1μm. Assume the wires on GaAs and on Si will be 1μm thick aluminum 1μm wide (250 *OHMS*/cm), spaced 1μm to the nearest-neighbor wires. Wires for both GaAs and Si will be capped with a dielectric of $\epsilon = 3.9$ on a field oxide of $\epsilon = 3.9$, 1μm thick. For Si the semiconductor surface is considered to be a ground plane, and for GaAs this surface is assumed to be an insulator with $\epsilon = 12.9$ with a ground plane far from the wires (consistent with wiring over semi-insulating GaAs). Computation of the wiring capacitance per unit length under these conditions[7] gives 1.91 pf/cm for Si and 1.56 pf/cm for GaAs. We will assume $g_m = 100$mS/mm for both technologies and unloaded gate delays of 20 ps for GaAs and 106 ps for Si n-MOS.

Using Eq.(8) the results of computing delays for Si and GaAs as a function of line length in circuit pitches n is shown in the upper and lower shaded regions of Fig. 5, respectively, assuming a fanout of 3 minimum-width gates. The boundaries of performance for minimum width and 10 x minimum width drivers are at the upper and lower edges of the two shaded regions, respectively. A clear performance advantage for GaAs over Si n-MOS is predicted for line lengths of less than 500 pitches. At 500 pitches and longer there is considerable overlap in delay for the two technologies. An estimate of the average line length per gate required by a given number of gates per chip can be determined using the theory of Heller et al.[10]

$$n = (6FO/Nf)\ G^{(2p-1)} \tag{9}$$

where N is the number of levels of interconnect, f the efficiency of utilization of these levels by the wires, G the number of gates per chip and p the Rent coefficient which we shall take as 2/3 representative of random logic. Assuming $f = .75$ and $N = 2$, $n = 12G^{1/3}$. The number of gates corresponding to selected n's are shown on the abscissa of Fig. 5. At 10,000 gates per chip the advantage of GaAs has been greatly diminished. Of course, we have neglected many other parameters such as power dissipation, threshold uniformity, etc. necessary to make a rigorous technology comparison. However, it is expected that complex GaAs circuits with transistors having g_m = 200-400 mS/mm will appear, this having been achieved with thinner active layers.[11] The parasitic capacitance of the gates would then be expected to be about 2-4 times higher, thus trading off power dissipation for ability to drive wires. The delay for g_m = 400 mS/mm technology is shown as the solid lines in Fig. 5 for minimum size drivers (the upper line) and for $w/\lambda = 10$ drivers (the lower line). The interconnect length at which overlapping performance with Si begins is ten times longer here and a clear performance advantage is predicted at all line lengths.

V. CONCLUSIONS

We have shown that the first moment of the impulse response of RC ladders is a good approximate measure of delay time in LSI circuits, and used it to show quantitatively some examples of the effects of resistance in the interconnect lines on delay. We did not consider other interconnect resistance effects on fine line chips, such as disturbs in power and ground rails, that are at least as important and troublesome as effects considered here. More rigorous models of the effect of interconnect resistance on delay are just now being taken into account in CAD tools.[12,13]

ACKNOWLEDGEMENTS

We would like to thank J. A. Cooper, Jr. for line capacitance computations and R. K. Curran, A. B. Glaser, S. Waaben, and O. Wing for valuable discussions.

FIG. 5 Si - GaAs COMPARISON

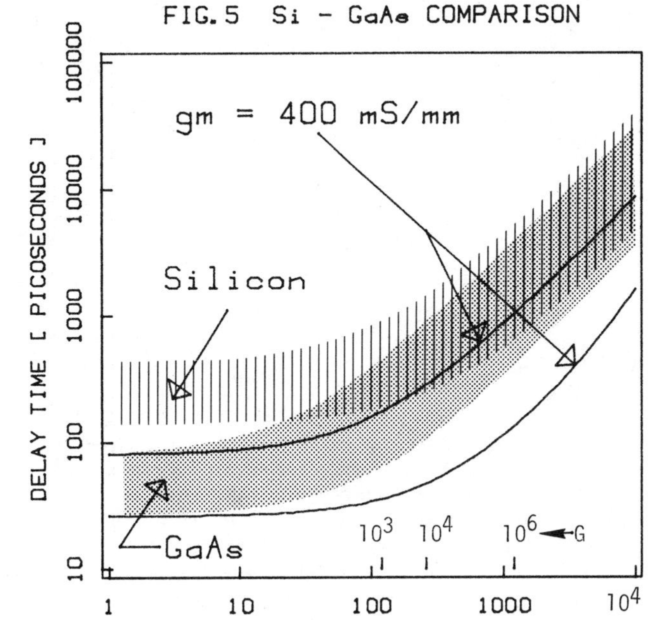

REFERENCES

[1] A. K. Sinha, J. A. Cooper, Jr. and H. J. Levinstein, "Speed limitations due to interconnect time constants in VLSI integrated circuits," IEEE Elect Device Lett., EDL-3, No. 4, pp.90-92, April 1982.

[2] W. C. Elmore, "The transient response of damped linear networks with particular regard to wideband amplifiers,", J. Appl. Phys., Vol 19, pp.55-64, January 1948.

[3] M. S. Ghausi and J. J. Kelly, "Introduction to Distributed-Parameter Networks," Holt, Rinehart and Winston, 1968.

[4] D. L. Carter and D. F. Guise, "Delay and rise time in arbitrary R-C ladder networks. I. Unbranched circuits," submitted to IEEE Trans. Circuits Sys.

[5] L. W. Nagel, "ADVICE for circuit simulation," ISCIS Conf., Houston, TX, 1980.

[6] J. Rubinstein, P. Penfield, Jr. and M. A. Horowitz, "Signal delay in RC tree networks," IEEE Trans. CAD, to be published.

[7] J. A. Cooper, Jr., private comm.

[8] M. R. Namordi, W. A. White, "GaAs E/D Circuits," Digest of Technical Papers, IEEE International Solid State Circuits Conference, p.38, New York, NY, Feb. 1983.

[9] M. Abe, T. Mimura, N. Yokokoyama, H. Ishikawa, "New Technology Towards GaAs LSI/VLSI for Computer Applications," IEEE Trans. on Electron Devices, Vol. ED-29, No. 7, pp.1088-1094, July 1982.

[10] W. R. Heller, W. F. Mikhail, W. E. Donath, "Prediction of Wiring Space Requirements for LSI," Design Automation and Fault Tolerant Computing," pp.117-144 (1978).

[11] H. Nishi et al., "Through-AlN Implantation for a High Transconductance Self-Aligned GaAs MESFET," IEEE Device Research Conference, Pgm. Digest, Paper IVA, June 1983.

[12] R. Putatunda, "Automatic calculation of delay in custom generated LSI/VLSI chips," IEEE International Conference on Circuits and Systems Proceedings, pp.193-196, September 1982.

[13] N. P. Jouppi, "TV: An nMOS timing analyzer," Proceedings of the Third Caltech VLSI Conference, pp71-85, March 1983.

THIN FILM INTERCONNECTION LINES FOR VLSI PACKAGING

A. Deutsch

C. W. Ho

IBM T. J. Watson Research Center

Yorktown Heights, New York

Abstract

This paper discusses the design of thin film lossy interconnection lines for future high performance computer package applications. The electrical characteristics of a triplate structure design are described and the effects of via loading and processing variation on the lines are analyzed. A simple approach is given for adjusting line resistances by changing the line widths within the wiring plane while maintaining a constant characteristic impedance so that multiple receiver nets can be accommodated. The high performance of the programmable design makes the thin film package extremely versatile for interconnecting high speed VLSI bipolar logic and memory chips on a multichip module.

I. INTRODUCTION

Future VLSI computer packages require high wiring density and improved performance in order to interconnect the increased number of chip input/output terminals and high speed devices. One approach being pursued is the use of thin film, photolithographically defined interchip wiring lines[1] The use of a thin film wiring layer can replace many thick film module wiring layers to reduce the lengths of interconnection wires so that performance improvements can be made.

It was shown before[2] that a triplate structure design for thin copper lines with polyimide dielectric (ε = 3.5) provides a high speed, high wiring density interchip package environment. Fig. 1 shows two possible implementations for a line center-to-center distance of 25.4 μm: one line per channel (Fig. 1a) with interconnecting vias on the same pitch with the lines, or two lines per channel (Fig. 2a) in which case the vias are on twice the line pitch. Both designs result in similar performance, however the two lines per channel configuration may be preferable because it allows space for larger diameter vias (vias are generally placed in between the lines). In Section V, a third design is discussed, (Fig. 1c) that allows even larger minimum features and thus less resistive lines.

II. ELECTRICAL CHARACTERISTICS OF TRIPLATE STRUCTURE

The design shown in Fig. 1a uses X and Y signal lines between two solid ground planes. Such a closed transmission line system when compared with a single ground plane case, yields a reduced noise coupling between lines which in turn allows the use of wider and thicker conductors and therefore less lossy lines. The following dimensions are used: w = 8 μm, $s_1 = s_2$ = 17.4 μm, $t_1 = t_2$ = 5.5 μm, and $h_1 = h_2 = h_3$ = 6 μm. The characteristic impedance is close to 50Ω (Z_o = 41Ω) and the use of low-dielectric polymer with ε = 3.5 results in propagation delay of τ = 65 ps/cm. With a resistivity of ρ = 1.8 μm • cm of the copper metallurgy, the conductors have a resistance of R_o = 4.09 Ω/cm. It was explained in Ref. 2 that such resistive lines can be used to propagate high speed pulses over long transmission lines as long as the lines do not have the usual terminating resistor but an open circuit that allows the signal to double and compensate for the attenuation.

The design in Fig. 1b has similar dimensions except for line separations. s_1 = 21.8 μm and s_2 = 13 μm. The electrical characteristics τ, Z_0 and R_0 are maintained the same, however vias that interconnect the two levels or connect to chip or module signal or power I/O's can be placed only in the wider separation s_1 as opposed to Fig. 1a where smaller vias were placed in both separations: s_1 and s_2. It will be shown in the next section that there is a very small increase (1.5%) in noise coupling due to reduction of s_2 in the two lines per channel design.

III. THE EFFECT OF VIAS ON THE ELECTRICAL CHARACTERISTICS

In order to ease the analysis of the three dimensional structures of Figs. 1a and b, measurements of self and mutual capacitances were made on a 500 times scaled-up model shown in Fig. 2. The effect of vias was analyzed and the results were combined with two-dimensional capacitance and inductance programs[3] and a circuit simulation program[4] in order to assess the change in coupled noise, characteristic impedance and propagation delay. The model in Fig. 2 incorporated both one and two lines per channel designs. Figs. 3a and b show the effect of vias on the self (C_{22}) and mutual (C_{12} and C_{23}) capacitances of the X-layer conductors for a via population ranging from 0-100% and for three cases: when the Y layer is fully populated (full load) with conductors, half populated (half load), and totally missing (no load). Similar effects were previously predicted in Ref. 5.

The self capacitance C_{22} is affected much less by the presence of vias than the mutual capacitances C_{12} and C_{23}. In the two lines per channel case, C_{22} and the mutual capacitance C_{23} between lines that have no vias in between are practically not influenced by the presence of vias, the only observed change is due to Y conductor loading. The mutual capacitance C_{12} however, is strongly influenced by vias in this design due to the larger spacing and wider vias.

In the thin film package with vias on 25.4 μm (Fig. 1a) or 50.8 μm (Fig. 1b) pitch it is estimated to have only a 15% via utilization and half Y loading. For these conditions, C_{22} increases only 7.7% for the one line per channel design and 6.8% for the

CH1935-6/83/0000/0222$01.00 © 1983 IEEE

two lines per channel. The mutual capacitance C_{12} decreases only by 28% for the first case while C_{12} decreases by 62% and C_{23} by 16.7% for the latter case.

The small change in self capacitance C_{22} results in even smaller change in characteristic impedance $Z_o = \sqrt{L_{22}/C_{22}}$ and propagation delay $\tau = \sqrt{L_{22}C_{22}}$ For the one line per channel design, Z_0 decreases only by 3.7% and τ increases by 3.8%; for the two lines per channel case, Z_0 decreases by 3.2% and τ increases by 3.3%.

The effect of the presence of vias on the noise that couples from two active lines into an adjacent quiet line was analyzed using the same method described in Ref. 2. The near-end-noise (NEN) is proportional to the sum $C_{12} / C_{22} + L_{12} / L_{22}$ of capacitive and inductive couplings[5] while the far-end-noise (FEN) is proportional to the difference of the two couplings. The decrease of C_{12} and increase of C_{22} results in decrease of NEN but increase in FEN. Fig. 4 shows the increase in FEN for 5 cm long lines for one and two lines per channel designs as a function of via loading.

In addition to the inhomogeneities caused by crossover lines and vias the loss in the lines causes continuous reflections that effect the coupled noise. In the case of short unterminated transmission lines and not very lossy (R_0 = 4.09 Ω/cm) sustained reflections are present both on the active and quiet lines. The superposition of NEN and FEN causes the peaks of the coupled noise not to occur at the starting end (for NEN) or the far end (for FEN) of the quiet line. Fig. 5 shows the noise coupling (NEN and FEN) for the one line and two lines per channel designs, for 15% via population and half Y loading as a function of coupled line length. Due to reflections, the worst coupled noise is obtained for short lines and the two lines per channel design has larger coupled noise than the one line per channel design by only about 1 - 1.5%. The worst coupled noise of 11% is very close to the 10% design objective and Section V shows how the coupled noise is reduced when the over-and-undershoots on the active signal line are eliminated by use of a diode clamp network at the end of the line.

Due to the low via utilization projected, the electrical characteristics, coupled noise, characteristic impedance and propagation delay, are very slightly altered by their presence.

IV. SENSITIVITY ANALYSIS

An analysis was made of the effect on the transmission line electrical characteristics of varying all the triplate structure dimensions, line width w, line thickness t, height of insulator h by ± 10% and a ± 7.5% variation of line resistivity (ρ). It was assumed that the line resistance R_0, capacitance C and inductance L are functions of the form

$$F = f_o + \frac{\partial f}{\partial \rho}\Delta\rho + \frac{\partial f}{\partial w}\Delta w + \frac{\partial f}{\partial t_1}\Delta t_1 + \frac{\partial f}{\partial t_2}\Delta t_2 + \frac{\partial f}{\partial h_1}\Delta h_1 + \frac{\partial f}{\partial h_2}\Delta h_2 + \frac{\partial f}{\partial h_3}\Delta h_3$$

where $\frac{\partial f}{\partial w}$ for example is

$$\frac{\partial f}{\partial w} = \frac{\partial f(w,\rho_0,t_{10},t_{20},h_{10},h_{20},h_{30})}{\partial w}$$

Since the variations are small, the $\partial f/\partial w$ term can be approximated by the slope of the regression line through the calculated points. The line dimensional variations were assumed independent and normally distributed.

The 3σ results obtained by statistically varying all the dimensions and the resistivity simultaneously are summarized in Fig. 6. The variation of the conductor width (w) and height (h) above ground have the strongest effect while the thickness (t) has only a second order type influence on the electrical characteristics. The largest variation was obtained for line resistance R_0(± 16.3%) and for coupled noise (+23.5%, -19.24 % which represents +1.9%, -1.57% of the signal swing). The line impedance Z_0 varied only by +5.9% and -4.95% and the propagation delay varied by +4.4% and -2.8%.

V. PROGRAMMABLE LINE WIDTH AND GROUND PLANE DESIGN

It is explained in Ref. 2 that for unterminated lossy transmission lines with lengths ℓ such that $R_0 \cdot \ell \leq 2\ Z_0$ high speed low distortion propagation can be maintained. Beyond such lengths the signals start exhibiting RC-type slowdown. Fig. 7a shows that for lines with R_0 = 4.09Ω/cm, distortionless propagation for a 1ns wide pulse can be maintained for lengths up to ℓ = 15 cm for the case of one driver and one receiver chip. When the resistance R_0 is dropped to 2.18Ω/cm, the same signal can be transmitted over lengths in excess of 24 cm as shown in Fig. 7b.

Fig. 1c shows a method of incorporating in the same triplate structure described earlier, a lower resistance line R_{02} = 2.18Ω/cm) with width w_2 = 15 μm. Two adjacent narrow lines (w_1 = 8 μm) are joined to form the new wide line and the characteristic impedance is maintained the same (Z_{02} = 42.2Ω) by replacing the solid ground planes with mesh sections with appropriate widths and separations as shown in top view in Fig. 8. This approach offers a simple transition from narrow to wide lines since all the vias are kept on the same grid.

The only concern that can be raised for such a programmable design is the possible increase in coupled noise when the narrow lines traverse under one of the slots in the ground planes. The width of the mesh (w_3) is designed to minimize any such increase and still maintain the same impedance for both the narrow and wide lines while lowering R_{02} as much as possible.

It is expected that at the high circuit density levels of a future VLSI multichip module package very few (about 5% of the nets) far-end multi-receiver nets (one driver chip placed far from a group of receiver chips) are required. The NEN and FEN coupled noise was analyzed for 4.2 cm long lines for the case with narrow 8 μm lines with 95% of the line length having solid ground planes and the rest of the 5% of the line length crossing over randomly distributed mesh ground plane sections of the type shown in Fig. 8. Figs. 9a and b show the coupled noise for the all solid ground plane case and Fig. 9c shows the minimal effect the presence of the ground slots have (the near-end coupled noise increases from 11% to 11.6%).

VI. MULTI-RECEIVER NETWORKS

In future high performance digital systems, it is expected that most of the nets on multi-chip modules are of the single load type (one driver and one receiver chip). A few percentage of the total number of nets however, can still require distributed or cluster type of interconnections.

Groups of 3 - 4 receiver chips close to a driver chip can be wired with the high resistance lines (R_0 = 4.09Ω/cm) and still achieve first incidence switching[2] of every receiver as shown in Fig. 10a and b.

In the case of far-end nets (receiver chips are far from the driver chip), the use of low resistance lines offers improved wirability. Figs. 10c and d show how a far-end distributed net can go from 14.9 cm to 18 cm long connections when the resistance is dropped from 4.09Ω/cm to 2.18 Ω/cm, and far end cluster nests up to 16.5 cm can be easily achieved as well (Fig. 10e). Due to the small number of such configurations, the mesh areas required for the wide lines of the design in Fig. 1c can be limited to avenues spaced at appropriate intervals on the module.

In areas of lower wiring demand, the 50.8 μm pitch design with mesh ground plane exclusively can be used within the same X and Y signal plane-pair as shown previously. In this case, wiring configurations can arise where the driver chip is in an area of the module with high resistance lines and is connected to a group of receiver chips with low resistance. Fig. 10f shows the signals on such a net where half of the initial wire segments (2.5 and 4.1 cm long) connecting the driver chip to receiver chips R_1 and R_3 have R_o = 4.09Ω/cm while the rest have R_o = 2.18Ω/cm. The signals arriving at the four receivers show minimal distortion.

VII. CONCLUSIONS

Thin-film lossy unterminated transmission lines have been shown to successfully satisfy the wiring requirements of future multi-chip module packages. Thin film lines offer high wiring density and reduction of power dissipation due to elimination of the power dissipated in terminating resistors. The electrical characteristics show small variation due to crossover lines, interconnecting vias and processing tolerance.

The triplate structure with programmable line width and solid or mesh ground planes offers a flexible design within the same X and Y wiring plane. The reduction in resistive drops enhances the package wirability and allows long point-to-point and multi-receiver nets.

ACKNOWLEDGEMENT

The authors are grateful to V.A. Ranieri for performing the extensive capacitance measurements of the scaled-up model and for coordinating the excellent machining work of the numerous vias and via-holes needed for the experiment. We also thank I. Feinberg and L. Wu for the fruitful collaboration and discussions that led to better understanding of critical packaging issues.

REFERENCES

[1] C. W. Ho, D. A. Chance, C. H. Bajorek, R. E. Acosta, "The Thin Film Module as a High-Performance Semiconductor Package", IBM J. Res. Develop., vol. 26, No. 3, pp. 286-296, May 1982.

[2] A. Deutsch, C. W. Ho, "Triplate Structure Design for Thin Film, Lossy, Unterminated Transmission Lines", 1981 IEEE Intern. Symposium on Circuits and Systems Proceed., vol. 1, pp. 1-5, April 1981.

[3] W. T. Weeks, "Calculation of Coefficients of Capacitance of Multiconductor Transmission Lines in the Presence of a Dielectric Interface", IEEE Trans. Microwave Theory Tech., vol. MTT-18, pp. 35-43, Jan. 1970.

[4] W. T. Weeks, A. J. Jimenes, G. W. Mahoney, D. Mehta, H. Qassenoadeh and T. R. Scott, "Algorithms for ASTAP-A Network Analysis Program", IEEE Transactions on Circuit Theory, vol. CT-20, pp. 628-634, Nov. 1973.

[5] C. S. Chang, G. Crowder and M. F. McAllister, "Crosstalk in Multilayer Ceramic Packaging", 1981 IEEE Intern. Symposium on Circuits and Systems Proceed., vol. 1, pp. 6 - 11, April 1981.

[6] Private Communications with L. Yuan and B. J. Rubin from IBM GTD Laboratory, Hopewell Junction, N.Y.

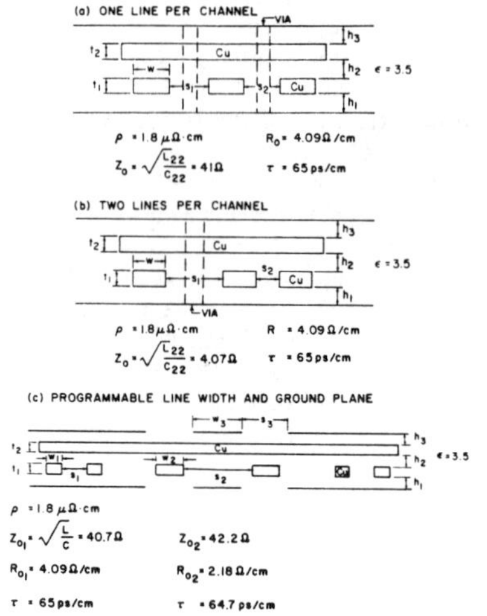

Fig. 1. Triplate Structure Designs: a) One line per channel, b) Two Lines per Channel, c) Programmable Line Width and Ground Plane.

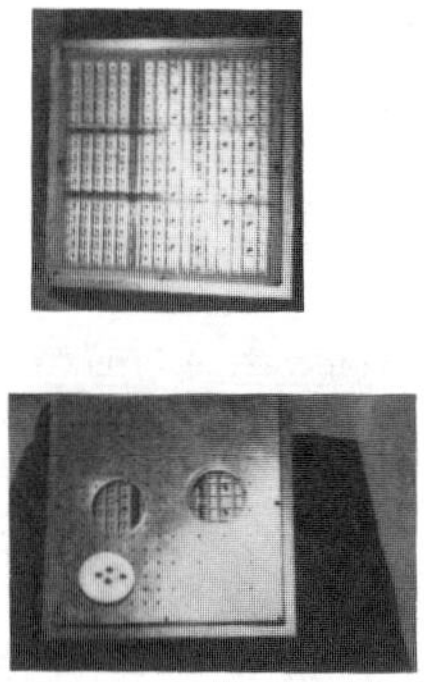

Fig. 2. Top View of 500-Times Scaled up Model with Vias, and with and without Top Ground Plane.

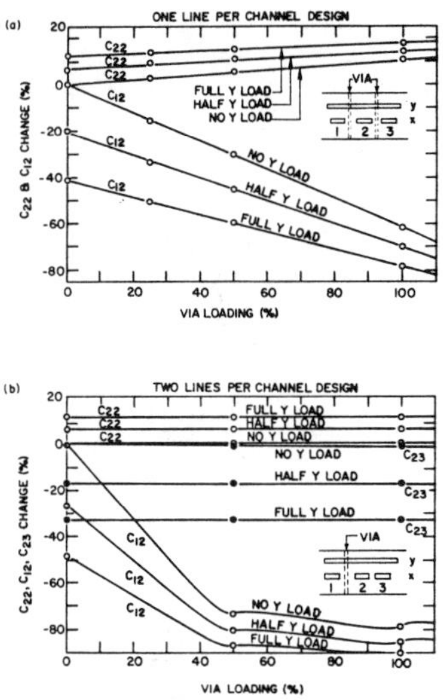

Fig. 3. Self and Mutual Capacitance Change as a Function of Crossover Lines and Via Loading for: a) One Line per Channel and b) Two Lines per Channel Design.

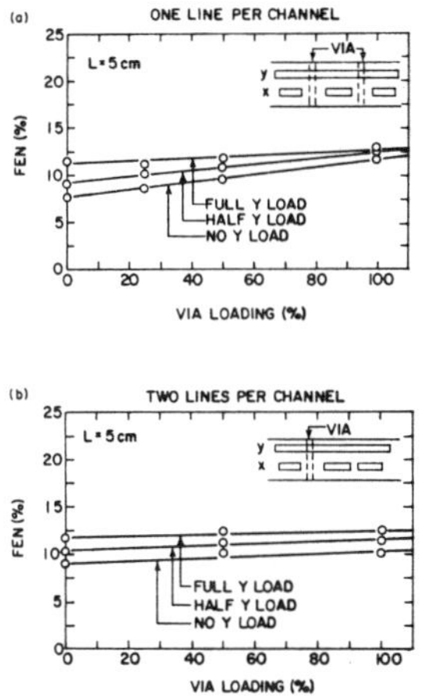

Fig. 4. Far-End Noise (FEN) as a Function of Crossover Lines and Via Loading for a) One Line per Channel and b) Two Lines per Channel Design.

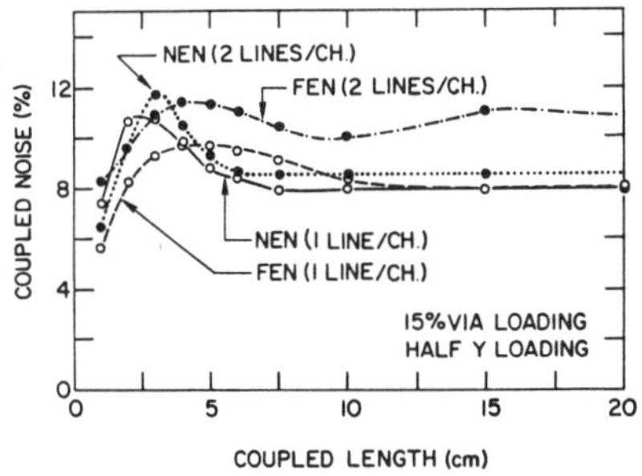

Fig. 5. Coupled Noise (NEN and FEN) as a Function of Coupled Length.

	L(cm)	MEAN	
FEN(%)	5	8.18	
ΔFEN(%)		-1.57	+1.9
Z_0 (Ω)	10	41.36	
ΔZ_0(%)		-4.95	+5.9
R_0(Ω/cm)	10	4.913	
ΔR_0(%)		-16.26	+16.27
τ(ps/cm)	20	71.99	
Δτ(%)		-2.79	+4.39

Fig. 6. Sensitivity Analysis Results

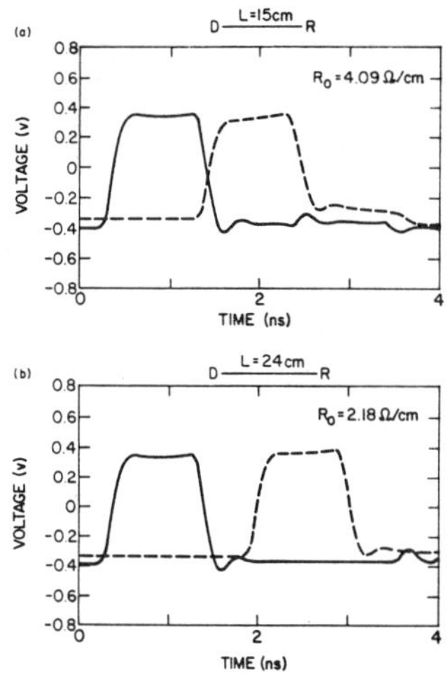

Fig. 7. Signal Response for Lossy Diode Clamped Line with a) Length ℓ = 15 cm and Resistance R_o = 4.09Ω/cm and b) ℓ = 24 cm and R_o = 2.18Ω/cm

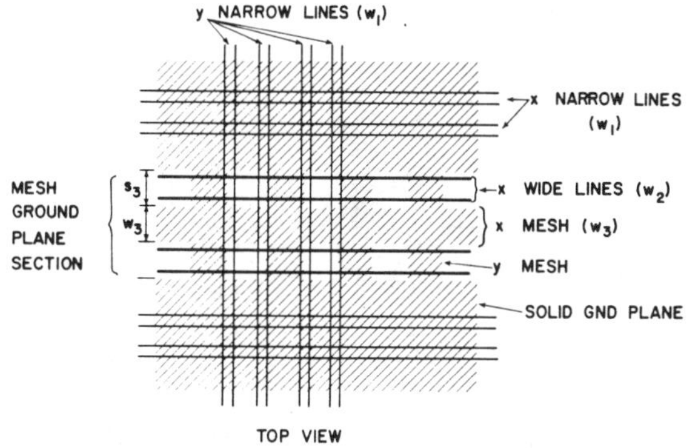

Fig. 8. Top View of Programmable Line Width and Ground Plane Design.

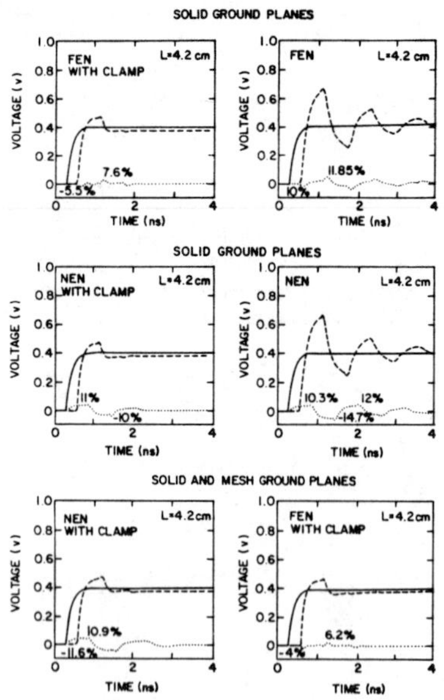

Fig. 9. Signal Response and Coupled Noise for a) 4.2 cm Long Lines with and without Diode Clamp and Triplate Structure and Solid Ground Planes and b) Triplate Structure with 45 Sections of Mesh Ground Planes.

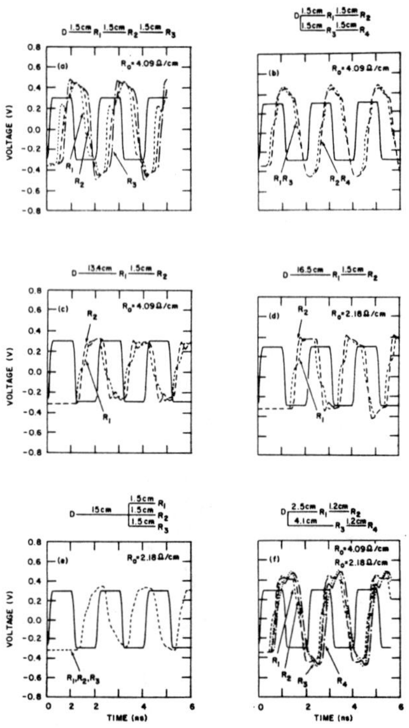

Fig. 10. Multi-Receiver Networks: a) Near-End Distributed with Three Receivers and Lines with R_o = 4.09Ω/cm, b) Near-End Cluster with Four Receivers and R_o = 4.09Ω/cm, c) Far-End Distributed with Two Receivers and R_o = 4.09Ω/cm, d) Far-End Distributed with Two Receivers and R_o = 2.18Ω/cm, e) Far-End Cluster with Three Receivers and R_o = 2.18Ω/cm, f) Distributed Cluster with R_o = 4.09Ω/cm and 2.18Ω/cm.

VLSI HIGH PERFORMANCE COMPUTERS

by Dr. John Foggiato

Amdahl Corporation

Abstract

Today's mainframe computers use LSI to achieve both high performance and reliability. These large systems, which operate at greater than 10 million instructions per second (MIPS), have individual logic circuits containing between 400 to 1000 gates. One must then evaluate the impact of VLSI which will be defined as circuits containing in excess of 10,000 gates. Further system capabilities will be enhanced performance, reliability and functionality of the large systems. VLSI will present a number of challenges which will be confronted by both system designers and the technologists.

Introduction

High performance systems encompass the more publicized supercomputers used in scientific environments and the commercially oriented mainframe. The important criteria is the measurement of performance, be it the response time in an interactive environment or throughput of large batches of computations. The benefit of better response time is system programmer or user productivity. On the other hand, high throughput performance for large batch jobs reduces costs through timely availability of data. In the typical data processing environment of today, both batch and interactive jobs are required.

The term VLSI has been used to denote the transition from small densities of circuitry on a chip to the realm of millions of components. The definition for this paper will deal with circuits containing greater than 10,000 gates and operating at clock frequencies of 20MHZ or higher. As our national project, VHSIC measures the transition from LSI to VLSI, the product of gate density and clock frequency exceeds 2×10^5 gate MHZ in a one square centimeter chip. Examples of available VLSI circuits containing a high density of components include the Motorola 68020 microprocessor and the Hewlett Packard chip used in the HP9000 Work Station Computer. However, these are MOS circuits. For the large mainframe computers, bipolar technology is needed to achieve high performance. Future technology trends will be reviewed, including reliability as an important criterion.

To illustrate the capabilities of VLSI, Figure 1 projects the system cycle times achievable in future systems. As can be seen, the gate delay of the VLSI circuits will primarily influence the performance capabilities of the system. The derivative of this shorter system cycle time is performance objectives of 50 to 70 MIPS in the 1990 time frame.

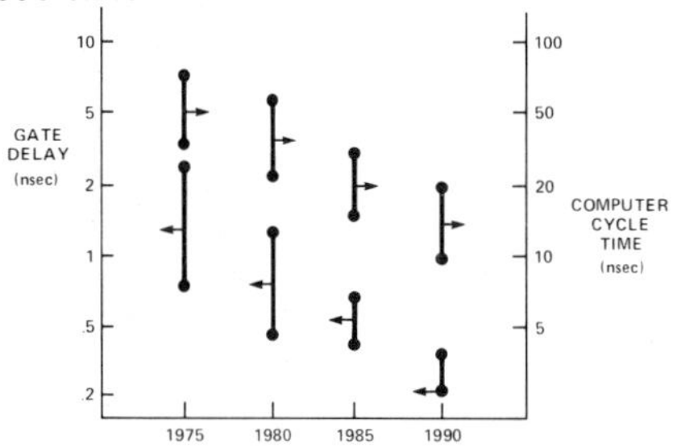

Figure 1

Computer cycle times and logic circuit gate delays

What are the high performance systems? IBM is well known as the developer of large systems with a specific architecture. Later, the Plug Compatible Manufacturers, initially Amdahl and now some others, supplied systems with better price/performance and more features than those offered by IBM. Also, there are the large systems supplied by non-IBM compatible suppliers, collectively known as the BUNCH. These have dominated the scientific environment with some use in the commercial sector. The scientific community continues to demand more performance and higher accuracy, thus there are the recent developments from Cray and CDC, followed closely by Sperry. For those wanting both scientific and commercial capabilities, attached array processors have been developed. What about the future? How will the 5th generation supercomputer conceived by the Japanese fit with the available products?

Why are continued enhancements to these systems required? In reviewing today's systems limitations, system availability and performance are dominant. Diagnostic capabilities have been designed into systems. The more desirable feature, however, is to have systems which can repair themselves - the so-called fault tolerant computers. Today, a number of techniques, such as parallel processors and software, are used to provide this capability. Other system limitations involve limited I/O capabilities (i.e.,

CH1935-6/83/0000/0227$01.00 © 1983 IEEE

channels and memory), thus limiting full utilization of the performance capability of the central processor. Futuristic systems will have to be more user friendly and provide extensive intercommunication capabilities. Overcoming these limitations in large systems will be reviewed. Architectural limitations encompass optimal implementation of instruction sets, and in this endeavor, VLSI will have the greatest impact.

As noted, VLSI will impact many areas. Performance enhancements are provided through faster operating speeds and closer packing of functional units. Reliability coupled with fault tolerant design will provide the user greater system availability. Both functional and physical architectures will be changed.

The physical size and operational requirements (power and cooling) will continue to decrease. New interfaces will be developed and some examples will be given. As such, stepping into the world of VLSI will provide the system user with numerous functional characteristics which are unavailable today due to development and implementation costs.

Another aspect of VLSI will be to provide attachable processors to supplement the larger central processor (CPU). Figure 2 represents a number of processor elements to enhance functional capabilities. The futuristic CPU, shown in the center, will operate at a high performance level. Integrated or attachable processors, more specifically application oriented, will supplement the central unit. For scientific environments, a vector or array processor provides computational capability, with the main CPU having a large data base. Interactive environments will operate through a transaction unit which will be heavily communications oriented. Data Base Management (DBM) will be conducted through a larger storage unit. For the large storage requirements of a large CPU, a special Input/Output (I/O) processor provides external communication links, as being done in all large systems today. For applications programs requiring fast access to large data bases, a high performance buffer memory would be available.

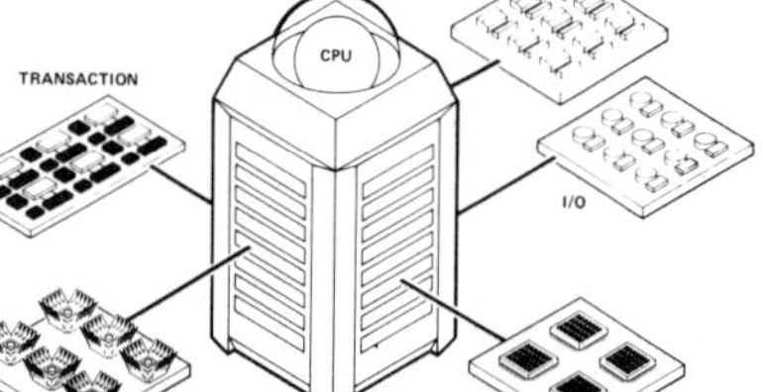

Figure 2

Examples of off-loaded processors

What will be the system of the future? Setting 1990 as the reference date, systems performing as single processors in the 50 to 70 MIPS range will be available. Memory channels and I/O capability will be enhanced by a factor of 6 to 10 over today's availability. More importantly, with the rapid advances in hardware, will the operating systems and applications software be able to fully utilize these capabilities? As such, the impact on the users will be substantial.

One can envision the potential of VLSI. What, however, are the challenges to realizing VLSI circuitry for these large systems? First comes the technology which encompasses semiconductors, packaging, design automation and testing. Various methodologies and needed resources will be described for each area. Once developed, how does one transfer the technology to manufacturing? Automation will become an important factor since this will be the only way to achieve manufacturing yields at minimal costs. Finally, the most important resource, personnel, will provide the greatest challenge. Not only training, but orientation and motivation will require the development of training facilities and new management methodologies. The subject of personnel will also include suggestions regarding untapped resources.

In summary, we should also bring into perspective the efforts of Japan and Europe. These will be summarized and compared to those within the U.S.

Hardware and Software Synergism: Key to VLSI

J.K. Hassan

IBM Corporation

The continuing advance of electronic technology has been a driving force behind the evolution of computing machinery, an evolution in which the unit cost falls while the unit capability rises. The constantly evolving ratio of cost to capability requires a frequent review of the design tradeoffs to be made. The revolutionary impact of VLSI technology on computer architecture, and design methodology by which large-scale embedded computer systems can now be realized much more easily. An embedded computer system can be defined as a computer system which controls a very large hardware comples. Any real-time control system fits into this category.

The difficulties in the design of large-scale embedded computer systems fall into two categories: First, the requirement on real-time control, criticality, high performance, and fault tolerance keeps stressing the available technology to its limits. Sophisticated computer architectures, such as pipelining, interleaved memory and parallel processing, have to be used in order to satisfy the imposed requirements. These complicated architectures usually lead to a very expensive design. Other difficulties in developing large-scale embedded computer systems are primarily caused by the largeness of the systems and the ever changing system environment. Despite these problems, difficulties in the design of large-scale embedded computer systems have been alleviated by the rapid advances in the following two areas: First, the emergence of VLSI technology and second, the advances in design methodologies.

The use of bit slice microprocessors provides a multicomputer system with dynamic reconfiguration capabilities. These capabilities allow reconfiguration, under software control, of the system hardware into a variable number of processors of different characteristics (such as word size and instruction set). There are many potential uses of such capabilities in the areas of multiprogramming, parallel processing, and fault tolerant systems. The reconfiguration capability introduces a number of issues concerning processor function modularization that do not arise in fixed architectures. These issues for potential processor modules are realizable with current VLSI technology. This architecture has only recently become feasible as a result of the great reduction in the ratio of cost to capability provided by VLSI bit slice microprocessor technology.

Due to the new synergism between architecture and VLSI hardware algorithms for pattern matching and string processing for VLSI implementation are available. The advent of very large scale integration (VLSI) technology has stimulated renewed interest in a fundamental issue in the context of LSI: What to put into a chip? Because they have a regular geometry based on the repetition of simple cells, memory devices are natural candidates for VLSI implementation. But there is always some communication and system overhead for utilizing a pure memory device in the total system architecture. An improvement in this situation can be made if both memory and logic can be combined to build a special-purpose processor that can take over some of the specialized functions of CPU, reduce system bottlenecks and simplify the overall communication and control structure. The basic problem seems to be the identification of tasks which can be built into special-purpose VLSI chips. Factors such as speed, power, cost reliability, custom design versus off-the-shelf and degree of risk must all be considered during the decision-making process which culminates in the choice of a specific integrated circuit technology. The feature which is called "chip architecture" must also be considered because of the effect it has in areas such as system performance, design flexibility and engineering changes. Chip architecture is defined from both an essential and an existential point of view. It consists of the integrated curcuit design and layout plus application-oriented features such as mask programmable versus software programmable logic function. The performance figure of merit for logic chips is the signal propagation delay through a gate from input to output, ranging from greater then 10ns for medium speed to less than 1ns for very high speed gates. For memory chips, the performance figure of merit is the read access time (the delay from a valid address at the input to valid data at the output). This ranges from greater than]00ns for medium speed to less than 10ns for very high speed memory.

The increasing demand for new data processing systems which have higher performance, smaller system size, lower cost and higher reliability are increasingly demonstrated. High speed LSIs are indispensable to meet these requirements. However, they disclose many problems as the integration level is increased.

Low energy, high performance VLSI technology, with an average of 2000 gates and a delay of 0.9

CH1935-6/83/0000/0229$01.00

nanoseconds per basic gate, utilizes a transistor array masterslice and low energy technology design activities and will facilitate the realization of systems based on VLSI are necessary. GAte arrays and other forms of integrated circuit chips continue to increase their level of integration and their speed-power performance. To cope with the very large scales in integration (VLSI) on chips, effective design strategies and computer-aids for design (CAD) are emerging. Array chips, design strategies and CAD all contribute to reducing the time, cost, and risk in VLSI chip design.

The ability of the semiconductor industry to produce chips with higher and higher circuit densities has created a challenge for the product designer: to utilize this capacity and still develop chips rapidly at reasonable cost. A multi-faceted approach to VLSI design that significantly reduces product development time and resource from those required with existing methods can be based on the use of PLA structures or macros. It consists of a hardware/software modeling technique, use of laser-personalizable PLAs for rapid modeling of PLA macros, and a method for repairing design errors (that may hide other errors) on the actual VLSI wafers with a laser tool. A two pass VLSI design is, therefore, highly probable.

Emerging technologies are Josephson and Gallium Arsenide. Superconducting technologies require many engineering considerations which have to be accomodated in order to exploit VLSI tools for the purpose of realizing an ultra-high performance computer. Different aspects of cryo-electronic technologies achieve the following objectives: 1) The limitations of known computer switching elements in order to justify investigating superconductive devices and the role they may play in those future ultra-high performance computers. 2) Superconductive devices, logic and memory circuits emphasizing the influence of their basic properties on maintaining their promising performance. 3) The practical considerations which lead to optimum designs and perhaps stimulate invention of new devices and circuits. 4) Building blocks for ultra-high performance computers, switching devices in general must satisfy simultaneously certain physical and engineering requirements. During the last few years there has been growing interest in the use of gallium arsenide for high speed digital integrated circuits. This is due to the intrinsic advantages of the high electron mobility of this material combined with the availability of a semi-insulating substrate. Very high switching speeds, approaching those of Josephson junction devices, have been demonstrated for GaAs logic employing short channel Schottky gate MESFETs. A reproductible ion implantation technology in GaAs has recently been developed. This technology does provide MESFETs with the uniformity in pinchoff voltage required for LSI circuits. Because of these technological advances, low pinchoff voltage depletion mode MESFET logic has been able to achieve low power dissipation and high circuit density with little sacrifice in speed, thereby overcoming many of the former objections to the feasibility of GaAs LSI circuits.

A typical software/hardware synergism is shown by the functions of a microprocessor and how to determine their internal dynamic behavior by means of electron beam testing techniques. The 8-bit microprocessor 8085 can be chosen as a typical representative for this purpose. During the 1970s; the instrument most commonly used in laboratories for this purpose was the mechanical probe. Some laboratories additionally applied electron beam testing techniques, which they used to supplement mechanical probing. If the two techniques are compared, the electron probe is found to have a number of advantages over the mechanical probe. However, the ten-to-twenty-fold higher cost of equipment, the distinctly more complicated measuring techniques and the unavailability, until recently, of the commercial equipment for the scanning electron microscope for electron beam testing have so far been obstacles to the acceptance of this otherwise more advantageous alternative. During the 1980s, it is to be expected that electron beam testing will be introduced as a standard method for internal probing and that the mechanical probe will only remain in use as a supplementary tool.

The complexity of VLSI circuits creates an enormous demand for efficient and effective layout techniques. Classically, the layout process has been divided into three subproblems, namely, partitioning (assigning logic to chips), placement (physically assigning circuits to physical locations on a chip), and interconnection (connecting the pads of the circuits so that pads associated with the same signal net are electrically common and isolated from the other nets). To help achieve goals, numerous design automation layout tools have been developed, and new ones will continue to be developed. The degree to which these tools are used is a function of the design style being employed. For polycell and masterslice layouts, automated placement and interconnection techniques have been successfully employed since the early 1970s. Most of the techniques employed are very similar to those used for PCB layout. For custom VLSI, there is a greater demand for very high circuit density. Automatic layout techniques are beginning to be used heavily. However, layout design faces two other major problems. One results from industry's ability to pack over 10,000 gate equivalents into a single chip. Besides this increase of complexity, today's micro-electronics technology has made a variety of processes--each with its own set of design rules--available for integration. Diversity has existed for a long time, but complexity raised the problem, since development of efficient systems for designing complex systems is costly and time-consuming. Layout design shares the complexity problem with any other design task. From the proliferation of different device technologies, layout design seems to suffer most heavily.

Software simulation technologies which are especially devoted for the design verification of VLSI logic circuits, such as gate array LSIs and building block LSIs are emerging as an important area in CAD is called macromodeling, macro-simulation and mixed-mode simulation.

Session 1-4c
Workstations II

Chairman

D. Stamm
Daisy, Inc.

THE ZYCAD LOGIC EVALUATOR AND ITS APPLICATION TO MODERN SYSTEM DESIGN

NICHOLAS VAN BRUNT

ZYCAD CORPORATION

ABSTRACT

The Zycad Logic Evaluator is a special purpose digital computer designed to perform gate level and device level logic simulation. It has sufficient capacity to simulate entire systems as large as 1.5 million gates and speed of up to 1ØØØ times conventional software logic simulators. It was designed to provide a dramatic improvement in design verification capabilities. The Logic Evaluator is also useful for fault simulation.

The design verification problem has been around since people began designing complex systems. The advent of VLSI has worsened the design verification problem by making changes in fabricated designs more difficult, but the problem has existed for a long time.

Software logic simulators have been around in various forms for 2Ø years. They have become mature as they were applied to real product design cycles, especially at companies designing large systems. The major deficiency with these simulators has always been their inability to simulate an entire system, at a level consistent with the real physical structure, fast enough to execute diagnostic programs.

Multilevel software simulation systems solve the capacity problem by discarding most of the detail of the system to produce more compact models. This solves the capacity problem, but introduces problems when interfacing the compacted models to more detailed models in other parts of the simulation. Since the compacted models are a simplification of the real physical design, it is often difficult to decide how much detail to include or exactly which features are important. Many errors and misunderstandings typically result, particularly in complex areas. It is precisely these areas that require careful, detailed, simulation. It becomes clear that the actual detailed interactions of the physical system cannot be accurately simulated if high level models are used to eliminate or "gloss over" the very detail that must be verified.

Hierarchical design techniques reduce many potential design errors, but are only useful where a good deal of pattern and repitition exist. Experienced designers know that it is in the control and interface areas between these orderly blocks that most design errors occur. This is caused by the number of unique or difficult to define conditions that must be satisfied.

Most structured design techniques are patterned after approaches that worked well in software design because hardware, unlike software, processes events and data concurrently. This adds an extra dimension of complexity to the design considerations that must be verified. (Examples will be given) The need to verify this concurrent operation makes it undesirable to simulate only pieces of a design at a time. At some point the entire design should be verified with as much detail as possible at a level very close to the actual physical structure if the last design errors are to be found. The Logic Evaluator was designed for this purpose.

The concurrency of hardware processors is the key to the Logic Evaluator's performance. The Logic Evaluator is an event-driven simulator. (An event is defined as a change of state on a gate output.) All events can have associated delays, which allows the modeling of arbitrarily complex timing relationships. There are many steps within the processing of a single event

CH1935-6/83/0000/0232$01.00 © 1983 IEEE

which can be done concurrently. Also, a large number of separate events can be processed simultaneously, if they occur at the same simulated time. In the case of a large simulation, many events occur at the same time, once they have been digitized into discrete sample times. The Logic Evaluator takes advantage of this natural concurrency through the use of multiple processors, pipelining, and multiple memories. The architecture is unique, and is optimized to only perform event simulation. No software process is used to process an event. The simulation algorithms are totally implemented in hardwired control logic, just as instructions are implemented in general purpose supercomputers. (The archetecture will be elaborated on and illustrations provided.)

The Zycad Logic Evaluator is a commercial product. First deliveries are occuring this spring. Examples of actual problems run on the Logic Evaluator will be provided to contrast it to existing simulators and illustrate its unique features.

SL-2000

AN INTEGRATED SYSTEM FOR CUSTOM IC DESIGN

W.M. VANCLEEMPUT

SILVAR-LISCO, PALO ALTO, CALIFORNIA 94304

1 Overview

The SL-2000 System combines schematic entry and logic simulation with a set of layout design tools for custom integrated circuits. SL-2000 is centered around a design database that fully supports hierarchical design. The layout tools support several IC implementation styles including:

-- standard cell

-- gate array (masterslice).

The overall structure of the SL-2000 System is illustrated in Figure 1.

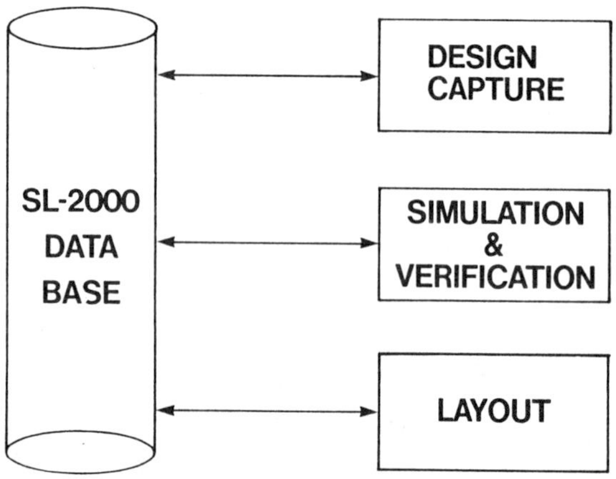

Figure 1: The structure of the SL-2000 system.

The SL-2000 database contains:

-- logic design information

-- simulation models

-- layout information.

This database is implemented in a distributed manner to optimize performance of each individual function.

SL-2000 fully supports hierarchical design concepts for top-down as well as bottom-up design methodologies.

1.1 Design Entry

Logic designs can be entered using any of the following methods:

-- alphanumerically, using the Structured Design Language (SDL)

-- interactively, with a graphics schematic editor, the Computer Aided Schematics System (CASS)

-- with a digitizing tablet.

Digitized schematic information can be edited with CASS. A separate netlist extraction program establishes a relationship between the graphical data and the underlying circuit connectivity.

1.2 Design Rule Verification

The netlist data obtained from the SDL description or from the netlist extraction is subsequently checked according to a number of typical design rules:

-- fan-in/fan-out

-- number of outputs in a net

-- existence of components in the library.

Beyond these simple rules, the user can calculate predicted reliability and testability measures for the design. A new design tool allows the designer to check for the violation of timing constraints with a timing verification subsystem (HELIX).

1.3 Logic Design Manipulation

In a hierarchical design system, two types of tools are necessary to modify the logic design database:

-- Hierarchical Design Expander (HIDEX) takes a high-level block diagram plus the definition of each of the blocks in terms of lower-level primitives. This information is used to create a netlist at a lower level of abstraction.

CH1935-6/83/0000/0234$01.00 © 1983 IEEE

-- Hierarchical Partitioning (HIPAR) changes the grouping of elements in order to accommodate the physical implementation medium. Consider the example of a 100,000 gate CPU originally designed for a 5,000 gate ECL gate array. At the critical decision point, however, the availability of this particular gate array is in question and the designer must use a 3,600 gate version of the same array. The impact of this decision on the logic design is small provided we know how to "repartition" the design to match the new set of physical constraints.

SL-2000 DESIGN CAPTURE
SDS: STRUCTURED DESIGN SUBSYSTEM

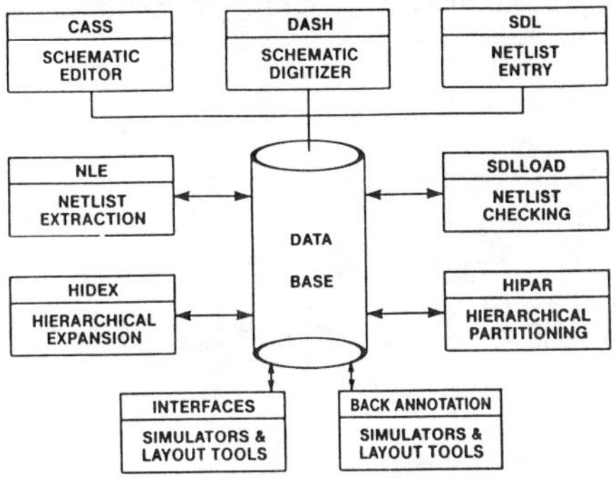

1.4 Simulation

The use of simulation in the design of a digital system is limited by two counteracting requirements:

-- accuracy

-- computing resources.

Depending on the stage of the design process and the nature of the implementation technology, a different set of trade-offs is required. The SL-2000 System contains a set of mutually-complementing simulators.

-- HELIX is a hierarchical multilevel, mixed-mode simulator. Its major areas of application are:

- architectural and register-level simulation

- functional fault simulation

- timing verification

- gate-level logic simulation.

Each of these capabilities can be implemented by building the appropriate model library.

-- BIMOS is a MOS-oriented logic-level simulator, capable of correctly simulating bidirectional transfer gates.

-- DIANA is a mixed-mode simulator with the following application areas:

- switched-capacitor network simulation

- transistor-level timing simulation

- logic level simulation using accurate timing macromodels.

-- SWAP is a switched capacitor network analysis program that works directly in the frequency domain. Among its special features are: sensitivity, noise and distortion analysis.

-- SPICE is a device-level simulator developed at the University of California at Berkeley.

An interactive logic analyzer (LOGAN) allows the user to interact with the schematic diagrams and the simulation results. This capability replaces the traditional practice of building breadboards and performing measurements.

SL-2000
SIMULATION/VERIFICATION

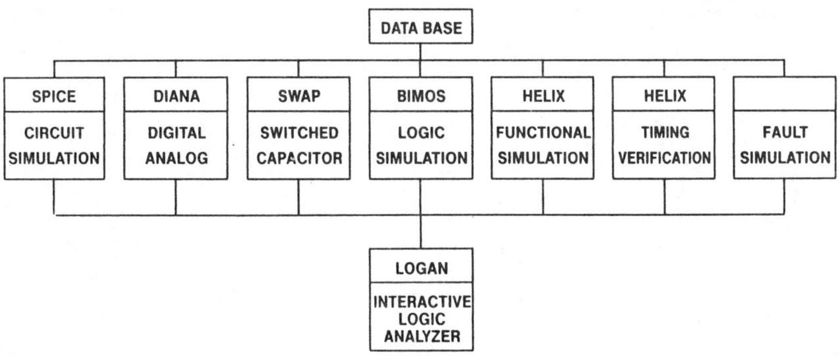

1.5 Layout Tools

The SL-2000 System supports several physical implementation strategies:

-- standard cell

-- gate array (masterslice).

SL-2000: LAYOUT CAPABILITIES

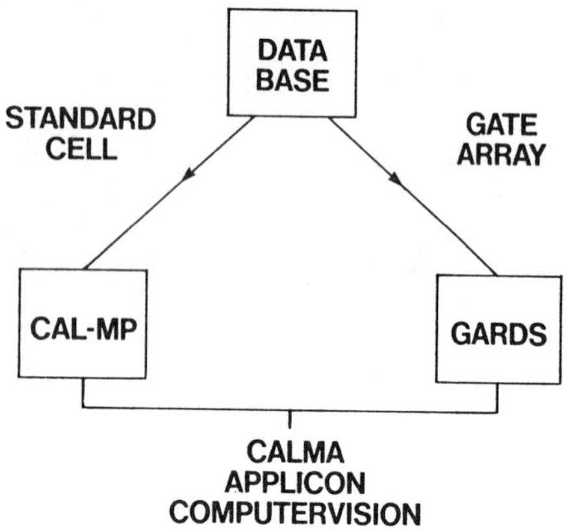

The CAL-MP layout software provides automatic place/route capability for standard cell designs. CAL-MP prefers row-based designs, but can also handle large functional blocks such as PLA's and ROM's.

CAL-MP: OVERVIEW

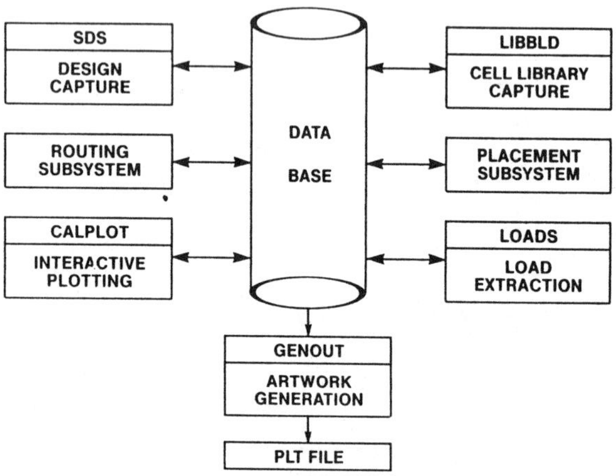

The GARDS Subsystem allows for automatic placement and routing of single-layer as well as double-layer gate arrays. While CAL-MP's routing algorithms ensure 100 percent routing completion by making the circuit area larger, GARDS cannot guarantee 100 percent routing completion. Therefore, an interactive routing editor is available to complete the layout. This editor enforces connectivity as well as geometric design rules.

The results of either layout subsystem can be output in a format that is compatible with most commercially available artwork editing systems such as APPLICON, CALMA and COMPUTERVISION.

GARDS: OVERVIEW

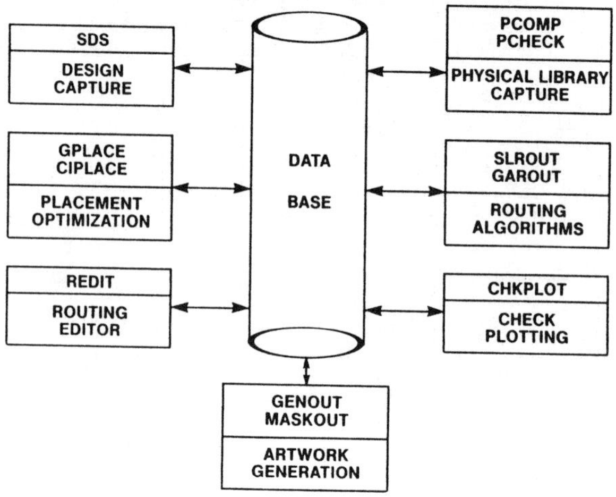

1.6 Implementation

The SL-2000 System has been implemented with the goal of maximizing its useful life. This is realized in the following ways:

-- technology-independence: all technology data are stored in the database rather than in the software

-- manufacturer-independence

-- CPU portability: high-level language implementation

-- graphics terminal-independence: using a device-independent graphics package

-- data portability allowing for partial completion of designs on one type of CPU; work can be resumed at a later time, using a different type of CPU or graphics terminal.

INTERACTIVE LOGIC SIMULATION IN A DISTRIBUTED ENGINEERING WORKSTATION ENVIRONMENT

Dan Holt and Steve Sapiro

CAE Systems, Inc.
Sunnyvale, California 94089

ABSTRACT

Implementation of an interactive logic simulator in a graphics-based engineering workstation environment is discussed. Both the user interface and the simulator/workstation interface of the system are described. The simulator is based on a nine level-strength node state model that supports switch, gate, and functional level primitives. Details of the simulation model and some new switch-level simulation implementation results are also presented.

1. INTRODUCTION

Computer-based engineering workstations are becoming an important aid in shortening the design cycle and lowering the design cost of digital circuits. In addition to design capture, many workstations are capable of local interactive design analysis and verification which not only provides immediate feedback but also reduces dependence on mainframe CAD systems. Accurate interactive logic simulation coupled with a graphics-based schematic editor and data-base is a powerful digital design verification tool. Since many designers think in terms of schematics and waveforms, design errors can be quickly located when simulation results are displayed directly on the circuit schematic. A highly interactive user interface, including edit of simulation input waveforms, interrupt, resume, and signal setting allows a designer to "debug" the hardware without incurring the expense of actually building it.

An accurate mixed switch, gate, vector simulation model encourages hierarchical verification and simplifies the user/data-base/simulator interface. An accurate model minimizes the manual alterations and intervention required to generate a simulation network description from the circuit schematic data-base.

2. ENVIRONMENT

The IDEAL logic simulator is an integrated member of the CAE-2000 workstation. As such, it shares a number of general purpose CAE-2000 tools such as the netlist extractor and waveform editor. In order to maintain a level of consistency, all simulation result displays, and the actual user interface of IDEAL are part of the CAE-2000 shell. IDEAL can be run in a batch background mode on the Apollo, in batch mode on an external host, or as a child process of the CAE-2000 shell for interactive simulation. Because of this multitasking, longer simulations can be run while the designer is performing other tasks on the workstation. When IDEAL reaches a breakpoint, or completes the simulation, it signals the designer with a message on the screen.

3. SIMULATION MODEL

IDEAL is a mixed switch, gate, and vector/functional level simulator with a fairly conventional time-based, event-driven, nine-state foundation. Vectors are multi-node data paths as described in [1, 2]; functional primitives are currently predefined high-level elements such as shift-registers, adders, and memories. All node and vector events are scheduled into and retrieved from a conventional time-wheel based event queue. Simulation proceeds by retrieving events from the time queue, updating the node or vector state, and evaluating all fanout devices for future events. All node and vector states are composed of three logic levels (0, 1, and X) and three logic strengths (high-z, resistive, and driving). In general, strengths are only used at MOS transmission gates and wire-ors where the the signal with the highest strength dominates. Assignable rise and fall delay, inertial (spike rejection) delay, charge decay, and input or wire delay are also supported at all primitives.

Since most schematic data bases contain a mixture of circuit (switch), logic gate, and higher-level functional devices, some form of mixed-level simulation is desirable. A conventional gate-level logic simulation foundation for mixed-level simulation is employed in IDEAL for the following reasons:

- o conventional logic simulation algorithms and data structures are highly developed and usually quite effective for modeling logic and higher-level devices.
- o adding switch capability to a conventional simulator is more straightforward than adding higher-level capability to a pure switch simulator, especially when assignable delays and non-MOS technologies need to be supported.

o the simulator will probably be used more for logic simulation than either switch or functional simulation.

The switch, gate, and functional devices may be freely interconnected through nine-state nodes and vectors. There is no need for special vector "merges" and "splits", and no need for special interfaces at switch devices. Each vector bit is nine-state, but the interpretation of the entire vector is undefined except at individual primitives. For example, the vector may be a coded real number at one primitive, and a 2's integer at another. A preprocessor automatically accounts for implicit wire-ors, with the only restriction that only full vectors can be implicitly wire-ored (though partial vectors can be explicitly wire-ored).

3.1 GATE-LEVEL PRIMITIVES

Gate-level primitives are single node output, up to four node or vector-bit input, unidirectional devices that employ a conventional yet highly-developed simulation algorithm. In addition to the basic logic gates, MOS transmission gates, and a limited number of latches are supported.

The nine level-strength state model [3] has proved quite effective for TTL and many complex MOS circuits. In general, signal strengths are ignored at most device inputs, and most device outputs have a fixed strength for each output level. Input signal strength is only used at wire-ors, resistors, and MOS transmission gates. The wired network of Figure 1 will produce a wire-and or wire-or characteristic depending upon the fixed strength that is assigned to the level-1 and level-0 outputs of the nand gates. A fixed driving output strength for logic level-0 will override resistive-1 at the wired-logic device producing a wire-and. A great deal of flexibility is possible because any of the 27 possible fixed output strengths for each output level can be assigned to the basic logic gates.

A number of transmission gate primitives have been implemented in an attempt to overcome many of the common problems encountered in MOS logic simulation. The additional primitives are not expected to be as accurate as the switch-level primitives, but they do retain the speed of unidirectional gate-level simulation. Two of the common problems are charge transfer logic and processing of unknowns at gate inputs. The default transmission gate type will not transmit high-z inputs to the output. For charge transfer logic, transmission gates may be flagged to transmit high-z inputs. Transmission gates can also be flagged as "stable" in response to unknowns at gate inputs. In Figure 2, the default transmission gate type will generate a driving-unknown output following a 1 to unknown event at the gate input because the strength of theoutput is unknown. The stable transmission gate types will generate a driving-0 because the output logic level is known even though the output strength is unknown. A network preprocessor will examine node capacitance to set the "transmit" flag, and will set the "stable" flag where appropriate.

Gate-level primitives in IDEAL use a "descriptor" based data structure [2], single-pass signal propagation and device evaluation, and evaluation by zoomtable lookup. The descriptor data structure requires 12 fixed bytes per device plus 4 additional bytes for each input. An additional 12 bytes per device is used for external data such as node and device location pointers. The zoomword is presented in Figure 3. The level and strength data is stored modulo-3. Since most devices ignore signal strength, the two least significant bytes are the pointer into the zoomtable. Wire-ors and transmission gates require three additional table lookups to access the strength data. A network preprocessor automatically expands high input count devices into an equivalent network of four input devices.

3.2 SWITCH PRIMITIVES

The switch primitive is required to overcome the limitations of unidirectional gate-level simulation of complex MOS transmission gate networks. The primary limitations overcome are bidirectional signal flow, simplistic charge sharing, and an overly pessimistic response to unknowns at gate inputs. Switch primitives are groups of bidirectional transmission gates interconnected through source and drain nodes identical to Bryant's "transistor groups" [4]. Such a static partition isolates bidirectional signal flow and transforms the network into a set of unidirectional switch devices. Unidirectional logic gates need not be expanded into transistors even if they are a source at a node included in the switch device. These devices are simply a pseudo input to the switch device.

Switch devices are event driven and evaluated by iteration through a table of "on" transistors. The transistor types include p and n-channel, resistors, and gated resistors. The node types are statically identified by a preprocessor as either large or small capacitance nodes. This data is coded into the node state with an additional strength (large-cap-high-z) resulting in 12 states for switch device evaluation. The evaluation algorithm is as follows:

[1] load a node state table with the high-z or large-cap-high-z strength of the current output level of each switch node. If the node has a pseudo input as described above, that state has precedence.

[2] load a table with the "on" transistors of the switch device. The table data includes a flag for resistive and the source/drain nodes of the transistor.

[3] iterate through the table of transistors evaluating new source and drain states by table lookup where the highest strength state will dominate. The node state table is updated

immediately. Iteration is complete when no node states are updated following a pass through the set of transistors.

[4] compare new node states with previous node states and schedule events as necessary.

The algorithm is very straightforward and easy to implement. It does not require any sort of unit-step mechanism, and does not require any sort of signal path propagation as in most switch simulators. When unknown gate inputs are present, the switch device is evaluated by first setting all unknown transistors "on" for solution with the above algorithm, followed by a solution with all unknowns "off." The two results are then compared, and any level differences set to unknown. An initial attempt at coding the unknown data within node states as in [5] was abandoned because it was inconsistent with the iterative table lookup algorithm. Treatment of capacitance is not as accurate as [6], and treatment of delay not as accurate as [5], but more accurate treatment of both may have a major impact on efficiency.

Switch devices use approximately twice the simulation time as a unidirectional gate-level version of the switch network. The probable causes of this are a two-pass evaluation method, and the higher number of events caused by the bidirectional activity within the switch device. A preprocessor option will automatically set all unidirectional transistors bidirectional, but because of the potential speed problem, many that are obviously undirectional (connected to VSS and VDD, etc.) are reset. A more in-depth analysis of clock assertions as in [7] would probably result in very few truly bidirectional transistors.

3.3 VECTOR/FUNCTIONAL PRIMITIVES

Vectors are multi-node data paths, and vector primitives are multi-vector or node input, multi-vector output devices. In addition to vector versions of all the gate-level primitives, many functional devices such as adders and shift-registers are supported. Most gate-level features such as nine-states, decay, and inertial delay are supported. The system allows direct interconnection with gate and switch devices, supports input concatenation, and allows implicit wire-oring of output vectors. Vector devices use a two-pass evaluation algorithm, and use a descriptor-based data structure that includes redundant input data storage.

4. INTERACTIVE FEATURES

The usefulness of a logic simulator depends on how quickly a designer can locate errors in the hardware. A highly interactive user interface that includes display of the schematic and formatting results into waveforms greatly improves the productivity of the user.

Simulation proceeds normally until an interactive interrupt is generated at the end of a simulation timepoint by one of the following:

- o user programmable interrupt specification that includes absolute or periodic timepoints and signal rising and falling edges.
- o asynchronous interrupt by a user "break" key.
- o interrupt on certain errors such as detection of zero-delay loops.

During an interrupt, input waveforms can be altered; signals can be set to temporary values; signals can be set to fixed values; RAM/ROM contents can be changed; and events can be scheduled. All simulation data displays can be altered, and pending events can be displayed. All simulation options such as decay simulation and trace mode can be altered. The simulation can be restarted from timepoint zero or any timepoint where a checkpoint/restart file was created. A limited number of device and node parameters such as delays and node decays can also be altered at advanced timepoints. The basic philosophy is that changing any simulation data that is "easy" to change internally is supported. Changing network topology at an advanced timepoint is not allowed.

5. CONCLUSIONS

A powerful and flexible interactive design verification tool has been developed. Accurate, interactive, mixed-level simulation greatly improves the designer's ability to debug his hardware and to experiment with alternative design configurations. The nine-state level-strength model is indeed fast and flexible for TTL and many complex MOS circuits. The switch simulation algorithm presented is reasonably accurate, fast, and easy to implement. It is also adaptable to concurrent fault simulation. Indeed, the authors believe that a distributed network of small computers may provide a solution to the sometimes overwhelming computational complexity of fault simulation.

6. REFERENCES

[1] M. Tokoro, et. al., "A Module Level Simulation Technique for Systems Composed of LSI's and MSI's," Proc. DAC78, pp. 418 - 427, June 1978.

[2] E. Ulrich, et. al., "High-Speed Concurrent Fault Simulation with Vectors and Scalars," Proc. DAC80, pp. 374 - 380, June 1980.

[3] Information Systems Design, "LOGIS User's Manual," Version 2.

[4] R. Bryant, "An Algorithm for MOS Logic Simulation," Lambda Magazine, pp. 46 - 53, Fourth Quarter 1980.

[5] M. Heydemann, G. Hachtel, and M. Lightner, "Implementation Issues in Multiple Delay Switch Level Simulation," Proc. IEEE ICCC82, pp. , Oct. 1982.

[6] G. Jordan and R. Apte, "Modeling of MOS Transistors in a Logic Simulator," Proc. IEEE ICCC82, pp. 431 - 434, Oct. 1982.

[7] R. Apte, N. Chang, and J. Abraham, "Logic Function Extraction for NMOS Circuits," Proc. IEEE ICCC82, pp. 324 - 327, Oct. 1982.

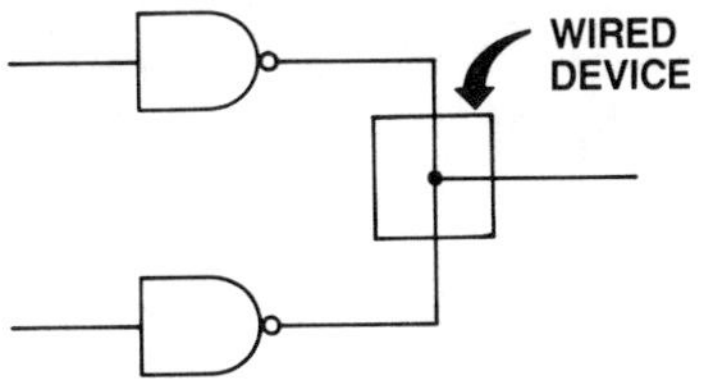

FIGURE 1. WIRED NET

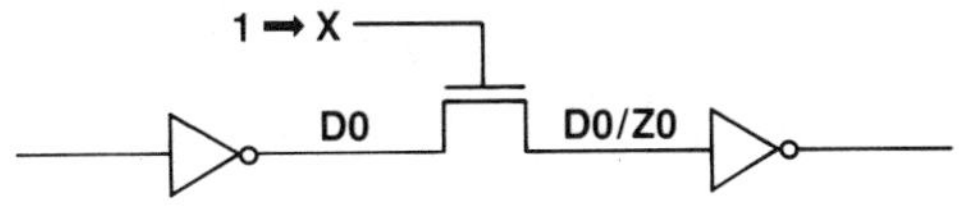

FIGURE 2. UNKNOWN GATE INPUTS

SCHEDULED EVENT	IO STRENGTH	SCHEDULED FLAG	TYPE	IO LEVEL

FIGURE 3. GATE ZOOMWORD

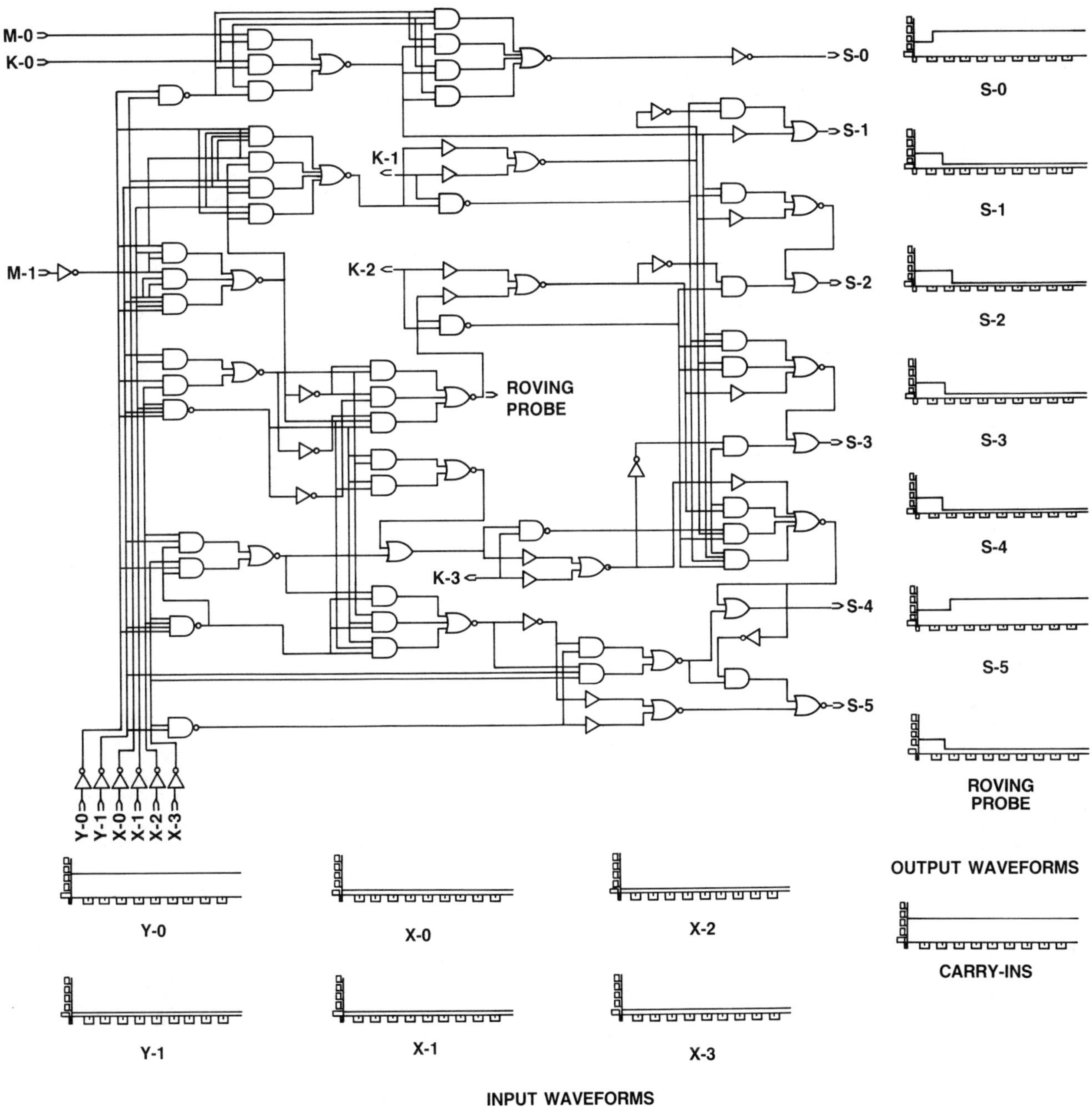

AN ENGINEERING WORK STATION FOR GATE ARRAY DESIGN

Michael Feuer - California Automated Design, Inc.
and
Stephen Swerling - Mentor Graphics Corporation

California Automated Design, Inc.
2620 Augustine Dr. #280, Santa Clara, CA 95051
Mentor Graphics Corp.
10100 S.W. Nimbus Ave, G7, Portland, OR 97223

The IDEA System is a set of design and analysis tools adapted for use on the Apollo engineering work station. The logic design systems include: a symbol editor, an interactive schematic editing system, a design traverser, a system for design expansion, a wide range of netlisters, an interactive mixed-mode simulator, a timing verifier derived from SCALD, and a circuit simulator called MSPICE. The design subsystems for gate arrays are: an automatic hierarchical layout system, an automatic routing system, and an interactive graphic editing system with advanced error-prevention features. The advanced nature of the IDEA System of engineering tools has resulted in the fastest semi-custom integrated circuit turn-around yet achieved.

Section I.

Introduction

Many of the advantages of gate array integrated circuits - simple, rapid design, and short fabrication cycles - require sophisticated computer tools for realization. These tools must support the evaluation of alternative approaches and prevent trivial errors in logic entry and layout. To the extent practical they must automate tedious design functions. Current trends toward increased densities and higher performance will increase the need for such tools. This paper describes a complete system for the design of gate array integrated circuits.

Section II provides a technology-independent overview of the design entry and analysis portions of the system. Section III describes aspects of the system which make it especially well-suited for gate array development. Section IV describes a portion of the system which transforms a design into actual gate array chips and feeds back physical layout information to the analysis tools.

Section II

Design Entry and Analysis

A. Overview

The IDEA System contains a set of technology-independent design and analysis tools. Users can configure the individual tools to their particular design situation - analog or digital, off-the-shelf TTL or customized IC's.

The generality and flexibility of the design tools in the IDEA System stem partly from the way hierarchical design is handled. A component symbol may represent anything from an electrical primitive (e.g., a resistor, transistor or gate), to a functional block with an arbitrary number of lower levels of hierarchy. There is no inherent system difference between two such component symbols. These symbols and their sheet descriptions can be created in any order. Any component symbol, including those which are described for simulation purposes by behavioral language models (see below) can be included on a higher-level sheet. Only data for the symbol is copied; complex components can, therefore, be used in multiple places without replicating lower hierarchical levels.

Properties in the IDEA System are used to convey the information required to describe the non-graphical aspects of a component. This includes pin numbers, part name, reference identifiers, rise and fall delays and simulation model identification.

The multi-tasking capability of the Apollo computer, on which the IDEA System is based, permits numerous schematic sheets to be edited simultaneously; text documentation can be prepared in one process while editing a schematic in another process.

B. Component Symbol Creation

If a user requires a component symbol not found in the supplied component libraries, a new symbol may be easily created with the Symbol Editor. This editor supports lines, arcs, rectangles, and pins and is used to define the default set of attached properties. The Symbol Editor includes a "case frame" concept - multiple variants of a symbol can be included on a single symbol definition sheet. A particular variant is selected at the time the component is activated for use during schematic editing.

C. Schematic Editing

The Network Editor is used to create graphical schematics in the IDEA System. In a top-down design environment, this editor is used to partition the system into its high-level blocks, and to

CH1935-6/83/0000/0242$01.00

make project assignments to sub-teams. In this editor, components are placed on an arbitrary sized (up to many D-size sheets in area) sheet, nets or busses are drawn, and properties are added or edited as needed. Sheet size may be changed at any time. Certain design checks are performed interactively, to prevent many classes of errors; other checks are run voluntarily by the user. Examples include: bus ripper checks, frame checks, out-of-date references, fan-out, and unconnected inputs. Two important constructs distinguish the Network Editor:

- Replication frames can be used to represent regular arrays of components such as registers and adders. Parameters are supported so that "n-Bit" modules can be defined. Replication frames can be nested to construct multi-dimensional arrays such as memories. Frames may also be used for conditional inclusion of a section of circuitry. These constructs can enhance a custom IC design significantly.

- Various ways of rotating a component or group of components alleviate the need for storing multiple versions of component symbols in libraries. Text is automatically adjusted during rotation to be readable from below or from the right.

D. Design Traversal/Review

Once a hierarchical design has been created, it is necessary to review elements in the hierarchy, to determine their status, and to move between them. The design Traverser provides read-only views of schematic sheets in a hierarchy, and, using component names and textual indentation, depicts any selected portion of a hierarchical design. This tool identifies missing (e.g., not yet designed) components, and can provide a user with simultaneous views of parent, peer and child sheets whilst editing a schematic. A component on a schematic can be selected, and its sheet(s) brought into view.

E. Design Expansion

Any part of a hierarchical design can be Expanded for analysis, netlisting, or physical design. Parameters can be set during design expansion, and either completely flat or hierarchical expansions can be performed. Incremental expansion is supported so isolated changes require only a short reprocessing run. Properties can be edited following design expansion for re-analysis with new propagation delays subsequent to physical design.

F. Netlisters

A wide range of Netlisters is provided for translating the information to the formats required by tools outside the IDEA System. Netlisters are available for analysis tools (e.g., TEGAS, SPICE, ILOGS) and physical layout systems (e.g., Applicon, Scicards, Redac, etc.).

G. Logic Simulation

A range of analysis tools is available as part of the IDEA System. One of the most significant is the interactive, mixed-mode, nine-state simulator developed for the IDEA System. This simulator, whose modeling capabilities are identical to the well-known ILOGS mainframe simulator, handles bi-directional transfer gates and wired-OR's. Models for circuitry can be written in a Pascal-based modeling language. Output data from the simulator can be displayed in either waveform or list form, or can be written to a file for later review. Results produced by the mainframe version of ILOGS, or any other simulator, can be similarly brought back to the IDEA System for analysis. During simulation, any desired schematic can be viewed and un-named nodes probed, so that any signal can be displayed. A sophisticated break-point system supports complex trigger definitions for stopping the simulator. Worst case timing analysis has not been included, for performance reasons; the timing verifier is used for this purpose.

H. Timing Verification

A second analysis tool is used for Timing Verification. The IDEA timing verifier is related to the Lawrence Livermore Laboratory public-domain SCALD program, but takes advantage of the same human interface capabilities described in the simulator section above. Various types of timing errors are flagged. These include setup and hold violations and conflict with user asserted values. Nodes with errors are identified on the schematic in view; a list of error explanations can be transcribed. The strengths of signals displayed as waveforms are encoded visibly, so that all information regarding such signals is available. Users may specify clock skew.

I. Circuit Simulation

The third major IDEA analysis tool supports analog circuit simulation. MSPICE is a set of input and output Pascal procedures which can easily be coupled to a public-domain version of the well-known UC Berkeley SPICE simulator, or to an in-house proprietary modified version, or to any analog simulator which can take advantage of MSPICE's human interface and post-analysis capabilities. The simulator itself need not actually run on the same machine as MSPICE, nor does it need to run on the same network of IDEA Apollo nodes. Facilities within MSPICE include probing of un-named nets, forcing of voltages or currents, interactively varying circuit parameter values, creating arbitrary graphs of currents, voltages, or computed signals (e.g., power), and running multiple iterations of an analysis while automatically varying parameters. A set of user-specified models can be supported with a model file. Time and frequency domain, and DC operating point modes are supported. Signals at a particular point in the analysis can be displayed graphically on the schematic.

Section III.

Gate Array Design

The capabilities of the IDEA System greatly facilitate gate array development:

- If it is known in advance that a particular block of yet-to-be-designed logic is to be a gate array device, a symbol can be created for inclusion on schematics, and if appropriate, a behavioral language model developed. This means that a significant amount of analysis can be performed before the logic internal to the device is actually designed. Conversely, if a decision to utilize a gate array is made after logic is designed, the appropriate section can be cut-and-pasted from its schematic diagram onto another, so that the semi-custom device can be represented by its own sheet or set of sheets.

- Since a "fully-flattened" set of graphical schematics may not be required for the gate array, Network Editor replication frames can be used to substantially reduce the graphical complexity of the logic schematics.

- Schematic symbol and physical layout libraries are available for a number of the popular gate array devices. Creating a new library involves copying the appropriate data from existing libraries, and adding any unique symbols with the appropriate editors.

- An engineer can proceed with a gate array design in several ways. If the gate array family has been predetermined, then the schematic symbol library for that device would be used. If, the particular device has not been selected at logic design time, advantage can be taken of a generic circuit symbol library. The complete design can be developed, analyzed, and modified, while device vendors and chip styles are still being selected. Once the selection has been made, it is a simple matter to force all references to components to point to a specific vendor's library. Detailed analysis can then be performed using the gate array vendor's supplied parameters.

- Once physical layout has been completed, actual path length data can be fed back to the design expander for more accurate resimulation or timing verification of the chip.

- Several vendors of gate arrays require particular net list formats. Netlisters for these targets are typically available on the IDEA System. If a proprietary netlister is required, the same facilities used in developing the supplied netlisters, plus fragments of existing programs can be used to quickly implement a program for a new target.

- As gate array chips with larger device counts become available, partitioning of an existing design into fewer physical chips can be accomplished by creating a symbol for the new device and an associated sheet containing only the earlier devices.

Section IV

Gate Array Physical Design

A. Overview

Physical design is the part of the integrated circuit design process which transforms the logical network into geometric patterns necessary for fabrication.

If the shapes that implement individual logic gates or macros are thought of as components, then physical design consists primarily of the placement of these components on a chip and the routing of metal interconnections among the component's pins. These functions are accomplished in the IDEA System by the CADI automatic placement, automatic routing, and interactive graphic editing subsystems.

The input network is extracted from the IDEA data base with a special netlister and entered into a physical design file for automatic placement and routing. Data about the location of placement sites, and the like is provided by selected gate array vendors and distributed with the IDEA System.

B. Automatic Placement

Automatic placement is accomplished by a series of hierarchical design algorithms which attempt to mimic the decisions which a human layout designer might make if confronted with similar choices. In this sense, automatic placement and routing are examples of applied artificial intelligence. In fact, typical game playing programs for chess, backgammon, and checkers are quite similar in concept to layout programs.

The automatic placement subsystem assigns blocks to sites on the gate array image in such a way as to minimize wiring congestion and length. The user may specify the pad or input/output buffer locations or leave them free for automatic assignment. Locations and orientations of selected macros may also be specified by the user. In addition, the user may suggest an initial placement of the macros. These capabilities are useful in processing engineering changes. Critical connections or nets may be assigned a weighting factor which will influence the placement programs to bring the affected macros closer together. The macros need not be of uniform size or shape but may consist of any combination of placement sites.

The CADI placement subsystem consists of the following sections running as a single batch job.

a. Input processor
b. Clustering
c. Global placement
d. Detailed placement
e. Pad assignment
f. Output processor

C. Automatic Routing

The automatic routing subsystem defines the interconnect paths or routes for a network of logic, using technology rules and specifications of a particular array family. Interconnects are constructed along a pre-defined grid structure, using standard widths and spacings.

A logic list of the network, along with component placement data, is used to determine a map of blockages for the chip with its macros in place. Any routes supplied by the user are also reflected in the map. Paths for each net are selected using a hierarchical scheme to attempt to minimize total length and then are modified if necessary to avoid area congestion. Internal connections are symbolic "from-to" paths following the track grid structure; geometrical form is supplied by an output processor using technology rules. The length of the routes on each level of metal is multiplied by vendor-supplied factors to yield interconnect capacitance values which are fed back to the IDEA data base for accurate simulation and timing verification.

Automatic routing achieves a high percentage of connections and uses available tracks efficiently. For gate array images of typical intrinsic routability, these programs will achieve 99-100 percent of the connections. Because of the look-ahead nature of the path determination and a structured constraint analysis with channel routing, connections do not simply fill up all space in increasingly desperate attempts to reach a target node. Consequently, missed connections are not only few in number, but also relatively straightforward to complete without major rework of existing routes using an interactive graphic editor.

For single-level metal gate arrays with diffused or polysilicon underpasses, a separate router is available. Working on new principles, this router has achieved unexpected success in this traditionally difficult field. Automatic completion rates of over 95 percent of the connections are typical.

The CADI two-level router consists of the following sections running as a single batch job:

a. Input processing - map generation
b. Global routing
c. Continuity assignment
d. Street routing
e. Maze running
f. Postprocessing

D. Interactive Graphic Editor

The interactive graphic editor is used for routing the interconnections on a gate array in an interactive mode using the color display and the touch pad of the Apollo computer. Other optional input/output devices as well as an interactive placement editor are now under test for a subsequent release.

The routing editor can be used to modify existing routes and to complete residual connections after automatic routing, or to specify the routing of critical nets before automatic routing. It is even possible to route a chip completely using only the interactive graphic editor.

This application has been very carefully engineered with respect to human factors. It provides users the ability to select colors and screen layout. It is menu driven and simple to learn. It keeps track of connectivity and prevents inadvertent creation of loops. It features error prevention rather than error detection, permitting the user to select only valid options during the session.

An interactive graphic editor of this type differs from a turnkey graphics system such as a Calma or Applicon in that no design rule or connectivity errors can be introduced during the design. Consequently there is no need for running design rules checking or electrical rules checking once the technology data base has been verified.

Section V.

Conclusion

This paper has presented a complete commercial gate array design tool based on a single user general purpose computer system. Its application software permits the entry and verification of engineering designs and the automatic and interactive layout of the chip. Such a tool, coupled with gate array fabrication capability, provides the fastest semi-custom integrated circuit turn-around time yet achieved.

Session 1-4d
Real-Time Systems

Chairman

M. Wei
American Bell

A CONTENT ADDRESSABLE SYSTOLIC ARRAY FOR SPARSE MATRIX COMPUTATION

Omar Wing
Department of Electrical Engineering
Columbia University
New York, New York 10027
(212) 280-3122 (Office)

ABSTRACT

A systolic array is proposed which is specifically designed to solve a system of sparse linear equations. The array consists of a number of processing elements connected in a ring. Each processing element has its own content addressable memory where the nonzero elements of the sparse matrix are stored. Matrix elements to which elementary operations are applied are extracted from the memory by content addressing. The system of equations is solved in a systolic fashion and the solution is obtained in NZ + 5N - 2 steps where NZ is the number of nonzero elements along and below the diagonal and N is the number of equations.

I. INTRODUCTION

Sparse matrix computations are characterized by non-numeric and numeric operations on unstructured data and hence are unsuited for processing on a conventional vector or array computer. In this paper we propose a computation structure which obtains the solution of a system of sparse linear equations by exploiting the sparsity and parallelism inherent in the problem. The structure is a content addressable systolic array which consists of a number of processing elements connected in a ring. Each processing element is connected to a content addressable memory which stores the matrix data. The equations are solved in a systolic fashion without central control as the data are pipelined around the ring, and the solution is obtained in NZ + 5n - 2 steps, where NZ is the number of nonzero elements along and below the diagonal and n is the number of equations.

In sparse matrix computation, matrix elements with zero value are not represented and all trivial operations (multiplications by zero; divisions of zero) are not done. The number of nonzero elements and their locations vary from row to row. If we want to realize the highly desirable feature of absence of central control, then each data item presented to a processor must carry with it the data value, its location in the matrix, and the instruction to be applied to it. In other words, the computation system should be "data driven."

II. PREVIOUS WORK

Parallel solution of sparse matrix equations has been considered by a number of authors [1-11]. The idea of using content addressable memories for sparse matrix computations is probably first reported in [11]. The proposed structure consists of a linear array of processing elements, each with its own content addressable memory, a RAM, and some local control. The array is not systolic or data driven, and broadcast of some data is required.

III. PARALLEL SOLUTION OF SPARCE MATRIX EQUATIONS

Consider solving Ax = b in which A is n x n. We assume the order in which the rows of A are to be eliminated has been determinated and thus the locations of all the fill-ins are known. As usual we obtain the LU decomposition of A and solve for x from the upper triangular system Ux = y, where y is solved from the lower triangular system Ly = b.

The solution algorithm is given below for reference.

```
Procedure LU;  (* LU decomposition *)
  begin
    for k:=1 to n-1 do (* sequentially *)
      begin
        z:=1./a(k,k);
        for j:=k+1 to n such that a(k,j) ≠ 0 do
             (* in parallel *)
           a(k,j):=a(k,j) *z; (divide
                operation *)
        for i:=k+1 to n such that a(i,k) ≠ 0 do
             (* sequentially *)
        for j:=k+1 to n such that a(k,j) ≠ 0 do
             (* in parallel *)
           a(i,j):=a(i,j) - a(i,k) * a(k,j);
                (* update *)
      end;
    end; (* end of LU *)
Procedure L; (* Solve Ly = b *)
    begin
      y(1):=b(1)/a(1,1);
      for k:=2 to n do (* sequentially *)
          for i:=k to n such that a(i,k) ≠ 0 do
               (* in parallel *)
            b(i):=b(i) - a(i,k-1)*y(k-1);
                (* update *)
          y(k):=b(k)/a(k,k);
      end; (* end of L *)
Procedure U; (* Solve Ux = y *)
     begin
       for i:=n to 1 do (* in parallel *)
           x(i):=y(i);
       for k:=n-1 to 1 do (* sequentially *)
       for i:=k to 1 such that a(i,k+1) ≠ 0 do
            (* parallel *)
        x(i):=x(i) - a(i,k+1)*x(k+1);
   end; (* end of U *)
```

Consider LU decomposition first. There are two types of operations at each pivoting step k: divide and update. The divide operation applies to all the nonzero elements in row k. For the update operation, for each nonzero element below the pivot a(k,k) pairing with each nonzero element in row k, there is an update operation. These divide and update operations can be done in parallel as noted in the algorithm. But in a systolic array, they should be staggered over the array. This suggests that we associate the column index with the space index (processor number), the row index with the time index of a minor cycle, and the pivot index with the time index of a major cycle.

IV. CONTENT ADDRESSABLE SYSTOLIC ARRAY

The array consists of n processors, interspersed with registers (latches) (see Fig. 1). Each processor is associated with a column of the matrix. We shall assume that the number of processors equals the number of columns for the time being; the contrary case will be considered later. Each processor is connected to a content addressable memory which contains the nonzero elements of a column. Each memory word has four fields: the value of an element, its row number, an operation bit, and an end-of-column bit. The operation bit indicates whether the divide or update operation is called for as a data item is moved along the array from processor to processor. The end-of-the-column bit is to indicate that the element is the last of a column.

We shall denote the 4-field word of a(i,j) by w(i,j). With reference to Fig. 1, the principle of operation of the systolic array can be explained as follows.

t = 1 (time step 1): PE-1 selects w(1,1) and places it in the output register.

t = 2: PE-2 uses the row number of w(1,1) as a key to select a(1,2); the operation bit indicates a divide operation and a(1,2) := a(1,2)/a(1,1); a copy is retained in PE-2; w(1,1) is placed in the output register; PE-1 selects w(2,1) and places it in the output register.

t = 3: PE-3 uses w(1,1) to select an element in column 3 with the same row number; CAM has no respondent; no operation is called for; w(1,1) is placed in the output register. PE-2 gets w(2,1) and the operation bit indicates an update; the row number of w(2,1) is used to select a(2,2); a(2,2) := a(2,2) - a(2,1)a(1,2); w(2,1) is placed in the output register. PE-1 selects w(4,1) and places it in the output register.

t = 4: PE-4 will do divide of a(4,1). PE-3 does nothing. PE-2 will update a(4,2) and recognizes a(4,1) as the last nonzero element of column 1; this signals the end of a pivoting step and completion of a major cycle.

PE-1 is now logically disconnected from the array. The second major cycle begins with PE-2 passing a(2,2) to PE-3, which does nothing and with PE-4 updating a(4,4) := a(4,4) -a(4,1)*a(1,4). It ends when PE-2 has sent a(6,2) onto PE-3, and the third major cycle begins. There are n-1 major cycles. When all are done, we have obtained the LU factors of A stored in their original locations in the content addressable memories.

The computation process is sketched in Fig. 2. If we assume that all the operations in a minor cycle can be done in one time step, then it takes NZ + n - 2 time steps to complete the LU decomposition, where NZ is the number of nonzero elements in L. If data broadcast is used as in [11], the completion time is NZ - 1 steps.

To solve Ly = b, we put b in a stack to the left of PE-1. At each minor cycle, an element of b, b(i), is popped and sent to every processor in turn. At the kth processor, b(i) is updated as b(i) := b(i) - a(i,k)*y(k). (See Procedure L.) The computation process is shown in Fig. 3. The final y vector is placed in a stack at the end of the array with y(1) in the bottom and y(n) on top. It takes at most 2n steps to compute y.

To solve Ux = y, elements of y are popped from the right hand side stack and sent to the processors one at a time, with y(n) first and y(1) last. At each processor, x(k) is updated (see Procedure U) in the same manner as forward elimination except data are now moving from the right to the left. After 2n steps, the solution is stored in the left hand side stack with x(1) on top and x(n) in the bottom.

In summary, to solve Ax = b, NZ + 5n -2 steps are required on a systolic array. By letting each data item carry not only the value of a matrix element but also its row number and the type of operation it calls for, central control is eliminated. Elements in each column are selected for divide or update operation by content addressing. By overlapping the parallel operations, we can achieve systolic processing and data broadcast is avoided. By placing vectors b and y in a stack, the same array is used to solve the triangular systems to obtain the final solution.

V. CONTENT ADDRESSABLE SYSTOLIC RING

The interesting and important instances of sparse matrix computation are those in which the order of A, n, exceeds the number of processors, p, by many folds. Several methods have been suggested to partition matrix A into pxp submatrices or nxp segments to fit into a system of p processors [11-13] We propose to partition A into $\lceil n/p \rceil$ nxp submatrices. The nonzero elements of column j are placed in the content addressable memory of processor k = j (mod p), with k = p if j(mod p) = 0. PE-p is now connected to PE-1 to form a systolic ring as shown in Fig. 4.

We shall denote the set of nonzero elements along and below the diagonal from row i to row j by D(i,j) and call the elements the active elements. In the case of n > p, LU decomposition proceeds as before except that the elements of D(1,p-1), which will be used in the divide and update operations for the first p-1 pivoting steps will be placed in a FIFO queue to the right of PE-p after each active element has passed it. When PE-(p-1) has sent the last nonzero element in column p-1 (the last element of D(1,p-1)) onto PE-p, the first item on the queue, w(1,1), is removed and returned to PE-1, where it will divide a(1,p+1) if it is nonzero. At the next time step, the second element on the queue is removed and sent to PE-1 where it will update a nonzero element in column p+1, if a(1,p+1) ≠ 0. At the same time, w(1,1) arrives at PE-2 to divide a(1,p+2) if not zero. After p such steps, w(1,1) enters the queue again, followed by other active elements. When the last element of D(1,p-1) arrives at PE-1 where it divides a(1,2p+1) and a second "super" cycle is started. The process is repeated $\lceil n/p \rceil$ times until all the columns are processed. This completes the elimination steps of the first p rows.

The next p rows are eliminated in the same way, with PE-1 emitting a(p+1,p+1), followed by the active elements of D(p,2p+1). The process is repeated until all the rows are eliminated.

The absence of central control should be noted. Forward solution of Ly = b and back substitution Ux = y are similarly processed.

VI. SUMMARY

We have presented a conceptual design of a content addressable systolic array specifically intended for solving a system of sparse linear equations. The array is characterized by its modular structure, local connectivity, absence of central control, simplicity of the processing elements, and systolic operation. Connected to each processing element is a content addressable memory where the nonzero elements of a column of the sparse matrix are stored. The matrix elements to be divided by a pivot or to be updated in the elimination process are extracted from each CAM by content addressing. Parallel operation is realized by staggering the elimination tasks among the processing elements; no scheduling is required.

REFERENCES

1. D.A. Calahan, "Parallel solution of sparse simultaneous linear equations," Proc. 11th Allerton Conf. Circuit and System Theory, October 1973, pp. 729-735.

2. ____________, "Exploiting parallism in design algorithms" in Basic Questions of Design Theory, W.R. Spillers, Ed. New York: Elsevier, 1974.

3. C. Pottle and Y.M. Wong, "Nonlinear circuit simulation on a parellel microcomputer system," Proc. 1976 IEEE Int. Symp. Circuits and System, IEEE 76H 1074-4 CAS. April 1976, pp. 394-397.

4. O. Wing and J.W. Huang, "A computation model of parallel solution of linear equations," IEEE Trans Computers, July 1980, pp. 632-638.

5. O. Wing and J.W. Huang, "A parallel triangulation process for sparse matrices," Proc. 1977 Int. Conf. Parallel Processing, August 1977, pp. 207-215.

6. J.W. Huang and O. Wing, "Optimal parallel triangulation of a sparse matrix," IEEE Trans. Circuits and Systems, September 1979, pp. 726-732.

7. O. Wing and J.W. Huang, "An experiment in parallel processing of Gaussian elimination of a sparse matrix," Proc. IEEE Int. Symp. Circuits and Systems, April 1976, pp. 263-266.

8. O. Wing and J.W. Huang, "On minimum completion time and optimal scheduling of parallel triangulation of a sparse matrix," IEEE Power Engineering Society Meeting, July 1978.

9. J.W. Huang, "Computation Models and the Scheduling of Parallel Solution of Linear Equations, Ph.D. Thesis, Columbia University, 1978.

10. J.W. Fawcett, "Solution of Large Sets of Linear Equations Using Cellular Arrays, Ph.D. Dissertation, Syracuse University, Syracuse, NY, June 1981.

11. R,M, Kieckhager and C. Pottle, "A processor array for factorization of unstructured sparse networks," Proc. Int. Conf. Circuits and Computers, 1982, pp. 380-383.

12. D. Heller, "Partitioning big matrices for small systolic array," in Proc. of USC Workshop on VLSI and Modern Signal Processing, S.Y. Kung, Ed. University of Souther California, 1982, p. 109.

13. P. Kuekes and R. Schreiber, "Fast Sparse Matrix computation," Digest COMPCON-81, 1981, pp. 251-253.

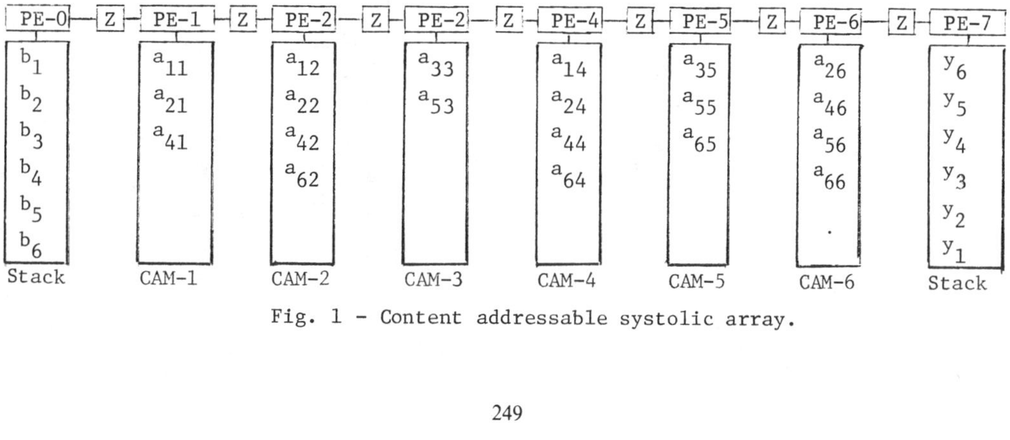

Fig. 1 - Content addressable systolic array.

Step	PE-1	PE-2	PE-3	PE-4	PE-5	PE-6
1	a_{11}	-	-	-	-	-
2	a_{21}	$a_{12} \leftarrow a_{12}/a_{11}$	-	-	-	-
3	a_{41}	$a_{22} \leftarrow a_{22}-a_{21}a_{12}$	-	-	-	-
4	-	$a_{42} \leftarrow a_{42}-a_{41}a_{12}$	-	$a_{14} \leftarrow a_{14}/a_{11}$	-	-
5	-	a_{22}	-	$a_{24} \leftarrow a_{24}-a_{21}a_{14}$	-	-
6	-	a_{42}	-	$a_{44} \leftarrow a_{44}-a_{41}a_{14}$	-	-
7	-	a_{62}	-	$a_{24} \leftarrow a_{24}/a_{22}$	-	-
8	-	-	-	$a_{44} \leftarrow a_{44}-a_{42}a_{24}$	-	-
9	-	-	a_{33}	$a_{64} \leftarrow a_{64}-a_{62}a_{24}$	-	$a_{26} \leftarrow a_{26}/a_{22}$
10	-	-	a_{53}	-	-	$a_{46} \leftarrow a_{46}-a_{42}a_{26}$
11	-	-	-	-	$a_{35} \leftarrow a_{35}/a_{33}$	$a_{66} \leftarrow a_{66}-a_{62}a_{26}$
12	-	-	-	a_{44}	$a_{55} \leftarrow a_{55}-a_{53}a_{35}$	-
13	-	-	-	a_{64}	-	-
14	-	-	-	-	-	$a_{46} \leftarrow a_{46}/a_{44}$
15	-	-	-	-	a_{55}	$a_{55} \leftarrow a_{66}-a_{64}a_{46}$
16	-	-	-	-	a_{65}	$a_{65} \leftarrow a_{56}/a_{55}$
17	-	-	-	-	-	$a_{66} \leftarrow a_{66}-a_{65}a_{56}$

Fig. 2 - Pipelined LU decomposition on systolic array

Step	PE-1	PE-2	PE-3	PE-4	PE-5	PE-6
1	$y_1 \leftarrow b_1/a_{11}$	-	-	-	-	-
2	$y_2 \leftarrow b_2-a_{21}y_1$	y_1	-	-	-	-
3	$y_3 \leftarrow b_3$	$y_2 \leftarrow y_2/a_{22}$	y_1	-	-	-
4	$y_4 \leftarrow b_4-a_{41}y_1$	y_3	y_2	y_1	-	-
5	$y_5 \leftarrow b_5$	$y_4 \leftarrow y_4-a_{42}y_2$	$y_3 \leftarrow y_3/a_{33}$	y_2	y_1	-
6	$y_6 \leftarrow b_6$	y_5	y_4	y_3	y_2	y_1
7	-	$y_6 \leftarrow y_6-a_{62}y_2$	$y_5 \leftarrow y_5-a_{53}y_3$	$y_4 \leftarrow y_4/a_{44}$	y_3	y_2
8	-	-	y_6	y_5	y_4	y_3
9	-	-	-	$y_6 \leftarrow y_6-a_{64}y_4$	$y_5 \leftarrow y_5/a_{55}$	y_4
10	-	-	-	-	$y_6 \leftarrow y_6-a_{65}y_5$	y_5
11	-	-	-	-	-	$y_6 \leftarrow y_6/a_{66}$

Fig. 3 - Systolic solution of Ly = b.

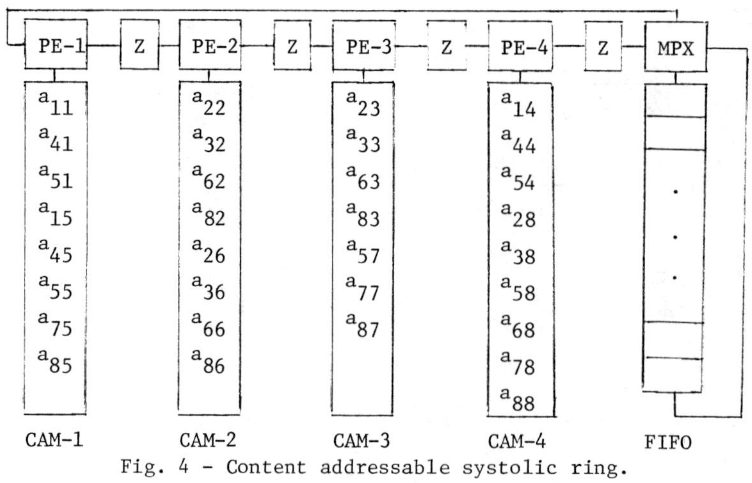

Fig. 4 - Content addressable systolic ring.

"Proceedings of the 1983 IEEE International Conference on Computer Design".

PERFORMANCE CONTROL OF TASKS IN A MULTIPROCESSING ENVIRONMENT

Howard A. Sholl

Computer Science and Engineering U-157
University of Connecticut
Storrs, Connecticut 06268

Abstract

In time-dependent environments which exist in process control and/or monitoring applications a designer may need to use multiple processors to insure the necessary task performance. This paper describes an investigation into the feasibility of using processing power multiplexing techniques to accomplish this intent. Processing power is adjusted by time division multiplexing (internal to a processor) and input data stream multiplexing (external to a processor). In the M/G/c queuing model studied, each software task is assigned a job queue, and its performance parameters of service rate capability, total system delay time mean and variance are considered. A profit function approach is used to provide a relative basis for distributing the processing power over the tasks, and a numerical optimization strategy developed and tested. The results demonstrate the feasibility of the approach and indicate directions for future work.

Introduction

Controlling the execution time performance of tasks in a real time multiprocessing environment is a problem a designer must resolve to achieve an effective system design. The intent of this work is to develop techniques which may be utilized in such an environment to assist a designer.

Application of multiple processors to the improvement of execution time performance has been considered in a variety of ways. Utilization of parallel control streams has been shown to offer realizable performance enhancement, but is problem dependent[1, 2, 3, 4, 5, 6]. Pipeline processing methods, in which a task is decomposed into a sequence of subprocesses, provide a means of increasing data rate and decreasing delay time, but the decomposition is problem dependent[7]. Processing power multiplexing, considered in this study, is also applicable. A processor may be shared among multiple processes by time multiplexing its computing time, if each process requires less than the power of a single processor. Multiple copies of a process may be utilized in multiple processors by multiplexing the input (output) data stream to (from) the desired processors, if the process requires more than the power of a single processor. This approach has the advantage of avoiding the problem-dependent property of algorithm decomposition inherent in the other two approaches and is thus better suited as a control mechanism. However, there remain some applications in which the input data stream may not be logically decomposable into jobs[7], or in which some algorithm decomposition may yet be needed to achieve the desired performance level.

The multiplexing approach is considered in this paper to evaluate its potential for achieving a means of controlling the task performance parameters of service time and response time mean and variance.

Approach

Performance Parameter Control

Let processing power, n_i, be defined as the amount (relative to an entire processor) of processor time assigned to a task i. Now control of the performance parameters can occur in the following manner:

Service time.

Let T_i = the reference service time on a dedicated processor.

Now, $ts_i = \dfrac{T_i\lceil n_i\rceil}{n_i}$ the actual service time with processing power n_i.

In this expression it is assumed that if $n_i > 1$ (more than one processor is required), then equally-powered copies of the task are used to form multiple servers of that task's job queue.

Also, the mean service rate capability is $\frac{n_i}{T_i}$. Clearly, processing power manipulation can control both service time (up to T_i) and service rate.

Response Time Mean and Variance

The response time (total time in the system) consists of queuing delay and service time. By controlling the service rate, the response time is also controllable, although not independently.

The variance in response time is also controllable since variations about the mean decrease as processing power is increased. Thus one can increase processing power to decrease the variance magnitude, and then add in a fixed delay, if necessary, to achieve a desired mean. Since during such a delay other processes may execute no effective processing power need be lost.

CH1935-6/83/0000/0251$01.00

Queuing Model

To evaluate the feasibility of this approach an M/G/c queuing model was used to fit the following operational context:

1. Task execution time is large compared with any required communication delays. This implies that tasks run as entire units on a processor (no task decomposition exists.)
2. Each task is assigned to an independent M/G/c queue where m is adjusted to fit a specified processing power assignment.
3. A number of independent tasks exist which must be controlled in a simultaneous/concurrent manner.
4. Implementation-level constraints are not imposed (since the intention is to evaluate the control capability first, and then later translate the results into specific implementation-level support requirements).

Processing Power Assignment

In order to assess the potential of this approach, the problem of optimally assigning processing power to a set of tasks operating in the above environment was addressed. Determination of an optimal assignment was aided by the definition of a profit function for each performance parameter of each task. (For this initial study only the performance parameters of service time mean and response time mean were investigated-response time variance is currently under investigation). Sample profit functions are shown below.
The profit function approach allows one to specify the relative importance of the performance parameters to be controlled over the task set, and how this importance varies with the performance parameters.

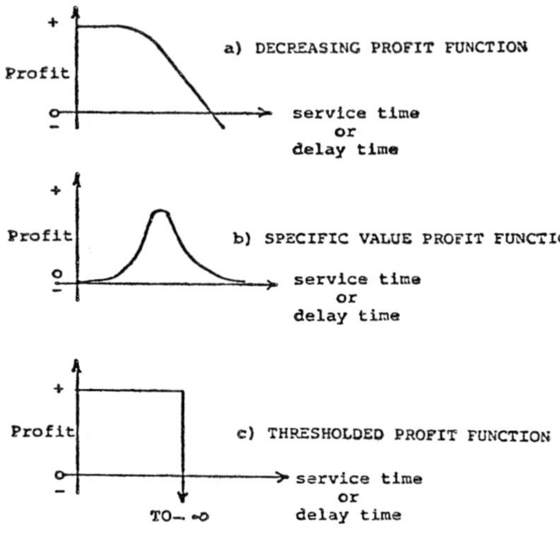

TYPICAL PROFIT FUNCTIONS

As noted above possible profit functions include: gradual deterioration (a); desirable operational region (b); or threshold (c). Now, given the specification of the tasks (mean, variance of service time), arrival rates, and profit functions, an optimization problem may be described which finds a distribution of an available amount of processing power which maximizes total profit. In effect, the solution mechanism for this optimization problem is the control mechanism desired.

Solution of Optimization Problem

Analytic Solution Approach

An analytic solution to the assignment problem was considered first, using the method of LaGrange multipliers.

Let $P_T = \sum_{i=1}^{r} (P_i + Q_i)$ be total profit

where P_i = service time profit
Q_i = delay time profit
r = task quantity

Find $\underline{n}$ = (n1, n2, ... nr) to maximize P_T subject to

$$\sum_{i=1}^{r} n_i = m$$ where m = the number of processors used

Using the method of LaGrange Multipliers, one can find the constrained maximum by maximizing

$P_T + \theta\,(m - \sum_{i=1}^{r} n_i)$, where θ is a LaGrange multiplier.

Equation System

$$\frac{\delta}{\delta n_i}\left(P_T + \theta\,(m - \sum_{i=1}^{r} n_i)\right) = 0 \qquad r + 1 \text{ equations}$$

$$\sum_{i=1}^{r} n_i = m \qquad n_i\,,\ \theta \text{ unknowns}$$

where
$$\frac{\delta P_T}{\delta n_i} = \frac{\delta P_i}{\delta n_i} + \frac{\delta Q_i}{\delta n_i}$$
$$\frac{\delta P_i}{\delta n_i} = \frac{dP_i}{dts_i} \cdot \frac{\delta ts_i}{\delta n_i}$$
$$\frac{\delta Q_i}{\delta n_i} = \frac{dQ_i}{dtd_i} \cdot \frac{\delta td_i}{\delta n_i}$$

While this approach is potentially solveable, it was found that an analytic solution was intractible even for small systems as the profit functions become varied and of high degree. Also, each system specification requires a different analytic solution. Because of these considerations a numerical optimization approach was selected.

Numerical Optimization Approach

A gradient type of optimization strategy with "optimal" step size was selected, in which solution adjustments were made to correct for threshold violation effects (profit function thresholds and traffic utilization limitations). Since an optimal solution was desired for a set of possible processor quantities, it was decided to establish first a minimal processing power solution, and use this as a basis (starting poit) for the optimization procedure for the next larger processor quantity. This approach was then followed by iterating through the set of desired total processing power values.

Minimum Processing Power Solution

Let ρ_j = the untilization of queue j. Assume, for generality, that up to s tasks are assigned to a queue.

Now,

$$\rho_j = \sum_{i=1}^{s} \frac{\lambda_i T_i}{n_i}$$

Let ρmax be a specified bound on utilization. The minimization problem is now:

For each queue, find a set of n_i, $1 \leq i \leq s$, values such that $m = \sum_{i=1}^{s} n_i$ is minimized under the constraint: $\rho MAX = \sum_{i=1}^{s} \frac{\lambda_i T_i}{n_i}$

using a LaGrange multiplier approach an analytic solution may be found:

θ = LaGrange multiplier

$$\frac{\delta}{\delta n_i} \left(\sum_{i=1}^{s} n_i + \theta \left(\sum_{i=1}^{s} \left(\frac{\lambda_i T_i}{n_i} - \rho_{MAX} \right) \right) \right) = 0$$

$$\rho_{MAX} = \sum_{i=1}^{s} \frac{\lambda_i T_i}{n_i} \quad \text{set of s+1 equations}$$

unknowns θ, $\{n_i\}$

Solution:

$$n_i = \frac{(\lambda_i T_i)^{1/2} \sum_{j=1}^{s} (\lambda_j T_j)^{1/2}}{\rho_{MAX}}$$

This result may now be adjusted to fit any violated profit function thresholds by establishing the threshold requirements for n_i and using the above result on only the "unthresholded" tasks.

"Optimal" Step Size Determination

It is desirable to establish a method of adjusting a solution for increased profit in a manner which can be assured of reaching an optimum point. There are two environments in which this must be done:

Adjusting the processing power assignment, $\underline{n}$, when m is fixed.

Adjusting the processing power assignment, $\underline{n}$, when m is to be increased.

For both cases a common theoretical basis may be established:

Given $\underline{n} = (n_1, n_2, ---, n_r)$

$$\sum_{i=1}^{r} n_i = m$$

$$P_T = \sum_{i=1}^{r} (P_i + Q_i), \text{ total profit}$$

Modify $\underline{n}$ to $\underline{n}^1$ where $n_i^1 = n_i (1+w_i)$

w_i = the desired change in n_i

Such that P_T is maximized subject to

$$\sum_{i=1}^{r} n_i (1+w_i) = m_R, \text{ where } m_R = \begin{cases} m & \text{case 1} \\ m+1 & \text{case 2} \end{cases}$$

Again using a LaGrangian approach, the following equation set may be established and solved:

$$\frac{\delta}{\delta n_i} \left(\sum_{i=1}^{r} (P_i + Q_i) + \theta \left(\sum_{i=1}^{r} n_i (1+w_i) - m_R \right) \right) = 0$$

$$\sum_{i=1}^{r} n_i (1+w_i) = m_R$$

unknowns $\{w_i\}$, θ

where:

$$\frac{\delta P_i}{\delta n_i} = \frac{dP_i}{dts_i} \cdot \frac{\delta ts_i}{\delta n_i}$$

$$\frac{\delta Q}{\delta n_i} = \frac{dQ_i}{dts_i} \cdot \frac{\delta td_i}{\delta n_i}$$

Solution:

$$1 + w_i = \frac{m_R \left(\frac{\delta P_i}{\delta n_i} + \frac{\delta Q_i}{\delta n_i} \right)}{\sum_{j=1}^{r} n_j \left(\frac{\delta P_j}{\delta n_j} + \frac{\delta Q_j}{\delta n_j} \right)}$$

In the implementation, the profit functions P,Q were defined tabularly, for generality; and the service time, ts, and response time, td, were defined analytically[8]. The result above provides an estimate of the total change required in the $n = (n_1, n_2, .. n_r)$ vector to achieve a local optimum. If the starting vector is close to an optimum point, convergence is fast; however, in the other extreme, it is possible for a divergent pattern to occur. Behaviorally, the approach is similar to Newtonian optimization methods. It was found desirable to include a stepsize parameter, s, to reduce the extent of the computed change, and this was sufficient to remove the divergent property over the test cases used. The stepsize parameter was included after the "optimal" step was computed:

Constraints

There were two types of constraints considered in the implementation:

utilization < ρ_{max}
service or response time < profit function threshold

These were included in the optimization process by first computing the "optimal" change, and then moving in that direction only as far as the thresholds would allow. An "optimal" change was then recomputed over the non-thresholded subset of tasks for that iteration.

Experimental Results

A variety of test cases were evaluated, using the types of profit functions described. It was found that the algorithm was properly able to find a local optimum, but that more than one local optimum point may exist, particularly with complex profit functions, such that the global optimum might be mssed. The algorithm can cope with mixtures of thresholded, specific value, and decreasing profit functions, as desired while finding a local optimum. In these cases it is necessary to start at alternative initial points in order to find other local optima. Clearly, within the modelling context, performance control is possible, and the method used can be viewed as a feasible design approach.

Conclusions/Further Work

This paper has reviewed the initial results of a study of the problem of achieving performance control of a set of real time tasks in a multiprocessor environment. Both a set of theoretical results that form the basis for optimization and an implementation were described. The solution algorithm provides a method of establishing a locally optimum assignment and provides results that may be used to maintain a high level of performance as the number of processors may vary. This type of result is useful for sytems which may need to adjust the processor quantity to maintain a desired level of reliability.

Also, the technique used to determine an "optimal" step change produced an analytic result, implying that it may be applicable (fast enough) to be used in a dynamic environment where performance monitoring and adjustment are achieved. In this instance, it is assumed that a local optimum exists and regular corrections are made as a system parameter (e.g. arrival rate) varies.

Further work is needed and planned in the following areas, before a complete evaluation is possible.

Variance Control

Control of the variance of the performance parameters is clearly desirable/necessary. This area is currently under investigation.

Queuing Structure Alternatives

The current model does not include a wide capability of queuing structure alternatives. Expansion of this area is desirable.

Implementation Details

The model ignores a number of implementation-level factors that will influence the results. Currently, a hardware/software system is being developed to aid the assessment of these factors.

Dynamic Application

The potential of the approach in a dynamic environment should be more thoroughly investigated.

References

1. M.J. Flynn and J.L. Hennessey, "Parallelism and Representation Problems in Distributed Systems," IEEETC, Dec. 1980, pp 1080-1086.

2. D.J. Kuck, et.al., "Measurement of Parallelism in Ordinary Fortran Programs," 1973 Sagamore Conference on Parallel Processors.

3. D.J. Kuck, "Parallel Processing of Ordinary Programs," in Advances in Computers, M. Rubinoff and M.C. Yovits, eds., Academic Press, 1976, pp 119-179.

4. M.J. Gonzalez and C.V. Ramamoorthy, "Recognition and Representation of parallel Processable Streams in Computed Programs", in Parallel Processor Systems, Technology and Applications, L.C. Hobbs, et.al. ets. Spartan Books, 1970, pp 335-375.

5. E.H. Riedmann, "Automated Translation of Sequential Programs into Parallel Programs", in Parallel Computers - Parallel Mathematics, M. Feilmeier, ed. North Holland, 1977.

6. M.C. Wei and H.A. Sholl, "An Expression Model for Extraction and Evaluation of Parallelism in Control Structures," IEEETC, Sept., 1982.

7. H.A. Sholl, et.al., "A Multi-Microprocessor System for Real Time Classification of Railroad Track Flaws," IEEETC, Oct., 1982.

8. A.O. Allen, "Probability, Statistics, and Queuing Theory." Academic Press, 1978.

A REAL-TIME MULTI-MICROCOMPUTER ARCHITECTURE EMPLOYING A FULLY PARALLEL CROSSBAR SWITCH*

W. C. McDonald and M. W. Beasley

System Development Corporation
Huntsville, AL 35805

ABSTRACT

This paper describes the architecture and organization of a multi-microcomputer system employing a fully parallel crossbar switch. The design provides a flexible system capable of application to a wide range of real-time control and data acquisition problems. The system logic has been partitioned into eight board types that permit easy configuration of a variety of systems employing different numbers of processing elements, I/O processors, and sharable memories.

INTRODUCTION

This paper describes the architecture and organization of a distributed multi-microcomputer system employing a fully parallel crossbar switch. The design provides a flexible system capable of supporting a range of real-time control and data acquisition applications. The system logic has been partitioned into eight board types that are used as basic hardware building blocks to configure systems to accommodate a range of throughput, I/O, and memory requirements. The building blocks are a single-board microcomputer, extended EPROM/real-time clock module, bus interface unit, shared memory, interrupt controller, flag memory, input/output, and clock and reset module.

The system is a tightly coupled multiprocessor system with access to shared memories provided through a fully parallel crossbar switch. One of the primary advantages of the system is the flexibility in adjusting the hardware complement to meet functional and performance requirements. For example, the number of processing elements can be varied to suit the load and perform new functions, sharable memory can be added to suit capacity, and I/O processors can be added to accommodate interfaces and data rates.

Many multi-microprocessor architectures have been proposed and built for a variety of real-time and high-reliability applications.[1] A few multi-minicomputer systems have been implemented using crossbars as an interconnection mechanism.[2,3,4] The architecture discussed in this paper differs from previous crossbar systems in several important ways. It is implemented with advanced 16-bit microprocessors that have a large address space and thus overcomes the inefficiency that occurs with small address space processors encountered in C.mmp.[5] Second, the system has been designed to maximize data accessibility by all processors, thus reducing system port-to-port times. Finally, the system logic has been partitioned into basic building blocks that provide considerable flexibility (within design limits) in configuring a system to meet functional and performance requirements.

SYSTEM ARCHITECTURE

The system is a tightly coupled multiprocessor architecture containing both private and shared memory. It can be configured with up to 7 processing elements and up to 12 sharable memories. High-speed input/output is handled by I/O processors. The number of I/O processors may vary, but there can be no more than seven I/O and main processors combined. The processors and the sharable memories are interconnected with a crossbar switch. The processing elements are single-board microcomputers, each with local private memory. In addition to the sharable memories, each processing element is provided access to an interprocessor interrupt controller that enables any processor to interrupt any other processor including I/O, flag memory for synchronization of access to shared memory, programmable interval timers for time-slicing and watchdog timing, a real-time clock for system time coordination, and extended EPROM for application and kernel operating system programs. Access to the interrupt controller and the flag memory is provided through the crossbar switch.

Each of the sharable memories, the flag memory, and the interprocessor interrupt controller are addressed by a separate segment number of the processor address space as shown in Figure 1. A bus interface unit, connected to each processing element (including the I/O processors), controls a memory bus to which all memory modules, the flag memory, and interrupt controller are connected. Each module monitors the memory segment portion of the address output by each processor to identify requests and contains decision logic to resolve contention and to select the processor request to be honored. The seven busses provide full interconnection, which coupled

*This work was sponsored by the Ballistic Missile Defense Advanced Technology Center under Contract DASG60-82-C-0019.

CH1935-6/83/0000/0255$01.00

with the selection logic on each module, provides the crossbar function. Simultaneous requests to unique memories cannot interfere with each other, and logic delays are not dependent on processor/memory combinations.

SEGMENT NUMBER	SUBSYSTEM ADDRESS
00	LOCAL EPROM
01	LOCAL RAM
02	MEMORY #1
03	MEMORY #2
04	MEMORY #3
05	MEMORY #4
06	MEMORY #5
07	MEMORY #6
08	MEMORY #7
09	MEMORY #8
0A	MEMORY #9
0B	MEMORY #10
0C	MEMORY #11
0D	MEMORY #12
0E	FLAG MEMORY
0F	INTERRUPT CONTROLLER
10	EXTENDED EPROM
11	REAL-TIME CLOCK
12	INVALID
•	•
•	•
•	•
7F	INVALID

Figure 1. Crossbar Segment Addressing

Processing Element

The processing elements are Zilog Z8001-based, single-board microcomputers each containing 64K bytes of private RAM, parity generation and checking logic, 8K bytes of EPROM, and two RS422 serial I/O ports. The serial I/O ports can be used for a control console to download programs from host computers, or to monitor system activity. The private RAM is implemented with 64K dynamic RAM chips. It is primarily for application program space and program stack space. The onboard EPROM provides space for a nucleus monitor.[6]

Memory Subsystem

The memory subsystem contains the memory bus interfaces, the arbitration unit, and the memory array. Each of the seven bus interfaces is connected to one of the processing element busses. The bus interfaces detect requests for the memory subsystem, latch the associated address (and data), and signal the arbitration unit that the request is pending. The arbitration unit is responsible for sequencing requests to the memory array on a first-come, first-served basis. As shown in Figure 2, each bus interface in the memory subsystem is associated with a 4-bit counter. The arrival of a request in a bus interface starts the counter, which is incremented every clock cycle, until the request is granted. At each clock pulse, the counter with the highest value represents the request that has been waiting the longest. The selection logic is PAL-based combinational logic that selects the counter with the highest value. Granting the request clears the associated counter. Ties are arbitrarily resolved in favor of the lowest-numbered interface. The memory array contains 64K bytes of memory (with parity) and is implemented with 55 ns, 16K by 1 NMOS static RAM chips. Static RAM is used instead of dynamic RAM to avoid conflicts between memory requests and memory refresh requests. RAM speed is chosen to allow multiple requests to be satisfied at a rate of one per clock period.

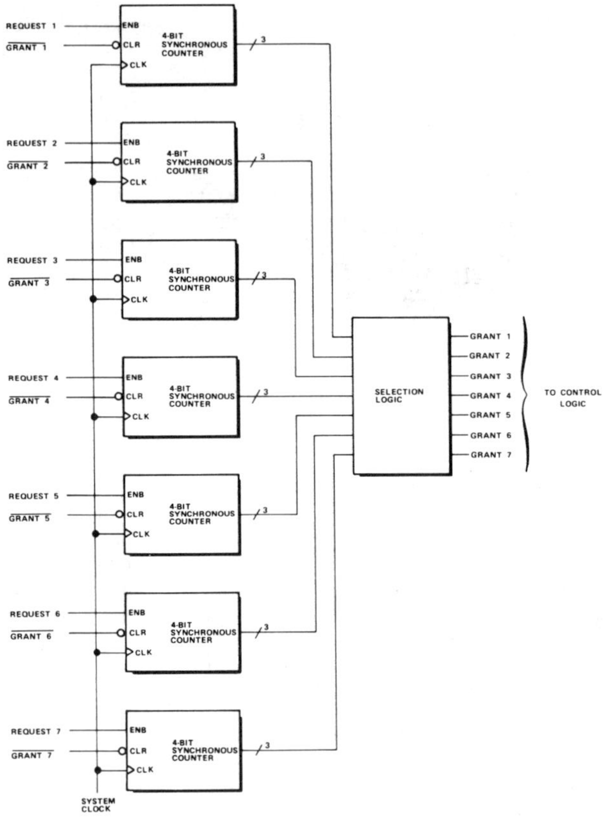

Figure 2. Arbitration Unit

Flag Memory

The Flag Memory provides 16,384 one-bit flags that can be used for synchronizing accesses to the sharable memories. These flags are provided because the Z8001 Test and Set (TSET) instruction does not guarantee indivisible access to a memory location. Each read request to a flag implements a true TSET operation. The value of the flag is returned to the processor and the flag is set to one. A write request to the flag will set it to zero. The flag memory uses the same bus interface, arbitration, control logic, and board layout as the sharable memory modules. Two bits at each address are provided for error checking.

Interrupt Controller

The Interrupt Controller provides the mechanism for any processor, including input/output processors, to interrupt any other processor. A processor is interrupted by writing data to a 16-bit register associated with it. The interrupted processor reads the data as part of its interrupt acknowledge sequence. Once the data are written to an interrupt register, further writes to that register are blocked until the interrupt acknowledge sequence has occurred. A processor attempting to interrupt must write to the appropriate interrupt register and then read its own status register. If the status is true, the interrupt request has succeeded. If the status is false, the interrupt request has failed and the processor must try again. Contention logic to resolve simultaneous interrupt requests is implemented in a 2910 microsequencer.

Memory Bus

Each processing element including the I/O processors is connected to a bus interface unit that drives the bus to which all memory modules, the flag memory, and the interrupt controller are connected. The bus interface unit also includes logic to detect and trap references to invalid segment addresses. On the memory end of the bus, another bus interface detects requests, latches the associated address and data, and signals the arbitration unit that a request is pending. The bus, illustrated in Figure 3, is bidirectional. It includes 18 bits for address and data as well as bits for request, grant, and interrupt request. Clock and reset signals have separate dedicated paths.

SIGNAL	DIRECTION
• INTERRUPT REQUEST	I
• REQUEST	O
• GRANT	I
• GROUNDS	–
• ADDRESS/DATA 17	B
• ADDRESS/DATA 16	B
• ADDRESS/DATA 15	B
• ADDRESS/DATA 14	B
• ADDRESS/DATA 13	B
• GROUNDS	–
• ADDRESS/DATA 12	B
• ADDRESS/DATA 11	B
• ADDRESS/DATA 10	B
• ADDRESS/DATA 9	B
• ADDRESS /DATA 8	B
• GROUNDS	–
• ADDRESS/DATA 7	B
• ADDRESS/DATA 6	B
• ADDRESS/DATA 5	B
• ADDRESS/DATA 4	B
• ADDRESS/DATA 3	B
• GROUNDS	–
• ADDRESS/DATA 2	B
• ADDRESS/DATA 1	B
• ADDRESS/DATA 0	B

I – INPUT TO MICROPROCESSOR INTERFACE
O – OUTPUT FROM MICROPROCESSOR INTERFACE
B – BIDIRECTIONAL

ADDITIONAL SIGNALS DISTRIBUTED TO ALL CIRCUIT CARDS IN THE CROSSBAR ARE CLOCK AND RESET. THESE SIGNALS HAVE SEPARATE DEDICATED PATHS.

Figure 3. Memory Bus

Input/Output Processor

Input/Output uses a separate processor board to support transfers between sharable memory and external devices. I/O processors use the same bus interface module and have the same performance and capabilities as the main processing elements. The maximum burst transfer rate for an I/O processor is 945K bytes/s. Actual transfer rate will depend on the protocol and error checking requirements for the specific application. To maintain maximum flexibility, detailed I/O protocol is left to the software. Logic to interface with an external device (e.g., another computer, an external bus, or a sensor) is implemented an an I/O interface board. Additional I/O processors may be included; however, the number of main processing elements and I/O processors combined cannot exceed seven.

Extended EPROM/Real-Time Clock Module

Extended EPROM and a real-time clock readout are provided on a single card for each processing element and I/O processor. The card provides up to 64K bytes (using 27128 s) of EPROM located in segment 10_{16}. There is one card per processor. A single, free-running, 31-bit real-time clock with a granularity of 2 μs is accessible to the associated processor by addressing segment 11_{16}. Since the clock is 31-bits wide, two accesses are needed to read it into the processor. To ensure coherence of clock data, the first access causes the clock to be copied into a latch associated with the processor. The least significant half of the latch is then copied from the latch onto the processor data lines. The second access copies the most significant half of the latch onto the processor data lines. Real-time clock readouts are driven from a single master clock and are synchronized within ±1 least significant bit. Bit 32 of the clock readout is used for power-up detection.

Interval Timers

An interrupting interval timer is provided for each processor including I/O processors. The interrupt period is selectable between 2 μs and 131 ms; the timer can be loaded, stopped, and started under program control. The interval timers are located on the interrupt controller board.

Clock and Reset Module

The entire system operates synchronously from a single system clock. The clock and reset module contains the master crystal oscillator and drivers for distributing the clock throughout the system. A slave clock module is required to drive processors in a second card cage. Delay lines are also provided to enable synchronization of the clocks throughout the system. A master reset is provided to reset all processors and the crossbar switch to a known state on power-up or when a nonrecoverable abnormal condition occurs.

PERFORMANCE CONSIDERATIONS

The system operates on a 6 MHz clock. A Z8001 requires three clock cycles (0.5 μs at 6 MHz) to perform a memory access. If a memory cannot meet the response requirements, the processor is placed in the WAIT state for additional clock cycles. Access to the onboard private memory does not require WAIT states. Access to extended EPROM requires one WAIT state for a total access time of four clock cycles. Contention-free access to a sharable memory (including Flag Memory) requires four WAIT states for a total access time of seven clock cycles (1.17 μs). When multiple requests are pending for the same memory, they are handled at one additional clock cycle per request. The Interrupt Controller may require more WAIT states depending on its work load.

IMPLEMENTATION CONSIDERATIONS

The system logic has been partitioned into eight board types: microcomputer, memory subsystem, interrupt controller, flag memory, bus interface, real-time clock/extended EPROM, clock and reset module, and I/O interface. Memory boards, the interrupt controller, and the flag memory are large (13 by 23 in) double-layer Multiwire(R)* boards. The remaining boards are double-height Eurocard format boards. Most of the logic is in the memory subsystem (approximately 320 dual in-line packages per board). The system is implemented in low-power Schottky TTL.

The primary implementation problems result from the number of wires and signals required. All busses are parallel; 21 signals are on each bus. To minimize crosstalk problems, all bus signals are differentially driven and transmitted over twisted pair cable. The microcomputer bus signals are daisy-chained to minimize the overall cable length.

STATUS AND SUMMARY

The system was first implemented as part of a distributed testbed for real-time application[7] with 6 processing elements, one I/O processor, and 12 sharable memories. Subsequently, the basic hardware building blocks have been used to configure a variety of system configurations for use in data acquisition and control applications where system response times are critical and high reliability with tolerance to faults is a necessity. The hardware building blocks have proven effective for configuring systems within the limits of the design--7 processors and 12 sharable memory modules for full interconnection. Systems may be configured with more processors by accepting less than full memory access from each processor.

*Multiwire(R) is a U.S. registered trademark of the Kollmengen Corporation.

REFERENCES

1. E. T. Fathi and M. Kriger, "Multiple Microprocessors Systems: What, Why, and When," Computer, March 1983.

2. S. H. Fuller and S. P. Harbison, "The C.mmp Multiprocessor," Technical Report, Carnegie-Mellon University, Computer Science Department, 1978.

3. W. A. Wulf, R. Levin, and S. P. Harbison, HYDRA/C.mmp: An Experimental Computer System, McGraw-Hill Book Company, 1981.

4. H. B. Baskin, B. R. Borgeson, and R. Roberts, "PRIME - A Modular Architecture for Terminal-Oriented Systems," Spring Joint Computer Conference, 1972.

5. W. A. Wulf and S. P. Harbison, "Reflections in a Pool of Processors: An Experience Report on C.mmp," 1978 National Computer Conference, AFIPS Press, 1978.

6. N. Vosbury and C. Bryant, "System Software for Experiments in Distributed Computing on a Distributed Testbed," Proceedings 3rd International Conference on Distributed Computing Systems, October 1982.

7. W. C. McDonald and R. W. Smith, "A Flexible Distributed Testbed for Real-Time Applications," Computer, October 1982.

The ITT 1240 Highly Reliable Distributed Control Architecture

R. A. Conroy
ITT Advanced Technology Center
Shelton, Connecticut

H. A. Malec
ITT Programming
Stratford, Connecticut

Abstract

ITT's experience with fully distributed control in the ITT 1240 Digital Exchange has validated it's fundamental innovative principles. The architectural concepts, and their software/hardware realizations, make the ITT 1240 system a highly reliable system in which data switching in its many forms (digitized voice, packet, message, etc.) is efficiently handled.

This paper contains an architectural overview of the fully distributed ITT 1240 system emphasizing the following points:

1. *The software architecture is fundamentally reliable because of the independence of functional modules and the rigor of interface mechanisms.*
2. *The hardware architecture is fundamentally reliable because the basic failure units are small and use LSI and VLSI componentry.*
3. *The architecture is inherently fault tolerant through the decoupling provided by message communication in the systems network.*
4. *Reliable performance is easily engineered to any level.*
5. *The field performance has reached steady state system effectiveness predictions within three months of initial cutovers.*

Introduction

The ITT 1240 architecture attains four major objectives: (a) to provide for a smooth modular growth in systems which can be extended to more than 100,000 terminations, (b) to provide a single architecture which satisfies many switching system requirements, (c) to respond to the advances in hardware technology by actually taking advantage of these for both performance improvement and cost reduction with minimal impact on the existing software, and (d) to limit the effects of service failures to small modular units.[1]

Figure 1 presents the top level organization of the ITT 1240 System. Control of the system is distributed throughout the functional modules which communicate among each other through the Digital Switching Network. Each module in Figure 1 is equipped with a standard Control Element consisting of a commercial microprocessor with its memory and a special functional unit called the Terminal Interface (TI) which interfaces each module with the digital network.

The TI allows the microprocessor in one module to communicate with the microprocessor in another module. The TI also acts as a transmission interface between the terminals and the network. The Digital Switching Network (DSN) is realized with a unique switch port using a custom LSI circuit. Sixteen switch ports constitute a switch element and the network is made up through repetitive use of these switch elements.

The interface between each module and the digital switching network is a pair of 4 megabit (32 channel PCM) links used for packet communication or virtual paths. As long as this interface is maintained, architectural or technological evolution within a module has no impact on the rest of the system. The ability of the control elements to communicate through the digital switching network obviates the needs for specialized communications links between the nodes, thereby facilitating the smooth, modular design growth from a system with a few terminations to one with more than 100,000 terminations through use of varying numbers of switch elements.

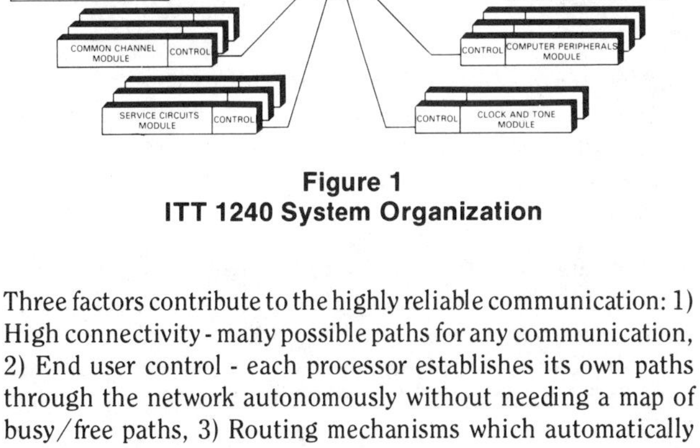

Figure 1
ITT 1240 System Organization

Three factors contribute to the highly reliable communication: 1) High connectivity - many possible paths for any communication, 2) End user control - each processor establishes its own paths through the network autonomously without needing a map of busy/free paths, 3) Routing mechanisms which automatically bypass failed sections of the network in most cases.

CH1935-6/83/0000/0259$01.00

Experience with the design and the operation of computer controlled communication systems has shown that the performance of the software package determines to a great extent the quality of the service to be offered. Therefore during system and software design, special emphasis has to be placed in order to reduce the number of faults still undetected when starting the actual system operation. To further increase the system reliability it is necessary to limit the propagation of corrupted data which may have been generated erroneously.

Software Design Approach

In ITT 1240[4,8], the answer to the above objectives are basic software modules that control the implementation of all functions; they are called 'Finite Message Machines', (FMM). The required independence of these FMMs is achieved by the fact that they communicate with one another only by messages, even in case the of FMMs within the same microprocessor or control element. Finite Message Machines interwork in a clearly defined way.

- Inter-FMM communication is by messages only, with each message represented by a block of data.
- Each FMM is capable of receiving a specific set of input messages, and of generating a specific set of output messages.

An FMM is made up of one or more processes, each of which can be in one of several states. While it is in any given state, it is capable of receiving only a specific set of messages from another FMM. In the dialing phase of a telephone connection, for example, the FMM would be in the state 'dialing'. Therefore, the input message may only be from the set that includes 'calling subscriber went on-hook, subscriber dialed digit n, and maximum interdigit time exceeded'. All other input messages to the FMM would be disregarded in that state. The FMM then processes the input message, generates the appropriate output message and only then goes to a new state.

The necessary communication between processes in the FMMs is supported by the operating system. Each control element contains a message handler and a network handler program within the control element operating system. Every message from one FMM to another FMM is routed and transported by the message handler program. The receiving FMM can either be in the same control element or in a distant control element and the sending process does not need to know the location of the receiving process.

The management of data in the system to provide secure, reliable and efficient data access and storage is achieved through use of a fully distributed data base control system. Data is viewed through the relational model and is kept physically separated from FMMs, so that FMMs are unaware of the location or storage format of the data.

The software is structured in layers with virtual machines at each layer; that is, software is arranged in functional layers with each layer making use of lower layers with independence from the detailed implementation of the lower layers. In this way changes do not affect higher layers. An important example is the device handler software module and its related hardware, which together form a virtual machine with a standard interface to the operational software; as a result, changes in hardware do not affect the operational software.

This modular software design with its clearly separated processes and message transfer has direct advantages for the reliability of the system.

Isolation of FMMs is a principle which considerably increases software reliability. No FMM explicitly uses data of another FMM, thereby eliminating errors due to misinterpretation of the data layout of other FMMs. Messages which are not accepted are discarded, thus preventing the propagation of errors. This gives the software a high degree of fault-tolerance, because most error situations can automatically be overcome by restarting an FMM.

The FMMs can be tested in a straightforward manner both during software design and system operation. In testing, the input message sequences are simulated and the generated output message sequence is checked. This enables a thorough and application-oriented testing and allows quick tracing down of residual bugs found later during operation. This leads to a reduction of the maintenance effort needed in the field.

These qualitative advantages for distributed control need to be translated into quantified advantages in evaluating the system effectiveness of a design. For software controlled, real-time communications systems, the prediction of the reliability and of the quality of service, even when using the modular FMM approach, is presently not as established as the more hardware oriented modeling techniques. However, it is important for system effectiveness calculations that a prediction be made for the software components of the system. This should take into account both transient and catastrophic software errors.

With the fully distributed control in ITT 1240, the probability of a software catastrophic error which affects the full system is extremely small. The more interesting parameter is the probability of a catastrophic software error which affects one control element and which causes that control element to be restarted or reloaded.

The actual frequency of these fault events can be derived from relevant field data or system analysis. The ratios that can be used as a first estimate are that the hardware transient faults occur at a rate approximately ten times the hardware catastrophic rate and software transient faults occur at about 100 to 500 times the software catastrophic rate. These ratios tend to be pessimistic.

Hardware Design Approach

As previously stated, the ITT 1240 Digital Exchange hardware architecture is based on distributed control with a variety of autonomous modules connected to a digital switching network[2,3]. The fundamental unit of processing power has a low failure rate because there are few components. It consists of two to four printed board assemblies — a processor, a terminal interface, and optional memory boards. The converse impact of needing many processing units is a basic structural property of fully distributed control, and its reliability is managed by the inherent re-configurability and common pool sparing features of the architecture.

Both standard and custom LSI circuits are used and deferred maintenance concepts are an integral part of the design. The design approach has also followed the concept of reliability performance measurement[6].

The system effectiveness effort has analyzed many design concepts throughout the system, one example of which is the single bit error correction and double bit error detection memory with deferred maintenance concept[7]. Table 1 presents a comparison of two studied repair strategies for the memory of the Auxiliary Control Elements (ACE).

Failure Mode	Repair Strategy: All parity failures deferred; total memory failure causes urgent repair	Repair Strategy: No repair action with parity failure; total memory failure causes urgent repair
	Contribution to memory board unavailability (minutes per year)	
2 failures in 1 word	0.2	11.4
2 failures in each of the 256 words	0	0.8
2 failures in each of the 64K words	0	0
Failure in common circuitry	3.1	3.1
Board removed for replacement	3.2	0
Total	6.5	15.7
Average number of boards replaced per year	8.4 boards/y	2.1 boards/y

Table 1

As can be seen, the deferred but later repaired parity failures lead to an ACE memory unavailability of 6.5 minutes per year, whereas the decisions to take no repair action at all in case of parity bit failures results in an increase in unavailability by a factor of 2.4. However, the average number of boards to be replaced per year drops in the latter case by a factor of 4 and could lead to a considerable reduction of the maintenance actions.

This latter alternative was finally chosen, because the increased memory board unavailability can be compensated by spare Auxiliary Control Elements which may be activated when needed; this has almost no effect on the system availability and on the quality of service. On the other hand the fault tolerance of the memory organization with its error-correction feature allows many failures to be deferred or even to be disregarded, and gives a direct reduction in maintenance costs.

Fault Tolerance and Deferred Maintenance

In non-repairable structures it is important to have fault tolerance for reliability, whereas deferred maintenance can be very appropriate for repairable structures. The previous example on memory errors shows how these two strategies can be effectively combined to provide an optimal solution. The objective of deferred maintenance is to limit the number of urgent service calls to a cost-effective level. This is accomplished by the deferring of certain non-urgent maintenance activities to the next urgent or semi-urgent service call.

Deferred maintenance concepts can be incorporated effectively into both simplex and redundant architectures in both common control and distributed control communications systems. The merits of partially distributed architectures versus fully distributed control architectures has been investigated. Figure 2 presents a general overview of the relative gain of implementing deferred maintenance techniques as a function of several different system control architectures in terms of system effectiveness.

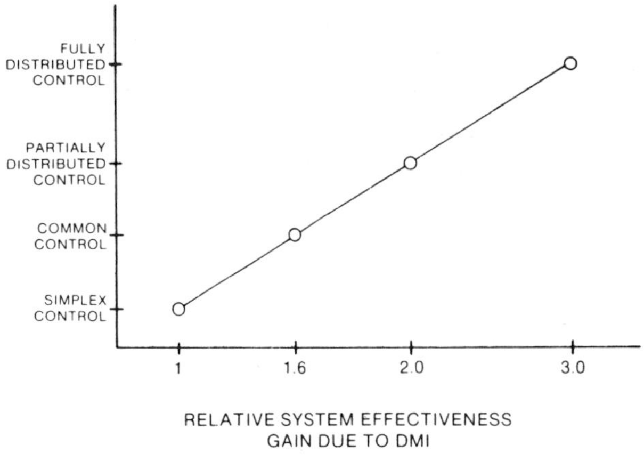

Figure 2
Deferred Maintenance Implementations (DMI)

In the future, further investigation of fault tolerant and deferred maintenance techniques for fully distributed communications control systems will continue due to control placement options. The most recent philosophy has been to place the control as close as possible to the user interface. This implementation reflects the power of a distributed configuration. Another concept of incorporating the control into the switching network itself, could also prevent the control from being overloaded and in addition lessen the probability of user loss of service.

Other advantages of this latter approach are the ease of software version updating and graceful degradation of system functions. As the control has been moved towards the system user interface, the user may have observed a lower availability of service if fault tolerant or deferred maintenance techniques were not incorporated.

System Effectiveness

Since the 1960s, ITT has applied its own stringent system effectiveness specifications to ensure that its designs meet foreseeable needs in all countries. These living requirements have guided the ITT 1240 development from the earliest feasibility design stages to the final fault-tolerant design. The system effectiveness objectives were established for different configurations as matrix goals[5].

Each configuration of the ITT 1240 (i.e., independent toll and local exchanges, supervised exchange, remote subscriber unit) has a set of objectives related to its position in national or international switching plans with each objective sub-divided, where appropriate to take account of the maintenance scenarios of (a) continuously attended exchange, (b) exchange attended during normal working hours only, and (c) unattended exchanges. System effectiveness requirements include specifications for the above variences for levels of unavailability, call effectiveness, and mean downtime[5].

The most feared event in central control processor systems is the simultaneous failure of the processors resulting in service failure for all customers in an exchange. This event is usually specified in terms of no more than several minutes average per year. The reliability of the distributed control of the ITT 1240 is such that total system unavailability is measured in terms of seconds per year.

Field Experience

At the time of writing, there were several ITT 1240 exchanges in operation in Europe. These first exchanges have demonstrated that reliability is realized in the field as has been predicted by the appropriate modeling. Additional early failures could occur and these initial operational failure rates may be adjusted later. The majority of the following data has been taken from two 1240 operational systems (a local and a toll system).

After eight months of operation, there were no microprocesor failures; this gives a point estimate of 209 FITS. The predicted failure rate used in the system effectiveness models was 300 FITS. The predicted 64K dynamic RAM device hard failure rate was 600 FITS. The field operational failure rate for this device during the first nine months of operation, with three recorded hard failures, was tenfold less than predicted.

The switch port device, which is a custom LSI with 3300 equivalent RAM bits (or 11500 transistors), had a predicted failure rate of 470 FITS in the system effectiveness models, using MIL-HDBK-217C, and later 215 FITS using MIL-HDBK-217D. There are two other equivalent custom LSI devices, the receive port and the transmit port. Taking the field operational data for these three device types, there were 4 hard failures in nine months, yielding a point estimate of 109 FITS.

One of the best benchmarks for software field performance is the frequency of processor fault-interrupts, restarts and reloads. The advantage of the distributed architecture has been dramatically demonstrated in the ITT 1240 as the system has reached predicted reliability steady state conditions for these undesireable events very rapidly within 3 months of initial cutover. After 9 months, it was operating better than original long term projections in the system effectiveness models. A significant contribution to this success is the effect of the distributed control architecture in reducing the impact of software faults, *e.g.* since each control element is responsible for a very small part of the total system, then any software fault will impact only a very small part of the total system.

Summary

Highly reliable communication between processors is absolutely essential to making fully distributed control work. The ITT 1240 manages this by using the same Digital Switch Network for processor communication as for the digital voice and data paths. Thus the ability to communicate between processors is guaranteed through multiple and redundant network paths.

Reliability is further enhanced through use of LSI, both custom and off-the-shelf, and a software structure which emphasizes functional separation and strict interface control. A well-planned combination of fault-tolerant design and deferred maintenance repair strategy results in a system effectiveness which exceeds the most stringent specifications.

ITT's experience with fully distributed control in the ITT 1240 has demonstrated that high reliability is inherent in the concept. The improvement in reliability is so dramatic that fundamental methods of thinking about reliability change. This design advantage has now been demonstrated in commercial application with the first half dozen ITT 1240 systems working in Europe.

References

1. M. Langenbach-Belz, A. Melis, and H. Verhille, "ITT 1240 Digital Exchange An Introduction", *Electrical Communication,* Volume 56, No. 2/3 1981, pp 114-125.
2. S. Das, K. Strunk, and F. Verstraete, "ITT 1240 Digital Exchange Hardware Description", *ibid,* pp 135-147.
3. J. Cotton, K. Giesken, A. Lawrence and D. Upp, "ITT 1240 Digital Exchange Digital Switching Network", *ibid,* pp 148-160.
4. L. Katzschner and F. Van den Brande, "ITT 1240 Digital Exchange Software Concepts and Implementation", *ibid,* pp 173-183.
5. J. Dutt and H. Malec, "ITT 1240 Digital Exchange System Effectiveness", *ibid,* pp 198-206.
6. J. Dutt and H. Malec, "Reliability Standards for Digital Switching Systems", *EOROCON '82,* 14-18 June 1982, Copenhagen, Denmark.
7. J. Dutt and H. Malec, "Fault-Tolerance Allowing Deferred Maintenance Techniques", *3rd International Conference on Reliability and Maintainability,* 18-21 October 1982, Toulouse, France.
8. R. Arranz, R. Conroy and L. Katzschner, "Structure of the Software for a Switching System with Distributed Control", *International Switching Symposium '81,* 21-25 September 1981, Montreal, Canada.

Session 2-1a
VLSI Technology III

Chairman

K. Kataoka
NTT

A HIGH SPEED 16-BIT EXPANDABLE MULTIPLIER USING 5000 GATE ECL GATE ARRAY

Toru Takahashi, Hiroshi Nakashiba, Toshio Nakamura and Kodo Kimura

NEC Corporation
1753 Shimonumabe, Nakahara-ku, Kawasaki 211 Japan

ABSTRACT

A High speed 16 bit expandable multiplier LSI with ECL-100K compatibility is developed using 5000 gate ECL gate array.
Utilization of the advanced bipolar process and four layers metalization realize the basic gate delay of 0.35 ns and producible chip size of 9.4 x 9.2 mm.
As a result, high speed multiplication time of 20 ns is measured for 16x16 bit datas. The multiplication time for 64 x 64 bit is expected as 74 ns.
Sophisticated CAD system and macro block library in gate array approach make enable the completely automatic layout and routing with 100% routability resulting maximum design efficiency.
The multiplier LSI with 164 signal pins is assembled in a 208 pin PGA package with heat sink using TAB assembly technology.

INTRODUCTION

This paper describes a high speed 16 bit expandable multiplier LSI which have been developed using 5000 gate ECL gate array.
Introduction of LSIs is indespensable for mainframe computers to improve the total performance. It will also promise increased reliability, decreasing cost and smaller hardware size.
Particularly, decreased media delay between chips is effective to improve the critical path delay which may determine the system performance.
Full custom design approach has used to be employed to develop these LSIs for inhouse mainframe computer application.
However, the average system development time tends to increase due to the complexity trend in the system configuration, in spite of the continuing decrease of the average product life cycle. Mainframe computer system is no exception. Design automation using CAD system seems to be an important approach to solve these above mentioned problems.
Sence up to 2000 gate ECL gate arrays are spread in various field where the speed performance is dominant[(1)], our goal is to develop a higher complexity ECL device for high speed digital application without sacrificing the sub-nanosecond delay performance.
On the other hand, demands for hardware logic engine is increased recently. Particularly, multiplier LSIs become widely used not only for scientific calculation in computers but also for signal and image processing applications. Consequently, a high speed 16 bit expandable multiplier LSI is developed using newly designed 5000 gate ECL gate arrays employing four layers metalization and advanced bipolar process technology.

5000 GATE ECL GATE ARRAY

Major features of the gate array are listed in Table 1.

Supply voltage	: VEE = -4.5V, VT = -2V
I/O level	: ECL-100K compatible
Power dissipation	: 7.45 W (typ.)
No. of pins	: 176 (Signal)
No. of cells	: 2176
No. of gates	: 5440
No. of metal layers	: 4 (Two for Customize)
No. of tracks	: 816(X) + 672(Y)
Circuit type	: Two Level Series Gated ECL
Chip size	: 9.4 mm x 9.2 mm
Package	: 208 pin PGA

Table 1. Characteristics of the ECL Gate Array

Advanced bipolar process with oxide isolation and self-aligned electrodes is employed in combined with poly silicon resistors which require no area in the silicon substrate.
The most unique aspect of the process is the introduction of the four layers metalization. This feature is effective to reduce the die size since chip size is primarily determined by the area for wiring tracks.
The first poly silicon layer is only used for intra-cell wiring and inter-connections to the 1-2 via holes.
The second and the third alminum layers are completely used for customization of the logic. The fourth gold layer is only used for power and GND bussing and reference voltage distribution.
Total of 672 tracks run in the second layer with 3um metal width and 6um pitch, 816 tracks run in the third layer with 4um metal width and 7 um pitch.
As a result, realizable chip size of 9.4 x 9.2 mm

CH1935-6/83/0000/0264$01.00 © 1983 IEEE

is obtained in spite of its high gate complexity. In order to decrease the total power consumption, termination voltage V_T(-2V) is used not only for output emitter follower termination, but also for internal emitter followers. In addition, power pins of VCCA for output emitter followers are provided separately from power pins of VCC for internal circuits so as to obtain stable operation against simultaneous output switching.
The 32 rows of 68 internal cells yield the equivallent of over 5000 gates. They are surrounded by the I/O buffers and reference voltage generators.
Output buffers which require relatively large area are only located at the left and right side of a chip resulting minimized chip size.
Each internal cell consists of 10 transistors and 10 resistors.
This number is relatively large compare to the traditional cell design approach but effective to organize the actual complex logic functions.
Figure 1 shows an example of the circuit schematic which performs the carry save adder function. Notice that only two cells are required to realize this complex function owing to the employment of the two level series gated ECL circuit and wired logic functions.

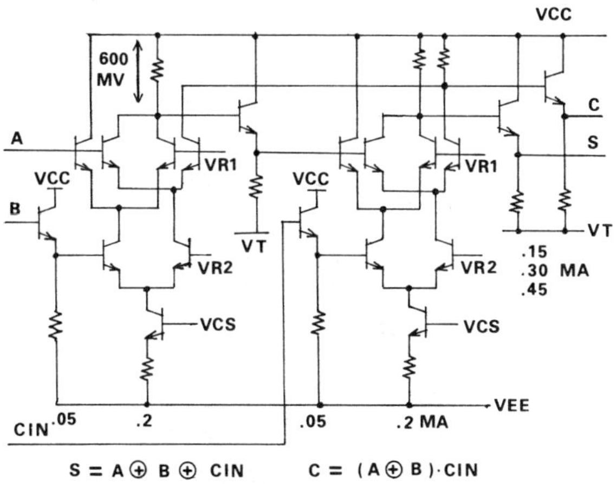

Figure 1 Circuit schematic of the carry save adder

Affect of the stray capacitance due to interconnections become more critical when the gate complexity is increased. To solve this problem, emitter follower configuration is employed for all the outputs of macro blocks.
These emitter followers are designed to have three kind of current values.
According to the loading capacitance value, the optimum current level is automatically selected by the CAD program so as to minimize the media delay.
The internal logic swing is designed as 600 mV typically which is kept constant against the temperature and supply voltage variations so that the stable operation is obtained with enough noise margin.
A ratio of each current level in a block is determined by computer simulation so that optimized speed performance is obtained within a limited power consumptions.
Notice that each cell corresponds to over three equivqllent gates in this example. Even when the over all function blocks are considered, a cell corresponds to 2.5 equivallent gates in average.
Since increased gate complexity requires more signal pins, the gate array LSI provides up to 176 signal pins including 88 output pins in maximum. Packaging technology will be discussed in the later section.

16 BIT MULTIPLIER LSI

Rapid progress in digital signal processing and scientific calculation in mainframe computers requires high speed multiplier LSI which can handle wide bit datas[2].
To fulfil the requirement, a high speed 16 bit expandable multiplier LSI has been developed using the above mentioned 5000 gate ECL gate array.
Figure 2 shows a block diagram of the multiplier LSI.

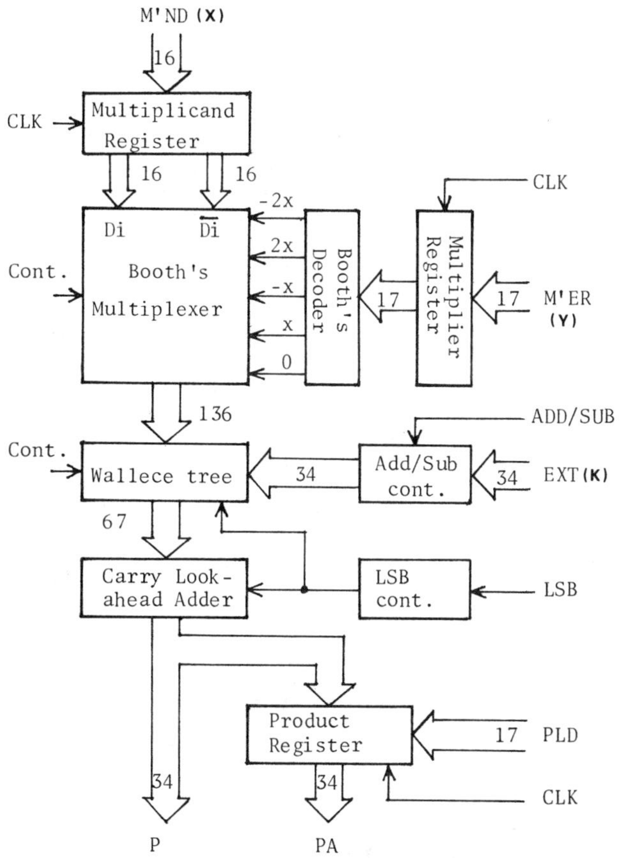

Figure 2 Block diagram of the multiplier LSI

The LSI provides three registers to store all the I/O datas resulting easy to data management such as pipelining.

In order to increase multiplication speed, and reduce the number of the partial product, modified Booth's algorithm is employed instead of the conventional sequential multiplication.
Modified Wallace tree, which postpone addition of the carries to a later stage, is also employed for final addition of the partial products.
Except to the conventional add functions, subtract functions, is also avairable not only for single mode but also for expansion mode offering flexible utilization. These extra inputs for expansion are applied to the later stage of the Wallace tree in order to decrease the delay time for the path since it determine the speed performance in expansion mode.
The final product result can be obtained through the 34 bit carry lookahead adder. The output register which provides additional inputs for pre-loaded data, is also provided to store the product result.
All the registers are designed to be operated with independent clock signals so that flexible and precise timing adjustment is allowed.
As a result, high speed and flexible 16 bit multiplier has been designed.
To implement this complex functions, total of 20 unique macro blocks are designed and used. Each macro block can be designed with minimum cell usage because a cell contains enough number of components as described in the previous section. Notice that a cell corresponds to approximately 2.8 equivallent gates in case of this multiplier LSI. The multiplier requires 1153 cells of 2176 total internal cells. Consequently, the total of 3460 equivallent gates can be implemented with only 53% the internal cell utilization.
Although this multiplier LSI requires 164 signal pins consist of 96 inputs and 68 outputs, the gate array with 176 I/O pins make enable this high pin count with 93% pin utilization.
To avoid saturation and mulfunction of emitter followers due to the simultaneous output switching, total of 12 VCCA pins are provided for the output emitter followers separated from VCC pins for internal logic.
The resistor value of 75 ohm is employed for the output terminator so as to reduce the total power consumptions. The multiplier LSI dissipates only 4.93 watts typically in spite of its high gate complexity and subnanosecond delay performance.
Since 58% of the total power is dissipated at the I/O buffers and reference voltage generators, internal blocks which performs the multiplier function dissipate only 2.1 watts.
Completely automatic design approach using CAD system is employed this multiplier LSI. First, logical function is verified by the inhouse developed logic simulator called LOGOS which can handle up to 10k gate complexity at a time and performs approximately ten times faster than commercial logic simulators. Next, automatic block placement and routing is performed using inhouse developed layout program named MASTER2[3] which can manage up to 20k gate complexity at a time.
A 100% completion routing is successfully obtained for this complex multiplier LSI.

The total CPU time required for the placement and routing procedure is 60 minutes on the mainframe system with 10 MIPS performance.
After the routing is completed, the appropriate current value is selected automatically for all the output emitter followers in macro blocks according to the actual loading capacitance values. Timing verification and delay time estimation are also performed by the precision delay simulator LOGOS-D. The actual wiring and loading capacitances are taken into account precisely here.
Figure 3 shows the distribution result of the net's length. The total net's count is 1131 and average net length is only 2.48 mm for the sake of macro block approach. The average number of fan-out is 2.52 per net.

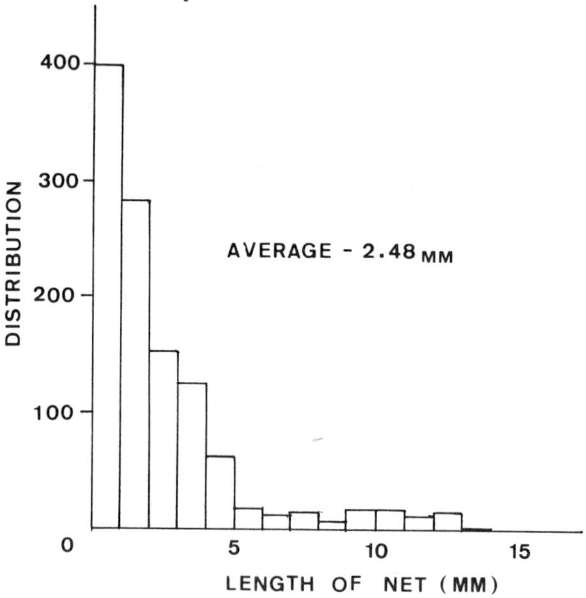

Figure 3 Distribution of wiring net's length

The photomicrograph of the multiplier LSI chip is shown in Figure 4.

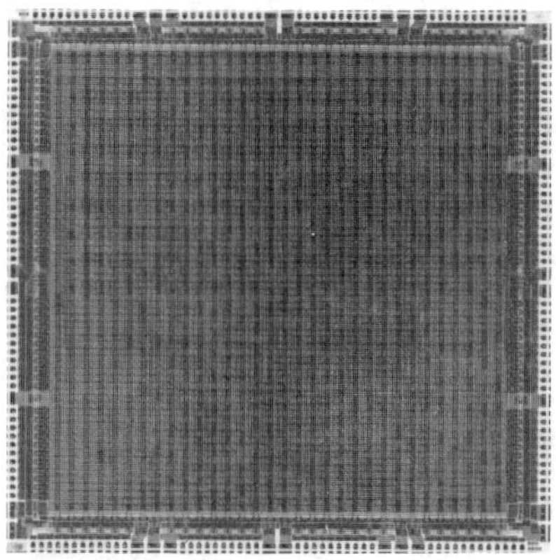

Figure 4 Photomicrograph of the multiplier LSI

The chip size is 9.4 x 9.2 mm. To realize large pin count within a limited chip size, batch bonding technology using a flexible tape as a chip carrier is introduced.
Each pad provides gold bump for thermal compression bonding for this assembly approach called TAB technology.
As a result, small pad size of 60 um square with fine pad pitch of 140 um is realized.
The LSI is assembled in a newly developed pin grid array package with 208 pins.
In addition, chip is directly mounted on a metal substrate where the heat-sink is attached on the opposite side. No material such as ceramic is existed between the chip die and the heat-sink material.
As a result, excellent heat transfer characteristic can be obtained using traditional forced air cooling method.
The thermal resistance from junction to air is measured as 5°C/W with 800 lfpm air flow.

PERFORMANCE

Multiplication time is primarily determined by the delay time from the clock input for the multiplier register to the product output (Tm).
The delay time from the expansion input to the product output is also important when the LSI is used in expansion mode (Te).
The multiplication for 16n x 16n bit datas can be roughly estimated using the following equation.

$$Tm + 3(n-1)Te$$

Figure 5 shows the switching waveforms represent the delay from the multiplier clock input to the product result.

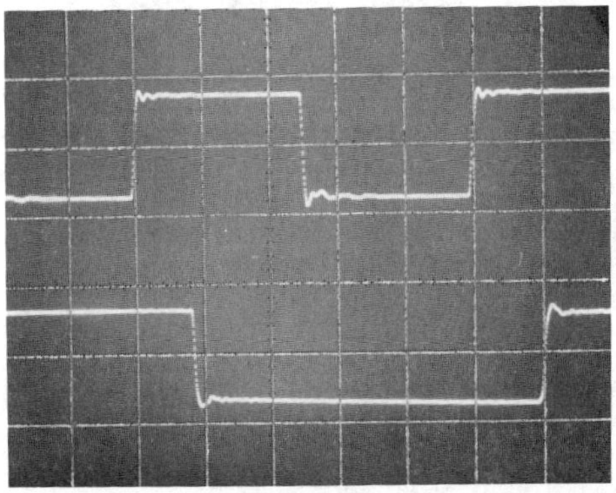

Figure 5 Switching waveforms of the multiplier LSI

This delay path through 24 current source stage which correspond to 41 equivallent gates. Delay time of 20 ns can be observed for this critical path which determine the multiplication time.
Notice that an equivallent gate delay of 0.49 ns is realized for this actual complex circuits.
In addition, delay time measurement by the gate chain circuit results a basic internal gate delay is 0.35 ns. The delay time of 6 ns is also measured from the expansion data input.
As a result, multiplication time of 20 ns can be expected for 16 x 16 bit datas.
For an example of expansion mode, 74 ns is expected for 64 x 64 bit datas.
The perfect samples have just been obtained, and evaluation in detail continues.
Major characteristics of the multiplier LSI are summarized in Table 2.

Gate complexity	:	3460
Pin count	:	164 (signal)
Power dissipation	:	4.93 W (typ.)
		8.66 W (max.)
No. of cells	:	1153
No. of nets	:	1131
No. of fanouts	:	2.52 (average per net)
Wire length	:	2.48 mm (average per net)
Multiplication	:	20 ns (16 X 16 bit)
Time	:	74 ns (64 X 64 bit)

Table 2. Characteristics of the multiplier LSI

CONCLUSION

The high speed 16 bit expandable multiplier LSI with ECL-100K compatibility has been developed using the newly designed ECL gate array with 5000 gate complexity.
Utilization of the advanced bipolar process and four layers metalization realize the basic internal gate delay of 0.35 ns and producible chip size of 9.4 x 9.2 mm.
As a result, high speed multiplication time of 20 ns is measured for 16 x 16 bit datas. The multiplication time for 64 x 64 bit datas is expected as 74 ns.
Sophisticated CAD system and macro block library in gate array approach make enable the completely automatic layout and routing with 100% routability resulting maximum design efficiency.
The multiplier LSI with 164 signal pins is assembled in a 208 pin PGA package with heat-sink using TAB assembly technology.

ACKNOWLEDGEMENT

The authors wish to thank S.Yawata for his helpful advise and encouragement; M.Yorikane for his processing support; Y.Umeki for his design support and electrical measurement.

REFERENCE

(1) Sato,F et al, "A Subnanosecond 2000-Gate Array with ECL-100K Compatibility", Proc. of CICC 83, pp.23-26, May,1983.
(2) Takahashi,T et al, "A High Speed Multiplier using Subnanosecond Bipolar VLSI Technologies", Dig. of Tech. Papers ESSCIRC 79, pp. 110-112, Sept, 1979.
(3) Kato,H et al, "An Automatic Layout System for Gate Arrays up to 20K Gate", Proc. of CICC 83, pp.19-22, May,1983

A CMOS Gate Array for Computer Applications

Isao OHKURA, Kazuhiro SAKASHITA, Kazumi SUGIZAKI,
Yoichi AKASAKA and Tatsuya ENOMOTO

Mitsubishi Electric Corp.
LSI Research and Development Lab.
4-1, Mizuhara, Itami, Hyogo 664, JAPAN

Abstract

An 8K gate CMOS gate array has been developed intending to apply it for large systems such as CPUs in computers. High density and high speed operation are accomplished by the gate isolation configuration and 2 um n-well CMOS fabrication technology. RAMs or registers are realized by utilizing the basic cells in small area owing to the gate isolation configuration. The CAD system contains facilities of cell reduction to accomplish a hierarchical logic design effectively. It also supports Level-Sensitive-Scan-Design (LSSD) and automatic test generation in order to attain high testability. As an example of application to CPUs, a 32b RALU is implemented in the 8K gate CMOS gate array. The RALU contains two 32b x 32word RAMs which equivalently construct a two-port RAM. The RALU occupies 86% of the internal basic cells and is realized by fully automatic layout procedures.

I. Introduction

While CMOS gate arrays become to be used for large systems due to low power consumption, these are requested to have additional features in order to get wider computer applications;

1) high complexity and high speed operation,
2) hierarchical design facilities for short design time,
3) precise logic and timing verifications,
4) effective RAM implementation.

An 8K gate CMOS gate array has been developed under the consideration of these points. The LSI features will be described in Section II. In Section III, the CAD system used for the LSI design will be presented. It allows the hierarchical logic design and the precise logic and timing verifications. The RAM implementation will be discussed in Section IV. Finally, 32b RALU implemented in the gate array LSI will be presented as an example of the applications.

II. LSI Features

The basic cell configuration based on the gate isolation and 2 um CMOS technology has attained the complexity of 8K gate and the typical gate delay of 1.5 ns under 2 F/O and 1 mm of aluminum wiring, which implies that STTL circuits are replaceable to the gate array.

Basic Cell Configuration

A basic cell contains a pair of a p-channel and n-channel transistor. The basic cells are arranged in a row closely without oxide isolation between them, resulting in a serial p-channel transistor chain and a n-channel one.

Standard cells, such as inverters, NAND gates, NOR gates, AND-NOR gates, OR-NAND gates, flip-flops and shift registers are constructed by using plural basic cells. The poly-Si gates of p-channel transistors and n-channel ones located on the right and the left ends of the cells are connected to VDD and GND lines respectively to set these transistors in the cut-off states. Compared with conventional basic cell configurations, the gate isolation enables to construct various logic cells without unused transistor pairs. As reported before[1][2], the gate isolation configuration attains 1.3 to 1.8 times higher complexity than the conventional 2 transistor pair basic cell configuration in which the basic cells are isolated by thick oxide region.

LSI Chip Structure

Figure 1 illustrates the LSI chip structures. 216 basic cells were arranged in a row and formed a block, and 112 blocks were arranged in 28 rows and 4 columns, and as a result, 24,192 basic cells were set in the LSI chip. The chip equivalently implements 8,064 2-input NAND gates. Each intercell wiring zone located between the rows, allows 40 tracks for holizontal wirings using the first aluminum layer, the pitch of which was 5.0 um. The vertical wiring pitch by the second aluminum was 8.5 um. The five pairs of VDD and GND aluminum lines ran vertically along the both side of each column. The chip had 200 I/O buffer cells around the basic cell matrix region. All these buffers can be used for either input buffer, output buffer, 3-state buffer or bidirectional buffer by changing the artwork data on the customized stages. The chip size was 9.89 x 9.80 mm.

III. CAD System

Various design tools have been developed to support logic design, timing verification, layout, test program generation and so on. These programs were constructed arround the data base which allows any nesting levels of hierarchical description. The fully automatic layout is executed or any manual modifications are not required. The facilities for hierarchical logic design and automatic layout make possible to design an LSI in a short time and to reduce design errors. The CAD system also support the scan design for improvement of test

CH1935-6/83/0000/0268$01.00

coverage. Test patterns are automatically generated by one dimensional path sensitization method.

Among others the following two features of the CAD system should be noted.

Hierachical Design and Cell Reduction

90 standard cells and 50 standard macros have been registered in the data base. The standard cells are the fundamentals for logic design and also layout elements of the LSI. The standard macros were constructed by standard cells and small macros. The macro library includes the commonly used logical blocks equivalent to TTL standard MSIs. The system designers can define the temporary macros in addition to the standard macros. When a standard cell in a macro becomes to be unused on the system construction, the layout interface program extracts the cell in order to increase the effective logic density. The facility is called cell reduction and is shown in figure 2.

Precise Logic and Timing Verifications

The logic simulator used in the system is a commercially purchased delay assignable simulator. The interface program assigns three kinds of delay time according to the progress of the LSI design, that is norminal delay, provisional delay and real delay. In the norminal delay mode which is used for logic verification, a nominal delay value is assigned to each kind of gates. The provisional / the real delay mode is for the delay analysis before / after the layout procedures, the delay of which is estimated from the fan-out number and the provisional / real wiring length. The delays of a gate in the last two modes are calculated for each input pin under the consideration of both the loading condition and the input waveform[3][4].

Moreover, any sequential circuits in the LSI are checked on the simulation results by the timing verifier "SIMCHK" in various timing margins such as set-up times, hold times, spikes and hazards. The timing margins are verified in the two different situations, that is for the skew error between two input pins and for the logic error caused by racing between two logic paths. These two facilities are effective for the precise delay analysis and the minute timing design.

IV. Implementation of RAM

To constract a large system by gate array LSIs, it is important that various configuration of RAM with high speed and low power consumption are easily implemented in the LSI. RAMs on internal basic cells have no constraint on the configuration or bring about no remain, while they seems to be inferior in the point of the complexity and the power delay product compared with the RAMs implemented by exclusive areas. The gate isolation has supressed the demerit of RAMs on basic cells, as described bellow.

Memory Circuit Configuration

Figure 3 shows a microphotograph of the memory cell and figure 4 shows the memory circuit for one column. The CMOS memory cell was constructed by six transistors. Two bit lines run toward the column direction and the sence amplifier detects the voltage difference between the two bit lines. The precharging of bit lines is executed by the inverters in the data input circuit. The gate isolation has realized one memory cell by four basic cells which is two times denser than the conventional oxide isolated configuration with two transistor pairs in the basic cell.

The standard cell was composed of 1b x n word memory cell, the input circuit, the sence amplifier and output circuit which correspond to one column. The block diagram of 4b x 32 word RAM circuit was shown in figure 5. The address decoder circuit is constructed by standard cells. The word lines are routed in the wiring regions between the cell arrays by automatic wiring program.

RAM Repertry

Table 1 shows the standard cells for RAM. Any configurations of RAM, bit counts per one word and word counts, are able to be implemented by using these cells. As shown in table 2 the standard macros were also prepared to shorten the design time. These macros are applicable to any smaller word length than the ones in the table without any useless cells owing to the cell reduction facility described before. The read access time of 20 ns, minimum write pulse width of 15 ns, cycle time of 20 ns were obtained. The power consumption was 140 mw/ 16 bit at VDD=5V and 5 MHz operation, in 16b x 32 word RAM.

V. An LSI Application

A 32b ALU with RAM was implemented in the 8K gate CMOS gate array. Figure 6 shows the block diagram. Two 32b x 32 word RAMs were included in the LSI and they correspond to the registers for the ALU or the two port RAM equivalently. The RALU has the same functions as the system constructed by eight Am 2901s [5] except the point of adopting two level synchronous circuits. The ALU and the control circuits was constructed with eight 4 bit-slice blocks. To attain high speed operation, three level carry look ahead generator was adopted to the ALU expansion.

The shift register latches (SRL) are used for all memory circuits, that is A latch, B latch and Q register except the RAMs, and they are controled by two different clocks which are never superposed each other. These SRLs are connected to become a shift register chain for scan design.

The 32b RALU occupies 20,773 basic cells or 86% of all. The layout procedures are completely performed automatically. The input pin count is 63 and output is 40, including 4 input pins and 1 output pin for LSSD.

VI. Conclusion

An 8K gate CMOS gate array with gate isolation configuration has been developed intending to apply it for large systems such as CPUs in computers. The LSI attains 1.5 ns/gate, and also can implement RAMs efficiently. A 32b RALU based on LSSD was realized in the LSI by the fully automatic layout.

Acknowledgement
The authers wish to thank Drs. H. Oka, Y Gamoh, and H. Nakata for their encouragements and Drs. K. Sato, S. Murai, and Y. Kuramitsu for their helpful discussion.

References

[1] Ohkura, I., et.al.,"Gate Isolation - A Novel Basic Cell Configuration for CMOS Gate Array", Proc. 1982 CICC, p.307.

[2] Sakashita, K., et.al.,"A 10K Gate CMOS Gate Array with Gate Isolation Configuration", Proc. 1983 CICC, p.14.

[3] Ohkura, I., et.al.,"A New Exact Delay Logic Simulation for ED MOS LSI", Proc. 1980 ICCC, p.953.

[4] Tokuda, T., et.al.,"Delay Time Modeling for ED MOS Logic LSI", IEEE Transactions on CAD/ICAS (will be published in July 1983).

[5] AMP Corp., Sunnyvale CA.,"Bipolar Microprocessor Logic and Interface Data Book",1981.

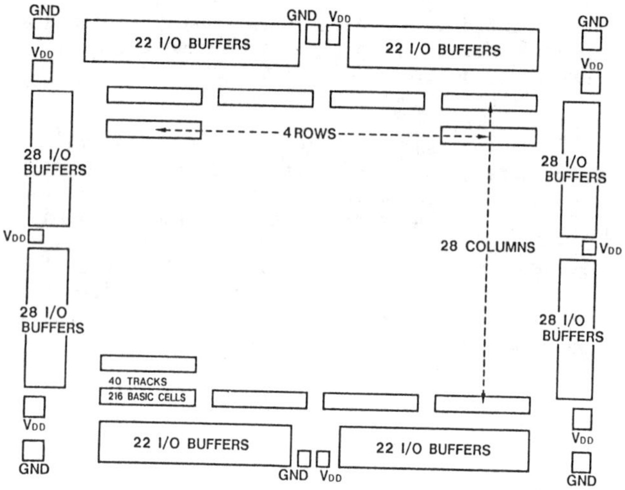

Figure 1. Chip Structure

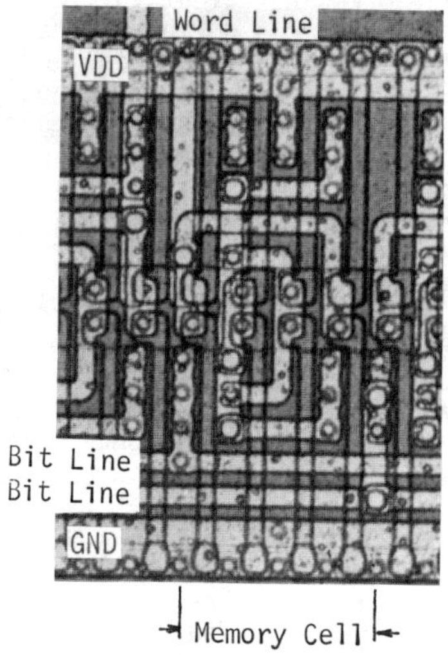

Figure 3. Memory Cell

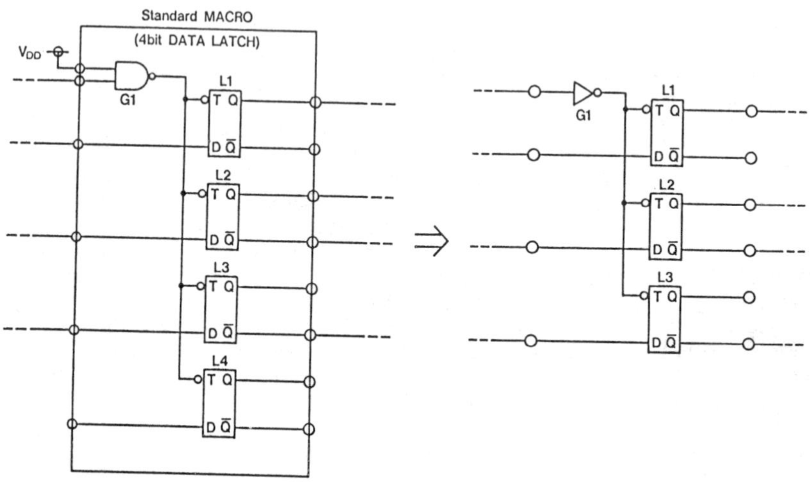

Figure 2. Cell Reduction Facility

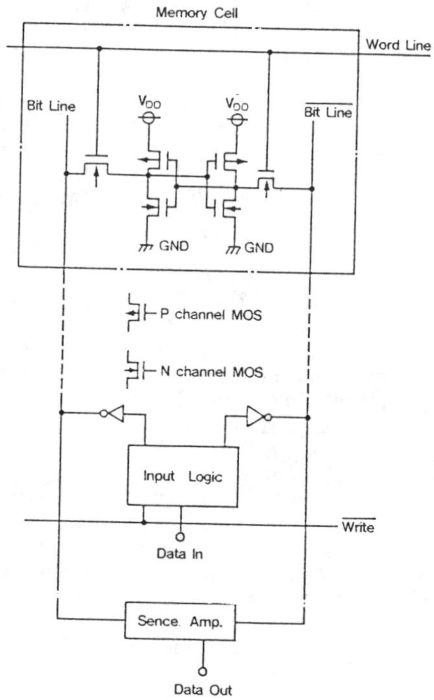

Figure 4. Memory Circuit (1 Column)

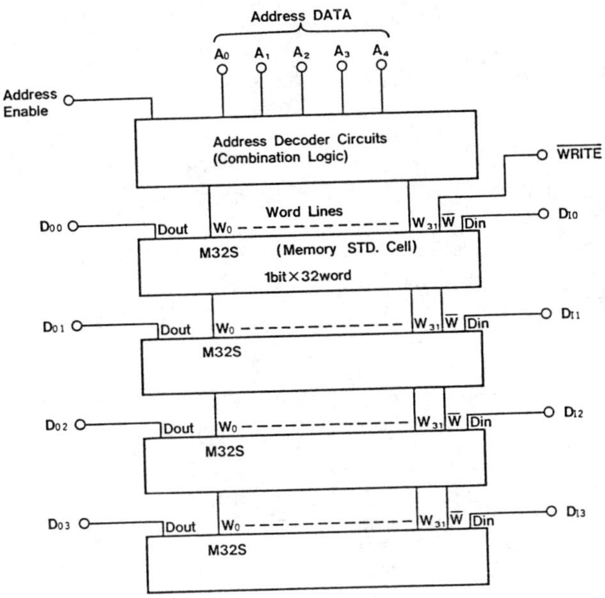

Figure 5. 4b x 32w Memory Circuit Configuration

Cell Name	Configuration	Basic Cell Count
M32S	1 bit x 32 word	147 BC
M16S	1 bit x 16 word	83 BC
M08S	1 bit x 8 word	51 BC

Table 1. Standard Cell Repertory for Memory Circuit

Configuration	Access Time	Basic Cell Count
8 bit x 128 word	30 ns	5,707 BC
16 bit x 64 word	30 ns	5,609 BC
32 bit x 32 word	25 ns	5,397 BC
32 bit x 16 word	20 ns	3,059 BC

Table 2. Standard Macro Repertory for Memory Circuit

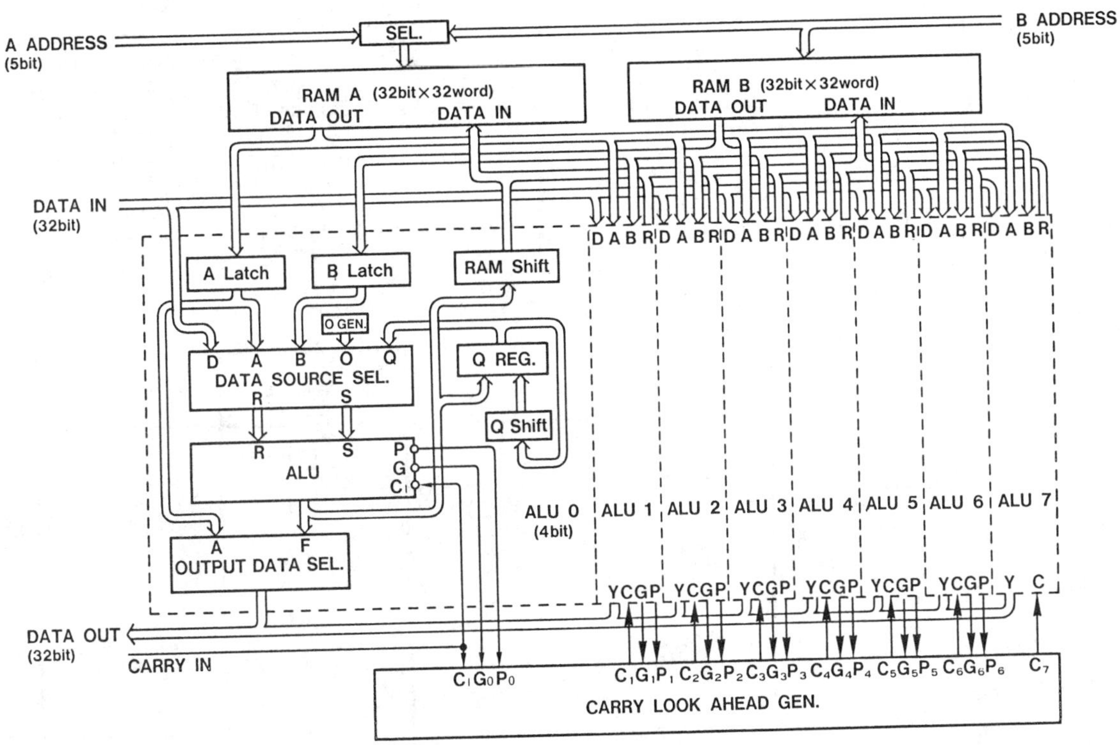

Figure 6. 32 bit RALU Block Diagram

STATUS AND PROSPECTS FOR BIPOLAR ECL GATE ARRAYS

*Mitsuo Usami *Satoshi Hososaka *Akio Anzai *Kanji Otsuka
**Akira Masaki ***Shingo Murata +Mitsuru Ura ++Mitsuo Nakagawa

*Device Development Center **Central Research Laboratory
***Kanagawa Works +Hitachi Research Laboratory ++Katsuta Works
Hitachi Ltd., Tokyo, Japan

ABSTRACT

The performance of high speed bipolar gate arrays using ECL circuit technology as of 1983 is reviewed especially in Hitachi. To keep their improvements, advanced fine pattern process of 1.5 μm photolithography, low power supply circuit technique of Non-Threshold Logic and low thermal resistance VLSI package using Silicon Carbide ceramics are discussed. These technologies enable to realize a high speed bipolar gate array chip with high gate density of 5000 gates. An attempt is made to prospect the future of bipolar gate arrays in the coming years.

INTRODUCTION

The first application, in Hitachi, of ECL gate array (1) to digital systems started in 1974 with 160 gates of 1 ns propagation delay time as shown in Table I. With the advances of integrated circuit technology and design automation systems, in 1978, we developed 550-gate ECL gate array (2) (3) employing V_{EE}= -4.5V and V_{TT}= -2V circuitry with a subnanosecond switching delay of 0.7 ns, which was used for Hitachi's very large computer systems HITAC M-200H. This LSI using oxide-isolation 3 μm-technology was the first high density ECL gate array to be used widely in mainframes at that time.

Two major high speed ECL gate arrays following the above LSI were developed in 1979-1980, which were speed-density hierarchy chips of 550-gate (0.35 ns/gate) and 1500-gate (0.45 ns/gate) ECL gate arrays (4) for M-280H, S-810 super computing systems and other Hitachi computer series. These LSIs' main technologies were minimum 2 μm photolithography, shallow diffusion with full ion implantation and planar three-layer metallization for obtaining higher density.

To review the progress of ECL gate arrays according to years, a figure of merit defined as (propagation delay time) / (number of gates in LSI) is useful as shown in Fig. 1. The past developments of ECL gate arrays are shown by a straight line, and the tpd/gate-complexity have been decreasing every year by a factor of 0.65. Fig. 2 shows the power-delay product of ECL circuits of gate arrays, which have been also improved annually by a factor of 0.75.

Table I ECL Gate Arrays Developed in Hitachi

	160 - Gate	550 - Gate I	550 - Gate II	1500 - Gate
Developed Year	1974	1978	1979	1980
Propagation Delay Time (ns)	1	0.7	0.35	0.45
Power Dissipation (W/Chip)	1.8	4	4	4
Number of Signals	46	88	88	88
Power Supply V_{EE} (V)	-5.2	-4.5	-4.5	-4.5
Photolithography (μm)	4	3	2	2
HITAC Computer Systems applying LSIs	M - 180	M - 200H	M - 280H S - 810	M - 280H S - 810

CH1935-6/83/0000/0272$01.00 © 1983 IEEE

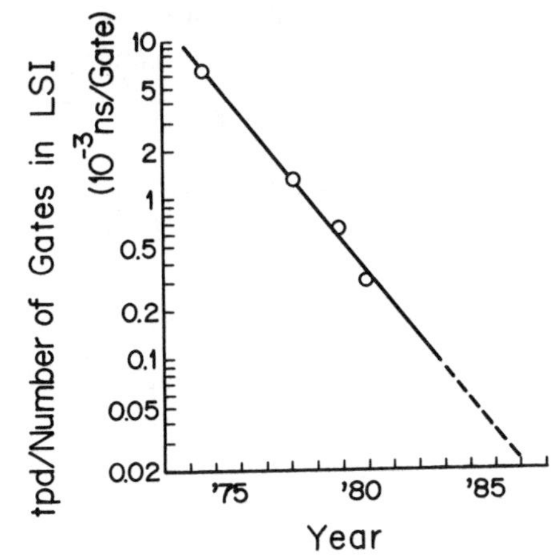

Fig. 1 Performance improvement of ECL gate arrays

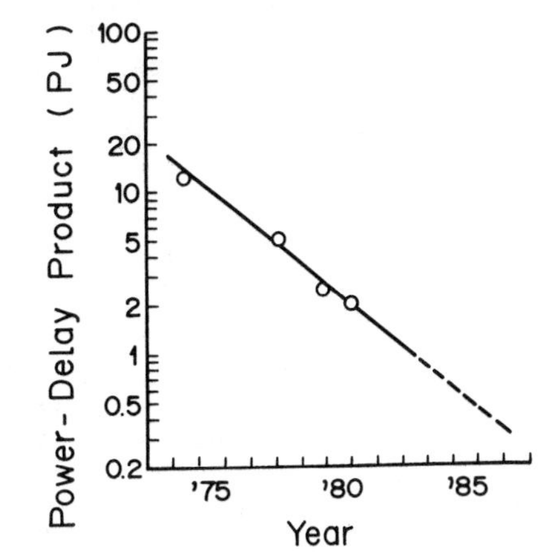

Fig. 2 Power-delay product of ECL gate

PRESENT TECHNOLOGIES

PROCESS

The high integration of 1500-gate and high performance of 0.45 ns have been enabled by the process progress such as a 2 μm-technology with a shallow diffusion and precise etching technique. Scaled oxide-isolation technology using a wafer stepper and fine dry etching techniques allows 1.5 μm groundrules, i.e., a small emitter and base contact sizes of 1.5 μm x 3 μm, and minimum line space of 1.5 μm. The transistor size of 1.5 μm-process is 38 % smaller than that of 2 μm-process. Moreover, a thin epitaxial layer of 1.2 μm and shallow junction depth using doped poly silicon technology provide small junction capacitances and high cut-off frequency as shown in Fig. 4.

CIRCUIT

Fig. 3 (a) shows the basic configuration of the ECL circuit using in ECL gate arrays. The ECL circuit provides both OR and NOR output performing an emitter-dotting wired OR and a collector-dotting wired AND function. Fig. 4 shows the simulation results of tpd vs I_{CS} using a 1.5 μm-technology process. It is proven that the ECL circuit reaches 0.25 ns at I_{CS} of 1 mA. This figure also shows

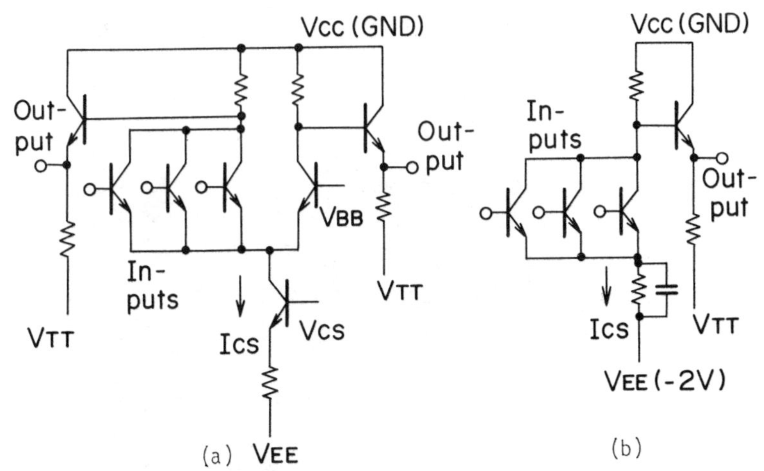

Fig. 3 (a) Conventional ECL circuit

(b) NTL circuit with an emitter-follower

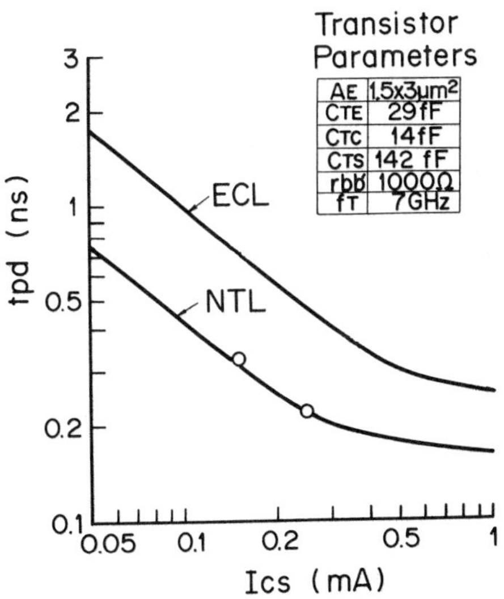

Fig. 4 Comparison of propagation delay time

gate delay of a Non-Threshold Logic (NTL) circuit (5) with an emitter-follower (Fig. 3 (b)), which is less than 50% of that of the ECL at the same I_{CS}. In order to reduce a gate power, the NTL circuit with a low supply voltage of -2V and a high speed operation is the same attractive circuit technology as series gating.

PACKAGE

A novel material of Silicon Carbide (SiC) ceramics (6) has adaptable characteristics to form a VLSI package, because that SiC ceramics has better thermal conductivity of 270 W/m°C than metal aluminum of 230 W/m°C and thermal expansion coefficient of 3.7 x 10^{-6}/°C equal to that of silicon chips of 3.5 - 4 x 10^{-6}/°C. The structure of the package is very simple, that is, a large size VLSI chip is mounted directly on a SiC ceramics substrate which provides good heat transfer properties from the chip to the surface of an aluminum heat sink attaching to the back side of the package. Fig. 5 shows the thermal resistance of a VLSI flat package applying SiC ceramics of 2 x 2 cm^2 is able to be reduce to less than 3°C/W in order to dissipate high power of 4W or more.

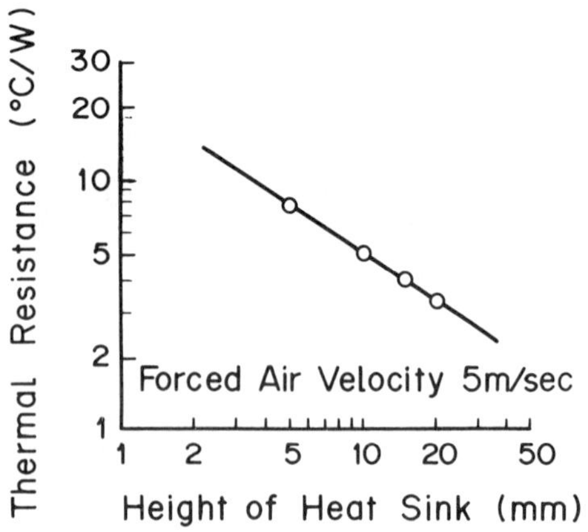

Fig. 5 Thermal resistance of a SiC ceramics package

A VLSI TEST CHIP

To evaluate the above mentioned technologies, a tentative 5000-gate gate array chip was fabricated as shown in Fig. 6. The features of this circuit test chip are summarized in Table II.

EVALUATION CIRCUITS

This chip consists of three sections as follows.

—Internal Reference Voltage Monitor/Control Section
This VLSI assumes six internal voltage levels design, i.e., two levels for internal NTL/ECL circuits and four levels for input and output ECL buffers. This section is only for margin tests.

—Ring-Oscillator Section
The propagation gate delays depending on gate types and loading conditions are measured by this section which was laid out regularly in the chip.

—Actual Path Logic Section
Actual path logic from computer system was placed to evaluate logic function operation.

Table II Features of the 5000-gate Gate Array Circuit Test Chip

Number of Circuits	5112
Internal Circuit	3-In NTL/ECL NOR Gate
Input Circuit	ECL OR/NOR Gate
Output Circuit	3-In ECL OR/NOR Gate
Power Supply	-3V, -1.8V
Signal Pins	120
Interface Level	ECL 100K
Emitter Size	1.5 x 3 μm^2 (Min.)
Pitch of Metal	5 μm (1st Layer)
	7.5 μm (2nd Layer)
Chip Size	9.3 x 9.3 mm^2
Material of Package	Silicon Carbide Ceramics
	(HITACERAM SC-101)
Thermal Resistance	<3°C/Watt

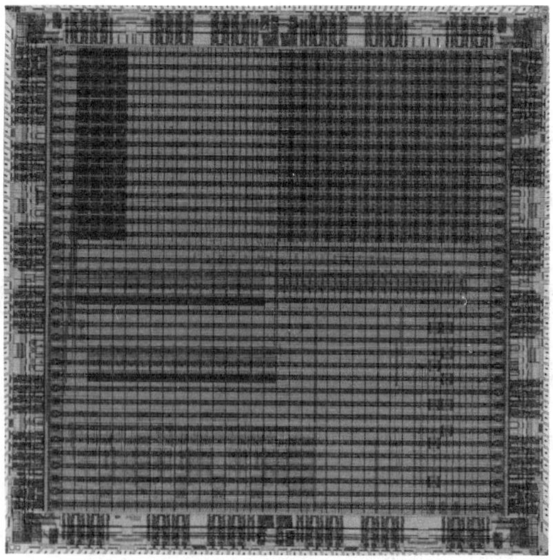

Fig. 6 Microphotograph of the gate array chip

RESULTS

Light loaded propagation delay time of the NTL gate with 3 fan-ins was 320 ps at ICS of 0.15mA and 220 ps at that of 0.25 mA at Ring Oscillation (Fig. 4). The tpd dependency on load capacitances was 230 ps/pF at I_{EF} of 0.3 mA as shown in Fig. 7.

A functional operation test by the actual logic showed ± 20 % operation range for designed power supply voltage of VEE = -3V.

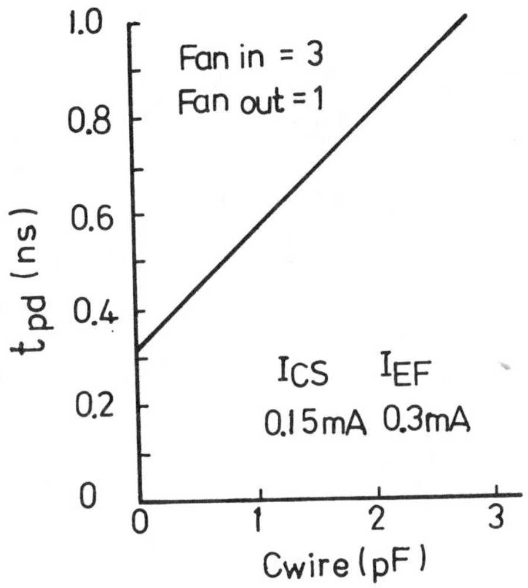

Fig. 7 Propagation delay time of an internal gate

FUTURE TRENDS

High performance bipolar gate arrays, today, go to faster switching time and higher density (7). By projecting the curves of the trends, one can estimate the status of VLSIs in the future; ultra high speed of less than 100 ps and very large scale integration of more than 10 thousand gates.

In order to achive these VLSIs, several technological breakthrough must be performed. In terms of process technology, finer pattern technique less than 1 μm (8), (9) as well as four layers metallization (10) will be indispensable. On the other hand, low voltage operation circuits will be very useful to reduce power-delay product. The new material of SiC ceramics will solve the thermal resistivity problem of VLSI packages and be used widely in several forms.

ACKNOWLEDGEMENTS

The Authors would like to thank M. Tanaka and T. Nitta, Device Development Center; T. Chiba, Central Research Laboratory; and K. Furumaya, Kanagawa Works of Hitachi Ltd. for their encouragement and advice.

REFERENCES

(1) A. Masaki, Y. Harada and T. Chiba, "200-Gate ECL Masterslice LSI", ISSCC Tech. Dig., Feb., 1974

(2) A. Hayasaka, K. Mitsusada and K. Ito, "Very-Large-Scale Integrated Circuits for Computers", Hitachi Review Vol. 3, 1980

(3) A. Anzai, T. Hirai, R. Takagi and T. Nitta, "Developments in Subnanosecond Logic LSIs", Compcon 78, Tech. Dig., Feb., 1978

(4) A. Anzai, Y. Oya, T. Nitta and T. Chiba, "The High-Speed Bipolar LSI Technology", High Speed Digital Technologies Conference, Tech. Dig., Jan., 1981

(5) H. Mukai, T. Sudo and H. Kindo, "NTL-LSI Circuit Design Consideration", Rev. Electr. Commun. Lab., NTT, Tokyo, Sept.-Oct., 1973

(6) S. Ogihara, K. Maeda, Y. Takeda, K. Nakamura and M. Ura, "High Thermal Conductivity and High Electrical Resistivity in SiC Ceramics Containing BeO as Additives", American Ceramics Society, Apr., 1983

(7) M. Suzuki, S. Horiguchi and T. Sudo, "A 5000 Gate Bipolar Masterslice LSI with 500 ps Loaded Gate Delay", ISSCC Tech. Dig., Feb., 1983

(8) N. Hanaoka and A. Anzai, "Perspective of Scaled Bipolar Devices", IEDM Tech. Dig., Dec., 1981

(9) H. Higuchi, A. Anzai, N. Honma and A. Hayasaka, "High Performance Bipolar LSIs; Their Present Status and Future", Symposium on VLSI Technology Dig. Tech. Papers Sept., 1982

(10) Klein, K., Miersch, E.F., Remshardt, R., Schettler, H., Schluz, U., and Zuhlke, R., "A Study on Bipolar VLSI Gate Arrays Assuming Four Layers of Metal", IEEE J. Solid-State Circuits, Vol SC-17, June, 1982

Session 2-1b
VLSI Verification

Chairman

H. Schichman
Bell Labs

DESIGN VERIFICATION FOR VERY LARGE DIGITAL NETWORKS BASED ON CONCURRENT SIMULATION AND CLOCK SUPPRESSION

E. Ulrich, M. Kearney, J. Tellier, S. Demba

Digital Equipment Corporation, 100 Minuteman Road, Andover, MA 01810

ABSTRACT

This paper presents a methodology for design verification of very large networks based on Concurrent Simulation and Clock Suppression. "Concurrent Case Simulation" consists of the concurrent injection and simulation of many input bit vectors into a given network model. This permits, relative to normal serial simulation, concurrent simulation of large numbers of cases or "good machines" at a very high speed. This speed advantage increases with the number of cases.

Clock Suppression is an auxiliary technique to avoid simulation slowdowns if networks with very large clock fanouts must be simulated.

Advantages of Concurrent Case Simulation are: (1) the fundamental ability to simulate many cases concurrently, "against" each other, rather than serially, (2) to observe differences between cases in a more effective and natural style than with serial simulation, (3) to run many more cases than would be possible with serial simulation, and (4) to achieve a speed advantage of 100:1 or better for concurrent simulation of 10,000 cases.

INTRODUCTION

This paper outlines Concurrent "good machine" Simulation and Clock Suppression, and their application to the problem of design verification/simulation. This establishes a methodology for "Concurrent Case Simulation."

Opportunities inherent in Concurrent Simulation [7] are concurrency, and speed based on concurrency. Thousands of individual "concurrent machines" are simulated at a very high speed because activity in the typical concurrent machine is largely identical to that of a single "reference machine". In other words, activity simulated for this reference machine represents identical activities occurring in most concurrent machines. In addition, distinct activities occurring in non-identical parts of concurrent machines need to be simulated.

Understanding the speed advantage of Concurrent Simulation is based on visualizing the time-consuming role of clocks (and other high-frequency, high-fanout signals) during normal (serial) simulations of a network, especially if the network is large. Most of the CPU time expended during a typical serial simulation (possibly 98% for a 100,000 gate network) is for clocks and clock-derived signals, and is typically expended again and again during many serial simulations of the same network. However, if N serial simulations are replaced with a single concurrent one, then N-1 re-simulations of the same routine activity are avoided. Thus, the speed advantage of Concurrent Simulation increases with both, the number of (concurrent) machines simulated and the clock fanout of the network under simulation.

In Concurrent "Good Simulation" (as originally suggested in [3]) concurrent machines are created and simulated by driving many distinct input bit vectors simultaneously into a network model. In Concurrent Fault Simulation [1,3,4,6,7,8] concurrent "bad" machines are created by injecting fault sources into the network model. Thus, the creation of concurrent machines is one difference between Concurrent Fault Simulation and Concurrent Good Simulation, but not a fundamental one. There are other minor differences, but a Concurrent Simulator of adequate generality can be used for either type of Concurrent Simulation.

The combined advantages of Concurrent Simulation (primarily with respect to normal serial one-machine-at-a-time simulation) and of Clock Suppression are:

1. A speed advantage of at least 100:1 when 10,000 input patterns are simulated simultaneously, increasing to 200:1 or better when scalar (gate) simulation is supported with vector simulation. Due to Clock Suppression, this speed prevails even for large networks.
2. All concurrent machines are advancing synchronously through simulated time, and are simultaneously available for observation. Furthermore, individual machines (i.e., more than one!) may be used as "relative observers", of other machines. This is a powerful feature which is impossible with serial simulation.
3. Accuracy beyond conventional standards is possible due to the speed of concurrent simulation. This is based on the ability to use detailed primitive models with high inherent (state and timing) accuracy. Primitives which can be used without the usual fear of simulation slow-downs include transistor models.
4. If a concurrent simulator is suitably implemented, the accuracies of the reference machine and of concurrent machines are identical. This makes simulation results repeatable, regardless of which machine is the reference machine.
5. The simulation of distinct cases of arithmetic or logical functions, and of macro- or micro-instructions, is a natural application for Concurrent Simulation. Thousands of cases can be simulated concurrently. Case comparisons, to find differences and similarities between related cases, are easily performed. The option to have several "relative observer" machines, each observing several other machines, is especially suitable for case simulation.
6. The possibility exists to "randomize" a given input sequence, thus creating a number of slightly different input sequences automatically. Simulating these input sequences, i.e., related cases, concurrently is likely to produce distinct behaviors in the various machines created. This method of simulating the consequences of slightly different input sequences has potential applications for automatic random machine initialization, automatic random test pattern generation, and automatic random timing verification.

I. THE CONCEPT OF CONCURRENT SIMULATION

Concurrent Simulation is based on the fact that the average small change to the structure or to input patterns of a digital network will produce proportionally small differences in states or behavior of the network. For example, the average stuck-at-fault in an average network is likely to produce a few faulty states in the fault's down-stream neighborhood rather than pervasive consequences throughout the network.

Useful concepts in Concurrent Simulation are the reference machine, implicit concurrent machines, and explicit concurrent machines. The reference machine exists continually in simulated time and consists of the entire network, of its input pattern, of all quiescent and active elements and their nodal states. Each Concurrent machine consists of a normally very large "implicit" part identical to its counterpart in the reference machine, and a normally very small "explicit" part distinct from its counterpart in the reference machine. Only the explicit part, and this is crucial for simulation speed, and storage efficiency, is explicitly represented and simulated. Figure 1 shows a conceptual viewpoint of reference machine R and a single concurrent machine.

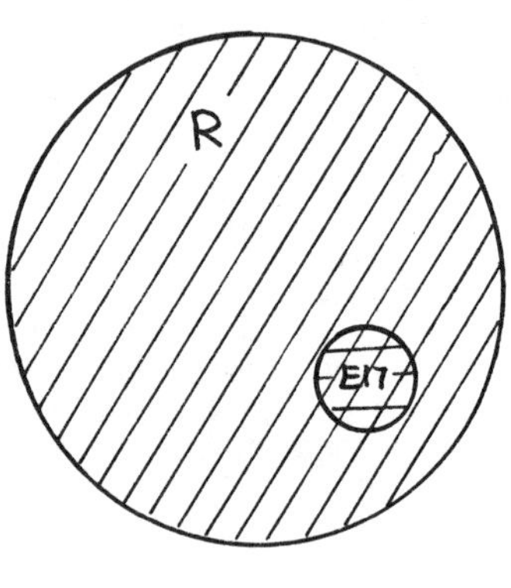

Figure 1: Reference machine R and the explicit part (E17) of concurrent machine C17 are shown.

A relevant fact for concurrent simulation is to distinguish quiescent and active network parts. The typical activity in any digital network usually remains below 5% [3,11], and this holds for concurrent machines.

As a simulation advances through time, small "elements" of the implicit concurrent machines become distinct from the reference machine and are "diverged" as "explicit elements". These elements are again "converged" (they disappear) if they become identical to their counterparts in the reference machine. The information which is diverged includes an identification number for the particular concurrent machine involved, input and output states of individual gates (or the state vectors representing registers and memory words), and timing information about the particular element involved. Most elements being diverged or converged are idle, but active elements are also diverged or converged.

Although concurrent fault simulation and concurrent good simulation are fundamentally the same, the sizes of the explicit concurrent machines created are different. During fault simulation a network is usually simulated subject to single-bit fault sources which tend to create "unintelligent" faulty machines, i.e., small explicit concurrent machines. These small machines resist being observed (detection), and typical fault simulations require hundreds or thousands of clock cycles to create temporarily large explicit machines which are detected.

In contrast, for concurrent good simulation the sources of concurrent machines are meaningful multi-bit input patterns (vectors) which quickly create relatively intelligent good machines, i.e., large explicit machines which are easily observed simply because they are large. These good simulations require very few clock cycles, i.e., only a few macro- or micro-instructions need to be simulated. Thus, although explicit good machines are larger than explicit faulty machines, the CPU time for concurrent good simulation is much less than that for concurrent fault simulation.

A very important practice in Concurrent Simulation is the "dropping" of individual concurrent machines after they have been observed. This has resulted in order-of-magnitude CPU time savings for fault simulation, and will also be helpful for good machine simulation.

II. CONCURRENT CASE SIMULATION

Case simulation is a methodology to verify a new digital design. It consists of simulating, observing, and comparing distinct but similar cases of arithmetic, logical, or control functions, or of doing this for individual macro- or micro-instructions. Simulating related cases concurrently rather than serially is an opportune application for Concurrent Good Simulation. For example, an adder circuit may be tested in a rather natural style by adding N distinct sets of numbers concurrently, and comparing the results. This is handled as N concurrent machines. Also, if the adder uses M memory locations to obtain data and R registers to store results, this would lead to N*(M+R) different cases.

The simulation of cases typically leads to successive simulations, evolving from simple cases to more complex ones. For example, for instruction simulation one may start off with "No-ops" and progress via "Add" and "Subtract" to "Multiply" or "Divide". This evolution will usually be accompanied by a gradual increase of the network under design, and by an increase in the number of cases. Some of the benefits of Concurrent Case Simulations are the following:

(1) The capacity for handling large numbers of cases.

(2) The ability to perform case comparisons DURING the simulation rather than AFTER performing two or more serial simulations.

(3) The concept of "relative observation". For example, the addition 0+0 may serve as an observer for the additions 1+1, 100+100, etc. By reporting only the difference between such related cases will it be possible to reduce the total simulation output. An even more powerful use of relative observation is to cause concurrent machines to compute identical results, and to report simulation outputs only if this fails to occur. For example, the additions 0+1000, 1+999, 2+998, etc. will normally produce identical results. Nothing will be reported if these identical results occur, but any case producing a result different than 1000 will be reported.

(4) The ability to "build up" the number of cases during successive simulation runs, and allowing the simulation/debugging of new "foreground" cases while still retaining and running old "background" cases to assure that they run as during previous simulations. This could be done with the help of case "signatures" which are generated for individual concurrent machine, are forwarded from one simulation run to the next, and are compared for equality. Of course, the number of background cases could gradually be reduced

from one simulation to the next, thus merely maintaining a small sample of the background cases.

III. IMPLEMENTATION HIGHLIGHTS

At this time there are approximately 25 Concurrent Fault Simulators in existence, including various versions at GTE Labs [6,7], 2 at Bell Labs, and one at DEC [5,8,9]. Most of the important implementation problems have been adequately solved, with most solutions applying equally to Concurrent Good Simulation and Concurrent Fault Simulation. The specific features mentioned here have been implemented in DEC's DECSIM Simulator. The DECSIM experience suggests that the Concurrent Algorithm itself provides only part of the efficiency, with the following implementation features contributing significantly to overall power and efficiency:

- Reliance on table-lookup methods, particularly "zoom" and "traverse" tables [9]. This gains speed and programming simplicity, i.e., replacement of many serial decisions with a few parallel ones, at the expense of table storage.
- The implementation strategy of list-traversal. This is based on the fact that (circular) linked lists are the natural data structures for concurrent simulation, and that the simulation overhead largely consists of the coordinated traversal of a number of lists. These traversals must be flexible and fast, and "multi-list traversal" has become an important implementation strategy.
- An opportunistic method of processing simultaneous signals emitted from "fraternal" concurrent elements [8], leading to overhead-sharing for the simulation of similar explicit concurrent machines. This method is crucial, leading to significant reductions in CPU time when concurrent machines are unusually active.
- The use of a relatively small fixed-size storage block to contain the required information for a typical explicit concurrent element. This storage block is generous for scalar information but just large enough to contain the information for a typical bit vector, e.g., 32 to 36 bits. Using of a fixed-size storage block avoids the complexities of garbage collection typical for list processing programs.
- Unusual emphasis on observation and statistics gathering. The objective is convenient and efficient observation, but also automatic program decision making (such as termination) based on statistics.

IV. CLOCK SUPPRESSION

Clock suppression [10] is a concept which can save large amounts of simulation CPU time for large networks with large clock fanouts. Clock suppression may be regarded as an extension of the method of selective tracing [11] of activity. While selective tracing avoids the unnecessary simulation of inactive regions of a network, clock suppression avoids the unnecessary simulation of routine activity, i.e., activity that provides no information except timing. Some background facts regarding clock suppression are the following:

For a typical network, the clocks and the data input paths should be viewed as distinct origins of activity. For large networks these origins produce activity (for a logic simulator) in sharply divergent proportions. For example, an instruction channel injects a single instruction at a time into a network and the resulting activity will be whatever is required to decode and execute this instruction. An instruction channel has a FIXED width of, say, 80 bits per micro-instruction, amplified by a typical wire fanout of approximately 2.5. The average information transfer rate will only be approximately 100 bits per micro-instruction because typically only 50% of these bits will change from one instruction to the next. The number of signals (events) produced in response to this should be normal and will rarely exceed 200. In contrast with the fixed channel width of the instruction stream, the clock channel width is proportional to network size and is exactly equal to the total clock fanout; i.e., it may be 50 for a 500-gate network. For each clock fanout point there will be two signal changes per clock cycle, and the average information transfer rate will be, assuming a fanout of 50, 100 bits per clock cycle. Thus, with an injection rate of one microinstruction per clock cycle and a clock fanout of 50, the ratio of clock-produced activity to instruction-produced activity should be roughly 100:100 or 1:1 for a network as small as 500 gates. For larger networks with clock fanouts of 500 or 5,000, this ratio would increase to 10:1 or 100:1, respectively. For networks with two or more clocks per micro-instruction these ratios would increase proportionally.

Of course, when an increase in network size is coupled with an increase in the instruction channel width, then these ratios would decrease.

Clock suppression consists of (a) suppressing the delivery of the clock itself and of (b) suppressing (by default) additional activity indirectly caused by the clock in further levels of logic.

One possible method of implementing clock suppressions is a mechanism where simple sequential clock-driven circuits such as transmission gates, flipflops, and registers (but no combinational logic!) remain DISCONNECTED from their clock source, but are briefly connected if and when such a circuit receives an input signal other than the clock.

V. SPEED

Optimally implemented, the efficiency (speed per individual machine) of Concurrent Simulation increases with the number of concurrent machines. This is based on the facts that each extra concurrent machine is largely identical to the reference machine, while also having a high probability of being partially identical to other explicit concurrent machines. The cost of simulating each additional implicit concurrent machine is always zero, while simulating each additional explicit concurrent machine is usually overhead-shared with that of simulating other nearly identical explicit concurrent machines.

An approximation for the effective concurrent simulation speed (with DECSIM fault simulation) is Sc=C*K*Sb, where C is the number of concurrent machines, K is the overhead for concurrent simulation (which typically ranges from 0.1 to 0.01), and Sb is the basic simulator speed in events/sec. For 10,000 concurrent faulty machines and an overhead of 0.05 we have Sc=500*Sb, a speedup of 500:1 over serial simulation.

The speed of concurrent fault simulation cannot be equaled for concurrent good machine simulation, i.e., explicit concurrent good machines will be larger than explicit concurrent faulty machines and will thus require more CPU-time. Analytical estimates and experiments indicate that typical explicit concurrent good machines are likely to be 5 times larger and proportionally more active than explicit faulty machines. Therefore, for 10,000 good machines we have approximately Sc=100*Sb, a speedup of 100:1 over serial simulation.

The DECSIM simulator basic speed (Sb) is roughly 5,000 events/sec. (on a VAX 11/780), and Clock

Suppression appears promising to approximately maintain this speed for typical networks, even for those with large clock fanouts. Therefore, the effective speed for 10,000 concurrent faulty machines is approximately 10,000*0.05*5,000=2,500,000 events per second, and the effective speed for 10,000 good machines is roughly 10,000*0.01*5,000=500,000 events per second. However, these speeds are for pure scalar (gate) level simulation, and are at least twice that for combined scalar/vector level simulation. Thus the effective speed for simulating 10,000 concurrent machines will range from one million to five million events per second. Given this speed, the concurrent simulation of 10,000 micro- or macro-instructions can be done in minutes on a VAX 11/780.

It is interesting to compare these simulation speeds with the fastest potential alternative, i.e., serial "simulation engines" such as ZYCAD or those proposed in [2]. Simulation engines are expected to achieve a speed of one million events per second, but large clock fanouts will probably deteriorate this speed at the same rate as that of software simulators. Therefore the combination of Concurrent Simulation/Clock Suppression is likely to be faster than engine simulation. Also, the use of a faster machine than the VAX 11/780 shifts the advantage further towards Concurrent Simulation/Clock Suppression.

The merits of these competing approaches become clearer by considering situations where few or many machines must be simulated. Hardware engines will be indeed substantially faster when only a few thousand relatively small machines are to be simulated, but speed is of very little importance in these cases. Concurrent Simulation/Clock Suppression becomes increasingly faster for very large numbers of large machines, i.e., when speed becomes vital.

CONCLUSIONS

Concurrent Simulation, which has been successful for fault simulation, was outlined as being applicable to "good machine" simulation. The advantages of this method are speed improvements ranging from negligible to extremely large, depending on the number of concurrent machines and on clock fanout.

Concurrent Case Simulation, as a methodology for design verification/simulation, was outlined. It was estimated that a minimal speed advantage of 100:1 would prevail for 10,000 concurrent cases. Using vector and scalar simulation in combination, an effective speed advantage of at least 200:1 for 10,000 cases would be typical. The effective simulation rate would be one million events per second or better, and the simulation of 10,000 macro- or micro-instructions would be done in minutes of CPU time on a VAX 11/780.

A more important advantage than speed is the fact that running cases concurrently, or against each other, establishes a powerful design verification methodology.

A question that may be asked is whether or not it is excessive to simulate thousands of cases. An answer to this is to consider simulation of the instruction set of a modern computer. A typical instruction set has 256 instructions, and simulating each instruction with at least 50 different parameters (data, registers, index registers, etc.) appears reasonable. Simulating 10,000 distinct cases is therefore not excessive.

A comparison of the speed of concurrent simulation vs. hardware simulation engines (for 10,000 machines) showed no clearcut advantage for either approach. However, simulation engines are indeed faster for small numbers of small machines and concurrent simulation is faster for large numbers of large machines. Hardware simulation engines, to surpass the speed of software, will need to incorporate concurrent algorithms and Clock Suppression into hardware. Of course, the superior flexibility of software over hardware remains an important point in favor of software simulators, this being similar to the superior flexibility of high-level languages over assembly languages. In other words "program" changes to a hardware simulator are clearly a difficult problem. Given the speed and flexibility of concurrent simulation, the need for simulation engines is questionable.

A very important side-effect benefit of Concurrent Simulation is the fact that a single simulator would be used for Design Simulation and Fault Simulation. The convenience gained and the extra work avoided by employing a single simulator and therefore using a single network model is substantial.

Finally, it should be said that many megabytes of primary memory are required to do the things described here.

ACKNOWLEDGEMENTS

The authors thank Jack Arabian for reviewing this paper.

REFERENCES

[1] Abramovici, M., et al. "Concurrent Fault Simulation and Functional Level Modeling." 1977 Design Automation Conference (DAC).

[2] Abramovici, M., et al. "A Logic Simulation Machine." 1982 DAC.

[3] Breuer, M. and A. Parker. "Digital System Simulation: Current Status and Future Trends." 1980 DAC.

[4] Breuer, M. and Friedman, A. "Diagnosis & Reliable Design of Digital Systems."

[5] Phillips, N. and Tellier, J. "Efficient Event Manipulation - The Key to Large Scale Simulation." 1978 IEEE Test Conference.

[6] Schuler, D. et al. "A Program for the Simulation and Concurrent Fault Simulation of Digital Circuits described with Gate and Functional Models." 1979 IEEE Test Conference.

[7] Ulrich, E. and T. Baker. "The Concurrent Simulation of Many Nearly Identical Digital Networks." 1973 DAC.

[8] Ulrich, E. et al. "High-speed Concurrent Fault Simulation with Vectors and Scalars." 1980 DAC.

[9] Ulrich, E. "Table-Lookup Techniques for Fast and Flexible Digital Logic Simulation." 1980 DAC.

[10] Ulrich, E. and D. Hebert. "Speed and Accuracy in Digital Network Simulation Based on Structural Modeling." 1982 DAC.

[11] Ulrich, E. "Time-Sequenced Logic Simulation Based on Circuit Delay and Selective Tracing of Activity." 1965 ACM National Conference.

A Formal Verification Method for Hierarchical VLSI Designs

Kyushik Son
Zen Kishimoto

Computer Science Laboratory
GTE Laboratories, Inc.
40 Sylvan Road
Waltham, Mass. 02254

Abstract

This paper presents a formal verification method that proves the functional correctness of a VLSI design by using a description of the design written in a Hardware Description Language (HDL). In this method, functional abstraction is conducted by using symbolic simulation to prove the correctness of inductive assertions of functionality. These inductive assertions are defined by the designer as the verification process proceeds through the intermediate levels of a hierarchically decomposed VLSI design.

I. Introduction

The advent of computer-aided simulation techniques has made it possible to verify an intended design by computer simulation, increasing the level of confidence in new logic designs. This logic simulation approach is based on the gate and functional level models of the elements in the logic circuit.

Logic simulation for verifying the functional correctness of a design depends on a set of inputs. These inputs are normally selected by a designer according to the specification of the intended logic. The verification is performed by manually comparing the simulation outputs to the expected outputs of the design.

Logic simulation, however, is inadequate for verifying the complete set of input and output relationships. The success of logic simulation as a design verification tool clearly depends on the completeness of the defined input and output relationships. With VLSI systems, logic simulation becomes too CPU intensive to verify the complete set of those relationships.

Several new approaches for formally verifying the functional correctness of VLSI designs have been proposed throughout industry and universities. Of these approaches, two are particularly noteworthy: logic comparison from functional abstraction and formal software verification methods.

The verification approach using logic comparison from functional abstraction can be considered a bottom-up method. It involves deriving the equivalent function from the structural description of a design and then performing a logic comparison between the abstracted function and the corresponding high level specification for the functional equivalence. The functional abstraction can be done either by using transformation rules for Boolean expressions or by defining the structural distance of two different descriptions.

Hardware verification based on formal software methods has become popular because the high level designs of VLSI systems are more frequently specified using HDLs that are similar to programming languages. Two of these methods, known as symbolic simulation and the inductive assertions approach, are particularly promising for the verification of VLSI designs.

Symbolic simulation is an extension of conventional logic simulation in which the simulation is performed using symbolic rather than actual input values. The results of the simulation will be symbolically expressed subfunctions which are partitioned according to the conditions required to perform each subfunction. The verification process using symbolic simulation thus involves confirming each subfunction expressed in terms of initial symbolic values with the intended logic.

The success of a symbolic simulation relies on two key factors: the complexity of the symbolic computations and the feasibility of the construction of the intended logic for proving the final symbolic results. To solve the problems related to these two factors, a symbolic expression simplifier should be further developed to reduce the complexity of symbolic computations. Also, another mechanism to prove the correctness of the symbolic results must be considered, rather than simply comparing two different implementations of the same logic.

The formal method of design verification using inductive assertions involves proving the functional correctness on the basis of assertions planted between statements in the description of the intended logic. Proving these assertions using the

CH1935-6/83/0000/0281$01.00

axioms for given programming constructs forms the kernel of this method. However, theorem provers used in program verification must be adapted to the unique features of hardware description languages. Particularly, the timing characteristics of hardware design should be incorporated within these theorem provers. So far, no generalized algorithm for this theorem proving technique has been developed.

To implement a verification process for a hierarchical design methodology, we consider both symbolic simulation to verify the data flow of the overall functionality and inductive assertions to validate the algorithmic description of hierarchically decomposed functional blocks.

The following section discusses a general verification process that uses both symbolic simulation and inductive assertions. In Section III, a procedure for validating the assertions planted within functional blocks will be discussed and illustrated with an example, showing the verification procedure for the modified Booth's multiplication algorithm.

II. Verification of Hierarchical VLSI Designs

In a hierarchical design methodology for VLSI systems, an entire system is designed using functional decomposition. The design can be functionally decomposed according to various factors. These can be, for example, a syntactic hierarchy, where ease of specification and readability are the main constraining factors, a geometric hierarchy, where the geometric constraints necessary to implement the logic design govern the decomposition, or a spatial hierarchy, where the spatial locality of related logic is the major consideration.

Whatever factors are applied, the consistency between hierarchical levels must be verified to prove the functional correctness of the entire system, and each level must be proven to be correctly implemented. Therefore, the hierarchically decomposed components must be interconnected according to the data-flow mechanism of the design, and each component should be algorithmically correct. Thus, the verification process for hierarchical designs must perform two distinct steps. The first is to verify the data-flow mechanism between functional blocks, and the second is to validate the algorithmic implementation of each functional block.

In the method proposed here, a symbolic simulation technique is used for the first step and an inductive assertion approach is used for the second step. In the following, the general procedure using these two methods will be described.

In symbolic simulation [1], symbols instead of actual values are assigned to input variables of the HDL description, and simulation is performed using these symbols. A symbolic simulation proceeds in the same way as a logic simulation, except for the handling of decision points in the HDL description. Since each variable is assigned initial symbols, the outcome of a decision point cannot be determined naturally. Hence, a particular path from a decision point must be selected by the user. This selection leads to a condition for selecting the outcome. Successive encounters of decision points along the path result in the accumulation of these outcome conditions. When the simulation is terminated, the accumulated conditions constitute the predicate of a subfunction implemented by following the path.

The results of the symbolic simulation along each path are stored in a symbolic simulation tree. The corresponding predicate and computation information of each path is stored at each leaf node of this tree.

Since the symbolic simulation supplies information about these subfunctions, it can be used to derive the conditions and computations constrained at the inputs and outputs of a functional block. These constraints are known as the input and output assertions.

The input and output assertions for each module corresponding to a functional block can be embedded at the intermediate levels of the hierarchical decomposition. These intermediate assertions are known as inductive assertions because they are used to inductively prove the functional correctness from inputs to outputs.

Let an inductive assertion A be injected at a location P in the module. When the result of executed conditions from the input assertions satisfies A at P, and when the result of executed conditions from A satisfies the output assertions, we conclude that the module is verified. This is because the module is proven to be correct from the entry to P and then from P to the exit, so by induction the module must be correct from the entry to the exit.

Our verification method starts with the modules corresponding to functional blocks in the lowest level of the hierarchical decomposition (i.e. the primitives). For these modules, there are no "call" statements invoking any other modules. Symbolic simulation is applied as in the verification process previously described.

After all the modules corresponding to terminal functional blocks are verified, any modules that only call already-verified modules can be selected for verification. The same verification process is applied to these modules, except for additionally verifying the "call" statements to other modules. When a call statement is encountered, outputs are computed using the input and output relationships of the called module: the call statement is substituted with those outputs. Since all the called modules are already verified and the input and output relationships are available for those modules, the substitution can be done for all the "call" statements. This process is repeated until the module corresponding to the entire description is verified.

Since symbolic simulation tends to produce lengthy and complex symbolic expressions, it is vital to have a good mechanism for simplifying the expressions. Moreover, in order to show that the result of a simulation satisfies an assertion, a good mechanism for validating the assertion by using symbolic expressions is also required. The next section describes such a mechanism.

III. Validation of the Assertions

The inductive assertions technique, known as Floyd's method [2], utilizes the input and output assertions to prove verification conditions. These verification conditions are derived by cutting the loops and by characterizing the behavior of the loop-free subfunction. Cutting the loop by inserting an appropriate set of inductive assertions is the essence of this technique, and it is not always easy to derive that set.

To validate the inductive assertions, we need a mechanism to determine the validity of the output assertions within the limit of the input assertions. This mechanism is commonly called a theorem prover [2].

Three different methods can be considered for proving the verification conditions for inductive assertions: induction, contradiction, and deduction. The inductive method validates the relationship for stepwise computations, which are often derived by a user during the proving steps. The contradictory method shows the unsatisfiability of the negated output assertions by applying the repeated rules backwards into the input assertions. The deductive method deduces a finite set of mappings from the input assertions to determine whether or not the output assertions are valid.

It is known, however, that the validity of a predicate calculus is generally unsolvable [2]. We therefore adapt a deductive method--aided by symbolic simulation--to simplify each function with respect to the initial input symbols. The simplification steps usually cannot be automated to build a complete deduction system. Thus, interactive interventions by a user will be conducted during the theorem proving steps to prove the output assertions.

The following example demonstrates the derivation of inductive assertions for an algorithmic description of a functional block and the inferencing needed to verify the inductive assertions. This example has been selected to show that the algorithmic validations are conducted by inductive assertions, rather than by symbolic simulation. However, the validation for a data flow description that can be characterized by cause-effect relationships is easily performed by symbolic simulation [3].

Example : The arithmetic logical unit (ALU) of a computer system is an example of a functional block which can be described with an HDL. To validate the algorithm and configuration implemented for the ALU, symbolic simulation can be applied to the descriptions of the ALU, and then all the possible paths through the logic must be verified individually. However, such an algorithmic description contains iterations of some statements, which may lead to indefinite looping due to the symbolic initial values. This indefinite looping can be avoided by using inductive assertions.

In this example, the modified Booth's algorithm [4] for a multiplier is described. This algorithm is used to produce partial products for the Monolithic Memories Inc. 8 x 8 multiplier (Part No.: 67558) [5].

The modified Booth's algorithm is a method for reducing the number of iterations to generate partial products by recoding the multiplier into a higher-radix number. Thus, the number of multiplier bits shifting over will be more than one bit.

Let the unsigned 8-bit multiplicand and multiplier be X and Y, respectively. Let Y<-1> denote an extra bit padded with a logic 0 at the low-order end of Y. Then, every three bits of Y<7:-1> for 0 ≤ i ≤ 8 will be grouped, resulting in five groups. Thus, five partial products will be calculated by the designated operation specified in Table I. Each time the next operation is performed, two bits of the partial product will be shifted to the right. Figure 1 shows the description of the modified Booth's algorithm for this multiplication, written in the Instruction Set Processor Simulator (ISPS) language [6].

Note that in the ISPS character set, the symbols ' and @ denote a binary constant and a concatenation operator, respectively.

Y<i+1>	Y<i>	Y<i-1>	Operation
0	0	0	0
0	0	1	+ X
0	1	0	+ X
0	1	1	+ 2X
1	0	0	- 2X
1	0	1	- X
1	1	0	- X
1	1	1	0

Table I. Specification of Modified Booth's algorithm.

To prove the functional correctness of this description, five inductive assertions are inserted--denoted with comment characters "!"--to specify the initial condition and the sufficient conditions for four distinct operations. The five assertions are designed to confirm the initial constraint, two additions and two subtractions during the cycles.

Let %A denote the initial value of the accumulator (A), if Y<-1>, and let ?A denote the previous contents of A, if Y<-1>=1. Initially, %A=0. If Y<-1>=1 at the current cycle, then -2X was performed at the previous cycle with Y<1:-1>='100. Thus, ?A=-2X.

The verification process requires that we validate the four assertions, assuming that the input assertion is correct. Depending on the three multiplier bits, four different operations will be performed. Thus, the proving procedure will be divided into four cases. The predicate at each decision point will be deduced by performing symbolic simulation. The three-bit pattern for Y will be derived by using the initial condition: A0.

The uniform right-shifting of two bits on A@Y will be interpreted from A0, such that the partial product of A@Y will be multiplied four times when another operation at the next cycle is performed onto A. The statement "IF i=10 => LEAVE MULT" will be interpreted from the statements "i=0" and "i=i+2", such that the loop will be repeated five times. In Figure 2, each assertion will be verified using the facts discussed above. Some assertions require several cycles to satisfy the predicates.

Hence, the design for the modified Booth's algorithm has been verified by validating those assertions which are sufficient to prove the partial functional correctness with respect to the input and output assertions. The validity of the output assertions has been determined from the input assertions, aided by interactive interpretations from a designer.

Concluding Remarks

A formal verification technique that uses both symbolic simulation and inductive assertions has been presented. Symbolic simulation was performed to abstract the functionality of the design from its HDL description and to verify the cause-effect data flow. Inductive assertions have been used to validate an algorithmic description of the functional block. An example applying the technique to the modified Booth's multiplication algorithm was demonstrated. This example has shown how to avoid exhaustive looping by using the inserted assertions to perform verification.

References

1. S.L. Hantler and J.C. King, "An Introduction to Proving the Correctness of Programs," Computing Surveys, Vol. 8, No. 3, September 1976, pp. 331-353.

2. Z. Manna, Mathematical Theory of Computation, McGraw Hill, New York, 1974.

3. Z. Kishimoto and K. Son, "Symbolic Execution for VLSI Design Verification," presented at IEEE Workshop on VLSI and Software Engineering, October 1982.

4. O.L. MacSorley, "High Speed Arithmetic in Binary Computers," Proceedings of the IRE, January 1961, pp.67-91.

5. S. Waser, "State-of-the-Art in High Speed Arithmetic Integrated Circuits," Computer Design, July 1978, pp.67-75.

6. M.R. Barbacci, G.E. Cattell, and D.P. Siewiorek, The ISPS Computer Description Language, ISPS Reference Manual, Carnegie-Mellon University, August 1979.


```
  MULT(REF X<7:0>, REF Y<7:0>) :=
! A0: ∀i(((X<i>=0) ∨ (X<i>=1))
!        ∧((Y<i>=0) ∨ (Y<i>=1)))
    Begin
    A = 0, Y<-1> = 0, i = 0
    REPEAT
      Begin {TC}
      Z = Y<0> + Y<-1> - 2 * Y<1> NEXT
      IF Z = 1  => A = A + X
! A1: ((Y<1:-1>='010) ⊃ (A=%A+X))
!     ∨((Y<1:-1>='001) ⊃ (A=?A+4X))
      IF Z = 2  => A = A + 2 * X
! A2: ((Y<1:-1>='011) ⊃ (A=?A+8X))
      IF Z = -1 => A = A - X
! A3: ((Y<1:-1>='110) ⊃ (A=%A+3X))
!     ∨((Y<1:-1>='001) ⊃ (A=?A+12X))
      IF Z = -2 => A = A - 2 * X
! A4: ((Y<1:-1>='100) ⊃ (A=%A+2X))
      i = i + 2 NEXT
      IF i = 10 => LEAVE MULT NEXT
      A@Y SR0 2
    End
  End
```

Figure 1: The modified Booth's method.

```
Case 1 for A1:

      (Y<0>+Y<-1>-2*Y<1>=1)
   => ((Y<0>+Y<-1>=1) ∧ (Y<1>=0))
   => (((Y<0>=1) ∧ (Y<-1>=0)) ∧ (Y<1>=0))
      ∨(((Y<0>=0) ∧ (Y<-1>=1)) ∧ (Y<1>=0))
   => ((Y<1:-1>='010) ∨ (Y<1:-1>='001))

    If (Y<1:-1>='010),
       then A = %A + X => A = X
    If (Y<1:-1>='001) => (Y<3:-1>='00100),
       then A = ?A + 4X => A = 2X.

Case 2 for A2:

      (Y<0>+Y<-1>-2*Y<1>=2)
   => (Y<1:-1>='011)

    If (Y<1:-1>='011) => (Y<3:-1>='01100),
       then A = ?A + 8X => A = 6X.

Case 3 for A3:

      (Y<0>+Y<-1>-2*Y<1>=-1)
   => ((Y<1:-1>='110) ∨ (Y<1:-1>='101))

    If (Y<1:-1>='110) => (Y<3:-1>='00110),
       then A = %A - X + 4X => A = 3X.
    If (Y<1:-1>='101) => (Y<5:-1>='0010100),
       then A = ?A - 4X + 16X => A = 10X.

Case 4 for A4:

      (Y<0>+Y<-1>-2*Y<1>=-2)
   => (Y<1:-1>='100)

    If (Y<1:-1>='100) => (Y<3:-1>='00100),
    then A = %A - 2X + 4X => A = 2X.
```

Figure 2: Validation of the assertions.

AN APPROACH TO MIXED LEVEL SIMULATION

M. A. d'Abreu, K. L. Cheong, C. T. Flanagan

Honeywell Corporate Solid State Laboratory
12001 State Highway 55
Plymouth, MN 55441
Phone: (612) 541-2303

ABSTRACT

The process of designing VLSI systems continues to increase in complexity. The VLSI system designer can handle this complexity only through abstraction of the design process. Designs are described at different levels of abstraction, where higher levels correspond to greater abstraction from the actual implementation of the VLSI system.

The design refinement process, i.e. the reduction of the abstraction in the description, is an iterative process and has to be supported by a simulator. High level descriptions are usually simulated at the register transfer level and the designer sees the data transformation that takes place only at the end of clock cycles. As more and more structure is incorporated into the design description, technology dependencies have to be taken into account. Simulators have to support the transition between different levels of description.

INTRODUCTION

This paper describes algorithms and data structures for a simulator that supports mixing of different levels and types of design descriptions. The simulator is being implemented in stages. The primary goal for this development is to establish a good foundation for various types of simulators and thus develop a mixed level simulator. The first version of the simulator allows the designer to simulate MOS designs at the switch level as well as a combination of MOS switches and higher level primitives such as gates, registers, memories, etc. The simulator allows the primitives to have different delays. This is allowed when the design is simulated in terms of MOSFETS, as well.

The simulator supports various technologies [1]. When the level of description reaches a stage where technology dependencies have to be considered, the designer has to specify the necessary parameters.

Some of the more interesting technologies, with respect to the simulator, are Current Mode Logic (CML), Integrated Schottky Logic (ISL), and MOS. CML provides the designer with a means to implement complex logic gates by allowing the current source of the differential gate to be controlled as shown in Figure-1. Reference [1] describes how this is handled by the simulator. The ISL gate, which inverts the logic level of the input signal, has a single input and multiple outputs. [2,3]. Complex logic functions are implemented by dotting the outputs of different gates. One of the outputs does not have a Schottky diode. When non-Schottky outputs of two gates are dotted, the output value at the other outputs will be the logical product of the complement of the inputs. Thus the non-Schottky outputs behave as bidirectional signals.

Proper choice of data structures for representing the network interconnection is important for a simulator to support CML, ISL, and MOS IC design. These technologies make extensive use of wired logic. Hence data descriptors for wired outputs are needed.

NETWORK REPRESENTATION

The network topology is modeled as a collection of element and node descriptors and their relationships. Every element in the network has a descriptor which consists of the element type, name key, and pointers to its input and output descriptors. The

CH1935-6/83/0000/0285$01.00

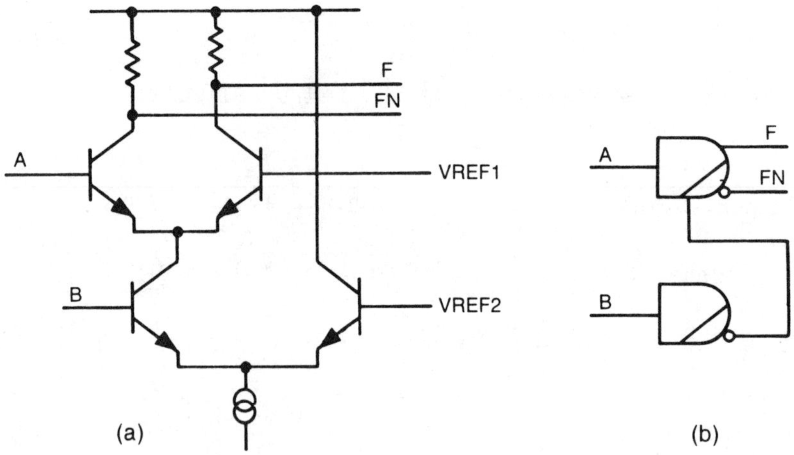

Figure-1 Series Gating to Control the Current Source.

element descriptor also contains other information that is needed during simulation. The output descriptor has a pointer to a node descriptor, in addition to other fields.

Node descriptors provide a flexible and powerful means to accurately describe relationships between various element descriptors. Nodes describe the interconnection of the elements. They can perform logical functions, can retain and decay charge, and can cause execution of user commands. Figure-2 shows the types of nodes currently defined.

Vector nodes are processed in their entirety; that is, if a component of the vector is active the state of the whole vector is propagated. The fact that some of the vector components have not changed state does not affect the efficiency of the simulator. An interface node is inserted whenever two different levels of network description are mixed. Interface nodes perform the required analog to digital and digital to analog conversion whenever these two levels of simulation are mixed. The conversion includes voltage and time scale mapping. When RTL and gate level descriptions are mixed, the conversion involves the mapping of logic states and discrete time points into logic levels and clock cycles respectively.

The timing wheel and the event lists form the core of the data structure needed by the simulator. Events are classified into four major categories: (i) events that propagate signal changes, (ii) events that cause the execution of user commands and control the simulator, (iii) events that propagate vectors, and (iv) events that control the evaluation of groups of MOS switches. The latter type of event is used during the MOS switch level simulation.

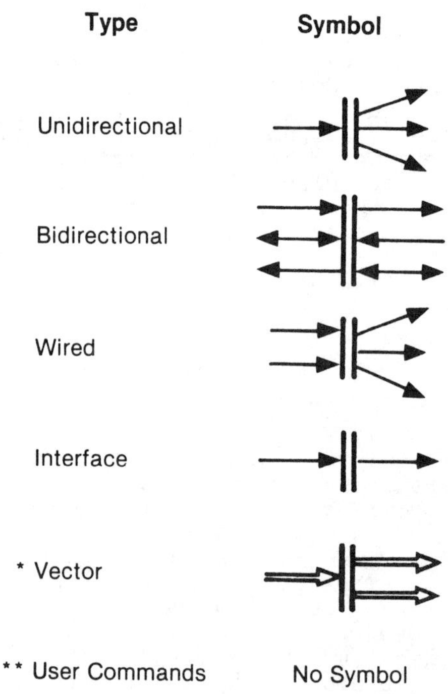

* Vector nodes can be wired and bidirectional.
** Command nodes are special class of other nodes.

Figure-2 Types of Nodes

SIMULATION ALGORITHM

The first version of the simulator allows mixing of gates and MOS switch networks. The ease of creating new node and event types is the feature that enables the simulator to mix various levels of simulation. The simulator works with the logical value together with the drive capacity of the signal. Prior to the start of simulation, static analysis, based on the technology, is performed on the network. In the case of MOS technology, MOS switches are partitioned into classes [4]. Checks are also made for the violation of logic design rules [1] thus the complexity of the simulation algorithm could be reduced.

```
Do until (max_time);
  • Do for all events at this time_step;
      - Fetch event;
      - Get event_type;
      - Goto label (event_type);
          event 1: Regular signal update;
          event 2-8: System Control, User commands;
          event 9: Switch group evaluation;
    End event_loop;
  • Process wired node stack;
  • Evaluate zero delay element stack;
  • Evaluate non zero delay element stack;
End max_time_loop;
```

Figure-3 Simulation Algorithm

Figure-3 gives a high-level description of the simulation algorithm; note that prior to the execution of this algorithm all node descriptors have been tagged according to their type. The most important steps in this algorithm in connection with this paper, are included in the processing that takes place under event-1 and event-9.

Figure-4 describes the activities that take place during the processing of event-1. After the output value is updated, the node descriptor for this signal is accessed. Processing of unidirectional nodes involves updating the input value for all node fanouts and then pushing the fanouts into the zero or non-zero delay stack. If the node is a wired node, then the node itself is stacked-up. After all the events for this time step are processed, the stacked wired nodes are evaluated. For each wired node in the stack, the new logic state is determined by the logic it performs, and then this value is propagated to all fanouts of this node. The fanouts are then pushed onto the appropriate evaluation stacks.

```
• Update value for signal;
• Get node descriptor for this signal;
• Goto label (node_type);
    Unidirectional: Process fanouts;
    Wired: Push node into wired stack;
    Bidirectional: Push node into bidirectional stack;
    Vector:
    Interface:
    User Command: Execute command, process
                  generic node;
• End;
```

Figure-4 Event 1 Processing

Bidirectional nodes are activated whenever there is a change in state at the data ports of a bidirectional element. The processing of the bidirectional stack involves the calculation of a new state for all bidirectional elements that are connected in a closed loop. The elements in a closed loop are grouped into classes prior to simulation. It must be pointed out that during the evaluation of a new state for a class, inactive elements are bypassed. If the control (gate) input to any switch changes state then an evaluation is scheduled for the class to which the switch belongs. The scheduling of an evaluation event (event-9) allows for detection of race conditions when multiple control (gate) inputs for a class switch simultaneously or within a minimum time window. Figure-5(a) shows the condition under which a class is evaluated due to a change in the data input, and Figure-5(b) shows the case for which an evaluation is scheduled for a class.

Initialization of the circuit is accomplished by performing a zero delay simulation of the entire circuit. A simulation in this mode does the required checks for oscillations, zero delay loops, and proper initialization of MOS latches.

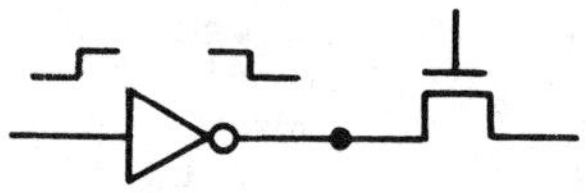

(a)

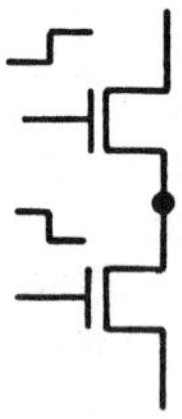

(b)

Figure-5 Evaluation for a Class of Switches

CONCLUSIONS

The approach for mixed level simulation described in this paper has been implemented and is currently under beta-test. The performance of the algorithms and user feedback have been very encouraging. The simulator is implemented in PL/1 and is very modular, thus making enhancements to the system easy.

REFERENCES

[1] M. A. d'Abreu, K. L. Cheong, C.T. Flanagan, J. Winkler, "A Simulation Technique for Current Mode Logic", Proceedings of the 1983 National Computer Graphics Conference, Chicago.

[2] "H2176 ISL Gate Array", Bipolar Product Data Sheet, Honeywell Solid State Electronics Division, June 1982.

[3] J. Lohstroh, "ISL, A Fast and Dense Low-power Logic Made in a Standard Schottky Process", 1978 European Solid-State Circuits Conference, Digest of Technical Papers, Sept 1978.

[4] R.E. Bryant, "MOSSIM - A Switch Level Simulator for MOS LSI", Proceedings of the 18th Design Automation Conference, Nashville Tennessee, June 1981.

Session 2-1c
Physical Design I

Chairman

K. W. Wu
Bell Labs

PARAMETERIZED PROCEDURAL MODULE GENERATION FOR PHYSICAL INTEGRATED CIRCUIT DESIGN

David H. Potter
John D. Tauke
Aris A. Yiannoulos

Bell Telephone Laboratories

ABSTRACT

This paper presents a general-purpose module generator using parameter-driven procedures to generate families of customized layout modules from seeds. A seed contains both layout geometries and control directives.

The module generator is quite general and is applicable to various layout styles and technologies. The seeds can be defined by non-programmers using graphical aids, and no program changes are involved. Furthermore, the seeds can be created at run time. This module generator is embedded in an advanced graphical editor, called GRED, that is part of a system of VLSI computer aids used at Bell Laboratories [1] [2].

1. INTRODUCTION

Frequently, there is a need for a family of layout modules where each module has some geometries modified to suit various circuit requirements like speed, power, fan-in or fan-out numbers, or fabrication technology. The motivation for this work was to automate the generation of such custom layout modules from a single seed.

A seed is a collection of layout geometries and control directives, and is capable of accepting parameters. Some geometrical elements will become the final layout geometries. Some others are used in conjunction with the directives to control the module generator. Figure 1 illustrates the creation and use of seeds for the production of customized modules.

Using GRED, the creator of a seed has the ability to see what the module being made actually looks like. The conventional features of the graphics system can be used, as well as the special features designed to aid in the insertion of the directives.

The module generator is a command interpreter that has access to the graphics processing capabilities of the graphic system into which it is embedded but is not technology specific. The technology specificity is imparted to the individual seeds which contain specialized procedures appropriate for families of layout modules and technologies. This separation is quite beneficial, as it provides the flexibility to allow user-definable seeds at run-time.

When a custom module is to be made the user invokes the module generator and specifies a seed, along with input parameters. Based on the parameters supplied, the seed directives cause arithmetic and logical evaluations and geometric operations to be performed for the creation of the module.

2. SEED STRUCTURE, CREATION AND USAGE

2.1 Seed Structure

A seed contains directives and geometries. Together they form a graphical programming language. The form and scope of the directives are similar to those of standard programming languages. The description of the geometries is in XYMASK [3], a standard graphics language used at Bell Laboratories.

The directives operate on variables of numeric, string or logical type, and on geometric objects. A complete set of arithmetic operations, logical flow control and an extensive mathematical function library are available. There are also means provided to prompt for parameters to be input, to read the reply, to use a previously defined value of a variable, or to assign a constant value to a variable. There are facilities provided to do error checking on input data and to report computed values or other information to the user.

The geometries are of two kinds, control geometries and target geometries.

Control geometries are used in conjunction with command directives to move or to delete target geometries. Target geometries, in their final form, will constitute the generated module.

The control geometries are border polygons and offset vectors. The border polygon selects those vertices to be moved or geometries to be deleted. The offset vector, in conjunction with input parameters, specifies the direction and magnitude of the move.

Geometric objects which have only some of their vertices in control geometries will undergo stretching. It should be noted that not only target geometries can be deformed, but also control geometries. For this reason, the order of move/stretch operations is important.

For additional flexibility, all of the above operations can be embedded in an IF-THEN-ELSE control structure. Execution of directives or inclusion of target geometries may be skipped, based on the evaluation of logical or arithmetic expressions.

The geometric definition of a resistor seed is shown in figure 2. The target geometries are shown solid and the control border polygon and offset vector are shown dotted.

2.2 Seed Creation

A seed could be written entirely with just a text editor. However, since most of the contents of a seed is of a graphical nature, the use of the graphical editing features in GRED is natural. GRED provides special commands for inserting directives. If the seed is created interactively in GRED, it can be tested and used immediately.

2.3 Seed Usage

The module generator is invoked by a single command in the GRED graphical editor. To generate a module, the user types:

IC <name> [parameters]

where <name> is the name of the seed which is invoked to generate a customized layout module and [parameters] are the input data required by that seed. Under control of seed directives, the parameters can be prompted for if not specified on the command line, or can be given default values.

When the parameters are defined the system will generate the specific module to suit the requirements specified by the parameters or return with an error message if those requirements cannot be met. GRED then prompts for the inclusion point at which the generated module will be placed. The module is also added to the job data base for further inclusion at other points.

3. APPLICATIONS

The initial application of the module generator was the creation of four seeds that can generate a set of bipolar components that otherwise would have to be stored in a comprehensive library. The components are transistors and diodes, resistors, capacitors and I/O pads. These seeds are easily design-rule and technology updatable.

An earlier library that was replaced by these four seeds contained some 650 components. Because of the drastic size reduction, maintenance is much easier. Figure 3 shows assorted resistors generated from the resistor seed, and figure 4 illustrates a variety of capacitors generated from the capacitor seed. Figure 5 exhibits some of the diverse devices that can be created from the transistor seed.

Other potential applications include a library of components for CMOS design, a series of biquad filters and op-amps and a set of components for an automated memory generator.

4. CONCLUSION

A parameter-driven module generation facility embedded in an interactive graphical editor has been presented. A seed can be created either with graphical aids or textually. Module generation seeds can be defined by non-programmers and do not require program changes. Seeds can be called up from libraries or can be created during interactive layout editing sessions. This is in contrast to schemes which require seeds to be written in a computer language and to be linked into the module generator. The present module generation facility is especially valuable in situations where many variations of basic components are needed. The approach presented is technology independent and can be applied to many layout styles.

Experience has shown that this technique is well suited for building flexible, complete libraries of small components. Current work indicates that it will also be applicable to building larger and more complex cells.

5. ACKNOWLEDGMENTS

The authors are indebted to Hermann Gummel and Kwok Wu for their encouragement, support and helpful suggestions. Y.C. Wu's contribution to the module generation project is also appreciated.

REFERENCES

[1] A. K. Bose, B. R. Chawla and H. K. Gummel, "A VLSI Design System", 1983 IEEE Proceedings of International Symposium on Circuits and Systems, pp. 734-739.

[2] "GRED - An Interactive Graphical Editor for Layout", to be published.

[3] B. R. Fowler, "XYMASK", Bell Laboratories Record, Vol. 47, No. 6, July, 1969, pp. 204-209.

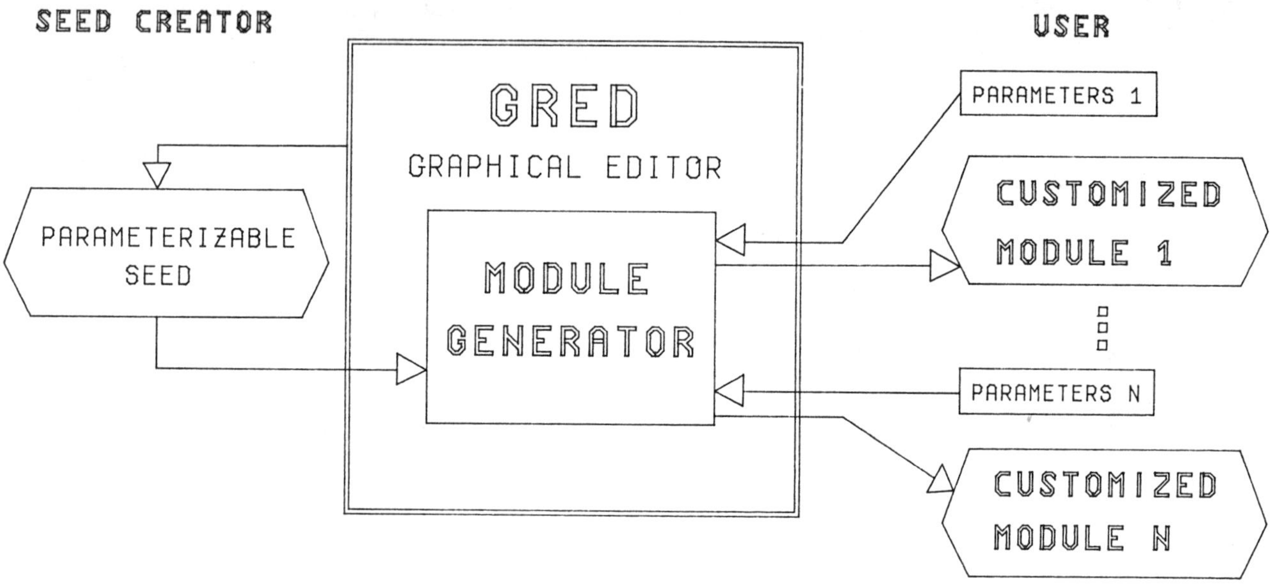

Figure 1 Seed Creation and Module Generation Process

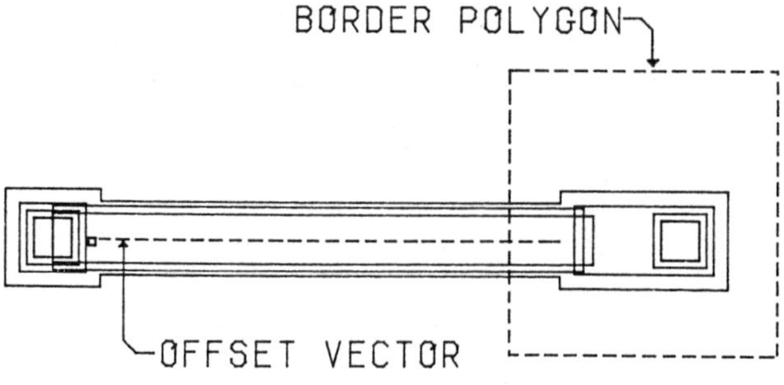

Figure 2 Resistor Seed

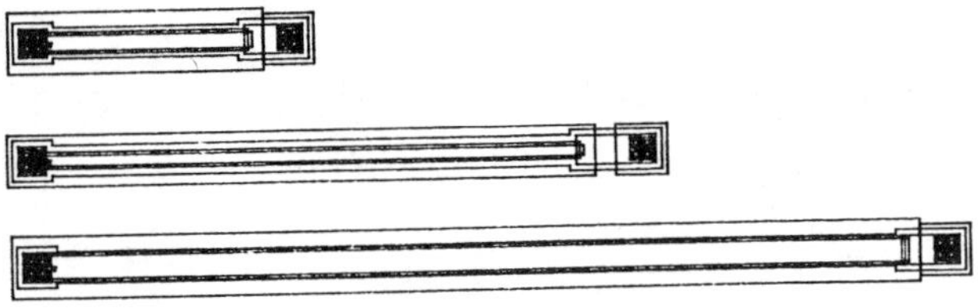

Figure 3 Customized Resistor Modules

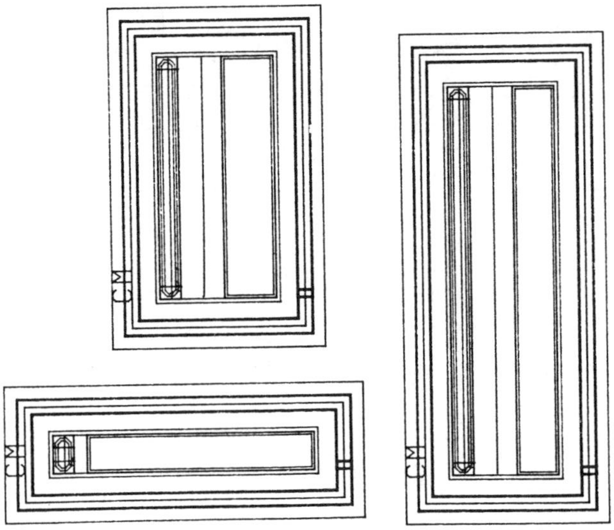

Figure 4 Customized Capacitor Modules

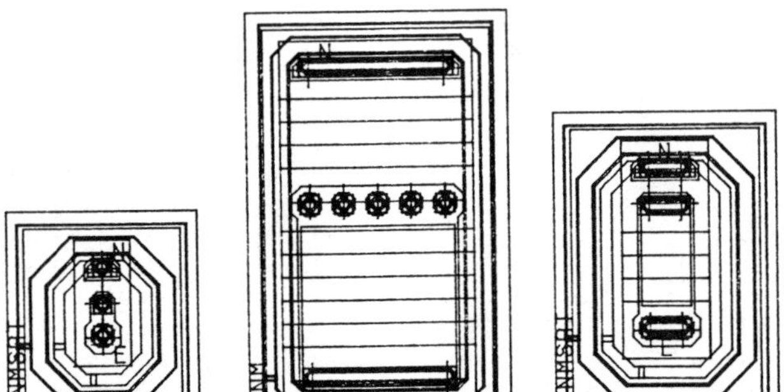

Figure 5 Customized Transistor Modules

VERIFICATION OF SCHEMATIC CONNECTIVITY WITH LAYOUT CONNECTIVITY

LINCOLN A. FAJARDO

Bell Laboratories
Allentown, Pa. 18103

ABSTRACT

This paper discusses a computer program to verify that the circuit connectivity specified by the designer in his or her schematics is the same as implemented in the circuit layout. This program is part of a system of computer aids used at Bell Laboratories for the design of VLSI circuits.

INTRODUCTION

In the IC development industry today there is a clear trend toward greater degrees of automation in every phase of the design process [1]. Many design engineers are using graphical design capture tools to aid them in the design definition stage. These tools produce connectivity lists which can be transmitted immediately to circuit simulators for the design verification stage. In some cases layout can proceed automatically, using algorithms to place geometries within some kind of constraining layout design style (polycell, gate matrix). In other applications, where performance or silicon area are critical issues, layout must still be done by hand using an interactive graphics layout tool. The disadvantage of using this technique is that it raises a vital question: Is the circuit that has been laid out the same as the circuit which had previously been simulated and verified? At Bell Laboratories the method being used to answer this question involves the automated comparison of the connectivity list used for circuit design verification with the list extracted from the completed layout. The following pages will describe the program XCOLA which was written to perform this automated comparison.

DESIGN ENVIRONMENT

Typically a design process begins with a designer capturing a schematic on SCHEMA[2]. SCHEMA is a computer program allows the designer to graphically manipulate device symbols and then to interconnect them. Once the schematic has been defined, SCHEMA is then asked to produce two outputs: schematic drawings (See Figure 1 for a brief example.) and a hierarchically structured connectivity list (See Table 1). The connectivity list is given immediately to a simulator to verify the correct behavior of the circuit.

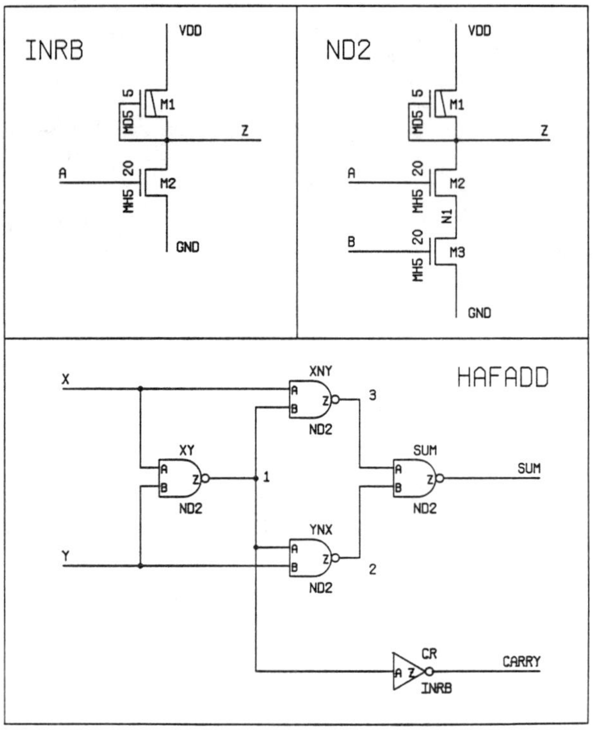

Figure 1. Schematic for Half-Adder

When the design is verified in the simulator, the schematic drawings are given to a draftsman who uses an interactive graphics layout tool to define the geometries that make up the layout. Nodes are labeled with the same names specified in the schematic. The completed layout is analyzed by the layout analysis program, HCAP [3] to check for design rule violations. Once the layout is made self consistent, HCAP is then asked to produce a connectivity list that also includes device parameters and parasitics. The stage is set for verifying that the schematic connectivity is consistent with the layout connectivity.

The designer then uses XCOLA (Expand and COmpare LSL or ADVICE [1]) to compare the two connectivity lists. XCOLA reads in the connectivity lists and expands them down to the lowest level devices. Because extra levels of hierarchy are often employed to facilitate layout, XCOLA first matches node names that have the full hierarchical path and then those whose hierarchical tokens are a subset of another node name. Ambiguities are flagged.

CH1935-6/83/0000/0294$01.00 © 1983 IEEE

Devices that are matched in detail, are then checked to see if their physical parametrization is within the specified tolerance of the schematic.

TOPOLOGICAL SOLUTION

A certain fraction of the errors detected by XCOLA are due to mislabeled nodes. Ideally the comparison program would perform the comparison without requiring extensive labeling of the schematic and layout. However, to try all possible topological permutations would require solution times that increase exponentially with the number of devices.

To reduce computation times, N. Kubo et. al. [4] have proposed an algorithm for partitioning connectivity into small "vertex" groups and only comparing groups with similar characteristics. M. Takashima et. al. [5] have applied this technique enhanced with some heuristic algorithms. It was found necessary to label input and output nodes to resolve "parallel circuit" ambiguities. Even with this approach considerable CPU time was required to partition an LSI circuit and then perform the matching. Topological solutions often require considerable ingenuity to decipher problems when there are several discrepancies.

XCOLA does not follow this approach. Instead, it relies on the labeling of node names for fast CPU times and clearly defined errors. Its been our experience that labeling nodes facilitates layout and subsequent edits. The number of nodes to be labeled can be reduced significantly if the design contains hierarchical repetitions.

XCOLA FEATURES

Running XCOLA is straight forward: The designer specifies the two connectivity files to be compared and the options. The connectivity lists can be hierarchically structured in either AIL (ADVICE Input Language) [6] or in LSL (Logic Simulator Language) [7] or a mixture of the two. Typically device level circuits are in AIL while the higher hierarchical levels are in LSL. To satisfy the needs of real users a number of options are needed:

1. Unconnected Terminals. In SCHEMA and HCAP unique nodes names beginning with special prefixes are assigned to unused terminals of a device. The designer can specify these prefixes so that all such node names are matched.

2. Interchangeable Terminals. Many devices have terminals that may be used interchangeably and thus may be connected differently in the schematic than in the layout. The designer can specify the prefix of such devices and which terminals are interchangeable.

3. Device Characteristics. The user can specify how closely the parameters specified on the schematic match those of the actual layout.

4. Ignore Parasitics. The user has the option of ignoring classes of generic devices. Typically this is used to ignore parasitics generated from the layout.

5. Estimate Parasitics. By means of a user defined subroutine the designer can estimate the capacitance on a node by knowing the number of devices on the node and their characteristics.

6. Estimate Area and Loads. The designer can request for polycell or device counts and node fanout audits.

7. Select Output. The user can specify how verbose the output from XCOLA should be.

In Table 2 we have simulated the output of HCAP for the circuit shown in Figure 1. We have included a misconnected MOS transistor and a transistor with a smaller width. In Table 3 we show a listing of the XCOLA output comparing the examples in Tables 1 and 2. The first part of the output shows the options used in making the comparison. All devices whose generic names begin with "C" are ignored (i.e. parasitic capacitance). For all transistors terminals 1 and 3 (source and drain) are interchangeable. And all transistors must have their width within 5% of the width specified in the schematic. Next is a listing of the subcircuits in each connectivity list and if they had a counterpart to be compared against. The second part of the output summarizes the results for each connectivity list. Since a list of matching devices was requested, these are listed next. Note the matching of devices whose source and drain have been interchanged. The transistor out of tolerance is shown next, followed by the one improperly connected.

PROGRAM DETAILS

The expansion of node names follow these rules: Nodes that are inputs or outputs of a subcircuit are replaced by connections to the subcircuit; Nodes that are declared globals (i.e. power, ground, high, low, etc...) are left unchanged. Internal nodes names are concatenated with the occurrence name of the subcircuit. This insures that the node names are unique and meaningful.

To reduce the number of detailed comparisons due to interchangeable terminals two hash words, g and h, are generated for each device:

$$g = \sum_{i=1}^{N} A(i)$$

$$h = \sum_{i=1}^{N} 4^{i-1} \sum_{j==i} A(j)$$

where N is the number of terminals of the device, A(i) is the program address of the node connected to terminal i of the device. The expression "j==i" means for all terminals j that have been declared equivalent to i. Devices that are connected to same nodes regardless of which terminal have the same g value. Devices with same nodes connected to the appropriate interchangeable terminal have the same h value. Only devices with matching hash words are compared in detail. For large circuits this hash method has a rejection of 95% or better. This prevents a large amount of time from being wasted going through all possible combination of interchangeable terminals for devices that do not match.

XCOLA APPLICATIONS

Since the schematic design may be completed far ahead of the layout, it is useful to estimate parasitics as described above for circuit simulation. After the layout is completed, the designer can verify that the correct connectivity was attained and that the devices were properly characterized. In addition using the estimated parasitics and the tolerance checking, the designer can quickly determine areas which paths may have to re-simulated.

Because of the iterative nature of the design process and the large numbers of persons working on the same project, an automatic way of validating the design/layout is needed. One approach used successfully is based on the date when the different data files were generated. It requires that the connectivity files be more recent than the schematic and layout files from which they are generated. Then the XCOLA comparison file and the simulation files must be more recent than the connectivity files.

CONCLUSIONS

XCOLA has been found to be a fast and efficient tool for detecting discrepancies between schematics and layouts. XCOLA has also been found to be useful before layout when the designer can invoke a special feature to estimate parasitic capacitance and thereby obtain a more realistic simulation.

REFERENCES

[1] A. K. Bose, B. R. Chawla, and H. K. Gummel, "A VLSI Design System", Proc. ISCAS Conference, 1983.

[2] C-C. Liaw, M. Tong, and L. A. Fajardo, "A System for High-Level Design Capture and Synthesis", to be published Bell System Technical Journal.

[3] P. A. Swartz, B. R. Chawla, T. R. Luczejko, K. Mednick, and H. K. Gummel, "HCAP - A Topological Analysis Program for IC Mask Artwork", Proc. IEEE International Conference on Computer Design 1983.

[4] N. Kubo, I. Shirakawa, and H. Ozaki, "A Fast Algorithm for Testing Graph Isomorphism", Proc. ISCAS Conference, pp 641-644, 1979.

[5] M. Takashima, T. Mitsuhasi, T. Chiba, K. Yoshida, "Programs for Verifying Circuit Connectivity of MOS/LSI Mask Artwork", Proc. 19th DAC, pp 544-550, 1982.

[6] L. W. Nagel, "A Computer Program to Simulate Semiconductor Circuits", ERL Memorandum ERL-M520, University of California, Berkeley, California, May 9, 1975.

[7] B. R. Chawla, H. K. Gummel, P. Kozak, "MOTIS, An MOS Timing Simulator", IEEE Transactions on Circuits and Systems, Vol. CAS-22, No. 12, pg 091, 1975.

TABLE 1

CONNECTIVITY LIST FOR HALF-ADDER FROM THE SCHEMATIC

```
.SUBCKT  INRB (A,Z)
.GLOBAL GND,VDD
  M1 (VDD,Z,Z) MD5  SW= 5
  M2 (Z,A,GND) MH5  SW= 20
.FINIS

.SUBCKT  ND2 (A,B,Z)
.GLOBAL GND,VDD
  M1 (VDD,Z,Z) MD5  SW= 5
  M2 (Z,A,N1) MH5  SW= 20
  M3 (N1,B,GND) MH5  SW= 20
.FINIS


NETWORK:
CKTNAME: HAFADD;
INPUTS: X,Y;
OUTPUTS: SUM,CARRY;

INRB:
      CR,(1),(CARRY);
ND2:
      SUM,(3,2),(SUM);
      XNY,(X,1),(3);
      XY,(X,Y),(1);
      YNX,(1,Y),(2);
```

TABLE 2

CONNECTIVITY LIST FOR THE HALF-ADDER FROM THE LAYOUT

```
.SUBCKT HAFADD(X, Y, SUM, CARRY)
M1 (VDD,SUM,SUM) MD5  SW=5
M2 (YNX.N1,1,2) MH5  SW=20
M3 (XNY.N1,1,GND) MH5  SW=20
M4 (XNY.N1,X,3) MH5  SW=20
M5 (VDD,CARRY,CARRY) MD5  SW=5
M6 (CARRY,1,GND) MH5  SW=18
M7 (SUM.N1,3,SUM) MH5  SW=20
M8 (SUM.N1,2,GND) MH5  SW=20
M9 (VDD,1,1) MD5  SW=5
M10 (XY.N1,1,X) MH5  SW=20
M11 (XY.N1,Y,GND) MH5  SW=20
M12 (VDD,2,2) MD5  SW=5
M13 (VDD,3,3) MD5  SW=5
M14 (YNX.N1,Y,GND) MH5  SW=20

C1 (SUM,GND) CAP 100 40
C2 (CARRY,GND) CAP 100 40
C3 (X,GND) CAP 100 40
C4 (Y,GND) CAP 100 40
C5 (1,GND) CAP 200 60
C6 (2,GND) CAP 150 50
C7 (3,GND) CAP 150 50
C8 (XY.N1,GND) CAP 150 50
C9 (XNY.N1,GND) CAP 150 50
C10 (YNX.N1,GND) CAP 150 50
C11 (SUM.N1,GND) CAP 150 50
.FINIS
```

TABLE 3
OUTPUT OF XCOLA

```
OPTIONS: EXPAND HIERARCHY; LIST MATCHING DEVICES;
   IGNORED DEVICES:
       C*
   DEVICES WITH EQUIVALENT TERMINALS
       M* 1 3
   DEVICES WITH TOLERANCE SPECIFICATIONS
       M* 0.05

INPUT FILES AND SUBCKTS READ IN
FILE 1, HAF.SCH:    INRB  ND2  HAFADD
FILE 2, HAF.HCP:    HAFADD

SUBCKTS NOT COMPARED:  INRB, ND2;  HAF.SCH

* * * * * * * *  SUBCKT COMPARED:   HAFADD    * *

HAF.SCH: 14 DEVS, 12 MATCH, 1 OUT TOL., 1 NOMATCH
HAF.HCP: 14 DEVS, 12 MATCH, 1 OUT TOL., 1 NOMATCH

. . . . . . MATCHED DEVICES WITHIN TOLERANCES . .

HAF.SCH:  MD5/5  CR.M1    ( VDD  CARRY  CARRY )
HAF.HCP:  MD5/5  M5    ( VDD  CARRY  CARRY )

HAF.SCH:  MD5/5  SUM.M1    ( VDD  SUM  SUM )
HAF.HCP:  MD5/5  M1    ( VDD  SUM  SUM )

HAF.SCH:  MD5/5  XNY.M1    ( VDD  3  3 )
HAF.HCP:  MD5/5  M13    ( VDD  3  3 )

HAF.SCH:  MD5/5  XY.M1    ( VDD  1  1 )
HAF.HCP:  MD5/5  M9    ( VDD  1  1 )

HAF.SCH:  MD5/5  YNX.M1    ( VDD  2  2 )
HAF.HCP:  MD5/5  M12    ( VDD  2  2 )

HAF.SCH:  MH5/20  SUM.M2    ( SUM  3  SUM.N1 )
HAF.HCP:  MH5/20  M7    ( SUM.N1  3  SUM )

HAF.SCH:  MH5/20  SUM.M3    ( SUM.N1  2  GND )
HAF.HCP:  MH5/20  M8    ( SUM.N1  2  GND )

HAF.SCH:  MH5/20  XNY.M2    ( 3  X  XNY.N1 )
HAF.HCP:  MH5/20  M4    ( XNY.N1  X  3 )

HAF.SCH:  MH5/20  XNY.M3    ( XNY.N1  1  GND )
HAF.HCP:  MH5/20  M3    ( XNY.N1  1  GND )

HAF.SCH:  MH5/20  XY.M3    ( XY.N1  Y  GND )
HAF.HCP:  MH5/20  M11    ( XY.N1  Y  GND )

HAF.SCH:  MH5/20  YNX.M2    ( 2  1  YNX.N1 )
HAF.HCP:  MH5/20  M2    ( YNX.N1  1  2 )

HAF.SCH:  MH5/20  YNX.M3    ( YNX.N1  Y  GND )
HAF.HCP:  MH5/20  M14    ( YNX.N1  Y  GND )

? ? ? ? ? ? ? DEVICES OUTSIDE TOLERANCES ? ? ? ? ?

HAF.SCH:  MH5/20  CR.M2    ( CARRY  1  GND )
HAF.HCP:  MH5/18  M6    ( CARRY  1  GND )

? ? ? ? ? ? ?   UNMATCHED DEVICES   ? ? ? ? ? ? ?

HAF.SCH:  MH5/20  XY.M2    ( 1  X  XY.N1 )
HAF.HCP:  MH5/20  M10    ( XY.N1  1  X )
```

HCAP - A TOPOLOGICAL ANALYSIS PROGRAM FOR IC MASK ARTWORK

P. A. Swartz, B. R. Chawla, T. R. Luczejko, K. Mednick, H. K. Gummel

Bell Laboratories, Murray Hill, N.J. 07974

Abstract

A layout analysis tool called HCAP is described. The program accepts an artwork description of a circuit and extracts connectivity and parasitic device information from the layout. The connectivity is used to verify circuit integrity and to provide input to circuit simulators. As a consequence of the layout analysis, HCAP is capable of sophisticated, selective design-rule checking. The information on circuit connectivity is used to filter out false alarms and expose real violations. A novel feature of the program allows the user to specify circuit blocks of arbitrary size and complexity as subcircuits which are not analyzed in detail. This permits rapid establishment of inter-block connectivity and parasitics when a detailed analysis of the entire circuit is not desirable or possible. HCAP has been adapted to many different technologies and design methodologies. Flexibility is maintained by keeping most of the technology and design style information in user accessible files external to the program.

Introduction

An important problem in IC design is verifying that the physical layout implements the design intent. A variety of issues come together here. The problem is concerned not only with the logical correctness of the layout but involves considerations of driving power and loading of circuit elements as well. In addition, the manufacturing process has an impact through design-rule requirements. This paper presents an approach to layout analysis and verification which has been in wide use at Bell Laboratories for several years for VLSI design. In the next section, we give a general overview of the HCAP program. Following that is a more detailed description of design-rule checking. After that is a section which illustrates a technique used for hiding part of the circuit from the program. Finally, we conclude by giving some statistics on HCAP.

Overview

With the emergence of VLSI circuitry, there have been numerous attacks on the problem of layout verification[1-6]. As with most of these, HCAP accepts a geometric description of the layout. The particular form used here is a mask description language called XYMASK[7]. It specifies geometries on given mask levels and very importantly, allows a circuit to be described hierarchically. The first part of the HCAP data structure is a compiled version of the input XYMASK and preserves any hierarchy present.

The actions that HCAP takes are controlled by the user through a command file. This allows custom tailoring of the program to a wide variety of technologies and design styles. As much as possible, factors which depend on a particular technology are kept out of the program and placed in this file. HCAP looks to this file for information about the mask levels being used, and beyond that for process and design-rule data. Furthermore, as the name suggests, there are commands in the file which cause HCAP to interpret the XYMASK and perform the appropriate operations for analysis of the layout. The processing differs depending on the technology.

Typically, as the first step in the layout analysis, the hierarchy is flattened and overlapping pieces merged. HCAP employs a working buffer for this purpose. Figures in the data structure can be accessed by mask level and read into the buffer where a geometry engine is used to consolidate the shapes. An example of a command line is

READ N5 OR* WRITE METAL

which reads in all geometries on level N5, merges them, and stores the results in a file called METAL. The set of routines which handle the ORing of figures can also perform other Boolean functions such as the AND and the NOT of two sets of geometries. A scan line algorithm is employed for the Boolean operations. Another important command is the GROW command which expands or contracts figures, depending on the argument given. This allows the user to properly account for processing changes, such as diffusion outgrowth. There are no restrictions on the figures except that they must be polygons.

After each set of figures is merged, it can be saved in a file for later processing. Commonly created files consist of all shapes on the metal, polysilicon, and diffusion levels. Several different mask levels may be combined into one file during the process of merging figures. Other important files are also formed at this stage. For example, the gate regions of transistors (Poly AND Thinox in MOS technologies) might be created.

The second major step in this process is to resolve the connectivity. An important concept is the notion of connectable levels. There can be any number but typically there are a few, such as Metal, Poly, and Diffusion. HCAP associates shapes in the connectable levels with nodes in the circuit. Initially, each shape is presumed to be electrically distinct. HCAP is informed through the command file how these levels are connected to each other. Wherever two or more levels overlap and a prescribed contact window is present, the program consolidates the electrical nodes. The user has the option of labeling nodes; unlabeled nodes will be given default names by HCAP.

Once the nodes are resolved, transistors, capacitors, and resistors can be identified and characterized. Their place in the circuit is recorded for later generation of descriptions for other programs. For transistors, details about channel length, width, and implant type are calculated. Special attention is paid to the capacitive elements at the source-gate-drain interface. For simple path-like resistors, HCAP will calculate a center line resistance and correct for corners and end effects. The designer may also label devices for easier referencing in the programs which process HCAP output.

After the devices have been characterized, the circuit is fairly well understood. The user can return to the geometry engine to obtain parasitics associated with each node. HCAP can report the areas and perimeters of shapes and from these calculate parallel plate and

CH1935-6/83/0000/0298$01.00 © 1983 IEEE

fringing capacitance. Through the command file, the user may select not only which types of capacitance to report but may also set minimum values below which the capacitance will be discarded.

A multitude of output forms can be gotten from HCAP. One of these is simply a picture of the mask such as is shown in Fig. 1. Here the various levels are contrasted by different shading. Since both connectivity and parasitics are known, HCAP can produce a circuit level description for a simulator such as ADVICE[8]. It also has the ability to recognize logic configurations and group transistors into logic gates for MOTIS[9] descriptions. As a by-product of the logic recognition, a schematic diagram of the circuit based on the logic grouping of the transistors can be produced. Another output which is available is a skeleton description of the circuit for placement and routing using LTX[10]. All of these forms of output correspond directly to the layout. A major benefit of HCAP is that changes to the layout can be quickly reflected in updated descriptions for other programs.

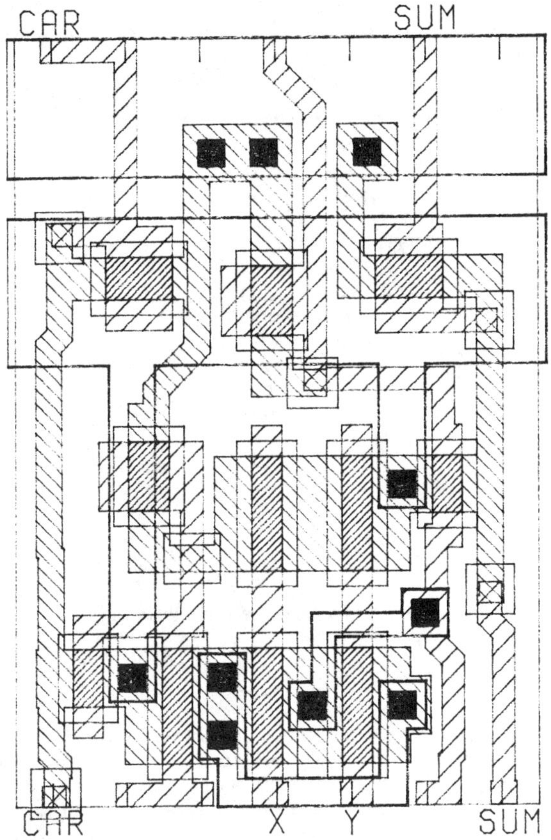

Figure 1

Design-rule Checking

Preliminary checks of the mask descriptions are performed as part of the layout analysis. However, these checks are cursory and designed only to reveal obvious inconsistencies in the artwork such as short circuits between the power and ground buses. The bulk of the mask checks are done after extraction of the layout is completed. There are two reasons for this. The first is that the checking task is facilitated by being able to work on the "cleaned up" artwork. Many of the rules are directed towards shapes in files produced during the initial merging phase.

The second important reason for doing DRC after layout analysis is that a great deal of information about the circuit has been derived and can be used to advantage. The most important information is the connectivity. Armed with this, HCAP is able to intelligently weed out real violations from false ones, e.g., clearance violations between two parts of the same node.

HCAP provides a preparation phase just before mask checking is performed. Here the user is free to create any special files which do not come automatically out of the circuit extraction. This is done by returning to the geometry engine and selectively eliminating or admitting shapes to the buffer. The full power of the geometric commands may be brought to bear. Geometries in the buffer can be expanded, trimmed, or in general, distorted in many ways so that a variety of special rules may be enforced for particular classes of figures. In cases where certain shapes, because of size or implant, behave differently under processing than others on the same mask level, this ability to discriminate is indispensable.

Following any preparatory operations, HCAP proceeds with the design-rule checking. The commands generally involve one or two files of shapes. Such things as clearance checks with or without regard to nodal information have already been mentioned. Internal clearance checks, for contact windows, are also available. HCAP has the ability to check figures for self-intersection, or notches and nubs which fall below a specified size. Other commands include minimum area and width checks.

The output from these checks is an XYMASK file. The XYMASK describes only the figure(s) that fail the tests. A post processor may be used to view the errors and make measurements on them. As shown in Fig. 2,

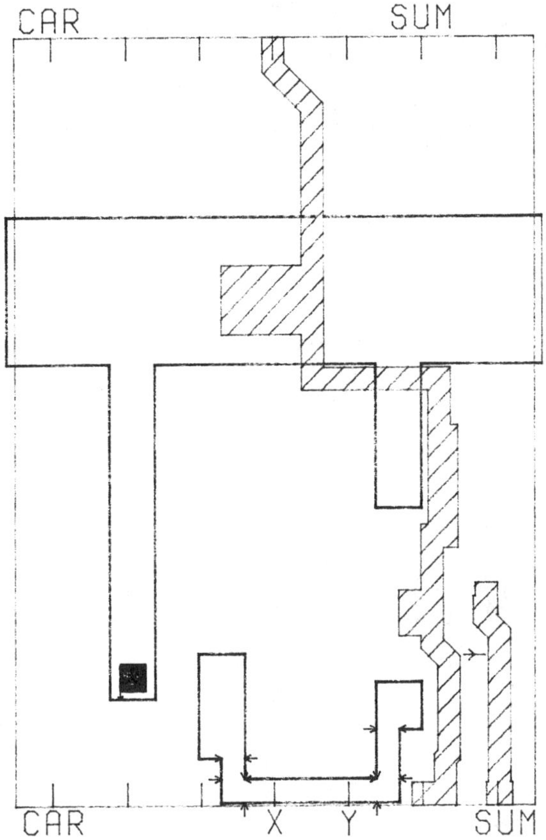

Figure 2. Mask errors: 1) Contact too close to edge; 2) Width too small; 3) Clearance violation

arrows indicate where the errors occurred and thus simplify the user's problem of determining where the errors are. On the left in Fig. 2 is displayed a contact which was not centered properly. A check for internal clearance of contacts to metal revealed that it was too close to the edge of the metal geometry shown. In the middle of Fig. 2, a geometry had a width smaller than the acceptable minimum and was flagged. On the right side of Fig. 2, a clearance violation was detected. Normally separate pictures are generated for different design-rule checks. We have combined them here to indicate the variety of checks which can be performed.

Black Box Analysis

Under some circumstances, it is not feasible to perform a full analysis of the layout. Bipolar technologies in particular present severe if not insurmountable problems for characterization programs. In MOS technologies, where the problems are considerably easier, it may simply be undesirable to analyze the full layout. Parts of the artwork may already have been examined or, alternatively, a compact or more accurate simulation model may exist for a collection of devices. A good example is an operational amplifier, where a full subnetwork description is not as useful.

HCAP provides a convenient way to deal with these situations. The part of the circuit which a designer does not wish HCAP to analyze is captured in a subcircuit. The subcircuit is then instantiated in the full layout at the appropriate places. Ordinarily, HCAP would fathom all levels of the layout hierarchy during processing. However, with the appropriate commands to the program and the proper decorations to the artwork, the user can hide the subcircuits from HCAP. We refer to these hidden subcircuits as "black boxes," since HCAP skips over them and has no knowledge of their interior.

If HCAP determines that a subcircuit is a black box, it proceeds no further down that chain of the layout hierarchy. The rest of the circuit is analyzed in full detail. After deriving the nodal information, HCAP associates terminals on the black box with signals in the circuit. Thus, the black boxes get "plugged in" to the rest of the circuit. Furthermore, the parasitics of the interconnections are obtainable as in a full-fledged extraction.

To illustrate how this mechanism works, we show a simple half adder circuit made of black boxes in Fig. 3.

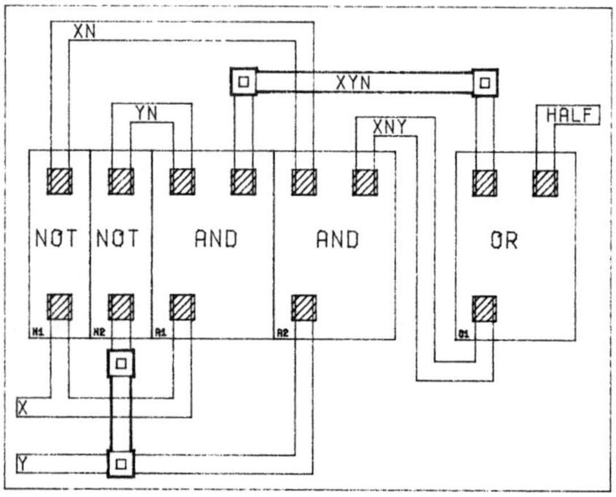

Figure 3. Half Adder circuit built with Black Boxes. Black Box terminals are shaded, poly is shown with light lines, and metal is shown with heavy lines. Electrical nodes have been labeled using XYMASK STEXT.

HCAP has identified the subcircuits named NOT, AND, and OR as black boxes and Fig. 3 shows the extent of HCAP's knowledge of them. The shaded areas represent the terminal locations within each black box. Signal lines from outside the black boxes run in and overlap the terminals. This indicates to HCAP how the subcircuits are wired to the rest of the circuit.

The output for this example is shown in Figure 4. The first half of the figure shows the logic description derived by HCAP. The second half lists the capacitance of the external wiring. With appropriate models for NOT, AND, and OR, the full circuit can be exercised on a simulator such as MOTIS or ADVICE. The models of the subcircuits could very well have been provided by HCAP in another analysis.

Through the use of black boxes, most of the difficulties of extracting bipolar circuits are contained in the detailed device modeling. This allows semi-automatic extraction of these layouts without large expenditures of computer time. The technique also permits quite large (essentially unlimited) and complicated layouts to be handled easily. In fact, the extraction of a large chip can be done in a hierarchical fashion, since the full analysis on a moderately sized module can be followed by an analysis of a larger circuit in which the original module is treated as a black box, and so forth.

```
OPTION:  MULTISOURCE=Y;
NETWORK:
CKTNAME:  HALF      ;
INPUTS:   X,Y;
OUTPUTS:  HALF;
 NOT:  N1,  (X),(XN)  ;
 NOT:  N2,  (Y),(YN)  ;
 AND:  A1,  (YN,X),(XYN)  ;
 AND:  A2,  (XN,Y),(XNY)  ;
 OR:  O1,  (XYN,XNY),(HALF)  ;

 OPTION:  ROUCAP:
 Y          =     0.371,
 XYN        =     0.338,
 XN         =     0.276,
 X          =     0.185,
 YN         =     0.108,
 XNY        =     0.246;
 HALF       =     0.066;
```

Figure 4. Output results of Black Box analysis.

Conclusion

The HCAP program has been in production use for several years. It has been used to inspect circuits in a variety of technologies. Among these are NMOS, CMOS, Depletion NMOS, and bipolar. Additionally, the program has been applied to printed circuit boards. It has been adapted to different design styles: standard cell, gate matrix, gate array, and custom designs. The suitability of HCAP to these many applications has been achieved by placing technology and design dependent information into a command file. With this file, each user may guide the HCAP processing in a way which suits the design at hand.

The processing speed depends a great deal on how much analysis is required and the number of design rules handled. For a detailed extraction of moderately sized circuits (a few hundred transistors) HCAP can process about 2 transistors/sec on a 1 MIP machine.

References

1. B. T. Preas, B. W. Lindsay, and C. W. Gwyn, "Automatic Circuit Analysis Based on Mask Information," Proc. 13th DAC, (1976), pp. 309-317.

2. H. S. Baird, "Fast Algorithms for LSI Artwork Analysis," Proc. 14th DAC, (1977), pp.303-311.

3. C.M.Baker and C. Terman, "Tools for Verifying Integrated Circuit Designs," Lambda Magazine, 4th Quarter, 1980, pp.22-30.

4. M. Hofmann and U. Lauther, "HEX: An Instruction-Driven Approach to Feature Extraction," Proc. 20th DAC, (1983), pp.331-336.

5. T. S. Szymanski and C. J. Van Wyk, "Space Efficient Algorithms for VLSI Artwork Analysis," Proc. 20th DAC, (1983), pp.734-739.

6. A. Gupta, "ACE: A Circuit Extractor," Proc. 20th DAC (1983), pp.721-725.

7. B.R. Fowler,"XYMASK," Bell Laboratories Record, Vol. 47,No. 6, July 1969,pp.204-209.

8. L. W. Nagel, "ADVICE for Circuit Simulation," Proc. 1980 International Symposium on Circuits and Systems, Houston, Texas, April 28,1980.

9. B. R. Chawla, H. K. Gummel, and P. Kozak, "MOTIS - An MOS Timing Simulator," IEEE Trans. on Circuits and Systems, Vol. CAS-22, Dec., 1975, pp. 901-910.

10. G. Persky, D. N. Deutsch, D. G. Schweikert, "LTX - A System for the Directed Automatic Design of LSI Circuits," Proc. 13th DAC, (1976), pp. 399-407.

MASK LAYOUT COMPARISON IN LSI DESIGN ENVIRONMENT

Yu-Chun Wu

Bell Laboratories
Allentown, PA 18103

ABSTRACT

A set of mask comparison programs is presented. These programs compare two versions of a layout and output the differences which then can be examined graphically with an interactive graphical editor or listed out textually. These programs are very useful for layout change control. Mask comparison is difficult for VLSI circuits because of the huge amount of data involved, and because there is no unique description of a layout. To overcome this, a variety of techniques are employed. Hierarchical, geometrical comparisons and merge operations are applied in this set of comparison programs to cope with various types of situations. This set of programs for layout change verification is part of a system of computer aids used at Bell Laboratories [1]. Due to the fast execution of these programs, frequent and comprehensive audit of the layout databases has now become feasible.

1. INTRODUCTION

The rapid growth in circuit density and complexity has created the need for an efficient tool for mask layout change verification. During the design cycle, it is often necessary to make changes to a mask layout database. The changes are made through various CAD tools and occasionally result in an error being introduced into the database. Such errors could be the result of hardware, software or human procedures.

Historically, layout change verification has been done in several ways. One is to plot the entire chip and manually locate and verify the changes. Another way is to plot the two versions of the chip in two colors on a level by level basis. For example, the old layout is plotted in red, and then it is overlaid with the new layout in green. Geometrical mismatches can be detected by visual examination of all plots. Black represents a match, red or green represent mismatches. These methods are time-consuming, tedious, error-prone and costly.

Another method is to use text comparison utility programs such as "DIFF" to find textual differences between two layouts represented in text format. While very efficient for special kinds of changes, such a method is not generally applicable. In particular, it does not work if there have been changes in the hierarchy.

COMXY, COMXZ and COMXM are a set of mask layout comparison programs which have been written to handle efficiently the typical design scenario, in which minor design changes are made with an interactive graphical editor. The programs are general enough however, to handle major changes at somewhat lower efficiency. COMXY, COMXZ and COMXM take input either in XYMASK [2], a mask geometry or artwork description language used at Bell Laboratories or in internal data structure format of GRED [1] [3]. GRED is an advanced hierarchical graphical editor developed at Bell Laboratories. The output file which contains only the differences between the two layouts is composed of two difference sets. One difference set contains all the geometries which are on one layout but are not on the other layout, and vice versa for the other difference set. The layout changes can be plotted on various plotting devices or viewed graphically by using the rich set of graphical commands in GRED. This new method is faster, more reliable and cost-effective in comparison with the traditional methods.

2. PROGRAM COMXY

In VLSI design, hierarchy is frequently employed to manage complexity. As do other mask layout languages, XYMASK also supports hierarchy. That is, a cell or symbol (called CLUMP in XYMASK) is a collection of geometric atomics such as rectangles, paths or polygons and can also include (or call) instances of other cells.

Hierarchy presents both opportunities and problems for mask comparison. Some approaches, among them COMXZ and COMXM described below flatten the hierarchy at the beginning. While such an approach eliminates certain problems, it introduces others and can lead to very long run times.

COMXY does mask comparison on a hierarchical basis. Not only are individual comparison steps performed on a hierarchical basis but so is the reporting of differences. This approach not only provides more meaningful results but also achieves better speed performance and requires less memory than approaches which flatten the whole hierarchy before comparison.

CH1935-6/83/0000/0302$01.00

2.1 Algorithms

The algorithms implemented in the hierarchical comparison program consist of both upward propagation and downward propagation techniques.

2.1.1 Upward Propagation

At the comparison processing stage, COMXY starts with the bottom-level cells, proceeds upward through all cells in the hierarchy tree, searching for mismatched cells, atomics or inclusion records. A flag U associated with each cells is used to mark the cells that contain mismatched records. Each cell having its U flag set will later be extracted as a diagnostic cell which contains only the mismatched records. If there are no placement changes for a diagnostic cell, only the mismatched records in that diagnostic cell will be reported.

The comparison criteria used are as follows:

For cell comparison, if a cell exists in one layout but not the other, the U flag associated with the cell is set.

For atomics comparison, in each cell, if an atomic record in one layout is found not to have a syntactically-identical atomic record in the other layout, the record is considered as a mismatched atomic record. The U flag associated with the cell is then set.

For inclusion record comparison, in each cell, if an inclusion record in one layout is found not to have a syntactically-identical inclusion record in the other layout, the record is considered a directly-mismatched inclusion record. However, even if a pair of syntactically-identical inclusion records are found, further checking is required to determine whether their descendant cells contain mismatched records. If the U flag has been set for one of the descendant cells, the inclusion records are considered indirectly-mismatched inclusion records. For both direct and indirect mismatches, the U flag associated with the cell will be set.

Figure 1 shows the upward propagation in a hierarchy tree. A mismatched atomic record RECTANGLE is found in cell D, this mismatch is propagated to the inclusion records in its ancestor cells A and B.

2.1.2 Downward Propagation

At the next stage, COMXY finds all the inclusion records which have been changed and extracts their descendant cells. The descendant cells extracted at this stage will be in their original form to represent the placement changes when examining the output file. COMXY starts with the top-level cells and proceeds downward through all cells in the hierarchy tree, searching for directly-mismatched inclusion records. For each directly-mismatched inclusion record, COMXY follows the inclusion chains downward and marks every affected cell by setting a flag D (for downward propagation). Each cell having its D flag set will later be output as a whole cell. If the cell in addition has the U flag set, then it will also be output as a diagnostic cell, i.e. as a cell containing only records that differ from their counterparts.

Figure 2 shows the downward propagation in a hierarchy tree. A cell C is relocated to location (7,7), thus all the cells under this inclusion record, cells C, E and F are output.

2.2 The Speed Performance

Program COMXY was coded in FORTRAN and runs on VAX 78Ø/VMS computers. The following table shows some statistics gathered from current chip design projects. The data to be compared were available in the GRED internal data structure format. Showing in the table are CPU time for comparison processing only and total CPU time which includes the input/output required for reading the two layouts and writing out the differences to a disk file, in addition to the time required for performing the actual comparison.

Number of XYMASK Lines	Approx. No. of Geometries	CPU Sec. for Comparison	Total CPU Sec.
81,ØØØ	76Ø,ØØØ	25	83
6Ø,ØØØ	36Ø,ØØØ	18	53
27,ØØØ	148,ØØØ	11	36
14,ØØØ	421,ØØØ	6	19

2.3 Potential False Alarms

Because a layout can be described in many different ways with a mask geometry description language such as XYMASK, the COMXY program may report differences which represent layout features that are functionally the same.

A false alarm can occur if the same geometry is described in two different ways. For example, a four sided polygon can be described as POLYGON, RECTANGLE (BOX) or PATH (WIRE). False alarms can also occur if the two layouts are described in different hierarchical form. Figure 3 shows an example of this kind. In the old layout, cell A consists of one inclusion record REPEAT which describes an array of four instances of cell B, with the array's lower left corner origin at (Ø,Ø) and replication interval 1ØØ units in the x-direction, and Ø units in the y-direction. In the new layout, the hierarchical structure is altered. It now consists of four INClude records and each of them places cell B individually. In spite of the different representations, the layout is the same.

Another comparison program COMXZ, which eliminates these potential false alarms is presented in the next section.

3. PROGRAM COMXZ

COMXZ is a geometrical comparison program that serves as a supplement to its companion program COMXY. Different from its companion program COMXY,

COMXZ performs the comparison without regard to the hierarchical structure. COMXZ uses a divide and conquer approach to partition the chip into windows, extracts geometries within each window, and converts each geometry into a canonical form (polygon) before the comparison.

For those cases in which COMXY would generate "false alarms" due to valid differences between the two XYMASK layouts, COMXZ eliminates the false alarms and extracts only the true geometrical differences between the two mask layouts.

3.1 Implementation

COMXZ flattens the whole hierarchy and compares the geometries on the chip plane. Because the amount of data to be compared could be very large, COMXZ applies a dynamic partitioning technique to stay within the available memory space and achieve better speed performance. COMXZ first estimates the number of geometries and the amount of memory space needed by traversing the hierarchy from bottom up. It then automatically partitions the chip into windows and performs the comparison within each window. During the geometry extracting process, if there is insufficient memory space, the current window will be partitioned recursively until all the geometries within the window fit into the available memory space.

Although COMXZ is superior to COMXY in that it compares the two layouts and extracts only the true differences, it is inferior to its companion program COMXY in speed. On the average, COMXZ is about 15-20 times slower than COMXY. Therefore, COMXZ is recommended to be used on the outputs of COMXY, which contains only the differences as a result of hierarchical comparison.

3.2 Exceptional Cases

Sometimes, the geometric figures in the database are changed because of figure-merging or figure-fracturing. Figure-merging consists of combining two or more geometric figures into a single figure. For example in figure 4, a U shaped polygon is formed after a boolean "OR" operation has been performed on three separate rectangles. Figure-fracturing occurs when a geometric figure is subdivided into several smaller figures. For example in figure 5, a C shaped polygon is subdivided into six smaller pieces during the EBES (Electron Beam Exposure System) database creation process. Merged and unmerged figures will be reported as mismatches by COMXZ.

The next comparison program, COMXM compares layouts with merged or fractured figures.

4. PROGRAM COMXM

COMXM is another geometric comparison program that compares two layouts and outputs the differences in XYMASK format. COMXM merges the figures on each mask level and then performs the comparison on the merged figures.

For each mask level of the two layouts, COMXM extracts all the geometries from the hierarchy and performs the boolean "OR" operation to merge them. The resultant merged figures are then compared.

Although COMXM eliminates the limitations that exist in COMXZ, the COMXZ program is faster. Therefore COMXM is recommended to be used on the output of COMXZ.

5. CONCLUSION

A comprehensive set of mask layout comparison programs has been presented. These programs are automatic, efficient, accurate and it minimizes the need for check-plots. It also reduces human errors because only the differences are presented to the designer. Most of all, tight coupling with a graphical editor (GRED) makes frequent layout comparison convenient.

COMXY is an intelligent hierarchical comparison program. It compares two physical layout databases using their hierarchical structures and generates the differences, if any, in XYMASK format. Furthermore, it preserves the original hierarchical structure in the output file so that the user can examine the differences in proper perspective. It is the fastest one among the three and is best suitable for comparing two layout databases with similar hierarchical structure.

COMXZ is a companion program to COMXY. It flattens the hierarchy, partitions the chip and performs the comparison on geometries which have been transformed into a canonical form (polygon). Therefore, it is suitable for comparing two databases with different hierarchical structures or different syntactical descriptions.

COMXM is a program which reconciles the differences between merged and fractured figures. Since it involves considerable processing, it should only be used as a last resort.

These three programs have been designed in such a way that they can be run in a "pipelining" fashion (see figure 6) with COMXY at the beginning followed by COMXZ and COMXM. The output from one program can be directly fed into the next program to achieve maximum throughput.

COMXY, COMXZ and COMXM have been used in IC design projects for layout change control. They have proven to be very effective and efficient tools for that purpose.

6. ACKNOWLEDGMENTS

The author is indebted to Hermann Gummel and Kwok Wu for their encouragement and support.

REFERENCES

[1] A. K. Bose, B. R. Chawla and H. K. Gummel, "A VLSI Design System", 1983 IEEE Proceedings of International Symposium on Circuits and Systems, pp. 734-739.

[2] B. R. Fowler, "XYMASK", Bell Laboratories Record, Vol. 47, No. 6, July, 1969, pp. 204-209.

[3] "GRED - an interactive graphical editor for layout", to be published.

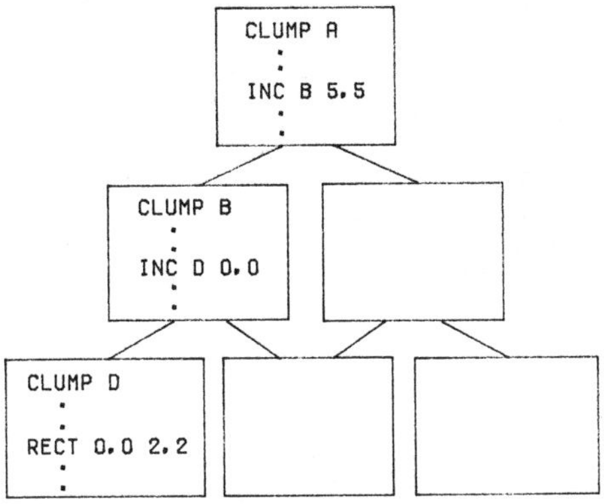

FIGURE 1 UPWARD PROPAGATION

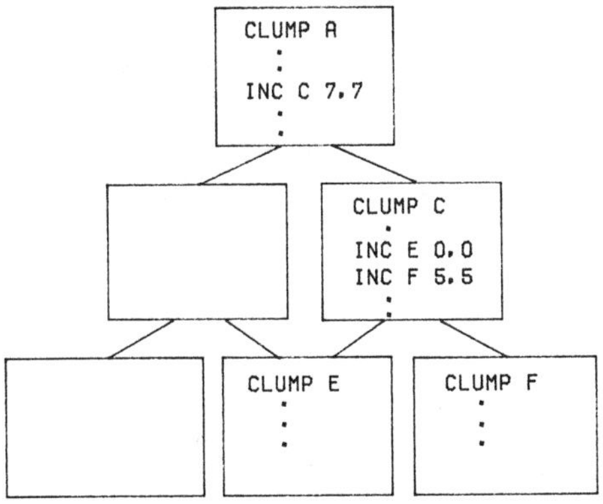

FIGURE 2 DOWNWARD PROPAGATION

```
OLD   CLUMP A
      REPEAT B 0.0 4 100.0
      END CLUMP
----------------------------
NEW   CLUMP A
      INC B 0.0
      INC B 100.0
      INC B 200.0
      INC B 300.0
      END CLUMP
```

FIGURE 3 HIERARCHICAL FALSE ALARM

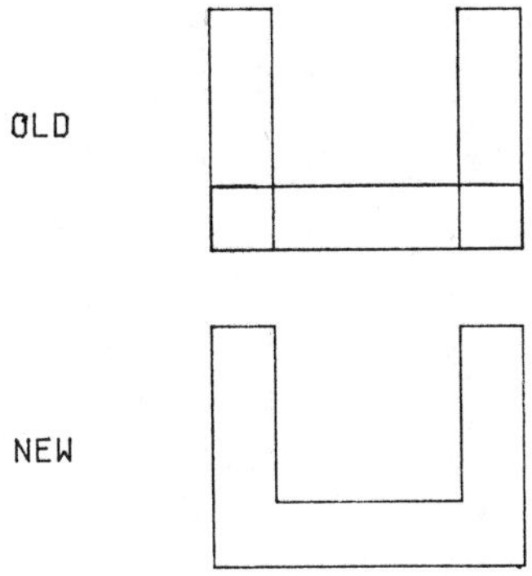

FIGURE 4 FIGURE-MERGING

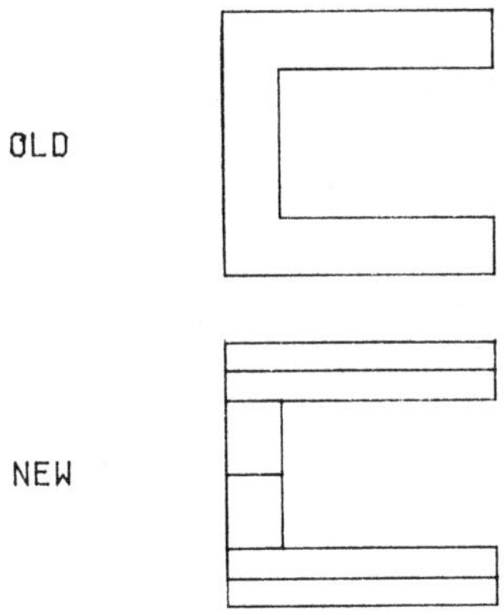

FIGURE 5 FIGURE-FRACTURING

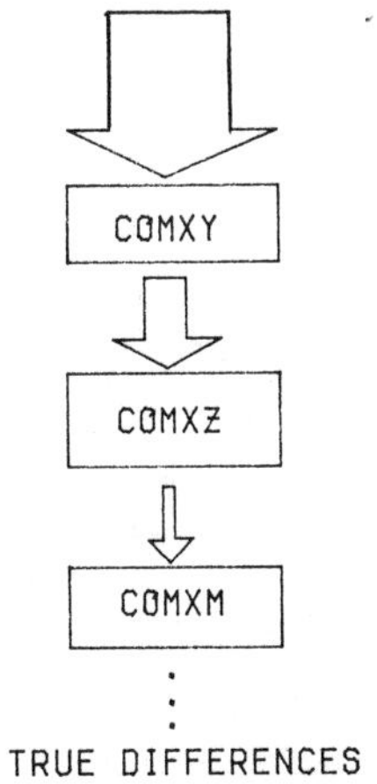

FIGURE 6 COMPARISON SYSTEM OVERVIEW

Session 2-1d
Chip Architecture

Chairman

E. Swartzlander
TRW

ALGORITHM TRANSFORMATIONS IMPROVING CONTROL PART IMPLEMENTATION

Monika Obrębska

Computer Architecture Group , IMAG BP 68

38402 St Martin d'Heres Cédex , FRANCE

ABSTRACT.

We present several basic methodologies used in the design of control parts of microprocessors. We discuss their classification with respect to the qualities of the corresponding design expressed in terms of the cost of hardware, the cost of design and the performance. We show how these different implementations may be obtain when modifying the description of the instructions interpretation algorithm. Lastly we study the evolution of the design efficiency versus the circuit (algorithm) complexity.

1.INTRODUCTION.

Technological improvements induce not only the increase of the total circuit area but also the increase of the transistor mean density. As a consequence the total number of transistors integrated on a chip is growing up and by the same time the total circuit complexity. This means mainly, in the microprocessor design domain, the increase of the instructions interpretation algorithm complexity. This is a crucial problem especially in control part design where using different methodologies may lead to quite different qualities of the layout.

2.PRINCIPLES OF INVESTIGATION.

In our work we were interested in the comparison of different control part design methodologies and in the establishment of the referencial classification which should help the designer in his decisions. This has been done in three stages.

First the basic principles of different design styles used by the IC manufacturers were explicited. Secondly each of these styles was applied to the same exemple (algorithm) in order to obtain a set of normalized solutions. At last the results were extended for all other general purpose microprocessors.

3.OUTLINE OF CONTROL PART DESIGN METHODOLOGIES.

As a matter of fact, five basic design styles were brought out by the microphotographic analysis of the internal architecture of existing circuits. These are : the random logic style used in MC6800 [14] and in MC 6809 [1], the single PLA style [6] used in INS SC/MP and INS 8070 [7], the microprogramming style used in MC 68000 [5],[9], HP 9000 [16], TI9900 [10] and 52752 FAST, the multiple PLAs style used in GI CP1600 [3], HP MC2 [4],and CDP 1802 [19], and the design using timing generators employed in I8080, I8085 [12], Z80 [11] and Z8000 [18] microprocessors. But each of these methodologies was used to built a different machine, executing a different algorithm and often implemented with different technology so they could not realy be compared.

4.ALGORITHM TRANSFORMATIONS.

The idea was to apply each of these methodologies to the MC 6800 microprocessor and to obtain the corresponding hardware realisation of the same algorithm [2],[17]. By doing so we found a set of comparable solutions each reflecting the efficiency of the specific method. We found also the necessary modifications of the algorithm description leading to the corresponding solutions.

In fact we shall remember that in any microprocessor the control part is an automaton supervising the data processing and the flow of informations. We can also say that the control automaton is executing the instructions interpretation algorithm. The behaviour of this automaton can be described in different manners, but in general we distinguish two basic forms of description.

The Moore Form where the automaton is defined as a quintuple

(S,I,O,d,l) and
S is a set of states
I is an input vocabulary
O is an output vocabulary
d : S x I → S is a transition function
l : S → O is an output function.

When this form of description is used it means that all the commands generated by the control automaton (output vocabulary) are depending only on the present state of the automaton.

The Mealy Form where the automaton is defined as a quintuple

(S',I,O,d,l) and
S' is a set of states
I is an input vocabulary
O is an output vocabulary
d : S' x I → S' is a transition function
l : S' x I → O is an output function.

This form of description implies that the commands generated by the control automaton are depending not only on the present state of the automaton but also on the present input vocabulary.

The behaviour of the same automaton may be often described in both Moore or Mealy Forms. It is obvious that the set S contains in such a case greater number of states than the set S'. As a consequence the implementation of the control automaton described in Moore Form would need more hardware. On the other hand the implementation of the Mealy Form is more difficult because the extraction of d and l functions is not always evident.

The efficiency of each design methodology was estimated in terms of the cost of hardware, the cost of design time and the performance. The cost of hardware was evaluated as a function of the total area of the circuit. The time and the ease of the design were estimated according to the percentage of structures which can be generated automatically like ROMs or PLAs in each solution. We called this percentage regularity factor. The performance was expressed by the cycle time of each control part. The ratio of the new cycle time over the MC 6800 original cycle time was called the speed factor (figure 5).

RANDOM LOGIC. The MC 6800 original control part is implemented with random logic (figure 1) and generates about 80 commands for the operation part. It is executing an interpretation algorithm which described in Moore Form has 167 different states. First we considered several solutions implementing the interpretation algorithm described in this form.

SINGLE PLA. The single PLA implementation (1 PLA) was obtained from the equations giving the list of activation states for each command and the list of the former states for each state. Notice that the d and l functions are computed in the single unit. In the layout proposed for this solution the OR matrix was optimized, but the global characteristics are poor.

MICROPROGRAMMING. In microprogramming approach each state of the interpretation algorithm is defined by its microaddress. Once the microinstruction format is defined, we obtain the microprogramme by a direct transformation of the interpretation algorithm. This is very simple in horizontal microprogramming but the qualities of the corresponding layout are not very good (uProg H). In vertical microprogramming the commands are coded allowing to decrease the ROM area (uProg V).

In these solutions the interpretation algorithm was described in two parts. One part describes the output function (in ROM) and the second part describes the transition function (implemented either with the random logic or with the PLA structure). The description of the decoding functions must be given to complete the vertical solution.

Among above three solutions, only the vertical microprogramming methodology is acceptable comparing to the random logic approach.

We saw that if we extract several instructions properties directly from the operation code then the MC 6800 interpretation algorithm can be easily described in Mealy Form. These global properties, such as operand register designation, addressing mode, fetch type, condition code modification etc..., are used to generate in the same algorithm state different commands for the operation part. In this way we obtained the description of the interpretation algorithm which has no more than 71

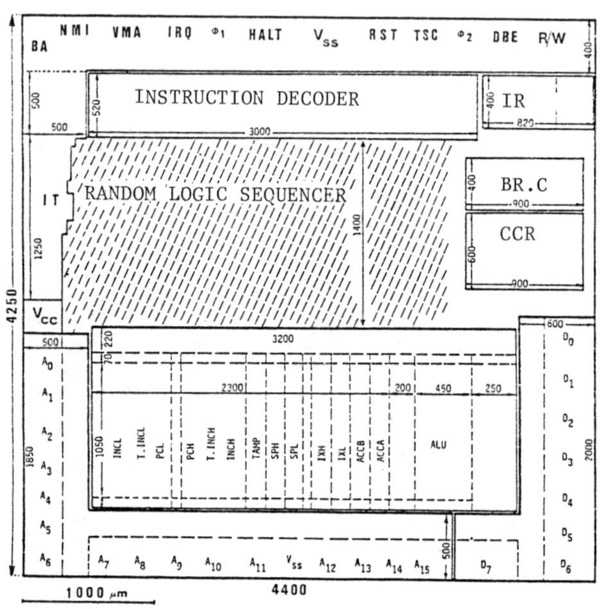

Figure 1. Original MC6800 layout.

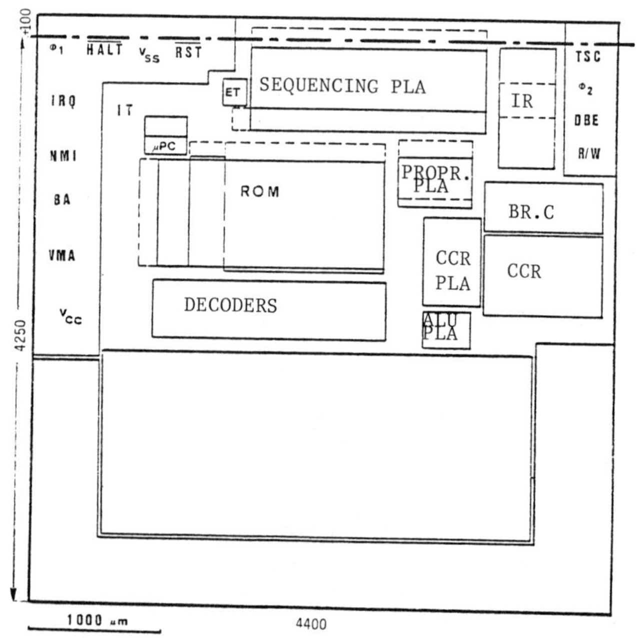

Figure 2. Layout for the solution with parametered microprogramming.

different states. By the same the realization of the sequencing (transition function) is simplified.

The implementation of the interpretation algorithm described in this form may be done using either the multiple PLAs methodology or the microprogramming methodology.

MULTIPLE PROPERTY PLAs. In this case we use basicly three PLAs : the property PLA, the sequencing PLA and the command PLA. The command PLA takes into account the algorithm state comming from the sequencing PLA and the conditions comming from the property PLA . The layout proposed for this solution prove that it is an interesting methodology: all its characteristics (m.Prop.PLAs) are better than those of the previous solutions.

PARAMETERED MICROPROGRAMMING. In this case we studied two different layouts where the coded commands are stored in a ROM but the sequencer is implemented with random logic (uProg.PA) or with a PLA (uProg.PP),(figure 2). The property PLA used here is the same as in the multiple PLAs solution. The special decoders are taking into account the property lines and the coded commands to generate the final signals for the operation part. These two solutions are very satisfying because they are easy to design, performant and not too costly.

MULTIPLE PARAMETER PLAs. Another kind of multiple PLAs design methodology can be based on parameters extraction i.e. the local properties of functional units. This properties are extracted from the operation code and translated directly into the sets of commands by a specialized parameter PLAs. This approach decreases the number of different states in the interpretation algorithm to 56. The layout proposed (figure 3) for this solution is very regular and the total circuit area is not increased (m.Para.PLAs).

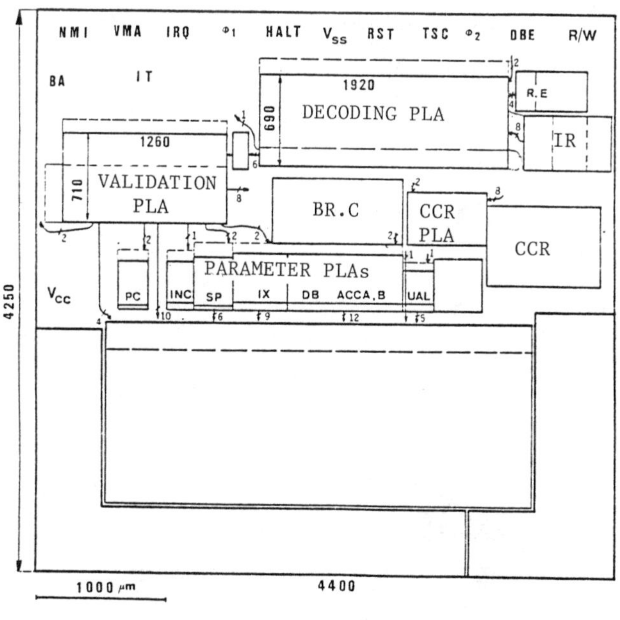

Figure 3. Layout with multiple PLAs.

TIMING GENERATORS. The methodology using timing generators is based on the fact that each command is characterized by the "moments" of its activation during the instruction interpretation. In existing microprocessors the interpretation algorithm is described at 1-st level of interpretation in terms of major cycles and at 2-d level of interpretation in terms of minor cycles. The layout proposed for such a solution is very satisfying (T Gen. 2Ls.). The major cycle automaton has 8 different states and the minor cycle automaton has 6 states.

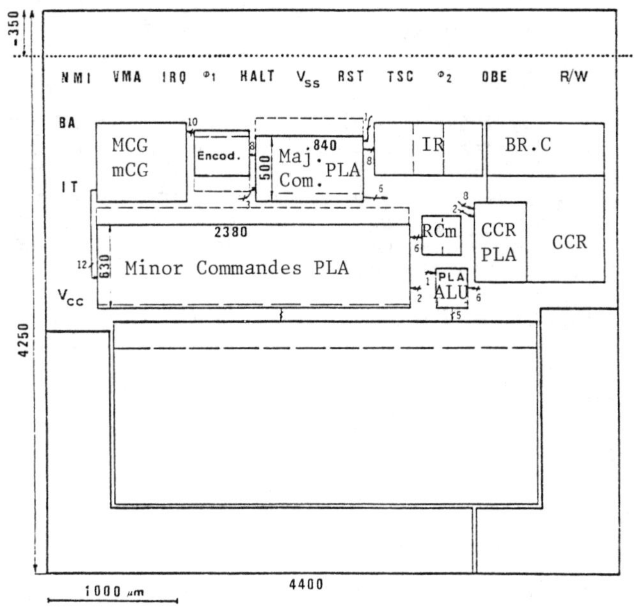

Figure 4. Layout for 2 levels timing generators.

We found that the decomposition and the parametrization of the interpretation algorithm allow to express it in the Moore form. This form of description may be implemented with several methodologies giving very good qualities of the design. The best solutions seem to be obtained for the most parametrized descriptions, where the number of algorithm states is minimized (figure 6). The main problem to solve now is the automatization of the algorithm analysis in order to transform the Mealy description into the desired Moor form.

	Area Increase %	Regularity Factor %	Speed Factor
MC 6800	0	15	1
1 PLA	40	53	2.08
uProg H	26	55	1.66
uProg V	15	49	1.60
m.Prop.PLAs	11	57	1.15
uProg. PP	2.3	45	1.06
uProg. PA	1.2	38	1.23
m.Para.PLAs	0	44	1.30 -> 1.17
T Gen. 1L.	-2	34	0.74
T Gen. 2Ls.	-8	35	0.74

Figure 5. Efficiency characteristics of different design methodologies.

5. GENERALISATION.

The complexity was defined as the number of states of the equivalent Moore automaton executing the interpretation algorithm. We made the assumption that all the characteristics of the algorithm are linear functions of the complexity. In this way, the curves showing the evolution of the total control part area versus complexity could be obtained. To validate these results, we have examined the MC 6809 microprocessor: its point (466, 18.12) plotted in figure 7 confirms our generalisation.

6. CONCLUSIONS.

Our study shows that the use of regular structures will be necessary to menage the complexity of the future microprocessor circuits [15]. Time and facility of the design are becoming more important that the total area of the circuit. The regular structures are much easier to design, modify and test than the random logic circuits. They must be used as optimised building blocs in the modular architecture approach to decrease the cost of the global layout. The appropriate CAD tools should be implemented to insure the automatic analysis of the interpretation algorithm [8],[13], the choice of the best solution and the generation of the corresponding masks.

REFERENCES.

[1] J.Abdo, F. Rodriguez, Analysis of the MC 6809 Master's Report, ENSIMAG, June 1981.

[2] F.Anceau, Architecture and Design of Von Neumann microprocessors, NATO Advanced Summer Institut July 1980.

[3] C.Bernard, Analysis of the CP1600 micropressor Master's Report, ENSIMAG, June 1979.

[4] C.Bernard, Analysis of the MC2 microprocessor IMAG Report, June 1980.

[5] A.Bosseboeuf, Analysis of the MC68000 microprocessor, IMAG Report,June 1980.

[6] J.P.Braun, Design and implementation of VLSI circuit with CMOS/SOS technology, Master's Report, CNAM, June 1978.

[7] H.Derantonian, Analysis of the INS807X microprocessors, IMAG Report, September 1981.

[8] H.Derantonian, PAMELA: a system for the generation of PLAs sequencers,to be published.

[9] M.Guitet, Microprogramming in MC68000 micropro cessor, Master's Report, ENSIMAG, June 1981.

[10] K.Guttag, Data compression shrinks microproces sor's control store, Electronics, January 1983

[11] A.Guyot, Comparison of the Z80 and I8085 microprocessors, IMAG Report, September 1979.

[12] V.S.R.Malladi, Analysis of INTEL microprocessors,IMAG Report, September 1979.

[13] S.Marine, F.Anceau, K.Jahidi, IRENE : a language for the description of integrated circuits IMAG Report , March 1983.

[14] M. Nemmour, Analysis of the MC 6800 execution, EDF/IMAG Report, May 1979.

[15] M. Obrebska, Comparative Survey of Different Design Methodologies for Control Parts of Microprocessors, Doctor Thesis, June 1982.

[16] D.Patterson, The microprogramming, Scientific American, May 1983.

[17] E.Presson, Sequencing in the MC 6800 microprocessor, NPL Report, IMAG, 1978.

[18] R.Reis, Analysis of the Z8000 microprocessor, IMAG Report, September 1980.

[19] R.Sarwat, Analysis of the CDP 1802 microprocessor, IMAG Report, June 1979.

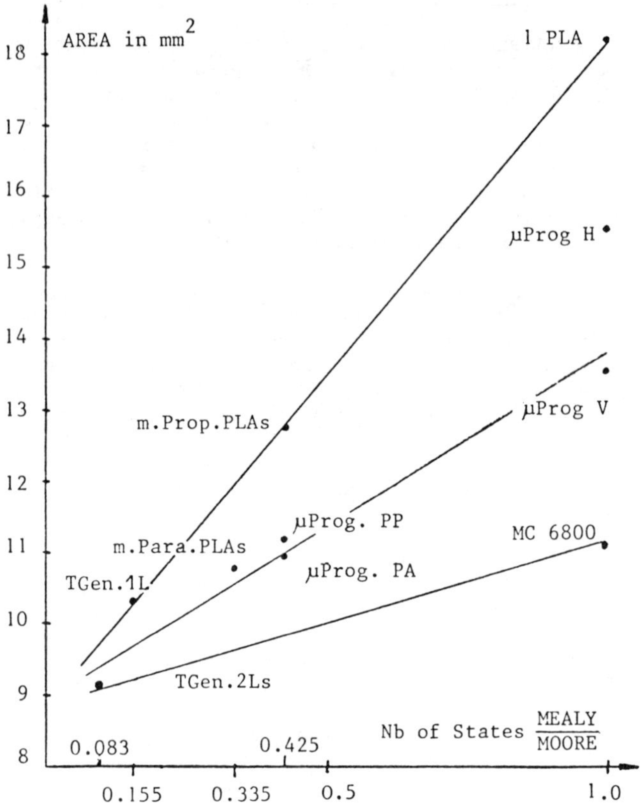

Figure 6. Algorithm Transformations improving control part area.

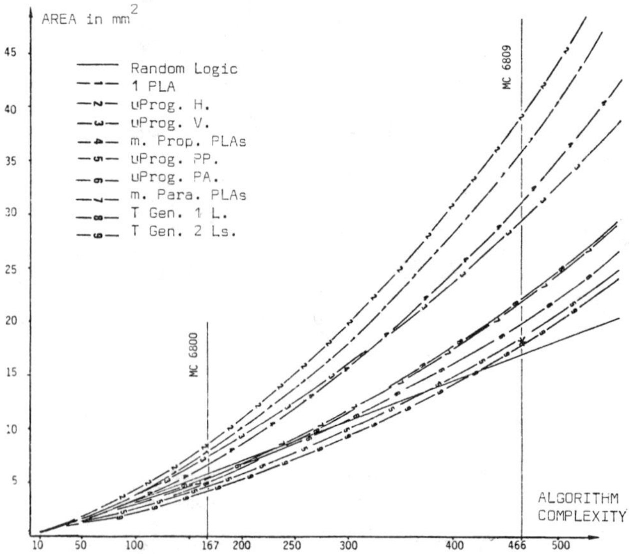

Figure 7. Area versus algorithm complexity.

A VLSI IMPLEMENTATION OF THE 32-BIT ECLIPSE ARCHITECTURE

Chandra R. Vora, Mark Hecker, Robert Murdoch*, Donald Wiser, Steve Hamilton, Jerry Wang*, Nabil Takla*

Data General Corporation, 4400 Computer Drive, Westboro, MA 01580
*Data General Corporation, 433 Mathilda Ave., Sunnyvale, CA 94086

Advanced NMOS processing with titanium disilicide to reduce line resistances in dimensions down to two micrometers makes possible a VLSI implementation of the Data General 32-bit ECLIPSE architecture. With a chip set codenamed microEAGLE containing approximately 300,000 devices, high performance mainframe capabilities will become available for small, cost effective products.

The five chips that comprise the microEAGLE chip set are the CPU, the microsequencer (uSEQ), the Floating Point Arithmetic unit (FPU), the System Input/Output unit (SIO), and the Burst Multiplexer Channel unit (BMC). The five chips can be configured in a low-cost single bus design (Figure 1) or a high performance dual bus design (Figure 1A). In the single bus design, the CPU, SIO and BMC chips contend for one bus. In the dual bus design, the SIO and BMC contend for one bus, while the CPU uses the other.

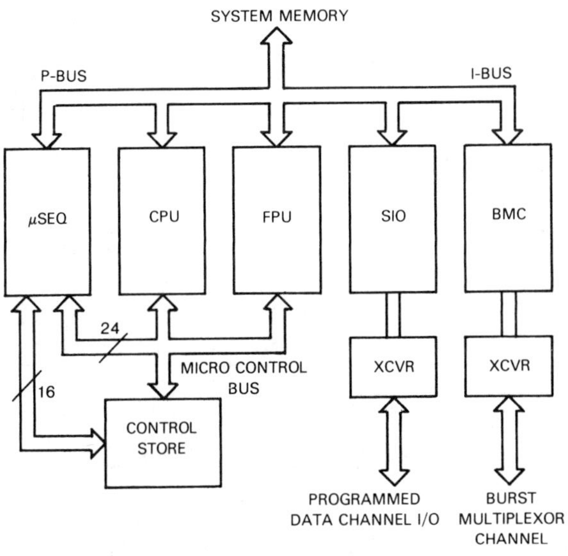

FIG. 1 A TYPICAL SYSTEM USING VLSI CHIP SET

The microEAGLE chip set runs the existing ECLIPSE MV/series instruction set and addresses 4 Gbytes of virtual memory with true demand paging techniques. It has the ability to save the complete microstate of an instruction in case of a page fault so that complex instructions can be resumed where the page fault occurred rather than at the beginning.

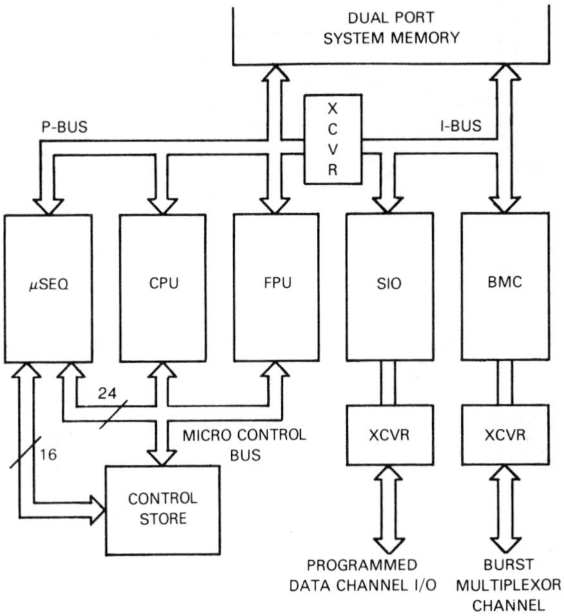

FIG. 1A A TYPICAL DUAL BUS SYSTEM USING VLSI CHIP SET

The Central Processing Unit Chip

The CPU (Figure 2) is a true 32-bit chip which performs all fixed point operations and addressing in the microEAGLE system. It contains an ALU, four accumulators, eight general registers, and twelve special purpose registers, each 32 bits wide. It consists of five major units: the fixed point arithmetic unit, the address translation unit, the autonomous fetch unit, the control unit, and the external interface.

The fixed point arithmetic unit consists of a dual port register file, an ALU and a shifter. There are two 32-bit busses between the register file and the ALU. One bus is multiplexed between memory data and one register file source, and the other bus is multiplexed between the second register file source and the ALU result. With a high

CH1935-6/83/0000/0311$01.00 © 1983 IEEE

degree of parallelism, the CPU is able to perform 32-bit register-to-register operations in a single cycle, and 32-bit register-to-memory arithmetic operations in three cycles.

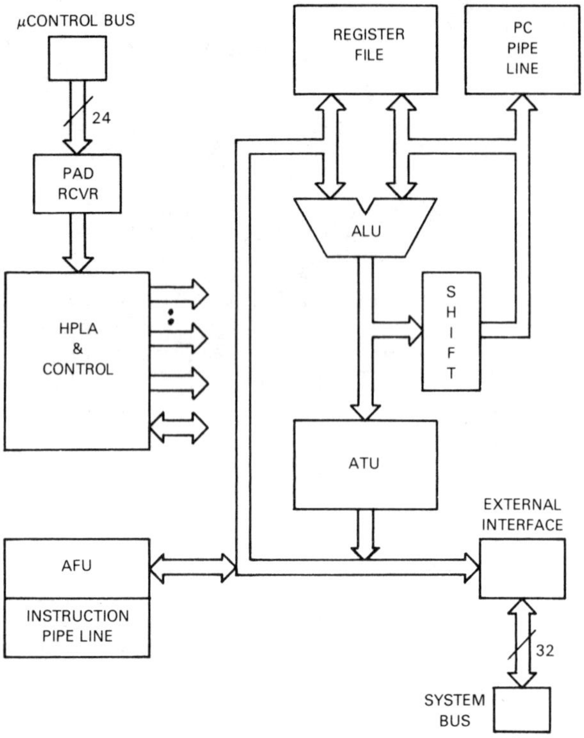

FIG. 2 CPU BLOCK DIAGRAM

An on-chip address translation unit (ATU) provides memory management in 2 Kbyte pages via 16 address translations in a fully associative cache, supporting a physical address space of 128 Mbytes. The on-chip design permits an address calculation and subsequent translation in one cycle. The ATU also supports the Long Address Translation (LAT) process. Performed by microcode, this process translates a logical address when the translation is not in the ATU cache and saves the context block during page faults when necessary.

The CPU has an autonomous instruction fetch unit (AFU) which permits simultaneous fetch, decode and execution. It consists of the instruction register (IR) pipeline, and the program counter pipeline. The IR pipe holds five words of the instruction stream, which can be used for displacements and immediates. The AFU initiates a prefetch operation whenever the ATU has a valid translation for the required memory address and the system bus is not being used for any other transactions.

The CPU interfaces with the other system elements via the system bus and the micro control bus. The system bus is a tri-state, bi-directional 32-bit wide address/data bus. When carrying a 26-bit address, the remaining six bits specify control information regarding bus transfers. In addition to system memory accesses, the bus is used to perform interchip transfers, and console ROM, ECO RAM, and microcode RAM accesses.

The micro control bus is a 24-bit wide bus that connects the uSEQ and external control store memories with the CPU and FPU.
The CPU has an autonomous sequencer that takes 24-bit vertical microinstructions from the micro control bus and performs one or more 64-bit horizontal microinstructions as specified and modified by the vertical. Exception conditions such as LATs and memory waits are handled in the horizontal microcode without interrupting the vertical microinstruction sequence.

The microSequencer Chip

The uSEQ chip (Figure 3) performs instruction decoding and vertical microcode sequencing. It contains an instruction queue, decode logic and microaddress generation logic.

Each double word of instructions fetched from memory by the AFU within the CPU is simultaneously loaded into the uSEQ chip. Both chips place the received double word into their respective instruction register

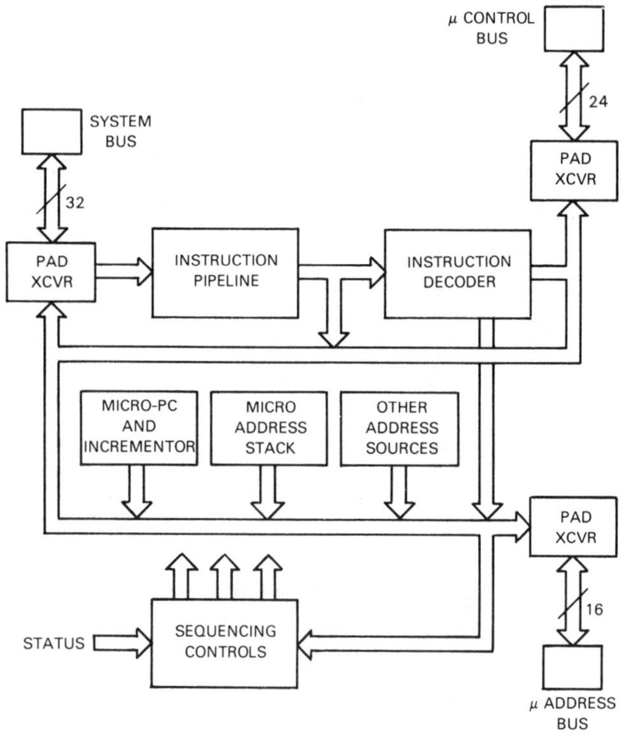

FIG. 3 MICRO SEQUENCER BLOCK DIAGRAM

pipelines. As the execution of each macro-instruction is completed, the uSEQ decodes the next instruction in the pipe. Control is provided by the CPU chip to allow proper maintenance of the uSEQ IR pipeline in the event of branching or the extraction of displacements or immediate operands from the pipe.

The macro-instruction opcode to be decoded is presented as input to a programmable logic array called the Starting PLA (SPLA). The SPLA provides the sequencing information and the first 24-bit vertical micro-instruction to be sent to the execution chips. The sequencing information typically specifies the address in the external control store of the next microinstruction to be executed. Having the first microinstruction on-chip eliminates the need of waiting to fill the microinstruction pipeline when decoding a new instruction. The entries in the external control store each consist of further sequencing information and another 24-bit microinstruction.

The sequencing information from the external control store is processed by the micro-sequencer to generate the next micro-instruction address. It is accomplished using sequencing logic within the uSEQ chip. This logic includes the micro-PC, six entry stack, decision logic, and an 8-bit counter. The uSEQ also hases the ability to save its state in the event of a page fault.

Conditional microinstruction aborts further enhance performance. The uSEQ is two micro-instructions ahead of the CPU and FPU, yet depends on them for the results of tests for branches. Using conditional aborts, the microcode can, for example, assume the branch chosen is a loop branch, thus optimizing the iterative looping case. If the alternate branch is taken, the currently executing microinstruction is aborted and the micro-instruction pipeline is flushed.

Vertical microcode is stored in either ROM or RAM-based control store. In order to accommodate engineering changes, the uSEQ chip contains an associative memory which can override the selection of up to eight addresses in the ROM control store and divert the control flow to the writable control store. Hardware is also provided to aid in the loading and verification of the writable control store.

Floating Point Arithmetic Unit Chip

The FPU chip performs the floating point instruction set of the ECLIPSE MV architecture. It contains 64-bit data paths, 64-bit accumulators and general registers, and independent mantissa and exponent ALUs. A 64-bit hex shifter prescales or normalizes operands in one cycle. The unit performs 4 bit/cycle multiply and 2 bit/cycle divide. The multiply hardware uses carry save adder techniques. These features result in enhanced floating point instruction performance shown in Table 1.

OPERATION	VLSI SYSTEM μS	ECLIPSE MV/8000 μS
ADD REG. + REG.	0.4	0.22
LOAD MEM. REG.	0.8	0.44
F.P. ADD 64-BIT	1.6	0.88
F.P. MULT. 64-BIT	7.6	3.96
F.P. DIV. 64-BIT	13.2	6.93
JUMP	1.2	0.66
PUSH 4 32-BIT REG.	4.0	2.42

TABLE: PERFORMANCE COMPARISON WITH ECLIPSE MV/8000 SYSTEM

There are three major functional blocks in the FPU: the execution unit, the external interface, and the control unit. The execution unit (Figure 4) consists of three principal sections: the sign logic, the mantissa ALU, and the exponent ALU. Two busses provide input to these three sections. The A bus supplies data from the register file to the ALUs, and the B bus supplies data from the register file and other functional blocks. The B bus is time multiplexed to act as a destination bus.

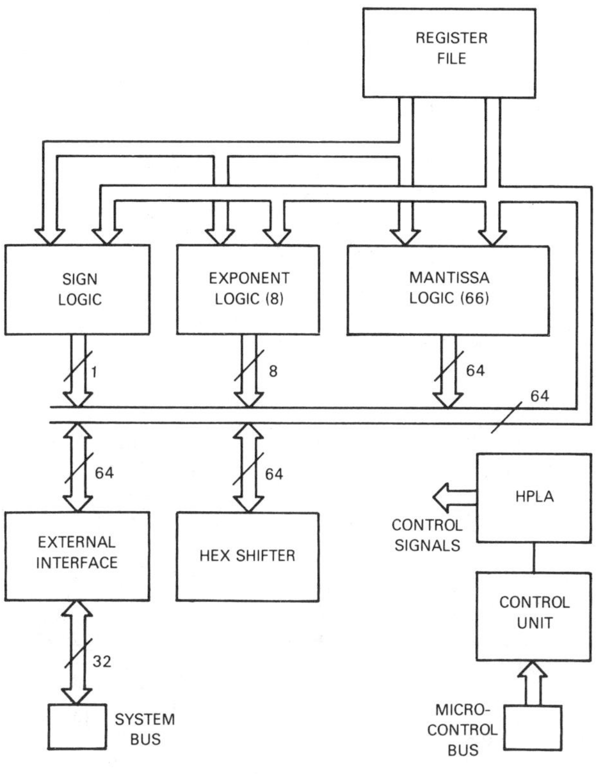

FIG. 4 FPU BLOCK DIAGRAM

The core of the execution unit is the carry save/propogate adder (CSPA). This is a 66-bit wide arithmetic unit which provides the combined functions of a carry-propagate adder and a carry-save adder. The data inputs to the CSPA contain the data latches and the data paths necessary to implement the multiply and divide algorithms. The shifter is used to shift the data on the B bus either left or right in groups of 4 bits at a time.

The external interface interacts with the system bus, sourcing data to, and receiving data from the other elements of the system. To optimize floating point data transfers, the external interface accesses system memory using addresses generated by the CPU. The external interface unit consists of a dual-ported 64-bit bus interface register, the floating point status register, state save paths and control logic.

The control unit receives microinstructions from the uSEQ chip via the microcontrol bus and generates control signals for the chip. The same two level microcode structure is used on the FPU as the one used on the CPU. The CPU and FPU share a single horizontal address space, with the CPU recognizing some addresses, and the FPU others, allowing CPU-oriented and FPU-oriented verticals to be mixed in the microcode stream. Some verticals are recognized by both chips, permitting concurrent operations (see Figure 5).

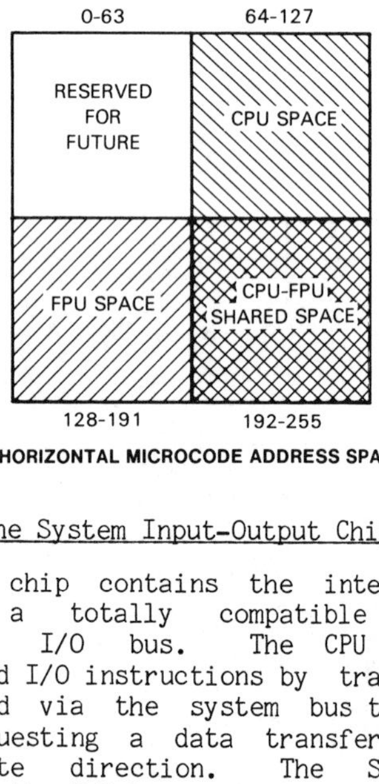

FIG. 5 HORIZONTAL MICROCODE ADDRESS SPACE

The System Input-Output Chip

The SIO chip contains the interface to generate a totally compatible ECLIPSE MV/series I/O bus. The CPU executes programmed I/O instructions by transmitting a command via the system bus to the SIO chip, requesting a data transfer in the appropriate direction. The SIO chip generates the I/O bus control signals needed to perform the transfer over the I/O Data bus.

In addition to the I/O functions performed by the SIO chip, it provides a number of standard I/O device such as, a real time clock, a programmable interval timer, a power fail monitor, and an asynchronous line interface. When programmed I/O commands are directed to these devices, the SIO chip logic recognizes the device codes and responds accordingly.

The Data Channel controller within the chip performs direct transfers of data between the I/O bus and system memory. The Data Channel map, which contains the logical to physical address translations for these transfers, resides in physical memory. Recently used address translations are cached on the SIO chip. The chip also provides data prefetching and caching for data channel output, and buffering for data channel input.

The SIO chip must be able to operate correctly at a variety of different system clock frequencies. It contains a 32-bit wide Configuration Register and a 16-bit Control Register to specify the operating parameters. Proper logical operations occur over a frequency range extending from 10.66 MHz to 26.66 MHz, allowing for production variations and future advances in technology.

Burst Multiplexor Channel Chip

The BMC chip contains the interface to generate a compatible ECLIPSE MV/Series BMC I/O bus. The ECLIPSE Burst Multiplexor Channel transfers bursts of data at high speed between system memory and an I/O device controller without CPU involvement. Bursts may vary in size from one to 256 16-bit words of data. The BMC chip transfers data at approximately 8 Mbytes/sec, multiplexing transfers from up to eight controllers.

In addition, data prefetching and caching for output, and buffering for input are performed. The BMC prefetches 4 words of data per device in anticipation of sequential memory accesses. The BMC also provides 8 words of data buffering to allow BMC bus activity to continue while the higher priority SIO chip performs memory transfers.

Conclusion

The chip set design is based on 2 micron design rules and uses a single 5V supply with TTL-compatible levels on I/O signals. The system clock uses eight-phase logic at 20 MHz. The chips are packaged in 84 to 100 pin packages. The microEAGLE chip set provides approximately 50 per cent of the performance of the ECLIPSE MV/8000 (see Table 1).

Acknowledgements:

The authors would like to acknowledge Thomas Morin, Andreas Meyer, William Baxter, Mark Travaglio, Galdino Rosas, Dan Blasinski.

RIDGE 32 ARCHITECTURE -- A RISC VARIATION

Ed Basart and Dave Folger

Ridge Computers
586 Weddell Drive
Sunnyvale, CA 94089

ABSTRACT

A recent computer architecture innovation is the Reduced Instruction Set Computer (RISC), exemplified by the RISC I [1] at the University of California, Berkeley, the MIPS machine [2] at Stanford, and the 801 [3] at IBM. RISC architecture contrasts with the traditional, complex instruction set machines like the DEC VAX-11 [4] and the IBM System/38 [5]. The Ridge 32, a high speed "graphics engine," is the first Reduced Instruction Set Computer in commercial production.

This paper shows how superior performance is designed into the Ridge 32 hardware, and how a small instruction set can out-perform more complex systems.

ARCHITECTURAL GOALS

L. J. Shustek [6] found that LOAD, CALL, and BRANCH instructions usually appear in compiled code more often, are executed more often, and take longer to execute than other instructions. It was also found that simple instructions are used most frequently in code generation, and that an operation performed with a sequence of simple instructions often executes faster than a corresponding sequence of supposedly more powerful complex instructions.

The Ridge 32 architecture was inspired by these observations, and by the need for a general register machine with a simple instruction set on which to implement high level langauges.

Ridge Computers was formed in May, 1980, to build a high-performance, moderately priced engineering and scientific workstation. The Ridge 32 functionality fulfills a need for on-line engineering analysis and high-speed graphics. Certain RISC design criteria were established for the Ridge 32:

- o The Ridge 32 must be a high performance computer on a workstation budget.
- The reduced instruction set must be simple and consistent to facilitate implementation of high level languages. (Ridge currently supports Pascal, FORTRAN, and C, with others planned.)

Additional Ridge criteria, which differ from the Berkeley RISC project, include the following:

- To satisfy scientific needs and support bit-mapped graphics, a complete set of instructions (including floating point, multiplication, division, and bit manipulation) must be performed in hardware.
- The Ridge 32 must not be constrained to one-chip VLSI architecture. A multi-chip design allows it to be compatible with a family of computers that will outlive its first implementation.
- To accomodate very large application programs, the Ridge 32 must have paged virtual memory and full 32-bit addressability.
- The Ridge 32 must have fast context switching supported by hardware to allow the use of many I/O devices, multiple processes, and multiple users.
- Simplicity and efficacy in design will be a higher priority than minimizing the instruction set.

INSTRUCTION SET DESIGN

Ridge supports 32-bit addresses, 8-, 16-, and 32-bit integers, and 32- and 64-bit real numbers.

The machine contains sixteen 32-bit registers. Any arithmetic or address operation can be performed on any register. They are not specialized, for example, for counting or indexing.

The use of multiple instruction formats distinguishes the Ridge 32 from other RISC

CH1935-6/83/0000/0315$01.00

machines. Half word, full word, and word-and-a-half instructions are supported. This allows full 32-bit address reach, aiding code generation for high level languages. A single 32-bit instruction format would have allowed only 16-bit addressability.

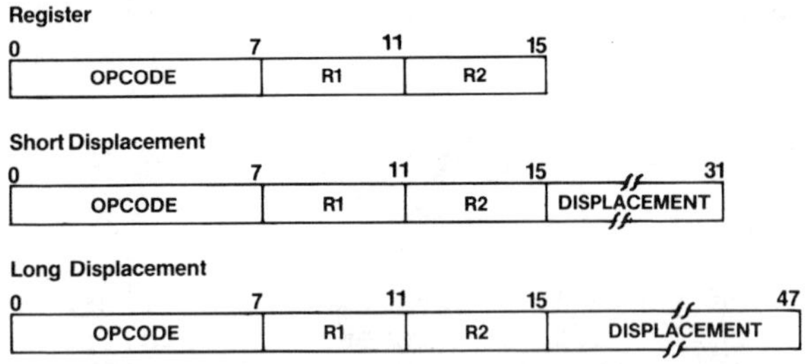

Fig. 1. Ridge 32 Instruction Formats

With just 16 address bits, accessing large arrays and inter-module calls would require the use of base registers, which complicates compilers and linkers, and results in extra instructions loading base registers. This is a generally recognized limitation on the IBM-370 series of computers that Ridge avoids by providing the long displacment instruction form.

Ridge performance benefits from the existence of half-word instructions, which reduce code size. Although the advent of 64K-bit memory chips have removed the restrictions on memory size, the problem today is now one of obtaining high enough band width for the instruction stream.

The Ridge opcode is always one byte long and contains all the information about instruction and operand type. This feature is probably the most important factor in the high performance of the Ridge 32 and the Berkeley RISC machine.

In contrast to the concise format of the Ridge instructions, the VAX has a larger instruction set with many addressing modes.

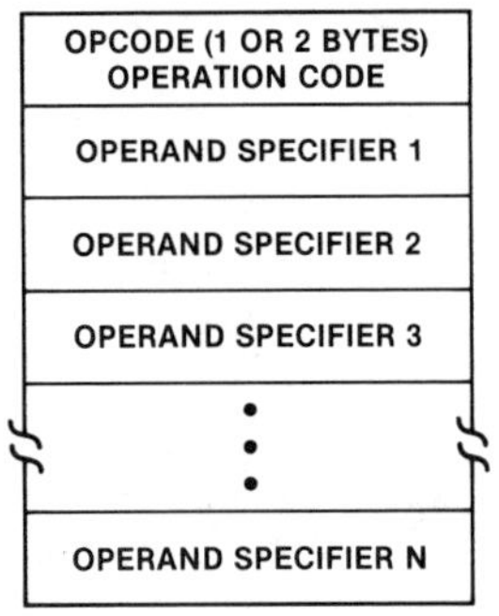

Fig. 2. VAX-11 Instruction format

The VAX requires one more clock cycle than the Ridge to execute any instruction. A VAX opcode contains a value indicating the number and type of operands that follow, and this value must be decoded in a clock cycle before the operands are fetched and executed. Ridge 32 instructions are fixed in format, so the operands can be fetched without decoding their type or quantity.

CONDITIONAL BRANCHES

The implementation of branch instructions is critical to the performance of pipelined machines. Without special handling, a conditional branch instruction empties the pipeline because the processor cannot prefetch the next instruction until the outcome of the branch has been determined. For this reason, branches can be among the slowest instructions on high performance machines. The Ridge 32 utilizes a technique to load the instructions into the pipe which are the most likely result of the branch, thus reducing the chance that the pipeline is loaded with instructions on the wrong path.

Example: Pascal REPEAT UNTIL loops usually execute more than once, so the compiler generates opcode with a branch instruction that is "predicted" to branch more often than not. (A bit is set in the opcode to designate that the specified target is predicted.) At run-time, the processor pre-fetches the instruction at the predicted target. When predicted correctly, the target instruction is already in the pipeline. When predicted incorrectly, which presumably occurs infrequently, the pipeline is cleared and the actual target instruction is loaded into the pipeline with only a 4 clock-cycle delay. With the processor pre-fetching the predicted target, the loop executes at the same speed as a linear section of code.

The IBM-370 has an architecture on which branch-prediction is difficult. It uses base registers to determine the branch target. Because the base register can be modified by instructions immediately before the branching, it is difficult to predict where a branch instruction will go if it is taken.

The Ridge 32's compare-and-branch instruction has several advantages over the traditional "branch-on-condition-code" type of instruction:

- Compare-and-branch does not check condition codes, so it saves one operation.
- Compare-and-branch does not destroy any register contents, thus preventing the

necessity of copying one of the values to compare.

- Because Ridge does not use the concept of condition codes, and does not have to spend time setting them, the instruction pipeline is simplified, and less process state information has to be stored on an interrupt or instruction abort.

IMPLEMENTATION TRADEOFFS

The Ridge 32 processor, pictured in Figure 3, was implemented with off-the-shelf Schottky bipolar logic. Future implementations of the Ridge architecture will use faster logic families and lower cost VLSI-based versions.

The Ridge processor consists of separate fetch and execute units that work simultaneously. The fetch unit always contains the next instruction to be executed, thus increasing the effective instruction stream bandwidth.

Ridge 32 performance is further increased by a cache into which all recently-executed instructions are stored. In the likely event of needing to execute those instructions repeatedly, they will be quickly accessible.

The fetch and execute units form a four-stage pipeline that allow overlapped instruction execution and shortens the overall instruction cycle. The pipeline stages are: 1) instruction decode, 2) operand fetch, 3) execute, and 4) store result. The first stage of the line begins as an instruction is fetched from the cache, or from memory, and then decoded. On the next machine cycle, a second instruction is decoded while the operands are fetched for the first instruction. The third machine cycle starts decoding a third instruction, while the second instruction's operands are fetched, and the first instruction is now executed. The fourth machine cycle stores the results of the first instruction while the second and third instructions advance to the next stage in the line. Each new machine cycle advances four instructions to the next stage of the pipeline; a new instruction enters and one instruction completes.

COST OF IMPLEMENTATION

The Ridge processor is fabricated on two 15 by 16-inch printed circuit boards, containing the equivalent of 478 16-pin DIP packages total.

Execute Board Function	% Area
Memory Interface	12
Barrel Shifter	12
ALU Control	10
ALU	8
Register File	7
Register File Decode	9
Mutiply and Divide	5
Control and Clocks	37
	100

Table 1 (continued next page)

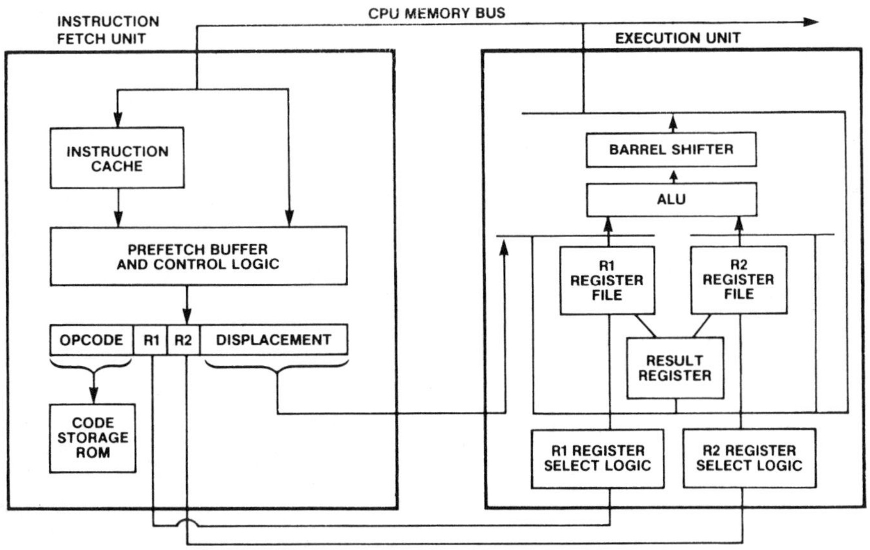

Fig. 3. The Ridge 32 Processor

Fetch Board Function	% Area
Micro-Sequencer & Control Store	21
Prefetch Control	21
Instruction Cache	14
Memory Interface, Prefetch Buffers	12
Half-Word Boundary Logic	6
Branch Prediction Logic	2
Branch Target Prefetch	4
Control and Clocks	20
	100

Table 1. Percentage of Board Area Taken by Functions

BENCHMARK RESULTS

Benchmarks verify the theoretical performance of the Ridge 32.

Machine	PUZZLE BENCHMARK Pascal, subscript version
IBM-3081	1.3 seconds
IBM-370/168	2.1
Ridge 32	2.2
RISC I	4.7
DEC VAX-11/780	11.2
DEC VAX-11/750	18.4
M68000	19.0

The Puzzle Benchmark is dominated by array references, testing, and branching. On the Puzzle, the Ridge achieves more than 4 MIPS. The results shown for the Berkeley RISC machine are for C, which performs somewhat better than Pascal. Although some of the Ridge vs. VAX performance difference can be attributed to the Pascal implementations, this benchmark shows Ridge's advantage in having simple architecture in basic integer testing operations.

Machine	WHETSTONE BENCHMARK	
Ridge 32	1,500	
IBM 4341 Group II	1,500	thousands of
VAX 11/780 with FPA	1,168	instructions
VAX 11/780 w/o FPA	751	per second
VAX 11/750 w/o FPA	331	
M68000 (8 MHz)	70	

The Whetstone Benchmark performs extensive floating point computations, procedure calls, integer arithmetic, and array references. Here, floating-point speed on the Ridge is roughly equivalent to the VAX-11/780 with the floating point accelerator. When the integer computations of the Whetstone are averaged in, the Ridge comes out noticeably faster than the VAX-11/780 with the accelerator.

Million Instructions Per Second ...	1.95
Frequency of Branch Instructions ...	11%
Branches Predicted Correctly	86%

Table 2. Ridge Pascal Compiler Performance

CONCLUSION

The Ridge 32 suggests the superior performance and commercial viability of the RISC concept. Simple architecture and a reduced instruction set allows the inclusion of virtual memory, floating point arithmetic, and bit instructions in an engineering workstation. Higher performance can be achieved through clever pipelining and a simple instruction set, without exotic technology. Future Ridge computer designs will demonstrate that the advantages of architectural simplicity apply to both lower cost and higher performance technologies.

Regstr. format	Operands	Comments	cycles of .125 ms
ADD	R1,R2	R1:=R1+R2 (32-bit register add)	1
ADDI	R1,value	R1:=R1+value (Add small constant)	1
LSL	R1,R2	shift R1 to the left (R2 times)	3
LSLI	R1,count	shift R1 to the left (cnt times)	1
Short format	**Operands**	**Comments**	**cycles of .125 ms**
LOAD	R1,offset	R1:=contents of (PC+offset)	5
LOAD	R1,R2,offset	R1:=contents of (R2+offset)	5
CALL	R1,offset	R1:=PC; PC:=PC+offset	2#
BR	offset	unconditnl branch (PC:=PC+offset)	2*
BR	R1=R2,offset	if R1=R2 then PC:=PC+offset	2*#
BRI	R1>value,ofs	if R1>value then PC:=PC+ofs	2*#
LOOP	R1, n,offset	R1:=R1+n; if R1<0 then branch	2*#
LADDR	R1, offset	R1:=offset	1
LADDR	R1,R2,offset	R1:=(contents of R2)+offset	1

* based on short offset
\# based on correctly predicted target

Table 3. Assembly Language Instructions in Three Simple Formats

REFERENCES

1. "A VLSI RISC", Patterson, D.A., Sequin, C.H., IEEE Computer, Sept. 1982
2. "The MIPS Machine", Hennessy, J., et al., Digest of Papers -- Compcon Spring 82, Feb. 1982.
3. "The 801 Minicomputer", Proc. Symposium on Architectural Support for Programming Languages and Operating Systems, March 1982.
4. "VAX-11/780: A Virtual Address Extension to the DEC PDP-11 Family", AFIPS Conf. Proc. Vol. 47, 1978 NCC
5. IBM System/38 Technical Developments, Utley, B.G., IBM publication GS80-0237, 1978
6. "Analysis and Performance of Computer Instruction Sets", Shustek, L.J., Ph.D. thesis, Stanford Linear Accelerator Report, Stanford University, May 1978

HIERARCHICAL TOP-DOWN DESIGN AND LAYOUT OF A VLSI SINGLE CHIP MEMORY MANGEMENT UNIT FOR BELLMAC*-32 CPU

P. W. Diodato, J. Kumar, C. L. Kung, J. H. Lee, M. C. Lega, Y. T. Oh, W. W. Troutman, and M. S. Tsay
Bell Telephone Laboratories, Inc.
Murray Hill, New Jersey

and

L. H. Blendinger, A. K. Goksel, R. A. Heaton, M. S. Kaplan, and P. V. LaMaster
Bell Telephone Laboratories, Inc.
Holmdel, New Jersey

1. ABSTRACT

A memory management unit for the BELLMAC*-32 Central Processing Unit (CPU) has been conceived, implemented in silicon, and proven to be fully functional in a period of only ten months. The memory management unit (MMU) is a single chip VLSI device consisting of approximately 100,000 transistors which are implemented in a 2.5 micron twin-tub CMOS technology. The MMU is capable of operating at an internal clock rate of 8.0 Mhz. An overview of the MMU architecture and the design methodology used is presented in this paper.

2. INTRODUCTION

The MMU [1] is one chip in a family of six BELLMAC VLSI chips fabricated over the past two years. The focus of the BELLMAC project can be described as a 2.19x3.65 inch module which is used to interconnect the six BELLMAC chips. This module is used in a variety of processor applications in the Bell System, all of which demand elaborate memory management capabilities. As a result of these demands, a comprehensive memory management VLSI chip has evolved. The MMU works in conjunction with the BELLMAC-32 CPU [2]. It supports paged and unpaged segments of memory, on-chip miss-processing, a multi-user environment, and a protection scheme involving four types of privileges for each of the four levels of accessibility.

3. MMU ARCHITECTURE

3.1 System Environment

There are two modes of operation for the MMU: 1) As a co-processor to the CPU, and 2) as a memory mapped peripheral for the CPU. In the peripheral mode, all of the internal registers and descriptor caches of the MMU can be addressed as physical memory locations. The MMU goes into the peripheral mode when external logic activates the chip select pin. The operating system makes extensive use of this mode for initialization of the MMU, flushing descriptor caches and handling faults. The peripheral mode also enhances the controllability and observability of the chip, thereby improving its testability.

3.2 Address Space

The address space of the BELLMAC environment is divided into four sections. These four sections are further divided into segments which may be paged or contiguous (unpaged). The MMU supports paged and contiguous segments simultaneously. A paged segment is divided into fixed 2048 byte portions of physical memory. Each contiguous segment has a granularity of eight bytes and a maximum size of 128K bytes.

* BELLMAC is a trademark of Western Electric.

3.3 Descriptor Caches

The simultaneous support of paged and contiguous segments is made possible by the three on-chip memories, referred to as descriptor caches. These caches store the starting address, also known as the physical base address, of each segment. The three caches are configured as a single 32-word segment-descriptor cache and a two-way set-associative 64-word page-descriptor cache. A single descriptor holds such information as the physical base address, the access permissions and the various utility bits used for swapping and virtual memory management tasks. These descriptors provide mapping information between virtual and physical addresses. For each contiguous segment there is a corresponding segment-descriptor word and for each paged-segment there is a corresponding page-descriptor word. The MMU monitors the system multiplexed address/data bus. If the 32-bit address presented on the bus is a virtual address then a search is made in the MMU descriptor caches for a match, also referred to as a hit. If a hit occurs, a check for segment length and access permission is performed. In parallel with this check, the MMU proceeds with the physical address calculation.

3.4 Address Translation

The address calculation from virtual to physical is done in one of the following two ways.

1. For contiguous segments, the physical address is computed by adding the segment base address to the segment offset field of the virtual address.

2. For paged segments the physical address is computed by concatenating the page address to the page offset field of the virtual address.

Details of address translation processing are illustrated in Figure 1a and 1b.

3.5 Miss-Processing

In the event that a particular descriptor is not resident in the MMU descriptor caches at the time that the CPU requests access to a particular segment of data, the MMU on-chip processor, known as the miss-processor, takes control. The miss-processor then fetches the appropriate descriptors from physical memory and loads them into the appropriate MMU cache, thus initiating the address translation. The miss-processing function of the MMU is independent of the CPU.

3.6 Demand Paging

If a page or segment is not found in physical memory, the MMU invokes a feature called demand paging. In this instance, the MMU instructs the CPU that a required page or segment is missing in the physical memory and must be read into the referenced location in physical memory. The operating system takes the responsibility for these activities and it invokes the necessary input/output routines that swap new data into the physical memory.

CH1935-6/83/0000/0319$01.00 © 1983 IEEE

4. MMU TOPOLOGY

4.1 Functional Description

The MMU is partitioned into two register transfer datapaths and several centrally located control blocks as shown in Figure 2. The lower datapath is used to determine if the virtual address is paged or unpaged, if the access permission is valid, and the length of the particular segment in question. The upper datapath contains the arithmetic logic necessary to do the actual address translation. These abstract functional blocks shown in Figure 2 are realized physically and illustrated in Figure 3. The five large blocks sandwiched between the two datapaths represent the control logic of the MMU. A large portion of this control is occupied by the Miss-PLA, labeled PLA1, which in some sense can be considered a stand alone processor in the MMU. In total there are four independent state machines in the MMU, two are implemented in PLA form (Fault-detection PLA and Miss-processing PLA) and two in Random logic, (translation controller and input/output controller).

4.2 Physical Layout

The topology for the MMU is the result of a carefully planned effort to organize the chip in a regular topology. This topology resulted in an efficient distribution of power, ground, and special clock lines [3,4]. Figure 4 shows the major segments of power (VDD) and ground (VSS). The two major clock lines of the MMU are not shown but they flank VDD and VSS. The centrally-located control allows for an efficient distribution of control lines to the datapaths. A narrow strip of random logic, referred to as resynchronization logic, is found on the inside boundary of each datapath. This resynchronization logic is used primarily to synchronize the control data input to each datapath as a function of the four phase digital timing scheme. The inherent synchronization, which is a result of the symmetric chip topology is a key asset in the timely completion of the electrical design.

4.3 Top Down Design

The details of the MMU topology have been planned for in advance of the final chip integration with the use of a top-down hierarchical design methodology [2]. This methodology allows for the iterated design, with imitation superblock boundaries, of the entire chip. The iterative attempts to construct the chip take place in parallel with the evolving architectural and logic design. Each attempted construction is used to forecast area and performance-limiting features of the design. Problems which are observed can be corrected at an appropriate level of the design since no aspect of the design is considered final. Obviously, the faster one iteration takes place, the more iterations can be completed in a reasonable time, hence a more complete forecasting of limiting features can be detected. The early forecasting of parameters was made possible by LTX2 [5], a graphics software tool which was used to construct the chip and verify its connectivity. The top-down design scheme, aided by LTX2, works well and the superblock level design of the chip can be done, with imitation modules, in a matter of weeks. The exact information forecast by LTX2 comes in the following three forms: 1) Area of the chip, 2) Plots of the chip with inter-superblock routing present, and 3) The calculation of parasitic capacitance on each inter-superblock conductor. The early forecasting of the chip area is a key aspect of the design methodology. It is helpful when ordering the chip package many months in advance of the final layout and in the determination of tradeoffs concerning area versus desired architectural features, the latter reason being critical in order to meet cost objectives.

The plots with inter-superblock routing are helpful in the planning of terminal positions on superblock boundaries and the guiding of special-purpose signals so that they reside on low resistive paths.

Finally, the calculation of parasitic capacitance per signal is very helpful as a supplement to the electrical timing simulation done by MOTIS [6].

5. SUPERBLOCK CONSTRUCTION

The construction of intra-superblock modules for the MMU was done primarily with a symbolic layout tool known as Gate Matrix [7]. An example of the Gate Matrix symbolic coding and corresponding XYMASK [8] geometrics for a small module of the MMU is shown in Figure 5. This type of layout is design rule independent, hence it can be ported or updated to a more efficient set of design rules quickly. This updatability is an important aspect of all VLSI chips assembled in the BELLMAC family.

A second method of XYMASK generation, which was used solely for the frame circuits of the chip, has a slightly different form of symbolic coding. The XYMASK generated by this coding was geared for the special features, such as guardbands, diodes, and current limiting resistors, which are found only in the frame. Such circuits are shown in Figure 6.

Finally, the PLA generation [9] was accomplished by a "C-like" programming language. The language converts a logical truth table and the ordered list of input/output signals for the PLA, into the XYMASK of the dynamic AND-OR planes with associated support logic necessary for the realization of that function specified in the truth table.

6. CIRCUIT DESIGN

The particular type of circuit implementation for any superblock in the MMU varied due to performance and area consideration. Some superblocks were realized purely with dynamic circuits, for example, the translation adder and various length comparators found in the datapaths. Other superblocks used conventional static CMOS, such as the random logic sections. Regardless of the circuit implementation, all inter-superblock conductors were synchronized with a four-phase, digital timing scheme. The accuracy of this scheme was assured by the use of a local clock decoder technique [2]. This technique assures an absolute reference of timing phase boundary at any point on the chip, thus eliminating any race conditions which could have resulted from clock skew induced from the physical layout.

7. FABRICATION

The fabrication of the MMU was done with a 2.5 micron twin-tub CMOS technology [10]. This technology allows for the independent optimization of N and P channel transistors, an N+ substrate which guards against static latch-up and a low resistance TaSi gate material. The electrical channel length of 1.5 microns (nominally) made it possible for the MMU to work at the desired internal clock rate of 8.0 Mhz at 70^oC.

8. CONCLUSION

The memory management architecture for the BELLMAC project was proposed early in the first quarter of 1982. Ten months later in the fourth quarter of the same year, a fully functional model was made which ran at the target frequency of 8.0 Mhz (at 70^oC). The functionality was verified by testing in an actual computer system. Some chips worked at frequencies as high as 15 Mhz. The dimensions of the chip in 2.5 micron CMOS design rules are 8890 by 9980 microns, which is within 5% of the original area estimate forecast by the top-down design. The MMU was implemented with the use of a hierarchical top-down design methodology. This methodology allows for the shortest possible design time, the ability to forecast limiting features of the design and the optimization of the chip for silicon area and electrical performance. The system requirements have been met and the MMU, mounted on the BELLMAC module, has been incorporated in a wide range of applications within the Bell System.

9. ACKNOWLEDGEMENTS

The authors of this paper wish to acknowledge the contributions of B. Cheng, V. Herrero, J. A. Fields, F. D. LaRocca, H. G. Nguyen, K. N. Wong and W. J. Yang. The authors also thank R. H. Krambeck, M. K. Maul, B. T. Murphy, H. C. So, and J. J. Molinelli, D. C. Fowlis and G. P. Sampson for their encouragement and counsel throughout the project.

10. REFERENCES

1. P.M. Lu, W.A. Dietrich, M.L. Fuccio, L.N. Goyal, C.J. Chen, D.E. Blahut, J.A. Fields, A.K. Goksel, F.D. LaRocca, "Architecture of a VLSI Map for BELLMAC-32 Microprocessor," IEEE Spring Compcon83, 1983, pp. 213-217.
2. D.E. Blahut, R.H. Krambeck, H.F.S. Law, H. Shichman, and H.C. So, "Hierarchical Design Methodology for a Single Chip 32 Bit Microcomputer," IEEE., ICCC 1982; pp. 16-20.
3. M. Shoji, "Electrical Design of the BELLMAC-32A Microprocessor," IEEE., ICCC 1982., pp. 112-115.
4. P.W. Diodato, R.L. Parson, W.W. Troutman, D.N. Deutsch, M.F. Jukl, M. Wiesel, "CAD Construction of a VLSI Memory Management Unit," IEEE., ICCAD 1983.
5. B.W. Colbry and J. Soukup, "Layout Aspects of VLSI Microprocessor Design," International Symp. on Circuits and Systems, Rome, Italy, May 1982, Vol. 3, pp. 1214-1228.
6. V.D. Agrawal, A.K. Bose, P. Kozak, H.N. Nham, and E. Pacas-Skewes, "Mixed Mode Simulation in the MOTIS System," Journal of Digital Systems, 1981, pp. 383-386.
7. A.D. Lopez and H.F.S. Law, "A Dense Gate Matrix Layout Method for MOS VLSI," IEEE Trans. Electron Devices, Vol. ED-27. No. 8, Aug. 1980, pp. 1671-1678.
8. B.R. Fowler, "XYMASK", Bell Laboratories Record 47, No. 6, July 1969, pp. 204-209.
9. H.F.S. Law and M. Shoji, "PLA Design for the BELLMAC-32A Microprocessor," IEEE., ICCC 1982., pp. 161-164.
10. L.C. Parrillo, R.S. Payne, R.E. Davis, G.W. Reutlinger and R.L. Fields, "Twin-Tub CMOS- A Technology for VLSI Circuits," IEEE., IEDM paper, 29.1, Dec. 1980, pp. 752-755.
11. P.W. Diodato, D.C. Fowlis, G.P. Sampson, "A Technology Updatable Microcomputer," IEEE., ICCAD 1983.

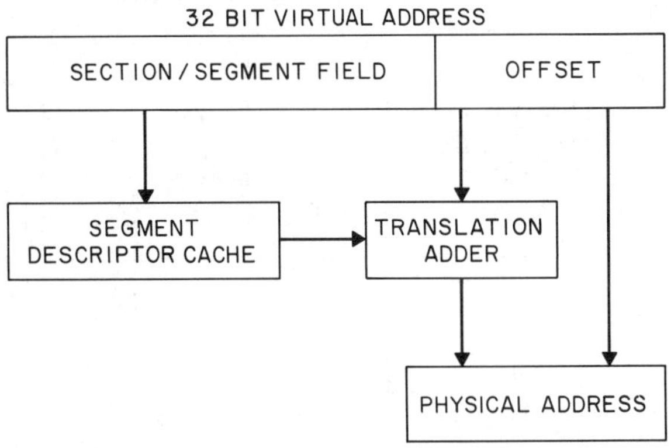

FIGURE 1A. CONTIGUOUS ADDRESS TRANSLATION

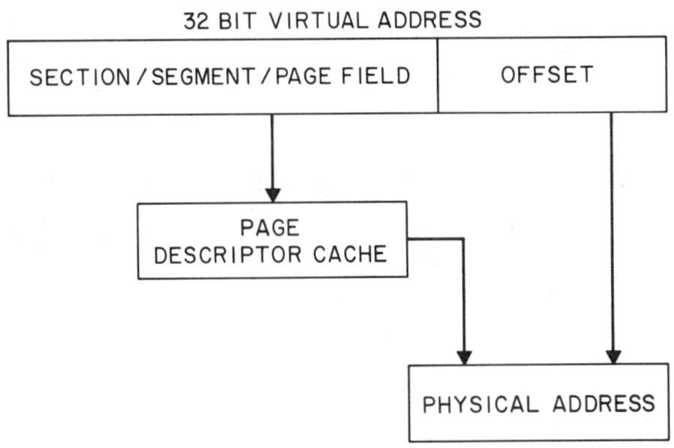

FIGURE 1B. PAGED ADDRESS TRANSLATION

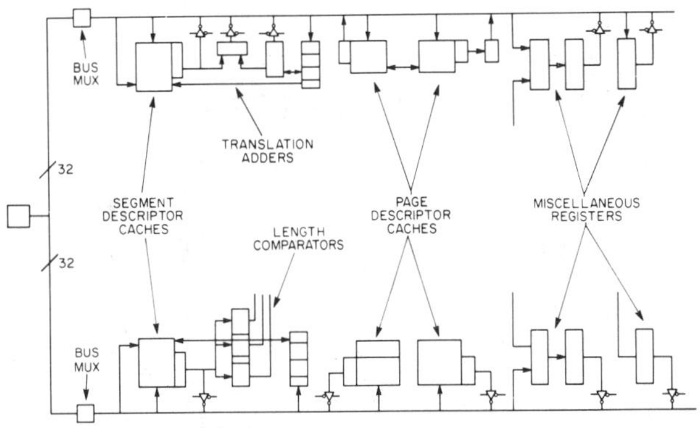

FIGURE 2. DUAL DATAPATH DATA FLOW DIAGRAM
(COURTESY OF REF. 4)

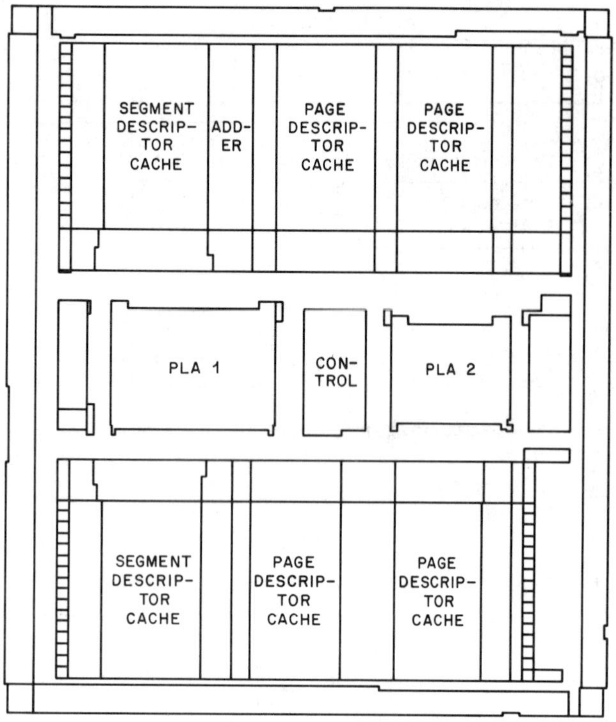

FIGURE 3. PHYSICAL REPRESENTATION OF DUAL DATAPATH TOPOLOGY (COURTESY OF REF. 4)

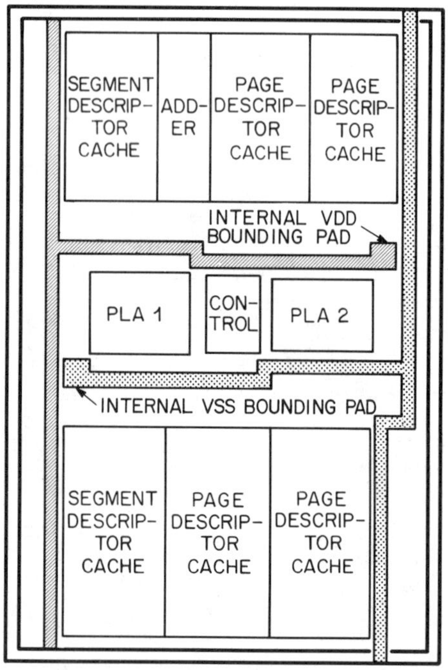

FIGURE 4. SIGNAL TOPOLOGY FOR VDD AND VSS

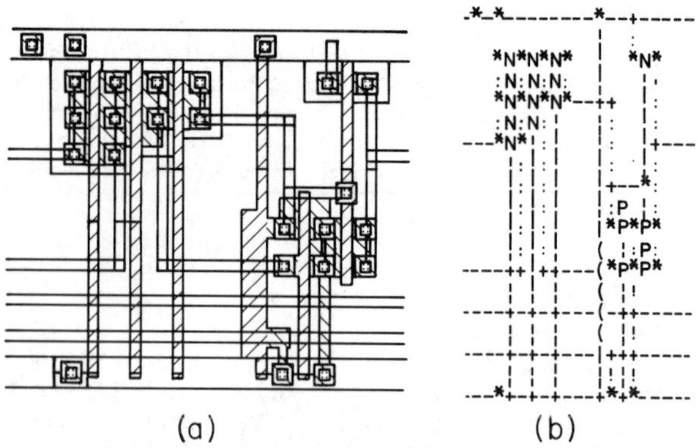

FIGURE 5. (a) XYMASK FEATURES AND (b) SYMBOLIC CODING OF GATE MATRIX MODULE (COURTESY OF REF. 4)

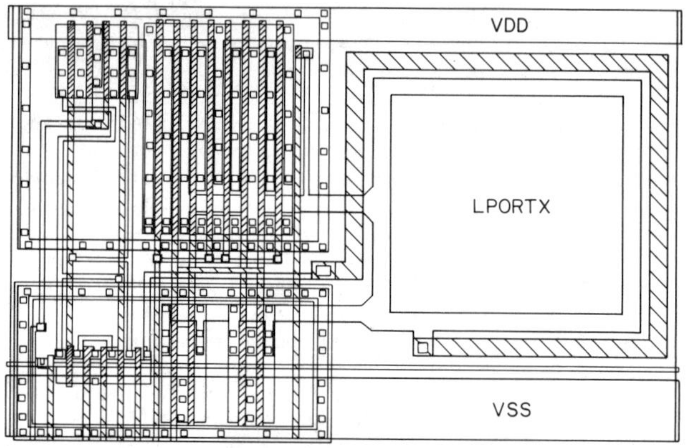

FIGURE 6. XYMASK FEATURES OF FRAME MODULES (COURTESY OF REF. 11)

FIGURE 7. 2.5 MICRON CHIP PHOTOGRAPH OF THE MMU

Session 2-2a
VLSI/Mass Storage

Chairman

T. A. Schwartz
CDC

STORAGE SYSTEMS ARCHITECTURAL OVERVIEW

Richard S. Wasserman

IBM Information Systems & Technology Group
Poughkeepsie, N.Y. 12602

ABSTRACT

This paper is a tutorial on storage hierarchies in computer systems. Price performance tradeoffs amongst storage technologies are discussed, leading to structural and architectural alternatives in organizing storage systems using multiple technologies.

CH1935-6/83/0000/0324$01.00

VLSI IN MAGNETIC DISK STORAGE

by J.R. Brand

Magnetic Peripherals, Inc.
Minneapolis, Minnesota 55435

Abstract

Market and technology forces are constantly pushing magnetic disk drives toward greater storage capacity and increased data-access bandwidth. With the reduction in the length of product development cycles due to increased competition and the steadily increasing consumer demand for product quality, the form and complexity of drive electronic systems are changing dramatically. As the circuit and logic designers of these systems push the gate (transistor) per die density to VLSI levels, the primary challenge is the smooth transition from LSI to VLSI. The orderly application of product-specific VLSI to the analog and digital subsystems of disk drives requires that this change be evolutionary in character rather than revolutionary. The familiar progression from semi-custom array methods through array macros and standard cells to structured custom is appearing in drive logic systems. This paper examines approaches presently being used to increase integration density in the drives of one major manufacturer with internal CAD and micro-electronics support facilities.

Introduction

Magnetic disk drive products span the range from mechanics-only flexible media units sold as component parts for inexpensive microcomputers to complex fixed-media rigid disk drives whose controllers connect directly to mainframes. Storage capacities run from less than 200 kilobytes to above one gigabyte; fundamental data/frequencies at read/write heads vary from a few hundred kilohertz to existing state-of-the-art 24 megahertz; and, drive-to-host interface is established at virtually any place in the data path from the head itself to high-bandwidth busses on minicomputer-like controllers directing multiple actuators which themselves position multiple heads simultaneously. As a consequence, drive electronics can be as simple as the head itself or as complex as tens of thousands of gates of mixed logic types selected for different speed-power tradeoffs.

At Magnetic Peripherals, Inc., (MPI) emphasis is placed primarily on the design and manufacture of high-performance, high per-box capacity disk drives for marketing by Control Data Corporation (CDC); included are full-electronics flexible media drives and Winchesters from 3.5" to 14". MPI is a majority-owned subsidiary of CDC, a major manufacturer of computer hardware and vendor of computer services. MPI has access to a wide variety of internal computer-aided design (CAD) and microelectronics resources as well as those available to any manufacturer from external vendors. This environment has accelerated the tendency toward VLSI application in MPI's products.

The following discussion provides an overview of methods available to electronics designers at MPI. Within the distinct areas of analog and digital design disciplines, selected methodologies are highlighted. As might be expected, the greatest potential for VLSI remains in the future; in disk drives, however, that particular future is much less than five years away. A brief summary of the most promising areas is presented.

Design: Environment and Selected Approaches

Target performance specifications, final package requirements, and knowledge of the idiosyncracies of the chosen magnetic components (heads and disks) enable the start of drive electronics design. This may be sufficient for an engineer to solve part of the design problem by using a pre-existing LSI component; e.g., an internally-developed multi-channel, preamplifier/writer chip is usable across the company's line of drives with thin-film read/write heads. The remainder of the data channel, servo system and control logic development is often unique relative to previous drives. The designer now considers the available selection of integrated circuit approaches within the company.

An MPI engineer can implement his design in a range of custom and semi-custom IC products and technologies. These rely on ECL, IIL, CMOS, and linear bipolar processing from both internal and external sources; VLSI complexity is possible. Design automation assistance is available and required on all internal semi-custom design; custom projects are supported by CDC's Micro-circuits Division. Corporate-wide efforts to

CH1935-6/83/0000/0325$01.00 © 1983 IEEE

facilitate access to computer-aided design tools are resulting in the specification and provision to the electronics community of a common, local-area-network-based and schematic-entry equipped engineering workstation.[1] In the midst of these resources for the product designer, only one is missing: time, the latitude for experiment on silicon.

Typical drive product development cycles are less than eighteen months and dropping quickly: the CDC 9270-6 "Cricket 1" 3.5" Winchester evolved from conception to prototype (discrete electronics) in approximately 90 days. IC technology selection is rarely final at the prototype stage; however, the emphasis from that point on is to complete the task reliably and quickly. This is the crux of the choice of the evolutionary path to high-density circuitry on silicon: analog and digital gate/device arrays with their supporting design systems "fit the schedule" and allow for reasonably straightforward changes almost at the last minute.

Digital: The VLSI-6000 CMOS Gate Array CDC is actively involved in evaluating Mead-Conway[2] techniques and tools for VLSI design. However, a two-layer metal CMOS array[3,4] conceived for mainframe applications and available for use in peripherals is currently the corporate form of VLSI. With full mainframe-based design automation support by CDC, this array of approximately 24,000 transistors from the Technology Development Division has many unique features including:

- 80-85 percent routability
- 172-pin pin-grid array package
- large selection of logic macros
- versatile I/O structure and pin assignment
- built-in maintenance system

In particular, the on-chip maintenance system[3] (OCMS) costs only twelve percent of the active die area; it strongly supports the designer of complex control logic in designing testability and maintainability into the drive. Finally, both the basic array with OCMS and the supporting design automation system appear capable of considerable extension.

For less complex designs, the engineer will have access to a family of smaller CMOS arrays selected from the product lines of certain vendors. The same design automation resources will be available as for the VLSI-6000.

Analog: The Analog Master Chip Array In support of disk and tape drive electronic design, CDC's Microcircuits Division has implemented two-layer metal, linear bipolar custom analog-digital circuits in the 1500-2000 transistor-equivalent range. The servo/motion control chip in the "Cricket 1" is one example of such an effort. The designs were structured, full-custom products of a team approach. In the limited context of bipolar analog circuitry, this emitter density per die can be classified as "VLSI" with only minor license.

Similarly, the Analog Master Chip (AMC) can be considered "analog VLSI". This two-layer metal, linear bipolar semi-custom transistor/resistor array family was developed by MPI and CDC Microcircuits Division for application to data and servo channel design. Supported by MPI's Technology Division in the "simulator-to-routed-silicon" design automation mode common to digital arrays, the AMC brings to the capable analog design engineer the ability to tackle complex on-chip analog electronics without extensive silicon experience. Features available include:

- specification by analog simulator code
- 28-40 pin I/O depending on die chosen
- 500-600 high-performance transistors, Schottky devices, and resistors
- low-noise amplifiers and high-current drivers
- supports CML and ECL logic and analog circuitry

AMC applications have included use in the data channels of 3.5", 9", and 14" drives. The resulting reduction in printed circuit board space devoted to analog electronics in these units was the critical contribution to meeting packaging requirements.

Trends In Drive Electronics

As the performance requirements begin to emerge for the next generation of rigid disk drives, virtually every area of support electronics will be a candidate for VLSI techniques. The driving consideration will be that the drives' "intelligence" must increase to the point where real-time compensation for mechanical, magnetic, and electronic imperfections considers all drive characteristics. As examples:

(1) Error correction to compensate for catastrophic loss of bursts of data due, e.g., to media imperfections. As data rates migrate to higher frequencies with increasing BPI and improved ECC techniques are used to reduce the controller's housekeeping load, special purpose bipolar signal processors will be required.

(2) Data path optimizations to make dynamic corrections for, e.g., the individual playback characteristics of different read heads as a common data channel is multiplexed from head-to-head. Variations in gains and time-domain behaviors in the channel electronics due, e.g., to temperature and age, can be tuned out by constantly measuring performance.

(3) Servo performance enhancement to allow each head nearly independent, dynamic adjustment to the servo information embedded in the data track of interest. With the improved functional density

possible in VLSI, the design approach will likely involve a dedicated central servo processor. It is interesting to note that, as the much-discussed 2000-plus track-per-inch region is approached, the servo processor will be seeking to locate data bit-cells whose dimensions are on the order of the processor's interconnect metallization!

Other areas, such as intelligent drive interfaces to a host computer or controller, remain for VLSI application. It is arguable that these systems, whose purpose is to mediate between some type of processor and streams of data and servo information are relatively less unique to disk drive technology than the above examples ; these functions are often just standardizing extension of the host processor into the drive.

A significant challenge in this coming generation of high-density integrated electronics is to approach the above problems so that the solution realizations can be rapidly migrated from the technology-flagship large-diameter (14") drives downward to physically smaller units. Solutions heavily dependent on custom and semi-custom bipolar circuitry (high-speed analog and ECL) generally carry severe size and power penalities in the context of drives embedded in equipment, for example. As much as possible, the use of CMOS digital and analog design should be investigated even though the actual circuit and system designs are increased in complexity.

Conclusion

Introduction into magnetic disk drive electronics of any level of VLSI requiring significant design methodology changes is tough: the rapid evolution of drive technology and requirements is a sufficient headache for any engineer. Array approaches have excelled and will continue to do so for some time: new, faster gates presented by a well-planned design automation system look just like new, faster gates! The simplicity and predictability of arrays keep the designer isolated from unforgiving semi-conductors.

The promise of the Mead-Conway style of VLSI design will remain just that until CAD tools appropriate to the complexity management problems are available. The M-C concepts are elegant, powerful and ultimately quite reasonable. Any manufacturer in the business of manipulating data cannot ignore the potential advantages of recognizing in silicon the inherent flow of the information. Fascinating results will be reported by those who are careful, fortunate, and able to pursue development uncoupled from agressive product schedules.

This ability to separate serious new-technique evaluation from near-term products is of major advantage to MPI in its approach to VLSI:

- The Analog Master Chip was developed in such an environment; when the concept solidified, development was accelerated to merge with a requirement for space reduction of the channel electronics in CDC's FSD 9" Winchester.

- MPI has had the opportunity to monitor and help steer CDC's CMOS gate array strategies and choose the appropriate time to target applications for the VLSI-6000.

- Finally, with the resources of its own Technology Division and CDC's Micro-circuits Division available, MPI is agressively developing processing technologies; design methodologies; computer-aided engineering workstations; and, prototype VLSI systems-on-silicon.

The reason for this activity is simple: the next generation of magnetic disk drives is almost here.

Acknowledgement and Notes

The internal development and application of high-performance, semi-custom analog integrated electronics for magnetic disk drives has resulted from the efforts of many. The author wishes to recognize the contributions and support of the engineering and management teams of the following organizations: Technology and Twin City Disk Divisions, Magnetic Peripherals, Inc.; and, Micro-circuits Division, Control Data Corporation.

References not available to the public have been listed below. The purpose is to identify contributions and provide historical perspective. No material has been used from these references which is not already public knowledge.

References

[1] Harris, T.R., "Integrating the engineering workstation into the product development process," Proceedings, Electro '83.

[2] Lesmeister, G. J., "VLSI Design Methodologies & Tools," 1982 Hardware Symposium, Control Data Corporation (internal report), Nov. 9, 1982.

[3] Bach, R.E., "A 6000-Gate CMOS Array," 1982 Hardware Symposium, Control Data Corporation (internal report), Nov. 9, 1982.

[4] Resnick, D.R., "Testability and maintainability with a new 6K gate array," VLSI Design, March/April 1983, pp. 34-38.

[5] Brand, J.R., and Hester, R.E., "The Analog Master Chip - A New LSI Array," 1982 Hardware Symposium, Control Data Corporation (internal report), Nov. 9, 1982.

MAGNETIC BUBBLE TECHNOLOGY: STATUS AND FUTURE*

by Mark H. Kryder and Donald E. Thomas

Department of Electrical Engineering
Carnegie-Mellon University
Pittsburgh, PA 15213

Abstract

Magnetic bubble technology today uses permalloy asymmetric chevron patterns on magnetic garnet materials to provide high density non-volatile storage on 1 Mbit chips. 4 Mbit chips, already announced, will soon be shipped in quantity. It is projected that new device technologies such as wide-gap permalloy and ion-implanted contiguous disk devices will lead to 16 Mbit devices in 1985 and 64 Mbit devices in 1988. The price per bit, presently around 20 millicents, is expected to decline to 4 millicents by 1985. The use of limited amounts of current-access technology on predominantly field-access memory chips is projected to simplify controller design and improve performance and functionality of bubble devices through the use of on-chip logic.

Introduction

Magnetic bubble technology was patented in 1967[1] brought to market in 1975[2] and today is being manufactured by several U.S., European, and Japanese firms. The standard chip capacity today is 1 Mbit, although 4 Mbit devices have been announced.[3]

Magnetic bubble memories offer non-volatile storage of information in a highly reliable solid state package which is resistant to harsh environments (vibration, extremes of temperature, radiation). They offer good granularity (1 Mbit/module) and therefore are advantageous over floppy or Winchester disks where reliability or cost/bit for a small number of bits are critical.

In this paper the characteristics of bubble memory technology today are reviewed and the bubble technologies of the future are projected. It is expected that bubble chip capacity will increase to 64 Mbit/chip and that future bubble devices will incorporate on-chip logic to improve bubble memory performance in systems.

Field Accessed Permalloy Bubble Memory Technology

Bubble memory today uses field-access permalloy asymmetric chevrons shown in Fig. 1 to propagate bubbles.[4] A rotating magnetic field applied in the plane of the film induces magnetic poles on the asymmetric chevrons causing the bubble domains to propagate. Bubble generation, SWAP, replication, and detection are achieved using similar permalloy structures with current-accessed conductor lines.[5] The basic cell of an asymmetric chevron propagator is 7w x 9w where w is the minimum lithographic feature in the pattern. Bubble domain size is about 2w. In volume production today, w is about 1μm and the basic cell is about 7μm x 9μm.

Bobeck has proposed a wide-gap version of permalloy technology in which the cell size is 4w x 4w.[6,7] This technology opens the path to higher densities and some groups have recently reported operational 4μm period devices based on it.[8] However, manufacturers have failed to adopt it since it was first announced in 1981.[6] High drive field requirements and critical design tolerances have slowed adoption of this technology for 4μm period devices.

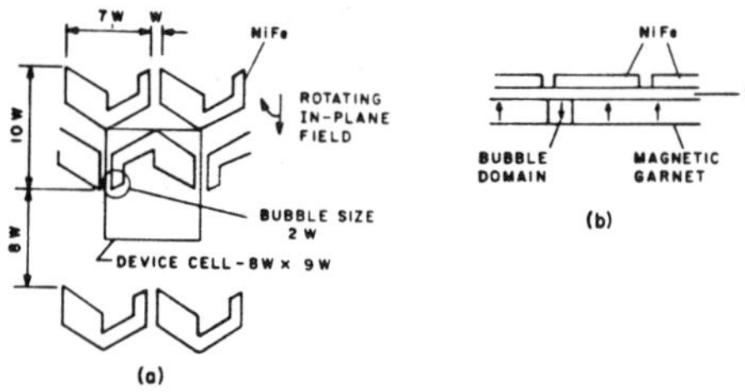

Fig. 1. Asymmetric chevron propagation elements for bubble domain memories a) Topview b) Cross-section

Ion Implanted Contiguous Disk Technology

A promising future bubble technology are the ion implanted contiguous disk devices[9] shown in Fig. 2. Here stress created by ion implantation induces an in-plane anisotropy in the surface layer of the bubble garnet.[9] Domain walls which form near pattern boundaries when an in-plane magnetic field is applied have poles (magnetic charge) on them and attract the bubble domains. When the in-plane field rotates, the charged walls circulate around the non-implanted area causing the bubble domains to propagate.[9] The cell size of this technology is, in the best case, 2w x 2w , although many groups are now using design rules of 4w x 4w. This ion implanted contiguous disk technology thus offers a factor of four to sixteen gain in density over the currently manufactured permalloy asymmetric chevron technology. This technology is being pursued by a majority of the industrial firms involved in bubble technology, but has not yet found its way into any manufactured products, because it would require some changes in manufacturing technologies and/or chip architecture. It remains the most promising high density technology and it appears that once the permalloy technology has been pushed to its limits (4-6μm period), ion implanted devices will take over.[10]

*This research was supported under NSF Grant ECS-8017700.

CH1935-6/83/0000/0328$01.00 © 1983 IEEE

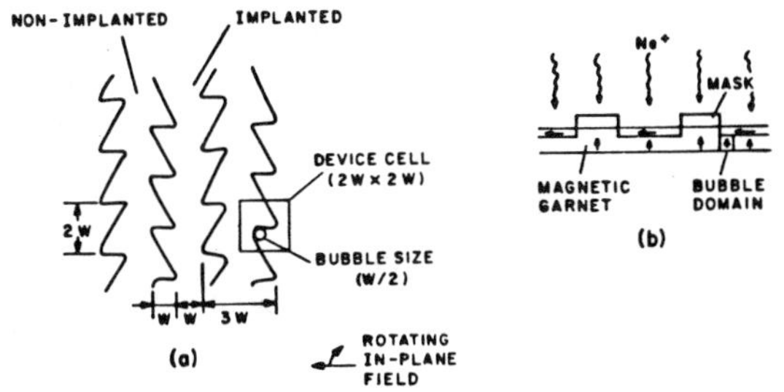

Fig. 2. An ion implanted contiguous disk shift register
a) Topview b) Cross-section

Projections for Bubble Memory Technology

Bubble memory technology has made great advances in density and chip capacity since its invention. Figure 3a shows the capacity of bubble chips as a function of the year in which they were introduced. The rate of progress has been rapaid and projections are that the chip capacity will quadruple every two to three years at least until 64Mbit devices are reached. This projection seems reasonable since several laboratories have reported permalloy devices with 4μm period, and ion-implanted contiguous disk propagation structures with periods as small as 1.5μm have been demonstrated.[11]

The price per bit for bubble memory has been declining as shown in Fig. 3b. The price per bit will surely come down as the chip capacity increases, but significant price decreases will also result from economies of volume production.

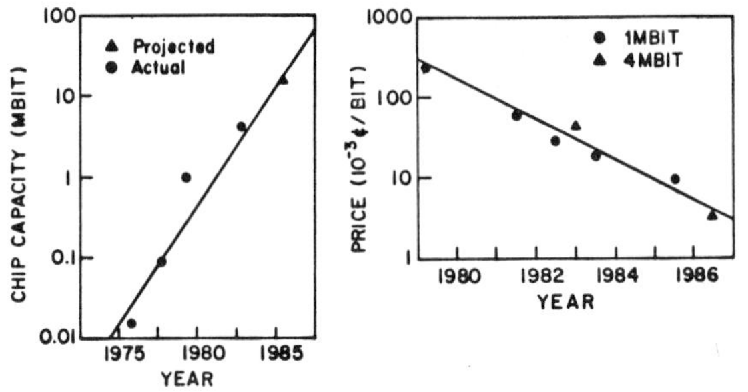

Fig. 3. a) Bubble memory chip capacity versus year b) Bubble memory price/bit versus year

The access time of bubble memory devices depends on chip architecture, chip capacity, and the frequency of the drive field or current used to access the data. All bubble memories manufactured have used major-minor loop or block-replicate chip organizations, shown in Fig. 4, in which data in parallel "minor loop" shift-registers are accessed by another shift register (major loop or output path) leading to the detector. Neglecting some small overhead for propagating and stretching the bubble to a detectable size, the access time to the first bit in a block of information is proportional to the number of bits in a minor loop, which is proportional to the square root of the chip capacity.

To avoid an increase in access time with increasing chip capacity, manufacturers have been putting multiple detectors on a chip and are raising drive frequencies. Since detectors occupy considerable space and since the associated sense electronics is moderately expensive, there is a limit to the number of detectors which a manufacturer can put on a chip.

The ability of manufacturers to increase drive frequency as bit density is increased is limited by the fact that required drive field amplitude increases with bit density. This is because bubble domain size scales approximately inversely with magnetization while drive field requirements tend to scale

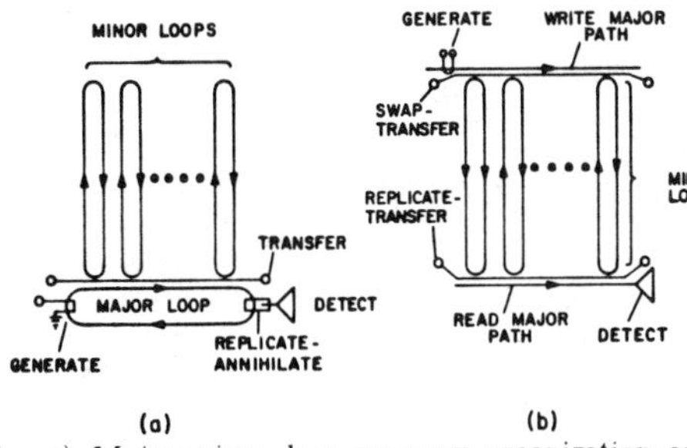

Fig. 4. a) Major-minor loop memory organization and
b) block-replicate chip organization

linearly with magnetization. Hence to increase bit density, higher magnetization materials are used, but this in turn requires higher drive fields. Furthermore, the use of higher frequency drive fields in a field-access device requires an increase in the volt-ampere product of the drivers supplying power to the coils. These considerations lead to the conclusion that the number of bits which may be driven in a module.

$$N = K(V\text{-}I)/f\ H^2\ d^2 \qquad (1)$$

where f is drive field frequency, H is drive field amplitude, V-I is volt-ampere product of the drivers, d is bubble diameter and K is a constant which depends on coil design.[12] This equation makes clear that higher data rates require higher power current drivers and that the tendency for drive field requirement to increase with decreasing bubble size limits the number of bits which can be placed in a module.

In spite of the trade-offs on bit density, access time, and data rate, it is expected that bubble memory bit density will advance to 64Mbit/chip by 1988 with no decrease in performance. Improved current drivers, the use of slightly higher drive field frequencies and the use of multiple detectors on individual chips will make this possible. Alternative architectures and/or current-access major path shift registers to be discussed below could provide significant improvements.

Current Accessed Bubble Devices

Both permalloy and ion-implanted devices are field-accessed in that a rotating in-plane magnetic field is used to propagate the bubbles. An alternative means of propagating bubbles is to use current-access shown in Fig. 5.[13] Two conductor sheets with apertures etched into them are used to carry currents. The currents flow around the apertures, creating poles at the ends of the apertures in analogy to the poles induced on permalloy bars by an in-plane field. The poles attract the bubble domains and, if the current sequence shown in Fig. 5 is used, the bubbles are caused to propagate.

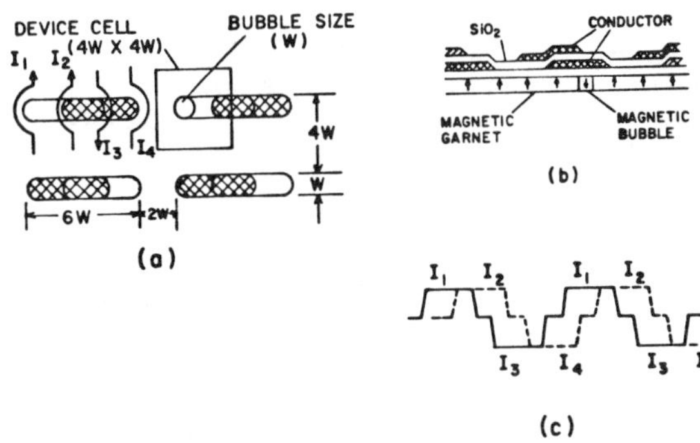

Fig. 5. Current-accessed bubble propagation elements a) Top-view b) Cross-section c) Current pulse sequence

The device cell of the current access technology is 4w x 4w and thus is four times more dense than the presently manufactured permalloy devices, but is four times less dense than the highest density ion implanted contiguous disk devices.

Current access makes possible higher data rates than field-access technology. Field-access data rates are presently around 100 kHz and are limited by the inductance of the coils used to provide the rotating field. The conductor sheets of the current-access devices present low inductance and can be driven by high-frequency pulses. Improvements in data rate from the present 100 kHz to 1-10 MHz range are possible.

In spite of the advantages of current access technology it suffers from a serious drawback. On-chip power dissipation is high and tends to increase inversely with the cell size.[14]

The power dissipation problem makes the use of current-access for memory unlikely, but it may be used for limited amounts of on-chip logic and for high data-rate read-out channels on otherwise field-access chips. By restricting the current-access to small regions of a chip, average power dissipation can be kept low. A current-access major path with shortened, but more numerous field-access minor loops could, for example, shorten access time and improve data rate.

Even without increasing the bit rate of the current-access major path, a shorter access time at the system level can be achieved with a current-access major path because it allows independent control of propagation in minor loops and in major paths. Independent control of minor loops and major paths makes it possible, during readout of data on the major path, to bring the next block of data in the minor loops to the replicate/transfer gate and hold it while the previously transferred data is read. Access to the second and succeeding blocks does not then involve any cycle time in the minor loops. Assuming the address of the second block is known when the first block is transferred to the read-out major path and assuming the read-out major path has as many bits as the minor loops, the second block can be accessed with no gap in data after the first block.

Magnetic Bubble Logic

Recent reviews of magnetic bubble logic, given by Chang[15] and Chen and Chang[16], illustrate an exciting variety of potential applications. Unfortunately, however, the commitment of manufacturers to conventional bubble memory and the large investment required to develop bubble logic functions to the point that they can be manufactured has thus kept bubble logic from being implemented in manufactured devices.

There are several features which distinguish bubble logic design styles from electronic logic design styles: 1) Sensing of on-chip information by off-chip processes is costly in terms of both time and area. This is because information must be propagated to a detection area and then sensed externally. The detection area requires considerable space. 2) Controlling of on-chip information by off-chip processes is also costly in terms of both time and area. 3) Algorithmic control of the processing must be distributed such as in dataflow architectures as opposed to centralized control such as microprogramming. Again, information must be propagated from where it is generated to where it is used and thus a central bubble logic controller would be slow. 4) Fan-out capability is severely limited because of the use of conservative logic. 5) Synchronization of events at the processing nodes in the data-flow is a critical and difficult problem because information is not easily held in place while waiting for "the other operand."

The first two items above provide motivation for use of magnetic bubble logic and the last three items are direct implications of that choice. The important point is that appropriate uses of magnetic logic can be found in functions where localized control can be generated and exerted. For instance, it is easy to use control information to divert streams of data toward processing nodes. Indeed, the faulty loop corrector described below shows a useful implementation of this style of design. However, it is very difficult to do a typical processing function such as "increment a variable and compare that variable to some value to test for the end of a control loop" in less than the order of n cycles where n is the bit width of the variables. Since all other information will have moved n positions in that time there cannot be a bit-by-bit control of the data stream. Thus applications of bubble logic should initially be limited to data-flow and management functions where localized control can be brought to bear on streams of information.

Current-access technology is the present choice for implementing bubble logic. There are sound physical reasons for this: 1) There are few constraints on geometry and orientation of the apertures in the current sheets as compared to the constraints in ion implanted contiguous disk devices; 2) The interaction fields between two bubbles in a logic gate on a current-access device are not shunted away as in permalloy devices or ion-implanted devices; 3) Simulation of current-access technology is considerably easier and more accurate than simulation of permalloy or contiguous disk devices which require knowledge of the micromagnetic behavior of permalloy elements or of charged walls; 4) Finally, current-access technology is not limited in data rate by the inductance of drive field coils.

Magnetic Bubble Faulty Loop Corrector

A recent illustrative example of a practical application of bubble logic is a new memory architecture which uses redundant minor loops to improve yield (as do presently manufactured devices), but into which serial streams of data with no gaps for the faulty loops can be written.[17] Figure 6 shows a block diagram of a faulty loop corrector chip. A variable delay function is located at the input and output of the major paths. Gaps in the data stream can be inserted or removed, respectively, by incrementing or decrementing the delay causing subsequent data to catch up or fall behind. These delays are controlled by a program, unique to each chip, that is loaded into the control loops at manufacture, much as faulty loop maps are loaded into contemporary chips.

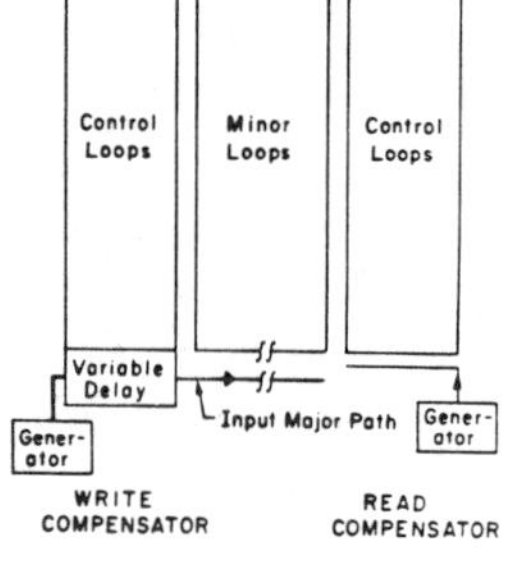

Fig. 6. A block diagram of a memory chip with invisible faulty loop correction

The faulty loop corrector employs bubble logic gates in current-access bubble technology. Figure 7 shows a diagram of a bubble logic gate used to implement the variable delay. The gate is a two-input, three-output device with one input line (x) controlling the flow of data in the other input (p) into one of two alternative outputs (r and q). The data path from p to r is easier for the bubble to traverse than the path from p to q. Data entering at p takes the easy path (from p to

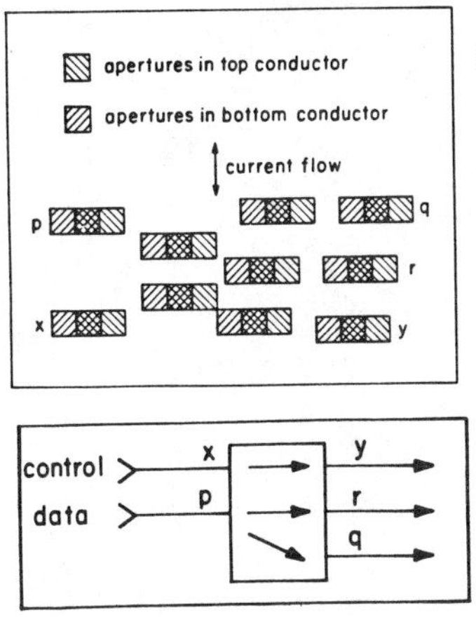

Fig. 7. Bubble interaction gate used in the faulty loop corrector

a) Diagram,

b) Schematic representation

r) if no control bubble is present at x. When a control bubble is present at x, the repulsion between the bubbles forces the data bubble along the hard path from p to q.

To implement the variable delay function, the bubble logic gate of Fig. 7 is used to determine whether data flowing at p takes a shorter path or is delayed. Figure 8 illustrates an implementation of a variable delay compensator. The shortest path through the compensator is straight through and will be achieved only if the control loops have no bubbles to deflect the data bubbles into the detours. To insert gaps into the data, the data must be deflected into the detours by bubbles in the control loops. By choosing the delays in the detours to be the Fibonacci numbers [1,1,2,3,5,8....($F_N=F_{N-1}+F_{N-2}$)], up to $F_{N+1}-1$ gaps may be inserted wherever desired in the data stream. Furthermore, any one of the control loops in the compensator may be faulty itself and the compensator will still operate. Thus the compensator itself has built in redundancy.

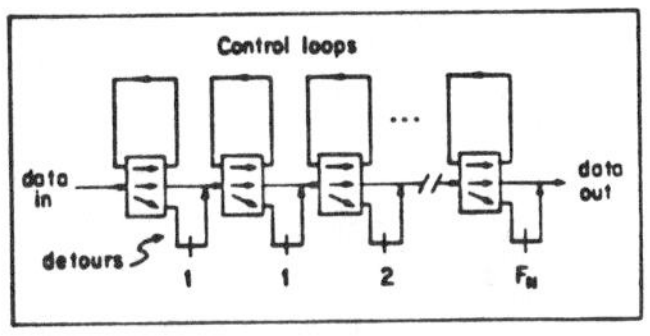

Fig. 8. Variable delay compensator used in the faulty loop corrector

An analysis of the impact of the faulty loop corrector on controller design showed that the on-chip faulty loop corrector, which required less than 2% area overhead on the bubble chip, reduced the number of gates required on the controller chip by 67% while the access time was decreased due to the ability to independently control major paths and minor loops.[17]

Conclusions

Permalloy bubble memory technology today in the form of 1Mbit to 4Mbit chips enjoys a special niche in the wide range of memory applications. When non-volatility is needed in a reliable non-mechanical system at moderate cost, bubble memories are the preferred choice. Applications include airborne avionics, flight recorders, robotics, numerical control machine tools, and portable terminals. Wide gap permalloy and/or ion implanted contiguous disk technology will make possible high-density chips with 16 Mbit and 64 Mbit capacity at approximately 2 - 3 year intervals. Slightly higher drive field frequencies and multiple detectors on a chip will keep access times in these devices in the 10msec to 50msec range.

Given the base of a growing market for field access technology, it is expected that bubble devices with improved performance and functionality will be made by using current access technology for higher speed logic on chips with field access memory. The on-chip logic may be used to take over some of the controller functions; however, future devices will most likely also make use of the logic capability to enhance the functionality of the chip. For example a chip with a built in memory hierarchy and cache for fast access might be constructed so that much faster access times are achieved in a system application.

References

1. U.S. Patent 3,460,116.

2. Electronic News, October 20, 1975.

3. D. Rose, "A 4 Megabit Bubble Chip," paper FA-2, Intermag Conf., Philadelphia, April, 1983.

4. A. H. Bobeck, and I. Danylchuck, IEEE Trans. Magnet., MAG-13, 1370 (1977).

5. P. I. Bonyhard and J. L. Smith, IEEE Trans. Magnet., MAG-12, 614 (1976).

6. A. H. Bobeck, et al., "Wide-Gap, Gap-Tolerant Permalloy Bubble Propagate Elements," paper EA-1, Magnetism and Magnetic Materials Conf., Nov., 1981.

7. A. H. Bobeck, "Wide Gap Permalloy Bubble Devices," paper FA-4, Intermag Conf., April, 1983.

8. T. Yanase, et al., "Design and Characteristics for a 4μm Period Permalloy Bubble Device," paper FA-1, Intermag Conf., April, 1983.

9. Y. S. Lin, G. S. Almasi, and G. E. Keefe, J. Appl. Phys, 48, 5201 (1977).

10. Mark H. Kryder, IEEE Trans. Magnet., MAG-17, 2385 (1981).

11. Y. Sugita, Hitachi Central Research Laboratories, Private Comunication.

12. Mark H. Kryder, IEEE Trans. Magnet., MAG-15, 1009 (1979).

13. A. H. Bobeck, et al., Bell Sys. Tech. J., 58, 1453 (1979).

14. Mark H. Kryder, IEEE Trans. Magnet., MAG-17, 2392 (1981).

15. H. Chang, "Magnetic Bubble Memory Technology," in Encyclopedia of Computer Science and Technology, edited by J. Belzer, 10, Marcel Dekker (1978).

16. T. C. Chen and H. Chang, "Magnetic Bubble Memory and Logic," Adv. in Comp., 17, 223 (1978).

17. R. L. Smith, et al., "A Major-Minor Loop Memory Organization with Invisible Faulty Loop Correction," paper GA-3, Intermag Conf., April, 1983.

SEMICONDUCTOR MEMORY FUNCTIONS IN VLSI-BASED SYSTEMS

Bruce Threewitt
Product Planning
Advanced Micro Devices, Inc.
Sunnyvale, California 94086

ABSTRACT

The need for storage functions permeates the typical digital system. The following discussion outlines some of these applications areas with special attention paid to those uses where semiconductor memory can be utilized. The rapid evolution of very dense semiconductor techniques heralds the penetration of even the mass storage applications by solid state memories. Manufacturers of these devices face several development challenges, including special problems in data integrity and security. The final section describes some of the possible techniques to be applied to the mass storage problem.

INTRODUCTION

The need for memory in digital systems is driven by the principle contained in O'Toole's Law which states that the job always expands to fill the available memory space, plus 10%. This precept implies that no maximum limit on the number of required memory bits in the system exists. Various types of memory functions are found in each of the main blocks of the system (Figure 1). The memories electrically near the CPU require very high performance and are, therefore, RAMs built with the fastest of the semiconductor technologies. Both the access time and the cycle time of such systems need to be as near zero as possible. The performance of this type of memory is usually obtained at the expense of the density (cost). Fortunately, these memories are used only as data accelerators (caches) and are usually small in size. In general a memory hierarchy exists in the system where performance and cost are the two exclusive parameters (Figure 2).

The main memory in Figure 1 is also usually RAM. Since a much larger amount of storage is needed in this function, a less expensive memory on a per-bit basis is required. The main memory need not be as fast since the cache raises the effective speed. The main memory still requires random access, but the transfers to a cache are largely serial. Thus, semiconductor memory circuit techniques that shorten the effective cycle time of the main memory are desirable, even at the expense of more complex memory controls.

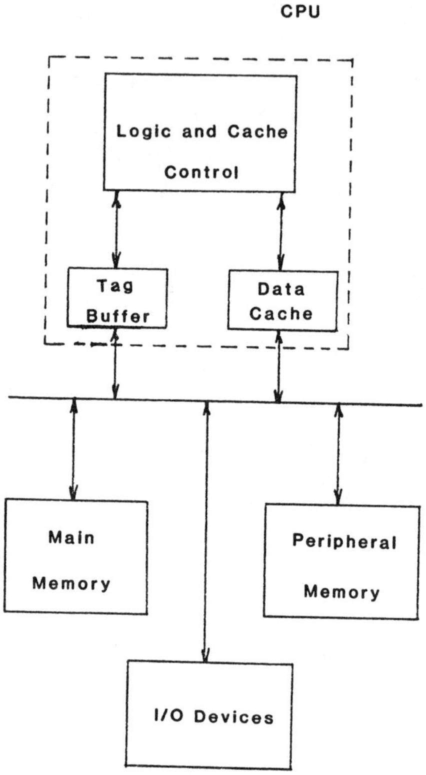

Figure 1 - System Storage Functions

Input-output devices such as printers and terminals are also using a large amount of memory for such functions as spooling, formatting, and preprocessing of local data in preparation for sending the

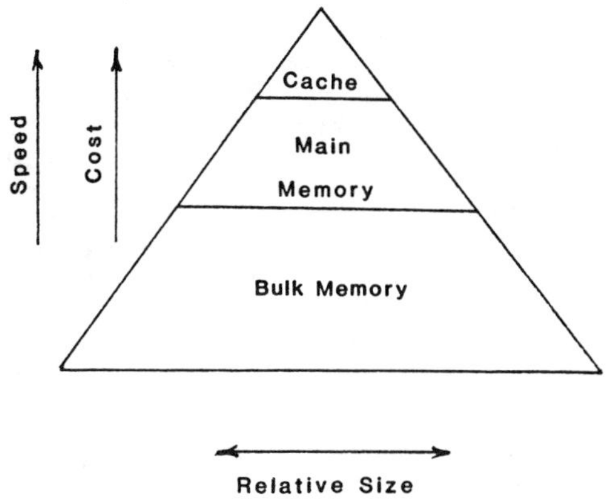

Figure 2 - Memory Hierarchy

data to the CPU. This recent increase in the amount of local memory used has increased the throughput of systems significantly by reducing the amount of data that must be sent to the CPU for processing. Performance requirements in this area are less stringent because these memories accept data in a word format on one side and supply it to a mechanical or human interface on the other side.

The peripheral memories have characteristically used magnetic techniques to gain the needed density and cost because this applications area resides in the lower portion of the hierarchy shown in Figure 2. These memories have been serial in nature with the concommitant reduction in performance. However, the serial data rates are relatively high. Until such time as the semiconductor memories grow sufficiently in density to begin replacing parts of the traditional peripheral memories, buffer applications to speed up the average performance and ease the interface will use semiconductor devices.

SEMICONDUCTOR SOLUTIONS

Semiconductor memories draw upon several device technologies and circuit techniques to tailor their features and performance to a particular application. Users should be aware of the array of options available with varying performance, cost, writability, and ease of use.

In random-access read/write memories, for example, speed and cost parameters determine whether bipolar (currently ECL) or MOS technologies are used. In recent years MOS techniques have narrowed and in some cases eliminated the performance gap between bipolar and MOS. Thus, TTL-compatible memories are now all designed using n-channel MOS (NMOS) or complementary MOS (CMOS) rather than bipolar technology because of the cost advantages offered by MOS.

RAMs are characterized by two types of circuit approaches: static, which requires no refreshing to retain data, and dynamic, which must be exercised on a periodic basis in order to retain data. The refresh time overhead amounts to about 3% for typical dynamic RAM (DRAM) products. Figure 3 shows the basic difference between the two approaches at the cell level. The cell complexity in terms of components per cell determines the feasible density at a given technology level and, therefore, the cost per bit. As Figure 3 implies, the DRAMs provide the storage function for the lowest cost per RAM bit. However, the refresh and address manipulations necessary to successfully use the DRAMs add to the system overhead. On the other hand static RAMs provide the higher performance both in terms of access time and cycle time. Coupled with their inherent ease of use, the static RAMs find many applications in digital systems, especially where relatively small amounts of local memory are desired.

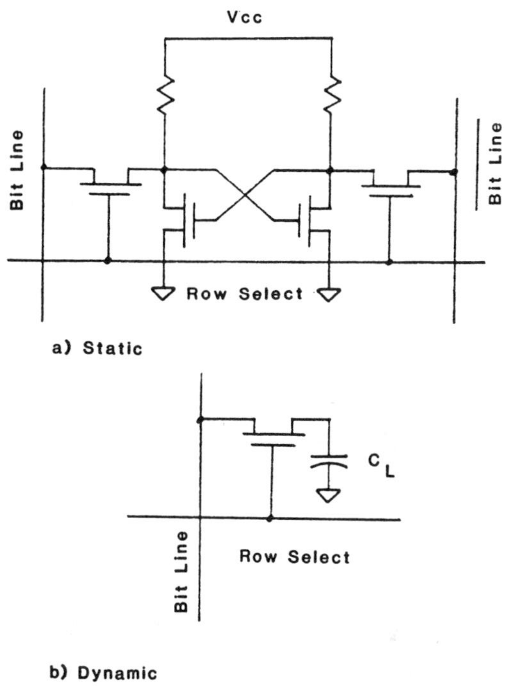

Figure 3 - RAM Cells

Not all of the applications implied by Figure 1 require a data writing capability, at least not on a regular basis. Several functions might be read-only or read-mostly. For these needs, a wide variety of electrically, fuse-, and mask-programmable memories are offered. These devices find their way into control stores for controllers or into program storage in local microprocessor applications. Additional flexibility for read-mostly memories has recently surfaced in the form of electrically erasable and reprogrammable MOS memories.

Several specialty functions are being added to semiconductor memories to solve specific system problems. These functions include dual-port memories, FIFOs for use in I/O interfaces, video-oriented memories, and the like. These specialty products help to reduce package count where relatively small amounts of memory are required.

PRODUCT DEVELOPMENT ISSUES

Semiconductor memories are the most component-intensive devices that integrated circuit vendors manufacture. For that reason they are commonly used to push the technology to the next level of complexity. A further advantage from a manufacturers point of view derives from the huge volume which the industry ships of these devices. That volume, of course, helps to reduce the cost to manufacture other products on the same production line. The drawback to these characteristics is that the memory products, which represent such a large and visible market, become commodity products with fierce competition and lowered profits. That reduced profit makes financing the needed growth through product and technology development very difficult. A benefit can be obtained by increasing the functionality of the memory products, provided that a measure of flexibility is retained. Thus, specialized memory-based system solutions are attractive to the integrated circuit vendors and to users alike.

Two of the more intractable problems arising from the increased density of semiconductor memories, data integrity and data security, have made recent strides toward solutions. The data integrity problem occurs because as density increases, the amount of energy stored in each cell tends to diminish. In effect the signal-to-noise level in the system is reduced, increasing the incidence of so-called "soft" errors. That term covers errors that are not caused by structural damage to the memory. The most widespread cause of soft errors is the natural alpha-particle-emitting materials used in the packaging of the memory devices.[1] As feasible chip sizes increase, the probability of a soft error increases.

To verify the electrical margins of a given memory device generally requires at least $n^{3/2}$ cycles where n is the number of bits in the memory. Thus, as density increases, the testability of the memory products is reduced, further aggravating quality or reliability problems.

The data integrity issues are being addressed in two ways: at the system level and at the device level. Use of VLSI logic products to do error detection and correction has lowered the cost for improved data integrity to the point where most systems can afford the added cost and slower speed. Chip designs are incorporating some of these features directly in the form of on-chip ECC and redundancy. This trend will continue as semiconductor memories move toward mass storage.

The data security problem arises from the use of large nonvolatile memories to transport software and data bases. Read-only semiconductor memories, for example, are currently large enough to contain whole operating systems on which much of the system's added value is based. To protect this data from being copied is a worthwhile goal. However, the hardware needed to accomplish this aim has only recently begun development. The use of low-cost standard RAMs and other memories has to some extent hampered this effort since users have been reluctant to pay for any extra features on the memory devices. Until some extra control can be added directly to the memory to validate accesses, this problem cannot be solved. A simplified block diagram of a protected memory is shown in Figure 4. Absolute security is a nebulous term and may not be attainable. The goal of security-assist hardware is to make the task of unauthorized copying too cumbersome for the "casual" copier.[2] Access might for example require some _a priori_ knowledge of the content of the data. Electrically copying the data then becomes a much more complex problem, and security is enhanced.[3]

These development issues will carry into the use of semiconductor memories in mass storage. In each case these issues are compounded by the increased density found in mass storage. The following section discusses some of the possible approaches under study by the industry for building semiconductor mass storage devices.

SEMICONDUCTOR MASS MEMORY

Several new approaches to drastically increase the density of semiconductor memory are under study by the industry. These techniques include major departures in packaging, wafer materials, and circuit structures. By merely scaling the lithography, semiconductor memory will evolve to the several-megabit-per-chip density. To participate in the mass memory arena, however, may require a more revolutionary approach.

Some of the more promising methods under consideration have merely dusted off some older magnetic techniques and applied current lithographic capabilities. One offering in this area was recently reported in the literature in which a garnate material is used as a magnetic medium for creating bubble-like domains, storing as many as 100 bits per site [4]. This type of memory is inherently serial and will not offer the performance of a RAM, but the density provides a solid state mass storage capability. Manufacturers do not like to depart from the silicon material base used in volume IC production. To overcome that bias requires major cost benefits. A convergence of silicon and magnetic techniques might eliminate that limitation. Other ways to store more than one bit per cell are also being studied. The data integrity problems mentioned earlier are especially relevant in this case.

An analogous effort is underway in the charge storage area as well.[5] This method involves a nonpatterned target array and a separate charge-gun mounted in a special package in close proximity. By using a nonpatterned target structure, the lithography challenges are significantly reduced so that larger chip sizes become feasible.

Until the density of semiconductor memories makes solid-state mass storage possible, performance remains the only incentive for users to incorporate these devices in their mass storage systems. Thus, in the interim these systems will contain a mixture of semiconductor and mechanical devices where the ICs will be used to enhance performance.

CONCLUSIONS

Semiconductor memories are evolving toward mass storage densities. Cost, data integrity, and data security issues will have to be resolved before solid-state mass memories become feasible. When those problems are solved, the performance enhancements that will result will have a drastic impact on system architectures. Magnetic, charge storage, and new circuit techniques are all under investigation. The added system functionality that these semiconductor products will offer will help the user to reduce package count, increase system reliability, and reduce cost while adding performance. Further, the manufacturers of the components will realize a better return on their development dollars.

REFERENCES

[1] T.C. May and M.H. Woods, "Alpha-particle-induced Soft Errors in Dynamic Memories," IEEE Trans. Electron Devices. vol. ED-26, pp. 2-9, Jan. 1979.

[2] R.W. Davis, "Using ... Authenticated EPROM," Intel Corp. Application Note 152, p. 2, Feb. 1983.

[3] M. Kaufman, "Address Translator," U.S. Patent 4,377,844, Mar. 22, 1983.

[4] "...One Gigabit of Memory on a Single Chip," Semiconductor International, p. 20, July 1983.

[5] M.H. Eklund, "...4-megabit Memory Chip," ICE, Corp. report, Issue 2-1, 1982.

Session 2-2b
VLSI Evaluation

Chairman

J. Molinelli
Bell Labs

THE DELTA-I SIMULTANEOUS SWITCHING PROBLEM

Gary Ditlow
IBM T.J. Watson Research Center
P.O. Box 218
Yorktown Heights, NY
(914) 945-1690

Al Brown
IBM Poughkeepsie Laboratory
D/D18 B/707
Poughkeepsie, NY
(914) 463-8109

Abstract

An important task in the design of high performance computers using VLSI technologies is the analysis of Delta-I noise. That is, noise which occurs when chip drivers switch large amounts of current across the inductance in the power distribution system. When too many drivers switch simultaneously, the noise immunity of the technology is reduced and as a consequence, the technology becomes very prone to false switching.

In this paper we describe the Delta-I problem from technology and system viewpoints and then introduce new algorithms based on Boolean Algebra techniques to identify potential Delta-I failures. We create a Boolean model for excessive switching activity in a combinational logic network and demonstrate how to solve Boolean equations for a set of input patterns that may lead to a Delta-I failure by simultaneously switching many chip outputs.

Introduction

A key question in any Delta-I analysis is what set of input conditions will cause a set of output drivers to switch simultaneously. These input conditions assume the outputs have been forced to a known state first. Hence, two sets of input conditions are required: one to initialize the outputs and the other to cause the outputs to change state. We call such conditions a two-step input sequence. The outputs may change from 0 to 1, from 1 to 0, or they may not change state at all. Only output transitions are of interest. In addition, a cancellation effect takes place when one output switches from 0 to 1 and another switches from 1 to 0.

The issue then is to find a two-step input sequence (if it exists) such that the number of outputs switching in the same direction (assuming a cancellation effect) is greater than some technology limit. Furthermore, those outputs which are identified must switch within a time window which technology has agreed to.

The three areas we focus on in this paper are technology, systems, and design automation. Technologists design innovative circuits and packages to minimize the effect of excessive simultaneous switching. Logic designers partition the chips in high performance computers so that the timing of their logic cause a small number of driver circuits to switch simultaneously. Design automation is also key in the analysis of Delta-I noise. We will present an algorithm which demonstrates the feasibility of a Boolean algebra based approach for automatically creating two-step input sequences which produce a Delta-I failure.

Technology Considerations

Delta-I noise occurs when chip drivers switch large amounts of current across the inductance in the power distribution system. Since the voltage across an inductor does not change instantaneously, a noise spike emerges on the power bus and propagates through quiet drivers to receivers. If the amplitude of the noise penetrates the threshold of the receivers while the clock is on, then a false signal can latch on the receiving chip (Figure 1). When too many drivers switch at the same time, the ground plane also shifts. If the disturbance is greater than the noise immunity, then the technology is subject to false switching. A first order estimate for the amplitude of the noise is given by the formula in Figure 2.

CH1935-6/83/0000/0337$01.00

Delta-I problems can manifest themselves as logic delay increases or as significantly reduced noise tolerances (Davidson82). High speed technologies are susceptible to this type of noise due to large driver currents, fast rise times, and the inductance in the power supplies and package. In high performance computers, the problems can be far too extensive and complex for manual analysis.

Innovative circuit designs attempt to control the driver currents and rise times. Circuits designed with active feedback limit dI/dt by using saturating transistors which turn off very slowly.

System Considerations

Designers of high performance computers as well as technologists specializing in packaging and circuit design, work together to engineer a computer and a technology so that the effect of Delta-I noise is reduced. There are a variety of design trade-offs which can be used to minimize the switching noise.

Balancing circuits by using a dual rail design tends to cause cancellation due to positive noise on a falling transition, and negative noise on a rising transition. The design trade-off is better noise immunity at the expense of more chip IO's.

Staggered switching using late clocks is another method of minimizing Delta-I noise. If n drivers switch in the same direction within a given time window, the noise on the power supply is n times the noise of one driver. By staggering the switching times of the drivers, the noise is spread out in time at the expense of a slower computer and more intricate clock distribution system.

Another approach is to partition the computer so that busses which typically switch together are on separate chips. Ideally, the chips should use different power supply distribution systems so that the noise on one weakly couples to the noise on the other. By introducing additional ground planes in the package, the effective inductance is reduced since there are additional paths to ground. Smaller inductance reduces the amplitude of Delta-I noise.

Engineering judgement based on years of experience is a practical methodology to follow for minimizing the effect of Delta-I noise. However, accurately predicting the amount of switching activity is too large for manual intervention. Therefore, the need exists for verification programs which discover if there are any two-step sequences for which a large number of drivers simultaneously switch.

Design Automation Considerations

To determine if a Delta-I problem exists, we must discover a two-step sequence which causes m-out-of-n outputs to switch simultaneously. By constructing a suitable Boolean model (Brown83), we can solve the Boolean equations to decide if such sequences exists to cause a Delta-I failure. The model which we construct, corresponds to the original combinational network with additional logic added to correctly model the simultaneous switching behavior of the outputs. The final form of the model consists of a single output with twice as many inputs as the original network. We double the number of inputs to model the two-step input sequence.

The outputs of each model are inputs to a decision function which counts the number of outputs switching in the same direction. Many different kinds of decision functions are possible. The simplest one is where we count the number of drivers switching from 1 to 0. A more elaborate model would consider cancellation effects by adding up the number of drivers switching from 1 to 0 and then subtracting the number of drivers switching from 0 to 1. The output of any decision function is input to a comparator circuit which compares the number of outputs switching to some technology limit. Figure 3 illustrates the steps involved in this model building phase.

The output of the comparator is a single output t. If we can satisfy t by finding an input condition x which causes t to be one, then the pattern x is a Delta-I test pattern.

Our formulation of the Delta-I problem is an instance of the well-known Satisfiability problem (Cook71,Garey79). Many techniques exist to solve such problems. The approach we use is a decomposition technique based on successive elimination(Rudeanu74,Hammer68) a variant of which was implemented by Bahnsen in verifying logic networks(Smith82). That is, given a Boolean function t we want to find input patterns x which satisfy t=1. We solve t for x by recursively expanding around an input at both zero and one. The expansion creates two new functions which are Boolean partial derivatives of t as defined by Brown (Brown70). The recursive evaluation continues until a partial derivative evaluates to zero or one at the leaf of the expansion tree. Only leaves of the tree which evaluate to one are of interest since they represent a two-step input sequence which result in simultaneously switching outputs. Figure 4 illustrates the input patterns that cause at least one output to switch from 1 to 0. The procedure in Figure 5 finds the two-step input sequence by solving a system of Boolean equations using the recursive procedure of successive elimination.

If we were interested in the existence of a two-step input sequence rather than the explicit determination of all such sequences, then a formulation developed by the first author using the Cutting Algorithm(Savir82) is an appropriate and polynomial-time algorithm. The Cutting Algorithm attempts to count the number of input sequences which cause the output t=1. The count is exact for trees, but for functions with reconvergent fanout, a lower and upper bound are computed. The Cutting Algorithm uses analytic techniques which examine the circuit topology to derive tight lower and upper bounds. A sufficient condition for the existence of a Delta-I test pattern is for the lower bound to be greater than zero.

Conclusions

The effect of Delta-I can be effectively managed by coordinating the combined efforts of technologists, systems designers, and design automation analysts. Input patterns to detect potential Delta-I failures are efficiently found by building a Boolean model of the problem and solving the corresponding Boolean equations for a set of two-step input sequences that will result in many of the outputs switching simultaneously.

Acknowledgments

Evan Davidson clarified many of the technology issues for us. George Katopis was helpful in the analysis of potential Delta-I problems. Stu Tucker identified engineering trade-offs which are appropriate in the design of large computers. Heidi Petersen helped in developing models of the decision functions.

References

(Brown70) A. Brown and H. Young, "Algebraic Logic Network Analysis (ALNA)," IBM TR001974, January 1970.

(Brown83) A. Brown and G. S. Ditlow, "Delta-I Failure Detection Technique," IBM Technical Disclosure Bulletin, Vol. 25, No. 9, February 1983, pp. 4870-4871.

(Cook71) S. A. Cook, "The complexity of theorem-proving procedures," Proc. 3rd Ann. ACM Symp. on Theory of Computing, pp. 151-158.

(Davidson82) E. E. Davidson, "Electrical Design of a High Speed Computer Package," IBM J. Res. Develop., Vol 26, No. 3, May 1982,

(Garey79) M. R. Garey, and D. S. Johnson, Computers and Intractability, A Guide to the Theory of NP-Completeness, W. H. Freeman and Co., 1979.

(Hammer68) P. L. Hammer and S. Rudeanu, Boolean Methods in Operations Research, Springer-Verlag, New York, 1968.

(Ho82) C. W. Ho, D. A. Chance, C. H. Bajorek, and R. E. Acosta, "The Thin-Film Module as a High-Performance Semiconductor Package," IBM J. Res. Develop., Vol 26, No. 3, May 1982, pp. 286-296.

(Katopis78) G. A. Katopis, "Methodology and Computer Aids for the Noise Tolerance of ECL Logic Gates," 12th Asilomar Conference, November, 1978.

(Rudeanu74), Boolean Functions and Equations, North-Holland Publishing Co., New York, 1974.

(Savir82) J. Savir, G. Ditlow, and P. Bardell, "Random Pattern Testability," IBM Research Report RC9643, Oct 10, 1982.

(Smith82) G. L. Smith, R. J. Bahnsen, and H. Halliwell, "Boolean Comparison of Hardware and Flowcharts," IBM J. of Research and Development, Vol 26, No 1, January 1982, pp. 106-116.

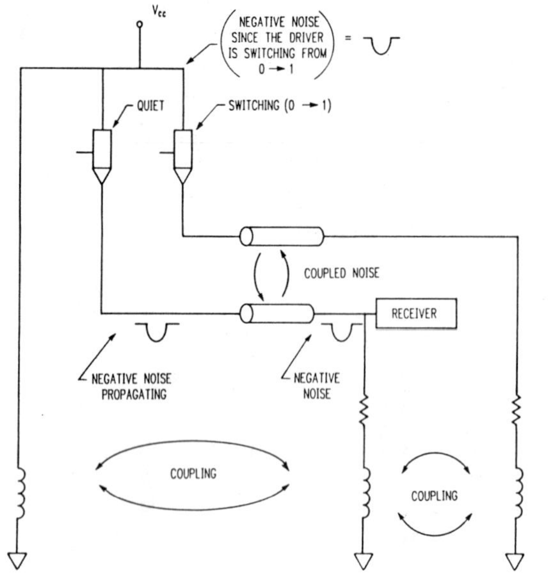

Figure 1

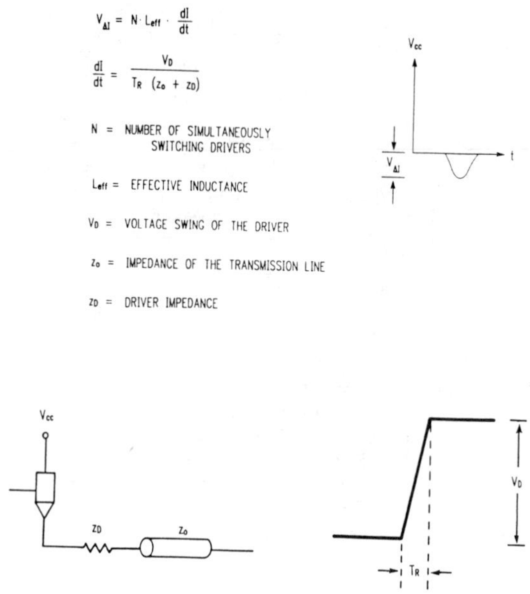

Figure 2

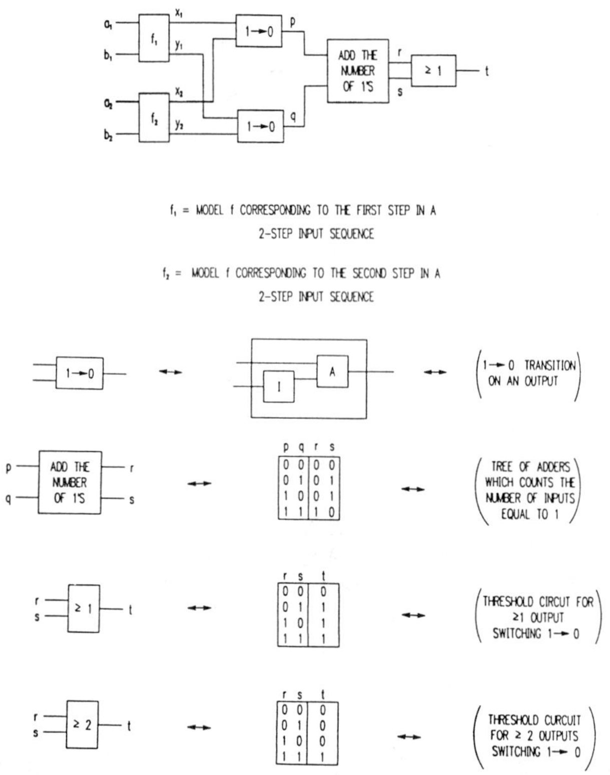

Figure 3

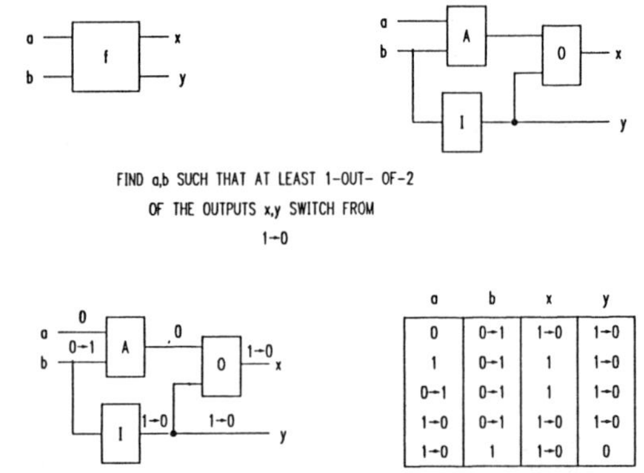

a	b	x	y
0	0→1	1→0	1→0
1	0→1	1	1→0
0→1	0→1	1	1→0
1→0	0→1	1→0	1→0
1→0	1	1→0	0

Figure 4

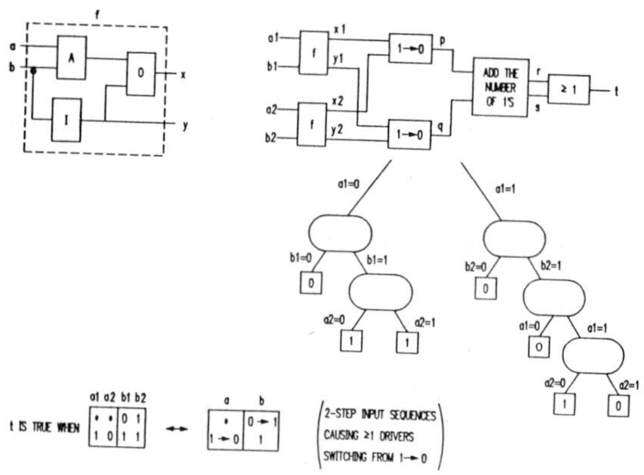

Figure 5

EVALUATING THE PERFORMANCE IMPACT OF MICROARCHITECTURAL FEATURES IN A VLSI PROCESSOR

RONALD M. GUFFIN

HEWLETT-PACKARD
CUPERTINO, CALIFORNIA, USA

ABSTRACT

This paper describes a method for evaluating the quantitative effect on processor performance of adding features to the microarchitecture. The general method involves assessing the amount of time saved for each proposed feature for each instruction in a mix of instructions appropriate to the application domain of interest. The overall effect on performance can then be determined by weighting the time saved for each instruction by its relative frequency of execution and adding the contribution from each instruction in the mix. To illustrate the method, eight different features ranging from variable length shifts to a compare condition code option are evaluated and their effect on performance is computed. This method not only provides a good way to justify the inclusion of a particular feature but also provides an accurate way to predict the overall level of performance that will be achieved.

INTRODUCTION

Continuing increases in practical complexity levels of VLSI processors provide opportunities to enhance performance through more efficient microarchitectures. While one approach might be to simply add all the features that can be thought of or would fit on the chip, a better approach would be to add those which provide the most improvement at the smallest cost. The problem that we will address here is to determine how much improvement each potential feature would contribute.

GENERAL APPROACH

In order to determine what quantitative effect each feature would have on performance, it is necessary to define what the processor is expected to be doing. A recent paper by Clark and Levy (1) describing instruction set usage on the VAX-11/780, showed that this usage can be quite variable in a general time sharing system. Typically some instructions were used more often than others, with the top 25 representing from about 65% to 93% of instruction executions, and from 62% to 98% of execution time. The top 25 instructions however varied widely depending on whether the processor was executing part of the operating system, a Whetstone benchmark (a Fortran benchmark), or a COBOL benchmark. The Whetstone top 25 included a lot of floating point, while the COBOL top 25 included a lot of character, numeric and packed decimal operations. Because instruction execution time varies widely, the ordering of instruction by frequency and time also varies for each operating domain. By determining how much time is saved for each instruction in a mix by adding a feature, it is possible to get a good approximation to the overall effect on performance for that domain. In the following, we will use a more or less typical mix, i.e. typical for o.s. kernel functions, to illustrate the general method.

CONSIDERATIONS IN ASSESSING PERFORMANCE EFFECTS

Processor performance is commonly measured in millions of instructions executed per second (MIPS), and due to the varying execution time of each instruction will relect an overall weighted average. Instantaneous MIPS can vary considerably, being highest in programs using primarily the simpler, faster instructions as in operating systems, and lowest in programs using a lot of the more complex and slower instructions like the decimal operations in COBOL. One problem with the MIPS metric is that it includes the effect of the technology used in implementation so that a processor with an inefficient microarchitecture but implemented with a fast type of logic (e.g. ECL) could have a higher MIPS measure than a more efficient one implemented with a slower technology. The effect of technology can be eliminated by using an Average Number of Clocks per Instruction (ANCI) metric instead. The following equation relates this with MIPS and the clock rate.

$$\text{Ave Clks per Instr} = \text{Clock Rate/MIPS}$$

Thus a 1 MIP machine with a 10 Mhz clock would average 10 clocks per instruction. Since the systems MIPS metric may also include the effects of IO waits, interrupt handling, etc., and since our interest here is in processor performance only, our metrics will assume 100% program execution.

In evaluating the performance impact of a particular feature, it is important to take into account the effect of memory access time. Figure 1 shows how the performance of 3 different hypothetical processors would vary as a function of this parameter. Towards the right where access times are high, each processor will be memory bound and performance differences will be small. Towards the left however where access times are low, such as with a local cache, performance differences can be substantial. Features whose beneficial effects would be masked by the long waits for memory data or instructions in the memory bound region and thus would not be worth including in the design, could be very worthwhile in the processor bound region on the left. Consequently any evaluation must take this factor into account. For our purposes, we will assume in the following that our processor will have a fast 1 clock access cache since these can now be provided inexpensively.

One other factor which needs to be considered is that adding a particular feature may affect the clock period. Obviously, if this feature would increase the clock period by 10% but decrease the average number of clocks per instruction by only 5%, then it should be left out. On the other hand if the figures were reversed then it may well be worth adding. The point is simply that the fact that an increase in clock period would be incurred is not sufficient in itself to reject a particular feature. Since accounting for this factor is beyond the scope of this paper, we will assume that none of the potential feature additions cause a reduction.

POTENTIAL PERFORMANCE ENHANCING FEATURES

Depending upon how simple the base line processor design is that is to be used to compare against, there are many approaches to improving performance. These may include any combination of the following:

1. Automatic support features, such as instruction prefetch, bounds checking, or micro-traps,
2. More powerful micro-operations, such as concatenating variable length shifts, three address operations, variable field operations, two condition jumps, or a special compare condition code setting option,
3. Increased parallelism, such as might be provided by increasing the number of bits in the microinstruction,
4. Increased resources, such as more scratch pad registers, result buses, or counters,etc,
5. Coprocessors, such as a floating point processor,
6. Increased pipelining, such as going from a 2 phase design to a 3 phase design.

While all of these are of interest and potentially valuable, we will limit the analysis to the first two categories. The instruction prefetch feature automatically keeps the next instruction buffer full as a background activity so that an explicit microinstruction doesn't have to be executed to do this fetch. The bounds checking feature provides hardware to automatically check addresses for validity on each access which avoids two explicit comparisons by microcode. Where memory access is more than two clocks, then the comparisons can be done while waiting, but for a fast access cache this would require extra time. Micro-traps are a mechanism whereby exceptional conditions are handled by hardware detection rather than a ucode test. One example of this in a stack architecture is the problem of keeping the hardware stack adjusted so that a stack operation can be performed on valid operands. This can be done by explicit tests or by microtrapping when a stack operation is attempted and one or both operands are invalid. Because the operands are normally valid a high percentage of the time, the latter approach can save a great deal of time. The variable length shifter refers to some form of barrel shifter which allows any length shift (typically 1 bit less than the data width) to be done in 1 clock rather than 1 bit per clock. Three address refers to the ability to perform an operation on the contents of two registers and put the result in a third. A two address machine would require the result to replace one of the original operands which sometimes requires an extra cycle to avoid. Variable length field operations refer to the capability to perform operations on an arbitrary field within a word, or extract or insert arbitrary fields. This is helpful in handling packed data such as operand descriptors or floating point operands. Two condition jumps would permit a branch to be performed based on the logical combination of two conditions. The compare condition code setting option would permit the condition code for compares to be set in one clock rather than to have to use several microinstructions to make use of the standard condition code setting logic.

PERFORMANCE IMPACT ANALYSIS

We will start with the assumption that our base line processor has none of the features of type 1 or 2 above, and that we want to determine how much performance would be gained by adding them to our design. The method used to determine this involves starting with a list of the top 26 instructions used (chosen because they happen to provide almost exactly 80% of the instruction executions), ordering these by frequency of use, and determining the average number of clocks required for each one in the base line processor. We then determine how many clocks we can save in each instruction for each feature. The resulting matrix is shown in Table 1. The instructions shown are from the HP3000 (2) and the frequency distribution is based on the work described by Blake (3). The microarchitecture has been described previously by Guffin (4).

It is obvious from the table that some features are more beneficial than others. The instruction prefetch feature affects nearly all of the instructions and while only saving, at most, one clock, adds up to a 10% improvement in performance. The variable length shift feature on the other hand only affects two instructions but saves a lot of clocks on each one and adds up to more than a 5% improvement. The three address feature only saves a fractional clock on a few instructions and so only adds up to a little over 1% improvement. This may seem surprising but most of the top 26 instructions are quite simple and opportunities to exploit this feature are rare. Usage of this feature is higher in the more complex instructions like floating point and decimal and if these are important then this feature may be useful. The fractional clocks shown in many cases are a consequence of different instruction options. For example, for memory reference instructions a certain percentage of executions use indirect addressing which takes more clocks than for direct addressing.

Another factor to consider in deciding whether a particular feature is worthwhile is the cost to implement it. For example, the variable length shifter might be relatively expensive, while the compare CC setting option, while only providing a little over 2% improvement, will require a few gates to implement. Another factor is the relative emphasis on performance and cost. The two condition jump feature would probably not be worthwhile in a low cost processor but might be included in a high performance design.

Collectively the eight features evaluated here reduce execution time by 41%. This would result in a performance improvement of:

$$\text{Performance Improvement} = 1/(1 - \text{Time Savings}) = 1/.59 = 1.69$$

or 69%. This of course neglects other degrading factors like cache miss penalties, etc., which should be accounted for in an actual evaluation. Processor efficiency is improved from an average of 8.6 clocks to 5 clocks per instruction.

CONCLUSIONS

This paper has evaluated the effect on performance of including a variety of microarchitectural features into a VLSI processor design. The results indicate that while some features are much more valuable than others, substantial performance improvements can be achieved by increasing the power of the microarchitecture. While the evaluation made here was for a typical mix of instructions, the effect on a Whetstone or COBOL mix can be made in a similar way. The results that were achieved here are, of course, only valid for the HP3000 architecture implemented on a particular microarchitecture. Nevertheless, the general evaluation technique should be quite useful for evaluating other microarchitectures. Hopefully, this will provide a more quantitative basis for feature selection in place of the rather subjective approach often used in many current processor designs.

REFERENCES

1. D. Clark and H. Levy, "Measurement and Analysis of Instruction Use in the VAX-11/780", Proc 9th Annual Computer Architecture Symposium, April 1982, p.9-17,

2. HP3000 Machine Instruction Set Reference Manual, Hewlett-Packard Co., Manual Part No. 30000-90022,

3. R. Blake, "Exploring a Stack Architecture", Computer, May 1977, p.30-38,

4. R. Guffin, "A Microprogramming Language-directed Microarchitecture", Proc 15th Annual Workshop on Microprogramming, Oct. 1982, p.44-49.

Table 1. Performance Impact of Various Microarchitectural Features

Instr	Freq %	Ave Base Line Clks	Freq x Clks	/---Clocks Saved / Instr Prefetch	Bounds Checkers	, Freq Micro Traps	x Clocks Var Len Shifts	Saved by Adding These Features---/ Three Addr	Var Len Field Ops	Two Cond Jumps	Compare/ CC Option
LOAD	19.2	7.92	152	1,19.2	1.72,33	1,19.2		.24,4.6			
BCC	11.6	9	104	.5,5.8							
STOR	6.7	9.38	63	1,6.7	3.24,22	1,6.7		.24,1.6			
CMPI	4.5	↳8	36	1,4.5		1,4.5			1,4.5		2,9
LDXI	4.5	3	14	1,4.5							
BR-P	3.3	5.1	17								
STAX	2.7	4	11	1,2.7		1,2.7					
DUP	2.5	5	13	1,2.5		2,5					
EXF	2.5	23	58	1,2.5		1,2.5	12,30		3,7.5		2,4.8
CMP	2.4	6	14	1,2.4		1,2.4					
LDX	2.4	5.92	14	1,2.4	1.72,4						
PLDA	1.9	5	10	1,1.9		1,1.9					
LDI	1.7	4	7	1,1.7		1,1.7					
BRO	1.6	7.8	12	.5,.8		1,1.6					
ZERO	1.5	4	6	1,1.5		1,1.5					
ANDI	1.5	4	6	1,1.5		1,1.5					
SUBM	1.4	8.36	12	1,1.4	1.72,2	1,1.4		.24,.3			
CMPM	1.2	10.36	12	1,1.2	1.72,2	1,1.2		.24,.3			2,2.4
LRA	1.1	4.84	5	1,1.1	.24,0	1,1.1					
LDXA	1.0	4	4	1,1.0		1,1.0					
ADDI	.98	4	4	1,.98		1,.98					
PCAL	.9	57	51					2,1.8	5.13,4.6	1.18,1	
EXIT	.9	40.67	37						3.77,3.4	2,1.8	
LDD	.83	9.38	8	1,.83	2.5,2	1,.83					
ADDM	.8	8.36	7	1,.8	1.72,1	1,.8		.24,.2			
LSR	.74	15	11	1,.74		1,.74	10,7.4		1,.74		
	80.3%		687	69	67	60	37	9	21	2.8	16
Performance Improvement				10%	9.8%	8.7%	5.4%	1.3%	3%	.4%	2.3%

Baseline Processor Efficiency = 687/80.3 = 8.6 Clocks per Instruction
Improved Processor Efficiency = 405/80.3 = 5 Clocks per Instruction

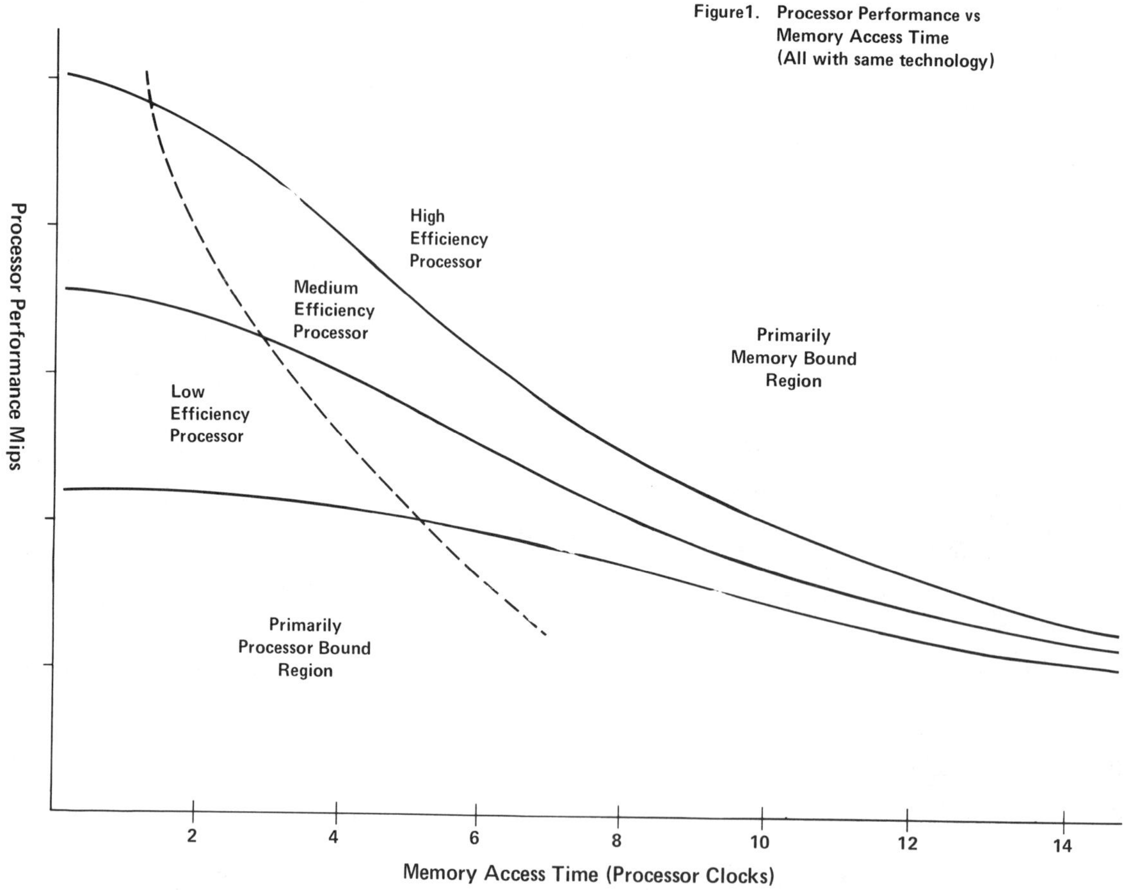

Figure1. Processor Performance vs Memory Access Time (All with same technology)

A Linear-Time Algorithm for Race Detection in Transistor Switch-Level Circuits†

Vijaya Ramachandran‡

Department of Electrical Engineering and Computer Science
Princeton University
Princeton, N.J. 08544

1. Introduction

In [R83a], we developed the notion of a well-designed circuit, and a design for a fast and reliable switch-level simulator for such circuits. In particular, the simulator handled race conditions. In this paper, we develop a linear-time algorithm for the detection of race condition in any *nonoscillating circuit* (i.e., a circuit that is acyclic within a clock phase), and use it as a basis for a fast switch-level simulator.

In Section 2, we derive a race detection algorithm for nonoscillating circuits which will stop if a *controlling node* (i.e., a gate of a transistor) ever settles at the undefined or unknown level x. In Section 3, we derive a more general algorithm to handle level x at controlling nodes. The algorithms we derive assume that all inputs that can change are gated through a clock, so that switching of external signals occurs only in the clocked mode.

2. The Basic Algorithm

We assume that the reader is familiar with the terminology and results of [R83a]. In it, we pointed out that under the LSV simulation, the transistor nodes are ordered so that the switching of a node in transistor group i does not influence any node in group $i-1$ or less. In addition, in this section, we assume that controlling nodes cannot have level x. This means that a race condition occurs at node n in transistor group i if and only if two different *sequences* of switchings of the gates of transistors whose source-drains are in this group, causes node n to settle at different final values. This immediately suggests an algorithm to detect the presence of a race condition at any node n in transistor group i: If there are k transistors in group i that changed state at some time during this phase, then, for each of the $k!$ different ways the gates of these transistors can switch, determine the level at node n. If the level is the same in all cases, then there is no race condition at node n, otherwise there is. (Actually, the number of different sequences of gate switchings will be greater than $k!$ if there are hazards. However, once the switching pattern of the gates is known, the number of different sequences is known, and is finite.)

The algorithm described above is clearly not fast. We will now present a race-detection algorithm that is linear in the number of transistors in the circuit.

†This research was supported in part by DARPA under ONR N00014-82-K-0549.

‡Author's address after August 1983: Coordinated Sciences Laboratory, University of Illinois at Urbana-Champaign, Urbana, Illinois 61801.

We first introduce the notion of a *switched node*.

Definition 1: A node n is a *switched node* at a given clock phase if

1) it changed its level during the phase; or
2) it had a hazard during the phase; or
3) its level during the phase was x (i.e., either an undefined level u or an unknown level w.)

Note that n is a switched node if and only if it can be at both level 0 and level 1 during this clock phase.

Next, we introduce the concept of a *switched graph*. Recall from [R83a] that, while preprocessing the circuit, we introduced an edge between the source and drain of each transistor to get the transistor groups. This gave us an undirected graph, whose connected components were the transistor groups. After we have simulated the nodes in transistor groups 1 to $i-1$, we form the switched graph of group i as follows:

Algorithm 1: Switched Graph

1) Retain edges for only those transistors whose gates are not switched nodes, and are at logical level one. Collapse each connected component of this graph into a single node whose level is the common level of all the nodes in the component, and whose strength is the strength of a strongest node in the component. Call each such node a *level-one-macronode*.
2) Introduce an edge for each transistor whose gate is a switched node. This graph may have multiple edges between vertices (because of step 1), but we are not concerned by that.

The graph obtained by steps 1 and 2 is called the *switched graph* of the ith transistor group for this clock phase.

Lemma 1: Every node in a level-one-macronode gets the same value at all points in the clock phase and this is the value obtained by treating the macronode as a regular node in the switch level circuit.
Proof Trivial.▪

In what follows, we will treat a level-one-macronode as a regular node in the circuit until the final updating of the individual nodes at the end of the clock phase.

Note that the switched nodes in step 2) of Algorithm 1 will satisfy either 1) or 2) of Definition 1 since we do not allow gates at level x. The switched graph is clearly a subgraph of transistor group i, and may contain several connected components. Let C be any connected component of the switched graph.

Theorem 1: If n_1 and n_2 are two input nodes in C having

CH1935-6/83/0000/0345$01.00

different initial levels, then there exists a sequence of switchings so that every node in C has an undefined level at some point in the sequence.

Proof Turn on all of the off transistors in C in any order. Clearly, this is part of a valid sequence of switchings, since every off transistor in the switched graph did switch on. At this point, n_1 is connected to n_2, and this results in the value u^i, which is transmitted to all nodes in C, since all transistors in C are on.▪

The proofs of Theorems 2 to 4 below are similar to that of Theorem 1, and can be found in [R83b].

Theorem 2: If n_1 is any input node in C with initial level l_1, and n_2 is any node in C with strength s_2 and initial level l_2, then there exists a switching sequence such that each of nodes n_1, n_2 can have value merge$(l_1^i, l_2^{s_2})$ at some point during the switching.

Theorem 3: If C contains a pullup node, then for any node in C that is not an input node, there exists a sequence of switchings such that the node is at level 1 at some time during the switching.

Theorem 4: Let D be the graph C with strong nodes removed, and let D_1 be a connected component of D. Then, for any nodes n_1, n_2 in D_1, there exists a switching sequence such that each of nodes n_1, n_2 can have value merge(l_1^w, l_2^w) at some point during the switching, where l_1 and l_2 are the initial levels of n_1 and n_2 respectively.

Theorem 5: A node n cannot take on any level other than the ones specified by Theorems 1 through 4 at any point during any sequence of switchings.

Proof The proof ([R83b]) is by a case-by-case analysis and will be omitted for lack of space.▪

Note From Theorems 1 through 5, we have the result that, to determine the different levels at which a node in the ith transistor group can be during a clock phase, the only information we require about the gate of each transistor in the group is whether the transistor could have been both on and off, whether it remained on throughout, or whether it remained off. The exact sequence of switchings of a transistor, or even the number of times a transistor switched, or the set of levels through which it switched does not affect the levels at any node in the group. In other words, the only information we require about a controlling node is whether it is a switched node, a node that stayed at level 1, or a node that stayed at level 0.

From the above results, we can formulate a simple algorithm to detect the presence of switching at a node (denoted by level x) in a switched component C.

Algorithm 2: Switched Nodes

1) If there are two input nodes at different levels, or if there is an input node at level x in C, then every node in C has level x; else
2) if C contains a pullup node and an input node at level 0, then all nodes in C are switched nodes except the input nodes; else
3) form D from C by deleting all strong nodes. Find the connected components of D. Each node in a connected component D_1 of D is a switched node if there is node in D_1 at initial level x; or if there are two nodes in D_1 at different initial levels; or if there is a strong node in C at a different level from every node in D_1.

Clearly, this algorithm can be implemented in time linear in the number of nodes by making one DFS to execute steps 1 and 2, and another to execute step 3.

Next, to identify the subset of nodes that have a race condition, we look at the subgraph of the switched graph that contains only edges that can be on at the end of the phase. This refers to transistors whose gates have final level 1. We call this graph the *on−graph* of transistor group i for this clock phase. We still treat level-one-macronodes as regular nodes. The on-graph may again contain many connected components. Let C_o be any connected component of the on-graph. From [R83a] we know that there can be no race condition at any node in a connected component that contains a strong node. So we only need look at connected components that contain only weak nodes for a possible race condition. For such connected components, we have the following theorem:

Theorem 6: Let C_o be a connected component of the on-graph that contains only weak nodes. For each node n in C_o, let S_n be the set of possible levels that node n could have taken during the sequence of switchings. Then, each node in C_o could be at any level in the intersection of the S_n's at the end of the phase.

Proof See [R83b].▪

Corollary: A node in C_o gets level x if and only if some node in C_o is a switched node, or equivalently, if and only if every node in C_o is a switched node.

We use the above theorem in the following algorithm for detecting races. The output from Algorithm 2 is taken as input. The algorithm computes the final level at each node, and indicates the presence of a race condition at a node by assigning it the level x.

Algorithm 3: Final Levels with Race Detection

1) Form the on-graph from C by retaining only edges for gates that have level one.
2) Make a DFS of the on-graph, keeping track of the levels of strong nodes encountered, and the presence of switched nodes.
3) If a connected component has a strong node, then there is no race condition in that component, and the final level of all nodes in the component is the merge of the levels of the strong nodes; else {all nodes are weak} if there is a switched node in the connected component, then all nodes in the component have level x, otherwise, all nodes in the component have the single common level.

We now have the tools to give the restricted simulation algorithm with race detection for a nonoscillating circuit whose transistor groups have been topologically sorted. We assume that we have pointers so that we can extract each transistor group by turn. The number of transistor groups in the circuit is assumed to be N.

Algorithm 4: Restricted Simulation Algorithm with Race Detection

1) Set input nodes to their new levels, and clocks to correspond to the new phase. Mark all input nodes that changed from their previous level.
2) For $i = 1, 2, \cdots, N$, process transistor group i:

a) Form the switched graph of transistor group i using Algorithm 1, using unmarked gates at logical level 1 for step 1, and marked gates for step 2.

b) Make a DFS of the switched graph to obtain the connected components. With each component C, use Algorithm 2 to detect the presence of switched nodes. Mark all controlling nodes that switched.

c) Apply Algorithm 3 to each component C to detect race conditions, and to find the final level at each node.

d) Decompose each level-one-macronode into its component nodes, and assign the final level of the macronode to each of the nodes. If the macronode is a switched node, mark each controlling node in it as a switched node.

e) If a controlling node gets final level x, then flag the error and stop.

3. The General Simulation Algorithm

In this section, we extend Algorithm 4 of the previous section to handle the case when the gate of a transistor can be at level x. We need to reexamine the results of the previous section under this additional feature. First, note that we can obtain the unknown level w in a new way. Consider the situation in Figure 1. The gate of the transistor (node a) has level x, so the transistor can be either on or off in the steady state. If a is at 1, then node c will be at level 0; if a is at 0, then c will be at level 1. Hence c can have two different levels, and so is at level w. Note, thus, that when we allow a race condition at a controlling node, we *can* have a race condition at a strong node, and hence Lemmas 1 and 2 of [R83a] do not hold in this case.

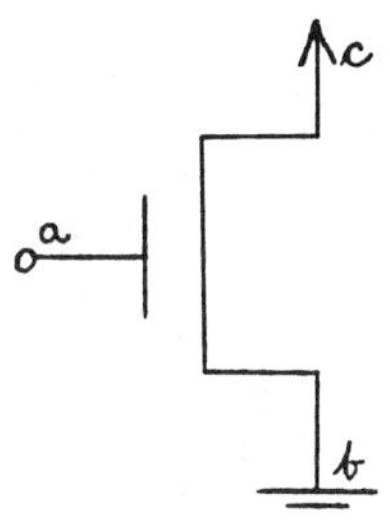

Figure 1

Fortunately, however, Theorems 1 to 5 of the previous section do hold for this case, since we have defined a switched node to be one that could have value x as well. This means that Algorithm 2 goes through without change.

The main addition we need to make is in the evaluation of the final levels. Now we may have transistors in the ith transistor group whose gates are at level x, so at the end of the clock phase, these transistors may be either on or off. This, of course, is what gives rise to race conditions in strong nodes. We will now develop the tools to evaluate the final levels at nodes when controlling nodes are allowed level x. First, we construct the *final graph* in Algorithm 5. The algorithm takes the output from Algorithm 3 as input.

Algorithm 5: Final Graph

1) Collapse each connected component of the on-graph into a single macronode that has the common final level of all nodes in that component, and the strength of a strongest node in the component. Treat the macronode as a regular node.

2) Introduce an edge for each transistor whose gate has level x. This graph may have multiple edges (because of step 1).

Note that Algorithm 5 is very similar to Algorithm 1, except that in step 1, we use transistors whose gates have final level 1, rather than transistors whose gates are at level 1 all through the clock phase; and in step 2 we use transistors whose gates have final level x, rather than transistors whose gates are switched nodes. We can modify Lemma 1 to obtain the following:

Lemma 2: Every node in a macronode of Algorithm 5 gets the same set of final levels, and this is the set obtained by treating the macronode as a regular node in the circuit.

Next, we have the interesting result that each of Theorems 1 to 5 translates in a very direct way into a corresponding theorem about the final levels of nodes in the final graph. We state the five new theorems below. Let C be a connected component of the final graph.

Theorem 7: If n_1 and n_2 are two input nodes in C at different levels, then there exists an assignment of levels from the set $\{0,1\}$ to gates with undefined levels (henceforth called *gate assignment*) so that every node in C has a final level x.

Theorem 8: If n_1 is any input node in C at level l_1, and n_2 is any node in C with strength s_2 and level l_2, then there exists a gate assignment such that each of nodes n_1, n_2 can have final value merge($l_1^i, l_2^{s_2}$).

Theorem 9: If C contains a pullup node, then for any node in C which is not an input node, there exists a gate assignment such that the node gets final level 1.

Theorem 10: Let D be the graph C with strong nodes removed, and let D_1 be a connected component of D. Then, for any two node n_1, n_2 in D_1, there exists a gate assignment such that each of nodes n_1, n_2 can have final value merge(v_1,v_2), where v_1 and v_2 are the values of n_1 and n_2 respectively.

Theorem 11: A node n cannot have a final level that is different from the ones specified by Theorems 7 through 10.

Since Theorems 7 through 11 have an exact correspondence with Theorems 1 through 5, the algorithm to determine the final values corresponds in a direct way to Algorithm 2. However, we will not need the full power of Algorithm 2 because we need to check for race conditions only at strong nodes. For weak nodes, the levels obtained in Algorithm 3 are the final levels, as shown in the following theorem:

Theorem 12: Let C_0 be a weak macronode in the final graph, whose level from Algorithm 3 is l. Then, this level does not change when Algorithm 2 is carried out on the final graph.

From Theorem 12, we have the result that, given the output from Algorithm 3, we only need check for new final levels at strong nodes. Let C be a connected component in the final graph.

Algorithm 6: Final Levels at Strong Nodes with Race Detection

1) If there are two input nodes at different levels, or if there is an input node at level x in C, then every strong node in C has level x; else

2) if C contains a pullup node and an input node at level 0, then all pullup nodes in C have level x.

We now give the general simulation algorithm for nonoscillating circuits.

Algorithm 7: Simulation Algorithm with Race Detection for Nonoscillating Circuits

1) Set input nodes to their new levels, and clocks to correspond to the new phase. Mark all input nodes that changed from their previous level.

2) For $i=1,2,\cdots,N$, process transistor group i:

   a) Form the switched graph of transistor group i using Algorithm 1, using unmarked gates at logical level 1 for step 1, and marked gates for step 2.

   b) Make a DFS of the switched graph to obtain the connected components. With each component C, use Algorithm 2 to detect the presence of switched nodes. Mark all controlling nodes that switched.

   c) Apply Algorithm 3 to each component C, to detect race conditions at weak nodes, and to find the final level at each node, assuming all transistors with x at their gates are off.

   d) Form the final graph of transistor group i using Algorithm 5.

   e) Apply Algorithm 6 to detect race conditions at strong nodes and their final set of levels.

   f) Decompose each macronode in the final graph and set the level of each component node equal to the final level of the macronode.

   g) Decompose each level-one-macronode into its component nodes, and assign the final level of the macronode to each of the nodes. If the macronode is a switched node, mark each controlling node in it as a switched node.

4. Conclusion

In this paper, we have given a switch-level simulation algorithm that detects race conditions in all nonoscillating circuits. The paper has been condensed due to space limitations. We refer the reader to [R83b] for a more complete presentation, for all proofs, and for an example. The running time of the algorithm is linear in the number of transistors, since it uses a DFS a constant number of times in each transistor group.

The significant contribution of this paper is the method we have introduced to identify any sort of switching at a node during a clock phase. Similar results have been obtained in the context of ternary simulation of logic-gate circuits (see e.g., [E65], [BY79]). These results were not applicable to race detection at weak nodes in switch-level circuits. However, Bryant [B83] incorporated ternary simulation in the switch-level simulator MOSSIM II, and found it to work well. The results of this paper prove why ternary simulation can be applied to general nonoscillating switch-level circuits. Since our algorithm is linear-time, it should find widespread application in the field.

In the presence of fanout from a node, Algorithm 7 can sometimes indicate the presence of race condition when in reality, the circuit has no race condition. This is because the algorithm looks at the state of each transistor gate in isolation. If a node at unknown level is gate to more than one transistor, the algorithm does not ensure that this node is at the same final level at all gates. The problem of correctly identifying race condition in a nonoscillating circuit in the presence of fanout can be shown to be NP-complete by a simple reduction from 3-SAT; it is, thus, highly unlikely that an efficient algorithm exists for this problem [GJ79].

Acknowledgements: I would like to acknowledge useful comments from Andrea LaPaugh and Ken Steiglitz. In particular, Ken Steiglitz pointed out the NP-completeness of the general race detection problem for nonoscillating circuits. I also wish to thank Randy Bryant at Caltech for drawing my attention to the work on ternary simulation.

References

[B83] R. E. Bryant, "Race detection in MOS circuits by ternary simulation," to appear in *VLSI 83*, F. Anceau ed., North-Holland Publishing Co., August 1983.

[BY79] J. A. Brzozowski and M. Yoeli, "On a ternary model of gate networks," *IEEE Transactions on Computers*, vol. C-28, no. 3, March 1979, pp. 178-183.

[E65] E. B. Eichelberger, "Hazard detection in combinational and sequential switching circuits," *IBM Journal of Research and Development*, vol. 8, 1965, pp.90-99.

[GJ79] M. R. Garey and D. S. Johnson, *Computers and Intractability: A Guide to the Theory of NP-Completeness*, Freeman, San Fransisco, 1979.

[R83a] V. Ramachandran, "An improved switch-level simulator for MOS Circuits," in *Proc. of ACM IEEE 20th Design Automation Conference*, Miami Beach, Florida, June 27-29, 1983, pp. 293-299.

[R83b] V. Ramachandran, *Studies in VLSI Layout and Simulation*, Ph. D. thesis, Department of Electrical Engineering and Computer Science, Princeton University, Princeton NJ 08544, August 1983.

Session 2-2c
Physical Design II

Chairman

C. Séquin
University of California

AN EFFICIENT DATASTRUCTURE AND ALGORITHM FOR VLSI ARTWORK VERIFICATION

J. T. Fokkema*

Department of Mining Engineering

T. G. R. van Leuken

Department of Electrical Engineering

Delft University of Technology

Delft, the Netherlands

Abstract

A polygon can be represented by a set of vertical line segments. A new algorithm is presented which produces these line segments using a mechanism called the stateruler. Line segments belonging to the same polygon obtain the same polygon-number. The input of the algorithm is a collection of rectangles. Experience with an implementation of this algorithm shows fast execution on a large number of rectangles. The algorithm is of O(n) complexity, where n is the number of rectangles.

1. Introduction

The basic element in a layout data structure for VLSI circuits is often a box(rectangle) and the description of a mask layer consists of a large collection of unordered boxes. Then verification is a difficult task because many boxes overlap. The input for a design-rule checker or a circuit extractor should be a representation that is efficient and without redundancy. The best way to do this is with a polygon representation.

A polygon is uniquely described by a set of connected line segments. There are two approaches to accomplish this goal : the bit-map approach [5] and the method that confronts the edges of the boxes with other edges[2,3]. Both methods have their disadvantages, the first requires a large memory allocation and the second has an algorithmic complexity of O(n**m) where m>1.

The method proposed in this paper is also based on edge confrontation, however the area of confrontation is restricted. The implementation of the method results in an one-pass algorithm. The new aspect of this algorithm is that it does not search for information, all required information is present in the area of interaction.

* The basic ideas of this research were formed during the stay of J.T. Fokkema at Stanford University under the support of the Netherlands Organization for the advancement of Pure Research (Z.W.O.). The organization is gratefully acknowledged.

In our case a polygon is described by start and stop occurrences. We use the term start occurrence for a line segment when the interior of the polygon is situated on the right side. A stop occurrence denotes a line segment where the interior is located on the left side.

2. Computational model

We consider a particular mask area as a two-dimensional integer space D2, where the horizontal and vertical values are denoted by the integers x and y, respectively. Then a mask box is uniquely described by the following integers: xl, the left horizontal value; xr the right horizontal value; yb, the bottom vertical value and yt, the top vertical value.

The mathematical model underlying the computational model for the generation of occurrences is the set-theoretic relation, which is a subset of the Cartesian product of a list of domains. A domain is a simple set of values. Then the Cartesian product of domains Dom(1)*Dom(2)* ... *Dom(n) is the set of all n tuples[6]:

$$\langle\ a[p](1),\ a[p](2),\ldots\ a[p](n)\ \rangle,$$

such that a[p](1) is in Dom(1), a[p](2) is in Dom(2) and so on. The integer n is called the arity of the tuple, p denotes the element number of the tuple in the set A and a[p](2) denotes the second component of a[p].

The collection of mask boxes can be represented by the set B of tuples of arity four such that

$$B = \{b[p]:\ \langle\ xl\ ,yb\ ,yt\ ,xr\ \rangle\},$$

and Dom(i) = D1.

where D1 is the one-dimensional integer space. We call a tuple set A ordered if there is at least one component a[p](q) such that for all p:

$$a[p](q)\ <\ a[p+1](q).$$

If K and L are both subsets of the ordered set A, then we use the notation

K <- L,

for the insert of L in K, such that K maintains its order. In the following section, an algorithm for occurrence generation will be discussed based on interactions between tuple sets.

3. The algorithm for occurrence generation

To explain our algorithm, we introduce the following ordered tuple sets.

The set of events E.

E = { e[n] : <b[q](1), b[q](2), b[q](3), b[q](4)>,
e[n](1) < e[n+1](1) or
e[n](1) = e[n+1](1) -> e[n](2) <= e[n+1](2)},

and Dom(i) = D1.

It is noted that E has the same elements as B, however ordered with respect to xl and yb.

The set of states S.

S={s[j]:<bottom, top, duration, group>,
s[j](1) < s[j+1](1) and s[j](2) < s[j+1](2) },

with Dom(i) = D1.

The duration and the group are the state variables of the set S.

The set of occurrences OC.

OC = { oc[m] : <X value , bottom value ,
top value , occurrence type, group>,
oc[m](1) < oc[m+1](1) or oc[m](1) = oc[m+1](1) ->
oc[m](2) < oc[m+1](2) and oc[m](3) < oc[m+1](3)},

with Dom(1), Dom(2), Dom(3), Dom(6) = D1,
and Dom(4) = { start occurrence, stop occurrence}.

The set of groups G.

G = { g[r] : <r, number>,
g[r](1) < g[r+1](1)},

with Dom(i) = D1.

The set of numbers N.

N = { n[t] : <t, G_connect, group number>,
n[t](1) < n[t+1](1) },

with Dom(i) = D1.

The sets S and OC are linked with the set G via the component group, which refers to a element number of G. The set G is linked with the set N via the component number, which refers to a element number in N. The component G_connect in the element definition of N denotes the number of elements of the set G that are connected to a particular element in N.

The algorithm we propose for the generation of occurrences strongly resembles the basic algorithm of a logic simulator. A particular box of E, say e[n], has to be confronted with the state of the mask at Xpresent where Xpresent equals e[n](1). The state of the mask for a particular Xpresent is given by the stateruler { Xpresent, S}. The stateruler is the vertical cross-section of the mask at Xpresent and the states s[j](3) indicate the x value where the fields s[j](2) - s[j](1) will terminate. Xpresent divides the mask area in a past and a future with respect to the x value. We notice that no information of the past is available, only information of the future is present in the stateruler. The stateruler allows a two-dimensional extension in the future of the one-dimensional stateruler. This is just the information needed for a correct insert of the event in the stateruler.

Let e[n] denote the event that has to be merged with the stateruler. e[n] partitions the set S of the stateruler in three subsets :

(i) S1, such that S1 = { S1[j] is in S and
S1[j](2) < e[n](2) }

(ii) S2, such that S2 = { s2[j] is in S and
(S2[j](1) or S2[j](2)) is in [e[n](2),
e[n](3)]}, where [... , ...]
denotes the closed interval.

(iii) S3, such that S3 = { S3[j] is in S and
S3[j](1) > e[n](3)}

To determine the merge position of e[n] in S all elements in S1 have to be visited. Because the imposed order of E every member of S1, S1[j], for which S1[j](3) = Xpresent is a stop occurrence. By generating the stop occurrences S1 is updated. The subset S2 indicates the area in the stateruler where the start occurrences are generated. The resulting modified set S2 merged with the set S3 determines the reduced stateruler, needed for the evaluation of the next event e[n+1] if e[n+1](1) = Xpresent. If e[n+1](1) does not equal Xpresent, each element of the reduced stateruler is tested for stop occurrences and updated, in the same way as S1. Then the next Xpresent is determined by the minimum value of the X of the next event and the smallest state of the set S. If the next value of Xpresent equals the minimum state, only stop occurrences are generated for that particular Xpresent. In the other case the procedure described above is repeated.

Program I shows an implementation of the main algorithm. In the first block the input boxes are ordered and the initialization takes place. The next block realizes the stop- and start-occurrence generation by the subroutines STOP and START respectively. The first call of STOP evaluates the subset S1, the call of START performs the confrontation of the event e[n] with the set S2 and the second call of STOP produces the stop occurrences of the reduced set S or the set S itself in the case that Xpresent equals the minimum state. The insert of the dummy event <max, 0, 0, max> flushes the stateruler, hence generates the last stop occurrences. In the last block the occurrences are counted.

ProgramI. : Occurrence Determination.

Input: The set of unordered boxes .
Output: The set of ordered occurrences .

```
Procedure OCCURRENCE(B) :
begin
 E <- Quicksort(B); Xpresent <- e[1](1);
 min <- e[1](4);n <- 1; j <- 1;max <- 0;
 key <- 1;
 while n<= last element number of E do
 begin
  max <- maximum(e[n](4),max);
  if n= last element number of E and key =1
  then begin
   E <- <max,0,0,max>;key <- 0
  end
  if e[n](1) = Xpresent
  then begin
   if n < last element number of E
   then begin
    STOP(j,{Xpresent,S},min, flag <- e[n](2),OC);
    START(j,{Xpresent,S},e[n], OC,G,N)
   end
   n <- n+1
  end
  else begin
   STOP(j,{Xpresent,S},min,flag<-S[last](2)+1,OC);
   Xpresent <- minimum(e[n](1), min);
   min <- e[n](4);j <- 1
  end
 end;
 COUNT(OC,G,N)
end
```

ProgramII. : Stop Occurrence Generation

Input: j,the actual element number of S; {Xpresent,S}, the stateruler; min,the smallest element of the set {S[1](3), and OC the set of occurrences.
Result: S and OC are updated by generating the stop occurrences; j and min are updated by stepping through S.

```
Procedure STOP(j,<Xpresent,S>,min,flag,OC)
begin
 if S is empty
 then return;
 while S[j](2) < flag do
 begin
  if S[j](3) = Xpresent
  then begin
   OC <- <Xpresent,S[j](1),
   S[j](2), STOP OCCURRENCE,S[j](4)>;
   delete S[j] from S
  end
  else begin
   min <- minimum(min,S[j](3)); j <- j+1
  end
 end
end
```

Program II shows the implementation of the subroutine STOP. When S[j](3)=Xpresent a stop occurrence is scheduled. The group of the stop occurrence is the same as the group of the deleted element S[j]. If there is no stop occurrence the minimum state is adapted.

ProgramIII. : Start Occurrence Generation

Input: j,the actual element number of S; {Xpresent,S}, the stateruler; e[n], the event n; OC, the set of occurrences;G, the set of groups and N the set of group numbers.
Result: S,OC,G and N are updated by generating the start occurrences.

```
Procedure START(j,{Xpresent,S},e[n],OC,G,N)
begin
 if S is empty
 then begin
  OC <- <e[n](1),e[n](2),e[n](3), START OCCURRENCE,
  last element of G+1>;
  S <- <e[n](2),e[n](3),e[n](4), last element of G+1>
  G <- <last element of G+1, last element of N+1>;
  N <- <last element of N+1,1,0>;
  return
 end
 k <- j; empty the sets F, O and O';
 if S[k](1) < e[n](2)
 then begin
  F <- <S[k](1),e[n](2),S[k](3), S[k](4)>;
  if S[k](2) > e[n](2)
  then
    S[k](1) <- e[n](2);
  else
    delete S[k] from S
 end
 while S[k](2) <= e[n](3) do
 begin
  O <- <S[k](1),S[k](2),S[k](3) <- maximum(e[n](4),
  S[k](3)), S[k](4)>; delete S[k] from S; k <- k+1
 end
 if( S[k](1) <= e[n](3)
 then begin
  F <- <e[n](3),S[k](2),S[k](3), S[k](4)>
  if (S[k](1) != e[n](3) )
  then
   O <- <S[k](1),e[n](3),S[k](3)<-maximum(e[n](4),
     S[k](3)), S[k](4)>; delete S[k] from S
 end;
 O' <- Construct the complement set O' of O with
 respect to the interval [e[n](2),e[n](3)]; F<-O;
 for each o'[l] in O' do
 begin
  F <- o'[l] resulting in the element f[p];
  f[p](3) <- e[n](4); determine f[p](4) ;
  if OC[last-1](4) = START OCCURRENCE and
     OC[last-1](2) = f[p](1)
  then
   merge START OCCURRENCES ;
  else
   OC <- <Xpresent,f[p](1), f[p](2),
   START OCCURRENCE, f[p](4)>
 end
 for each f[p] in F such that f[p](1) = f[p-1](2)
 and f[p](3) = f[p-1](3) do
 begin
  f[p](1) <- f[p-1](1); delete f[p-1] from F
 end
 S <- F
end
```

In program III the implementation of the start occurrence generation is described. The subset S2 is determined and deleted from S. S2 is further partioned in the boundary set F and the overlap set O (the true projection domain of the event on S). The states of the elements of the set O are modified according the overlap rule that the new element state is the maximum value of the old element state and the event state(the xr value). Next the complement set O' with respect to the event field e[n](3) - e[n](2) is determined. The elements in this set form the start occurrences and the state of all elements equal the xr value of the event. O is inserted in F and successively all elements of O' are inserted in F. Now the environment of the start occurrence is known and the group can be determined. Finally the set F is reduced so that no connected fields have the same state. The set F is inserted back into the set S. Figure 2 shows an event confrontation with the stateruler and the corresponding sets.

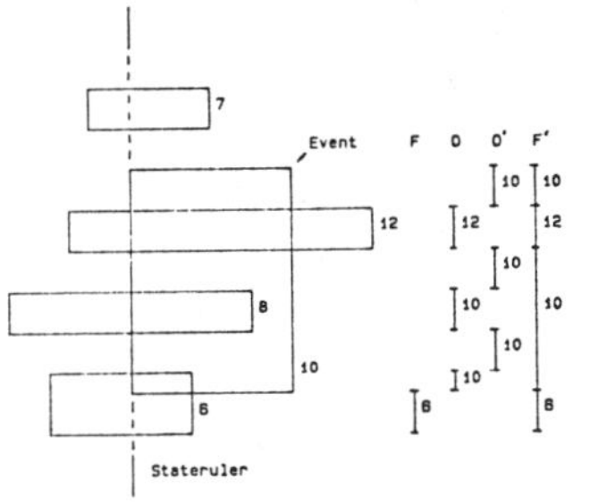

Figure 1. Event confrontation for a particular case with the corresponding sets F, O, O'and F. The numbers shown refer to the states of the fields.

ProgramIV. : Group Counting

Input: OC,the set of occurrences; G ,the set of groups and N,the set of group numbers.
Result: the set OC is updated with respect to the group numbers.

```
procedure COUNT(OC,G,N)
begin
 n <- 1;
 for each OC[k] in OC do
 begin
  if N[G[OC[k](6)](2)](3) = 0
  then begin
   N[G[OC[k](6)](2)](3) <- n;
   OC[k](6) <- n; n <- n+1
  end
  else OC[k](6) <-N[G[OC[k](6)](2)](3)
 end
end
```

In program IV the counting of the occurrences is performed. Since the group structure is already embedded via the links of OC, G and N, the counting is a trivial matter.

4. Results

The algorithm described above is implemented in the program language C and is running under the UNIX * operating system on a VAX-11/750.

---

* Unix is a Trademark of Bell laboratories.

We investigated the linear-time complexity of the presented algorithm. We chose a random set of boxes as input to guarantee a constant design style and mask property. Table I shows the results for an increasing number of boxes and clearly illustrates the linearity of the time complexity.

TABLE 1. Computer runtime for a random set of boxes.

Number of random boxes	CPU time in sec.
100	1.8
1000	16.9
10000	188.8
50000	963
100000	2063

5. Conclusion

The new algorithm presented in this paper is efficient both in runtime and space. It generates a database suitable as input for the plotter, the pattern generator, the design rule checker and the circuit extractor.

We already have implemented a design rule checker, a circuit extractor and a program that performs Boolean operations on masks. Both programs use the occurrence sets as input. The working is also based on the stateruler formalism. Hence they also have a linear-time complexity. The programs only differ from the described algorithm in the nature and number of state variables of the fields in the stateruler. This implies that the confrontation of the events (now recruited from the set of occurrences) with the stateruler changes in every application.

References

[1] H.S. Baird: "Design of a Family of Algorithms for Large Scale Integrated Circuits Mask Artwork Analysis", M.S. Thesis, Dept. of Computer Science, Rutgers University, New Brunswick, New Jersey, May 1976.

[2] H.S. Baird: "Fast Algorithms for LSI Artwork Analysis", Proc. 14th DA Conf., New Orleans, June 1977, pp. 303-311.

[3] U. Lauther: "An O(nlogn) Algorithm for Boolean Mask Operations", Proc. 18th DA Conf., Nashville, Tenn., June 1981, pp. 555-562.

[4] P. Losleben, K. Thompson: "Topological Analysis for VLSI Circuits", Proc. 16th DA Conf., June 1979, San Diego, pp. 461-473.

[5] J.A. Wilmore: "A Hierarchical Bit Map Format for the Representation of IC Mask Data", Proc. 17th DA Conf., June 1980, Minneapolis, pp. 585-590.

[6] A.V. Aho, J.E. Hopcroft, S.D. Ullman: "The Design and Analysis of Computer Algorithms", Reading, MA: Addison-Wesley, 1974.

X-Y POLYGON, X-Y CONVEX POLYGON, AND VLSI LAYOUT

Delfin Y. Montuno
Alain Fournier

Computer Systems Research Group
Department of Computer Science
University of Toronto
Toronto, Ontario
CANADA M5S 1A4

ABSTRACT In VLSI layout, placement of function modules and routing of wires to interconnect them are some of the basic design problems. The most common geometric shape of building block for function modules in VLSI layout is rectangular. The realization of the layout could benefit from a larger class of shapes for building blocks of function modules, especially if area compaction is desirable. In this paper, we extend the concept of rectilinear polygons to x-y polygons and introduce the idea of x-y convexity. Next, we present algorithms that find x-y convex hull(s) in Θ(n) time for an x-y polygon, Θ(nlogn) time for a set of points, Θ(plogp+n) time for a set of nonintersecting x-y convex polygons, and O(nlogp) time for a set of intersecting x-y polygons, where n is the total number of vertices and p is the number of x-y polygons. The space (i.e. storage) complexity for all the algorithms is Θ(n).

I. Introduction

In VLSI layout, placement of function modules and routing of wires to interconnect them are some of the basic design problems. The most common geometric shape of building block for function modules in VLSI layout is rectangular. The basic function modules, numbered in hundred of thousands in a chip, are usually hierarchically grouped in higher and higher level function modules, thus in larger and larger rectangles. The geometric shapes used for interconnection are rectilinear polygons, i.e. polygons with only horizontal and vertical edges and sometimes with rounded corners. Two general strategies used in layout are placement first and then interconnection[2,3] and graph layout first and then embedding[4]. In both strategies, the realization of the layout could benefit from a larger class of shapes for building blocks of function modules, especially if area compaction is desired.

In this paper, we present a more general set of building blocks called **x-y polygons** and a set of algorithms for computing **x-y convex polygons**. We define an x-y polygon to be either a point, a connected rectilinear line segments, a rectilinear polygon, or a connected union of the latter two types of polygons (see Fig. 1). An x-y convex polygon is an x-y polygon such that any line parallel to either x- or y- axis intersects it no more than once (see Fig. 2). With the extended concepts of x-y polygons and x-y convexity, we will then present algorithms that find the x-y convex hull(s) in Θ(n)* time for an x-y polygon, Θ(nlogn) time for a set of points, Θ(plogp+n) time for a set of nonintersecting x-y convex polygons, and O(nlogp) time for a set of intersecting x-y polygons, where n is the total number of vertices and p is the number of x-y polygons. Note that alternative definitions of rectilinear convex hulls for point sets and algorithms to compute them are given by Ottmann, Soisalon-Soininen, and Wood[6].

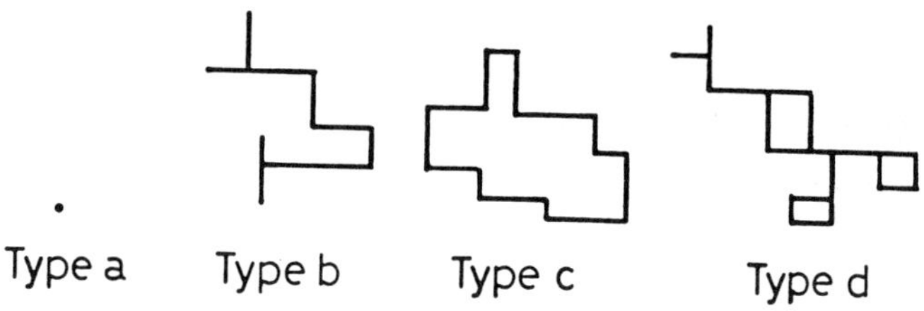

Fig. 1 Types of x-y polygon.

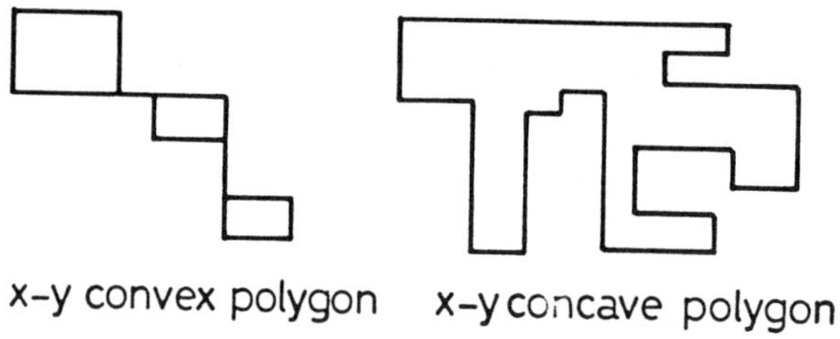

Fig. 2 Kinds of x-y polygon.

The computed x-y convex hull is given in terms of the input coordinates (in clockwise order).

*f(n) = O(g(n)) iff there exist two positive constants c and m such that |f(n)| ≤ c |g(n)| for all n > m. f(n) = Ω(g(n)) iff there exist two positive constants c and m such that |f(n)| ≥ c |g(n)| for all n > m. f(n) = Θ(g(n)) iff f(n) = O(g(n)) and f(n) = Ω(g(n)).

Without such restriction, there could be an infinite number of them. Even with such restriction, there could be an exponential number of them. However, we can always find one of them uniquely and generate others if needed. In all our algorithms for finding the x-y convex hull, only comparison between coordinates of input is used. Though new vertices could be output, they contain no new coordinate values. Therefore, a decision tree model is adequate as our model of computation[1].

Other than being useful for compaction purposes, x-y convex hull algorithms can also be used in path finding between two given points. A typical example is as follows:

> Given an x-y polygon and two points, one on its leftmost side and the other on its rightmost side, the shortest rectilinear path between them is either the top or bottom x-y convex hull of the x-y polygon. These two paths can be obtained in O(n) time by finding its x-y convex hull (see Fig. 3).

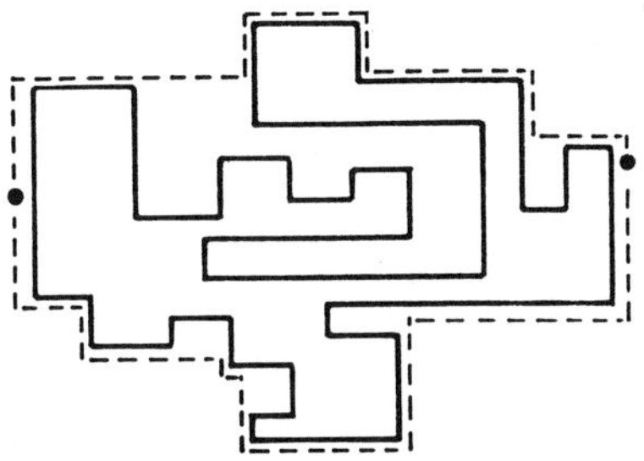

Fig. 3 Shortest rectilinear path between 2 extreme points on leftmost and rightmost sides of an x-y polygon.

Finding a path in a densely or sparsely packed set of x-y convex polygons using x-y convex hull algorithms is also an interesting and worthwhile research problem.

In the next section, we outline the algorithms. In Section 3, we analyze the algorithms and present their time and space complexity. Finally, we conclude with some open problems.

II. Algorithms

The algorithms for finding x-y convex hull of an x-y polygon, a set of points, or a set of x-y polygons consist of the following 3 main steps. A more detailed and rigorous presentation of the algorithms is also available[5].

1. Find the top left, top right, bottom right, and bottom left x-y convex hulls (see Fig. 4).
2. Find the top and bottom x-y convex hulls by finding the union the top left and top right x-y convex hulls and bottom left and bottom right x-y convex hulls, respectively.
3. Find the minimum y x-y convex hull by merging the top and bottom x-y convex hulls, at the same time, count and list other possible places in the minimum y x-y convex hull that can be modified to obtain other x-y convex hulls (see Fig. 5).

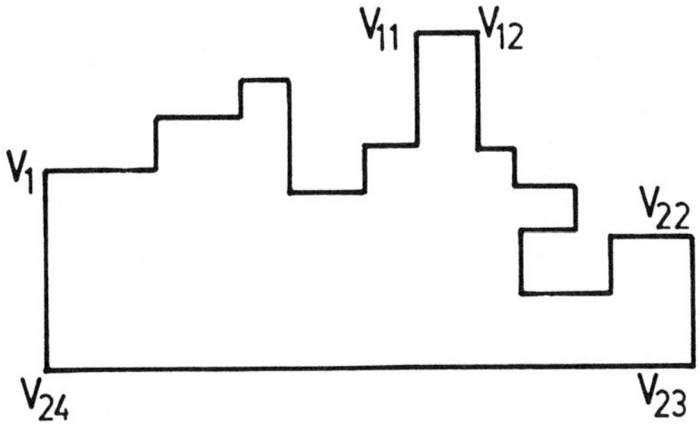

Top left = $(V_1, \ldots, V_{11})$
Top right = $(V_{12}, \ldots, V_{22})$
Bottom right = (V_{23})
Bottom left = (V_{24})

Fig. 4 Sides of an x-y polygon.

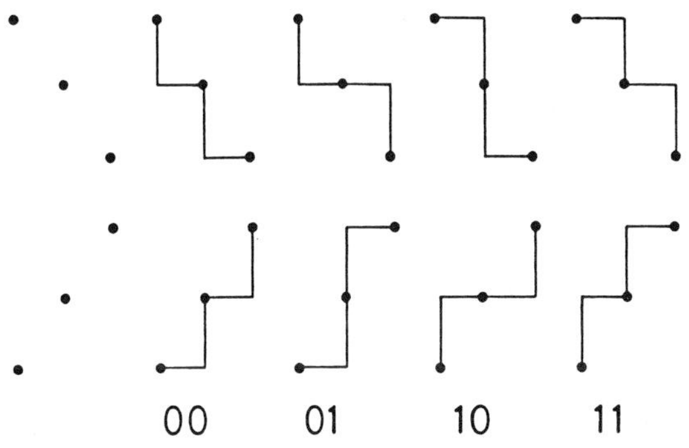

Fig. 5 X-y convex hulls of 2 sets of 3 points with different arrangements.

For step 1, an algorithm for the top left x-y convex hull suffices because the top right, bottom right, and bottom left x-y convex hulls can be found by applying the same algorithm with x coordinates replaced with (maxX-x) coordinates, x and y coordinates interchanged, or y coordinates replaced with (maxY-y) coordinates for $\pi/2$, π, and $3\pi/2$ rotation, respectively, where maxX and maxY are the maximum x and y coordinates of the minimum bounding x-y rectangle (these rotations can be done explicitly once in O(n) time, where n is the number of vertices or implicitly by replacing appropriate coordinates as required).

For Step 2, an algorithm for merging the top left and the top right x-y convex hulls suffices again because of the symmetry of the cases.

For step 3, we give an algorithm for merging the top and bottom x-y convex hulls, which we show to take O(n) time. We can, however, for the sake of efficiency, merge in constant time if the input consists of only one x-y polygon.

Note that we are not concerned in this paper with optimizing the constant factor of the algorithm but with demonstrating achievable asymptotic complexity.

X-y Polygon Given an x-y polygon, its top-left x-y convex hull is found by gathering its x-y increasing convex vertices in scanning the polygon from left to right until the left uppermost (first maximum y from the left) x-y convex vertex is reached; then in the right to left scan, remove convex vertices that are under some x-y convex vertices (see Fig. 6). Finally the resulting top left x-y convex hull is determined from this ordered list of convex vertices. Since each vertex is scanned a constant number of times, the time complexity of the algorithm is O(n).

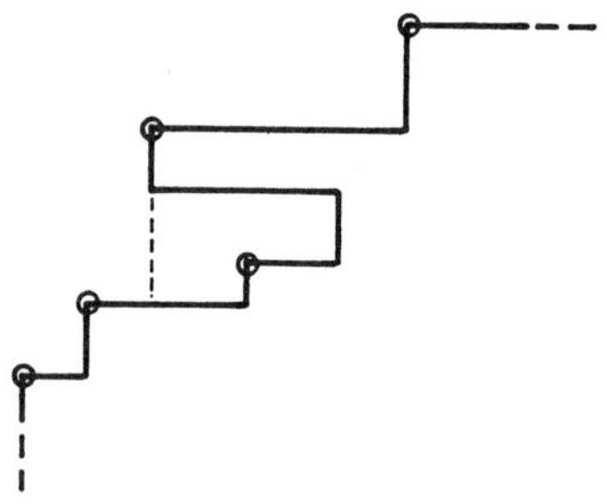

Fig. 6 X-y increasing convex vertices.

Point Set Given a set of points, we first sort them by increasing x value and then decreasing y value. Then proceed to find its top-left x-y convex hull as given for a simple x-y polygon. Since the points have to be sorted, the time complexity of the algorithm is O(nlogn).

A Set of Non-intersecting X-y Convex Polygons Given a set of non-intersecting x-y convex polygons, we first sort them using their respective upper leftmost (first minimum x from the top) vertices as key as in the point set. We then scan the polygons from left to right gathering x-y increasing convex vertices until the left uppermost x-y convex vertex of the set is reached. Since the ordering of the polygons on the scan line at a given position is fixed, we can always choose the polygon on the topmost position of the scan line. Changes could occur on the ordering on a scan line only when the upper leftmost vertex of a polygon currently to the right of the scan line is encountered or the last vertex of the topmost polygon is reached. In the latter case, we have to choose between the next uppermost polygon currently intersecting the scan line and the nearest polygon currently to the right of the scan line, choosing the one on the scan line if it is to the left and/or above the other. In the former case, we have to continue with the top polygon if it is above the newly encountered polygon and replace it if it is below. A straightforward implementation of the algorithm would require O(nlogp) time, where n is total number of vertices and p is the number of polygons. However, using a heap to maintain the ordering on the scan line and exploiting x-y convexity and non-intersecting condition to limit the number of updating, we are able to obtain an algorithm of O(plogp + n) time[5].

A Set of Intersecting X-y Polygons Given a set of intersecting x-y polygons, we first find the x-y convex polygon of each x-y polygon of the set and then build a heap using the coordinates of the upper left-most vertices as key such that the root contains the vertex of with minimum x and maximum y value. Since the x-y convex hull can be built from convex vertices, only convex vertices are stored in the heap. A convex vertex of the top-left x-y convex hull is obtained from the root of the heap; its content is then updated to the next adjacent convex vertex and the heap is heapified again to satisfy the heap condition. This process is continued until the left uppermost convex vertex of the set is reached. The resulting top left x-y convex hull is then determined from this ordered list of convex vertices. Since the heap contains at most p nodes at any given time and a total of n nodes are inserted into and deleted from the heap, the algorithm requires at least O(nlogp) time.

Merging Top-left and Top-right X-y Convex Hulls Merging the top-left and top-right x-y convex hulls requires only a list concatenation.

Merging Top and Bottom X-y Convex Hulls Merging top and bottom x-y convex hulls requires a scanning of the convex hulls from left to right and determining a choice of alternative convex hull edges. This choice occurs when two points or two rectangles do not overlap on both their x and y projections. This implies that there could be an exponential number of x-y convex hulls, each having the minimum area requirement, if we are to enumerate all of them (see Fig. 5). However, we can always find one of them and store only the locations of possible changes for identifying other x-y convex hulls.

III. Analysis of x-y Convex Hull Algorithms

Analysis The algorithms for finding the top left x-y convex hull of an x-y polygon, a set of points, and a set of nonintersecting x-y convex polygons can be shown to be optimal. Optimality for an x-y polygon is obvious. For a set of points or a set of nonintersecting x-y convex polygons, we show that the algorithm is optimal by proving that sorting is reducible to finding the x-y convex hull of a set of points or a set of nonintersecting x-y convex polygons. Element uniqueness is also reducible to the same problems.

Since the proof for both are similar, we only give the proof for that of a point set. We also leave reducing element uniqueness to above problems to the reader.

Given a set of distinct numbers X_i, $1 \le i \le n$, transform each X_i, in a total of O(n) time for all X_i's, to a point (X_i, X_i) in the x-y plane. Find the x-y convex hull of such set of points. The order of the convex vertices on the top left convex hull from the left to the right is the sorted order of X_i's. Therefore, sorting is reducible to x-y convex hull algorithm of a point set and the algorithm is optimal.

The time complexity of the algorithm for a set of nonintersecting x-y convex polygons is actually O(plogp+n). The term O(plogp) is necessary because the algorithm must be able to sort the upper leftmost vertices of p x-y convex polygons; the term O(n) is present because the number of vertices output could be O(n). The O(plogp + n) time also resulted from the O(p) insertions and deletions of p x-y convex polygons into and from the heap of maximum depth O(logp) and from O(n) time in scanning vertices for updates. If $n \le kp$, then only O(plogp) is necessary, though k could be a very large constant.

The algorithm for finding the top left x-y convex hull of set of intersecting x-y polygons takes O(nlogp) time. The strategy is to find first the x-y convex hulls of all x-y polygons in O(n) time. Next, by x-y convexity a vertical line intersects the horizontal edges of the top left x-y convex hulls at most once for each x-y convex polygon. We could then order the x-y polygons on this vertical line. Since these x-y convex hulls could possibly intersect one another, there is, thus, no guarantee that the order of these convex polygons would be preserved as we sweep the vertical line from left to right. We also know that the vertical line intersects at most p horizontal edges at a time. In the worst case, we can delete or insert in O(logp) time. Since there are at most O(n) such operations, the time complexity is O(nlogp) which is greater than just sorting the upper leftmost vertices of all x-y polygons, but much less than O(nlogn) time, which is the time to sort all vertices.

Space Complexity It is straightforward to show that all the algorithms use O(n) space.

IV. Conclusion

We define the concept of x-y convexity and extend the concept of rectilinear polygons to x-y polygons. We also present various x-y convex hull algorithms for various kinds of objects. An interactive graphics system is currently being implemented that allows input of arbitrary x-y polygon(s) and display of its (their) x-y convex hull.

Some of the open problems are detecting and finding intersection of a set of x-y polygons, dynamic compaction of a set of rigid x-y polygons, and efficient routing around x-y polygons.

References

[1] A.V. Aho, J.E. Hopcroft, and J.D. Ullman, *The Design and Analysis of Computer Algorithms*. Addison-Wesley Publishing Company, 1974.

[2] D. Dolev, K. Karplus, A. Siegel, A. Strong, and J. Ullman, Optimal wiring between rectangles. 13th Symp. Theory of Computation, (May, 1981), pp. 312-317.

[3] A.S. LaPaugh, A polynomial-time algorithm for optimal routing around a rectangle. Proceedings of the Twenty-first Annual IEEE Symposium on Foundations of Computer Science, (Oct. 1980), pp. 282-293.

[4] C.E. Leiserson, Area-Efficient graph layout. Proceedings of the Twenty-first Annual IEEE Symposium on Foundations of Computer Science, (Oct. 1980), pp. 270-281.

[5] D.Y. Montuno and A. Fournier, Finding x-y Convex Hull of a Set of x-y Polygons. University of Toronto Tech. Rept. CSRG-148, (Nov. 1982).

[6] Th. Ottmann, E. Soisalon-Soininen, and D. Wood, On the definition and computation of rectilinear convex hulls. University of Waterloo Tech. Rept. CS-83-07, (April,1983).

An Efficient Algorithm for Boolean Mask Operations

W. E. Donath and C. K. Wong

IBM T. J. Watson Research Center
Yorktown Heights, NY 10598

Abstract:

An efficient algorithm to find the intersection and complementation between two polygon sets is proposed. It is founded on the well-known scanline technique. It runs in O(N log N) time, where N is the total number of intersection edges. Furthermore, problems related to round-off errors are dealt with explicitly.

I. Introduction

We consider the problem of calculating Boolean mask combinations between two polygon sets. The operations of interest are intersection, complementation, and union. They arise in several aspects of the design process, such as design rule checking, connectivity checking, and so on. Because of the large number of elements in the layers of a design, such operations require a large amount of computation. Considerable effort has been devoted to the development of efficient algorithms (see ref. 2-4, 6, 7, 9, 10).

In reference 7, Lauther used the idea of scan lines, first conceived by Shamos and Hoey (ref. 8) and further developed by Bentley and Ottman (ref. 5), to determine the intersection of polygons in two mask layers. In this approach, two separate scan lines (OLDSCANLINE, NEWSCANLINE) were used. Continuous updating of both lines was necessary. The author further assumed that for a given technology the number of different x-coordinates was $O(\sqrt{N})$, where N is the number of edges in the completely intersected set (i.e. output set), and that the average number of edges crossing a vertical cut through the layout was also $O(\sqrt{N})$. Consequently, the time for maintaining the scanlines was O(N).

In the present paper, the same scanning approach is adopted. However, we only need one scanline and update is performed only after an intersection is found. Thus, the maintainance is truly O(N log N), without Lauther's assumption. Furthermore, we deal explicitly with some of the round-off problems raised by Lauther.

The typical situation of design rule checking programs is that the amount of data is very large, and space complexity for the in-core operations is important. Our situation is different, since we work in an environment where a lot of preprocessing has already been done. We typically work with a few nets at a time, and all data are in core. Therefore space complexity of the order of O(N) was adequate, while time complexity was still very important; also, it would be easily possible to modify our implementation using one of the concepts Lauther has pioneered (i.e. separation of corner vertices and intersection vertices) to obtain a space complexity of O(S), where S is the maximum edges crossing any scanline.

Except for the highly unusual situation where many edges meet in a single point, the implementation of our algorithm runs in O(N log N) time; this special case could be handled without affecting this time if we modified our implementation appropriately.

II. The Algorithm

The algorithm developed here produces the intersection and complementation between two sets of polygons; computation time is O(N log N). The polygon edges may run in arbitrary directions, and inside shapes are handled. The algorithm performs implicitly a union on the polygons in the two sets. There are two advances we make on the closely related work of Lauther (ref. 7): for one, we avoid the necessity for updating the scan line - which makes the algorithm O(N log N) in the sense of Lauther even if the chip consists of a large number of parallel lines starting and stopping in some order and secondly, we concern ourselves directly with the problem of truncation met when we have non-rectilinear lines.

Each polygon is defined in terms of its corners and the *edges* connecting adjacent corners; the set of *vertices* consists of the corners and intersections between edges of different polygons; *segments* are parts of the edges, and may have associated with it several edges, which run either collinear or approximately collinearly. The coordinate grid is defined on the set of integers; notice that an edge therefore does not run in a straight line, but rather in a stepwise fashion - however, we assume that the exact step function is not critical, but can be defined to suit the convenience of the computations. The explicit consideration of truncation leads to several special situations; in particular, we generate intersections of edges such that before the intersection the edges are treated as collinear, while afterward they diverge and secondly, we find that it is possible that when an edge is broken in an intersection, the two parts of that edge have opposite directions in the scan procedure (see figs. 1 and 2).

The algorithm uses essentially the concepts developed by Shamos and Hoey (ref. 8) for finding all the intersections of a set of edges and is also similar to that developed by Lauther (ref. 7); the scan line is parallel to the x-axis and moves in the positive y direction. The scan is implemented by scanning through all the corner vertices of the polygons and those vertices generated by the intersection of the scan lines - in a sorted order.

To describe the algorithm in somewhat more detail, we consider the program in terms of three subprograms - the vertex list program, the edge segment program, and the main program.

CH1935-6/83/0000/0358$01.00

The *vertex list program* maintains the vertices in a heap (ref. 1) and allows the following operations:

1) Identify the FIRST vertex - i.e. that vertex with smallest y and, if several exist, at that y with smallest x.
2) ENTER a vertex
3) REMOVE any vertex from the heap.

In a heap structure, the ENTER and REMOVE commands can be executed in log n compute time, while the FIRST command requires constant time. The operation of a heap is given in ref. 1; note that we allow removal of any vertex from the heap, which requires maintainence of an index into its heap position for all vertices in the heap. In essence, Lauther (ref. 7) maintains the set of vertices in two different structures - one for the corners of the polygons, the other for the intersections of the edges. Thereby he obtains relatively modest core requirement: we have no limitation on core, and therefore we maintain the two types of vertices together. It would be easily possible to modify the algorithm so as to use the Lauther scheme and obtain the same space complexity.

The *edge segment program* maintains its data in a serial order with a balanced tree program (see ref. 1), and its data consist of line segments x_1, y_1, x_2, y_2, each of which may have several edges associated with it. All line segments in the tree structure either cross, end, or start at the current scan line. The segments are directed: x_1, y_1 is a point inside the scanned area, while x_2, y_2 extends into the area to be scanned. The operations on the lines are ENTER a segment, REMOVE a segment, or RIGHT_NEIGHBOR or LEFT_NEIGHBOR of some proposed segment. The x_1, y_1 value of the segment is the location at which the segment starts; x_2, y_2 may be either the nearest final point of any edge associated with the segment, or may be the point at which two nearly collinear edges associated with the segment may separate (see fig.(1)). The program also maintains a chain of pointers on the segment list in their order in the tree. Two segments A and B are compared as follows:

1) Determine which of the two segments has the later (i.e. larger) starting y value, and find for both segments their x coordinate at the y value.
2) If a segment has after step 1 a lower x value, then it is to the left of the other segment and the comparison is finished.
3) If the two segments have identical x values in step 1, then they must start at the same y value, in which case that segment with the lower final x value is to the left of the other. (If two segments have the same x value after step 1, and should they start at different y values, then the segment that starts earlier intersects the segment that starts later: the earlier segment should have been picked up when the later segment started out and had been broken)

Horizontal segments can be handled by this program just like any other edge - there is no special distinction. Note, however, that its x value is always its starting x value.

The *edge segment program* also updates with each insertion and deletion of a segment the intersection vertices that are created or removed. For each insertion of a segment, at most 2 such vertices are created and 1 may be removed, while for each removal of a segment 2 such vertices may be removed and 1 created.

Input to the program is a list of edges of the polygons - each of which represents a transition in the nature of the planar space. Associations between the edges, vertices, and segments are maintained as follows:

1) For each starting point of any edge, a vertex is created, which has associated with it the edges of which it is the starting point.
2) For each segment, there is a chain of edges, which run collinear with that segment. Each edge has a pointer to that segment with which it is associated (a null value on the pointer indicates that the edge is not crossing the scan line).
3) Pointers are provided to allow one to determine for each segment the intersections that it may have with its right and left neighbor.
4) Each vertex, that is created by an intersection, has two edges associated with it. The corresponding segments can be recovered from the edge to segment pointer.
5) It is possible for a polygon edge to change its direction for a segment as the scan line proceeds along (see fig. 2); suppose an edge is very flat - i.e. has a large negative increment in the x direction and a small positive increment in the y direction. Truncation after an intersection may cause the first segment to have all of the y increment, but the second segment may have no y increment at all - in which case the second segment must be reversed. Provisions were made in the data structure to handle such reversed segments, and care was taken to choose the intersection vertex so that the first segment always gets some y increment under such circumstances.

We are now ready to describe the *main program* in pseudo code:

1) Input the edges of the polygons and generate all the starting verteces. Initialize the edge segment program.
2) This step collects all edges passing through the x,y of the FIRST vertex on the heap. (Should the heap be empty, then the program has finished and exits to a procedure for collecting the polygons from the set of output vertices). One collects all vertices with the same x,y on the heap and removes same from heap. One collects all the edges associated with those vertices, the segments associated with these edges, and the edges associated with these segments. Lastly, one uses the LEFT_NEIGHBOR and RIGHT_NEIGHBOR functions of the edge segment program to collect all the other segments intersecting x,y. All segments collected here are removed from the edge segment list. One takes care that duplicate occurrences of edges are removed. It is in principle possible that a large number E of edges intersect at one point - and the algorithm could be made to run E lg E in that number of edges; however, the implementation used here runs E^2, since normally E would be relatively small.
3) The edges are sorted on the direction, with which they pass through x,y , and continuation segments are generated where appropriate. Neighboring outgoing edges are bundled on the basis of full or near collinearity. It is possible that two edges are almost collinear, in which case the outgoing segment would contain both edges up to a point, where it would be thought that the two edges have separated sufficiently to be treated separately.

4) Information associated with the segments, in particular the segment which is the LEFT_NEIGHBOR of the leftmost segment in the edge list, is used to find for each of the arcs defined by the edges whether the arc region is covered by polygon 1, by polygon 2, by both, or by neither. Corresponding vertices of the output polygons are now created, and chained to the vertices created at previous x,y intersections.

5) New segments are created for the outgoing edges and entered into the edge segment list. The program now repeats Step 2.

As has already been previously indicated, the algorithm is N log N - with one possible reservation - i.e. if many edges intersect at one point. However, this dependency is implementation dependent rather than fundamental to the method employed here.

III. Experimental Results

The algorithm was implemented in a PL/I program; the intersection and the two complementations between two polygon sets was performed. For experimental results, the two polygon sets consisted of m by n arrays of either squares or diamonds, whereby the two polygon sets were slightly displaced from one another. For each element of the array, there is an intersection with the corresponding element in the other array, so that there are 12 intersection edges per element of the array. Experimental computing times (adjusted for IBM 370/168 processor) were as follows:

Polygon Shape	Size of Array	Computing Time	Time per Vertex
Square	2 by 2	0.032 sec	0.7 msec
Square	16 by 32	6.3	1.0
Square	30 by 30	11.7	1.1
Diamond	2 by 2	0.044	0.9
Diamond	16 by 32	8.0	1.3
Diamond	30 by 30	15.1	1.4

If n is the number of intersection verteces, we can find for the computing time the formulae (in msec/vertex)

$$0.38 + 0.050 \log_2 n \text{ for rectilinear polygons and}$$
$$0.60 + 0.057 \log_2 n \text{ for arbitrary polygons}$$

Fig. 2 Reversal of Edge Direction due to truncation on intersection

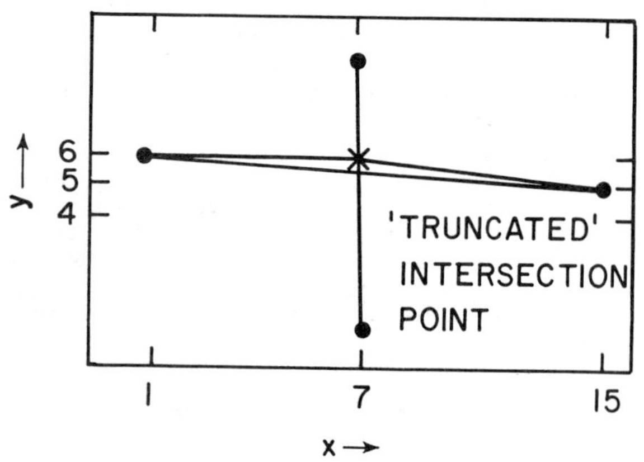

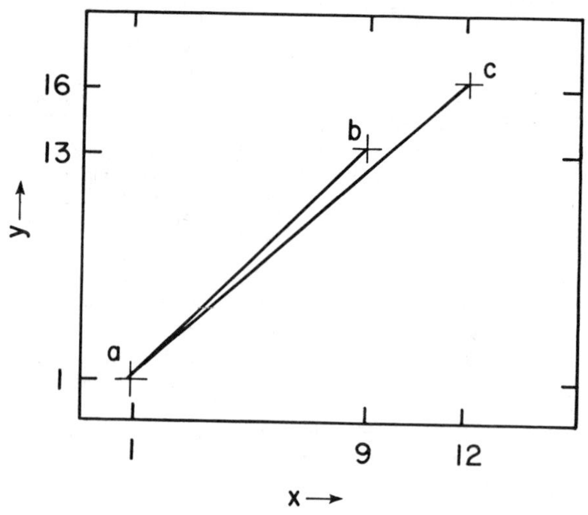

Fig. 1 Near Collinearity of two edges. Edge a-c may well be approximated by path a-b-c

References

1) A.V. Aho, J. E. Hopcroft, S. D. Ullmann: 'The Design and Analysis of Computer Algorithms,' Reading, MA: Addison-Wesley 1975

2) D. Alexander: 'A Technology Independent Design Rule Checker,' NCA Corporation, 383 Oakhead Parkway, Sunnyvale, Calif.

3) H. S. Baird, 'Design of a Family of Algorithms for LSI Artwork Analysis,' M.S. Thesis, Dept. of Computer Science, Rutgers University, New Brunswick, N.J. May 1975

4) H. S. Baird, 'Fast Algorithms for Artwork Analysis,' Proc. 14th DA Conf. 1977 pp.303-311

5) J. L. Bentley and T. A. Ottmann: 'Algorithms for Reporting and Counting Geometric Intersections,' IEEE Trans. Comp., Vol. 6-28, No. 9, Sept. 1979, pp. 643-647

6) U. Lauther, 'Simple But Fast Algorithms for Connectivity Extraction and Comparison in Cell Based VLSI Designs,' Proc. ECCTD 80 (1980) Warsaw, pp. 508-514.

7) U. Lauther, 'An O(N log N) Algorithm for Boolean Mask Operations,' Proc. 18th DA COnf., 1981, pp. 555-562.

8) I. Shamos, D. Hoey: 'Geometric Intersection Problem,' Proc. 17th Ann. Conf. Foundations of Computer Science, Oct. 1976, pp. 208-215

9) L. Szanto: 'Network Recognition fo a MOS Integrated Circuit from its Masks,' Teska Ekectronics 9(1976), pp. 67-75.

10) M. Yamin, 'Derivation of All Figures Formed by the Intersection of Generalized Polygons,' The Bell System Technical Journal, Vol. 51, No. 7, Sept. 1972, pp. 1595-1610

Session 2-2d
VLSI Logic Design I

Chairman

J. Foggiatto
Amdahl Corporation

POSSIBILITIES FOR APPLYING CARROLL CELLS TO LOGIC DESIGN EFFORTS

Gary R. Gladden
Honeywell BSD
Technical Center
2332 McGaw Avenue
Irvine, Ca. 92714
(714) 557-6940

ABSTRACT

A unique method of diagrams devised by Lewis Carroll in the 1800's, present an intriguing set of possiblities for future study. Carroll's diagrams are rectilinear diagrams containing up to 1024 (1K) compartments (herein called Carroll Cells). This paper traces the fundamental issues and describes the use of VENN diagrams in analysis of combinatorial logic to that diagram's limitations. Carroll Cells are then described as well as their usefulness for evaluating first-order logical expressions without tables of combinations. Finally, having defined Carroll Cells, the paper suggests the practicality of utilizing the Carroll Cell geometry in a VLSI "chip" architecture for directly executing combinatorial logic. Moreover, it is suggested that under further examination, Carroll Cells could replace logic gates and ultimately blur the distinction between memory and processors and usher in a new computer architecture.

INTRODUCTION

Logic design describes the interconnection of logic elements in digital computer hardware so that a desire function is performed. These logic elements are called as "gates". Logic gates consist of the following types[1].

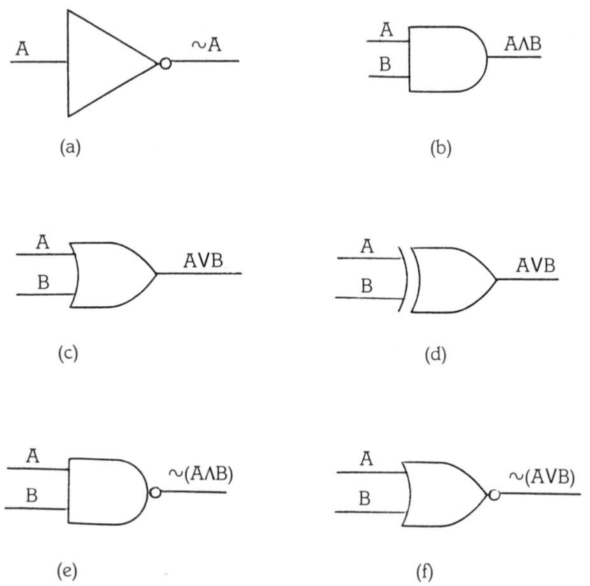

(a) INVERTER (b) AND (c) OR (d) XOR (e) NAND (f) NOR

FIGURE 1

The "value" of the inputs and outputs to these gates are binary (either 1 or 0) values. When evaluating the resultant outputs of these gates, a table of combinations are arrived at by applying the rules of first-order logic. Thus the tables of combinations associated with the gates in Figure 1 are shown in Figure 2.

(a) INVERTER

IN	OUT
0	1
1	0

(b) AND GATE

IN		OUT
A	B	
0	0	0
0	1	0
1	1	1

(c) OR GATE

IN		OUT
A	B	
0	0	0
0	1	1
1	0	1
1	1	1

(d) XOR (EXCLUSIVE OR)

IN		OUT
A	B	
0	0	0
0	1	1
1	0	1
1	1	0

(e) NAND (NOT AND)

IN		OUT
A	B	
0	0	1
0	1	1
1	0	1
1	1	0

(f) NOR (NOT OR)

IN		OUT
A	B	
0	0	1
0	1	0
1	0	0
1	1	0

FIGURE 2

All logic networks therefore can be described in terms of a combination of elementary gates.

VLSI then concerns itself with devising complex logic network designs utilizing these gates and reducing the networks to smaller physical sizes.

FUNDAMENTAL ISSUES

The rules for translating gate inputs to logical outputs follow the rules of first-order logic. In the following discussion first order logic will consist of the following symbols:

"NOT" ~

"AND" Λ

"OR" V

As well as the quantifiers "all" ∀ and "some" ∃.

Now one may describe the tabular inputs from Figure 2.

(a) The negation (inversion of A is A.

(b) AΛB reads {∃x| x∈A AND x∈B}

(c) AVB reads {∃x| x∈A OR x∈B}
or both

(d) Same as above only x∈ both A and B

(e) ~(AΛB) ~{∃x| x∈A AND x∈B}

(f) ~(AVB) ~{∃x| x∈A OR x∈B}

VENN DIAGRAMS

An adjunct to the "truth" table evaluation of logical relations as characterized above, a method of diagrams is commonly used to illustrate the "belonging" of some variable to the class of A and B. These diagrams are called VENN diagrams after their inventor.

A depiction of two of the above examples (b) and (f) are in Figure 3

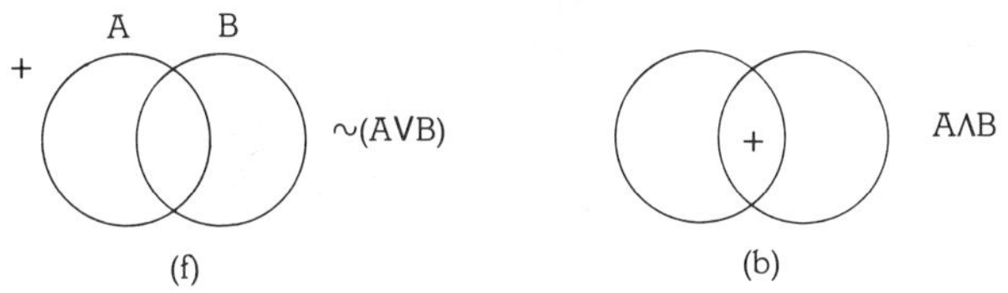

FIGURE 3

Similar VENN Diagrams can be produced for various other relationships. VENN'S diagrams have two shortcomings however. One, the lack of a "universe of discourse" or an area of existence outside the (in this example) class of (A and B Figure 3 (f) for example). Second, classes of five elements are the limit for diagramatic representation. Beyond that the curvelinear representations become too confusing. The VENN diagram for 5 elements is drawn below.

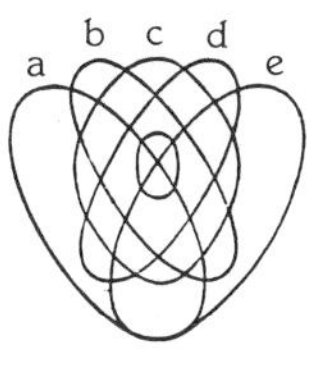

VENN diagrams then are appropriate for depicting relationships in the first order logic of up to 5 variable classes.

CARROLLS METHOD

Earlier we mentioned that miniaturizing the elementary logic gates forming a logic network was the substance of VLSI endeavors. We would like to discuss a method described by Lewis Carroll (the author of "Through the Looking Glass") in an antique little book [2] which can overcome the aforementioned shortcomings of VENN diagrams. They have an identifiable "universe of discourse" class and admit up to 10 variables or 1024 logical combinations (2^{10} = 1024 or 1K commonly). Furthermore, and even more enticing is the fact that they use a rectilinear geometry. We will discuss the implication of this geometry later.

Suppose we were dealing with the examples above; 2 classes named A and B. For 2 terms a bilateral box is drawn in which the North half is assigned 'A', the West 'B' and the East '~B', the South '~A, as in Figure 4.

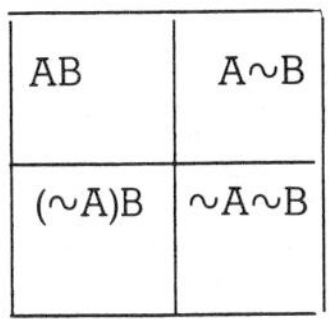

FIGURE 4

Applying the previous examples of the AND and NOR gates to a bilaterial diagram we have

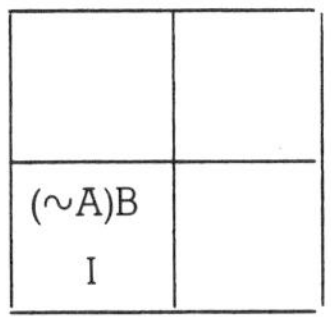

(b) AND GATE

"I" indicating at least one element exists and that it is a member of both A and B

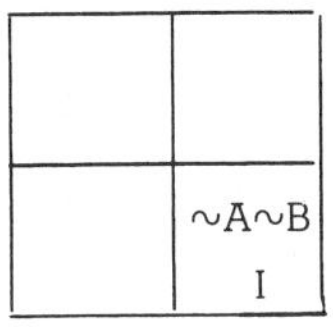

(f) NOR GATE

'I' indicates the existence of an element in a cell. The 'I' is used because it may cross two borders thus indicating existence in two quadrants simultaneously. A further notation is the symbol 'O' which indicates no member of a particular class exists (a particular cell is empty).

The bilaterial diagram and its larger family of diagrams I have named Carroll Cells in honor of their inventor.

TRILATERAL AND BEYOND

While in logic combination (or truth tables) and VENN DIAGRAMS, the representation of input and output values grow unwieldy for terms larger than two, the Carroll Cell method remains manageable.

For three classes (A, B, and C for example) we draw what Carroll described as a trilateral diagram. Figure 5 is such a diagram. Notice that for 3 classes, 8 individual cells are the result ($2^3 = 8$).

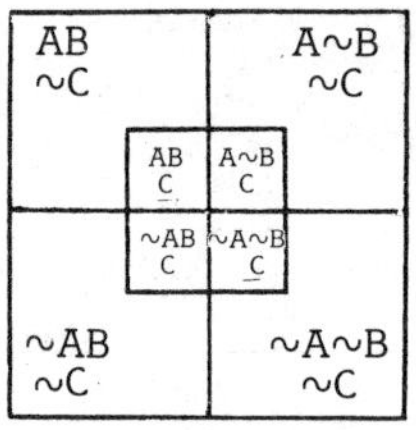

FIGURE 5

The location of the cells are described in the following table.

ABC	=	NORTH WEST QUARTER, INNER PORTION
AB~C	=	NORTH WEST QUARTER, OUTER PORTION
A~BC	=	NORTH EAST QUARTER, INNER PORTION
A~B~C	=	NORTH EAST QUARTER, OUTER PORTION
~ABC	=	SOUTH WEST QUARTER, INNER PORTION
~AB~C	=	SOUTH WEST QUARTER, OUTER PORTION
~A~BC	=	SOUTH EAST QUARTER, INNER PORTION
~A~B~C	=	SOUTH EAST QUARTER, OUTER PORTION

TABLE 1

This rectilinear Carroll Cell representation varies its geometrics for the number of variable classes until it reaches a limit (this has not been verified) of 2^{10} or 1024 unique cells. The following figures show the progression of diagrams up to the one containing 1024 cells:

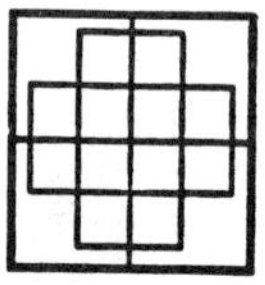

4 VARIABLES = 16 CELLS

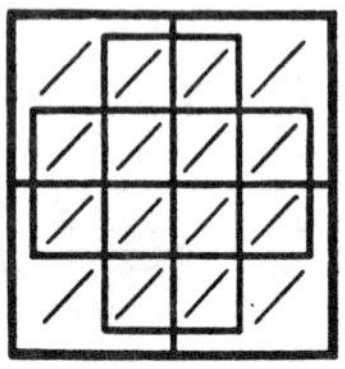

32 CELLS

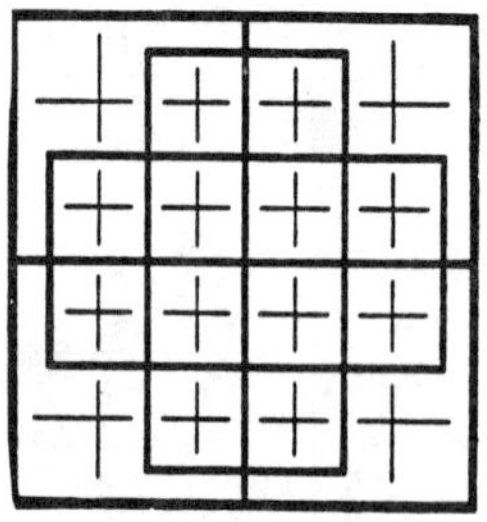

64 CELLS

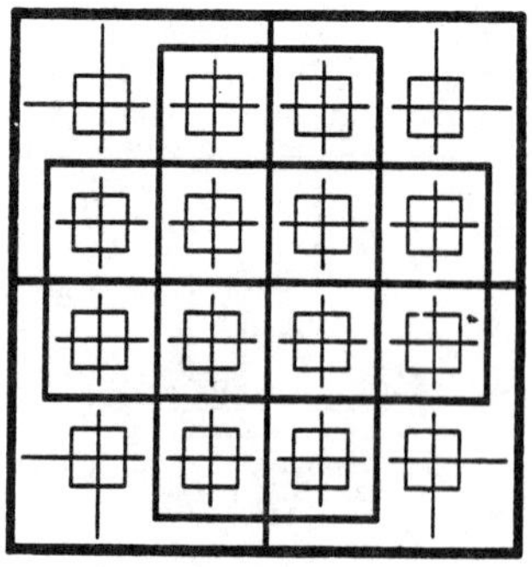

512 CELLS

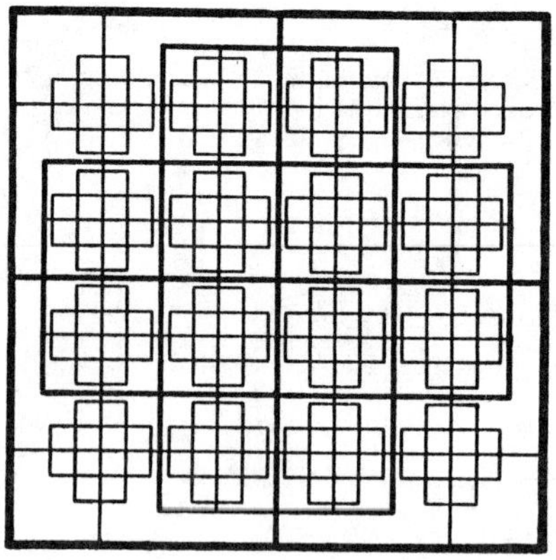

1024 CELLS

TABLES TO DIAGRAMS

We have seen previously that the behavior of logic gates are called combinatorial logic networks. That is, output values are dependent upon input values at any instant of time and are associated through given logical operations (symbolized by gates). We have seen furthermore than gate symbology can be mapped onto a diagram representation called VENN DIAGRAMS. For a large number of inputs, however, the VENN DIAGRAMS become unuseable. The Carroll Cell, however can be economically used for up to 10 inputs variables.

Suppose that instead of evaluating tabular combinatorial logic representation we were to evaluate the logic under study by the Carroll Cell Diagram? Suppose further that this cell being rectilinear were able to be miniaturized to a dimension commensurate with current VLSI? With some appropriate method for attaching input lines and output sensors or detectors (to sense when a certain cell was occupied) could the supposed Carroll Cell function as a monolithic (that is non-logic element specific) executor of inputs and emmitter of outputs according to the rules of combinatorial (or first order logic)? Again, the pursuit of these suppositions may or may not be profitable, but a VLSI geometry based on Carroll Cells is intriguing.

A CARROLL CELL GEOMETRIC LOGIC

Assume we wish to design the following simple network:

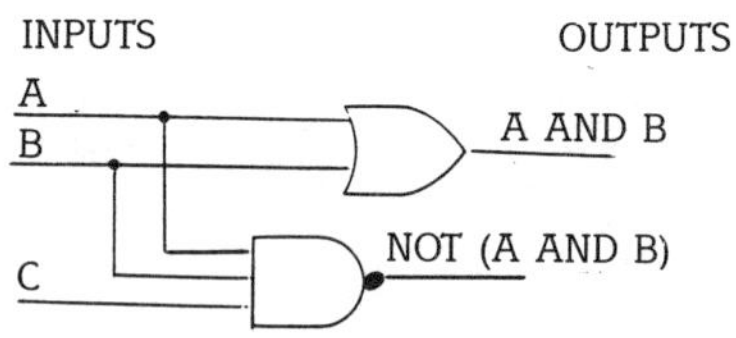

FIGURE 6

The equivalent Carroll Cell would display the following results:

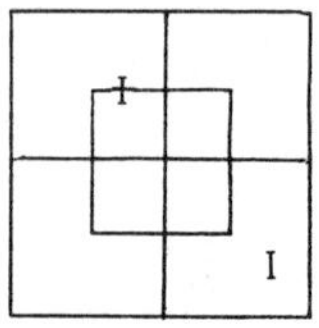

FIGURE 7

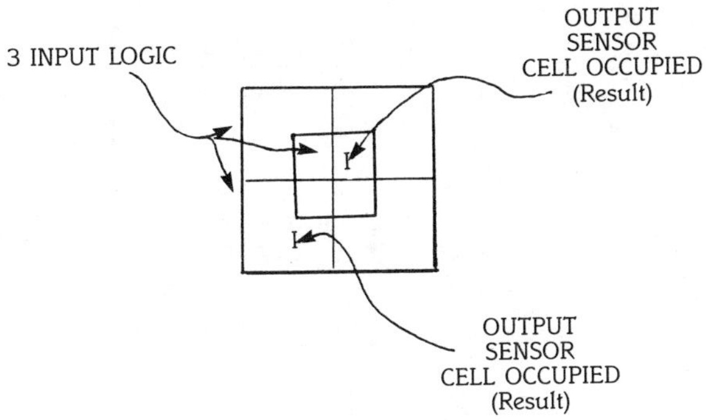

FIGURE 8

Any multiple input logic network could with proper electrical interconneciton and location sensors be represented and evaluated in a Carroll Cell geometry. In addition, since the occupancy of a particular cell could be (in our new technology) detected for a given set of input values, when inputs changed and with current (or last-state) occupancy sensed the new output values could be sensed.

Therefore it seems likely that Carroll Cells could also represent "sequential logic" networks.

FURTHER STUDY

So far we have supposed a great deal and certainly the fruitfulness of pursuing these representations may be questioned. It might be rightly assumed that these rectilinear diagrams devised by Lewis Carroll are only pedantic diversions. Carroll's Diagrams on the other hand, however, present a clever expression of a logical analysis method.

Because logic design is fundamental to our industry would not it be more than just fanciful to pursue this diagram method and perhaps reduce it to a VLSI network? Its form is simple and perhaps such a form would lend itself to a very dense VLSI.

Perhaps experimentation would show how several 1024 Carroll Cell "chips" could be stacked upon one another or laid side by side to form an exceptionally dense computer.

Furthermore, future study might indicate that this Cell structure is all that is needed for processor or memory tasks. As quickly as inputs (logic values) entered the geometry they would be evaluated, occupied cells sensed and new inputs evaluated. Separate storage and processing would become unnecessary.

It seems certain that no more tedious evaluations by hand or computer or incompletely specified combinatorial tables would be required. Each Carroll Cell would evaluate inputs regardless of their states and only significant cells sensed for their outputs. Furthermore, the Carroll Cell could represent all combinatorial networks.

Other courses of study might be the applicability of Carroll Cells to robot vision, certain cells corresponding to picture elements, or to data flow machine architectures. An impact in the field of Cellular Automata may be in the offing.

CONCLUSIONS

Lewis Carroll's method of logical inference has not had much exposure perhaps due to the obscurity of the reference, or to the antiquated method of diagrams.

Carroll Cells, however, are a logical examination tool embodying the first-order logic on which switching networks are based. If further study could result in a miniaturization of Carroll Cell geometry and exploration of its uniqueness, it might prove to have numerous distinct advantages over combinatorial logic circuitry of today. Perhaps it might point to a whole new architecture.

BIBLIOGRAPHY

[1] E.J. McCLUSKEY. Logic Design, *Encyclopedia of Computer Science*, edited by Anthony Ralston.

[2] LEWIS CARROLL. Symbolic Logic Part I, 4th Ed., Published 1897, Republished 1955, Berkeley Enterprises, Inc., N.Y., N.Y.

Optimal Clocking Schemes for High Speed Digital Systems

Stephen H. Unger
Computer Science Department
Columbia University, N.Y., N.Y. 10027
(worked on this project while on
sabbatical leave at IBM T. J. Watson
Research Center, Yorktown Heights, N. Y.)

Chung-Jen Tan
IBM Data Systems Division
Kingston, N.Y. 12401

ABSTRACT

The clock period of a digital system determines the rate at which data is processed, and so should be minimized, consistent with reliable operation. Based on a worst case analysis, clocking schemes for high-performance digital systems are analyzed. These are 1 and 2-phase systems using simple latches, and 1-phase systems using edge-triggered flip-flops. Within these categories (any of which may be preferable), it is shown how optimal trade-offs between clock period and propagation delay tolerances can be made by appropriately choosing the parameters of the clocking system as functions of technology parameters.

1.0 INTRODUCTION

Almost all digital systems operate synchronously, with elementary actions governed by pulses generated by a central clock and distributed to various subcircuits thru a clock distribution network. This study is limited to systems in which clock pulses are fed <u>only</u> to clocked storage devices. Storage elements considered are simple latches <1>, which propagate the D-inputs while C=1, and which hold their outputs constant while C=0, and edge-triggered D-flip-flops <1> (ETDFF's), whose states change to conform to the D-signals present when the clock pulse goes on.

Our object is to determine the trade-offs between the clock pulse rate and the tightness of the constraints required on the logic delays. It is assumed that maximum and minimum values are given for the parameters describing the operation of the storage elements, the clock skew and pulse width variation, and the maximum delay, D, thru the logic. The D-value is a worst case estimate incorporating physical factors based on the lengths of the longest paths considered necessary to accomplish the tasks to be performed by the system. Functions are derived, on a worst case basis, specifying optimal clock parameters as functions of both the aforementioned terms and the lower bound on the shortest delay, d, thru the logic.

A central issue is the treatment of clock skew, S. It is shown that the effects of skew may be taken into account by lengthening the clock period or by widening the clock pulses. The latter solution permits faster operation, provided that the resulting tighter lower bounds on the logic delays can be reliably satisfied. A continuum of intermediate solutions exists.

1.1 Skew and Pulse Width Variation

<u>Clock skew</u>, S, is uncontrolled, random deviation from the nominal timing of the leading edge of a particular clock pulse at a particular latch input. Maximum skew is treated as an additive factor, fixed for a given system. Note that a specified edge of a particular clock-pulse may be delayed by as much as S at one node, and advanced by as much as S at another node. The <u>maximum</u> deviation in timing between any 2 nodes is thus 2S. <u>Clock pulse width variation</u>, Wt, is defined as the ratio of the maximum to the minimum values of pulse width. It represents an uncontrolled, random deviation from the nominal value of the width of clock pulses, exclusive of effects due to skew.

1.2 Basic Constraints

For a synchronous system to operate properly, certain timing malfunctions must be prevented:

1. the clock period must be long enough to allow changes during one period to traverse the latch-logic loop in time to control latch states during the next period.
2. clock pulse widths must allow the slowest latches time to respond to data input changes.
3. a state change of one or more latches cannot cause additional latch state changes (of the same latch or of other latches) during the same period (the multi-stepping problem.)
4. An output change from one latch must not be able to penetrate the logic in time to abort a change in another latch state during the same period (critical race).
5. spaces between clock pulses must be adequate to allow latches to "forget" previous false inputs that were too short to change their states (the recovery time problem).

1.3 Notation

The clock parameters are the period p, the width w, and, for 2-phase systems, T12, the phase lag of C2 behind C1. Latches are characterized by propagation delays Lpc and Lpd from the C and D-inputs respectively, and the set times Lsc and Lsd measured from changes at the C and D-inputs respectively. For the 2-phase case, subscripts 1 and 2 are used to distinguish between the phases or between the first and second latches of each pair.

Capital letters and small letters are used to designate maximum and minimum values, respectively, of the various parameters. Note that if C persists for less than lsd after a D-change, that D-change is guaranteed not to affect the state of the latch. There are latch designs where lsd is negative; a change in D occurring in some interval after the trailing edge of the clock pulse might affect the output or state of the latch.)

2.0 1-PHASE SYSTEMS WITH LATCHES

The systems discussed here employ simple polarity hold latches controlled by the clock C, as shown in Fig.1. Only a skeletal argument is given; details are supplied elsewhere <2>.

2.1 The Basic Relations

Define Gi as the maximum lag of the D-input to a latch behind the ith C-pulse to that latch, in the absence of skew. This quantity may be negative. We begin with:

$$Gi = Max(Lpc, Gi-1 + Lpd) + D - p.$$

Let the upper bound on Gi be G, the D-lag, solve for p and bring G inside the brackets to obtain:

$$p = Max(Lpc - G, Lpd) + D. \quad \text{(Basic constraint on p.)} \tag{2.1}$$

The minimum time it takes from the leading edge of C for a signal to propagate thru a latch, thru the logic, and to set a receiving latch is given by the left side of the inequality below. The worst case effect of clock skew might extend the interval during which a sequence of 2 or more steps might occur to W + 2S. Hence, in order to avoid multi-stepping, it is necessary to satisfy:

$$lpc + d + lsd > W + 2S.$$

Solving for d and using equality, since one goal is to minimize the lower bound d, we have:

$$d = W + 2S - lpc - lsd \quad \text{(basic constraint on d).} \tag{2.2}$$

The following constraint ensures that the slowest latch can be switched under the combined worst-case conditions of skew, clock pulse width variation, and propagation delay distribution in the logic and latches:

$$w > Max(Lsc, G + 2S + Lsd) \quad \text{(basic constraint on w).} \tag{2.3}$$

Given that W = (Wt)w and that the pulse width should not be greater than necessary, the previous 2 expressions yield:

$$d = (Wt)Max(Lsc, G + 2S + Lsd) + 2S - lpc - lsd. \tag{2.4}$$

2.2 Deriving the Optimal Trade-Offs

The D-lag should be reduced at least enough so that p is dependent on it, i.e.:

$$G < Lpc - Lpd. \tag{2.5}$$

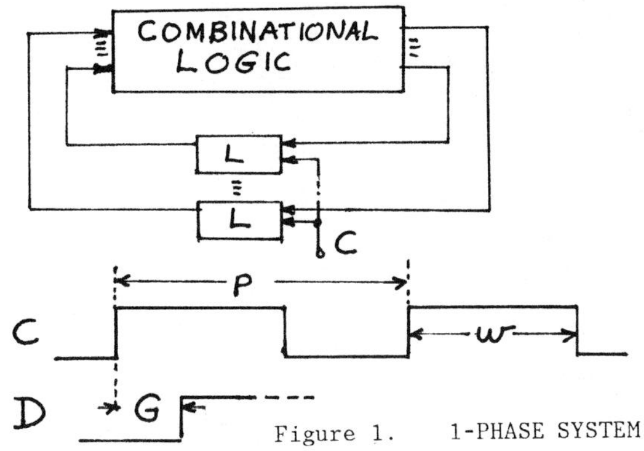

Figure 1. 1-PHASE SYSTEM

This reduces (2.1) to:

$$p = Lpc + D - G. \tag{2.6}$$

If G is maximized, then (2.6) indicates that the minimum value of p is Lpd + D, clearly irreducible, as it corresponds to the maximum loop delay.

G should be constrained to exceed the point where reducing it would increase p (see (2.4)). i.e.:

$$Lpc - Lpd > G > Lsc - Lsd - 2S. \tag{2.7}$$

This reduces (2.4) to:

$$d = 2(Wt + 1)S + (Wt)(G + Lsd) - lpc - lsd. \tag{2.8}$$

There is no point in reducing G at the cost of increasing p beyond the point where d becomes negative. This determines J, another lower bound on G, found by solving (2.8) for G and setting d = 0:

$$J = (-2(Wt + 1)S - (Wt)Lsd + lpc + lsd)/Wt.$$

Combining with (2.7) gives us:

$$Lpc - Lpd > G > Max(Lsc - Lsd - 2S, J). \quad \text{(bounds on G)} \tag{2.9}$$

Now we can determine p as a function of d by eliminating G from (2.8) and (2.6):

$$p = Lpc - (d - 2(Wt + 1)S - (Wt)Lsd + lpc + lsd)/Wt + D. \tag{2.10}$$

This gives minimum p-values for d in the range:

$$2(Wt + 1)S + (Wt)(Lpc - Lpd + Lsd) - lpc - lsd > d > Max(0, (Wt)Lsc + 2S - lpc - lsd). \tag{2.11}$$

From (2.3) and (2.8) we can derive:

$$w = Max(Lsc, 2S + (d - 2(Wt + 1)S + lpc + lsd)/Wt). \tag{2.12}$$

Relations (2.10), (2.11), and (2.12) enable us to specify optimal values of p and w, given the largest tolerable value of d.

3.0 2-PHASE SYSTEMS

In this case a pair of latches, L1 and L2, appears in each feedback path, controlled by clock pulses

C1 and C2, respectively as in Figure 2.

3.1 The Basic Relations

Four parameters are needed to specify the clocks for a 2-phase system: the period p, clock pulse widths w1 and w2, and the phase lag T12 of the C2-pulse behind the C1-pulse. (Sometimes it is advantageous to replace T12 with the overlap of the leading edge of the C2-pulse over the end of the C1-pulse.)

As in the 1-phase case, the maximum lag, G1, of a D-input to a latch behind the corresponding C1-input to the same latch, in the absence of skew, is used as an intermediate variable. The analog to (2.1) is:

$$p = \mathrm{Max}(G1 + L1pd + L2pd,\ L1pc + L2pd,\ T12 + L2pc) + D - G1 \text{ (basic constraint on p)}. \quad (3.1)$$

To prevent the possibility of multiple transitions during one clock period, it is necessary to ensure that no D-input to an L1-latch during a particular C1-pulse can propagate thru that latch, thru the succeeding L2-latch, and thru the logic in time to disturb the state of any other (or the same) L1-latch. The effects of worst-case clock skew must be taken into account:

$$T12 + l2pc + d + l1sd > W1 + 2S.$$

This leads to the minimum value of d:

$$d = W1 + 2S - T12 - l2pc - l1sd \text{ (basic constraint on d)}. \quad (3.2)$$

(In (3.2), the subexpression, W1 + 2S - T12, describes the maximum overlap of a C1-pulse and the succeeding C2-pulse. Satisfying (3.2) ensures that multi-stepping cannot occur.)

The minimum width of the C1 clock pulse must be adequate to switch the slowest latch, even when the clock skew is such as maximally to increase the lag of the D-input to an L1-latch beyond the no-skew value given by G1:

$$w1 = \mathrm{Max}(L1sc,\ G1 + L1sd + 2S) \text{ (basic constraint on w)}. \quad (3.3)$$

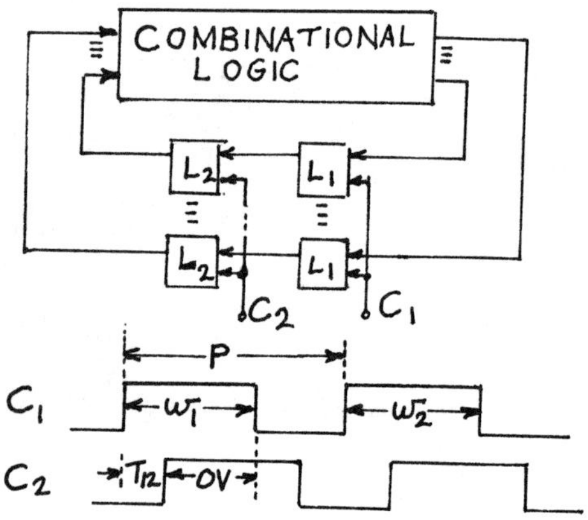

Figure 2. 2-PHASE SYSTEM

Substituting Wt times the right side of (3.3) for the W1 term of (3.2) yields:

$$d = (Wt)\mathrm{Max}(L1sc,\ G1 + L1sd + 2S) + 2S - T12 - l2pc - l1sd. \quad (3.4)$$

3.2 Simplifying the Basic Relations

Assuming that S, Wt, and the latch parameters are fixed by the technology, and that D is fixed by the system requirements and the technology, we now determine the clock parameters so as to minimize p and d. This is done via appropriate choices of the D-lag G1 (which determines w1), and T12. The G1, T12 pair determines p and d. Since we can't minimize p and d simultaneously, we develop a function specifying the minimum p-value for each attainable value of d, and associated functions that specify T12, w1, and the bounds on C2's width for each such point. First (3.1) and (3.4), are simplified to:

$$p = T12 + L2pc + D - G1, \text{ and} \quad (3.5)$$

$$d = (Wt)(G1 + L1sd) + 2(Wt + 1)S - T12 - l2pc - l1sd. \quad (3.6)$$

Next we derive the following bounds on G1 and T12:

$$L1sc - L1sd - 2S < G1 < T12 + L2pc - L1pd - L2pd \text{ (bounds on G1)}, \quad (3.7)$$

$$\text{and } L2pd + L1pc - L2pc < T12 < (Wt)(G1 + L1sd) + 2(Wt + 1)S - l2pc - l1sd \text{ (bounds on T12)}. \quad (3.8)$$

These basic relations can be manipulated to find the optimum trajectory of p as a function of d. In general, this consists of a 2-section, piecewise linear curve. The minimum value of p is:

$$p = D + L1pd + L2pd. \quad (3.9)$$

The expression for the segment of the p vs d curve terminating with this value of p is:

$$p = D + (L2pd + L1pc)(1 - 1/Wt) - (d + 2(Wt + 1)S - L2pc - (Wt)L1sd - l2pc - l1sd)/Wt. \quad (3.10)$$

This expression is valid over the d-range:

$$\mathrm{Max}(0,\ (Wt)L1sc - L2pd - L1pc + L2pc + 2S - l2pc - l1sd \le d \le Wt(L1pc - L1pd) - L2pd - L1pc - L2pc + (Wt)L1sd + 2(Wt + 1)S - l2pc - l1sd. \quad (3.11)$$

In this range, T12 is held fixed at:

$$T12 = L2pd + L1pc - L2pc \quad (3.12)$$

and the value of w1 is set at:

$$w1 = (d + L2pd + L1pc - L2pc - 2S + l2pc + l1sd)/Wt. \quad (3.13)$$

For d ranging between 0 and the lower bound above, w1 is fixed at L1sc, and T12 is specified by:

$$T12 = -d + (Wt)(L1sc - L1sd - 2S). \quad (3.14)$$

The minimum value of p is:

$$p = -d + (Wt - 1)(L1sc - L1sd - 2S) + L2pc + D + (Wt)L1sd + 2(Wt + 1)S - l2pc - l1sd. \quad (3.15)$$

Bounds on the C2-pulse width are generally not critical. They can be found <2> by considering constraints analogous to those used C1.

4.0 ONE-PHASE SYSTEMS WITH ETFF'S

Here the latches in Fig. 1 are replaced by positive-edge-triggered D-FF's (ETFF's). We define the set-up time U as the minimum time that a D-signal must be established at a FF input prior to the leading edge of the clock input, and the hold time H as the minimum time that the D-input to a FF must remain unchanged after the leading edge of the clock input. The timing constraints are:

$$p > Lpc + D + U + 2S, \quad (4.1)$$

to ensure that there is always time for the D-input to a FF to be set up prior to the leading edge of the next clock pulse.

$$w > wm; \quad (4.2)$$

minimum clock pulse width from the FF design.

$$lpc + d > 2S + H \quad (4.3)$$

to prevent a critical race in which a fast FF might stop a slow FF from completing a state change. It can be shown <2> that this constraint dominates the one derivable from the multi-stepping problem. From (4.3) we deduce the constraint on d:

$$d > 2S + H - lpc. \quad (4.4)$$

There is no width compensated system using ETFF's to compensate for skew; this must be done by increasing p, since it is only the leading edge of the clock pulse that counts.

5.0 CONCLUSIONS

It has been shown here that, for both 1 and 2-phase systems, faster operation is possible by appropriately extending the clock pulse widths to counter the effects of clock skew. We refer to this concept as width compensation. This is in contrast to the more usual data-first systems, in which D-inputs to clocked state devices must be set up prior to the leading edges of the clock pulses. The penalty for the faster operation of the width-compensated systems is more stringent lower bounds on the permissible short-path logic delays.

The choice of system type, as well as the choice of clocking parameters depends upon the technology parameters. No one mode of operation is uniformly superior to all others; each may be optimal for some particular set of circumstances. Where the propagation delays can be held within very tight limits, and the clock skew and width tolerance are small, a 1-phase system with simple latches is likely to result in fastest operation. But if these factors are such that this system would require delay pads associated with each latch, then a 2-phase system would usually be faster. If clock skews are high and short-path delays cannot be tightly controlled, then edge-triggered FF's may be advantageous. A basic advantage of the 2-phase system is its compatibility with LSSD operation <3> (where all state devices in the system may be chained together as a shift register to facilitate testing). Edge-triggered systems can probably be adapted to LSSD.

Substantially higher data rates for digital systems are obtainable by narrowing the bounds within which the delays thru the logic are constrained and by making clock generation and distribution more precise. Our results also apply to synchronous pipelined systems <4><5><6>, a special case with no actual feedback.

REFERENCES

1.Engineering Staff, Texas Instruments, Inc., The TTL Data Book for Design Engineers, Second Edition, 1981, Texas Instruments, Inc.

2.Unger, S. H., Tan, C. J., "Clocking Schemes for High Speed Digital Systems",IBM Research Division Report, RC 9754, 12/13/82.

3.Eichelberger, E. B. and Williams, T. W., "A Logic Design Structure for LSI Testability", Proc. of the 14th Design Automation Conference, New Orleans, LA, June 1977, pp.462-468.

4.M. J. Flynn, T. G. Hallin, "Pipelining of Arithmetic Functions", IEEE Transactions on Computers, 8/72, pp. 880-886.

5.G. K. Fawcett, "Maximal Clocking Rates for Pipelined Digital Systems", MS Thesis, EE Dept., U. Illinois, Urbana, 1974.

6.Peter M. Kogge, The Architecture of Pipelined Computers, 1981, McGraw-Hill.

Acknowledgment

The comments and criticisms of Klim Maling have been most helpful. We also profited from stimulating discussions with Vijay Kumar, Gregory Buchanan, Dale Kurschner, Gerald Maley, and Patrick Meehan.

ON THE IMPLEMENTATION AND USE OF SYSTOLIC ARRAY PROCESSORS

H. T. Kung

Department of Computer Science, Carnegie-Mellon University
Pittsburgh, Pennsylvania 15213

In recent years many systolic algorithms have been designed and several prototypes of systolic array hardware have been constructed. Major efforts now started in attempting to use systolic array processors in large, real-life applications; practical issues on the implementation and use of systolic array processors in systems have begun to receive substantial attention. This paper first examines various implementation issues and alternatives, and then identifies some work that is essential to the eventual, wide use of systolic array processors.

INTRODUCTION

Much of the time consuming "front-end processing," that deals with large amounts of data obtained directly from sensors in areas such as vision and radar, is inherently regular and parallel. By exploiting this regularity and parallelism systolic architectures, when properly implemented, can achieve high performance with low cost.

In a systolic architecture, a systolic array in particular, data are passed from processing element (PE) to processing element during a computation. By ensuring that once a data item is brought out from the memory it can be used effectively at each PE it passes, a systolic array can achieve a high computation throughput with a modest memory bandwidth. As we will see later in this section, recent research has shown that this is possible for a wide class of compute-bound computations where multiple operations are performed on each data item in a repetitive manner. Being able to use each input data item a number of times is just one of many advantages of a systolic architecture. Other advantages, including modular expansability, simple and regular data and control flows, use of simple and uniform cells, elimination of global broadcasting and fan-in, and (possibly) fast response time, are also characteristic.[1] These properties are highly desirable for VLSI implementations; indeed the advances in VLSI technology have been a major motivation for the systolic architecture research.

Mathematical algorithms implementable with systolic architectures are called *systolic algorithms*. A systolic algorithm has many systolic *cells*, which can operate in parallel, and each of them is to be implemented by a PE in a systolic architecture. A large number of systolic algorithms are known today; extensive references can be found in a separate paper.[2] The following is a partial list of problems amenable to systolic solutions, which should indicate the breadth of the systolic approach.

- *Signal and image processing*:
  - FIR and IIR filtering and 1-D convolution;
  - 2-D convolution and correlation;
  - discrete Fourier transform;
  - interpolation;
  - 1-D and 2-D median filtering; and
  - geometric warping.
- *Matrix arithmetic*:
  - matrix-vector multiplication;
  - matrix-matrix multiplication;
  - matrix triangularization (solution of linear systems, matrix inversion);
  - solution of triangular linear systems;
  - solution of Toeplitz linear systems;
  - QR-decomposition (least squares computations, covariance matrix inversion);
  - singular value decomposition; and
  - eigenvalue problems.
- *Non-numeric applications*:
  - data structures—stack and queue, priority queue, searching, and sorting;
  - graph algorithms—transitive closure, minimum spanning trees, and connected components;
  - geometric algorithms—convex hull generation;
  - language recognition—string matching and regular expression;
  - dynamic programming;
  - polynomial manipulation—multiplication, division, and greatest common divisor;
  - integer greatest common divisors; and
  - relational data-base operation.

Designing algorithms for systolic arrays constitutes the very beginning phase of the total effort of constructing a systolic device for application use. Activities following this phase include at least architectural specification of the systolic array processor, processor implementation, integration to the host system and application software development. This paper focuses on issues beyond the algorithm design phase. In the presentation, we shall assume that the reader is familiar with the basic systolic literature[1, 3] and knows at least a few simple systolic algorithms.

IMPLEMENTATION ALTERNATIVES

Many alternatives exist for the implementation of systolic array processors at both chip and board levels. We discuss them along two dimensions: flexibility and interconnection topology.

*This research was supported in part by the Defense Advanced Research Projects Agency (DoD), ARPA Order No. 3597, monitored by the Air Force Avionics Laboratory under Contract F33615-81-K-1539.

CH1935-6/83/0000/0370$01.00 © 1983 IEEE

FLEXIBILITY

1. *Single-purpose systolic array.*[4, 5, 6, 7] The most straightforward way to implement a systolic algorithm is to build a special-purpose systolic array processor just for that algorithm. This approach is reasonable if one or more of the following conditions hold: (1) the performance of the processor is of ultimate importance and the use of the processor is well-understood; (2) the processor will be used in large quantities despite the fact that it is single-purpose; and (3) the design and implementation cost of the processor is low—this is the case for those systolic arrays consisting of only few types of very simple cells like some of the pattern matching and correlation arrays.

2. *Multi-purpose systolic array.*[8, 9] This type of systolic array processor can implement a predefined set of systolic algorithms. The approach is based on the observation that many systolic algorithms like those for convolution and matrix multiply can be executed on systolic arrays of very similar structures. Therefore a processor array, with little overheads for providing the necessary flexibility, can perform a number of functions under some simple controls.

3. *Non-programmable building-block.*[10] Some common functions such as multiplier-accumulation are performed by cells of a large number of systolic arrays. Therefore it is possible to construct building-block processors capable of executing a few predefined, commonly used functions, and then connect them to form a variety of systolic array processors of different sizes and shapes. A good building-block should meet the needs from current as well as future systolic algorithms.

4. *Programmable building-block.*[11, 12] In this approach building-blocks, by programming, can implement a large family of systolic cells. Of course the programmable approach is not as efficient as the non-programmable one, because of the overheads for supporting the programmability. Nevertheless it fulfills the need of implementing those systolic arrays, which each are not important enough for warranting individual, custom hardware implementation, but in aggregate can be implemented cost-effectively by some programmable building-blocks. Moreover in some systolic designs, the instruction that the cell executes in a cycle depends, in a complicated way, on the inputs to the cell and the cell's state during that cycle. A notable example is the the systolic array for computing greatest common divisors.[13] The programmable approach seems to be the only effective way to implement those complicated, data-dependent systolic arrays.

5. *Programmable systolic arrays.*[14, 15] These are systolic array processors where a fixed number of programmable PEs are connected together in certain manner, possibly with other control circuits. This approach is more flexible than the multi-purpose systolic array approach in the sense that the PEs are programmable and sometimes even their interconnections can be configured by software control before a computation starts.

INTERCONNECTION TOPOLOGIES

Systolic arrays typically call for simple and regular array interconnections among their PEs. We classify interconnections for n-PE arrays into three classes according to the number of computations performed for each I/O operation:

1. *2-D (two-dimensional) systolic arrays,* such as square and hexagonal arrays, for which at each cycle, $O(n)$ PEs perform computations while $O(\sqrt{n})$ boundary PEs perform I/O, and thus the computation over I/O ratio is $O(\sqrt{n})$. Examples are systolic arrays for matrix operations.[3, 16]

2. *Degenerated 2-D systolic arrays,* that is, one-dimensional arrays for which at each cycle, $O(n)$ PEs perform computations and $O(n)$ PEs perform I/O, and thus the computation over I/O ratio is $O(1)$. Examples are systolic arrays for solution of triangular linear system[3] and orthogonal transformations.[17]

3. *Linear or 1-D systolic arrays,* for which at each cycle, $O(n)$ PEs perform computations but only the two end PEs perform I/O, and thus the computation over I/O ratio is $O(n)$. Examples are various systolic convolution arrays.[1]

For situations where the communication bandwidth between the host system and the systolic array is a bottleneck, linear arrays are preferable to 2-D arrays and degenerated 2-D arrays are most undesirable. Sometimes for a given problem there could be both linear and 2-D systolic array solutions. For example, two-dimensional convolution can be performed by a linear systolic array[18, 19] or a 2-D systolic array.[5] In this case we can choose a linear solution to solve the communication bandwidth problem. In addition linear arrays (and degenerated 2-D arrays) have the additional advantage that they can always be safely synchronized by a global clock.[20] To avoid the excessive I/O bandwidth required by degenerated 2-D arrays, it may sometimes be cost-effective to associate each PE with a memory, where data required by the PE can be cached.[8, 9]

In general some "design-time" rather than "run-time" flexibility at the interconnection level should always be provided not only for implementing different interconnections, but also for system maintenance, debugging and fault-tolerance. The idea is that by modifying interconnections faulty PEs can be detected, and replaced or avoided during a computation. The usefulness of run-time reconfigurability for systolic array processors is not clear, since typically interconnections do not change during a computation.

WHAT FURTHER WORK IS NEEDED?

From the preceding section, we see that much progress has been made in the past several years in the design and implementation of systolic array prototypes. But, as in the development of any other computer system, there is a large gap between prototypes and usable systems. In this section, we identify some work that is essential to bring systolic array processors to a wide user community.

SYSTOLIC BUILDING-BLOCKS

The building-block approach is most natural to the implementation of systolic array processors, because the same PE is replicated many times for each individual array and many systolic arrays appear to share PEs of similar structures. Potentially, systolic building-blocks can be used in large quantities and therefore they can be made available at low cost. However, it is not easy to specify an efficient and easy-to-use building-block. Consider for example one of the obvious non-programmable, building-block candidates—a multiplier-accumulator chip augmented by some systolic I/O ports. This chip will indeed be able to implement many systolic arrays in signal and image processing, but only in their theoretical versions. The problem becomes non-obvious when practical issues are considered, such as:

- Support for both real and complex arithmetic? Complex arithmetic is fundamental to signal processing, but real arithmetic is useful for other applications. For VLSI implementation where the pin bandwidth limitation is a major concern, the complex multiplier-accumulation is very different from the real counterpart—the former performs twice as many real arithmetic operations per I/O as the latter.
- Support for modular expansion of data word widths?
- Support for systolic data streams moving at multiple speeds and in multiple directions? Support for multiplexing arithmetic units of the chip to implement multiple systolic cells ? On-chip buffering registers, delay registers, FIFOs and bidirectional ports must now be considered.
- Mechanism for shipping out from each cell the accumulated results, without interrupting an ongoing systolic array computation? Support for systolic controls moving from cell to cell

together with data?

The design space for programmable systolic building-blocks such as the CMU Programmable Systolic Chip (PSC)[11, 21] is even larger. Successful building-blocks can be determined only through extensive experiments and careful requirement analyses of various classes of systolic algorithms. The challenge here is to find ways to speed up this timing consuming learning process.

SYSTEM SUPPORT

After having identified the "right" systolic building-blocks and built efficient systolic array processors, we still need substantial system support to make them easy to use.

- *Development systems.* We need a basic design and test environment for the development of systems incorporating systolic array processors. The problem here is not unlike those faced by earlier microprocessors. In the case of microprocessors, moderate amounts of memory, I/O (ports, terminals, etc.) and permanent storage were sufficient in the beginning. In contrast, due to their high performance, systolic systems typically require a complete microcomputer with a large memory, a complex I/O controller (simple DMA facilities cannot handle complex access patterns required by many systolic arrays) and fast I/O devices (disks, frame-buffer, ADC/DAC subsystems).
- *Host interface facilities.* Interfacing a systolic array processor to a host system requires certain hardware and software interface facilities. The controller between the host and the systolic array has to handle potentially high data rates, complex access patterns and tight timing constraints. In addition it is necessary to provide drivers (to interface the systolic array to the host operating system), decomposition routines (to decompose the algorithm or problem to match the size of the given systolic array), as well as software debugging and hardware testing tools.
- *Programming aids.* For most of the systolic array processors being implemented, programming tools are little more than assemblers and simulators. Any widespread use of a systolic array processor demands availability of efficient compilers for high level languages. Users should be able to program a systolic array processor by just providing the mathematical definition of a systolic algorithm. This compiler is not necessarily more difficult to develop than a conventional compiler, because both the data flow and timing information of a computation will be specified in user programs.

ALGORITHMS REFLECTING IMPLEMENTATION NEEDS

In spite of the large number of systolic algorithms already designed, a major effort is still needed in developing systolic algorithms that are convenient for implementation. Currently, when implementing a systolic algorithm on a systolic processor, we often find that the algorithm must be substantially modified and in some cases totally new algorithms have to be invented. It is useful to know the following experiences:

- *Systolic algorithms using linear arrays.* For implementation convenience, systolic algorithms using linear arrays are superior than those using 2-D arrays—it is not an accident that most of the systolic algorithms that have been implemented use linear arrays. Besides for reasons noted earlier that linear systolic arrays enjoy the advantages of low I/O bandwidth requirements and ease of synchronization by clocking, we find it is convenient to balance the computation over I/O ratio for algorithms using linear arrays. As illustrated in Figure 1, the ratio increases linearly with the number of systolic cells allocated to each PE. For example, suppose that in a systolic algorithm the boundary cell (denoted by the circle in the figure) takes k times more operations than the internal cells (denoted by squares). In implementation we can achieve a balancing in speed between the two types of cells by simply allocating k rather than one internal cell to each PE.

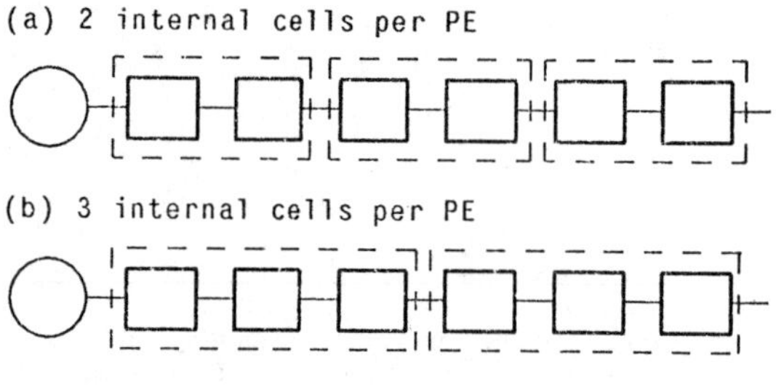

Figure 1: Balancing I/O and computation for linear array: (a) 3 cell operations and (b) 2 cell operations per I/O.

- *Systolic algorithms for pipelined units.* Systolic array processors often employ PEs which are pipelined units themselves. This is typically true for floating-point processing elements, which could have four to ten pipeline stages. For performance, it is important that one can take advantage of this second level of pipelining in a systolic implementation.[19] This can always be achieved for systolic algorithms without cycles, i.e., without feedback, but without decreasing throughput, it is impossible for algorithms with cycles. We know that certain problems such as convolution and graph connectivity can be solved by systolic algorithms either with or without feedback loops.[1, 22]
- *Problem decomposition.* In practice, given problems are often larger than what a given systolic array can handle in one pass. In this case an original problem must be decomposed into subproblems, and the systolic array then processes these subproblems one at a time. Some systolic algorithms can be easily modified to work on subproblems with relatively small I/O and control overheads, but some can not. [23, 24] The design of a systolic algorithm is really not complete unless its decomposition methods are provided, although in many cases partitioning schemes are self-evident once the basic algorithm is presented.

Being simple and regular is not everything about systolic algorithm design; for implementation convenience, other issues such as balancing computation with I/O, use of pipelined processing elements and problem decomposition are crucial.

CONCLUSIONS

Among the many implementation alternatives stated, the building-block approach (programmable or non-programmable) is likely to be most influential to the implementation and use of systolic array processors, although we do expect very high performance special-purpose systolic processors will continue to be built. Commercial vendors' interest in this area represents the best solution to the problem of providing large numbers of reliable building-blocks for system use at reasonably low costs. Several on-going research projects[10, 14, 21] should help provide some initial information on the specification of such building-blocks. However, the best building-blocks probably cannot be determined until extensive programming experiences on the use of systolic array processors have been accumulated. Development of catalogs or application notes on programming examples for various existing systolic machines should be a useful starting point.

To make systolic array processors available for general user communities, we face many conventional system support and implementation problems. There are really no magic solutions to make these problems go away but to solve each of them. The issue here is not how to solve these problems since their solutions in general do exist, but to recognize their importance to the success of the total approach by investing proper efforts on them.

ACKNOWLEDGMENTS

The author had many useful discussions with A. Nowatzyk of CMU on various topics of this paper.

References

1. Kung, H.T., "Why Systolic Architectures?," *Computer Magazine*, Vol. 15, No. 1, January 1982, pp. 37-46.

2. Fisher, A.L. and Kung, H.T., "Special-Purpose VLSI Architectures: General Discussions and a Case Study," *Proceedings of the USC Workshop on VLSI and Modern Signal Processing*, November 1982.

3. Kung, H.T. and Leiserson, C.E., "Systolic Arrays (for VLSI)," *Sparse Matrix Proceedings 1978*, Duff, I. S. and Stewart, G. W., eds., SIAM, 1979, pp. 256-282.

4. Corry, A. and Patel, K., "A CMOS/SOS VLSI Correlator," *Proceedings of 1983 International Symposium on VLSI Technology, Systems and Applications*, 1983, pp. 134-137.

5. Kung, H.T. and Song, S.W., "A Systolic 2-D Convolution Chip," *Multicomputers and Image Processing: Algorithms and Programs*, Preston, K., Jr. and Uhr, L., ed., Academic Press, 1982, pp. 373-384.

6. Foster, M.J. and Kung, H.T., "The Design of Special-Purpose VLSI Chips," *Computer Magazine*, Vol. 13, No. 1, January 1980, pp. 26-40.

7. Tamura, P.N., Haugen, P.R. and Betz, B.K., "Time Integrating Digital Correlator," *Proceedings of SPIE Symposium, Vol. 431, Real-Time Signal Processing VI*, August 1983.

8. Blackmer, J., Frank, G. and Kuekes, P., "A 200 Million Operations per Second (MOPS) Systolic Processor," *Proceedings of SPIE Symposium, Vol. 298, Real-Time Signal Processing IV*, August 1981, pp. 10-18.

9. Yen, D.W.L. and Kulkarni, A.V., "Systolic Processing and an Implementation for Signal and Image Processing," *IEEE Transactions on Computers*, Vol. C-31, No. 10, October 1982, pp. 1000-1009.

10. Kuekes, P. and Shen, J., "One-Gigaflop VLSI Systolic Processor," *Proceedings of SPIE Symposium, Vol. 431, Real-Time Signal Processing VI*, August 1983.

11. Fisher, A.L., Kung, H.T., Monier, L.M., Walker, H. and Dohi, Y., "Design of the PSC: A Programmable Systolic Chip," *Proceedings of the Third Caltech Conference on Very Large Scale Integration*, Bryant, R., ed., Computer Science Press, Inc., March 1983, pp. 287-302.

12. Sorasen, O. and Solberg, B., "VLSI Implemented Systolic Array Processor," *Proceedings of VLSI 83 International Conference*, Trondheim, Norway, August 1983.

13. Brent, R.P. and Kung, H.T., "Systolic VLSI Arrays for Linear-Time GCD Computation," *Proceedings of VLSI 83 International Conference*, August 1983.

14. Bromley, K., Symanski, J.J., Speiser, J.M., and Whitehouse, H.J., "Systolic Array Processor Developments," *VLSI Systems and Computations*, Kung, H.T., Sproull, R.F., and Steele, G.L., Jr., eds., Computer Science Press, Inc., October 1981, pp. 273-284.

15. Symanski, J.J., "Systolic Array Processor Implementation," *Proceedings of SPIE Symposium, Vol. 298, Real-Time Signal Processing IV*, Society of Photo-Optical Instrumentation, August 1981, pp. 27-32.

16. Gentleman, W.M. and Kung, H.T., "Matrix Triangularization by Systolic Arrays," *Proceedings of SPIE Symposium, Vol. 298, Real-Time Signal Processing IV*, Society of Photo-Optical Instrumentation Engineers, August 1981, pp. 19-26.

17. Heller, D.E. and Ipsen, I.C.F., "Systolic Networks for Orthogonal Equivalence Transformations and Their Applications," *Proceedings of Conference on Advanced Research in VLSI*, MIT, January 1982, pp. 113-122.

18. Kung, H.T. and Picard, R.L., "Hardware Pipelines for Multi-Dimensional Convolution and Resampling," *Proceedings of the 1981 IEEE Computer Society Workshop on Computer Architecture for Pattern Analysis and Image Database Management*, IEEE Computer Society Press, November 1981, pp. 273-278.

19. Kung, H.T., Ruane, L.M., and Yen, D.W.L., "Two-Level Pipelined Systolic Array for Multidimensional Convolution," *Image and Vision Computing*, Vol. 1, No. 1, February 1983, pp. 30-36.

20. Fisher, A.L. and Kung, H.T., "Synchronizing Large VLSI Processor Arrays," *Proceedings of the 10th Annual International Symposium on Computer Architecture*, June 1983, pp. 54-58.

21. Fisher, A.L., Kung, H.T., Monier, L.M. and Dohi, Y., "Architecture of the PSC: A Programmable Systolic Chip," *Proceedings of the 10th Annual International Symposium on Computer Architecture*, June 1983, pp. 48-53.

22. Tchuente, M. and Melkemi, L., "Systolic Arrays for Connectivity Problems and Triangularization for Band Matrices," Tech. report R.R. No. 366, IMAG, Institut National Polytechnique de Grenoble, March 1983.

23. Hong, J.-W. and Kung, H.T., "I/O Complexity: The Red-Blue Pebble Game," *Proceedings of the Thirteenth Annual ACM Symposium on Theory of Computing*, ACM SIGACT, May 1981, pp. 326-333, To appear in *Journal of the ACM*.

24. Kung, H.T. and Yu, S.Q., "Integrating High-Performance Special-Purpose Devices into a System," *VLSI Architecture*, Randel, B. and Treleaven, P.C., eds., Prentice/Hall International, 1983, pp. 205-211.

Session 2-3a
CMOS I

Chairman

S. Hu
Wang Labs

ADVANCED CMOS PROCESS FOR LSI CIRCUIT DEVELOPMENT

C. S. Wang, S. A. Saller, K. L. Chen,
D. B. Scott, P. L. Shah, and R. D. Davies

Semiconductor Group/CMOS Product Customer Center
Texas Instruments Incorporated
Dallas, Texas 75265

ABSTRACT

CMOS circuits are now being used for more highly integrated designs due primarily to the inherent low standby power advantage. One of the main limitations to the use of CMOS in LSI circuitry is the increased process complexity inherent in CMOS processes. In this paper a twin-well CMOS process is described which requires 9 masking operation (only two more than the comparable NMOS process). Modeling and characterization of active and parasitic device leakage dependence on process variations are discussed. Additionally characterization results which show the sensitivity of P to N channel spacing with alignment are also discussed. Finally, the effect device performance variations, resulting from process variations, on circuit performance are shown to be within acceptable limits.

INTRODUCTION

A key advantage of the use of CMOS for LSI and VLSI circuits is the ability to obtain high performance circuits that have low standby power requirements. In this work a CMOS technology is described that allows for these circuit enhancements without the traditionally expected increase in process complexity[1,2]. The process features 2 micron polysilicon linewidths, a twin-well structure and the use of P/P+ epitaxial substrate to avoid latch-up[3].

SECTION II: PROCESS FLOW

A single poly/single metal, twin-well CMOS process using 9 masking steps and featuring 2 micron poly and 350 Å gate oxide on P/P+ epi substrates has been developed. This process uses a single mask to define both N and P tanks, a process unique to twin-well. The oxide grown over the N-well during the first local oxidation acts as a self-aligned mask for the subsequent P-well implant.

After forming the second local oxidation mask, a blanket channel stop can be performed to adjust the parasitic thick field threshold voltage. This process both saves a masking level and avoids the conventional double coat resist process commonly found in many CMOS processes and thus resulting in a more manufacturable process.

Both threshold voltage implants are performed on the pre-gate oxide. This avoids possible contamination of the gate oxide due to resist processing prior to poly deposition by allowing the poly to be deposited immediately after gate oxidation.

After $POCl_3$ doping, the 2 micron poly gates are patterned and anisotropically etched. The N-channel S/D implant and anneal are performed prior to the P-channel S/D formation, thus minimizing the diffusion of boron and hence P-channel junction depth. Densified plasma oxide is used as a diffusion barrier to the phosphosilicate glass (PSG), which is reflowed to improve metal step coverage.

A single step dry oxide etch which produces 60-65 degree sloped walls is employed to open contact windows in the interlevel oxide. This simplifies the contact process to a single step while offering the step coverage advantage inherent in sloped profiles.

Silicon doped aluminum is deposited to a thickness of 1 micron and anisotropically etched to form a 7 micron minimum pitch interconnect. A layer of silicon nitride is deposited and patterned to form a protective overcoat for the completed circuit. Fig. 1 shows a cross-section of a final structure without the protective overcoat. Table 1 contains a list of the major process parameters.

SECTION III: DEVICE CHARACTERIZATION, ACTIVE LEAKAGE SOURCES

In order to maintain the CMOS advantage of low standby power, the threshold voltage and its variation as a function of channel length must be comprehended. The subthreshold current can be described as[4]

$$I_d = I_{ref} (e^{[(V_g - V_t)/(NkT/q)]}), \qquad (1)$$

in which Vt is the threshold voltage when I_d=1 micro-amp. Thus, for zero gate voltage, the subthreshold leakage current is given by

$$I_d = I_{ref} (e^{[-V_t/(NkT/q)]}). \qquad (2)$$

Observing (2), the leakage current is exponentially dependent upon the threshold voltage. In fact, for a shift in threshold voltage from 1.0 to 0.6 V, a corresponding four order of magnitude increase in subthreshold leakage has been typical. Thus for most cases, the threshold voltage must be adjusted high enough such that the subthreshold voltage is not the dominant leakage factor when compared with other leakages sources.

As shown in Fig. 2 and Fig. 3 for patterned gate lengths of 2 microns, the issue of subthreshold leakage becomes critical. Any variations in threshold voltage causing higher than anticipated leakage currents can be compounded by short channel effects. In order to predict the leakage current due to process variations, the functional dependence of V_t with channel length must be known. It has been shown[5] that

$$V_t = V_{to} - A\,(L_{eff})^{(-B)} \qquad (3)$$

where, for N-channel devices fabricated with this process A=0.25 and B=1.28, and A=0.12 and B=1.78 for the P-channel devices.

SECTION IV: DEVICE CHARACTERIZATION, PARASITIC LEAKAGE SOURCES

The cross-section in Fig. 4A shows the two parasitic field devices which are unique to CMOS. The P-moat and the P-substrate form the source and drain respectively of the P-channel parasitic field device, whose channel length is denoted by L_P. The N-channel parasitic field devices has the N-tank as the drain, the N-moat as the source and channel length, L_N. Also shown in Fig. 4B are metal gate field threshold voltages of these parasitic devices as a function of the moat to tank boundary spacings.

Observing Fig. 4B it can be seen that a total N-moat to P-moat spacing of 7 micron can be achieved without causing a significant lowering of the field threshold voltage. A histogram of the N-channel field threshold voltage measured across a slice is shown in Fig. 5 for a 4 micron N-moat to tank edge spacing. The results clearly indicate that the alignment registration between the tank and moat lithography levels is such that for a 4 micron spacing significant variations in the N-channel field threshold voltage were not observed. However, at 3 micron spacing an unacceptable variation in the field V_t was observed (Fig. 6).

SECTION V: ACTIVE DEVICE CHARACTERIZATION AND CIRCUIT PERFORMANCE

The characterization of this process has been concerned not only with the performance of the device themselves but also their performance in a circuit. For the discrete device performance, the devices were evaluated in terms of their current carrying capability at nominal operating conditions. The drain current was measured with V_{ds} = 5.0 volts and V_{gs} = V_t+4.0 volts. Thus, the effect of any threshold variations was normalized out. Fig. 7 shows a plot of this device drive current as a function of the effective channel length for the P-channel devices in this process.

The behavior of these devices in an actual circuit was evaluated using a long inverter chain with 215 inverters in series. The devices in this inverter chain had an aspect ratio of 58 microns/2 micron. The average per stage propagation delay observed was 0.63 nanoseconds and a power-delay product of 43.62 fJ was measured. To evaluate the effect of process variations on circuit performance the delay time was measured for a number of circuits. A histogram of the delay per stage illustrates the effect of process variations on circuit performance (Fig. 8).

SECTION VI: CONCLUSIONS

The characterization of leakage sources is of paramount importance and their modeling has been shown to be possible without the use of two-dimensional simulations. The modeling and subsequent characterization of leakage sources has lead to the choice of optimized process parameters (V_t doses) in order to minimize the active device leakage. Parasitic threshold has also been measured and shown to be very alignment sensitive. This variation can be minimized by proper choice of design rules to avoid the alignment sensitive region.

ACKNOWLEDGEMENTS

The authors would like to thank the engineering and manufacturing staffs of Dallas MOS1 for their assistance in fabricating these devices. Thanks also to Michael Maberry, Laurence Schultz, Helen Simon, Delores Williams and Michael Woodruff for their help in taking the characterization data.

REFERENCES

1. L.C. Parrillo et al., "Twin-Tub CMOS-A Technlogy for VLSI Circuits," IEDM Digest of Technical Papers, pp. 752-755, Dec. 1980.

2. D.B. Scott et al., "Considerations for Scaled CMOS Source/Drains," IEDM Digest of Technical Papers, pp. 538-541, Dec. 1981.

3. R.S. Payne et al., "The Elimination of Latchup in Bulk CMOS," IEDM Digest of Technical Papers, pp. 248-251, Dec. 1980.

4. R.M. Swanson and J.D. Meindl, "Ion Implanted Complementary MOS Transistors in Low Voltage Circuits," IEEE J. Solid State Circuits, Vol. SC-7, p. 146, 1972.

5. H. Masuda et al., "Characteristics and Limitations of Scaled-Down MOSFET's Due to Two-Dimensional Field Effects," IEEE Trans. Electron Devices, Vol. ED-26, pp. 980-996, June 1979.

CROSS-SECTION OF THE TWIN-WELL DEVICE STRUCTURE

FIG. 1

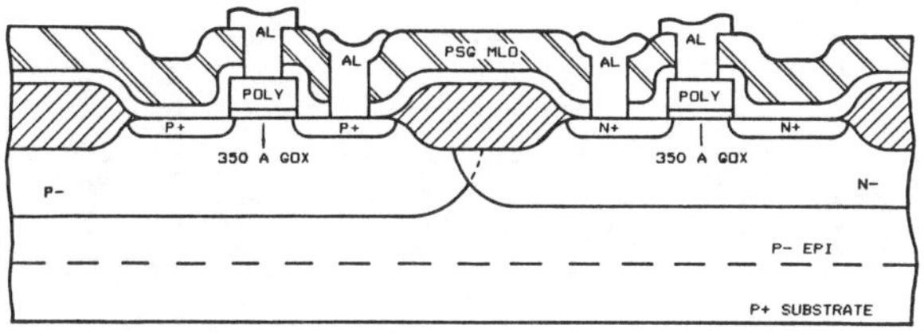

THRESHOLD VOLTAGE VS. ELECTRICAL CHANNEL LENGTH

FIG. 2

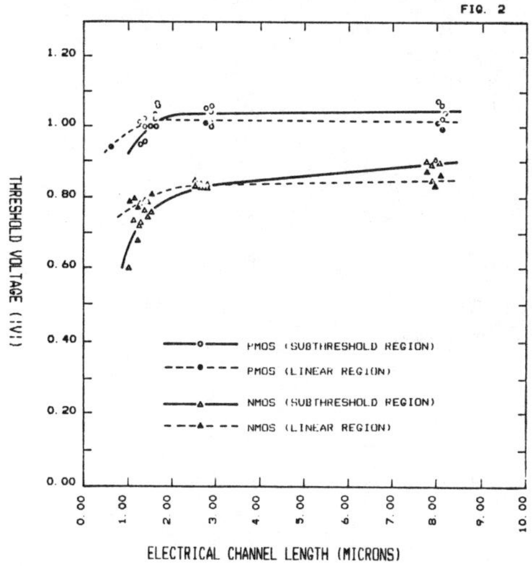

NMOS LEAKAGE CURRENT VS. ELECTRICAL CHANNEL LENGTH

FIG. 3A

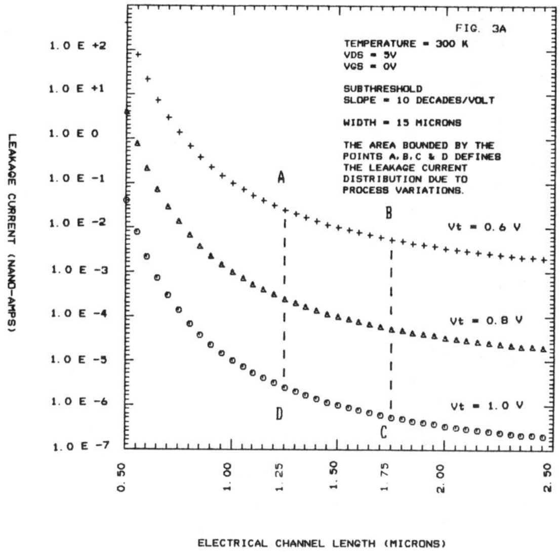

PMOS LEAKAGE CURRENT VS. ELECTRICAL CHANNEL LENGTH

FIG. 3B

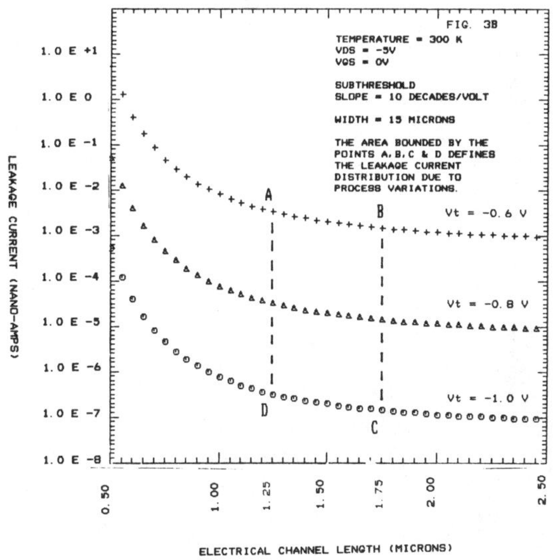

PARASITIC THRESHOLD VOLTAGE VS MOAT-TO-TANK SPACING

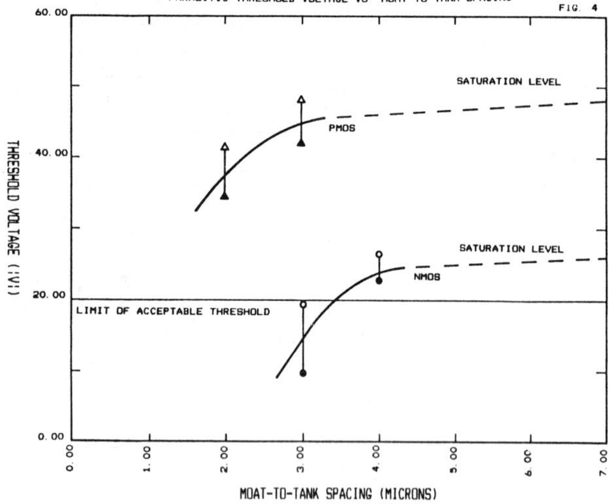

NMOS PARASITIC THRESHOLD VOLTAGE AT 4 MICRON SPACING

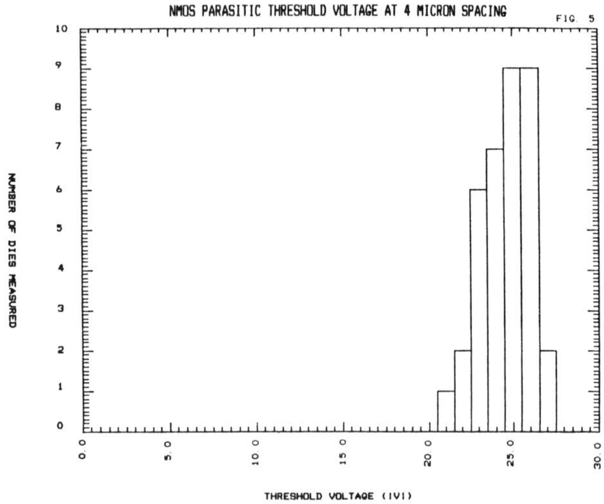

NMOS PARASITIC THRESHOLD VOLTAGE AT 3 MICRON SPACING

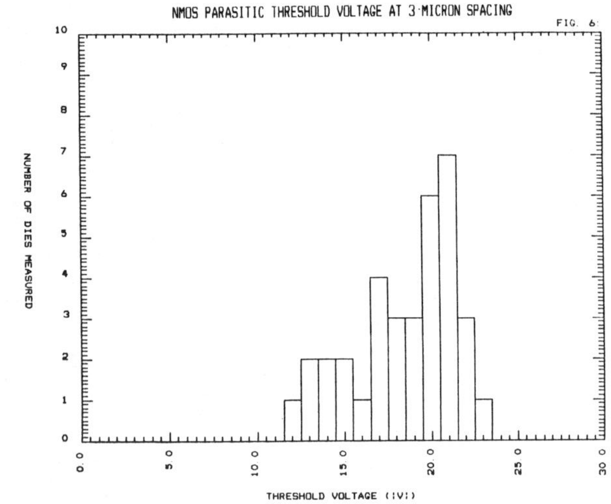

CROSS-SECTION OF THE PARASITIC FIELD DEVICE STRUCTURE

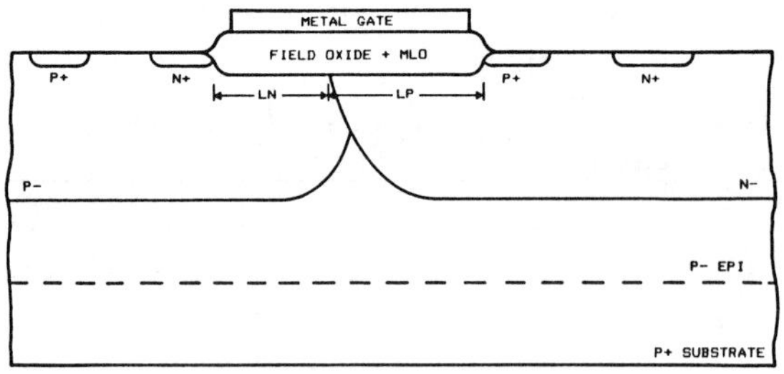

PMOS INVERSE DRIVE CURRENT VS. ELECTRICAL CHANNEL LENGTH

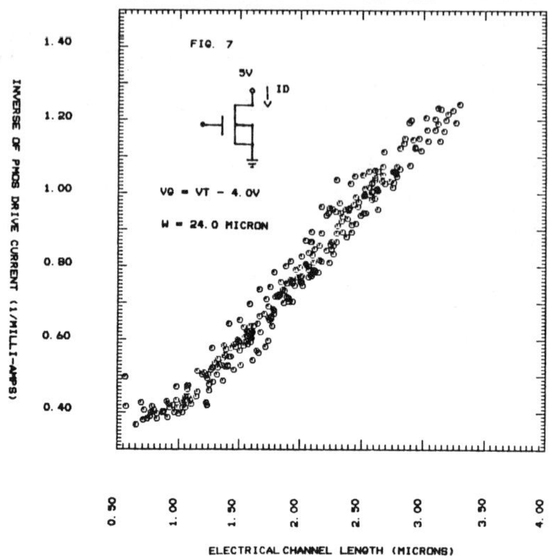

DISTRIBUTION OF INVERTER DELAY TIME PER STAGE

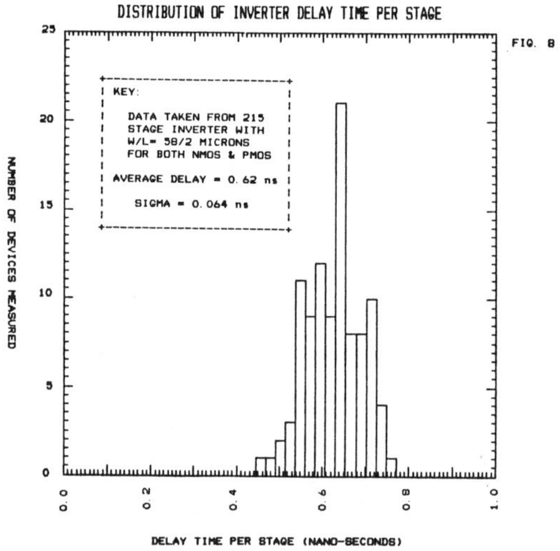

MAJOR PROCESS PARAMETERS

TABLE 1

N-TANK DEPTH	3.2 MICRON
N-TANK SURFACE CONCENTRATION	2.2E16 ATOMS/CC
P-TANK DEPTH	3.5 MICRON
P-TANK SURFACE CONCENTRATION	7.0E15 ATTOMS/CC
FIELD OXIDE THICKNESS	8500 ANGSTROMS
GATE OXIDE THICKNESS	350 ANGSTROMS
POLY SHEET RESISTANCE OUTSIDE TANK	17 - 20 OHMS/SQ
POLY SHEET RESISTANCE INSIDE TANK	19 - 23 OHMS/SQ
P+ SHEET RESISTANCE	150 - 200 OHMS/SQ
P+ JUNCTION DEPTH	0.4 MICRON
N+ SHEET RESISTANCE	20 - 35 OHMS/SQ
N+ JUNCTION DEPTH	0.35 MICRON

CMOS VS. NMOS COMPARISONS IN DYNAMIC RAM DESIGN

Paul Madland, Joseph Schutz
Richard Green, Roger Kung

Intel Corporation
3585 S.W. 198th Avenue, Aloha, Oregon

ABSTRACT

This paper will explore circuit design tradeoffs and system implications of high performance CMOS processes when applied to dynamic RAM's with comparisons to NMOS. Included are discussions on consequences of wells for soft error rates, refresh time enhancements and increased internal decoupling. A comparison is made of CMOS and NMOS circuits for clock generation, address buffers, sensing and decoding. The advantages of CMOS for static column operation is also discussed.

INTRODUCTION

In recent history, CMOS technology has proven to be a viable alternative to NMOS for use in dynamic RAMs [1,2]. A number of benefits of this technology include (1) use of junction barriers to block collection of thermally generated and alpha particle generated carriers (2) automatic power down of most logic allowing standby power to be set by junction and device leakage (3) dampened supply noise due to chip generated current spikes (4) simple active power reducing techniques (5) complementary cross coupled sense amplifiers and (6) simple implementation of fully static column operation including high gain differential amplifiers and static column decoders. These advantages will be discussed in two basic categories. Process and well related items and circuit related items.

WELL TECHNOLOGY

Alpha Immunity of Well Technologies

Since 1978, when alpha particle induced soft errors in DRAM's were first reported [3], scaling of cell size in VLSI memory has been limited by soft error rate (SER) considerations. Initial work to reduce SER focused on controlling the alpha flux through the use of improved package materials (with low alpha activity) and die coats. Cell size and stored charge can be scaled further without die coat if the charge collection efficiency of the memory array can be reduced. Several protection barriers have been reported [4] including buried grids [5], EPI substrates, p+ to p− barriers [6], and buried p-p junction barriers, of these the most effective is the p-n junction that is inherent in CMOS technology.

The charge collection efficiency for a 32K p-channel array inside an n-well was measured directly using a thorium foil source. This data is shown in figure 1 along with collection efficiency curves for similar arrays with p-type reflection barrier and no barrier. Comparison of collection efficiencies for the three cases indicates the following relative soft error rates (for a stored charge of 0.5 million holes):

N-well over p-substrate	SER = 1x
P-well over p-substrate	SER = 12x
No well (p-substrate)	SER = 330x

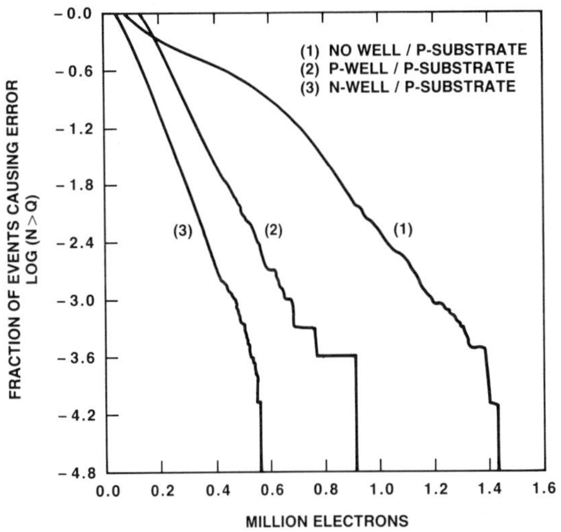

FIGURE 1. CHARGE COLLECTION EFFICIENCY

Accelerated testing with thorium foil on a 64K CMOS DRAM confirms the excellent soft error protection provided by the n-well (< 10 FITs for 75ff cell at 1 μS cycle time with VCC = 5 volts). Detailed analysis indicates that greater than 95 percent of soft errors resulted from the unprotected n+ junction on the bit line. CMOS DRAM's normally employ complimentary bit line sensing (described in a later section) and require a small amount of n+ junction area attached to the bit lines for precharge and restore operations. This area becomes the dominant charge collector.

CH1935-6/83/0000/0379$01.00 © 1983 IEEE

Extended Refresh and Low Power Operation

One of the attractive features of an n-well CMOS technology is the superior refresh characteristics over its n-channel counter part. This extended refresh feature coupled with the inherent low power operation offered by CMOS can not only reduce system overhead due to refresh cycles but also make battery back-up operation possible.

Refresh characteristics of NMOS DRAMs are usually limited by thermally generated electron hole pairs from defect sites within a diffusion length of the memory cells. The minority carriers are collected by the memory cells degrading their refresh times. CMOS has numerous advantages which dramatically reduce this leakage current. Among these are: (1) The thermally generated currents in the substrate are collected by the reversed biased well-substrate junction and thus have no influence on the memory cell. (2) The thermally generated minority carriers collected by the memory cells are generated in the well whose depth is substantially less than the diffusion length. Also the thermally generated minority carriers in the well are shared by the memory cell junction depletion region and the well-substrate junction depletion region. (3) The doping gradient of the well results in a retarding electric field that reduces the collection of the thermally generated minority carriers by the memory cells.

Internal Capacitive Supply Decoupling

Low defect process lines and implementation of redundancy has allowed larger and larger chips to be produced. Speed requirements of dynamic RAMs has also increased for new generations. This combination has increased the capacitance which must be moved each cycle and increases the speed with which it moves creating power supply noise. Damping these effects is usually accomplished by increasing the decoupling capacitances and reducing the inductance and resistance from the capacitor to the chip. Dual bonds, twin supply lead packages and hybrid capacitors bonded under the die have all been considered to reduce the problem. CMOS cannot remove the high current spikes required to move all the bit lines each time a word line change is required. The single largest difference is in the on chip decoupling of CMOS.

For the same die area, CMOS DRAMs have significantly more VCC to VSS capacitance than their NMOS counterparts. This is due to the capacitance of the wells to substrate. This is especially true of CMOS devices that use a well in the memory array. Although EPI is normally considered in CMOS as a method to reduce SCR latch up, it can also aid in increasing the well to substrate capacitance. EPI will enhance this capacitance as the well depth approaches the EPI thickness. As reverse bias is applied, the well depletion region extends into the highly doped region of the EPI next to the high concentration substrate. At normal supply voltages of 5.0 volts the increase can be 1.5 to 2 times. In addition to the capacitance increase, the low substrate resistance improves the frequency response of the well capacitance.

CIRCUIT TRADEOFFS

Clock Generators

CMOS shows its greatest advantage in basic clock generation. N-channel clock generators when built for minimum current draw, have inherent time delays associated with the precharge and release of the boot strap capacitance. Figure 2a shows a typical n-channel clock generator. For HMOS III with effective channel lengths of 1.2μm this delay is about 1.3 to 2 nS. When delays less than this are required, the stage must be charged with an earlier stage requiring more current. Although other clock generators types attempt to reduce this problem they are typically slightly slower, require more chip area and require higher node voltages adding restrictions to the process and layout [7,8].

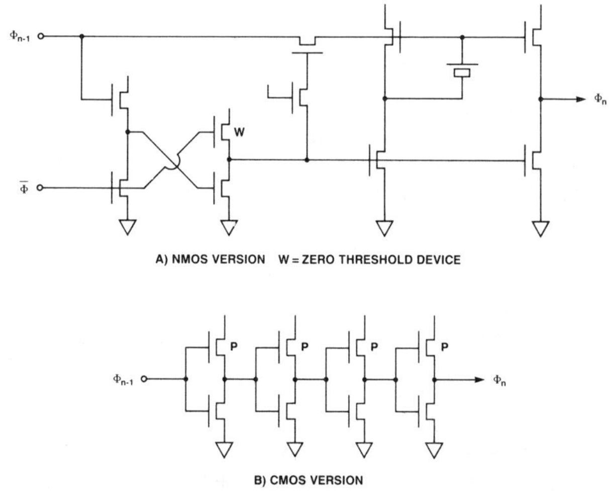

FIGURE 2. CLOCK GENERATORS

CMOS clocks are simply strings of inverters arranged for optimum speed to a given load. The number of stages and device sizes are chosen based on the load driven. Figure 2b shows a comparable circuit to the n-channel clock of the same figure. Two inverter clock delays can be generated and can be as small as 0.5nS.

Table 1 shows a comparison of basic NMOS and CMOS clocks. The comparison is made using identical effective channel lengths of 1.2μm, 250Å gate oxide and a load of 4.0pF. In addition to the higher efficiency of CMOS in driving capacitive loads, CMOS also gives higher current drive per input capacitance compared to n-channel clock generation. This is extremely beneficial in the TTL to MCS level conversion and amplification to the first large clocks. Because of the high CMOS gain, small inverters may be used for input stages then amplified on chip for only a small cost of delay. This allows for low input capacitance and small DC current draw of the input stage with TTL levels applied.

NMOS VS. CMOS CLOCK COMPARISON
(4pF capacitive load and 5.0 volt supply)

	CMOS	NMOS
Rise time	1.4nS	1.7nS
Rise delay	1.8nS	1.3nS
Fall time	0.8nS	2.0nS
Fall delay	1.7nS	2.8nS
Supply charge	37.7pC	61.6pC
Input charge	.15pC	10.5pC
Output charge supplied	20.0pC	20.0pC
Total/output charge ratio	1.89	3.08
Input/output charge ratio	0.008	0.525

All delays are 50% points and all rise and falls are 10% to 90%. All charge in coulombs per cycle.

TABLE 1

Simplified Address Buffer

Static input buffers are easily implemented in CMOS. A comparison of the number of devices in a CMOS static address buffer vs. an n-channel address buffer has been reported [1]. These circuits are not only smaller but remove the clock timing margins required of most NMOS designs thus yielding smaller delay.

Select vs. Deselect Type Decoders

Internal current reducing techniques are also easily implemented with CMOS. In n-channel DRAMs the second largest current problem (next to the charge and discharge of the bit lines) is the charge and discharge of the row decoders. N-channel designs typically charge all the decoders including the decoder pass gates to the VCC supply. Then the address lines become valid discharging all but one decoder. Figure 3a shows a sample n-channel decoder. Figure 3b shows a possible select type decoder in which only the selecting decoder is required to move.

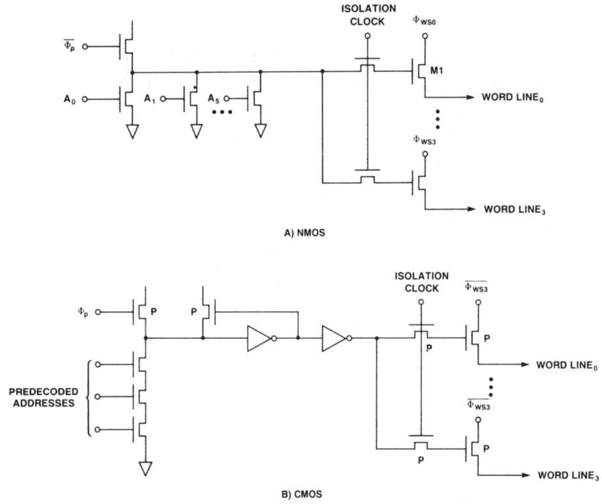

FIGURE 3. ROW DECODERS

Note that predecoded addresses may be used in CMOS limiting the number of stacked devices to 3 for a one of 256 decode. For a 256K with two row decoders, there can be as many as 4096 devices the size of the word line drive devices (M1). Not having to charge and discharge these devices each cycle represents a significant current savings as well as reduction in current spikes.

Complimentary Bit Line Sensing

Active bit line restore functions in n-channel use from six to eight devices and suffer from their non-cross coupled nature. Thus it is necessary to sense and amplify the bit line signal significantly before restore can occur. A two device cross coupled restore is possible in CMOS (figure 4 contains both the sensing devices and the restore devices). The bitlines need only be amplified sufficiently to be immune to the imbalances of the restore latch before the opposite polarity latch turns on. The single disadvantage of the dual polarity cross couple is that some of the bit line junction must be outside the alpha protective well. In some instances this exposure will dominate the soft error rate as explained earlier.

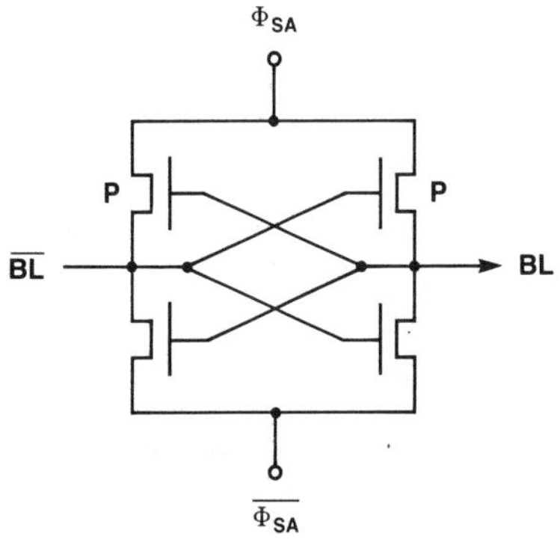

FIGURE 4. CMOS SENSE AMPLIFIER AND RESTORE

Static Column Operation

The heart of the back end of the dynamic RAM is the column decoder. The clocking and power requirements of the decoder do the most to determine the cycle time and power consumption of static mode operation. To implement static access from stable external addresses the combination of column decoder and address buffer have to perform together to appear static to the user. This can be accomplished by implementing purely static circuitry or with clock assisted circuitry. The purely static implementation would have static address buffers and static column decoders, the clocked version would have static address buffers, from which an address change pulse would be derived to precharge the column decoder for the next selection [8]. The implementation of the purely static approach is the most direct in either CMOS or NMOS.

In the purely static NMOS version the address buffers and the selected bank of column decoders would all be consuming D.C. power. This implies that all column decoders except the selected one are conducting current, as well as the current dissipation of the address buffers. Additionally the normal n-channel DRAM process does not require depletion or zero threshold transistors. This additional process complexity makes the process more complex without the advantages of CMOS.

To implement a NMOS clocked version the address buffers would be used to derive an address change pulse. This would precharge the column decoder and reclock the selection of the I/O pass transistors. This technique removes the need to dissipate D.C. current in the column decoder. This approach should have lower power dissipation than the purely static n-channel version especially at the 256K level where more column decoding is required.

CMOS technology permits the best possible approach to this problem. The elegant solution alone justifies the use of CMOS for advanced high density memories. CMOS permits the use of static buffers which only dissipate power while transitioning, and static decoding which uses power only to charge the selected decoder and discharge the previously selected decoder.

Two very similar 64K DRAMs are previously reported. One is NMOS [9] and the other is CMOS [1], both implement the static column operation, both utilize purely static circuitry, a summary of performance follows:

	CMOS	NMOS
Static mode current Measured at 60nS cycle	20mA	58mA
Minimum column access time	35nS	35nS
Speed-power product	3.5μJ	10.2μJ

The above comparison clearly shows the power advantage of CMOS as well as its comparable speed.

CONCLUSIONS

CMOS technology can be used to solve some of the more significant problems of soft error rate, high power and chip complexity in DRAM designs.

Additional benefits emerge including the capability of battery backup and extended refresh operation, low active power in peripheral circuits and extremely fast column data rates. These factors should make CMOS a dominate force in future DRAM development.

ACKNOWLEDGEMENTS

The authors would like to thank Amr Mohsen and Ronald Chwang for their comments and contributions.

REFERENCES

[1] R. Chwang, et. al., "A 70nS High Density CMOS DRAM," ISSCC Digest of Technical Papers, p. 56, 57; Feb. 1983.

[2] K. Shimohigashi, et. al., "An n-Well CMOS Dynamic RAM," IEEE Trans. Electron Devices, Vol ED-29, No. 4, pp. 714-718, April 1982.

[3] T.C. May and M.H. Woods, "Alpha-Particle-Induced Soft Errors in Dynamic Memories," IEEE Trans. Electron Devices, Vol ED-26, No. 1, pp. 2-9, Jan. 1979.

[4] G.A. Sai-Halasz, M.R. Wordeman, and R.H. Dennard, "Alpha-Particle-Induced Soft Error Rate in VLSI Circuits," IEEE Trans. Electron Devices, Vol ED-29, No. 4, pp. 725-731, April 1982.

[5] M.R. Wordeman, R.H. Dennard, and G.A. Sai-Halasz, "A Buried n-grid for Protection Against Radiation Induced Charge Collection in Electronic Circuits," IEDM Digest, pp. 40-43, 1981.

[6] S.W. Fu, A.M. Mohsen, T.C. May, "Alpha Particle Induced Charge Collection Measurements and the Effectiveness of Reflecting Barriers on VLSI Memories," IEDM Digest, pp. 632-635, 1982.

[7] John Y. Chan, et. al., "A 100nS 5V Only 64Kx1 MOS Dynamic RAM," IEEE Journal of Solid State Circuits, Vol. SC-15, No. 5, pp. 839-846, October 1980.

[8] S. Sheffield Eaton, et. al., "Circuit Advances Propel 64K RAM Across the 100-nS Barrier," ELECTRONICS, pp. 132-136, 24 March 1982.

[9] Kim C. Hardy and Rahul Sud, "A Fault-Tolerant 30nS/375mW 16Kx1 NMOS Static RAM," IEEE Journal of Solid-State Circuits, Vol. SC-16, No. 5, pp. 435-443, October 1981.

[10] Fumio Baba, et. al., "A 35nS 64K Static Column DRAM," ISSCC Digest of Technical Papers, pp. 64-65, 23 Feb. 1983.

THE EFFECT OF CELL DESIGN ON CMOS GATE ARRAYS

Michael P. Gagliardi
Project Engineer, Design Engineering
General Electric Company
Microelectronics Center
Research Triangle Park, North Carolina 27709

ABSTRACT

A two man-year study was undertaken to determine the optimal gate array family to complement the routing software and three micron (two level metal) CMOS process available at General Electric-Intersil. The duration of the study was approximately four months, requiring as many as six engineers and two layout specialists. The study considered internal cell structures, I/O capabilities, process technology, buss architectures, and most importantly, routability issues. This paper concerns itself with internal cell structures and routability only.

INTRODUCTION

In designing Gate Array masters for customization, there exist two extremes in the design process. The first extreme would be to limit the performance and density of the array and insure 100% auto-routability. The complement to this approach would be to design the array as one would a custom circuit, purely for performance and transistor density. Neither of these approaches provide the competitive edge necessary for today's growing semi-custom market. The first, although a goal for any array family, would not permit the gate counts felt necessary to handle the market needs. The chips would be excessively large, and therefore, expensive. The second would permit more cost effective circuits in terms of die size, assuming the circuits could be built. On the other hand, there would be little or no automation to the design process. This would result in more engineering support needed, and a longer design cycle leading to more expensive chips.

The intent of this paper is to present a design methodology undertaken at the General Electric Microelectronics Center, offering both auto-routability as well as high gate counts for reasonable die sizes.

CELL SELECTION FACTORS

As a first step in any study, it is essential to define a means of measurement based on the goals of the study. In this study, the measurement criteria were separated into the following five categories:

1. The number of two-input gates versus die size (GD/W)
2. Combinational versus sequential logic measured in cell area
3. Performance
4. Layout rule dependency versus die size (GD/W)
5. Routability of the array

The first of these issues relates the gate count of a given array to the projected good die per wafer (GD/W). This data accounts for the different amounts of area consumed by the various cell types. The cells requiring more area will demand larger die sizes for equivalent gate counts, resulting in lower yields, fewer number of die sites, and therefore, higher die cost.

The ability to handle flip/flop intensive logic is also important, in that it provides a way to rank the cells by the type of logic most commonly used. Although some of the cells are fairly straight forward in design, many were designed to handle specific logic functions. For instance, one cell type may have the ability to be easily configured into a D-type flip/flop, while causing channel congestion when configured as a combinational logic block. Therefore, a cell that would permit a small cell count and chip area when used to implement a flip/flop intensive application would result in a larger cell count and chip area when used to implement a combinational intensive application.

Inversely proportional to the projected GD/W is the performance rating of the cell types. In general, if the cell size is laid out for minimum area, the transistor widths become limited. In this study, all transistors were sized at their maximum limitation based on metal interconnect required and availability of routing channels over the active area regions. Another point considered was that although larger device sizes lead to higher performance ratings, they require more cell area. Larger cell area leads to longer interconnect paths for the routing metal, and hence, larger loading on those devices which leads to degradation in performance.

Another issue important to the choice of the cell was dependence on design rules. It became evident that designing a cell to excel in one of the previously mentioned categories can be disasterous should a determination to increase routing channel grids be made. That is, a particular cell can be laid out to a certain X and Y grid that makes it very desirable, but if the grid is increased for

CH1935-6/83/0000/0383$01.00 © 1983 IEEE

other reasons, it may be much less attractive. This is primarily due to an inefficient utilization of area on a larger grid scale.

Finally, the issue of routability plays a major role in the cell selection. The various cell types utilize the two layers of metal quite differently based on their respective layouts. Many of the cells can and do cause routing congestion because of the desire to keep down the number of contact locations required for macro configurations (metal defining logic blocks). Data based on macro layouts provided insight on the number of open vertical and horizontal routing channels, as well as the number of terminal locations for each cell type. This data was then used to predict the overall number of routing channels required for the given array sizes and thus, die size.

REVIEW OF CELL TYPES

Fifteen unique cell types were studied, which varied in transistor count from 4 to 16 transistors per cell. The layouts differed in the number of N-channel and P-channel devices with common gates, as well as the locations of the common gate devices. Some of the cells left all the gates uncommitted (Figure 1) while others only shared gates between certain devices (Figures 2 & 3). Shared source and drain regions were varied from combinations of two transistors to the "sea of gates" approach (Figure 1), where all the devices in a column are repeated with shared source and drains. All the cell types were chosen for their uniqueness from a much larger sample. The General Electric-Intersil cell (Figure 4) contains three common gate pairs and separated source/drain areas.

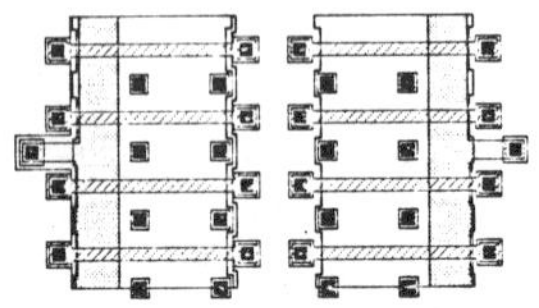

Figure 1: Sea of Gates

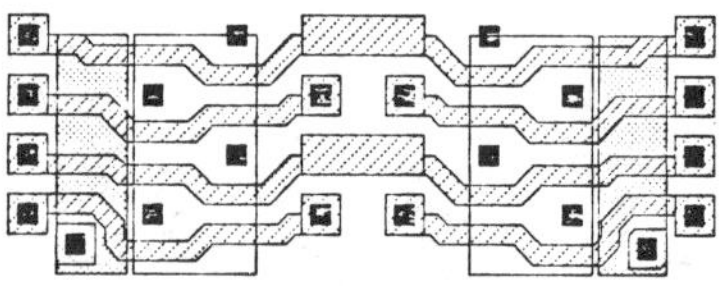

Figure 2: Common and Shared Gates

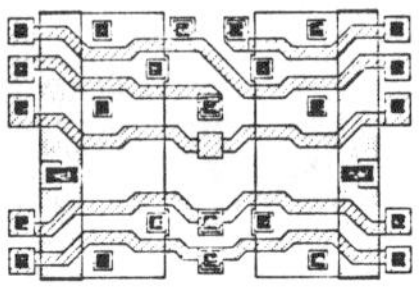

Figure 3: Separated Source/Drains

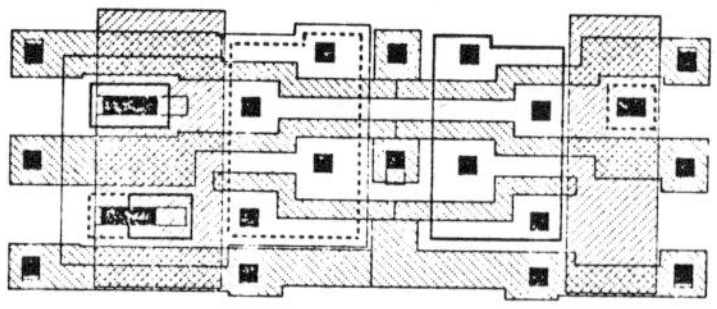

Figure 4: MEC 3u Gate Array Cell

LAYOUT OF CELL TYPES

Once the number of cell types was limited to 15, it was necessary to layout the cells utilizing the three micron layout rules. After several iterations of each cell, the data for chip areas was obtained. This avoided any chance for error due to guesswork on horizontal and vertical grid pitch. Figure 5 illustrates the layout of a D-type flip/flop with reset.

Data was also taken to determine cell areas and rankings for cells assuming larger grid scales. For example, a particular cell may be laid out with staggered gates allowing an eight micron horizontal grid. If in fact, processing constraints do not permit a grid of less than nine microns, then the cell would be rated at the larger grid. This type of analysis varied the rankings of the cells and allowed for a comparison based on process tolerances and reliability issues.

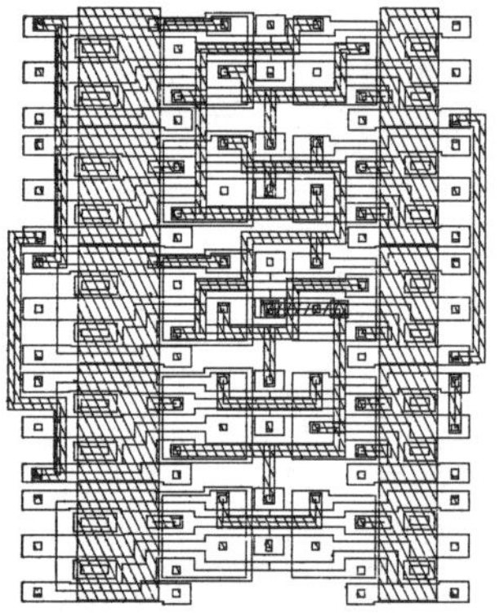

Figure 5: D Flip/Flop with Reset

MACRO DATA BASE

Each of the cell types, with their unique gate and source/drain combinations, are more suited to particular types of logic functions. It is, therefore, necessary to determine the logic types most commonly used in the portion of the marketplace for which the gate array is intended. Once the primary logic functions are determined, then data can be obtained to support those cell types best fitting that market.

For the study, macro counts were taken from several circuits built on previous gate arrays. The macros were then separated into random logic and flip/flop intensive logic. For the flip/flop intensive logic, the macros consisted of delay type elements with combinational logic included to realize the desired flip/flop. The random logic was divided into the specific types of logic functions, and a sample representing at least 75% of those macro types was used in the study.

Next, a weighting system was developed for the combinational logic elements. This was done to provide those elements most often used, with a weight proportional to the frequency of their use. The weight was then used in equation (1) to determine the average area utilized for combinational logic (i.e., an exclusive OR and a 1x inverter require different amounts of cell area, but are not used in the same quantities).

$$Y = \Sigma\, (W_i * [\#\ CELLS]_i * CELL\ AREA) \quad (1)$$

where: Y = Average Combinational Area on a Chip; W_i = % of Times Utilized in Survey; $[\#\ CELLS]_i$ = # of Cells Taken up by Macro; CELL AREA = Used to Determine Chip Area

ROUTING CHANNEL DETERMINATION

The number of routing channels necessary for auto-routable cells is dependent upon the types of logic (% of combinational versus sequential) to be put on the array as well as the physical aspects of the cells. For the study, software was written by the Advanced Methodologies Group of the Microelectronics Center, to account for both of these dependencies. The program was developed around equation 2, which has been shown to provide accurate channel counts for the routing software developed at General Electric-Intersil.

$$K = (H * V) / (G ** R) \quad (2)$$

where; H = number of horizontal tracks
V = number of vertical tracks
G = number of three input gates
R = constant; $1 < R < 1.5$ for the study, R = 1.01 and is used to influence the added routing space as a function of G
K = route constant derived empirically for each cell type

H and V were determined from equation (2) using a K factor derived by members of the CAD group. The initial K factors were determined largely through extensive knowledge of the routing software and applied to the inter- as well as intra-cellular tracks. A similar equation relating the H/V ratio to the number of terminals on the vertical and horizontal sides of the cell type and the free channels for the macros was utilized. With this equation, an optimum K factor was selected and was then verified through actual routing of several chips.

CELL PERFORMANCE

After the intercellular tracks (channels) were defined for each cell type, the transistors were again laid out to maximize device width. For cells requiring more routing tracks, the device widths were increased proportionally. The widths for the cell types in the study ranged from 20.0 microns to 51.5 microns.

Although wider device widths typically yield better performance, they yield larger die sizes as well as increased loading for similar fanouts. With a larger array, longer interconnect paths are expected, so it was necessary to define a means of measuring performance based on die sizes. Figure 6 illustrates a logical representation of the circuit used to analyze the cell types. Figure 7 illustrates the method for relating the interconnect capacitance to the projected die size. By simulating each of the circuits, data was taken for pair delays and propagational delays under loaded and unloaded conditions.

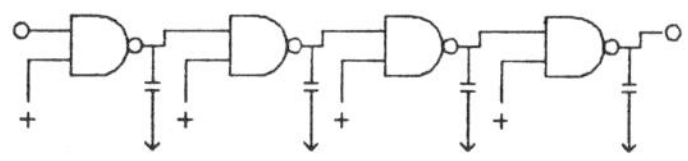

Figure 6: Logical Representation

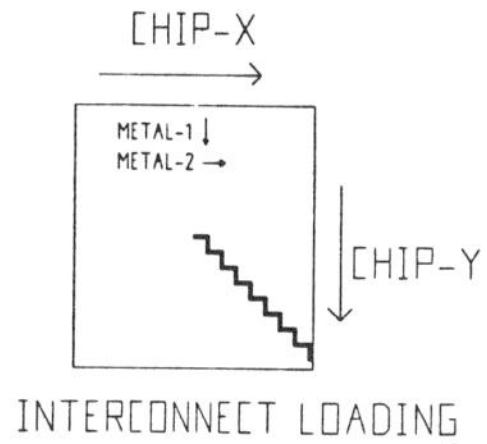

Figure 7: Interconnect Capacitance

CONCLUSION

A design methodology for internal gate array cells is proposed. What are believed to be the dominant issues of cell design are listed and a means of collecting relevant data is explained. This methodology has led to the introduction of a new generation of gate arrays at General Electric-Intersil. The new family of arrays were developed in 3 micron, 2 layer metal, CMOS technology. The arrays range in size from 400 gates to approximately 15,000 gates and contain pad counts ranging from 40 to 200 pads.

ACKNOWLEDGEMENTS

The author would like to thank M. Bates, M. Keith, J. Watters, and W. Wolf for the long hours and great efforts they put forth to help develop this methodology and carry it to conclusion. The author would also like to thank D. Hightower and T. Weis for the software and CAD support for this effort, as well as C. Gregory and T. Jochum for their technical assistance and good judgement on many of the issues.

SYSTEM IMPLICATIONS OF CHMOS DYNAMIC RAMS

Joseph P. Altnether

Intel Corporation, Hillsboro, Oregon

ABSTRACT

Building a DRAM on CHMOS technology permits the incorporation of a new architecture. Because of the lower power, static column circuits have been used to increase the RAM performance. The technology and the architecture combine to provide four key features at the system level: high data bandwidth, usable speed, low power and high reliability. This paper will explain the high bandwidth operation, and show the advantages at the system level. Usable speed will be demonstrated in an iAPX 286 system design, and comparisons of low power and high reliability will show the improved system level performance.

INTRODUCTION

The applications of Dynamic RAMs (DRAMs) are changing. Widespread use of microprocessors has increased the variety of applications. With these new applications have come new memory requirements. Graphics, for example, require high data bandwidth, while portable systems need low power. These features are not optimized in current NMOS DRAMs. As a result, solutions to the application problems until now, have ranged from non-existant to CMOS and Fast Static RAMs. CHMOS DRAM technology however, allows new features not previously practical with NMOS. A DRAM built with CHMOS technology, provides more than only low power at the system level. The operating parameters of a CHMOS DRAM are compared to a second generation NMOS DRAM in Table 1. Comparisons of CHMOS DRAM parameters with those of the NMOS DRAM show a two-fold increase in performance. In addition to faster access, of significance is the RAS to CAS delay or t_{RCD} window. This is particularly useful in high speed system design. Active power consumption is reduced by a factor of two, while standby power consumption is reduced by a factor of 50 at CMOS levels. Additionally, the soft error rate has been decreased by two orders of magnitude.

Because of the lower power, new architectures can be incorporated that were not feasible with NMOS. NMOS DRAMs use dynamic or clocked circuits to reduce power consumption. CHMOS DRAMs with inherent low power, however, can incorporate static circuits. An example is in the column decode circuitry. Here, static circuits reduce the number

Table 1. Selected Parameters

PARAMETER	NMOS DRAM	CHMOS DRAM
Access	150 ns	100 ns
Cycle	260 ns	160 ns
t_{RP}	100 ns	50 ns
Active Current	55 mA@ 260 ns	37 mA@ 160 ns 22 mA@ 260 ns
Standby Current	5 mA	2 mA@ TTL levels 100 uA@ CMOS levels
SER	0.1%/1k-hr	0.001%/1k-hr
t_{RCD}	30-65 ns	25-80 ns
t_{CAC}	85 ns	20 ns
Page Mode Cycle	125 ns	65 ns

of clocks by a factor of three and drastically reduces the CAS precharge time. Because the column address decoder circuits are static, access time on a row of data is equivalent to a Static RAM. This permits not only high speed serial data streaming, but also easier system design for high speed single access. The CHMOS technology and the Static Column Circuits offer four advantages at the system level: high data bandwidth, usable speed, low power and high reliability.

HIGH BANDWIDTH

Two distinct types of static column operation are possible: Static Column Mode and Ripplemode™. As shown in Figure 1, Static Column Mode has standard DRAM row address and RAS operation. Access begins exactly like an NMOS DRAM with row addresses latched into the DRAM by RAS. The 256 bits on the selected row are read into the internal sense amplifiers. During read cycles, CAS serves only as an Output Enable. Typically, access from column addresses is 30 ns. and from CAS, 10 ns. Column addresses are not latched by CAS (cause and effect). Access is from valid column

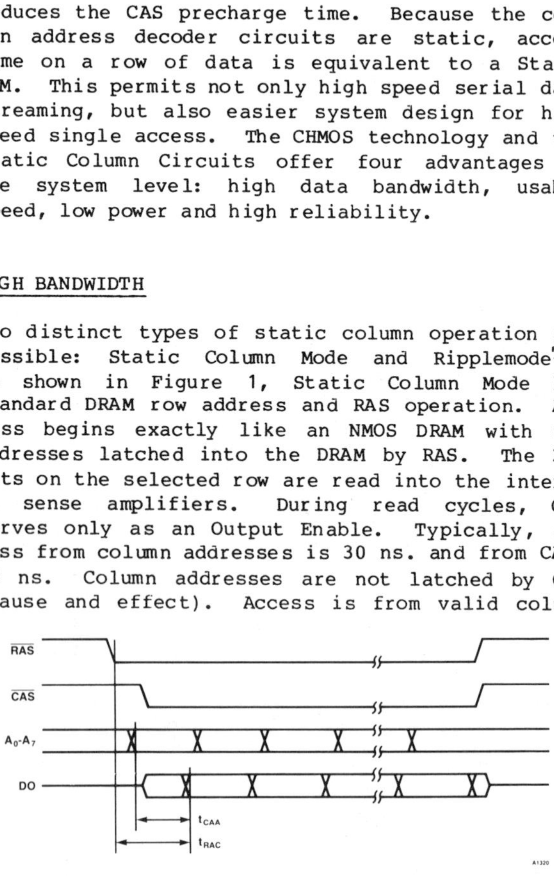

Figure 1. CHMOS DRAM Static Column

CH1935-6/83/0000/0386$01.00 © 1983 IEEE

addresses, not CAS. All 256 bits in the row can be accessed merely by changing addresses. In practice, it is possible to permanently enable CAS, or ground it. Cycle and access time after the first access is typically 50 ns. Continuous high speed data streaming is slightly less than this because the first access is from RAS at 100 ns and row precharge time, t_{RP} of 50 ns must be included to calculate the continuous data rate.

$$t_{CY} = \frac{t_{RAC} + 255*(t_{CAA}) + t_{RP} + 2t_T}{256} \quad \text{(EQ. 1)}$$

where t_{CY} = Maximum continuous cycle time
t_{RAC} = Row address access
t_{CAA} = Column address access
t_{RP} = RAS precharge time
t_T = Rise/Fall time

Substituting:

$$t_{CY} = \frac{100 + 255*65 + 50 + 10}{256} \text{ ns}$$

$$t_{CY} = 65.37 \text{ ns}$$

Data can be continuously streamed from the CHMOS DRAM at 15.29 MHz. Performance of an NMOS DRAM is considerably less. On a current second generation NMOS DRAM, for example, t_{RAC} is 150 ns, t_{CAA} is 125 ns and t_{RP} is 120 ns. For the NMOS DRAM, then:

$$t_{CY} = \frac{150 + 255*(125) + 120 + 10}{256} \text{ ns}$$

t_{CY} = 125.52 ns or 7.97 MHz

The CHMOS DRAM has a twofold increase in performance over the NMOS DRAM.

While operating at the same device speeds, Ripplemode operation differs slightly. It also has static column decoders but column addresses enter the DRAM via flow-through latches. As shown in Figure 2, CAS latches addresses and serves at an Output Enable. When CAS is inactive (high), the address latch allows addresses to flow through to the column decoders. Activating CAS latches the addresses and enables the D_{OUT}. Column access begins as soon as CAS is inactive and CAS precharge time is included in the address access time. Latching the addresses at the device allows system addresses to change and set up for the next cycle. This address pipelining provides an additional 15% increase in the system data rate.

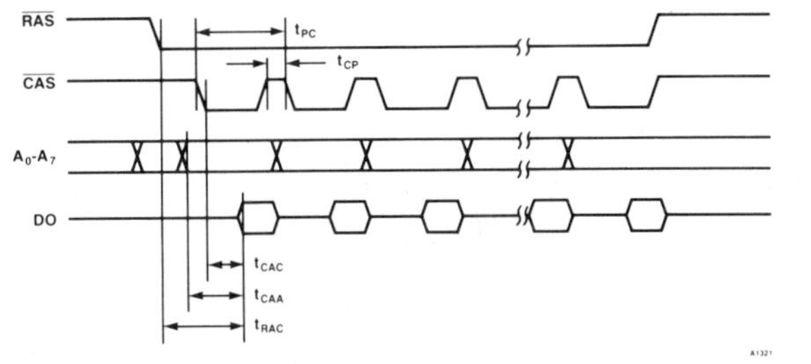

Figure 2. CHMOS DRAM Ripplemode Operation

USABLE SPEED

Static column decoders provide system level advantages, not only in data streaming, but also in driving a single access as well. One of the major difficulties in DRAM system design has been meeting the minimum cycle time. DRAM cycle timings are shown in Figure 3. At the system level, Row addresses must be valid prior to RAS and held for t_{RAH}. Next, addresses must be switched to column addresses and be valid prior to CAS activation minus any skew associated with CAS. The critical parameter is the RAS to CAS delay, t_{RCD}. If the system can meet the minimum t_{RCD}, access is from RAS or t_{RAC}. Violating t_{RCD} increases the access time to t_{RCD} (actual) plus t_{CAC}. The t_{RCD} term is violated if the system cannot switch from row to column addresses quick enough. Thus, the RAS to CAS delay, t_{RCD}, has contradictory requirements. It must be as short as possible for fast access, yet as wide as possible for ease of system design.

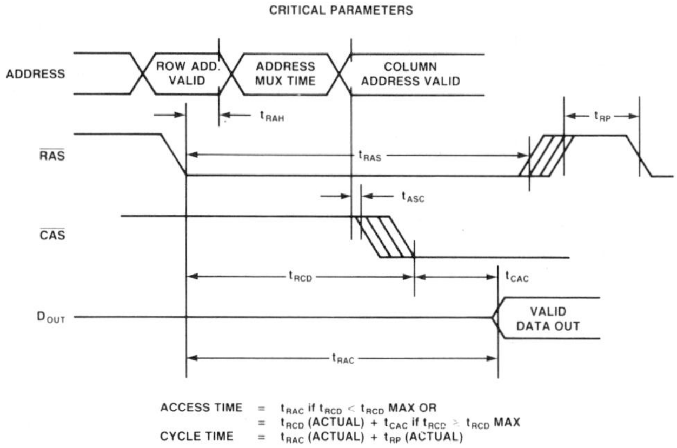

Figure 3. NMOS DRAM Timings

The CHMOS DRAM eliminates this problem by accessing from valid addresses rather than CAS, as shown in Figure 4. This removes CAS and its associated timings from the critical timing path. With t_{CAC} of 15 ns and t_{RAC} of 100 ns, the t_{RCD} window is 85 ns wide. This is a 50% improvement over NMOS DRAMs. Eliminating CAS from the critical timing path and relaxing t_{RCD} simplifies designing high speed systems. Systems that could previously operate only with high speed static RAMs can now obtain the DRAM benefits of low power, high density and low cost as well. Systems like the iAPX 286, shown in Figure 5, can operate

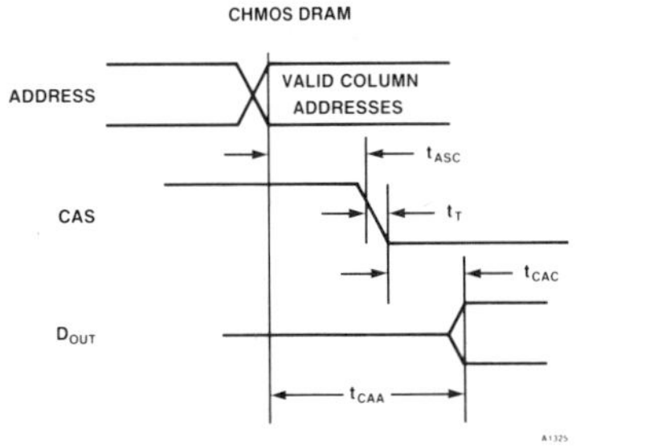

Figure 4. CHMOS DRAM Access Timing

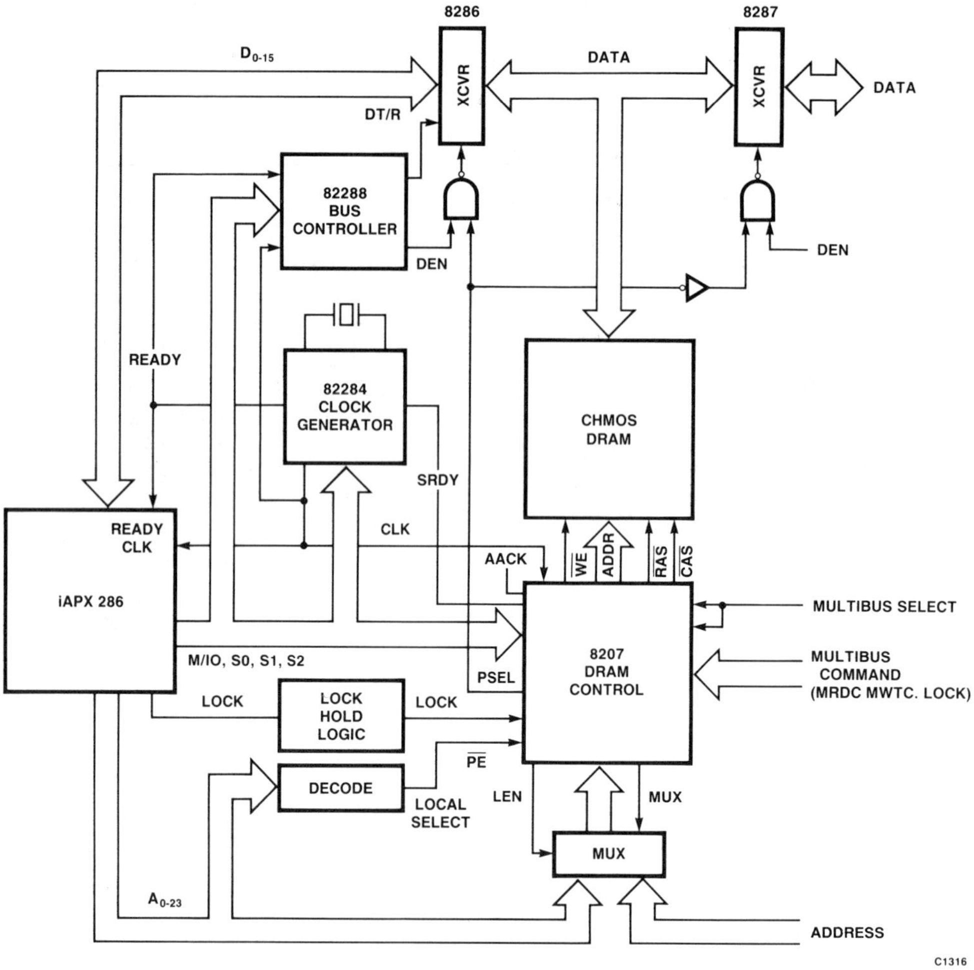

Figure 5. iAPX 286 System with CHMOS DRAMs

at higher speeds. In this system , the critical timing parameter is t_{CAC}. The fast t_{CAC} of the CHMOS DRAM allows this system to operate at 8 MHz without WAIT states. Using any other DRAM would force the injection of WAIT states during memory cycles with a corresponding reduction in system performance. The CHMOS DRAM not only simplifies system design but also improves performance.

LOW POWER

The two previously discussed advantages are a function of both the technology and the architecture. The technology in itself offers several advantages at the system level: low power and high reliability. A DRAM memory system has three power components: active, standby, and refresh. The memory system power is expressed by:

$$P = V(M*I_A + K*I_S + N*I_R) \quad \text{(EQ. 2)}$$

where:

- P = Memory power
- V = Voltage (5.5 volts, worst case)
- M = Number of active devices
- I_A = Active current
- K = Number of devices in standby
- I_S = Standby current
- N = Total number of devices
- I_R = Refresh current

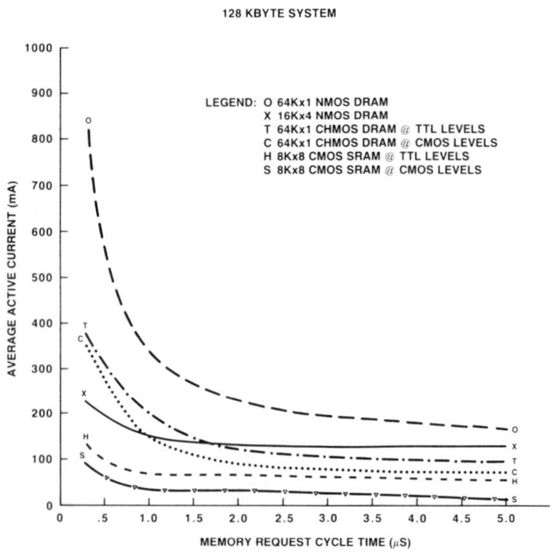

Figure 6. System Current as a Function of Duty Cycle

As shown in Table 1, the CHMOS DRAM active and standby currents are reduced by a factor of two. This reduces the system power by 50%. System power can be further reduced by driving RAS to a CMOS level (V_{CC} - .5v). In this case, the standby current is reduced to 100 uA. This 5000% reduction is significant because in large systems, the majority of the memory devices are in standby. When memory duty cycle is considered as in Figure 6, system power is reduced by an additional factor of two or three. The low operating power is useful in portable operating systems.

Low standby power is derived from two elements, the inherent low power of CHMOS and the capability to extend the refresh period from 4 to 64 ms. The refresh period is a function of leakage current and stored charge. The CHMOS DRAM has extremely low leakage and high storage capacitance, which relates to stored charge. Storage capacitance of the CHMOS DRAM is .125 pF as compared to .060 pF to .080 pF of NMOS DRAMS. The results of this are shown in Figure 7. In this graph, average standby current is plotted against the refresh period. NMOS DRAMs curves end at 4 ms because NMOS DRAMs do not specify refresh periods greater than 4 ms.

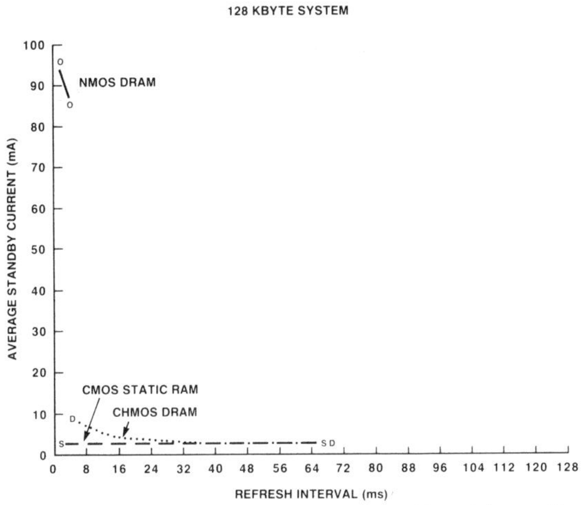

Figure 7. Standby Current as a Function of Refresh Period

Assuming the refresh period could be increased, the average standby current would asymptotically approach the device standby current. In the case of 128K byte system, 16 NMOS devices drawing 5 mA each, would consume 80 mA in the limit. Compared to the standard 4 ms refresh period, the power saving is less than 10%. CHMOS DRAMs, on the other hand, approach 1.6 mA at the same limit, providing a six-fold decrease in power consumption. CHMOS DRAMS have an active power reduction of 100% and a standby power reduction of 600%. This extremely low standby power is useful in battery back-up applications.

HIGH RELIABILITY

System reliability is quantified as Mean-Time-Between-Failures (MTBF). As shown in Equation 3, it is an inverse function of the number of devices used in the system and the device failure rate.

$$MTBF = \frac{1}{fN} \qquad \text{(EQ. 3)}$$

where:

f = Device failure rate (%/1k-hr)

N = Number of devices used

Two events have occurred to reduce the system reliability. First, in spite of the increased density of RAMs, more RAMs are employed per system. Secondly, the higher density of DRAMs has reduced the storage area for individual data bits. Along with this size reduction is a corresponding reduction in stored charge. At about the 16K density, the difference between a stored "1" and a stored "0" is about one million electrons. The stored charge is small enough to be affected by ionizing radiation in the silicon. Alpha particles, the most common type of radiation, are a result of the residual radioactivity in the package. As they travel through the silicon, they generate hole-electron pairs. Typically, over one million hole-electron pairs are generated. If the hole-electron pairs are created near a node sensitive to one million electrons, data could be changed.

Data would remain incorrect until re- written by the system, hence the term "soft errors." While soft errors occur five times more often than hard errors, their effect on the system reliability is minimized by error correction cir- cuits (ECC). ECC does increase the system reli- ability, but for a price. A 16-bit word requires six additional ECC bits, increasing the memory size by 37%. This larger memory requires more P.C. board area, larger power supplies and more cooling.

More importantly, the system performance is reduced. ECC adds 50 ns to the memory access time. A 100 ns RAM is a 150 ns RAM at the system level, adding 50% to the access time. Cycle time is significantly increased, especially when writing one byte into a two-byte word. First, the word must be fetched and corrected. Next, the new byte is merged into the word, ECC bits are generated and the entire 22 bits written into the memory. Adding to this are the cycle precharge time and any timing skews. Consequently, a RAM with a 200 ns cycle time cannot operate faster than 335 ns in this environment. This is a 68% increase in cycle time.

The better alternative is to reduce the device error rate and possibly eliminate ECC. The CHMOS DRAM has a soft error rate two orders of magnitude less than equivalent NMOS DRAMs. The system implications of this are shown in Figure 8, which plots system size versus system reliability. The NMOS curves were generated using measured .002%/1k-hr HER and .08%/1k-hr SER, while the CHMOS curves use the same .002%/1k-hr HER but a .001%/1k-hr SER. These curves compare system reliability of NMOS and CHMOS DRAMs with and without ECC. The MTBF of an uncorrected 256K byte NMOS system is 10,000 hours or one year. Comparatively, a similar size system using CHMOS DRAMs has an MTBF of 1 million hours or 114 years. The MTBF of the non-corrected CHMOS system compares

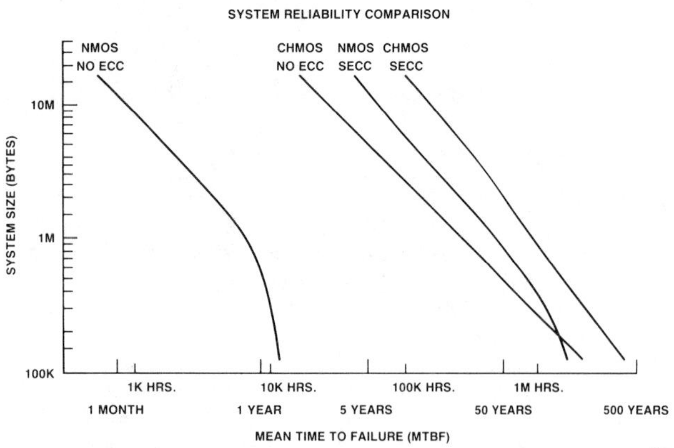

Figure 8. System MTBF Versus System Size

favorably with the NMOS system with ECC. It is significant to point out that below 256K bytes the non-corrected CHMOS system is more reliable than the corrected NMOS system! In this region, better reliability is achieved using CHMOS DRAMs without ECC and system speed is improved due to the elimination of ECC.

CONCLUSIONS

By combining new architectures with CHMOS technology, DRAM performance is dramatically improved. The CHMOS DRAM can meet a variety of application requirements unmatched by NMOS DRAMs. These range from high data bandwidth for graphics and telecommunications, to low power for portable and remote systems, to usable speed for high speed applications and to high reliability for sensitive applications. Applications requiring any or all of these features can benefit from the performance of CHMOS DRAMs.

REFERENCES

1. R. Chwang, et al, "A 70 ns High Density CMOS DRAM," ISSCC Digest of Technical Papers, p. 56-57, February, 1983

2. T. C. May and H. C. Woods, "Alpha-Particle-Induced Soft Errors in Dynamic RAMs," IEEE Trans. Electron Devices, Vol ED-26, No 1, pp. 2-9, January, 1979

3. W. Righter and J. Altnether, "Designing Memory Systems for Microprocessors using the Intel 2164A and 2118 Dynamic RAMs," Application Note AP-133, Intel Corporation, April, 1982

Session 2-3b
VLSI Simulation I

Chairmen

A. E. Ruehli
IBM Research

D. Zein
IBM

THE WAVEFORM BOUNDING APPROACH TO TIMING ANALYSIS OF DIGITAL MOS IC'S.

John L. Wyatt, Jr., Charles Zukowski, Lance A. Glasser, Paul Bassett, and Paul Penfield, Jr.

Department of Electrical Engineering and Computer Science
Massachusetts Institute of Technology
Cambridge, Massachusetts 02139

ABSTRACT

The waveform bounding approach to fast timing analysis of MOS VLSI circuits is discussed. The idea is to compute rigorous closed-form expressions giving upper and lower bounds for transient voltage waveforms, rather than exact values. The goal is to enable rapid computation without sacrificing user confidence in the results.

I. Background and Objectives

Existing approaches to timing analysis and simulation of digital integrated circuits fall, roughly speaking, into three classes:

i) Methods such as SPICE2 [1] and ASTAP [2], based on essentially exact numerical solution of the network's differential equations, are accurate and reliable. But even with the increase in speed afforded by the waveform relaxation method [3], exact numerical solution is too slow for the needs of the VLSI era.

ii) Specialized MOS timing simulators like MOTIS-C [4] and SPLICE [5] rely on table lookup of device characteristics for speed, and save additional time by terminating a Newton-Raphson or similar iteration before convergence is reached. SPLICE is in addition a mixed-mode circuit, timing and logic simulator and uses a selective trace algorithm to exploit latency. In both these programs the termination of an iterative step prior to convergence saves time at the cost of accuracy and, in some instances, of numerical stability [6]. The improvement in speed over SPICE2 is typically one to two orders of magnitude for SPLICE [5] and about two orders of magnitude for MOTIS-C [7].

iii) More recently, some researchers are exploring an alternate approach to timing analysis and simulation based on a radically simplified electrical description of the network. RSIM [8], CRYSTAL [9], and TV [10,11] fall at the far end of the speed-accuracy tradeoff curve from SPICE2. A MOSFET is typically represented in these programs by an extremely simplified model: a linear resistor in series with a switch. And a polysilicon or diffusion line is represented by a lumped capacitance in RSIM, or by a delay in TV obtained by simply averaging the upper and lower delay bounds obtained by Rubinstein, Penfield, and Horowitz [12]. These programs are potentially very fast and have a number of attractive user-oriented features. The drawback, of course, is that there are no absolute known limits to the error in their total delay estimates. The user can never be sure the answers they give are close enough.

The objective of the waveform bounding approach to timing analysis and simulation is to combine the computational speed that results from avoiding the numerical solution of differential equations with the user confidence in the result that comes from rigorous uncertainty bounds. Our attack on the timing analysis problem is based on a careful fundamental study of the differential equations describing the dynamics of gates, pass transistors, interconnect, and the standard digital circuits constructed from them.

In addition to the MIT group working on this project, Mark Horowitz [12,13] is currently completing a dissertation on MOS timing analysis at Stanford.

II. Response Bounds for Interconnect

2.1) Linear Interconnect Models

This section summarizes the results obtained in [12]. In this work an MOS signal distribution network as shown in Fig. 1 is modelled as a branched linear RC line, i.e., an RC tree, as in Fig. 2.

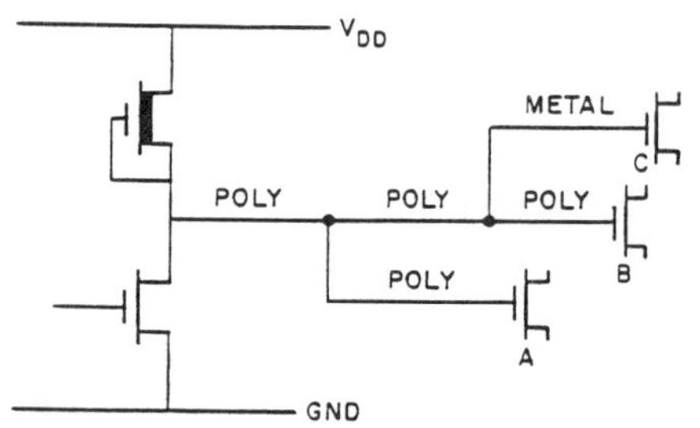

Figure 1. Typical MOS signal-distribution network. The inverter is shown driving three gates.

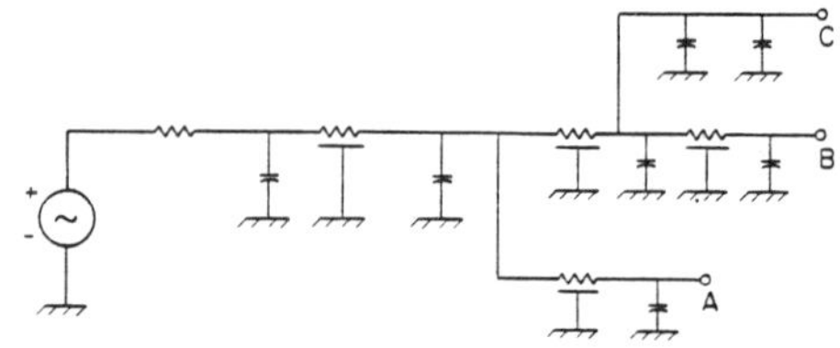

Figure 2. The linear RC tree shown above is a model for the network of Fig. 1. The voltage source is a unit step at time t = 0.

For any two nodes in the network, $R_{\ell m}$ is defined as

----

the sum of the resistances along the route consisting of the intersection of the path from the input to node ℓ with the path from the input to node m, as illustrated in Fig. 3. The three time

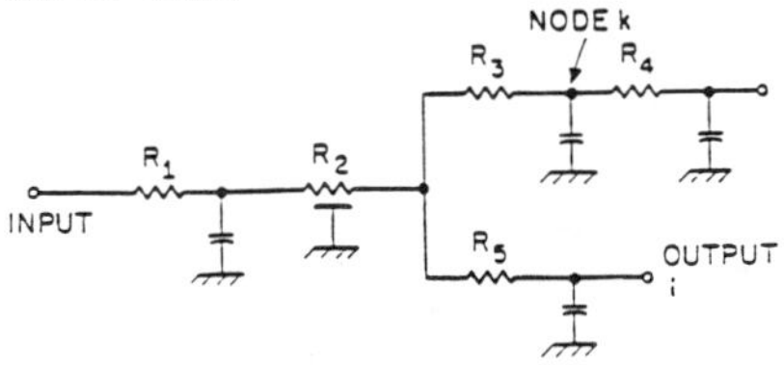

Figure 3. Illustration of resistance terms. For this network, $R_{ki} = R_1 + R_2$, $R_{kk} = R_1 + R_2 + R_3$, and $R_{ii} = R_1 + R_2 + R_5$.

constants used to derive response bounds are

$$T_P \triangleq \sum_k R_{kk} C_k \tag{1}$$

$$T_{Di} \triangleq \sum_k R_{ki} C_k \tag{2}$$

$$T_{Ri} \triangleq \sum_k (R_{ki}^2 C_k)/R_{ii}, \tag{3}$$

where the summations are taken over all nodes of the network. The derivation in [12] shows that $\underline{v}_i(t) \leq v_i(t) \leq \bar{v}_i(t)$, for all $t \geq 0$, where $v_i(t)$ is the actual zero state step response at any terminal node i, and the bounds $\underline{v}_i(t)$ and $\bar{v}_i(t)$ are given by

$$\underline{v}_i(t) \triangleq \begin{cases} 0, & 0 \leq t < T_{Di} - T_{Ri} \\ 1 - \dfrac{T_{Di}}{t+T_{Ri}}, & T_{Di}-T_{Ri} \leq t < T_P-T_{Ri} \\ 1 - \dfrac{T_{Di}}{T_P} \exp[(T_P-T_{Ri}-t)/T_P], & T_P-T_{Ri} \leq t \end{cases} \tag{4}$$

$$\bar{v}_i(t) \triangleq \begin{cases} 1 + \dfrac{t-T_{Di}}{T_P}, & 0 \leq t < T_{Di}-T_{Ri} \\ 1 - \dfrac{T_{Ri}}{T_P} \exp[(T_{Di}-T_{Ri}-t)/T_{Ri}], & T_{Di}-T_{Ri} \leq t, \end{cases} \tag{5}$$

as illustrated in Fig. 4.

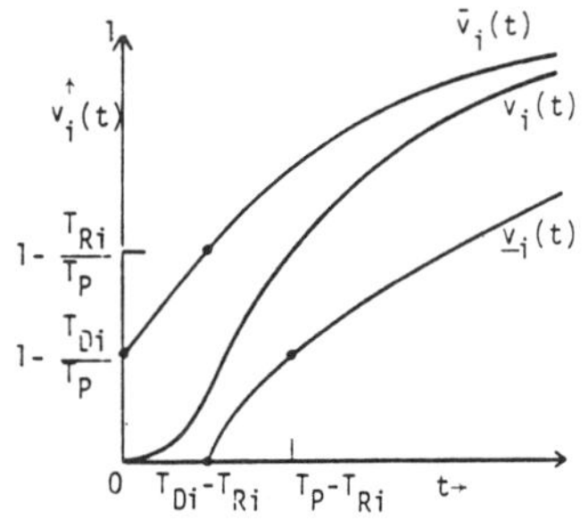

Figure 4. Form of the bounds with the distance from the exact solution exaggerated for clarity.

The time required to compute these bounds grows only linearly with the number of elements in the network. Recent applications of this result include [10,11,14,15 16]. The ultimate goal of this portion of the project is to derive a hierarchy of such bounds, permitting the user to trade off accuracy for computation time.

2.2) Nonlinearities Affecting Interconnect

The linear circuit model in Fig. 2 fails to incorporate three types of nonlinearities present in Fig. 1 or related circuits: the nonlinear output resistance of the inverter, the nonlinear gate-to-channel capacitance of the MOSFET loads, and the nonlinear capacitance from any diffusion line to substrate. This section describes recent work [17-20] that allows the bounds for linear networks [12] to be applied to RC lines incorporating such nonlinearities. (Further research is needed for branched lines, i.e. RC trees.)

Using the notation and sign conventions illustrated in Fig. 5, the

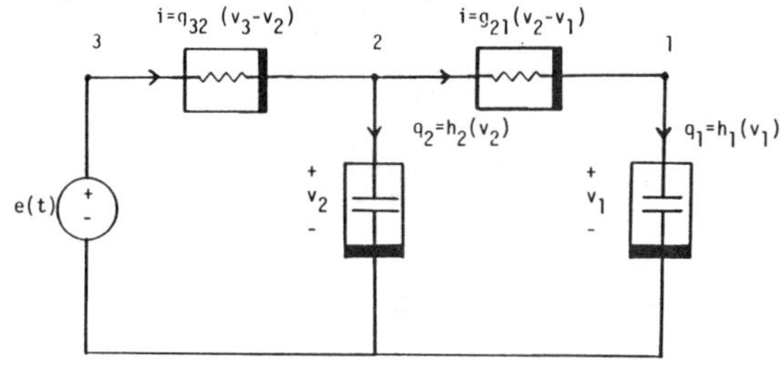

Figure 5. Two-capacitor example of a nonlinear, nonuniform RC line.

state equations for any nonuniform, nonlinear lumped RC line with N capacitors can be written in the form

$$\dot{v}_j = \frac{1}{C_j(v_j)} \left[g_{j+1,j}(v_{j+1}-v_j) - g_{j,j-1}(v_j-v_{j-1}) \right], \quad 1 \leq j \leq N, \tag{6}$$

where $g_{10} \triangleq 0$, $v_{n+1} \triangleq e$, and the capacitor constitutive relations $q_j = h_j(v_j)$ are continuously differentiable with $C_j(v_j) \triangleq h_j'(v_j) > 0$ everywhere. We assume the resistor curves are continuously differentiable, strictly increasing, and pass through the origin.

Lemma 1 [19]

Consider any nonlinear, nonuniform RC line. At any instant during an "up" transition (i.e. $e \geq 0$, $\dot{e} \geq 0$) from equilibrium,

$$v_j(t) \geq 0, \; v_j(t) \leq e, \; v_j(t) \leq v_{j+1}(t)$$
$$\text{and } \dot{v}_j(t) \geq 0, \; 1 \leq j \leq N.$$

Lemma 1 is proved in [19]. Using it, we give a proof in [20] of the

Monotone Response Theorem for Nonlinear, Nonuniform RC Lines. Given a nonlinear RC line as described above. Suppose that (because of circuit parameter uncertainty, the use of linearized models for nonlinear elements, replacing the exact input by input bounds, etc.,) we do one or more of the following:

a) overestimate the input e(t),

b) underestimate one or more *R*'s,

c) underestimate one or more *C*'s.

The resulting circuit model will then necessarily overestimate the output $v_j(t)$ at each instant t during "up" transitions (i.e., during transitions where $e \geq 0$, $\dot{e} \geq 0$ throughout.)

A similar result holds for "down" transitions and estimate errors of the opposite sign. Using part a) of the assumptions, this theorem allows us to computationally propagate upper and lower signal bounds through the network. Using parts b) and c), it allows us to replace a nonlinear line by two linear ones, one strictly faster and one strictly slower, to which the linear network bounds (4,5) in turn apply. We have not yet succeeded in finding a generalization of this result that will apply to nonlinear RC trees.

III. An Approach to Waveform Bounding for MOS Logic Gates

The results reported here apply to MOS device models of the form

$$i_D = \frac{W}{L} f(v_{GB}, v_{DB}, v_{SB}), \tag{7}$$

where D,G,S and B refer to drain, gate, source and substrate, respectively. For specificity we consider only n-channel devices in this paper. No special algebraic form for f is assumed, only that f is continuously differentiable and satisfies the natural monotonicity conditions

$$\frac{\partial f}{\partial v_{GB}} \geq 0, \ \frac{\partial f}{\partial v_{DB}} \geq 0, \ \frac{\partial f}{\partial v_{SB}} \leq 0 \tag{8}$$

everywhere, Thus a wide variety of device models are allowed, with the exception that (7) does not allow for short-channel effects.

Our approach will be to reduce a multiple-input logic gate by steps to an "equivalent bounding inverter" and then to find bounds for the response of this inverter.

3.1) Reduction of Series-Parallel Transistor Network to "Equivalent Bounding Transistor"

We have developed a method for reducing any series-parallel transistor network to a single "equivalent bounding transistor." Using the technique recursively, one can replace the pullup or pulldown network of a multiple-input gate by a single transistor and have rigorous bounds for the error produced by this simplification.

For example, a parallel connection of N transistors, all identical except for widths, lengths and gate voltages, satisfies

$$i = \hat{f}(\vec{v}_{GB}, v_{DB}, v_{SB}) \triangleq \sum_{j=1}^{N} \frac{W_j}{L_j} f(v_{GB_j}, v_{DB}, v_{SB}), \tag{9}$$

where $\vec{v}_{GB}$ is the vector of gate voltages. We have proven that, because of the assumptions (8), there exist W_{eq}, L_{eq} independent of $\vec{v}_{GB}$, and $\underline{v}_{GB}$ and $\bar{v}_{GB}$ that depend on $\vec{v}_{GB}$, such that (9) can be replaced by the simpler bounds

$$\frac{W_{eq}}{L_{eq}} f(\underline{v}_{GB}, v_{DB}, v_{SB}) \leq i \leq \frac{W_{eq}}{L_{eq}} f(\bar{v}_{GB}, v_{DB}, v_{SB}), \tag{10}$$

for all $v_{DB} \geq v_{SB}$, describing a single transistor with a range of gate voltages. The function f is the same throughout (9) and (10). Figure 6 illustrates this process for N = 2.

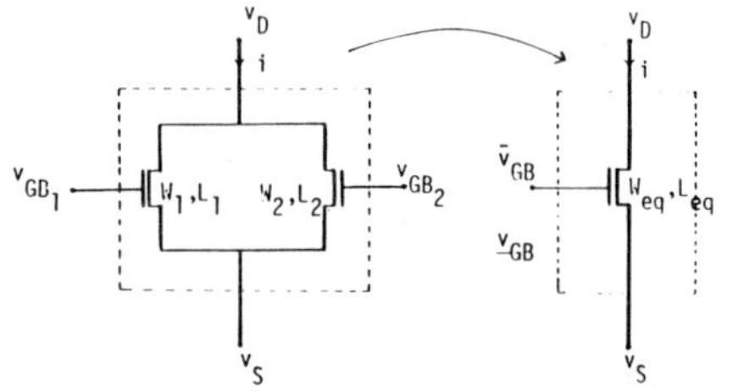

Figure 6. Replacing a parallel transistor network by an "equivalent bounding transistor". The cost of this simplification is that the exact value of i for the network on the left is replaced by a range of values in the simpler model corresponding to $\underline{v}_{GB} \leq v_{GB} \leq \bar{v}_{GB}$.

3.2) Reducing a Multiple-Input Gate to an "Equivalent Bounding Inverter"

A gate can be modelled as an "equivalent bounding inverter" by performing the reduction outlined in section 3.1 on both the pullup and pulldown networks, reducing each to a single transistor. Initial trials, comparing SPICE2 simulations of the original network with simulations of the "equivalent bounding inverter" indicate that the resulting bounds for $i_{out}(v_{out})$ differ from the exact values by only about $\pm$ 10% for practical circuits.

3.3) Bounding the Response of an Inverter and Load to Input Transitions

When applied to some multiple-input gates, the reduction procedure described in the previous two subsections may yield an inverter in which the pullup gate is externally driven. But for simplicity we consider here only the case of a standard NMOS depletion - load inverter as in Fig. 7.

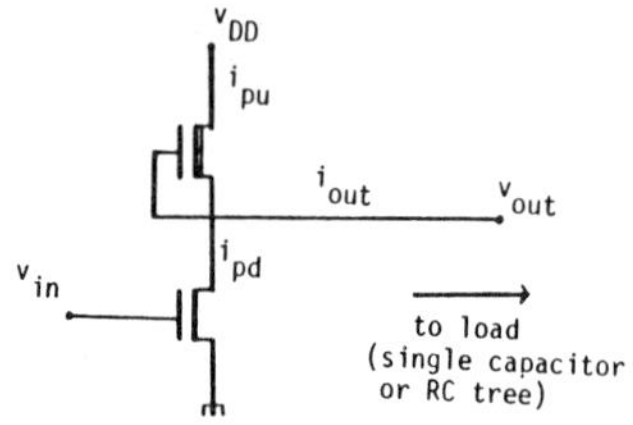

Figure 7. Depletion-load inverter.

To bound the response time of the loaded inverter we need simple bounds on the function $i_{out}(v_{out}, v_{in})$, which is the difference of the pullup and pulldown currents:

$$i_{out}(v_{out}, v_{in}) = i_{pu}(v_{out}) - i_{pd}(v_{out}, v_{in}). \tag{11}$$

Simple linear bounds on both the pullup and pulldown currents are shown in Fig. 8. The resulting bounds for the output curve $i_{out}(v_{out})$ depend on $v_{in}(t)$.

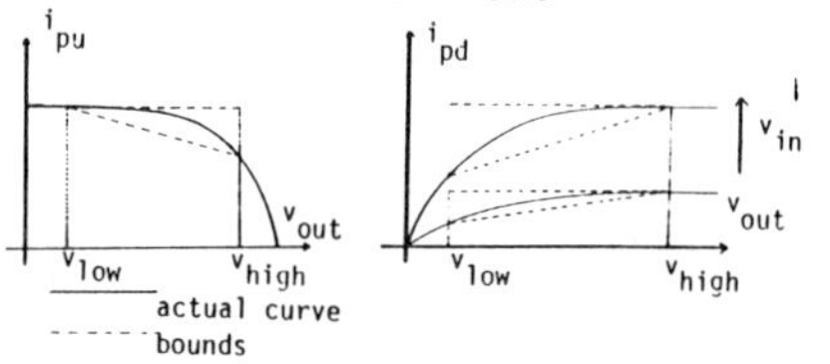

Figure 8. Simple linear bounds on the pullup and pulldown currents. The latter depend on v_{in}, and hence on t.

Initial simulations using this approach indicate that the delay bounds for these simplified models differ from the delays obtained from SPICE simulations by about $\pm$ 15%.

IV. Further Work in Progress

Much work remains to be done before the theoretical basis for the waveform bounding approach to timing analysis is complete. Among the larger remaining problems are:

1. extending the Penfield-Rubinstein bounds to incorporate time-varying source resistances, such as those modelling the pulldown current in Fig. 8,

2. finding bounds for the response of an RC tree containing pass transistors,

3. investigating the tolerance in the bounds obtained so far and finding tighter ones where necessary, and

4. incorporating effects of the Miller capacitance into bounds.

ACKNOWLEDGMENT

This project was supported by the Motorola Research Grant Fund, an IBM Fellowship, and NSF Contract No. ECS8118160.

REFERENCES

[1] L.W. Nagel, "SPICE2, A Computer Program to Simulate Semiconductor Circuits," Memorandum No. ERL M520, Electronics Research Laboratory, U.C. Berkeley, Berkeley, CA, 94720, May, 1975.

[2] W.T. Weeks, A.J. Jimenez, G.W. Mahoney, D. Mehta, H. Qassemzadeh, and T.R. Scott, "Algorithms for ASTAP-A Network Analysis Program," IEEE Trans. Circuit Theory, vol. CT-20, pp. 628-634, Nov. 1973.

[3] E. Lelarasmee, A.E. Ruehli, and A.L. Sangiovanni-Vincentelli, "The Waveform Relaxation Method for Time-Domain Analysis of Large Scale Integrated Circuits," IEEE Trans. CAD, vol. CAD-1, no. 3, July 1982, pp. 131-145.

[4] S.P. Fan, M.Y. Hsueh, A.R. Newton, and D.O. Pederson, "MOTIS-C: A New Circuit Simulator for MOS LSI Circuits," Proc. 1977 IEEE Int. Symp. on Circuits and Systems, pp. 700-703, April, 1977.

[5] A.R. Newton, "Techniques for the Simulation of Large Scale Integrated Circuits," IEEE Trans. Circuits and Systems, vol. CAS-26, no. 9, pp. 741-749, Sept. 1979.

[6] G. De Micheli and A. Sangiovanni-Vincentelli, "Characterization of Integration Algorithms for the Timing Analysis of MOS VLSI Circuits," Int. Jour. of Circuit Theory and Appl., vol. 10, pp. 299-309, 1982.

[7] M.Y. Hsueh, A.R. Newton, and D.O. Pederson, "New Approaches to Modelling and Electrical Simulation of LSI Logic Circuits," Journees d'Electronique, pp. 403-413, Lausanne, Switzerland, 1977.

[8] C.J. Terman, "User's Guide to NET, PRESIM, and RNL/NL," private communication, available from the author.

[9] J. Ousterhut, "CRYSTAL: A Timing Analyzer for NMOS VLSI Circuits," Proc. 3rd Caltech Conf. on VLSI, March 1983, pp. 57-69.

[10] N. Jouppi, "TV: An NMOS Timing Analyzer," Proc. 3rd Caltech Conf. on VLSI, March 1983, pp. 71-85.

[11] N. Jouppi, "Timing Analysis for NMOS VLSI," ACM IEEE 20th Design Autom. Conf. Proc., June 1983, pp. 411-418.

[12] J. Rubinstein, P. Penfield, Jr., and M.H. Horowitz, "Signal Delay in RC Tree Networks," to appear in IEEE Trans. CAD.

[13] M. Horowitz, "Timing Models for MOS Pass Networks," Proc. 1983 IEEE Int. Symp. Circuits and Systems, pp. 198-201.

[14] R. Putatunda, "Auto Delay: A Program for Automatic Calculation of Delay in LSI/VLSI Chips," ACM IEEE 19th Design Autom. Conf. Proc., pp. 616-621, 1982.

[15] R. Putatunda, "Automatic Calculation of Delay in Custom Generated LSI/VLSI Chips," Proc. IEEE Int. Conf. on Circuits and Computers, Sept. 1982, pp. 193-196.

[16] E. Tamura, K. Ogawa, and T. Nakano, "Path Delay Analysis for Hierarchical Building Block Layout System," ACM IEEE 20th Design Autom. Conf. Proc., June 1983, pp. 403-410.

[17] J.L. Wyatt, Jr., "Inequality Theorems for Nonlinear Differential Equations," VLSI Memo 82-127, Dept. of Elect. Eng. and Comp. Sci., MIT, Cambridge,MA, 02139, Nov. 1982. All VLSI Memos are available from the Microsystems Program Office, Room 36-575, MIT.

[18] J.L. Wyatt, Jr., "Monotone Behavior of Nonlinear RC Meshes," VLSI Memo 82-128, Nov. 1982.

[19] J.L. Wyatt, Jr., and P. Bassett, "Spatial Monotonicity and Positive Invariance for Nonlinear RC Lines and Trees," VLSI Memo 83-140, April 1983.

[20] J.L. Wyatt, Jr., "Partial Ordering and Monotone Sensitivity for Nonlinear RC Meshes and Lines," VLSI Memo 83-145, June, 1983.

This work was supported by DoD under ARPA order number 3972, issued by the Office of Naval Research under Contract N00014-80-C-0622.

ROBUSTNESS ASPECTS OF AN ADAPTIVE WAVEFORM RELAXATION SCHEME

F. Odeh*
A. E. Ruehli
IBM Thomas J. Watson Research Center
Yorktown Heights, New York 10598

C. H. Carlin
Ecole Polytechnique Federal de Lausanne
Switzerland

Abstract

The waveform relaxation technique can be employed in the time solution of large scale MOSFET transistor circuits. In this paper an enhancement to the technique is given based on the quasi one-way property of MOSFET circuits which makes the analysis of independent subportions possible. This result can be employed in an adaptive scheme. Further, we prove the robustness of the approach as well as the convergence of the discrete system which is solved numerically.

1. Introduction

The waveform relaxation (WR) approach is a promising technique for the analysis of large scale MOSFET circuits[1]. The major application of this approach is the time waveform analysis of VLSI circuits. Mainly conventional general purpose circuit analysis programs like SPICE[2] are designed to handle a multitude of different circuits. Hence, it is possible to design special algorithms and special purpose programs for specific tasks which perform substantially better than conventional ones. The large scale circuit analysis field has a long history which is not reviewed here but the reader is referred to two recent survey papers[3, 4] which contain extensive references. The implementation of the WR method shows an improvement in computation speed between a factor of 10 to 100[1, 5] above a general purpose circuit analysis tool like SPICE.

This short paper is concerned with two robustness aspects which relate to the implementation of the WR method. The first is that the behavior of the resultant WR waveforms is basically unchanged if we ignore the effect of distant local weak feedback paths. This device is appropriate for reducing the storage requirements of the method. The second aspect pertains to the convergence of the method under numerical discretization which is clearly a necessary step for its implementation.

In section 2, we examine the error induced by ignoring local feedback paths which is the same as breaking the circuit into parts. Here, we exploit a special property exhibited by many realistic MOSFET circuits namely their unilaterality or one-way behavior[6]. We show that quasi one-way circuits which have relatively small local feedback can be analyzed by considering portions of the circuit only. We consider only very simple configurations for simplicity. The extension to general circuits is of an algorithmic nature and will be reported on in the future.

In section 3, we show that the WR method converges for small time steps h when the circuit equations are discretized by means of implicit multistep methods which possess certain well known stability properties.

2. Impact of Small Coupling

The waveform relaxation (WR) method for solving a smooth system or circuit

$$\begin{aligned} F(y,\dot{y}) &= 0 \\ y(0) &= y_0 \end{aligned} \tag{1}$$

where a dot , • , denotes differentiation with respect to time and y and u usually represent nodal voltages as defined in[1].

The circuit is decomposed into m subcircuits

$$\begin{aligned} F_i(y_i, \dot{y}_i, \tilde{u}_i) &= 0 \\ y_i(0) &= y_{0i} \end{aligned} \qquad 1 \leq i \leq m \tag{2}$$

where the $\tilde{u}_i$ are the decoupling inputs (loading currents and/or voltages y_j, $\dot{y}_j$ $j \neq i$). If one assumes that the equations (2) are solvable w.r.t. the dotted variable, i.e., that (2) is reducible to

$$\begin{aligned} \dot{x}_i &= \tilde{f}_i(x_i; \tilde{u}_i) \\ x_i(0) &= \quad x_{i0} \end{aligned} \tag{3}$$

then a relaxation such as Gauss-Seidel or Gauss-Jacobi yields a WR algorithm whose canonical form is given by

$$\begin{aligned} \dot{x}^s &= f(x^s; x^{s-1}, \dot{x}^{s-1}) \\ x(0) &= x_0 \end{aligned} \tag{4}$$

Here superscripts refer to the relaxation sweep number and f is a smooth function of its arguments. In[1] the convergence of the iteration (4) was shown under suitable assumptions on the behavior of f in terms of its arguments especially the third. The function f takes into account the effect of all feedback paths on each specific subcircuits. However, there are many examples of circuits, e.g. combinational ones, which are basically one-way and where only nearest neighbors interact with weak feedback. For example, the chain of circuits

*This research was sponsored in part by the Air Force Office of Scientific Research (AFSC), United States Air Force, under Contract No. F44620-77-C-0088. The United States Government is authorized to reproduce and distribute reprints for government purposes notwithstanding any copyright notation hereon.

symbolically given in Figure 1, and in greater detail in Figure 2 (where the fifth circuit is not shown) are one-way if one neglects the small Miller feedback capacitances $c_{12}, c_{23}, \ldots$. It is intuitively clear that if these feedbacks are small, then a capacitive feedback $c_{n,n+1}$ should have an exponentially small effect on the mth circuit if $n - m$ is large enough. This section is concerned with a more quantitative description of this intuitive picture.

For simplicity, we first illustrate the above remarks in the case of solving the linear algebraic system

$$Ax = b \tag{5}$$

where A has an upper-Hessenberg form $A = T + E$ where T is a lower triangular invertible matrix and E has zero entries everywhere except for $E_{i,i+1}$ which are assumed small $O(\varepsilon)$. These non-zero entries may be arranged into a vector $(v_1, v_2, \ldots)$ which may be again denoted by E, slightly abusing the notation. For a fixed given N, decompose E into $E_1 + E_2$ where $E_1 = (v_1, \ldots, v_{N-2})$ and E_2 is the remainder. Noting that v_i corresponds to feedback from variable x_{i+1}, then, in order to ignore the feedback from variables x_j, $j \geq N$ one simply has to equate E_2 to zero. Let y denote the solution to the truncated system

$$A_1 y \equiv (T + E_1)y = b \tag{6}$$

and define the error

$$e = x - y \tag{7}$$

then one has

Lemma 1

The components of the error e satisfy the estimates

$$\begin{aligned} e_j &= O(\varepsilon^{N-j}) \qquad i \leq j \leq N-1 &(8)\\ &= O(\varepsilon) \qquad\qquad j \geq N &(9)\end{aligned}$$

Proof

Since $A - A_1 = O(\varepsilon)$, the estimate (9) follows from regular perturbation theory. To show (8), consider the error equation

$$(T + E)e = -E_2 y. \tag{10}$$

Hence

$$-e = (I + LE)^{-1}(LE_2 y) \tag{11}$$

where L, the inverse of T, is bounded and is also lower triangular.

The vector $E_2 y$ has zero components in the first $(N-2)$ entries, hence the same is true for $r = LE_2 y$. Since $E = O(\varepsilon)$ we may expand $(I + LE)^{-1} = \sum_{k=0} (-1)^k (LE)^k$. Then, substituting in (11) one obtains

$$-e = \sum_{k=0} (-1)^k (LE)^k r. \tag{12}$$

Since the matrix E corresponds to a (forward) shift and r_i vanishes for $1 \leq i \leq N-2$, the expansion (12) for e_j, $i \leq j \leq N-1$, starts with $k = (N-1) - j$. Using the facts (i) that the expansion is convergent in any norm, for small ε, and hence is of the same order as its first term and (ii) that both E and r are $O(\varepsilon)$ we immediately obtain (8).

The same principle applies to the solution of initial-value problems for smooth differential system. Consider for example the circuits chain of Figure 2. Then in terms of the node voltages $(v_1, \ldots, v_4)$, the nodal equations are

$$\begin{aligned}
(c_1 + c_{12})\dot{v}_1 + I_1(v_1,u) - c_{12}\dot{v}_2 &= 0\\
-c_{12}\dot{v}_1 + (c_2 + c_{12} + c_{23})\dot{v}_2 + I_2(v_1,v_2) - c_{23}\dot{v}_3 &= 0\\
-c_{23}\dot{v}_2 + (c_3 + c_{23} + c_{34})\dot{v}_3 + I_3(v_2,v_3) - c_{34}\dot{v}_4 &= 0\\
-c_{34}\dot{v}_3 + (c_{34} + c_4)\dot{v}_4 + I_4(v_3,v_3) &= 0
\end{aligned} \tag{13}$$

where the currents $I(v)$ include both transistor currents for the active and the load device and u is the input to (13). The capacitive coupling $c_{i,i+1}$ is assumed to be $O(\varepsilon)$ and it is included in the matrix E in the previous discussion. Hence, ignoring c_{34} for example should have, by analogy, an effect on v_1 which is $O(\varepsilon^{4-1}) = O(\varepsilon^3)$. To show the validity of such an estimate we employ a continuation method. Write (13) in the form

$$F(v;u) = 0 \tag{14}$$

where we may assume that

$$F(0;0) = 0. \tag{15}$$

Imbed (14) in a smooth family of problems

$$F(v(p);\, u(p)) = 0. \tag{16}$$

Such that $p = 0$ corresponds to (15) and $p = 1$ to the actual problem (14). One approach is to change u from zero to its value in a linear fashion. Then, with $\partial v = \dfrac{\partial v}{\partial p}$, one has

$$v = \int_0^1 (\partial v)dp$$

and it is sufficient to show that (∂v) behaves in the manner described above when the feedback is cut. But (∂v) satisfies a *linear* system of differential equations given by

$$F_v(v(p);u(p))\partial v = -F_u(\partial u). \tag{17}$$

For example, the first component of (17) reads

$$(c_1 + c_{12})(\partial\dot{v})_1 + J_1(p)(\partial v)_1 - c_{12}(\partial\dot{v})_2 = -J_2(\partial u) \tag{18}$$

where J_1, J_2 are resp., the derivative of I_1 w.r.t. v and u and are assumed globally smooth and bounded. Hence (17) represents a small perturbation on a triangular system of smooth linear ordinary differential equations with given initial data and may be inverted in the same manner as (5), e.g., the solution of (18) may be written as

$$(\partial v)_1(t) = \int_0^t G_1(t,\tau)[c_{12}(\partial\dot{v}(\tau))_2 - J_2(\partial u(\tau))]d\tau \tag{19}$$

where G_1 is the smooth Green's function which inverts the linear differential operator in (18). The proof of Lemma 1, with F_v substituting for A, then shows that the error committed in $(\partial v)_1$ by ignoring c_{34} is $O(\varepsilon^3)$ as desired.

Another way of showing the above result is by linearization around the solution y to the problem (14) when the feedback E_2 is set equal to zero. If $e = v - y$, then corresponding to the linear error equation (10) one has the nonlinear equation

$$G(e) \equiv F(y + e) - F(y) = -E_2 y \tag{20}$$

where E_2y again has its first $(N-2)$ components equal to zero. One can solve (20) by a variety of linear methods, e.g., by a modified Newton method starting with $e_0 = 0$, then $e_1 = e_0 + \delta e_0$ where δe_0 satisfies

$$F'(y)(\delta e_0) = -E_2 y. \tag{21}$$

The above argument shows that

$$(\delta e_0)_j = O(\varepsilon^{N-j}) \tag{22}$$

where $N = 4$ in the special example. Continuing the iteration and making use of the remainder form of Taylor s expansion one finds that $(\delta e)_n$, which is defined by

$$F'(y)(\delta e_n) = -[F(y+e_n) - F(y) + E_2 y] \tag{23}$$

also satisfies the estimate (22). Since the iteration is convergent for small ε [in fact uniformly in t over bounded intervals] the desired estimate follows again.

Next, we give an illustration of how the small coupling can be exploited to analyze the circuits in the inverter chain without considering all circuits simultaneously. Here, we assume that we are able to import the portion of the circuits which are analyzed together from disk into main storage which insures that it can be processed rapidly. Based on the insight gained from the analytical results, we propose the following analysis sequence for the subcircuits in Figure 1.: 1,2,3,4,, 1,2,3,4,, 1,2,3,4 until the voltage waveforms for circuit 1 only converge by comparing waveforms for the entire analysis time interval (25). Now we can discard subcircuit 1 from active storage and we store waveform 1 on disk. Also we fetch circuit 5 from disk and now we analyze the circuits as before in the sequence 2,3,4,5,, 2,3,4,5 . . . .. Hence, this approach lets us analyze infinitely many subcircuits if we assume infinite disk storage. The number of subcircuits which must be analyzed in one sequence is determined by the error tolerance ε_T due to the truncation of the chain. Let d be the smallest integer such that

$$\varepsilon^d < \varepsilon_T. \tag{24}$$

Then the sequence of circuits includes is 1,2, . . . , d. In the above example we assume that $d = 4$. This is a consequence of Remark 1.

<u>Remark 1</u>

Since the linearized operators F_v in (17) [or $F'(y)$ in (23)] are almost triangular, a Gauss-Seidel relaxation for solving either (20) or (17) also converges. However, in the initial relaxation sweeps, the effect of the $(K+1)$ feedback $c_{K,K+1}$ on, for example, the first subsystem is not felt. To take it into account at least $(K+1)$ sweeps have to be performed.

<u>Remark 2</u>

The monitoring of the convergence in the canonical waveform relaxation method[1] is accomplished by monitoring the convergence in the circuit or

$$\max_{t\in[0,T]} \| y^s(t) - y^{s-1}(t) \| < \varepsilon_w \tag{25}$$

for all nodes. Else, the "transportation" of a voltage by the process discussed in Remark 1 can result in a non monotonic decrease of the error (25) with respect to the iterations due to local feedback. Here, the local ε feedback attenuates the voltage feedback and local monitoring as discussed above is possible.

<u>3. Multistep Methods and Wave-form Relaxation</u>

In[1] the convergence of the relaxation (WR) scheme (4) was given under basically- the assumptions that f is smooth, with Lipschitz constants K_1, K_2 w.r.t. its first two arguments and contractive, with constant $r<1$, w.r.t. the third. This section is concerned with the convergence of the discretization of (4) by means of a k-step multistep formula on a uniform mesh with time step h

$$\sum_{j=0}^{k} a_j x_{n-j} - h \sum_{j=0}^{k} b_j \dot{x}_{n-j} = 0 \tag{26}$$

or, in an obvious notation

$$ax - hb\dot{x} = 0. \tag{27}$$

If (ρ,σ) are the standard multistep polynomials, then $\rho(z) = z^k a(z^{-1})$ and $\sigma(z) = z^k b(z^{-1})$ where z corresponds to forward shift by $\Delta t = h$. We assume that the method is *implicit*, and

$$\begin{array}{ll} \text{(i)} & \text{consistent: } \rho(1)=0,\ \rho'(1)=\sigma(1)=1 \\ \text{(ii)} & \sigma(z) \text{ and } \tilde{\rho}(z) \equiv \rho(z)(z-1)^{-1} \text{ have} \\ & \text{roots strictly inside the unit disk.} \end{array} \tag{28}$$

The discretized form of (4) takes the form

$$\frac{1}{h}\frac{a}{b}x^s = f\left(x^s;\, x^{s-1},\, \frac{1}{h}\frac{a}{h}x^{s-1}\right) \tag{29}$$

where a/b above is a convolution operator. Regarding (29) as a mapping between sequences, we show that this mapping is a contraction in some appropriate norm. Let the sequences p, q be defined by

$$p = \frac{1}{h}\frac{a}{b}x^{s-1},\quad q = \frac{1}{h}\frac{a}{b}x^s. \tag{30}$$

Then x^{s-1} is defined by the linear difference equation

$$\begin{array}{ll} ax^{s-1}(n) = hbp \equiv hw(n) & n \geq k \\ \text{with } \{x_0, x_1, \ldots, x_{k-1}\} \text{ given} & \end{array} \tag{31}$$

The solution to (31) is a convolution

$$x^{s-1}(n) = h \sum_{m=k}^{n} H(n-m)w(n) + hc(n). \tag{32}$$

In (32) $c(n)$ is determined by the fixed initial data and H is the one-sided Green's function given by

$$H(t) = \sum_{i=1}^{k} (z_i)^{t+k-1} / \rho'(z_i) \tag{33}$$

where $\{z_i\}$ are the (simple) roots of ρ. The representation (32, 33) may be easily obtained by taking the z-transform of (31) as was done, for example, in[7]. If one defines the operator S by

$$(Sp)_n \equiv x^{s-1}(n) = hc(n) + h \sum_{m=k}^{n} H(n-m)bp(m) \tag{34}$$

then the iteration (29) reads

$$q = f(Sq;\, Sp,\, p). \tag{35}$$

To show that (29) defines a contraction, introduce the norm

$$\| p \| = \sup_{n \leq N} \beta^{-n} | p_n | \tag{36}$$

where β is to be chosen later. For any two bounded sequences p_1, p_2 we first estimate the distance between their images Sp_1, Sp_2. From (34) and with $w = bp$

$$\| Sp_1 - Sp_2 \| = h \sup \beta^{-n} | \sum^{n} [H_1(n-m) + H_2(n-m)] [w_1(m) - w_2(m)] | \tag{37}$$

where H_1 corresponds to the root of ρ at $z = 1$, and H_2 is the exponentially decreasing kernel corresponding to the other roots. Estimating separately the effect of H_1, H_2, which we may call S_1, S_2 and using the accuracy condition $\rho'(1) = \sigma(1)$ one has

$$\begin{aligned} \| S_1 p_1 - S_1 p_2 \| &= h \sup \beta^{-n} \sum^{n} \beta^{m} \beta^{-m} | p_1(m) - p_2(m) | \\ &\leq h \| p_1 - p_2 \| \sup \beta^{-n} \frac{(\beta^{n+1} - 1)}{(\beta - 1)} \\ &= O(h) \| p_1 - p_2 \| \text{ for } \beta < 1. \end{aligned} \tag{38}$$

A similar, but easier, estimate holds for the H_2 term in (37) [the β-weighting is in fact unnecessary in estimating S_2 since $H_2(t)$ is exponentially small]. Hence one has

$$\| S(p_1 - p_2) \| = O(h) \| p_1 - p_2 \| . \tag{39}$$

Now, from the assumed Lipschitz continuity and contractivity of f one has

$$\| q_1 - q_2 \| \leq K_1 \| Sq_1 - Sq_2 \| + K_2 \| Sp_1 - Sp_2 \| + r \| p_1 - p_2 \| . \tag{40}$$

Combining (39, 40) one obtains

$$\| q_1 - q_2 \| (1 - O(hK_1))^{-1} \leq (r + O(K_2 h)) \| p_1 - p_2 \| .$$

Hence, taking h small enough, (29) defines a contraction in the norm (36) which proves the result. An easy sufficient condition is that both $hK_1 < 1$ and $hK_2 < (1 - r)$. However, one should remark that if K_2 is of moderate size but K_1 is large,, as in the case of stiff subsystems, the first of the above conditions indicates that h should be taken very small, i.e., $h < K_1^{-1}$. A more refined analysis (to be published) shows that the use of A-stable methods circumvents this difficulty.

Conclusion

We have shown that the cutting of the circuit into subcircuits with small coupling of order ε adds robustness as well as other advantages to the waveform relaxation method. Further, the bounds on the discretization which result from the numerical integration scheme indicate that a restriction exists on the maximum size of the time step h.

Acknowledgements

We held discussions on the subject with many researchers in the field, at several universities, as well as IBM Yorktown. Their contributions are acknowledged here without mentioning any names since they are too many to acknowledge appropriately.

References

1. E. Lelarasmee, A. E. Ruehli, and A. L. Sangiovanni-Vincentelli, "The waveform relaxation method for time-domain analysis of large-scale integrated circuits," IEEE Trans. CAD Integ. Circ. Sys., vol. CAD-1, pp. 131-145, July 1982.
2. L. W. Nagel, "SPICE2: a computer program to simulate semiconductor circuits," Univ. of California, Berkeley, ERL Memo ERL-M520, May 1975.
3. G. D. Hachtel and A. L. Sangiovanni-Vincentelli, "A survey of third generation simulation techniques," Proc. IEEE, vol. 69, pp. 1264-1280, Oct. 1981.
4. A. E. Ruehli and G. S. Ditlow, "Circuit analysis, logic simulation and design verification for VLSI," Proc. IEEE, vol. 71, pp. 34-48, Jan. 1983.
5. J. White and A. L. Sangiovanni-Vincentelli, "Relax2: "A modified waveform relaxation approach to the simulation of MOS digital circuits," Conf. Proc. IEEE ISCAS, vol. 2, pp. 756-759, Newport Beach, CA, May 1982.
6. A. E. Ruehli, A. L. Sangiovanni-Vincentelli, and N. G. Rabbat, "Time analysis of large scale circuits containing one-way macromodels," IEEE Trans. Circuits Syst., CAS-29, pp. 185-189, March 1982.
7. F. Odeh and W. Liniger, "Nonlinear fixed-h stability of linear multistep formulas," J. Math. Anal. App., 61, pp. 691-712, 1977.

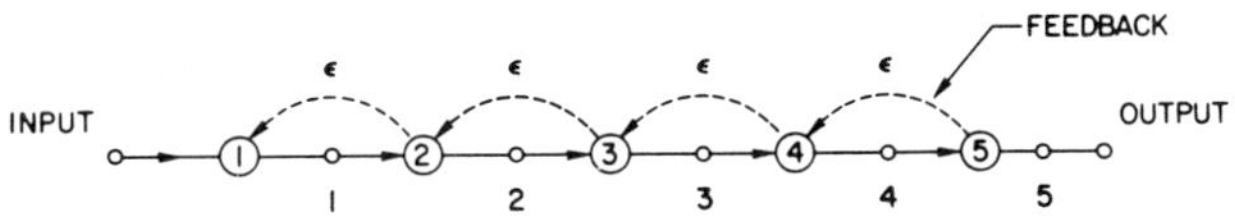

Fig. 1 Chain of inverters with weak local feedback.

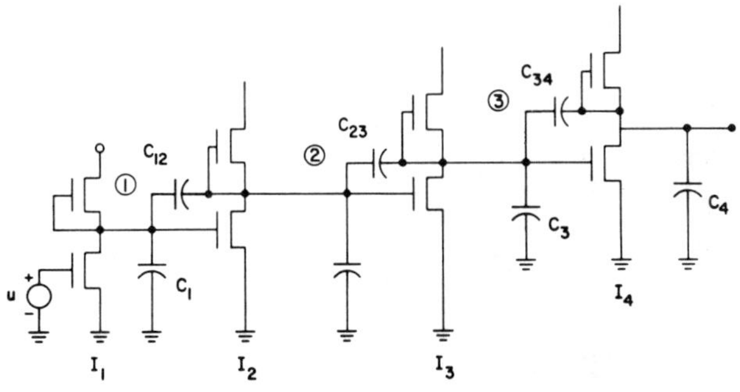

Fig. 2 Example circuit

BLOSSOM: An Algorithm and Architecture for the Solution of Large-Scale Linear System

Fu-Hwa Ko and
Alberto Sangiovanni-Vincentelli

Department of Electrical Engineering and Computer Science
U.C. Berkeley

Abstract

Block LU factorization and other partitioned matrix algorithms are used in a VLSI computing system, BLOSSOM, to solve very large-scale matrix problems. With only one type of VLSI arithmetic processing unit needed for submatrix computations, a reconfigurable processor array is used. Since numerical properties are vital to matrix computations, we use neighbor pivoting on submatrices with scalar elements and partial pivoting on full matrix with submatrices as elements. Natural topological structures of certain kinds of operand matrices are also exploited to help reducing pivoting costs. BLOSSOM is designed such that submatrices of different sizes can be efficiently processed even when the number of processing units is less than that of the submatrix elements. Comparisons with other special hardware schemes are provided as a reference.

1. Introduction

The solution of Large-scale linear System of algebraic Equations(LSE) is needed in the analysis and simulation of many engineering systems. Finite-element analysis, circuit simulation, and power system analysis are few examples. In these applications, sparse matrix techniques have been used extensively to speed up the solution process. The time complexity of these techniques has been experimentally estimated to be $O(n^{\alpha})$, where $1.2 \le \alpha \le 1.5$ and n is the number of equations. When analyzing very large scale systems(more than 10,000 equations), the computational complexity of these techniques makes the solution process very expensive and the use of large main-frames such as the IBM3081 indispensable.

New architectures, in particular vector computers such as the CRAY 1, have inspired the design of new algorithms to exploit parallelism in the solution process. An important example is the program CLASSIE[14] for the simulation of electronic circuits. Along these lines, peripheral array processors, such as the FPS164, can also be used in conjunction with hosts such as the VAX11/780 to speed up the solution process. However, this speedup is not enough to cope with the problems to be solved in the VLSI era.

The advent of VLSI technology has made the cost-effective design of special purpose machines possible. Examples of these machines are the Yorktown Simulation Engine(YSE) for logic solution[3] and Systolic Arrays [11]. Special purpose machines have also been proposed for the solution of LSE[7, 9, 5, 15]. Most of these machines limit the size of the operand matrix. When no size limit is imposed, the operand matrix has to be partitioned into submatrices of equal sizes. Only Johnsson in[7] and Pottle in[9] treated the related numerical properties and matrix sparsity is exploited only in[9]. However special matrix structures, such as the Bordered Block Diagonal Form (BBDF) or the Bordered Block Triangular Form(BBTF), commonly expected in engineering problem, are not exploited in [9]. In this paper, we propose a new algorithm-architecture BLOSSOM for the solution of LSE.

The paper is organized as follows. In Section 2 we propose a variant of block LU decomposition with block neighbor pivoting to ensure numerical stability and accuracy. A parallel-pipeline architecture, described in Section 3, is designed to implement the block LU decomposition. This architecture supports other matrix operations used as subprocedures by block LU decomposition such as the multiplication and the inversion of submatrices. In Section 4, we describe the hardware implementation of these matrix operations. Finally, a comparison with other hardware schemes is listed in table 1.

2. Block LU Factorization

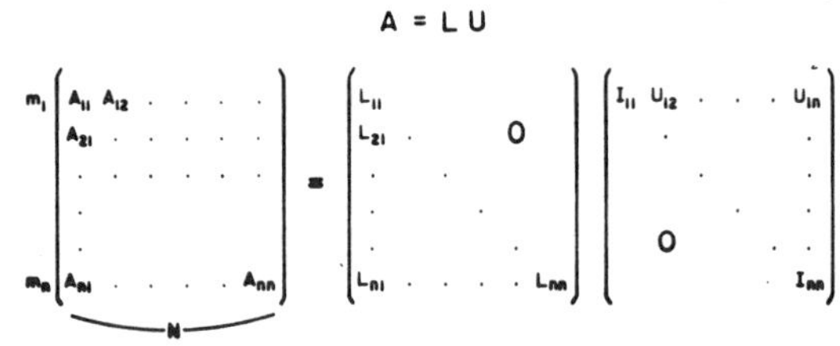

Fig.1 Block LU decomposition

Let $A \in \mathbb{R}^{N \times N}$ be partitioned into n^2 submatrices as shown in Fig.1, where $1 \le i \le n$ and $\sum_{i=1}^{n} m_i = N$. Note $A_{ii} \in \mathbb{R}^{m_i \times m_i}$. Let L_{ij} and U_{ji} denote respectively the submatrices of the block L,U factors of A in the (i,j) and (j,i) positions. The following algorithm performs block LU decomposition:

Algorithm 1.1. Block LU-decomposition:

(1) $i=1$;

(2) Compute $\tilde{L}_{ii} = A_{ii} - \sum_{k=1}^{i-1} L_{ik} U_{ki}$;

(3) Compute the inverse of $\tilde{L}_{ii}$, denoted by $\tilde{L}_{ii}^{-1}$;

(4) If $i=n$, stop.

(5) Compute for all k, $i < k \le n$:

$$L_{ki} = A_{ki} - \sum_{l=1}^{k-1} L_{kl} U_{li};$$

$$U_{ik} = \tilde{L}_{ii}^{-1}[A_{ik} - \sum_{j=1}^{k-1} L_{ij} U_{jk}];$$

(6) $i=i+1$; go to (2);

Note that the structure of this algorithm is the same as Crout's method except that each step is based on submatrices instead of scalar elements. Thus, the block L,U factors are not triangular matrices but rather block triangular matrices and the multipliers $\tilde{L}_{ii}$ in step 3 are computed by means of a matrix inversion instead of scalar inversion.

For the algorithm to be well-defined, it is necessary that L_{ii} be nonsingular. George[6] computed the block LU factorization under the condition that the operand matrix is diagonally dominant. However, Bunch[1] pointed out that diagonal dominance of A is not sufficient to guarantee that the algorithm is well defined. Bunch conjectured that *block diagonal dominance** is a sufficient condition. The following theorem states that strictly block diagonal dominance guarantees the nonsingularity of the diagonal submatrices of each reduced matrix of A, which implies the existence of block L,U factors. The proof can be found in [10].

Existence Theorem : Let A be a nonsingular $N \times N$ matrix partitioned as in Fig.1. If A is strictly block row diagonally dominant, then each reduced matrix of A in block Gaussian elimination is strictly block (row) diagonally dominant.

Remark: The uniqueness of block L,U factors is proved in [10].

Sparsity has been exploited extensively in scalar LSE solution algorithms to reduce computational complexity. This concept is used block-wise in BLOSSOM. Operations performed by BLOSSOM involve nonzero blocks only. To exploit the sparsity further, concurrent subtasks of block LU decomposition on matrices of special topological structures such as BBDF, BBTF can be executed in parallel by BLOSSOM. These subtasks are, for example, inversion of diagonal submatrices on the block diagonal of a matrix in BBDF, forming of L factors on the i'th block column, etc.

Dahlquist et al[2] indicated that a compact scalar Gaussian elimination algorithm can not incorporate complete or diagonal pivoting easily because of the order in which L,U factors are computed. This limitation also applies to block LU decomposition. When partial pivoting is used in block LU decomposition, square blocks in the same block column of the present pivot are checked to find the one with smallest-norm inverse. We add partial pivoting to algorithm 1.1 as follows: after L_{ki}'s are computed (including $\tilde{L}_{ii}$), compute their inverses. Compare the norms of these inverses to determine a new pivot, and then compute U_{ik}'s. The hardware implementation of block LU decomposition is thus based on algorithm 1.1 with partial pivoting added.

3. System Architecture

The BLOSSOM system consists of a host-interface, an executive control unit(ECU), a sequencer, and a reconfigurable processor array as depicted in Fig.2. The following is a behavioral description of each functional block.

3.1. Host-Interface

The host machine acts as an intelligent interface between the user and BLOSSOM. A request for a LSE solution issued by the user is sent to BLOSSOM by the host machine. If BLOSSOM is not being occupied, data will be loaded into BLOSSOM, freeing the host for other duties.

BLOSSOM is intended to be capable of being attached to different kinds of host. For this reason, the host-interface consists of two parts. The first part is dedicated to accommodate different host machines by providing facilities such as converting integer numbers into floating point numbers and matching different word lengths. The second part initiates block LU decomposition as described in the sequel.

The data path and control path are generally kept separate in most processor designs. In BLOSSOM, the operation set is very small and the amount of data is very large. Hence control signals are embedded into the data stream. Since the host-interface recognizes only partitioned matrices and vectors, the host provides a data separator for each submatrix and vector segment in the data stream. The host-interface generates a proper data representation each time a separator is encountered. This data representation and the instruction words are then sent to ECU, while data

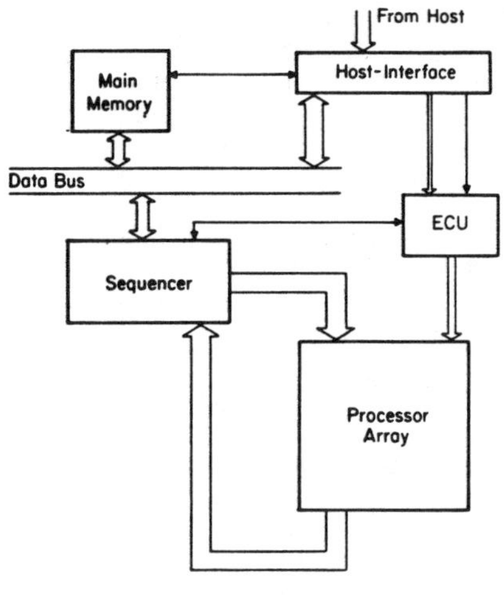

Fig.2 Block Diagram of System Architecture of BLOSSOM

are sent to the main memory.

3.2. Executive Control Unit(ECU)

The ECU decodes host-interface instructions on matrices into sequencer instructions on submatrices and vector segments. Thus each task requested by the host is partitioned into several subtasks and each subtask is carried out by the processor array under supervision of the sequencer.

The set of sequencer instructions corresponding to each host instruction is stored in a control memory of ECU, and the status of the sequencer is monitored by the ECU. The ECU also generates control signals indicating whether processors in adjacent rows (columns) should be connected. We denote them by RECONr.i (RECONc.j).

The subtasks under execution by the sequencer are monitored by the ECU. Whenever a subtask is completed, the ECU can perform one of the following actions on the next subtask on the queue: execute it immediately, wait until more processors become available, or wait until all processors become available. The choice among these possible actions is determined by the sizes of the subtask and the available processor subarrays.

3.3. Sequencer

The ECU generates concurrent requests for the sequencer, which in turn initiates and monitors actions in the processor array. The control logic in the sequencer is then divided into four parts: the Task Control Unit (TSU), the Memory Management Unit (MMU), the Sequencing Unit (SU), and the Feedback Buffer (FB), as shown in Fig.3.

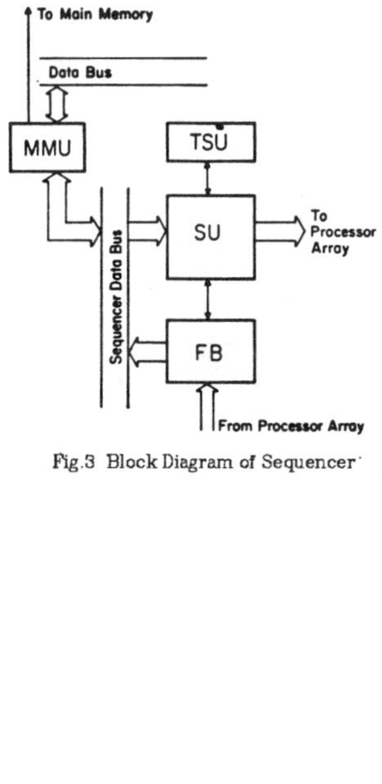

Fig.3 Block Diagram of Sequencer

---

* A is block (row) diagonally dominant if A_{ii} is nonsingular and $\sum_{j=1, j\neq i}^{n} \|A_{ii}^{-1}\| \|A_{ij}\| \leq 1$ for $1 \leq i \leq n$, where $\|\cdot\|$ denotes an operator norm. A is strictly block (row) diagonally dominant if the strict inequality holds for each i.[4]

3.3.1. Task Control Unit & Memory Management Unit

The TSU interfaces with the ECU and maintains the status registers. Each instruction received from the ECU is directly added (instead of decoded) into the data stream that is loaded by the MMU and sequenced by the SU.

The MMU manages the main memory system. Since the amount of data of each operand matrix can be very large, virtual addressing scheme is used. The operand address received from the ECU is the starting virtual address of a submatrix or vector segment. The MMU computes the ending virtual address from it and translates both virtual addresses to physical addresses. The MMU is required to fetch data for concurrent tasks from the multiport memory. The host-interface would try to allocate potentially parallel data objects into different parts of the memory hierarchy to be fetched through different ports. However, memory conflicts are possible when pivoting is employed. If conflicts happen, the MMU would inform the TSU. The latter can either reject the subtask and return it to ECU, or keep the subtask in its queue till the conflict disappears.

3.3.2. Sequencing Unit & Feedback Buffer

The SU adds different delays to data loaded by the MMU before sending them to different processor rows or columns according to the requirements of the computing algorithms. The computation of the number of these delays is done in the SU under the control of the TSU. Since the delay is uniform from processor column to column and from processor row to row in each subtask, only one computation is needed for each subtask. The computations for different subtasks are done sequentially to limit the hardware complexity.

The FB is used when the data output from the processor array is resent to the processor array again. To the SU, the FB is a multiport buffer between it and the processor array. The FB has the capability to reverse the existing delay relationship among a row of data.

3.4. Processor Array

The processor array consists of processor elements, local data links, local control links, SU data registers, FB data registers, and the ECU RECON buses, as shown in Fig.4. The local data links carry multiplexed data words. There is one ECU RECON bus per processor row and one per processor column.

Each subprocedure used by block LU decomposition has a corresponding microprogram stored in each processor, which is activated when that subprocedure is initiated by the sequencer. These subprocedures assume that all segments of a row (or column) of an operand are stored in one processor row (or column). A register file in each processor is used as a queue to enable such a storage scheme. Each processor has four data buses connecting to its neighbors which are named "data ports". The processors also make individual decisions on whether connections to neighboring processors are maintained or not by invoking proper microprograms.

4. Partitioned Matrix Operations

The procedures employed by block LU decomposition may use any or all forms of matrix, submatrix, vector, vector segment, and scalar for their operands. A procedure is implemented by a string of commands issued by its controlling module, and is labeled by its controlling module. For example, block LU decomposition is ECU controlled. Due to the present space limitation in this paper, only submatrix inversion is described in details here. Other operations are listed for reference only.

4.1. Pipelined Inversion of a Submatrix

Let $P \in R^{p \times p}$ be the operand matrix and $I \in R^{p \times p}$ be the unity matrix. The pipelined inversion is essentially Gaussian elimination applied to the augmented matrix $[P|I]$ resulting in $[I|P^{-1}]$. The procedure is listed below:

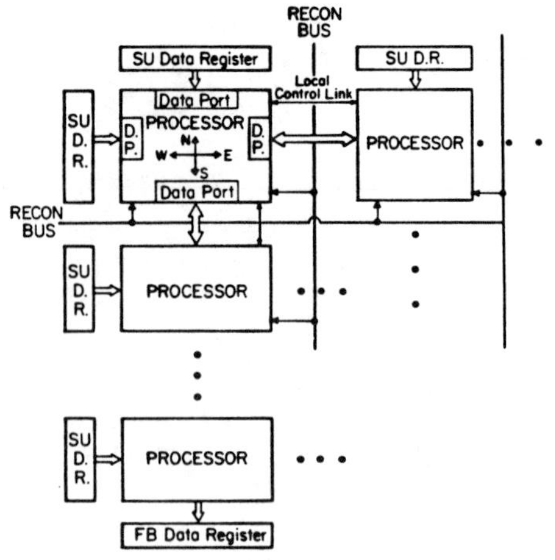

Fig.4 Processor Array

for $j=1$ to p {
$B_{jj} = \frac{1}{B_{jj}}$;
$B_{js} = B_{js} B_{jj}$, for *all* $s=1$ *to* p;
if $i \neq j$
then $P_{ik} = P_{ik} - P_{ij} P_{jk}$, for *all* $k,i=1$ *to* p;
$I_{ik} = I_{ik} - P_{ij} I_{jk}$, for *all* $k,i=1$ *to* p;
}

Note that $B_{xy} = [P_{xy} \mid I_{xy}]$, and P_{xy}, I_{xy} are scalars.

Two important features are added to the hardware implementation of this procedure: "folding of operand submatrix" and "neighbor pivoting". The inverting process is partitioned into two phases. The first phase is triangularization with neighbor pivoting. Backward substitution is the second. The two phases are executed as two contiguous steps following a one-time loading of data.

When neighbor pivoting is used, the leading element of the next adjacent row is reduced to zero by adding a multiple of the current row. When this multiplying factor exceeds unity, the former row becomes the current row and the latter has its leading element reduced to zero. Thus this factor is kept fractional, which suggests that this triangularization process is numerically stable as confirmed by empirical results[15, 13].

Let Pj<k> denote the k'th segment of the j'th row of "P" and Pj<k>g denote the g'th nonzero element of Pj<k>. The same notation applies to the unity matrix "I" and every operation applied to Pj is applied to Ij as well. We describe next the implementation of the algorithm on a per-processor-row basis. The following symbols are used: Pe denotes the current pivot row on that particular processor row, Pi the row that is currently processed, Po the row that has its leading element reduced to zero and sent to the next processor row, "pr" the processor row index, "ino" the iteration number of the whole operand in the processor array, "$q \times q$" the size of the processor array, and "h" the number of segments for any one column (or row) of P after folding.

Algorithm 4.2.1 Triangularization:

(1) $ino=1$;
(2) If Pe<ino>1 < Pi<ino>1 then go to (6);
(3) $m = \frac{Pi\text{<}ino\text{>}1}{Pe\text{<}ino\text{>}1}$;
(4) $Po \leftarrow Pi - mPe$;

(5) go to (9);

(6) $m = \frac{Pe<ino>1}{Pi<ino>1}$;

(7) $Po \leftarrow Pe - mPi$;

(8) $Pe \leftarrow Pi$;

(9) $\#pi = \#pi + 1$;

(10) If $\#pi < (p - q(ino - 1))$ then go to (14);

(11) $\#pi = 1$; $Pe \leftarrow Pi$;

(12) $ino = ino + 1$;

(13) $nu = pr + q(ino - 1)$; $Pnu = Pe$;

(14) If $ino \le h$ then go to (2),

(15) stop;

In this algorithm, each element of a segment of a row of "P" is processed in a way that it lags or leads its neighbor-column elements by one step as in most algorithms used in systolic arrays. The multiplier "m" is generated from the diagonal position and transmitted to every column. Since the "I" matrix has different zero-nonzero topological structure from the "P" matrix, the algorithm has two distinct phases for "P" and "I". The "I" phase can significantly lag the progress of "P" matrix in time, but the operations are essentially the same. A similar backward substitution algorithm can be found in [10].

4.2. Other Matrix Operations

The operation set of BLOSSOM also includes: system solution and matrix multiplication (both controlled by the ECU), submatrix multiplication, submatrix addition, submatrix norm computation, and submatrix-vector multiplication (all controlled by the SU). Detailed description of these operations are listed in [10].

5. Concluding Remarks

BLOSSOM differs from other special purpose machines by using block LU decomposition and by exploiting matrix sparsity commonly expected in engineering problems. Table 1 lists the comparison of BLOSSOM with other hardware schemes proposed in [12, 7, 9, 8, 5], where the complexity of $O(Nwh^2)$ for a LSE solution by BLOSSOM happens in the case of $N \times N$ BBDF matrices under mild conditions[10]. With the architecture designed to exploit the parallelism of the block LU decomposition algorithm, BLOSSOM is efficient in solving problems encountered in the VLSI era. Current work on BLOSSOM also involves its simulation, its performance evaluation, and the expansion of its operation set.

6. Acknowledgement

The authors wish to thank Mr. Giovanni De Micheli for the helpful editorial comment. This work was sponsored in part by the Naval Electronic Systems Command contract (DARPA N00039-83-C-0107).

References

1. James R. Bunch, "Block Methods for Solving sparse Linear Systems," in *Proceedings of the Symposium on Sparse Matrix Computations*, September 1975.
2. Germund Dahlquist, Ake Björck, and Ned Anderson, *Numerical Methods*, Prentice-Hall, 1974.
3. M.M. Denneau, "The Yorktown Simulation Engine," in *Proceeding of the 19th Design Automation Conference*, pp. 55-59, 1982.
4. David G. Feingold and Richard S. Varga , "Block diagonally dominant matrices and generalizationa of the Gershgorin circle theorem," *Pacific J. of Math.* , pp. 1241-1250., 12(1962).
5. Ron J. Gal-Ezer, K. S. Arun, and D.V. Bhaskar Rao, and Sun-Yuan Kung, "Wavefront Array Processor: Language, Architecture, and Applications," *IEEE Trans. on Computers*, vol. C-31, no. 11, November 1982.
6. Alan George, "On block elimination for sparse linear systems," *SIAM Numerical Analysis*, pp. 585-603, 1974.
7. Lennart Johnsson, *A Computational Array for the QR Method*, MIT Conference in Advanced Research in VLSI, January 1982..
8. Kai Hwang and Yeng-Heng Cheng, "Partitioned Matrix Algorithms for VLSI Arithmatic Systems," *IEEE Transactions on Computer*, vol. C-31, December 1982.
9. R.M. Kieckhafer and Christopher Pottle , "A Processor Array for Factorization of Unstructured Sparse Matrices," in *ICCC Confer. Proc.* , 1982.
10. Fu-Hwa Ko and Alberto Sangiovanni-Vicentelli, *BLOSSOM: An Algorithm and Architecture for the Solution of Large-scale Linear System.* Initial Report
11. H.T. Kung, "Why Systolic Architectures?," *IEEE Computer*, January 1982.
12. H.T. Kung and C.E. Leiserson, "Algorithms for VLSI Processor Arrays," in *Symposium on Sparse Matrix Computations and Their Applications*, 1978.
13. L.D. Rogers, *Optimal Paging Strategies and Stability Considerations for Solving Large Linear Systems*, 1973. Ph.D. thesis, University of Waterloo, Department of Computer Science.
14. Andrei Vladimirescu, *LSI Circuit Simulation on Vector Computers*, October 1982. Memo. No. UCB/ERL M82/75.
15. W.M. Gentleman and H.T. Kung, "Matrix Triangularization by Systolic Arrays," in *Proc. SPIE Symp.*, vol. 298, SPIE 1981.

Table I. Comparison of hardware schemes on LSE solution

Attributes	Systolic Array *by H.T. Kung et al* *: LU on banded matrix	Systolic Array *by H.T. Kung et al* *: Ortho. Triangularization	Computational Array for QR *by Johnsson*	LU of unstructured sparse matrix *by Pottle*	Partitioned matrix algo. *by Hwang and Cheng*	BLOSSOM	WAP *by S.Y. Kung et al*
LSE Soln. Algorithm.	Needs FE & BS facilities.	Needs BS facility.		Needs FE and BS facilities.	Needs extra special hardware	Offered	
Size limit of operand	Window size ≤ sz(PA)*	sz(PA)	Computational window ≤ sz(PA)	No limit if segmented.	No limit. Intended for any size.		Not for all sizes unless expanded.
Pivoting Strategy	no	NeP	no need	None on the fly	None	NeP and ParP**.	None specified
Interprocessor Comm.	Local. Sync.	Pipeline.	Local. Pipeline. Some broadcasting. Sync.		Not specified.	Local. Pipeline.	Handshake mechanism.
Complexity	$3n + \min(r,s)$†	$3n$	$rn+3n+an$	nl††	$O(N)$ or $O(\frac{N^2}{m})$‡	$O(\frac{N^3}{x^2})^{(1)}$, $O(Nwh)^{(2)}$,	n
Number of processors	rs	$\frac{n^2+n}{2}$	$r(3(r+s)+2)$	$\ge d^{\&}_{max}$	$O(N^2)$ or $O(Nm)$	x^2	n^2
Feasible operand prop.	Has L,U factors.		Any matrix within size limit.	Needs to be symbolicly factorized.	Pivotal block in each step has L,U.	Block diagonal dominance.	Same as Hwang's if extended.

* sz(PA) denotes size of processor array.($n \times n$ in most cases)

** NeP denotes neighbor pivoting. ParP denotes partial pivoting.

† $r+s-1$ is the bandwidth of a banded matrix.

†† l denotes the mean degree of columns of a sparse matrix.

‡ N denotes size of operand matrix. m denotes submatrix size.

(1) matrix is full. (2) w is the original border width. h is a quantity related to x and the sizes of operand submatrices.

& d_{max} denotes the maximum number of nonzeros in any row of the operand.

Session 2-3c
Artificial Intelligence

Chairmen

P. W. Horstmann
Syracuse University

S. Goto
NEC

ON INTELLIGENT CAD SYSTEMS FOR VLSI DESIGN

S. Chen

Department of Computer and Information Sciences,
Department of Mathematics, University of Florida
Gainesville, Florida 32611

ABSTRACT

Some issues of intelligent CAD and knowledge-based expert systems for VLSI design are discussed. A simple distributed expert system for VLSI design is suggested. At the data level of the expert system, a semantic model generalizing relational data base for hierarchical VLSI design environment is proposed.

1. INTRODUCTION

Since the beginning of computers, the possibility of designing new and more powerful computer systems by existing systems have been looking into. Whether full automation can ever be reached or human intervention is still necessary is a much debated question. At present, full automation seems to be unrealistic. With the advances in artificial intelligence and knowledge-based expert systems, the concept and the notion of intelligent CAD systems or expert systems for VLSI design have begun to attract attentions among scientists of various disciplines. Even though a truly intelligent expert system has yet to be established in the near future, some partially intelligent CAD systems have been accomplished. See references in (1), (8),(18), (20),(21),(23).

In this paper, we shall formalize and discuss some issues of knowledge-based expert systems for VLSI design. Because of the ever progressing and changing VLSI technology, these systems must be modular, extendable and complex. We propose a simple but efficient distributed expert system for this purpose. Several modules of knowledge bases or algorithm packages are interconnected by a ring or bus network. A global controller integrates possibly different views from different modules or design databases into a blackboard. The complete CAD system should contain at least the following features:

1. behavior optimization,
2. hardware synthesis,
3. silicon compiler or other approaches,
4. system simulation,
5. verification and testing,
6. interactive user interface

An expert system consists of three levels:

1. data level (declarative knowledge, database),
2. knowledge-base level (program, operator),
3. control level (control strategy, state space search)

The data level of VLSI design is very complex. In fact, data structures are so large that they are often large files. The first requirement for a successful CAD system will be to design a systematic and efficient data representation scheme. We propose a semantic geometric model which generalizes the frame system (Minsky), relational database (Codd), generalized relational database (Su and others) and semantic network for representing data and knowledge. The database has to be distributive and different views are integrated. The knowledge and control levels are more simpler than ordinary expert systems for human interfaces still play an important role in present day VLSI CAD systems. The distribution is over the ring or bus structure of several modules of algorithm packages implemented in hardware or software. Thus the control strategy is quite easy. Only few papers are concerned with artificial intelligence techniques applied to VLSI design (20), (21), (26).

2. KNOWLEDGE-BASED EXPERT SYSTEMS

There are three levels of an expert system. The data level represents declarative knowledge. The knowledge-base level supplies domain-specific problem solving knowledge and contains procedural knowledge to manipulate the data for a problem and to solve the problem. The control level gives control strategies of procedural knowledge.

Declarative knowledge is given by formulas in first order logic, frames or semantic networks. Frames are data structures in which all knowledge about a particular cell, chip or PCB is stored. Frames contain fields or slots for various information data and values. Hierarchically, frames have pointers to other frames. This representation is quite convenient for VLSI data files. Semantic networks (semantic nets) are graphs whose nodes represent cells, modules or chips while labeled arcs represent relations (connections) among them. Again, semantic networks support hierarchical data and are suitable for VLSI applications.

The procedural knowledge at the knowledge-base level can be given as computer programs called pattern invoked programs or as operators mapping state to state in the state space. The state space search from the initial state to the goal state is performed by these operators. Another description of this level is the concept of production systems. If certain conditions hold, then primitive actions are taken to change states. Procedural knowledge is represented using production rules.

At the control level, control strategies are needed for state space search. Data-driven search and goal-driven search are two common methods. A basic technique is backtracking by exploring one path as far as possible. If the path deadends or procedure backtracks to a previous state, a new operator is chosen.

State space search usually can be extensive. For the immense complexity of VLSI devices, a truly intelligent expert system is not feasible. We like to emphasize that CAD does not require optimal solutions but near optimal solutions. Thus, our system will focus more on the data level and how to simplify data representation than the knowledge-base and control levels. For VLSI design, we only have few operators or programs in the knowledge-base level so that the control level is rather simple.

3. DISTRIBUTED EXPERT SYSTEMS

Distribution and parallelism are used in expert systems (7), (13), (22). The simple problem reduction method decomposes or partitions the problem into subproblems. Iteratively, the problem is partitioned into primitive ones. Each primitive problem can be solved on a component expert system. A more general environment is a collection of cooperating expert systems communicating among them over an interconnection network. Smith (22) proposed a contract net for distributed problem solving. There are a great deal of communications of that system. We propose an efficient distributed system with a ring or bus interconnection network whose protocol and control are rather standard and simple (5). A global controller or a metaplanner controls the communication and problem solving globally. It is interfaced with users who are VLSI designers. A blackboard integrates various different views from the component expert systems and is controlled by the global controller. Each component expert system is concerned with one specific problem domain. For instance, in automated placement and routing, each stage of initial placement, dissection, loose routing, fine routing and adjustment may be one component system or the whole layout strategy is run on one system. In the case of silicon compiler, one may have a language translator and a hardware generator component systems together with a feedback system for size, power and performance evaluation and design adjustment. Other important factors, such as test generation, rule verification, system simulation, hardware synthesis, can be implemented on several component systems connected by a ring or a bus. The communication network will send the data from one component to another in the order of design procedures.

The hierarchical planning and problem solving tactics proposed by Sacerdoti (19) offers a general approach which is applicable to VLSI hierarchical design environment. Plans are represented as a partial ordered sequence. A sequence may be modified to complete the design.

4. VLSI DATABASES

VLSI data structures are extremely complex. We focus our attention to the data level of the expert system. In the hierarchical environment, lower level data attributes should not need to be stored. Otherwise, the data types become very cumbersome. One can use some algorithms to compute those unstored data and only store those algorithms. This is one aspect of an intelligent system. By using more regular and library modules, the data structure is further simplified. The reason that regular structures such as memory achieve much higher and denser capacity is the design complexity mentioned here. Manipulating hierarchical data wisely will increase the VLSI device capacity. The structural approach will simplify the data types and clarify these algorithms used in CAD.

Equivalent logic types should be grouped into equivalent classes. Each logic type is given a name which will be the key for accessing the data. A VLSI device is partitioned into subdevices which are further partitioned into components. Thus, a device with a certain name will have pointers toward characteristics, such as wiring capacitance, clock time, power voltage, pin type and many others, and component logic types, together with their characteristics. This complex network will be simpler with this structural analysis. Within a fixed technology domain, this can be achieved, but will have to be updated constantly.

Symbolic and graphical design approach is merged with the hierarchical approach. Data types become two dimensional. For the future, we may speculate three dimensional data. Implemented in CAD graphics environment, data representation is generalized to image database. We have generalized the semantic data model of Su (24) and others to a symbolic database model. For simplicity, we shall explain these ideas in terms of the semantic data model of (24) and its extension. A generalized relational database model is proposed with several associations and operations. The formal properties of these associations and operations are studied. In this formal study, consistency, dependancy and management can be treated in a systematic way. Design rule checking and verification are made easier as consistency verification of the database. By associations-aggregation, interaction and composition, we may produce new design by existing designs. Summarization is useful to give descriptive and statistical data and information. Wiring, connectivity and boundary components are expressed in terms of these associations. Furthermore, in this structural environment, new devices of similar types may be produced simply by update, modification of existing design databases.

This data model may be imbedded into a universal semantic model proposed by D. Scott. It is an infinite dimensional algebraic structure with three Boolean values- true, false and undefined. Several simulation models (2), (4), (10), (12), (16) are multi-valued logics (finite-dimensional algebraic structures) with two Boolean values-true and false, and are used in test generation and digital logic simulation. Our new approach includes all previous results. It is not surprising that when data structures are analysed carefully, many operators (algorithms, programs or artificial intellience systems) on them as operands become simpler. We hope that our suggestions will be useful to the VLSI community and actually be implemented.

5. CONCLUSION

A systematic approach toward intelligent CAD systems (or expert systems) for VLSI design is proposed. There are two new points. A modular and distributed expert system for VLSI design is suggested, together with a general data representation scheme. We also note that due to human interface CAD systems for VLSI need not be so complicated as usual artificial intelligence systems.

6. REFERENCES

1. J. P. Avenier, Digitizing, layout, rule checking- the everyday tasks of chip designers, Proc. IEEE, Vol. 71, No.1, Jan. 1983, 49-56.

2. M. A. Breuer and R. L. Harrison, Procedures for eliminating static and dynamic hazards in test generation, IEEE Trans. Computers, Vol. C-25 No.6, June 1976, 630-636.

3. B. Chandrasekaran et al., An approach to medical diagnosis based on conceptual structures, Proc. 6th International Joint Con. Art. Intell., 1979, 134-142.

4. S. G. Chappell, C. H. Elmendorf and L.D. Schmidt, LAMP: logic-circuit simulator, The Bell System Tech. J., Vol. 53 No. 8, Oct. 1974, 1451-1476.

5. S. Chen, Distributed Expert Systems, in preparation.

6. J. A. Darringer, W. H. Joyner, L. Berman and L. Trevillyan, Logic synthesis through local transformations, IBM J. Res. Develop., Vol. 25 No. 4, July 1981, 272-279.

7. L. D. Erman and V. R. Lesser, A multi-level organization for problem solving using many diverse cooperating sources of knowledge, Proc. 4th International Joint Conf. Art. Intell., 1975, 483-490.

8. M. Feuer, VLSI design automation: an introduction, Proc. IEEE, Vol. 71 No. 1, Jan. 1983, 5-9.

9. S. J. Hong and R. Nair, Wire-routing machines-new tools for VLSI physical design, Proc. IEEE, Vol. 71 No. 1, Jan. 1983, 57-65.

10. Y. H. Jea and S. A. Szygenda, Boolean vector algebraic structures: properties and applications to digital logic simulation, Computer Aided Designs, Vol. 11 No. 6, Nov. 1979, 315-323.

11. C. Y. Lee, An algorithm for path connections and its applications, IRE Trans. Electron. Comp., Vol. EC-10, Sept. 1961, 346-365.

12. S. C. Lee, Vector Boolean algebra and calculus, IEEE Trans. Comp., Vol. C-25 No. 9, Sept. 1976, 865-874.

13. D. Lenat, BEINGS: knowledge as interacting experts, Proc. 4th International Joint Conf. Art. Intell., 1975, 126-133.

14. C. Mead and L. Conway, Introduction to VLSI Systems, Addison-Wesley, 1980.

15. E. F. Moore, Shortest path through a maze, Annals Computing Lab. Harvard University, Cambridge MA, Vol. 30, 1959, 285-292.

16. P. Muth, A nine-valued circuit model for test generation, IEEE Trans. Comp., Vol. C-25 No. 6, June 1976, 630-636.

17. D. S. Nau, Exper computer systems, Computer, 16, 1983, 63-85.

18. C. Niessen, Hierarchical design methodologies and tools for VLSI chips, Proc. IEEE, Vol. 71 No. 1, Jan. 1983, 66-75.

19. E. D. Sacerdoti, Problem solving tactics, Proc. 6th International Joint Conf. Art. Intell., 1979, 1077-1085.

20. C. H. Sequin, Managing VLSI complexity: an outlook, Proc. IEEE, Vol. 71 No. 1, Jan. 1983, 149-166.

21. S. G. Shiva, Automatic hardware synthesis, Proc. IEEE, Vol.71 No.1, Jan. 1983, 76-87.

22. R. G. Smith, A framework for distributed problem solving, Proc. 6th International Joint Conf. Art. Intell., 1979, 836-841.

23. J. Soukup, Circuit layout, Proc. IEEE, Vol. 69, No. 10, Oct. 1981, 1281-1304.

24. S. Su, SAM: a semantic association model for corporate and scientific/statistical databases, CIS Tech. Report, No. 8182-6, University of Florida, June, 1982.

25. T. Sudo, T. Ohtsuki and S. Goto, CAD systems for VLSI in Japan, Proc. IEEE, Vol. 71 No. 1, Jan. 1983, 129-143.

26. G. J. Sussman et al., Design aids for digital integrated systems, an artificial intelligence approach, Proc. IEEE Int. Conf. Circuits and Computers, 1980, 612-615.

27. S. Trimberger, J. A. Rowson, C. R. Lang and J. P. Gray, A structured design methodology and associated software tools, IEEE Trans. Circuits Sys., Vol. CAS-28, July 1981, 618-634.

SLED/SDL: A FLEXIBLE SYMBOLIC VLSI DESIGN SYSTEM

Gregory A. Haynes and Robert C. Pettengill *

Texas Instruments Incorporated
Dallas, Texas 75265

* E-Systems Incorporated
Dallas, Texas 75266

Abstract

A VLSI design system using a Lisp based procedural design language has been written. Technology independence, adaptability to different symbolic design methodologies, user customization, and ease-of-use are the major design goals. The system supports both integrated circuit layout, schematic, and logic diagram capture. Extensive use of hierarchy, symbolic representation and careful design of the graphics display software allow the system to be used with low-cost graphics terminals.

1. System Overview

Symbolic and procedural techniques in computer-aided design systems have the potential to provide significant improvements in managing the complexity of VLSI circuit design. SLED/SDL is a system developed within a working VLSI design group which applies ideas from the Design Procedure Language (DPL) from MIT [1,2] to the problem of graphical symbolic design. Flexibility and extensibility were emphasized so that this system could serve as a working "breadboard" to test new design methodologies and artificial intelligence based synthesis tools. The initial implementation is on a DEC VAX 11/780 using VT125 terminals and the Franz Lisp language [3]. The use of hierarchy, bounding boxes [4], and symbolic representation, along with careful design of the graphics display software, provides adequate graphics performance with a low-cost display.

This paper describes the two major components of the design system: SDL, a symbolic design language based on the Design Procedure Language, and SLED, a Symbolic Layout EDitor, which manipulates SDL definitions to produce geometric layouts, schematics, and logic diagrams. The goal has been to combine the powerful procedural definition capabilities found in text oriented design languages with the graphical editing characteristics of symbolic layout systems.

The large address space of the VAX makes it possible to include the data required for a design (even for a very large circuit) in a single data structure which resides in virtual memory. Designs are constructed from <u>geometric primitives</u> including rectangles, arbitrary closed polygons, wires, lines, and points. Documentation is provided by means of text strings that can be placed in much the same way as the geometric primitives. The basic structural unit is called a <u>cell</u> or <u>symbol</u>, and hierarchy is provided by allowing cells to be composed of other cells (with the only restriction being that recursive composition is not allowed). The data structure is uniform in that geometries or primitive symbols are represented as cells which have the restriction that other cells cannot be placed within them. Cells that contain other cells are called <u>composite</u> cells. A cell placement is referred to as an <u>instance</u> of the cell.

The primary reason for combining the design information in a single data structure is to make it easier for the designer to edit the design. In general, much of the information about connectivity, certain circuit parameters, and signal names needs to be entered only once, and if the data is changed, it is only necessary to change it in one place, rather than in multiple files or documents. This can reduce the number of errors and the amount of conflicting information that might otherwise occur.

Several features make the SLED editor easy to learn and easy to use. Since both composite symbols and primitive symbols are described in a uniform manner, it was easy to define a small set of commands that modify any graphic object in a uniform way. Editors that provide different command sets for different types of symbols are much more complex for the user. In addition, the same editor is used for entering logic diagrams and for all phases of editing layouts. The user interface was made modeless by allowing any command to be given at any time. Not only does this eliminate the need for special menus of legal commands, but it allows a more natural style of design.

In SLED, the design data may consist of a logic diagram (Fig. 1) and a layout (Fig. 2). Other arbitrary collections of data, referred to as <u>abstractions</u>, may be incorporated in the future into SLED. Each abstraction is divided into <u>partitions</u>, which currently include the structure partition and the symbolic partition. Each cell (except for the geometric primitives) may have a logic abstraction, a layout abstraction, or both, and each abstraction generally has a structure partition and may have a symbolic partition. Both the structure partition and the symbolic partition may contain placements of geometric primitives, but placements of composite cells are restricted to be in the structure partition. The logic diagram and the layout have separate hierarchical decomposi-

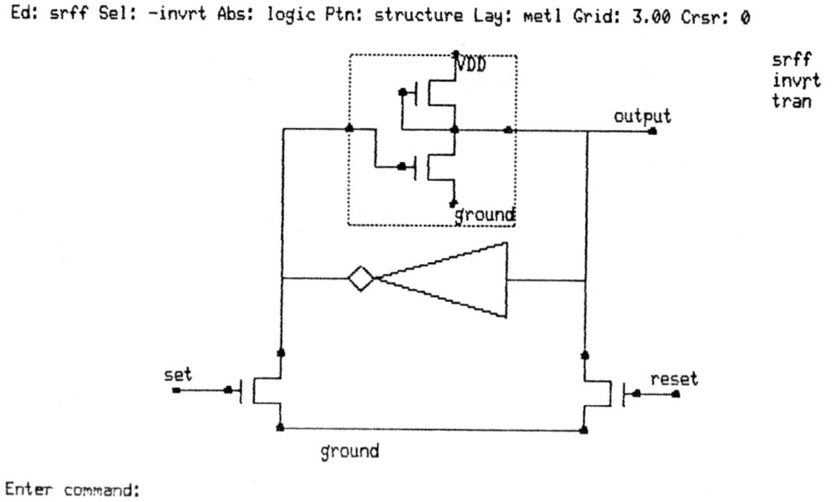

Fig. 1. A SLED logic abstraction example.

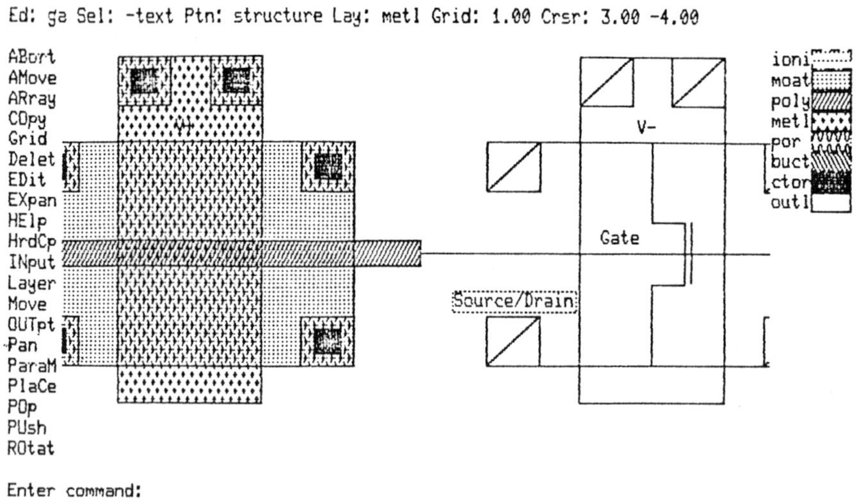

Fig. 2. A SLED layout example with SYMBOLIC and STRUCTURE visible.

tions. The designer is not forced to maintain identical physical and logical hierarchies as in many other design systems.

In order to take advantage of existing operating system, source control, and database management tools, each cell or symbol in the design is stored externally in a separate file. Since the symbolic partition of a cell contains only geometric primitives and not placements of composite cells, each file contains all the information needed for SLED to display the symbolic view of the cell. As cells are expanded to show their structure, the sub-cells may be loaded into the internal data structure automatically as needed. A search path mechanism allows the user to specify the order in which directories are searched for the occurrence of the cell.

2. SDL - Symbolic Design Language

The Symbolic Design Language, SDL, is based on the Design Procedure Language and the RSA chip assembler [5]. The two types of files where SDL is used are technology files and symbol definition files.

The technology file (Fig. 3) performs two functions: It contains the technology dependent design rule information for a set of designs, and it describes the technology dependent display information for the SLED graphics editor.

Symbol definition files (Fig. 4) describe the design symbols or cells and how they combine to make up the design. Each file contains a single def-sym function which describes the symbol. The

```
;;; Sample CMOS SLED Technology Initialization File
(setq *scale* 0.25)        ; lambda in real user units
(setq *pxPerGrid* 12)      ; use 12 for GIGI
(setq *default-grid* 0.5)  ; cursor grid in microns
(setq *directory-path*  ; look in current dir first
  '("ud: cad.layout " "ud: cad.sled ") )
;;;; Layer Definition ;;;;;;;;;;;;;;;;;;;;;;;;;;;
;;;; specify color, line type, and stipple pattern.
(def-layers
  (IONI 1 1 0  2  o)  ; Vt ion implant, yellow
  (MOAT 1 0 0  1  b)  ; source/drain diffusion, red
  (POLY 0 0 1  1  k)  ; poly silicon gate, blue
  (OUTL 1 1 1  1  z)  ; symbolic outlines
  ... )
;;;; Design Rules ;;;;;;;;;;;;;;;;;;;;;;;;;;;;;;;;
(def-rules
  (IONI width 6.0)
  (MOAT width 3.0)
  (CTOR width 2.0) ... )
(def-rule MOAT space 4.0)
(def-rule MOAT space POLY 1.0)
(def-rule MOAT overhang POLY 2.0)
(def-rule MOAT overlap CTOR 1.0)
;;;; set geometry defaults for the given layers
;;;; specify width, height, angle, etc.
(set-layer-primitive-defaults
  (poly rect ( (w 5.0) (h 5.0) (a 0) ) )
  (poly wire ( (w 2.0) ) )
  (ctor rect ( (w 2.0) (h 2.0) (a 0) ) )
  (ctor wire ( (w 2.0) ) ) ... )
;;; User customization of SLED in Lisp may follow
```

Fig. 3. An example SDL technology file.

```
; CONTENTS OF THE FILE TRAN.SDL
(def-sym tran ()   ; no parameters for this symbol
  ( (layout)
    (symbolic
      (moat
        (line ((0. 3.) (0. 2.) (-1. 2.) (-1. -2.)
               (0. -2.) (0. -3.) ) )
        (point ( (0. 3.) ) () 1 "Drain")
        (point ( (0. -3.) ) () 1 "Source") )
      (poly
        (line ( (-4. 0.) (4. 0.) ) )
        (point ( (-4. 0.) ) () 1 "Gate") ) )
    (structure
      (moat
        (cpr ((0. 0.)) ((w 4.0) (h 6.0) (a 0)) 1
          "Source/Drain") )
      (poly
        (rect ((-4. -1.) (4. 1.)) () 1 "GatePoly")
        )))
  ( (logic)
    (symbolic
      ... )
    (structure
      ... ) ))

; CONTENTS OF THE FILE TRANARRAY.SDL
(def-sym tranarray ()
  ((layout)
    (structure
      (parts        ; a 5 X 2 array of tran symbols
        (tran ((0. 0.)) () 1 "TArray1"
            ( (5 2) (8.0 6.0) ) ) ))))
```

Fig. 4. Example SDL symbol definitions.

def-sym function contains placements of other symbols in the form of Lisp function calls to the placed symbols. In addition, the symbol calls may have arguments which are used to parameterize the design. This symbol definition explicitly supports:

* Multiple views of the design with separate hierarchies. These are called design abstractions.

* Symbolic representations of every symbol abstraction. These are called symbol partitions.

Each symbol definition consists of a list of abstractions in which that symbol is defined. Currently supported abstractions are LAYOUT and LOGIC. Each abstraction consists of a list of partitions. Layout partitions include: symbolic, structure, interconnect, and compaction. Each partition consists of a list of layers. Layers allow display attributes to be set for the SLED editor. Each layer consists of a list of component parts. Composite part symbols are placed on a special layer called "parts". Primitive symbols inherit display attributes from their layers. Part or symbol calls are a list which consists of the symbol name, the location of the symbol in the super symbol (this is a list for some primitives), a list of parameter overrides for procedural symbols, a coded transformation for specifying rotations or reflections, an instance name (unique to this super symbol), an array transform for repeated parts, and finally information about which partitions (or symbolic representations) should be displayed in the SLED editor for this part.

3. SLED - Symbolic Layout Editor

The operation of the editor can be explained in terms of the edit cell, which is the cell currently being modified, and the selected instance, which is the placement to which the modification applies when a command such as DELETE, MOVE, ROTATE, or ARRAY is given. The selected instance may be a placement of a primitive or a composite cell. All the commands that apply to the selected instance operate uniformly on placements of primitives or composite cells. For example, selecting and moving a rectangle is just the same as selecting and moving a composite cell, both in the editor commands and in the software which implements these commands. The only objects which may be selected are those placed one level deep in the hierarchy of the edit cell. Detecting repeatedly on the same point cycles through a circular list of symbols whose bounding box includes the point. The use of the bounding box for the select function results in a simple implementation and in practice seems to be very convenient for the user as well. Edit in place is accomplished by means of the PUSH command, which makes the selected cell become the edit cell. The POP command restores the most recent cell from which a PUSH was given to the status of edit cell.

Two commands are available for making changes in the physical placements of cells. The MOVE command moves the selected cell placement by a specified displacement. The AREA MOVE command is used to move by a specified displacement the individual points that lie within a rectangular region. This single command can be used to lengthen rectangles, move parts of wires, move symbols, or stretch entire sections of a cell.

A HELP command is available which searches the SLED User Manual and displays the appropriate sections of text. After the user has finished reading this text, control returns to the point in the editor at which help was requested, and editing can be resumed.

In a logic diagram, a cell is usually defined to include terminals which are the points at which the cell is connected by wires to other cells. All interconnection is done with the WIRE command, which places wires that are constrained to begin and end on terminals. After the interconnections are placed, it is possible to generate a netlist description of the circuit for logic simulation. Wires that are connected to terminals make use of a generalization of the method of representing points that make up the wire. This feature allows cells to be moved without breaking the connections made by the wires.

4. Summary

We believe that we have made significant progress in creating a flexible, easy-to-use graphical VLSI design system. This system provides a high level of support for symbolic and procedural methods as well as traditional geometric drawing. We have shown that a low cost display can be used very effectively for symbolic design. The rapid decline in the cost of personal workstations with bit mapped displays should allow SLED to be implemented with support for multiple window views of different design abstractions. The macro expansion for procedural symbols has been only partially implemented and the lack of graphical tools for editing of procedural definitions is a significant area where extensions to SLED should be made. SLED and its prototypes have been used for schematic entry, geometric and symbolic layout in both NMOS and CMOS technologies. In addition, placement and routing software is being developed to allow SLED to be used in standard cell designs. The ability to implement the Symbolic Design Language as macro expansion extensions to Lisp, rather than writing our own interpreter and parsers, had a major role in the rapid implementation of this design system.

5. References

1. J. Batali and A. Hartheimer, "The Design Procedure Language Manual", MIT Artificial Intelligence Lab. A.I. Memo No. 598, September 1980.

2. G.J. Sussman, J. Holloway, and T.F. Knight, Jr., "Computer Aided Evolutionary Design for Digital Integrated Systems", MIT Artificial Intelligence Laboratory, AI Memo No. 526, May 1979.

3. J.K. Foderaro and K.L. Sklower, "The FRANZ LISP Manual", University of California at Berkeley, April 1982.

4. K.H. Keller and A.R.Newton, personal communication.

5. R.L. Rivest, "Documentation of the RSA Chip Assembler", MIT Dept. of Electrical Engineering and Computer Science, VLSI Memo No. 80-20, July 1980.

Designing a High-Level Silicon Compiler

Philip E. Agre

MIT Artificial Intelligence Laboratory

Abstract. This talk describes some of the issues bearing on the organization of a silicon compiler, a program that takes as input some code in a high-level language and produces as output an integrated circuit that executes the procedure described by that code. I have written a program called SC that takes as input a (small) program in Scheme (a lexically scoped and tail-recursive dialect of Lisp) and produces as output a layout specification for an NMOS chip that communicates with the Scheme-81 chip and its list-structured memory over the Scheme-81 chip's memory bus, computing the desired function whenever it is provided with an appropriate set of arguments.

SC is organized as a sequence of modules that span the great abstract gap between the program's input and output. Adjacent modules communicate using intermediate representations of the program being compiled. The first few modules correspond to portions of conventional compilers and the last few are taken directly from existing IC layout tools; the interesting portions of SC are those that mediate the conversion between the "compiler" representations, which emphasize the temporal organization of the process described by the program, and the "silicon" representations, which emphasize the physical organization of the chip that will realize this process.

The SC project began as an attempt to capture in a program a part of the design method employed by the designers of the Scheme-81 chip, starting with a program describing the chip's behavior and constructing the chip by tailoring a register machine for efficient implementation of that particular program. My thesis is that the best way to write a silicon compiler is to start with a simple and powerful design method that is employed by real designers to make real chips and to use artificial intelligence techniques to capture that design method in a program.

CH1935-6/83/0000/0413$01.00

Session 2-3d
Digital Systems I

Chairmen

A. M. Sauer
Siemens

S. Hirschhorn
GTE Labs

NEEDS FOR HIGH-LEVEL DESIGN TOOLS

Monika Gonauser and Anton M. Sauer

SIEMENS AG, Corporate Information Technology
Munich, West-Germany

ABSTRACT

Advances in technology and increasing complexity of systems are intensifying the needs for an integrated System Design Environment (SDE). In this paper the authors discuss the structure of a computer system, and the relation of such a structure to the design process. The design activities in and between each level of system design are outlined. Finally, steps toward realization are briefly printed out.

MOTIVATION

The increasing complexity of systems makes the design of systems like computers more and more difficult. Advances in technology, as the basis of each real system, have an increasing influence on the architecture of a system, especially a computer system. New and more complicated computer structures are possible. And therefore, we have to think about design methodologies which meet all requirements and necessities.

The process of designing and deeveloping computer systems is the transformation of the designers' original ideas into a concrete implementation. This process is actually influenced by some technical or behavioural constraints, or by the history of a computer product. The more complex the design of a system gets the greater is the need for computer aided design tools. To do the design task in a cost effective way, the design activities need to be supported and managed by a computer based System Design Environment (SDE).

The goal of SDE is to assist the designer in exploring alternatives and deriving more detailed descriptions from less detailed specification, reviewing and undoing design decisions and tradeoffs, insuring that changes made in one portion will be manifest in the entire design, documenting the design activity, and improving the designers productivity.
If we think about the needs for High-Level Design Tools we have to consider that
- today almost no system assists designers during all design phases,
- there is no system with integrated tools,
- there is no management system, which allows to integrate the work of different parts of a project.

HIGH-LEVEL DESIGN

It is common practice to structure the design process of a computer system roughly into the following 3 steps:

- design and development of architecture and general circuit diagram,
- design of boards and integrated circuits,
- technology process and production.

It's possible to separate this 3 steps, and to define interfaces between each step. And therefore, separated tools in each step can be developed and also used in the design process. But, today, there is no integration.

The steps in the top-down design of a computer system are in a relation to the system structure, and therefore to the levels of system design (Fig. 1) /1/.

In the first step the structure of a system, consisting of hardware and software is analyzed in a top-down manner. At the highest level, which represents the users point of view, the purpose, the attributes and the behaviour of the system are described. Successive refinements of the structural analysis finally lead to the detailed level of circuits, which is of

CH1935-6/83/0000/0415$01.00 © 1983 IEEE

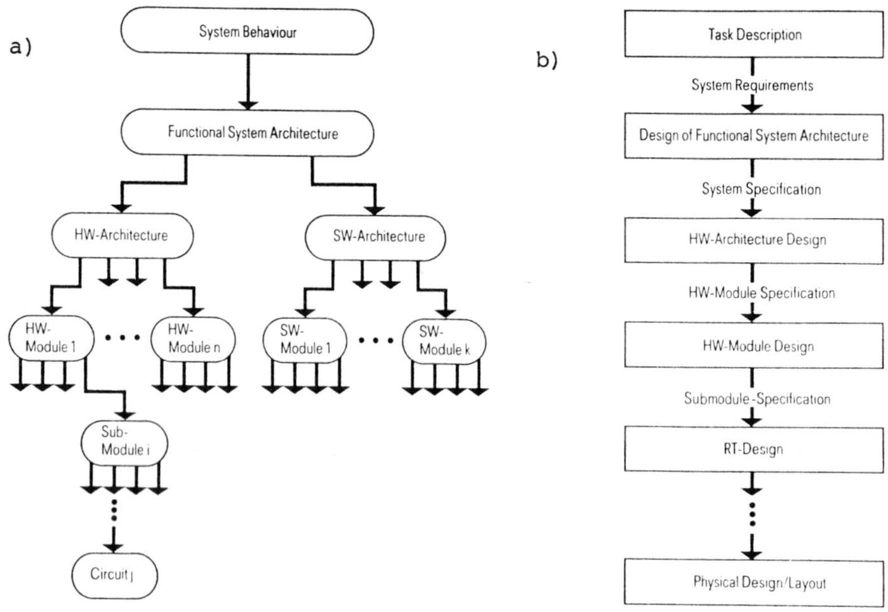

Fig. 1: Relationships between
a) system structure and
b) system design /1/

interest to the designer and manufacturer only (Fig. 1a).

Having discussed a method to describe a system structure at different levels of abstraction and to partition the system within these levels, the next issue is to look at the corresponding steps in designing such a system.

The static structure of the design method represents the framework for the actual design process (Fig. 1b). At all the abstraction levels of Fig. 1a it contains a body for design activities and orders them in a hierarchical manner. It does not show, in which sequence and how often these bodies are "instanciated" by the designer. Besides that, the static structure shows the interfaces between these bodies. In a top-down design process, which is widely accepted as a general design methodology, the result of the activity within each level is a document containing the requirements or specifications for the next deeper level of design. The proposed top-down methodology does not prohibit a bottom-up approach for iterations or for more "local" design steps (The software branch is not discussed here).

Existing tools are predominantly supporting the design process on gate and circuit level /2,3/. But they are not suited to support the designer on system level design, as we suggested in Fig. 1b. Though no such integrated system is known to be in productive use today, a number of techniques have been devised, developed and used to partially support the design process on the system level (for example /4,5,6/). Examples for practical use of this hierarchical method are given in /7/.

DESIGN ACTIVITIES

There is a dynamic flow of design activities between and within the design levels of Fig. 1b.

As stated before, the top-down-approach as a general guideline for the design process has proven its advantages. But a more general design method must take into account influences from lower levels to higher ones, which means for instance, that - if specifications are not met - iterations of the design steps must be possible in a clear and easy manner.

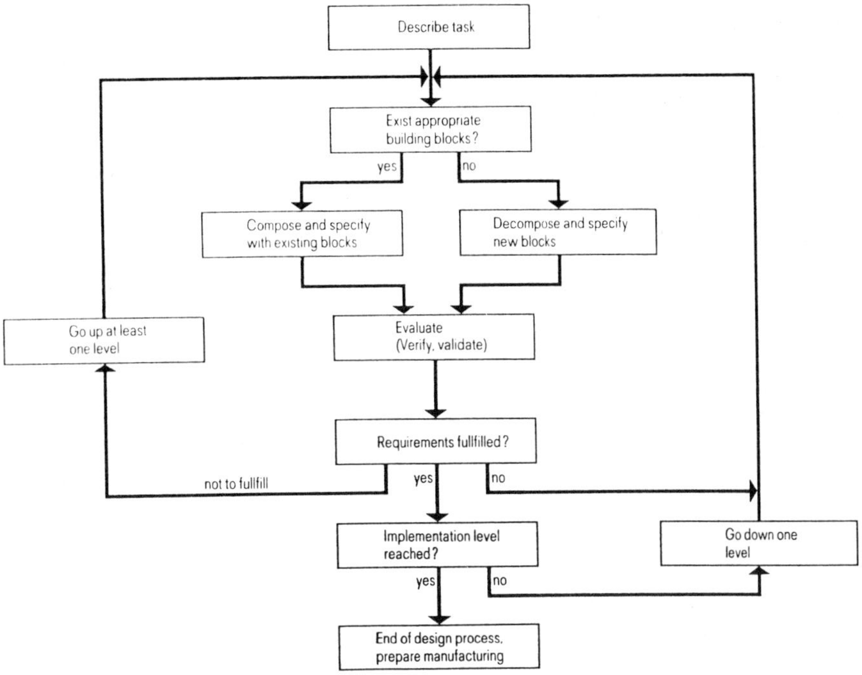

Fig. 2: Design activities /1/

Design is a non-monotonic process. At each step previous design details might be invalidated. Fig. 2 /1/ shows a simplified flow chart of the design process. It demonstrates the need for short transitions between design levels in order to easily iterate on parts of the design and, if necessary to change specifications or requirements at higher levels.

The design within one level is possible in two ways. Depending upon the availability of appropriate building blocks a bottom-up ("compose") or a top-down ("decompose") approach is carried out.

Important is the evaluation of the resulting substructure. It has to be checked, whether the requirements or specifications, set up in the next higher design level, are met. Furthermore various design alternatives, which are known or have to be worked out in further iterations, should be evaluated and compared. Proven methods for verification are simulations as well as analytical methods. Of increasing interest is the area of testability analysis and test pattern generation.

Further work has to be done to verify and improve the dicussed design methodology. But the above discussions pointed out important facts, which should be kept in mind for the development of a computer aided System Design Environment (SDE).

STEPS TOWARDS REALIZATION

The partitioning of the design process, especially the register-transfer-level, demonstrates:

- A large amount of work has to be done, before a register-transfer-model of the system can be developed.

- Many design decisions can be made at high levels of abstraction, that means at an early stage in the project. Dead-end solutions can be detected earlier and the designer's effort can be concentrated to worthy alternatives.

- The design decisions can base on simulated or analytically verified results at nearly all levels of system design.

A deeper look into the design flow underlines the need for a flexible CAD/CAE system covering all levels of system design with no restrictions to change design levels in a fast and conform way, and with a uniform user-interface.

Workstations which are available since last year support some aspects of the design process in a convenient way. System Design Environment will be done by a combination of engineering workstations and mainframes machines. Workstations support schematic capture, layout, communication with each others and mainframes in the background. A mainframe is used for analysis, scheduling, material resource planing, and design management. This combination of personal workstations and mainframes will improve product quality. The engineers can easily try several approaches, do more computer analysis, produce more consistend documentation, and can better control their design.

CONCLUSION

In order to specify, design, and develop levels and tools for High-level System Design, we started a common project between Carnegie-Mellon University in Pittsburgh, Pa. and the Research Labs of Siemens in Princeton, N.J. and Munich, Germany. The first result of this joint effort was a proposal for research and development /8/, outlining the overall system and the stages of design. The other papers of this session present the current work at these 3 places. The goal is to develop and integrate tools in a System Design Environment which increases the designers productivity, allows more efficient management and results in design of computer systems of better performance, reliability and testability.

REFERENCES

1. Gonauser, M.; Kober, R.; Wenderoth,W.: A Methodology for Design of Digital Systems and Requirements for a Computer Aided System Design Environment. IFIP Working Conference on Methodology for Computer System Design, Sept. 15 - 17, 1983, Lille France

2. McWilliams, T.M.; Widdoes, Jr., L.C.: SCALD: Structured Computer-Aided Logic Design. Proceedings of the 15th Design Automation Conf., June 1978, New York, pp. 271 - 277

3. Director, S.W.; Parker, A.C.; Siewiorek, D.P.; Thomas, D.E.: A Design Methodology and Computer Aids for Digital VLSI System. IEEE Transactions on Circuits and Systems CAS-28, July 1981, pp. 634-645

4. Foulk, P.W. et al.: AIDS - an Integrated Design System for Digital Hardware. Proc. IEEE, Vol. 127, Pt. E, Nr. 2, March 1980

5. Ellenberger, D.J.; Ng, Y.W.: AIDE - A Tool for Computer Architecture Design. Proc. 18th Design Automation Conference, 1981, pp. 796 - 803

6. Newmann, P.S.: Towards an Integrated Development Environment. IBM Systems Journal, Vol. 21, Nr. 1, 1982, pp. 81 - 107

7. Kober, R.; Wenderoth, W.: Problems and Practical Experience in High-Level Design. Proc. of the 1983 IEEE Conference on Computer Design

8. Siewiorek, D.P.; Giuse, D.; Birmingham, W.P.: Proposal for Research on Demeter: a Design Methodology and Environment. Carnegie Mellon University Pittsburgh, Pa., 22. Jan. 1983

A Perspective on the Levels and Methodologies in Digital System Design

Patrick P. Fasang, Peter Ulrich, Michael Whelan

Siemens Corporate Research and Support, Inc.
Research and Technology Laboratories
Princeton, New Jersey

ABSTRACT

The design activities performed by a digital system designer range from converting marketing information into design specifications, to final implementation of a chip, board, or system. In this paper the authors share their views on the high levels of design which include hardware requirements, architecture, module (subsystem) and register transfer level. In each design level, different design activities are performed which should be supported by tools. The increasing complexity in digital system design requires an environment with integrated tools to support the design. The basic features of such a system are described.

I. Design Process

High levels in the design process of a digital system are first described. These levels range from system design to register transfer level specification. Here emphasis is laid on hardware design. Levels are defined based on the different activities that are performed during that stage of the design. The description is done in a top down fashion, although the design process in reality more resembles a middle out procedure.

Figure 1 shows the logical partitioning of the design process in terms of levels of abstraction. The highest level shown is that of system requirements. System requirements describe what the system as a whole is meant to do, without regard to any partitioning of functionality between hardware and software. The subsequent system design process partitions the functionality of the system between hardware and software, thus yielding both hardware and software requirement specifications. In this discussion, we shall only consider the hardware design.

II. Hardware Requirements

The hardware development process starts with a hardware requirements specification which contains quantitative, qualitative and comparative information on the features of the hardware system. In the case of a general purpose computer, hardware requirements could specify, e. g.,

Performance	(e.g. >3 MFLOPS, Whetstone)
Memory size	(range from - to)
Configurations	(minimal - maximal configuration)
System availability	(MTTF, MTTR)
Price/Performance	(as compared to competition)
Compatibility	(with existing software)
Standards	(e.g. Corporate, FCC, UL)

This is meant to be representative of the kind of information which would be found in a hardware requirement specification. Actual specifications may be incomplete, vague, ambiguous, and possibly inconsistent.

The hardware requirements definition is an activity usually consisting of discussions between designers, marketing personnel and possibly technology representatives. These discussions lead to a system specification which satisfies the marketing requirements and is feasible to design and to manufacture. In addition, the required features must

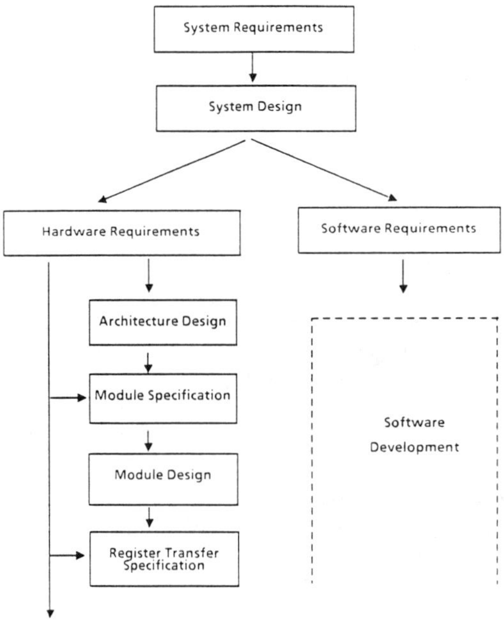

Fig. 1 Information flow in the design process

CH1935-6/83/0000/0419$01.00 © 1983 IEEE

be consistent and balanced. Examples would be the relationships (processing power - memory, - I/O bandwidth) and (price-performance).

Most of this requirement information is related directly to the hardware architecture requirements; however, some aspects, such as the technology to be used, may directly apply to lower levels. In most cases it is important that lower level issues be discussed and agreed upon at an early stage. This is because bringing a product successfully to market requires the cooperation of marketing, design, development, and technology personnel. Since the lead time for developing new technologies and setting up and scheduling manufacturing facilities may be long, it is imperative that each group know, and agree to, what is expected of it.

III. Hardware Architecture Level

Design activities at the hardware architecture level start with requirements which are deduced from the decisions made during the hardware requirements specification activity. The architecture design considers different system configurations (in the sense of maximal and minimal configurations) and models (different products). On this level, the functionality of the system, as described in the requirements, is partitioned into subsystems.

The design deals with system performance, parallelism of operation, redundancy, and reliability. The result of the architecture design is a specification of the types of subsystems to be used (like processor, memory, I/O controllers, etc.), their interconnection and requirements for their operation, e.g. memory size, I/O bandwidth and service rates. The data flow between the subsystems is given, but usually not the width of the busses transferring the information. However, the interconnection of two subsystems implies that the same interconnect protocol (both in terms of hardware and software) has to be used by both. The specification of this protocol is part of the module level design phase.

Techniques used to model the system at this level include block diagrams containing functional blocks, their interconnections, and data rates between them, PMS [Sie74] like descriptions, or Graph Models of Behavior [Raz79] descriptions. These models may also serve the purpose of documentation.

At this level, analysis and simulation tools are used to estimate system performance [Ale82] and reliability [Sie82]. The exact form of, and parameters associated with, these models may not be known accurately, and simplifying assumptions and approximations must be made. As the design proceeds to lower levels, the form and parameters of the higher level models can be better determined, and the higher level models may have to be reexamined in the light of more accurate information. This leads to information flowing from lower levels to higher levels in the design process (data extraction in Fig.2).

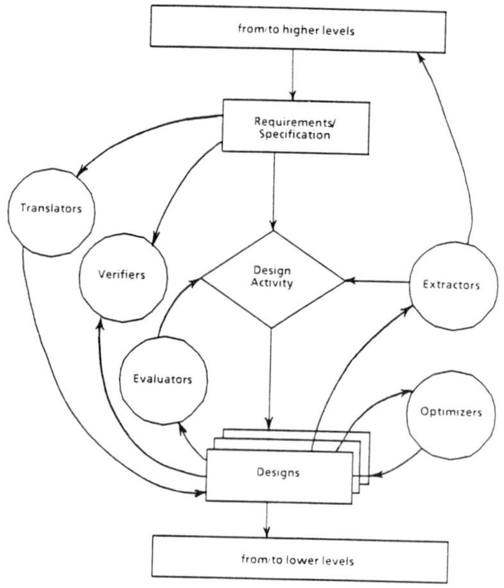

Fig. 2 Tool classes

IV. Module (Subsystem) Level

Module level design is based on the module specification, i.e. on the requirements for the blocks or subsystems that come out of the architecture design. The module design consists of designing each of the subsystems.

Problems to be solved in the module design phase are specification of the subsystems in terms of data and control portions, and control flow. In the case of a microprogrammed system this includes microinstruction format and microinstruction flowcharts. The design of each module includes synchronization mechanisms, data path widths, communication types and speeds (e.g. serial, parallel, synchronous, asynchronous), and access mechanisms. Modeling low level parallel operations, as in the case of pipelining, is done on this level.

The result of the module level design is a specification of the behavior of each module and specification of the manner in which the modules are to be connected together.

Simulation is used to verify the correctness and viability of designs by examining the dynamic behavior of the models. In addition, interconnections between modules may be tested for static inconsistency, e.g. the number, and type, of connections between modules.

V. Register Transfer Level

The register transfer (RT) level design describes in details the behavior of a subsystem ,as specified at the module level, in the terms of register transfer primitives and procedural control.

There is no clear distinction between module and register transfer design. From subsystem to RT design, the model undergoes refinement in both structure and timing. Therefor, although the same language may be used to describe models at both the module and register transfer levels, there is a difference in the implied semantics associated with the descriptions. At the register transfer level, there is an implied connection between the elements in the RT language description, particularily the structure of the module and the realization of that module.

The result of the register transfer level design is a description of the system in terms of primitive elements such as adders, registers, decoders etc.. Microprogram development could be done on this level. The primitive elements which are appropriate to use in a RT description are technology dependent (in the broad sense of gates per chip, as opposed to the narrow sense of CMOS vs. NMOS) and can be expected to increase in complexity with time.

VI. Design Activities

There are several design activities which are common to all the design levels. The design in each level starts with the requirements specification for the design at that level. This is obtained from the design at the previous (higher) level with some input from the general system/hardware requirements being possible. The design at each level creates a model that represents the system to be developed and constitutes requirements for the next lower level.

The design activity at each level is itself an iterative process that consists of

- building a model of the system to be designed,
- changing the model to meet the requirements,
- optimizing the model,
- designing different models for different configurations.

The designer works using intuition, experience, and knowledge of trends and previous designs.

As each design at a particular level is proposed, it must be checked for functional correctness and compared to the less detailed model that was available at the previous level. It will also be necessary to check if the design meets the goals set up in the appropriate requirements specification. Often it is only possible to answer these questions with the increased knowledge gained from the lower level design process. This is because at higher levels, simplifying assumptions and approximations had to be made. This leads to an information flow from lower to higher design levels that may cause models at the higher levels to be reexamined and changed. If models at the higher level are changed, then design activities which made use of these models may have be to reexamined.

To support the design activity there are several classes of tools, the purposes of which are common to all levels. These tool classes and their relationship to the design process at a particular level are illustrated in Figure 2. This does not imply the existence of tools of each type at all levels. These tool classes are

Verification/validation tools. They serve to check whether the design satisfies the requirements and performs correctly. The result is a qualitative answer.

Evaluation tools. They yield quantitative information such as price and reliability.

Translation tools. They support the creation of a lower level design from a higher level one. An example is a synthesizer that (semi)automatically generates a lower level design, or special parts thereof, from a higher level description.

Optimization tools. They manipulate a design at one level in order to enhance attributes associated with it; e.g., performance, price, testability, or power consumption. Obviously these are, in part, antagonistic goals.

Data extraction tools. They translate relevant data from lower level designs to higher levels of abstraction. This is necessary when for analysis (e.g. of performance) at a higher level, the lower level model yields additional information on parameters associated with the higher level model.

This classification is based on the purposes for which the tools are used as opposed to the manner in which the tools operate. In practice, a single tool may be used for several purposes, e.g., a simulator may be used in one instance for verification (e.g. to check for deadlock in protocols), or for evaluation (e.g. bus performance). This categorization is not complete, it only covers the most usual usage of tools. Tools such as input generators (e.g. generating the external stimulus to be aplied to a model during a simulation) do not fit neatly into this scheme since they may depend on models at several successive levels of abstraction.

VII. System Design Environment

Because of the growing complexity of system design, it has become apparent that the designer must be supported by automatic tools rather than manual design procedures. A system design environment (SDE) is meant to serve this purpose. Such an SDE is not merely a collection of tools but the tools are integrated in such a way that they work together and user interaction with these tools is via a uniform user interface.

The SDE (Figure 3) is centered around a database which provides access to all the data. This includes mechanisms for data management, administration, and recovery. The database provides uniform access to input, output, and intermediate files. This obviates the need for tools to perform their own data management functions (as is normally the case today).

Integration also means that the SDE user interface provides the user with uniform access to all tools, hiding the specialized handling that is connected with each tool.

In additon to tools, a SDE has to contain libraries of design data and modules, and documentation aids.

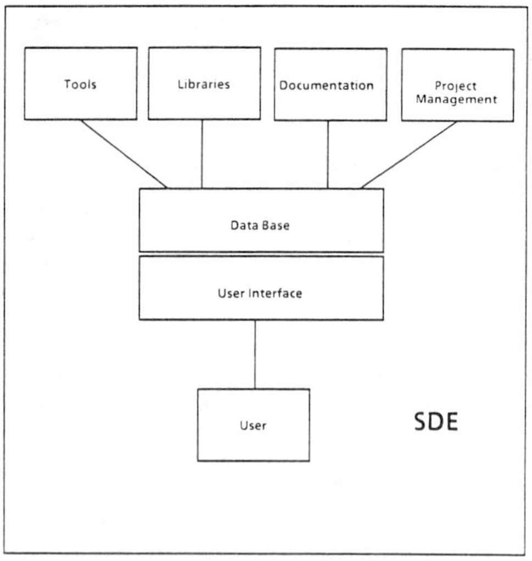

Fig. 3 Overall view of SDE

In order to facilitate integration, wherever possible, common model descriptions are used for several tools. This implies the use of tool independent languages for describing the structure, behavior, and attributes associated with a model. Such an approach alleviates some of the problems associated with maintaining consistency between different models of a system associated with different tools, since where possible, a single tool independent description would be used, with each tool extracting the needed information from the description.

Over and above activities directly related to the designer, there are organizational aspects that greatly influence the design procedures. The large amount of design data generated requires that there be provisions for effective data storage and project management facilities. Larger design projects may span several years and involve many different designers. In order to support projects of this scale, management tools are necessary. The functions supported by these tools would be scheduling of design activities and communication between designers.

VIII. Conclusion

A perspective on the levels and activities in digital system design is given. The design process is described and broken into levels. In the context of system design , three levels are considered, namely, system design, architecture, and module. These are levels in which very small numbers of tools exist. The different types of generic tools which are needed to perform the design task at a given level are identified, and their functions are explained. A system design environment (SDE) is a setting which consists of tools and an environment. The tools are integrated in such a way that they work together and complement each other. The environment, with its database and user interface, enables the designer to properly and conveneniently use the tools.

References

[Ale82] "Performance Modelling in the Design Process" W. Alexander, R. Brice AFIPS, Proceedings of the NCC, Volume 51, 1982, pp 257-262

[Raz79] "Evaluation Methods in SARA - The Graph Model of Behavior" R. Razouk, M. Vernon, G. Estrin Conference on Simulation, Measurement and Modelling of Computer Systems, 1979 pp 189-206

[Sie74] "Introducing PMS" D. P. Siewiorek, IEEE Computer, December 1974

[Sie82] "The Theory and Practice of Reliable System Design" D. P. Siewiorek, R. S. Swarz Digital Press, 1982

DEMETER - A DESIGN METHODOLOGY AND ENVIRONMENT *

Dario Giuse, Daniel P. Siewiorek, William P. Birmingham

Carnegie-Mellon University
Pittsburgh, PA 15213

ABSTRACT: The design of large digital systems is a complex activity that involves several levels of abstraction. The higher levels of design have not received enough attention in the past; as a result, some of the main issues are poorly understood. The DEMETER Project tries to provide a framework for digital design and to define an integrated environment that will support the whole range of design activities.
This paper describes D0, the preliminary phase of the DEMETER Project. D0 is an experiment in communication among existing tools in a specific design domain (Reliability Analysis). The selected tools execute on different computer systems and use widely different data representations. D0 provides a unifying framework that allows isolated tools to cooperate in a useful way and makes the underlying structure invisible to the user.

INTRODUCTION

Computer assistance in the form of more or less independent CAD tools has been used in the past twenty years for the design of digital systems. It is now apparent that a new approach is needed, since isolated tools lack many of the capabilities required in the management of a large project.

Only an integrated Design Methodology and Environment, spanning all the levels of design from the definition of the task down to the circuit level, can help manage the growing complexity of digital design and increase the productivity of the design process.

The DEMETER Project (DEsign METhodology and EnviRonment) [4] was initiated at Carnegie-Mellon University to investigate the higher levels of design. The goal of the DEMETER project is to define the attributes of a design methodology and environment for digital systems; the design environment will support the full array of design activities (simulation, analysis, optimization, synthesis, etc.). The system will insure that changes made in one portion will be manifest in the entire design. It will also be possible to review and undo design decisions and tradeoffs.

By concentrating mainly on the design stages above the Register Transfer level, the DEMETER project should eventually improve the designer's productivity by one to two orders of magnitude.

* This research was supported by the Semiconductor Research Corporation - CMU Center for Computer-Aided Design, and by Siemens Corporation.

THE D0 SYSTEM

In order to obtain a better insight into the complexity involved in the DEMETER system, we decided to start our research with a limited subset of the problem. This first phase has been named D0, for DEMETER phase 0.

This phase will not include the creation of a highly polished system, but rather will emphasize the exploration (and possibly the rejection) of several strategies. The experience gained from this simple, "throw-away" system will be important in the subsequent phases of the project.

In order to concentrate exclusively on the most important issues, only off-the-shelf software tools were used in D0. In particular, efficiency considerations were considered irrelevant at this stage.

Characteristics of the D0 system

The most important characteristics of the D0 system are:

- D0 ties together existing tools that operate in a **well-defined domain**, specifically Reliability Analysis.
- The tools are written in **different languages** and run on **different machines** under different operating systems.
- The **input/output formats** for the various tools are widely different.
- A **commercial data base** system is used.
- **Standard language-processing tools** are utilized to convert between different input/output formats.

The following goals have been intentionally excluded from the first implementation effort:

- Uniform user interface.
- Version control policies.
- Change propagation mechanisms.

Fig. 1 illustrates the configuration of the D0 system. The main tools have been shown as darker boxes, while small circles indicate that a translation between different formats is required. Small boxes indicate the main formats.

CH1935-6/83/0000/0423$01.00

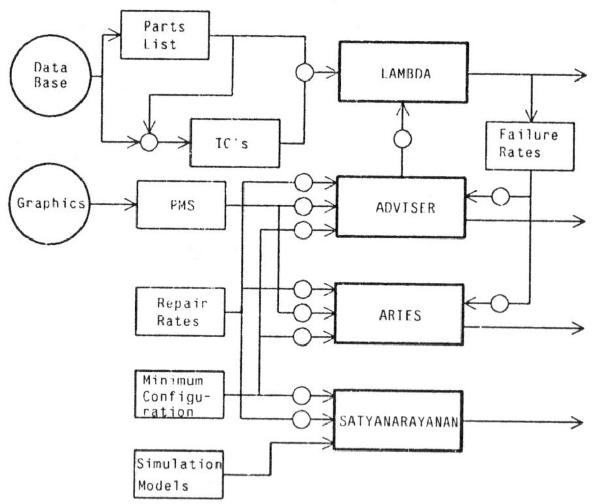

Figure 1. The D0 configuration

The tools

Four tools in the Reliability Analysis area and a Data Base system are connected by D0. These tools were chosen as examples of typical stand-alone systems. The tools are:

- **Lambda**, a MIL-217D Failure Rate Analysis program written at CMU by Steve Elkind [10]. Lambda is based on a parts-count reliability model.
- **Adviser**, a PMS-level program for symbolic reliability evaluation written at CMU by Vittal Kini [2]. Given a PMS-level structure and the failure rates for individual modules, Adviser computes the symbolic reliability formula.
- **Aries**, an interactive Markov reliability estimation system written at UCLA by S.V.Makam, A.Avizienis, and G.Grusas [1]. Aries is able to deal with systems whose states are enumerated by the user, including systems with repair.
- **M. Satyanarayan's** as yet unnamed simulation program [8]. Given a structure and the description of the individual modules, the program evaluates the performance of the whole system.

D0 tries to integrate these tools in a homogeneous environment. A user of the D0 system is practically unaware of the nature and details of the various tools and translations being executed.

IMPLEMENTATION

It was decided to keep each tool in its native environment and completely unmodified. This is a pragmatic approach to the world of computer-aided design, where very large tools are available that required tens of man-years to write. It is our view that rewriting existing tools is generally unfeasible.

The different computers are connected by a local network. Only existing connections were used, and no special-purpose communication software was written. This approach is similar to that of the Designer's Workbench system [5].

The physical and logical configuration of the system is indicated in Fig. 2. The logical configuration differs from the physical one in that only the Main Host is logically connected to all the other machines.

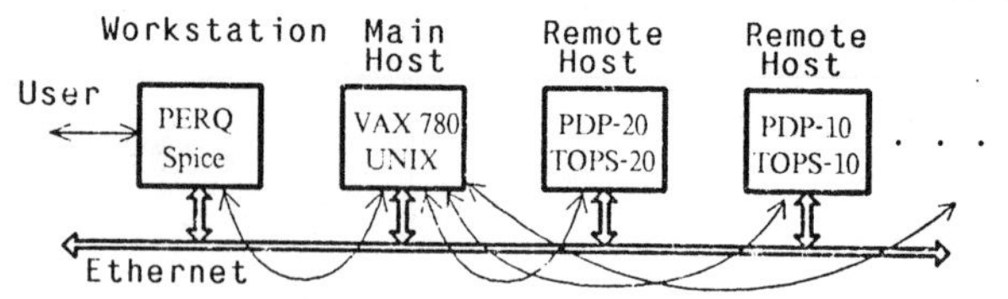

Figure 2. Physical and logical configuration of the D0 system.

The Main Host

The present implementation of the D0 system uses as the Main Host a VAX computer running the UNIX operating system. UNIX proved a very convenient framework because of the following features:

- The Shell provides a powerful job-control mechanism with full-blown control structures such as conditionals and loops.
- The I/O redirection and pipelining mechanisms allow the connection of different programs in a simple and effective way.
- UNIX supports a wide variety of program-generation and language-translation tools.
- Different data base systems are available under UNIX.

Remote Tools

In order to simplify the communication with remote hosts, remote tools are always executed as batch jobs. This also applies to tools that were originally created for interactive use.

A remote tool is executed in a standardized way. An input file and a file of commands are first sent to the remote host; the tool is executed generating an output file; the output file is retrieved onto the main host. The whole process is completely controlled by the Main Host.

Format Translations

Three general-purpose UNIX language processing tools were investigated for the task of format translations: YACC [9], LEX [7], and AWK [3].

Yacc is a compiler compiler; its input is a description of the input grammar in BNF format. It is possible to specify semantic actions to be executed for each grammar rule being recognized. The output is a C program that will parse any file in the input language.

Lex is a lexical analyzer generator. Its input is a table of regular expressions and a set of actions; the output is a C program that will perform lexical analysis and partition an input stream into tokens.

Awk is a pattern scanning and processing language. The input is a set of patterns and actions. When a pattern is recognized in the input, the corresponding action is executed. Unlike Yacc and Lex, Awk does not generate an executable program. Instead, Awk interprets the input rules directly while scanning the input files.

After preliminary work with the three programs, Awk was chosen as the translation tool of choice. The considerable expressiveness of the Awk language makes the translators easy to write and maintain. The debugging time is also reduced, since Awk is an interpreter and does not require any compilation. Awk is particularly effective for the formats used by current generation CAD tools, whose outputs are never very complex syntactically since they are heavily oriented towards records and fields.

Data Base

A standard Unix data base system, INGRES [6], was investigated. INGRES is a relational data base that supports views, protections and integrity constraints. The system provides full recovery and consistency control, and has adequate support for very large data bases.

An interesting feature of INGRES is that all the query commands may be embedded in C programs. It is thus possible to execute any data base operation completely under program control.

RESULTS

At the time the present paper was written, two tools and the data base had been integrated into the D0 system. The two tools are LAMBDA and ADVISER. The data base system is INGRES. The information flow in a typical run is briefly described below.

This experimental system emulates a portion of a typical DEMETER scenario: a designer wants to quickly evaluate the reliability of a system at the PMS (Processor, Memory, Switch) level before further expanding the design. D0 provides a system that automatically invokes tools as appropriate and displays the results in an easily understood form. The final result of each iteration is an automatically generated diagram that depicts the estimated reliability of the system at various times.

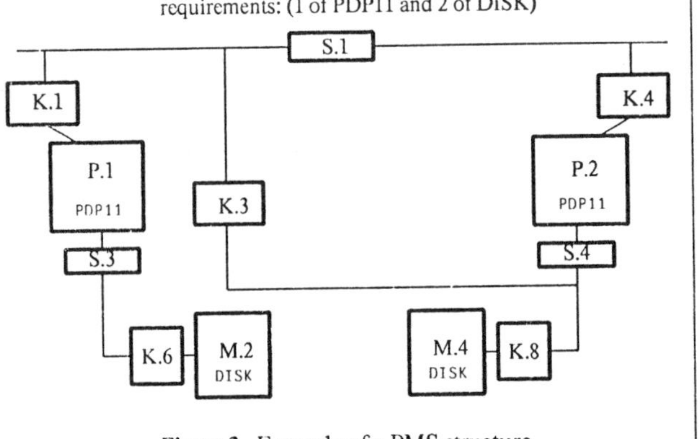

Figure 3. Example of a PMS structure

Fig. 3 shows an example of a redundant, dual processor system. Two PDP-11's are connected via a Dynabus. This structure was given as input to the D0 system. The output, shown in Fig. 4, plots the probability that the system will be functioning (ordinates) versus the time in hours (abscissae; logarithmic scale).

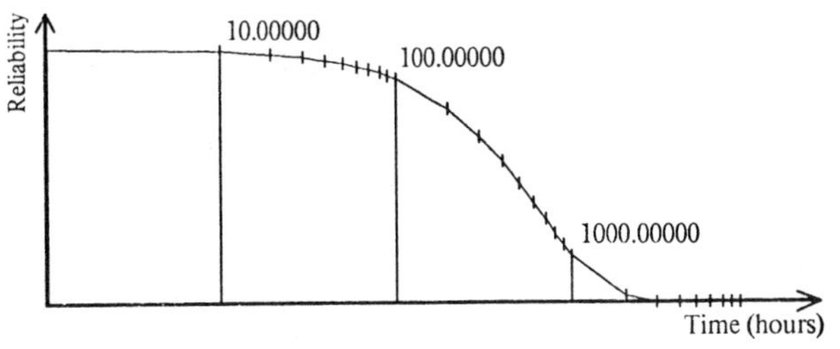

Figure 4. System Reliability

The designer may control one or more of three system attributes:

- The graphical representation of the PMS-level structure of the system. The structure is manipulated through a graphics editor running on the work-station.
- The description of the operating conditions of the digital system, such as temperature, environment, etc.
- The description of modules in terms of their submodules. This description is hierarchical, down to primitive components.

Varying any one of these descriptions causes D0 to automatically invoke all the necessary tools, taking care of the communication among the various machines.

If, for instance, the PMS diagram is modified, a symbolic description of the PMS structure is extracted from the diagram and merged with the failure rates for individual modules computed by Lambda. The resulting file is sent to the appropriate host, where Adviser is run as a remote job. The output is brought back to the Main Host, and the Reliability Formula is evaluated for different values of time, yielding different points of the Reliability curve. A diagram of the curve is generated and displayed on the work-station.

The typical elapsed time for a change in the PMS structure, in the case of a simple system with a dozen blocks, is under two minutes. The typical elapsed time for changes in all of the three input representations at the same time is under 4 minutes.

The overhead time (including translations and communication among machines through the Ethernet) is typically less than 30 percent of the total, and tends to decrease for more complex structures where the processing time of the tools becomes by far the dominating factor.

Manpower requirements

The present configuration of D0 required three man-months for its implementation. About one third of this time was spent in the analysis of the communication problems and in the creation of the simple communication module used by the Main Host. Because of the use of powerful language processing tools the problem of translating between different formats occupied less than one third of the total time. The remaining time was spent on debugging, maintenance of the system, etc.

The only new programs written were a 20-line C program that takes the user's account and password for the remote hosts and a 96-line C interface to the INGRES data base. The system consists of about 300 lines of Shell command language and Awk input language. This figure should be compared with a total of about 95,000 lines of code that comprise the three tools (Lambda, Adviser and Ingres).

Standard tools

Using only standard tools has proved an enormous advantage in this phase of the project, as can be seen from the figures in the preceding section. No modification of programs was required on either the Main Host or the workstation.

The strongest asset was the availability of the UNIX system, with its many efficient and specialized tools. The program-generator tools made it possible to leave the rest of the system unchanged, since translation from one format to another was made very simple. Writing a translator between two different formats using Awk, for instance, only takes a few hours. Simple translations can often be prepared in less than 30 minutes.

Data Base

The requirements for the data base system in an environment such as DEMETER are quite heavy. The data base must provide complete support for many designers for long periods of time. All the previous versions of the digital system being designed should be stored in the data base.

One of the main goals of the DEMETER project is thus to determine whether any existing data base system may be suitable for such a demanding application. Suitability must be measured both in terms of response time and of size of information that may be stored.

The experience gathered in this field is not yet sufficient to provide a clear answer, but a few points are worth mentioning:

- A program-level interface to the data base is essential. The whole functionality of the data base should be usable from within a program.
- It does not seem possible to store all the information about a design into the data base itself. The data base may instead function as a catalog being used to store *pointers* to the information. The information may then be stored in a traditional file system.

Error recovery

The problem of error recovery has turned out to be a difficult one. While error recovery in a complex environment is always delicate, a multi-host environment such as D0 represents a particular challenge.

As an example, consider the Adviser program. More than 20 different failure modes have been observed while executing Adviser. These modes include problems with the communication between the Main Host and the PDP-20, errors in the input file, and plain program bugs.

The tools react differently to bad input, and in a few cases they may just crash in the middle of execution. In such cases the connection is typically *not* aborted, i.e. the communication program is left hanging without any event that signals an abnormal termination.

The number and variety of error conditions that may occur suggest that some "understanding" of the different steps is necessary. Such knowledge can hardly be captured in an algorithmic program or in a Shell script. From the experience with the D0 system it appears that full error recovery in a multi-host environment can only be achieved via an expert system.

CONCLUSIONS

D0 is an experiment in integration. Originally incompatible tools are integrated under a uniform system that hides unnecessary details. All the interaction with the system takes place through a single work-station.

D0 demonstrates that the use of standard tools and methodologies can significantly reduce the time required to integrate separate tools into one environment. Experience with format translation and multi-host communication has also emerged from the experiment.

Results from the D0 experiment will constitute the basis for the next phase of the DEMETER project. It is our hope that the project will help improve our understanding of the whole design process, eventually leading to one to two orders of magnitude increase in designer's productivity.

REFERENCES

[1] Srinivas V. Makam, Algirdas Avizienis, Gintaras Grusas, *"ARIES 82 Users' Guide"*, Technical Report, UCLA, August 1982.

[2] V. Kini, D.P. Siewiorek, *"Automatic Generation of Symbolic Reliability Functions for Processor-Memory-Switch Structures"*, IEEE Trans. Comp., pp. 752-771, August 1982.

[3] Alfred V. Aho, Brian W. Kernighan, Peter J. Weinberger, *"Awk - A Pattern Scanning and Processing Language"*, Technical Report, Bell Laboratories, September 1978.

[4] Daniel P. Siewiorek, Dario Giuse, William P. Birmingham, *"Proposal for research on DEMETER - a Design Methodology and Environment"*, Technical Report, SRC-CMU Center for Computer-Aided Design, Carnegie-Mellon University, January 1983.

[5] L.A.O'Neilly, J.R.Breiland, R.A.Friedenson, P.H.McDonald, T.J.Thompson, *"The Designer's Workbench"*, The Bell System Technical Journal, pp. 1757-1825, November 1980.

[6] R. Epstein, *"A Tutorial on INGRES"*, Technical Report, UCLA, December 1977.

[7] M. E. Lesk and E. Schmidt, *"Lex - A Lexical Analyzer Generator"*, Technical Report, Bell Laboratories, October 1975.

[8] M. Satyanarayanan, *"A Methodology for Evaluating Storage Systems and its Application to a Network File System"*, PhD Thesis, Carnegie-Mellon University, September 1982.

[9] Stephen C. Johnson, *"Yacc: Yet Another Compiler-Compiler"*, Technical Report, Bell Laboratories, July 1978.

[10] Steven A. Elkind, *"Lambda - A MIL-217 System Failure Rate Analysis Program"*, Technical Report, Carnegie-Mellon University, June 1982.

PROBLEMS AND PRACTICAL EXPERIENCE IN HIGH-LEVEL DESIGN

Rudolf Kober and Wolfgang Wenderoth

SIEMENS AG, Corporate Information Technology
Munich, West-Germany

ABSTRACT Progress in technology and new complex architectures are the reason for difficulties and deficiencies in present-day computer system design. Based on a survey of typical design projects of general and special purpose computers, the most significant problems are discussed. It turns out that the design process is mostly lacking in formalized or formal methods for description, verification, and analysis in particular at the higher levels of design, whereas the subject of design gives rise to increasing difficulties stemming from a strong tendency to parallelism and concurrency. Three examples of already designed systems are used to illustrate the design process with emphasis on the higher levels of design, and to demonstrate the need and, particularly, the use of high-level design tools.

1. Introduction

In the past, research in the area of computer aided design and design automation has resulted in a good knowledge of the design process and a substantial number of most valuable design tools. Presently in this respect, the advent of VLSI and the trend to parallel and distributed architectures has two major effects: first, existing tools reach their limitation e.g. in processable gate count and CPU time; second, a gap of new quality tools for architecture level design emerges. This necessitates research on design methodologies and computer aided tools in order to develop a new generation of design systems. The basis needed for this research is a thorough understanding of the design process and related problems. For this reason a survey was made on typical design projects of minis, mainframes, switching systems, data link control computers, and process control computers. In chapter 2 an characterization of the design process is given as found in the survey. Chapter 3 addresses the prevailing problems and deficiencies. In chapter 4 three examples of already finished designs are used to verify these problems, and to demonstrate the need and, partially, the use of highlevel design tools. The intention is to motivate further research towards an integrated design environment /1,2/.

2. Design Process Characterization

Computer design is largely determined by the nature of the target system. With mini and mainframe computers, one important constraint, besides price/performance requirements, is compatibility with existent computer systems to allow continued use of existent user software and existent operating systems. That leaves only limited freedom in architectural design. With special purpose computers, e.g. switching systems, additional requirements such as high reliability or availability lead to completely new architectures. Typically, these systems are parallel in nature.

Though the design may differ substantially, a common methodical abstraction is possible for the design process. Usually, the design process is carried out on several levels. Figure 1 shows a representation of hierarchical design levels abstracted from the design processes which have been surveyed. Typically, design starts on high levels with a small team of chief designers, and ends on low levels, where possibly several hundred engineers are involved. Characteristic high-level design activities are described in chapter 4, whereas a more detailed description can be found in /2/.

Though, the design levels are hierarchical, the computer system design process is not a purely top-down but rather an iterative process involving feedback loops within and between levels.

3. Problems in High-Level Design

Design and testing have an increasing share in the life cycle of present-day computer systems. Correction of design errors is getting more and more costly. The underlying problems and difficulties are caused by the complexity of design object, by methodical deficiencies in the design process, by a tool gap, and by designers attitude. In the following an overview of typical problems, as found in the survey, is given, which is fairly representative but by no means exhaustive.

Design

The growing use of custom or semi-custom VLSI chips creates two problem classes. First, designs should be correct, because correcting errors can lead to a redesign of a whole chip thus causing high costs and delays in the order of months. Second, testability is decreasing unless provisions are made in the original design, which reduces the chip area functionally used.

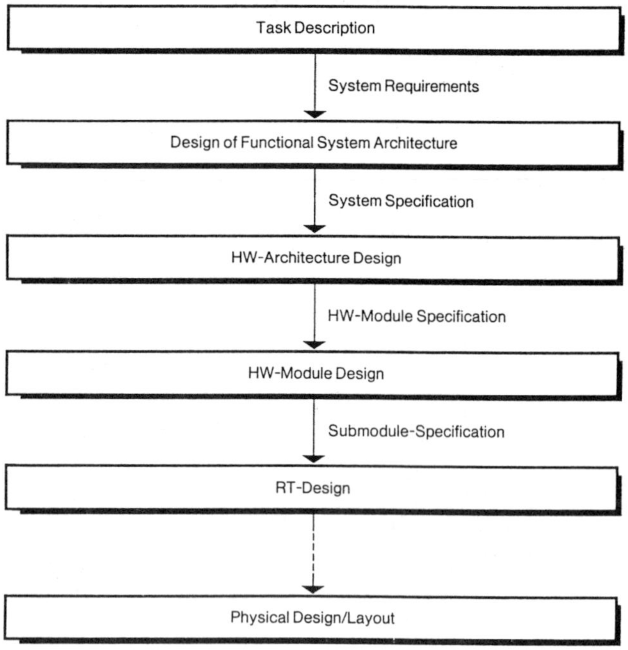

Figure 1: Levels of system design

On the architecture levels, system design uses increasingly parallelism and concurrency to meet performance and reliability requirements. Related problems are for instance:

* synchronization and communication schemes of distributed processes and systems,
* interconnection systems in multiprocessor systems,
* complex pipelining mechanisms in mainframe computers, and
* organization and size of cache memory.

To cope with these and similiar design problems, almost no design tools are in practical use today, although this area has been a field of research at universities since the begin of the seventies /3/,/4/.

Representation

In general, present-day design methods are lacking in formal computer aids for both design and management. Reasons are, that no formal or formalized languages are established on the system design levels, and a variety of mutually unrelated languages exist on the lower design levels. Where languages have been introduced in the design process, they are rather used to describe designs after completion than being a real design aid.

Formal languages similar to higher programming languages have the advantage of being machine readable and processable. But due to a lack of graphic representation, they have their drawbacks when a system structure has to be described. Many attempts to introduce description languages have failed to meet with any success due to a lack of acceptance by designers and management. The reasons for this are the amount of time required to learn these languages, the lack of compatibility with the design process (e.g. no graphic descriptive media), and the lack of intelligibility and clarity.

Documentation

Documentation is a general problem in hardware and software design projects. Very often a design is documented after completion, the design documentation itself is sometimes outdated when completed. Usually, system requirements and higher level designs are documented in prose with supplementary illustrations. Natural language documentation tends to be incomplete and ambiguous and aggravates communication especially in larger projects.

Verification and Validation

In computer system design, the verification and validation problem is concerned with two aspects: to verify the functional correctness and to validate the design with respect to performance requirements.

Usually, simulation tools or if not available manual examinations are used for verification. Tools for functional simulation are predominantly established on gate and circuit levels, and are in the process of introduction on the transfer level. On higher levels of design, no appropriate tools exist today. Most of these tools are self-contained with their own specific description language and user interface. The user has to learn a variety of totally different languages and conventions and has to manually transform representations and results if he wants to use them as inputs for other tools.

With the advent of VLSI many of todays simulation tools are no longer able to meet the requirements. Though it is extremely helpful to simulate a complete system e.g. on the gate level, cost and feasability reasons are prohibitive. What is needed are concepts for hierarchical and multilevel simulation. Work in these areas, which tackles the problem for all levels of design, is currently done at universities /5/ but is still in a research stage.

Results created by functional simulation are as good as the load data are. There is no proof of correctness. Approaches for formal specification which allows to prove correctness of a design are still in a research stage and are primarily applicable to software design.

To do performance evaluation for a new design, it is still common practice today, to measure prototypes, or to measure on predecessor models and extrapolate. This is acceptable, as long as the architecture does not change very much. On designs with new architectures, traffic and/or functional simulation is necessary yet on architecture levels to gain confidence in the performance characteristics and to detect bottlenecks e.g. in interconnect systems of multiprocessor systems. The problems with performance simulation lie in the modelling of system and system load. Modelling difficulties are the reason why simulations, when performed, are done by specialists rather then by the designer themselves. Even though, this does not solve the problem of establishing a representative system load e.g. benchmarks, instruction mixes, and load and dialogue types.

Though, many designers are convinced of the usefulness of high-level computer simulation, they rather do "manual" simulation e.g. to verify a pipeline mechanism. Typical arguments are that under time pressure, which is part of their every day job, they cannot afford to spend extra time to learn simulation tools. Consequently, appropriate verification and validation tools have to be developed which are easy to learn and thus acceptable to the designers, and which are elements of an integrated environment.

4. Examples of computer architecture design

Examples for already designed systems are given to illustrate the requirements for an integrated system design environment.

Low-end data processing system

The first example is the design of a lower-end DP-system for an existing line of mainframes using existing standard microprocessor-components. It had to be compatible with an existing operating system, that means with an existing hardware/software-interface. Therefore the functional system architecture was fixed, and nearly no design activities had to be carried out in that level (refer to figure 1).
The further design was carried out in two phases:

In the first one primarily the performance, which can be achieved by the given standard processors, was studied. The basic problem is the following: Though standard microprocessors have a reasonable throughput with their native instruction set, much performance is lost when emulating a different instruction set. After rough concepts concerning the hardware architecture the main objective was, to study one part of the system in detail, namely the processor block, without spending much effort for the rest of the system. Different standard microprocessors were evaluated. The "microprograms" for the most important instructions were coded, and the execution times were calculated "by hand".

The resulting performance was rather low, but not low enough to stop the project. Performance improvements by using an array of microprocessors in pipelined and parallel manner and the expected further progresses in VLSI-technology motivated the second, more detailed design phase. Now on the hardware architecture level a more precise block diagram was worked out, and the interface and behaviour for each block was specified.
An important activity was "balancing" the system. The memory bandwidth, bus transfer rate, I/O date rate, hit rate of the cache (if necessary) and processor performance were optimized by analytical models. The parameters for these calculations, for instance the number of memory accesses per instruction, were measured on existing fore-runner systems under typical loads.

In the next step the blocks were refined; as an example we look again at the processor block.
Important questions during its final design were:

* Pipelining, number of pipeline stages.
* Parallel processing, number of processors.
* Cost/performance as function of processor count.
* Local or global memories.
* Number and type of buses.
* Control strategy (synchronization, arbitration, pipeline conflict resolution,...)

These problems had to be evaluated under time pressure, so no simulations were carried out. Too much time would have been needed to build simulation models. Decisions were made on the basis of experience and some rough analytical calculations.

Safe data processing systems

The second example is the development of a safe data processing system with non-stop processing capabilities. The main aim was to achieve high availability without making the user responsible for programming checkpoints etc. Fewer restrictions than in the first example about the operating system and programming languages led to more design alternatives and decisions at higher levels of system design.

During the design of the functional system architecture different approaches for error detection and correction were discussed:

* Full hardware support for both, for instance by using master-checker configurations for error detection and duplication of these units for error corection.
* Mixed hard-/software approach, with fewer hardware for error detection, supported by software selftests, and software concepts for error correction (checkpointing).

The problem is evident: Just having started the system design, a decision has to be made leading to different implementations of both hard- and software. A detailed analysis of both alternatives, which means designing and evaluating them, would have cost too much time and manpower. So expert knowledge in that field was put together, rough estimates for the necessary design effort and the resulting manufacturing costs of the hardware, and some special-purpose simulations were made. They led to a decision for a compromise, and the functional architecture could be worked out.
Similar problems arose in the next stages of the design. Some alternatives, mainly influenced by the hardware architecture design, were for instance:

* Multiprocessor versus multicomputer
* Hardware/software-interface (intermediate language versus assembler)
* Processor architecture (stack versus register machine)

They all had strong influence on the corresponding software design level and of course on all following levels of refinement. The lack of multi-level simulators and of easy-to-use tools to evaluate and optimize the MTBF and availability of the system again made it necessary to implement ad-hoc, special purpose programs. Nevertheless major problems had to be solved "by hand". Even if the designers are rather confident about their decisions, nobody can be absolutely shure, whether they were the best ones.

Attempts were made to improve documentation and version control of the hardware design by using formal description methodes. Existing Hardware Description languages (HDL) were not ready-to-use in our environment, so we decided to use ADA. The function of the I/O-sytem was specified down to the hardware modules.
Although it had benefits in reducing misunderstandings between the designers, it showed many needs for improvements:

* ADA is not well suited to express asynchronous problems (e.g. interrupts).
* Automatic generation of simulation models out of the HDL-description was not possible.
* Cross-checks between HDL-representation and realization had to be carried out manually.

Multicomputer system

The last example, a parallel processing system for speeding up time-consuming calculations, was a research projcet with only few requirements. A main motivation was to study computer structures, which can make intensive use of identical VLSI-components.

In contrast to the two other examples, "unconventional" structures like data-flow architectures were considered and compared to arrays of von Neumann-modules. The impact of these high-level-discussions on hardware and software is enormous: operating system, compiler and hardware implementation are totally different for both approaches. On the other side, only few expert knowledge was available in that time, and no simulation experience for that kind of problems was established. So other rational arguments, like availability of microprocessors with conventional architecture and with increasing development support, the limited budget for that project and the lower risk led to the decision for the array of von Neumann-models.
After that general design decision again many further alternatives with similar importance had to be investigated, for instance:

* computation scheme
  synchronous (clear organization, high efficiency for limited class of application areas, easy expandability to some hundreds or thousands of modules)
  versus
  asynchronous (complex organization, more overhead, wider class of applications, limited expandability).
* Central versus distributed control.
* Communication mechanism
  synchronous versus asynchronous,
  common memory versus mailboxes in local memories.

The tradeoffs were evaluated partially by using simulation methods. So the computation scheme was selected on the basis of typical applications, which were transformed to synchronous and asynchronous simulation models. It turned out, that enough applications fitted into the clear synchronous mode of operation. Analytical models helped in the calculation of processing power and in evaluating the overhead for coordination and communication. Of course some important decisions had to be made on the basis of experience and rough estimates due to the lack of established evaluation and simulation tools for these high-level decisions.

5. Requirements for a System Design Environment

The above examples have shown the need for a system design environment, which integrates formal and graphic description methods, version control, multi-level-simulation, tools for reliability and costs calculations, and an data base for expert knowledge. Integrated means:

* Consistency between requirements, formal and graphic representation, simulation models and documentation must be garanteed automatically.
* An easy-to-learn, common user-interface for all parts and tools of the systems.
* Changes between representation levels (design levels) and between tool environments (design - simulate/verify - modify) must be possible easily and fast.

6. Conclusion

An overview of representative design problems was given. Some typical problems have been verified on three concrete examples of computer system design. It turns out that many difficulties stem from the lack of appropriate and efficient tools. Great benefit on the architecture levels can result from the use of functional and performance simulation. The results of the survey and the experience made during the design of the three examples urgently demand the development of an integrated system design environment.

REFERENCES

1. M. Gonauser, A.M. Sauer, Needs for High-Level Design Tools, Proc. of the 1983 IEEE Conference on Computer Design.

2. M. Gonauser, R. Kober, W. Wenderoth, A Methodology for design of digital systems and requirements for a computer aided system design environment, IFIP WG. 10.0, Sept. 1983.

3. G. Estrin, A Methodology for design of digital systems - supported by SARA at the age of one, AFIPS Proc. of the NCC 79, Vol. 47.

4. F. Heath, C. Rose, The case for integrated hardware/software design, with CAD implications, IEEE Computer Conference Digest, Sept. 1972.

5. R. Piloty, D. Borrione, The CONLAN Project: Status and future plans, Proc. 19th DA Conference, June 1982.

Session 2-4a
CMOS II

Chairmen

W. R. Young
Harris Semiconductor

R. I. Kung
Intel Corporation

Panelists

W. R. Young, *Harris Semiconductor*
R. I. Kung, *Intel Corporation*
E. J. D'Escoubet, *Solid State Scientific*
C. R. Gregory, *General Electronic*
D. B. Scott, *Texas Instruments, Inc.*
T. J. Rogers, *Cypress Semiconductor*
T. Frederiksen, *National Semiconductor*

AN OVERVIEW OF CMOS TECHNOLOGIES

W. R. Young

Harris Semiconductor
P. O. Box 883
Melbourne, FL 32901

Abstract

There are many CMOS processes in use or under development: P well, N well, twin well, dielectrically isolated, silicon-on-sapphire, silicon-on-insulator, high temperature, and radiation hardened. Each process optimizes some aspect of the devices performance, manufacturability or cost; however, no single process offers a complete solution for all applications. Techiques have been developed to minimze or eliminate latch-up in CMOS devices to allow the successful use of these technologies. Current circuit design techniques have lead to improved circuit performance and increased packing densities.

Introduction

CMOS is one of the oldest MOS technologies with development beginning in the mid-60's. Originally CMOS was considered only as a low power technology. Modern CMOS technologies are designed to provide not only low power but also high performance and cost effective solutions to VLSI circuit applications. There are several different processing approaches being developed for CMOS: P well, N well, twin well, dielectrically isolated, silicon-on-sapphire, silicon-on-insulator, high temperature, and radiation hardened CMOS processes. Each process will be discussed considering its primary advantages and disadvantages. The P well process, being the original CMOS process, will be discussed first and serve as a reference for comparing others. Bulk CMOS processes will be discussed first followed by processes using some form of dielectric isolation. The control of SCR induced latch-up in CMOS processes will then be discussed. Finally, modern VLSI CMOS design techniques will be reviewed.

CMOS Processes

P Well Processes

The P well CMOS processes are the oldest[1]. These start with a N- wafer into which a P- well is diffused. Most recently developed processes employ some form of side wall oxide isolation. Here, the active device and low resistance regions are delineated; field region doping, if used, is performed; and the field regions are oxidized. Next the gate oxide is grown; threshold adjustment implants, if employed, are performed; and the gate material is deposited and defined. The source-drain and low resistance regions are now doped. Then the gate material to metalization insulator is formed; contact areas are opened; and interconnect metal is deposited and patterned. Finally the circuit is passivated and the passivation is removed from the bond pad area to complete the process.

These processes offer three types of transistors: PMOS transistors with common N substrate bodies, NMOS transistors with P well bodies and a vertical common collector substrate NPN transistors. The starting material is the body of the PMOS transistors and the collector of the NPN transistors. The N substrate is connected to the most positive voltage on the chip. This positive voltage provides a common voltage for all PMOS transistor bodies and a common collector supply for the NPN transistors. The substrate NPN transistor can provide positive drive capability to enhance the positive capability of the PMOS transistors. This device has been used to great advantage in CMOS PROM's. The NMOS device, being fabricated in an isolated P well, allows the body to be connected to either the device's source or to a potential equal or more negative than the source. This flexibility in body bias control is helpful in maximizing the NMOS devices performance in many design applications.

The proponents of P well technologies cite the following advantages over N well technologies:

1. Proven reliability and history of manufacture.
2. Reduced alpha particle sensitivity.
3. Closer matched NMOS and PMOS performance.
4. High performance NPN availability.

N Well Processes

N well processes are relatively new[2-3]. These start with a P- wafer into which a N- well is diffused. Once the N well has been diffused, the subsequent processing is similar to that for P well technologies.

This process offers three types of transistors: NMOS transistors with common P substrate bodies, PMOS transistors with N well bodies and a vertical common collector substrate PNP transistors. The starting material is the body of the NMOS transistors and the collector of the NPN transistors. The P substrate is connected to the most negative voltage on the chip. This negative voltage provides a common voltage for all NMOS transistor bodies and

CH1935-6/83/0000/0432$01.00 © 1983 IEEE

a common collector supply for the PNP transistors. The substrate PNP transistor can provide current sink capability to enhance the drive capability of the NMOS transistors. The PNP is used primarily in input protection circuits. The PMOS device being fabricated in an isolated N well allows the body to be connected to either the device's source or to a potential equal or more positive than the source. This flexibility in body bias control is helpful in maximizing the PMOS devices performance in many design applications.

The proponents of N well technologies cite the following advantages over P well technologies:

1. Compatibility with existing with NMOS processes.
2. Substrate quality, availability and cost.
3. Improved NMOS performance.
4. Reduced latch-up sensitivity.

The primary disadvantages of this process are increased performance difference between the NMOS and PMOS transistors and the loss of the vertical NPN transistor of the P well process which can be used to aid the performance of the PMOS transistor. The P well process starts with a N substrate which optimizes the PMOS transistor; the P well must be about an order of magnitude heavier doped for good process control. This increased doping level of the well degrades the NMOS transistors optimum performance; however, the higher mobility of the electron relative to the hole still allows the P well NMOS transistor to out perform the PMOS transistor. The N well process reverses the doping levels; thereby, increasing the performance differences.

Twin Well Processes

The twin well processes are relatively new. These start with high resistively wafers. There are several choices for starting material: high resistivity P- wafers, high resistivity N- wafers, high resistivity P- epi on P+ substrates, or high resisitivity N- epi on N+ substrates[4]. The epi approach offers reduced latch-up sensitivity and improved substrate bipolar transistors. Once the substrate has been selected, the process begins by diffusing both a N- well and a P- well. After the wells have been diffused, the subsequent processing is similar to that for P well technologies.

The primary advantage of the twin well approach over conventional bulk CMOS methods is the optimization of both the NMOS and PMOS transistors. The disadvantages are the availability of the starting material and increased process complexity.

Radiation Hardened Bulk CMOS

Radiation hardened bulk CMOS processes require solving the following problems:

1. Elimination of latch-up.
2. Elimination of radiation induced reduction of field threshold voltages.
3. Control of radiation induced transistor threshold voltage shifts.

Latch-up is eliminated by using a N- epi on a N+ substrate. The N+ substrate lowers the resistance of the parasitic base to emitter resistor in the lateral PNP transistor of the SCR to the point where the SCR is prevented from latching. Radiation tends to lower the thresholds of NMOS devices and raise the thresholds of PMOS devices; therefore, only NMOS field thresholds must be controlled. The lowering of field thresholds can be prevented by the use of heavily doped guardbands to prevent parasitic field devices from turning on. P+ guardbands are used in the N channel field regions to prevent NMOS field devices from turning on. Threshold shifts induced in regular CMOS devices are minimized by the use of special radiation hardened gate oxidation processes. These processes are designed to reduce the radiation induced charge trapping in the gate oxide.

Harris Semiconductor has developed a radiation hardened bulk CMOS process. This process is similar to a conventional P well process using a N- epi on N+ substrate except that a P+ guardband is defined and diffused after the P well diffusion and a special radiation hardened gate oxidation process is employed in place of the normal gate oxidation process. Using this process, 4K RAMS employing a six transistor memory cell have been fabricated. These memories have a worst case access times of less than 300 ns over the full military voltage and temperature range. The parts are latch-up free and are radiation hardened to greater than $2X10^5$ RAD-Si total dose.

Silicon-On-Sapphire

Silicon-on-sapphire is one of the oldest methods of developing a dielectrically isolated technology. A thin epitaxial layer of silicon is grown on a sapphire substrate. Devices are isolated by removing the silicon between devices either by etching or by local oxidation of the silicon. Sapphire is aluminum oxide (AL_2O_3); therefore, silicon-on-sapphire requires the use of low temperature processing to minimize the autodoping of the silicon layer by the aluminum in the substrate. In order to minimize the high temperature processing steps, ion implantation is used for all doping operations.

This technology offers a theoretical performance advantage over bulk technologies due to reduced parasitic capacitances to the substrate and improved packing densities offered by the silicon-on-sapphire. Performance is limited by reduced carrier mobilities compared to those of bulk devices. This is caused by stress in the silicon layer, induced by differences in the temperature coefficiences between the silicon and sapphire. Lattice mismatch between the silicon and the sapphire substrate cause high defect densities in the silicon layer. These defects cause increased leakages in silicon-on-sapphire devices. As interconnect geometries are scaled to submicron dimensions the substrate terminates fewer field lines and the capacitance is determined to a larger extent by fringing fields and coupling to adjacent lines. This reduces the effective capacitance difference between bulk and silicon-on-sapphire interconnect. Packing density of silicon-on sapphire is theoretically better than bulk silicon because spacing is not required for device to well spacing. Bulk processes require spacing for the well to surround the contained device and separation from the well to a non-contained device. The silicon-on-sapphire device only require minimum device isolation space reguardless of the device type. Layout density is not determined solely by device spacing but also by device geometries and interconnects; therefore, neither technology is inherently denser.

The availability and cost of starting substrates and the lack of a demonstrated performance advantages have prevented this technology from developing any large commercial markets. This technology is still being pursued because of its ability to be radiation hardened. Dielectric isolation prevents latch-up and eliminates radiation induced field threshold reduction problems associated with bulk processes. Two additional problems not common to bulk processes must be solved in silicon-on-sapphire in order to obtain radiation hardened devices[5]. The first is minimization of the non-saturation of the drain current versus drain voltage in the N- channel device. This effect is caused by the floating body employed in silicon-on-sapphire designs and avalanche multiplication which lowers the device threshold as a function of applied drain bias. The second effect is back-channel leakage caused by charge trapped at the body to substrate isolation dielectric interface. These effects have been controlled to allow fabrication of devices which are hardened in excess of 10^6 rads total ionizing dose.

Dielectrically Isolated CMOS

Dielectric isolation is the process in which each active area of single crystal silicon is surrounded on the sides and bottom by an insulating layer of silicon dioxide and the structure is imbedded in polycrystalline silicon for mechanical strength. The process begins with a conventional N- type silicon wafer. The side of the wafer which will become the bottom of the single crystal islands is etched to form the sidewall pattern. Next silicon dioxide and polycrystalline silicon are grown to fill the etched "moats". The opposite side of the wafer is then polished until the insulating sidewalls appear at the surface of the wafer. The wafers are now ready for conventional silicon intergrated circuit fabrication techniques. Dielectric isolation has the following advantages:

1. The isolation eliminates the parasitic SCR's which might create latch-up.
2. Leakage currents to the substrate are greatly reduced under high temperature conditions.
3. Bipolar, JFET, or MOS devices can be fabricated on the same chip.
4. The bodies of MOS devices are not inherently connected together as in junction isolated structures. This allows the bodies to float, be connected to fixed potentials, be shorted to the device's source or be modulated as required for optimum circuit performance.
5. Both vertical NPN and PNP transistors can be fabricated on the same chip. These transistors exhibit lower collector to substrate capacitance than their junction isolated equivalent devices. The vertical structure and reduced parasitic capacitances enhance the device's high frequency performance.

Harris Semiconductor is currently delivering standard products designed using a dielectric isolated linear silicon gate CMOS process. This process has been optimized for linear circuits and contains both vertical NPN and PNP transistors, non-selfaligned CMOS transistors and NiCr resistors. A family of high speed analog switches and multiplexers and a 12 bit multiplying digital-to-analog converter have been developed using this process.

High Temperature CMOS

CMOS circuits which can operate at temperature in excess of 350°C have recently been demonstrated by Harris Semiconductor[6]. The process employed is similar to a conventional self-aligned silicon gate P well process except that dielectric isolation is used to separate PMOS devices from the NMOS devices. The use of dielectric isolation prevents SRC action and eliminates P well to N substrate leakage. The thresholds in the process are adjusted to keep the devices enhancement at the elevated temperatures. Thin dielectric isolated islands, 10u, were used to improve the packing density compared to conventional dielectric isolation processes and to minimize the silicon volume thus reducing leakage currents. The primary limitation is the leakage current of the PMOS drain. The leakage current characteristics of the PMOS P+N- diode was 225 ua compared to an equal geometry NMOS N+P- diode leakage current of 89 ua at 10 volts reverse bias at 325°C. These differences in leakage currents are caused by the different doping levels in the two diodes. It is expected that adjustments in the process and device geometries will allow operation above 400°C.

Silicon-On-Insulator

Silicon-on-insulator may be one approach for achieving higher density and performance in future CMOS structures. There are two different approaches being investigated. One approach[7] implants an insulating layer below the surface of the silicon wafer. If the insulating layer is deep enough, either an epitaxial layer can be grown or the residual layer can be used directly. This approach requires the availability of high current implanters to implant the nitrogen or oxyen required for the insulating layer. Once the implant is complete the implant damage must be annealed. The second approach[8] recrystallizes a polycrystalline layer deposited on an oxidized silicon wafer. A laser or graphic heater is moved over the surface locally melting the polycrystalline material and the material recrystallizes into single crystal material. Several approaches are being investigated to control crystal orientation of the recrystallized material. Once the material is annealed in the first approach or is recrystallized in the second approach the conventional CMOS processing can be employed.

Silicon-on-insulator material potentially offers material costs approaching those of bulk silicon wafers, device mobilities of bulk silicon and the performance advantages of a dielectrically isolated substrate. Experimental results to date have demonstrated various aspects of this technology; however, a commercially viable process has yet to be demonstrated.

SCR Control

The basic CMOS structure contains a parasitic SCR formed by the PMOS source (P+), PMOS body (N-), NMOS body (P-), and the NMOS source (N+). If the NMOS or PMOS source to body junction becomes forward biased then the SCR can be fired causing latch-up. The P well process contains a high performance vertical NPN transistor formed by the NMOS N+ source region, the P- well and the N- substrate. A lateral PNP transistor is formed by the PMOS P+ source region, the N- substrate and the P- well. Similarly, the N well process contains a high performance vertical PNP transistor formed by the PMOS P+ source region, the N- well and the P- substrate. A lateral NPN transistor is formed by the NMOS N+ source region, the P- substrate and the N- well.

Two approaches have been used to control the SCR structure in CMOS processes. The first method is to eliminate the four layer structure by means of dielectric isolation. Three approaches to this form of isolation are currently used: dielectric isolation, silicon-on-sapphire or silicon-on-insulator. These isolation techniques separate the PMOS and NMOS devices by a dielectric material such as silicon dioxide, silicon nitride or sapphire; thus, preventing the formation of the four layer structure. The second approach is to control the SCR in junction isolated structures.

Several approaches to controlling SCR action in junction isolated devices have been employed. The SCR cannot cause latch-up as long as the product of the NPN and the PNP is less than one. One approach is to lower the beta of the transistors to the point where latch-up is prevented. The doping levels are determined by the MOS transistor characteristics required; therefore, they cannot be changed in order to lower the bipolar transistor beta. These betas can be lowered by reducing the carrier life time by gold doping or neutron damage but these methods cause increased device leakages. Another method is to increase the base width of the bipolar transistors. The vertical transistors base width cannot be increased without greatly affecting the packing density of the process and only a modest beta reduction is achieved because of the light doping in the well. The base width of the lateral device can be increased by increasing the spacing between the substrate isolated device and the nearest well. This approach is used only on input and output devices because of the large spacing required to lower the lateral beta to a safe level. Guardrings have been used around wells to collect carriers injected laterally in the substrate and prevent them from reaching the well and causing latch-up. These guardrings are the same diffusions as the well diffusion but contain no diffusions of a different doping type. The guardrings are connected to the same potential as the well it is protecting. Another approach is to prevent the NPN and PNP transistors form turning on by connecting the bipolar's base to its emitter. The use of P- epi on P+ or N- epi on N+ as a starting substrate for N well or P well process respectively has successfully lowered the value of the base to emitter resistor in the lateral transistor to the point where latch-up does not occur[9-11].

CMOS Circuit Design

VLSI CMOS circuits use the best of both the conventional CMOS structure and the best of the NMOS structure. The series connections in a CMOS structure cause the area of the series devices to grow as the square of the number of inputs in order to maintain the same output impedance. A second problem with the conventional CMOS structure is the requirement for each input to drive both a NMOS and a PMOS transistor. One method of overcoming these difficulties is the use of CMOS domino circuits as proposed by Bell Laboratories[12]. This logic uses a single clock to precharge the output of a dynamic NMOS logic stack high while opening the connection to ground. The output of this gate then drives only a CMOS buffer. The buffer supplies the required drive capability for subsequent loads. This logic is dynamic and requires all inputs which are to be low, to complete their transition during precharge. At the end of the precharge cycle, the output precharge path is opened and the connection to ground is enabled. The output buffer makes at most a single transition and remains there until the next precharge. RAM cells employed in CMOS circuits may be of the six transistor type or a four NMOS transistor cell using high resistance poly load resistors. ROM, EAROM and EEROM cells are similar to those used in NMOS circuits. The basic NAND and NOR capability of CMOS allows the elimination of many inverters. VLSI CMOS circuits may employ as many as two or three times as many NMOS as PMOS devices and achieve comparable packing density to NMOS circuits.

References

1. M. R. Gulett, "The CMOS Controversy," VLSI DESIGN Forth Quarter 1981, pp. 50-51.
2. R. L. Maddox, "Reverse CMOS Processing," Solid State Technology Feb. 1981, pp. 128-131.
3. T. Ohzone, H. Shimura, K. Tsuji and T. Hirao, "Silicon-Gate N-Well CMOS Process by Full Ion-Implantation Technology," IEEE Trans. Electron Devices, Vol. ED-27, No. 9, Sept. 1980, pp. 1789-1795.
4. L. C. Parrillo, et al., "Twin-Tub CMOS-A Technology For VLSI Circuits," IEDM 1980, pp. 752-755.
5. B. L. Buchanan, D. A. Neamen and W. M. Shedd, "SOS Device Radiation Effects and Hardening," IEEE Trans. Electron Devices, Vol. ED-25 No. 8, Aug. 1978, pp. 959-970.
6. J. D. Beasom, R. D. Moore, G. Mohammed, and B. L. Draper, "350°C CMOS Logic Process," IEDM 1981, pp. 350-353.
7. J. Gosch, "Buried-Nitride C-MOS May Double Device Density," Electronics Jan. 27, 1983, pp. 70-71.
8. H. W. Lam, "Laser-Recrystallized Silicon-On-Oxide— The Ideal Silicon-On-Insulator Structure for VLSI?," IEDM 1980, pp. 556-558.
9. W. D. Raburn and J. E. Schroeder, "Limitations On Bulk CMOS VLSI Due To Parasitic SCR Paths," UGI Symp. 1981, pp. VII 44-51.
10. W. D. Raburn,"A Model For The Parasitic SCR in Bulk Silicon," IEDM 1980, pp. 252-255.
11. R. S. Payne, W. N. Grant and W. J. Bertram, "Elimination Of Latch Up In Bulk CMOS," IEDM 1980, pp. 248-251.
12. R. H. Krambeck, C. M. Lee and H. S. Law, "High-Speed Compact Circuits With CMOS," IEEE J Of SSC, Vol. SC-17, No. 3, June 1982, pp. 614-619.

Session 2-4b
VLSI Simulation II

Chairmen

A. E. Ruehli
IBM Research

J. Jess
Einhoven University

RSIM — A LOGIC-LEVEL TIMING SIMULATOR

Christopher J. Terman

M.I.T. Laboratory for Computer Science
545 Technology Square, Room 418
Cambridge, MA 02139

Abstract

RSIM is an event-driven, logic-level simulator that incorporates a linear transistor model useful in simulating digital MOS circuits. Use of a linear model allows a designer to determine both the functional behavior and approximate timing characteristics of large circuits (10K - 50K transistors) with relatively modest expenditures of computer time. This paper describes the linear model, presents a simulation algorithm based on this model, discusses the accuracy of the simulator's predictions, and reports on a year's industrial experience with RSIM.

Introduction

What can one do for a designer who wants feedback on the operation of a large digital circuit? Traditionally there are several options:

- gate-level simulation.[1-3] Can accommodate large circuits but models are inadequate for switching logic found in MOS designs.
- switch-level simulation.[4,5] Simple model accurately simulates many large digital MOS designs, but provides no information on performance of the circuit.
- timing simulation.[6-8] Usable performance predictions but circuit size limited to <10K devices.
- circuit analysis.[9-11] Most accuracy but only small circuits can be simulated. Even with the newer relaxation techniques,[12,13] circuits still limited to a few thousand devices.

This paper discusses RSIM, a logic-level simulator that uses a simple linear transistor model: the result falls somewhere between switch-level and timing simulation. Rather than perform a detailed simulation of each transistor's operation, RSIM uses the linear model to directly predict the logic state of each node and to estimate transition times if the nodes change state. The net effect is to trade some accuracy in the predictions for an increase in simulation speed. When the linear model is conservatively calibrated, its predictions can be used to identify potential problem circuits in need of more accurate analysis. A large portion of most circuits pass the scrutiny of RSIM and so the expense associated with detailed simulation of the whole circuit is avoided.

In addition to serving as the basis for simulation, the linear model can be used in timing analysis and might serve to quickly generate initial waveforms for a relaxation-based circuit analysis program.

A Linear Model for MOS Networks

The transistor model in RSIM can be quite simple since it is only used to predict the final logic state of a node and the length of time each state transition takes. As an example of how the model works, consider a simple inverter: one can think of the effective resistance of its component devices at any moment as

$$R_{eff:pullup} = \frac{i_{ds:pullup}}{v_{ds:pullup}} \qquad R_{eff:pulldown} = \frac{i_{ds:pulldown}}{v_{ds:pulldown}}$$

Although the effective resistances of the transistors change as their terminal voltages vary, it might be possible to use "average channel resistances" to characterize the transistors' behavior. The other salient feature of a transistor's operation is its switch-like behavior: with certain voltages on a transistor's terminal nodes it makes no connection at all between its source and drain terminals — the transistor is "off". As the relative terminal voltages change, the transistor turns "on", conducting current between its source and drain terminals. Of course, the transistor is more "on" at some times than others, but distinctions between different "on" states will be ignored for simplicity.

It is possible to build on the two observations made above to construct a linear transistor model for RSIM:

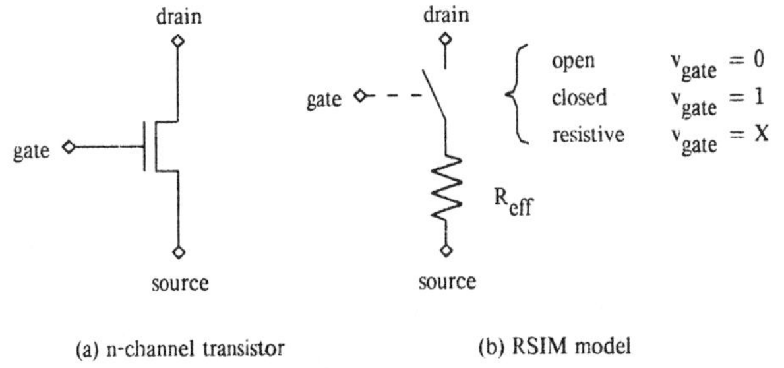

Figure 1. *RSIM model for an n-channel transistor*

It is easy to tabulate what sort of connection exists between the source and drain terminals as a function of the gate voltage:

$$R_{ds} = \begin{cases} R_{eff} & \textit{switch closed} \quad (v_{gate}=1) \\ \infty & \textit{switch open} \quad (v_{gate}=0) \\ [R_{eff}, \infty] & \textit{switch resistive} \quad (v_{gate}=X) \end{cases}$$

Note that uncertainty about the state of the switch leads naturally to a resistance interval describing the source-drain connection. In fact,

CH1935-6/83/0000/0437$01.00 © 1983 IEEE

all the network calculations use interval arithmetic, and the bounds of the resulting intervals are used when converting voltages to logic states, etc.; no other mechanisms are needed to deal successfully with X states in the network. Models for other types of transistors differ in how the state of the switch is determined from v_{gate}:

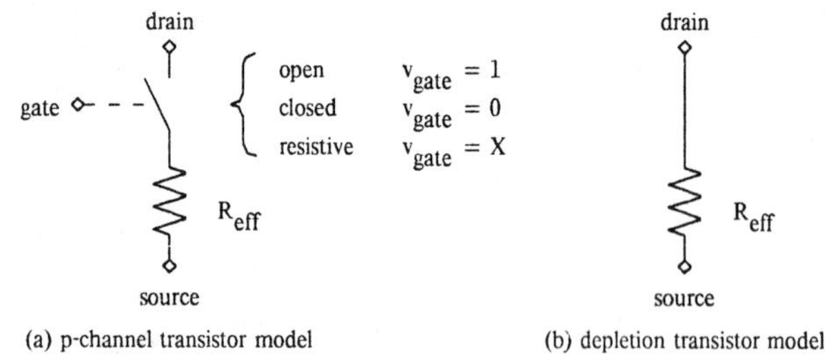

Figure 2. *RSIM models for p-channel and depletion transistors*

The effective resistance R_{eff} is determined separately for each transistor and depends on

width,length dimensions of the active transistor area. Various non-linear effects make R_{eff} a more complicated function of the transistor geometry than just length/width.

type Most MOS circuits contain more than one type of transistor. The different types are distinguished by different values for their threshold voltage — since the current conducted by a transistor is a function of its threshold voltage and hence its type, the modeling resistance also depends on the transistor type.

context Accuracy in choosing the effective resistance can be improved by distinguishing several contexts in which a transistor may appear: for example, an enhancement transistor can be used as a pulldown or source-follower in addition to its default "pass" gate configuration. Surprisingly few contexts need to be recognized to encompass a large portion of digital MOS designs.

The determination of R_{eff} is made once for each transistor and does not depend on any dynamic properties of the circuit to be simulated; during simulation the only device information RSIM uses about a transistor is its effective resistance.

Voltages in this model are quantized into one of three values; this corresponds to our intuition for *digital* logic and greatly simplifies the simulation calculations. If all node voltages are normalized to fall in the range [0, 1], then the possible quantized values are

0 logic low -- voltages in the range $[0, v_{low}]$;

1 logic high -- voltages in the range $[v_{high}, 1]$;

X intermediate or unknown voltages.

where v_{low} and v_{high} are the predetermined logic thresholds.

How is the value of a node determined? RSIM characterizes the effect of the network on a particular node by the Thevenin circuit equivalent for all pieces of the network that directly influence the value of the given node.

V_{thev} a voltage interval $[V_-, V_+]$ in the range [0, 1] specifying the possible voltages the output node may have.

R_{drive} a resistance interval $[R_-, R_+]$ in the range $[0, \infty]$.

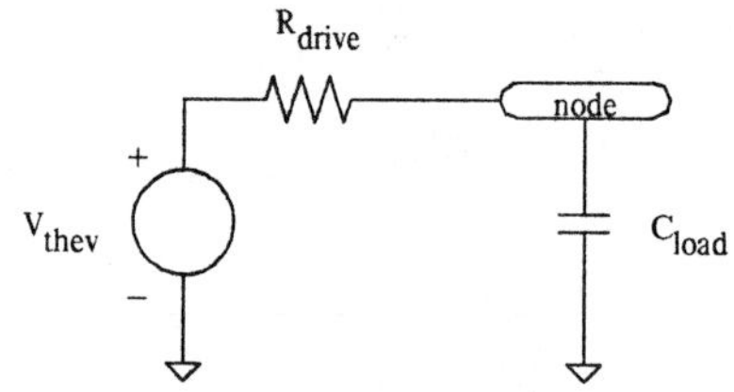

Figure 3. *Equivalent circuit for a network node*

V_{thev} and R_{drive} are, in general, intervals since the equivalent transistor resistances from which they are derived might themselves lie in an interval. A node's final value is determined by comparing V_{thev} with the low and high logic thresholds and choosing the appropriate logic state. If the calculation of a node's final value yields a result different from the node's current value, a transition has been discovered and the simulator must predict how long it will take for the node to reach its new value. Given the model shown in Figure 3, an obvious choice for the transition time is $R_{drive}C_{load}$. With suitable definitions of R_{drive} and C_{load}, this is the approach adopted by RSIM.

Actually RSIM uses not one, but three effective resistances for each transistor. To see why, recall that RSIM is trying to predict the transition time and final voltage, as shown in Figure 4.

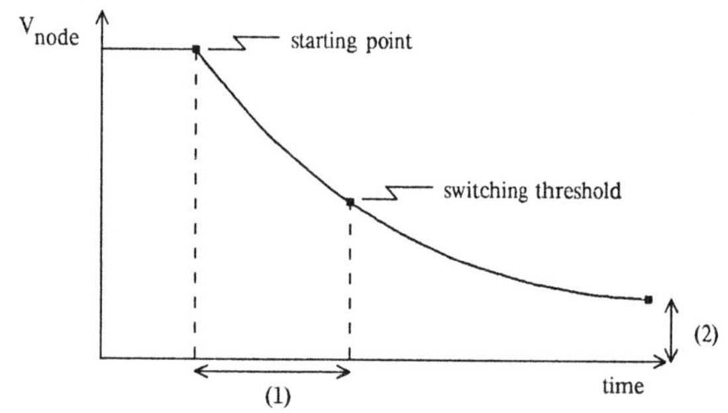

Figure 4. *R_{eff} used to predict (1) transition time and (2) final voltage*

One would like to calibrate the model to give accurate predictions for both parameters, but that is impossible with a single set of resistances. To solve this problem RSIM uses three resistances for each transistor:

R_{static} used when calculating V_{thev}, the final voltage.

R_{dynlow} used when calculating R_{drive} for high-to-low transitions.

$R_{dynhigh}$ used when calculating R_{drive} for low-to-high transitions.

Two "dynamic" resistances are used so that the asymmetric behavior of "pass" devices can be accurately predicted. All three values (V_{thev} and two versions of R_{drive}) are calculated simultaneously, so the extra overhead introduced by multiple resistances is not very large.

The RSIM Simulation Algorithm

Basic to the operation of RSIM is the notion of an *event*. An event specifies (i) a node in the network, (ii) a new logic state, and (iii) a time at which the node's value is to be changed to the new logic state. RSIM maintains a list of events, sorted by time, that tells what processing remains to be done. Whenever the user changes an input, an event is added to the list; when the list is empty, the network has "settled" and RSIM waits for further input.

When started on an initial event list, RSIM sequentially processes events from the list, stopping (1) when the list is empty, (2) when a node the user is tracing changes value, or (3) when the specified amount of simulated time has elapsed. Processing an event entails

(a) removing the event from the event list.

(b) changing the node's state to reflect its new value.

(c) calculating any consequences, *i.e.*, new events, resulting from the node's new value. First all nodes that might be affected by the change are found and marked — this requires a tree-walk of the network starting at the source and drain nodes of transistors for which the changing node is the gate. The tree-walk follows source/drain connections, stopping at input nodes or non-conducting transistors. For each marked node two calculations are made: (1) a charge-sharing calculation that models changes of state due to charging/discharging of the node capacitances and (2) a final-value calculation that determines the node's ultimate state.

As seen in step (c) the network is naturally partitioned into stages, each stage consisting of nodes "shorted" together by source-drain connections. The values for all nodes in a stage are recalculated whenever nodes are added or removed from the stage because of a transistor turning on or off. Since nodes are only added to the event list when their values change, portions of the circuit unaffected by the current set of changes are not re-evaluated — the algorithm is event driven.

A node can have two events pending:

(1) a charge-sharing event describing an immediate change in the node's state due to the redistribution of charge among the capacitors for nodes in the current stage. This type of event is generated only when a node is added to a stage (*i.e.*, when a transistor turns on).

(2) a final-value event describing what the final, driven state of the node will be. This type of event is generated only when $R_{drive} < \infty$.

The charge-sharing calculation models what happens when two or more charged nodes in different logic states are connected. In this case, all the connected nodes will reach the same logic state; this state is determined by the relative capacitances and initial logic states of the nodes in the stage. For example, if a large (high capacitance) node such as a data bus were connected by a pass transistor to a small node such as the input to a register cell, then the small node would "share" the charge of the large node as its final value regardless of the charge it had initially. In such cases, the charge-sharing value is determined from

$$V_{share:min} = \frac{\sum_{1\ nodes} c_i}{\sum_{all\ nodes} c_i} \qquad V_{share:max} = \frac{\sum_{1\ nodes} c_i + \sum_{X\ nodes} c_i}{\sum_{all\ nodes} c_i}$$

where the sums are over nodes in the current stage. Since nodes at logic state X contribute an undetermined amount of charge to the result, V_{share} is an interval whose bounds represent the worst case assumptions about the actual values of X nodes. This interval is compared with the logic thresholds when calculating the charge-sharing value:

$$Charge-sharing\ value = \begin{cases} 0 & V_{share:max} \le v_{low} \\ 1 & V_{share:min} \ge v_{high} \\ X & otherwise \end{cases}$$

If $R_{drive} < \infty$, and the node is not an input, the final state of a driven node is calculated from the V_{thev} interval $[V_-, V_+]$:

$$Final\ value = \begin{cases} 0 & V_+ \le v_{low} \\ 1 & V_- \ge v_{high} \\ X & otherwise \end{cases}$$

If this value differs from the charge-sharing value then the appropriate event is scheduled $R_{eff}C_{eff}$ seconds in the future where

$$R_{eff} = \begin{cases} R_{drive:high} & final\ state = 1 \\ R_{drive:low} & final\ state = 0 \\ min\ of\ above & final\ state = X \end{cases}$$

$$C_{eff} = \begin{cases} \sum_{0\ and\ X\ nodes} c_i & final\ state = 1 \\ \sum_{1\ and\ X\ nodes} c_i & final\ state = 0 \\ min\ of\ above & final\ state = X \end{cases}$$

where the sums are over nodes in the current stage. The lumped capacitance represented by C_{eff} is overly conservative; more accurate models are available[14,15] and should be incorporated into RSIM.

Experience with RSIM

RSIM has been in use at both university and industrial environments for about 18 months. During that time it has simulated several hundred designs, ranging in size from very small to approximately 40K transistors. Because RSIM was fast enough to simulate whole circuits, it often uncovered circuit flaws that had fallen between the cracks during simulation of the individual components. The trend has been to think of RSIM as a companion to circuit analysis, using it for all logic-level verification and preliminary timing analysis, and then resorting to circuit analysis for those paths identified as critical by RSIM.

The simulation algorithm is embedded in a Lisp-like command language[16] that has been used to write quite elaborate programs to drive the simulation and process the results. Since programs to prepare simulation input are much less tedious to construct than the input itself, designers have been able to conduct more exhaustive tests than they were able to do using earlier simulators. For example, it is a simple matter to take a set of test vectors used to drive an RTL simulation, use those vectors as input for an RSIM run, and compare the predicted outcomes, all under program control.

With careful calibration, RSIM's predictions for combinational logic are within 30% of those of SPICE. For circuits relying on analog behavior (sense amps, bootstrapped nodes, etc.) or chains of "pass" devices, the predictions are less accurate. To compensate, several "escape" mechanisms exist: it is possible to specify the logic thresholds and transition times for individual nodes so that one can incorporate the results of more detailed simulation into RSIM. Usually this mechanism need be invoked for only a few critical nodes (*e.g.*, clock driver outputs). Another approach is to identify problem subcircuits and replace them with logically equivalent circuits that can be simulated easily; a network preprocessor[17] that performs subcircuit matching and replacement is available and has been used to good effect. Finally, it is possible to calibrate the linear model so as to guarantee that the predictions are an upper bound for the actual transition times.[18] Calibrated in this manner, RSIM is a fairly reliable "filter" for detecting circuits in need of more careful analysis.

Acknowledgement

This research was supported by the Defense Advanced Research Projects Agency (DoD), Contract No. N00014-83-K-0125.

References

[1] Alia, G., P. Ciompi, and E. Martinelli, "LSI Components Modelling in a Three-Valued Functional Simulation," *Proceedings of the 15th Design Automation Conference*, June 1978.

[2] J. Jephson, R. McQuarrie, and R. Vogelsberg, "A Three-Value Computer Design Verification System," *IBM Systems Journal*, No. 3, 1969.

[3] Szygenda, S. and E. Thompson, "Digital Logic Simulation in a Time-Based, Table-Driven Environment," *IEEE Computer*, Vol. 8, March 1975.

[4] Bryant, R., *A Switch-level Simulation Model for Integrated Logic Circuits*, M.I.T. Laboratory for Computer Science TR-259, 1981.

[5] Baker, C. and C. Terman, "Tools for Verifying Integrated Circuit Designs," *Lambda*, Fourth Quarter, 1980.

[6] Chawla, B. R., H. K. Gummel, and P. Kozak, "MOTIS — An MOS Timing Simulator," *IEEE Transactions on Circuits & Systems*, Vol. CAS-22, No. 13, December 1975.

[7] Newton, A. R., *Timing, Logic and Mixed-mode Simulation for Large MOS Integrated Circuits*, NATO Advanced Study Institute on Computer Design Aids for VLSI Circuits, Sogesta-Urbino, Italy, July/August 1980.

[8] Nham, H. N., and A. K. Bose, "A Multiple Delay Simulator for MOS LSI Circuits," *Proceedings of the 17th Design Automation Conference*, 1980.

[9] Nagel, L., *SPICE2: A Computer Program to Simulate Semiconductor Circuits*, Memo No. ERL-M520, University of California, Berkeley, May 1975.

[10] Sangiovanni-Vincentelli, A. L., *Circuit Simulation*, NATO Advanced Study Institute on Computer Design Aids for VLSI Circuits, Sogesta-Urbino, Italy, July/August 1980.

[11] Weeks, W. T. et al., "Algorithms for ASTAP — A Network Analysis Program," *IEEE Transaction on Circuit Theory*, Vol. CT-20, November 1973.

[12] Lelarasmee, E., A. E. Ruehli, and A. L. Sangiovanni-Vincentelli, *The Waveform Relaxation Method for Time Domain Analysis of Large Scale Integrated Circuits*, Memo No. UCB/ERL-M81/75, University of California, Berkeley, June 1981.

[13] Newton, A. R. and A. L. Sangiovanni-Vincentelli, *Relaxation-based Electrical Simulation*, University of California, Berkeley, 1983.

[14] Horowitz, M., "Timing Models for MOS Pass Networks", *Proceedings of the IEEE International Symposium on Circuits and Systems*, 1983.

[15] Penfield, P. and J. Rubinstein, *Signal Delay in RC Tree Networks*, M.I.T. VLSI Memo No. 81-40, January 1981.

[16] Terman, C., *User's Guide to NET, PRESIM, and RNL*, M.I.T. Laboratory for Computer Science, September 1982.

[17] Iler, J., *A VLSI Circuit Recognizer for Enhancing Simulator Accuracy*, MS Thesis, M.I.T. Department of Electrical Engineering and Computer Science, January 1983.

[18] Terman, C., *Simulation Tools for Digital LSI Design*, PhD Thesis in preparation, M.I.T. Department of Electrical Engineering and Computer Science, 1983.

MULTI-LEVEL SIMULATION FOR VLSI

Karl J. Lieberherr

Department of Electrical Engineering and Computer Science
Princeton University, Princeton, New Jersey 08544
and
Laboratory for Computer Science, Massachusetts Institute of Technology
Cambridge, Massachusetts 02139

Abstract: We propose a powerful, unifying notation (called Zeus) for describing circuits to a multi-level simulator. At the highest level, Zeus introduces a novel circuit model for which we describe a fast simulator. Although this simulator is simpler and faster than previous simulators, it is able to diagnose many common design mistakes.

1. Introduction

A hardware/software system designer relies heavily on hardware simulation. There are many applications: Partial verification at various stages of the design (design simulation), fault simulation, debugging of software before hardware is ready, assessing the impact of a change, allowing measurement, evaluation, and identifications of bottlenecks etc.

The goal of design simulation is to detect errors in a circuit description by exercising it on a set of carefully chosen inputs. An error is a short proof which shows that a circuit description does not satisfy a specification (which might be a circuit description). If the circuit is small enough, then all possible inputs can be simulated, thereby proving the correctness of the circuit with respect to a given model. Usually the circuits are too big and therefore simulation only helps for detecting errors and not for proving correctness.

Recently there have been several efforts for formally proving the correctness of circuits. There are two basic approaches. The first one provides a proof that a given circuit description satisfies a specification (see e.g. [Gordon (1981), Moskowski (1983)]). The second method uses correctness preserving transformations [Leiserson (1983), Lam/Mostow (1983)]. Since the correctness problem is co-NP-hard in most cases, it is unlikely that simulation is replaced by formal verification. Correctness proofs might be very useful if they are developed during the design and therefore help to guide the design process.

Precise simulations of electrical behavior require enormous amounts of computation. Therefore the designer usually compromises by using incomplete and/or imprecise simulations, simulating precisely those small regions of the design that depend on critical electrical behavior, while simulating the bulk of the design at higher levels.

The simulation levels can roughly be divided into 1. register transfer level, 2. logic level, 3. switch level and 4. circuit level. Unfortunately a different notation is often used for describing the input at the four levels. In this paper we propose the Zeus language [Lieberherr/Knudsen (1983a,b)] as input language to simulators for all four levels. (Zeus has been slightly extended.) Zeus is powerful enough to serve as unifying input language for many more design tools, like silicon compilers. The advantages of a common input language for a large variety of design tools are obvious. Probably the most important one is that the user of the tools has to learn just one notation. (See e.g. [Ackley (1982)] for detailed motivation.)

The Zeus language, at the highest level, is a technology independent hardware description language. It is designed for describing chips from the register-transfer level down to the layout level in a hierarchical way.

2. Zeus as simulator input

Because of space restrictions we refer the reader to [Lieberherr/Knudsen (1983b)] for an introduction to Zeus. The syntax is given in the appendix.

At the highest level Zeus describes a circuit as a hierarchical network of unidirectional switches, registers and boolean gates. In the following we summarize the features of Zeus which are relevant to simulation at the highest level.

There are several restrictions on the networks which can be expressed in Zeus at the highest level. These restrictions facilitate simulation and testing and are widely believed to make the design of hardware safer.

Zeus enforces regular design in the sense of [Roth (1980)]. This means that all feedback has to be clocked. Non-clocked feedback is hidden in predefined components.

Zeus has two basic signal types: boolean and multiplex. A signal of type boolean is assigned exactly once. A signal of type multiplex is assigned conditionally several times, however at most one assignment may be 0 or 1 at simulation time.

A conditional statement represents a unidirectional switch. It appears to be a good design practice to use switches predominantly in a unidirec-

CH1935-6/83/0000/0441$01.00 © 1983 IEEE

tional way (see e.g. [Jouppi (1983)]). Switches which are used bidirectionally are hidden in predefined components.

Every Zeus type can have a sequence of attributes which are denoted by keywords and enclosed between "{" and "}". The purpose of these attributes is to give special information to various design automation tools. Each tool (e.g. a simulator) recognizes certain attributes and ignores others. For example,

TYPE
  HorizontalMetalWire = boolean { layer = metal, direction = horizontal }

defines a horizontal metal wire. The attributes are accumulative, e.g.

TYPE
  x(n,m) = HorizontalMetalWire { length = n, width = m }

defines a horizontal metal wire of length n and width m.

We sketch now the input language for the various levels.

2.1. Register transfer level

This level corresponds to the highest level of Zeus and has all features described above. During the early design phase it is convenient to use a general purpose programming language for behavioral description. We use Modula-2 [Wirth (1982)] because of its simplicity and elegance and since the Zeus syntax has a Modula-2 flavor anyway.

In a component type definition there may be a definition/implementation module pair. The definition module must export a procedure which defines the behaviour of the component. A synchronous component type definition is described by a Modula-2 procedure specifying the actions which are performed for each clock cycle. An asynchronous component type is described by a Modula-2 procedure specifying the behaviour from start to completion. We omit here the interface between Zeus types and Modula-2 types.

2.2. Switch level

A switch level description specifies a network of interconnected transistors. R. Bryant has proposed a model at this level which allows a fast simulation [Bryant (1981)]. Improvements to Bryant's simulation for interesting restricted classes of circuits are proposed in [Ramachandran (1983)]. In order to use Zeus at this level we drop the functional part of Zeus and add predefined transistor component types, depending on the technology.

In the following we outline how to use Zeus for describing MOSSIM-II like input [Bryant (1982)]. We introduce three predefined components: Ntrans (n-type transistor), Ptrans (p-type transistor), Dtrans (depletion mode transistor). Each signal of basic type can have an attribute called "capacitance" (size in MOSSIM II). Each transistor has an attribute called "conductance" (strength in MOSSIM II) which indicates the conductance of the transistor relative to other transistors which may form part of a ratioed path.

2.3. Circuit level

One possibility to describe the input to a circuit level simulator is to give a layout level description. It is then necessary to have a program which extracts the input required by simulators like SPICE [Nagel (1975)].

Zeus is suitable for describing chips at the layout level in any two-dimensional technology since each assignment statement and signal of basic type can have a direction, layer, length and width attribute. Each signal of basic type is called a wire. E.g. x:=y {layer= metal, direction= horizontal} means that the wires x and y are connected by a horizontal piece of metal. Contact cuts are inserted when necessary.

For mixed-level operation the simulators need to have a common interface which defines how signal values at different levels are related. A flexible level translation is proposed in [Infante (1983)].

3. Simulation algorithm

We describe how to simulate a synchronous Zeus hardware description at the highest level. The basic simulation algorithm is a topological sorting algorithm, once the Zeus description has been translated into a node-labeled directed multigraph. This simulation does not provide delay information.

A Zeus hardware description is translated into a node-labeled directed multi-graph as follows. The predefined function components AND, OR, NAND, NOR, XOR, NOT and EQUAL are translated into one node which is labeled with the corresponding name of the function. For example, the expression AND(a,b), where a and b are signals of basic type, is translated into an AND-node with two incoming edges from nodes a and b and one outgoing edge. The predefined component REG is translated into a node labeled REG.

Each instantiated signal of basic type is translated into a node labeled SIGNAL. An instantiated component is translated into a graph recursively. If the component has n IN parameters and m OUT parameters of basic type, then it is translated into a graph with n ingoing and m outgoing edges. A similar rule holds for function components.

An assignment statement of the form a:=b, where a and b are of basic type, introduces an edge from node b to node a. The assignment statement a==b introduces a new signal in the class of signals which are equivalent to a (or b). An assignment statement of the form a1:=f(b1,b2, ... ,bn) where a1 is a signal of basic type, introduces a graph for f and the exiting edge is directed towards node a1. These rules generalize to assignments involving structured signal variables. The connection statement is considered as a sequence of assignment statements.

The IF-THEN-ELSE statement which is used for describing switches can be reformulated as a sequence of IF statements of the form

IF b1 THEN s1 END

For the translation of this statement to a graph we distinguish several cases.

a) If s1 is an assignment statement s:=e then we introduce a node, labelled IF, which has two entering edges from nodes b1 and e. The exiting edge goes to signal s.

b) If s1 is a connection statement it can easily be rewritten as a sequence of assignment statements for which rule a) applies.

c) If s1 is a conditional statement of the form
IF b2 THEN s2 END
then we introduce an IF node and apply the generation rule recursively.

d) If s1 is a replication statement we replicate the application of the generation rule.

e) If s1 is a conditional generation statement then we apply the generation rule recursively to the description which is selected. Note that in an instantiated component the values of all boolean expressions occuring in conditional generation statements are known.

The replication and conditional generation statement are handled in the way described above.

This completes the description of the node-labeled directed graph associated with a Zeus hardware description. According to the rules of Zeus this graph is free of cycles if we remove all nodes which are labeled REG. This graph is topologically sorted for guiding the evaluation for various input values.

To simulate an instantiated component s we evaluate the graph corresponding to s as follows. Let $p_1, p_2, \cdots, p_n$ be a topologically sorted sequence of all nodes in the graph corresponding to s. Each node p_i has a label which indicates whether the node is a processing node (labels AND, OR, NAND, NOR, XOR, NOT, EQUAL and IF) or a signal forwarding or storage node (labels SIGNAL, REG).

We perform a three value simulation with values 0,1 and NOINFL. NOINFL is a high impedance value often denoted by Z. A fourth value UNDEFINED is not needed in our simulation since it is considered illegal. It is a good design practice to bind unused pins either to 0 or 1 and therefore also all internal signals of a component should have value 0,1 or NOINFL.

If p_i is a signal node corresponding to an IN parameter, then p_i is assigned the value (0 or 1) assigned to the IN parameter.

If p_i is a signal node corresponding to the out pin of a register of type REG, then p_i is assigned the bit value stored in the register. If p_i is a SIGNAL node which has several predecessors, then p_i is assigned the value 0(1) as long as exactly one predecessor has value 0(1). If all predecessors have value NOINFL, then p_i is assigned value NOINFL. Otherwise an error message is printed.

This Zeus simulator runs in linear time in the number of basic signals. It allows a faster implementation than the switch-level simulators by Bryant [Bryant (1982)] and the improved version by Ramachandran [Ramachandran (1983)]. Nevertheless the Zeus simulator catches many design errors.

4. Zeus syntax

We give the complete syntax of Zeus. There are a few changes to the syntax given in [Lieberherr/ Knudsen (1983b)].

Hardware= {CompilationUnit} {declaration}.

CompilationUnit= see Modula-2 syntax .

declaration= constDeclaration| typeDeclaration| signalDeclaration| moduleDeclaration.

constDeclaration= CONST {ident "=" constant ";"}.

ident= letter {letter|digit}.

constant= ConstExpression| structConstExpression.

ConstExpression= SimpleConstExpr [relation SimpleConstExpr].

relation= "="| "<>"| "<"| "<="| ">"| ">=".

SimpleConstExpr= ["+"| "-"] ConstTerm {AddOperator ConstTerm}.

AddOperator= "+"| "-"| OR.

ConstTerm= ConstFactor {MulOperator ConstFactor}.

MulOperator= "*"| DIV| MOD| AND.

ConstFactor= number| "(" ConstExpression ")"| NOT ConstFactor| ident ["(" ConstExpression {"," ConstExpression} ")"].

number= digit {digit}.

structConstExpression= ConstExpression| "(" structConstExpression {"," structConstExpression }")"| BIN "(" ConstExpression "," ConstExpression ")".

typeDeclaration= TYPE {ident ["(" idlist ")"] "=" type ";" }.

type= (arrayDeclaration| componentDeclaration| ident ["(" ConstExpressionList ")"]) ["{" AttributeList "}"] .

AttributeList= Attribute { ";" Attribute } .

Attribute= ident "=" ConstExpression .

arrayDeclaration= ARRAY "[" ConstExpression ".." ConstExpression "]" OF type.

componentDeclaration= COMPONENT "(" [fparams {"," fparams }] ")" ["{" layoutStatementList "}"] [[":" type] IS [USES idlist ";"] {declaration} ["{" layoutStatementList "}"] [DefinitionModule IMPLEMENTATION ProgramModule] /comment: Modula-2/ [BEGIN StatementSequence END]].

fparams= [IN| OUT] fieldlist.

fieldlist= idlist ":" type.

idlist= ident {"," ident }.

StatementSequence= statement {";" statement}.

statement= [assignment| connection| conditional| replication| condGeneration| result| parallel| sequential| with].

assignment= signal (":="| "==") expression.

signal= (ident {"[" (ConstExpression [".." ConstExpression] "]"| NUM "(" signal ")") "]"| "." ident [".." ident]})| "*".

expression= signal ["{" AttributeList "}"] | ident ["(" ConstExpressionList ")"] [expression]| structConstExpression| "*" [":" ConstExpression]|

"(" expression {"," expression }")".
connection= signal [expression].
conditional= IF expression THEN StatementSequence {ELSIF expression THEN StatementSequence} [ELSE StatementSequence] END.
replication= FOR ident ":=" ConstExpression (TO| DOWNTO) ConstExpression DO [SEQUENTIALLY] StatementSequence END.
condGeneration= WHEN ConstExpression THEN StatementSequence {OTHERWISEWHEN ConstExpression THEN StatementSequence} [OTHERWISE StatementSequence] END.
result= RESULT expression.
parallel= PARALLEL StatementSequence END.
sequential= SEQUENTIAL StatementSequence END.
with= WITH signal DO StatementSequence END.
ConstExpressionList= ConstExpression {"," ConstExpression }.
signalDeclaration= SIGNAL {idlist ":" type ";"}.
moduleDeclaration= MODULE ident ";" { import } [export] { constDeclaration | typeDeclaration } END ident.
import= [FROM ident] IMPORT idlist.
export= EXPORT [QUALIFIED] idlist.
layoutStatementList= LayoutStatement {";" layoutStatement }.
layoutStatement= [basic| order| align| replication| boundary| condGeneration| with].
basic= (signal ["{" AttributeList "}"])| (signal "=" type).
order= ORDER directionOfSeparation layoutStatementList END.
directionOfSeparation= ident.
align= ALIGN alignmentline layoutStatementList END.
alignmentline= ident.
replication= FOR ident ":=" ConstExpression (TO| DOWNTO) ConstExpression DO layoutStatementList END.
boundary= (TOP| RIGHT| BOTTOM| LEFT) layoutStatementList.
condGeneration= WHEN ConstExpression THEN layoutStatementList {OTHERWISEWHEN ConstExpression THEN layoutStatementList} [OTHERWISE layoutStatementList] END.
with= WITH signal DO layoutStatementList END.

Acknowledgement

This research has been partially supported by grants MCS80-04490 from the American NSF and 82.001.82 from the Swiss NSF.

References

Ackley1982a. D. Ackley J. Carnegie E. B. Hassler, "Hardware Description System Development," *ICCC, New York*, pp. 608-611 (1982).

Bryant1982a. R. Bryant M. Schuster D. Whiting, "MOSSIM II: A switch-level Simulator for MOS LSI (User Manual)," *Technical Report No. 5033, California Institute of Technology*, (1982).

Bryant1981a. R.E. Bryant, "A switch-level model for integrated logic circuits," *VLSI Conference, Edinburgh*, pp. 329-340 (1981).

Gordon1981a. M.J.C. Gordon, "Register transfer systems and their behaviour," pp. 23-36 in *Computer hardware description languages and their applications*, ed. M. Breuer, R. Hartenstein,North Holland (1981).

Infante1983a. B. Infante M. Bales E. Lock, "MADL: A language for describing mixed behaviour and structure," pp. 115-126 in *Computer Hardware Description Languages*, ed. T. Uehara M. Barbacci,North-Holland (1983).

Jouppi1983a. N.P. Jouppi, "TV: An nMOS Timing Analyzer," pp. 71-86 in *Third Caltech Conference on Very Large Scale Integration*, ed. R. Bryant,Computer Science Press (1983).

Lam1983a. M. Lam J. Mostow, "A transformational model of VLSI systolic design," pp. 65-78 in *Computer hardware description languages and their applications*, ed. T. Uehara and M. Barbacci,North Holland (1983).

Leiserson1983a. C.E. Leiserson F.M. Rose J.D. Saxe, "Optimizing synchronous circuitry by retiming," pp. 87-116 in *Third Caltech Conference on Very Large Scale Integration*, ed. R. Bryant,Computer Science Press (1983).

Lieberherr1983a. Karl J. Lieberherr Svend E. Knudsen, "Zeus: A hardware description language for VLSI," *Tech. Report 51, Institut für Informatik, ETH Zentrum, CH-8092 Zurich, Switzerland*, (Feb. 1983).

Lieberherr1983b. Karl J. Lieberherr Svend E. Knudsen, "Zeus: A hardware description language for VLSI," *Proceedings of the ACM/IEEE 20th Design Automation Conference*, (June 27-29, 1983).

Moskowski1983a. B. Moskowski, "A temporal logic for multi-level reasoning about hardware," pp. 79-90 in *Computer hardware description languages and their applications*, ed. T. Uehara and M. Barbacci,North Holland (1983).

Nagel1975a. L.N. Nagel, "Spice2: A computer program to simulate semiconductor circuits," *Univ. of Calif. at Berkeley, Electronics Research Laboratory, Memorandum No. ERL-M520*, (May 1975).

Ramachandran1983a. V. Ramachandran, "An improved switch-level simulator for MOS circuits," *Proceedings of the ACM/IEEE 20th Design Automation Conference*, (June 27-29, 1983).

Roth1980a. P. Roth, "Computer Logic, testing and verification," *Computer Science Press*, (1980).

Wirth1982a. N. Wirth, *Programming in Modula-2*, Springer (1982).

A TABLE-DRIVEN DELAY-OPERATOR APPROACH TO TIMING SIMULATION OF MOS VLSI CIRCUITS*

by V. B. Rao, T. N. Trick and I. N. Hajj

Department of Electrical Engineering
and the Coordinated Science Laboratory
University of Illinois
Urbana, Illinois 61801

Abstract

A switch-level simulator for MOS circuits is described which includes a table-driven delay operator for timing analysis. The delay table is constructed by simulating several different circuit primitives on a circuit simulator under different load and signal conditions. The simulator provides timing information fairly close to that obtained by conventional circuit simulators, but at speeds approximately two orders of magnitude faster.

I. Introduction

In this paper we describe a simulator which attempts to reconstruct the digital-equivalent of the analog waveforms in an MOS circuit. It uses 3 states (0,u,1) to describe signal values. The signal waveforms are then described by a sequence of transitions between signal-states and time-points at which these transitions take place. A delay-operator is used to account for the circuit-dynamics by delaying the transitions suitably. The characteristics of this operator are stored in tables as a function of some parameters so that a delay value can be obtained by table-lookup techniques. This simulator attempts to overcome most of the shortcomings of the multiple-delay simulator of Nham and Bose [1] which have been pointed out in [2].

In the next section we explain some notations and definitions which are used to describe the algorithms used in this simulator. Section III deals with the partitioning and analysis-sequencing of the various partitioned blocks of the network. In Section IV we describe the basic analysis procedures and in Section V we describe the delay-operators. Finally, in Section VI we present some aspects of the computer implementation and some examples.

II. Notation and Definitions

The ternary algebra used has three states '0', 'u' and '1' with '0' and '1' representing low and high values of the signal respectively while 'u' represents the intermediate state or the unknown state. A transition is defined as a change in the ternary state of a signal occurring at a certain time instant. Only 4 out of the possible 6 transitions are allowed, namely the $0 \rightarrow u$, $u \rightarrow 1$, $1 \rightarrow u$ and $u \rightarrow 0$ which are denoted by types 1,2,3 and 4 respectively. A minimum resolvable time 'h_{min}' is is chosen to discretise the time interval of interest $[t_o,t_f]$ so that transition-time points can be represented by integers $\{0,1,2,\ldots,N\}$ where $N=(t_f-t_o)/h_{min}$. Thus any transition can be represented by an ordered pair (m,n) with 'm' denoting transition-type i.e. 1,2,3 or 4 and 'n' denoting transition time. Signal waveforms which are constant in the interval $[t_o,t_f]$ are represented as (4,-1) for constant '0', (1,-1) or (3,-1) for constant 'u' and (2,-1) for constant '1'. Thus any ternary signal waveform could be represented by a <u>sequence</u> of transitions

$$S=(m_1,n_1),(m_2,n_2),\ldots,(m_k,n_k) \text{ where } m_i \in \{1,2,3,4\} \text{ and } n_i \in \{-1,0,1,2,..N\}$$

An analog-waveform $x_a(t)$ is converted to its 'ternary-equivalent' x(t) as follows: Let V_{SS} and V_{DD} denote the minimum and maximum values of $x_a(t)$. Two threshold limits V_L and V_H are selected such that

$$V_{SS} < V_L < V_H < V_{DD}. \text{ Then } x(t)=\begin{cases}0, \text{if } V_{SS} \le x_a(t) < V_L \\ u, \text{if } V_L \le x_a(t) \le V_H \\ 1, \text{if } V_H < x_a(t) \le V_{DD}\end{cases}$$

The ternary-signal-waveform x(t) can then be represented by a sequence of ternary-transitions. In this simulator V_{SS}=0V, V_L=1V, V_H=3V and V_{DD}=5V to represent voltage waveforms of NMOS circuits.

The aim of this simulator is to construct a sequence of transitions $S=(m_1,n_1),(m_2,n_2),\ldots,(m_k,n_k)$ for each node of interest so that it approximates the 'ternary-equivalent' of the analog waveform $S'=(m_1',n_1'),(m_2',n_2'),\ldots,(m_k',n_k')$ such that $m_i = m_i'$

$$\text{and} |n_i-n_i'| \le \text{ntol} \quad \forall i=1,2,\ldots,k$$

where 'ntol' is a tolerance parameter that can be chosen according to level of timing accuracy desired. Typical values of h_{min} and ntol are 0.01ns and 0.1ns respectively so that the constructed ternary waveform S, has its transition times within 0.1ns of the corresponding ones in S'.

An NMOS network is assumed to be composed of (i) input sources (SRC), (ii) multi-functional blocks (MFB) and (iii) pass-transistor blocks (PTB).

An MFB is an inverter-like configuration which consists of a depletion-type load device whose source node is the output node which could be pulled down to ground by a collection of enhancement-type driver transistors or pulled up to supply voltage (V_{DD}) by the load device. The inputs to the MFB are the signals at the gate terminal of the driver transistors. The output of an MFB is thus a B-ternary function of its inputs. The only nodes of interest in an MFB are

*This work was supported by the IBM corporation and by the Joint Services Electronics Program (U.S. Army, U.S. Navy and U. S. Air Force).

CH1935-6/83/0000/0445$01.00

its output node and input nodes which are collectively called external nodes while the rest of its nodes are called internal nodes.

The rest of the MOS transistors (i.e. the ones not included in MFB's) are termed pass-transistors. A PTB comprises of a collection of pass-transistors with a DC-path between the drain-nodes and source-nodes of these transistors. A PTB is in fact a maximal set of such pass-transistors. All the nodes of a PTB are usually nodes of interest.

The nodes of interest are further divided into 3 classes namely i) source nodes of strength = 2, ii) pull-up nodes which are output nodes of MFB's of strength = 1 and iii) normal nodes which are of strength = 0.

III. Partitioning and Analysis Sequencing

The entire NMOS network is partitioned into blocks which are either SRC's, MFB's or PTB's. First the primary-input-sources are identified as SRC's directly from the information available in the input-deck. The load-devices, which are depletion type devices are next identified. The transistors encountered tracing along all DC paths from the source node of the depletion device to ground constitute an MFB. For each MFB there is a list of external nodes (which is the output node along with all input nodes), internal nodes and information about the interconnection of driver-transistors. After all the MFB's have been identified, the remaining enhancement-devices are labelled as pass-transistors. Also for each MFB, the set of internal nodes is examined to check if it is also an external node for some other MFB. If this situation occurs then the node in question is deleted from the internal node-list of the first MFB and transistors encountered in all paths between the node and output of the MFB or the node and ground are deleted from the MFB and are labelled as pass-transistors. The pass-transistor-blocks PTB's are then formed by grouping all pass-transistors together which have DC-paths between their drain and source nodes.

After partitioning, the next step is to decide the order in which the various partitioned blocks have to be analysed. This ordering scheme is termed analysis sequencing [3]. First, a subroutine GRAPH constructs a directed graph G with each block of the circuit, i.e., SRC, MFB or PTB being represented as a vertex. If two blocks share a common external node then an edge is introduced between the two corresponding vertices with a direction assigned in order of priority as: 1) all edges must be directed away from SRC's, 2) if the common node is an output node of an MFB then the edge is directed away from the MFB, 3) if the common node is an input node of an MFB then the edge is directed towards the MFB. After the graph is constructed, a modified version of depth-first-search, first proposed by Tarjan [8], is performed on the graph G. The root for starting the procedure is one of the SCR vertices. Strongly-connected-components (SCC's) are identified by this method. For each SCCj, the minimum number of edges is found which, if deleted, would destroy all cycles in SCCj. Since this problem is in general NP-complete [9], a heuristic algorithm is used to find an improvement given any set of such edges.

IV. Switch-Level Simulation

In the graph G, first sequences are constructed for the SRC nodes from the given input data. Next the MFB's PTB's, and SCC's are processed in the order in which they occur in the graph. The output of MFB's in G are computed from the input sequences by eliminating internal nodes from the connection graph for that MFB. If no pass-transistors are connected to the output of the MFB, then the output sequence is passed through a delay operator (described in the next section) and then a filter which arranges the sequence in chronological order and eliminates incompatible transitions.

In the case of pass-transistors, an algorithm based on the node strengths is used to evaluate the response of pass-transistor chains. In the case of more than one pass-transistor, the delay is computed by using a combination of the delay tables described in the next section and a procedure based on the Elmore time constant [5].

Finally, strongly connected components, such as, sequential circuits, are processed using the waveform relaxation technique in which the time window encompasses one complete transition. Because of the window size, very few iterations are required for convergence.

V. Delay-Operator

The delay-operator works on the sequences of transitions produced by the simulator described in the previous section and delays each transition time with reasonable accuracy. It consists of 5 categories. Each category can be further divided into 2 types depending on the nature of the input waveforms.

Category 1: A simple inverter driving a lumped grounded capacitance C_1. The input V_{in} is the signal at the gate node of the driver, and the output is the voltage across the capacitor.
Type '0': - V_{in} rising from 0V to 5V.
Type '1': - V_{in} falling from 5V to 0V.

Category 2: A pass-transistor whose drain node is connected to a constant DC voltage source and the gate node is driven by a rising pulse from 0V to 5V. The source node which is the output node is connected to a lumped capacitor C_2.
Type '0': - DC source at 0V.
Type '1': - DC source at 5V.

Category 3: A pass-transistor whose drain node is connected to a voltage source V_{in} and the gate node is held fixed at 5V. The output node is the source node connected to a lumped capacitor C_2.
Type '0': - V_{in} rising from 0 to 5V.
Type '1': - V_{in} falling from 5V to 0.

Category 4: A simple inverter driving a pass transistor. A lumped capacitance C_1 is connected to the output of the inverter and a grounded capacitor C_2 is connected to the source node of the pass transistor which is also the output for this configuration. A rising signal from 0 to 5V is applied at the gate node of the pass-transistor while the gate of the driver transistor of the inverter is held fixed.

Type '0': - gate of driver at 0V.
Type '1': - gate of driver at 5V.

Category 5: Same configuration as category 4 except that the gate of the pass-transistor is held fixed at 5V while the input signal in applied to the gate of the driver transistor.
Type '0': input signal rising from 0 to 5V
Type '1': input signal falling from 5V to 0.

Standard devices:
Load: $W = 5\mu$, $L = 10\mu$, $V_{TL} = -3.0V$
Driver: $W = 10\mu$, $L = 5\mu$, $V_{TD} = +1.0V$
Pass-transistor: $W = 10\mu$, $L = 10\mu$, $V_{TP} = +1.0V$.

We define the resistance of a transistor as L/W. Thus $R_L=2, R_D=1/2, R_P=1$, for the standard devices. To compute the characteristics of the delay-operator standard devices were chosen and C_1 was chosen to be 0.01pF for categories 1, 2 and 3. In categories 4 and 5, the pass-transistors were chosen as standard devices and capacitance C_2 was fixed at 0.1pF. The values of C_1/C_2 were varied from 0.01 to 10. R_D/R_P was varied from 0.01 to 5.0, while R_L/R_D was fixed at 4 in each case. These configurations were simulated on SPICE2 [6] with a level = 1 model for the devices for different values of rise time (fall time) on the input signal in each case. Two threshold limits of 1V and 3V were chosen. Let t_1 and t_2 denote the first and second threshold crossing times respectively in the input signal waveform and t_1' and t_2' the corresponding times in the output waveform. The delay operator stores the value of $t_1'-t_1$ and $t_2'-t_1'$ for each case simulated on SPICE2. We will use Δt_o as a generic name for these time-differences. In each category we have 4 Δt_o's (2 for each type) for each set of the circuit and device parameters. From the nature of the differential equations governing the behavior of the output, it can be shown that

$$\Delta t_o = \begin{cases} f(t_2-t_1, R_DC_1) & \text{in Category (1)} \\ f(t_2-t_1, R_pC_2) & \text{in Categories (2) and (3)} \\ f(t_2-t_1, R_DC_1, R_pC_1, R_pC_2) & \text{in Categories (4) and (5).} \end{cases}$$

where f is a generic name used to denote 'function of'.

By scaling the time variable in the differential equation, one can handle a wide range of impedance values not stored in the table.

$$\frac{\Delta t_o}{k} = \begin{cases} f\left(\frac{t_2-t_1}{k}, \frac{R_DC_1}{k}\right) & \text{for Cat. (1)} \\ f\left(\frac{t_2-t_1}{k}, \frac{R_pC_2}{k}\right) & \text{for Cat. (2) and (3)} \\ f\left(\frac{t_2-t_1}{k}, \frac{R_DC_1}{k}, \frac{R_pC_1}{k}, \frac{R_pC_2}{k}\right) & \text{for Cat. (4) and (5)} \end{cases}$$

Then k can be chosen to make $\frac{R_DC_1}{k} = 0.005pF$ in Cat. (1), $\frac{R_pC_2}{k} = 0.01pF$ for Cat. (2) and (3) and $\frac{R_pC_2}{k}=0.1pF$ for Cat. (4) and (5) which are the standard values of R_DC_1 for Cat. (1) and R_pC_2 for Cat. (2), (3), (4) and (5) respectively.

Thus, $\Delta t_o = kf\left(\frac{t_2-t_1}{k}\right)$ for Cat. (1), (2) and (3)

$= kf\left(\frac{t_2-t_1}{k}, \frac{R_DC_1}{K}, \frac{R_pC_1}{k}\right)$ for Cat. (4) and (5)

Hence, the entries in the tables for storing Δt_o are dependent on only one parameter, namely t_2-t_1 in Cat. (1), (2) and (3), and on 3 parameters, namely t_2-t_1, C_1/C_2 and R_D/R_p and R_D/R_p for categories (4) and (5).

VI. Implementation and Examples

The algorithms described in the previous sections have been implemented in a computer-program written in FORTRAN which runs on a VAX 11/780 computer. The circuit-description is in a SPICE-like input format. A number of NMOS circuits have been implemented by this program. A few examples are given below.

Example 1 is a one-bit-full-adder circuit with part of the logic realized using pass-transistors as shown in Fig. 1(a). Nodes (1) and (2) are the input bits with node (3) as the carry-in bit. Node (4) is one carry-out bit while node (5) is the sum bit. The ternary-waveforms at nodes (4) and (5) produced by our simulator are shown as dotted lines in Fig. 1(b). For comparison purposes the waveforms in solid lines are the ones produced by SPICE2 [6]. This circuit took 0.45 CPU-seconds for partitioning and 0.58 CPU-seconds for the actual analysis. Total job-time for SPICE2 was 119.32 CPU-seconds.

Example 2 is an SR-Flip Flop circuit shown in Fig. 2(a). Node (1) is an enable bit which is held at logic level '1'. The waveforms at the remaining nodes are shown in Fig. 2(b) with dotted lines representing those produced by our simulator while the solid lines represent SPICE2 waveforms. This example is presented to illustrate the ability of our simulator to handle feedback loops using waveform-relaxation type methods. This circuit required 0.38 CPU-seconds for partitioning and 1.3 CPU-seconds for analysis of the various blocks. The number of iterations for convergence of the waveforms within the feedback loop was found to be 3. SPICE2 on the other hand requires 90.37 CPU-seconds of total job-time.

To compare computational-complexities, i.e., the rate of growth of CPU-time vs. the number of transistors, we simulated a standard two-input NAND gate realization of a full adder circuit. Both partitioning and analysis CPU-seconds taken by our simulator for one-bit to four-bit adders is shown in the following table. The total job-times taken by SLATE [7] and SPICE2 [6] are also listed for comparison. We observe that the analysis time is fairly linear and about 60 times faster than SLATE and about 200 times faster than SPICE2G.1. The partitioning time on the other hand shows a growth exponent of about 1.7. Efforts to reduce partitioning time are being investigated.

Table 1

Adder Bits	Number of Devices	CPU Time(s): Switch-level Simulator Partitioning	Switch-level Simulator Analysis	SLATE	SPICE 2G.1
1	33	1.37	0.77	61.1	184.0
2	66	3.97	1.98	133.2	371.7
3	99	9.73	2.94	-	-
4	132	14.42	3.92	252.9	767.0

VII. References

[1] Nham, H. N. and A. K. Bose: "A Multiple Delay Simulator for MOS LSI circuits," Proceedings of the 17th Design Automation Conference, June 1980, pp. 610-617.

[2] Rao, V. B. and T. N. Trick: "Accurate Multiple Delay Calculations for MOS Circuits," IEEE Int. Conf. on Circuits and Computers, Sept. 1982, pp. 439-442.

[3] Wei, Y. P.: "Large-Scale Circuit Simulation," Ph.D. Thesis, University of Illinois, Urbana, 1983.

[4] Rao, Vasant B.: "Algorithms for a Multiple-Delay Simulator," M.S. Thesis, University of Illinois Urbana, 1982.

[5] Horowitz, M.: "Timing Models for MOS Pass Networks," IEEE International Symposium on Circuits and Systems Proceedings, May 1983, pp. 198-201.

[6] Nagel, L. W.: "SPICE2: A Computer Program to Simulate Semiconductor Circuits," ERL Memo., No. ERL-M520, University of California, Berkeley, May 1975.

[7] Yang, P.: "An Investigation of Ordering, Tearing and Latency Algorithms for the Time-Domain Simulation of Large-Circuits," Ph.D. Thesis, University of Illinois, Urbana, August 1980.

[8] Tarjan, R. : "Depth-First Search and Linear Graph Algorithms," SIAM Journal of Computing, Vol. 1, No. 2, pp. 146-160, June 1972.

[9] Even, S.: "Graph Algorithms," Computer Science Press, Inc., 1979.

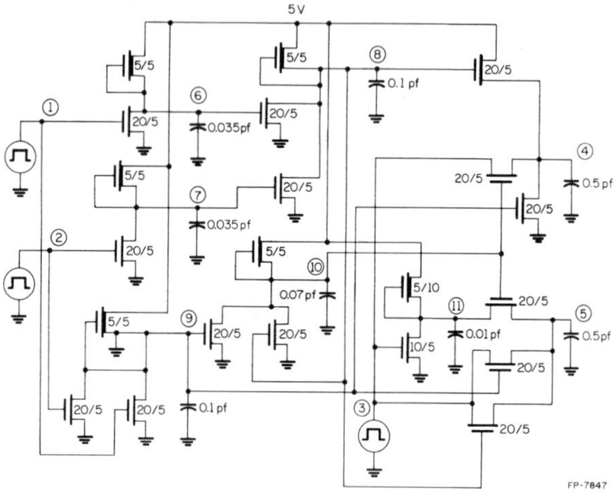

One-Bit Full Adder

Fig. 1(a)

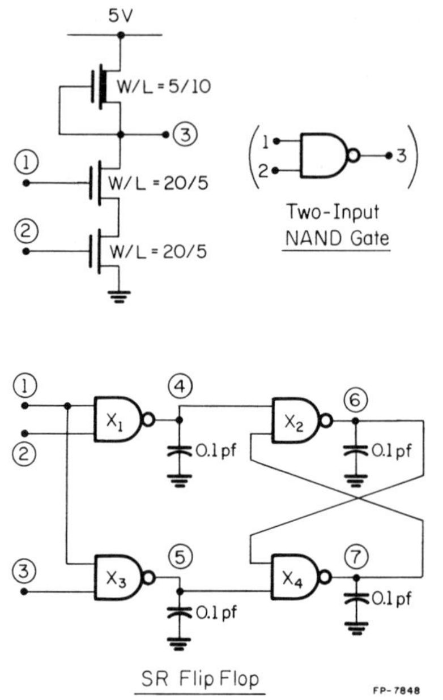

Fig. 2(a)

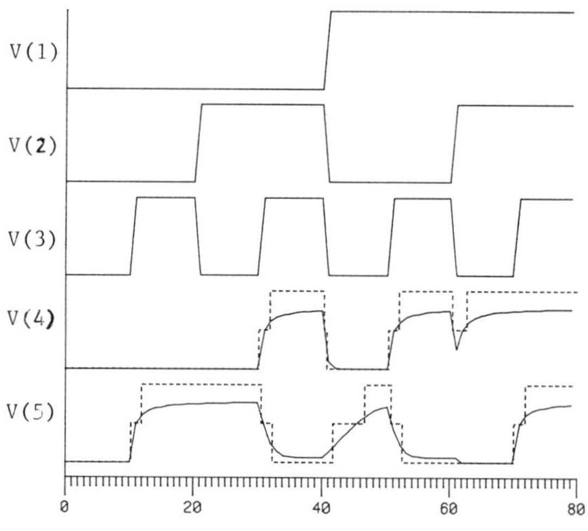

Fig. 1(b)

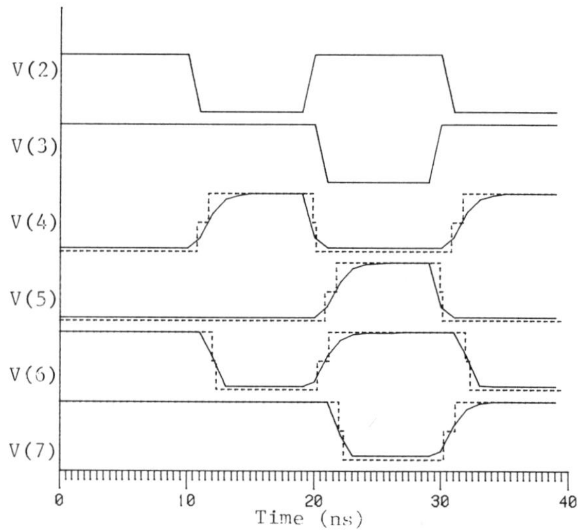

Fig. 2(b)

Session 2-4c
Artificial Intelligence

Chairmen

S. Goto
NEC

P. W. Horstmann
Syracuse University

Design of 3D VLSI Chip Using Artificial Intelligence

Doug Lenat

Stanford University
Stanford, California

This paper will not be published in the proceedings, but will be presented at the conference, at the author's request.

A Rule-based Routing System

Tomoyuki Fujita and Satoshi Goto

NEC Corporation
1-1, Miyazaki 4-chome, Miyamae-ku
Kawasaki, Kanagawa, 213 Japan

Abstract

This paper presents a new interactive routing system based on artificial intelligence technique. In general, an expert designer carries out the design by exploiting his own specific knowledge, called "rule of thumb". The knowledge specifies a certain operation in accordance with a specific situation, which is represented in a form of the design objects, their properties, and logical relations among them. This system accepts the designer's knowledge in clausal form of the first order predicate logic, similar to Prolog language. The system applies the clausal form design logic into the physical design by concatenating the knowledge based on predicate calculus. In order to solve a large scale problem, SPARK interpreter is introduced. SPARK interpreter deals with the logical expression of a specific situation, especially for one which includes procedural relations, and adapts its inference result to a physical design strategy. This rule-based system has been developped to solve the practical large scale routing problem.

Introduction

A routing problem is to determine a connecting path for each given net. From a computational point of view, a routing problem is considered to be a hard combinatorial problem. It is too difficult for computer programs to attain 100% routing with deterministic algorithms. In most practical cases, a large portion of the design time is spent on manual design for unconnected nets whose paths cannot be found by automatic programs. In these years, an integrated CAD system, which fully integrated the automatic and interactive functions, has been developped to reduce the design time [1,2]. However, a large part of the design activity still depends upon the designer's ability. Therefore, assistance from a more powerful computer program is considerably required to solve the large scale design problem. Recently, several new CAD programs, based on artificial intelligence technique, have been reported [3,4,5] . However, it seems that there remains some problems to solve the large scale real problems. This paper proposes a new CAD system for use in practical routing design problems. The system is focused on applying the designer's specific knowledge into the design process.

In the design process, an essential difference between human beings and a computer program is considered to be an intutive ability for reaching an appropriate goal. The ability is considered to be based on a mechanism for applying a "rule of thumb". The rule is a set of knowledge acquired through the expert's design experience and quickly leads to an appropriate design goal. In order to achieve fast and high quality design, a CAD system should have a mechanism with the designer's knowledge.

As a primaly step to make a computer an expert design assistant, this paper presents a new interactive routing system which treats with the designer's specific knowledge for a practical design problem.

System Overview

Figure 1 shows a system configuration. The system includes two kinds of databases, knowledge base and design database. The knowledge base stores a certain amount of design knowledge about specific routing situations: properties of design objects, relations among the objects, and heuristics for guiding problem solving. The design database, which is a conventional database, contains physical information about the design objectives.

The input language of the system is a non-procedural one, similar to Prolog [6] . The designer's specific knowledge is described by this language. The SPARK interpreter is to compile the language and make inference on the knowledge. The inference result is transformed into a certain procedure sequence. The procedure sequences implies how to manipulate physical data on the design database and which design primitives should be executed. The procedure sequences are executed by Design Executer, which realizes high speed processing.

Through the design process, SPARK interpreter evaluates the logical expressions for which knowledge should be applied to the problem. According to SPARK's results, the high speed physical design is performed. These design mechanisms enabled to design a practical routing problem with a reasonable design time.

CH1935-6/83/0000/0451$01.00

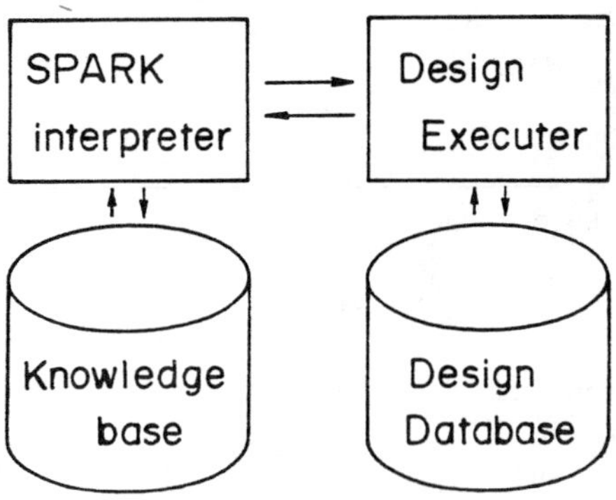

Fig.1 System Configuration

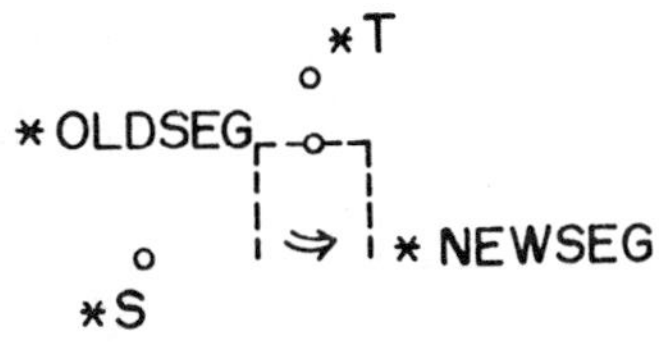


```
*S : (1,4)        *OLD_SEG : (3,5),(3,7),(4,7)
*T : (4,8)        *NEW_SEG : (5,5),(5,7),(4,7)
                  *BLK_SEG : (3,5),(3,7)
```

Fig.3 Specific Routing Situation

Design Knowledge Representation

The proposed system provides a new routing strategy based on artificial intelligence technique. The system is focused on the use of an expert designer's knowledge. The knowledge is expressed as a Horn clause in the first order predicate logic. Terminologies and notations in predicate logic are based on [7].

First of all, the designer has to describe a routing logic with his knowledge in a predicate logic form. The logic is stored in the knowledge base. This process is the same as if he explains the situation to a low level assistant.

Consider a simple case for embedding overflow wires as shown in Fig.2, where the connecting path between S and T cannot be found. The designer can explain the situation by using variables and their properties in stead of the individual design objects. In the routing problem, the variables are considered to be wiring segments, pins, blocks, ... etc. The variables are represented as capital letters headed by '*'. The example is shown in Fig.3. The routing logic is stated as follows.

A connecting path (*S,*T) is blocked by a wiring segment *BLK_SEG. The segment is a portion of *OLD_SEG. The (*S,*T) can be connected by rerouting the path *OLD_SEG along the path *NEW_SEG.

In this case, the designer may describe the situation in a clausal form as

```
MIRROR_ALTERNATIVE (*S,*T,*OLD_SEG,*NEW_SEG)
      <= LINE_BLOCK (*BLK_SEG,*S,*T,*OLD_SEG),
         ALTERNATIVE (*NEW_SEG,*OLD SEG,*BLK_SEG),
         PASS_THROUGH (*S,*T,*BLK_SEG).
```

This clause means that the designer defined the situation MIRROR_ALTERNATIVE by three stages: LINE_BLOCK, ALTERNATIVE, PASS_THROUGH. The predicate LINE_BLOCK states the pair (*S,*T) is blocked by a segment *BLK_SEG. The next predicate ALTERNATIVE indicates that an alternative path *NEW_SEG is a mirror image of the path *OLD_SEG. The last predicate PASS_THROUGH denotes a condition wherein the connection path for (*S,*T) must be routed along *BLK_SEG.

In a similar manner, the designer can explain each situation in more detail step by step. At last,

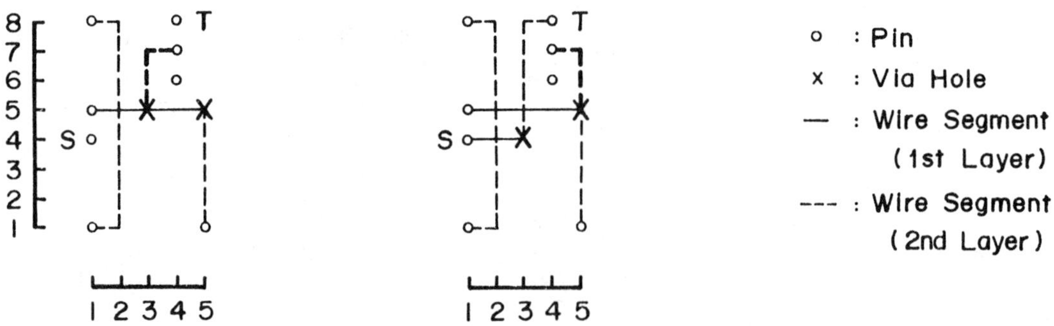

(a) Pair (S,T) cannot be connected (b) Pair (S,T) can be routed

Fig.2 Routing Example

the designer may obtain the expression as follows.

```
LINE_BLOCK (*BLK_SEG,*S,*T,*OLD_SEG)
  <= UNCONNECTED (*S,*T),
     BLOCKING (*BLK_SEG,*S,*T,),
     DEL_UNIT (*OLD_SEG,*BLK_SEG).

ALTERNATIVE (*NEW_SEG,*OLD_SEG,*BLK_SEG)
  <= MIRROR (*SKETCH,*OLD_SEG,*PAIR_ID),
     WIRE_ROUTE (*NEW_SEG,*SKETCH,*PAIR_ID),
     DISJOINT (*NEW_SEG,*BLK_SEG).

PASS_THROUGH (*S,*T,*BLK_SEG)
  <= PATH (*CONNECT_PATH,*S,*T),
     INTERSECT (*BLK_SEG,*CONNECT_PATH).
```

In the above expression, each predicate corresponds to a certain design primitive. Therefore, the system can proceed with the design by applying this knowledge.

Sophisticated Interaction

As a result of expressing specific situations, logical relations among the design objects are obtained. Then, the system provides a sophisticated interaction for the designer. Followings are the several sophisticated interaction examples.

(1) <= MIRROR_ALTERNATIVE (S, T,*OLDSEG,*NEWSEG).

> The system finds wiring segments *OLDSEG to be modified and finds its alternative path *NEWSEG to be rerouted by specifying a pair of pins (S,T).

(2) <= MIRROR_ALTERNATIVE (*S,*T, OLDSEG,*NEWSEG).

> The system finds a pair of pins (*S,*T) to be connected through the rule and finds an alternative path *NEWSEG for the rerouting, by specifying certain wire segments OLDSEG to be modified.

(3) <= MIRROR_ALTERNATIVE (*S,*T,*OLDSEG, NEWSEG).

> The system finds an unconnected pair (*S,*T) and wiring segments *OLDSEG such that the connection (*S,*T) can be completed by rerouting *OLDSEG along with the given path NEWSEG.

(4) <= MIRROR_ALTERNATIVE (*S,*T,*OLDSEG,*NEWSEG).

> The system finds all pairs (*S,*T) to be connected, wires *OLDSEG to be modified for the connection, and its alternative path *NEWSEG for the rerouting, since nothing is specified.

The above example shows that a single rule can be applied to four different design interactions. More design interactions can be provided by considering the combinations of the instances. Therefore, the designer can proceed the routing by exploiting the knowledge from the several view points. By accumulating more knowledge or rules, the designer complete the more complex routing in a quick manner.

SPARK Interpreter

Spark interpreter is a new Prolog-like interpreter to solve problems which involves objects and their relationships between objects, especially for procedural relationships. In the practical design problems, there are many kinds of procedural or directional relationships among the design objects. A procedural relation R(x1,x2,...,xn,y1,y2,..,ym) means that a set of items y1,y2,...,ym can be derived from another set of items x1,x2,...,xn through a procedure R. The conventional Prolog-like interpreter has a problem to treat with the procedural or directional relations, implemented as built-in predicates. The problem is that the resolution is dependent on the order of the procedural predicates.

In the SPARK, a relation can be defined by several procedures without worrying about any redundancy. For example, an arithmetic relation X=Y+Z can be definded by arithmetic procedures X<=Y+Z, Y<=X-Z, Z<=X-Y. SPARK interpreter selects out the most suitable procedure which meets the variable dependency with the other procedures. Then, SPARK interpreter finds out a procedure sequence in an optimal order trying any redundant searches.

For example, the first and second interactions in the previous section are resulted in the following procedure sequences.

```
(1) Mirror_alternative ( S, T,*OLD_SEG,*NEW_SEG) :
      Check_if_unconnected        (S,T),
      Find_blocking          (*BLK SEG,S,T),
      Get_delete_unit        (*OLD_SEG,BLK_SEG),
      Get_mirror_image       (*SKETCH,OLD_SEG,*PAIR_ID),
      Find_wire_route        (*NEW_SEG,SKETCH,PAIR_ID),
      Check_if_disjoint      ( NEW_SEG,BLK_SEG),
      Find_path              (*CONNECT_PATH,S,T),
      Check_if_intersect     ( BLK_SEG,CONNECT_PATH).

(2) Mirror_alternative (*S,*T, OLD_SEG,*NEW_SEG) :
      Divide_delete_unit     ( OLD_SEG,*BLK_SEG,*S,*T),
      Find_unconnected_pair       (*S,*T),
      Check_if_blocked       ( BLK_SEG,S,T),
      Get_mirror_image       (*SKETCH,OLD_SEG,*PAIR_ID),
      Find_wire_route        (*NEW_SEG,SKETCH,PAIR_ID),
      Check_if_disjoint      ( NEW_SEG,BLK_SEG),
      Find_path              (*CONNECT_PATH,S,T),
      Check_if_intersect     ( BLK_SEG,CONNECT_PATH).
```

On the executing stage, the system executes the above procedure sequence one by one. If the execution of a certain procedure is failed, the system backtracks the procedures until reaching a procedure which provides another output.

SPARK interpreter has several features over the conventional Prolog-like interpreter as follows.

(1) A user need not mention about the orderings of procedural predicates. Therefore, a designer can easily express the logical relations including procedural or directional relations.
(2) High speed resolution can be achieved by performing a dependency directed search among the variables. Then, the system can deal with a large scale practical design problem.

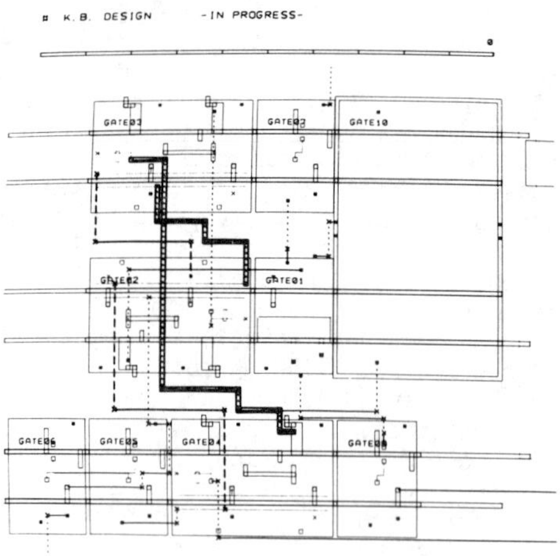

Fig.4 An Inference Result

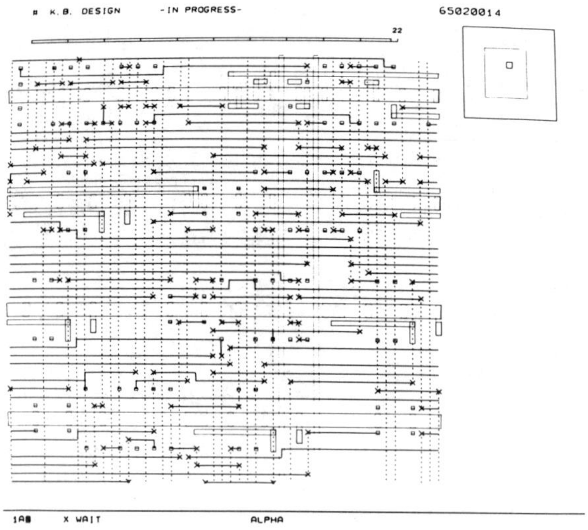

Fig.5 Practical Routing Problem

System Features

The proposed rule-based routing system has several features over the conventional CAD systems in the practical design process. Followings are particular ones.

(1) The system can consult with the design strategies from the stored specific routing knowledge: which wires should be modified, which pair can be connected, which unrouted area could be used as a rerouting path.

(2) By specifying a certain instance as a design goal, the system can show which operations should be performed to reach the goal and can give an explanation what kind of the operations should be performed.

(3) The system helps the designer to attain a high quality design under the practical constraints. The system can deal with signal delay time or signal interference by taking account of wiring length, topological shape of a wiring path, wiring branch restrictions, ... etc. Fig.4 shows an example of the design results. Figure 4 indicates two pairs of wiring segments by bold lines which contribute to reducing the number of via holes by local rerouting in a pairwise manner.

(4) The proposed rule-based system achieves an acceptable interaction speed for the practical routing problem. The inference mechanism was evaluated on the practical routing problem. Fig.5 shows a display screen view the real routing problem. The problem consists of about 10,000 wiring segments on 400 X 500 grid chip. The response time for an interaction is from 3 seconds to 3 minutes by using NEC MS/50 mini-computer.

Conclusion

This paper presents a new interactive routing system based on artificial intelligence technique. A designer can store his own specific knowledge in the system and use them in the design process.

The system is under development to realize a fully automatic routing system.

Acknowledgement

The authors would like to thank Dr. T. Mikami and Dr. M.Naniwada for their continuous encouragement. They are grately indebted to CAD group members in NEC Corporation, especially Mr. H.Mori and Miss K.Mitsumoto, for their valuable suggestions.

References

[1] H.Mori, T.Fujita, M.Annaka, S.Goto and T. Ohtsuki;"Advanced Interactive Layout Design System for Printed Wiring Boards", in "Hardware and Software Concepts in VLSI", ed. by G. Rabbat, pp.495-523, Van Nostrand Reinhold Company Inc., 1983.

[2] S.Goto, T.Matsuda, K.Takamizawa, T.Fujita, H. Mizumura, H.Nakamura and K.Kitajima; "LAMBDA, an integrated master-slice LSI CAD system", INTEGRATION, vol.1, no.1, pp.53-69, North Holland Pub. Co., 1983.

[3] G.L.Steel,Jr. and G.J.Sussman; "Constraints", Artificial Intelligence, Vol.14, pp.1-39,1980.

[4] J.McDermott ;"Domain Knowledge and the Design Process", Proc. of the 18th Design Automation Conference, pp.580-588, 1981.

[5] T.J.Kowalski and D.E.Thomas; "The VLSI design automation assistant: learning to walk", Proc. of the International Symposium on Circuit And Systems, pp.186-190, 1983.

[6] W.F.Clocksin and C.S.Mellish; "Programming in Prolog", Springer-Verlag, 1981.

[7] R.Kowalski ; "Logic for problem solving", Elsevier North Holland Inc., 1979.

Generating Distinguishing Tests based on Hierarchical Models and Symptom Information

Mark Shirley and Randall Davis

MIT Artificial Intelligence Laboratory
545 Technology Sq., Cambridge, MA 02139
(617) 253-5848

Abstract: Given a failing piece of digital hardware and a list of components that might be responsible for its failure, we wish to distinguish which of the components is faulty. We want to design tests that use the working portions of the device as a probe for observing the components that might be failing.

This paper presents our methodology and an example of its operation, demonstrating that (1) we use the symptoms of failure to tailor the test generation process (path sensitization) to a particular faulty device; (2) we use a hierarchical device description to focus effort on those portions of the device that might be faulty; and (3) we reason about properties of signals as well as their values.

1. Introduction

Consider the problem of diagnosing a piece of digital hardware malfunctioning in the field. Two aspects of the problem are especially relevant to our work. First, the problem is one of diagnosis rather than validation. Since the device has passed the usual tests during manufacture, we don't have to verify that every subpart works, we have to find the source of the malfunction. Second, the presence of the malfunction is manifest in a specific set of symptoms.

One common approach to the problem --- symptom-directed diagnosis --- calls for the repeated execution of three steps: (1) determining the set of **candidate parts**, i.e., parts whose failure could account for the observed misbehavior; (2) generating a test vector which distinguishes between the candidate parts; and (3) interpreting the test results, thereby reducing the number of candidates requiring further consideration. This process is continued until all of the remaining candidates lie within an appropriately small portion of the device, e.g., a field replaceable unit.

A program capable of Step 1, candidate generation, is described in [2]. Here, we describe a second program capable of carrying out Steps 2 and 3: generating test vectors guaranteed to reduce the set of candidates and producing interpretations of the results of those tests.

In the most general terms, the problem we address is the generation of tests to distinguish between plausible causes of failure. More specifically, we examine how the standard approach of path sensitization can benefit from knowledge of the candidate parts and from using a more elaborate vocabulary for describing signals.

2. An Overview of our Approach

Important features of our method include:

The Candidate Parts: Knowledge of the candidate parts allows us to generate a test whose result depends only on parts known to be good and exactly one of the possibly faulty parts. We say that the test is **isolated** from the effects of the remaining candidates. Traditional tests designed before the device has failed cannot, of course, take advantage of symptom information. We argue that as a result they are often unable to distinguish which candidate is faulty, and we claim that it is computationally intractable to design unambiguous tests in the absence of symptom information.

A Hierarchical Device Model: Our device descriptions are hierarchical. For example, a 4-bit adder is composed of 4 bit-slices, which are in turn composed of gates, etc. The tests we produce distinguish between candidates at high levels of the hierarchy first, i.e., between large portions of the circuit before single gates.

A General Fault Model: A fault is any deviation from correct device behavior -- not only those deviations that can be accounted for by specific fault mechanisms (e.g. stuck at faults).

A New Vocabulary for Describing Logic Values: In exploring the kind of reasoning done by experienced engineers, we find that while they often reason about the *value* of a signal, an important component of their problem solving involves reasoning about *properties* of signals as well (e.g., sensitivity to a faulty part). We have developed a vocabulary for describing several useful properties. We have also developed the machinery for reasoning about them and propagating them in much the same manner as numeric values are propagated in the traditional approach.

Fault Simulation: Our test generation method is guaranteed to produce a test which determines whether or not a particular candidate is working. We subsequently use fault simulation to discover possible "extra distinguishing power" that comes as a serendipitous result of the test.

The Simplifying Assumptions: We assume a single point of failure and that the failure is nonintermittent. We also assume that device inputs act as inputs, outputs act as outputs and that the device description accurately reflects its structure (i.e., there are no bridge faults.)

Due to space limitations, this paper emphasizes only points 1 and 2. (See [5] for more details.) In Section 3 we describe the traditional technology for test generation in the absence of symptoms. Building on that, we introduce extensions to take advantage of knowledge about candidate parts. We then present an example that illustrates our approach and conclude with some thoughts about possible improvements.

3. Path Sensitization: The Traditional Technology

We test a component by controlling its inputs and comparing its output against the correct value. This is easily accomplished for any component whose inputs and outputs are directly accessible. When inputs are not directly accessible, we have to use a portion of the circuit to propagate signals to the component being tested. Correspondingly, when component outputs are not directly accessible, we have to use another portion of the circuit to propagate the outputs to a place where they can be observed. This process of setting up inputs and observing outputs is called **path sensitization**. The basic steps of the process are:[1]

1. **Focus selection**: Select a part to test (we call this part the **focus** of the test). Hypothesize a way in which the focus could be malfunctioning (e.g., output stuck-at-0).
2. **Fault sensitization**: Select inputs for the focus that will cause different values to appear at the output depending on the condition of the focus. Propagate the choice of

[1] See [4] or [1] for a more complete description.

focus inputs back to the primary inputs. Backward propagation (often called line justification) ensures that the focus will receive the desired inputs.

3. **Error propagation**: Propagate the focus output forward to an observable output without loosing information. The route chosen is called the **sensitive path**, because it is sensitive to a fault in the focus. Error propagation is a combination of path selection (i.e., choosing a route for the sensitive path) and backward propagation (preparing the parts on the sensitive path to transmit the focus output.)

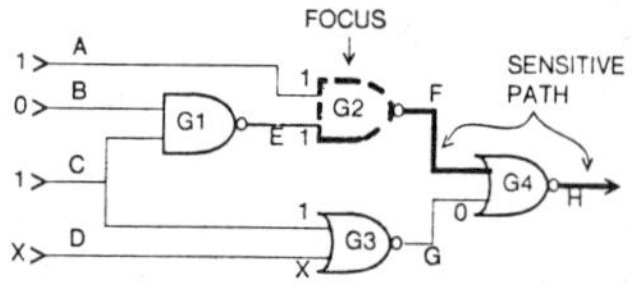

Figure 1: A simple example

Figure 1 shows a simple example. Assume that in Step 1 we select G2 as the focus and hypothesize that its output (F) is stuck at 1.

In Step 2 we set both inputs of G2 to 1. This will establish different values at the output depending on the condition of the focus. Next, we propagate the inputs of the focus back to primary inputs. For A this is trivial. Ensuring that E = 1 requires a little more work. We have three ways to accomplish it: [*i*] B = 0, [*ii*] C = 0 and [*iii*] B = 0 and C = 0.

A simple heuristic -- where possible, reduce the amount of work to do -- suggests that we try [*i*] or [*ii*] (which require one assignment) before [*iii*] (which requires two). This leaves us with a choice between two methods that appear equally good; the traditional technology has no reason to choose one over the other. We arbitrarily choose the first, setting B = 0 and leaving C unspecified. The portion of the test vector we have generated so far --- A = 1, B = 0 --- will ensure that both inputs of G2 are 1.

In Step 3, we route the output produced by G2 to an observable output (H). This requires that G = 0, which in turn is accomplished by setting C = 1. The finished test vector is:[2]

[A = 1, B = 0, C = 1, D = X]

3.1 A Problem with the Existing Technology

The problem with the traditional approach is that we made choices in the absence of any information about which parts other than G2 might be faulty. Consider what happens if we want to use our test in a situation where B might be stuck at 1. That is, as before, we want to determine whether F is stuck at 1, but now need to allow for the additional difficulty that B might be faulty instead. The test we generated above won't work: the output at H is 0 if F is stuck-at-1 *or* if F is good and B is stuck-at-1.

What can we do? We might try designing tests that anticipate such contingencies, i.e., design tests that cover all possibilities. In the example above, for instance, we could generate three tests, one corresponding to each of the choices for ensuring E = 1. But in general this means exponentially more tests and soon becomes computationally intractable.

We suggest instead that knowledge about which parts might be faulty can be used to create more robust tests. In the current example, for instance, if we had known that B might be a problem, we could have ensured E = 1 by setting C = 0, rather than by relying on B. This eventually results in [A = 1, B = X, C = 0, D = 1], a test vector capable of determining unambiguously the status of F, even if B is stuck-at-1.

As this simple example shows, when we have an idea of which components might be faulty, we can generate more informative test vectors. *Our second test is guaranteed to yield information about F because it relies only on working parts.*

[2] X indicates a don't care value

This brings us to one central feature of our method. When tests are designed before device failure, there is no way to decide which parts to rely on, hence there is no good means of avoiding ambiguous test results of the sort shown above.

However, if we know which parts are potentially faulty and which parts are good, we can design tests that rely only on the good components. Such tests are guaranteed to yield useful results, indicating the condition of the focus unambiguously.[3]

4. Extensions to the Traditional Technology

Knowing which parts might be faulty allows us to use a number of powerful techniques. The utility of these techniques depends upon the existence of a **candidate generator** (described in [2])

The candidate generator is given a structural and behavioral description of a device and the failure symptoms, and it deduces from that the list of candidate parts and their inputs and outputs.

Two of the techniques made possible by having knowledge of the candidate parts are:

- **Avoid candidates**: During error propagation, we use a number of heuristics to route the sensitive path away from candidate parts.
- **Use tunneling**: If a candidate proves to be impossible to avoid, we "tunnel under" it. The trick is to remember that all non-primitive components are implemented as networks of simpler parts. It may be that only *some* of those parts could have been responsible for the candidate's apparent failure. We therefore drop down a level in the functional hierarchy, run the candidate generator on the newly exposed portion of the circuit, and attempt error propagation through the lower-level network.[4]

5. An Example

The following example illustrates how symptom information can be used to design effective tests. In particular:

- We design a test that relies only on parts known to be working. The test is isolated from the effects of the candidates and is therefore guaranteed to provide information about the focus.
- We perform path sensitization at multiple levels of the device description hierarchy.

In this paper we discuss isolation and tunneling: the complete example appears in [5].

The circuit in Figure 2 comes from [3]. Figure 2 also shows a symptom case in which the circuit doesn't behave as expected. Inputs and observed outputs are indicated in the figure; expected outputs are included for comparison. A fault is indicated by the discrepancy between the expected value at F of 4 and the observed value of 2.

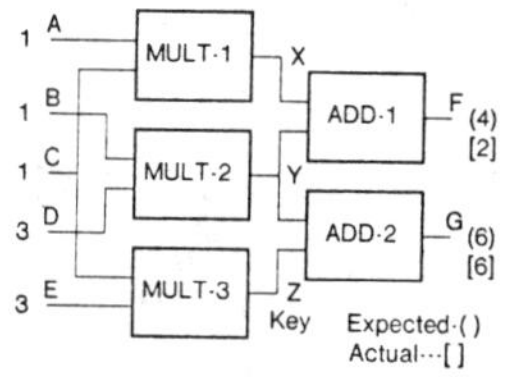

Figure 2: A circuit and symptoms

[3] This argument applies to diagnostic tests rather than "GO / NOGO" tests, which are designed to detect as many potential faults as possible. For fault detection, tests similar to the ambiguous one above are desirable. That test will detect faults at both F and B, making it unnecessary to design a separate test for B.

[4] Running the candidate generator at the lower level of detail may reveal that no *single* part in the candidate's substructure could have caused the original problem. If so, then the candidate could not in fact have caused the original fault and can be eliminated from the list of suspects. Tunneling under, therefore, may reveal that the way above was clear after all.

5.1 Candidate Generation

In the current example the candidate generator determines that the adder could have malfunctioned by taking correct inputs of 1 and 3 and producing an incorrect output of 2, or, alternatively, the multiplier could have malfunctioned with inputs of 1 and 1 but an output of -1, again resulting in the observed 2 at F. None of the other devices at this level of the device description hierarchy can account for the observed symptoms.[5]

5.2 Path Sensitization

The task now is to design a test that determines whether ADD-1 or MULT-1 is faulty. We attempt to test one of these candidates while avoiding the other. If the candidate we test fails, it is the problem; if it works, then it can be eliminated as a suspect (in this case, leaving only the other candidate.)

Focus selection: A simple heuristic method attempts to select the candidate that will be "easiest" to test. In this case it selects the multiplier. If it turns out to be impossible to test the multiplier, we will begin again with the adder as the focus.

Fault sensitization: We need inputs that will force MULT-1 to produce an incorrect output if it is faulty and a different, correct output otherwise. The candidate generator provides those values: if we set the inputs of MULT-1 to 1 and 1, and MULT-1 is in fact the single failure in this circuit, its output will be -1. If MULT-1 is working correctly, its output will of course be 1. If (as we assume) the fault is nonintermittent, we can count on MULT-1 to produce this behavior any time both of its inputs are 1.

We set A = 1 and C = 1, then propagate those values back to primary inputs (a trivial operation in this case.) MULT-1's output is now sensitive to the presence of a fault in MULT-1, i.e., the value there is guaranteed to be one thing if it is broken and another if it is functioning correctly. We thus refer to the value at the output of MULT-1 as a **sensitive** value.

Error propagation: We begin to propagate the sensitive value from the focus to an an output where it can be observed. Here, we immediately encounter a dilemma. There is no path from MULT-1 to an observable output (F or G) that does not pass through the candidate ADD-1.

Tunneling: Since there is no way to avoid the blocking candidate, ADD-1, we attempt to tunnel through it, i.e., we drop down a level in the functional hierarchy and attempt error propagation through the adder. Running the candidate generator on the adder's substructure yields SLICE-0 and SLICE-1 as the only subcomponents of ADD-1 that can fully account for the observed symptoms. We replace ADD-1 with the two new candidates, yielding a revised candidate set of MULT-1, SLICE-1 and SLICE-0.

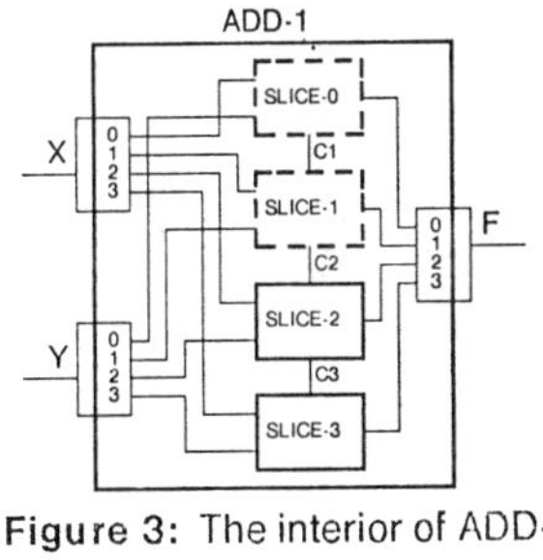

Figure 3: The interior of ADD-1

Figure 3 shows the substructure of ADD-1 with the new candidates marked by dashed lines.[6] The rectangles along the border of the adder are input / output ports, representing the shift in level of abstraction from the 4-bit datapaths (e.g., X) to the single-bit wires (e.g., X0, X1, X2 and X3).

[5] Simplifying assumptions about the character of the fault are used in this process. These assumptions and what to do in the event they prove invalid are discussed in [2].

[6] ADD-1 is a 4-bit carry chain adder, but in order to follow the example in [3], we have to discard the carry out of SLICE-3. The example shown in the figure does this.

Note that with the shift in level, the datapaths X, Y, and F are broken down into their constituent wires. So too must the sensitive value on X be brought down to the lower level. Since the output of MULT-1 is either 1 or -1 (i.e., either 0001 or 1111 in 4-bit 2's complement), the top three bits carry information about whether or not MULT-1 is working. These bits carry the same information for our purposes, so error propagation through ADD-1 could begin at any one of X1, X2, or X3.

Candidate avoidance: We choose to begin at X3, noting that the further the path is from candidates, the more likely it is that the path can be protected from their effects.[7] In this case, we need to get information through SLICE-3 unaffected by any malfunction in SLICE-0 or SLICE-1. Since the carry chain connects all four slices, it is not trivial to ensure that SLICE-3 is truly isolated from errors in the lower-order slices.

Propagation through SLICE-3: To push the sensitive value from X3 through SLICE-3 to F3, we reason as follows. One of the slice's inputs, X3, is sensitive to the status of MULT-1. We must choose the slice's remaining inputs (Y3 and C3) so that its sum output is also sensitive. That is, we must preserve the information present at one input when transmitting it to the output. If we can do this for every device on the sensitive path, then we will have created a chain of devices that transmits information from the beginning of the path (the focus) to the end (an observable output).

What values should we put on the other inputs to ensure that the output is also sensitive? The sum output of SLICE-3 is functionally a three-input exclusive-or (XOR) of its inputs. In this case all that is necessary is that the values of the remaining inputs be *known*. If we observe the output and know two inputs of an XOR gate, we can infer the third input. And if we can unambiguously infer the value of the sensitive input from the output, then the output preserves information about it, i.e., the output is sensitive. Note that we do *not* have to determine specific values just yet. It is only necessary that they be known when we attempt to invert the device's function, i.e., when we interpret the test results.

What does it mean for the values to be "known"? It means that they can be predicted from the test vector. But this in turn means that the values cannot depend upon any of the candidates, since, in general, a candidate's behavior is unpredictable. As a result, during path sensitization, we only need impose the condition that these values not depend upon any candidate. We call values that display this property **isolated** values, since they are isolated from the effects of all candidates.

We can ensure that a signal is isolated if its value doesn't depend on any candidates, i.e. because no candidates lie upstream. Thus if we could ensure that no candidates lie upstream of both Y3 and C3, their values would be isolated, hence "known", and thus a sensitive path would have been established through SLICE-3 *without choosing specific values for these inputs*.

For Y3 this is easy: our system determines that there are no candidates driving Y3, so it is isolated. Isolating C3 is more difficult however, because two candidates, SLICE-0 and SLICE-1, lie upstream. In this case we try specific values and propagate them back to primary inputs, taking care to avoid candidates on the way. In this case a value of 0 for C3 propagates successfully, setting the lower bits of Y (Y2, Y1 and Y0) to 0.[8]

[7] We use a number of heuristics to guide the selection of a sensitive path. Some of them are for avoiding candidates while others take into account features of the device description network (e.g. take the shortest route to an output).

[8] There is some interesting reasoning necessary to conclude that these assignments will ensure that C3 = 0. Again, space limitations preclude us from presenting that reasoning here. See [5].

The reasoning above makes important use of our vocabulary for describing properties of logic values. The notion of "isolated" captures an important *property* of the value on Y3 and reduces the amount of search necessary: we were able to establish a sensitive path without assigning a specific value to Y3, reducing the need for backtracking. Each such term we can identify and add to the vocabulary aids similarly in reducing search.

Traditional path sensitization algorithms (e.g., the D-algorithm) use a vocabulary that describes one property of a value: sensitivity to the presence of a fault. Our vocabulary adds additional terms like "isolated" and "unchanged" (for details, see [5]) and hence often performs less search.

5.3 Finishing off the Test Vector

The assignments to the individual bits of Y constrain its value to be either 0 or 8; the choice doesn't affect whether the test will work or not, so we arbitrarily choose Y = 0. We propagate that value backward to the primary inputs by setting B = 0, completing the path sensitization process.

Once the path sensitization process has succeeded, we finish the test vector by selecting values for the unspecified inputs. Constants appeared at A, B, and C (1, 0, and 1 respectively), so these assignments are established. Nothing appeared at D and E (i.e., these are "don't care's"). Arbitrarily setting D and E to 0 yields:[9]

$$[A = 1, B = 0, C = 1, D = 0 \; E = 0]$$

This test works by isolating SLICE-3 from the effects of candidates SLICE-1 and SLICE-0, and then routing the sensitive signal through the high-order slice. Figure 4 shows the sensitive path through ADD-1. The output reached was F3, the high-order bit of F. The sensitive value propagated to F3 indicates that, after applying the test vector, observing a 1 means that MULT-1 is faulty and observing a 0 means that it is working properly. Thus the test is guaranteed to distinguish whether the multiplier is working.

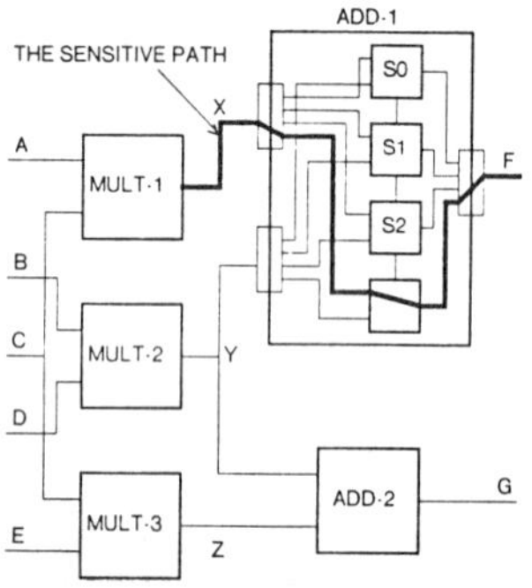

Figure 4: The sensitive path

Note that the circuit was never examined at the gate level. The hierarchical device description helped to focus the energies of the test generator on only those portions of the circuit where the fault might lie, and it was never necessary to consider individual gates. This is a marked advantage for testing very large circuits.

6. Relationship to Previous Work

In [4], Roth describes a modification of the D-algorithm for generating tests to distinguish between two potential failures. The method uses the D-algorithm to propagate sensitive signals forward from both potential failures at the same time. Our method represents an advance over the distinguishing D-algorithm in several areas: (1) we use a hierarchical device description rather than a single gate-level one; (2) our fault model is based on detection of discrepancies rather than deviations that can be accounted for by particular fault mechanisms (e.g. stuck at faults); and (3) our method handles any number of candidates in the same fashion, namely we choose one to test and isolate the test from the rest.

In [3], Genesereth describes a system that uses a hierarchical device description. In the example from the paper, the DART system produces a test which is not guaranteed to distinguish between the candidates. Unlike this system, we augment path sensitization with knowledge of known candidates and a hierarchical description to generate tests that are guaranteed to distinguish between candidates.

7. Implementation Status and Future Directions

The current implementation has been run on several examples including the one shown in Section 5. It was written in Franz Lisp and runs under Berkeley Unix on a VAX 11/780. Important areas of future work include test generation for devices containing state, and explicit representation of how parts are used in a test.

8. Conclusion

We have presented a test generation methodology for distinguishing among candidate parts, i.e., those parts which could have caused an observed misbehavior. Through repeated applications of these tests we can determine which part is faulty.

We generate the tests by augmenting the existing technique of path sensitization to take advantage of knowledge of the candidates parts. We choose one of the candidates to be the focus of the test and avoid letting the result of the test rely upon the functioning of other candidate. During path selection, if we cannot avoid a candidate at one level of the device description hierarchy, we attempt to "tunnel through" the candidate, attempting to avoid the new candidates at the next lower level.

We describe logic signals in terms of several properties important for tests which distingish between candidate parts. Thus the test generator describes logic signals in terms of isolated and unchanged values as well as the usual sensitive values. The importance of this vocabulary lies in its ability to reduce the number of arbitrary choices the test generator must make, thereby usually reducing the amount of search necessary to find an appropriate test vector.

References

1. Breuer, M. A. and A. D. Friedman. *Diagnosis and Reliable Design of Digital Systems.* Computer Science Press, Inc., Rockville, Maryland, 1976.

2. Davis, R. Diagnosis via Causal Reasoning: Paths of Interaction and the Locality Principle. Proceedings of the National Conference on Artificial Intelligence, August, 1983.

3. Genesereth, M. R. Diagnosis Using Hierarchical Design Models. Proceedings of the National Conference on Artificial Intelligence, AAAI, August, 1982, pp. 278-283.

4. Roth, J. P. *Computer Logic, Testing, and Verification.* Computer Science Press, Inc., Rockville, Maryland, 1980.

5. Shirley, M. H. Digital Test Generation from Hierarchical Models and Failure Symptoms. Master Th., MIT Artificial Intelligence Lab., May, 1983.

[9] We are currently investigating whether a second pass of the path sensitization algorithm can select "good" values for the unspecified inputs.

DEPENDENCY MAINTENANCE IN THE DESIGN PROCESS

John Batali

The Artificial Intelligence Laboratory
Massachusetts Institute of Technology
Cambridge, Massachusetts 02139

ABSTRACT I argue that to contribute to a design effort, an individual must understand more than just the object being designed. Aspects of the design *process*, especially the alternatives that were explored and why they were rejected or accepted, must also be understood. I present a simple account of the design process, paying particular attention to the aspects of that process in which particular design *decisions* are made. Decisions depend on the goals which must be satisfied by the design as well as states of affairs which may have been influenced by other decisions. The dependencies among these decisions ought to be represented explicitly. Such a representation is a tool for managing a project, can serve as documentation of the design process, and is also an important part of the interface between human and computer members of the design team.

INTRODUCTION

In this paper, I present an account of the design process. This account is meant to be accurate, as far as it goes, but is also meant to suggest the capabilities of a kind of design tool. Most existing design tools deal with the *object* being designed. However an important aspect of contributing to the design of some object is an understanding of how the design came to be what it is. So my account of the design process will be presented in order to suggest how tools could be developed to help understand the design *process*, rather than just its results.

The suggestions in this paper come from some recent work in artificial intelligence in which details of a program are explicitly represented by the program in order, for example, that the program can recover correctly from errors. Programs of this sort also are more reasonable to deal with in that they can provide reasons for why a certain step was taken. My account of the design process is based on my experience in designing VLSI circuits[1,2] as well as observing the design process in other domains. Very little discussed here is specific to the domain of VLSI design.

This paper describes research done at the Artificial Intelligence Laboratory at the Massachusetts Institute of Technology. Support for the Laboratory's VLSI research is provided in part by the Advanced Research Projects Agency of the Department of Defense under Office of Naval research Contract number N00014-80-C-0622 and in part under the Advanced Research Projects Agency under office of Naval Research contract N000-14-80-C-0505.

During the period of the herein described research, the author was, sequentially, an IBM VLSI fellow, a research assistant at the AI lab, and an employee of International Thinking Machines.

FIVE THESES

(1) For a member of the design team, understanding a design is more than simply understanding how the designed object works. In order to contribute intelligently to the design process, one must understand not only the object being designed, but also one must be familiar with details of the process that led to the design.

(2) The design *process* can be modeled as a structure of decisions, each of which can or does contribute an aspect of either the finished design or the design process itself.

(3) Each decision depends on and is influenced by the results of other decisions in the process. Other influences on decisions come from factors external to the design process. Understanding the design process is enhanced with the explicit characterization of these influences and dependencies.

(4) Changes in external factors may change the validity of decisions made in the design. Also the design may contain flaws or it may be necessary to modify it for some reason. The dependency structure can be used to locate the decisions which need to be reassessed in such situations.

(5) This analysis is just as applicable to higher level decisions as to lower level ones. The dependency structure is just as important to the leaders of the design effort as it is to the people who actually do the work. In particular, automated design tools ought to have access to, and modify, the structure.

UNDERSTANDING THE DESIGN

What must an individual understand in order to contribute to the design effort? Certainly he must understand the portion of the object on which he is working. He must understand what it is supposed to do. He must understand how it works or how it is supposed to work. This understanding is also important to any user of the object being designed inasmuch as details of the object are important to its use. This "object level" understanding of the design is also of interest to the engineering student and the industrial spy.

But to actually contribute to the design process, one must understand more than just the object level design. One must also understand the design process. For any detail of the object, there is often a long sequence of decisions and alternatives behind it. In some cases, a detail of the design is not completely understandable simply in the context of the final design. Many decisions are arbitrary in terms of the finished product, or are at least not obvious. However, when situated in the entire process, the decision often seems natural and correct.

For example, consider the issue of placement and routing. The existence and path of a particular wire can be understood in terms of the locations of its termini, as well as the obstacles it most avoid in its path. However where do *those* facts come from? Well, the modules were placed in some locations, presumably with some thought as to their routability. In many cases, though, the precise location of modules, and thus the absolute constraints on the paths of wires, is relatively arbitrary in that they could be changed without significant effect on the performance of the object. However the decisions *were* made, and the router must either live with them or determine that they are wrong for some reason or other. In any case, to understand why a certain wire must go in a certain path involves more than just understanding the physical constraints on the path. One must also understand why those particular constraints exist, and this involves understanding the decisions that introduced the constraints.

Consider a new engineer joining a project midway through its course. He must be given a description of the goals of the project, of course, and also the state of the current design. But he also must be given a picture of what went into the design. He must be given some idea of the false paths that were taken and why they failed. Often a novice to a project will make suggestions that seemed reasonable to the rest of the team also, and were explored. The reasons that these suggestions failed must be presented to the new member of the design team.

At any point in the process there may be one or more "current versions" of the design. But there are other paths that could have been or actually were taken. These paths may have involved much work and may have even produced results that, in the current state of the design, are not useful. For some reason it was discovered, after some effort, that a particular path was not good. In some cases, though, it turns out that the decision to surrender a path was made incorrectly, or the reasons for its inadequacy turn out later to be invalid. It would be nice to be able to find and use as much of the work done on these alternative paths as possible.

In general, changes to external factors, as well as internal bugs and changes of mind, can change the reasons for selecting one detail over another. In some cases, this occurs often (different technologies, changes in microcode details) and we want to be able to produce the new design with as little fuss as possible. Much of the work in the design process consists in paths parallel to the actual design which may be useful in other situations. It is important to be able to recognize which work is relevant to the design, both to be able to manage the work effectively, and also to correctly use the results.

A MODEL OF THE DESIGN PROCESS

In this section I present a simple model of the design process. The model is developed in such a way that will make it clear that the relevant parts of the model could be explicitly represented in a design system. Certainly the model won't capture all the important parts of the design process, but the dependencies and their relationships that are captured here can be very useful at many stages in the design.

We shall describe the initial state of a design process as presenting the designer with some set of *goals*. The whole point of the design process can be described as the satisfaction of this set of goals, or alternatively, discovering why these goals cannot be satisfied. The goals in a typical design may be more or less specific: "Design a non-inverting PLA superbuffer driver that can fit on a 10 micron pitch." "Make the best damn microprocessor ever."

Some goals related only to aspects of the finished product. To make the notion of a goal more general, we will allow other sorts of goals whose content depends not only on the final result, but on some other decisions that constrain the current design. For example, the PLA driver above had a requirement of a 10 micron pitch. That requirement is presumably not a requirement of the final design (who cares how big the PLA is in the final chip?) but is a requirement generated by the fact that the PLA terms are intended to pitch on 10 microns and the rest of the PLA design depends on that fact. The introduction of these constraint goals, which are introduced as a result of other design decisions, allows us to speak of all portions of the design process as aiming at the satisfaction, or proving the impossibility of, goals.

The design process proceeds by the making of *decisions*. A decision satisfies one or a number of goals. A decision is influenced by:

* Facts over which the designer at this level has no control: the diffusion resistance, the microcode, etc.

* The existence of the goals that it satisfied -- it makes no sense to satisfy a goal that is not relevant to the design.

* The results of other decisions. These results may constrain the available set of options.

There is no clear distinction among these influences. The separation is merely one to aid understanding among those who would have to examine the decision. "Facts" are simply those states of affairs which are more difficult to do anything about. Some goals -- the top-level design goals -- also can't be ignored, but goals and results of other decisions can be negated by changing some earlier decision.

The *result* of a decision is the state of affairs that comes to be after the decision is made. The result of a decision will change the design in some ways to satisfy the goals mentioned in the decision. A result will also modify the design process by introducing new goals and constraints. As in the above example, the decision to make the PLA pitch on 10 microns influences all goals involved in building the PLA's pieces. The decision to place a certain module in a particular location not only adds the fact of that location to the design, but also thereby introduces specific goals to the router.

USING THE DEPENDENCY GRAPH

The simple model thus presented captures some of the important relationships involved in the design process. Rather than filling it out more completely, I will argue that it can and should be represented explicitly at all stages of the design process. Some discussion of the techniques for representing such structures are presented in[3,4]. The general idea is that decisions and their results be represented as nodes in a graph structure, with the various sorts of dependency linkages be represented by the edges of the graph. One can then trace the results on which a certain decision depends, and one can trace the consequences of a certain result.

Just having the dependency structure for a design represented explicitly can be a useful documentation tool. A new member of the design team can read the history of the design and come to a more complete understanding of the design than any look at the partially finished product.

An even more important reason for maintaining the dependency graph is its utility in modifications to the design. Modifications can take several forms, and can be motivated in several ways. Consider the case where there is some inconsistent set of goals or subgoals. In this case, someone in the design process would announce that the goals he has been given cannot be satisfied. With the explicitly represented dependency graph we can locate the decisions that led to the failure. The technique is called "dependency based backtracking.[5]" The idea is to locate the set of goals or facts that are unsatisfiable. For each fact and goal in this set, we locate, from the dependency graph, the goals and facts on which the current ones depend. This process will visit all the decision nodes in the dependency graph which contributed to the current failure. The design team can choose to reject one or more of the decisions that led to the failure. The team must them find some other way to satisfy the goals that were previously satisfied by the rejected decisions.

There is no reason to expunge the nodes which depended on the rejected decisions. They still follow from the decisions, it is just that the decision is rejected -- its as if it hadn't been made. We must annotate the dependency graph to indicate which of he set of decisions therein have actually been accepted as contributing to the current design. Thus the dependency graph actually represents not only the process that led to the current design, but several other paths that were followed and rejected for some reason or another. Again, understanding such paths is important for the understanding of the design process. It is important to realize why an otherwise attractive proposal didn't prove fruitful.

Another reason for the importance of keeping these false paths represented is for the later modification of the design. Later design failures, or new goals, or changes in external factors may render invalid the reasons that led to the rejection of some path, or the acceptance of others. It may be, for example, that a path which had been rejected now becomes acceptable. By following the dependencies of a new factor through the dependency graph, the designer can tell which decisions might need to be reassessed. And if the results of the decisions that are currently invalid are available, all the work done to generate the results of that decision need not be done again -- the results are stored at precisely the place in the graph where they are relevant.

LIMITATIONS OF THE MODEL

There are two major limitations to the model here presented. The first is that it is not generative -- the model provides no account of *how* the design decisions are made. It is entirely a tool of documentation and information management. In the design of large systems with many contributing workers, however, such a tool becomes crucially important. Even an individual designer, working on a large project, needs some way to organize the multitude of separate, yet interacting decisions he must make.

The other limitation of the model is that many decisions in the design process seem not to be explicitly representable. In many cases the factors influencing a decision are things like "gut feelings" and late night weariness. And it is even more difficult for a designer to be able to determine the effects of a design decision, or even the subgoals it introduces.

Such limitations are important and are indeed major shortcomings of the suggestions I am exploring. But the dependency tree would be a useful tool even if it were very incomplete. Any information in the tree is useful as it represents at least an attempt to explicitly represent the design process. If the information in the tree is inaccurate, using the tree as described above can locate inaccuracies in the tree as well as in the design. Of course, an important aspect of the usefulness of such a tool is the interface it presents to the designer. Unfortunately I will not explore such issues in this paper.

WHO SHOULD USE IT

The design dependency graph is an important tool at all levels of the design process. For the top manager, the graph can be used to asses both the progress of the design and its methods. The manager can explore some alternatives to the currently accepted ideas, he can also explore the consequences of changes to external factors or influences. The dependency graph is thus very closely related in form and content to the electronic spread sheet, except that includes more complicated relationships among factors.

The design engineer can use the tree to explore the constraints and consequences of the goals he is assigned to satisfy. He can also use the graph to represent his on progress on the satisfaction of those goals.

And finally, if computers are to truly be considered a part of the design team, rather than just tools, they must also be able to manipulate and make use of the design graph. A design program must be able to take as its input a representation of its goals and their place in the graph. The tool may use the the graph in its operations, these methods can be used to back up and other sorts of recovery from errors.

TWO EXAMPLES

I will take as examples of this sort of "responsible program" two of my efforts in design tools.

In the design of the SCHEME-81[2] chip we decided, rather than use a special purpose router, to instead develop a small set of more specific routers for use in particular situations. For example, a "channel router" was developed to be used when the human designer decided that it was applicable. In some cases, however, the channel router would not do a perfectly acceptable job. Such cases arose when certain kinds of pathological cases arose in the routing -- certain kinds of cycles for example. Rather than attempt to solve this problem completely, it was decided to let the program make use of a very simple algorithm, and be able to note its failures. It would then report its failures to the designer, along with what it could determine about the causes for these failures. These "causes" were factors over which the router had no control but which could be modified by the designer -- such as the order of input lines to a subcircuit. The router was thus able to suggest modifications to the design that would make it able to do its job.

Another example comes from the author's PLA Generator Generator program PRECIOUS. This system constructs a PLA Generator program which is a DPL[6] program to lay out a certain kind of PLA according to a user's specifications. The specifications are in terms of goals which the program tries to satisfy. A simple way to write such a program is the "optimistic" approach -- one simply looks, for each goal mentioned by the user, for a way to satisfy the goal. This is done by annotating each part in a library with the kinds of goals that it can satisfy.

The process continues until each goal is satisfied. When this works, and the result is acceptable to the user, the process is finished. However if there is some problem, the program determines which of the "blind choices" it made was at fault, and restarts the design from that decision. This "dependency based backtracking" control structure is possible only with the explicit representation of the decisions and dependencies involved in the design.

REFERENCES

[1] John Batali, 1981 **A Vision Chip**, Master's Thesis: MIT Department of Electrical Engineering and Computer Science.

[2] John Batali, Edmund Goodhue, Chris Hanson, Howie Shrobe, Richard M. Stallman, Gerald Jay Sussman, 1982 "The Scheme-81 Architecture -- System and Chip," **Proceedings of the 1982 Conference on Advanced Research in VLSI at MIT.**

[3] John Doyle, 1978 **Truth Maintenance Systems for Problem Solving**, MIT AI Technical Report 419.

[4] David McAllester, 1982 **Reasoning Utility Package User's Manual**, MIT AI Memo 667.

[5] Gerald Sussman & Richard Stallman, 1975 "Heuristic Techniques in Computer-Aided Circuit Analysis," **IEEE Transactions on Circuits and Systems, CAS-22.**

[6] John Batali & Anne Hartheimer, 1980 **The Design Procedure Language Manual**, MIT AI Memo 598.

Session 2-4d
Digital Systems II

Chairman

D. P. Siewiorek
Carnegie-Mellon University

ANALYSIS OF COMPUTER PERFORMANCE AND RELIABILITY

Kishor S. Trivedi

Department of Computer Science
Duke University, Durham, N.C. 27706

ABSTRACT The need for increased reliability and computing power coupled with advances in technology has given rise to sizeable and complex computer systems. The users and designers of such systems need tools to evaluate the effectiveness of such systems. Current approaches to reliability evaluation, performance evaluation, performability evaluation, and optimization of systems with multiple objectives are discussed.

1. Introduction

There are three general techniques for system evaluation: testing, simulation, and analytic modeling. The testing approach will produce the most accurate results but will tend to be rather expensive. Further, if the system being evaluated is not yet available for experimentation, the testing method will not be applicable. In any case, the use of testing techniques implies the need for experimental designs and statistical analysis of observed data [1].

If the system is not available for experimentation, either a simulation model or an analytic model is used for its evaluation. Simulation models can be more accurate than analytic models but less accurate than the testing method. Like testing, simulation models also require a careful design of (simulation) experiments and statistical analysis of output data [1,2]. Simulation models can be either driven by an event trace or can be driven by distributions of random times to the occurrence of events, in which case, procedures to generate random variates are needed [1,2].

Simulation models can be expensive both to develop and to run, while the alternative of analytic models appears quite attractive. A large body of analytic models for computer system evaluation is available [1,2,3]. Often, models of complex systems are not analytically tractable, in which case either approximate methods [1,2,4] or hybrid (analytic combined with simulation [5] or analytic combined with testing [6] and such other combinations) methods are needed.

If a model is used for system evaluation, the modeler may have decided to ignore certain features of the system (its structure, its workload, its fault-occurrence behavior, or its fault-handling behavior). This may have been done to simplify the model in order to make it tractable or because the modeler felt that these features do not have a significant influence on system effectiveness. It is therefore important to validate the model against data collected by testing on the system itself, or if this is not possible, to analyze the effects of the ignored features on the output of the model.

There are at least two types of errors (ignored features) in system validation via modeling: modeling errors and errors in the solution process. One type of modeling errors are structural errors (missing or extra states and missing or extra arcs). It is sometimes possible to show that the model structure corresponds to actual system behavior [7]. Most stochastic models assume that certain random variables are statistically independent. If in fact, this assumption does not hold, we would like to be able to analyze its effect on the model outputs. Markov models are quite common in stochastic evaluation of systems, which imposes a severe restriction of exponential holding times in system states. It is then desirable to be able to analyze the effect of violation of this assumption. Further, it is possible for the system to start operation in a state other than that which is assumed by the model. Some reliability prediction packages include the analysis of such uncertainty [5]. The fourth class of modeling errors relate to variations in the input parameters. It is possible to analyze the effect of such parametric errors in reliability prediction models [5] and in queueing models for system performance evaluation [8].

Errors in the solution process can be classified into approximation errors and numerical errors. The approximation

CH1935-6/83/0000/0464$01.00

errors arise while solving complex models where an exact solution is either impossible or computationally expensive. Courtois [9] has analyzed approximation errors in queueing theoretic models. A detailed account of approximation errors in discrete optimization models is available in [10]. Finally, truncation and roundoff are two types of numerical errors.

2. Reliability Evaluation

Lifetesting is the most common technique for the reliability evaluation of individual components [1,6]. However, when applied to complex fault-tolerant computer systems it may become prohibitively expensive if not impossible [3,11]. Simulation models have been used in reliability evaluation in at least two different ways. To assess the effectiveness of various fault-handling mechanisms, a common technique is to use fault-injection in a simulation model of the system. Observations of system response to repeated fault injections are then analyzed to obtain estimates of parameters such as fault detectability [1,6]. It is also possible to use distribution-driven simulation models for reliability prediction [6].

Analytic models of many different types have been discussed for reliability evaluation. Combinatorial methods [1] have been used in the CARE package [12]. Markov methods [1] form the basis of the ARIES package [13]. The CARE II model uses Markov renewal methods [14] while the CARE III model uses a combination of non-homogeneous Markov and Markov renewal techniques [15].

Fault-tolerant systems used in life-critical applications are expected to provide ultra-high reliability. Reliability prediction models for such systems need to be able to deal with a large number of states, need to allow non-exponential distributions, need to allow the user to specify the fault-handling behavior in detail, and need to provide the changes in the predicted reliability due to variations in input parameters.

The problem of large state space is effectively handled using decomposition/aggregation [16]. The use of decomposition also allows us to employ different techniques for the solution of submodels. For example, HARP package uses simulation to evaluate the fault-handling submodel while it uses non-homogeneous Markov techniques to solve the reliability model [5].

To allow nonexponential distributions, several general techniques are available. The method of supplementary variables [17] and the Markov renewal approach [18] are effective for relatively small problems. Non-homogeneous Markov techniques have been effectively utilized to model non-Poisson failure processes in nonrepairable systems [1,5,15]. The Coxian method of stages is yet another approach to "Markovize" a non-Markovian problem [1,19]. It has the disadvantage of expanding the state space, however. Distribution-driven simulation allows the modeler to use quite general distributions and hence allows more realistic models. HARP uses distribution-driven simulation to evaluate fault-handling submodels.

A common approach to reliability analysis is to assume that the fault-handling mechanisms can be characterized by a single parameter known as coverage. However the estimation of coverage parameter can be quite elusive. Furthermore, the reliability is known to be strongly dependent on coverage. It is therefore desirable to provide mechanisms to the user to specify more elemental behavior of the fault-handling mechanisms. Stiffler has proposed several coverage models earlier and shown that it is possible to calculate coverage using Markov renewal techniques. More complex models of coverage are currently under investigation and stochastic Petri nets are used for this purpose [5].

The effect of variations in failure rates and coverage parameters on the predicted reliability can be bounded by analytical methods [5]. Effects of various other potential errors in the modeling process and in the solution technique need to be classified, analyzed, and bounded.

3. Performance Evaluation

Common techniques for performance evaluation are measurement, simulation, and analytic modeling. The first two techniques require the use of statistical methods to design experiments and to analyze output data. For a treatment of measurement techniques see [20] while for a discussion of simulation methods see [2].

A model for performance evaluation requires three sets of inputs: workload characteristics, hardware configuration, and software configuration. Thus, the resource requirements and traffic parameters of individual programs as well as the mix of programs needs to be specified. The types of hardware resources, their interconnections, and capacities need to be specified. As regards the software configuration, scheduling algorithms, file/data base allocations and such other parameters need to be specified. The outputs of such models consist of throughput, average response time, and queue lengths.

Many different analytical techniques have been used for the solution of performance models. The fastest techniques are available for product form queueing networks [1,2]. Models not belonging to this class can sometimes be solved using Markov chain techniques. More often, however, this either leads to excessive computation times or makes too many assumptions for the results to be useful. Various approximation techniques have been used to solve large problems [2]. These include decomposition [4,9], iterative [4], and diffusion approximations [2]. It is often necessary to remove the "exponential-distribution" assumption of Markov models. Solution of such non-Markovian models can be obtained by the method of supplementary variables [17], Markov renewal techniques [18], or by the method of embedded chains [1].

Many packages are available for the specification and solution of performance models [2]. Current efforts involve extending the range of applicability of tractable analytic models for the evaluation of parallel and distributed systems [4] and the investigation of new, user-based measures of system performance [21].

4. Performability Evaluation

The separation of reliability and performance evaluation of a computer system is no longer appropriate for systems with graceful degradation capabilities. Combined evaluation of performance and reliability of such systems is an active area of research [22,23,24].

Many of the efforts rely on the fact that times between failures are several orders of magnitude larger than the time to complete the execution of a program. Thus the state probabilities of reliability model can be computed without regard to program behavior and for each reliability-model-state, performance can be evaluated. Using these two separate evaluations, composite measures can be obtained. Although such decomposition approaches will generally be adequate, exact methods are nevertheless interesting [23].

5. Optimization

While comparing a set of alternatives a designer may wish to study possible tradeoffs. This may be done either by a sensitivity analysis or by an overall optimization involving cost, performance, and reliability attributes [25,26,27]. These problems are generally discrete optimization problems and hence tend to be computationally intensive. If the number of control variables is small, then exhaustive search may be feasible. For medium-sized problems branch-and-bound techniques are appropriate. For large problems an approximate solution is virtually the only feasible method of solution. Two distinct approaches to obtain an approximate solution have been used. One approach is to remove the integer constraints (relaxation), solve the continuous optimization problem, discretize the continuous solution, and obtain bounds on the discretization error [28]. By contrast, another approach is to develop a simple algorithm to directly obtain a suboptimal solution to the discrete optimization problem and then to derive bounds on the performance of the approximation with respect to the exact discrete optimum [10,26].

6. Conclusion

We have given a brief overview of current approaches to reliability evaluation, performance evaluation, and design optimization for computer systems. Due to increased complexity of computer systems, we expect an emphasis on the use of hybrid models, approximate solution techniques, and error analysis.

References

[1] Trivedi, K., Probability and Statistics with Reliability, Queueing, and Computer Science Applications, Prentice-Hall, 1982.

[2] Lavenberg, S. (ed.), Modeler's Handbook, Academic Press, 1983.

[3] Geist, R., and Trivedi, K., "Ultra-High Reliability Prediction for Fault-Tolerant Computer Systems," to appear, IEEE Trans. Comp.

[4] Heidelberger, P. and Trivedi, K., "Analytic Queueing Models for Programs with Internal Concurrency," IEEE Trans. Computers, Jan. 1983.

[6] Geist, R., Trivedi, K., Dugan, J. and Smotherman, M., "Design of the Hybrid Automated Reliability Predictor," Proc. Digital Avionics Systems Conference, Seattle, 1983.

[6] Siewiorek, D., and Swarz, R., The Theory and Practice of Reliable System Design, Digital Press, 1982.

[7] Wensley, J., et al., "SIFT: Design and Analysis of a Fault-Tolerant Computer for Aircraft Control," Proc. of the IEEE, October 1978.

[8] Suri, R., "Robustness of Queueing Network Formulae," JACM, July 1983.

[9] Courtois, P., Decomposability: Queueing and Computer Science Applications, Academic Press, 1977.

[10] Coffman, E., Computer and Job/Shop Scheduling, John Wiley, 1976.

[11] Trivedi, K., Gault, J., and Clary, J., "A Validation Prototype of System Reliability in Life-Critical Applications," Proc. Annual Tech. Symp. on Pathways to System Integrity, June 1980.

[12] Mathur, F. and Avizienis, A., "Reliability Analysis and Architecture of a Hybrid-Redundant Digital System: Generalized Triple Modular Redundancy with Repair," Proc. AFIPS SJCC, 1970.

[13] Ng, Y. and Avizienis, A., "ARIES - An Automated Reliability Estimation System," Proc. Annual Reliability and Maintainability Symposium, January 1977.

[14] Stiffler, J., et al., "An Engineering Treatise of the CARE II Dual Mode and Coverage Models," Final Report, NASA Contract L-18084A, November 1975.

[15] Stiffler, J., Bryant, L., and Guccione, L., "CARE III Final Report, Phase I," NASA Contract Report 159122, November 1979.

[16] Trivedi, K. and Geist, R., "Decomposition in the Reliability Analysis of Fault-Tolerant Systems," to appear, IEEE Transactions on Rel.

[17] Cox, D., "The Analysis of Non-Markovian Stochastic Processes by the Inclusion of Supplementary Variables," Proc. Cambridge Phil. Soc., Vol 51, 1955.

[18] Ross, S., Applied Probability Models with Optimization Applications," Holden-Day, Inc., 1970.

[19] Cox, D., "The Use of Complex Probabilities in the Theory of Stochastic Processes," Proc. Cambridge Philosophical Society, 1955.

[20] Ferrari, D., Computer System Performance Evaluation, Prentice-Hall, Englewood Cliffs, NJ., 1978.

[21] Geist, R. and Trivedi, K., "The Integration of User Perception in the Heterogeneous M/M/2 Queue," Proc. IFIP Symposium, PERFORMANCE '83, College Park, Md.

[22] Meyer, J. et al, "Performability Evaluation of the SIFT Computer," IEEE Trans. Comput., June 1980.

[23] Baccelli, F. and Trivedi, K., "Analysis of M/G/2 Standby Redundant System," Proc. IFIP Symposium, PERFORMANCE '83, College Park, Md.

[24] Trivedi, K. and Yu, P., "Reliability and Performance Analysis of a Ringnet," IBM Research Report RC-9792, January 1983.

[25] Geist, R. and Trivedi, K., "Optimal Design of Multilevel Storage Hierarchies," IEEE Trans. Comp., March 1982.

[26] Bannister, J. and Trivedi, K., "Task Allocation in Fault-Tolerant Distributed Systems," to appear, Acta Informatica.

[27] Trivedi, K., "Designing Linear Storage Hierarchies so as to Maximize Reliability Subject to Cost and Performance Constraints," Proc. 1980 Int. Conf. on Comp. Arch.

[28] Wagner, R. and Trivedi, K., "Hardware Configuration Selection Through Discretizing a Continuous Variable Solution," Proc. IFIP International Conference on Performance Modeling, Toronto, 1980.

System speed, space and power estimation using a higher level design notation

Paul J. Drongowski

Department of Computer Engineering & Science
Case Western Reserve University
Cleveland, Ohio 44106

ABSTRACT

Practical systems design must take into account the constraints imposed by the implementation technology. In the case of VLSI systems, technology dependent considerations include circuit size, routing complexity, wire delays, off-chip communications, power consumption and heat dissipation. These considerations affect the logical and physical partitioning of the system, eventually affecting the control programming of the system as well. In the worst case, if any constraints are violated, the design may not be feasible and it must be modified or abandoned. This paper states a set of practical requirements which must be satisfied by a design language and its mechanical aids.

1. Introduction.

As circuit complexity increases, the precise specification of higher level system functions becomes increasingly important. Reliable and functionally correct VLSI hardware systems cannot be developed using ad hoc techniques from incomplete or ambiguous product descriptions. Rigorous specification at the architectural level provides a firm basis from which a detailed design can be synthesized. Because systems must fit into a particular design context, they must satisfy many externally imposed functional and physical design constraints. Since many of these constraints depend upon the implementation technology or lower level components, feedback about the expected execution speed, space and power is essential for making good architectural (and organizational) choices.

This paper outlines some practical issues in architectural and organizational level design. We are developing a graphical hardware description language for organizational design which satisfies these requirements. Preliminary results are indicated and our future efforts in this area are outlined.

2. Problem definition.

There are many issues in design that go beyond the construction of a system which computes a correct result. Real systems must be designed into a _context_ that will impose temporal, structural and physical constraints upon it. This section presents and explores the major issues in high level (i.e., above logic) hardware design.

Architecture is the abstract, visible behavior of a computing engine. This level of specification is concerned with the description of _what_ is computed (and how the engine communicates with its environment) rather than _how_ the result is obtained. The engine can be recursively decomposed into subsystems, each possessing its own architectural specification. The internal structure, behavior, and physical characteristics of the engine -- the algorithm and machine for computing the result -- is its _organization_. As an aggregate, the subsystems implement the visible behavior of the engine and specify how the result is obtained. Any level in a design can be characterized by its architectural (external) and organizational (internal) specifications. Similar distinctions have been drawn by other researchers such as intent versus behavior in [Est77] and the behavioral versus functional/structural levels of [Nes82].

Architecture requires a very high level description of behavior. This description must capture:

* Visible behavior in its entirety, including external interconnections, signalling protocols (plus sequencing) and critical timing constraints,

* Externally imposed physical constraints such as die size, power consumption, heat dissipation (cooling) and voltage rails.

The architectural specification must be complete enough to verify that an implementation of the computing engine will be compatible with its design context and to provide a base from which the organizational description can be derived. Ideally, the specification should not be tainted by details about structure or layout which are reserved for the organizational level. If the architectural specification is not over-constrained, the design engineer or synthesis system will have additional flexibility during the translation from architecture to organization.

The architecture to organization mapping is one to many. The nature of this mapping is a major research problem which we will not address in this paper. Instead, we will focus on machine organization.

We are interested in a particular set of constraints which deal with speed, space and power (S/S/P) trade-offs at the organizational level. Execution speed is an important quality since results must be produced in a timely manner as the design context requires them. In VLSI, space is critical since excessively large dies may reduce production yields below an acceptable economic threshold. Power simply cannot be ignored because special cooling or packaging provisions may render a design infeasible with respect to its electrical or mechanical context. Further, power can be traded against space to obtain higher execution speeds.

The organizational specification describes the algorithm and resources required to implement an architecture. Following the work of [vanC79] and [Row80], we have separated the organizational design into three domains:

<u>Structural</u>. The structural design defines a set of resource components, such as storage, control and combinatorial circuit elements, and their interconnections. A subset of the components have direct, visible counterparts in the architectural specification.

<u>Behavioral</u>. The behavioral design specifies the sequence of control operations on the resource components which implements the intended system (architectural) behavior. The operations are the control signals that must be asserted to evoke the work to be performed by those components. Control operations may be stored as microcode in a PLA or ROM, or implemented in random logic.

<u>Physical</u>. The physical design describes the spatial arrangement of the geometric form for the resource components and the wire paths that interconnect them. Other aspects of the physical design include component delay and set-up times, power consumption and heat dissipation, and other technology dependent information.

The translation from the architectural specification to the organizational one includes all of the issues involved in traditional physical design automation (DA) plus the more contemporary topic of mechanical program synthesis (algorithm design) from software engineering.

The interrelationships between the behavioral, structural and physical (B/S/P) domains are quite complex and make the evaluation of organizational alternatives a difficult task. Exploration of the design space requires the description and comparison of several different organizations. To objectively compare alternatives, an S/S/P value must be calculated for each alternative. Traditionally, the S/S/P values have been found by simulating the designs at both the functional and electrical levels, and by measuring or estimating the floor plans.

Engineers, however, do not usually enumerate all possible designs (or even a few) and select among the best. By intuition or experience, usually just one organization looks feasible or "best." As the system is refined to its primitive components, the engineer will test the design against its S/S/P performance criteria. If the design fails to satisfy one or more of the criteria, it must be changed. Hence, exploration of the design space proceeds by enumerating a single design and then manipulating the design until it satisfies the performance criteria.

The interrelationships of the B/S/P domains makes improvement a difficult design task. If the specified system does not execute fast enough, for example, the engineer may modify the design in one or more of the following ways:

* Increase parallelism by adding new datapaths and operators.
* Change the placement of existing components to reduce wire lengths along the critical delay path.

* Move critical delay path wires to a lower capacitance material.

* Allocate more power to critical path components.

In some cases, changes can be confined to a single domain. Changes in the structural domain almost always upset the behavioral or physical domains because the behavioral control code and layouts must change according to the new structure.

To examine as many different alternatives as possible by moving quickly through the design space, excessive detailed design must be avoided. Any effort that is required to refine a design to a level of detail which is S/S/P analyzable will increase the amount of effort to be discarded when the design is modified. For example, if a design must be fully routed (an expensive operation) before delays along the critical path can be estimated, much of this work will be discarded when component arrangements are changed to reduce critical delay path wire capacitances.

This requirement calls for a very high level hardware description language for architectural design which is S/S/P analyzable. In this language, designs are expressed in terms of functional building blocks such as register files, ALU's, shifters, etc. Tools for delay analysis exist today, but they operate on gate and transistor level descriptions. An architect simply cannot afford to refine a design to gates or transistors before getting feedback about S/S/P. With today's short project schedules, a language and its CAD system should lead the engineer to a good design sooner and without unproductive excursions through the design space.

As more functionality is borne by firmware or software, architectural design aids must provide for trades between these elements and dedicated hardware components. Again, conventional delay analysis tools are insufficient to estimate the execution speed of microcoded functions. Simulation is often used to obtain these estimates. However, test data must be generated and turnaround times are long. Since there are critical paths through the microcode which must satisfy externally imposed time constraints, the design tool should provide some measure of delays through firmware as well as the hardware. Improvements along the critical delay path in the hardware will reduce the system clock period (for synchronous systems) thereby reducing the microinstruction cycle time. The design system should track these improvements from the critical hardware path through critical paths in the microcode.

3. Research.

To meet the organizational level requirements outlined above, we are devising a graphical hardware description language, called "Gdl." In Gdl, the system structure (data, control and condition paths) is specified hierarchically in block diagrams. These diagrams identify the major system components, their interfaces and interconnections. System behavior is described in a collection of control graphs which are an abstract representation of the control events that are needed to route and compute results through the data and control paths. The physical arrangement of the components are captured in floor plans. All of these representations are familiar to design engineers. We have merely formalized the syntax and semantics of the block diagrams, etc. into a language.

Gdl is based upon a graphical notation for the design of nMOS VLSI systems which are fully self-timed and have restricted interconnection topologies. As part of this earlier work, an analytical procedure was developed to estimate the execution speed of designs expressed in this notation. An experimental implementation of the procedure demonstrated that analysis can be performed with sufficient speed to support interactive design [Dro82]. (The application of other circuit analyzers on large nMOS designs has shown that static analysis is a promising alternative to detailed electrical or logical simulation [Jou83, Ous83].) We are extending the original method to include more conventional control organizations (e.g., microcode.) An implementation of the new procedure will be included in a suite of CAD tools to support design with Gdl along with an editor, verification and simulation tools. We intend to apply results from the microanalysis of programs to microcode, thereby extending performance estimation beyond critical path analysis of combinatorial logic to include the system software [Coh 82, Weg75].

In the course of writing this report, we have come to realize how little is known about true architectural level design. Most of our design languages are procedural and have distinct biases toward one particular style of implementation or another [Dro79]. These issues have been clouded by the imprecise or ambiguous use of common terms such as "architecture," "organization," "behavior," "functional," etc. A set of standard definitions for these terms would be a genuine contribution to the discipline of digital systems CAD.

Acknowledgements.

Special thanks are extended to Chuck Rose, J.R. Bammi, Randy Harr and Sundar Iyengar, for their comments and insight.

References.

[Coh82] Jacques Cohen, Computer-assisted microanalysis of programs, CACM, Vol. 25, No. 10, October 1982, pg. 724-733.

[Dro79] Drongowski, P.J. and Rose, C.W., Application of hardware description languages to microprogramming: Method, practice and limitations, MICRO-12, ACM SIGMICRO and IEEE Computer Society, November 1979.

[Dro82] Paul J. Drongowski, An engineering aid for the design of nMOS VLSI systems, Ph.D. dissertation, Computer Science Department, University of Utah, August 1982.

[Est77] Gerald Estrin, Modeling for synthesis -- The gap between intent and behavior, Proc. of the Symposium on Design Automation and Microprocessors, ACM SIGDA and IEEE Computer Society, Palo Alto, California, February 1977, pg. 54-59.

[Gra83] John J. Granacki and Alice C. Parker, The effect of register transfer design tradeoffs on chip area and performance, Twentieth Design Automation Conference, ACM SIGDA and IEEE Computer Society, June 1983, pg. 419-424.

[Hit82] Robert B. Hitchcock, Sr., Timing verification and the timing analysis program, Nineteenth Design Automation Conference, ACM SIGDA and IEEE Computer Society, June 1982, pg. 594-604.

[Jou83] Norman P. Jouppi, Timing analysis for nMOS VLSI, Twentieth Design Automation Conference, ACM SIGDA and IEEE Computer Society, June 1983, pg. 411-418.

[McS82] David W. McSweeney, Timing verification of VLSI logic circuits, MIT Conference on Advanced Research in VLSI, MIT, January 1982.

[Nes82] J.A. Nestor and D.E. Thomas, Defining and implementing a multilevel design representation with simulation applications, Design Research Center, Carnegie-Mellon University, Technical Report DRC-18-56-82, December 1982.

[Ous83] John Ousterhout, Using Crystal for timing analysis, User's Manual, Computer Science Division, EECS, University of California at Berkeley, 1983.

[Par79] Frederic I. Parke, An introduction to the N.mPc design environment, Sixteenth Design Automation Conference, ACM SIGDA and IEEE Computer Society, June 1979, pg. 513-519.

[Ros83] Charles W. Rose, Greg M. Ordy and Frederic I. Parke, N.mPc: A retrospective, Twentieth Design Automation Conference, ACM SIGDA and IEEE Computer Society, June 1983.

[Row80] James A. Rowson, Understanding hierarchical design, Ph.D. dissertation, California Institute of Technology, Pasadena, California, April 1980.

[Sas81] Tohru Sasaki, et al., Hierarchical design verification for large digital systems, Eighteenth Design Automation Conference, ACM SIGDA and IEEE Computer Society, June 1981, pg. 105-112.

[Ste82] Mark Stefik, et al., The partitioning of concerns in digital systems designs, MIT Conference on Advanced Research in VLSI, MIT, January 1982.

[Tam83] Eiji Tamura, et al., Path delay analysis for hierarchical building block layout system, Twentieth Design Automation Conference, ACM SIGDA and IEEE Computer Society, June 1983, pg. 403-410.

[vanC79] William M. vanCleemput, Hierarchical design for VLSI: Problems and advantages, Proc. of the 1st Caltech Conference on VLSI, Caltech, January 1979.

[Weg75] Ben Wegbreit, Mechanical program analysis, CACM, Vol. 18, No. 9, September 1975, pg. 528-539.

DEVELOPING VLSI SYSTEMS WITH A SILICON ENGINEERING ENVIRONMENT

Walt Scacchi
Computer Science Department
University of Southern California
Los Angeles, CA 90089-0782

INTRODUCTION

The next generation of advanced VLSI systems will lead changes the work habits, structure, and operation of VLSI system development organizations [9,32]. Organizing system development activities with available staff and resources is a precarious endeavor; local reorganization is continual in response to changing circumstances. When the organization of the project no longer matches the tasks needed for system development, then costly problems such as work slippages or mis-timed circuit logic appear. Subsequently, we find that problems of producing VLSI systems in a CAD environment are primarily organizational -- how to organize available staff and computing resources to perform technology analysis, system functional specification, floor plan and cell design, layout, simulation and test, fabrication, and so forth. As such, solutions must also be of an organizational character if they are to be most effective. Thus, our approach is to bring together a set of automated tools, development methodologies, project management strategies together with staff and computing resources into an organizational form we call the system factory [30].

The system factory approach consists of applying recent results in software engineering and the sociology of computing work to the engineering of VLSI systems [6,19,27,28,31]. These results outline organizational problems in system development, as well as organizational strategies for dealing with them. Specifically, problems in computing system development can often be traced to:

-- Computing infrastructure difficulties [19], e.g., are the local computing resources adequate to the demands upon them? We find that an adequate infrastructure for VLSI system development must integrate a computer-aided development environment with a system engineering methodology and set of strategies for managing system development.

-- Division-of-labor difficulties [21,23,28,31], e.g., are the problems addressed by systems in development suitable for automated support, or do they intrinsically require human discretion? The purpose for dividing work among engineers and computing systems is to achieve greater production efficiency. Large systems are produced by more than one designer. Techniques for articulating, decomposing, transforming, and evaluating system development tasks are thus part of this division process.

-- Innovation-handling difficulties [27], e.g., are start-up and shake-down problems recognized and dealt with as such, or are they perceived as something else--inadequate computing technology, for example? How should software aids and engineering design systems be configured to support both current and future VLSI system development activities? New computing environments must be evolvable, portable to other processors, and not significantly restrict system design methodology.

-- Error-handling procedures and their adequacy [31], e.g., are software bugs eliminated fixing the system, or are they handled in practice by users performing "workarounds"? To minimize the propagation of work errors into system design, testing and evaluation techniques must be performed as early as possible during system development, as well as throughout development.

Examining the patterns of these difficulties allows us to develop both analytical models of the problems arising in CAD work, and diagnostic and corrective procedures for managing them. Elsewhere, we identify more than ten corrective strategies for managing problems that appear that are applicable in different VLSI system design organizations [31].

The remainder of this paper outlines our approach to support the development of VLSI systems throughout their life cycle, a set of software tools, and a development methodology that can integrate both.

SUPPORTING THE VLSI SYSTEM LIFE CYCLE

The "system life cycle" denotes the set of formal activities that occur in producing a system. These activities span from the conception of a system through its routine use and evolution. Ideally, we would like to support these activities with automated tools and an engineering methodology for their use [18]. While comprehensive life cycle support is the goal, we do not yet understand what is required to provide this to project managers and system engineers alike. Nonetheless, we can describe our current understanding.

The development and evolution of a VLSI or software system follow a similar, but distinct course. The activities (or "stages") of the life cycle for VLSI systems include:

-- Initiation and Acquisition: making a commitment to acquire or develop a new VLSI system.

-- Requirements Analysis: determining the performance and production characteristics of a VLSI system to be developed with available staff and resources.

-- Selection and Partition: determining whether the system's functional components (e.g., cells) are available, and choosing between alternative divisions of hardware and software components.

-- System Specification: describing what computational functions the desired system is to perform.

-- Architectural Design: dividing, placing, and routing system functions across computational modules into a circuit floor plan.

-- Detailed Design: designing algorithms and circuit cells that realize the computational modules in the floor plan.

-- Implementation: layout, mask making, and fabrication of designed components in preparation for testing and installation in another system.

-- Testing: verifying that the implemented system fulfills system specification and design while validating its performance and production characteristics.

-- Documentation: providing a written record of what the system does and how to use it.

-- Use: operating the developed system under a variety of idiosyncratic circumstances.

-- Evaluation: evaluating the operation, performance, and applicability of the system in light of changing production or performance characteristics.

-- Evolution: enhancing, tuning, repairing, and converting a developed system to maintain its useful operation.

Overall, there is a high degree of concurrency across the activities occurring during a VLSI (or software) system's life cycle. Each life cycle activity requires a different kind of description of the system being developed: system specification, architectural design, detailed design, implementation (layout), and test plans. A complete and consistent description is the desired product of each activity, although these descriptions will often be incrementally constructed and revised. Developing these descriptions using machine-processable languages appears to be a key to providing automated life cycle support. However, three higher-order activities remain [18]:

- articulating the various system descriptions,

- establishing a mapping or set of transformations

between successive system descriptions, and

- evaluating the completeness, consistency, and correctness of the system descriptions vis-a-vis one another.

These activities are by necessity incremental, iterative, and ongoing reflecting many system sub-life cycles. But as we improve our ability to engineer this cycling, a growing array of resources must be provided, new tools will be necessary, new forms or subdivisions of work will emerge, and a more complex organizational and technological infrastructure will appear. With this in mind, we constructed an experimental environment and methodology supporting VLSI system development throughout its life cycle.

A SILICON ENGINEERING ENVIRONMENT

Among the new breed of software tools for VLSI is the "silicon compiler" [2,37]. The idea for using a silicon compiler follows an approach that makes VLSI system development more like simple software development -- "programming". As such, a silicon compiler takes in a silicon program (a high-level circuit description) and translates it into a executable form (a layout), assuming there are no bugs in the program. But, as builders of large software systems realized, the development of complex software systems entails much more than writing and compiling programs; it requires an integrated approach combining software environment, development methodology, and project management strategies [6,18,28,29].

Our silicon engineering environment (SEE) represents such a comprehensive of automated software tools that support the VLSI development life cycle. The SEE tools include:

-- Language-oriented editor generator: Given a language definition, the generator produces a syntax-directed editor for the language. Such an editor prevents the construction of system descriptions with either syntax or compilation errors, by virtue of its "knowledge" of the language being edited. A user constructs code fragments in the language through individual editing commands. The editor produces as output an intermediate representation[1] and a processable system description. If this description is stated in a system specification language (e.g., ISPS [3]), then the description can serve as input to a functional simulator or other prototyping facility [cf. 12]. If the description being constructed is a silicon design, then then a layout without syntax errors (ie, design rule violations) can be produced [cf. 13,14,21]. If a layout code generator is interfaced to this tool, then a simple silicon assembler can be built.

-- Parser Generator: A parser generator takes in a formal definition of a language's syntax, as does the editor generator, and generates a set of parsing tables and driver for that language [1].

-- Module Interconnection and Interface compiler: This compiler processes a system description expressed in a module interconnection language. It supports silicon composition at a "floor plan" level [33,34]. System modules are the compilation unit. The compiler checks the consistency of module interfaces, and signals, and timing sequences. It outputs a module configuration (architectural design) that can be interfaced to a system assembler, or to a graphic display.

-- Relational Data Base Management System: This is a single user, data base management system that supports the relational data model [5]. This RDBMS can be used in either stand-alone or interfaced to other tools to provide file services and data base management. This RDBMS supports a "Query-By-Example" user-friendly screen interface [38] and could be extended to querying graphic layouts "by pictorial example" [10]. In addition, a separate "Office Procedures By Example" interface is implemented which connects the RDBMS system to electronic mail and other office automation facilities [5,38].

-- Configuration Management System: This is a special-purpose file management system that interfaces a file system to a system composition facility [5]. As a file system, it can manage the configuration of files across different versions, consistent with the file contents.

-- Language-Oriented Testing System: This testing system takes in compilable code in a language defined externally. Given a program, it generates assertions about the code's behavior (ie, static analysis) which are added to the code locally, then checks them against the program's execution (dynamic analysis) with selected test data inputs [35]. Test data sets can be generated through either extension to a syntax-directed editor, or through the testing system itself.

[1] this representation is in the form of an attributed syntax tree, rather than "CIF" which is a lower-level form.

-- Flow Analyzer: The flow analyzer processes the intermediate representation that describes system behavior. This tool is usually interfaced to a syntax-directed editor or parser to provide its input. It checks for the completeness of the code (control-flow analysis) as well as the consistency (data-flow analysis) of the use of timing signals and other variables [24,25]. It can also be interfaced to either a code generator to support code optimization [25], or to a RDBMS for traceability analysis or program explanation [7].

-- Shell Processor: The shell processor provides the same functionality as does the one operating in the UNIX(tm) programmers workbench. It allows users to build shells containing programs that can communicate via "pipes" [20].

-- Macro-Preprocessor: The macro-preprocessor supports a user's desire to extend the surface form of a language-based system description [e.g., 36], without altering the internals of the target compiler. It also supports a conditional compilation facility [11].

-- General-Purpose User Interface: This "front-end" allows a user to run multiple programs/tools through a dual-window display manager and command processor [8]. It also maintains a log file of all user commands.

-- Documentation Facility: This facility supports the definition of program or documentation "templates". The data base routines interpreting these templates help to maintain the configuration of structured system development documents. A report generation subsystem allows any subset of template elements to be selected and assembled into a formatted document.

-- Layout Code Generator: The role of this tool is to translate our high-level intermediate representation into a low-level form (e.g., CIF NMOS). To achieve greater flexibility and host processor independence, a table-driven code generator-generator [17] is under consideration but not yet developed.

A key feature of this SEE is its use of software tool modules. The tools can be used independent of each other, or configured together as tool sets (ie, tool assemblies). For example, if we wanted to construct a simple silicon compiler, we would first define a silicon programming language, then interconnect a language-oriented editor or parser, flow analyzer, semantic analyzer (hand-built), and layout code generator. As such, the ability to configure an integrated tool set to support different VLSI life cycle activities is under user control and facilitated through common interfaces and intermediate representations.

These tools represent more than 40K lines of PASCAL targeted to both DECsystem-10 and VAX-VMS systems, and about 4K pages of development documentation. However, it is fair to say that none exist in a production-quality form. Rather, they represent second generation prototypes (except for the layout code generator). But the tools are sufficiently developed to support experimentation and classroom investigations in the use of silicon engineering environments.

SUPPORTING VLSI SYSTEM DEVELOPMENT WITH A SEE

Language-based development methodologies are emerging throughout the VLSI community. Language-based methodologies and tools exist which support particular life cycle activities: for example, ISPS and RTL-like languages are used for describing a system's functional specification [3,25]; SAGA Input Language [34] and Linked Module Abstraction [33] are used to describe and generate system architectural designs; and, Design Procedure Language [4] and ALI [21] are used to describe detailed circuit designs that can be translated into CIF NMOS layouts. Further, verification of circuit functions, logic and timing behavior, and layouts based on language-based (ie, "syntax-directed") testing tools and methodologies are appearing, similar to those supporting software verification [cf. 16,22]. However, a set of language-based techniques and tools are yet to be brought together to support the complete system life cycle. But, using a language-based formalism of some kind, VLSI development methodologies and procedural support tools can be integrated to more effectively support system engineering.

We intend to develop VLSI systems by following a development methodology that utilizes language-based system descriptions for each life cycle activity with a language-directed SEE. The tools we choose are, in general, "language-independent": they can be specialized to a system description language (e.g., extended PASCAL) of the engineer's choice so long as it can be defined within the tool's language range. This means it is possible to generate a silicon tool set that can process alternative language-based descriptions for a particular system engineering application. Similarly, it is possible to specialize a set of tools to support a single graphic or text language, as well as to specialize a family of tools supporting system descriptions occurring at each life cycle stage.

Most of the tools are table-driven and can be specialized to support different language-based VLSI descriptions through language initialization procedures. This initialization is done in the same manner as giving a formal language definition to a common table-driven parser generator to produce a parser[2] for that language [1]. Using this table-driven approach, VLSI system life cycle products such as system requirements, specifications, architectural designs, detailed designs, testing plans, executable layouts, and structured diagrams can be processed as long as they can be described via the language definition formalism. Subsequently, these kinds of descriptions are amenable to automated manipulation supporting system evolution [cf. 7].

Tools in the SEE can be specialized to "process" the various system descriptions. This processing includes support for syntactically correct description construction, parsing the description into a testable intermediate data flow form [cf. 24,25], flow analyzing and optimizing this form, functionally simulating system behavior derived from this form, and querying static or simulated relationships expressed in a description.

PROGRESS TO DATE

One of the principal goals of the system factory project is to develop and gain experience in using an advanced SEE, development methodology, and project management strategies [30]. All of our work was performed by more than 50 Computer Science graduate students here at USC. The prototype SEE described above was developed and redeveloped through two complete life cycles (generations) as of late 1982. The SEE actually represents a <u>language-independent software engineering environment</u> that supports silicon life cycle engineering through the use of hardware and circuit description languages [29]. One environment can through extension and careful choice of languages, support the development of both software and VLSI systems [18]. Thus, our current efforts focus on comparing, developing, and evaluating languages, and language-based design methodologies supporting VLSI development.

In parallel to VLSI language investigations, we are planning to redevelop the SEE tools. Two reasons underlie this. First, we are moving the SEE from a mainframe-based configuration to a distributed network of personal workstations and VAX-750 midiframe. This move will allow us to investigate improved graphics, local processing, and networking facilities to support the SEE. Second, we want the SEE to further support an iterative, incremental, and ongoing mode of VLSI system development. We envision a VLSI development process whereby activities performed later in a system's life cycle (e.g., design verification, testing, rules checking, and simulation) are incrementally performed as the system design emerges. This is yet another strategy currently under investigation in the realm of software engineering [15,26].

Finally, we have to date not attempted to produce any VLSI chips. Our efforts were directed to developing an advanced capability supporting computer-aided, life cycle engineering of VLSI systems. Getting hastily produced, small circuit layouts fabricated was not a motivation. However, we do hope to begin to produce interesting VLSI circuits using the SEE, language-based life cycle methodology, and project management strategies next year.

CONCLUSIONS

This paper describes components of the system factory approach to VLSI system development. Supporting VLSI system development throughout its life cycle is critical. Currently available tools such as silicon compilers provide only partial support. A silicon engineering environment was presented which is targeted to provide life cycle support. In order to do this, a language-based VLSI system development methodology rooted in the use of the SEE was described. Finally, the progress made in pursuing this approach to VLSI system development should convince the reader that language-based environments and methodologies represent an emerging thrust in VLSI system research.

---

[2] We use an augmented-BNF notation to describe the syntax and operational semantics of a language as is common in the construction of software compilers [1].

ACKNOWLEDGEMENTS

Preparation of this report was supported by a grant from the USC Faculty Research and Innovation Fund.

REFERENCES

1. Aho, A.V., and J.D. Ullman; <u>Principles of Compiler Design</u>, Addison-Wesley, Reading, MA (1979)

2. Ayers, R.; <u>VLSI: Silicon Compilation and the Art of Automatic Microchip Design</u>, Prentice-Hall, New Jersey, (1983).

3. Barbacci, M.; "Instruction Set Processor Specifications (ISPS): The Notation and Its Applications," <u>IEEE Trans. Computers</u>, Vol. C-30(1), 24-40, (1981)

4. Batali, J., N. Mayle, H. Shrobe, G. Sussman, and D. Weise; "The DPL/Daedalus Design Environment," in J. Gray (ed.), <u>VLSI 81</u>, Academic Press, 183-192, (1981)

5. Bennett, J.; "A Database Management System for Design Engineers," <u>Proc. ACM/IEEE 19th Design Automation Conf.</u>, 268-273, (1982)

6. Boehm, B.; "Seven Basic Principles for Software Engineering," <u>J. Software and Systems</u>, Vol. 3(1), 3-24, (1983)

7. Brown, J.C. and D.B. Johnson; "FAST: A Second Generation Program Analysis System," <u>Proc. 3rd. Intern. Conf. Software Engineering</u>, 200-206, (1978)

8. Carlson, E.D. and W. Metz; "A Design for Table-Driven Display Generation and Management Systems," IBM report RJ2770(35325), IBM Research Laboratory, San Jose, CA (1980)

9. CAS; "Special Issue on Computer-Aided Design for VLSI," <u>IEEE Trans. Circuits and Systems</u>, Vol. CAS-28(7), (1982).

10. Chang, N.S. and K.S. Fu; "Query By Pictorial Example," <u>IEEE Trans. Software Engineering</u>, Vol. SE-6(6), 519-524, (1980)

11. Comer, D.; "MAP: A PASCAL Macro Preprocessor for Large Program Development," <u>Software - Practice and Experience</u>, Vol. 9, 203-209, (1979)

12. Davis, A.M.; "The Design of a Family of Application-Oriented Requirements Languages," <u>Computer</u>, Vol. 15(5), 21-28, (1982)

13. Davis, T., and J. Clark; "SILT: A VLSI Design Language," technical report 226, Computer Systems Laboratory, Stanford Univ., Stanford, CA (1982a)

14. Davis, T. and J. Clark; "YALE User's Guide: A SILT-Based VLSI Layout Editor," technical report 233, Computer Systems Laboratory, Stanford, Univ., Stanford, CA (1982b)

15. Feiler, P.H. and R. Medina-Mora; "An Incremental Programming Environment," <u>IEEE Trans. Software Engineering</u>., Vol. SE-7(5), 472-482, (1981)

16. Foster, M.; "Syntax-Directed Verification of Circuit Functions," in H.T. Kung, B. Sproull, and G. Steele (eds.), <u>VLSI Systems and Computations</u>, Computer Science Press, Rockville, MD, 196-202, (1981)

17. Ganapathi, M., C. Fischer, and J. Hennessy; "Automatic Compiler Code Generation," technical report 225, Computer Systems Laboratory, Stanford Univ., Stanford, CA (1982)

18. Katz, R., W. Scacchi, and P. Subrahmanyam; "Development Environments for VLSI and Software Engineering", <u>Journal of Systems and Software</u>, (to appear)

19. Kling R. and W. Scacchi; "The Web of Computing: Computing Technology as Social Organization," in <u>Advances in Computers</u>, Vol. 21, 3-85, Academic Press, New York, (1982)

20. Kernighan, B.W. and J.R. Mashey; "The UNIX Programming Environment," Software - Practice and Experience, Vol. 9, 1-15, (1979)

21. Lipton, R., S. North, R. Sedgewick, J. Valdes, G. Vijayan; "ALI: a Procedural Layout Language to Describe VLSI Layouts," Proc. ACM/IEEE 19th Design Automation Conf. 467-473, (1982)

22. Moriconi, M.; "A Designer/Verifier's Assistant," IEEE Trans. Software Engineering, Vol. SE-5(4), 387-401, (1979)

23. Mudge, C.; "Components of a Silicon Compiler," in J. Gray (ed.), VLSI 81, Academic Press, NY (1981)

24. Munchnick, S.S. and N.D. Jones; Program Flow Analysis, Prentice-Hall, New York, (1981)

25. Parker, A. and L. Hafer; "Automating the Design of Testable Hardware", in J. Gray (ed.), VLSI 81, Academic Press, 357-363, (1981)

26. Ryder, B.; "Incremental Data Flow Analysis," Proc. Tenth Principles of Programming Languages Conf., 167-176, (1983)

27. Scacchi, W.; The Process of Innovation in Computing: A Study of the Social Dynamics of Computing, Ph.D. dissertation, Dept. of Information and Computer Science, University of California, Irvine, CA (1981)

28. Scacchi, W.; Managing Software Engineering Projects: A Social Analysis," TR 82-116, Computer Science Dept., University of Southern California, Los Angeles, CA (1982)

29. Scacchi, W.; "A Language-Independent Software Engineering Environment," Proc. IEEE Workshop on VLSI and Software Engineering, Rye Town, NY, (October, 1982) (to appear).

30. Scacchi, W.; "Developing a System Factory: A Case Study," technical report, Computer Science Dept, University of Southern California, Los Angeles, CA (in preparation)

31. Scacchi, W., L. Gasser, and E.M. Gerson; "Problems and Strategies for Organizing Computer-Aided Design Work," IEEE Intern. Conf. Computer-Aided Design, San Jose, CA (September, 1983), (to appear)

32. Sequin, C.H.; "Managing VLSI Complexity: An Outlook," Proceedings IEEE, Vol. 71(1), 149-166, (1983).

33. Stefik, M., D. Bobrow, A. Bell, H. Brown, L. Conway, and C. Tong; "The Partitioning of Concerns in Digital Systems Design," Proc. of the Conf. on Advanced Research in VLSI, 43-52, (1982)

34. Szepieniec, A.A.; "SAGA: An Experimental Silicon Assembler," Proc. ACM/IEEE 19th Design Automation Conf., 365-370, (1982)

35. Voges, U., et al.; "SADAT--An Automated Testing Tool," IEEE Trans. Software Engineering, Vol. SE-6(3), 286-290, (1980)

36. Wakerly, J.; "PASCAL Extensions for Describing Computer Instruction Sets," Computer Architecture News, Vol. 8(7), 15-23, (1980)

37. Werner, J.; "The Silicon Compiler: Panacea, Wishful Thinking, or Old Hat?", VLSI Design, Vol. 3(5), 46-52, (1982)

38. Zloof, M.M.; "QBE/OBE: A Language for Office and Business Automation," Computer, Vol. 14(5), 13-32, (1981)

A Production VLSI Design CAD System

Kris Vorm

General Electric, Aerospace Controls System Department

1. ABSTRACT

The design and efficient use of a VLSI CAD system today is, at best, a three dimensional problem. Software cost, software functionality and design methodology issues are interwoven to produce a nearly fluid and everchanging garden from which to harvest a CAD system solution. New software functions, new hardware abilities, and technology advances often cause a CAD system to become obsolete before its learning curve and design methodology have been surmounted. Our VLSI CAD system is based on three levels of simulation (register, gate, and transistor), gate array and standard cell automatic placement and routing, parallel fault simulation, static testability analysis, geometrical design and electrical rules checking, and a CAD executive which controls the operation of each program in the system and performs some document control functions. The CAD system was constructed with a "buy-if-you-can", "build-if-you-must" philosophy. The system is library driven and has a graphical engineering work station front end.

2. INTRODUCTION

This paper will present the General Electric, ACSD (Aerospace Controls Systems Department) VLSI CAD system and its use in the design of 1.25 micron CMOS custom chips. The paper is divided into four sections, a software description of the CAD system from a structural standpoint, a design methodology description, a description of each program in the system, and a summary.

Our VLSI CAD system was assembled by LSI chip designers who had more experience in the area of CMOS chip design than in computer science. The particular CAD programs were chosen both for their functionality and their availability. Issues such as host computer compatability, software maintenance, on-going enhancement, and user training dominated the CAD program selection decisions. We found that the programs, wonderful as they were, required significant amounts of integration "glue" and design methodology attention. The glue software proved to be more expensive than the main CAD programs in the system. Even after the glue was in place, choosing a design methodology which fit the CAD system, the department's product design time schedule, CMOS technology's flexibility requirements, and the available manpower experience level was far from automatic.

3. CAD System Architecture Description

Our VLSI CAD programs are surrounded and controlled by a CAD Executive which runs a VAX 11/780, several Apollo Domain engineering work stations, and remotely, a CALMA GDS II graphics system. The VAX performs all large simulations and heavy number crunching jobs, while the Apollos act as front end work stations for block diagram capture, register transfer level simulation, small gate level simulation, and test description language processing. Calma's GDS II graphics system forms the back-end of the CAD system for physical design. In general we run small jobs on the small machines and large jobs on the large machines, so, several CAD programs reside on both the Apollo and the VAX computers.

The CAD software architecture was built around an internally written CAD executive which performs the user interface, calls and executes the CAD programs themselves and provides some project security and document control functions. Two entry mediums are available, block diagram capture on the Apollo, or direct textual input on a VAX terminal. The design's structure (netlist and instantiation list) is captured in a hardware description language called HDL. All netlist driven programs derive their inputs from this description. The graphical entry programs available on the Apollo, however, provide their own netlist generation capability, eliminating the need for HDL, so, an HDL description for each graphically entered design is produced for documentation purposes after the design modification cycle is complete.

Three levels of simulation are supported: register transfer, gate level, and transistor level. There is a need for

CH1935-6/83/0000/0476$01.00 © 1983 IEEE

a multi-level design description language with suitable simulation abilities, but no such language has passed the economic, functional, and integration tests. A simulation based register level to gate level description comparison program was written as a short term solution until a multi-level language becomes available in early 1984.

Two home-grown placement and routing programs are connected in parallel with identical capabilities. Both are library driven channel routers, one for standard cell designs and the other for General Electric's (Intersil) gate array designs. Both routers provide run length information to an internally written program called RC-Delay, which calculates the added media delay for each signal path and adjusts the circuit's gate level description for final simulation.

4. Design Methodology

This VLSI CAD system is used for diverse design areas from radar signal processing to small appliance controllers. Each area produces exceptions and special cases from the originally intended use of the CAD programs. This section describes the design methodology used by ACSD in the design of VLSI CMOS chips.

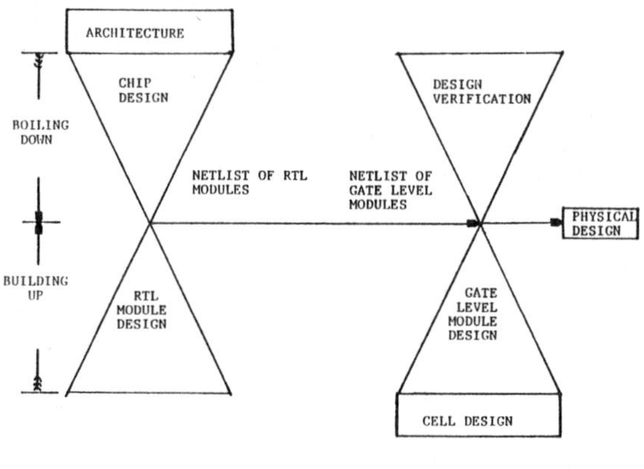

FIG. 1: VLSI DESIGN FLOW

Figure 1 shows the design flow in six pieces. The left hand butterfly contains the architecture, chip design, and register level macro design tasks, and the right hand butterfly shows the design verification, gate level macro, and cell design tasks. Fabrication and testing are not included in this discussion.

The architecture analysis task deals with system specification, throughput analysis, fault tolerance issues, and other pre-design aspects in the overall system. This analysis may have been performed during a previous LSI implementation. As the system is partitioned into smaller subsystems the design flows into the upper half of the left hand butterfly.

Top-down chip design is performed using ISPS and the PMS interconnection language. In the early design stages, many chips look like programmer's flowcharts or algorithms diagrams. A high level ISPSmodel can easily be generated to simulate these algorithms. The designer's task is to translate his design from flowchart to register transfer logic form. As logic blocks begin to take shape, the work station's block diagram capture software is used to show the structure of the chip. Blocks of ISPS code are written to correspond to the logic boxes on the block diagram. These boxes are then interconnected automatically using PMS, and presented to the designer for ISPS simulation. Simulation results are compared to those from a higher level description to insure that the system's function has not changed during the boiling down, or partitioning process. The target in this process is two fold, a chip description in the form of a netlist of modules, and a standard ISPS library of available, logic blocks which corresponds to previously tested digital circuits, such as adders, multipliers, barrell-shifters, and small controllers. If the designer partitions the chip into a logic block which is not in the ISPS library, he must prepare a specification for it and initiate the RTL module design task in the lower half of the left hand butterfly to design and enter it into the ISPS library.

The job of the RTL module designers is to insure that the logic within a module can be constructed from available gate level cells and macros. When completed he deposits the new module into the ISPS library.

The chip design task is complete when the center point of the left hand butterfly in figure 1 is reached. The design, at this point, is a PMS description, automatically generated from the captured block diagram. It defines a netlist of interconnected logic blocks which can be built from existing cells and macros. Simulation has been performed at each design level in this boiling down process to insure correctness of the logic function.

The transition from the register level chip description to the gate level chip description is made with the push of a button. This transformation usually requires a logic synthesizer, an RTL to gate level translator, or manual recoding into a new language. ACSD's CAD system utilizes the work station's ability to generate hardware descriptions in any target language from the captured graphical block diagram. The gate level description language is substituted for the register level language. Figure 1 shows this transformation with the arrow connecting the two butterflies. The logical sameness of the two libraries is mandatory and is insured through the use of the ISPS to Tegas comparator program which simulates both module descriptions in their respective simulators, and compares their results at a common module boundary.

The design is now in the gate level verification realm, which includes timing analysis, testability analysis, and test vector coverage grading. (upper right hand butterfly in Figure 1) Timing analysis is presently performed by simulation until a static timing analysis program can be located and incorporated. Our CMOS circuits are designed to have symmetrical rise and fall times, which reduces the need for worst case simulation, but it is still useful during media delay effects simulation. The digital circuit description includes three delay quantities, the intrinsic propragation delay of each logic cell, the additional delay due to the input capacitance of each fanout, and the delay caused by the cell to cell interconnections, called media delay. Our CAD system allows the designer to model the circuit using only the cell delays from the schematic library. He next executes the DECAL program which automatically calculates and inserts the circuit's fanout delays. A statistical analysis of interconnect length, based on previous layouts, shows a bell shaped distribution of run length. Minimum and maximum delays are generated to reflect the three sigma below and above on the media delay curve. Upon simulation, the circuit will be exercised with short media delays and with long media delays, which produces a more accurate circuit model.

Static testability analysis is performed to allow test point and logic modification to the gate level macros to correct any testability trouble spots.

Finally, the circuit netlist is flattened by the gate level language compiler, and written out to a file which is made up of several linked lists, containing all of the information in the source description. This file is easily machine readable and can provide a netlist to any hierarchical level. A home-grown translator program extracts the netlist and reformats it for the automatic placement and routing programs, described below. Exact metal and poly lengths are reported by the router program and used in an RC Delay extraction program. This home-grown program converts the extracted geometrical information to delay values, and re-inserts them into the gate level description. Final simulation is then accomplished with all three delay parameters present.

5. CAD Tool Description

This section describes many of the individual programs in ACSD's VLSI CAD systems.

Register Transfer Simulation - Carnegie Mellon's ISPS register transfer language provides a flexible and robust hardware description and architecture evaluation capability at minimal cost. The ISPS interactive simulator has been greatly enhanced to include several useful and time saving I/O features. The PMS structure-only language interconnects ISPS or other PMS descriptions to create a larger PMS hardware description for architecture evaluation.

Gate Level Simulation - Comsat General's TEXSIM gate level simulator family comprises the heart of the VLSI CAD system; TEGAS 5 for bipolar circuits, TEXSIM for CMOS circuits, and TEXMOS for CMOS or NMOS circuits. TEXSIM is a four valued switch level simulator with several added features, such as the ability to retrofit added media delays to the circuit model without recompiling the source. TEGAS and TEXSIM share the same compiler, but TEXSIM allows the inclusion of several bidirectional MOS primitives. TEXMOS is a sixteen valued simulator capable of modelling passive pullups and bidirectional MOS transfer gate circuits. TEXMOS also contains a Timing Constraint Language used to describe the allowable setup and hold time parameters in the circuit. The ability to model signal decay after a specified time is also included in TEXMOS. TEXSIM and TEXMOS will be combined with TEXSIM B (behavioral) to form one simulation package in early 1984. TEGAS 5 and TEXMOS are executed on the VAX 11/780, while TEXSIM is executed on both the Apollo work station and the VAX.

Transistor Level Simulation - Berkeley's SPICE program's transistor model must be modified to match the particular CMOS process to be used in the fabrication of a chip. In many cases, silicon foundry companies are not willing to divulge the details of their process, which makes transistor level simulation very difficult. Several cell libraries, custom I/O circuits, and many analog circuits have been successfully simulated at ACSD, using Spice. An internal program called CELCAP was devised to extract a Spice circuit model directly from the Calma geometrical artwork. In the past, a designer measured the area of each polygon to calculate the amount of capacitance to assign in the Spice circuit description. This was obviously very time consuming and error prone. Celcap is now used on all transistor layouts for digital cells, custom macros, and analog designs.

Automatic Placement and Routing - VLSI chips with 20k to 30k gates are layed out using macro blocks, each of which might be a standard cell layout of a portion of the overall chip. The penalty for using present automated placement and interconnect programs is area inefficiency. More efficient routing is certainly an important issue, but designs must continue in the meantime. A new, internally written, four-sided block router will be installed in the CAD system before the end of 1983. Many router restrictions are emerging as the geometries shrink. For example, parallel runs may produce crosstalk, and one bus wire may not run directly under another to minimize capacitive coupling. Efficiency is no longer the only problem facing the automatic channel router today. Presently, our CAD system has the ability to interconnect blocks of standard cells as if they themselves were standard cells. This automation produces a correct layout, without manual intervention. The engineering work station is a prime candidate for an interactive placement floorplanning and routing program, since it has the necessary CPU and graphics capabilities.

Engineering Work Station - Block diagram capture, RTL simulation, and TEXSIM simulation are performed on an Apollo engineering work station, running Comsat General's TEGASTATION software. Hierarchial block diagram capture has become an important function in VLSI design. The higher level diagram has become the central design point, from which submodules are assigned to individuals or teams of designers. The work station's graphics capabilities are also a welcome sight in the engineering environment. TEXSIM's simulation results are viewed graphically on the Apollo, using CGIS's TSCOPE program, a logic analyzer mode output post processor.

Circuit Libraries - As each circuit block is described, it is stored in the work station library. Eventually an ISPS and a TEXSIM description exist for the same block. A program named CIT, (Comparator ISPS to Tegas) was written to logically compare the two models at their module boundaries. CIT was written as a short term solution, pending a mature static timing analysis program and a multi-level, simulatable, HDL. This step will be removed when CGIS's behavioral level language, TEXSIM B, is included in the CAD tool set, since TEXSIM and TEXSIM B modules will be compatible and simulatable in a single simulator. CIT is a subsystem which envokes both ISPS and TEXSIM, runs the simulations on identical input stimuli, performs signal and pin name mapping, and finally compares the simulator outputs to report any differences. The outputs are strobed in the same way an actual device under test would be strobed on a tester. This comparison relies on the use of good test vectors, an expensive commodity which is hard to acquire. Static timing analysis programs will be selected and incorporated into the CAD system by early 1984.

Physical Design - In the standard cell approach, the interconnect capacitance between the cells adds a different amount of delay to each signal path in the fanout. This "media delay" is calculated from the CALMA artwork by an internal program named RC Delay, and presented to the TEXSIM simulator to prove the circuit still meets the timing specifications. TEXSIM's TCF file is used to numerically add these delays to the circuit description, allowing delays to be changed without recompiling the source code. The circuit can then be resimulated to verify its worst case delay operation including the interconnect capacitance loading effects.

Testability Analysis - COPTR which stands for Controllability, Observability, Predictability, and Testability Reports, provides a tool to analyze the testability of a chip without simulation. COPTR creates a table containing relative merit numbers, assigned for sequential controllability to the "one" and "zero" state, combinational controllability to the "one" and "zero" state, etc., for each circuit node. The COPTR table is used in two places; first, it is used by the designer identify trouble spots in the design before simulation. Second, the COPTR table is used by Comsat General's new automatic test vector generation program, ATG, to resolve path sensitizing conflicts during test vector generation. This added "intelligence" results in better test vectors for more complex circuits. In general, automatic test vector generation may not survive the move into VLSI, except when additional fault coverage is needed on an LSI size circuit.

6. Summary and Conclusions

ACSD has designed and tested a VLSI CAD system for use in 1.25 micron CMOS VLSI chip designs. The simulator programs and the computer equipment where purchased, while the CAD Executive, Automatic placement and routing, media delay extraction, parasitic capacitance extraction, and dozens of "glue" translators were written internally. CMOS chips are designed more sensibly, with lower risk, due to advanced features, including engineering work station front ends, a central CAD executive, and multiple levels of simulation. Our libraries are verified automatically through the use of comparitor programs and central data bases. Our software reflects the high quality expected when experienced designers from diverse areas of a large corporation combine their talents to produce a usable CAD system.

Session 2-5a
VLSI Design

Chairman

J. Jess
Einhoven University

"A Hierarchical Layout to Circuit Extractor using a Finite State Approach"

J. Annevelink, P. Dewilde, J.T. Fokkema

Delft University of Technology

ABSTRACT

The paper describes a layout-to-circuit extraction program that is based on an hierarchical description of the layout data and uses a new, finite state scanning algorithm to recognize the boundaries of the elements that are part of the target description. The method is independent of the target description which may be a symbolic description, an electrical circuit, or a switch-level description. The method used is illustrated by an implementation for NMOS-circuits.

Introduction

The growing complexity of IC-designs makes the development of hierarchical CAD-tools necessary ([1], [2], [6]). This paper describes an hierarchical layout to circuit extractor. The program exploits the hierarchy and repetition present in the layout description of an IC. By doing this the run_time and memory requirements are not longer a function of the number of layout features of the complete IC, but only of the number of features present in the individual models.

The algorithms developed are general in that they allow the extractor to recognize any kind of electrical function. The electrical functions are expressed in element conditions (EC's), which can be compared with the Boolean expressions used to derive new functional layers in other types of extractors ([3], [4], [5], [6]). The EC's form the basis of the boundary recognition algorithm (BRA), on which the extraction algorithm is based.

Method

The extractor operates on a description of the layout as sorted edge_files. These edge_files are derived from the model description by hierarchically instancing it ([7]). The hierarchical model instance is independent of other model instances. This is possible because models may be interconnected only via so_called terminals. A terminal is like an ordinary rectangle, the difference is that it can have a name and that it shows a place where layout features of different models may overlap. Terminals are represented by edge_files as if they are ordinary polygons. Thus the edge_files represent individual process masks, but also convey electrical interconnect information. A polygon can be represented by a set of vertical edges. Since we consider paraxial layouts only the vertical edges suffice to describe the polygon. An edge is represented by a record consisting of five parameters, resp. the X or horizontal value, the Y or vertical values, the edge_type and the group_number. Basically we consider two edge_types, START and STOP. We speak of a START edge when the interior of the edge is situated on the right side of the edge. Conversely we speak of a STOP edge when the interior is on the left.
The group_number of an edge relates the edges belonging to the same polygon. For a terminal edge_file it also relates a polygon with a terminal.

From the edge_files the extractor makes stateruler_profiles. A stateruler_profile is a vertical transverse of the layout. It consists of a linked list of fields, characterized by a bottom and top value and a set of layer_states. The bottom and top values of the field are non_overlapping, while the profile is sorted on the bottom value. The layer states are a set of Boolean variables, giving the past and future state of each layer. (Note that the number of

CH1935-6/83/0000/0481$01.00

layers is equal to the number of edge_files) The state of a layer is TRUE when the indicated layer is present, or FALSE when the layer is absent. The past state of a layer is the state of the layer between the previous profile and the current profile, the future state is the state of the layer between the current profile and the next profile. We say that a field or a layer has changed state if the past and future states are not equal.

The stateruler profiles are scanned by the boundary recognition algorithm. For each field that has changed state this algorithm determines whether the field forms a boundary edge of an element. If the field is part of the boundary of more than one element we have found an interconnection between the elements. In general this will result in the definition of a network node.

If a boundary edge is recognized the boundary recognition algorithm performs an appropriate action to update the description of the network element from which the edge is part. The element that must be updated depends on the environment of the field in the stateruler_profile. The connection between the stateruler fields and the elements is made with edge_descriptors. The elements themselves are represented with element descriptors, linked together in element lists.

We conclude that by scanning the entire layout, making successive stateruler profiles and scanning these profiles as well, the extractor determines all network elements, as well as their connectedness.

The Boundary Recognition Algorithm

The boundary recognition algorithm (BRA) is based on the fact that network elements are formed when a certain combination of layers (process masks) is present over a certain area in the layout. This is expressed by the element condition (EC). The EC states the layers that must be present or absent to form a network element. The EC takes the form of a Boolean expression, the variables representing individual layer_states. If the layer is present the variable is true, and when the EC is true the element it describes is present. For example, the EC of a MOS enhancement transistor will be : "polysilicon_state and diffusion_state and not depletion_implant_state and not buried_implant_state and not crossing_implant_state".

To find the boundary edges of an element we will have to compare the future and past layerstates of a stateruler field as described in Section 2. If the future layerstates satisfy a particular EC, and the past don't, we have found a boundary edge, here a START edge, of the element described by the EC. Conversely we will find a STOP edge if the past layerstates satisfy the EC and the future layerstates don't.

The BRA determines whether an EC is satisfied by the past and/or future layerstates and sets a number of flags to show the edges that are found.

Structure of the Extractor

A simplified algorithm of the routine 'extract' is given in Figure 1. To make a full extraction of a model this routine should be called once for every model called in it. From the algorithm we see that a stateruler_profile is made by repeatedly calling select_edge and insert_edge.

```
procedure extract(model_name)
   model_name : STRING of char ;
begin
   edge_status  : integer ;
   edge         :
      record of
         y_low, y_high : integer ;
         group_no      : integer ;
         edge_type     : integer ;
         layer_code    : STRING of char ;
      end ;
   edge_pos      : integer ;
   sr_pos        : integer ;
            [ x_value of stateruler_profile ]

   Initialize local variables ;
   select_edge(edge_status, edge, edge_pos) ;
   sr-pos = edge_pos ;
   while( edge_status is not EOF) do
      begin
         repeat [ make a stateruler_profile ]
            insert_edge(edge, sr_pos) ;
            select_edge(edge_status, edge,
                        edge_pos) ;
         until(edge_status is EOF or
               edge_pos is not sr_pos)
         extract_profile(sr_pos) ;
         sr_pos = edge_pos ;
      end
end [ procedure extract ]
```

Figure 1 : Algorithm of routine 'extract'.

Each profile in turn is scanned by the routine extract_profile. This routine implements the boundary recognition algorithm, and calls the network_element update routines. The network_element update routines create and update the network data structure. The network data structure is given in Figure 2.

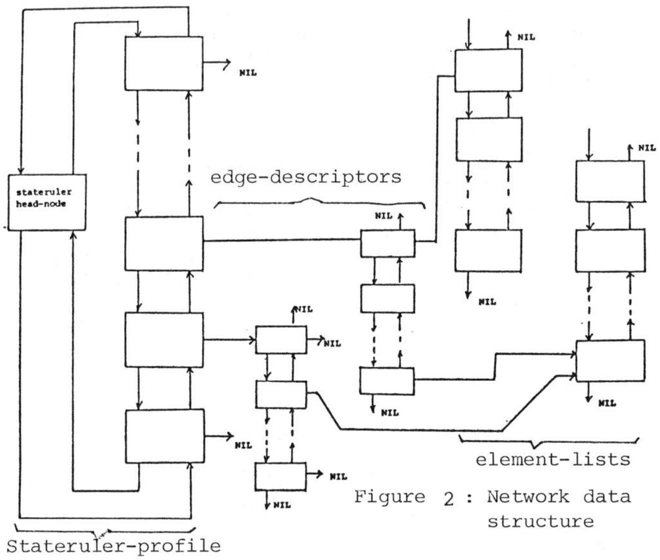

Figure 2 : Network data structure

A stateruler field contains a pointer to a list of edge_descriptors. An edge_descriptor has a type, e.g. poly_interconnection or enhancement_transistor, and contains a pointer to an element descriptor of the indicated type. An edge_descriptor also contains information concerning the boundary edge which caused it to be created. From this information the network_update routines determine the geometrical parameters of the element. The network itself is represented with element lists, consisting of element_descriptors. For each type of element the extractor recognizes there is a seperate element list.

Besides creating and updating the network data structures the update routines also create network nodes. The network nodes are part of a datastructure that reflects the topological structure of the layout. To this end the network nodes form classes and groups. A group contains all the network nodes that originate from a certain polygon in the layout description. Groups are divided in classes as a result of the intersection of layers. For instance in a MOS-process a diffusion group is divided into classes by the intersections with the polysilicon layer. We remark that the information necessary for determining the locality of elements is contained in the structure of the node_trees. This can be of importance for relaxation type simulators.

Implementation and Results

The algorithms discussed in the previous sections have been used to make a program that converts a hierarchical layout description into a network description that contains the same hierarchy.
For the current implementation the network description contains transistors and network references only. An example is the network description of a simple 4 bit register model, consisting of 4 latches as given in Figure 3. This network description can be used as input for a simulator, e.g. a switch-level simulator ([8]).

```
network register(in, out, phi1, phi2, vss, vdd)
  terminal in[1 .. 4], out[1 .. 4],
           phi1, phi2, vss, vdd ;
begin
  reg[1] := latch(in[1], out[1], phi1,
                  phi2, vss, vdd) ;
  reg[2] := latch(in[2], out[2], phi1,
                  phi2, vss, vdd) ;
  reg[3] := latch(in[3], out[3], phi1,
                  phi2, vss, vdd) ;
  reg[4] := latch(in[4], out[4], phi1,
                  phi2, vss, vdd) ;
end
```

```
network latch(input, output, phi1, phi2, vss, vdd)
   terminal input, output, phi1, phi2, vss, vdd ;
begin
   nenh 8       vss       6 ;
   nenh 9       vss       output ;
   nenh 5       vss       8 ;
   nenh phi1    input     5 ;
   nenh phi2    5         6 ;
   nenh phi2    8         9 ;
   dep  output  output    vdd ;
   dep  8       vdd       8 ;
   dep  6       6         vdd ;
end
```

Figure 3 : Example of a network description.

To show the correspondence with the layout description we have included a schematic representation of it as Figure 4. Figure 4 shows that the layout description of the 4 bit register model contains a reference to a latch model. This model reference is given an 'instance_name', in this case 'reg4'. The instance name is used to distinguish different references to the same model. The terminals given in the layout description can be found back in the network description too.

The extractor optionally determines the W and L of the transistors, as well as the node capacitances. The user can also specify that the extractor includes the geometric position of the transistors in the network description.

```
model register
 : 4 bit register
term metal xl xr yb yt input copyx dx 3 ;
                          |
                          |

                          |
term metal xl xr yb yt top vdd ;

mod_ref < reg4 > latch copyx dx 3 ;

   other primitives ;

model_end [ model register ]
```

```
model latch
 : S/R flipflop
term metal xl xr yb yt input ;
                   |
                   |

                   |
term metal xl xr yb yt vdd ;

   other primitives ;

model_end [ model latch ]
```

Figure 4 : Schematic layout description of register model and latch.

Table 1 shows the runtimes and memory requirements for an example. The example is a random counter consisting of of two submodels, i.e. an 8 bit shift-register with parallel-in shift select and a modulo-two feedback model. The shift-register in turn consists of 8 identical latches and parallel-in shift select models. The modulo-two feedback consists of 3 identical models making the ex-or function of two input signals.
The time necesarry to extract each of the individual models is given in Table 1. The number of elements mentioned in this table are respectively the number of transistors defined in the model and the number of terminals defined in the model and the models referenced therein..

circuit	elements	CPU-time
latch	10 + 10	2.4
select	7 + 8	1.8
mod2_fb	6 + 8	1.7
feedback	0 + 40	2.0
shift_reg	0 + 220	8.8
rand_cnt	0 + 60	3.3

Table 1 : Hierarchical extraction of random-counter.

If we compare the results we see that most of the time is taken by the extraction of the shift-register model. This model contains a large number of terminals, due to the fact that the latch and select models are repeated eigth times.

The extractor is written in C, and runs under the UNIX operating system on a VAX 11-750 minicomputer.

Conclusions

There are a number of conclusions and remarks we can make after having developed and implemented the algorithms discussed above. The most important are :

1. The models need to be extracted only once. That way we exploit the hierarchy and repetition present in the layout description of a model.

2. The hierarchy in the network description is the same as that in the layout description. This facilitates verification and interpretation of results obtained.

3. The algorithms developed are linear with respect to the number of elements extracted.

4. The characterization of elements by EC's enables us to extract any kind of electrical function, as long as it is based on the presence of a particular combination of masks.

5. By considering element boundaries interconnections are created implicitly whenever two boundary edges coincide.

The algorithms developed so far consider orthogonal rectangles only. Future work will be directed towards generalizing the algorithms, so that we will be able to extract 45 degree edges also.

References

[1] Haydamack, W.J., Griffin, D.J., "VLSI Design Strategies and Tools", Hewlett Packard Journal, June 1981

[2] Baker, C.M., Terman, C., "Tools for Verifying Integrated Circuit Designs", LAMBDA, 4th Quarter 1980, pp. 22 - 30.

[3] Baird, H.S., "Fast Algorithms for LSI Artwork Analysis", Proc. 14th DAC, 1977, pp. 303 - 311

[4] Scheffer, L., Apte, R., LSI Design and Verification Using Topology Extraction, Proc. 12th Asilomar Conf, 1978, pp. 149 - 153.

[5] Noice, D., Newkirk, J., Mathews, R., "A Polygon Package for Analyzing Integrated Circuit Designs", VLSI Design, 3th Quarter 1981.

[6] Newell, M., Fitzpatrick, D.T., "Exploitation of Hierarchy in Analyses of Integrated Circuit Artwork", IEEE Trans. on CAD, VOL. CAD-1, No. 4, Oct. 1982.

[7] Annevelink, J., Leuken, R. van , Fokkema, J., Hierarchical Verification of Integrated Circuit Artwork based on a Stateruler Scan Algorithm, Internal report, 1983, Network Theory Section, Delft University of Technology (to appear)

[8] Bryant, R.E., "An Algorithm for MOS Logic Simulation", LAMBDA, 4th Quarter 1980, pp. 46 - 53.

VLSI-LAYOUT DESIGN REQUIRES A NEW LANGUAGE

Kees van Berkel

Philips Research Laboratories, P.O.B. 80.000, 5600 JA Eindhoven, The Netherlands

ABSTRACT

IC layouts are commonly described in a layout description language (LDL) an, often human readable, data format. It is argued that current LDLs are hardly adequate to cope with VLSI layouts and more recent layout-design techniques, such as floor planning and symbolic layout. VLDL has been designed to overcome these drawbacks. Key features of VLDL are the concept of a **layout module**, providing for locality of consideration and the mix of symbolic and geometric layout. VLDL is implemented in a VLSI-layout design system named Voila.

INTRODUCTION

IBM's GL/1 [1], Philips' CMSK [2], Calma's Canonical Format [3], and Caltech's CIF [4] are examples of layout description languages. They all serve to record mask features and to interface between designers, CAD programs and mask-making facilities. Since these languages are essentially 2-D drawing languages, they have a high degree of technology independence and design-style independence. These two facts have resulted in a life-time of some of these languages already exceeding a dozen years.

With the advent of VLSI layouts the forementioned languages start showing inadequacies. This is primarily due to their inability to support **modular** layout design in order to handle the complexity of VLSI. Moreover, widely accepted design techniques, such as symbolic layout and floor planning, are outside the scope of these languages. Besides that, relations between a layout and its circuit/network can only be established through circuit-extraction programs, which are to be applied **after** completion of that part of the layout. Many CAD programs also suffer from the lack of structure in LDL descriptions of VLSI layouts. Layout-verification programs such as design-rule checkers and connectivity checkers, as well as mask-processing programs such as mergers, growers and preprocessors for pattern generators are becoming exceedingly greedy computer consumers.

Note that the subroutine-like facility in current LDLs, for example CIF's "symbol", does not aid modular design. A symbol provides neither means of **abstraction**, nor **locality** of consideration: a wire of global bus may be declared in the same symbol as a contact hole in of a 1-bit shift-register cell. This subroutine-like facility is merely a notational shorthand, thus reducing the size of a layout text.

We are convinced that VLSI layout requires a new language to overcome these drawbacks. VLDL (Voila Layout Description Language) is such a new language, and will be presented here. VLDL features a powerful **module** concept that allows for **hierarchical** layout design, in order to cope with the complexity. Modules can be used for top-down layout design (often called "floor planning") as well as bottom-up design. Different modules of layout can be designed either symbolically or geometrically, whereas the correspondence of a layout module to its circuit/network module can be expressed within the language.

Modular, hierarchical **layout** design and verification is only one aspect of hierarchical design methodologies and tools for VLSI **circuits**. For an introduction to this more global subject one is referred to Niessen [5].

The embedding of VLDL in the layout-design environment Voila is also discussed. The Voila VAX 780 based work station features, amongst others, fast picture generation, on-line incremental design-rule checking, and on-line incremental connectivity checking. We strongly believe that the module concept will increase the applicability and effectiveness of CAD tools such as routers, PLA generators, optimizers, libraries, etc.

Finally, some promising layout-language constructs, of which the applicability and implementation are not yet fully understood, such as parametrizing of layout modules, expressions denoting values of coordinates, and conditional placements, will be subject of consideration.

REQUIREMENTS FOR A VLSI LAYOUT DESCRIPTION LANGUAGE

Most of the geometry-description facilities of current LDLs should be supported by VLDL. However, at one point the freedom of expression should be restricted: the boundaries of all geometric primitives should consist of line segments that have angles of multiples of 45 degrees with the x-axis. In symbolic layouts, as will be discussed later, these angles should even be locked to multiples of 90 degrees. Here we make a concession to design flexibility and layout efficiency. Nevertheless, we are convinced that otherwise efficiency, functionality and reliability of VLSI CAD programs would be significantly less.

A next requirement is related to the complexity of VLSI layouts. VLDL should provide language constructs that allow abstraction of layout data and locality of consideration. These constructs should support modular, hierarchical layout design, top down as well as bottum up, modular layout verification and modular mask-processing.

Symbolic layout is a generic name for a layout design method with a considerable degree of abstraction from technological details and a reduced degree in expressive flexibility. Implementations of this notion differ widely [7] - [9], and numerous others. A disadvantage of most implementations is the strict separation of symbolic and geometric layout. VLDL should support symbolic layout and should allow for mixing the two forms of layout representation, i.e. it must be possible to place a geometric-layout module (of any complexity) in a symbolic-layout module.

Furthermore, the relations between a layout and the network or circuit it implements should be made explicit in VLDL such, that the consistency between the two can be verified on a modular basis.

DRAWING PRELIMINARIES

In this section VLDL's capabilities at the drawing level are briefly demonstrated. A VLDL layout consists of three essential ingredients: the **declaration of a layout space**, **geometries** and **attributes** that may give these geometries additional meaning.

A layout-space declaration may look like:

```
masks = (metal, poly, diff, oxide);
unit  = 0.5; grid = 0.25
```

This means: all geometries must define areas in one of the listed masks, all coordinates are given in units of 0.5 micron, whereas coordinates are rounded to the nearest multiple of 0.25 micron.

Geometries are constructed out of **geometric primitives**. VLDL knows six types of geometric primitives: dot, orthogonal rectangle, slanted rectangle, wire, polygon and text (Cf. Figure 1).

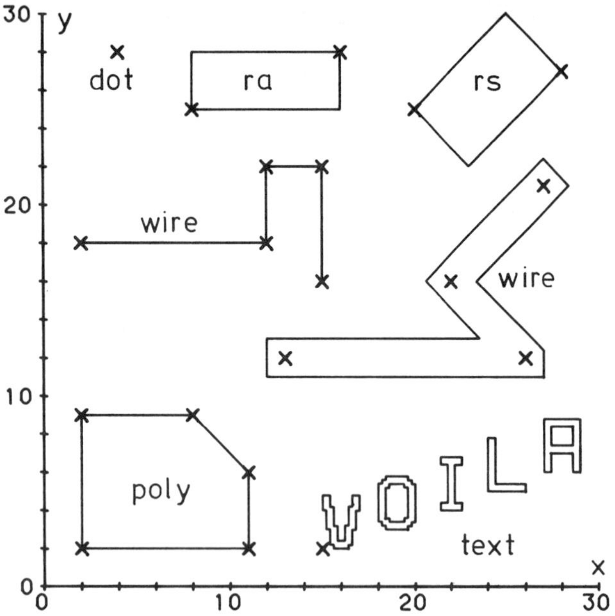

Figure 1.
Examples of geometric primitives of VLDL. The points marked by 'x' are part of the definition of that geometric primitive.

```
dot   (metal;  4, 28);
ra    (metal;  8, 25; 16, 28);
rs    (poly;  20, 25; 28, 27);
wire  (metal;  2, 18; 12, 18; 12, 22; 15, 22; 15, 16);
wire  (diff;    w = 2; 13, 10; 26, 10; 22, 14; 27, 19);
poly  (metal;  2,  2;  2,  9;  8,  9; 11,  6; 11,  2);
text  'VOILA'(poly;  15,  2; size = 3;  3,  1)
```

- A **dot** is defined by a mask and a point. It has no dimensions, and is only meaningful in a symbolic layout (Cf. the section Symbolic Layout).
- An orthogonal rectangle (**ra**) or slanted rectangle (**rs**) is defined by a mask and two opposing corner points.
- A **wire** is defined by a mask, a width (default value = 0; this default value is only meaningful in symbolic layouts), and a list of points that specify its centre line. Acute angles in a wire are allowed, but are blunted.
- A polygon (**poly**) is defined by a mask and a list of all corner points. Again acute angles are permitted as well as self loops. However, its orientation must be either consistently clock wise or anti-clock wise.
- A **text** is defined by a string, a mask, the position of the left--lower corner of the left-most character of the string, the size of a character and a vector specifying the relative position of two succeeding characters (default vector = size, 0).

Obviously there is quite some redundancy in having 6 different types of geometric primitives, since a polygon alone would suffice. However, depending on the geometry at hand, a particular (set of) geometric primitive(s) may yield a considerably more efficient and readable layout text.

Geometries may be **structured** by using **mask lists**, **compounds**, and **arrays** (Cf. Figure 2). A mask list is named list of masks, with optionally an offset specified for each mask. An example of a **mask-list declaration**:

masklist mop = metal, oxide, − 1.5, poly, 0.5.

A geometric primitive containing "mop" instead of a mask defines three geometries, one in metal, one in oxide shrinked by 1.5 units and one in poly, grown by 0.5 units.

A compound is a list of geometries enclosed by **begin** and **end**. An array is a geometry defined by an **array prefix** followed by another geometry that may be an array itself (a multi-dimensional array), resulting in a row of copies of the latter geometry. An array prefix specifies the index range, e.g. 0 . . 7, and a vector relating two successive array elements.

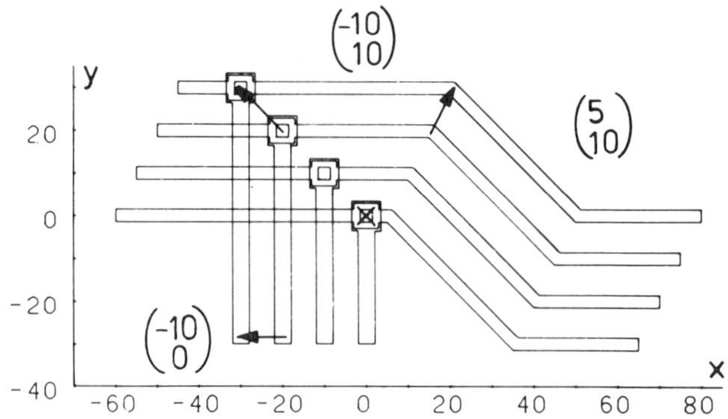

Figure 2.
An example of structured geometries, using a masklist, arrays and a compound. 'x' denotes the origin of the plane.

```
masklist mop = metal, oxide, −1.5, poly, 0.5;
array (0. . 3; 5, 10) of .
   wire (metal; w = 3; −60, 0; 6, 0; 36, −30; 66, −30);
array (0. . 3; −10, 10) of begin
   ra (mop; −3, −3; 3, 3);
   wire (poly; w = 4; 0, −28; 0, 0)
end;
array(0. . 3: − 10, 0)of  wire (poly; w = 4; 0, −28; 0, 0)
```

MODULAR LAYOUT DESIGN

Layout complexity can effectively be managed by application of the VLDL **module** concept. This concept is implemented by the VLDL constructs **module definition** and **module instance**. A module definition specifies an atonomous piece of layout. A module instance is an exact copy of that module definition and can be placed inside another module definiton, optionally after a rotation over a multiple of 90 degrees and/or a reflection in either x- or y-axis. A module may be instantiated any number of times, resulting in that number of congruent instances. A module definition contains a number of module instances (sub-modules) and/or geometries, that specify, amongst other things, interconnect tracks. A layout, being a module instance itself, can therefore be regarded as a **hierarchy** of modules. A module may denote a via, a load transistor, a gate, an adder, a pla, a data path, or any other layout entity.

The basic idea behind the notion module is **locality**:

Ideally design, verification and mask processing of a module should be executed without knowledge of the internal structure of its sub-modules, and vice versa, knowledge about the environment of each instance of the module should not be required either.

This locality is implemented by **assigning** a specific area to a module, the module's **domain**. The module's internals, "contents", must fit inside this domain, whereas other layout elements may not intrude it from the outside. The simplest domain consists of the module's bounding box. In general such a simple domain will lead to unnecessary waste of silicon area and a loss of flexibility at the next higher level in the module hierarchy. A VLDL domain may therefore have different shapes in different masks, may be non-rectangular and non-connected. This generality provides, for instance, a way to pass (partially) unused layers to the module's environment. A domain is an auxiliary quantity and will not be printed on silicon.

The afore mentioned locality is achieved if **modular design rules** are satisfied. In general a layout must comply with a set of geometric design rules, specifying the:
- minimum separation between two geometries (separation rules)
- minimum width of a geometry
- minimum overlap of one geometry over another, etc.

The contants of a layout module must satisfy these geometric layout rules. Furthermore, the separation rules are supplemented by **modular design rules**, in order to specify the separation between module's geometries and sub-module's geometries, without reference to the sub-module's contents. One of the separation rules may be:
*) two metal tracks must be separated by at least g micron.
Then the related modular design rules are:
1) the domains-in-metal of sub-modules may not overlap.
2) a metal track and a domain-in-metal of a sub-module must have a mutual separation of g/2 micron.
3) a metal track must be separated by at least g/2 micron from the module's domain-in-metal's boundary.
4) the domains-in-metal of all sub-modules must be completely covered by the domain-in-metal of the module.

Figure 3 illustrates these modular design rules.

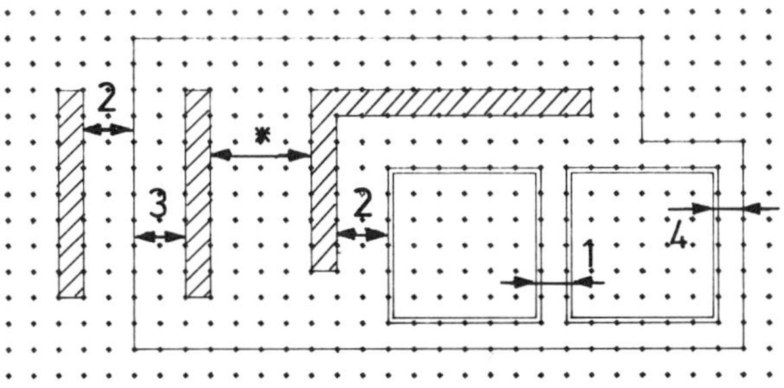

Figure 3.
Modular-design rules regarding domains. The shaded areas denote metal tracks, the not-shaded polygons the domains of a module and two sub-modules. The numbers correspond with the numbers of the design rules in the text. g = 4 grids.

Rules 1-4 provide a powerful **isolation** between modules. Too powerful, obviously, since **communication** between modules is inhibited in this way. The VLDL primitive for communication is the **terminal**. The simplest terminal would be the contact point located on the domain boundary. Again, such a simple terminal may lead to unnecessary silicon waste and to a loss of flexibility at the module's environment. A better way of looking at terminals is **not** to regard them as **locations** of interfacing the module's internals to the module's environment, but as two-dimensional **mask features** that are **shared** by the module's internals and the module's environment. Thus a wire, a contact hole, or any mask feature declared as terminal is considered to be part of the module in which it is declared, as well as part of the environment of an instance of that module. Note that this is feasible, because at both sides of a module's boundary we express ourselves in terms of wires and contact holes, i.e. in terms of objects represented by the same formalism. A VLDL terminal may therefore be defined by any geometric primitive and may be located anywhere relative to the domain boundary. There are additional modular design rules that specify if and how a terminal may be approached from the module's inside as well as from the outside. These additional modular design rules are actually relaxations of the rules around domains that were introduced above. Instead of going into the details of these modified rules, we refer to Figure 4. The domain together with a set of terminals will be referred to as the module's **abstraction** or **periphery**.

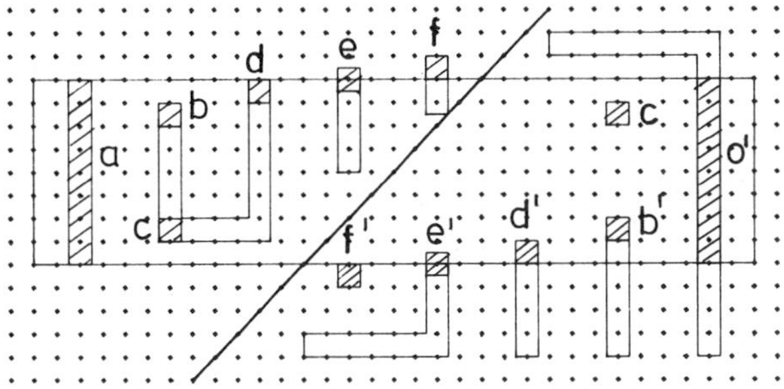

Figure 4.
Modular-design rules regarding terminals, by example. To the left of the diagonal a view is shown of a part of a module with its **formal** terminals (shaded), and some internal wires. To the right of it a view is shown of another part of the module, but now with its **actual** terminals (shaded) and some external wires. g = 2 grids. Notes: — terminal (a) crosses the module. An external wire is allowed to cross the module over (a'). — terminals (b) and (c) are g/2 inside the domain's boundary. Nevertheless, they still can be connected from the outside (b'), while, if not connected, a troublesome "proximity effect" towards the module environment can be avoided (c'). — terminals (e) and (f) partially extend over the domain's boundary, allowing actual terminals (busses!) to be shared by module instances (Cf. Fig. 5). — terminals that are located more than g/2 units inside the domain's boundary cannot be connected from the outside.

At this point it may be illustrative to mention a striking analogy with the procedure concept in a programming language like Pascal (Cf. table 1).

VLDL concept	VLDL example	Pascal concept	Pascal example
geometric prim.	**ra** in metal	variable	i: integer
compound	**begin ra** (. .) ra (. .) **end**	record	r: **record** i, j: integer **end**
array	**array** (. . .) **of ra** (. .)	array	a: **array** (. .) **of** integer
module definition	(Cf. Fig. 5)	procedure dcl	**procedure** swap (. .);
terminals	,,	formal var parameters	**var** i, j: integer
contents	,,	procedure body	**begin** k:=i; i:=j; j:=k **end**
domain	,,	scope of a var	scope of k
module instance	,,	procedure call	swap (a[1], a[n])

Table 1. The analogy between VLDL and Pascal.

The powerful procedure concept together with the technique of abstraction contributed significantly to the structuring and design of large programs. We anticipate an analogous result of the layout-module concept.

Advantages of the module formalism are numerous:

The periphery of a module defines a concise and precise **interface** between modules, and can therefore effectively be

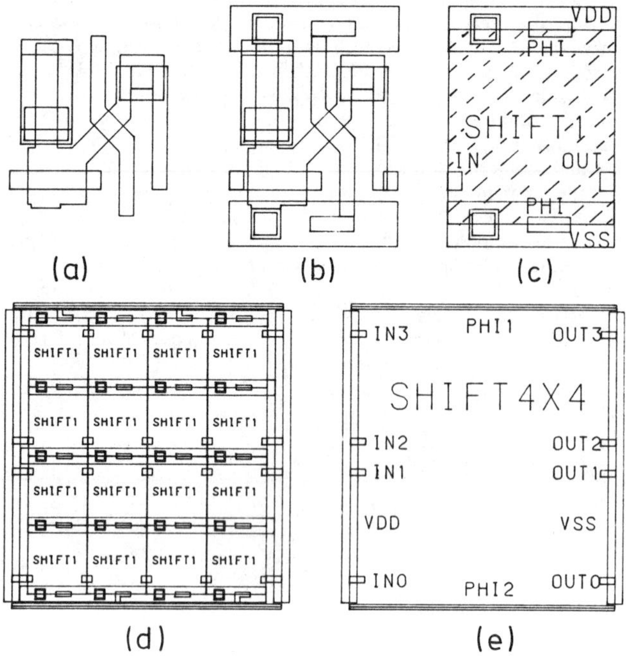

Figure 5.
A small-scale, realistic example using layout modules.
(a) the internals of a single-bit shift-register module "shiftl".
(b) the internals of shiftl, together with its terminals.
(c) the abstraction of shiftl, consisting of a domain in all layers, except metal (shaded rectangle), a set terminals (all other rectangles) and a so-called **drawing part**. A drawing part is a VLDL graphical comment feature, here consisting of the texts. Note that there is no domain in metal, allowing a metal wire to cross the module horizontally.
(d) module "shift4x4" consists of 4 rows of 4 shiftl modules each. Note that the horizontal vdd and vss tracks are shared between shiftl modules, in agreement with the modular--design rules.
(e) the abstraction of shift4x4.

used for splitting up a layout design into a number of more or less independent design masks. This can be done:

- in a **top-down** fashion, i.e. by decomposition of modules. Thus the area of a module may be partitioned into domains for the sub-modules with optional routing "channels" between them. Refining of this rough partitioning consists of global wiring, placing of sub-module terminals, and successively updating of the domains when more is known about the internals of the sub-modules. In other words, the concepts domain and terminals form the formalization of the commonly used notion **floor plan**.
- In a **bottum-up** fashion, i.e. by composition of modules. Library modules, silicon-compiled modules and parts of previously designed layouts can be used without considering possible interactions between their internals and the environment in which they are placed.

- Verification, i.e. design-rule checking, connectivity checking (Cf. the section "Connectivity") and layout extraction can be done modularly, and can therefore be done much less CPU intensive (estimates range from one to two orders of magnitude!). Reasons for this are:
  - Verification programs require a CPU time that grows faster than linear with the number of layout features.
  - A module definition needs to be checked, and not every instance of it.
  - Silicon-compiled modules and previously checked modules (such as library modules) can be excluded from checking.

  For reasonable module complexity, these checks can be done efficiently after every interactive update in an incremental way.
  Error reporting is also more concise, since an error is reported once in a module definition, and not once in each instance.
- The flexibility in the concepts are likely to result in more powerful hierarchical layout-synthesis and layout-optimization tools, such as routers, placers and compactors.
- Post processing, including merging, growing and logical combining of masks can be done modularly, saving considerable CPU time, as well as memory and disc requirements.
- Ideally, the pattern generator should also take advantage of the modular structure. Thus, a single flash should contain all features in one mask of a module definition, as opposed to a single rectangle or trapezium. This would result in a considerable more efficient pattern generation in the case of multiple module instantiation ("criss-cross flashing").

The modular approach also has its **price**: the domain and terminals need to be **specified** and **modified** during layout refinement. Although the number of rectangle equivalents involved may be relatively small (especially in the case of higher level modules), and some automation may be considered, there is a point to make here. A simple domain, for instance an identical rectangle for each mask, make the afore mentioned advantages maximal. However, the price in terms of waste of nano acres of silicon surface may be considerable. Refinement of the domain will reduce the silicon waste (zero in the limit), but will decrease the degree of abstraction and the related advantages, since the domain will reflect more of the module's internal structure.
Apparently, there is a trade off involved, of which the result depends on, amongst other things, available design effort/time, production quantities and circuit speed. The module concept makes this trade off explicit.

The concepts domain and terminal can be found in earlier literature [10]-[11]. In comparison with VLDL a number of differences can be noted:

- VLDL allows sharing of busses, for instance sharing of a power--supply track by two adjacent modules.
- In VLDL a terminal may be placed some distance inside the domain boundary, such that an external wire can be run as close to the domain boundary as elsewhere, even if there is no actual connection made to this terminal.
- In CMOS, a p- (or n-) well may entirely be declared as a VLDL terminal; at a higher-level module this well can be merged with other wells, saving area.
- (partial) free masks may be passed to the environment of a VLDL module. This makes it possible for an external wire to cross the module there.

SYMBOLIC LAYOUT

Symbolic layout has two aspects:

1. **layout representation**. This usually defines a set of symbols (vias, transistors, cells, etc.), a grid (if present), the allowed degrees of freedom (for instance orthogonal only), the wire widths, etc.

2. **layout transformation**. A symbolic layout is transformed into a "physical" or "geometric" layout, often including optimization, like compaction, merging of symbols, etc.

VLDL supports symbolic layout, primarily with respect to the layout representation aspect, although it specifies some minimum requirements for layout transformation. Compaction and other forms of optimization are therefore not regarded as part of VLDL, although it is easily conceivable that programs compact VLDL symbolic layouts.

The representation aspects of symbolic layout in VLDL, together with their intended use are discussed along with an example of a piece of symbolic layout and the corresponding physical layout (Cf. Figure 6). A geometric primitive in a symbolic layout specifies a physical area that includes all points within a distance of w/2 from any point of that geometric primitive, where w is the minimum width for that layer. This allows minimum-width wires to be specified by their center lines only, and it also inhibits violations of the minimum-width rules. Thus, the geometric primitive dot results a square with dimensions w*w. Of the geometric primitives wire and the polygon only the Manhattan variants are allowed; the geometric primitive slanted rectangle is not allowed in VLDL symbolic layout.

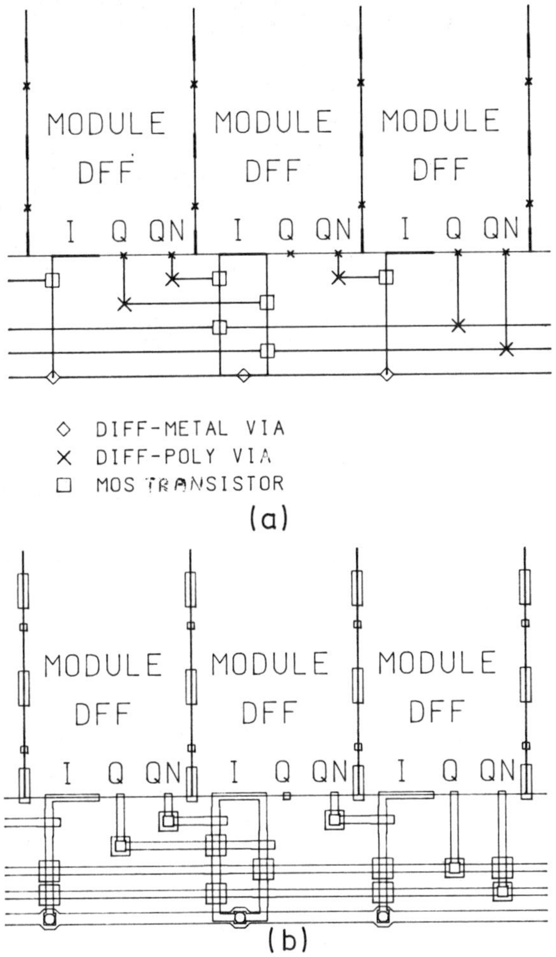

Figure 6.
An example of a symbolic layout.
(a) a piece of symbolic layout from a pseudo-random generator.
(b) the geometric equivalent of (a).

Symbols, such as standard transistors and vias are implemented using layout modules; the graphic appearance being defined in their drawing parts. A clever chosen abstraction may serve to hide the more exotic masks (for instance, contact windows, depletion, arsenic, overglass cut), and related design rules. Moreover, using this technique any geometric-layout module can be "symbolized". Thus a flip-flop, PLA or 2k ROM module can also be used as a symbol in a symbolic layout!

The aforementioned facilities can be used to **represent**, for instance, STICKS-type of layouts (Williams [8]). An optionally definable grid allows for other types of symbolic layout.

Note that in Figure 6 (not as part of VLDL) during transformation to geometric layout, diffusion-poly under passes are generated at those places where not explicitly a transistor has been specified.

CONNECTIVITY

A layout can be regarded as the implementation of a **network** (or **circuit**), where a network is a set of connected components, such as transistors and/or gates. These networks are often described in a **Network Description Language**, or NDL (e.g. PHILSIM [2]). The more successful NDLs contain an NDL **module** construct, where an NDL module is a set of connected sub-modules and/or components. The connections, often called **signals** are either **local** to that module, or are listed as the module's parameters in order to establish communication with the environment (**global** signals, Cf. procedure parameters). The consistency of a layout with the corresponding network can be verified. A program that does so is called a **connectivity checker**.

Both from a design and verification point of view it would be convenient if there is a correspondence on a modular basis, i.e. if there is a VLDL module for each NDL module, and if there is a terminal for each global signal, Unfortunately, this simple principle may yield inefficient layouts [12].

Consider the NDL and VLDL module hierarchies. The above expressed ideal, i.e. the proposed injection from the NDL module set into the VLDL module set is usually **almost** achieved. For this reason VLDL requires that the two module hierarchies denote equivalent structures. This means that if a VLDL module is the implementation of a particular NDL module, they should have the same name. Also their sub-modules must either correspond directly (by name), or after some, hopefully limited, expansion of one or more of the sub-modules (Cf. Figure 7).

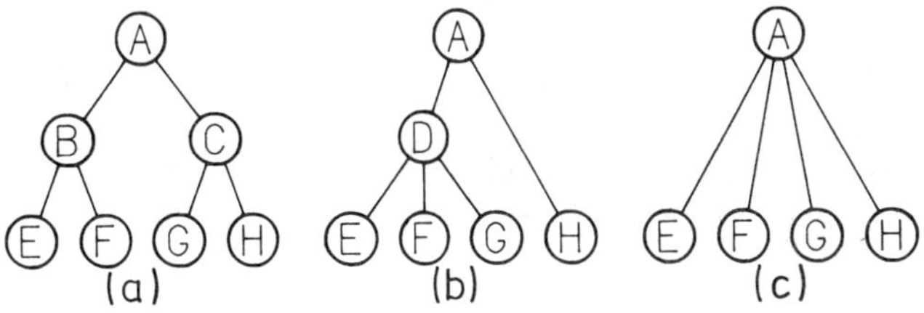

Figure 7.
Different hierarchies defining equivalent structures. Module hierarchies (a), (b) and (c) are equivalent after the expansion of modules B, C and D.

VLDL offers three facilities to specify a more general relation between global signals and terminals.

- Firstly, it can be specified that a global signal corresponds with a **set** of terminals, whereas these terminals are mutually connected within the module. The environment of the module may take advantage of this, for instance to provide for alternative terminals, or to pass power supply to an abutting module.
- Secondly, it can be specified a that a global signal corresponds with a **set** of terminals, whereas these terminals must be mutually connected in the module's environment, so that the module's internals may relay on their electrical equivalence. This allows for area-efficient distribution of global resources, such as power supply and clocks.
- Thirdly, it can be specified that (sets of) terminals are **logical equivalent**. Logical equivalent here means that wires to the terminals can be arbitrarily exchanged without altering the function of the circuit.

Thus, the NDL-VLDL verification can be done predominantly modular, without making concessions to layout efficiency.

VOILA, SOME NOTES ON A VLDL IMPLEMENTATION

Voila is a layout design environment, primarily intended for digital VLSI layouts for which design costs and elapsed design time are more important criteria for optimization than transistor density. However, the area of application is not clearly bounded, and larger parts of the system can also be used for other kinds of circuit layouts. Voila is an acronym for VLSI Oriented Interactive Layout Aid. It is also the French word for "there it is" or "that's it", a conotation reflecting its graphics capabilities.

A number of the more important components of Voila, of which the implementation is well on its way, are briefly discussed below. The core of Voila is a data structure named VIC (Voila Internal Code), implemented in Pascal. The VIC has semantics very close to VLDL, but is equipped with additional data, such as pointers, bounding boxes and hashed tables, in order to accomplish fast, selective data access. **Translators** from VLDL to VIC and vice versa link the VIC to the VLDL. A **display program** scans the VIC and paints a layout on a graphics display. An average picture (say, one of 1000 rectangle equivalents out of a layout of 200,000 rectangle equivalents) requires about 1 second of VAX/780 cpu time, elapsed time depending on the graphics hardware at hand. This also applies to LSI sized modules, since the contents of sub--modules need to be displayed. An **editor** drives the display program and manipulates symbolic as wel geometric layout in the VIC. The editor may also invoke the **incremental design-rule checker** and **connectivity checker** after every symbolic layout modification. For the latter NDL data are stored in a data structure similar to the VIC. Using this network data structure, some **interactive placement**, **routing** and **PLA generation** are planned. A designer may offer a symbolic-layout module for **transformation** to a geometric-layout module. Programs for **batch design-rule checks** (including checks for geometric design rules) as well as for **batch connectivity checks** are also part of Voila. Modular **circuit extraction** including circuit parasitics is a subject of further study.

Mask processing operations in Voila include **merge**, **plot**, **boolean combination**, **grow** and **pattern generation** of masks, all working on a module-by-module basis. The system is planned to be operational by spring 1984.

CONSIDERED VLDL EXTENSIONS

In the current VLDL definition modules have a static nature, i.e. all instances of a particular module are congruent. In the case of slightly different modules this yields duplication of VLDL code, since each of those modules may have a large common part. **Parametrizable modules** are a solution to this. Thus it would be convenient if expressions with parameters are allowed for coordinate values, as opposed to literal constants, and if parameters could be passed to modules during instantiation. Other extensions such as conditional geometries ("if <condition> **then** <geometry>"), **case** statements, **while** loops, etc. can be considered as well. We still regard it as a subject of further study, whether VLDL should be selectively extended this way, or a complete programming language such as Pascal should be embedded in VLDL (resulting in VLDL*). At this moment the latter solution seems preferable, since then also module generators (e.g. a PLA generator), routers, etc. could be completely written in VLDL*.

One drawback all these extensions have in common: modular correctness is not anymore an attribute of a module definition but one of a module instance, since it can only be verified for a particular set of module parameters.

ACKNOWLEDGMENTS

The author gratefully acknowledges the members of the Voila team, who contributed significantly to the expressed ideas. The reviewers of drafts of this paper are thanked for their useful comments and suggestions for improvements.

REFERENCES

[1] Lambert, D.R. (1981), "Graphics Language/One", Proc. 18th Design Automation Conf., pp. 713-719.

[2] Niessen, C. (1977), "Computer-Aided Design of LSI Circuits", Philips Technical Review 37, No. 11/12, pp. 278-290.

[3] Calma (1979), "GDS-II Reference Manual", Calma Computing Systems Inc.

[4] Sproull, R.F., and R.F. Lyon (1980), "The Caltech Intermediate Form for LSI Layout Description", Chapter 4.5 of "Introduction to VLSI Systems"/C. Mead and L. Conway, Addison-Wesley Publ. Company.

[5] Niessen, C. (1983), "Hierarchical Design Methodologies and Tools for VLSI Chips", Proc. of the IEEE, Vol. 71, No. 1, pp. 66-75.

[6] Lattin, B. (1979) "VLSI Design Methodology, the Problem of the 80's for Microprocessor Design", Proc. Caltech Conf. on VLSI.

[7] Gibson, D. and S. Nance (1976), "SLIC-Symbolic Layout of Integrated Circuits", Proc. 13th Design Automation Conf., pp. 434-440.

[8] Williams, J. (1978), "STICKS – A Graphical Compiler for High Level LSI Design", Proc. of the NCC, May 1978, pp. 289-295.

[9] Weste, N. and B. Ackland (1981), "A Pragmatic Approach to Topological Symbolic IC Design", Proc. VLSI 1981, Edinburgh, pp. 117-129.

[10] Sequin, C.H. and A.R. Newton (1982), "Description of STIF 1.0", from "Design Methodologies for VLSI circuits"/ edited by P.G. Jespers a.o., pp. 147-171, Sijthoff and Noordhoff, The Netherlands.

[11] Whitney, T. and C. Mead (1983), "Pooh, a Uniform Representation for Circuit Level Designs", to appear in the Proc. of VLSI 83, Trondheim.

[12] VanCleemput, W.M. (1979), "Hierarchical Design for VLSI: Problems and Advantages", Proc. Caltech Conf. on VLSI 1979, pp. 259-273.

DESIGN METHODOLOGIES APPLICABLE TO MICROPROCESSORS

Prathima Agrawal

Bell Laboratories
Murray Hill, New Jersey 07974

Abstract — **Microprocessors are treated as specialized VLSI chips and the design methodologies of some of the complex microprocessors that have been built are reviewed. The relationship between instruction set and control complexity, as it pertains to the design methodology adopted, is examined. Further, the importance of CAD tools for microprocessors and the direction in which they are headed are presented. The need for reducing human and computer resource intensive verification steps and for enhancing tools capable of efficient forward translations in the design process is emphasized. Although microprocessor is used as a paradigm, this paper is applicable to any VLSI design methodology.**

I. INTRODUCTION

A microprocessor is the result of a unique combination of advanced VLSI technology and system design. So far, undoubtedly microprocessors represent the largest and most complex VLSI parts ever to be realized on a single piece of silicon. Examples are the 146,000 transistor site CMOS Bellmac-32A* [1] and the 450,000 transistor NMOS FOCUS Chip [2] by Hewlett-Packard. These chips have encountered and solved special VLSI system related problems. Logic, circuit and layout methodologies and tools used for VLSI chips have been stretched to their limits, while designing these microprocessors. In addition, microprocessors experience special problems encountered by mainframe computer designers. These include optimizing the architectural design for a target performance and cost. The methodology adopted is concerned with guaranteeing that the proposed architecture embodies the instruction set correctly and predicting the performance for a given cost.

As with any system, competitive pressures demand that microprocessors with complexity often reaching those of mainframe machines are designed at reduced costs. Thus the challenge faced by the designers is to design and verify the microprocessor chip in the shortest possible design time. Due to the monolithic nature of the chip, design errors are costly and are time consuming to fix. Hence, there has been a trend to automate the design process as much as possible such that human errors are reduced. However, fully automated designs which result in very short design times may be less efficient than manual designs in terms of good floor plans and layout. (An example of such a microprocessor is given in [3].)

The compromise between the two extremes of complete automation and fully manual designs has been to combine human ingenuity with the speed of present day computers. Much of the mechanical chores are provided by computers and as is well known such a design discipline has come to be known as CAD or Computed Aided Design. Almost every microprocessor that is designed so far falls into this category [1,4,5]. The important human resource intensive steps in the design methodology are architecture, logic design, layout and verification. Although powerful simulators are available for verification purposes, the process of test generation is entirely manual. Most of the functional testing has been done to date using *ad hoc* procedures such as writing test programs that exhaustively test the processor for every instruction and every address mode, etc. They are *ad hoc* as they are specially generated for each microprocessor and there is no guarantee of complete functional coverage. For device testing higher structural fault coverage may be possible through design for testability [6]. However, microprocessor manufacturers have not exploited these for the area penalties that may be introduced by these methods. Attempts, however, have been made to focus some attention on the issue of design for testability in the HP chip [5].

The requirements on the design methodology for microprocessors is to set up a highly pipelined and parallel scheme of design and verification. Its performance is measured by the design interval and resources spent and its quality is measured by the number of errors in realizing the required design on silicon. In this paper, a walk through of the important steps in designing a microprocessor will be done by drawing examples from currently available microprocessors.

II. DESIGN STEPS FOR A MICROPROCESSOR

As with any system design, the design of a microprocessor goes through several stages. These can be coarsely divided, as shown in Fig. 1, into design specifications, functional design, logic, circuit and layout design and finally chip fabrication and testing steps.

The most time consuming, less algorithmic and entirely manual step is the design specification phase. Here, the important factors considered are the nature of the application for the microprocessor, the performance goals to meet the application, the instruction set to be supported in hardware, the internal architecture and the external protocol to arrive at the performance goals, the extent of the competitive market place for the microprocessor, the product cost when developed to be competitive and the technology choice to produce a cost effective product. It is important to give particular attention to the overall methodology that will be adopted for, this is what determines design cost, and the cost and quality of product.

At the functional level of design, a functional behavior of the system is conceived to meet the design specifications that were generated in the beginning. In essence, this level defines the high level topological structure which accommodates the instruction set of the microprocessor. Although the partitioning of the design steps, defined in this paper is by no means absolute, this stage can accomplish such decisions as the structure of the RALU, number of stages in the pipeline, the number and approximate size of the control elements such as PLAs, ROMs etc., whether an instruction cache is to be included, if so its size and location on the chip. Usually a register transfer level model is used to describe the design at this stage [1,4,5].

The next step in the design process is the logic design phase where the RTL or equivalent description of the functional design is transformed to a logic design description where primitives such as NANDs, NORs, FFs, multiplexers, decoders, etc. are used. Logic design is basically a manual step as the CAD tools available in this area have not replaced the decision making ability and efficiency of logic designers. Efficiency is an important attribute for a VLSI part where chip area determines the speed and the cost.

At the circuit design level, each logic element is further decomposed into MOS transistor level primitives. For example, a 2-input NAND gate at the logic level translates to a 4 transistor configuration in CMOS technology. Once the configuration of the transistors is finalized, layout or the interconnection process can begin. The main styles of layout are polycell, gate matrix [7] and hand layout. For large chips, such as microprocessors, usually a combination of pre-laid out blocks (Polycells) as well as customized layouts for larger functional blocks is beneficial. Very sophisticated layout techniques are used to produce the final layout. An example of such a layout system is the LTX2 System [8]. The output of these programs is a mask level database which is used in the physical chip fabrication.

* Bellmac is a trademark of Western Electric Company.

CH1935-6/83/0000/0491$01.00

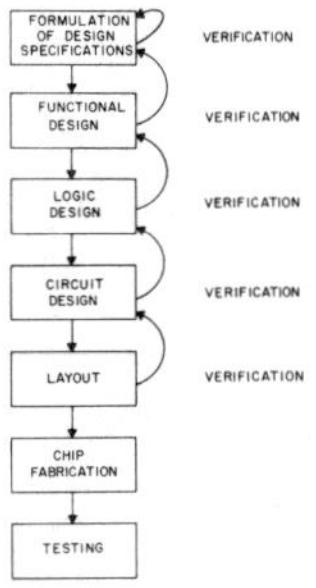

Fig. 1 Microprocessor Design Process.

The fabricated chip is tested using any of the specialized IC testers like Sentry or Takeda. The tests and their expected responses are developed at either the logic or circuit levels. A fault simulator at the logic or MOS level assesses the quality of tests.

The above methodology is adopted with some variation by different microprocessor designers. In this paper, however, the emphasis will be on the most crucial areas of design such as the instruction set design, control design and the overall verification methodology.

III. INSTRUCTION SET DESIGN

Sophistication in IC technology has made it possible for designers to conceive microprocessor designs with capabilities equivalent to those of mini or midi computers. One of the major requirements of such a provision is the ability to support high level languages and operating systems. From a cost-performance point of view, it is important for a microprocessor designer to determine the extent to which such support has to be provided in hardware. This software hardware trade off decision with respect to supporting one or more high level languages has enormous impact on the complexity of a microprocessor architecture, its implementation and the design tools that are needed to support the design activity.

A variety of instruction sets are adopted by various microprocessors that are commercially available [9] and also those that are designed at universities. The spectrum covers the simple instruction sets of RISC [10] and MIPS [11] microprocessors to the complex instruction sets of Bellmac, Motorola 68000 [12] and HP microprocessors. The choice of an instruction set is bounded in terms of complexity at both ends. On one end, it has to be at least as complex as to support high level languages without unduly increasing the complexity of compilers which translate the users program into instructions that can be efficiently executed by microprocessor hardware. At the other end, the complexity is constrained by the forces applied, by the technology at hand, such as area limits from yield and execution speed points of view. The commonly used metric for assessing the extent of complexity of instruction set is the cost to performance ratio.

While advocates of RISCy architectures emphasize the need for keeping instruction sets simple, it is important to remember that the meaning of "simple" instruction set is not absolute. It is intimately related to the technology available. As technology improves, "new" instruction sets which are relatively complex that effectively exploit the benefits of the new technology can be defined. The CAD tools that are provided should enable an architect to evaluate the effect of instruction set complexity on the cost performance ratio without neglecting the effect of technology. The impact of instruction set complexity on the efficient implementation of specialized architectural features, such as the number of stages in the pipelined control, degree of parallelism required to support an instruction set at a given cost and performance, needs to be evaluated.

Instruction set design is concerned with identifying those instructions that can be executed in hardware and supporting the rest in software. Once this identification is done, the architecture has to be designed so that the performance is optimized at a given cost. The methodology that is used for evaluating the relationship between the complexity of instruction sets and cost performance ratio should be carefully laid out. The result of this evaluation should have reasonable accuracy so that design decisions can be made up front and that there are few surprises when the chip is available.

The decision on what composes an instruction set should be arrived at by reviewing the potential application environment for the microprocessor. The instruction set should be rich enough to provide those instructions and addressing modes that are frequently used in these applications. Since, in general, no new high level language is invented for use by the microprocessor application, such statistics can easily be obtained by measuring the relative frequency of occurrence of instructions, addressing modes and data types in non-VLSI processors. This statistical study influences the optimization of the architecture.

The cost of a chip is determined from packaging and wafer yield considerations. As the size of a chip increases, the cost of implementation per device reduces due to the increased amount of integration. This advantage, however, is leveled off at certain chip size due to the drop in yield as shown in Fig. 2. Such data help in estimating the cost of a chip since the number of transistors can be estimated from the microprocessor data-width and the instruction set complexity. The advantage of using more complex instruction sets is in terms of increased performance. As shown in Fig. 3, the performance too levels off due to practical considerations like circuit delays, etc. The type of data shown in Figures 2 and 3 can be used for accomplishing a cost-performance tradeoff for a given technology.

The most commonly used design tool at this level is a simulator. A model for the proposed microprocessor architecture is generated using a specialized simulation language such as SIMULA [4] or a high level programming language such as C [1]. The model consists of a high level description of functional blocks as ALU's, registers, instruction queues, caches, etc. These models can provide an accurate behavior of the microprocessor at the clock cycle level. A number of pre-selected benchmark programs provide the performance predictions in terms of number of machine cycles needed to execute each instruction.

While the above type of simulators provide a fairly accurate behavioral description for the microprocessors, they have very little input from the technology which is used for implementation. Thus, factors such as the difference in the routing delays for different bit fields of the same word are not used in the model. Such simulators tend to exhibit a monotonically increasing performance with increased pipeline complexity, or the same performance for different data widths, etc. Due to these technology related issues that are not adequately modeled by the architectural simulators, the microprocessors tend to be optimistic with respect to a given performance. Worst case performance studies are generally undertaken by using the memory access time and cache hit ratios as variables. It should be remembered that although these factors influence the performance of a microprocessor, these are external to the chip itself.

As mentioned earlier, for every realizable technology, there is an upper limit on the level of complexity (translated to the number and interconnection of transistors) that can be supported. This is because, as the physical dimensions of the chip become larger, the yield of good chips decreases nonlinearly [13]. With increase in complexity, the ratio of the routing area to device area increases. This has an effect on the overall performance as the routing delays will start to be significant. The ability to accurately forecast the chip area even before detailed logic design and timing design are accomplished, as was done for the Bellmac-32A microprocessor [1], is a key factor in a successful design methodology. The tools that are available (a hierarchical chip router such as LTX2) for these purposes provide valuable feedback to the designers in the realization of the proposed architectures.

IV. DESIGN METHODOLOGY FOR CONTROL SECTIONS

Once the instruction set has been defined and performance and area projections are made, the next step is the functional design phase. Here, the details of the building blocks of the architecture, that

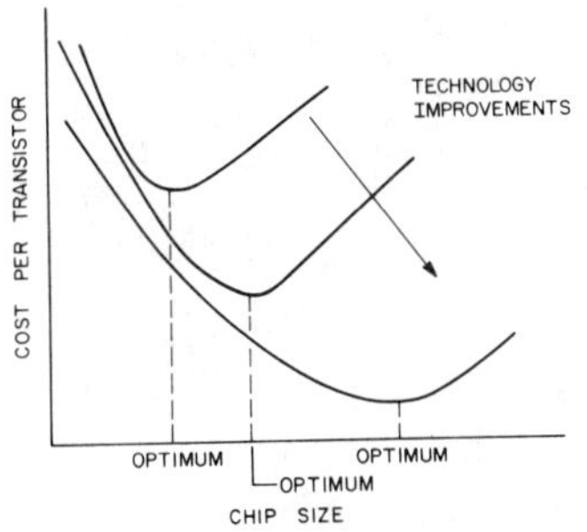

Fig. 2 Dependence of Device Cost on Chip Size and Technology.

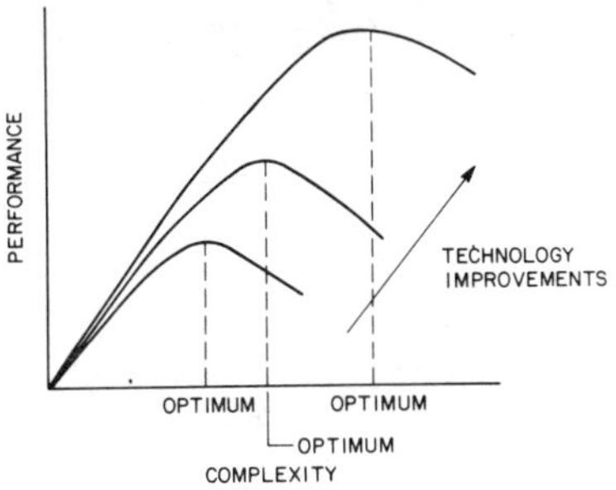

Fig. 3 Dependence of Performance on Chip Complexity and Technology.

accommodate a given instruction set, are designed. While the execution unit lends itself to be neatly partitioned into functional blocks such as ALU's, registers, barrel switches, etc., control units have comparatively very little functional regularity. Since, mainly control determines the speed of the microprocessor utmost care is needed in its design.

Control sections vary in nature and complexity from microprocessor to microprocessor. The trend, however, has been to provide a large degree of pipelining and parallelism so that a variety of operations take place within the control unit concurrently. This results in increased performance through the efficient utilization of resources such as execution unit, buses etc. In [14], an excellent account, of the way a typical control unit is designed, is given. Whether control sections are microprogrammed [5,12], PLA based [1] or combinational logic based [10], the methodology used for design verification of the control unit has been ad hoc and indirect. Typically, a register transfer level description of the control section is tied with a similar level description of the execution and I/O units. The resulting simulation model executes a machine language program and the output of the execution unit is monitored for the correct functioning of the chip. This method is used for indirectly debugging the control design.

The aim of the design verification effort for microprocessors has been to design a set of test programs which verify the behavior of the chip by extensive sequencing through the state space of the control. For a complex microprocessor, this space is large and there exist no measurable way of guaranteeing 100 percent functional correctness. This is a tricky problem since one never knows when the design has been tested completely. Although, design verification problem is not peculiar to microprocessors alone, much work remains to be done in terms of CAD tools for effective design verification.

As mentioned before, control section greatly influences the speed, area and performance of a microprocessor chip. Provision of CAD tools in accurately predicting these two valuable parameters significantly alters the total design time and reduces the chances of many design iterations. Although control section is less regular than the execution unit, use of regular structures such as PLAs and microprogrammed control schemes lend themselves to early projections of speed and area fairly accurately. For example, PLA synthesis tools that automatically translate high level control descriptions to connectivity tables can be used to predict speed and area along with the technology information. Bellmac 32A [1] made extensive use of such facilities.

Notice from the previous discussion that significant amount of information about the final product (area, speed and functionality) is obtained even before detailed logic or layout design has begun. This ability, made possible by the powerful CAD tools, encourages parallelism in the methodology. This also supports bidirectional feedback among functional and physical designs. The next section touches upon some of the advances, in CAD tools, which have been motivated by the challenges demanded by microprocessor designs.

V. CAD AND MICROPROCESSORS

Almost every microprocessor design today uses CAD tools. So much so that new CAD development has become an integral part of the design process. With increasing complexity of microprocessors, either existing tools have to be extended in their ability or new tools and methodologies have to be invented to meet the challenges offered by these complex designs.

CAD tool systems for microprocessors are a combination of general purpose VLSI tools and specific ones generated for the particular microprocessor design. The general VLSI design tools are logic simulators, MOS timing simulators, circuit analysis programs, physical design tools such as placement programs, routers, design rule checkers, layout analysis programs, etc. A detailed summary of such general purpose chip design tools at Bell Laboratories is given in [15]. References [4] and [5] also provide insights into CAD tools that were used for microprocessors at Intel and HP, respectively. Once a logic description is available, these well established tools can be used to help advance the VLSI design into manufacture. The problem, however, is the availability of effective tools for pre-logic phases of design. Although abundant material exists in the literature about logic synthesizers that derive logic description from a functional description [16,17], so far very little evidence, of practical applications of these in microprocessor design, is seen. Thus, pre-logic design is dependent, almost entirely, on human designers.

Due to the lack of efficient logic synthesizers, a great deal of simulation at the logic and timing levels needs to be done to ascertain the correct behavior of the models at this level. For large VLSI chips like microprocessors, not only is simulation expensive and time consuming, but very often the available simulators may not be able to support the size and complexity of these chips [5]. As an alternative to simulation, analysis programs that exhaustively verify one or more important properties of the design are beginning to be used. Examples of such programs are timing analyzers [18]-[20]. Such a program can evaluate every synchronous path in the circuit within minutes of CPU time as compared to days and months of simulation time.

Very few general purpose aids exist for the pre-logic levels of design. Thus simulation at the functional level has been done for specific problems using programs tailored to the design at hand. There are almost no analysis aids available at the functional level. As mentioned earlier, a number of test programs have to be written for verifying the correctness of the functional model. For microprocessors, this in general is a manual and tedious process with no metrics available for functional coverage. Several test generation algorithms, for microprocessors, have been proposed [21,22]. Practical application of these algorithms for large microprocessors needs to be evaluated.

VI. TRENDS IN MICROPROCESSOR DESIGN

The problems related to testing a microprocessor design has been mentioned earlier in this paper. As the microprocessors become more and more complex, there has been a strong commitment towards design for testability. As any large VLSI chip, microprocessors suffer from the limitations on testing imposed by the restricted number of I/O pins, the largeness of the state space and the tedium of wafer level probing. The HP microprocessor [5] has some features which make it testable. These features include the microcoded self test, diagnostic interface port and the provision of microprobe pads to cover transistors near critical control and data nodes. This represents a three step design for testability. Texas Instruments researchers [23,24] have proposed, a bit sliced architecture with data and control monitors and a built-in test for on-chip memory, to enhance testability. The penalty in area is predicted to be about 4-percent for a 16 bit microprocessor and 2.5-percent for a 32 bit microprocessor. Special problems of testing VLSI microprocessors have received much attention as current research interests [25]-[26].

A topic which has generated much curiosity and interest with respect to microprocessors is that of silicon compilers [27]. These are programs that automatically produce layouts for parts of or entire microprocessors starting from a set of high level functional specifications. The intermediate logic and transistor levels of design will be transparent to the user. Silicon compilers, in some cases, may provide the connection between major levels of design automatically as illustrated in Fig. 4. This hopefully, will reduce the criticality of the verification that is needed at every level. The verification effort shown in dotted lines will depend on the quality of the translations.

Future CAD effort for VLSI chips in general and microprocessors in particular should eliminate people intensive steps in the methodology and focus on providing intelligent tools for generating successively lower level models of the design. The resulting product should be optimized for area and performance. Such forward translation mechanisms will reduce the need for extensive verification steps.

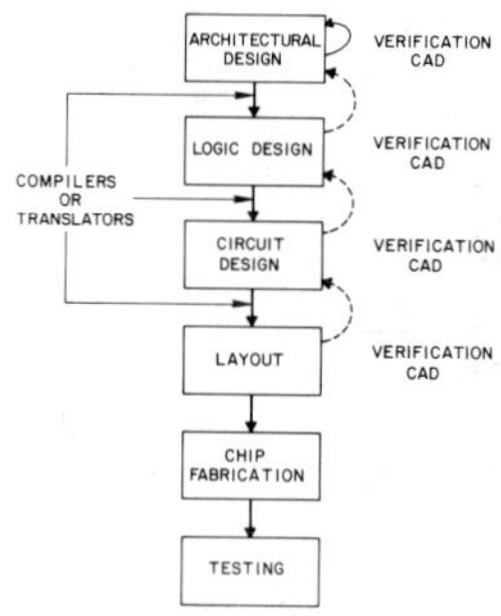

Fig. 4 Automated Design of Microprocessors.

VII. CONCLUSION

The complexity of microprocessors that will be designed in the future will increase with the availability of super MOS technology. Commercial microprocessors with one micron design rules have already been built. CAD tools that emphasize faster design turn around times and higher design accuracy will be needed to quickly adopt to these changing technologies. The design philosophy will shift from the dependency on specialized designers to more sophisticated design methodologies. These will enable the microprocessor designer to specify the design at a much higher level that is machine understandable and translatable to the mask levels. The increased automation and the embedding of testability features will remove the drudgery of elaborate design verification and testing steps.

REFERENCES

[1] D. E. Blahut et al., "Hierarchical Design Methodology for a Single Chip 32 Bit Microprocessor," *Proc. IEEE Int. Conf. on Circuits and Computers*, New York, NY, Sept. 28 – Oct. 1, 1982, pp. 16-20.

[2] J. M. Mikkelson et al., "An NMOS VLSI Process for Fabrication of a 32-bit CPU Chip," *IEEE J. Solid State Circuits*, Oct. 1981, pp. 542-47.

[3] J. Werner, "Computer-Aided Design and Design Automation for ICs in Japan," *VLSI Design*, May/June 1982, pp. 14-28.

[4] W. W. Lattin et al., "A Methodology for VLSI Chip Design," *LAMBDA*, Vol. 2, No. 2, 2nd Quarter 1981, pp. 34-44.

[5] M. Canepa et al., "VLSI in FOCUS: Designing a 32-bit CPU Chip," *VLSI Design*, Jan./Feb. 1983, pp. 20-24.

[6] E. B. Eichelberger and T. W. Williams, "A Logic Design Structure for LSI Testability," *J. Design Automation and Fault Tolerant Computing*, Vol. 2, May 1978, pp. 165-78.

[7] A. D. Lopez and H-F. S. Law, "A Dense Gate Matrix Method for MOS VLSI," *IEEE Trans. Electron Devices*, Vol. ED-27, Aug. 1980, pp. 1671-75.

[8] B. W. Colbry and J. Soukup, "Layout Aspects of the VLSI Microprocessor Design," 1982 *Proc. Int. Symp. on Circuits and Systems*, Rome, Italy, pp. 1214-28.

[9] A. Gupta and H. D. Tong, "An Architectural Comparison of 32-bit Microprocessors," *IEEE MICRO*, Feb. 1983, pp. 9-22.

[10] D. A. Patterson and C. H. Sequin, "A VLSI RISC," *IEEE Computer*, Sept. 1982, pp. 8-21.

[11] J. Hennessy et al., "The MIPS Machine," *Proc. COMPCON*, Feb. 1982, pp. 2-7.

[12] E. Stritter and T. Gunter, "A Microprocessor Architecture for a Changing World: The Motorola 68000," *IEEE Computer*, Feb. 1979, pp. 43-52.

[13] G. E. Moore, "What Level of LSI is Best for You?," *Electronics*, Vol. 43, February 1970, pp. 126-30.

[14] N. Tredennick, "How to Flowchart for Hardware," *IEEE Computer*, Vol. 14, No. 2, Dec. 1981, pp. 87-102.

[15] A. K. Bose et al., "A VLSI Design System," *Proc. IEEE Int. Symp. on Circuits and Systems*, Newport Beach, CA, May 2-4, 1983, pp. 734-39.

[16] J. A. Darringer and W. H. Joyner, "A New Look at Logic Synthesis," in *Proc. 17th Design Automation Conference.*, 1980, pp. 543-49.

[17] S. G. Shiva, "Automatic Hardware Synthesis," *Proc. of the IEEE*, Vol. 71, No. 1, Jan. 1983, pp. 76-87.

[18] V. D. Agrawal, "Synchronous Path Analysis in MOS Circuit Simulator," *Proc. 19th Design Automation Conf.*, Las Vegas, Nevada, June 1982, pp. 629-35.

[19] J. K. Ousterhout, "Crystal: A Timing Analyzer for NMOS VLSI Circuits," *Third Caltech Conference on Very Large Scale Integration*, Pasadena, CA, March 21-23, 1982, pp. 57-69.

[20] N. P. Jouppi, "TV: An NMOS Timing Analyzer," *Third Caltech Conference on Very Large Scale Integration*, Pasadena, CA, March 21-23, 1982, pp. 71-85.

[21] S. M. Thatte and J. A. Abraham, "Test Generation for Microprocessors," IEEE *Trans. Computers*, Vol. C-29, June 1980, pp. 429-44.

[22] K-W. Lai and D. P. Siewiorek, "Functional Testing of Digital Systems," *Proc. 20th Design Automation Conf.*, Miami, FL, June 27-29, 1983.

[23] S. M. Thatte et al., "An Architecture for Testable VLSI Processors," *Proc. IEEE Test Conf.*, Philadelphia, PA, Nov. 1982, pp. 484-92.

[24] T. Sridhar and S. M. Thatte, "Concurrent Checking of Program Flow in VLSI Processors," *Proc. IEEE Test Conf.*, Philadelphia, PA, Nov. 1982, pp. 191-99.

[25] R. Parthasarathi et al., "A Testable Design of General Purpose Microprocessors," *12th Int. Symp. on Fault-Tolerant Computing*, Santa Monica, CA, June 22-24, 1982, *Digest of Papers*, pp. 117-24.

[26] S. K. Jain and A. K. Susskind, "Test Strategy for Microprocessors," *Proc. 20th Design Automation Conf.*, Miami, FL, June 27-29, 1983.

[27] N. Bergmann, "A Case Study of the F.I.R.S.T. Silicon Compiler," *Third Caltech Conference on Very Large Scale Integration*, Pasadena, CA, March 21-23, 1982; pp. 413-30.

MACRO PLACEMENT BY MONTE CARLO ANNEALING

D. W. Jepsen and C. D. Gelatt Jr.

IBM T. J. Watson Research Center, Yorktown Heights, NY 10598

ABSTRACT: The method of optimization by simulated annealing (S. Kirkpatrick, C.D. Gelatt, and M.P. Vecchi, Optimization by Simulated Annealing, Science 220, 671-80 (1983)) has been adapted to the placement of macros on chips in custom VLSI. The present program places rectangles of vastly different sizes and aspect ratios. These are bordered by free space for wiring, and the resulting larger rectangles are placed in compact arrangements on a chip face. In intermediate stages of the placement process, the macros are moved around the surface of the chip allowing overlapping configurations, with the motion regulated by an optimization function taking wirability and overlap into account. Special macros and special requirements, such as short critical paths, are easily included in the formalism.

We have adapted the method of optimization by simulated annealing (S. Kirkpatrick, C. D. Gelatt, and M. P. Vecchi, Optimization by Simulated Annealing, *Science* **220**, 671-80 (1983)) to the placement of macros on chips in custom VLSI. This method has been used very successfully for placing circuits in masterslice arrays by Kirkpatrick and Gelatt, as well as for chips on cards and TCM's on boards. The present work differs from this previous work in that the objects to be place can be of vastly different sizes and shapes. At the present time, the program is restricted to rectangular objects of differing sizes and aspect ratios but generalization to more irregular shapes should not be difficult.

The application of the method can be broken into several parts. First one must define the set of configurations of the system out of which the optimal configuration is sought. In our case a configuration is specifiied by the locations of all of the macros on the chip as well as their rotations and reflections from standard orientations. For the purposes of the method, we include configurations in which macros overlap. In this way, macros will be able to move through each other during the optimization process to cluster with other macros to which they are strongly connected.

Next one must define a penalty function taking a value on each of these configurations which measures the degree to which a given configuration falls short of requirements. In the present case, we wish that the resulting configuration will be wirable and have no overlapping macros. Other features, such as short critical paths, could be introduced in future work. Presently we draw horizontal and vertical cuts across the area of the chip and count the number of nets crossing each of these cuts. We also measure the length of overlap of each macro with these cuts. We take a linear combination of these two quantities for each cut, square this sum, and sum these squares for all cuts to obtain the wiring penalty. One choice of linear combination gives the square of the chip width necessary to contain the wires and macros crossing a cut as the contribution of that cut to the penalty. We find that the macro width contribution to the penalty must be given a smaller weighting than this to reduce the unphysical correlations of this measure between widely separated areas on the chip but which lie on the same cut. A macro will move away from the cut at one edge because many macros are overlapping it at the other edge. On the other hand, some contribution from macro coverage of the cuts must be included in the penalty function to reduce congestion.

The penalty for overlap can be just the area of overlap, but a better measure is obtained if one weights the interiors of macros more than the edges. Then a small macro on top of a large macro will tend to move toward the edge. The strategy used in combining the penalty concerned with wirability with that for overlap will be mentioned later.

Next we must select a set of moves which can produce all possible configurations of the system in a rapid convenient way. Allowing overlap makes the definition of moves much easier. A move consists of a random length translation of a macro in either direction horizontally or vertically, a 90 degree rotation in either direction, or a vertical or horizontal reflection. Special macros, called IO objects, move around the edge of the chip with their pads facing out, and are also allowed to reflect in a center line perpendicular to the edge.

For the crudest method of optimization in this framework we would consider random moves in turn, carrying out those which reduce the penalty function and skipping those which cause it to increase. This procedure can only work when there exists a path to the desired minimum for which all steps are downhill. In general, reaching a configuration of reduced penalty requires coordinated motions of several macros. Performing such a motion in separate moves requires many steps which increase the penalty function.

To be able to reach the desired more remote, deeper minima simulated thermal equilibrium is used. The penalty function is treated as the energy of a given configuration, and the movement of macros is treated as random motion produced by contact with a thermal heat bath. This is accomplished by the Metropolis algorithm(Metropolis, Rosenbluth, Rosenbluth, Teller, and Teller, J. Chem. Phys. **21** 1087 (1953)). In this procedure one chooses moves at random and makes all moves which are downhill. One also makes or does not make uphill moves in a random fashion determined by the probability of this penalty increase in a "thermal" distribution.

According to Boltzmann's law, the configurations which are most probable at equilibrium are those with the lowest penalty function. The width of the distribution about the point of lowest penalty is narrowest for the lowest temperatures, but the rate at which equilibrium is achieved is very slow for low temperatures. Consequently we begin the simulation at a very high temperature in order to achieve equilibrium, and then slowly lower the temperature to maintain equilibrium and sharpen up the distribution about the state of lowest penalty. This is completely analogous to a physical annealing procedure.

In principle this procedure is sure to bring the system to the state of lowest penalty if carried out sufficiently slowly. In practice, the approach toward the optimal state becomes sluggish as the system becomes "cold", so that one settles for the configuration which results from the slowest annealing process feasible in the computer time allotted. The method produces good solutions in quite reasonable computation times if the step size is chosen large in the early stages of the process and progressively reduced as the more desirable configurations are reached. We also give greater weight to the wirability penalty at the beginning of the run and weight the overlap penalty more strongly at the end.

In the example to be presented the sum of the areas of the macros is 60% of the area of the chip. We introduce a border of free space around each macro with a width proportional to the corresponding thickness of the macro so that the total areas of the macros with their borders is 90% of the area of the chip. In the placing procedure we actually move the larger objects consisting of the macros and their border areas around the chip. In this way we use the striking ability of the method for packing different sized and shaped objects into a given area, at the same time reserving space for wiring around each macro.

The initial configuration used has all macros in the lower left corner. After some 500 trials the macros are spread randomly over the chip as shown in Fig. 1. After the full annealing procedure the configuration shown in Fig. 2 is obtained. When the macros are shown without their borders (Fig. 3), we find that suitable channels for wiring have been maintained about the larger macros. In subsequent work we have kept larger borders around the smaller macros to increase wiring space around them.

The set of macros shown in our figures were obtained from an actual IBM chip design, which had been placed (in a different arrangement) before our results were obtained. This other placement was simplified by combining some functionally related macros into supermacros. Thus it is interesting to see if our placement method put the separate parts of these supermacros near each other on the chip. In Fig. 3 each set of macros which were combined into a particular supermacro are shown with a common number. Large unnumbered macros were also supermacros in the other placement. Most of the supermacros appear as physically contiguous groups of the smaller macros in our placement. A few of them have been broken up. There is a group of small macros in the top middle section of the chip which forms a fairly compact mass in our placement, but which arise from pieces of several of the supermacros used in the previous placement. We believe that this collection of macros could have been profitably combined into a supermacro rather than parceled among the supermacros actually used.

Thus this algorithm can do grouping and partitioning as well as placement. Runs can also be made on the subsets of macros thus obtained to place them properly in supermacros of given shape. Finally the program can be used to place the supermacros. We see that the method is adaptable to hierarchical design.

At the present time we are working to include other factors in the evaluation of placements, in particular the constraints on placement which arise from power bussing. The framework is inherently simple, which makes it easy to add improvements and modifications for special situations.

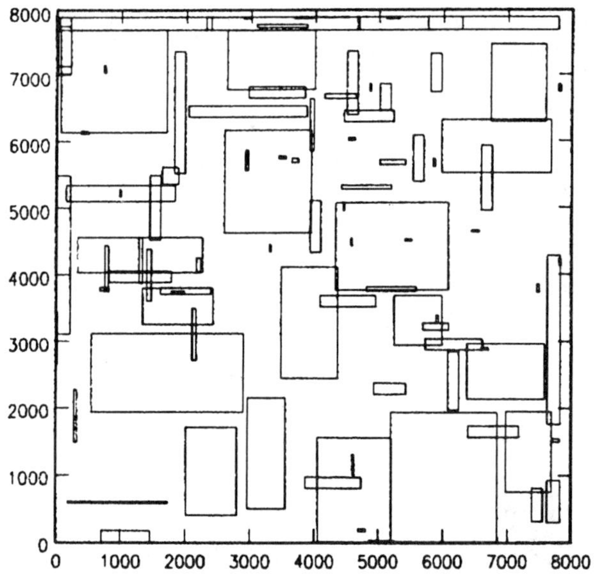

Fig. 1 Intermediate random configuration

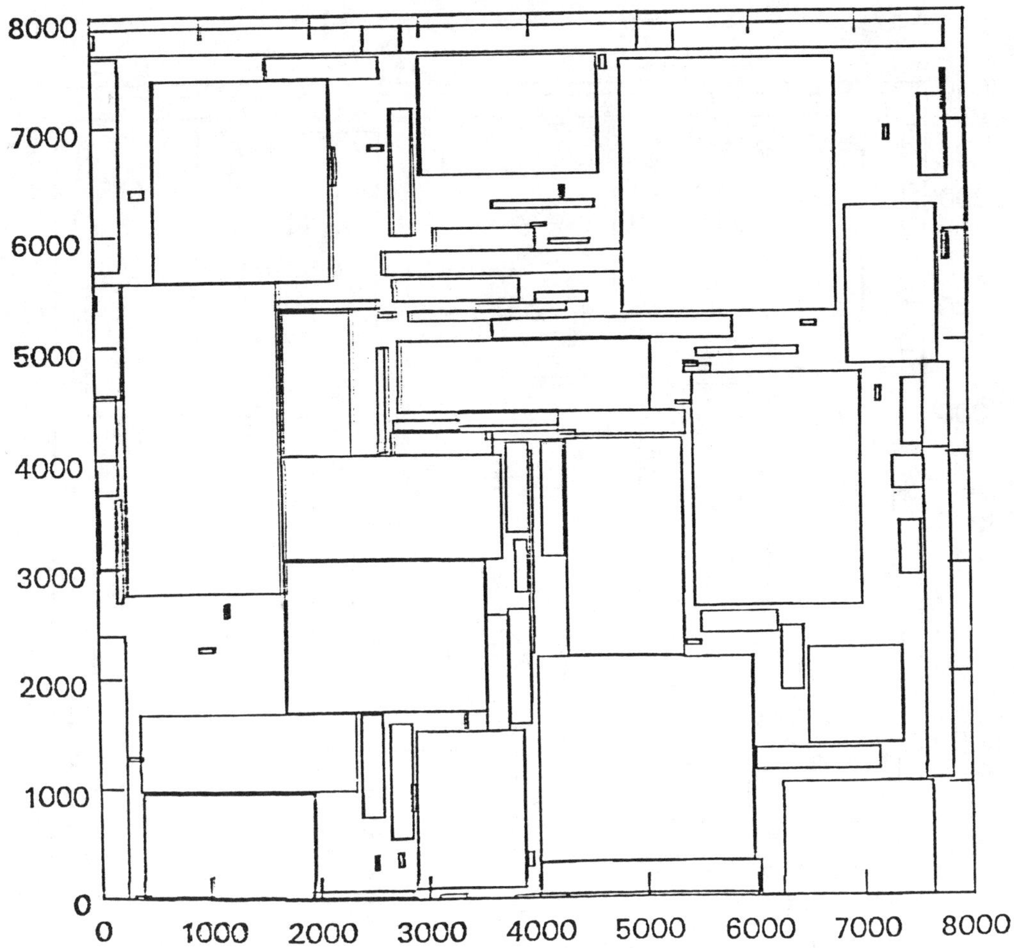

Fig. 2 Final configuration

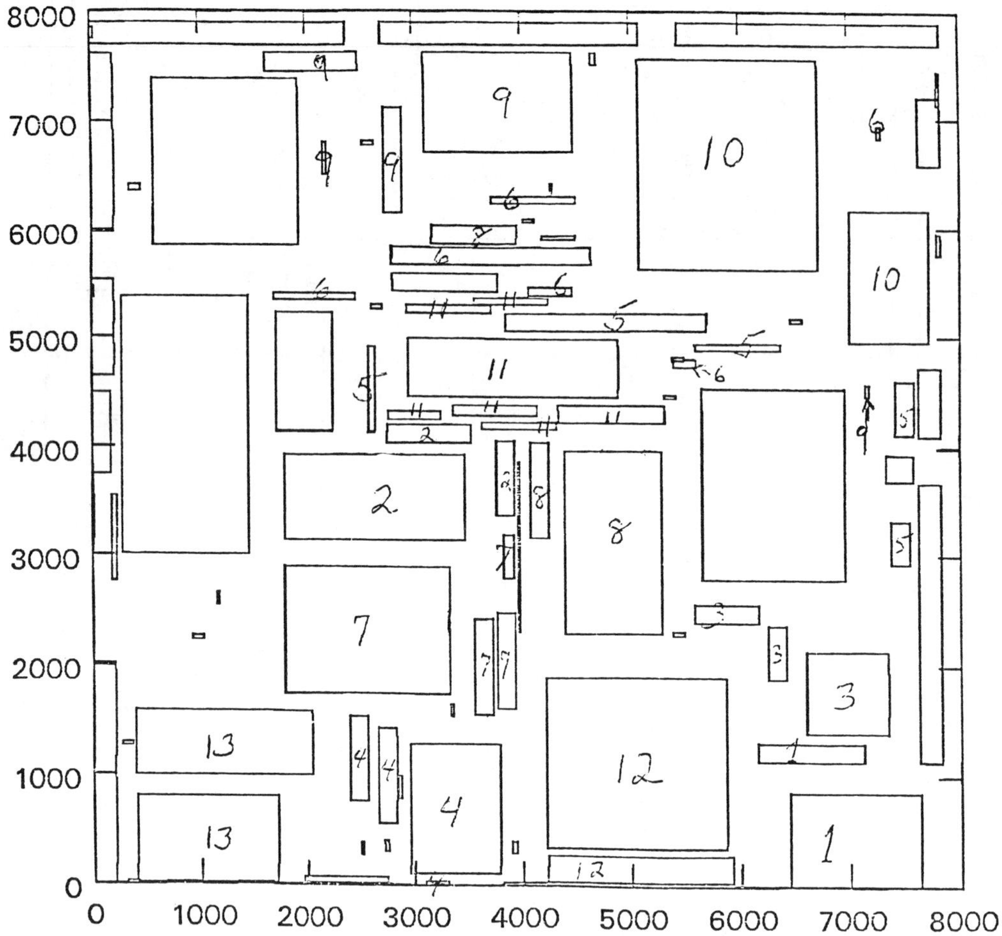

Fig. 3 Final configuration with border areas removed

EFFICIENT FLOORPLAN OPTIMIZATION

Ralph H.J.M. Otten

Thomas J. Watson Research Center
IBM corporation
P.O.Box 218
Yorktown Heights, NY 10598
United States of America

and

Eindhoven University of Technology
Department of Electrical Engineering
P.O.Box 513
5600 MB Eindhoven
The Netherlands

Abstract

This paper addresses the problem of assembling a number of rectangles, with constraints on size and shape, within one enveloping rectangle of which the dimensions are the arguments of an objective function to be minimized. The relative positions of the rectangles are given in the form of a slicing tree, or, equivalently, a series-parallel polar graph. This problem is an important part of floorplan design in a top-down layout design system. An algorithm for solving this problem in polynomial time is presented.

1. Introduction

The topology of rectangle dissections can be represented by polar graphs [1]. Given the rectangles, and given the topology, to find the smallest rectangle containing a configuration of the given rectangles without overlap is NP-hard [2,6]. However, under a restriction to series-parallel graphs, a restriction desirable for many other reasons [3,4], several of these *building block problems* have been shown to be solvable in polynomial time. If the only constraint on the individual rectangles is to have a given area, the problem can be solved for any given outer aspect ratio by simple depth first visit [5]. (Strictly speaking two depth-first visits, one for determining the areas of the dissected rectangles, and one for finding the dimensions and positions of all the rectangles.) By approximating the minimum area constraint by a piecewise linear bounding function (definitions will be given in the next section), the smallest enveloping rectangle and the associated dissection can be found efficiently for a mixture of minimum area rectangles and rectangles with a given minimum width [7]. This is important for realizing wiring space requirements if the channel width can be estimated with sufficient accuracy. Stockmeyer [6] gave an algorithm for finding the optimum orientations of rectangles with given dimensions in a given topology.

In this paper the solution to a more general problem is presented. In fact the restriction on the shape constraints to be imposed on the individual rectangles, is extremely weak. It essentially says that allowing one dimension to be bigger cannot force the other dimension to be bigger. Among the objectives that can be minimized by the algorithm to be presented in section 4, are the mostly used functions such as area and perimeter with or without restrictions on the aspect ratio.

Before presenting the algorithm a terminology is established in section 2, and in section 3 piecewise linear bounding functions are introduced and the way to combine them is explained.

2. Definitions

A *rectangle dissection* is a geometrical configuration consisting of a finite number of non-overlapping rectangles that together completely cover an enveloping rectangle.

A *slice* is either an undissected rectangle or a rectangle dissected into rectangles by parallel cutting lines.

A *slicing structure* is a rectangle dissection that can be obtained by recursively dividing rectangles into smaller rectangles by parallel lines.

The slices obtained by dissecting a slice over all its parallel cutting lines are called the *children* or *child slices* of the dissected slice and *siblings* or *sibling slices* of each other. By maintaining a consistent ordering among sibling slices (e.g. left to right, and top to bottom) each of the sibling slices except one has a unique next sibling separated from it by exactly one cutting line. The dissected slice is called the *parent* or *parent slice* of the obtained slices. An undissected rectangle in a slicing structure is called a *cell*. A slice containing more than one cell is called a *compound*.

The *topology* of a slicing structure is a tree with each slice represented by a node, and arcs from each node representing a parent to each node representing one of its children, leaving the parent's node in accordance with the ordering of the children.

The *longitudinal dimension* of a slice is the dimension it inherits from its parent slice. The other dimension is its *lattitudinal dimension*. The lattitudinal dimension of the common ancestor slice is the length of the side it transmits to its child slices. The other dimension is its longitudinal dimension.

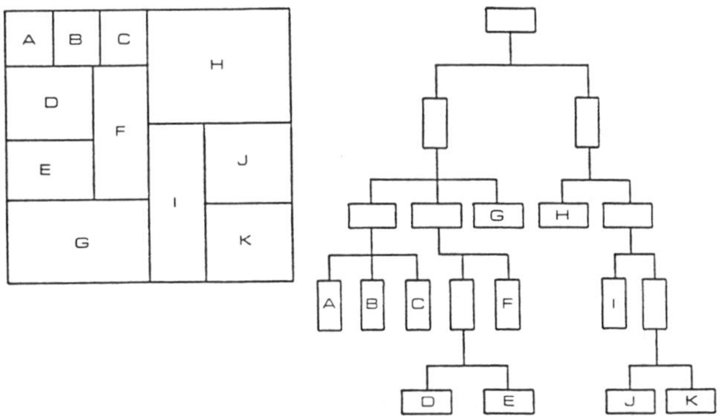

Figure 1. A slicing structure and its topology.
In this figure the leaves are labeled with the character in the corresponding undissected rectangle.

CH1935-6/83/0000/0499$01.00 © 1983 IEEE

A *bounding function* is a right-continuous, non-increasing, positive function of one variable defined for all real values not smaller than a given positive constant.

The *bounded area* of a bounding function f is the set of pairs of real numbers (x,y) such that f(x) is defined and $y \geq f(x)$.

The *inverse* f^{-1} of a bounding function f is a bounding function defining a bounded area with exactly those (y,x) for which (x,y) is in the bounded area of f.

The *shape constraint* of a rectangle is a bounding function of its longitudinal dimension. The bounded area of that function is the set of all permissible pairs of longitudinal and lattitudinal dimensions. (Figures 2, 3 and 4)

A *contour score* c is a function of two variables, defined for a convex subset Γ of the pairs of positive real numbers, which is quasi-concave and monotonously non-decreasing in its two arguments, i.e.

$$\forall_{x_0,x_1,y_0,y_1 \in \Gamma}[x_1 \geq x_0 \wedge y_1 \geq y_0 \Rightarrow c(x_1,y_1) \geq c(x_0,y_0)]$$

and

$$\forall_{x_1,x_2 \in \Gamma} \forall_{0 \leq \lambda \leq 1}[c(x_1) \leq c(x_2) \Rightarrow c(x_1) \leq c((1-\lambda)x_1 + \lambda x_2)]$$

3. Representation and manipulation of shape constraints

The input data for the algorithm is the topology of the slicing structure, shape constraints for the cells, and possibly some additional constraints on the shape of the enveloping rectangle. Two extreme examples of shape constraints are those of inset cells and those of flexible cells. An inset cell is a cell of which the detailed layout is stored somewhere, for example in a library of often occurring standard cells, or previously designed and stored cell layouts specific for one or a few circuits to be integrated. For this kind of cells the exact dimensions and pin positions are known. Sometimes even the orientation is given. In the latter case the algorithm only can assign a position to such a cell. If the orientation is not given, the algorithm has to determine the orientation that leads to the optimum geometry of the slicing structure. Since shape constraints are defined as bounding functions of the longitudinal dimensions, the shape constraint of a non-square inset cell with free orientation will be different from an inset cell with the same shape but with a fixed orientation. (Figure 2).

Often the detailed layout of a cell is not known. This is the case for cells that are further decomposed on an hierarchically lower level. The shape of such a cell depends on the geometry (floorplan) of this module which, in a top-down approach, is to be determined later. However, a fairly accurate estimate of the area the module will require, can be derived from the sizes of the elementary modules it contains, giving some additional area for wiring space. Sometimes a technology and style dependent algorithm will determine the inner layout of a cell. These cells often are quite flexible as far as their shape is concerned. Their area can be estimated from the functional specification and the rules of the target technology. Both kinds of cells, the further decomposed modules and the algorithmically specified cells, have a hyperbola of constant area as shape constraint. Very oblong rectangles, even if they satisfy the minimum area requirement, are not desirable, because they tend to increase the wiring space, and are difficult to realize. This should be reflected in the shape constraint by putting a lower bound on each of the dimensions. (Figure 3)

Inset cells have piecewise linear shape constraints. Such constraints can be conveniently represented by a sequential list of their breakpoints. This is not the case for flexible cells, and possibly other cell types occurring in practice. Of course, any shape constraint can be approximated by a piecewise linear bounding function with arbitrary accuracy. From the discussion of flexible cells in the preceding paragraph it is clear that a piecewise linear approximation with three breakpoints suffices considering the limited accuracy of any area estimation for the given examples. (Figure 4)

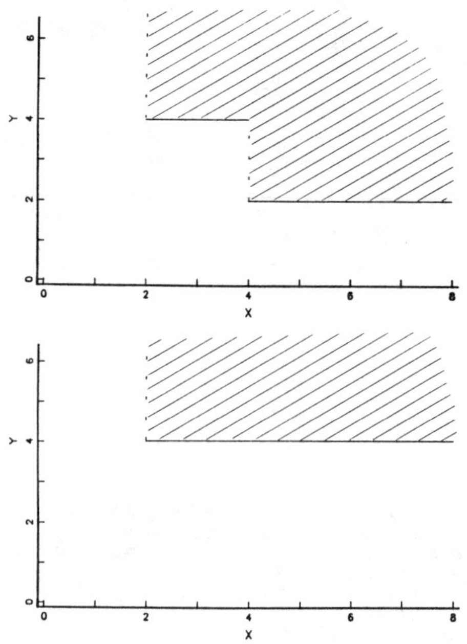

Figure 2. The shape constraint of an inset cell. If the orientation is free, the shape constraint consist of two pieces: one starting at (x,y) and the other at (y,x). For fixed orientation the shape constraint is a constant equal to the minimum lattitudinal dimension over the interval of allowed longitudinal dimensions. The shaded region indicates the bounded area.

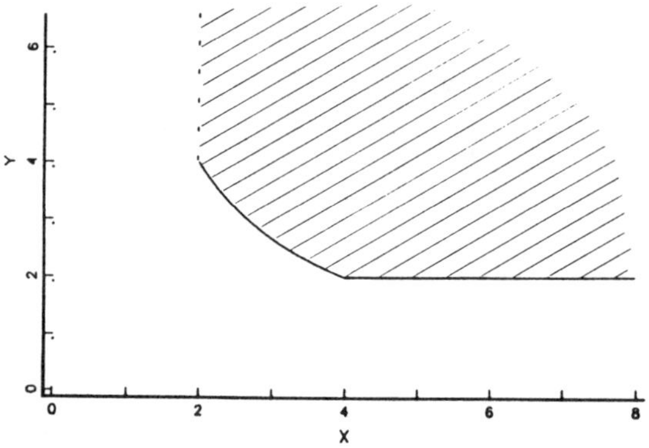

Figure 3. The shape constraint of a cell requiring a minimum area and minimum dimensions. The shaded region indicates the bounded area of the shape constraints.

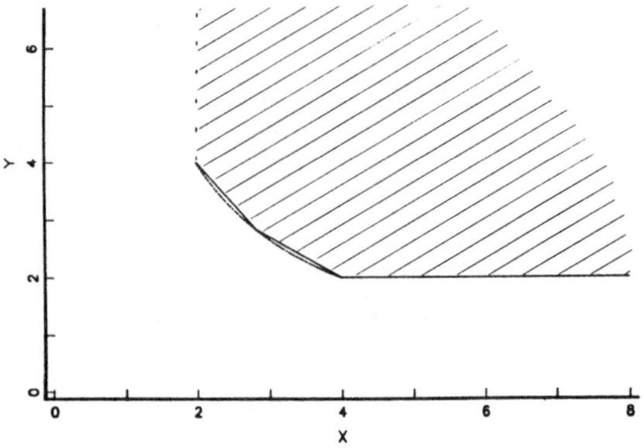

Figure 4. A piecewise linear approximation of the shape constraint of figure 3. The bounded area is indicated by shading.

Cells for which more breakpoints are desirable might exist. However, constraints not satisfying the requirements for shape constraint are not very probable. It would imply that allowing one dimension to be bigger would force the other dimension to increase too. So, an algorithm that accepts any piecewise linear shape constraint for a cell can satisfactorily solve all practical cases.

Part of the wiring will be allocated in spaces between the cells. For a more accurate floorplan design the size of these wiring areas can be estimated, and accounted for in the next floorplan calculations. Most methods for estimating the wiring space requirements deliver *channel widths* instead of areas. A minimum width constraint can be easily translated into a piecewise linear shape constraint. This shape constraint will be the same as that of an inset cell with fixed orientation, except that the function is defined over all non-negative real longitudinal dimensions. The constant value of this shape constraint is equal to the estimated width.

In many practical cases all areas bounded by the shape constraints will be convex. Of the above mentioned examples only the inset cells with free orientation define non-convex bounded areas. If all shape constraints define convex bounded areas previously published methods can be used [4,7]. If all cells are inset cells with free orientation Stockmeyer's algorithm [6] will find the optimum in polynomially bounded time. In the next section an algorithm will be described that can determine the optimum geometry for cells with piecewise linear shape constraints defining convex and non-convex bounded areas in $\mathcal{O}$(m.d.b)-time where m is the number of modules, d is the longest path from the root to a leaf in the structure tree, and b is the maximum number of breakpoints in a cell shape constraint.

The shape constraint of a compound slice can be derived from the shape constraints of its child slices. In the final configuration these child slices have to have the same longitudinal dimension, which is the lattitudinal dimension of their parent. The inverse of the compound's shape constraint is only defined on the intersection of the intervals on which the shape constraints of its children are defined. Its smallest possible longitudinal dimension for a given feasible lattitudinal dimension x is the sum of the values of the shape constraints of the children at x. So, the shape constraint of a compound is obtained by the addition of the shape constraints of its children in the interval where they are all defined, and determining the inverse of the resulting bounding function. These operations are easy for piecewise linear shape constraints, represented by a list of their breakpoints ordered according to the respective longitudinal dimensions. For each breakpoint of any child of which the first coordinate x is in the mentioned intersection, the shape constraints of all the children have to be evaluated and added. If the result is y, then (y,x) is a breakpoint of the parent's shape constraint. Ordering all these new breakpoints according to the y-values yields a consistent representation for the shape constraint of the corresponding compound slice.

4. The algorithm

The ability to obtain the shape constraints of a slice by adding the shape constraints of the child slices and inverting the result, enables us to obtain the shape constraint of the enveloping rectangle. The bounded area of that shape constraint is the set of all possible outer dimensions of the total configuration. A contour score always assumes its minimum value over the bounded area at the boundary determined by associated shape constraint. This is a consequence of the monotonicity of shape constraints and contour scores. For piecewise linear shape constraints that minimum will be assumed at at least one of its breakpoints, because of the quasi-concavity of the contour score. So, to find an optimum pair of dimensions for the common ancestor slice the contour score only has to be evaluated at the breakpoints of its shape constraint in the convex set of permissable pairs.

Given the longitudinal dimension of a slice and its shape constraint, its lattitudinal dimension can be found by evaluating its shape constraint for the given longitudinal dimension. After deriving the shape constraint for the common ancestor and determining a dimension pair for which the contour score assumes a minimum, the longitudinal dimensions of its children are known. So, for each of them the lattitudinal dimension, which in turn is the longitudinal dimension of its children, can be calculated. Continueing in this way will finally yield the dimensions of all slices in the configuration. If the shape constraint has a zero right derivative at the point where it has to be evaluated, some slack area will be included, i.e. the slice can be realized in a smaller rectangle without affecting its environment. In order to have the wiring channels connecting to other wiring channels at both ends this slack should be taken up by slices containing only one cell.

The algorithm consists of three parts:

1. Visit the nodes of the topology in depth-first order, and just before returning to the parent determine the shape constraint by adding the shape constraints of its children and inverting the result.
2. Evaluate the contour score for each of the breakpoints of the shape constraint of the common ancestor, and select a dimension pair for which the smallest value of the contour score has been found.
3. Visit the nodes of the topology in depth-first order, and before going to any of its children determine the lattitudinal dimension by evaluating its shape constraint for the inherited longitudinal dimension.

Clearly, when in the first step the same procedure has been applied to all children of a certain slice the shape constraints of these child slices are known and combining them in the way described yields the shape constraint of their parent. The process will end with determining the shape constraint of the common ancestor slice, and then the shape constraints of all slices are known. As explained in the beginning of this section the contour score will be evaluated at each of its breakpoints. The dimensions associated with the minimum value will become the dimensions of the enveloping rectangle. This means that after completing the second step the longitudinal dimension of the primogenitive (and all the other children) of the common ancestor is known. Together with the shape constraints this enough information to begin the process of the third step. At the beginning of a visit to a node in the structure tree, representing a certain slice, the lattitudinal dimension of that slice can be determined by evaluating its shape constraint at the value of the dimension that it inherits from its parent slice.

So, completing all three steps yields the dimensions of all slices in an optimum configuration for the given topology and cell shape constraints. To determine the position coordinates of the slices from these dimensions and the topology is straight forward. Also easy is to determine what orientation the inset cells can have in this optimum configuration.

5. An example

Figure 5 shows a result of the algorithm, a floorplan with 149 slices (56 cells, 38 compounds, and 55 wiring channels). The cells where a mixture of inset cells and flexible cells. The flexible cells had a limit of 2 on the aspect ratio. The optimization was for smallest area without restrictions on the aspect ratio or the outer dimensions. The topology was generated by the methods of [5]. The figure shows some non-used areas, a consequence of the presence of inset cells and aspect ratio constraints.

6. Acknowledgements

The author would like to thank Larry Stockmeyer and Phil Wolfe for several helpful discussions, especially the latter who introduced the idea of bottom up combination of slightly more restricted shape constraints.

7. References

1. Brooks, R.L.; Smith, C.A.B.; Stone, A.H.; Tutte, W.T.: *The dissection of rectangles into squares*, Duke Math. J., 7, 1940, pp. 312-340.

2. LaPaugh, A.S.: *Algorithms for integrated circuit layout: an analytic approach*, PhD-thesis, MIT, Laboratory for Computer Science, Cambridge, Mass, 1980.

3. Otten, R.H.J.M.: *Vlsi layout*, **Hardware and software concepts in vlsi**, edited by G.Rabbat, Van Nostrand Reinhold, New York, 1983, pp. 406-451.
   Also as: *Structured layout design*, EUT Report 80-E-111, Eindhoven University of Technology, Eindhoven, The Netherlands, 1980.

4. Otten, R.H.J.M.: *Layout structures*, Research Report RC 9657, IBM Thomas J. Watson Research Center, Yorktown Heights, NY, 1982.
   Also in: **Proceedings of the Large Scale Systems Symposium**, Virginia Beach, Virg., 1982, pp. 349-353.

5. Otten, R.H.J.M.: *Automatic floorplan design*, Research Report RC 9656, IBM Thomas J. Watson Research Center, Yorktown Heights, NY, 1982.
   Also in: **Proceedings of the 19th Design Automation Conference** Las Vegas, Nev, 1982.
   Also in: **Vlsi technologies in graphics**, edited by H. Fuchs, IEEE Computer Society, 1983.

6. Stockmeyer, L.J.: *Optimal orientations of cells in slicing floorplan design*, Research Report RJ 3731, IBM Research Laboratory, San Jose, Cal, 1982.

7. Otten, R.H.J.M.; Wolfe, P.S.: *Algorithm for wiring space assignment in slicing floorplans*, IBM Technical Disclosure Y0882-0682, 1982.

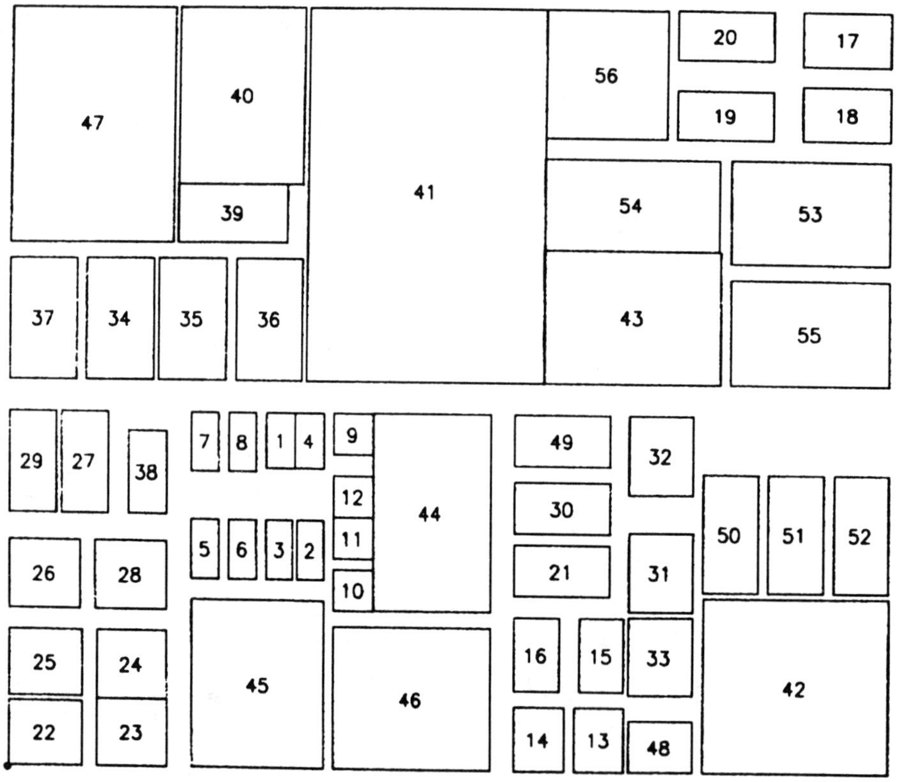

Figure 5. A result of the algorithm presented in section 4.

STATE ANALYSIS OF LARGE PIECEWISE LINEAR SYSTEMS*

J.Th.J. van Eijndhoven, J.A.G. Jess

Eindhoven University of Technology
Eindhoven, the Netherlands

INTRODUCTION AND SUMMARY

A multimode multilevel simulation program based on the piecewise linear modelling technique is currently under development. At the time of writing the program has been used for the state analysis of multilevel systems up to 320 interconnected elementary piecewise linear models with an equivalent of 512 boundary hyperplanes on a PDP11/60 minicomputer. This paper describes some aspects of latency and sparsity, which characterise the algorithms used in the program.

HIERARCHICALLY STRUCTURED PIECEWISE LINEAR SYSTEMS

A simple (piecewise linear) system is described by

$$\left.\begin{array}{l} 0 = Ax + Bi + g \\ v = Cx + Di + h \\ v \geq 0, \quad i \geq 0, \quad v^t.i = 0 \end{array}\right\} \qquad (1)$$

All entities in this equation are real vectors or matrices respectively. In particular x represents the port variables through which the system communicates with its environment. The vectors v,i represent internal variables. Their values characterise the region of operation. For instance $v \geq 0$, $i = 0$ characterises the region defined by $Cx + h \geq 0$ in the x-space. In this region the operation of the system with respect to its ports is given by the terminal constraint equation $0 = Ax + g$.

Simple systems can be connected together to establish compounds. Of course compounds can again be connected together to form higher level compounds. Thus we obtain a hierarchically structured description of a large piecewise linear system. The description can be symbolized by a directed tree graph (V,E). A vertex in the tree graph is a compound or a simple system. A directed arc from vertex $\sigma\epsilon V$ to vertex $\tau\epsilon V$ in the tree graph (notation $[\sigma,\tau\rangle \in E$) implies that τ is a subsystem of σ. The vertices with no outgoing edges are the "leaves" of the tree graph. All leaves are simple systems.

The tree has exactly one vertex with only outgoing edges. This vertex is called the "root". Any other vertex may be seen as the root of some subtree. Assume $\tau\epsilon V$ is the root of some subtree then the subtree is denoted by $(V(\tau),E(\tau))$. We agree upon $\tau\epsilon V(\tau)$. All vertices in $V(\tau)$ are the descendants of τ. The vertex set $\Sigma(\tau) = \{\sigma\epsilon V(\tau) \,|\, [\tau,\sigma\rangle \in E(\tau)\}$ is the set of sons of τ. Vertex τ is the father of all vertices in $\Sigma(\tau)$.

*This work was supported by the Netherlands Organization for the advancement of pure research (Z.W.O.)

With any vertex $\tau\epsilon V$ we associate a vector of terminal variables x^τ. Furthermore there are constraints of the form $0 = A^\tau x^\tau + g^\tau$ that bind the terminal variables as a consequence of the physical working of the compound system.

We have not bothered about which variable in x is an input and which is an output variable. In fact the number of output variables in x^τ is equal to the number of rows of A^τ. Let this number be n^τ. Then any nonsingular $n^\tau \times n^\tau$ submatrix of A^τ defines an admissible set of output variables. In fact the columns of such a submatrix define the output variables. Assume a choice being made then $x^\tau = [(x^\tau_{out})^t, (x^\tau_{in})^t]^t$ supposed to indicate the partition of x^τ into input and output vector respectively.

Since A^τ changes from region to region an admissible choice for one region may not be an admissible choice for some other region. Consider the example of Fig. 1. In the regions $x_1 < 1$ and $x_1 > 1$ the terminal constraints are

$$[0] = [0,1]\begin{bmatrix}x_1\\x_2\end{bmatrix} + [0]$$

and

$$[0] = [0,1]\begin{bmatrix}x_1\\x_2\end{bmatrix} + [-1]$$

respectively and x_2 is an admissible output variable. In the region $0 < x_2 < 1$ we have

$$[0] = [1,0]\begin{bmatrix}x_1\\x_2\end{bmatrix} + [-1]$$

and x_1 is an admissible output variable.

Let us consider a compound system τ and consider the subsystem associated with the vertex set $\{\tau\}\cup\Sigma(\tau)$. This stands for a network of interconnected systems $1,\ldots,k\epsilon\Sigma(\tau)$ with terminal variables x^τ and internal variables equal to the terminal variables $x^1,\ldots,x^k$. Given appropriate excitations it must be possible to solve this network. More than that, since within a region all constraint equations are linear, the network can be solved by Gaussian elimination. The network equations are as follows:

$$\begin{array}{l} A^1x^1 + g^1 = 0 \\ A^kx^k + g^k = 0 \end{array} \qquad (2)$$

$$S^\tau x^\tau + S^1x^1 + \ldots + S^kx^k + s^\tau = 0 \qquad (3)$$

In fact this is a tableau set up. Eq.'s (2) stand for the branch constitutive relations. Eq.'s 3 are topological and describe the interconnection of the subsystems and the relation to the terminal variables x^τ. If m^k is the length of the vector x^k then the number of rows in eq.(2) is

$$\sum_{i=1}^{k} (m^i - n^i) + n^\tau$$

and n^τ must be the rank of A^τ.Moreover any set of n^τ linearly independent columns of A selects a set of admissible output variables from x^τ.The remaining coordinates of x^τ are then input variables.
Fig. 2 is an attempt to render some insight into the structure of the equations (2) and (3).

ALGORITHMS FOR THE SOLUTION OF PIECEWISE LINEAR NETWORKS

Various algorithms have been published to deal with large piecewise linear systems.Among those the derivatives of the algorithms of Katzelenson [6] , Lemke [7], Dantzig-Cottle [8] and van der Panne [9] constitute a family.Any of those algorithms requires some of the following steps at some instance of its execution:

(1) Given an admissible excitation vector x_{in}^τ where τ is the root. Compute x_{out}^τ and the terminal vector x^σ of all subsystems $\sigma\epsilon V/\{\tau\}$
(2) Given an admissible incremental excitation vector Δx_{in}^τ. Compute Δx_{out}^τ and the incremental terminal vectors Δx^σ of all subsystems $\sigma\epsilon V(\tau)/\{\tau\}$
(3) Given the terminal constraint equations of all (sub)systems assume that the terminal constraint of some leaf σ has been changed after x^σ has penetrated a hyperplane, compute the new terminal constraints for all subsystems.
(4) Given a "perturbation term" $b^\sigma \Delta i^\sigma$ in the terminal constraint equation of some leaf σ and a constant admissible excitation vector x_{in}^τ of the root τ compute $\Delta x_{out}^\tau/\Delta i^\sigma$ and $\Delta x^\xi/\Delta i^\sigma$ for all subsystems $\xi\epsilon V/\{\tau\}$.

For the second and the third problem programs are indicated in the Appendix. The fourth problem requires a merge of the procedures "trace" and "dispatch". First incremental Δg^τ vectors have to be computed starting at leaf "σ" (where $\Delta g^\sigma = b^\sigma$) following the unique path up to the root applying a modification of the "trace"-procedure.This process comes to a standstill if for some τ on that path $\Delta g^\tau = 0$. Then we recompute Δx^σ for all $\sigma\epsilon V(\tau)/\{\tau\}$ using a modification of the dispatch procedure.

The actual algorithms all have their specific way to determine the trajectory along which the solution is sought. Along this trajectory there occur crossings of boundary hyperplanes. We call these crossings "events". In fact any of the algorithms maintains some sort of an event table in a typical way. In the last section we demonstrate the way this is done by the Katzenelson algorithm.

LATENCY, BLOCKDECOMPOSITION, SPARSITY

Latency indicates itself in the algorithms "dispatch" and "trace". In "dispatch" we make the recursive call dispatch$(\Delta x^\sigma,\sigma)$ dependent on whether $\Delta x^\sigma \neq 0$ since else no events will be generated in subtree $(V(\sigma),E(\sigma))$. The example in the next section will give an indication of the increase of computational efficiency.
In "trace" the correcting of terminal constraints is stopped at vertex τ if A^τ,g^τ is not changed. This means that a change of the terminal behaviour of some leaf is not "visible" at the terminals of and any other compound further up in the hierarchy . Again recomputation of incremental port vectors is restricted to the subtree $(V(\tau),E(\tau))$, where "dispatch" may cut further paths. The equations in Fig.2 are tableaus and thus inherently sparse. Their processing is crucial for the efficiency. The most general approach is to see the submatrix

$$[(T_{11}^\tau)^t,(T_{21}^\tau)^t]^t$$

as a bipartite graph and consider its canonical decomposition into irreducible components [5] which in fact yields the unique finest block triangular decomposition. Since the matrix has more rows than columns the decomposition must end up with a "tail", that is a rectangular block in the lower right corner of the (appropriately permuted) matrix. From this tail we discard n^τ rows applying sparse pivoting, such that the remaining square block can be processed optimally. This way we obtain a new partition

$$[(\hat{T}_{11}^\tau)^t,(\hat{T}_{21}^\tau)^t]^t$$

such that $\hat{T}_{11}^\tau$ is square and block triangular. The problem with this approach is that the structure of the terminal constraint matrices A is not fixed.This would in the worst case imply that the block triangular decomposition related to compound τ has to be done over every time one of the matrices $A^\sigma(\sigma\epsilon\Sigma(\tau))$ changes.
A better approach is the following: the leaves are generally library entries and most closely related to physical building blocks. From there we may obtain a clue which entries in the A of some leaf may be nonzero or large and which others may be zero or small in all regions of operation. Now before going into the block triangular decomposition of those compounds whose sons are all leaves set all coefficients of the first kind to one and all of the second kind to zero. The decomposition will force the ones into the square blocks and the zeros into upper triangular offblock area as much as possible. In proceeding bottom up to some higher level component σ the block structure of the associated tableau may yield an indication about the coefficients in A^σ. (Our example will give some insight with respect to this point). Otherwise we assume A^σ to be full.
This way of processing has some remarkable properties:

- if we assume that all A-matrices are full, the rows of any A-matrix will always be together in one block (this will be proven elsewhere;
- if for some A^σ a square $n^\sigma \times n^\sigma$ nonsingular submatrix of A^σ forms a block by itself this means that there is a sort of a unilateral signal flow through the respective subsystem . This flow is from the associated admissible inputs to the respective admissible outputs and may or may not fit with physics (an example of the latter kind is an operational amplifier: mostly its admissible output will be assigned to its physical input);
- refinement of some block with A^σ into smaller blocks can only be obtained by observing the zero nonzero structure of A^σ;
- if the rows of $A^\sigma 1$ and $A^\sigma 2$ are together in one block they share a common feedback loop;
- if pivots become zero in some block the searching of a new pivot sequence may be restricted to the block in question.

EXAMPLE

As an example we consider a matrix arrangement of 25 threshold elements. For simplicity the example is kept somewhat artificial, but it shows all essential features and all steps can be verified by hand. The threshold element is formally described by:

$$0 = \begin{bmatrix}-1,1\end{bmatrix}\begin{bmatrix}y\\x\end{bmatrix} + \begin{bmatrix}1,-1\end{bmatrix}\begin{bmatrix}i_1\\i_2\end{bmatrix} + \begin{bmatrix}-1\end{bmatrix}$$

$$\begin{bmatrix}v_1\\v_2\end{bmatrix} = \begin{bmatrix}0 & 1\\0 & -1\end{bmatrix}\begin{bmatrix}y\\x\end{bmatrix} + \begin{bmatrix}1 & 0\\0 & 1\end{bmatrix}\begin{bmatrix}i_1\\i_2\end{bmatrix} + \begin{bmatrix}-1\\2\end{bmatrix}$$

$$v_1, v_2, i_1, i_2 \geq 0 \quad v_1 \cdot i_1 + v_2 \cdot i_2 = 0$$

The first two lines of the matrix are shown in Fig. 3. The block triangular decomposition $[T_{11}^t, T_{21}^t]^t$ is shown in Fig. 4. As is expected from inspection the decomposition yields a triangular matrix. Table 1 describes the essential steps of the Katzelenson algorithm.It shows clearly that most operations have a fairly local character. All the activity in the system is on and above the diagonal of the matrix network. Finally Fig. 5 shows the function $x_{45}(\lambda)$ in the range $\lambda \in [0,3]$.This function is sort of representative for all the others.

CONCLUSION

This paper has the intention to show that piecewise linear modelling techniques can be adapted to large multilevel systems. Unilateral signalflow such as it often occurs in logic networks can be automaticcally detected using block decomposition techniques. This is important in view of the possible application of waveform relaxation, when the program is extended to handle time series analysis.
The example given shows that the simulation program can be easily made to operate fairly locally in the network it analyses, very much in the sense of an eventdriven logic simulator.

APPENDIX

```
begin comment program for case 2;
    procedure dispatch (Δx^τ,τ);
    begin if τ not leaf than
        begin for all σ∈Σ(τ) compute Δx^σ using T^τ_11 Δx^τ = -T^τ_12 Δx^τ;
              for any σ∈Σ(τ) do if Δx^σ ≠ 0 then
              dispatch (Δx^σ,σ);
        end
    end
    τ = root; compute Δx^τ_out from A^τ Δx^τ = 0;
    dispatch (Δx^τ,σ);
end

begin comment program for case 3;
    procedure trace (σ);
    begin if new A^σ,g^σ ≠ old A^σ,g^σ and σ not root then
        begin τ: = father of σ;
            compute A^τ: = T^τ_22 - T^τ_21 (T^τ_11)^-1 T^τ_12,
                    g^τ: = w^τ_2 - T^τ_21 (T^τ_11)^-1 w^τ_1;
            trace (τ);
        end
    end
    σ: = leaf where a penetration of a boundary hyperplane occured;
    trace (σ);
end
```

REFERENCES

1 W.M.G. van Bokhoven, "Piecewise linear modelling and analysis", Kluwer, Deventer, the Netherlands, 1981.

2 J.Th.J. van Eijndhoven, J.A.G. Jess, "The solution of large piecewise linear systems", Proc. of the IEEE ISCAS, pp.597-600, May 1982.

3 J.A.G. Jess, "Piecewise linear simulation : datastructures algorithms and modelling", Proc. of the IEEE LSSS, pp.354-359, Oct. 1982.

4 J.A.M. Brooijmans, J.Th.J. van Eijndhoven, J.A.G. Jess, H.O.Koopmans, A. Martis, "A new multilevel simulation program for large systems based on piecewise linear modelling", Proc. of the ECCTD, Stuttgart, Germany, September 1983.

5 A.L. Dulmage, N.S. Mendelsohn, "Coverings on bipartite graphs", Can.J.Math., vol.10, pp 517-534, 1958.

6 J. Katzenelson, "An algorithm for solving nonlinear resistive networks", Bell System Technical Journal, vol. 44, Oct.1965, pp. 1605-1620.

7 C.E.Lemke, "On complementary pivot theory", in: Mathematics of the decision sciences, Part I, Eds. G.B.Dantzig and A.F.Veinott Jr., New York, Academic Press, 1950.

8 R.W. Cottle, G.B. Dantzig", Complementary pivot theory of mathematical programming", Lin.Algebra and its applications, Vol. 1, 1968, pp. 103-125.

9 C. van der Panne, "A complementary variant of Lemke's method for the linear complementarity problem", Math.Programming 7 (1974), pp.283-310.

λ value of event	change of matrix entries of term. constraints	change of increments	eventlist (new events underlined)
λ = 0	initialization	initialization	1.0
λ = 1	$a_{11}, a_{12}, a_{13}, a_{14}, a_{15}$	$\Delta x_{21}, \Delta x_{22}, \Delta x_{23}, \Delta x_{24}, \Delta x_{25}$	1.2, 1.25, 1.33, 1.5, 2.0
λ = 1.2	a_{25}	Δx_{35}	1.25, 1.33, 1.4, 1.5, 2.0
λ = 1.25	a_{24}	$\Delta x_{34}, \Delta x_{35}$	1.33, 1.4, 1.5, 2.0
λ = 1.33	a_{23}, a_{35}	$\Delta x_{33}, \Delta x_{34}, \Delta x_{35}, \Delta x_{45}$	1.4, 1.4286, 1.5, 1.667, 2.0
λ = 1.4	a_{25}	$\Delta x_{35}, \Delta x_{45}$	1.4286, 1.5, 1.667, 2.0
λ = 1.4286	a_{34}, a_{35}, a_{45}	Δx_{44}	1.5, 1.667, 2.0
λ = 1.5	a_{22}, a_{24}	$\Delta x_{32}, \Delta x_{33}, \Delta x_{34}, \Delta x_{35}, \Delta x_{44}, \Delta x_{45}, \Delta x_{55}$	1.6, 1.667, 2.0
λ = 1.6	$a_{33}, a_{34}, a_{44}, a_{45}, a_{55}$	Δx_{43}	1.667, 2.0
λ = 1.667	a_{23}	$\Delta x_{33}, \Delta x_{34}, \Delta x_{35}, \Delta x_{43}, \Delta x_{44}, \Delta x_{45}, \Delta x_{55}$	2.0
λ = 2.0	$a_{11}, a_{12}, a_{13}, a_{14}, a_{15}, a_{22}, a_{32}, a_{33}, a_{43}, a_{44}, a_{54}, a_{55}$	$\Delta x_{21}, \Delta x_{22}, \Delta x_{23}, \Delta x_{24}, \Delta x_{25}, \Delta x_{32}, \Delta x_{33}, \Delta x_{34}, \Delta x_{35}, \Delta x_{43}, \Delta x_{44}, \Delta x_{45}, \Delta x_{54}, \Delta x_{55}$	3.0
λ = 3	none	$\Delta x_{11}, \Delta x_{12}, \Delta x_{13}, \Delta x_{14}, \Delta x_{15}$	empty

Table 1 : Katzenelson algorithm applied to the matrix threshold network

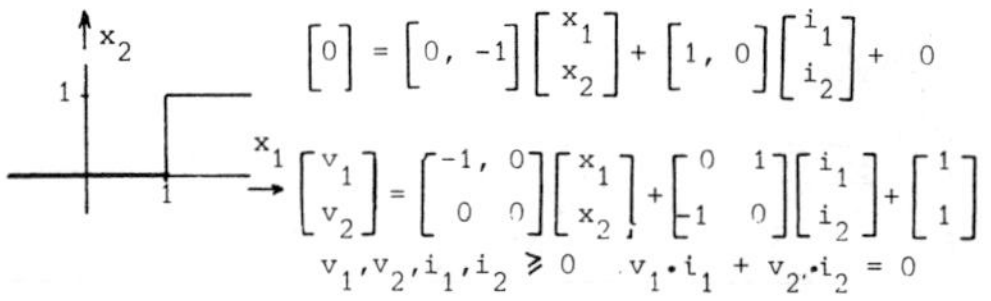

Fig. 1 : Example of a simple system, where the admissible output variable depends on the region of operation

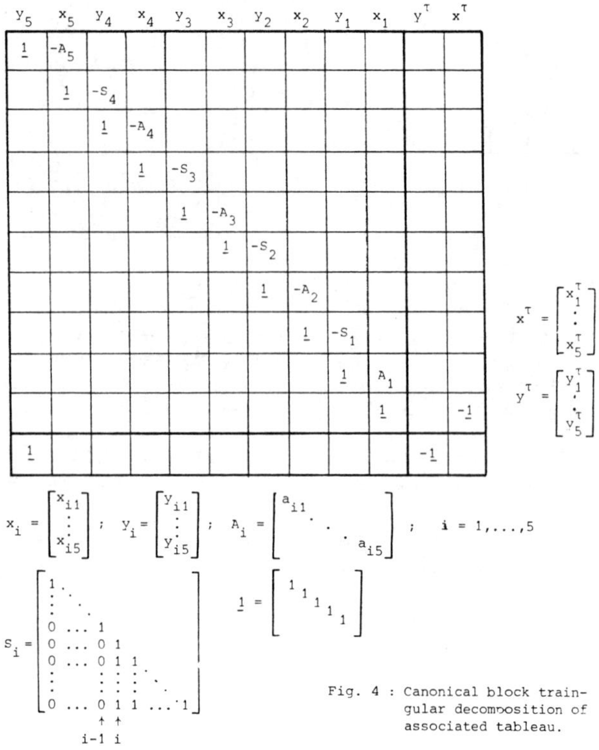

Fig. 4 : Canonical block traingular decomposition of associated tableau.

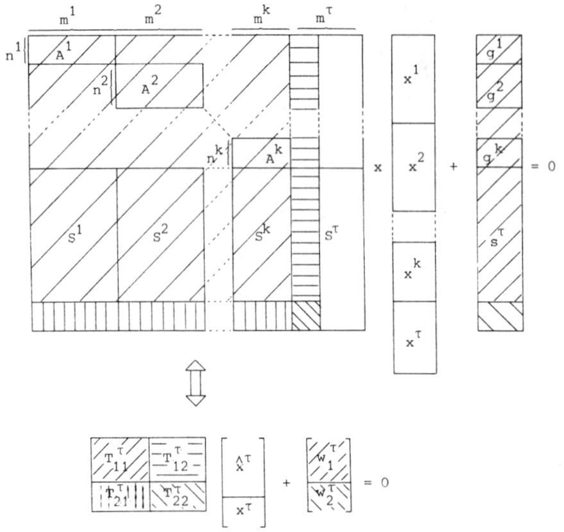

Fig. 2 : Tableau associated with compound "τ"

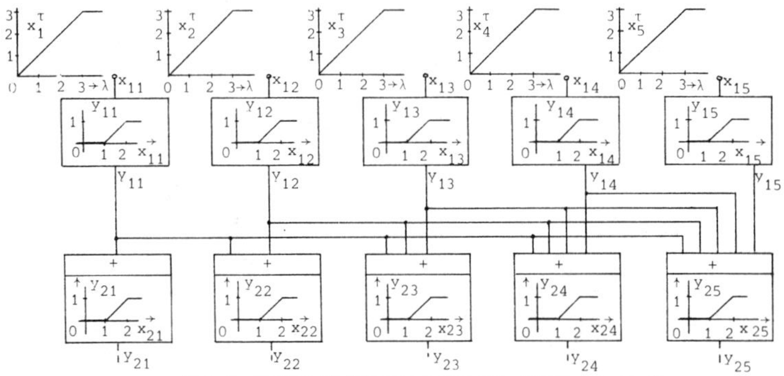

Fig. 3: Matrix of threshold elements (first two rows)

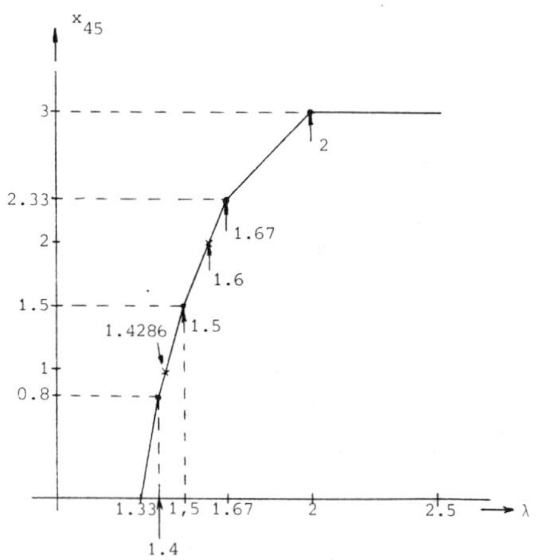

Fig. 5: x_{45} as a function of λ under the execution of the Katzenelson algorithm.

Session 3-1a
Josephson I

Chairman

J. Haddad
IBM

CIRCUIT DESIGN FOR A FULLY-DECODED HIGH-SPEED JOSEPHSON 1K-BIT RAM

A. Ishida, M. Yamamoto, K. Miyahara, and Y. Yamauchi

Atsugi Electrical Communication Laboratory, NTT.
1839 Ono, Atsugi, Kanagawa, Japan

ABSTRACT---Circuit design for a fully decoded high speed Josephson RAM, employing an NDRO ring memory cell and a current steering loop logic, is described. A cell configuration with a buffer circuit and a newly-designed single shot pulse generator brought a simple and high speed RAM control circuit including an on-chip timing generator. The circuit designed here was experimentally evaluated by integrating it in a 1K-bit RAM by means of 5-μm Pb-alloy technology. Read access time of 3.3 ns was obtained with the associated power dissipation of as low as 2 mW.

1. INTRODUCTION

Nondestructive readout (NDRO) Josephson ring memory cells and dc-powered current steering loop logic have been extensively investigated to realize a high speed fully-decoded RAM.(1)(2) However, two serious problems have never been solved: the improvement in operating margin of cell-to-driver interconnection and the best configuration of a high speed on-chip timing generator for random access memory operation.

In this paper, a new configuration of the memory cell with a wide interconnection margin and an edge-triggered single shot pulse generator which greatly simplifies the on-chip timing design are proposed. Design of a fully-decoded 1K-bit RAM with an on-chip timing generator, employing them, are discussed. Also, the circuit fabricated by 5-μm Pb-alloy technology is tested and evaluated experimentally.

2. MEMORY CELL DESIGN

The memory cell studied here is basically a quantized ring cell which stores double magnetic flux quanta.(1)(2) The flux is induced within a superconducting loop by switching the write gate; the presence of the stored flux is sensed by the interferometer sense gate. One design approach for this type of memory cell is minimization of the cell occupied area.(2)

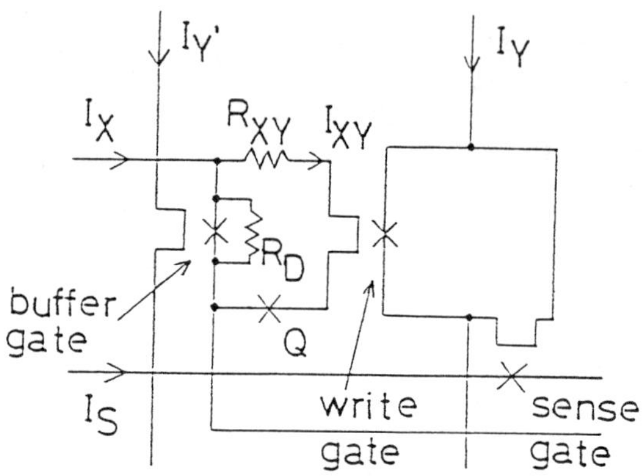

Fig. 1 Circuit configuration of NDRO ring cell with a buffer circuit.

This approach, however, causes difference in operating current level between four cell-driving signals, which results in a narrow cell-to-driver interconnection margin as well as complexity in cell driving circuits.

The approach presented here is how to realize a wide margin interconnection and a simple configuration for RAM control circuit. One solution is to design all cell driving current levels equal to each other. For this purpose, an additional gate, which is called a buffer gate,(3) was employed in a manner as shown in Fig. 1. The buffer gate consists of a double-junction interferometer with a single control input Iy', being biased by an X-selection line Ix. The output Ixy from the buffer gate is fed to the write gate of three-junction interferometer. As a result, cell driving with the equal current level and a wide cell-to-driver interconnection margin of 28% were successfully achieved with a slight increase in the cell occupied area.

With this three gates configuration, however, switching of the buffer gate must not only generate a large output current Ixy exceeding the write gate threshold, but also completely reset the Ix line of a X-driver loop. For this problem, a switched resistor configuration was designed : the small resistance Rxy as a buffer gate load is dominant in the stage of Ixy output, and the large one Rd is dominant in the stage of Ix resetting. Switching of the dominant resistor is performed using a single Josephson junction Q which operates as a superconducting switch.

3. DESIGN FOR TIMING CONTROL

In the present design, all cell drving signals are generated by a current steering loop logic.(2) The readout scheme is substantially represented in Fig. 2. For reading, the memory cell is selected through coincident driving of the X-selection sense line Is and the Y-selection line Iy. When the flux is stored in the selected cell, the sense gate switches to transfer the current Is back to the Is-driver branch. The back-transferred current resets the sense bus loop, and the readout data are finally picked up by the data gate.

In this readout scheme, the sense line Is, sense bus Ibus, and Y-selection line Iy should be driven in that order to avoid a latch-up failure.(2) Namely, the sense bus must be driven after the sense line has been driven completely because incomplete Is driving leaves a control current to the sense bus gates behind, which leads to a latch-up of the sense bus loop. Similarly, the resetting of sense line Is caused by Iy driving must be performed after complete Ibus driving.

To fullfill these requirements, a positive and a negative edge triggered single shot pulse generators, called POS and NOS, consisting of two cross-coupled logic loops, were newly designed.(4) The POS circuit generates a single shot pulse P at the rising edge of the input. On the other hand, the NOS circuit generates a delayed output $\tilde{N}$ at the rising edge and a single shot pulse N at the falling edge. The circuit configuration and operation principle for NOS are shown in Fig. 3.

The time sequence for readout operation is obtained as follows: The external chip select input CS activates a CS buffer, the output of which triggers a POS-X circuit via A4 address latch and a buffer circuit BUF-X. The single shot pulse P from POS-X drives the Is line and NOS. The output $\tilde{N}$ from NOS drives the Ibus loop, and N drives the Iy line via BUF-Y and POS-Y. In this way, a driving sequence Is-Ibus-Iy is achieved.

In a similar way, a driving sequence of the order Ix-Iy-Iy' for writing "1" and that of the order Ix-Iy' for writing "0" were also obtained. A RAM control circuit based on the timing design stated above is schematically shown in Fig. 4.

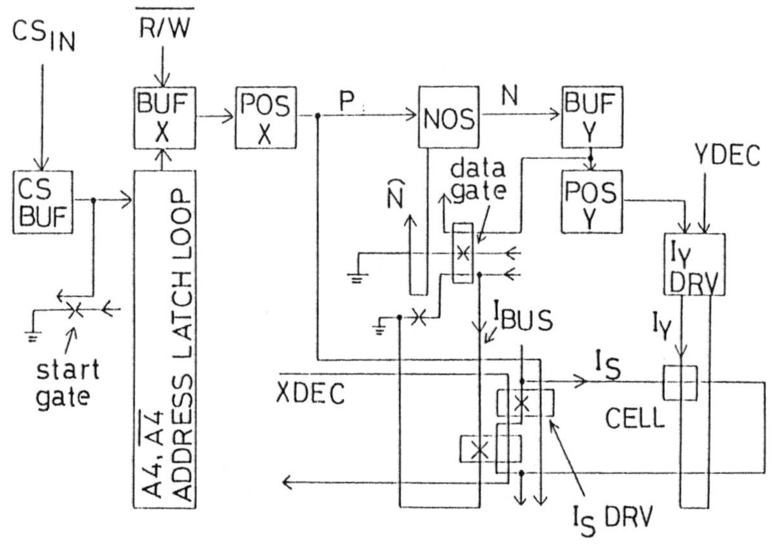

Fig. 2 Schematic drawing of the read timing generation. The gates consist of double junction interferometer.

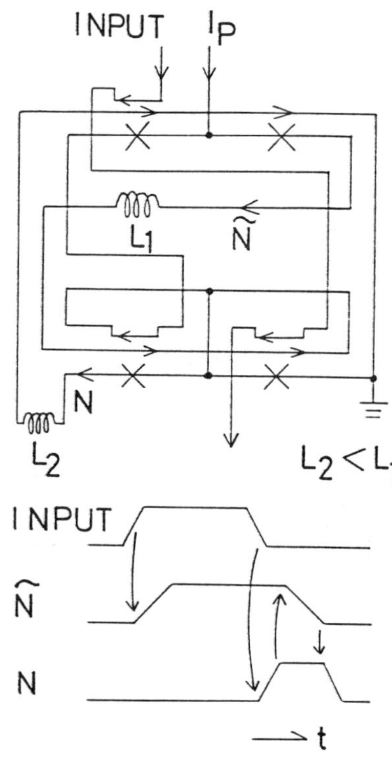

Fig. 3 Configuration and operation principle of the negative edge triggered single shot pulse generator, NOS.

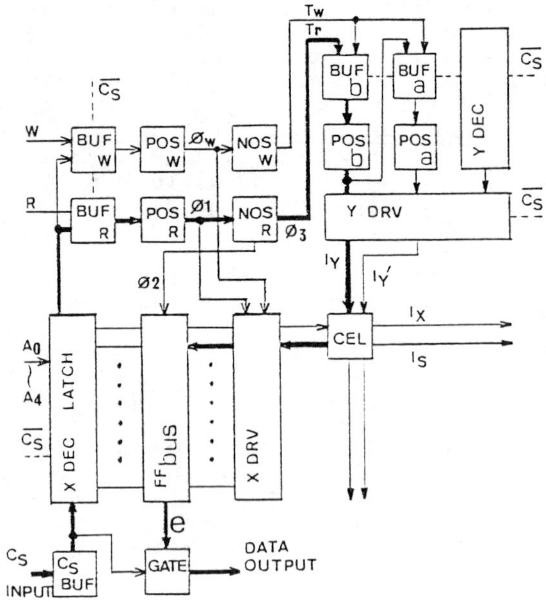

Fig. 4 Schematic representation of the RAM control circuit designed here.

4. DESIGN FOR ADDRESS DECODING

For address decoding, single-rail external address inputs, Ain, must be converted into dual-rail signals of true and complement, A and $\bar{A}$. For this purpose, an address buffer circuit was designed by employing an interferometer gate with an offset input as shown in Fig. 5. A pair of current steering logic loops are provided for generating a true and a complement outputs. An address input is fed to both the

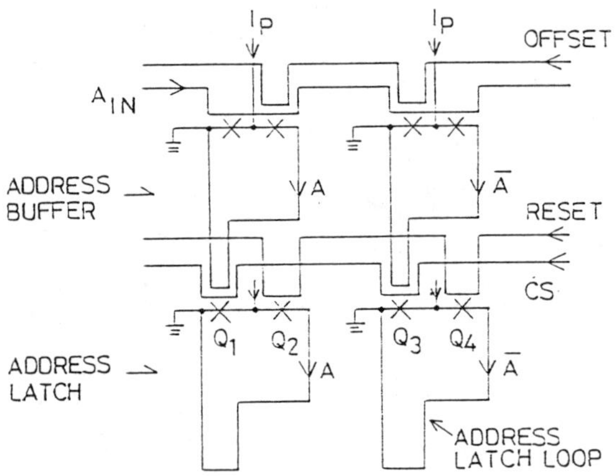

Fig. 5 Configuration of address buffer circuit. Q1 and Q2 are 2-input AND gates.

the set and reset gates of the two loops, but an offset input is fed to the A-loop reset gate and the $\bar{A}$-loop set gate, only, so as to cancel the external address input. With this pair loop configuration, single-to-dual rail conversion with a wide operating margin of more than 30% was attained.

A pair of address latch loops are provided not only to latch the buffer output but also to make an appropriate timing for supplying it to the decoder. The latch is set at the time of the CS input and reset by the RESET input.

A dual rail loop logic was introduced to construct a 5-to-32 bit decoder. The basic unit of the present dual rail decoder is shown in Fig. 6. Two output loops are provided for true and complement outputs, which enables us a wide margin inter-stage connection.

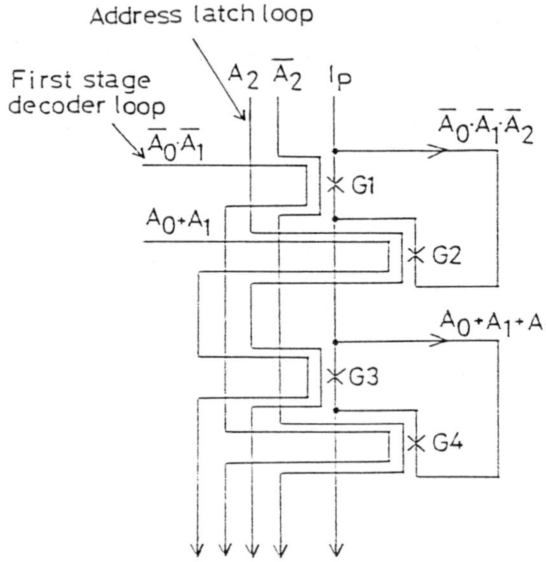

Fig. 6 Basic unit of the duail rail loop decoder. G1 and G4 are 2-input AND, and G2 and G3 2-input OR gates.

5. EXPERIMENTAL EVALUATION

A 1K RAM chip was fabricated by using 5-μm Pb-alloy technology. A microphotograph of the chip is shown in Fig. 7. Chip and cell sizes are 9.3 mm x 9.3 mm and 160 μm x 160 μm, respectively. Random access read/write operation was successfully achieved for several bits of 1K-bit array. To estimate the access time, the time delay from the CS input to the data output was measured by a special monitor gate through a single output cable. A read access time of 3.3 ns was obtained, typically, as shown in Fig. 8. Associated power dissipation was about 2 mW, which was main-

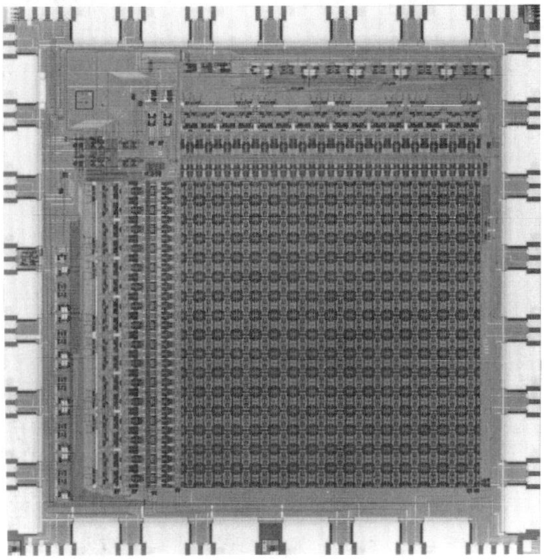

Fig. 7 Microphotograph of the fabricated 1K-bit RAM chip.

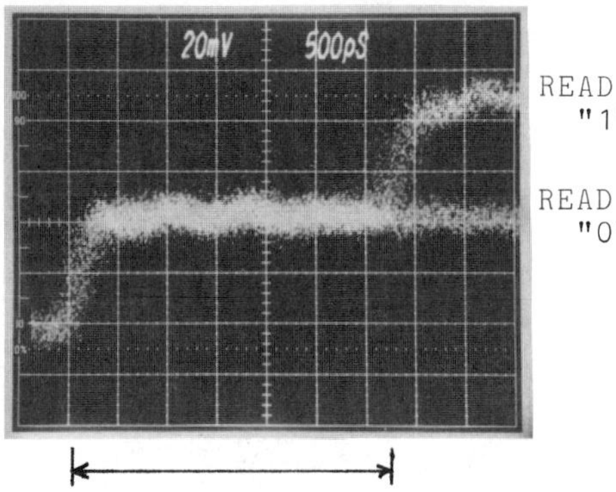

Fig. 8 Experimental estimation of read access time. Superimposed output signals for reading "1" and "0" are shown.

ly consumed at the source resistors for supplying dc power current to the logic loops.

The access time estimated here is not very fast. This is mainly due to a heavy inductance loads for address latches and POX-X. Increase in load driving power is indispensable. One of the promising solution is the higher voltage driving with twice gap-voltage which is easily achieved by a series connection of two interferometer gates.(2) With this driver configuration applied to the address latches and POS-X, read access time less than 2 ns can be expected.

6. SUMMARY

Circuit design for a fully-decoded Josephson RAM has been studied. A new configuration for a wide margin NDRO ring cell, an on-chip read/write timing generator, and an address decoding circuit were proposed. They were experimentally evaluated being integrated in a 1K-bit RAM. Read access time of 3.3 ns was obtained with the associated power dissipation of as low as 2 mW.

ACKNOWLEDGEMENTS

The authors thank M. Fujimoto and T. Ikegami for their encouragements throughout this work; K. Kuroda, J. Nakano, Y. Mimura and F. Yanagawa for their contribution to chip fabrication, and T. Yamamoto for her assistance in circuit design.

REFERENCES

(1) W. H. Henkels and J. H. Greiner,"Experimental single flux quantum NRDO Josephson memory cell," IEEE J. Solid State Circuits, vol. SC-14, p. 794, 1979

(2) S. M. Faris, W. H. Henkels, E. A. Valsamakis and H. H. Zappe," Basic design of a Josephson technology cache memory," IBM J. Res. Develop., vol. 24, p. 143, 1980

(3) W. H. Hekels,"Low-current level widest margin Josephson NDRO cell," IBM Tech. Disc. Bull., vol. 21, p. 412, 1978

(4) Y. Yamauchi and A. Ishida'"An up-transition edge triggered single shot pulse generator with Josephson Devices," IEEE J. Solid State Circuits, vol. SC-18, 1983 (to be published)

POWER SYSTEM DESIGN IN A JOSEPHSON PACKAGE

Patrick C. Arnett
IBM Thomas J. Watson Research Center
P. O. Box 218
Yorktown Heights, New York 10598

Abstract- The use of ac power for Josephson circuits and dc power for semiconductor circuits is a direct consequence of the fact that in current use Josephson circuits are latching and semiconductor circuits are non-latching. Ac power is a master clock for the Josephson circuits which are polarity independent and, therefore, can use each half cycle of the ac power as a complete logic cycle. Ac power distribution through thin film transformers has the advantage that currents do not increase as the system grows in size. Device switching disturbs are minimized by local voltage regulation with large area Josephson junctions.

Introduction

This paper reviews the characteristics of a power system design for Josephson latching logic circuits [1,2] and describes some of the tests of this design [3]. Latching logic circuits place a special requirement on the power supply system, namely, that the circuits must be reset between machine cycles. This is accomplished by using an ac power supply. In the transition from positive to negative half cycles the current supplied to the logic circuits goes to zero, thereby allowing the circuits to reset. The sinusoidal waveforms delivered to each chip are clipped to a nearly trapezoidal shape by regulators (large area Josephson junctions) attached to local power busses. Logic operations only occur during the active portion of the cycle while the signal is near the peak of the trapezoid. Supply resistors between the regulated power bus and the logic circuits define the operating current level. Current is supplied to individual chips through an integrated thin film transformer tree where at each transformer a single input branches into multiple outputs. Since ac power is the master clock, logic currents everywhere in the system must be nearly equal in phase and amplitude.

Development of this power scheme included optimization of the regulator [4] and transformer [5] designs. These were followed by system tests - at first mapping a 16 logic chip power system on a single chip for processing simplicity [6] and, finally, testing a power system in a package consisting of logic chips mounted on cards plugged into boards [7].

General Design

The logic circuit environment is shown by the schematic in Fig. 1(a). Several logic circuits fed by resistors are connected in parallel with a regulator junction to a power bus. Fig. 1(b) shows that the regulator nonlinear I-V characteristic acts to transform the incoming sinusoidal current waveform into the trapezoidal voltage waveform required for the logic circuits. The design of the regulator is determined by considerations of duty cycle, voltage regulation, disturbs, skew, and heating.

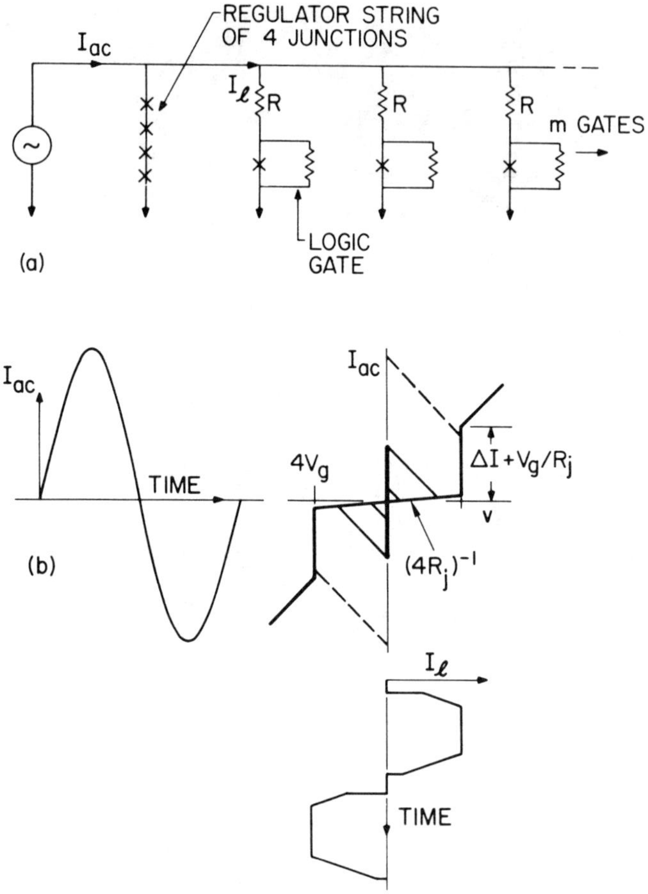

Fig. 1 a) Circuit diagram showing regulator junctions, current defining resistors, and logic gates.
b) Illustration of the effect of the nonlinear regulator characteristic to transform the sinusoidal current waveform into a regulated voltage waveform for the logic circuits. [1]

The duty cycle [8] is the fraction of the cycle that the voltage remains at the peak of the trapezoidal signal and corresponds to the fractional time spent on the current step of the regulator I-V curve. In the limit that the subgap current is negligible the duty cycle is given by

$$\gamma = \frac{2}{\pi}\cos^{-1}\frac{I_L}{I_T}$$

where I_L and I_T are the current to the logic circuits and the amplitude of the sinusoidal current delivered to the regulator and logic circuit combination, respectively. Thus, to achieve a 70% duty cycle more than twice the current delivered to the logic load is shunted through the regulator.

Voltage regulation refers to the flatness of the peak of the trapezoidal voltage waveform. It is defined by

$$Reg = 2\frac{V_{max} - V_{onset}}{V_{max} + V_{onset}}$$

It is influenced by the steepness of the current step in the regulator characteristic. Typical target values are ± 4%.

The regulator can be very effective in reducing the disturbs felt by one logic circuit when another circuit on the same power bus switches [9]. The corresponding effect in semiconductors is referred to as the the ΔI problem. The regulator acts as a clamp on the supply voltage. Thus, if all logic circuits star from the regulator (local regulation) a very small disturb (< 2%) is felt even if all circuits but one fire simultaneously.

Regulators can introduce skew in the power supply waveforms if their $I_m(0)$'s are large enough that they switch onto the current step rather than to the subgap. In this case, a variation in $I_m(0)$ would not only cause skew but also affect the duty cycle. $I_m(0)$ can be suppressed geometrically by making long regulators. The condition that the regulator switch to the subgap region is

$I_m(0) < I_{SG} + I_L/n$

where I_{SG} is the maximum subgap current and n is the number of regulator junctions in series. Typically, this requires that the junction be longer than 6 λ_j, or about 60 μm.

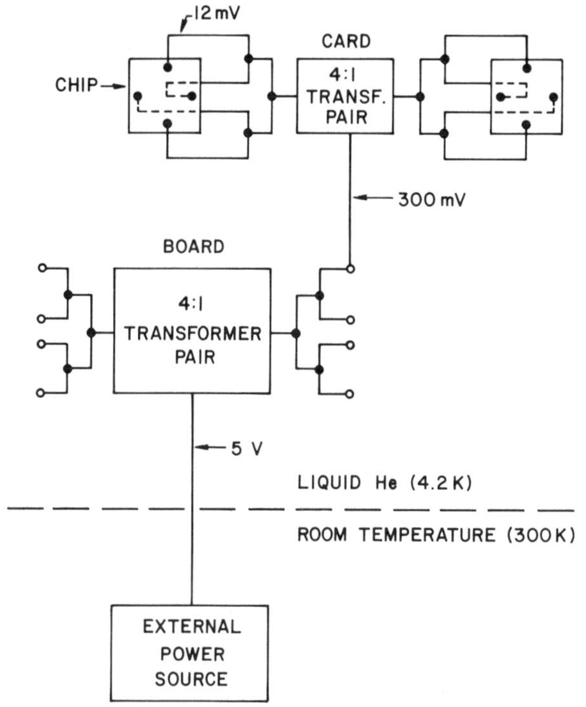

Fig. 2 Circuit schematic of a power distribution system through transformers. It is characteristic of the system that currents remain nearly constant throughout the system while the voltages increase from the chips to the room temperature source.

Heating will adversely affect the operation of the regulator. As the temperature rises the subgap current increases and the differential resistance on the current step increases. This leads to a lower duty cycle and poorer regulation. In planar junctions heating is determined by the current density. For current densities of 1000 A/cm^2 the temperature of the regulator rises about 0.4K above the substrate temperature which is considered acceptable.

Each regulator cell (regulator and logic circuits) should have an identical electrical position in the power system to every other cell. Thus, on a logic chip the input power fans out in a treelike network of power lines that maintain the electrical equivalence of each regulator cell. Since all logic circuits and regulators are supplied in parallel the currents at the input to the chip build rapidly to practical limits set by the critical currents of the lines. That is, to ensure that power lines are not driven normal by the required current levels the lines must be made wider, and wide power lines reduce the number of channels available for logic wiring. A safe limit for Pb-alloy lines is to keep the current/width below 7 mA/μm.

To supply many logic chips from the same room temperature source without having excessively high currents thin film transformers are used (Fig. 2). The idea is to maintain approximately constant currents throughout the power system by a transformer with a multiple turn primary and branching secondaries (Fig. 3). The transformer itself has been optimized in the placement of the two transformer secondaries in a ground plane hole to give the maximum current transfer from primary to secondary while keeping the electrical length of the primary winding below a quarter wavelength (exceeding a quarter wavelength causes a rapid decrease in current transfer). The optimum design has the spacing between secondaries three times the distance from a secondary to the edge of the hole

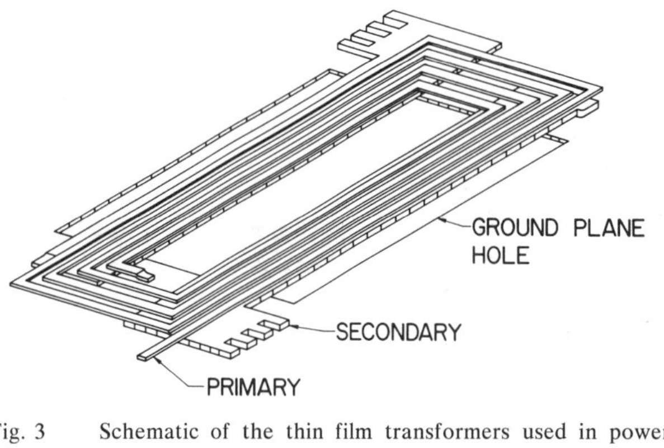

Fig. 3 Schematic of the thin film transformers used in power distribution. The primary winding wraps around four times over two secondaries before going to ground. Each secondary fans out four ways. The ground plane hole ensures a high coupling efficiency (>90%). [2]

and the hole width three times the secondary width. The secondary width is determined by the critical current characteristics of the windings.

Power System Test

The two transformer level power system was tested in the package environment shown in Fig. 4. The package elements consisted of chips, cards, feet with micropins, boards with microsockets, wiring modules with micropins, and a cable attaching adaptor with micropins [10]. The package elements combine thin film fabrication [11] with interconnections. The path of the ac power is down the I/O cable [12] onto the adaptor to the second level transformer primary. The transformer secondaries fan out to micropins [13] which are connected through mercury filled sockets to micropins on the feet. From the feet the power passes by a right angle solder fillet [14] onto the card to the primary of the first level transformer (Fig. 5). Finally, the first level transformer secondaries are connected to the circuits on chip by solder pads [15]. For redundancy, the four quadrants of the chip are identical with only the circuits in one quadrant active at any one time.

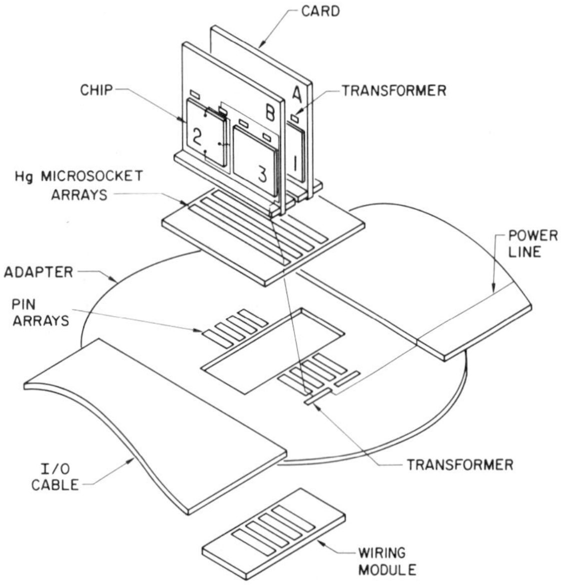

Fig. 4 Exploded schematic of the package assembly. The path followed by the power through the package components is indicated. [3]

The important characteristics for the power system are (1) the frequency response of the current supplied to the chips, (2) the maximum current amplitude available for circuit operation, and (3) delivering currents with nearly equal amplitudes and phases to all chips. Electrical simulations were used to design the power system. The behavior of the voltage at various points in the power system are shown in Fig. 6. As expected the voltage (as normalized to the logic circuit supply voltage) jumps in crossing each transformer. The voltages also show a maximum at between 200 to 250 MHz. The currents as normalized to the chip current (Fig. 7) are much harder to predict and reflect the standing wave resonances in various parts of the power system. From Fig. 7 one can see how the

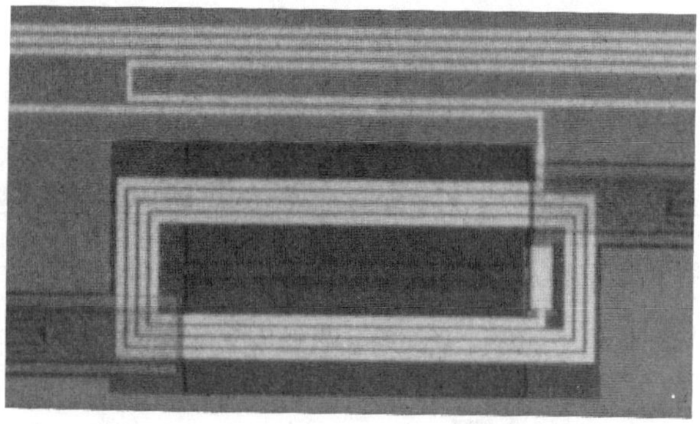

Fig. 5 Micrograph of one of four transformers on a card. Primary lines connect to right angle fillets at the base of the card. Power lines from the transformer secondaries connect to the chip solder pads.

supply current (x symbol) must be varied to maintain circuit operation over the range 10-400 MHz.

It is more convenient experimentally to measure the chip current as a function of frequency keeping the supply current constant (Fig. 8). For the frequency range 50-400 MHz the agreement among different chip quadrants and with the simulation is within the range expected when considering line width and insulation thickness variations inherent in the process. However, the maximum current delivered to the chips was less than required for the logic circuit experiments. To boost the current to the active quadrant the currents to the inactive chip quadrants were eliminated by removing their power solder pads. The successful result of this operation is shown in Fig. 9. The currents between 100-200 MHz, which are of interest for the logic experiment, are greatly enhanced.

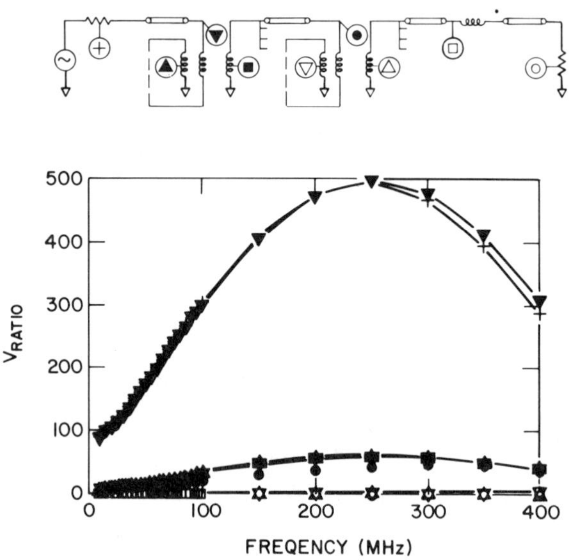

Fig. 6 An abreviated schematic of the power system indicating where voltages are monitored in a simulation, and the frequency dependence of those voltages as normalized to the logic bus voltage.

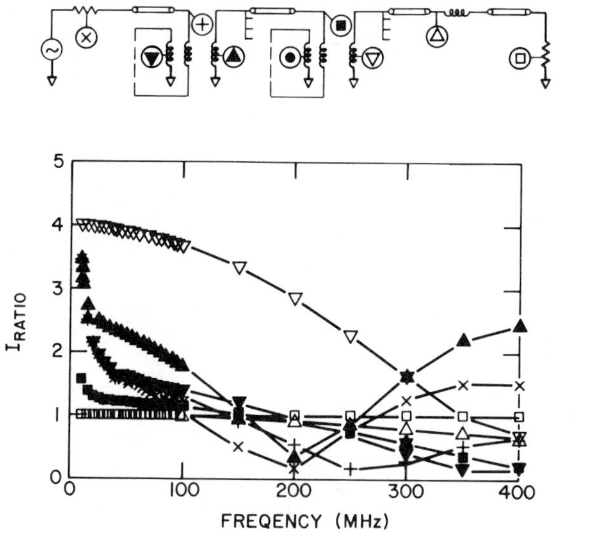

Fig. 7 An abreviated schematic of the power system indicating where currents are monitored in a simulation, and the frequency dependence of those currents as normalized to the logic load current.

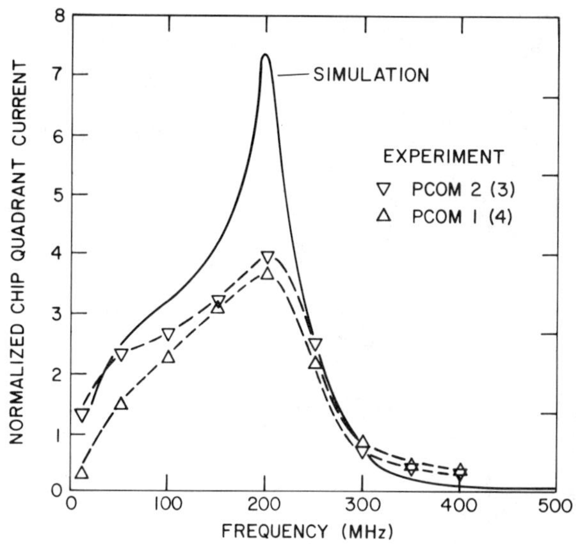

Fig. 9 A comparison of simulation and experiment for the case which has three quadrants from the same secondary with the active quadrant disconnected. [3]

Conclusions

The design of an ac power system for Josephson logic circuits uses thin film transformers to distribute current to many chips at the same current level required for one chip. The ac power signals are the clock for the Josephson circuits. Regulator junctions on-chip are effective in providing well defined operating levels for the circuits and in minimizing switching disturbs.

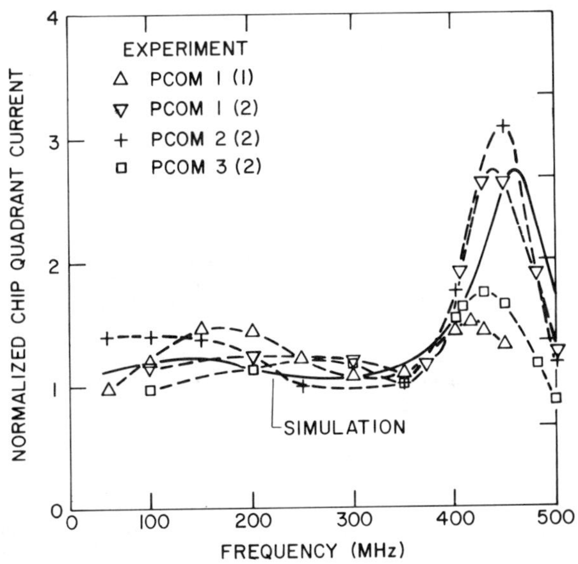

Fig. 8 A plot of chip quadrant current amplitudes at four locations on a card as a function of frequency. The quadrant currents are normalized to the input current from the 50 ohm source. [3]

References

[1] P. C. Arnett and D. J. Herrell, IEEE Trans. Magn., Vol. MAG-15, No. 1, 554 (1979).

[2] P. C. Arnett and D. J. Herrell, IEEE Trans. Microwave Theory Tech. 28, No. 5, 500 (1980).

[3] P. C. Arnett, IEEE Trans. Magn. MAG-19, NO. 3, 1250 (1983).

[4] M. B. Ketchen, Tech. Digest IEDM, 489 (1979).

[5] P. C. Arnett, Tech. Digest IEDM, 485 (1979).

[6] P. C. Arnett, IEEE Trans. Magn. MAG-16, NO. 5, 1233 (1980).

[7] M. B. Ketchen, et. al., IEEE Electrion Device Letters, Vol. EDL-2, No. 10, 262 (1981).

[8] C. J. Anderson and M. B. Ketchen, IEEE Trans. Magn., Vol. MAG-17, No.1, 595 (1981).

[9] M. B. Ketchen, Proc. IEEE ICCC80, 874 (1980).

[10] S. K. Lahiri, et. al., IEEE Trans. Comp., Hybrids, and Manufac. Technol., Vol. CHMT-5, No. 2, 271 (1982).

[11] A. Bright, J. H. Greiner, S. Klepner, A. J. Warnecke, and R. H. Wang, J. Vac. Sci. Technol. B1(1), 77 (1983).

[12] P. A. Moskowitz, R. W. Guernsey, and J. W. Stasiak, Cryogenic Advances in Engineering 27, Plenum Press, New York.

[13] S. K. Lahiri, P. Geldermans, G. Kolb, J. Sokolowski, and M. J. Palmer, IEEE Trans. Comp., Hybrids, and Manufac. Technol., Vol. CHMT-5, No. 1, 166 (1982).

[14] K. R. Grebe, C. Y. Ting, and D. P. Waldman, J. Electrochem. Soc. Extend Abstracts 80-1, 213 (1980).

[15] C. Y. Ting, K. R. Grebe, and D. P. Waldman, J. Electrochem. Soc. Extend Abstracts 80-1, 210 (1980).

JOSEPHSON COMPUTER TECHNOLOGY - AN OVERVIEW

Hans H. Zappe

IBM T. J. Watson Research Center
Yorktown Heights, New York 10598

The performance of computing systems has grown dramatically during the last 35 years. First, the original electro-mechanical machines were replaced with systems built in vacuum technology and the computing speed was increased from a few operations per second to about 10^4. Subsequently, the development of semiconductor computers further increased that capability by over three orders of magnitude. Progress is slowing now since semiconductor performance limits no longer seem as far away as they used to be.[1] Yet some users insist that there is need for a system performance improvement of one to two orders of magnitude in this decade alone.[1]. For this reason it is legitimate to ask whether the next generation of computers may not have to involve yet another technological change.

Superconductivity holds the promise to fulfill these needs. It is a radically new approach which requires operation in liquid helium at 4K, but which has technological advantages sufficiently broad so that many of the fundamental hardware problems facing a machine designer today can in principle be pushed far into the future. This does not mean that performance improvements in machines built with room temperature technologies have come to a halt. Improvements will be made but these are expected to become expensive and more of an incremental nature than has been the case in the past.

The main reason for the difficulties we presently encounter, is that the size of the computing system must be commensurate with the distance a signal can travel in a time interval corresponding to the cycle time of the machine. This implies that as the computing power of a machine with given architecture increases, its cycle time and its physical size have to decrease in the same proportion. Since the system size has to shrink, the power dissipation of the circuits should decrease at least as the square and possibly as the cube of the linear system size, otherwise the power density dissipated in the machine gets out of bound. Conversely, as means are found to extract the thermal energies, frightfully large power supply currents have to be fed into a progressively miniaturized system. This means that the operating voltages and therefore the switching energies of the devices have to decrease, and that to keep the signal to thermal noise ratio constant, the operating temperature has to decrease as well.[2]

At least equally important are problems which surface in package design. Since the signal lines must progressively shrink, their resistance increases faster than can be compensated for by the decrease in line length. As a result, transmission line properties of the signal wires in the package degrade rapidly and limit the distance over which a signal can be transmitted in a reasonably short time.

Josephson Technology has the potential to solve these problems. [2],[3],[4] Circuits built in conventional 2.5 μm technologies switch in tens of picoseconds at power levels of μW. Thus one deals not only with the required speed but also with the necessary low power dissipation. The second major advantage is that superconducting lines have no dc resistance and permit near lossless dispersionless transmission of signals up to frequencies of the order of 10^{12} Hz. This is important not only on chip but also in the package where one can use lines with widths near the resolution limit of the technology and line thicknesses of only a few thousand Å. Lastly, because of the lack of line resistance and because of the low power dissipation of the circuits, terminated transmission lines can be used throughout the system. This means that in a *well designed* miniaturized package there is essentially no difference between driving a signal on chip or through the package from chip to chip. Signals are transmitted in one pass at the speed of light in the chosen dielectric, from one circuit to the next.

At IBM, significant progress was made to develop such a superconducting technology. A logic family with an average logic gate delay of 44ps and a power level of 5 μW was experimentally evaluated.[5] A chip designed with these circuits, containing shallow logic networks between 11 register levels was operated at an unprecedented cycle time of 670ps with a total chip power dissipation of 350 μW[6] which is less than the dissipation of a single state of the art semiconductor circuit. This particular logic family is presently being redesigned more conservatively to increase operating margins. Because of the low power dissipation of the circuits, supply voltages can be regulated on chip. This eliminates the so-called delta I problem which in semi-conductor designs limits the number of simultaneously switching drivers. Furthermore, power is provided in the form of a sinusoidal voltage in Josephson technology. At this meeting, recent results on power supply design are presented.[7]

Cache memory circuits have been tested and were found to operate as predicted by circuit models. In an experiment, elements interconnected into circuits of varying complexity were investigated. This experimental memory chip also contained the cross-section of a fully decoded 1K bit NDRO, RAM array. The results of this investigation are being presented for the first time at this conference.[8]

To ensure that the high speed of these devices is maintained on the system level, a very small, dense package with microscopic pluggable connectors was developed.[9] To test chip operation in this

CH1935-6/83/0000/0516$01.00

package environment a cross sectional model was successfully built and tested. It represents a 1 bit critical path through a signal processor and is composed of four chips on two cards of the quasi three dimensional package. Signals emanating from the chips were successfully transmitted through the microscopic connectors provided to permit communication between cards. This experimental sub-system showed proper operation down to a cycle time of 3.7 nS.[10]

At IBM, most work in Josephson Technology was based on lead alloy superconductors. Devices built with these materials were susceptible to failure during the required, repeated temperature cycling from room temperature to that of liquid helium. Although, compositional and structural changes in these alloyed metals resulted in considerable improvement of cyclability, other more subtle instabilities of the device threshold made a change in base materials imperative. For this reason a Josephson Technology using niobium as a superconductor is now being developed. Niobium is a much stronger material than lead and a niobium junction hardly ever failed during thermal cycling. But being a high melting point material it has drawbacks. It is a strong getter and high quality junctions are difficult to to make. In addition its oxide has a nearly four times larger dielectric constant than the oxide grown on lead based tunnel junction materials. This results in an increase of circuit delay by roughly a factor of two.

One solution consists in decreasing the junction area. However, if the junction area is delineated by a photolithographic process, a smaller size junction may be subject to large area variations and therefore to large variations in device threshold. For this reason, the reduction in junction area is achieved in a manner in which it can be well controlled. The tunnel barrier is formed on the tapered edge of a thin niobium film and the junction is completed with a subsequently deposited lead based counterelectrode. Since the thickness of the niobium base electrode is only 0.2 μm, edge devices hold the promise of a nearly seven fold decrease in active junction area. This so called edge junction technology[11] is now being developed for all future circuit work.

Much progress has been made in Josephson Technology during a decade of determined effort. However, considerably more work is needed before one can expect to see a Josephson system compete with or overtake semiconductor machines. The biggest problem is that in Josephson Technology every aspect of a computing system has to be developed without the ability of taking advantage of existing hardware except in the tools required to fabricate the parts. It is necessary to invent and develop logic, memory, a process which permits the fabrication if these circuits out of appropriate superconducting materials, and a new quasi-three dimensional miniaturized package which can withstand the temperature excursions from room temperature to that of liquid helium.

The cross-sectional model showed, on a small scale, that the various elements can be assembled and that Josephson circuits, within the package, can communicate among themselves and with external room temperature hardware. However, one must realize that the test vehicle was assembled and operated in a laboratory environment. What remains to be shown is that chips and package parts can be manufactured again and again with an acceptable yield. IBM is attempting to prove this point in a development line. Technological problems center mainly around the reproducibility of the device threshold, but many other technical question still need to be resolved as well.

Whether Josephson Technology will be a technical and commercial success remains an open question. Only after a Josephson prototype machine is fully operated, will it be possible to ascertain that at least a first solution was found to all problems encountered in the design of a cryogenic computer. Construction of a prototype is planned, at IBM, although the name prototype does not imply the first of a series of machines. It is intended simply as a vehicle to demonstrate the technology.

[1] Worlton, J., "A Philosophy of Supercomputing," Los Alamos Scientific Laboratory Report, LA-8849-MS, UC-32, June 1981.

[2] Zappe, H. H., "Josephson Computer Technology," Nato Advanced Study Institute Series, Plenum Press, to be published July 1983.

[3] Anacker, W., "Computing at 4 Degrees Kelvin," IEEE Spectrum, pp. 26-37, May 1979.

[4] Special Issue on Josephson Computer Technology, IBM Journal of Research and Development, Vol. 24, No. 24, March 1980.

[5] Gheewala, T., "Josephson-Logic Devices and Circuits, IEEE Trans. on Electr. Devices, Vol. ED-27, No. 10, October 1980.

[6] Mukherjee, A., "High Performance Data Processing in Josephson Technology," IEEE Electr. Device Lett., Vol. EDL-3, No. 2, February 1982.

[7] Arnett, P. C., "Power Design in a Josephson Package," this issue.

[8] W. H. Henkels, K. H. Brown, T. V. Rajeevakumar, L. Geppert, J. W. Allan, Y. H. Lee, and J. T. Yeh, "Experiments on a Cross-section of a Josephson RAM Chip," this issue.

[9] Syamal K. Lahiri, H. Randall Bickford, Pieter Geldermans, K. R. Grebe, Paul A. Moskowitz, Menachem Natan, M. J. Palmer, Sampath Purushothaman, John Sokolowski, Bernard J. C. van der Hoeven, Member, IEEE, David P. Waldman, Run-Han Wang, C. T. Wu, and Tadashi Yogi, "Packaging Technology for Josephson Integrated Circuits," IEEE Trans. Comp., Hybrids, Mfg. Tech., Vol CHMT-5, No. 2, June 1982.

[10] M. B. Ketchen, Member, IEEE, B. J. van der Hoeven, Member, IEEE, J. Matisoo, Fellow, IEEE, J. H. Greiner, D. J. Herrell, Member, IEEE, R. H. Wang, R. W. Guernsey, C. J. Anderson, P. C. Arnett, S. Bermon, H. R. Bickford, A. A. Bright, P. Geldermans, T. R. Gheewala, Member, IEEE, K. R. Grebe, H. C. Jones, M. Klein, Member, IEEE, S. P. Klepner, P. A. Moskowitz, M. Natan, S. Purushothaman, J. Sokolowski, J. W. Stasiak, D. P. Walkman, A. J. Warnecke, C. T. Wu, and T. Yogi, "A Josephson Technology System Level Experiment," IEEE Electr. Device Lett., Vol. EDL-2, No. 10, October 1981.

[11] Broom, R. F., Oosenbrug, A., Walter, W., "Josephson Junctions of Small Area Formed on the Edges of Niobium Films, App. Phys. Lett., Vol. 37, pp.237-239, July 1980.

Session 3-1b
Design Specs

Chairman

J. L. Halvorsen
Evans and Sutherland

LAYOUT GENERATION OF PLA BASED CIRCUITS FROM A REGISTER TRANSFER DESCRIPTION

A.C. de Graaf

R. Nouta

Delft University of Technology
Department of Electrical Engineering
Mekelweg 4, 2628 CD Delft
The Netherlands

Abstract

This paper presents a package of C-programs, to generate a mask layout description of decoders and finite state machines in PLA-form. Two levels of input, a register transfer language and a self defined PLA-description language,are described, as well as translation between these descriptions. Issues are: user friendliness and easy updating for new design rules and other layout language.

1. Introduction

Many PLA layout generating systems [1,2] have been published during the last five years. Some of these try to give the designer a tool, that generates a PLA automatically from a minimum of input description. The result may be a PLA that does not correspond to the one the designer expected, because the program has made instant decisions. This paper describes a layout generating system, that is capable of generating the geometrical artwork of a decoder or finite state machine in PLA-form from two levels of description. At the highest level one can describe the function of a circuit with the Register Transfer Language (RTL) Digital Design Language (DDL)[3]. With a RTL the functions in a PLA can be described easily but some geometrical information has been dropped. Therefore it is necessary to add this information in some way. For this reason a second lower level language was introduced, which was called PLA Design Language (PLADL). With PLADL one can describe the composing blocks of the PLA artwork, at the same time one has options to customize the PLA. In figure 1 the block diagram of the system is given. The encircled parts represent text files, the rectangled parts represent programs or a cluster of programs.

Start point is a text file containing the DDL description of a decoder or a finite state machine. This description is translated by the Delft Logic Translator (DLT) into a PLADL-description. At this level the designer can alter parameters with respect to speed, clocking, and location of the planes in the PLA-area. The PLADL description is input for the PLA -layout generating cluster of programs MKPLA. With a PLADL source, MKPLA uses three additional files:

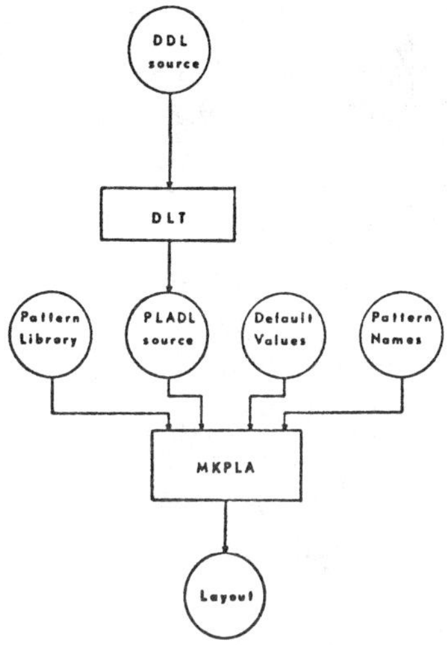

Figure 1. Block diagram of the PLA generating system.

1. Pattern Library, in which the predefined parts of a PLA are stored.
2. Pattern Names, which stores the names and bounding box information of the patterns defined in the Pattern Library.
3. Default Values, which are taken if the designer does not alter the parameters mentioned above.

In section 2 we will be concerned with the syntax definition of the two input languages DDL and PLADL.
In section 3 we will concentrate on the data structure and algorithm of the translating program DLT.
All programs are coded in the C-language[4], while with the translation extensive use has been made of the program generating programs LEX[5] and YACC[6] with the translating parts of the system.

2. Syntax definition of the two input languages.

In the PLA generating system two languages are used:

1. Digital Design Language (DDL), a high level register transfer language.
2. PLA design language (PLADL), a self-defined PLA-layout language.

PLADL is the input language of the PLA-layout generating cluster of programs MKPLA. Between DDL and PLADL a translator called Delft Logic Translator (DLT) is written. Next paragraphs describe the languages.

2.1 A DDL subset.

A DDL subset was sufficient to define both decoders and finite state machines. The syntax is shown in table 1. Each item between curly braces is a syntactic construct, which is defined by a sentence of syntactic constructs, tokens and delimiter characters. The vertical bar denotes a choice of a number of sentences, which define the same syntactic construct. The tokens and delimiter character must be literally present in the input text. Spaces in the syntax definition have no particular meaning, but are inserted for readableness.

```
{pla circuit}    ::= {automaton def} ¦ {operator def}
{operator def}   ::= <op> {OP header} : {OP block} .
{OP header}      ::= {OPerator} ¦ {OP header} , {OPerator}
{OPerator}       ::= {identifier}
{OP block}       ::= {TErminal decl}{BOolean decl}
{automaton def}  ::= <au> {AU header} : {AU block} .
{AU header}      ::= {identifier}
{AU block}       ::= {TErminal decl}{IDentifier}{STate decl}
{TErminal decl}  ::= <te> {TErminal list} .
{IDentifier decl}::= <id> {ID def list} .
{BOolean decl}   ::= <bo> {Boolean eqns} .
{STate decl}     ::= <st> {STate list} .
{TErminal list}  ::= {TErminal}¦{TErminal list} , {TErminal}
{TErminal}       ::= {identifier}
{ID def list}    ::= {IDentifier} = []
                   ¦ {ID def list} , {IDentifier} = []
{IDentifier}     ::= {identifier}
{BOolean eqns}   ::= {OPerator} = {Boolean exp}
                   ¦ {BOolean eqns} , {OPerator} = {Boolean exp}
{Boolean exp}    ::= {product}¦{Boolean exp} + {product}
{product}        ::= {complement}¦{product} * {complement}
{complement}     ::= {TErminal}¦{TErminal} '
{STate list}     ::= {STate def} . ¦{STate list}{STate def} .
{STate def }     ::= {STate identifier} : {STate action}
{STate identfier}::= {identifier}
{STate action}   ::= {unconditional} ¦ {conditionals}
{unconditional}  ::= -> {STate identifier}
                   ¦ {unconditional}{ID list}
{ID list}        ::= {IDentifier}¦{ID list} , {IDentifier}
{conditionals}   ::= {IF statement} .
                   ¦ {conditionals} , {IF statement} .
{if statement}   ::= if {simple boolean exp} then {unconditional}
                   ¦ {if statement}{else part}
{else part}      ::=
                   ¦ else {unconditional}
{simple boolean
             exp} ::= {product} ¦ {sum}
{sum}                 ::= {complement} + {complement}
                   ¦ {sum} + {complement} + {complement}
{integer}        ::= {digit}¦{integer}{digit}
{identifier}     ::= {letter}¦{identifier}{letter}
                   ¦ {identifier}{digit}
{digit}          ::= 0 ¦ 1 ¦ 2 ¦ 3 ¦ 4 ¦
                     5 ¦ 6 ¦ 7 ¦ 8 ¦ 9
{letter}         ::= a ¦ b ¦ c ¦ d ¦ e ¦ f ¦ g ¦ h ¦
                     i ¦ j ¦ k ¦ l ¦ m ¦ n ¦ o ¦ p ¦
                     q ¦ r ¦ s ¦ t ¦ u ¦ v ¦ w ¦ x ¦
                     y ¦ z
```

TABLE 1. Syntax of DDL subset

Dietmeyer and Duley[3] distinguish two facility types, simple and complex facility types. Complex facility types define a block of hardware. Like a subroutine in software, a complex facility declaration can contain local declarations, which can be simple facility declarations. Simple facility declarations define the basic pieces of a circuit as wires,registers,logic and memory. In our situation we need two complex facility types:

1. <OP> the operator type. Within an operator declaration a combinational circuit is defined. Each operator maps into one output in the PLA. Two local declarations are allowed:
   - A terminal declaration, which are physically the inputs of the PLA.
   - A boolean declaration defines the relation between terminals and operators.
2. <AU> the automaton type. The automaton declaration must contain a TErminal declaration, an IDentifier declaration and a STate declaration. Within the state declaration part the finite automaton is defined. During each state a state action is executed. A state action can be either unconditional or conditional.

2.2 PLADL a low level PLA Design Language.

PLA's have a number of characteristic parameters:

1. the mask layout language in which the PLA is described.
2. the name of the pattern to be generated.
3. the number of inputs.
4. the number of outputs.
5. the number of productterms.
6. the personality matrix.
7. the number of feedback terms (in case of finite state machines)

These parameters must be present in a specification of a PLA. Further optional parameters with respect to speed,clocking and relative position of the pieces a PLA is built of, can be specified. Optional parameters are:

a. The horizontal position of the AND and OR planes.
b. The vertical position of the inputs and outputs.
c. The length of the pullup transistors.
d. The width of the enhancement transistors in the planes.
e. The interval between extra groundlines is placed in the planes, if needed at all.
f. The type of inputs and outputs.

The PLA is thought of being constructed from a number of hardware blocks, which correspond to the above mentioned essential pieces of a PLA. Within each block parameters can be set to shape the PLA to the designers wishes. Table 2 shows the syntax of the PLA Design Language. The description format is the same as in the DDL subset syntax description.

```
{pla definition}  ::= {block list} $
{block list}      ::= {block spec}|{block list} ; {block list}
{block spec}      ::= {IO name} : {IO parameterlist}
                    | format : {format type}
                    | name : {identifier}
                    | terms : number = {integer}
                    | pullupsxand : ldep = {integer}
                    | pullupsxor : ldep = {integer}
                    | feedback : number = {integer}
                    | {array def}
{format type}     ::= cmsk | ldld
{IO name}         ::= inputs | outputs
{IO parameterlist}::= {parameter ref}
                    | {IO parameterlist} , {parameter ref}
{parameter ref}         ::= hpos = {hpos parameter}
                    | vpos = {vpos parameter}
                    | ck   = {clock spec}
                    | {parameter} = {integer}
{hpos parameter}  ::= left | right
{vpos parameter}  ::= top  | down
{clock spec}      ::= ac | an |{clock array}
{clock array}     ::= {clockbit}|{clock array}{clockbit}
{clockbit}        ::= N | C
{parameter}       ::= number | ldep | wenh | type | giv
{array def}       ::= array : {matrix}|{array def}{matrix}
{matrix}          ::= I | O | .
{integer}         ::= {digit}|{integer}{digit}
{identifier}      ::= {letter}|{identifier}{letter}
                    | {identifier}{digit}
{digit}           ::= 0 | 1 | 2 | 3 | 4 |
                      5 | 6 | 7 | 8 | 9
{letter}          ::= a | b | c | d | e | f | g | h |
                      i | j | k | l | m | n | o | p |
                      q | r | s | t | u | v | w | x |
                      y | z
```

TABLE 2. Syntax description of PLADL.

Explanation of optional parameter names with their possible values.

- **hpos** denotes the relative horizontal position of a block of inputs or outputs in the PLA-area.
- **vpos** denotes the relative vertical position of a block of inputs or outputs.
- **number** denotes the respective amount of inputs, outputs, productterms or feedback paths.
- **ldep** denotes the length in microns of pullup transistors in the specified block.
- **wenh** denotes the width of enhancement transistor in the planes.
- **type** specifies the type of inputs or outputs.
- **giv** specifies the number of productterms or outputs that are between two groundlines.
- **ck** denotes the clocking specification on inputs and outputs.

3. Delft Logic Translator.(DLT)

DLT translates a DDL-subset source into the input language for the layout generating program MKPLA. It consists of two parts, first a parser which fills a datastructure, that contains all information required to generate a PLADL description of a PLA. The second part interprets the data structure and generates the description.
The data structure is built from a number of linked lists, which have a strong connection to the basic facility types in DDL. There are for instance lists of terminals, identifiers and productterms. The DDL inputfile is parsed two times. The first time tables of C-structures with terminal, identifier, state or operator information are filled. Each terminal is characterized by its name and place number with respect to the other terminals in the layout. The terminal with the lowest place number is positioned most left in the layout. The same holds for identifiers, operators and states, only with states the place number indicates instead of the horizontal the vertical position of the productterms in the layout. In the second parse stored facility information is chained such that it defines a decoder or finite state machine. The information about a decoder is put in the table of operators. An outline of the data structure for operators is shown in figure 2.

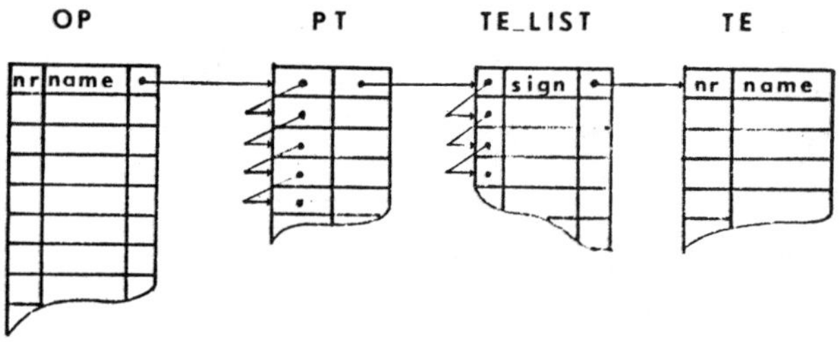

Figure 2. Diagram of the datastructure for OPerator declarations.

Each operator is an output in the PLA to be generated. In the Boolean declaration part a sum of productterms is chained to each operator. Because of the fixed form of the Boolean expressions, a sum of productterms is implemented as a linked list of productterms, a productterm as a list of terminals. Remember that during the first parse, the OP and TE tables are filled. Every operator or terminal is only one time present in the related table. A product is a list of references to the terminal table, stored in structure TE.LIST. To each reference a sign is assigned to mark whether the signal or its complement is meant. A sum of productterms is put in the linked list PT.
Information about a finite state machine is put in a table of states. Each state is characterized by:

- a state number, which in binary form is the state vector part of the productterms.
- a state name which denotes a state in a DDL program.
- a pointer to a list of actions, executed during this state.

The most complex form actions can have is a list of IF-THEN-ELSE constructions. Structure ACTION in the data structure is deducted from this complex form.
The members of structure ACTION are:

- a pointer to the Boolean expression in a IF-THEN-ELSE construction.
- two pointers to structure OUTPUT, which keeps information about the output settings and the next state.
- a pointer to the next action in the list of actions.

Structure OUTPUT fixes the output settings and the next state.

The last structure is the one that characterizes a Boolean expression, called BOOLEXP. Remember Boolean expressions allowed in IF-THEN-ELSE constructions are simple products or sums of terminals.
In figure 3 the connections between the structures is illustrated.

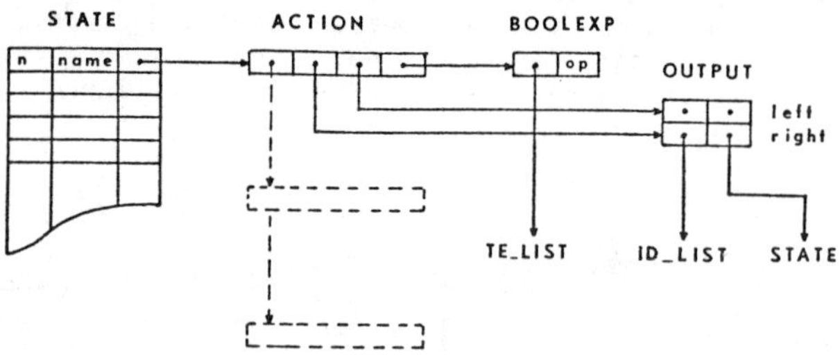

Figure 3. Diagram of the datastructure for the AUtomaton declaration.

From the datastructure a PLADL description of either a FSM or a decoder is extracted by the following algorithm. This algorithm is written in pseudo-C.

The algorithm:

```
COMMENT: if the state table is not empty , then an automaton is
         defined and a finite state machine PLADL description
         will be given. If the operator table is not empty, a
         decoder PLADL description is generated.
         variables: i,k,array_row : current counters.
                    cnt.te        : number of terminals.
                    sign          : normal      = 0.
                                    complement = 1.
                    STATE[]       : state table.
                    POP[]         : operator table.
BEGIN

i=0;
array_row = 0;
IF ( STATE[i] not empty )
{
   compute( number of feedback paths);
   i = 0;
   WHILE ( STATE[i] not empty)
   {
      current_action = STATE[i]->action;
      WHILE ( current_action )
      {
        IF ( current_action->boolean_expression )
        /* is the action a conditional one ? */
        {
           IF ( current_action->LEFT )
           /* is the condition TRUE */
           {
              compute( personality of current array_row );
              array_row = array_row + 1;
           }
           IF ( current_action->RIGHT )
           /* is the condition FALSE */
           {
              compute( personality of current array_row );
              array_row = array_row + 1;
           }
        }
        ELSE
        IF ( current_action->LEFT )
        /* is the action an unconditional one */
        {
           compute( personality of array_row );
           array_row = array_row + 1;
        }
        current_action = current_action->next_action;
      }
      i = i + 1;
   }
   printout(PLADL description of the FSM);
}
IF (POP[0] is not empty)
{
  fill(PLA_array with '.');
  i=0;k=0;
  WHILE ( POP[k] is not empty )
  {
    current_productterm = POP[k]->first_productterm;
    WHILE ( current_productterm not empty )
    {
      current_terminal = current_productterm->first_terminal;
      WHILE ( current_terminal not empty )
      {
        SWITCH ( current_terminal->sign )
        {
          case 0 : PLA_array[i][current_terminal->number]='I';
          case 1 : PLA_array[i][current_terminal->number]='O';
        }
        current_terminal = current_terminal->next_terminal;
      }
      current_productterm = current_productterm->next_productterm;
      PLA_array[i][cnt.te + k] = 'I';
      i = i + 1;
    }
    k = k + 1;
  }
remove( rows from PLA_array which define the same productterm );
printout( PLADL description of the decoder );
}
END
```

Conclusion.

A PLA layout generating system has been presented. The package of programs is running on our VAX 11/750. Especially the feature to define PLA's on two levels of description seems to be convenient. A designer can alter a number of optional parameters, to match speed, clocking and placing of inputs outputs.
In the future, minimization techniques will be applied to the PLA-array. We think of a logic minimization program package such as R.K. Brayton and others presented with ESPRESSO [7]. We also want to implement folding, but in such aa way, the designer desides which columns or rows should be folded.

References.

[1] L.A. Glasser,P. Penfield,"An interactive PLA generator as an archetype for a new VLSI design methodology.",ICCC 80,pp 608-611, Oct 1980.

[2] S. Kang,W.M. van Cleemput,"Automatic PLA synthesis from a DDL-P description.",18th DAC,pp 391-397,Nashville 1981.

[3] M.A. Breuer,D. Dietmeyer,J.R. Duley,"Digital Design Automation" Pitman 1976,pp 117-218.

[4] B. Kernigham,D. Ritchie,"The C Programming Language", Prentice-Hall Englewood Cliffs,ISBN 0-13-110163-3.

[5] M.E. Lesk,E. Schmidt,"Lex - A Lexical Analyzer Generator", Comp. Sci. Tech. Rep. No.39,Bell Laboratories, Murray Hill, October 1975.

[6] S.C. Johnson,Yacc: Yet Another Compiler Compiler, Computing Science Technical Report No. 32,Bell Laboratories, Murray Hill, 1975.

[7] R.K. Brayton,G.D. Hachtel,L.A. Hemachandra,A.R. Newton,A.L.M. Sangiovanni-Vincentelli,"ESPRESSO: An APL Program Package for Partitioned Logic Minimalization, Procc. ISCAS pp. 42-48, Rome 1982.

CONTROLLING VLSI COMPLEXITY USING A HIGH-LEVEL LANGUAGE FOR DESIGN DESCRIPTION

J.P. Gray, I. Buchanan and P.S. Robertson

Lattice Logic Ltd., 6 Albany Lane, Edinburgh EH1 3QP, Scotland

ABSTRACT

The control of complexity is a central issue in VLSI design and languages have a vital role to play. In this paper we present such a language, MODEL, and concentrate on two aspects of its development: programming-style-driven language design and debugging-method-driven compiler operation.

1. INTRODUCTION

In VLSI design there are primarily two sources of complexity: inherent complexity associated with the size of a design, and indirect complexity associated with the development of multiple descriptions of hardware.

It is important to recognise both in the development of design methodologies and the construction of support tools for particular design styles. In a sense, the first source of complexity is more easily treated. It is widely recognised that there is no alternative to structuring to contain inherent complexity. It therefore follows that there is no alternative to parameterisation and hierarchy in tool support.

There are two consequences of this observation: the first is that the features of parameterisation and hierarchy are most easily provided for in a programming language notation, and the second is that "good" structuring in a design should lead to higher performance with specific tools. There is more than an intuitive basis for this, as the time and space complexity bounds of tree-structured algorithms provide a strong incentive for structuring to increase tool efficiency.

The control of indirect complexity is more complicated. Different representations of structural, behavioural and physical data at different levels of design abstraction can lead to severe consistency-checking problems. There is a long history of attempts to finesse these problems by inventing notations that partially combine more than one description in a single representation, and compilers that generate one description from another using knowledge of a design style [1]. To date, these techniques have both confused the designer and produced inefficient hardware implementations.

A key decision in the work described in this paper was to work from the simplest possible design description. The MODEL notation was invented to support only structured design. This removes potential ambiguities in the interpretation of design descriptions. The production of behavioural and physical descriptions is seen as design-style dependent. In the case of specific styles it may be possible to automate the production of these descriptions [2], thus providing a simple and unambiguous design environment for hardware.

2. NOTIONS OF PROGRAMMING STYLE

The objective of this section is to discuss both high-level programming languages and hardware description languages, and to note the similarities between them.

Constructing a correct working program consists of two major activities: writing and debugging. Writing a program is a creative activity channelled along the lines of logical thought. Unfortunately, programs rarely work first time and therefore a coherent debugging strategy is required.

In fact, it is valuable to realise that a correct program can be completed faster if a strategy which integrates writing and debugging the program on a module-by-module basis is followed. Also, the way a program is written can have a substantial effect on the ease of debugging it.

Structured programming is one style which has gained widespread acceptance. Its main features are modularity and hierarchy, with top-down design and bottom-up construction. The style of programming is to construct reasonably small modules, each with a well-defined interface to the outside world. Thus each module can be tested in isolation. With bottom-up construction methods, higher-level modules make calls on existing, tested lower-level modules and can therefore themselves be verified more accurately.

The use of a high-level language is not mandatory in structured programming but it makes the job much easier. High-level languages already contain the constructs required to express structured programming concepts. These are: procedures with

CH1935-6/83/0000/0523$01.00 © 1983 IEEE

well-defined interfaces, the ability to construct hierarchical and recursive programs, and useful, concise control structures such as loops and conditionals.

It is not impossible to write structured programs in assembly languages, but it does require enormous self-discipline and an understanding of the importance of good programming style. High-level languages do not guarantee well-structured programs, but do lead the programmer into using the correct methodology.

A methodology for debugging a program is naturally tied in with the methodology for writing it. Debugging a poorly-written program can be extremely difficult; for example, the cause of a problem can be hidden some distance from where it comes to light, and the lack of structure can make tracking it down difficult. A structured program, written in a high-level language and backed up with a reasonable diagnostic system, can provide many advantages in debugging a program. Firstly, the high-level language compiler can build in run-time checking for errors such as array bound faults, division by zero and unassigned variables. In addition, the hierarchical nature of the design can be exploited in debugging by the run-time system producing a traceback through the hierarchy, showing on which line of the program the error occurred and the value of each variable in each routine through the traceback. Thus, errors can be tracked down with a well-defined methodology. There is no need for an online "fiddling" system such as DDT, or wholesale dumps of memory.

Many HDLs closely resemble assembly languages in their paucity of syntax and semantics. They have limited usefulness because it is extremely difficult to write a correct description of any significant length or complexity. Problems similar to those of debugging programs occur in debugging a design description. Often, HDL compilers perform no diagnostic functions and the designer is left to unravel the program with very little help. However, extending the diagnostic ideas of a high-level language compiler into an HDL compiler gives many advantages. Naturally, the errors detected may be different, eg dangling signal, illegal wired-or or incompatible signal vectors. As with a programming language, a design language compiler can provide diagnostic traceback facilities when an error is detected in a design. Thus a designer can have the same environment for debugging as a programmer.

Both programming and design environments benefit from the use of high-level languages in the quality of self-documentation provided naturally through the use of the language and through the actions of the compiler. High-level languages encourage the use of meaningful names for modules and variables or signals. This means that a program or design is more readily understood and is easier to maintain.

3. DESIGN OF MODEL

The fundamental structuring feature in the MODEL language is the concept of a PART. A part is a module with one or more input signals and one or more output signals. Parts can also have numeric parameters. The definition of a part specifies its internal structure in terms of instances of simpler parts and their interconnections. An instance of a part is the use of that part within the definition of a more complex part.

MODEL supports two data types: signals and integers. Signals may appear in one dimensional vectors and are the basic objects to be manipulated in the language. Signals or vectors may be simply connected together or may appear as input or output parameters to a part.

The major control structures of high-level languages are duplicated in MODEL, namely loops and conditionals. These allow parts to be more than just simple interconnections of other modules. A part can be described to fulfil a range of tasks e.g. an n-bit shift register or an n-bit counter.

A more complete description of MODEL appears in [3]. A few examples of MODEL programs are included in this paper to illustrate its salient points.

Example 1: An n-bit multiplexer.

```
PART mux (n) [data(1:n),ctl(1:n)] -> output
   SIGNAL d(1:n)
   INTEGER i
   FOR i=1:n CYCLE
      nand [data(i),ctl(i)] -> d(i)
   REPEAT
   nand [d] -> output
END
```

Example 2: A binary decoder.

```
PART decoder [a,b,latch] -> p(0:3)
   SIGNAL abar,bbar
   INTEGER i
   not [a] -> abar
   not [b] -> bbar
   FOR i=0:3 CYCLE
      nor [IF i&1=0 THEN a ELSE abar ENDIF,
           IF i&2=0 THEN b ELSE bbar ENDIF,
           latch] -> p(i)
   REPEAT
END
```

Example 3: A flip-flop

```
PART inner [p,q,r,s] -> a,abar
   nand [p,q,dbar] -> d
   nand [r,s,d]    -> dbar
END

PART flip flop [reset,clear,clock,d] -> q,qbar
   SIGNAL aout,bout(1:2)
   inner [reset,bout(2),clear,clock] -> --,aout
   inner [aout,clock,clear,d]        -> bout
   inner [reset,aout,clear,bout(1)]  -> q,qbar
END
```

Since the initial design of the compiler was undertaken, several enhancements have been included. The compiler now looks after optimisations made possible by VDD, GND or don't care used as inputs to parts. For example, a two-input NAND gate may be reduced to an inverter if one of the inputs is VDD. Pads proved to be more complex than originally anticipated and so a GENERAL PAD has been introduced into the language to allow a more flexible description of peripheral cells.

4. THE COMPILER

4.1 Description

The MODEL compiler operates in three phases:

Phase 1. The first phase is similar to the normal compilation process for high-level languages. A non-backtracking context-sensitive parser is used to generate an analysis record from which an intermediate code is assembled. During this phase a listing is produced along with "static" error messages, that is, those which are dependant on the static program text.

Phase 2. The second phase "executes" the intermediate code produced by phase 1. As the execution proceeds, "values" are created and combined; "values" are wires and primitive components (stages). Unlike the execution of conventional programs the values are never destroyed; instead, they become associated with a record of the control path through the program. It is this path that defines the hierarchy of the design. During this phase, "dynamic" errors are reported, ie those which are dependant on the execution path of the program.

Phase 3. The final phase of the compilation performs three main functions. Firstly it optimises the circuit produced by phase 2 by removing redundant logic. Secondly, it reduces to one wire groups of wires which are joined together. Thirdly, it outputs the Intermediate Design Language (IDL) file in the format required by the rest of the design system.

One particular problem which the third phase has to address is that of defining the locality of each wire. In order that the physical design subsystem can optimise placement and routing, the IDL file must define wires as close to their use as possible, hence phase 3 must move the definitions of signals to the lowest point in the hierarchy which is common to all uses.

4.2 Diagnostics

Phase 1. In order to simplify error detection and recovery, the MODEL language has been defined on a statement-by-statement basis; that is, the syntax does not define compound statements. The result of this is that, on detecting an error, the compiler can simply skip to the end of the statement (line) and continue from there.

Phase 2. As the exact location of "execution time" errors includes the path to the faulty statement, the second phase must keep track of the local execution history in source program terms. On detecting the error, the compiler can then identify the error line, the part containing it, the part from which that part was invoked and so on.

A MODEL program describing an adder together with diagnostic output produced by the compiler is shown below.

```
Source file: SIA1:[JEEM.ICCD]ADDERBUG.MOD;3
compiled on 27-JUN-1983 at 11:22:03

     Lattice Logic MODEL compiler Version: 3.4.0

     1 {tests and,nor,nand,not,xor}
     2
     3 {n-bit binary full adder with fast carry,}
     4 {similar to TI part SN74LS283           }
     5
     6 Constant bits=4
     7 Input Pad  a(1:bits), b(1:bits), cin
     8 Output Pad s(1:bits), cout
     9
    10 Part look (n) [p(1:n),g(0:n)] -> carry
    11    Signal temp(1:n)
    12    Integer j
    13    and[p(j:n),g(j-1)] -> temp(j) For j=1:n
    14    nor[g(n),temp(1:n)] -> carry
    15 End
    16
    17 Part slice (n) [a,b,p(1:n-1),
    18                 g(0:n-1),cin] ->
    19                 sum,cout,prop,gen
    20    nand[a,b] -> prop
    21    nor[a,b] -> gen
    22    xor[cin,and[prop,not[gen]]] -> sum
    23    look (n) [p(1:n-1),prop,
    24              g(0:n-1),gen] -> cout
    25 End
    26
    27 Part adder (n) [a(n:1 By -1),
    28                 b(n:1 By -1),cin] ->
    29                 cout,s(n:1 By -1)
    30    Integer j
    31    Signal c(0:bits),p(1:bits),g(0:bits)
    32    not[cin] -> g(0)
    33    not[g(0)] -> c(0)
    34    c(n) -> cout
    35    For j=1:n Cycle
    36       slice(j)[a(j),b(j),p(1:j),
    37                g(0:j-1),c(j-1)] ->
    38                s(j),c(j),p(j),g(j)
    39    Repeat
    40 End
    41
    42 adder (bits)[a(bits:1 By -1),
    43              b(bits:1 By -1),cin] ->
    44              cout,s(bits:1 By -1)
    45
    46 Endoffile
 1 fault detected

*Model fails -- Too many inputs

  Executing line 17 in slice
            n = 1

Called from line 36 in adder
            n = 4
            j = 1

Called from line 42
         bits = 4

 1 fault detected
```

5. SUMMARY

The MODEL compiler has been in use for about two years. It exhibits approximately linear time and space performance for reasonably structured designs, and rates of debugged code production are between 100 and 500 gates per day.

The MODEL language has been modified with use; these modifications have been largely syntactic alterations to improve the consistency of the grammar. One clear benefit of the MODEL language and compiler over many other HDL systems, eg an embedded language, is that diagnostics can be dealt with effectively by the compiler and this therefore leads to a shorter debugging cycle.

REFERENCES

[1] T. Vehara, M. Barbacci
Proceedings of 6th International Symposium on Computer Hardware Description Languages
Carnegie-Mellon University, Pittsburgh
May 1983

[2] J.P. Gray, I. Buchanan, P.S. Robertson
"Designing Gate Arrays using a Silicon Compiler"
Proceedings of 19th Design Automation Conference
IEEE Catalog No. 82 CH 1759-0

[3] MODEL Language Reference Manual
Appendix 1 of "Designing with Gate Arrays"
Lattice Logic, Edinburgh, Scotland

PERSONAL COMPUTERS AS KNOWLEDGE BASE DEVELOPMENT ENVIRONMENT

David Y. Y. Yun

Computer Science and Engineering Dept.
Southern Methodist University
Dallas, Texas 75275

and

IBM Research Center
P.O. Box 218,
Yorktown Heights, NY 10598

Abstract

"Intelligent Workstations" for engineering design are reaching the level of daily working environments of practicing engineers. Such useful tools as interactive graphics and program editing/debugging can be carried out on these tools most conveniently. Furthermore, pressing engineering needs arising from VLSI design are calling for new and better tools to improve the productivity of engineers. Unfortunately, the true local intelligence, such as program control structure and reasoning on symbolic data/knowledge, still tends to be weak even on the more powerful of the marketed workstations. We discuss a LISP based programming environment on a personal computer that is not only capable of assisting program (knowledge base, expert system) development, but also carry out local decision support inferencing and remote network access/transfer of data/knowledge.

1. Introduction

In a recent survey article on "sorting out the CAE Workstations", Jerry Werner, editor-in-chief of VLSI Design points out that "software is the key". The important issue is the integration of application software with the memory and processing power of the hardware and with the graphics and screen manipulation through the operating systems. It is increasingly recognized that the limitation in achieving high density, high functionality of VLSI chips is not physical or technological. Rather, it is the number of available design and process engineers, and their productivity, that is the problem. Specifically, it is the techniques and the "tools" used by them that need drastic improvement.

The complexity of achieving sub-micron chip features has been extrapolated from the experienced designers of LSI or MSI levels. The combinatorial explosion of details that face an engineer is a major limitation in realizing future needs, particularly those low volume, special purpose VLSI chips. Our premise, as shared by numerous practitioners of this field, is that the body of knowledge for VLSI design is composed largely of judgements and heuristics of the "experts". Evidence of this is represented by the fact that current successes in VLSI chip design are mainly engineering accomplishments, not scientific nor technological breakthroughs. The needs are design tools that assist in the exploitation of the hierarchical organizations and constraint properties during the design process.

Analogous problems can be found readily in the design and development of large software packages, a number of years back. Programmers, especially those with "experience", were thought to be in short supply. Numerous tools, such as high-level languages and compilers, as well as techniques, such as structured programming and schematic (hierarchical) diagrams, were needed and developed. The search for solutions to the problems in VLSI has followed a path not dissimilar to the past. Several fields of computer science have been found to have direct bearing on this particular area: data structure and algorithm design, computational complexity, programming languages, simulation, statistical methods and analysis, software engineering, and graphics. Perhaps due to the engineering nature and needs of the VLSI problems, the approaches encompassed by "Knowledge Engineering", the art of developing Knowledge Based Expert Systems, seem to have become the focus of significant amount of attention. The major ingredients of KBES, such as symbolic representation of data, heuristic judgement, expert knowledge, logical inferencing, interactive interface, hierarchical organization, and easy modification, do seem to match some essential needs of VLSI design. Several additional features of ES's, including careful expertise extraction, domain restriction, logic deduction tracking, natural language interface, and conducive system environment, also seem to be highly desirable for assiting a design engineer faced with a complex process.

We will give an assessment of some existing Knowledge Base Development Environments (KBDE) and extract a set of common desirable criteria in the next section. Then we discuss the attempt of developing such a KBDE on the IBM Personal Computer. Finally, we present the lessons drawn from our attempt so far and some targets for the future.

2. Experiences with Several KBDE's

Expert systems can be given a slightly more general definition to include all programs that use explicitly represented knowledge and computational inference procedures to solve problems that are normally expected of a recognized human expert. Such a system can be called simply a "Knowledge Based System", which is more abundant than known "Expert Systems". Chess programs, block-world geometry programs, robot arm manipulation, and computer algebra systems can all be included in this category. In Fact, most computer programs reflect a certain level of "human expertise". It is really the program

CH1935-6/83/0000/0527$01.00 © 1983 IEEE

organization, the separation of logic and data, that distinguishes a traditional program from an expert system.

Computer algebra systems, such as [MACSYMA] and [SCRATCHPAD], are acknowledged as knowledge based systems since they perform at skill levels even surpassing human experts (mathematicians). Practitioners in this field have gained valuable experience from the design and implementation of these mathematical KBSs. Symbolic data representation, knowledge (both algorithmic and heuristic) specification, modular and hierarchical organization, high-level transformation, and convenient user interface have always been major considerations underlying the design of these systems.

On the other hand, valuable experience has also been gained recently by our recent efforts of developing expert systems more in line with Feigenbaum's original specification [Feigenbaum 80]. We have installed and used two of the well known knowledge engineering tools, EMYCIN and OPS5. The project of developing a specific expert system using OPS5 is still on-going. EMYCIN was extensively exercised by a project at IBM Research, which we now briefly discuss.

The motivation stems from the realization that there exists a significant gap between the abundance of sophisticated facilities offered any computing environment and the users who usually only make mundane utilizations. Unfortunately, these "casual" users demand more friendly and intelligent interface from the system and application software. An expert system serving as an "intelligent agent" of a body of knowledge contained in the existing facilities must be valuable for improving the productivity of these users.

What exists in a modern, interactive operating system are some "static" help (tell, ?, etc.) "menus". However, menus are often too complex for novice users (going through screen after screen of text explaining the choices, with numerous foreign terminologies and/or meaningless acronyms) and too slow and tedious for an experienced user (knowing exactly what should be done, but must step through the menu sequence anyway).

The goal of the Dynamic HELP Project was to develop an KBES serving as an on-line consultant to users of the computer, who are reaching into the realm of unfamiliar facilities to get a quick response to their own computing needs. The domain of concern was initially the collection of the VM/CMS command modules and EXEC's that have associated TELL description files to serve as the source of knowledge. The domain of the resulting CMS-HELP ES was, in fact, restricted to those commands falling in the general paradigm of operating on files while passing them from a source to a sink. The knowledge about these commands was coded as rules in EMYCIN. CMS-HELP assists CMS users by reasoning about its own KB and giving advice about the use of a sequence of commands to accomplish a user's task. It performs at a level of competence comparable to a computing center consultant who carries on a conversation on the telephone to help a user working at a terminal.

Finally, the fundamental underlying environment for developing most of these KBS has been provided in various LISP environments such as MACLISP/EMACS, INTERLISP, YKTLISP/LISPEDIT, and LISP machines. OPS5 and MACSYMA are written in MACLISP in a programming environment that is greatly enhanced by the EMACS editor. EMYCIN and CMS-HELP are written in INTERLISP, which has the reputation of being one of the most friendly program development environments. OPS5 has recently been transported to YKTLISP, within the last 6 months. An expert system project using OPS5 and a new computer algebra system called NEWSPAD (for NEW ScratchPAD) are being developed in YKTLISP, and its associated edit/debug facilities.

A set of desirable features for a knowledge base development environment has emerged from these experiences. This resulted in the following list of criteria, which we adhered to in the design of our KBDE on the PC:

(1) an editor that "understands" the programming language, e.g. LISP expressions and programs can be "pretty printed" with parenthesis balancing features.

(2) an editor that "communicates" with the programming language, e.g. programs can be passed from the editor to the language for trial execution.

(3) a language that "interacts" with the programmer, e.g. offering tracing of functions and variables, allowing a program to be stepped through, and allowing change of variable value or redefinition of programs with continued execution.

(4) a system that "remembers" changes made by developers, in that a record of changes made is kept so as to allow a system regeneration, which is the result of most up-to-date work.

(5) a knowledge base that "organizes" well, in that rules can be collected into modular sets, which can be hierarchically or horizontally connected.

(6) a language that "adapts" to the programming style of the system developer, e.g. allowing a mixture of functional programming, a la LISP, with the declarative, non-procedural style of PROLOG.

(7) an inference mechanism that "conforms" to the intent of the system designer, in that deductions can be made either by forward or by backward chaining and controls can be set and modified dynamicly.

(8) a system that "tracks" the reasoning logic, in that reasoning sequence and branching can be monitored by the user.

(9) a system that "transports" easily, in that the rules, programs, data, and support functions can be made compatible with other dialects (of LISP, say) and portable to other system environments.

3. A KB Development Environment on the PC

With these criteria in mind, we carried out the design and implementation of MULE, the MULisp Extension, environment for the IBM personal computer; and WHIM, a symbolic programming and declarative logic inferencing environment built on (and with) MULE. The underlying LISP is the muLISP/muSTAR developed by The Soft Warehouse of Honolulu, Hawaii [MULISP 80]. Since MULE is embedded in LISP, it supports symbol and list manipulation capabilities and dynamic storage allocation (e.g. garbage collection). The environment provides a full screen program editor, a relatively complete set of interactive trace/debug facilities, and a logic deduction mechanism that is more flexible than some of the well known KE tools, such as EMYCIN and OPS5.

The overall system performance has been perceived to surpass that of a normally loaded time-sharing machine running LISP. The potential applications for the programming environment are numerous and include: computer algebra, expert systems (rule based or otherwise), natural language DBMS query ststems, etc. Although the environment is similar to several mainframe systems, it is new for microcomputers. The logic programming and inference extensions to LISP are based on other AI languages: PROLOG, CONNIVER, LOGLISP, and OPS5, modified by the experience gained with the CMS-HELP/EMYCIN, TEIRESIAS, and CENTAUR expert systems.

3.1 The MULE Programming Environment

The muSTAR editor assigns full screen functions to the various control keys allowing fast program changes. This feature makes the system easy to use because the same functions work on all in-core objects: programs, values and properties. Since the editor is written in muLISP itself, it is flexible and easy to modify, e.g. for dynamic reconfiguration of control keys. We have made significant extensions to the original muSTAR editor to enable pretty-printing, multiple levels of editing, and invocation of LISP from within the editor. Because this co-residancy, the edit/debug context switch can be rapidly done. Many additional functions have been added to enrich the available editing tools (e.g. parenthesis balancing and block moves). MULE also features macros and read-macros, which muLISP lacks. In summary, the MULE interactive programming environment behaves like a "baby INTERLISP".

There are four groups of top level commands that enable these features of MULE:

Editor Options: Editor commands invoke the full-screen editor to add or modify function definitions.

LISP Level Options: The three LISP level options are (1) the NREAD-eval-print loop that supports full read macros including prefix quote, (2) the read-eval-print loop that has no macros, and (3) eval-quote LISP.

Debugging Options: Debugging is simplified by tracing variables and the execution of functions. At any point, variable values may be reassigned and function execution resumed.

System Level Options: There are five system level options that (1) queries the user for the name of a source file and reads the file, name.LIB, into memory; (2) queries the user for a name and opens a file where the source codes on SAVELIST are saved. (3) another similar query opens the file where the source code for each object on a list with file-tag names are saved. (4) switches the floppy disk drive currently in use. (5) exits MULE, returning the user to the DOS level.

3.2 The WHIM Deduction System

The WHIM declarative (applicative) language extensions allow a non-procedural programming style similar to, but more general than, that provided by PROLOG. In our system, imperative LISP functions may be called from within declarative code, much like LOGLISP. Thus, users may mix the two styles to fit different aspects of a problem. The inference control scheme is explicit and more powerful: meta-rules and a partial ordering on goals allow incremental additions and modifications of the deductive process.

The WHIM deductive system extends MULE with facilities for managing knowledge bases. WHIM has two major and related parts, the Deductive Manager (DM) and the Knowledge Base (KB), as sketched here:

```
*-------------------------------------------------*
|                                                 |
|  Deductive Manager:                             |
|       Manages Knowledge Base,                   |
|       Generates Inferences and Explanations.    |
|       Knowledge Independent.                    |
|                                                 |
|  *------------------------------------------*   |
|  |                                          |   |
|  |    Knowledge Base:                       |   |
|  |         Contains Facts, Rules            |   |
|  |         and Reasoning Control Knowledge, |   |
|  |         plus Justifications.             |   |
|  |         Incrementally Additive.          |   |
|  |                                          |   |
|  *------------------------------------------*   |
|                                                 |
*-------------------------------------------------*
```

3.2.1 The Knowledge Base

Facts are represented using a form of first order predicate calculus, such as (ON ball table) and (RABBIT bugsy). Such clauses may contain variables, which are strings immediately preceded by a "?", e.g. (ON ?x TABLE) or (NOT (RABBIT ?y)).

All inference knowledge in the system is stored as rules. The two logical parts of a rule are the antecedent list and consequence list of clauses. Rules may be used in both forward and backward chaining manners. How a rule is to be used is not a static decision. Rather, it is a dynamic process in WHIM, controled by several mechanisms directly representable in the KB. A forward chaining rule has its antecedent clause "active", and a backward chaining rule has the consequence clause active. Bi-directional rules have both clauses active. Any active clause must contain only a single variable.

Any number of rules may be collected into a "rule set", which is defined with the DM command RULESET. Rules and rule sets may be members of more than one rule set. Such a nested rule-set forms a tree, which may be refered to as a group by the set name. These rule sets can then be manipulated by DM commands, such as PREFER and SWITCH.

Meta-rules allow an explicit and extensible control mechanism for deciding which rule to chain off for a given goal, if more than one unify. A meta-rule is simply a rule which refers to other rules instead of to the problem domain. Thus, meta-rules are not concerned with validity of rules, but with their applicability. An example of a meta-rule in English is

If the goal is to determine if X is an animal
And X is in the KB
Then set ruleset GENERALIZE-TO-ANIMALS forward.

A typical meta-goal is a PREFER command, which selects a preferable rule or ruleset for application. Meta-rules may also be forward or backward chaining, and may be included in rule sets.

A partial ordering on goals may be specified in the system. Such explicit control is essential to allow incremental addition to the KB. This feature is the result of recognizing the deficiencies in PROLOG (and similarly in OPS5), where subgoals are guaranteed to be evaluated in the same order they were written.

3.2.2 The Deductive Manager

The DM (sometimes known as inference engine) does not control the actual inferences made since all control information is explicitly represented in the KB. Instead, the DM provides a set of utilities to manage the KB (including a data dependency model of justifications), execute the selected inference rules, and generate explanations of action when requested.

A DM/KB interface is provided to allow DM commands to act on the KB. These commands are classified into the following catagories: Creation (used to build a KB), Normal (used in routine inference rules), Debug (used to understand system behavior), Meta (consequences of meta-rules).

The internal representation for the KB was chosen so that rapid matching between arbitrary fact, rule, and goal clauses can be made. One useful mechanism for this purpose is discrimination nets. It is used to provide a general (and fast) selection of possible matches, which may then be checked with the more exact (and slower) unification algorithm (see [Charniak 79]). The unification algorithm used is described in [Nilsson 80].

Justifications for the validity of each fact are also stored in the KB under the property JUST. Every fact entered by a user level ASSERT has the justification PREMISE. Facts can also be generated by rules. This justification information can be and is used to generate the pseudo-English explanations.

4. Assessment and Future Objectives

The ultimate objective of our efforts is to develop a KBDE or a "Knowledge Engineering Tool". Our system represents the mere basics for a knowledge editor. More complex facilities will be necessary to design, construct, execute, and fine tune large KBs. One such future facility is a batch handler, which offered by EMYCIN that provides for re-running old cases on a new KB. Another is a deduction browser that would allow a knowledge engineer to graphically browse through the sequence of rule calls created by the deduction process.

An important application file supported by MULE is UPLOAD.LIB. It contains the muLISP compatibility package for porting muLISP code or rule sets to other dialects of LISP, specifically now YKTLISP. This is merely a first attempt of tranporting LISP programs, rule sets, and knowledge bases, between workstations or with a central host machine. The longer term objective is to achieve a network of compatible workstations, each capable of interactive knowledge base development and local deductive reasoning, together with a central repository of data/knowledge on a host machine, which controls the flow of individual knowledge bases much like a librarian. Then, any knowledge base (deposited in the central bank) compatible with the inference mechanism and the language can be accessed, modified, reorganized, tested, and used for local reasoning. The network is not only intended for separate development of knowledge bases, but also for potential parallel processing of multiple reasoning paths. In addition to its librarian functions, the central processor also serves as reserved processor for the workstations (when additional compute power is needed) and controls the parallel processes by assigning tasks to the workstations.

It is our view that the PC is still not the most suitable vehicle for such a large scale knowledge base development. It is lacking in memory, speed, and quality graphics. However, the PC environment has proved to support program development and local reasoning quite admirably. It is satisfactory, as well as often more satisfying than time sharing, for small scale AI/KB applications. Furthermore, it certainly was an invaluable vehicle for our efforts and provides a guideline for our design of a better distributive AI/KB system. One outstanding potential lies in the combination of intelligent (knowledge base and logic inferencing) functions with a more powerful engineering workstation, such as the general purpose APOLLO DN600 or the more specialized VLSI design station CAE 2000. Such a locally intelligent engineering workstation should be easily realizable and will already significantly improve engineering productivity, even before the networking and central knowledge base librarian could be fully developed.

References

[Charniak 79] Charniak, E., Artificial Intelligence Programming, Wiley, 1979.

[Feigenbaum 80] Feigenbaum, E. A., "Knowledge Engineering: The Applied Side of Artificial Intelligence", Computer Science Department Report, Stanford University, 1980.

[MACSYMA] The MACSYMA Group, MACSYMA Reference Manual, MIT Laboratory for Computer Science, Cambridge, Mass.

[MULISP 80] The Soft Warehouse, muLISP/muSTAR-80 Artificial Intelligence Development System Reference Manual, Honolulu, Hawaii.

[Nilsson 80] Nilsson, N. J., Principles of Artificial Intelligence, Tioga Publishing Co., Palo Alto, Calif., 1980.

[SCRATCHPAD] Griesmer, J. H., Jenks, R. D., and Yun, D.Y.Y.: SCRATCHPAD User's Manual, IBM Research Publication RA70, June 1975.

[Werner 83] Werner, J., "Sorting Out the CAE Workstations", VLSI Design March/April, 1983, pp. 46-55.

Session 3-1c
CAD and VLSI I

Chairman

I. Shirakawa
Osaka University

SHARPS: A CAD SYSTEM FOR VLSI

T. Chiba, N. Kubo, T. Kambe, T. Inufushi, I. Nishioka, and S. Kimura

Engineering Center, Sharp Corporation
Tenri, Nara 632 Japan

Abstract

In this paper, a hierarchical CAD system SHARPS for VLSI, which is still partly under development, is outlined, highlighting the concepts and architectures. Several implementation results are also touched on.

Introduction

The increasing scale of circuits integration has changed VLSI design drastically, and a CAD system has to cover all the design phases of Fig.1, to guarantee short turn around time, and to permit any combination of CAD tools.

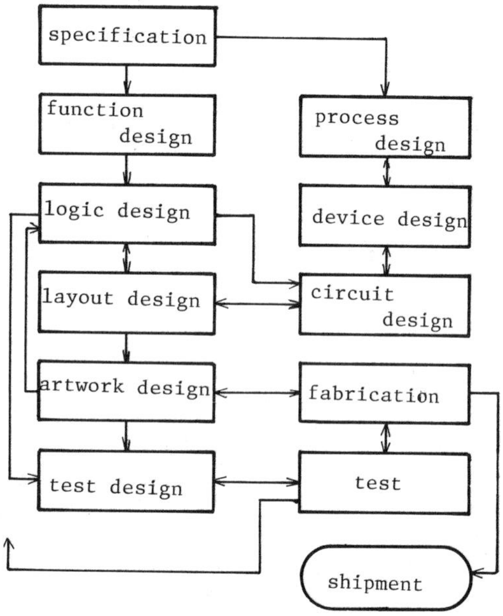

Fig. 1 Outline of VLSI design flow.

To fulfill these requirements, a total LSI design system is urgently required, which is constructed on the basis of
1) a total CAD data base to support consistently all areas of the design flow (for example, some revision of logic design yields corresponding data modifications of layout),
2) good automatic facilities which guarantee quick turn around time and a design of high quality,
3) interactive design facilities which closely combine cleverness of a designer with efficiency and accuracy of a computer, and
4) a network linkage among designers to assure a project team approach to VLSI design.

To achieve these demands for a total CAD system, a variety of softwares are to be linked as illustrated in Fig.2, by which maintainability of each software, establishment of a common data access method, and verification among data of different design phases are easily realized. On the other hand, in terms of hardware specifications the followings are necessary;
a) an engineering work station which is composed of a CPU, a raster scan type color graphics display, and mass storage, is to be used for interaction of all design phases,
b) a special purpose computer such as a logic simulator machine,
c) a high speed computer to cover time consuming batch type jobs such as DRC, ERC, placement, routing, etc., and
d) network facilities to realize the architecture of Fig.2.

Under these software and hardware environments, a total CAD system can be constructed as will be outlined in the following.

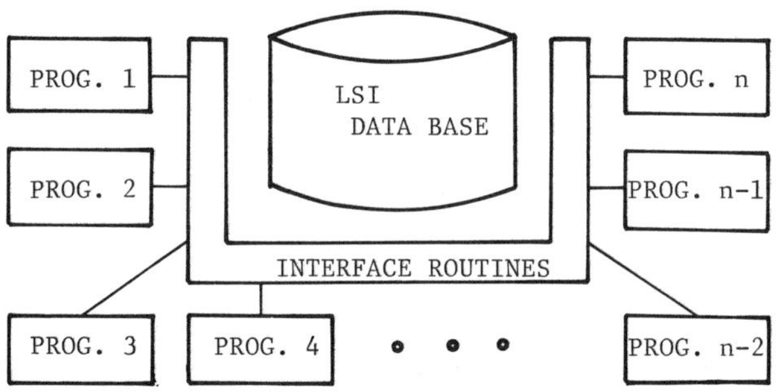

Fig. 2 Conceptual structure of software.

Outline of the system

In 1979 we started to develop our LSI design system SHARPS (Sharp Hierarchical Automatic Routing and Placement System)[1,2,3,4], which is mainly for general cell, standard cell (or polycell), and masterslice LSI's. Distinctive features of SHARPS are listed below.

A: Work Station

In general, LSI design phases such as function design, logic design, layout design, test design, etc., are iteratively executed in an order which is determined at a designer's choice. Therefore, to attain a quick turn around time, those overhead induced by these iterations of the design phases has to be eliminated by introducing an engineering work station with which a designer can perform a high

CH1935-6/83/0000/0532$01.00

quality design. This engineering work station is composed of
i) a color raster scan type graphics display with up to 8 MB segment memory,
ii) a microprocessor of 32 bits architecture with up to 8 MB RAM,
iii) an 80 MB Winchester type disk storage, and
iV) a monochrome console display.
Any interaction of all these design phases is performed through the use of this engineering work station independently of other computers in order to guarantee a quick response.

B: Computer Network

Assigning various kinds of jobs (such as logic simulation, interactive layout pattern editing, design rule check, PG tape generation, etc.) to their adequate computers (such as hardware simulators, engineering work stations, a main frame computer (IBM 3081/G), and super minicomputers (VAX 11/780)), an efficient use of a total design system is achieved. These computers are connected through the use of a local area network of an Eather Net type (10 Mb/s), to enable several designers to share the total VLSI design phases as illustrated in Fig.3.

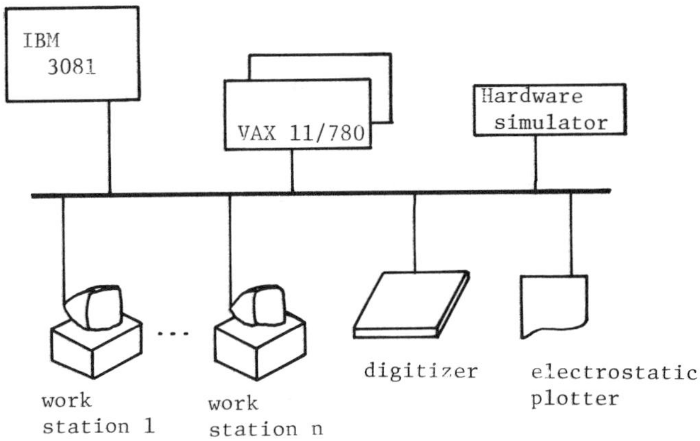

Fig. 3 Hardware configuration.

C: Total CAD Data Base

Any functional block at each hierarchy level has various kinds of data (such as for logic, layout pattern, circuits, etc.), but conventionally data in the block are treated differently and stored in separate files. Therefore consistency check for all these data is difficult. To cope with this barrier, in our CAD data base they are treated as different aspects of the same block. All programs are provided around this data base so that designers can choose any design flow at his own will. This architecture provides a user with any design approach and also guarantees maintainability of programs.

D: Utility Programs

This system is equipped with a wide range of automatic and interactive programs, which will be briefly described in the next section.

Design flow of the system

In the following, along with the design flow of Fig.1 we describe each design phase of SHARPS, taking the case of polycell design approach.

(I) Cell Design

Semiconductor process design and device design are bases of LSI development. Primitive devices are stored in a device library, and basic cells are constructed using these primitive devices. A circuit diagram of a cell is inputted through the use of a graphics editor as shown in Fig.4. In these design phases, process, device, and circuit simulators are used, and data (device parameters, process parameters, etc.) thus obtained are transferred to and from each other.

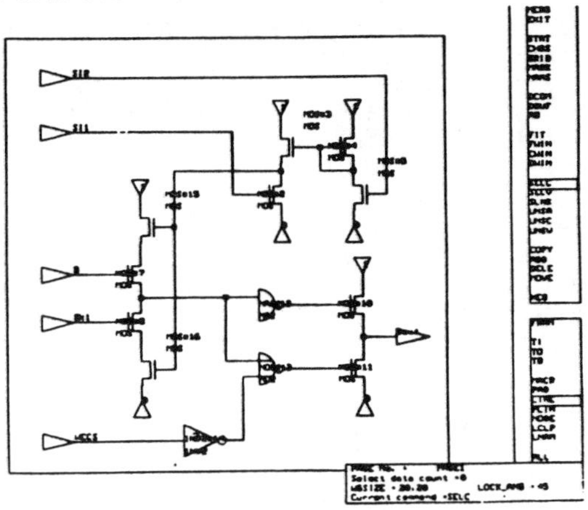

Fig. 4 Example of a circuits diagram displayed on a CRT.

(II) Logic Design

The first stage of logic design is to input a logic circuits diagram of an LSI. Elements of the diagram are boolean gate, flip-flops, and macros (a macro is defined by a combination of gates, flip-flops, and other macros). Logic design is implemented at engineering work stations, on which users can design and edit the circuits diagram interactively as shown in Fig.5.

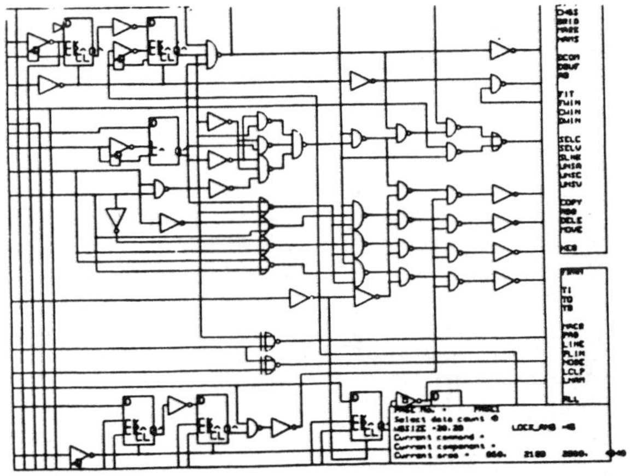

Fig. 5 Example of a logic diagram displayed on a CRT.

Moreover, an interactive simulation program supports users for logic verification of the circuits effectively. Features of this part of the system are summarized as follows:

(1) The hierarchical design can be efficiently performed by means of a facility to define macros.
(2) Several automatic routing facilities simplifies editing operations (for example, when the position of a gate is modified, the nets attached to it are automatically shifted so as to keep the interconnection).
(3) An interactive simulation program checks circuits behavior effectively.

(III) Layout Design

After verifying the correctness of inputted circuits by the simulator, layout design is performed in three phases; cell assignment, placement, and routing.

Cell assignment: Given a logic circuits and a set of layout cells in a library, the cell assignment problem is defined as follows: Find an exact cover of the circuits by a least number of cells. As readily seen, the problem contains two difficult subproblems; the subgraph isomorphism problem and minimum cover problem. Hence, a heuristic program[5] which consists of initial assignment and iterative improvement stages, is constructed, and experiments shows that the results are comparable to manual assignment.

Placement: The placement program first assigns sets of cells to rows, then determines positions of cells in each row. Given a set of cells and a number γ of rows, the set of cells are partitioned into γ classes. An initial solution is obtained by assigning these classes to γ rows. Then iterative improvements named a shuffling operation[2] are conducted so as to minimize the total number of nets which cross over rows.

The scheme for determining positions of cells in each row is the same as the one stated above. It should be noted that a hierarchical placement scheme is also implemented by introducing macros as well as cells into the shuffling stages, and hence higher performance has been provided.

Routing: Two routing programs are provided, one for a general cell or polycell LSI, and the other for a masterslice LSI. Both of them consist of two routing phases; global routing and detailed routing.

At the global routing phase, routing paths for each net are sought as a sequence of chennel segments. In case of a masterslice LSI, rerouting schemes are adopted to avoid local routing congestion and to get higher wirability.

At the detailed routing phase, line segments in each channel are generated. A dog-leg and a grid free routing schemes are adopted for more compact layout.

(IV) Artwork Design

After the automatic layout design, artwork data of an LSI are generated. For those LSI's manually designed, the system is provided with artwork design facilities. In such cases, designers first draw symbols (transistors, contacts, line segments, etc.) registered previously in a symbol library, then the layout drawing is digitized. Finally, the system transforms the inputted drawing into artwork data by referring to the symbol library.

Besides this program, SHARPS has two more tools for artwork design. One is a program[6] for random logic portion of MOS LSI, which automatically generates an array of ratioless MOS transistor circuits for a given set of boolean functions to be realized. Another is a compaction program for symbolic design: Given a rough symbolic layout drawing inputted through the use of a graphics display, the program automatically compacts the drawing and keeps design rules.

(V) Design Verification

After layout design, timing simulation and delay time checking programs are executed. They check setup and hold time of a flip-flop, delay time of critical paths, etc., taking account of propagation delay of routing length.

Other verification tools for artwork design are listed below.
(1) Geometrical design rule check (DRC): Checking a space between adjacent polygons, the minimum width of rectangle, etc.
(2) Electrical design rule check (ERC): Checking isolated polygons, short/open circuits, open gate transistors, etc.
(3) Generation of input data for simulation: After extracting from artwork patterns circuits elements (resistors, capasitors, etc.) and their interconnection information, input data to circuit simulators are generated.

(VI) Interactive Facilities

One of main features of SHARPS is that interactive facilities support users in all design phases. Since these facilities are implemented at a work station, short response time and high performance are always guaranteed. The interactive facilities are listed below.
(1) Logic circuits editing program (see (I)).
(2) Interactive routing program: This program is a depth-first maze router to be applied to two points specified by a user. When the batch type routing program fails in 100 percent routing, a user modifies portions of routing patterns and specifies two points of an incomplete net so as to be interconnected by this program[3].
(3) Artwork editing program: This program is for editing composite layout patterns and is also for displaying results of ERC program.

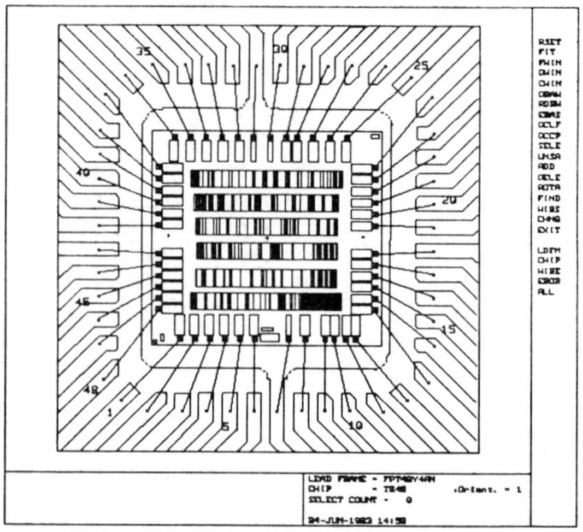

Fig. 6 Example of a package check program

(4) Package check program: In the layout design of polycell and general cell LSI's, positions of pads have to be checked whether bonding wires between pads and leads satisfy specified constraints (see Fig.6).

Experimental results

The system operates on the hardware of Fig.3. Since 1981, over fifty LSI's have been developed using the automatic layout facilities of the system. Fig.7 shows implementation results of a general cell LSI of a 8 bits microprocessor with 3,000 gates and a masterslice LSI with 700 gates, respectively.

Concluding remarks

Those advantages obtained by the total CAD system SHARPS are as follows.
a) Various types of design approaches to general cell, polycell, and masterslice LSI's can be easily realized by this software architecture.
b) Any flexible interaction among programs around the data base can be easily attempted to attain a high quality design, according to design environments. For example, in order to design a cell registered in a library, the process, device, and circuits simulators are to be dedicatedly combined through the use of the total CAD data base.
c) Each program can be changed or improved independently of other programs according as the design subject varies.

Acknowledgement

The authors would like to express their thanks to Prof. I. Shirakawa of Osaka University for his valuable criticism. They also acknowledge the encouragement received from Prof. Emeritus H. Ozaki of Osaka University and Dr. T. Nakamura of Sharp Co. Thanks are also due to our colleagues for their invaluable cooperation for programming.

References

[1] T. Chiba, N. Okuda, T. Kambe, I. Nishioka, T. Inufushi, and S. Kimura,"SHARPS: A hierarchical layout system for VLSI," Proc 18th Design Automation Conference, 1981.
[2] T. Kambe, T. Chiba, S. Kimura, T. Inufushi, N. Okuda, and I. Nishioka,"A placement algorithm for polycell LSI and its evaluation," Proc. 19th Design Automation Conference, 1982.
[3] T. Chiba, N. Okuda, T. Kambe, I. Nishioka, T. Inufushi, and S. Kimura,"A general cell layout system for VLSI," Proc. IEEE International symposium on Circuits and Systems, 1982.
[4] S. Kimura, N. Kubo, T. Chiba, I. Nishioka,"An automatic routing scheme for general cell LSI," IEEE Trans. on CAD (to appear in October 1983).
[5] S. Yoshida, T. Chiba, T. Kawata, and I. Shirakawa,"A logic assignment scheme in LSI design,"Monograph of IECE, CAS 82-147, 1983.
[6] I. Shirakawa, N. Okuda, T. Harada, S. Tani, and H. Ozaki,"A layout system for the random logic portion of an MOS LSI chip," IEEE Trans. on Computers, vol. C-30, No. 8, 1981.

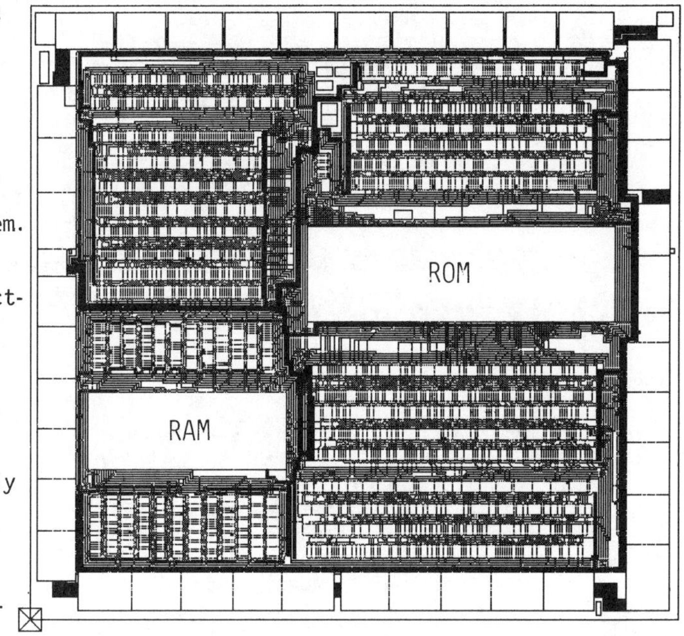

(a) General cell layout with four hierarchical levels.

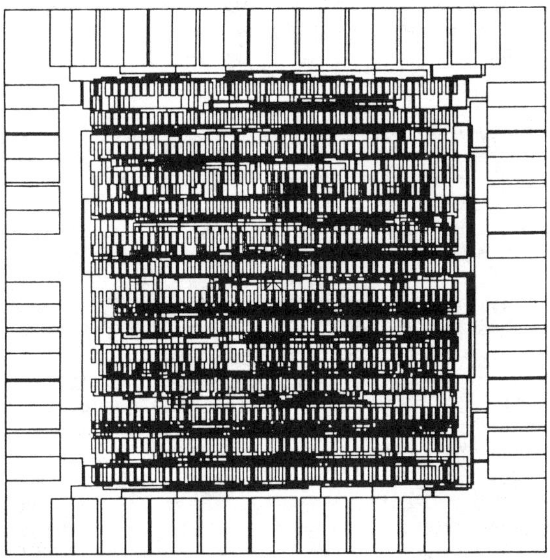

(b) Masterslice layout.

Fig. 7 Implementation results.

PERFORMANCE ANALYSIS OF AUTOMATIC PLACEMENT AND ROUTING FOR LARGE-SCALE CMOS MASTERSLICES

Hidekazu Terai*, Michiyoshi Hayase*, Tatsuki Ishii*, Chihei Miura*, Yasushi Ogawa*, Tokinori Kozawa*, Yasuo Sato**, Hisao Sasaki**, Tohru Hiyama***

* Central Research Laboratory, Hitachi Ltd. (Kokubunji,Tokyo,Japan)
** Device Development Center, Hitachi Ltd. (Kodaira,Tokyo,Japan)
*** Kanagawa Works, Hitachi Ltd. (Hadano,Kanagawa,Japan)

ABSTRACT

Automatic placement and routing algorithms for 6K gate CMOS masterslices and the results of experimental analyses are described. Three strategies for improving wirability are presented, and their effectiveness evaluated. The wiring potential of the channel router(CR) and the combination of the CR and the maze router(MR) are discussed from the perspective of processing time, wirability and chip size. It is estimated that CR+MR is 1.2 times more powerful than CR alone for reducing masterslice chip area. Placement and routing time are proportional to the number of cells, at 1.4 power and 1.2 power. Finally, wiring length is compared with respect to both design methods which partition the chip and those that do not.

1. Introduction

The number and variety of VLSIs that have to be designed has increased dramatically with the widening in variety of electronic equipment implementing them. To meet these design needs, the adoption of masterslice VLSIs is most effective. As the production volume of these LSIs is comparatively low, primarily because of their extensive variety, it is very important to reduce the design cost. For the manual wiring of unconnected pins, we estimate from experience that it requires about 1--2 person hours for each pin-to-pin connection. Therefore, improved wirability and less CPU time consumption need to be realized through automatic placement and routing programs. In this paper, we describe automatic placement and routing procedures which enable high speed processing and superior wirability, and also present the results of evaluative and experimental analyses.

2. Overview of CMOS Masterslice

Figure 1 shows the cell and chip structure of our CMOS masterslice LSI [1]. As shown in this figure, a 'Basic Cell (BCL)' is composed of 6 MOS FETs, and this acts as a 3-input NAND gate. These BCLs are arranged in the x direction to form masterslices on the chip.

The chip has 20 double cell rows and includes 4000 BCLs in total. The number of wiring tracks between rows on the first level is 35, and 4 tracks above a BCL are available on the second level.

The chip has 112 I/O pads and 8 power pads on the peripheral. Each of the 112 buffers is able to work as an input, output or bi-directional buffer. A good match between the I/O interface signal level and LS-TTLs is assured, thus making it possible to easily connect them to the conventional digital systems.

We define a 'Placement cell (PCL)' as a minimum logic unit. It is composed of one to about 50 BCLs, which is assembled horizontally and connected with each other using 1st and 2nd level metal layers. With regard to intra-cell track overflow, wiring tracks in the channel between the BCL rows are used for intra-cell connection. The height of a PCL is equal to 13 horizontal tracks. The width of the PCL is variable. These PCLs are pre-designed, and their electrical, logical and physical specifications are registered in the cell library accompanied by their pattern data.

A designer performs his logic design using these PCLs, and then inputs the schematic diagram to make the master logic file.

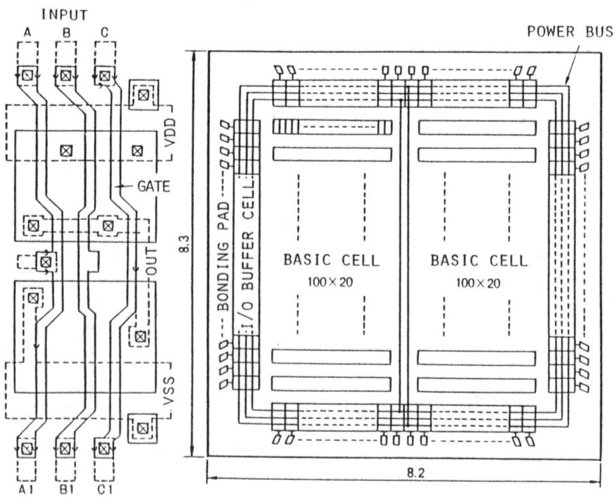

FIG.1 CELL & CHIP STRUCTURE OF CMOS MASTERSLICE

3. Organization of VLSI Layout System

The overall configuration of our DA system for logic VLSIs is shown in Figure 2. This system can be utilized not only for masterslice LSIs but also for fully customized LSIs, and it is based on building block methods [2,3,4]. Therefore, a designer can optimally choose either a masterslice approach or a building block approach when constructing the chip.

This system also allows a layout designer to partition the whole chip into sub-chips and perform the layout concurrently. Splitting and merging of physical design files is possible. Such an approach is effective in reducing the CPU resources required for placement and routing processes, as well as the design period.

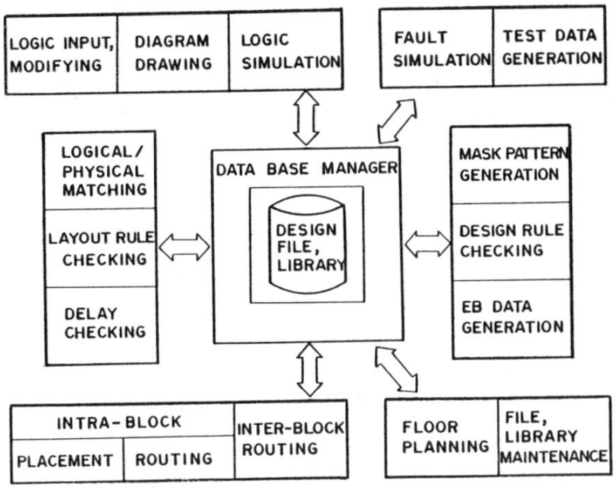

FIG.2 DA SYSTEM CONFIGURATION

4. Automatic Layout Algorithms

Figure 3 shows the functional diagram of our automatic placement and routing program. The placement procedure [4] is composed of two major components, that is initial placement procedures and improvement procedures. The routing procedure also has two main processes, global wiring and detailed wiring.

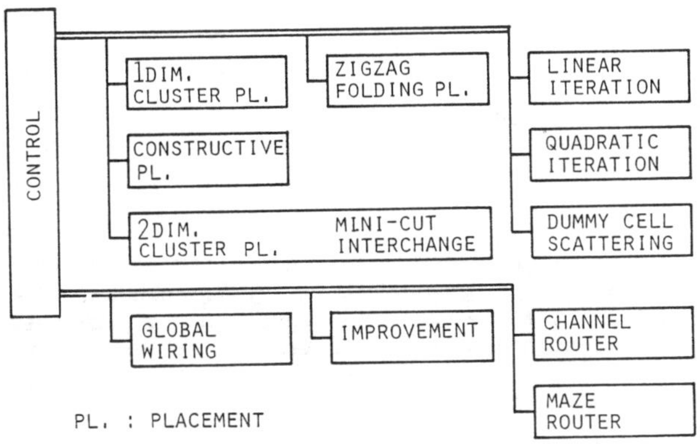

FIG.3 PROGRAM STRUCTURE

4.1 Initial Placement

Initial placement is the process whereby each logic unit (a PCL) is mapped onto the BCL rows.

Our mapping procedure is executed as a combination of bottom up and top down processings, and involves:

(1) Clustering the logic units.
(2) Forming the binary tree of clusters.
(3) Disassembling the tree from the top and mapping the node of the tree onto a partial domain of the chip. This domain is generated by partitioning the chip into smaller parts as the cluster tree is disassembled. Partitioning order is controlled by the parameters shown in Fig. 4.
(4) Interchanging the clusters across the cut line of the domain. An interchange is possible not only between the clusters but also between the PCLs.

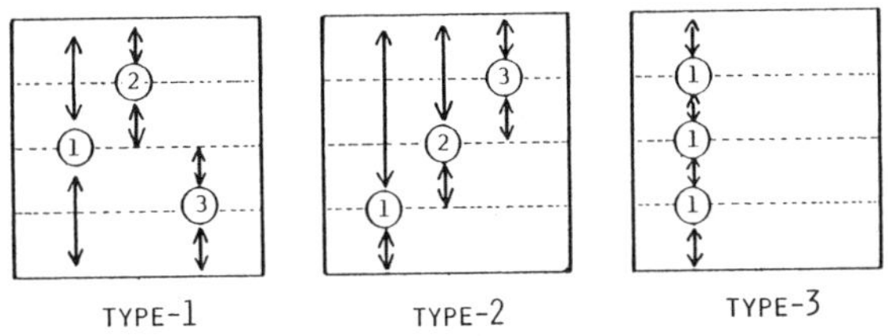

FIG.4 PARTITIONING ORDER

By mapping the logic units onto the BCL rows, a two-dimensional PCL array is obtained. Iterative placement improvement is accomplished at the next stage.

4.2 Improvement Procedure

In our program, there are two kinds of improvement procedures: linear iteration and quadratic iteration. The user can choose to use either of them separately or together.

(1) Linear iteration

The PCLs mapped onto each BCL row are transfered within that row only, using the Net Balance Placement algorithm [2]. Its objective function can result in a minimization of the trunk length along the row. This function is well-matched to the channel router.

(2) Quadratic iteration

Pairwise interchange is adopted. As the width of each PCL is variable, a simple one-to-one interchange results in irregularities in the mapped BCL row lengths. We thus employ an interchange among groups of PCLs whose lengths are the same. The objective function of this procedure is to reduce local wiring congestion.

When the utility of the BCLs is less than 100%, wirability can be effectively improved by scattering unused BCLs on the chip. Our program has a facility for scattering unused BCLs under a well-controlled strategy.

4.3 Global Wiring

The main task of global wiring is to determine the x coordinates where the vertical wiring path runs across the BCL rows. After completing this process for all of the nets, detailed wiring is executed to solve routing problems within every channel between adjacent BCL rows.

The second task of global wiring is to determine which side of the PCL terminals should be used. This problem results from the structure of a CMOS cell. The basic solution to this problem is as described here:

(1) The distribution of the trunk overlappings in the upper and lower channels is calculated as shown in Figure 5.

(2) The side selected is that at which the peak value of the distribution is smaller.

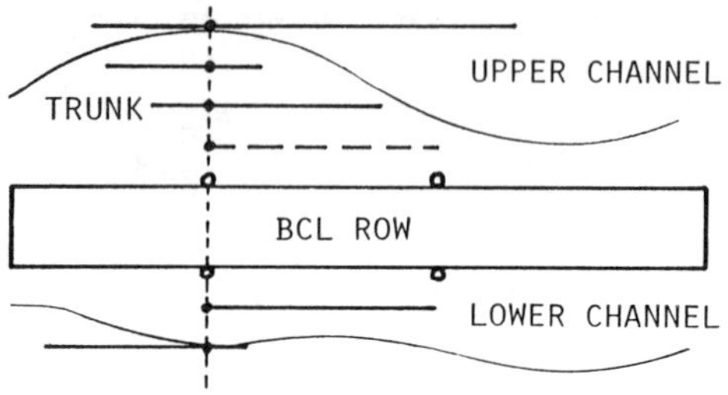

FIG.5 TRUNK DENSITY DISTRIBUTION

4.4 Detailed Wiring

Our routing program has two kinds of algorithms, a channel router(CR) and a maze router(MR). CR can perform routing in less CPU time than MR. On the other hand, MR has superior wiring potential. We combine these two algorithms to inter-balance their effects. That is, the major part of the wiring is done by CR, and the few nets which overflow the channels are connected by MR.

5. Performance Analysis

5.1 Some Strategies for Improving Wirability

After placement and global wiring have been completed, the number of required horizontal tracks is then estimated. If overflows are forecasted and congestion is localized, some strategies for improving wirability are then considered. We have verified the effects of the following three strategies.

S-1: Alteration of crossing points in the vertical wiring path.

S-2: Alteration of PCL sequences in a BCL row.

S-3: Alteration of trunk assignment in channels.

Their effectiveness is shown in Table 1. The experiment demonstrated that:

(1) S-1 reduced nonconnections by 98 (cf. Items 1 and 2);

(2) where S-1 was embedded, the reduction effect of S-2 alone is 8 (cf. Items 2 and 3) or 6 (cf. Items 4 and 5);

(3) on the other hand, S-3 reduced nonconnections by 42 (cf. Items 2 and 4) or 40 (cf. Items 3 and 5).

Therefore, the alteration of crossing points and trunk assignment is thought to be more powerful than that of PCL sequences.

TABLE.1 STRATEGIES FOR IMPROVING WIRABILITY

STRATEGY	ILLUSTRATION	ITEM 1	ITEM 2	ITEM 3	ITEM 4	ITEM 5
S-1 : CROSSING POINT	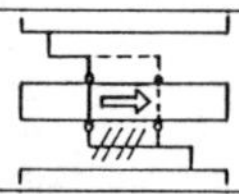	—	○	○	○	○
S-2 : PCL SEQUENCES	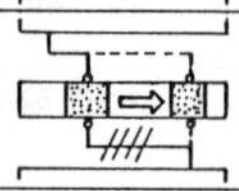	—	—	○	—	○
S-3 : TRUNK ASSIGNMENT	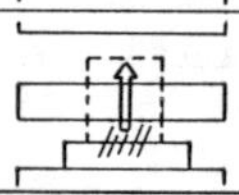	—	—	—	○	○
NO. OF UNCONNECTED PIN-TO-PINS		146	48	40	6	0

(///: CONGESTED REGION, —: NOT ADOPTED, ○: ADOPTED)

5.2 Effect of Combining Routing Algorithms

To confirm the effectiveness of combining CR and MR, we have experimented with automatic placement and routing for a chip in which the number of horizontal tracks is reduced to 74% of that shown in Fig. 1. Figure 6 shows the results. The nonconnections left by CR are perfectly cleaned up. Moreover, the increase in routing time is fairly small. Effectiveness is evident.

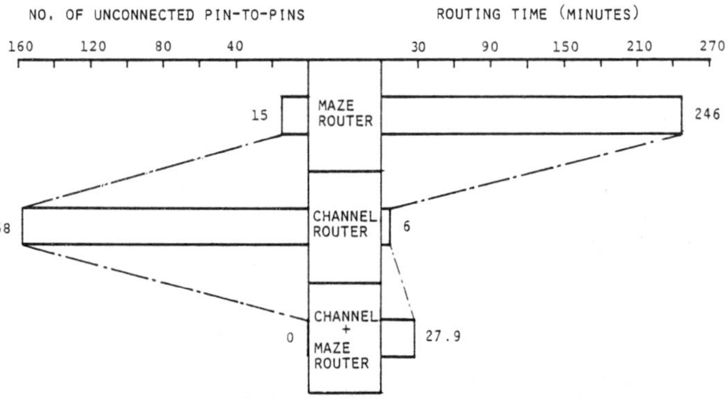

FIG.6 EFFECT OF COMBINING ROUTING ALGORITHMS

5.3 Processing Time Analysis

Figure 7 shows the processing time for the placement and routing algorithms. The lower section of the figure shows the time for the CMOS masterslice discussed in this paper. The upper part shows the routing time for our old MR. The routing time is improved approximately 100 times compared with conventional data, and placement and routing time are proportional to the number of PCLs, at 1.4 power and 1.2 power, respectively.

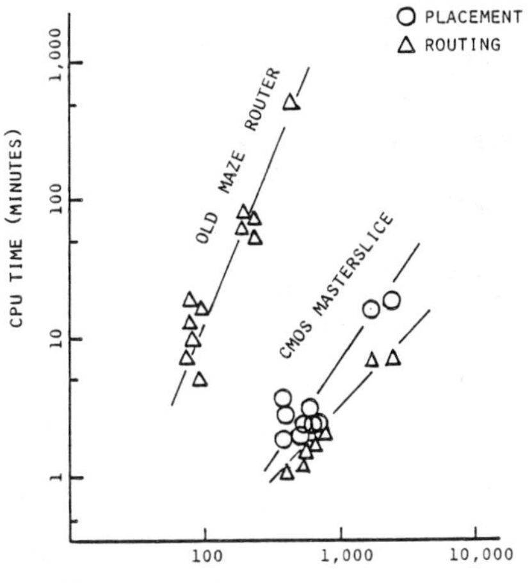

FIG.7 PROCESSING TIME

5.4 Wirability Analysis

We have also examined the potential for routing algorithms. Figure 8 shows the experimental relation between the number of nonconnections and the number of horizontal tracks. This figure shows the result of applying CR only, as well as that of combining CR and MR. The number of nonconnections decreases exponentially in relation to the increase in tracks, and the value finally reaches 0 for a certain number of tracks. This critical point corresponds to the router potential. For LSI-2, the potential for CR alone is 30 tracks and that for CR+MR is 23.

Here, we define S[CR] and S[CR+MR] as chip size fully routed by the channel router, and the combination of channel router and maze router. S[CR+MR]/S[CR]=(C+23)*N*W/(C+30)*N*W, where C,N, and W are the height of a BCL, the number of BCL rows, and the length of a BCL row, respectively. As C is equal to 13 in our case, this ratio is approximately 0.84. This means that the chip size as routed by CR+MR becomes 84% of that with CR alone. In other words, CR+MR is 1.2 times more powerful than CR.

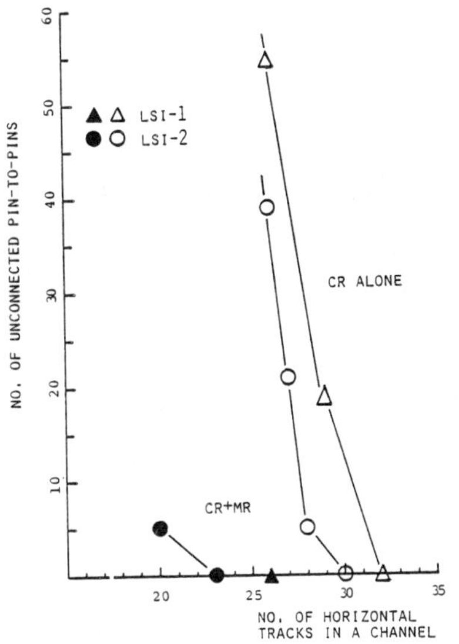

FIG.8 WIRABILITY VS. NUMBER OF HORIZONTAL TRACKS

5.5 Net Length Analysis

Figure 9 shows the net length distribution of a chip laid out automatically using two kinds of design methods. The solid line represents a chip in which the sub-chips are merged into a full chip after each partitioned portion (in this case a chip is partitioned into 4 sub-chips) was separately laid out. The broken line represents a chip laid out in a single effort without partitioning. While 97% of the nets were routed within the length of one side of the chip in the partitioned case, the value was 96% without partitioning. On the other hand, the maximum net length in the non-partitioned case is longer than that in the partitioned case. Thus it can be concluded that the partitionined layout method is superior to the non-partitionined layout from the viewpoint of delay management.

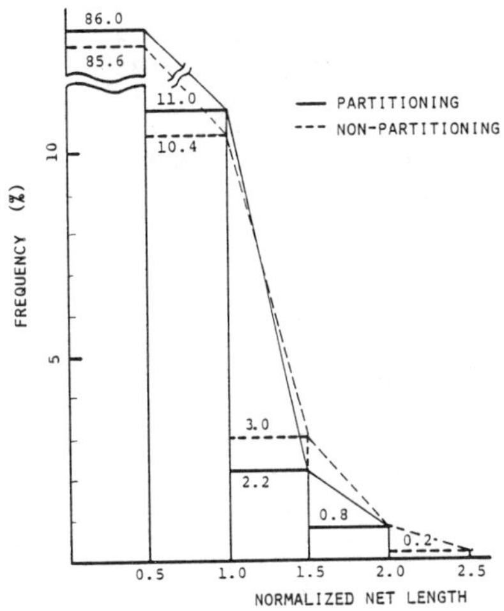

FIG.9 NET LENGTH DISTRIBUTION

6. Conclusions

We have described the functions and performance of an automatic placement and routing program, using a CMOS masterslice as an example. The combination of CR and MR is more poweful than CR alone with regard to wirabitily and processing speed. By now, dozens of VLSIs have been fabricated using this program. Wiring has been performed 100% automatically in every case.

7. References

1) T. Itoh, et al. ;A 6000 Gate CMOS Gate Array, ISSCC'82.
2) H. Terai, et al. ; Automatic Placement and Routing Program for Logic VLSI Design based on Hierarchical Layout Method, ICCC'82.
3) Y. Ohno, et al. ; Integrated Design Automation System for Custom and Gate Array VLSI Design, ICCC'82.
4) T. Kozawa, et al. ; Automatic Placement Algorithms for High Packing Density VLSI, DAC'83.

CIRCUIT PARAMETER EXTRACTION FOR HIERARCHICAL

AND EXACT VERIFICATION OF LSI ARTWORKS

Kazumichi Sakaguchi, Mikio Ikeda, Masayoshi Ogawa, and Koji Sato

LSI Research and Development Laboratory
Mitsubishi Electric Corporation
Itami, 664 Japan

ABSTRACT

A circuit parameter extraction program CCAP is presented, with an interactive editor of circuit simulator input. CCAP extracts properties of the external connection terminals, as well as circuit networks and parameters, to be utilized in hierarchical verifications. Procedures to extract circuit networks and parameters can be defined by the user to allow for program flexibility under various technologies.

Introduction

This paper describes an LSI mask artwork analysis program, CCAP. CCAP is a tool which can be used for confirming LSI circuit performance, not for design rule checking or verification of logical functions. It extracts the circuit network and circuit parameters from an LSI mask artwork, and provides circuit simulator input for the confirmation of circuit behavior. This program system has the following functions:

1) interfacing with the circuit simulator(SPICE2); the user can easily edit input data for circuit simulation by using a graphic terminal,

2) extracting properties of the external connection terminals, to be utilized in the "hierarchical" verification process[1],

3) allowing the user to describe the procedure of circuit extraction using operations on geometrical figures or specific commands in order to have program flexibility under various changeable technologies[2].

In the remainder of this paper, these functions are explained briefly.

System Overview

Figure 1 shows the flow of the CCAP system. Mask artwork designed in a graphic design system and the control data that describe the device structure and the parameter extraction procedure are put into the program. CCAP can handle any hierarchical level mask artwork (e.g. chip level, functional block level, gate level), for which it is not necessary to give node designations, such as character strings or numbers, to the mask artwork.

The program recognize circuit elements and wirings and extract the circuit network and the circuit parameters from the mask artwork in accordance with the extraction procedure described by the user. All MOS devices and basic bipolar devices can be extracted as circuit elements. Variation of the device structure is absorbed by the procedure description capability. The extracted network and circuit parameters are stored in an external memory for simulation and printout usage[3].

By using an interactive graphic terminal, the user can identify signal nets in the extracted network and edit the input data for circuit simulation semi-automatically.

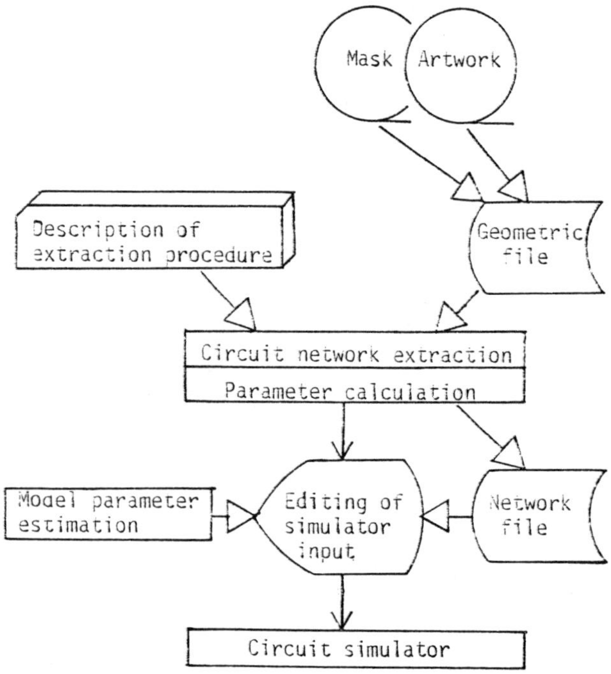

Fig. 1. System flow

Description of Extraction Procedures

In order to avoid frequent program renewals which must be carried out whenever new device technologies are put into use, the CCAP system was designed so that the user can describe the device structure definition and the circuit parameter extraction procedure as execution controls. As a matter of course, once the procedure description for the paticular device being used is prepared in the technology file, each user does not have to write it every time the program runs.

The control data are:

1) designation of which mask layers to use in the process, as well as layer type specifications (proper conductor, diffusion conductor, inter-layer connection, Power/Gnd etc.,) for each layer,

2) procedure to extract circuit elements, (to be described using specific operations on shapes),

3) definitions of inter-layer connections, i.e., the relative positioning of two figures in different layers to be considered as contacting each other, (We call a pattern element in network such as a polygon, a rectangle, or a path a "figure" in this paper.)

4) what kind of circuit parameters (e.g., gate capacitance, junction capacitance) to extract, together with their values for unit length or area.

Fig. 2 is an example, showing the MOS transistor extraction procedure. Here, "MOST" is a reserved command which means extraction of MOS transistors using results of the preceding four operations, i.e., (candidates of) gate, source/drain, channel dope and back gate.

```
CHANL= FIELD AND POLY
AF   = FIELD AND NOT POLY
GATE = CHANL CONTAIN POLY
SD   = CHANL TOUCH AF
CDOP = CHANL CONTAIN DOPE
BG   = CHANL CONTAIN WELL

MOST ( GATE, SD, CDOP, BG )
```

Fig. 2 An example of description to extract MOS transistors.

Calcululation of Capacitance

If conductors without contact are within short distances of each other, electrostatic capacitances are formed. In general, a conductor has a possibility of forming capacitance with other conductors in any direction of three dimensional space. In current technologies, however, vertical capacitance (i.e., capacitance between two vertically adjacent conductor figures) is usually ten times greater than lateral capacitance, since any two adjacent conductors are closer vertically than horizontally, and facing areas are vertically wider than lateral ones, as shown in Fig. 3. For this reason, vertical capacitances are the main consideration in this system.

Vertical capacitances are calculated by simply multiplying the overlapping areas of two conductors in different layers by the capacitance per unit area, since these layers are parallel. However, concerning junction capacitances between two adjacent diffused layers, the lateral capacitance has the same general value as the vertical capacitance and the above hypothesis that lateral capacitance is negligible does not stand up, because an area that is diffused in the silicon wafer is surrounded by other diffusion areas. (the substratum is also considered as a diffusion area.) Therefore, the program also calculates the lateral side-wall capacitances of diffusion areas in proportion to the edge length of the diffusion area.

Thus, the program has the capability of calculating all parasitic capacitances between any two conductor layers, i.e., the diffusion layers and the metallic layers. The substratum is implied as the lowermost layer.

Since there can be formed N(N-1)/2 kinds of capacitance from N conductor layers, the user specifies the set of layer pairs that contribute the capacitances to be calculated. Although each capacitance is formed by the conductors of two contributing layers, if any other conductor exists between these two layers, it obstructs formation of capacitance by shielding effect. Considering this phenomenon, areas of the conductor between the two contributing layers are subtracted from the overlapping area (Fig. 4).

In the case of more than one diffusion layer, however, areas of a lower diffusion layer are absorbed by areas of the upper diffusion layers as shown in Fig. 5. To allow for this phenomenon, figures of the upper areas are previously subtracted from those of the lower areas. For example, in order to determine the side-wall capacitance formed by two diffusion conductors P1 and P2 in Fig. 5, figures P2 and P3 are subtracted from P1 (call the result P1') and figure P3 is subtracted from P2 (call P2'). Then, the edges that are shared by (i.e., AND of) these resulting figures P1' and P2' are extracted (denoted by (a) in Fig. 5), and the total length of them is multiplied by the capacitance per unit edge length. The remaining side walls (denoted by (d)) which result from the difference in diffusion depth are ignored in the present version of CCAP.

The obtained capacitances are summed up along the pre-traced equi-potential wiring net for data compression.

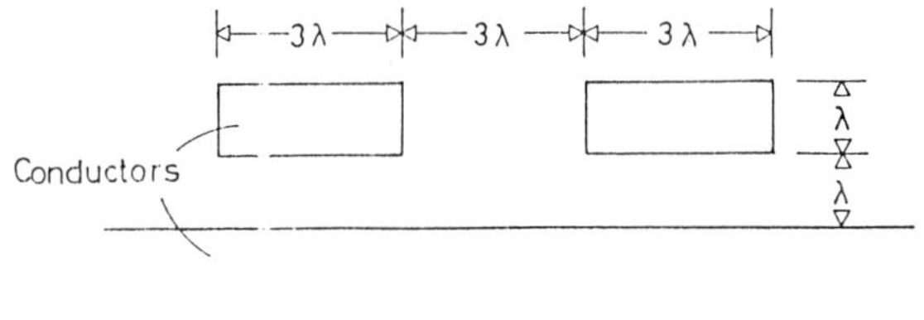

Fig. 3 Dimensions among conductors.

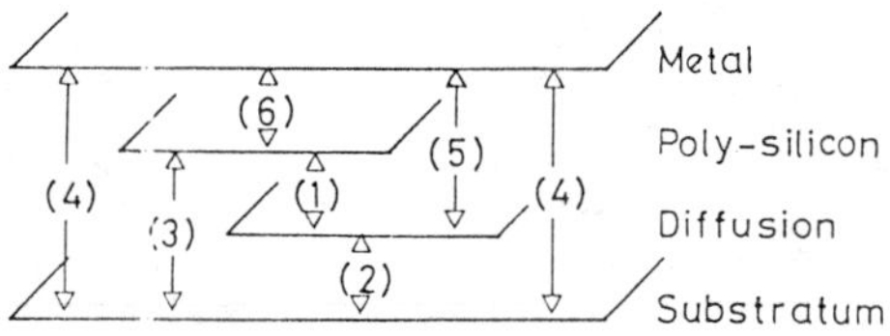

(1) Gate capacitance.
(2) Junction capacitance.
(3) Poly-Si substratum capacitance.
(4) Metal substratum capacitances.
(5) Metal diffusion capacitance.
(6) Metal poly-Si capacitance.

Fig. 4 An example of a conductor configuration and resultant capacitances.

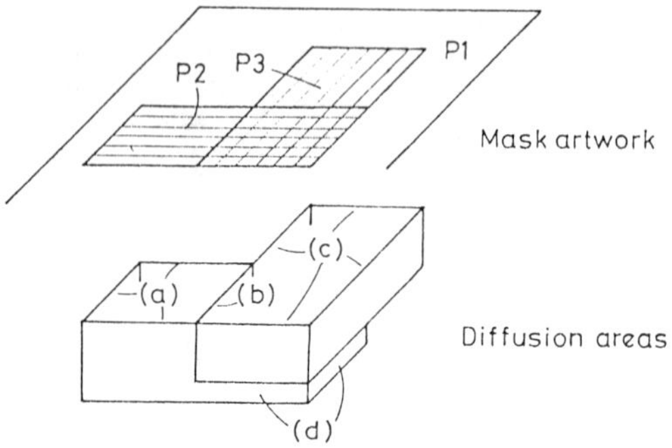

Fig. 5 An example of diffusion areas. Here, (a), (b), and (c) show the side wall edges between P1-P2, P2-P3, and P3-P1 respectively.

Operations on Figures

Various operations on figures were prepared for the realization of the program functions. Most of them are opened to users for description of extraction procedures. Those operations are as followes:

Data File Preparation

The external representation of an LSI mask artwork is converted into an internal on-line file, whose structure was designed so as to execute required operations at high speed. In this structure, a polygonal figure is represented by the set of component trapeziums, and the minimum enclosing rectangle is attached (Fig. 6). The enclosing rectangles are arranged according to their coordinates, and figure can be easily and rapidly searched for the execution of operations. Moreover, the adoption of "slit" method, in which polygons are sliced by horizontal slits, also contribute to the speed of operation because the complexity of the process is reduced from two-dimensional to linear.

Logical Operations

Double-operand logical operations AND, OR, AND-Not, and Exclusive-OR are available, which are used for extraction of regions forming transistors, capacitances, etc. An operation with three or more operands is implemented as a sequence of these double-operand operations. (When the user writes the procedure, he must express his desire in terms of such a sequence.) Since the full- and null-set are allowed as operands, any Boolean function can be realized.

Extraction of Inclusion and Overlapping Relations

By this routine, pairs of figures that hold inclusion and/or overlapping relations are extracted. These pairs are used for recognizing inter-layer connections and transistors (bipolar). To use this routine, the user must specify three arguments: two layer designations and the spacing distance. (The spacing is allowed to be negative.) When the spacing is positive, the routine extracts all figure pairs such that the figure of the first layer is properly included in the figure of the second layer and such that the least distance between them is more than the specified spacing distance. If the spacing is zero, pairs such that the inner figure touches the outer one are also extracted. In the case of negative spacing, the overlapping pairs are further added. The "coincidence" operation which detects exactly overlaping figure pairs is also available. This is implemented by utilizing the abobe zero-spacing operation two times.

Extraction of Touching Pairs

Those pairs of figures that abut are extracted. This operation is used to extract source/drain terminals which abut the channels of MOS transistors.

Area Calculation

The area calculation and the edge-length calculation routines (the latter is explained below) are used for obtaining transistor parameters and parasitic capacitances. As arguments of this routine, two layers and a logical operation must be specified. The area of each figure in the figure set which is extracted by performing the specified logical operation on the specified two layers is calculated. In order not to lose sight of the interconnection relations to the circuit nodes/elements to be extracted, the ancestor figures of each extracted figure are retained.

Edge-length Calculation

This routine is similar to the above area calculation routine, except that the output data are lengths of figure edges instead of areas. Logical operations on figures are performed as shown in Fig. 7. This routine is used for calculating those parameters that are proportional to edge length (e.g., lateral capacitance, L and W of a transistor).

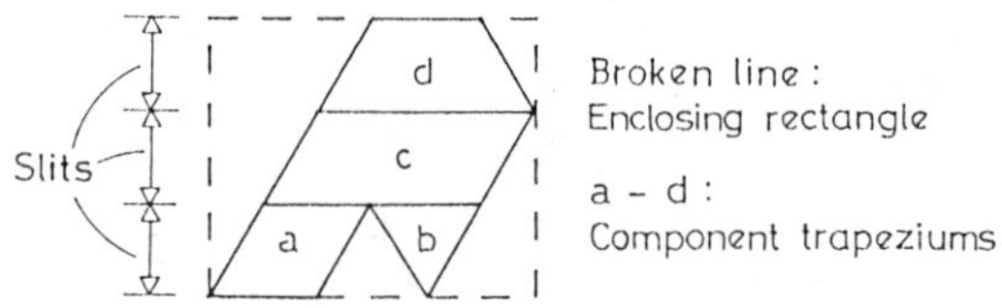

Fig. 6 Representation of a polygonal figure.

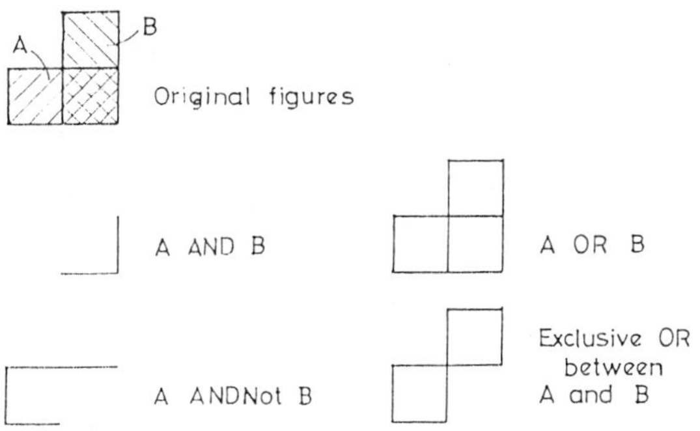

Fig. 7 Logical operations in the edge-length calculation routine.

The Printout

The printout data of the CCAP system consists of the extracted circuit network, circuit parameters, and the properties of the terminals to be connected with the external circuits. This last item enables the user to do hierarchical and/or divided mask artwork verification. For any single terminal, CCAP prints out the following:

1) location, width and layer number ,
2) signal type (power, ground, input, output, other),
3) circuit parameters of the wiring net connecting with this terminal.

As for 2) above, the present version can recognize signal type for ED-MOS, CMOS, and basic bipolar circuits. In ED-MOS circuits, for example, a signal line that connects with both gate and source terminals of a depletion transistor is regarded as an output line.

These terminal data can be utilized for checking consistency between two divided artworks of a circuit or between lower-level cells and a higher-level cell of an artwork having a nesting structure. The latter case we call hierarchical verification. Although it is performed manually at present, automated hierarchical verification capability is one of the future plans of this system.

Interface with Circuit Simulator

For the precise verification of circuit behavior, it is necessary to use a circuit simulator together with the data of accurate circuit parameter values extracted from the artwork design. Because of this, interfacing with the circuit simulator being used is very important for a system such as the one presented here.

For circuit simulation, we must prepare the following input data:

1) description of the circuit interconnection and parameter values (C, R, etc.) for each element,
2) parameter values of the transistor models used in the circuit,
3) control data such as I/O node designations, wave forms, step size, analysis types etc.

As for 2) above, the system is provided with a program which estimate the transistor parameter values by means of regression analysis from the static characteristics.

Also in the CCAP system, the user can interactively edit the above three kinds of data using a graphic terminal. By simple commands, the 1) and 2) are automatically edited in accodance with the input format of SPICE2. As for 3), since the interface program displays the schematic diagram of the extracted network (geometrically similar to the original artwork) on the terminal, the user can easily identify nodes and elements. He specifies input (i.e., control) nodes and output (i.e., observation) nodes by moving an electric pen on the tablet. Other control data are edited manually using the system editor.

Conclusion

A circuit parameter extraction system CCAP was reported in this paper.

Although CCAP is equipped with utilities for interfacing with SPICE2, there is a limitation of circuit scale that a circuit simulator can handle. So, our future plan includes the development of:

1) logic simulator interfacing utilities,
2) simple estimation method for wiring delay and power dissipation.

Furthermore, as a means of hierarchical verification, development of tools for

3) electrical rule checking using information on interfacing terminals, and
4) hierarchical performance checking

are also planned.

Acknowledgment

The authors would like to thank Dr. T. Yahara, Mr. M. Iwatsuki for their instructive discussions, and Dr. H. Oka and Dr. Y. Gamoh for their encouragement.

References

[1] L. V. Corbin, "Custom VLSI Electrical Rule Checking in an Intelligent Terminal," Proc. 18th DA Conf., pp. 696-701 (June 1981)

[2] B. T. Preas et al, "Automatic Circuit Analysis Based on Mask Information," Proc. 13th DA Conf., pp 309-317 (June 1976)

[3] T. Mitsuhashi et al, "An Integrated Mask Artwork Analysis System," Proc. 17th DA Conf., pp. 277-284 (June 1980)

[4] K. Sato et al, "MILD - A Cell-Based Layout System for MOS-LSI," Proc. 18th DA Conf., pp. 828-836 (June 1981)

ADULTS-C : An Automatic Contact-hole Generating System for VLSI Symbolic Layout

Tokuya Fujioka, Nobuyuki Nishiguchi,
Toshihiro Okamoto and Hiroshi Kawanishi

NEC Corporation, Kawasaki, 211 Japan

Abstract

An automatic contact-hole generating system, ADULTS-C, has been developed for symbolic layout of VLSIs. The system has the following features. First, it automatically generates suitable contact-holes in layout patterns at contact-hole reference points. Second, it can be adapted to any process technologies, such as NMOS, CMOS and bipolar. Finally, it can be used hierarchically. In this paper, the system overview and some experimental results are described.

1. Introduction

Recently design complexities of VLSIs are increasing exponentially according to the rapid evolution of process technologies. However, VLSI layout designs, especially for mass production, are in most cases done still by the handcrafted approach. Most layouts are drawn manually on mylar sheets and digitized by CAD turn-key graphics. The manual layout technique is effective in realizing high packing density, which provides lower product manufacturing costs. The disadvantages of this technique are its very long design time and the high possibility of human error inducement, such as design rule violations and electrical rule violations. Therefore, layout verification systems have become more important and many useful systems have been developed to date, including a design rule check program(1), electrical rule check programs(2, 3) and a logic comparison program(4). These systems are effectively used to reduce verification time for the geometrical layout patterns, but are not able to support initial layout design, which means the transformation of circuit designs into geometrical layout patterns. Under these circumstances, so-called symbolic layout tools(5, 6, 7, 8,) that produce the geometrical layout patterns from topological layout design of the chip are used and they reduce the possibility of numerous human errors. ADULTS-C is a kind of symbolic layout tools and automatically generates suitable contact-holes in layout patterns.

In section 2 the system overview is presented, and in section 3 details of the system are described. Experimental results are given in section 4.

2. System Overview

The features of the system are as follows:

1) For given reference points by designers, it automatically generates suitable contact-holes in the same location.
2) It can be adapted to any process technologies, such as NMOS, CMOS and bipolar.
3) It can be used hierarchically.

The system consists of two processes as shown in the flow chart in Fig. 1. In the preprocess, input layout patterns from CAD turn-key graphics are converted to the database. In the next process, a suitable contact hole is generated for each reference point according to the layout patterns surrounding the reference point. And actual layout patterns corresponding to input data are prepared for post process, namely the fabrication process like pattern generator and electron beam machines.

3. Details of the System

3.1 Types of contact-holes

Layout pattern data of VLSIs consist of many layers. For example in MOS technology, there are three layers, which are metal, polysilicon and diffusion. There exist many types of contact-holes according to the pairs of layers, and each detailed layout patterns for each type of contact-hole is stored as a symbolized cell in CAD turn-key system. Layout designers should select a suitable symbolized cell for each contact-hole reference point. This work is difficult and tedious. Fig.2 shows a symbolized contact cell and corresponding actual layout patterns.

Examples of types of contact-holes in CMOS silicon gate technology are as followings:

1) metal-diffusion contact
2) metal-polysilicon contact
3) diffusion-polysilicon contact (generally called buried contact)
4) p-well contact
5) n-well contact
6) substrate contact

P-well, n-well and substrate contact are used for connecting source lines with p-well, n-well

CH1935-6/83/0000/0544$01.00

and substrate, respectively. If other process technologies are used, different types of contact-hole are necessary. But it is very important that the type of contact-hole can be determined by the relations between layers under certain conditions described in the next section. The system can automatically generate the correct type of contact-hole at each reference point.

3.2 Contact-hole generation algorithm

For a reference point of a contact-hole, a Contact-Image, shown in Fig.3, is defined as a square area around the reference point, and it denotes the search area for deciding a type of contact-hole. The suitable contact-hole at each reference point is generated by judging the relation between layers of layout patterns contained in the Contact-Image. The size of a Contact-Image is variable and depends on each layout design rule. If there are only two layers, which are metal and diffusion, in a Contact-Image, its type can be easily determined to be the metal-diffusion contact type.

When the number of layers of the layout patterns contained in a Contact-Image is more than three, a priority table is used for deciding the most suitable type of contact-hole. Table 1 shows an example of the sequence of priorities for each layer in a CMOS silicon gate technology. Both diffusion and polysilicon layers are given the highest priority. If there are three layers, which are metal, diffusion and polysilicon, in a Contact-Image, a buried contact (diffusion-polysilicon contact) is generated according to the priorities because the diffusion and polysilicon priorities are higher than the metal one in Table 1. The system generates a type of contact-hole to connect between layers having higher priority.

Preparing for Contact-Image and the sequence of priorities for layers, the most suitable contact-hole can be automatically generated from the complicated relation between layers at each reference point in the layout patterns.

Steps for the contact-hole generation algorithm are as follows:

- Step 1) Sort the set of reference points of contact-holes (The first key is a y-coordinate and the second key is an x-coordinate).
- Step 2) Select and delete an element from the set. If the set is empty, stop.
- Step 3) Judge the relation between layers of layout patterns contained in the Contact-Image.
- Step 4) Generate the most suitable contact type according to given priority table.
- Step 5) Go to Step 2

3.3 Examples of generation

Fig.4 shows examples of generating the type of contact-hole. Fig.4(a) indicates the input layout patterns and Contact-Image at each reference point. This point(•) is only used to indicate reference information of a contact-hole. There are two layers, metal and diffusion, in the Contact-Image at C1 point. Thus, the program can easily judge that a suitable contact-hole at this point is the metal-diffusion contact, and generates the correct contact-hole cell. It is a little more complicated with regard to the C2 location. Four layers, where are metal, diffusion, polysilicon and p-well, are contained in the Contact-Image at this point. Four different types of the contact-hole are possible on that point. They are metal-diffusion contact, metal-polysilicon contact, buried contact and p-well contact. The system automatically selects the buried contact as the most suitable contact type at this location by considering the priorities for each layer, shown in Table 1. (Notice that diffusion and polysilicon layers do not touch each other in the Contact-Image.) In the same way, the p-well contact is generated at C3 location. Fig.4 (b) shows actual layout patterns after generating a suitable contact-hole at each reference point.

4. Experimental Results

The program size of ADULTS-C system is about 6K steps in Fortran. Fig.5 shows the relation between the number of contact-holes to be generated and CPU time using the NEC ACOS 1000 computer. CPU time is nearly proportional to the number of contact-holes.

Fig.6 shows the experimental result. The input geometrical layout patterns are shown in Fig.6 (a). The symbol (⊠) in this figure shows the reference point of the contact-holes. Fig.6 (b) shows actual layout patterns after contact-holes are generated by ADULTS-C system.

5. Conclusion

A new system for an automatic contact-hole generation which supports symbolic layout of VLSIs has been described. By using the system, the layout design time has been reduced by about 10%. CPU time for a chip layout containing about 50,000 contact-holes is less than 20 minutes. The ADULTS-C system has been proved to be practically useful and efficient in layout design of VLSIs.

6. Acknowledgement

The authors would like to express their appreciation and thanks to Dr. T.Yanagawa, Mr. T.Mukawa and Mr. F.Tsukuda of NEC Corporation for their encouragement and helpful discussions.

Reference

(1) D. Alexander, "A Technology Independent Design Rule Checker", Proc. 3rd USA-Japan Computer Conf., PP.412-416(1978)
(2) T. Fujioka, et al, "An Electrical Rule Check System of LSI Artwork Data: SIMPLE PALMS", Monograph of Technical Group on Design Technology of Electronics Equipment of Inform. Process. Soc. Japan, 10-2(1981) (Japanese)
(3) NCA Corp. : VERC User Guide (1981)
(4) H. Kawanishi, et al, "An Automatic Layout-Logic Verification Algorithm for VLSI", Proc. ECCTD'80, Vol.1, PP.436-447(1980)
(5) J.D. Williams, "STICKS-A Graphical Compiler for High-Level LSI Design", Proc. AFIPS Conf. Vol.47, PP289-259(1978)
(6) M.Y. Hsueh, "Symbolic Layout and Compaction of Integrated Circuits", ERL Memo. NO. UCB/ERL M79/80, University of California, Berkeley. (1979)
(7) N. Weste, "Virtual Grid Symbolic Layout", Proc. 18th DA Conf., PP.225-233(1981)
(8) D. Johannsen, "Bristle Blocks; A Silicon Compiler", Proc. 16th DA Conf., PP.310-313 (1979)

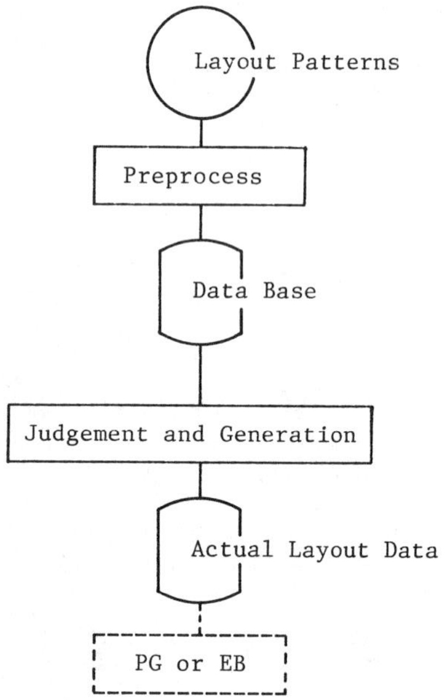

Fig. 1 A Flow Chart of ADULTS-C

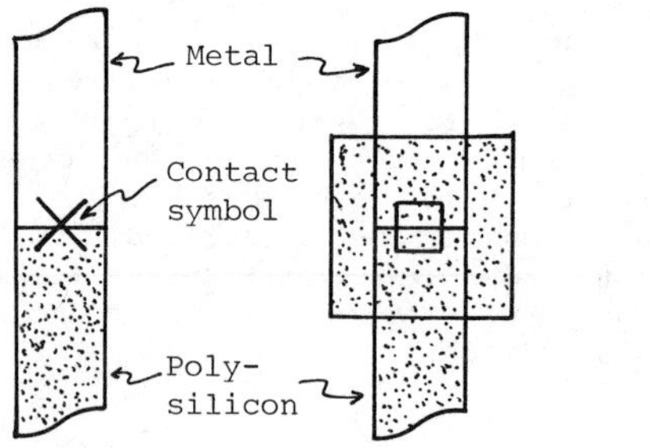

(a) Symbol (b) Corresponding layout patterns

Fig. 2 Contact Expression

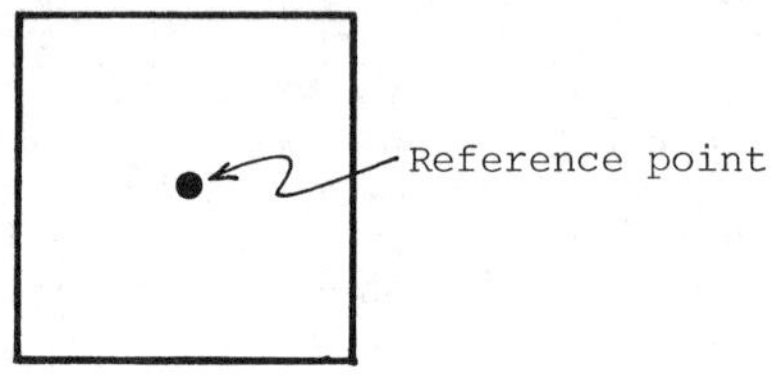

Fig. 3 Contact-Image

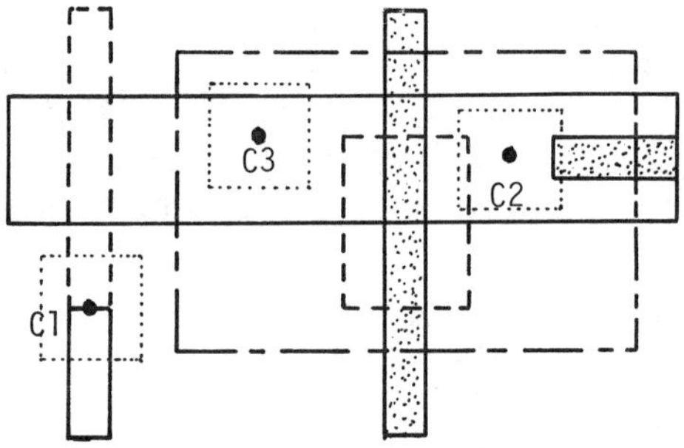

(a) Input data and Contact-Images

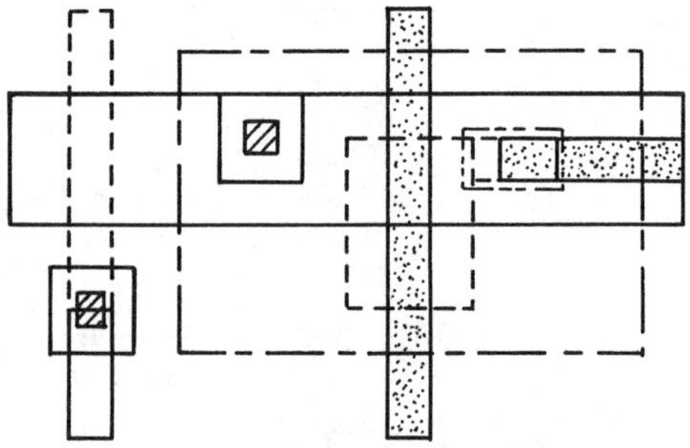

(b) Output data

:Diffusion :Metal
:Polysilicon :P-well
:Contact-Image

Fig. 4 Examples of Generation

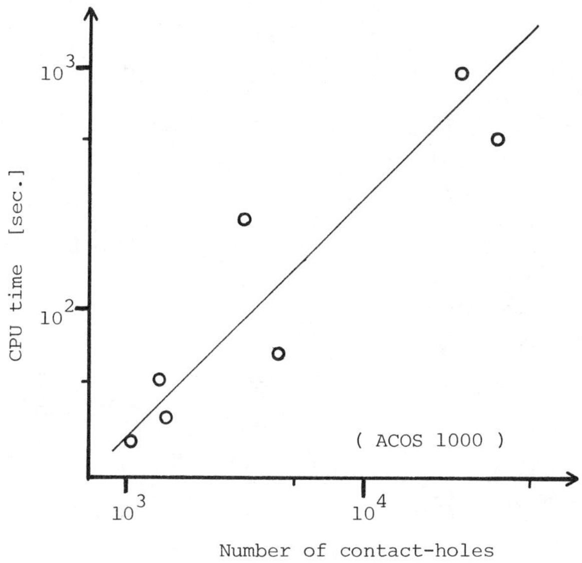

Fig. 5 CPU Time

Table 1. Priority for layers in CMOS silicon gate technology

Priority	Layer
1	Diffusion, Polysilicon
2	Metal
3	P-well, N-well
4	Substrate

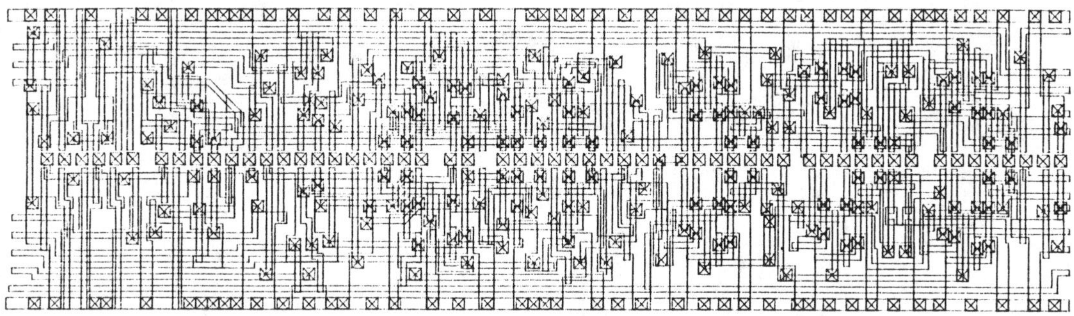

(a) Input layout patterns

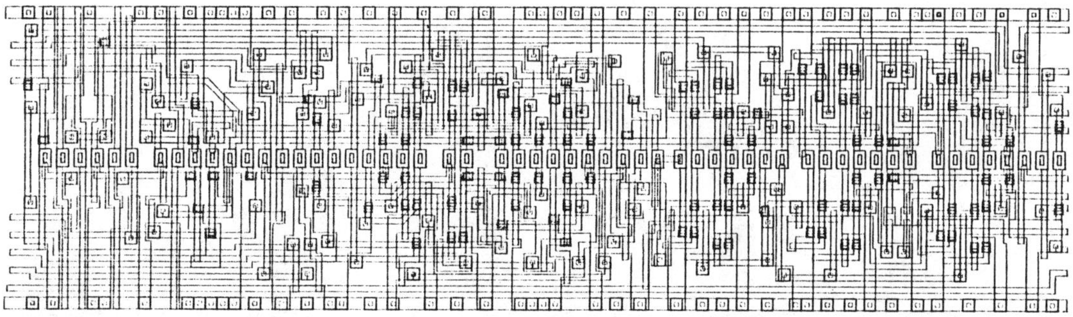

(b) Actual layout patterns

Fig. 6 Experimental Result

Session 3-1d
VLSI Architecture I

Chairman

K. S. Fu
Purdue University

VLSI ARCHITECTURES FOR MINIMUM-DISTANCE CLASSIFICATION†

Hsi-Ho Liu
Department of Electrical and Computer Engineering
University of Miami
Coral Gables, FL 33124

K. S. Fu
School of Electrical Engineering
Purdue University
West Lafayette, IN 47907

ABSTRACT

Minimum-distance classification is an important and useful technique in statistical pattern recognition and many other applications. It can be extended to syntactic approach as well; only string case is considered here. VLSI architectures using systolic arrays are presented in this paper to carry out these operations. It is shown that constant time complexity can be achieved regardless the size of reference samples.

I. Introduction

Minimum-distance classifiers (MDC) constitute an important subclass of linear classifiers in statistical pattern classification. Minimum-distance classification is also widely used in clustering analysis where the test samples are assigned to the nearest clusters according to their distances from the cluster centers. The same idea can be extended to syntactic approach where each pattern is represented by a string, a tree or a graph provided that the distances between the patterns can be computed. The computation of a MDC is time consuming especially when it is used repeatedly. The computation time for straightforward comparisons is proportional to N where N is the number of reference vectors. Special data structures and search algorithms have been proposed [1] where a time complexity of O(N log N) can be achieved. However, since the computations of MDC's are rather regular and simple, they can be performed by special VLSI architectures. A constant time complexity is attainable when parallel computation and pipelined data flow are fully exploited. In this paper we propose a VLSI MDC for statistical approach which deals with feature vectors and a VLSI MDC for syntactic approach which deals with strings. Both of them use systolic arrays [2] in order to gain maximum speedup. These VLSI systems are supposed to be used as special-purpose processors as shown in Fig. 1.

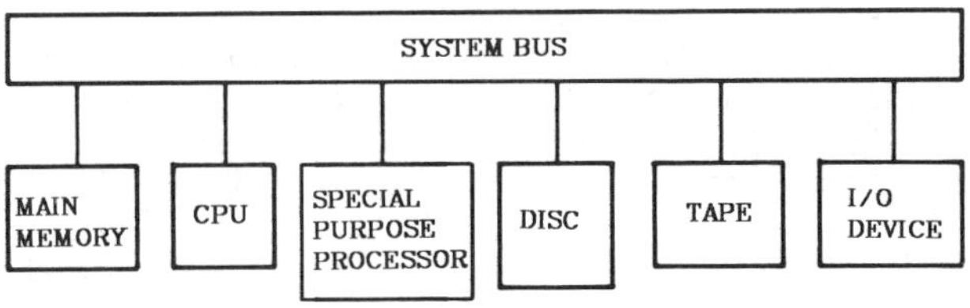

Figure 1 The special-purpose processor is attached to a host computer as a peripheral processor.

†This work was supported by the NSF Grant ECS 80-16580.

II. Systolic Array for Minimum-Distance Classification of Feature Vectors

Consider the classification problem of ℓ classes $C_1, C_2, \ldots, C_\ell$ where each pattern class C_j has a k-dimensional reference vector $M^j=[m_1^j, m_2^j, \ldots, m_k^j]$. A minimum-distance classification with respect to $M^1, M^2, \ldots, M^\ell$ is to classify the input vector $X^i=[x_1^i, x_2^i, \ldots, x_k^i]$ to class C_i if $d(X^i,M^j)$ is the minimum, where $d(X^i,M^j)$ is the distance defined between X^i and M^j. The distance computation can be formulated as a pseudo matrix multiplication problem $\underline{Y} = \underline{M} \circ \underline{X}$, i.e.,

$$\begin{bmatrix} M^1 \\ M^2 \\ \vdots \\ M^\ell \end{bmatrix} \circ [X^1 X^2 \cdots X^n] = \begin{bmatrix} m_1^1 & m_2^1 & \ldots & m_k^1 \\ m_1^2 & m_2^2 & \ldots & m_k^2 \\ \vdots & \vdots & & \vdots \\ m_1^\ell & m_2^\ell & \ldots & m_k^\ell \end{bmatrix} \circ \begin{bmatrix} x_1^1 & x_1^2 & \ldots & x_1^n \\ x_2^1 & x_2^2 & \ldots & x_2^n \\ \vdots & \vdots & & \vdots \\ x_k^1 & x_k^2 & \ldots & x_k^n \end{bmatrix}$$

$$= \begin{bmatrix} y_1^1 & y_1^2 & \ldots & y_1^n \\ y_2^1 & y_2^2 & \ldots & y_2^n \\ \vdots & \vdots & & \vdots \\ y_\ell^1 & y_\ell^2 & \ldots & y_\ell^n \end{bmatrix}$$

where $y_j^i = d(X^i,M^j)$ is the distance between vectors X^i and M^j, ℓ is the number of classes and n is the number of input vectors. All elements in the above matrices are real numbers. If we consider X^i and M^j as two points in the feature space, then $d_p(X^i,M^j) = (|x_1^i - m_1^j|^p + |x_2^i - m_2^j|^p + \ldots + |x_k^i - m_k^j|^p)^{1/p}$ and $d_\infty(X^i,M^j) = \max(|x_1^i - m_1^j|, |x_2^i - m_2^j|, \ldots, |x_k^i - m_k^j|)$ under L_p-metric. For simplicity we shall consider only the case of p=1, i.e., the absolute-value distance. The generalizations to p=2, i.e., the Euclidean distance, and p=∞ are given in Section V.

In the next step we shall transform matrix $\underline{Y}$ into an array of complex elements $Y' = (Y^1, Y^2, \ldots, Y^n)$ where $y^i = (y_q^i i, q^i)$, $y_q^i i = \min(y_1^i, y_2^i, \ldots, y_\ell^i)$. This simply means selecting the minimum distance for the input vector X^i.

Kung and Leiserson [3] proposed a hexagonal systolic array for multiplication of two n×n band matrices. Since the matrices $\underline{M}$ and $\underline{X}$ in the present study are neither square nor banded and the dimension n of $\underline{X}$ may be very large, therefore a hexagonal array is impractical due to the large number of processors required. Instead we use a square array for the pseudo matrix multiplication.

We shall need two different types of processing elements, one called 'compute' processor, the other called 'compare' processor. The former is the constitutive element for distance computation (or equivalently for pseudo matrix multiplication), the latter contributes to distance comparison. A processor array of 1 by k which performs the distance computation is shown in Fig. 2. Data flow and operations performed by each processor are given in Fig. 3. In order to minimize the overhead for communication and under the pin limitation data are

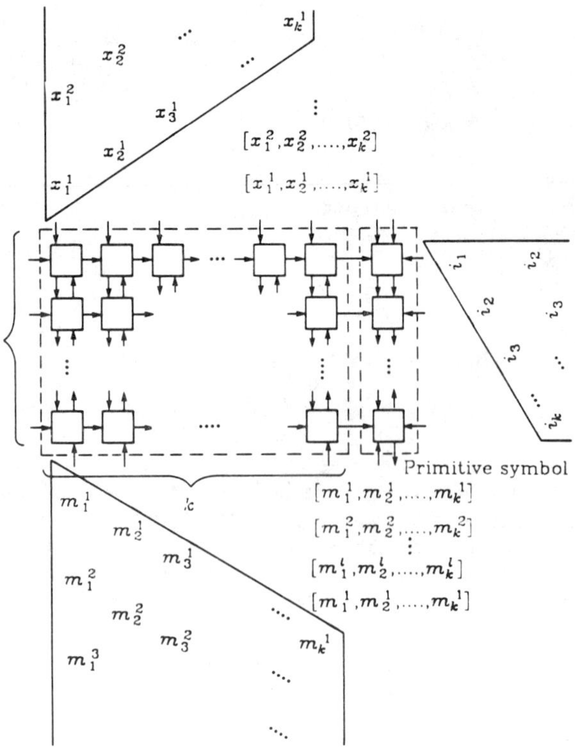

Figure 2 Processor arrays and data movement for primitive recognition.

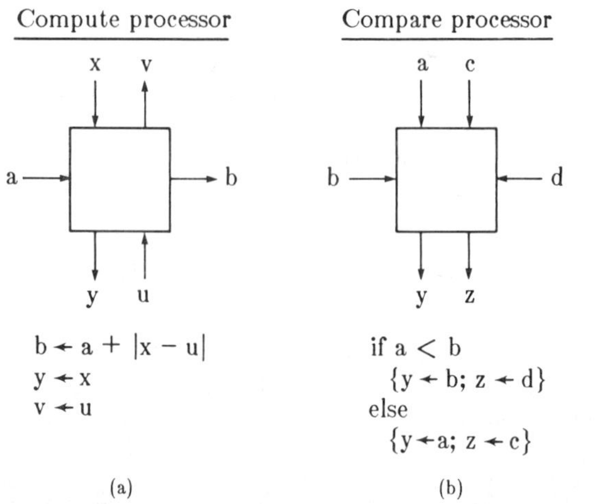

Figure 3 Data flow and operations of each (a) 'compute' processor and (b) 'compare' processor.

transferred serially between processors and to/from the host computer. Note that the data elements must be properly skewed as shown in Fig. 2. Since the input and reference data streams move in opposite directions, their data elements must be separated by one unit time as shown in Fig. 2; otherwise, some data will be passing instead of meeting each other. After completing computation of the corresponding components between the reference vector and the input vector, the partial sum moves to the right and enters the 'compare' processors, i.e., the rightmost column in Fig. 2. It can be seen from Fig. 2 that input vector X^1 is compared with the reference vectors in the sequence $M^1, M^2, \ldots, M^\ell$; X^2 is compared with the reference vectors $M^2, M^3, \ldots, M^\ell, M^1$, and so forth.

With a uniprocessor, the MDC needs $O(\ell \times k)$ computations and $O(\ell-1)$ comparisons to classify an input vector regardless the number of input vectors involved. With the processor array and pipelined data flow in Fig. 2 the classification result of the first input vector is available after $\ell \times k+1$ unit times, then it takes only 2 unit times to get a new result. A unit time is defined as the time for each processor to complete its operation cycle. Therefore the speedup is $\ell \times (k+1)/2$.

III. Systolic Array for Minimum-Distance Classification of String Patterns

In syntactic pattern recognition the patterns are represented by strings, trees or graphs [4]. The minimum distance classification in syntactic approach is similar to that of statistical approach except in distance computation. Distance computation is also the bottleneck in a MDC for syntactic approach. The time complexity for distance computation is $O(m \times n)$ where m and n are the lengths of the two strings. Therefore, distance computation is the area which needs speedup most. Since string patterns are very popular and tree patterns can be converted into strings, we will focus our discussion on the string patterns in this paper.

A number of string-to-string distance measures have been proposed. One type of measurement is based on the idea of insertion, deletion and substitution of string elements. They appear to be very useful in syntactic pattern recognition [4], information storage and retrieval, and database applications. The family of Levenshtein distance [5] are the most well-known examples. Other types of measurements use (time) warping functions to match the two patterns. These are found most often in speech recognition. An integrated circuit chip using systolic array for spoken work recognition has been constructed [6]. We shall propose in the following VLSI systolic arrays for the computation of Levenshtein distance and the like.

(a) Levenshtein distance

It is well-known that Levenshtein distance can be computed by dynamic programming. Therefore it can be implemented by VLSI architectures. For Levenshtein distance each insertion, deletion and substitution of string element is counted as one error transformation. The distance between two strings is defined as the minimum number of error transformations that is required to change one string into the other. We have developed a square processor array for this purpose. The structure of the processor array and data flow are shown in Fig. 4. The reference strings enter from the left, the test strings enter from the bottom. Both of them must be properly skewed as shown in Fig. 4 so that the string elements will arrive at the right processor at the right time. The control and the selection of minimum distance can be carried out by the host computer or by a special processor.

Each processor element $P_{i,j}$ only computes the partial sum

$$S_{i,j} = \min \begin{Bmatrix} S_{i-1,j}+1 \\ S_{i-1,j-1}+S(a_i,b_j) \\ S_{i,j-1}+1 \end{Bmatrix}$$

where $S(a_i,b_j) = 1$ if $a_i \neq b_j$; $S(a_i,b_j) = 0$ otherwise. The computation is divided into three stages. Fig. 5 shows the internal

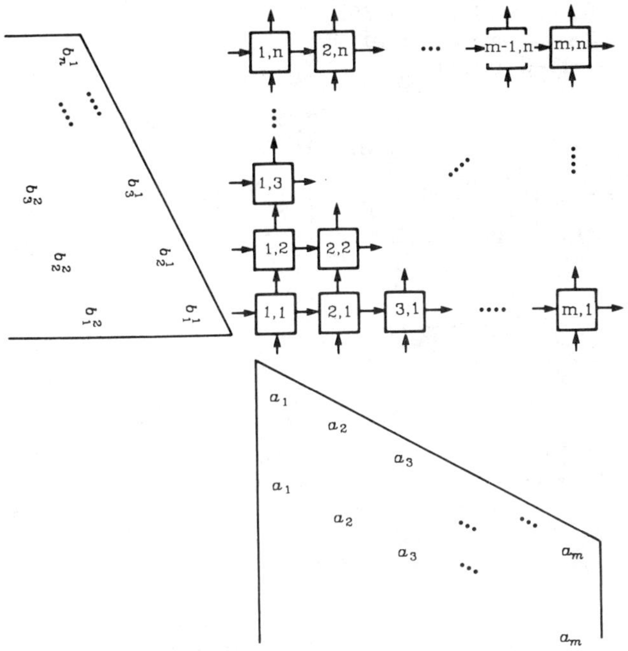

Figure 4 Processor array and data movement for computing Levenshtein distance.

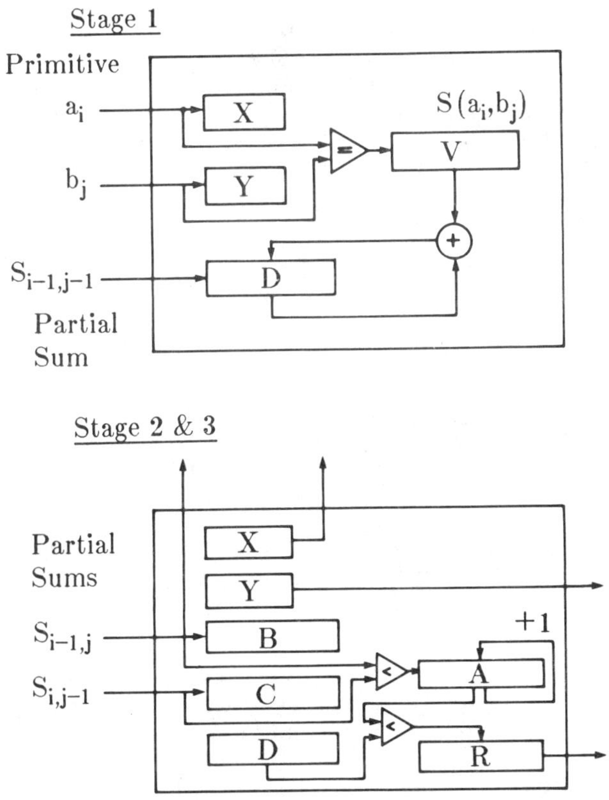

Figure 5 Internal structure and register transfer of PE $P_{i,j}$ at stage 1, 2 and 3.

structure and the operations of processor element $P_{i,j}$ at stage 1, 2 and 3. All the processors at the same diagonal (from upper left corner to lower right corner) in Fig. 4 perform the same operations. This format will move forward one step every 18 clock cycles. Details can be found in [7].

(b) Weighted Levenshtein Distance

Weighted Levenshtein distance in which different weights are associated with different error transformations is usually more favorable in practical applications. The major difference between this implementation and that in (a) is the storage problem, i.e., weights must be stored in each processor. We use a programmable logic array (PLA) to store the weight table. The internal structure of the processor element is given in Fig. 6. The computation of the partial sum is also divided

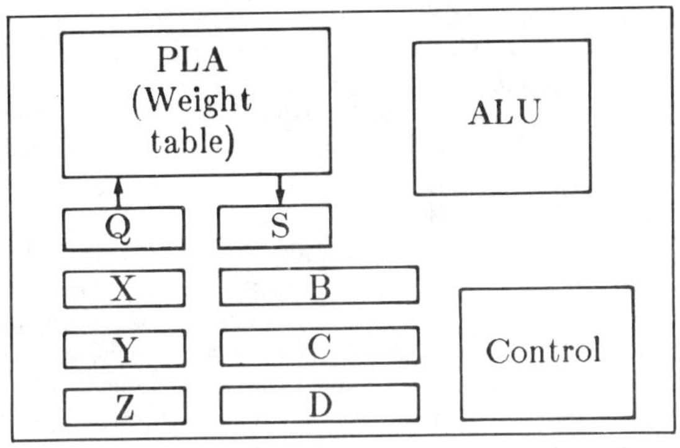

Figure 6 Internal structure of the PE for weighted string distance computation.

into 3 stages. However, it takes longer time in each stage due to the complexity of weight involvement.

Suppose that both the reference and the test strings have length ℓ. With a uniprocessor, the matching process between one input string and one reference string takes $O(\ell \times \ell)$ unit operations. A unit operation is defined as the computation of partial sum at location (i,j), $1 \leq i,j \leq \ell$. With the array processor and data flow of Fig. 4 it only takes 3 unit times to find the distance between a pair of strings.

IV. Simulation and Performance Evaluation

Simulations have been conducted for the systolic arrays described in sections II and III. The data from [7] were used to verify the correctness of the VLSI systems, where the results were positive. We also compare the real speedup with the estimated speedup. The speedup is defined in terms of the time to produce an output, i.e., as the ratio of time used by a sequential computer to the time used by a processor array. In our simulations, $\ell=13$, k=2, string length is 20, and only substitution transformation is considered. The results are given in Table 1. From Table 1 we can see that the real speedup does not match the estimated speedup. This is due to the fact that we only simulated the function, and data transfer was not executed in parallel. We should keep this in mind that a VLSI implementation usually partitions the original operation into distributed suboperations and uses different hardware technology, therefore the exact speedup can not be easily determined until the system is really built and tested.

Table 1 Computation time of sequential algorithm, simulated computation time for VLSI arrays using sequential computer, real speedups, estimated speedups and speedup ratio.

	Sequential algorithm	Simulated VLSI arrays	Real speedup	Estimated speedup	Speedup ratio
Feature vector classification	0.005 (sec.)	0.3 (ms)	17.4	19.5	89%
String Matching	0.07 (sec.)	15 (ms)	4.7	6.7	70%

V. Concluding Remarks

In Section II we only consider the computation of L_1-metric where the 'compute' processor performs subtraction, find absolute value value and then performs addition. For L_2-metric the 'compute' processor should subtract, multiply and then add. For L_∞-metric the 'compute' processor should perform subtraction, find absolute value and then performs comparison. We can see that these modifications are easy and do not change the overall data flow pattern.

In string distance computation we assume that both reference and input strings have equal length. This is not always true. In fact the length of strings can vary considerably. There is a need of compromise between efficiency and flexibility of the design. If an input string with length smaller than the (processor) array size, zeros or blanks can be appended to the string to make its length equal to that of the reference string. On the other hand, if an input string with length longer than the array size, it is better to handle this case separately, for example, by a host computer. Although partitioned or recursive approaches may be applied, it will interrupt the regular sequence of execution and degrade the performance of the system.

REFERENCES

[1] D. T. Lee, "On k-nearest neighbor Voronoi diagrams in the plane," *IEEE Trans. Comput.*, vol. C-31, no. 6, June 1982, pp. 478-487.

[2] H. T. Kung, "Let's design algorithms for VLSI systems," *Proc. Caltech Conf. on VLSI,* Pasadena, CA, pp. 65-90, Jan. 1979.

[3] H. T. Kung and C. E. Leiserson, "Systolic arrays (for VLSI)," in *Introduction to VLSI Systems* by C. A. Mead and L. A. Conway, Sect. 8.3, Addison-Wesley, Reading, Mass., 1980.

[4] K. S. Fu, *Syntactic Pattern Recognition and Applications,* Prentice-Hall, Inc., 1982.

[5] V. I. Levenshtein, "Binary codes capable of correcting deletion, insertions, and reversals," *Sov. Phys. Dokl.*, vol. 10, pp. 707-710, 1966.

[6] B. Ackland, N. Weste, and D. J. Burr, "An integrated multiprocessing array for time warping pattern matching," *Proc. 8th Ann. Symp. on Comput. Archit.*, Minneapolis, MN, pp. 197-215, May 1981.

[7] H. H. Liu and K. S. Fu "A syntactic approach and VLSI architectures for seismic signal classification," *Tech. Rept.*, TR-EE 83-5, Electrical Engineering, Purdue University, West Lafayette, IN, Jan. 1983.

AN EFFICIENT VLSI STRUCTURE FOR TWO-DIMENSIONAL DATA PROCESSING

G. R. Nudd, J. G. Nash, S. S. Narayan
Hughes Research Laboratories
Malibu, California 90265

and

A. K. Jain
University of California
Davis, California 95616

Abstract

We describe a VLSI array architecture consisting of identical processing elements which can be configured to perform a variety of low-level image analysis operations. Presently each processor consists of a custom designed nMOS chip, but potential exists to build single chip arrays with hundreds of elements.

I. Introduction

The rapid developments in microelectronics technology and VLSI have already provided significantly increased throughputs over conventional general purpose computing machines. It is likely that we have seen just the beginning of this trend, and in the near future relatively inexpensive processors which are fully programmable and capable of achieving throughputs of 100 to 1,000-MIPS (millions of instructions per second) will be available. It can be anticipated that a principal beneficiary of these developments will be image analysis and pattern recognition. The interest in this field is widespread and the range of potential applications very large, both for military systems and in the commercial world. Further, the throughput requirements even for relatively simple or 'low-level' operations is extraordinarily high. Hence it is probably reasonable to assume that the market for an easily programmable architecture that can perform a variety of operations such as enhancement, feature extraction and graphics, for example, will be sufficient to warrant a special purpose design. This is particularly true if the resulting hardware can be easily interfaced to the host computers presently in use and can thus take advantage of existing code development.

The aim of the work reported here is to develop a single programmable VLSI element which can be configured in variable sized arrays of identical modules to achieve the necessary throughput. Considerable effort has been expended to determine the basic computational operations required and the optimum VLSI structure in terms of regularity and data flow. At the low and intermediate level much of the processing is on local blocks of data in kernels of sizes from 3 x 3 to 32 x 32 pixels. This implies that the data can be processed in parallel stream format using the architectures based on the systolic [1] or wavefront [2] array concepts.

Research supported in part by a National Science Foundation Grant No. ECS8016581 and in part by ONR grant N00014-81-K-0191 under a Special Research Opportunity (SRO) program, and the Army Research Office under grant DAAG29-82-K-0077.

II. VLSI Processor Development

A variety of systolic arrays have been devised for applications including convolution, pattern matching, beam-forming and matrix algebra. Our interest is in developing a single VLSI processor which, with minor modifications, can be used over the full range of image and pattern recognition applications. To achieve this we have studied as wide a range of applications as possible. The results of this indicate that the bus oriented architecture with the primitive operations as indicated in Figure 1 would serve as an optimum element. The essential functions for the vast majority of applications are a high speed multiplication and accumulation capability, with perhaps some data manipulation such as a LIFO stack. The addition of a square root or division function can be useful at the borders of the array for some matrix operations such as inversion or singular value decomposition. For this reason our initial design has concentrated on the functions shown in the solid lines. Our Multiplication Oriented Processor (MOP)[3] is such an architecture with essentially a bit slice design using carry-save arithmetic to allow rapid redesign for variable internal word size. It consists of a 28-bit fixed point processor with a carry save multiplier, Manchester adder, a 28-bit wide LIFO stack and a 28 x 16-bit register for data storage.

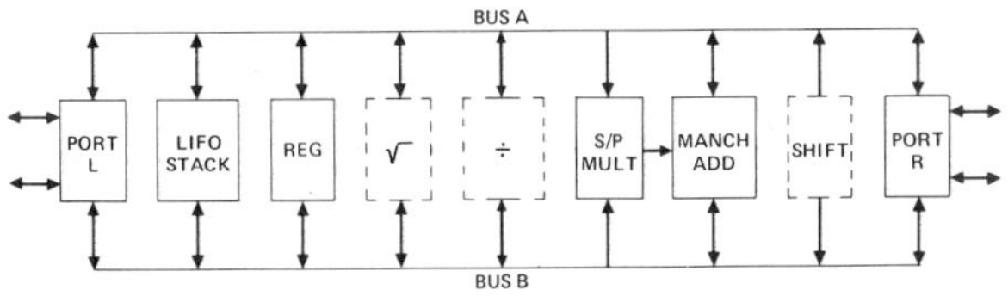

Fig. 1. Schematic of Optimum Processor Element

CH1935-6/83/0000/0553$01.00 © 1983 IEEE

The full processor is controlled by a 16-bit control word provided from the control store shown.

To provide the bandwidth and I/O flexibility required for array architectures we have included two bidirectional, tri-state, parallel ports with latches sufficient to receive or transmit two words per clock cycle. Facilities have also been provided to enable 2-D arrays to be configured with nearest neighbor north, east, south and west communication. In order to increase the speed of the MOP we have embedded the necessary control for all the arithmetic operation in hardware which provides for considerable speed advantage over a micro-coded approach. The multiplier uses a radix-4 Booth's algorithm and a three-stage pipeline to allow high speed partial product accumulation.

The initial version of the processor built in nMOS technology using 5 μm design rules is shown in Figure 2. The device, with approximately 15,000 transistors, operates from an internal clock frequency of 16 MHz for the multiplier and completes a full multiplication and accumulator operation in approximately 1 μsec. Our present estimates indicate that by decreasing the feature size to 1 μm we can reduce this cycle time by a factor of approximately five, and increase the integration to include a 6 x 6 array on a single chip.

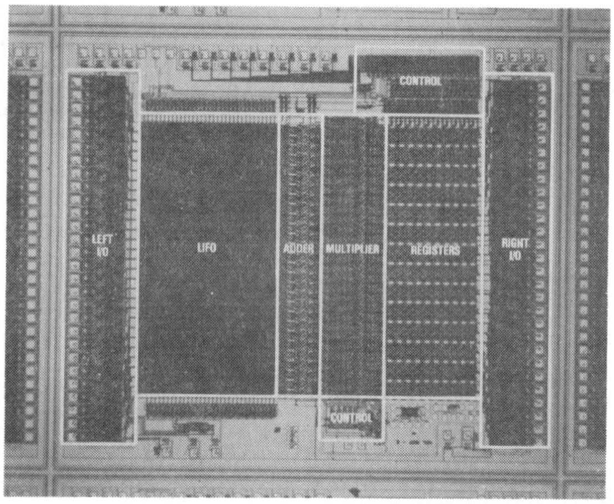

Fig. 2. Photograph of MOPS chip and functional elements

III. Toeplitz System Solver (TOPSS-28) Development

The facilities provided on the MOP chip, we believe, will allow any four-point tesselation to be configured, and hence two-dimensional convolutions, matrix multiplications and inversions, and general purpose linear system solution can be performed. Techniques for these more general operations have been discussed in a previous paper [4]. However, our initial interest has focused on the demonstration of a real-time Toeplitz System Solver, TOPSS-28. The reasons for this derive from the very widespread occurrence of the Toeplitz formulation and the relative simplicity of the array. It has been shown that a one-dimensional array is sufficient for solution of Toeplitz systems

$$\mathbf{H}\,X = \mathbf{Y} \qquad (1)$$

by using the Levinson algorithm [2]

$$\mathbf{X} = \mathbf{U}^{-1}D[\mathbf{U}^{+}]^{-1}\mathbf{Y} \qquad (2)$$

where $\mathbf{U}$ is an upper triangular matrix and D a diagonal matrix with elements u , u . . . u . To perform this operation in our TOPSS-28 processor the following successive steps are necessary:

Generation of U^{+}

First Back Substitution $\phi = [\mathbf{U}^{+}]^{-1}\mathbf{C}$ (3)

Second Back Substitution $\mathbf{X} = \mathbf{U}^{-1}\,\phi$.

The essential arithmetic elements in this process are high speed multiplication and accumulation to form the LU decomposition, and a LIFO data reflection to perform the transpose. The linear array configuration for this is shown in Figure 3.

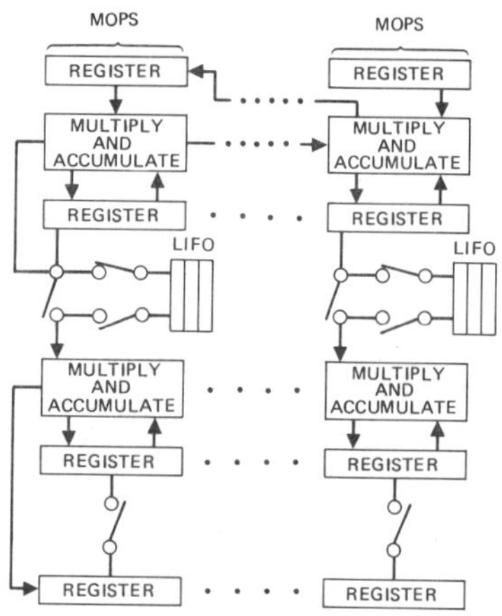

Fig. 3. Schematic of linear array of MOPS for Toeplitz System Solution (TOPSS-28)

IV. Application of Toeplitz Systems to Image Analysis

Toeplitz and nearly Toeplitz systems occur in a large number of image analysis and pattern recognition problems. For example, one particular sub-class involves the inversion of a large banded Toeplitz matrix of the form

$$H = \begin{bmatrix} h_0 & h_1 & \cdots & h_p & & \\ h_1 & \ddots & & & \ddots & \\ \vdots & & \ddots & & & h_p \\ h_p & & & \ddots & & \vdots \\ & \ddots & & & \ddots & h_1 \\ & & h_p & \cdots & h_1 & h_0 \end{bmatrix}_{N \times N} \quad (4)$$

where $N>>p$. Its application in boundary representation, image deblurring and restoration is described below.

a) Image Boundary Representation and Analysis

Increasingly important in the application of image analysis to robotics and industrial testing is the use of image boundaries for identification of objects by their shapes. In our recent studies [5] where several different methods such as Fourier descriptors, AR models and spline functions have been analyzed and simulated, we have found that the spline representation is most efficient in terms of the required number of parameters. For example, the B-spline functions are useful for data compression and the rapid synthesis of the object boundaries when characterized by a few so-called control points as indicated in Figure 4.

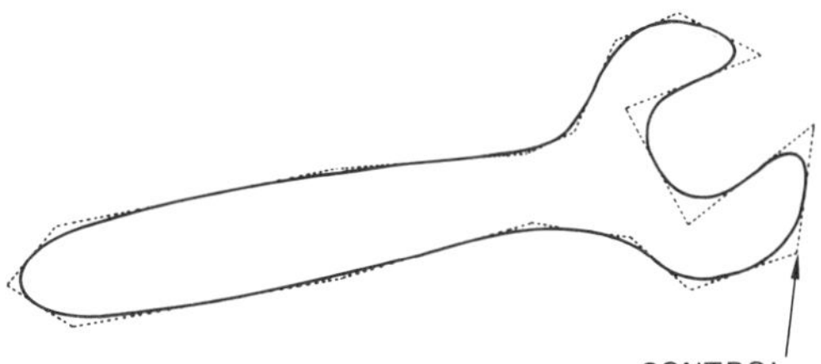

Fig. 4. B-spline fit of two-dimensional shape

In addition to their use in analyzing operations, such as rotation and scaling, spline functions are useful in "smoothing out" hand-drawn digitized curves in a computer graphics system to improve the resolution of the system.

The mathematical formulation of the spline functions can be developed in terms of the boundary parameter t. Then if $x(t)$, $y(t)$ denote the boundary addresses, the spline representation can be written

$$x(t) = \sum_{i=0}^{n} p_i N_{ir}(t) \qquad (5a)$$

$$y(t) = \sum_{i=0}^{n} q_i N_{ir}(t) \qquad (5b)$$

where p , q are the coordinates of the control points and N (t), i=0, 1, . . . are the normalized B-splines of order r, defined recursively as

$$N_{i,r}(t) = \frac{(t-t_i)N_{i,r-1}(t)}{t_{i+r-1}-t_i} + \frac{(t_{i+r}-t)N_{i+1,r-1}(t)}{t_{i+r}-t_{i+1}} \qquad (6a)$$

$$N_{i,1}(t) \triangleq \begin{cases} 1, & t_i<t<_{i+1} \\ 0, & \text{otherwise} \end{cases} \qquad (6b)$$

where t , t . . ., t are called the knot values. For closed boundaries, the corresponding B-splines are periodic. For uniform periodic B-splines

$$t_i = i \qquad (7a)$$

and

$$N_{i,r}(t) = N_{0,r}[(t-i+n+1)\mathrm{mod}(n+1)], \qquad (7b)$$

where the parameter r controls the order of continuity of the boundary. In image graphics often r is taken to be 3 or 4.

When the control points are given, Eqs. (5a) and (5b) can generate the appropriate boundary. The number of control points (n+1) necessary to reproduce a given boundary, which is usually much less than (1/10 to 1/10000 of) the number of boundary points (m) needed to maintain the resolution of the curve, is a function of its shape. It is well known [5,6] that if the boundary curves $x(t)$, $y(t)$ are sampled uniformly, and if the number of boundary points is equal to the number of control points, then the coordinate (p ,q) can be determined by inverting a banded Toeplitz matrix. Typically in image processing applications, since $(n+1) << m$, the problem of control point determination is to solve a rectangular banded Toeplitz system of the form

$$\mathbf{N}[\mathbf{P},Q] = [x,\mathbf{y}] \qquad (8)$$

where P, Q are (n+1)x1 vectors of control points p , q and **x**,**y** are mx1 vectors of boundary addresses, and **N** is an mx(n+1) Toeplitz matrix. Since Eq.(8) is a rectangular system of matrix equations, a unique solution for P, Q does not exist. However, the least squares solution is given by

$$[\hat{\mathbf{P}},\hat{Q}] = [\mathbf{N}^T\mathbf{N}]^{-1}\mathbf{N}^T[x,\mathbf{y}] \triangleq \mathbf{N}^-[x,\mathbf{y}] \qquad (9)$$

where $\mathbf{N}^- \triangleq [\mathbf{N}^T\mathbf{N}]^{-1}\mathbf{N}^T$ is the pseudo-inverse of **N**. For periodic B-splines it can be shown [5] that the (n+1)x(n+1) matrix $\mathbf{N}^T\mathbf{N}$ is (a) Toeplitz $\forall n$, (b) banded Toeplitz of n>r. In the case of open boundaries (i.e., open B-splines), $\mathbf{N}^T\mathbf{N}$ is nearly Toeplitz or nearly banded Toeplitz.

Until now these problems have been performed on general purpose machines and are prohibitively slow for real-time future graphics systems. The TOPSS-28 system has been designed to perform the computationally intensive elements of this calculation, such as the necessary inversion and matrix multiplication for real-time applications.

b) AR and MA Models in Image Processing

Another application of the Hughes TOPSS-28 processor is in the calculation of filter weights and coefficients for enhancement and data compression.

It can be shown that the covariance matrix of a pth order moving average process

$$x(n) = \varepsilon n - \sum_{k=1}^{p} b(k)\varepsilon(n-k), \quad n = 1 \ldots, N \qquad (10)$$

where (n) is a white noise sequence, is a banded Toeplitz matrix. Also, for a pth order autoregressive process

$$x(n) - \sum_{k=1}^{p} a(k)x(n-k) = \varepsilon(n), \quad n = 1 \ldots, N \qquad (11)$$

(n) = white noise sequence

the inverse of the covariance matrix of x(n) is nearly banded Toeplitz.

In many 2-D signal and image processing problems, such as data compression, noise smoothing and deblurring, each scan line of the image or a one dimensional transform of it is represented by a low order AR process [7]. Algorithms for processing such data often require inverting and/or finding eigenvalues and eigenvectors of banded Toeplitz matrices.

c) Image Restoration

In restoration of images blurred by spatially invariant point spread functions (PSFs), banded Toeplitz matrices arise with $p \ll N$ and N of the order of 16 to 1000. Typically, one is given observations:

$$\begin{aligned} v(m,n) &= h(m,n)*u(m,n)+\eta(m,n) \\ \eta(m,n) &= \text{additive noise} \qquad (12) \\ h(m,n) &= \text{point spread function (PSF)} \end{aligned}$$

and is required to estimate u(m,n). In a large number of image processing applications, the PSF is much narrower than the size of the image so that h(m,n) can be modeled as an FIR filter. Image restoration solutions such as Wiener filter, pseudo-inverse filters, etc., have been shown [8] to be equivalent to solving banded Toeplitz or nearly banded Toeplitz systems of equations.

V. Conclusions and Systems Issues

A practical architecture for image analysis and pattern recognition might prove to be a sequential host machine for the high level and control operations and a series of high speed attached processors for the low and intermediate level operations. This has the advantage of making use of existing code developed on the commercial machines while attaining real-time operation through the high speed primitives. Our analysis indicates that an array of high speed multiplication/accumulate processors can perform a wide variety of the bottleneck operations, including restoration, feature extraction, and graphics. As an initial step to this we have discussed the one-dimensional Toeplitz array which is widely useful. As a subset of this several situations exist where one is required to invert banded Toeplitz or nearly Toeplitz systems. Examples are in extrapolation of discrete bandlimited sequences, spectral estimation and adaptive image restoration.

A significant issue in the development of any custom VLSI hardware is matching the problem size to the array hardware. Frequently the kernel of the problem to be solved is larger than the available hardware and block techniques must be developed to resolve this issue. For Toeplitz systems this problem has been solved by using a circulant deconvolution of size N x N and a much smaller Toeplitz solver of size 2p x 2p (N>>2p).

REFERENCES

1. H. T. Kung, "Special-Purpose Devices for Signal and Image Processing: An Opportunity in VLSI," Proc. of Soc. of Photo-Optical Inst. Eng. (SPIE), pp. 76-84, July 1980.

2. S. Y. Kung and Y. H. Hu, "A Highly Concurrent Algorithm and Pipelined Architecture for Solving Toeplitz Systems," IEEE Trans. ASSP,, Vol. ASSP-31, pp 66-76, Feb. 1983.

3. J. G. Nash, G. R. Nudd and S. Hansen, "Concurrent VLSI Architectures for Toeplitz Linear Systems Solution," Gov. Microcircuit App. Conf., Orlando, Florida, November 2-4, 1982.

4. J. G. Nash, G. R. Nudd, and S. Hansen, "VLSI Processor Arrays for Matrix Manipulation," pp. 367-378, VLSI Systems and Computations, edited by H. T. Kung, B. Sproull and Guy Steele, Computer Science Press, 1981.

5. A. K. Jain, D. Paglieroni and S. Inamoto, "Image Boundary Representation Algorithms and Their Application in Vision Systems," Signal and Image Processing Lab., Tech. Report, Department of Electrical and Computer Engineering, University of California, Davis, 1983.

6. B. A. Barsky and D. P. Greenberg, "Determining a Set of B-Spline Control Vertices to Generate an Interpolating Surface," Comp. Graphics and Image Processing, Vol. 14, pp. 203-226, 1980.

7. A. K. Jain, "Advances in Mathematical Models for Image Processing," Proc. IEEE, Vol. 69, No. 5, pp. 502-528, May 1981.

8. A. K. Jain, "An Operator Factorization Method for Restoration of Blurred Images," IEEE Trans. Computers, Vol. C-26, No. 11, pp. 1061-1071.

SYSTEM DESIGN OF IMAGE RESAMPLERS FOR ELECTRONIC PUBLISHING

Zoltan Stroll and Sang-Chul Kang

Xerox Corporation
El Segundo, California

Abstract

The advent of VLSI is enabling the rapid development of the "electronic darkroom." Functions such as image scaling, rotation, spatial filtering, cropping, halftoning and tonal rendition modifications were traditionally performed photographically. This paper is a study of VLSI based systems to perform these functions electronically.

1. Introduction To The "Electronic Darkroom"

With the advent of the invasion of the office and publishing by electronics, new product opportunities have opened, and many have begun to be exploited. These include word, text, and graphics processing among a myriad of other applications. Whether locally or remotely generated or updated, data is stored, mailed, distributed, printed, edited, etc. Conspicuously absent from this battleground has been the handling of image data in a graphics composition/page makeup environment due to the prohibitive costs involved with scanning, storing, processing, and printing of scanned imagery. But this last bastion is now under attack and will fall in the near future to the onslaught of advances in very large scale integrated electronics, novel storage and maturing scanning and marking technologies.

A set of functions exist that are better handled by operating on the electronic master [1]. Included (but not limited to) are reduction, magnification, rotation, filtering, parametric and nonparametric cropping, halftoning/TRC modification, etc. The functions above can be selectively applied to the electronic master without obliterating it and the results interactively evaluated for adequacy. The first four can be accomplished by resampling the electronic master to the re-sized, rotated coordinate system using a loadable resampling function kernel.

The impetus to image resampling was provided during the last decade by NASA needs in removing satellite and scanning system errors from the LANDSAT imagery beamed back to earth [2,3]. Procedures were developed to estimate the error distribution profile followed by an image resampling process to warp the image back into the correct shape in a variety of coordinate systems. LANDSAT image data resampling requires a high degree of spatial and radiometric accuracy, a small amount of rotation (skew of up to 1.5°) and a reasonable processing rate (2 to 10 μsec/resampled pixel).

Theoretically, convolution with the sinc function provides ideal resampling [4]; unfortunately this requires processing a very large number of pixels. A truncated sinc function has been accepted as the quality resampling standard, with bilinear and nearest neighbor as standby options. The resampling process involves four nearest neighboring pixels (two on each side) in each direction, hence a single pass interpolation requires 16 multiply and add operations per resampled pixel, while a two pass system requires 4 multiply and add operations in each pass. Implementations are based on special purpose or specially programmed floating/fixed point signal processors.

The introduction of resampling in commercial applications has been lagging behind defense/space requirements; however, demand is now growing due to the increasing sophistication and technological advances that are being achieved in advanced systems products targeted at the office, communications and publishing industries. The commercial image data resampling requirements will have somewhat different characteristics than the ones formerly described. Sufficient quality can be obtained by using a 2-point (4-point two dimensional) resampling process with a loadable kernel function to provide certain filtering capabilities. However, the dynamic range requirements for rotation and scaling are much greater, necessitating the judicious use of system resources such as mass storage, main processor data buffering, private data storage and arithmetic operators. While pixel processing rates at the hardware level can be feasibly accomplished at .5 to 1 μsec/pixel, mass storage access may restrict the overall rate by a factor of from four to eight. Though the resampling process could also accommodate the removal of scanning system nonlinearities, it is anticipated that these functions will be performed in situ at the scanner.

Image cropping is accomplished through the dual functions of parametric cropping of the edges, along with the superposition of a composition bitmap for nonparametric cropping. The composition bitmap is at the CRT display resolution and size, and arbitrary resolution ratios among scanner, display and marker must be handled. Scanned image decompression and decoding as well as halftoning for black/white display or marking are also accommodated.

2. Overview Of System Design Concepts

This study investigates two approaches to image resampling: a single pass and a two pass method. At the heart of the single pass resampler is the four point two dimensional interpolator.

The original data scanlines are retrieved from disk, buffered in main memory, then segments along the new scan direction are shipped to the resampling hardware. Here the four point interpolator resamples the data into complete new scanlines, subsequently shipping them back to the memory for buffering, then storage on disk.

The original scanned data on disk is assumed to be in gray format. The single pass resampler can accommodate decoding of along-scan encoded data in the same operation only if no rotation is performed. This restriction does not pose much difficulty since the first operation would usually consist of decoding and conversion to CRT resolution in order to initiate interactive image manipulation. Thereafter, much of the interactive operation would be performed on the CRT resolution display master. The former is not a requirement, though processing the display master would be considerably faster than always returning to the full resolution gray master. For finished copy generation, of course, all of the processing parameters are applied to the full resolution master.

CH1935-6/83/0000/0557$01.00 © 1983 IEEE

The single pass resampler, in addition to scaling, rotation and filtering, accommodates all the "electronic darkroom" functions such as the application of the composition bitmap, parametric cropping and halftoning. The composition bitmap is aligned with the image before rotation. The mask is superimposed on the data before resampling. Cropping parameters are specified in terms of the resampled grid units and directions. The halftone cells are likewise aligned with the resampled grid.

The two pass resampler is so named because the image data passes twice through the resampling hardware, once for along line resampling, the second time for across line resampling, with intermediate storage on disk. The original data is retrieved from disk, buffered through main memory and shipped to the resampler hardware where along line resampling is performed by interpolating between two adjacent pixels on the scanline. Along line cropping and masking are also performed at this time. Subsequently, the data is returned in along line segments, each segment internally transposed to the across line direction. The buffered data is stored on disk in a format that allows efficient access to it in the across line segment order for the second pass. In the second pass, the segments are collected in across line sequence, and passed to the resampler for across line resampling in an analogous fashion to the along line process described above. Across line cropping and halftoning are also performed at this time. Then the segments, optionally transposed internally for rotation angles of 45° or less, are returned to be buffered through main memory and stored on disk. Unlike the one pass resampler, the two pass resampler does not restrict rotation while decoding.

3. Resampling Algorithms

The image resampling process consists of a coordinate transformation that determines the location of the resampled pixel in the original image space and an interpolation of neighboring pixels to obtain the resampled pixel value at the desired location. The coordinate transformation which is involved in rotation and scaling can be described in matrix notation as follows: let the original coordinates be x, y and the resampled coordinates be X, Y. Then $[X\ Y\ 1] = [x\ y\ 1]\ \mathbf{M}$ and post-multiplication with $\mathbf{M}^{-1}$ yields $[x\ y\ 1] = [X\ Y\ 1]\ \mathbf{M}^{-1}$, where

$$\mathbf{M} = \begin{bmatrix} a & d & 0 \\ b & e & 0 \\ c & f & 1 \end{bmatrix} \quad \text{and} \quad \mathbf{M}^{-1} = \begin{bmatrix} \Delta X_x & \Delta Y_x & 0 \\ \Delta X_y & \Delta Y_y & 0 \\ SP_x & SP_y & 1 \end{bmatrix} \tag{1}$$

are the forward and reverse coordinate transformation matrices, respectively, and $\Delta X_x = e/D$, $\Delta Y_x = -d/D$, $\Delta X_y = -b/d$, $\Delta Y_y = a/D$, $SP_x = (bf-ce)/D$, $SP_y = (cd-af)/D$, and $D = ae-bd$. The geometrical significance of these parameters can be seen in Figure 1.

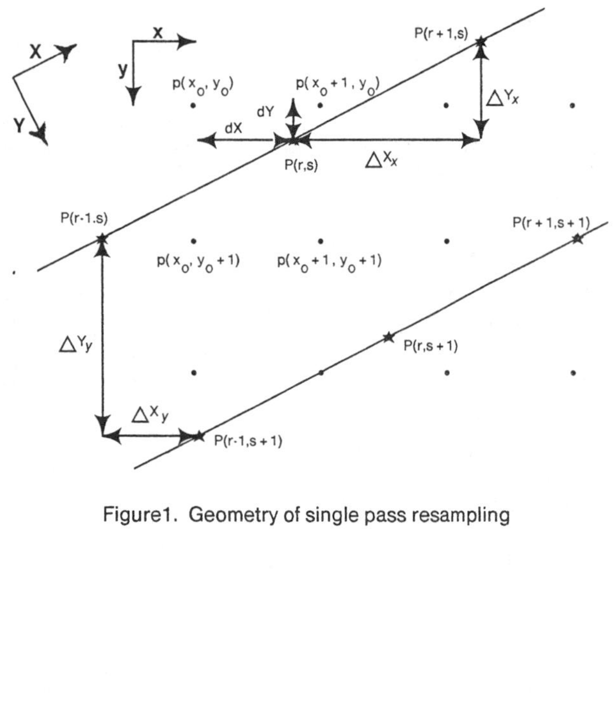

Figure1. Geometry of single pass resampling

Let an array of m x n pixels in the [x y] coordinate system, with pixel values at (i,j) denoted by p(i,j), be defined for the integers $0 \leq i \leq n-1$, $0 \leq j \leq m-1$ to represent the original image. The pixel value p(x,y) at (x,y) is obtained by convolution with the two dimensional interpolation kernel function W(x,y)

$$p(x,y) = \sum_{i=0}^{n-1} \sum_{j=0}^{m-1} p(i,j) W(x-i, y-j). \tag{2}$$

The above formula is simplified using an orthogonally separable interpolation function for which $W(x,y) = W(x)W(y)$ and interpolating only l neighbors. That is,

$$p(x,y) = \sum_{i=0}^{l-1} \sum_{j=0}^{l-1} p(x_i, y_j) W(x-x_i) W(y-y_j) \tag{3}$$

where x_i and y_j are one of the l nearest integers of x and y, respectively; p(x,y) has a "background" value for the pixels undefined in the given pixel array. Then the resampled pixel at the grid point (r,s) in the [X Y] coordinate system is $P(r,s) = p(x_r, y_s)$ where $[x_r\ \ y_s\ \ 1] = [r\ \ s\ \ 1]\ \mathbf{M}^{-1}$, $x_r = x_o + dX$, $y_s = y_o + dY$, $l = 2$ and

$$\begin{aligned} p(x_r, y_s) = {} & p(x_o, y_o) W(dX) W(dY) \\ & + p(x_o+1, y_o) W(dX-1) W(dY) \\ & + p(x_o, y_o+1) W(dX) W(dY-1) \\ & + p(x_o+1, y_o+1) W(dX-1) W(dY-1) \ . \end{aligned} \tag{4}$$

The above forms the basis of a one pass resampler. The mechanization involves only registers and adders to allow indexing through the original pixel space along the resampled scanline. Moreover, judicious rounding of parameters in $\mathbf{M}^{-1}$ and the use of limited significance integer values yields a system capable of meeting required spatial accuracy. When the interpolation function contains a limited number of discrete values, multiplication can practically be implemented by table lookup. Since the pixel space is closed, the multiplication can be iterated via the same lookup process.

An alternate approach to performing the coordinate transformation forms the foundation to the two pass resampler as follows: let $[X\ Y\ 1] = [x\ y\ 1]\ \mathbf{M} = [x\ y\ 1]\ \mathbf{M}_x\ \mathbf{M}_y$,

$$\text{where } \mathbf{M}_x = \begin{bmatrix} a & 0 & 0 \\ b & 1 & 0 \\ c & 0 & 1 \end{bmatrix} \quad \text{and} \quad \mathbf{M}_y = \begin{bmatrix} 1 & d' & 0 \\ 0 & e' & 0 \\ 0 & f' & 1 \end{bmatrix}, \quad \begin{aligned} d' &= d/a \\ e' &= e-bd' \\ f' &= f-cd'. \end{aligned} \tag{5}$$

Then $[x\ y\ 1] = [X\ Y\ 1]\ \mathbf{M}_y^{-1}\ \mathbf{M}_x^{-1} = [X\ Y\ 1]\ \mathbf{P}\ \mathbf{M}_{y'}^{-1}\ \mathbf{P}\ \mathbf{M}_x^{-1}$

$$= [X\ Y\ 1]\ \mathbf{P} \begin{bmatrix} \Delta X_2 & 0 & 0 \\ \Delta Y_2 & 1 & 0 \\ SP_2 & 0 & 1 \end{bmatrix} \mathbf{P} \begin{bmatrix} \Delta X_1 & 0 & 0 \\ \Delta Y_1 & 1 & 0 \\ SP_1 & 0 & 1 \end{bmatrix} \tag{6}$$

$$\text{where } \mathbf{P} = \begin{bmatrix} 0 & 1 & 0 \\ 1 & 0 & 0 \\ 0 & 0 & 1 \end{bmatrix} \quad \text{and} \quad \begin{aligned} \Delta X_1 &= 1/a \\ \Delta Y_1 &= -b/a \\ SP_1 &= -c/a \end{aligned} \quad \begin{aligned} \Delta X_2 &= 1/e' \\ \Delta Y_2 &= -d'/e' \\ SP_2 &= -f'/e'. \end{aligned} \tag{7}$$

The geometric significance of the parameters above can be seen in Figure 2. The two pass resampling process can be described as follows. In the first pass, an intermediate image is determined in the [X y] coordinate system as a rectangular array of m x n' pixels. The (x,y) coordinates for the new pixels defined in the [X y] coordinate system are obtained using $[x\ \ y\ \ 1] = [X\ \ y\ \ 1]\ \mathbf{M}_x^{-1}$. Note that the first pass resampling parameters, ΔX_1, ΔY_1 and SP_1, are defined by $\mathbf{M}_x^{-1}$. Each pixel value at (X,y) is obtained interpolating original pixels around (x,y). The intermediate image is then transposed in order to resample in the across scanline direction afterwards. This process is represented in terms of a multiplication by a permutation matrix **P**, i.e. [X y 1] = [y X 1] **P**. The second pass can be accomplished similarily. The final image is defined in the [X Y] coordinate system as an array of m' x n' pixels. The (y,X) coordinates for the pixels defined in the [X Y] coordinate system are obtained using [y X 1] = [Y X 1] $\mathbf{M}_{y'}^{-1}$ and [Y X 1] = [X Y 1] **P**. The final pixel values at (X,Y) are obtained interpolating pixels around (X,y). The second pass resampling parameters, ΔX_2, ΔY_2, SP_2, are defined by $\mathbf{M}_{y'}^{-1}$.

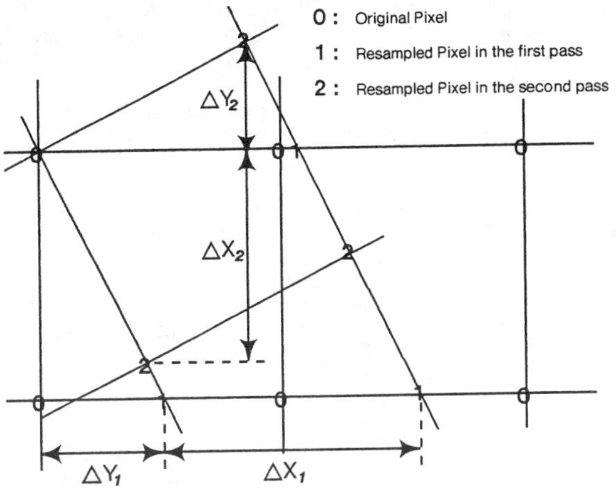

Figure 2. Geometry of two pass resampling

One point should be noted in two pass resampling. When the component "a" of the matrix **M** has a small magnitude, the resampling parameters, ΔX_1 (=1/a) and ΔY_1 (=-b/a) have large magnitudes. This would cause the pixels in the final image to be obtained interpolating original pixels which are not close to the resampling positions. This effect can be alleviated if the transformation matrix is permuted before resampling when the component "d" of the matrix M has a larger magnitude than the component "a". Thus, when |d/a| > 1, the resampling parameters are derived as follows:

$$[X\ Y\ 1] = [x\ y\ 1]\ \mathbf{M} = [x\ y\ 1] \begin{bmatrix} d & a & 0 \\ e & b & 0 \\ f & c & 1 \end{bmatrix} \mathbf{P}$$

$$= [x\ \ y\ \ 1]\ \mathbf{N}_x\ \mathbf{N}_y\ \mathbf{P},$$

$$\text{where } \mathbf{N}_x = \begin{bmatrix} d & 0 & 0 \\ e & 1 & 0 \\ f & 0 & 1 \end{bmatrix} \text{ and } \mathbf{N}_y = \begin{bmatrix} 1 & a' & 0 \\ 0 & b' & 0 \\ 0 & c' & 1 \end{bmatrix},\quad a' = a/d,\ b' = b\text{-}ea',\ c' = c\text{-}fa'. \tag{8}$$

The old coordinate system [x y] can be obtained from [X Y] as follows:

$$[x\ y\ 1] = [X\ Y\ 1]\ \mathbf{P}\ \mathbf{N}_y^{-1}\ \mathbf{N}_x^{-1} = [X\ Y\ 1]\ \mathbf{N}_{y'}^{-1}\ \mathbf{P}\ \mathbf{N}_x^{-1} \tag{9}$$

$$= [X\ Y\ 1] \begin{bmatrix} \Delta X_2 & 0 & 0 \\ \Delta Y_2 & 1 & 0 \\ SP_2 & 0 & 1 \end{bmatrix} \mathbf{P} \begin{bmatrix} \Delta X_1 & 0 & 0 \\ \Delta Y_1 & 1 & 0 \\ SP_1 & 0 & 1 \end{bmatrix},\quad \begin{matrix} \Delta X_1 = 1/d & \Delta X_2 = 1/b' \\ \Delta Y_1 = -e/d & \Delta Y_2 = -a'/b' \\ SP_1 = -f/d & SP_2 = -c'/b'. \end{matrix}$$

In the above case, the intermediate image is defined in the [Y y] coordinate system as a rectangular array of m' x m pixels. The coordinates are obtained using [x y 1] = [Y y 1] $\mathbf{N}_x^{-1}$. This image is then processed by using [y Y 1] = [X Y 1] $\mathbf{N}_{y'}^{-1}$. It is not required to transpose the image after second pass resampling since the obtained image is already in scanline order. Mechanization aspects mentioned in the one pass operation remain valid for the two pass resampler.

4. System Architecture and Performance

The architectural block diagram of the image resampling system is shown in Figure 3. The resampling subsystem is embedded in the classical computer system model. The programmable microprocessor performs the overall process control, particularly as related to setting up disk, DMA and resampler data flows.

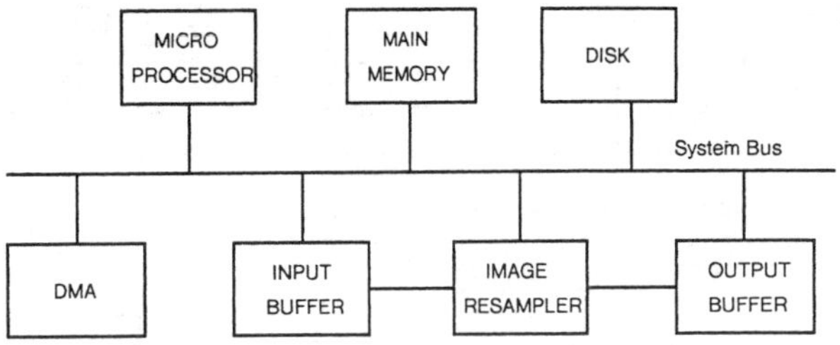

Figure 3. Resampling system block diagram

Let the resampled image be composed of a number of small blocks, Nblk. The following describes the one pass process which finds the resampled image by obtaining Nblk's in sequence:

0. Read first L scanlines of the original image into the input buffer of main memory. L is determined by the size of the input buffer available.
1. Identify next Nblk to be obtained in the resampled scan direction.
2. Find the original block of data, Oblk, containing the Nblk identified in Step 1. If the Oblk is not available in the input buffer, then read in another L scanlines of the original image.
3. Obtain Nblk by resampling Oblk. Store Nblk in the output buffer.
4. Repeat 1-3 until Nblksize (height of Nblk) scanlines of the resampled image are stored in the output buffer.
5. Write output buffer to disk file.
6. Repeat 1-5 until the image is complete.

The two pass process flow is as follows:

First Pass

1.1 Read H scanlines of the original image into the input buffer. H is determined by the buffer size.
1.2 Resample H scanlines and transpose blocks (of size H x H).
1.3 Write transposed blocks to the intermediate disk file.
1.4 Repeat 1.1 - 1.3 until the image is complete.

Second Pass

2.1 Read H scanlines of the intermediate image.
2.2 Resample H scanlines and transpose blocks if required.
2.3 Write transposed blocks to the disk.
2.4 Repeat 2.1 - 2.3 until the image is complete.

If transposition is performed in the second pass, then the stored image is in a multiplexed scanline format. The following steps can be used to demultiplex the scanlines of the final image:

3.1 Read H scanlines from the image obtained at the end of the second pass.
3.2 Demultiplex scanlines.
3.3 Write in the output file.
3.4 Repeat 3.1 - 3.3 until the image is complete.

The resampler subsystem functional flow diagram is shown in Figure 4 and that of the chip is shown in Figure 5. The chip is useful in both the one pass and two pass operation contexts.

The performance of the resampling system is determined by the subsystem and disk access performance. The resampling subsystem is estimated to be capable of generating resampled pixels at a rate in excess of one megapixel per second. This component has a minor effect on the overall throughput when compared with the disk access performance. Figure 6 graphs

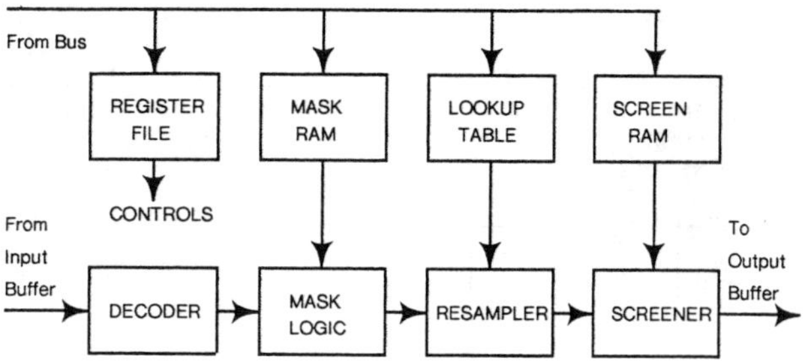

Figure 4. Resampler subsystem functional flow diagram

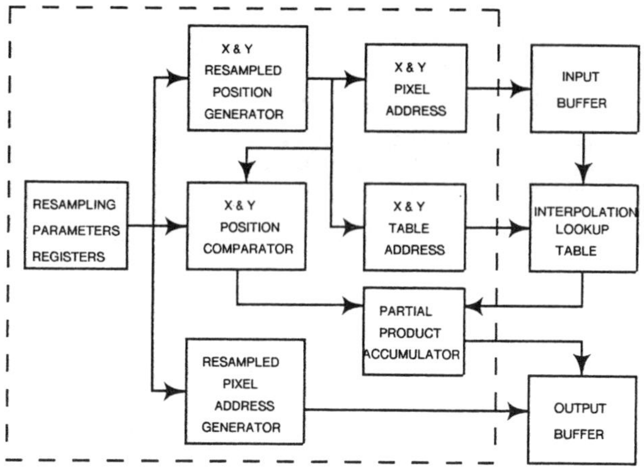

Figure 5. Resampler chip functional flow diagram

the disk transfer time as a function of rotation angle and available buffer memory space for a full page display oriented operation. The parameters used are as follows:

Image Size: 8"x10", 80 pixels/inch
(640 scanlines, 800 pixels/scanline)
Scaling Factor: 1
Transfer Rate: 35.4 Mbits/sec (formatted)
Inter File Seek Time: 30 msec
Intra File Seek Time: 10 msec
Rotation Latency: 8.33 msec.

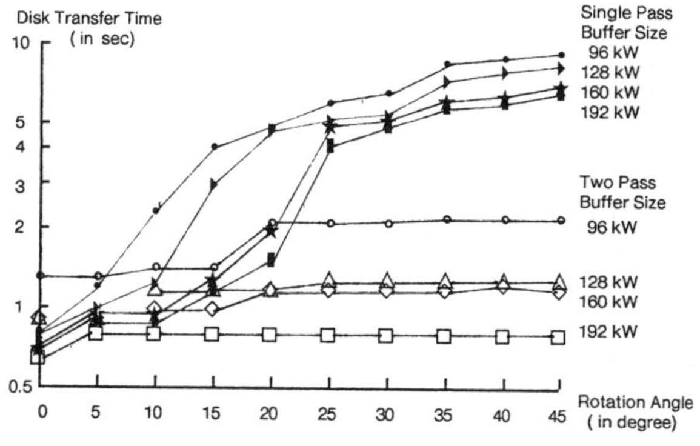

Figure 6. Disk transfer performance

It can be seen that the performance of the one pass system decreases with increasing rotation angles, while that of the two pass system is "constant." The one pass system is better at small rotation angles, whereas the two pass system is better at large angles. In both cases, the performance is between very good and adequate response time for the user performing interactive processing at the CRT display.

5. Tradeoffs and Conclusions

The functional difference between the single pass and two pass algorithms is that the single pass can accommodate a nonseparable kernel function, while the two pass cannot. Images have been synthetically resampled both ways using a separable four value approximation of the bilinear interpolation function. The resampled images were recorded on film in the gray domain and the results were found quite satisfactory, particularly when compared to the results obtained by resampling with the nearest neighbor kernel. While in terms of computational complexity, the single pass algorithm exhibits n^2 growth and the two pass behaves linearly with n, where n is the number of pixels to be interpolated along a scanline. From the performance point of view, the single pass has a systematic advantage by virtue of the ability to handle the data only once and compute in parallel, as opposed to a two stage systematic "pipeline" computation as in the two pass. The complexity and performance issues are related in that they can be traded off against each other in an implementation sense equalizing the benefits. The two approaches exhibit differing artifacts. The single pass approach is characterized by a static error in the desired versus actual angle of rotation and scale factor while being able to handle square grids on input and resampled output. Meanwhile, the two pass approach exhibits unequal static along/across line rotation angle and scale factor errors, resulting in the inability to create a perfectly square resampled grid, as well as occasionally interpolating over a parallelogram rather than square area. It must be emphasized that these static errors are under a fraction of a percent, expected to be quite adequate for commercial applications.

The system performance implications provide the most crucial distinctions. The features here have been gauged in terms of system performance and resource requirements. The major distinctions originate from disk handling. The single pass system is suitable to small angle rotation, up to about 5°; for larger angles, the performance rapidly deteriorates. In juxtaposition, the two pass system is suitable for arbitrary angle rotation as its performance is roughly "constant" with angle of rotation. While the performance generally improves with the addition of main memory, the performance curve shapes remain similar.

There are several disk related implications. The two pass operation requires additional storage for the intermediate formatted images; it also is sensitive to disk formatting. Since segments must be easily accessible, the disk sector size should ideally be compatible with this. Extrapolating the performance profiles based on disk handling, it would appear that a lower data rate disk would have a lesser impact on the two pass performance as the primary factor is head movement, not data rate. Similarly, it would appear that the two pass would benefit most from a fixed head disk.

From the foregoing it can be seen that the technology exists for building commercially feasible post scan image processing systems. The rapid technological advances in electronics will be soon enabling even more powerful and sophisticated systems.

References

[1] Z. Stroll, S. Kang, S. Miller, V. Kadakia, "Post Scan Image Processing System Study," Xerox Internal Report, Accession No. S81-00026, August 1981.

[2] R. H. Caron, S. S. Rifman, K. W. Simon, "Application of Advanced Signal Processing Techniques to the Rectification & Registration of Spaceboard Imagery." Technology Transfer Conference Proceedings, University of Houston, September 1974.

[3] K. W. Simon, "Digital Image Reconstruction & Resampling for Geometric Manipulation," Symposium on Machine Processing of Remotely Sensed Data Proceedings, W. LaFayette, Indiana, June 1975.

[4] W. K. Pratt, Digital Image Processing. New York: Wiley, 1978.

SELF-CHECKING VLSI BUILDING BLOCKS FOR FAULT-TOLERANT MULTICOMPUTERS

Yuval Tamir and Carlo H. Séquin

Computer Science Division
University of California, Berkeley, CA 94720

Abstract

The use of self-checking nodes and links for implementing fault-tolerant VLSI multicomputers is proposed. The system is composed of a large number of VLSI computers interconnected by high-speed dedicated links. Hardware that performs error detection is combined with system-level protocols that handle error recovery and fault treatment.

The self-checking nodes notify the rest of the system when their output is erroneous. In order to achieve high fault coverage, error detection is accomplished by duplication and matching. The critical circuit in this scheme is a comparator which must not be susceptible to faults that can remain undetected and later mask the failure of the functional modules. With both NMOS and CMOS technologies it is possible to implement a *self-testing* comparator that will produce an error indication if it incurs any single physical defect.

Introduction

High reliability and high performance are primary goals of most computer systems. There are fundamental limits to the increases in reliability and performance that are achievable by improvements in technology alone. The limits on performance can be overcome by exploiting parallelism while the limits on reliability can be overcome by using *fault tolerance* techniques.

Parallelism can be exploited by a system that consists of a large number of *computation nodes*, each able to execute a subtask of the problem being solved. A possible architecture for such a system, which is compatible with the constraints of VLSI, is to interconnect these computation nodes by high-speed dedicated links and *communication nodes* that provide hardware support for communication functions such as message routing. Each computation node is connected to one of the communication nodes. A communication node has several *ports* through which it is connected to computation nodes and other communication nodes.[9] We call such a system a *multicomputer*. The nodes and links are components (building blocks) that can be used to construct multicomputers with a wide range of sizes and topologies.

System *failure* occurs when the multicomputer doesn't perform according to its specifications at its interface with the "outside world." System failure is often the result of a failure of one of its components. Fault tolerance techniques attempt to prevent component failure from leading to system failure.[1] A multicomputer is especially well suited for reliability enhancement using fault tolerance techniques since it is partitioned into independent and "intelligent" components (the computation and communication nodes). Fault-free components can adapt to changes in faulty components and continue their operation in a way that leads to correct system output despite the fault.

A brief overview of techniques for implementing fault tolerance in a multicomputer is presented and the considerations that lead to the choice of an approach based on self-checking components are discussed. *Duplication and matching* is shown to be an effective practical technique for implementing nodes that are self-checking with respect to any likely fault.

Implementation of Highly Reliable Multicomputers

The reliability of any system can be enhanced by increasing the reliability of its components through *fault prevention*[1] techniques such as specialized design methodologies, stringent quality control, and extensive validation and testing. These techniques typically result in more complex designs, greater cost, and lower performance. Furthermore, the effectiveness of these techniques is limited by our inability to exhaustively test complex VLSI chips.[8]

Alternatively, the reliability of the components can be increased by employing *fault tolerance* techniques. These techniques attempt to ensure that each component will continue to perform according to its specifications despite faults. Unfortunately, no component can tolerate an unbounded number of faults. The contamination of the system by incorrect output from a faulty component can be prevented only if, at some stage, other system components find out about the failure of the component and physically or logically isolate it from the rest of the system.

At the system level, software (protocols) can be used to detect and recover from the failure of components. For example, identical tasks may be assigned to three nodes and a "majority vote" taken on the results. One of the problems with this approach is that if the results conflict, it may be very costly or impossible to locate the cause of the discrepancy. Additional problems are the high overhead in computation resources and communication bandwidth and difficulties in effectively handling transient faults.

If a node fails due to a transient fault, it should be reset to a "sane state" and remain active rather than be removed from the system. If neighboring nodes are responsible for detecting such a failure, they must be given the authority to initiate the reset. However, this authority also allows a *failed* node to reset operational neighbors. In order to prevent this situation, each node must be responsible for its own reset. Hence, the node should include a mechanism to detect its own *erroneous states*[1] and initiate the reset.

Some of the deficiencies with the aforementioned techniques can be overcome by implementing fault tolerance in a VLSI multicomputer using hardware *error detection* in conjunction with system level protocols which perform *error recovery* and *fault treatment*.[1] Errors caused by faults in the communication links are detected through the use of error-detecting codes. All nodes are self-checking and signal to the rest of the system when their output is incorrect so that it will not be accepted as correct. In addition, failed nodes attempt to reset themselves and reestablish a sane state. The immediate neighbors are informed whenever a node fails. If the node doesn't reset itself or fails too often, the neighbors can logically remove it from the system by refusing to communicate with it. The diagnostic status information is distributed throughout the system so that, eventually, no fault-free node will attempt to use the faulty component.

Self-Checking Nodes

For all likely faults, a self-checking component must either produce the "correct" outputs or somehow indicate that the outputs are incorrect. A component that satisfies this requirement is said to be *fault secure*.[10] If the component is not guaranteed to produce an error indication immediately following the first fault, it is possible for several faults to exist in the component simultaneously without any indication to the rest of the system. Even if the component is fault secure with respect to any single fault, several faults together may lead to the failure of the self-check mechanism and, eventually, to incorrect outputs from the component being accepted as correct by the rest of the system. In order to prevent this situation, the component must be *self-testing*.[10] In the presence of one or more faults, a self-testing component produces an error indication before additional faults can occur and lead to the failure of the self-check mechanism. Components which are fault-secure and self-testing are said to be *totally self-checking*[10] (*TSC*).

Error detecting/correcting codes can be used to implement *TSC* nodes. Redundant information is carried by busses, memories, and registers in order to detect (and possibly correct) errors.[10] Unfortunately, different coding schemes must be used for different parts of the node. The resulting increase in the complexity of the design and of design verification and testing may lead to a circuit in which failure modes that are more difficult to predict and "tolerate" are more likely to occur.

An alternative is to construct the *TSC* computation or communication node using two identical, *independent* modules, each performing the function of the node. Inputs from neighbor nodes are fed to both modules. If the modules operate synchronously, their outputs should always be identical. Except for the nearly-impossible case where both modules produce identical *incorrect* output, an error can be detected by a comparator which is part of the node. The output of the comparator is connected to neighboring nodes through dedicated wires. The output from one of the two modules is the "functional" output from the node (Fig. 1). A "no-match" signal from the comparator is used locally as a reset signal and is also sent to all neighbors as a failure indicator. Similar failure indicators from the neighbors cause an interrupt and invoke system-level routines that handle the node failure.

Implementing the *TSC* property in a component using duplication and matching may appear wasteful since it more than doubles the required hardware. However, this scheme becomes more attractive when issues such as design complexity, fault coverage,

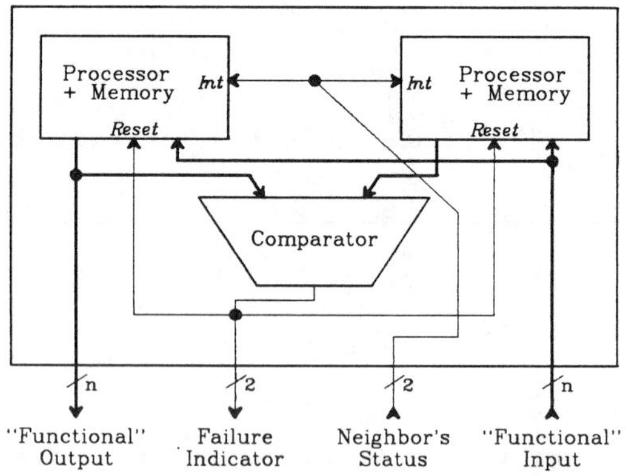

Fig. 1: *A Self-Checking Computation Node*

reliability prediction, and the ability to recover from transient faults are taken into account. Traditional fault models are not valid for VLSI.[5,12] As a result, low-cost error detection schemes, which are based on these models, may no longer be adequate. With duplication and matching, errors are detected as long as the comparator remains functional and the two modules produce different outputs the first time one or both of them fail. Since a faulty comparator can mask faulty functional modules, faults in the comparator must not go undetected, *i.e.*, the comparator must be self-testing. Thus a detailed analysis of the effects of all likely faults on the comparator is required.

Physical Defects and Logical Faults in VLSI

The design of self-checking circuits requires an understanding of the physical defects which commonly occur in VLSI and of the resulting logical faults. In the past the stuck-at fault model has been widely used to model, at the logical level, the effects of physical defects in circuits. This model does not cover many of the possible defects in VLSI.[3,5,12] The fabrication flaws and physical processes that can cause malfunction of NMOS and CMOS VLSI circuits are summarized in this section.

VLSI chip failures may be caused by design or fabrication flaws, may be due entirely to environmental factors, or are the end result of a degenerative process due to operational and environmental stresses but partially attributable to design or manufacturing defects.[4,10] Fabrication defects in MOS chips consist mainly of shorts and opens in each interconnection level, shorts between different levels, and large imperfections such as scratches across the chip.[5] Other fabrication defects include incorrect dosage of ion implants, contact windows that fail to open, misplaced or defective bonds, and penetration of the package by contaminants.[4] During the operation of the chip, faults may be caused by electromigration, corrosion, electrical breakdown of oxide, cracks due to thermal expansion, power supply fluctuation, and ionizing or electromagnetic radiation.[4]

At the logical level, most of the faults can be represented in a circuit model that consists of a network of switches, loads (for NMOS), and interconnection lines which directly correspond to the transistors and interconnections in the actual circuit.[5] Most of the physical defects, such as opens and shorts, can be represented in this model in an obvious way.[3] A "switch" may be permanently on or permanently off, corresponding to a gate input stuck-at-1 or 0, respectively. Shorted NMOS loads (pullups) are equivalent to an output line s-a-1. Disconnected gate inputs are usually equivalent to s-a-0 or s-a-1 faults.

Some physical defects have a more complex effect on the circuit. In NMOS, incorrect dosage of ion implants may cause a threshold shift in a load transistor. This can result in an output voltage that lies between the voltages assigned to logic 0 and logic 1. If the fanout from the gate is greater than one, some of the gates connected to its output may "interpret" it as logic 1 while others will interpret it as logic 0. If, at some point in time (clock cycle), the line is supposed to be a logic 1 but is interpreted by some of the gates as logic 0, we call it a *weak 1* fault. Conversely, if the line is supposed to be a logic 0 but is interpreted by some of the gates as logic 1, we call it a *weak 0* fault. A single physical defect, resulting in a single weak 0 or weak 1 fault, has the same effect as multiple s-a-1 or s-a-0 faults, respectively.

In CMOS, a transistor which is permanently off or a break in a line can result in a high impedance state where the output of a combinational logic gate is dependent on the previous output rather than the current input.[12] Such a fault (called a *stuck-open* fault) may escape detection even if all possible input vectors are used to test the circuit.[12]

Implementation of Self-Testing Comparators in VLSI

The duplication and matching scheme relies entirely on a self-testing comparator to detect faults in the functional modules. Implementing such a comparator requires knowledge of how different faults will affect the circuit. Fortunately, a comparator is a simple circuit that can be implemented with a regular structure and is therefore amenable to thorough analysis. Hence, we can have confidence in our ability to predict the likely physical defects, develop a valid fault model, and prove that the implementation we propose is indeed self-testing.

We assume that physical defects in the node occur one at a time. A fault that is the result of a single physical defect is called a *single fault*. It is assumed that there is a negligible probability that the time interval between the occurrence of successive single defects in the comparator or between a single defect in the comparator and an arbitrary collection of defects in the functional modules, is less then some value T. In order to ensure that faults in the comparator will not mask future faults in the functional units, during normal operation, the comparator must "test itself" for any single fault in less than time T.

Single Stuck-At Faults

As a first step to constructing a comparator which is self-testing with respect to *any* single fault, we will discuss the implementation of a comparator which is self-testing with respect to any single *stuck-at* fault.

In this context "two-rail" codes prove useful. They consist of all words (bit vectors) such that a specified half of the word is the complement of the other half. If the output of one of the modules in a self-checking node is complemented, a two-rail code checker can serve as a "comparator" that checks the validity of the output. Such a checker, which is self-testing with respect to any single stuck-at fault, can be implemented as a two level NOR-NOR PLA (Fig. 2).[13,2] The output from the checker is a two-bit two-rail code that is 01 or 10 (*code output*) if the input is a two-rail code word (*code input*), and 00 or 11 (*noncode output*) otherwise (*noncode input*). It can be shown that if any single stuck-at fault exists in the checker, there is an input two-rail code word that results in a 00 or 11 output, thereby "detecting" the fault.[13]

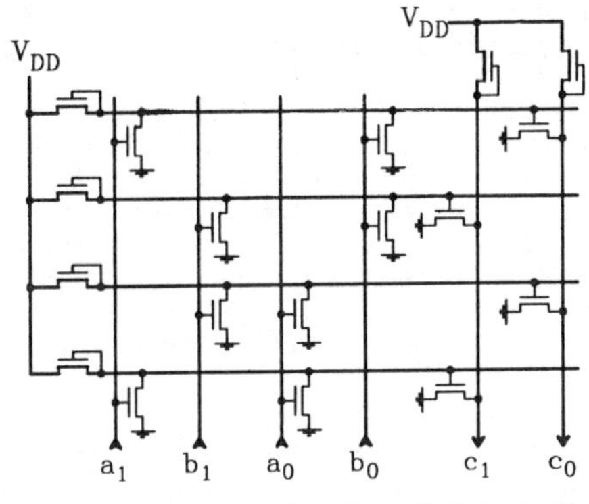

Fig. 2: *An NMOS Self-Testing Two-Rail Code Checker*

The requirement that the checker must be self-testing with respect to any single stuck-at fault poses severe constraints on its implementation. It can be shown that *any* two level AND-OR (or NOR-NOR) implementation for an input of $2n$ bits (n bits from each module) must use 2^n product terms, one for each code input.[11] If the output from each module is, say, 16 bits, this implementation is impractical since it requires $2^{16} = 65536$ product terms. Furthermore, all possible (2^n) code words must appear at the checker's inputs for it to perform a complete self-test.

Several small self-testing two-rail code checkers can be used as "cells" for constructing a self-testing checker for a wide input word (Fig. 3).[10,6] While the self-testing property is preserved, the number of input patterns required for a complete self-test is dependent only on the size of the largest "cell."[6]

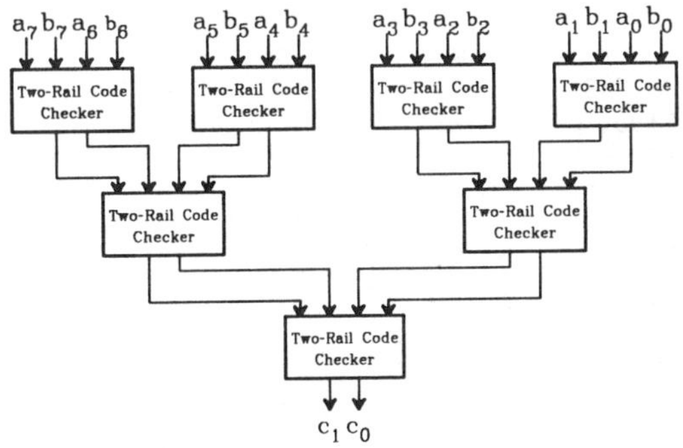

Fig. 3: *A Self-Testing Two-Rail Code Checker Tree*

Other Single Faults

The faults that commonly occur in a MOS PLA are stuck-at faults, shorts between adjacent lines, breaks in lines, and contact faults which include missing or extra devices at crosspoints.[13,7] In addition, weak 0/1 faults can occur on the input or product term lines. Fortunately, it turns out that a straightforward NOR-NOR PLA implementation of the checker discussed above is self-testing with respect to any one of the aforementioned single faults. The rest of this section contains an informal "proof" of this claim; a more formal proof will be presented elsewhere.[11] Faults in the input lines, product term lines, output lines, AND array crosspoints, and OR array crosspoints, are considered separately.

Any single stuck-at fault or short in the input lines will cause one or more 0's to change to 1's *or* one or more 1's to change to 0's (but not both) for some code input. It can be shown that such an error (called a *unidirectional error*[7]) on the input lines results in noncode output.[13] A break in the input line outside the AND array is equivalent to the line stuck-at-0 or stuck-at-1. A break in the middle of the AND array affects only some product terms. For an affected product term, if the break is equivalent to a stuck-at-1, the one code input that is supposed to select this product term won't, and a noncode output will result. If the break is equivalent to a stuck-at-0, there exists a code input that results in a noncode output since it selects two product term lines each of which is connected to a different output line.[11]

An extra device in the AND array is equivalent to the corresponding product term stuck-at-0. The code input that is supposed to select that product term results in a noncode output. If there is a missing device in the AND array, there exists a code input that produces a noncode output since it selects two product term lines, each of which is connected to a different output line.[11]

An extra device in the OR array means that one of the product terms is connected to both outputs. A missing device in the OR array is equivalent to the corresponding product term stuck-at-0. In either case, the code input that selects the relevant product term will result in a noncode output.

If the output lines are shorted, their values are equal and that is a noncode output. If one of the lines has a stuck-at fault, there exists a code input that causes the other line to have the same value so the output is noncode. For some code input, a break in one of the output lines is equivalent to a stuck-at-1 or stuck-at-0 fault on that line.

A stuck-at-0 fault on a product term line will result in a noncode output if the input is the code word that is supposed to select that product term line. A stuck-at-1 fault on a product term line will result in a noncode output to any input that selects a product term line that is connected to the other output line. A break in a product term line is equivalent to a stuck-at fault on that line since each product term line is connected to only one output line. A short between two product term lines will result in a noncode output if the input selects either one of these lines.[11]

Product term lines are not susceptible to weak 0/1 faults since each product term line is connected to only one output line (fanout of one) so that a weak 0/1 fault is equivalent to a *single* stuck-at fault. Input lines have a fanout greater than one and are thus susceptible to weak 0/1 faults. A weak 1 fault on an input line is equivalent to one or more missing devices in the AND array. Each product term that is connected to a "missing device" will be selected by an input code word that also selects a product term line that is connected to the *other* output line.[11] Thus, a noncode output will result. A weak 0 fault on an input line is equivalent to one or more product term lines which are stuck-at-0. Any code input that is supposed to select one of these product terms will result in a noncode output.

In CMOS chips PLAs are usually implemented in dynamic "pseudo NMOS."[12] All product term and output lines are precharged during every clock cycle before being selectively discharged according to the input. Therefore no state is preserved from one cycle to the next and the circuit is combinational despite any opens in the precharge or discharge paths.[11] Hence the PLA used in CMOS chips is only susceptible to the same faults as the traditional static PLA used in NMOS chips.

This analysis shows that for all single faults in our fault model, there exists a code input that results in a noncode output from the proposed two-rail code checker PLA. Thus, the checker is self-testing with respect to any likely single fault. Based on this result, it can be shown that the checker constructed as a tree of smaller self-testing checkers (Fig. 3) is also self-testing with respect to any likely single fault.[11]

Summary and Conclusions

We have presented an approach to increasing the reliability of future high-end systems beyond what is possible with technological solutions alone. The system consists of computation nodes and communication nodes, interconnected by high-speed dedicated links. These components are relied upon to detect errors while system level protocols are used for error recovery and reconfiguration.

The use of duplication and matching for implementing the self-checking nodes allows us to restrict a detailed analysis of the impact of all possible faults to the comparator, which is a relatively simple circuit. We have shown that the self-testing comparator, which is the backbone of our approach, can be implemented with NMOS and CMOS technologies.

Acknowledgements

We would like to thank Richard Fujimoto, Manolis Katevenis, and Robert Sherburne for reviewing a draft of this paper.

This research was supported by the State of California MICRO program and the Defense Advance Research Projects Agency (DoD), ARPA Order No. 3803, monitored by Naval Electronic System Command under Contract No. N00039-81-K-0251.

References

1. T. Anderson and P. A. Lee, "Fault Tolerance Terminology Proposals," *FTCS12*, pp. 29-33 (June 1982).
2. W. C. Carter and P. R. Schneider , "Design of Dynamically Checked Computers," *IFIPS Proc.*, pp. 878-883 (August 1968).
3. B. Courtois, "Failure Mechanisms, Fault Hypotheses and Analytical Testing of LSI-NMOS (HMOS) Circuits," pp. 341-350 in *VLSI 81*, ed. J. Gray, Academic Press (1981).
4. E. A. Doyle, "How Parts Fail," *IEEE Spectrum* **18**(10) pp. 36-43 (October 1981).
5. J. Galiay, Y. Crouzet, and M. Vergniault, "Physical Versus Logical Fault Models MOS LSI Circuits: Impact on Their Testability," *IEEE-TC* **C-29**(6) pp. 527-531 (June 1980).
6. J. Khakbaz and E. J. McCluskey, "Concurrent Error Detection and Testing for Large PLA's," *IEEE JSSC* **SC-17**(2) pp. 386-394 (April 1982).
7. G. P. Mak, J. A. Abraham, and E. S. Davidson, "The Design of PLAs with Concurrent Error Detection," *FTCS12*, pp. 303-310 (June 1982).
8. R. A. Rasmussen, "Automated Testing of LSI," *Computer* **15**(3) pp. 69-78 (March 1982).
9. C. H. Séquin and R. M. Fujimoto, "X-Tree and Y-Components," in *Proc. Advanced Course on VLSI Architecture*, Univ. of Bristol, England, ed. P.C. Treleaven, Prentice Hall (1982).
10. D. P. Siewiorek and R. S. Swarz, *The Theory and Practice of Reliable System Design*, Digital Press (1982).
11. Y. Tamir, "Fault Tolerance for VLSI Multicomputers," Ph.D. Dissertation (in preparation).
12. R. L. Wadsack, "Fault Modeling and Logic Simulation of CMOS and MOS Integrated Circuits," *BSTJ* **57**(5) pp. 1449-1474 (May-June 1978).
13. S. L. Wang and A. Avizienis, "The Design of Totally Self Checking Circuits Using Programmable Logic Arrays," *FTCS9*, pp. 173-180 (June 1979).

Session 3-2a
Josephson II

Chairman

L. Terman
IBM Research

A NOVEL APPROACH FOR DESIGNING JOSEPHSON PLA

by M.Morisue, K.Isaji, H.Fukuzawa and Z.H.Shaikh

Department of Electronic Engineering, Saitama University, Urawa, 338
Japan

ABSTRACT This paper describes a computer-aided approach for designing a very high performance logic system using Josephson PLA. A novel method to realize logical function with Josephson PLA is carried out successfully.

In this paper we describes two type of Josephson PLA, one of which is CIL PLA and the other is interferometer PLA. The principle of these devices is described and simulation results are verified. Especially the speed power product is reduced to 84fJ/bit for the adder.

Introduction

The Josephson PLA has a lot of flexibility and a capability of ultrafast realization of any logic system by only changing the connection of array. Therefore, Josephson PLA is treated here for VLSI technology to obtain fast and high density logical system.

So far very less work has been reported to realize the potentials of JPLA. Farris suggested the possibility of PLA but did not give the details of the circuit [1]. Therefore we propose a novel and acceptable model of JPLA for applications.

In this paper two types of the PLA [2] are explained, one of them is constructed by using CIL gates [3] and the other is interferometer gates [4]. In former, the output current of CIL gate at one stage is directly applied to the input of next CIL gate in series to perform a AND operation. In latter, the interferometer is used to perform a AND operation at each mesh point. The principle of these PLAs are based on mesh type decoder [2].

As an example of application of these PLAs, the simulation of 25x14 matrix PLA employing CIL gates is made with speed of 17ps/bit. The simulation results of 4bit adder using interferometer PLA shows the speed of 28ps/bit and power consumption of 3μw/bit.

JPLA using CIL

The fundamental construction of the PLA using CIL is shown in Fig.1. In this circuit the symbol ○ indicates a Josephson junction. The input signals of X_n are supplied to the one input terminal of CIL gates through the resistance R_n. When a gate current Y is applied to the CIL gate, the gates turn to the voltage state and produce a transmitting signal. If no input signal X_n is present, the transmitting signal stops there. Thus the function of PLA is achieved. The AND operation of CIL gate is explained by threshold characteristics as in Fig.2. The solid line indicates dynamic curve and dotted line static one. The shorter is the rise time of operating signal, the closer is the curve to the solid line. When input signal X(control current) and Y(gate current) are applied, the initial point O(superconducting state) moves to A(voltage state), hence CIL gates are switched and have achieved AND operation. The directionality of signal transmitting towards output terminal of PLA is obtained by introducing the decoupling coil into the output line of CIL gate. Fig.3 shows the basic model for PLA to achieve a 4bit adder, where the symbol ○ indicates the CIL gate. This is consisted of 25x14 matrix AND operation circuit. The set of differential equations for simulation are realized from the model shown in Fig.4(a). Using the equivalent circuit of Josephson junction as shown in Fig.4(b), these equations become

$$I_1=I_2+i_1+i_2 \tag{1}$$

$$\frac{dI_2}{dt}=\frac{-I}{L_2}(v_2-v_1+r_2I_2) \tag{2}$$

$$\frac{d\theta_{11}}{dt}=\frac{2e}{\hbar}v_{11},\quad \frac{d\theta_{12}}{dt}=\frac{2e}{\hbar}v_{12} \tag{3}$$

$$\frac{dv_{11}}{dt}=\frac{1}{C_{11}}(i_1+\frac{(v_1-v_2)}{R_D}+I_a-\frac{v_{11}}{v_{R11}}-I_{01}\sin\theta_{11}) \tag{4}$$

$$\frac{dv_{12}}{dt}=\frac{1}{C_{12}}(i_2+\frac{(v_2-v_1)}{R_D}-\frac{v_{12}}{v_{R12}}-I_{02}\sin\theta_{12}) \tag{5}$$

$$\frac{di_1}{dt}=\frac{L_{12}}{L_{11}L_{12}+L_{11}L_L+L_{12}L_L}[r_1(I_1-i_1-i_2)+L_L\frac{dI_1}{dt}+v_2-v_{11}-\frac{L_L}{L_{12}}(v_{11}-v_{12})] \tag{6}$$

$$\frac{di_2}{dt}=\frac{L_{11}}{L_{11}L_{12}+L_{11}L_L+L_{12}L_L}[r_1(I_1-i_1-i_2)+L_L\frac{dI_1}{dt}+v_2-v_{12}-\frac{L_L}{L_{11}}(v_{12}-v_{11})] \tag{7}$$

where θ_{mn} is the difference of phase angle of Josephson junction, v_{nm} is the voltage across the junction of J_{nm}, C_{nm} is the capacitance of the junction J_{nm}, v_n is the voltage at the central point of the junction J_{nm}.

For simulation the critical current and capacitance of the junction in the left hand of the CIL gate are 50μA and 0.2pF, and the right hand is three times of the left. Fig.5 shows the results of simulation when the input signals, $a_1=a_3=a_6=a_9=a_{12}=1$ and others at 0, are applied to PLA of Fig.3. It is seen from these results that the propagating delay time is about 17 ps/gate and power consumption is about 7 μw/gate.

CH1935-6/83/0000/0566$01.00

JPLA using interferometer

The most acceptable JPLA for application can be constructed with interferometer. As an example, a 4bit adder is described. Fig.6 shows the model of 4bit adder, which consists of driver and PLA. The driver outputs are NAND(X+Y), Equivalence(X=Y) and OR(X+Y) for two inputs of X and Y. Fig.7 shows the driver circuit using interferometers. The operation of 4bit adder is represented in mathematical terms from these circuit as follows

$$S1=(X_1=Y_1)\cdot C_0+(X_1+Y_1)\cdot \overline{C_0}$$

$$S2=(X_2=Y_2)\cdot X_1Y_1+(X_2=Y_2)\cdot (X_1+Y_1)\cdot C_0 +(X_2\oplus Y_2)\cdot \overline{X}_1\overline{Y}_1+(X_2\oplus Y_2)\cdot (\overline{X}_1+\overline{Y}_1)\cdot C_0$$

....

$$C4=X_4Y_4+(X_4+Y_4)\cdot X_3Y_3+(X_4+Y_4)\cdot (X_3+Y_3)\cdot X_2Y_2 +(X_4+Y_4)\cdot (X_3+Y_3)\cdot (X_2+Y_2)\cdot X_1Y_1 +(X_4+Y_4)\cdot (X_3+Y_3)\cdot (X_2+Y_2)\cdot (X_1+Y_1)\cdot C_0 \quad (8)$$

Fig.8 shows the model of PLA. The threshold characteristics of the interferometer are shown in Fig.9. A bias current is supplied to the interferometer and the operating point moves to O. A control pulse from driver moves the operating point O to A in zero-resistance state. At this stage when clock pulse to the interferometer is applied, the operating point moves to B and produces voltage and gate current propagates to the next stage. When no presence of signal from the driver, the operating point moves to C to stay in superconducting state even if the signal from the previous stage is applied. A high speed configuration of this PLA is achieved because the connection between interferometers are directly coupled.

Fig.10 shows the results of simulation for the model of 4bit adder as shown in Fig.6. All inputs for the adder are at level 1, and the adder performs the operation of (1111)+(1111)+(1). It is clearly seen that the adder can achieve the add operation with the high speed of 28ps/bit and power consumption of 3μw/bit.

Conclusion

We have proposed a novel Josephson PLA. The procedure to design the JPLA using CIL gate and interferometer are described. As an example of JPLA, a 4bit adder using interferometer is successfully simulated. The presented model can be expanded to a large model with numbers of bit to add and may perform any other logical function with high speed and reliability.

Acknowledgements

The authors are grateful to professor Y. Komamiya for his continuous encouragement and to Mrs. Kaminuma and Mr. Iizuka for their contributions to prepare this paper.

REFERENCES

[1] S.M.Faris, "Josephson LSI circuits", Proc. of ICCC 80, Vol 2, pp. 1196-1201, 1980.

[2] M.Morisue and K.Isaji, "A decoder using Josephson junction", 1982 Applied Superconductivity Conference(Preliminary), pp.52, Nov. 1982.

[3] T.R.Gheewala, "Design of 2.5-Micrometer Josephson Current Injection Logic (CIL)", IBM Jour. of R & D, Vol 24, No. 2, pp. 130-142, 1980.

[4] P.Gueret, A.Moser and P.Wolf, "Investigations for a Josephson Computer Main Memory with Single-Flux-Quantum Cells", IBM Jour. of R & D, Vol 24, No. 2, pp. 155-166, 1980.

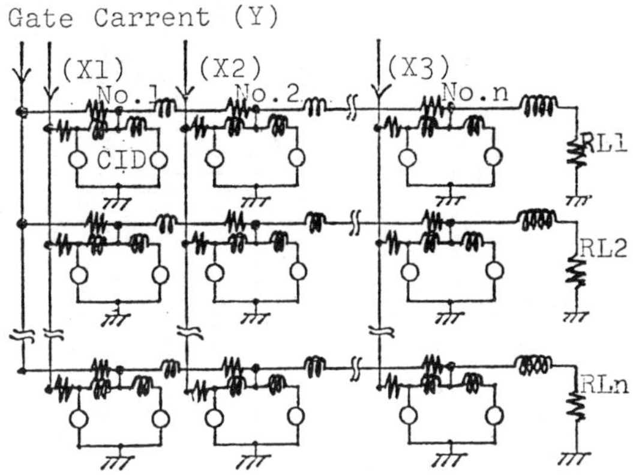

Fig. 1 Model of Josephson PLA using CIL

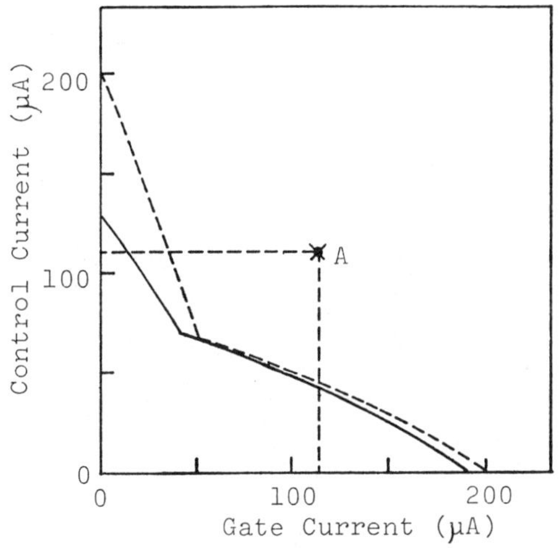

Fig. 2 Dynamic threshold characteristics of CIL gate

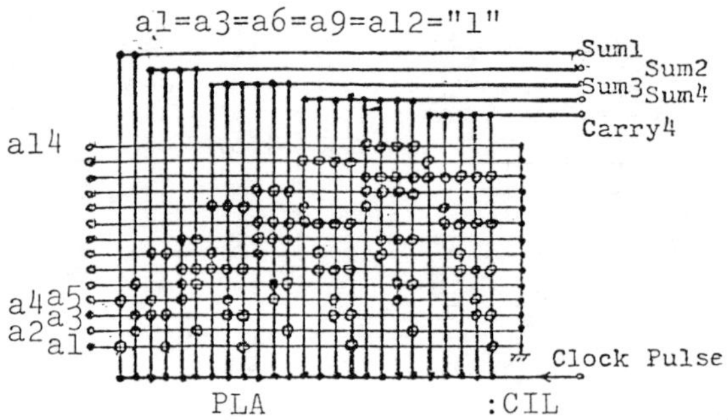

Fig. 3 Model of PLA using CIL gates

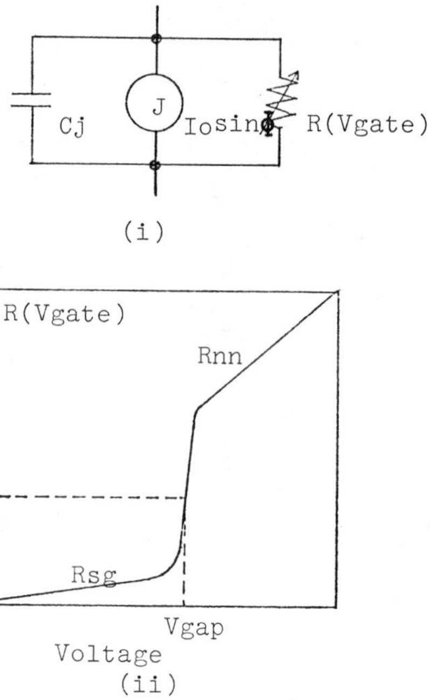

Fig. 4(b) Equivalent circuit of Josephson junction

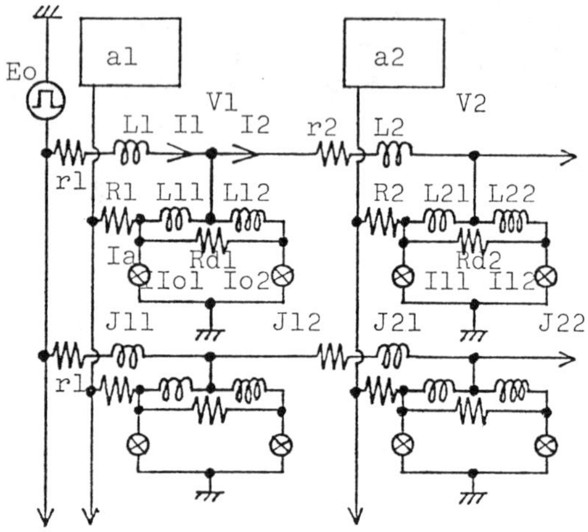

Fig. 4(a) Circuit of JPLA using CIL gates for simulation

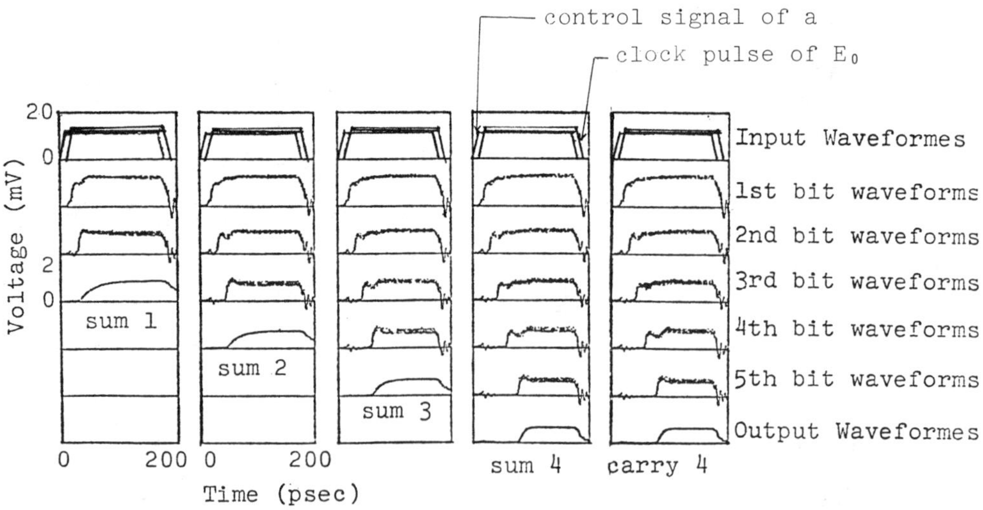

Fig. 5 Results of simulation show how the output voltage waveforms are obtained

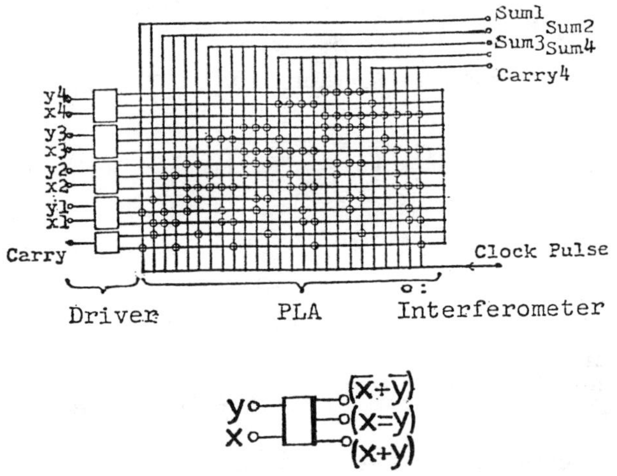

Fig. 6 Model of 4bit Adder

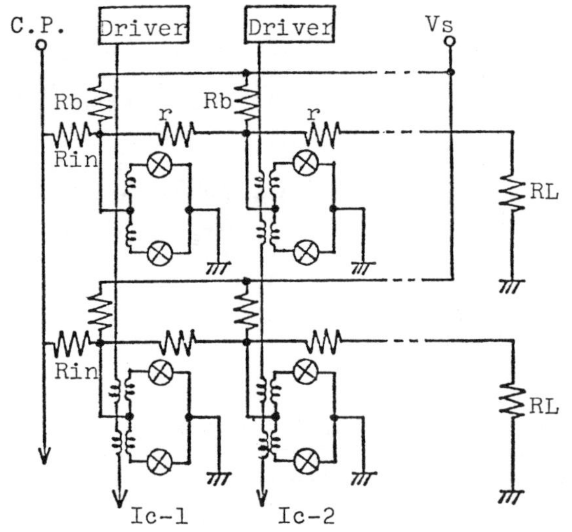

Fig. 8 Model of PLA using Interferometer

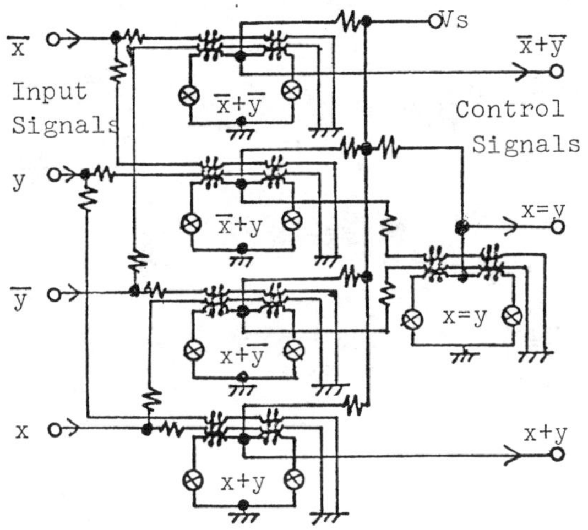

Fig. 7 PLA Driver Circuit using Interferometer

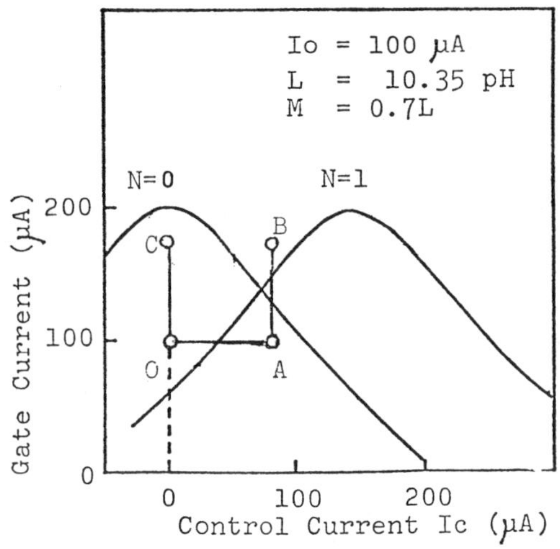

Fig. 9 Threshold characteristics of Interferometer

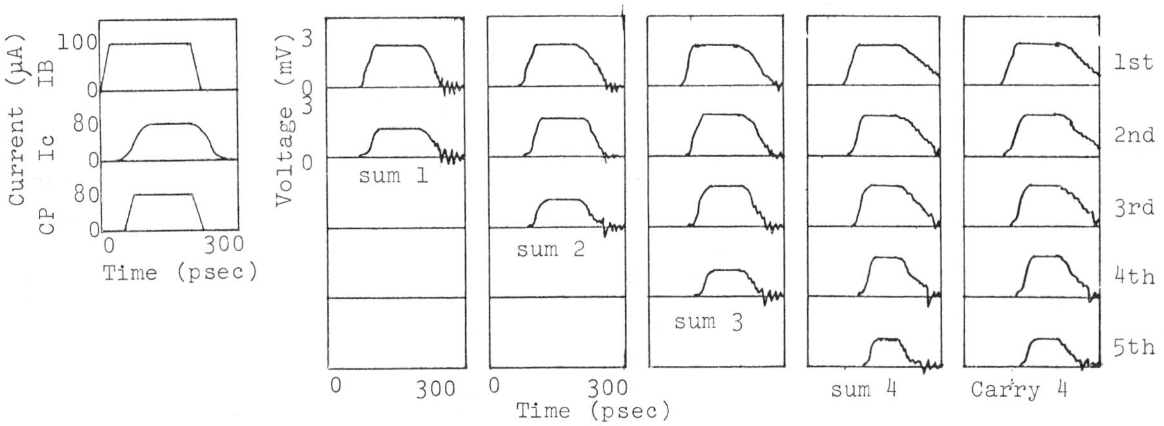

Fig. 10 Results of simulation

EXPERIMENTS ON A CROSS-SECTION OF A JOSEPHSON RAM CHIP

W.H. Henkels, K.H. Brown, T.V. Rajeevakumar, L. Geppert, J.W. Allan, Y.H. Lee and J.T. Yeh

IBM T.J. Watson Research Center, Yorktown Heights, NY

Abstract

A test vehicle intended to demonstrate individual and collective operation of all components of a Josephson NDRO RAM chip was designed, fabricated, and tested.

The experiments utilized a Pb-alloy technology incorporating a 2.5 μm minimum linewidth and 1.25 μm minimum overlay registration. The 6.35 mm x 6.35 mm test chip consisted of a cross-section of a 1k bit array, a cross-section of timing and reset circuits, individual sub-components, individual gates, and electrical parameter test sites.

A synthesis of all the results is consistent with a sub-nanosecond access time for 4k bits on a 7.1 mm chip.

Introduction

High speed memory circuits for the Josephson technology have been under development for several years.[1] Most initial experiments dealt with the properties of individual memory cells. It is only recently that attention has turned toward the development of complete VLSI memory chips.[2,3,4] Most recently the operation of a fully decoded 1K RAM built in a 5 μm Pb-alloy technology has been reported.[5]

The present paper reports the results of experiments on a 2.5 μm Pb-alloy technology 1Kx4 bit design. The objectives were to demonstrate individual and collective operation of all components of the design. The circuits were found to operate as predicted by circuit models, and a cross-section of a fully decoded 1K array was partially operated. The results are consistent with the design objective of a 4k bit chip with a sub-nanosecond access time. The intent of this paper is to give an overview of results; many details will be reserved for a future extended article.

Design & Fabrication

Much of the basic component design has already been described in a previous publication.[4] The present design differs primarily in some details of the sense bus, and in the decoder damping. Also, all double gates in the decoder address loops have been replaced by single gates.

The memory cell layout is shown in Fig.1(a), and its electrical equivalent circuit in Fig.1(b). The cell consists of a superconductive storage loop designed to store two quantum units of magnetic flux ($2\Phi_0$) for a binary "1" and zero flux for a "0". The cell vertical structure and operation are nearly identical to that of a non-optimized $1\Phi_0$ experimental cell previously reported.[6] The layout of the present design differs somewhat, in order to minimize array-line to storage loop coupling and to optimize the margins of the two-junction interferometer sense gate. For the present vertical structure a $2\Phi_0$ design is preferred over a similarly optimized $1\Phi_0$ design because it allows a smaller cell size (~2x) and supplies a more easily detected current to the sense bus. The cells are bit-organized in an array, and thus require both X and Y decoders. Reading requires activation of S and Y array lines; writing a "1" requires X,Y' and Y; writing a "0" requires only X and Y'.

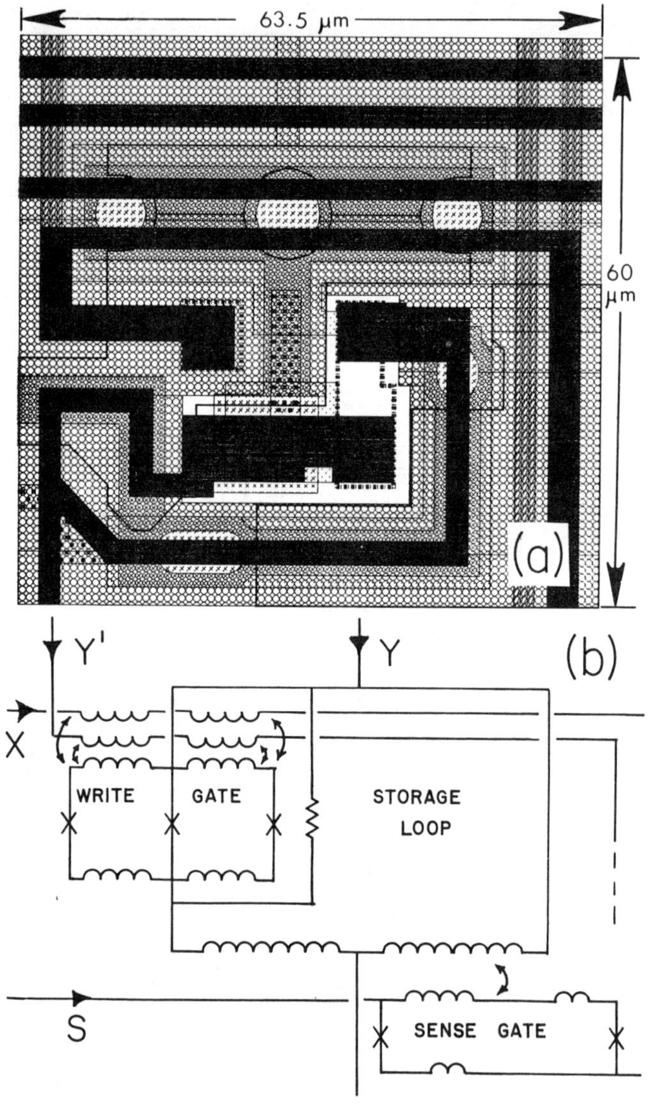

Fig.1. (a) Layout of the memory cell. (b) Electrical equivalent circuit of the cell.

Timing and resetting functions are controlled by triangular-shaped pulses which propagate along matched transmission lines of characteristic impedance 15 Ω. The pulses are initiated by self-resetting pulse generators. This timing scheme differs significantly from schemes which employ current transfer into superconducting loops[2,5] and is aimed at achieving very high speeds. It is readily extended to smaller dimensions ; its primary cost is that it requires good control of fabrication tolerances.

Fig.2 shows the experimental chip created to evaluate the design. The 6.35mm chip consists of circuits of varying complexity. At the low end of the hierarchy are individual gates and electrical parameter test sites. At the next level of complexity are individual components: isolated cells, a sense driver circuit, a decoder address loop, individual decoder loops, a sense bus loop, pulse generators, and separate timing and reset circuits. At the highest level of complexity is a cross-section of a 1K array. The upper photograph in Fig.2 shows some details of the array. The array is a cross-section in that 1) intentionally, not all cells are accessible, 2) all driver supply strings are individually powerable, and 3) timing and resetting for the array are externally controlled. Also, Josephson samplers[7] and monitor gates, as well as some bypass lines have been inserted, for monitoring and troubleshooting.

Fabrication was done with a liftoff process Pb-alloy technology,[8] incorporating a minimum linewidth of 2.5 μm and minimum overlay registration of 1.25 μm. The target Josephson current density and resistor sheet resistivity were 850 A/cm^2 and 2Ω/square respectively. Junction sizes in the design ranged from 2.5 μm to 7.1 μm in diameter. The chief fabrication challenges are the long lengths of 2.5 μm lines on a 5 μm pitch and the large area of insulation required over a complex topology.

At the time the design of the experimental chip was ~90% complete, the Josephson program at IBM changed over to a niobium edge-junction technology,[9] in order to improve the long term reliability of the junctions. The decision was made to complete the Pb-planar design, fabrication, and evaluation, because most of the circuit concepts are independent of the particular technology. However, because of the change of emphasis to Nb edge-junctions only a single fabrication run was made. Thus the results presented here correspond to a single design iteration and a single fabrication run.

PARAMETER	EXPERIMENT	NOMINAL DESIGN
R_d	3.3 Ω	2.9 Ω
L_{loop}	32.4 pH	33.1 pH
K	0.6	0.5
I_Y^{nom} $(2\Phi_o)$	0.23 mA	0.25 mA
S-line-to-cell coupling	4.5%	$\lesssim$ 1%
Y'-line-to-cell coupling	1%	1%
X-line-to-cell coupling	<< 1.5%	< 1%

Table I. Comparison of measured memory cell parameters with the nominal design.

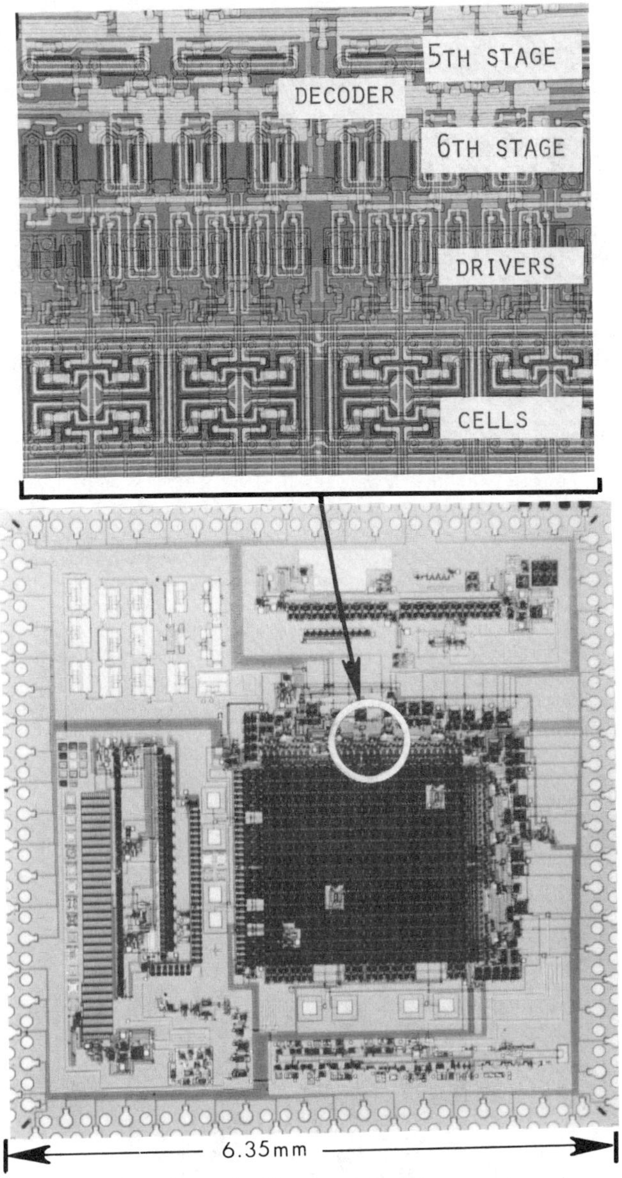

Fig.2. Experimental chip containing components, sub-circuits, and a 1k bit array cross-section. Upper photograph shows more detail of the array.

Experimental Results

Gates and Parameters

Measured gate characteristics were close to their nominal design specifications. The gate parameters of most importance are the mutual inductance between control line and gate, L_m, and the maximum Josephson current with no control current, $I_m(0)$. For the thirteen different gate designs required, the measured L_m's differed from their nominal design values by as little as <1%, for the write gate and decoder gate, to ~20% for one of the large interface gates. Other significant inductances deviated from their target values by ~0-10%. Experimentally required current levels were as close as -4% to the nominal design for some gates and -27% low in the worst cases. The measured on-chip spreads in $I_m(0)$ were typically <±6%. Resistors were within 15% of their target.

Measured memory cell parameters were also close to their nominal design values. Table I summarizes a comparison, where

R_d is the cell damping resistance, L_{loop} is the storage loop inductance, k is the loop inductance symmetry factor (k=0.5 means equal branch inductances), and I_Y^{nom} is the midrange Y operating current for $2\Phi_0$ operation. This degree of agreement was achieved without a prior iteration. Clearly with further iterations k and the S-coupling could be made to converge closer to the design target.

Circuits

The proper functional operation of all individual components was achieved. Fig.3 shows the waveforms for proper slow-speed functional operation of an isolated memory cell. Not shown here, but also observed, was proper operation under all half-select conditions. Through knowledge of the sense gate threshold characteristic the operating regions for 1,2,3 and $4\Phi_0$ operation were measured as a function of cell damping. The size of the observed regions agreed well with dynamic simulations of the cell write operation.

In the simplest test of the decoding principle a chain of two decoder loops driven by a pulse generator was shown to operate properly with the dc power varied by ±30%. In the most complex test, proper functioning of a full 6-bit decoder of an array was observed for several addresses. In this case the operation was monitored by individual monitor gates at the decoder outputs.

The sense bus was operated in the array environment, with an S-driver current range of ±22% and sense bus current range of ±25%. Proper functioning implied proper collective operation of the decoder, S-driver and S-array line current transfer, sense bus transformer, and the sense bus gate. The waveform of the sense current was measured on an isolated sense line via Josephson sampling. Current transfer into the line is shown in Fig.4(a); transfer out in Fig.4(b). Fig.4(c) shows a simulation of the nominal design. The agreement between the nominal simulation and the experiment is good. The peak amplitude of the oscillation is ±5%, and the time constant of the damping of the oscillation 0.5ns in both cases. The period of the oscillation and the 10-90% rise- and fall times of the experimental waveform were 100ps; the

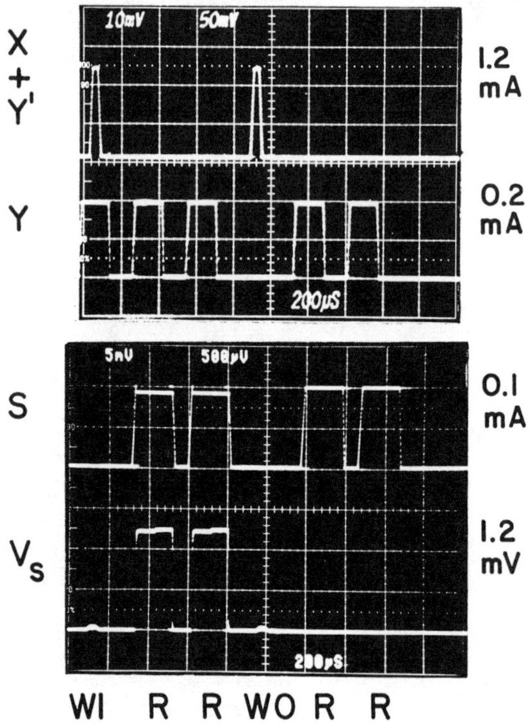

Fig.3. Waveforms showing proper quasi-static operation of an isolated memory cell.

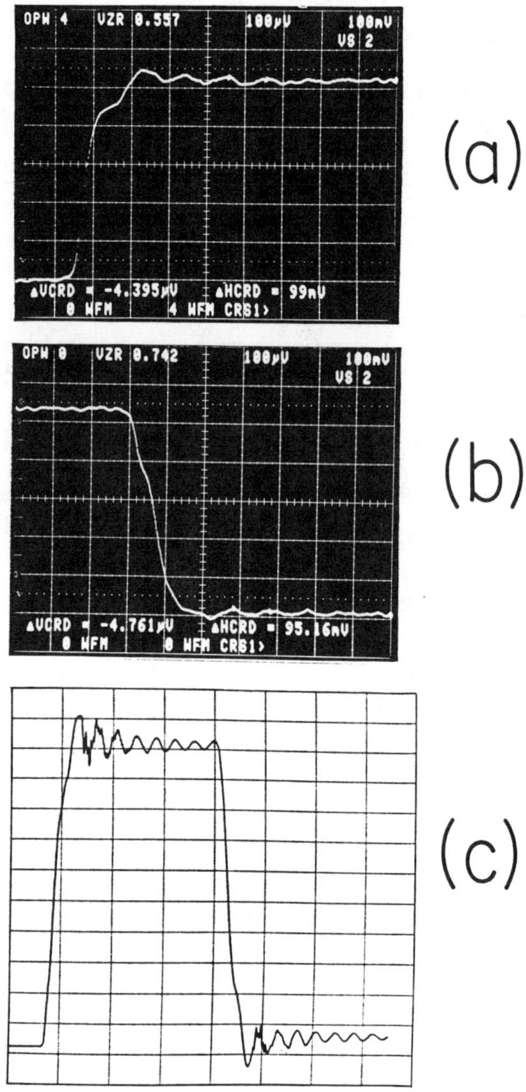

Fig.4. (a) & (b) Experimental sense line current waveform: vertical 29 μA/div.; horizontal 100 ps/div. (c) Simulation of the nominal design: vertical 20 μA/div.; horizontal 200 ps/div.

corresponding numbers in the simulation were ~80ps and 90ps. The chief difference from nominal is the amount of current transferred, which was ~9% low and is most probably due to differences between the experimental and nominal damping resistance in this first pass design and fabrication.

Similar agreement was observed for decoder address loop waveforms. In that case the current transferred was ~5-10% higher than nominal.

Longest Asynchronous Path Delay

In the course of these experiments no perfect arrays were found; no chips had X and Y decoders both functional. Consequently a complete access time measurement could not be made. However, it was found possible to partially operate some arrays and possible to measure the longest asynchronous delay through an array, for several addresses. This function requires operation of a logic-to-memory interface pulse generator, the X-decoder, a sense driver and sense line, the sense bus, and an output gate. A schematic of the measurement is shown in Fig.5. A decode start pulse simultaneously triggers the X-interface and the decode start monitor gate, A. Traversal of the X-signal through a decoded sense line eventually switches the memory-to-logic output gate, B. The difference in the arrival times at A and B is monitored by the output monitor gate. A magnification of a portion of the output waveform is shown in the photograph in Fig.5, which for this path indicates a delay of 230ps. Several precautions were taken to insure that the uncertainty in this measurement is (conservatively) $\sim\pm 50$ps. For five different addresses the mid-current-range value of this delay ranged from 180ps to 240ps. This delay is consistent with simulations of the interface, decoding, sense line, and sense bus delays.

Synthesis

The functional operation of all timing and reset circuits including the self-resetting pulse generators was observed in circuits separate from the array. The sampled waveforms and delays measured for the pulses and delay line were found to be in reasonably good agreement with the nominal design.

The longest asynchronous path delay discussed above, ~240ps, is the largest component of access time for the 1K array. Additional components, and simulation estimates, or separate experimental measurements of their values follow: 1) the time required for the interface gate to reach threshold, ~30ps, based on simulation; 2) the first stage address loop delay, ~80ps, based on measurements; 3) current transfer out of the sense line, ~150ps, based on the S-waveform measurements; and 4) S-Y timing safety required to prevent latch-up of the sense line, ~0-200ps. Item 4) was not measureable in our experiments, but at worst is the time required for the sense driver to reset to zero voltage after completion of current transfer into the sense line. Thus a synthesis of the experiments yields an access time of 700ps for the 1K array. Based on the size of the 1K layout and estimates for the space required for dc power, timing and reset circuits, we estimate that a 4x1k bit memory would fit on a 7.1mm square chip. For this size chip an additional ~100ps wiring delay would be required to route the address and data input lines to the four arrays. Thus a 4x1k bit, 7.1mm chip would have an access time of ~800ps.

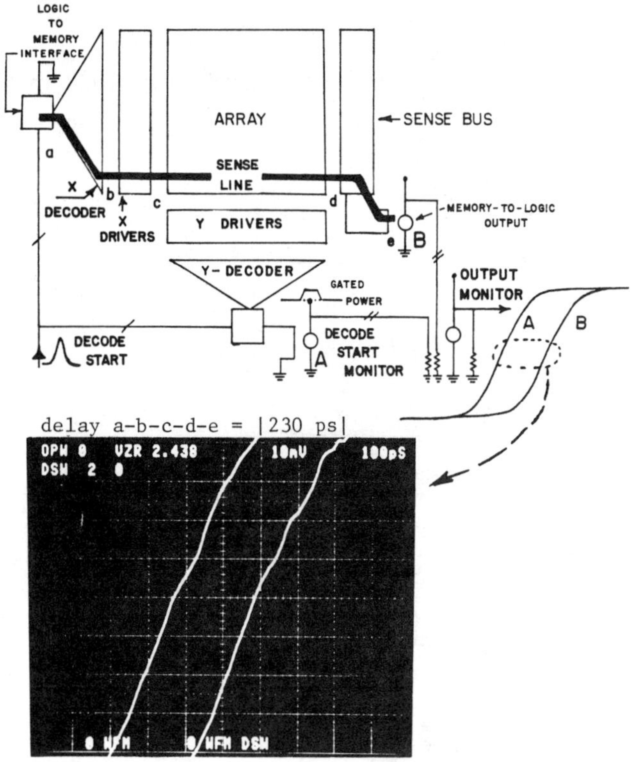

Fig.5. Schematic of the longest asynchronous path delay measurement. The photograph displays the measured delay for one address.

Acknowledgements

The authors wish to acknowledge and thank several people instrumental to the success of these experiments: C.J. Anderson for valuable assistance in the delay and waveform measurements; D.J. Herrell and H.H. Zappe for considerable technical input, guidance, and support; W.H. Chang for help in wiring and checking the layouts; J.H. Greiner for advice on the fabrication; E.E. Castellani, S. Renick, T. Ross, and R. Bickford for technical assistance in the fabrication; W.G. Thorn, Jr. and J.F. Smith for help in shipping the mask data; and J.C. Logue for his support.

References

[1] H.H. Zappe, "Memory-cell design in Josephson technology", IEEE Trans. on Electron Devices, vol. ED-27, no.10, pp. 1870-1882, October 1980.

[2] R.F. Broom, P. Gueret, W. Kotyczka, T.O. Mohr, A. Moser, A. Oosenbrug, and P. Wolf, "Model for a 15ns 16K RAM with Josephson junctions", IEEE J. Solid-State Circuits, vol. SC-14, p. 690, 1979.

[3] W.H. Henkels, "Fundamental criteria for the design of high-performance Josephson nondestructive readout random access memory cells and experimental confirmation", J. Appl. Phys., vol. 50, no. 12, pp. 8143-8168, December 1979.

[4] S.M. Faris, W.H. Henkels, E.A. Valsamakis, and H.H. Zappe, "Basic design of a Josephson technology cache memory", IBM J. Research and Development, vol.24, no.2, pp. 143-154, March 1980.

[5] M. Yamamoto, Y. Yamauchi, K. Miyahara, K. Kuroda, F. Yanagawa, and A. Ishida, "An experimental nanosecond Josephson 1K RAM using 5 μm Pb-alloy technology", IEEE Electron Device Letters, vol. EDL-4, no. 5, pp. 150-152, May 1983.

[6] W.H. Henkels and J.H. Greiner, "Experimental single flux quantum NDRO Josephson memory cell", IEEE J. Solid-State Circuits, vol. SC-14, no. 5, pp.794-796, October 1979.

[7] Sadeg M. Faris, "Generation and measurement of ultrashort current pulses with Josephson devices," Appl. Phys. Lett., vol. 36. no. 12, pp. 1005-1007, June 1980. D. B. Tuckerman, "A Josephson ultrahigh-resolution sampling system", Appl. Phys. Lett., vol. 36, no. 12, pp. 1008-1010, June 1980.

[8] J.H. Greiner, C.J. Kircher, S.P. Klepner, S.K. Lahiri, A.J. Warnecke, S. Basavaiah, E.T. Yen, J.M. Baker, P.R. Brosious, H.-C.W. Huang, M. Murakami, and I. Ames, "Fabrication process for Josephson integrated circuits", IBM J. Research and Development, vol. 24, no. 2, pp. 195- 205, March 1980.

[9] R.F. Broom, A. Oosenbrug, and W. Walter, "Josephson junctions of small area formed on the edges of niobium films", Appl. Phys. Lett., vol. 37, p. 237, 1980.

Miniaturization of Josephson Current Injection (CIL) Logic Circuits

H. Ko and T. Van Duzer

Department of Electrical Engineering and Computer Science
Electronics Research Laboratory, University of California
University of California
Berkeley, CA 94720

ABSTRACT

Projections of the performance of current injection logic (CIL) fabricated with 0.2 μm minimum feature is presented. A computer model is used to optimize the performance of the circuits. The results showed that the gate delay and the propagation delay are improved by almost an order of magnitude over 2.5 μm circuits, but the crossing delay improved less and the overall speed improvement is a factor of three. The number of logic gates on a chip is increased by over an order of magnitude. Gate count is further increase by a factor of about two by addition of a top ground plane, with negligible effects on speed. The calculations are made assuming lead-alloy technology but the principal results apply to niobium circuits as well.

1. Introduction

Josephson logic circuits fabricated with 2.5 μm minimum features have been shown to have switching speeds on the order of 10 ps, with a power dissipation of less than 10 μW. The power-delay product is at least two orders of magnitude less than current semiconductor devices, even those with 0.3 to 0.5 μm minimum features. Projections of the performance of Josephson circuits indicate that improvements can be made by scaling down the device sizes. Miniaturization of Josephson logic technology involves scaling down an entire logic family. We present here a projection of the performance of the miniaturized current injection logic family (CIL) with 0.2 μm minimum features.

2. CIL Circuits

The CIL logic family consists of two basic devices [1]: a three-junction magnetically coupled interferometer and an two-junction asymmetric current-injection gate. The equivalent circuit of an interferometer is shown in Fig. 1. This device is biased at a gate current $I_g < I_m$, where I_m is the critical current of the interferometer with zero applied control current. When a control current I_c is applied, the interferometer switches from the superconducting state to the voltage state, and the gate current is directed into the load resistor R_L. This device is usually made with two control lines and can be used as a two-input OR gate. The equivalent circuit of the other device, the injection gate, is shown in Fig. 2. Both $I_A < I_{01} + I_{02}$ and $I_B < I_{01} + I_{02}$ must be present for the gate to switch into the voltage state. Thus the injection gate functions as a two-input AND gate, and usually employs interferometers for input isolation. This requires the critical currents to satisfy the condition

$$I_m = I_{01} + I_{02}$$

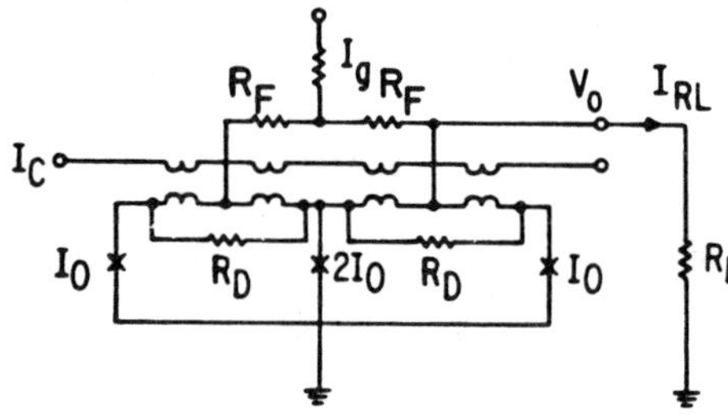

Fig. 1 A three-junction, magnetically coupled interferometer. [After Gheewala]

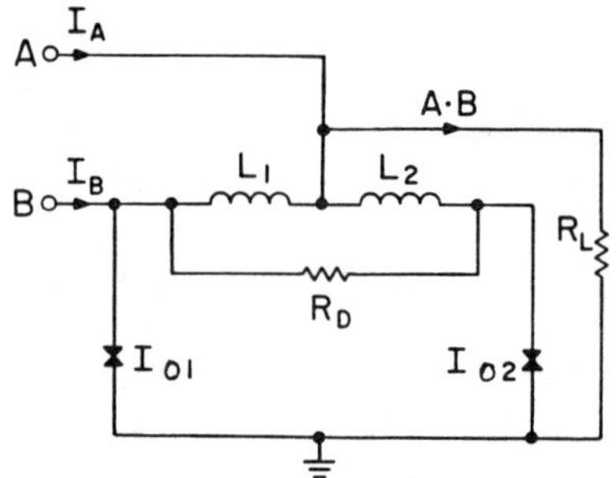

Fig. 2 A two-junction current injection gate. [After Gheewala]

Supported by AFOSR(JSEP) F49620-79-C-0178

2. CIL Circuits

The CIL logic family consists of two basic devices [1]: a three-junction magnetically coupled interferometer and an two-junction asymmetric current-injection gate. The equivalent circuit of an interferometer is shown in Fig. 1. This device is biased at a gate current $I_g < I_m$, where I_m is the critical current of the interferometer with zero applied control current. When a control current I_c is applied, the interferometer switches from the superconducting state to the voltage state, and the gate current is directed into the load resistor R_L. This device is usually made with two control lines and can be used as a two-input OR gate. The equivalent circuit of the other device, the injection gate, is shown in Fig. 2. Both $I_A < I_{01} + I_{02}$ and $I_B < I_{01} + I_{02}$ must be present for the gate to switch into the voltage state. Thus the injection gate functions as a two-input AND gate, and usually employs interferometers for input isolation. This requires the critical currents to satisfy the condition

$$I_m = I_{01} + I_{02}$$

There are a number of constraints to consider in miniaturization, and the following must be satisfied by CIL logic :

1. To insure adequate output signal when a CIL gate switches, the following must be satisfied :

   $$I_m R_L < V_g \tag{1a}$$

   where V_g is the gap voltage (V_g = 2.5 mV for lead technology). For high-speed operation R_L must be equal to the characteristic impedance Z_0 of the transmission line. Therefore, the constraint can be written

   $$I_m Z_0 < V_g. \tag{1b}$$

2. Critical current I_m of a gate must be > 40 - 50 μA to insure against noise switching.
3. Critical current density J_c of a junction must be < $10^4\ A/cm^2$ because of the unsuitability of the I-V characteristics of junctions with higher J_c's for latching logic applications [2]. (constraints 2 and 3 limit the minimum junction size to 0.12 μm^2).
4. To keep gains and margins constant, the product of the inductance L and critical current I_m must remain constant. For the interferometer

   $$L_i I_m = \Phi_i \tag{2}$$

   where L_i is the loop inductance of an interferometer loop, and Φ_i is chosen to be 2.4 mA–pH to optimize the device switching characteristics. For the injection gate the optimum condition are

   $$(L_1 + L_2) I_{02} = \Phi_0$$

   $$L_1 I_{01} = L_2 I_{02} \quad \text{and} \quad \frac{I_{02}}{I_{01}} = 3$$

   where Φ_0 = 2.07 mA–pH is a flux quantum.

These constraints combine to give relationships between circuit parameters such as Z_0, I_m, L, and J_c.

The total logic delay t_D of a CIL gate has four components,

$$t_D = t_0 + t_r + t_x + t_p$$

The turn-on delay t_0 depends only on the capacitance C and I_m ; for a given overdrive, $t_0 \propto (C/I_m)^{1/2}$ [3]. Since the capacitance of a junction is a very slowly varying function of the critical current density J_c [4], t_0 is approximately proportional to $[J_c]^{-1/2}$. The risetime t_r is proportional to the time constant CZ_0. For a given operating voltage V < V_g, $Z_0 \propto [I_m]^{-1}$ (constraint 1). Therefore, t_r is proportional to $[J_c]^{-1}$. For a critical current density of $10^4\ A/cm^2$, the gate delay $t_g = t_0 + t_r$ for a 2-input AND gate biased at 0.7 I_m is about 5 ps.

Each fan out causes a delay $2L_x\ /\ Z_0$ where $2L_x$ is the crossing inductance of an interferometer control line. For a series fan out of m, the crossing delay is

$$t_x = \frac{m 2L_x}{Z_0}. \tag{3}$$

For a fixed operating voltage, (1b) and (2) can be combined to give

$$L_i = \frac{\Phi_i Z_0}{V}. \tag{4}$$

Combining (4) and (3) gives

$$t_x = \left[\frac{m 2L_x}{L_i}\right]\frac{\Phi_i}{V}.$$

The crossing delay is proportional to ratio of the crossing inductance of the control line and the inductance of the interferometer. Since the control line and the inductor lengths are equal, t_x is proportional to the ratio of the inductances per unit length.

The propagation delay is $t_p = l\tau_p$, where l is the average path length in the circuit and τ_p is the delay per unit length of the transmission line. For a superconducting transmission line τ_p is approximated by the formula [5]

$$\tau_p = \frac{1}{c}\left[\varepsilon(1+\frac{\lambda_t}{d_0})\right]^{1/2} \tag{5}$$

where

$$\lambda_t = \lambda_1 \coth\frac{t_1}{\lambda_1} + \lambda_2 \coth\frac{t_2}{\lambda_2}$$

Here ε is the dielectric constant of the insulator, d_0 is the thickness of the insulator, t_1 and t_2 the thicknesses of the superconductors , and λ_1 and λ_2 their penetration depths.

3. Optimization

To optimize the performance of the circuits, the junction should be made with the highest possible J_c. At the limit of $10^4\ A/cm^2$ an interferometer with I_m = 50 μA would have junction size of 0.12 μm^2, which is larger than the minimum feature of 0.2 x 0.2 μm. Other considerations must be taken into account in optimizing scaled circuits; these cause minimum junction currents and sizes to be larger than this. The line widths of the interferometers and control lines should be made as small as possible, since the inductance per unit length of a superconducting strip line is inversely proportional to its width, and size reduction requires the highest possible inductance per unit length. Also, λ_t should be kept close to $\lambda_1 + \lambda_2$ by using sufficiently thick films, since any significant increase in λ_t would increase the propagation delay and the characteristic impedance, which is approximated by the formula [5]

$$Z_0 = \frac{1}{K(W,d_0)\,W}\left[\frac{\mu_0}{\varepsilon\varepsilon_0}d_0(d_0+\lambda_t)\right]^{1/2} \quad (6)$$

where W is the width of the transmission line, $K(W,d_0)$ is a factor which corrects for the fringe fields.

The determination of other parameters is not as straight-forward, since it involves trade-offs between various aspects of circuit performance. Increasing the width of the transmission lines would lower their characteristic impedance and allow a higher I_m (constraint 1), which implies a lower loop inductance L (constraint 4) and the size of the device is reduced. However, this increase causes greater mismatch between transmission lines and control lines, which requires larger space-consuming capacitors to correct [6], and of course wider lines take up more space on the chip. A thinner insulator (smaller d_0) lowers the transmission line impedance but also lowers their propagation speed, and increasing the dielectric constant produces similar results.

4. Model

To determine the optimum value of the parameters W, d_0, and ε, a computer model of a chip has been devised. In this model the chip is divided into cells. Each cell contains a logic gate which consists of N_i interferometers, and N_c injection gates, with the gate bordered by n transmission lines (Fig. 3). The relationships between the size of the devices and the size of the transmission lines are given by the circuit constraints. Since the gate delay only depends on the critical current density of the junctions, and the crossing delay only depends on the ratio of the inductances per unit length, they are uneffected by the parameters being varied; therefore, the parameters that give the minimum propagation delay will minimize the logic delay. The average propagation delay is assumed to be

$$t_p = mA_c^{1/2}\tau_p \quad (7)$$

where A_c is the total cell area, and m the average number of fanouts. Two separate optimizations are carried out, one to minimize the area A_c, and the other to minimize the delay $A_c^{1/2}\tau_p$.

The cell area is calculated in the following way. The area occupied by the devices is

$$A_{dev} = N_i A_i + N_c A_c \quad (8)$$

where A_i is the area of an interferometer, and A_c is the area of an injection gate. We assume the shape of this area is approximately that of a square; therefore, the area of the cell is

$$A_c = \left[A_{dev}^{1/2} + 2nW\right]^2 \quad (9)$$

The separation between the lines is assumed to be equal to the width of the lines W. This assumption is valid for W > 0.5 μm. For the optimized 0.2 μm circuit, results show that W should indeed be > 0.5 μm.

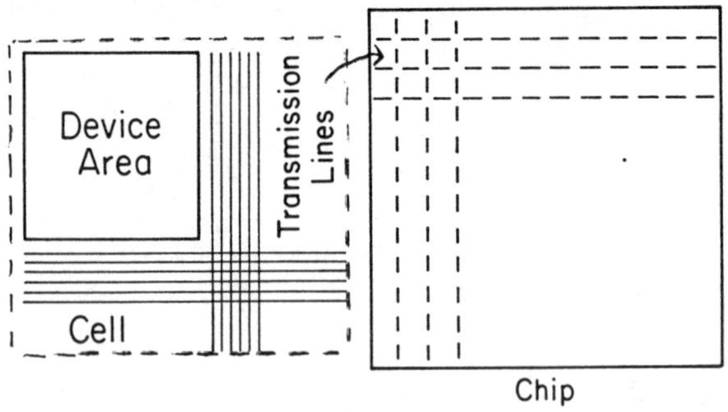

Fig. 3 Model of the chip and cell.

Only interferometers over ground-plane holes are considered since miniaturization requires the highest possible loop inductance and flux trapping probability is expected to decrease sufficiently with device size. The circuit constraints allow areas A_i and A_c to be written in terms of W, d_0, and ε :

$$A_i = \delta_i\,[\,l_i + 2w_e\,]^2 + 2A_{cap} \quad (10)$$

where l_i is the length of the inductor, w_e is the transverse distance between the inductor and the surrounding ground plane. A geometrical factor δ_i is taken to be the ratio between the area and the length squared of the 5 μm interferometers [8].

A capacitance of

$$C_x = \frac{2L_x}{Z_0^2}$$

is needed to match a transmission line to a control line [6]. The area of the matching capacitor is

$$A_{cap} = \frac{C_x d_1}{\varepsilon\varepsilon_1} \quad (11)$$

where ε_1 is the dielectric constant and d_1 the thickness of the dielectric material. For an interferometer with two control lines, an area of $2A_{cap}$ is needed. The length of the inductor is

$$l_i = \frac{L_i}{L'_i} \quad (12)$$

where L'_i is the inductance per unit length of the inductor. Substituting (6) into (4) and combining with (12) gives the result

$$l_i = \frac{\Phi_i}{L'_i VK(W)W}\left[\frac{\mu_0}{\varepsilon\varepsilon_0}d_0(d_0+\lambda_t)\right]^{1/2} \quad (13)$$

The area of the matching capacitor can be written

$$A_{cap} = \frac{2d_1 L_x}{\varepsilon\varepsilon_1 Z_0^2} \quad (14)$$

with

$$L_x = l_x L'_x \tag{15}$$

where $l_x = l_i$ is the length of the control line, and L'_x is its inductance per unit length. Substituting (13) and (6) into (14) gives

$$A_{cap} = \frac{2d_1\Phi_i K(W,d_0) WL'_x}{\varepsilon\varepsilon_1 VL'_i}\left[\frac{\mu_0}{\varepsilon\varepsilon_0}d_0(d_0+\lambda_t)\right]^{-1/2} \tag{16}$$

A similar argument gives for the injection gate

$$A_c = \delta_c\,[\,l_c + 2w_e\,]^2. \tag{17}$$

Where

$$l_c = \frac{4}{3}\frac{\Phi_0}{L'_c VK(W)W}\left[\frac{\mu_0}{\varepsilon\varepsilon_0}d_0(d_0+\lambda_t)\right]^{1/2} \tag{18}$$

is the length of the injection gate, δ_c is taken to be the ratio between the area and the length squared of the 5 μm device, and L'_c is the inductance per unit length of the injection gate inductor.

The parameters that can be varied in this model are ε, d_0, and W. The current IBM process includes two layers of dielectric, Nb_2O_5 and SiO, between the transmission line and the ground plane. The effective dielectric constant can be defined as

$$\varepsilon = d_0\left[\frac{d_1}{\varepsilon_1}+\frac{d_2}{\varepsilon_2}\right]^{-1} \tag{19}$$

where d_1 and d_2 are the thicknesses of the Nb_2O_5 and SiO layers respectively, and ε_1 and ε_2 their dielectric constants, and $d_0 = d_1 + d_2$ is the total thickness of the insulator. The dielectric constant of Nb_2O_5 is 29.4 and that of SiO is 5.6. The capacitors are formed by a metal layer over the Nb_2O_5 layer. For this process it is more convenient to consider the variations of d_1 and d_2 instead of ε and d_0.

Our model cell consists of two interferometers, and two injection gates bordered by 10 transmission lines on each side [9]. The operating voltage V is set at 2 mV. The parameters L', L'_x, and L'_c are calculated with a computer model based on the work of W. H. Chang [10, 11]. The inductance per unit length of the inductors increases as w_e decreases. Simulations show that the area of the devices is minimum for w_e = 3 μm.

5. Results

To minimize area and delay, the thickness d_1 of the Nb_2O_5 should be as small as possible. Although a decrease in d_1 increases the transmission line impedance Z_0, which requires the size of the devices to increase, computer calculations show that the reduction of the size of the capacitors and of the transmission delay per unit length τ_p more than compensate for this increase for the range of d_1 in current devices. The Nb_2O_5 layer is 0.035 μm thick in the current IBM Pb alloy process. This is the value of d_1 that we will use in our calculations. For minimum area, d_2 should also be as small as possible to reduce Z_0. The area of a cell A_c calculated using eqs. (8),(9), (10),(13),(16),(17),(18), and (19) is plotted vs. W and shown in Fig. 4. The curves are the loci of points of constant d_2. For $d_2 = 0.1\,\mu m$ the cell area is minimum at $W = 1.0\mu m$, having the value 2913 μm^2. The other circuit parameters are also functions of d_2 and W; at the conditions for minimum area, they are: $\varepsilon = 7.1$, $Z_0 = 22.1\,\Omega$, $L_c = 22.8\,pH$, $L_i = 24.5\,pH$, $L_x = 34.8\,pH$, $I_m = 90.6\,\mu A$, and $\tau_p = 0.015\,ps/\mu m$. for a fanout of two, $t_x = 6.3\,ps$, $t_p = 1.7\,ps$, and $t_D = 13.0\,ps$.

The alternate miniaturization is that of delay. Using eq. (5), and the above-mentioned set of equations, calculations of t_p are made and the results are plotted in Fig. 4. Curves of constant d_2 are plotted. For $d_1 = 0.035\,\mu m$, t_p is minimized at W = 1.2 μm and $d_2 = 0.32\,\mu m$. The other circuit parameters are: $\varepsilon = 6.09$, $Z_0 = 31.8\,\Omega$, $L_c = 32.9\,pH$, $L_i = 38.2\,pH$, $L_x = 50.2\,pH$, $I_m = 62.8\,\mu A$, $\tau_p = 0.011\,ps/\mu m$, and $A_c = 4757\,\mu m^2$. For a fanout of 2, $t_x = 6.3\,ps$, $t_p = 1.5\,ps$, and $t_D = 12.8\,ps$.

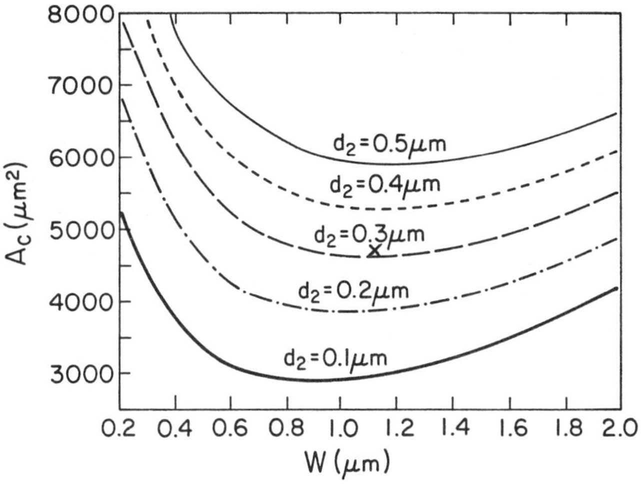

Fig. 4 Cell area vs. the width of the transmission line. The 'X' marks the point of minimum delay.

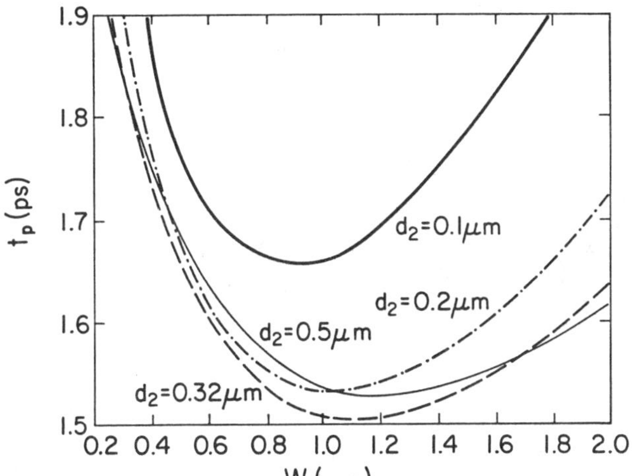

Fig. 5 Average propagation delay vs. the width of the transmission line,

Both the area A_c and delay t_p are increased when W is decreased below 1.0 μm because a decrease in W increases Z_0, and requires the currents to be reduced (constraint 1). A reduction in the critical current implies an increase in inductance and device size (constraint 4). Since τ_p is independent of W, the propagation delay is increased proportionally to device size.

6. Concluding Remarks

Miniaturized junctions can have a higher critical current density J_c, which significantly reduces the gate delay. Reducing the widths of the inductors increases their inductances per unit length and reduces device size; hence, the propagation delay is reduced and device count increased. The speed of the circuits does not vary significantly with the parameters W and d_2. This is because the gate and crossing delays, which accounts for about 90% of the total delay, are unchanged. However, the area of a logic gate can vary by a factor of two. The correct choice of these parameters can double the number of gates on a given chip.

A comparison with Gheewala's 2.5 μm CIL circuits shows that the gate delay is reduced from 28 *ps* to 5 *ps* by the change to 0.2 μm features. The propagation delay is reduced from 6 ps to 1.5 ps. The crossing delay show less improvement; for a fanout of two, t_x is reduced from 8.7 *ps* to 6.4 *ps*. The overall speed of CIL circuits is improved by about a factor of three. The improvement in area is much greater. A chip 6.35 mm on a side would hold about 1000 2.5-μm CIL logic gates while the same size chip can hold over 13,000 0.2-μm gates.

The performance of these circuits can be improved with an additional ground plane over the transmission lines [9]. Ignoring fringe field effects, this addition doubles the characteristic impedance of the transmission lines Z_0 for fixed values of W, d_0 and ε. Our estimates then indicate that the area of the circuits can be reduced by a factor of two with this change, but there is negligible effect on speed.

The calculations were made for lead-alloy technology, as was used in Gheewala's experiments. It should be possible to use higher current density in niobium junctions to offset its higher specific capacitance; by reducing the junction size, the critical current and capacitance will remain the same as in the lead-alloy case so gate delay is unchanged. The penetration depths in the two cases are almost equal so there is a negligible change of Z_0 and propagation delay. Therefore, the principal results found for lead-alloy technology apply also to niobium.

ACKNOWLEDGEMENTS

The authors wish to express their thanks to H. Zappe and M. Ketchen at the IBM Thomas J. Watson Research Laboratories for their suggestions and advice, and to J. Spargo for many helpful discussions.

REFERENCES

[1] T. R. Gheewala, IBM J. Res. Develop., 24, 30 (1981).

[2] Y. Tarutani and T. Van Duzer, IEEE Trans. Electron Devices, ED-30, 34 (1983).

[3] E. P. Harris, IEEE Trans. Magn. MAG-15, 562 (1979).

[4] J. H. Magerlein, IEEE Trans. Magn. MAG-17, 286 (1981).

[5] W. H. Chang, J. Appl. Phys. 50, 8129 (1979).

[6] H. H. Zappe, IEEE Trans. Magn. MAG-13, 41 (1977).

[7] H. H. Zappe, (unpublished)

[8] The 5 micron devices are the ones used in the Berkeley A/D conversion project. They are similar to the IBM devices.

[9] H. H. Zappe, Private Communications.

[10] S. Dhong, PH. D. Dissertation, EECS, University of California, Berkeley, CA, (1983).

[11] W. H. Chang. IEEE Trans. Magn. MAG-17, 764 (1981).

Session 3-2b
The IBM Machine

Chairman

R. L. Russo
Stanford University

Introduction to the IBM Los Gatos Logic Simulation Machine

John K. Howard, Richard L. Malm, Larry M. Warren

IBM Los Gatos Laboratory, Los Gatos, California

ABSTRACT

The IBM Logic Simulation Machine, also known as the LSM, is a high speed hardware Logic Simulator that simulates up to 64,512 logic expressions at a rate of 640 x 10^6 expressions per second. Two LSMs have been in use at IBM's Los Gatos Laboratory since 1980 where they have revolutionized logic simulation and design. This paper introduces the LSM, discusses the history of its development, and describes other logic simulators that have been built or proposed.

INTRODUCTION

The Logic Simulation Machine (LSM) is a special purpose computer designed to perform logic simulation at speeds heretofore unattainable. Evaluating a maximum of 64,512 logic expressions at a rate of up to 640 million expressions per second, the LSM represents a significant advance in simulation technology. The LSM is used to simulate logic designs in an interactive fashion, allowing a designer to simulate a design, make a change, and re-simulate — all within minutes.

The LSM was conceived in 1977 by John Cocke of IBM's Thomas J. Watson Research Center in Yorktown Heights, New York, by Richard L. Malm of Los Gatos Laboratory, and by John Schedletsky, also of Yorktown. The LSM was built and enhanced at Los Gatos by a hardware and software team headed by Richard Malm. The LSM has been used regularly at Los Gatos since 1980 and represents, we believe, the first major use of a hardware logic simulator for VLSI design efforts.

IBM's Los Gatos Laboratory is a small development lab located at the southern end of Santa Clara Valley (also known as "Silicon Valley") in California. At Los Gatos, a VLSI design group is engaged in VLSI design, development of advanced design tools, and development of sophisticated testers.

This is the first of four integrated papers[1 2 3] that describe the LSM. This paper introduces the LSM by presenting a short history of logic design and the problems presented by VLSI design in particular, by discussing conceptually the design of the LSM and major decisions made during its development, and finally, by examining some other logic simulators. The second and third papers present an overview of the hardware and software that comprise the full LSM system. The fourth paper discusses Los Gatos' experience with the LSM: what has been learned, how the LSM is used, how the LSM has changed logic design and simulation at Los Gatos.

Figure 1. The IBM Los Gatos Logic Simulation Machine

LOGIC SIMULATION

The need for logic simulators has long been recognized. If the accuracy of a design can be verified before a real physical device is built, much time and money is saved. In the competitive field of VLSI design, wasted time can mean the loss of a competitive edge.

If a simulator can closely simulate a design as presented by the designer, design errors can be found before the chip is manufactured. The turnaround time for a design that repeatedly needs to be altered and rebuilt is just too long for present needs.

Of course, neither the problem nor the solution are new. As early as the 1970's, some modelling or simulation method was needed. The form simulation took then was the building of hardware prototypes, simple models that were supposed to function like

CH1935-6/83/0000/0580$01.00 © 1983 IEEE

the real design. Prototypes proved too difficult to maintain as design changes were made and usually fell behind in representing the current state of the design.

The next step in the development of simulators was the introduction of software simulators—large computer programs that run on general purpose computers and simulate the functions of a design. As noted by Barto and Syzgenda[4], the Von Neumann architecture was developed originally to solve differential equations and is really not suited to tasks such as simulation at all. The massive data structures that need to be moved, maintained and updated require complex and large computer programs. In turn, these programs are excessively slow and therefore not adequate for current simulation needs.

Still, software logic simulators are the primary simulation tools used in the industry today. To abrogate the problems inherent with software simulators, parts of a design are often simulated separately so that the size of the problem—and the time required to simulate it—are reduced. This ignores the connection between the pieces. Also, the long simulation times often prohibit thorough verification of a design, and design errors can be extremely expensive.

Obviously, a better solution to the simulation problem is needed.

THE LSM SYSTEM

The IBM Logic Simulation Machine was developed in response to the limitations of the software simulators. Specifically, it was felt that a simulator capable of meeting current and future needs should:

- Be fast enough to allow interactive use.
- Have a large capacity to accommodate large designs.
- Give meaningful and precise output in an easy to read form.
- Allow for easy input of designs and design changes.
- Provide a simple method of stimulating the model.
- Be flexible and general in what it will simulate.

The LSM design was initiated and guided by these requirements. This resulted in the following design features:

- Evaluation rate of up to 640 million gates per second. Typical large simulations take just a few minutes.
- Current capacity of 64,512 gates, with expansion possible.
- Graphic output at user terminal of user selected data.
- Easy interface to logic entry systems.
- Test patterns are described by Pascal algorithms or tabular data.
- Delay, and dotted logic are supported. Function tables are programmable.

The LSM is a high-speed, highly specialized computer designed for and dedicated to logic simulation. It is composed of 64 processors operating in parallel with a 100 nanosecond clock. One processor controls simulation, and loads the LSM memories. The other 63 processors simulate random logic by evaluating 5 input random logic gates. Signals are four valued: logical 1 and 0, undefined and uninitialized.

The LSM is supported by an extensive software package residing on a general purpose host computer. An IBM System 370, Model 3081, is currently being used at Los Gatos. The software handles all preprocessing and post-processing of design models and simulation data and provides the primary user interface to the LSM. LSM software is written in Pascal and IBM's System Product Interpreter[5], a control language.

The software does not perform any actual simulation; its purpose is solely to provide a straight-forward and easy implementation of the LSM hardware. Figure 2 shows the relationship between the LSM and the supporting software. Both the hardware and software are discussed fully in the second and third papers of this series.

OTHER HARDWARE LOGIC SIMULATORS

The LSM was not the first hardware logic simulator built. In June of 1970, the Boeing Company put into use what it called the **Boeing Computer Simulator**[6] Recognizing the limitations of software simulators, Boeing developed their simulator to aid the design of digital equipment used in space programs.

The Boeing Computer Simulator was an event driven hardware logic simulator with a maximum capacity of 48,000 gates. Boeing also developed an input language, the "Register Level Simulation Language," RLSL, with which a model was described. Their simulator offered Boeing tremendous logic design advantages at that time.

Both the LSM and the Boeing Computer Simulator are hardware devices, but beyond that common trait, their differences are many. A significant difference is that the Boeing simulator was an event driven simulator and the LSM is not. This difference sets the LSM aside from many proposed and actual hardware and software simulators.

An <u>event driven simulator</u> monitors all the gates in a design and from one cycle of simulation time to the next evaluates new outputs for only those gates that had a change of inputs. Many sophisticated software simulators use this simulation technique, and the concept is a good one: why evaluate gates for which the outputs will remain the same? Theoretically, better performance might be achieved with an event-driven simulator.

The disadvantage in building a hardware event-driven simulator is in the increased complexity and cost due in large part to the large amount of data that needs to be passed among various parts of the simulator. This soon results in switching problems and communication bottlenecks.

LSM	Software
simulation	define model define test patterns start simulation view sim. results change model

Figure 2. Functions of the LSM and its Supporting Software

It was realized early that the LSM could provide more than sufficient simulation speed by relying on fast hardware and parallel processing techniques. This assumption has been more than borne out by experience with the LSM. A highly parallel design also makes the system extremely modular, allowing low entry costs and easy system expansion.

There are other differences between the Boeing Computer Simulator and the LSM:

1. BCS had non-programmable logic functions, implemented by logic cards. LSM functions are fully programmable.

2. BCS test patterns were interpreted on a host computer as the simulation was run. LSM test patterns are compiled and loaded into the LSM as machine instructions. Then the simulation runs.

3. Nets to be examined had to be described in the BCS test patterns. With the LSM, nets can be chosen interactively and separately from test patterns.

Still other differences between the LSM are largely a result of the BCS being built in 1970 when many of today's automated design facilities were not available. Boeing developed a modelling language because at that time sophisticated logic entry systems did not exist. The LSM is able to take advantage of these logic entry systems, compiling a design description that will later be used for production so that it becomes the simulation model. The LSM uses interactive software with CRT terminals; the BCS used punched card input and line printer output.

Nevertheless, Boeing was a pioneer in hardware simulator technology and the Boeing Computer Simulator was certainly a great advance beyond the simulators available in the early 1970's.

Other hardware simulation methods have been proposed. R. Barto and S. A. Syzgenda have presented a detailed description of a possible simulator architecture.[7] They describe an event driven simulator capable of simulating up to 32,768 gates.

According to the estimates given, Barto's and Syzgenda's simulator would perform 197,000 gate evaluations per second. A straight rate comparison is not, however, an accurate comparison of speeds between an event-driven simulator and the LSM. The event simulator assumes that only a percentage of the total gates in a design are active at any given time, and thus gains its efficiency through simulating only those active gates.

Even so, the LSM evaluates gates at such a high speed that event driven simulation is not necessary.

One of the latest entries into the hardware simulation field has been made by **Zycad.**[8] Zycad offers two hardware simulators, the LE-1001 and the LE-1002. Both are event-driven simulators. The most powerful of the two, the LE-1002, simulates up to 65,535 three-input gates at a rate of approximately 1 million events per second.

IBM has continued development of hardware logic simulators with the **Yorktown Simulation Engine**, also known as the **YSE**. The YSE is a second generation LSM, with enhancements added as a result of experience with the LSM. It provides larger capacity and higher speed than the LSM, simulating up to 1 million gates at a rate of 3 billion gate simulations per second. The YSE has been developed and built at the IBM Thomas J. Watson Research Center, Yorktown Heights, New York. It was presented at the IEEE 19th Design Automation Conference in 1982 and described in three papers. [9] [10] [11]

Where appropriate in the accompanying three papers, differences between the YSE and the LSM will be discussed. Presentation of the LSM in these papers gives a good historical perspective to the YSE, and because the LSM has been used extensively by IBM for over 2 years, that experience can be discussed.

CONCLUSION

The Logic Simulation Machine represents a new era in logic simulation and design. It is much faster and less cumbersome than other current simulation techniques, and IBM's use of the two LSMs at Los Gatos Laboratory has proven very successful. The LSMs are now used by IBM locations other than Los Gatos via IBM's computer communication network.

The LSM's speed frees the designer to spend more time creating and less time checking. Reworking due to design errors can be virtually eliminated. "What if" experimentation becomes possible with little time and hardware commitment. The power of a high speed hardware logic simulator like the LSM takes simulators beyond the verification role they have played in the past and makes them an integral part of VLSI design development.

ACKNOWLEDGEMENTS

As mentioned, the concept of the LSM was created by John Cocke (Yorktown), Richard Malm (Los Gatos), and John Schedletsky (Yorktown). Many others at Los Gatos and Yorktown contributed to the LSM project, and are not listed here. Richard Malm supervised development and implementation of the two LSMs at Los Gatos.

The authors thank Roy Russo of IBM for arranging Los Gatos Laboratory's participation in this conference and for his help with the papers and presentations given here.

REFERENCES

[1] Ted Burggraff, Al Love, Richard Malm, and Ann Rudy, "The IBM Los Gatos Logic Simulation Machine Hardware," Proceedings of the IEEE International Conference on Computer Design: VLSI in Computers , Rye, New York, October 31 - November 3, 1983.

[2] Jack Kohn, Richard Malm, Chuck Meiley, and Frank Nemec, "The IBM Los Gatos Logic Simulation Machine Software," Proceedings of the IEEE International Conference on Computer Design: VLSI in Computers , Rye, New York, October 31 - November 3, 1983.

[3] John K. Howard, Jack Kohn, Richard Malm, and Ann Rudy, "Using the IBM Los Gatos Logic Simulation Machine," Proceedings of the IEEE International Conference on Computer Design: VLSI in Computers , Rye, New York, October 31 - November 3, 1983.

[4] R. Barto and S. A. Szygenda, "A Computer Architecture for Digital Logic Simulation," Electronic Engineering , September 1980, 35-66.

[5] System Product Interpreter , International Business Machines Corporation, product number 5664-167.

[6] A. W. Van Ausdal, "Use of the Boeing Computer Simulator for Logic Design Confirmation and Failure Diagnostic Programs," Advanced Astronautics Science , 29 (June 1971), 573-594.

[7] Barto

[8] Zycad Logic Evaluator General Description , Zycad Corporation, November 1982, Order no. AA-040.

[9] Gregory F. Pfister, "The Yorktown Simulation Engine: Introduction," ACM IEEE 19th Design Automation Conference , June 1982, Paper 7.1, pp. 51-54.

[10] Monty M. Denneau, "The Yorktown Simulation Engine," ACM IEEE 19th Design Automation Conference , June 1982, Paper 7.2, pp. 55-59.

[11] E. Kronstadt and G. Pfister, "Software Support for the Yorktown Simulation Engine," ACM IEEE 19th Design Automation Conference , June 1982, Paper 7.3, pp. 60-64.

The IBM Los Gatos Logic Simulation Machine Hardware

Ted Burggraff, Al Love, Richard Malm, Ann Rudy

IBM Los Gatos Laboratory, Los Gatos, California

ABSTRACT

The IBM Logic Simulation Machine, also known as the LSM, is a high speed hardware Logic Simulator that simulates 64,512 logic expressions at a rate of 640 X 10^6 expressions per second. The LSM is a highly parallel system, composed of 64 processors operating simultaneously, each simulating part of a logic design. This paper presents the LSM architecture and discusses the decisions and changes made during development of the LSM.

INTRODUCTION AND OVERVIEW

The LSM is a complete computer architecture designed to perform logic simulation at high speed. It is not, however an exceedingly complex machine. It gains its speed through a highly parallel and modular architecture. The LSM's capacity can be increased incrementally with no speed degradation, and in most cases the addition of capacity can actually reduce simulation time by providing more processors among which a design can be spread.

Use of the LSM is made easy by an extensive software support package. The software package and other simulation facility considerations are discussed in three accompanying papers[1 2 3]. This paper describes only the LSM hardware architecture.

The LSM architecture provides specifically for:

- Simulation of Random Logic, PLAs, Memory.
- Variable delay simulation. Each gate can have a unique rise and fall delay time, from 1 to 256 times the basic unit of delay.
- Dotted Logic Gates with outputs tied together (dotted) can be simulated.
- Net activity trace The hardware can determine which gates in a design have changed value during a simulation. This may help determine the coverage of test patterns being used to stimulate the model.
- Conditional processing based on "events." When specified values for specified nets occur, simulation can be stopped.

Simulation in the LSM takes place in terms of a simulation time step which is one basic cycle through the LSM. During each simulation time step, all logic gates in the design are simulated, thus it takes many simulation time steps to simulate one real machine cycle.

As currently configured, the LSM can simulate 64,512 logic gates at a rate of 640 X 10^6 gates per second. This configuration consists of 64 processors, an inter-processor switch, and a host and an interface computer. The processors are distributed in the following manner:

1 required Control processor
x Logic Processors
y Array Processors
$x + y \leq 63$

The LSM is modular and flexible, allowing for low entry costs and incremental expansion of the system. The system can be configured to meet the needs of the simulation at hand, alteration of the configuration being simply a matter of plugging in or disconnecting boards.

The LSM can be extended indefinitely by adding more processors and a larger switch, and thus, can grow to accommodate future larger designs. There is no theoretical limit to the capacity it can obtain.

Figure 1 shows an overview of the LSM system. Each of the components is fully described below.

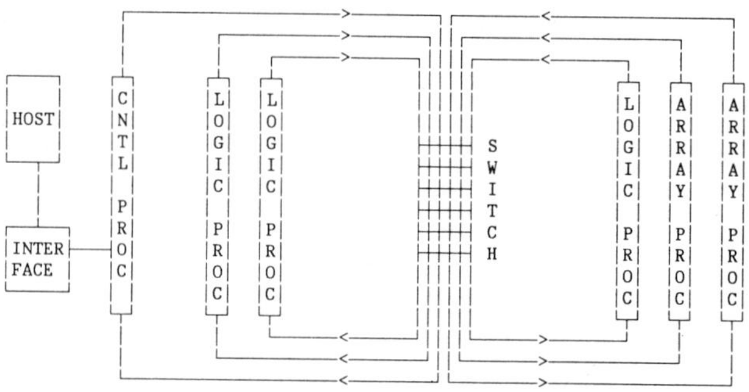

Figure 1. LSM Architecture, 6 Processor Configuration

Physically, the 64 processor LSM fits with its power supplies on six 19 inch square by 5 feet high racks. Each rack draws 190 amps at 5 volts. The total system contains approximately 42,000 TTL modules, 136,000 nets, and 612,000 wire points.

LOGIC PROCESSOR

The Logic Processor is the fundamental unit of the LSM and performs the actual simulation of a design. Sixty-three identical logic processors are incorporated into the LSM configuration. They operate in parallel, each stepping through its memories and simultaneously simulating a part of the total design.

The logic processor contains an instruction memory, a data memory, a logic unit, and a delay value memory. Figure 2 shows the basic structure of a logic processor.

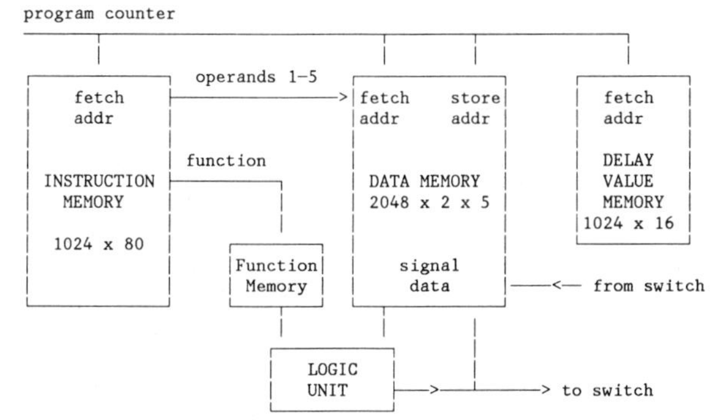

Figure 2. Overview of a Logic Processor: showing one of the two alternating data memories

The **instruction memory** contains 1024 80 bit instruction words. Each instruction word contains a function opcode, five input operand addresses, and other control values. The LSM executes only five input functions; any gate in a logic design with more than five inputs is decomposed into a sequence of five input functions.

Two **data memories** contain 2048 2 bit words each. The input operands addressed by the instruction word are fetched from the input data memory and results of the function evaluation are written to the output data memory. The two memories alternate as input and output memories. There are actually five copies each of these memories, each copy containing identical data, to facilitate the operand fetch. Addresses 0 through 1023 of the data memory contain signal data values calculated in the processor in which the data memory resides. Addresses 1024 through 2047 contain output from other processors.

The **function memory** contains 1024 64 bit words. Each bit in a word is an entry in a 64 entry truth table describing the output of a six input function. (One extra input is allowed for internal chaining of five input functions.) This means that the LSM can simulate up to 1024 native logic functions in each logic processor. The truth tables are fully programmable.

Simulation occurs as follows. The model and the test patterns containing model stimuli have already been loaded into the LSM processors. The model is distributed among the processors as determined by the LSM software and the number of instructions per logic processor (up to 1024) is known. Let this number be n.

An instruction is fetched from the instruction memory. The operands are fetched from the input data memory as addressed by the instruction. The function code and the operands are sent to the **logic unit** where the function is evaluated. The result (output) is sent to the inter-processor switch and results from other processors are received from the switch. The data are latched into the output data memory.

One fetch, evaluate, store cycle is called an instruction step. Instruction steps are executed in a seven stage pipeline, one instruction following another in the pipeline and one completing execution every 100 nanoseconds.

The required number (up to 1024) of instruction steps is executed, in this case n of them, and this makes up the simulation time step. Thus, the length of a simulation time step is not fixed for all simulations but depends on the design being simulated, the number of processors in the LSM configuration, and the distribution of the design in those processors.

The **delay value memory** contains rise and fall delay times associated with each instruction in the instruction memory. Thus, it contains 1024 words, each of which is 16 bits long. Eight bits each are used to represent the rise and fall delay time which range from 1 to 256 units of delay. A simulation time step is the basic unit of delay.

The minimum delay in the LSM is one unit, or one simulation time step, since results of the instructions in one time step are not available until the following time step. This is equivalent to a unit-delay simulator.

When more than one delay unit is specified for an instruction in the instruction memory, the output of the logic unit is not written to the output memory but is retained until the specified number of time steps have elapsed. The count for delay and the value being delayed are held in intermediate memories.

The instruction, output data, and delay memories are all accessed sequentially by the same program counter. No loops or branches occur. The input data and function memory are addressed by fields in the instruction word.

The input data and output data memories alternate in function each simulation time step, so that the memory that was written with new output values during the previous simulation time step is read during the current one. Thus arises the inherent unit delay propagation of output values in the LSM.

Also, because output values are always available for input the next simulation time step, no special ordering of instructions in the instruction memory is required to provide correct time sequencing.

The logic processor is implemented in low-power Schottky TTL on a 16 inch by 20 inch Multiwire board. Eleven boards are mounted vertically on one 19 inch square rack.

ARRAY PROCESSOR

The Array Processor is a special-purpose processor for the simulation of memory arrays such as RAM. These arrays can be simulated with the basic logic processor, but would soon exhaust the LSM's capacity. For example, to simulate a RAM, one LSM instruction would be required for each bit in the memory. This is clearly an inefficient use of the LSM.

The array processor was developed specifically to meet this need and greatly extends the abilities of the LSM. Up to 45 different memory arrays can be simulated on one array processor. The maximum size for any one array is 64K by 36 bits. Memories larger than 64K long or 36 bits wide must be broken into the basic 64K, 36 bit size. The total memory available on a single array processor is 512K 36 bit words.

The array processor was designed to look like a logic processor to the rest of the LSM system. Each array processor occupies one logic processor slot and communicates with the inter-processor switch in the same manner as a logic processor. Only its internal functions are different. The array processor may be substituted for a logic processor anywhere in the LSM configuration.

Arrays are defined by the user on the host computer using the LSM software. All read and write ports and address, data and control lines are defined. Each read and write port must be defined separately though they may share common inputs and controls. Then the array model is loaded into the array processor.

Once the array is defined and the model is loaded, address, data and control values are routed to the array processor and it performs the necessary reads and writes to simulate the modelled array.

CONTROL PROCESSOR

The control processor controls operation of the LSM. It starts, stops, and interrupts simulation; it reads and writes logic processor memories for simulation and diagnostics; and it communicates with the host computer through the interface computer, passing the results of simulation and loading new data into the LSM.

The control processor, like the array processor, is designed to look like another logic processor to the rest of the LSM system, so it passes data to and reads data from the logic and array processors through the same switching mechanism. The control processor occupies position 0 in the LSM rack configuration.

The control processor contains an **event mask memory** that provides for conditional simulation based on the results of instruction evaluation in other processors. The event mask memory contains 1024 4 bit words. Each bit corresponds to one of the four possible signal values and is turned on or off to indicate whether the occurrence of that value is to signal an event.

For example, if a certain instruction should never have an undefined output, an event can be triggered and simulation stopped if the output is undefined.

INTER-PROCESSOR SWITCH

The inter-processor switch is a 64 by 64 crosspoint switch that allows any processor (logic, array, control) to communicate with any other in the LSM. During each instruction step, all processors send simulation data to the switch and access the switch to obtain data needed from other processors. Any number of processors may access the output from any one processor simultaneously.

The switch is controlled by the **switch select memories**. There are 64 switch select memories in the interprocessor switch—one associated with each processor in the LSM. Each memory is a 1024 word 6 bit memory, and is stepped through sequentially by the same program counter used to access the logic processor memories. Each location in the memory contains a 6 bit address indicating the processor from which data is needed. During an instruction step, the address contained in the switch select memory selects a position in the interprocessor switch and routes the data to the memory's associated processor.

The switch select memories are loaded when the design model is loaded into the processor memories. The LSM software has determined where in the LSM the different pieces of a design will reside and loads the switch select memories so that proper communication takes place.

HOST AND INTERFACE COMPUTERS

External to the LSM system are two general purpose computers that handle user related functions.

The **host computer** provides the general purpose processing power necessary to take full advantage of the LSM. On the host computer, the user describes logic designs, creates test patterns, loads the LSM, starts simulation, and then views the simulation results. Diagnostic simulations and programs are also run from the host computer.

The **interface computer** links the LSM to the host computer, providing the necessary communication between the two and aiding data transfers. Simulations can also be loaded and run directly from the interface computer and many of the more detailed diagnostics are performed in this way.

At Los Gatos, the host computer is an IBM System 370 Model 3081; the interface computer is an IBM Series 1.

THE LSM AND THE YSE

A second generation LSM, called the Yorktown Simulation Engine, is being developed at the IBM Thomas J. Watson Research Center, Yorktown Heights, New York. The YSE has been described fully in three papers[4,5,6] presented at the 19th Design Automation Conference.

The LSM and the YSE are based on the same architectural concept, and many of the enhancements made to the YSE are a result of experience with the LSM. The most significant difference between the two simulators is the increased speed and capacity of the YSE. The YSE provides simulation of up to 1 million gates at a rate of 3 billion gates per second, compared to 64,000 gates at 640 million per second with the LSM. To date, the LSM has proven sufficient for most logic simulations, but for large VLSI simulation the increased speed and capacity provided by the YSE will be needed.

The YSE provides the larger capacity in two ways. First, each logic processor can simulate 4096 gates, as compared to 1024 on an LSM logic processor. Second, while the LSM now supports a maximum of 64 processors, the YSE allows up to 256 processors, all operating in parallel. This greatly increased parallelism provides the higher rate of gate evaluation in the YSE.

It should be noted that evaluating 4K gates on each processor rather than 1K does cause a longer simulation time step since the length of the time step is a function of the number of instructions evaluated.

This time increase is thought to be made up by the inclusion of rank order simulation in the YSE. Rank order simulation means, in effect, that a circuit can be simulated with zero delay. Instructions can use the output of other instructions within the same simulation time step. Note that rank order simulation forces chronological ordering of instructions in the instruction memories. This ordering must be handled by the software so that the memories are loaded correctly. The LSM hardware supports rank order simulation, but this feature has not yet been supported by the software.

The YSE does not directly provide the same variable delay capabilities of the LSM. To add delay to a design in the YSE, instructions must be used that simply pass the output along, or counters must be simulated that create delay for each net. The LSM allows the user to define unique rise and fall delay times for each gate output.

The YSE and the LSM handle function evaluation differently. LSM equations have 5 two bit inputs; YSE equations have 4 two bit inputs. The LSM provides for 1024 different functions; the YSE provides 256, but also provides 16 generalized De Morgan functions on each input, giving it the power to implement the equivalent of the LSM's 1024 functions. For example, the YSE does not need to have both AND and NOR functions as primitives since it can invert the inputs to the primitive AND to get a NOR.

The YSE offers more flexible use of the four signal states output from the function unit. One of the four states may, for example, be assigned to represent high impedance. The LSM's use of the four states is restricted in that the uninitialized state is determined by the function unit and its use is not programmable.

The YSE does not support net activity trace.

ACKNOWLEDGEMENTS

The LSM was conceived in 1977 by John Cocke of IBM's Thomas J. Watson Research Center in Yorktown Heights, New York, by Richard L. Malm of Los Gatos Laboratory, and by John Schedletsky, also of Yorktown.

Richard Malm supervised development of the LSM at Los Gatos Laboratory. Warren Begas, Ted Burggraff, Jack Kohn, Al Love, Gary McGill, Greg Pfister, Jack Powers, Ann Rudy, and John Shedletsky all contributed to the hardware design. Technical support was provided by Mai Nguyen, Bobby Sloan, Don Whiteman, and Ron Williams. The LSM was developed at Los Gatos Laboratory under the management of John K. Howard and Larry Warren.

REFERENCES

[1] John K. Howard, Richard L. Malm, and Larry M. Warren, "Introduction to the IBM Los Gatos Logic Simulation Machine" Proceedings of the IEEE International Conference on Computer Design: VLSI in Computers , Rye, New York, October 31 - November 3, 1983.

[2] Jack Kohn, Richard Malm, Chuck Meiley, and Frank Nemec, "The IBM Los Gatos Logic Simulation Machine Software," Proceedings of the IEEE International Conference on Computer Design: VLSI in Computers , Rye, New York, October 31 - November 3, 1983.

[3] John K. Howard, Jack Kohn, Richard Malm, and Ann Rudy, "Using the IBM Los Gatos Logic Simulation Machine," Proceedings of the IEEE International Conference on Computer Design: VLSI in Computers , Rye, New York, October 31 - November 3, 1983.

[4] Gregory F. Pfister, "The Yorktown Simulation Engine: Introduction," ACM IEEE 19th Design Automation Conference , June 1982, Paper 7.1, pp. 51-54.

[5] Monty M. Denneau, "The Yorktown Simulation Engine," ACM IEEE 19th Design Automation Conference, June 1982, Paper 7.2, pp. 55-59.

[6] E. Kronstadt and G. Pfister, "Software Support for the Yorktown Simulation Engine," ACM IEEE 19th Design Automation Conference , June 1982, Paper 7.3, pp. 60-64.

The IBM Los Gatos Logic Simulation Machine Software

Jack Kohn, Richard Malm, Chuck Meiley, Frank Nemec

IBM Los Gatos Laboratory, Los Gatos, California

ABSTRACT

The LSM is a high-speed, special-purpose computer designed to perform gate-level logic simulation at speeds that allow interactive use. It can simulate 64,512 gates at a rate of 640 X 10^6 gates per second. User oriented software running on a general purpose computer must provide an interface to the LSM that allows designers to fully utilize this simulator. This paper describes the LSM software and the requirements that directed its development.

INTRODUCTION

The Logic Simulation Machine (LSM), with its high speed and large capacity, presents many new possibilities to the logic designer. It is, though, precisely the power and speed of the LSM that proves to be the biggest challenge: how shall it be interfaced to the user? This question is best answered by understanding the implications of using such a simulator. Some of these implications are listed here.

- Designs may be simulated as whole systems, with all parts of a system communicating in the simulated realm as they will in the final product.
- Stimulation and testing of a design may be performed incrementally, with the user increasing the number of inputs stimulated and imposing known faults.
- Many simulations may be run in a short time.
- Only a portion of the output needs to be captured. Additional information can be gathered in subsequent simulations.
- Quick changes to a design are feasible, thus it can be debugged interactively. The user can fix errors as they are found.
- Instantly verifiable experimentation ("What if?" thinking) becomes possible.
- Logic simulation becomes part of the design process rather than being used solely for design verification.
- Users can feel more confident of the accuracy of their work.
- Simulation can be an enjoyable process.

An introduction to the LSM system, and a description of the LSM architecture are provided by the first and second papers of this series[1,2]. Use of the LSM within IBM is discussed in a fourth paper[3].

LSM SOFTWARE REQUIREMENTS

The LSM software has been designed to achieve the possibilities presented by the hardware, and in that way really was guided by the LSM itself. The following three rules determined the development of the LSM software:

1. It must be interactive.
2. It must take full advantage of the LSM's speed, capacity, and special features.
3. It must be easy to use.

Given these rules, the software must perform certain functions that the LSM itself cannot. Essentially, once the LSM is loaded with a model and with the stimuli to that model, it simulates the design. The software must perform all the preprocessing and postprocessing of data so that it may be loaded and read to and from the LSM.

Model descriptions, test patterns, and simulation parameters must all be converted to LSM machine code; this is done by the software. The software must make the simulation output intelligible and accessible to the user.

Figure 1 shows the organization of the LSM software. Each major portion of the software architecture is described fully in the following sections. LSM software is written in IBM PASCALVS[4], an IBM version of standard PASCAL, and in IBM System Product Interpreter, a control language .[5] The full LSM software package contains approximately 30,000 lines of code.

USER MENUS

The LSM menus provide the most basic interaction between the user and the LSM. Used in their fullest forms, the menus lead the user through the processes necessary to complete a simulation. As the user becomes more experienced with the LSM, he or she may choose to disable or bypass different sections of the menus. On the other hand a user may choose automatic mode which simply runs whatever process needs to be performed. The menus generally are used to set up parameters for the

CH1935-6/83/0000/0588$01.00 © 1983 IEEE

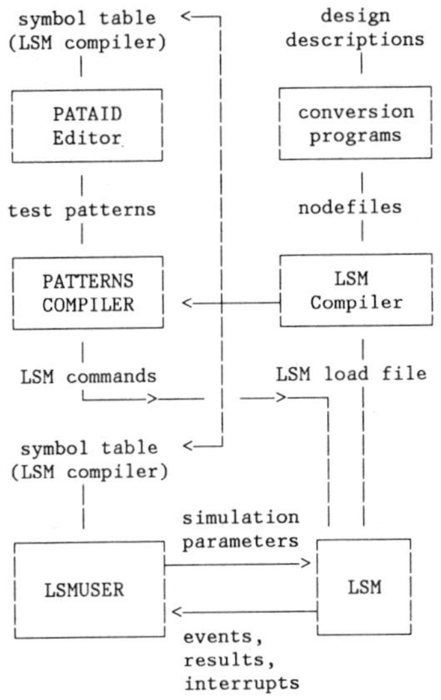

Figure 1. Software Architecture

execution of PASCAL programs that do the major processing. The user enters needed parameters in appropriate fields of the selected menu.

LSM COMPILER

The LSM compiler is the core of the system software, providing the link between design model descriptions and simulation. The compiler performs the following functions:

1. read user definitions of primitive functions.
2. read the input model from one or more nodefiles (defined below).
3. issue error messages for invalid logic.
4. partition the logic among the available processors.
5. schedule the logic in each processor and determine interprocessor switch memory data.
6. assemble instructions and create the LSM load file.
7. create a symbol table for use by other LSM software.

Input to the LSM compiler is always in the form of **nodefiles**. A nodefile is a low level logic description. Each record in the file describes a logic block which may have any number of inputs and outputs. The record describes the primitive function, the number and name of inputs to it, and the number and name of outputs from it.

The LSM compiler accepts one or more nodefiles as input and produces an LSM load file. When a design is described by more than one nodefile, the compiler resolves the interconnections between logic described in the different files. This is done on a net name basis.

An **LSM load file** consists of data to be loaded into the instruction memory, function memory, data memory, and delay value memory of each processor. For each array processor, the LSM load file contains the data to be loaded into the various array processor memories. The load file also contains data for the control processor and for the switch select memories of the interprocessor switch.

The compiler must also **distribute the design** among the processors in the LSM configuration, and it must **schedule instructions** (order them) within each processor. The compiler must provide for necessary communication among the processors, providing the interconnections between the logic that has been split among them.

Sometimes communication conflicts arise. During simulation, one processor may receive data from only one other at each instruction evaluation step. If the inputs to an instruction in a processor must be obtained from other processors, the instructions that create those signals must be scheduled or ordered so that all the inputs are accessible for the instruction evaluation. This necessary ordering must be determined by the compiler and is called the **scheduling problem**.

The scheduling problem is always resolved by either moving or copying instructions in the instruction memories of the logic processors so that conflicts are avoided. In some cases, grouping the nets in the processors according to their interconnection in the design may help avoid scheduling conflicts.

The LSM compiler is written in approximately 4400 lines of Pascal code. It processes 2000 input lines per second of CPU time on an IBM System 370/3081 host computer.

LOGIC DESCRIPTIONS

Logic designs are created at a high level using Computer Aided Design tools. Designers enter graphic descriptions of their design that are easy to read and change. These descriptions must then be converted to the nodefiles that the LSM compiler recognizes. Several interfaces between meaningful logic description techniques and nodefiles are in use at Los Gatos.

Some elements of a logic design must be described in special formats that provide the detailed information necessary for simulation. Arrays, delay propagation on nets and PLAs are all described in special formats. These formats are easy to read descriptions of the element being described. For example, the description of delay propagation is simply a list of nets and their associated rise and fall propagation delay times. Array descriptions detail the control, address and data lines of array data structures. PLAs are described through the use of a specialized PLA editor.

Ultimately, these descriptions are all converted to nodefiles (lists of logic blocks, input, outputs) and are included with the nodefiles that describe the remainder of the logic design. These are then compiled to LSM machine code by the LSM compiler.

Any number of nodefiles may be combined to describe a logic design. Logic that is incorporated frequently into larger designs may be described separately, converted to a nodefile and included in any number of models to be simulated. This again gives the designer extreme flexibility in creating a design.

PATTERNS COMPILER

Test patterns are sequences of stimuli applied to a model to simulate the real operation and must be described by the user. A test patterns facility was developed to help the user create these patterns. The user may accomplish the following tasks with the test patterns facility:

- set, read and compare values of nets
- define, start, and stop clocks
- control the running of the logic model cycle by cycle
- initialize an array (such as RAM) to a certain value
- load an array from a data file
- set an individual word in an array to a given value
- dump array contents to a data file
- define an "event" condition on a net which halts simulation if it ever occurs

The test patterns set up the simulation and define what will occur during simulation; they do not monitor simulation results and therefore cannot take decisive actions based upon events.

Test patterns can be described either in **tabular form** or **algorithmically**. With the tabular form, called preprocessor statements, the user can describe the patterns by creating a symbolic picture of the waveforms applied to nets during simulation. The user creates a table of net names versus time and shows the application of different signal values by representing a logical 1 by a +, 0 by a -, and so on.

Preprocessor statements are the simplest way to describe test patterns, but they are less powerful than the algorithmic descriptions. The algorithmic description of test patterns is a Pascal driven facility. The user describes with Pascal procedure calls how the model is to be stimulated, and those called procedures create the proper LSM instructions.

The preprocessor and Pascal patterns compiler is written in PASCALVS. The compiler contains approximately 6,500 lines of code.

THREE SIMULATION MODES

Simulation occurs in one of three possible modes recognized by the LSM software. The three modes are compare mode, strobe mode, and debug mode, and each mode provides a different level of detail in the simulation results.

Debug mode provides the greatest level of detail. When a simulation is run in debug mode, results are captured at every logic level within the model cycle, providing a literal trace of net behavior. Note that results are captured only for nets the user specifies using the net selection software described in the next section. Simulation can also be "windowed" in time, meaning that no results will be captured before time x or after time y, where x and y are determined by the user. This allows the user to concentrate on one time slice in the simulation, on only certain chosen nets, but with the greatest level of detail.

Strobe mode is similar to debug mode except that results are captured only once during each cycle. The user may specify at what level in the cycle results are to be captured, and he or she explicitly determines the nets for which data is collected.

Compare mode requires that the user predict in the test patterns what values certain nets will take on at certain times. If the simulation results differ from those predictions, data for those nets are gathered and may be used for the basis of more detailed simulation with either strobe or debug mode. The time at which simulated values deviated from predicted values is indicated specifically when the user views simulation results. Data collection is minimized since the user specifies compares on specific nets in the design. The user may also choose a time span for data collection.

SELECTING AND DISPLAYING NETS

An interactive program called LSMUSER lets the user select nets for simulation, run the simulation, and display the results. LSMUSER is the main user interface during simulation runs.

A special-purpose, full-screen editor is implemented within LSMUSER to aid the LSM user in selecting the simulation data to be captured and displayed. The editor features many functions typical of full screen text editors plus functions unique to simulation.

By limiting the number of nets for which simulation data is captured, the user greatly reduces the amount of data to be analyzed and can concentrate specifically on problems. LSMUSER allows the user to be very specific in describing for what nets data are to be captured and which nets will be displayed after simulation. The user builds lists describing these nets of interest.

The LSMUSER editor provides the following capabilities:

- nets can be chosen on the basis of their function in a design. For example, the user can specify all the primary inputs in a design.
- nets containing a specific substring of characters can be chosen.
- Nets may be grouped together and moved as a unit.
- Nets may be formed into a register. A register is one net containing one or more component nets. Registers may be "dissolved" into the separate component nets at any time.
- Fault values can be applied to nets.

- Predecessors and successors of nets can be found automatically.

After selecting nets for simulation, the user runs the simulation and then displays the results. Results may be displayed graphically as waveforms, or in textual form in hex, in binary or in symbolic representation (+ = logical 1, * = undefined, etc.).

When the results are displayed graphically, the user can change the time scale on his or her display terminal, thus either expanding or contracting the time span shown on the screen. The user may change which nets are being displayed at any time as long as the desired data were collected during simulation.

LSMUSER is written in PASCALVS. It contains 16,000 lines of code. LSMUSER accesses the symbol tables created by the LSM compiler in order to perform design dependent functions.

COMMUNICATIONS SOFTWARE

The LSM is attached to a host computer via an interface computer. At Los Gatos, the host is an IBM System 370/3081; the interface computer is an IBM Series 1. The LSM software runs under IBM VM/370. LSM software on the host computer communicates with the interface computer via a program called CADCOM (Concurrent Automatic Design Communications Facility). CADCOM enhances IBM's standard communication facility.

At any given time, many users may be trying to use one of the two LSM systems at Los Gatos Laboratory. These are not only local Los Gatos users but include remote IBM users that communicate with the LSM over an IBM computer network. (Remote users have their own software installations and ship the LSM load files across the network.)

These multiple users may all use the LSM software simultaneously to set up their simulations, but only one user at a time can actually run a simulation. The possible conflicts are resolved by a queueing facility program called LSMQ.

DIAGNOSTIC SOFTWARE

All diagnostics other than parity checks are implemented in software. Most diagnostics programs are simply simulations with known results. Each one of these simulations was designed to test specific LSM functions.

Other diagnostics take the form of memory initializations and reads. The data then obtained can be compared with correct data. Still other diagnostics are run from the interface computer rather than from the host computer. Typically these are the most detailed of the diagnostic programs.

Certain diagnostics are run automatically whenever the LSM is not in use. These provide a constant check on the accuracy and reliability of the LSM. When the LSM is requested for simulation, the diagnostic programs release control of the LSM and simulation can proceed. If errors are found by the automatic diagnostic programs, the problem is identified and logged. Pertinent memory contents and other data are also saved.

CONCLUSION

The LSM software has proven vital in the utilization of the LSM. Because users interact with the LSM through the LSM software, the quality of the software determines to a great extent the users' satisfaction with the LSM.

Development of new techniques and improvements to initial concepts has been a continuing process in development of the LSM software. User experience, innovation and requests have provided the software development team at Los Gatos with many of the ideas incorporated into the current software. As sure as there are more LSM users, there will be additional ideas for software enhancements.

ACKNOWLEDGEMENTS

The Logic Simulation Machine was invented by John Cocke (IBM Thomas J. Watson Research Center, Yorktown Heights, N.Y.), Richard Malm (IBM Los Gatos Laboratory, Los Gatos, CA), and John Schedletsky (Yorktown).

Jeff Barber, Wally Bass, Warren Begas (Burlington), Ted Burggraff, Daniel Burrier, Stan Head, Tim Kemp, Jack Kohn, Eric Kronstadt (Yorktown), Jean Lesser (Yorktown), Al Love, Chuck Meiley, John Moussouris (Yorktown), Frank Nemec, Greg Pfister (Yorktown), Jack Powers (San Jose Research) , Don Rozenberg (Yorktown), David A. Smith (Yorktown), and Gary Willis all contributed to the development of the LSM software system.

REFERENCES

[1] John K. Howard, Richard L. Malm, and Larry M. Warren, "Introduction to the IBM Los Gatos Logic Simulation Machine" Proceedings of the IEEE International Conference on Computer Design: VLSI in Computers , Rye, New York, October 31 - November 3, 1983.

[2] Ted Burggraff, Al Love, Richard Malm, and Ann Rudy, "The IBM Los Gatos Logic Simulation Machine Hardware," Proceedings of the IEEE International Conference on Computer Design: VLSI in Computers , Rye, New York, October 31 - November 3, 1983.

[3] John K. Howard, Jack Kohn, Richard Malm, and Ann Rudy, "Using the IBM Los Gatos Logic Simulation Machine," Proceedings of the IEEE International Conference on Computer Design: VLSI in Computers , Rye, New York, October 31 - November 3, 1983.

[4] Pascal/VS Language Reference Manual , International Business Machines Corporation, , Second Edition(April 1981), document number SH20-6168-1.

[5] System Product Interpreter , International Business Machines Corporation, product number 5664-167.

Using the IBM Los Gatos Logic Simulation Machine

John K. Howard, Jack Kohn, Richard Malm, Ann Rudy

IBM Los Gatos Laboratory, Los Gatos, California

ABSTRACT

The Logic Simulation Machine is a special-purpose computer designed to perform simulation of logic designs. The LSM is an interactive simulator that can simulate large systems in a matter of minutes, and thus takes its place in the VLSI design process as an important tool. It has, in fact, altered the way designers perceive logic simulation.

This paper describes the role of the LSM in the design process as experienced at IBM's Los Gatos Laboratory where two LSMs have been used since 1980. This paper also presents some examples of LSM use.

INTRODUCTION

This paper is the fourth of four integrated papers describing the Logic Simulation Machine. The first paper[1], provides an introduction to the LSM; the second[2] and third[3] papers discuss the hardware and software that comprise the LSM system. The intent of this paper is to show how the LSM is used in a real design environment.

The LSM is a high speed, highly parallel, special purpose computer with extensive software support. The LSM simulates 63K gates at a rate of 640 million gate evaluations per second. These descriptions are all accurate but are of somewhat secondary importance to the user who wants to use the LSM as a tool. "How do I use it?" "What can it do for me?" "How does it fit into what I am used to doing?" These questions are of primary importance to the prospective user of the LSM or to anyone contemplating using such a logic simulator in their design efforts.

THE LSM IN THE DESIGN PROCESS

Conventional attitudes toward logic simulators are not applicable to the LSM. The simulator is no longer used as an after-the-fact design verification tool. It has become an integral part of the design effort.

This has been accomplished at Los Gatos Laboratory by tying the LSM, through software, directly to the tools used to create logic designs. Interfaces between the standard Computer Aided Design (CAD) tools and the LSM have been provided. For example, design schematics are created with graphic entry devices. The designer describes a design with pictures and symbols that represent the various components. He or she uses a combination of text entry and light pen commands to describe the interconnections and functions of the different modules in a design.

With the LSM it is possible to go directly from the graphical description of a design through an automated conversion process to simulation of the design. This capability is provided by several software interfaces that are outside the scope of the LSM software. The crucial factor is the speed with which these conversions occur. If they are fast enough, the designer can readily shift back and forth between design and simulation. Thus the simulator moves upward in the hierarchy of design processes. It no longer remains at the tail end as a separate process. Figure 1 attempts to show the place of the LSM in the design hierarchy.

As indicated in the figure, there is a circular path between the creation and simulation of a design. This is called the **Experimentation/Verification loop** (XV loop). The viability of the XV loop is the most significant result of LSM use.

There has always been an iterative design-simulate process; indeed this is the reason for the existence of simulators in the first place. Simulators are intended to help designers locate design errors so they may be eliminated before a physical device is built. The difference here is that for the first time the XV loop is a fast and easy path. It is an individual process; one person can simulate and redesign with very little effort. Full simulation of a design on the LSM can virtually eliminate re-designing and rebuilding to correct design errors found after a physical prototype is built. Experimentation is possible more than with previous simulators simply because it costs little time or money to try a new idea. This concept may seem contrary to a conservative design outlook—one that presumes a designer knows exactly what he or she wants to do at all times—but experimentation results from creativity, and creativity fosters technological advances.

Users at Los Gatos Laboratory use a Tektronix 618 graphics display and an IBM 3277 terminal to per-

CH1935-6/83/0000/0592$01.00 © 1983 IEEE

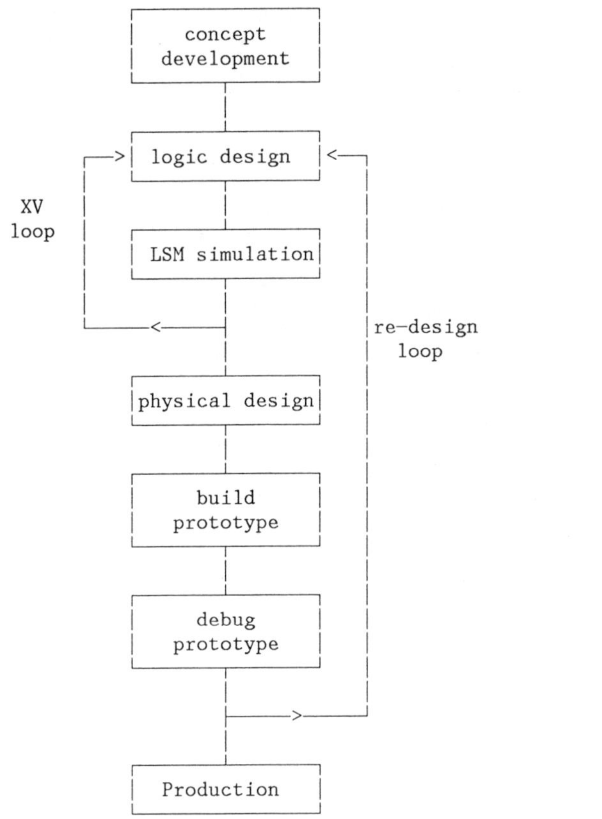

Figure 1. The LSM in the VLSI Design Process: the XV loop enhances creativity and virtually eliminates the need to rebuild prototypes

form simulations. This combination is designed around the needs of most LSM users. Simulation results may be viewed graphically as wave forms or as symbolic displays on the 618, while textual entry controlling that display is entered via the 3277. Symbolic signal output, in hex or binary, can be displayed on the 3277 or 618 displays.

The user sets simulation parameters, selects the nets to be simulated and displayed and runs the simulation from the 3277 terminal.

EXAMPLES AND COMPARISONS

The LSM has been used extensively since 1980 by designers at Los Gatos Laboratory and other IBM locations. Three examples of its use are given here.

First, the LSM was used to simulate a device controller chip being designed at Los Gatos. This chip contained a total of 52,748 transistors distributed as follows:

random logic	19820
PLA's	9047
on chip RAM	23881

22,000 LSM equations and 23 RAM ports (simulated by one array processor) were required to model the chip.

In the first stage of testing, the main functional islands of the chip (clock generator, ALU, etc.) were tested and debugged. Then, the full design was simulated, and logic errors were corrected. These simulations were all accomplished with test patterns. In a five month period, 1480 simulations were run on the LSM, approximately 15 per day. Only 64.5 hours of elapsed time were required on an IBM System 370 4341. It is estimated that simulation of this magnitude with a comparable software simulator would have taken more than ten years.

After the logic was debugged with test patterns, a more exhaustive testing was performed by simulating the real operation of the controller. 3280 lines of the chip's assembly language code were simulated. Each instruction class was tested with all possible addressing mode combinations, and each instruction was simulated at least one time. In addition, by using delay simulation, the chip was debugged in stages, with each progressive simulation operating at nearer to the simulated full speed of the chip. Delay was applied so that the chip was simulated at 1/4, 1/2, 3/4, full and intermediate speeds. At each speed, design errors were detected and corrected. Full regression testing (testing of all 3280 lines of machine code at all speeds) could be completed in less than one day's time. Each simulation took an average of 10 minutes.

Thus, the LSM became a true design tool, used to refine the design as it evolved. Design and simulation occurred alternately.

The LSM has also been used by designers at IBM locations other than Los Gatos. Remote users pass simulation data and parameters through an IBM computer communications network and run on one of the two Los Gatos LSMs. One such simulation involved a 5500 cell LSI system composed of four LSI chips. The following results were achieved.

7000 LSM equations, distributed among 26 logic processors, were required to model the system. The model was exercised with 1351 machine instructions. Total simulation time, including the time to load the LSM (14 seconds) and dump the LSM memories after simulation (5 seconds), was 48 seconds. One IBM software simulator running on an IBM 3081 computer took 70 times as long to simulate the same system. If the actual simulation time of 29 seconds on the LSM (48 total - 14 load - 5 dump) is used for comparison, the LSM is 100 times faster than the software simulator.

Simulation in this case was more at the machine level since the design included more than one chip. The test patterns used to exercise the model were the actual system diagnostics that will be sent along with the design for production. This means that no special test patterns were required.

This simulation was run in a highly detailed compare mode, in which most of the nets were compared. Simulation in either debug or strobe with a selection of nets would reduce the simulation time further still, making the difference between the LSM and software simulator even greater. Also, as models grow in size, the performance of software simulators is degraded, while more processors can be added to the LSM and the increased parallelism can actually improve simulation speed.

The LSM also was instrumental in the design of a chip tester at Los Gatos. The tester contains 12 million bits of RAM memory, 64 modules of 1000 gates each, and several cards of VTL modules. Using the LSM, the tester development group simulated its design at the chip, card, board and system levels. Many runs were performed at each level of simulation.

The result was that no design errors were encountered when the machine was built. It is estimated that the LSM reduced by an order of magnitude the time required to bring-up and debug the tester, simply because its architecture had been debugged during simulation.

CONCLUSION

Results like those described in the previous section are typical of simulation with the LSM. If anything, they provide conservative estimates of the potential for LSM simulation.

The LSM has become more than just a design verification tool and now assumes the role of an interactive design tool. It is used along with development of a logic design, providing easy and quick checking at all stages, and allowing experimentation with new ideas. Designers end up performing much more thorough checking than in the past because the verification can be performed as the design progresses. Ultimately, better logic designs are produced in less time and at less cost.

ACKNOWLEDGEMENTS

The LSM was invented by John Cocke of IBM's Thomas J. Watson Research Center, Yorktown Heights, New York, by Richard Malm of Los Gatos Laboratory, and by John Schedletsky of Yorktown. The hardware and software was developed at Los Gatos Laboratory under the direction of Richard Malm, John K. Howard and Larry M. Warren. Many persons at Yorktown and Los Gatos contributed to the LSM project and are not listed here. For more specific acknowledgement of these persons, see the preceding three papers in this series.

Scott Dahl of IBM Rochester, Minnesota, provided the LSM to software simulator comparison figures for the 5500 cell system described above.

REFERENCES

[1] John K. Howard, Richard L. Malm, and Larry M. Warren, "Introduction to the IBM Los Gatos Logic Simulation Machine" Proceedings of the IEEE International Conference on Computer Design: VLSI in Computers , Rye, New York, October 31 - November 3, 1983.

[2] Ted Burggraff, Al Love, Richard Malm, and Ann Rudy, "The IBM Los Gatos Logic Simulation Machine Hardware," Proceedings of the IEEE International Conference on Computer Design: VLSI in Computers , Rye, New York, October 31 - November 3, 1983.

[3] Jack Kohn, Richard Malm, Chuck Meiley, and Frank Nemec, "The IBM Los Gatos Logic Simulation Machine Software," Proceedings of the IEEE International Conference on Computer Design: VLSI in Computers , Rye, New York, October 31 - November 3, 1983.

Session 3-2c
CAD and VLSI II

Chairman

I. Shirakawa
Osaka University

PLACEMENT AND ROUTING ALGORITHMS FOR GATE ARRAY LSI

Shuji Tsukiyama, Ikuo Harada*, Masahiro Fukui**, and Isao Shirakawa

Department of Electronic Engineering, Osaka University
Suita, Osaka 565, Japan

ABSTRACT: This paper outlines an automatic layout system for gate array LSI's, which is constructed of a number of phases; clustering, rough placement, detailed placement, global router, pin assignment, and detailed router. This system is distinctive in that all layout procedures are organically combined to attain the ultimate object of realizing 100% interconnection by means of distributing all interconnection requirements over channels evenly.

1. INTRODUCTION

With the progress of the microelectronics technologies, the circuits which used to be composed of several printed circuit boards can be realized by means of a single LSI chip. Thus, the amount of design for a variety of LSI's, each produced in small quantity, is increasing rapidly. In order to make the design and production of such LSI's easier and less costly, a gate array LSI is introduced and numbers of layout methods have been developed[1-7].

In a gate array LSI chip, a certain number of cells (usually 1,000 through 5,000), each with a function of NAND or NOR gate, are preformed regularly, and the wiring space (channels) is predetermined. To attain 100% interconnection on such LSI chips, of primary importance is to distribute pairwise interruped interconnection requirements over cahnnels evenly so as to cut down peaks of localized interconnection requirements. To this end, all the layout procedures are to be organically combined so that features of one may be incorporated into those of another.

In this paper, we outline basic concepts of a series of layout procedures adopted in an automatic layout system for gate array LSI's.

2. SYSTEM OVERVIEW

Fig. 1 illustrates a layout model of the gate array LSI of our interest, where cells are arranged in a two-dimensional array of cell-rows and cell-columns. Each cell has a function of a logic gate such as NAND or NOR with pairs of electrically common terminals on its top and bottom sides, as shown in the figure. The area exclusive of these cells and surrounded by I/O cells is used for routing, which is divided into horizontal and vertical channels. The number of tracks available in a channel is designated as the capacity of the channel. Each portion of a vertical channel which lies between two adjacent cells on a cell-row is regarded as a feed-through cell[5].

Fig. 2 shows a general flow of our layout system. The first three phases constitute the placement process, and the remaining ones the routing process. Clustering is to put together each class of gates and macro-gates† which are connected strongly, into a cluster, by means of an affinity function. Rough Placement is to determine to which cell-row each gate (or macro-gate) is to be assigned and to which side of I/O cells each external pin is to be assigned. If the number of cell-columns is comparatively large, then it divides all cell-columns into a number of classes and determines not only to which cell-row but also to which part of the cell-row (e.g., right part, left part, etc.) each gate is to be assigned. Detailed Placement assigns each gate to a cell within a cell-row (or a part of a cell-row), and allocates each penetrating net†† of a cell-row to a vertical channel segment crossing the cell-row. Global Router is to determine, for each net, a set of those channel segments within which the interconnection route of the net is to be realized. Pin Assignment is to determine the position (the vertical track) of each floating signal pin†*. Detailed Router is to complete the final wire patterns for all interconnection requirements within each channel by means of a dogleg channel router.

3. CLUSTERING

In order to place those gates which are strongly connected each other or contained in a macro-gate, closely in an LSI chip, we classify a set of such gates into a cluster, and then we determine the position of each cluster. To classify a set of gates into a cluster, we introduce an affinity function CON(C) for a set C of gates defined as

$$CON(C) \triangleq \Sigma_{g \in C}|N(g)| \,/\, |C| \cdot |\cup_{g \in C} N(g)|,$$

where N(g) denotes a set of nets connceting to gate g in C. With the use of this affinity function, the clustering phase is conducted as follows.

1°: Each macro-gate is regarded as a cluster, and each gate not belonging to any macro-gate

---

* presently with Atsugi Electrical Communication Lab., NTT, Atsugi, Japan.

** presently with Central Res. Lab., Matsushita Electric Industrial Co., Ltd., Moriguchi, Japan.

† A macro-gate is a set of gates which are specified by a user to be laid out in a group.

†† A penetrating net of a cell-row is a net with terminals above and below of the cell-row but none in the cell-row.

†* A floating signal pin is a fictitious pin indicating a portion of a net which is to be connected to one of the logically equivalent terminals of a cell or is to be assigned to a feed-through cell with more than one tracks available.

constitutes a cluster by itself. Then, compute $CON(C_i \cup C_j)$ for each pair of distinct clusters C_i and C_j.

2°: Find a pair of clusters C_x and C_y such that $CON(C_x \cup C_y)$ is maximum and $|C_x \cup C_y|$ is minimum.

3°: If $CON(C_x \cup C_y)$ is less than the specified threshold value T (in our system $0.5 \le T \le 0.6$), then terminate. Otherwise, combine C_x and C_y into a new cluster and update CON values. Then, return to 2°.

Due to this clustering phase, not only the total wiring length but also the total processing time for placement procedures are reduced.

4. ROUGH PLACEMENT

The rough placement phase is composed of two procedures; the first one called Cluster-Partition determines to which part of the chip each cluster is to be allocated, and the second one called Cut-Placement determines to which cell-row each gate is to be placed.

In procedure Cluster-Partition, firstly a set of nets is partitioned into four subsets as shown in Fig. 3(a). The nets in subsets T, B, R, and L are to be interconnected around the top, bottom, right-hand, and left-hand sides of the chip, respectively. Concerning this partition, the following conditions are to be satisfied.

(i) Let s(T) be the number of gates contained in the clusters which are connected to nets in T; and let s(B), s(R), and s(L) be the numbers similarly defined. Then, these numbers must be approximately the same.

(ii) The number of those nets in each of subsets T, B, R, and L which are incident with the external pins is to be smaller than that of I/O cells on the corresponding side of the chip.

Secondly, according to this partition, a set of clusters and a set of external pins are classified into 9 blocks and 4 blocks, respectively, as shown in Fig. 3(b): Here, BLK(T), BLK(B), BLK(R), and BLK(L) are blocks consisting of clusters connected only to nets in T, B, R, and L, respectively; and EXT(T), EXT(B), EXT(R), and EXT(L) are blocks consisting of external pins similarly defined. BLK(LT) denotes a block consisting of clusters connected both to nets in L and to nets in B but not to others; and BLK(RT), BLK(LB), and BLK(RB) are blocks defined similarly. BLK(M) is a block consisting of the remaining clusters.

Finally, based on these blocks, a set of clusters and a set of external pins are partitioned into 4 and 8 subsets, respectively. This partition is conducted through two steps; the horizontal cut and the vertical cut. Fig. 3(c) shows schematically the vertical cut to be applied after the horizontal cut, where it should be noted that blocks EXT(L), EXT(R), BLK(L), BLK(M), and BLK(R) have been partitioned by the horizontal cut as shown, and that the shaded blocks are not partitioned this time.

In the case where the number of cell-columns is greater than the double of the maximum number of gates in a macro-gate, this procedure is again applied to each quater part of the chip, so that the chip is divided into 16 regions through the horizontal and vertical cuts.

In procedure Cut-Placement, an algorithm of the so-called min-cut palcement[8] is employed to determine to which cell-row each gate is to be assigned, where the horizontal cut lines are generated in a "bisection" manner[9], and the gates in a macro-gate are regarded tentatively as a single gate so that those gates should be assigned to the same cell-row. During the iterative generation of new horizontal cut lines in a bisection manner, if two cut lines are newly generated just above and just below an old cut line, then along the old cut line a set of gates which are currently specified to be located between the new cut lines is repartitioned to obtain a better placement.

A similar procedure is applied to the vertical cut lines, until the number of cell-columns between two consecutive vertical cut lines becomes less than the double of the maximum number of gates in a macro-gate.

It should be added here that the algorithm with "INTERMIX"[10] is used for partitioning a set of nets, clusters, or gates in this rough placement phase.

5. DETAILED PLACEMENT

Detailed Placement is strongly interrelated with Global Router, and determines the final placement of gates by taking a density estimation in each horizontal channel into account, as outlined in the following.

1° Placement of an Initial Cell-Row: Seek a cell-row, say the j-th cell-row, such that the number of nets incident with gates on the cell-row is greatest. Determine the placement of gates and penetrating nets of the j-th cell-row*† so that the peak of congestion for interconnection requirements within the cell-row may be minimized, and the interconnection length expected for each penetrating net may be reduced (see [11]).

2° Iterative Placement of Each Cell-Row: Let r be the number of cell-row. For each $(i,k) \leftarrow (j+1,j)$ until $(r,r-1)$ with the increment of both i and k by one, and then for each $(i,k) \leftarrow (j-1,j)$ until $(1,2)$ with the decrement of both i and k by one, determine the placement of gates and penetrating nets of the i-th cell-row so that the peak of congestion for interconnection requirements on the horizontal channel between the i-th and k-th cell-rows may be minimized, and the interconnection length expected for each penetrating net may be reduced.

3° Placement of I/O cells: Allocate external pins to the top and bottom I/O cell-rows and then to the left and right I/O cell-columns in a manner similar to 2°.

4° Improvement: Improve the placement of gates, external pins, and penetrating nets so as to minimize the maximum density of interconnection requirements in each horizontal channel, by means of a pairwise interchange.

6. GLOBAL ROUTER

At the termination of the placement process, all penetrating nets have been assigned to vertical channel segments, but those nets connecting to adjacent cell-rows have not been assigned, even if the use of feed-through cells makes the interconnection

*† The placement (or assignment) of penetrating net of a cell-row indicates to determine a vertical channel segment through which the interconnection of the net is to cross over the cell-row.

length shorter (see Fig. 4). Thus, this phase first assigns such nets to unsaturated feed-through cells so as to derive shorter interconnection.

Then, this phase constructs for each net N a rough sketch S(N), as illustrated in Fig. 5, which suggests a set of candidates for those channel segments along which the interconnection for N is to pass. On the basis of such a rough sketch S(N), the global router determines for each net N an interconnection pattern P(N), which is composed of a set of channel segments within which the interconnection of N is to be realized. In the process of constructing such P(N) from S(N), redundant portions of channel segments at which the interconnection requirements are most congested are to be deleted with top priority. By using rough sketches, we can take all interrelated interconnection requirements into account at once. Owing to limited space, the detailes of this phase are omitted (see [11]).

At the final stage of this phase, a rerouting scheme is employed so that those channel segments in P(N) for net N at which the congestion of interconnection requirements exceeds a certain number, can be replaced by a set of other channel segments of less congestion.

7. PIN ASSIGNMENT

According to our cell model, each cell has a set of logically equivalent (input) terminals. Thus, at the stage when the interconnection pattern P(N) for each net N has been determined, there arises a problem of which one of logically equivalent terminals is to be used for the net to be connected to any one of such terminals. Moreover, there also arises a problem of which vertical track of a feed-through cell is to be used for the net assigned to the feed-through cell with more than one tracks.

This phase of Pin Assignment is to solve these problem with the use of a heuristic algorithm which intends to satisfy the following criteria:

(i) The total number of cycles in the so-called vertical constraint graph[12] for all horizontal channels is to be minmized.

(ii) The peak of congestion of interconnection requirements in each horizontal channel does not exceed some specified number.

(iii) The number of vertical constraints[12] in each horizontal channel is to be minimized.

The heuristic algorithm reduces a set of vertical tracks to which each floating signal pin can be assigned, by paying attention to vertical constraint graphs and congestions of all horizontal channels, until each floating signal pin has a unique vertical track to be assigned.

At the termination of this phase, the whole routing problem of an LSI chip is divided into the channel routing problems of each horizontal channel and the left and right vertical channels.

8. DETAILED ROUTER

Given a set of interconnection requirements in each channel, this phase of Detailed Router is to generate final wire patterns with the use of horizontal and vertical tracks within the channel. For simplicity, we assume that a given channel is a horizontal one henceforth.

The algorithm used in this phase repeats the process of introducing a dogleg[12] of a net N at a vertical track t for a pair of N and t which are selected in a heuristic manner, until the vertical constraint graph of the channel is changed into an acyclic one. Then, in order to reduce the length of the longest directed path in the vertical constraint graph to such an extent as a quater of the channel capacity, another set of doglegs are generated with the use of a different heuristic algorithm.

In this dogleg generation, at each vertical track at most one dogleg is admitted for a net which does not contain a terminal located at the vertical track, and the concept of the feasible region of a net[13] is used in such a way that a dogleg is to be generated for net N at vertical track t so that both the feasible regions of subnets N' and N" of N seperated at t can be expanded as much as possible. The feasible region of a net is calculated from the vertical constraint graph and the zone representation[12], and implies a set of candidates of tracks to which the unique horizontal wire segment of the net can be laid out.

After these processes, detailed wire patterns are generated with the use of a merging procedure, which is modified slightly from the one proposed in [12] by using the concept of the feasible region of a net. Owing to limited space, the details are omitted here.

9. CONCLUSION

We have outlined an automatic layout system for gate array LSI's, stressing mainly basic concepts of a series of layout procedures. Owing to limited space, we only brief an experimental result: Given a circuit of 899 gates, 927 nets, and 2771 terminals including 99 external pins, to be realized on a (15 cell-rows) x (60 cell-columns) = (900 cells) array, the system attain 100% interconnection, where the maximum densities of all channels at the final stage of Global Router and at the initial stage of Detailed Router are 11 and 12, respectively, for which the number of horizontal tracks used for interconnection does not exceed 13 in any horizontal channel.

ACKNOWLEDGMENT: The authors would like to express their appliciation to Prof. Emeritus H. Ozaki of Osaka Univ., for his continuous encouragement, and to Dr. T. Chiba, Sharp Co., Dr. H. Kawanishi, Nippon Electric Co., and Dr. K. Sato, Mitsubishi Electric Co., for providing us with useful information and data of LSI examples in practice. They also wish to thank T. Kimoto and T. Ikeda, Osaka Univ., for their assistance in the programming of this system and the collection of implementation results.

This work was supported in part by the Grant in Aid for Scientific Research of the Ministry of Education, Science, and Culture of Japan under Grant: B-57460117 (1982,1983).

REFERENCES

[1] R. Kamikawa, et al., Proc. 13th DA Conf., pp. 245-250 (1976).

[2] K. Ueda, et al., IEEE J. Solid-State Circuits, vol. SC-13, no. 5, pp. 716-721 (1978).

[3] H. Yoshizawa, et al., Proc. IEEE ISCAS, pp. 470-473 (1979).

[4] H. Shiraishi, F. Hirose, Proc. 17th DA Conf., pp. 458-464 (1980).

[5] G. Persky, et al., Proc 13th DA Conf., pp. 399-407 (1976).
[6] K. A. Chen, et al., Proc. 14th DA Conf., pp. 298-302 (1977).
[7] I. Harada, et al., Mono. IECE of Japan, CAS82-149, pp.61-66 (1983) (in Japanese).
[8] M. Breuer, J. DA & FTC, vol. 1, no. 4, pp. 343-362 (1977).
[9] L. I. Corrigan, Proc. 17th DA Conf., pp.406-413 (1979).
[10] M. Burstein, Proc. IEEE ISCAS, pp. 477-480 (1982).
[11] S. Tsukiyama, et al., IEEE Trans. CAD, (Oct. 1983) (to appear).
[12] T. Yoshimura, E. S. Kuh, IEEE Trans. CAD, vol. CAD-1, no. 1, pp. 25-33 (1982).
[13] M. Fukui, et al., Mono IECE of Japan, CAS 81-121, pp. 91-98 (1982) (in Japanese).

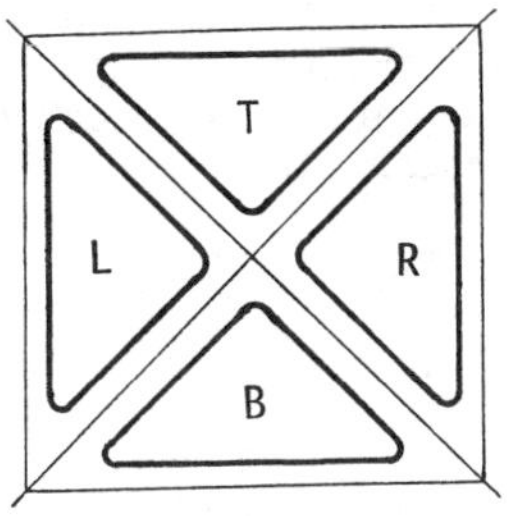

Fig. 3(a)
A partition of a set of nets.

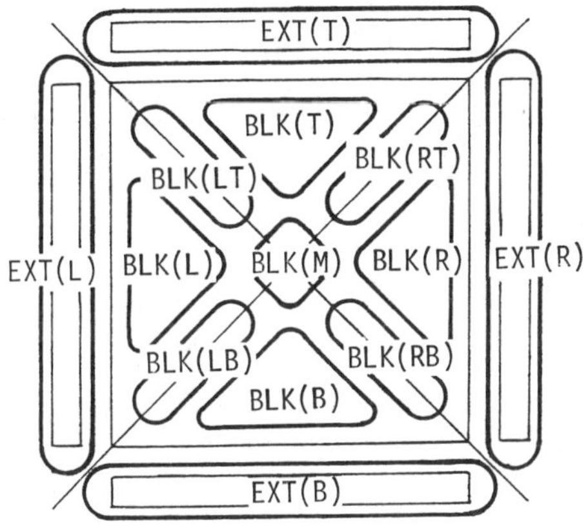

Fig. 3(b)
Blocks.

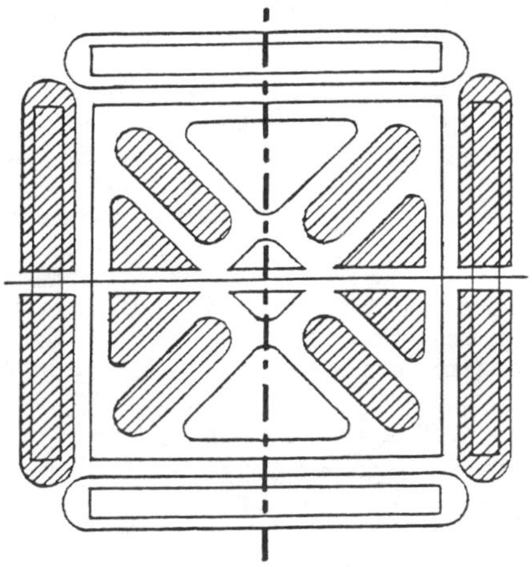

Fig. 3(c)
A vertical cut.

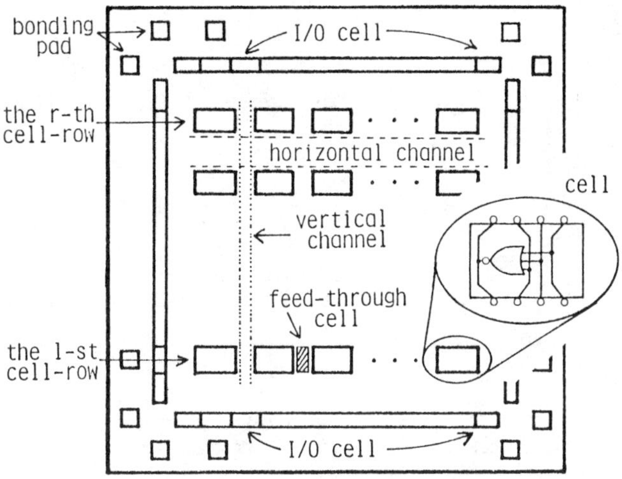

Fig. 1 A gate array LSI chip.

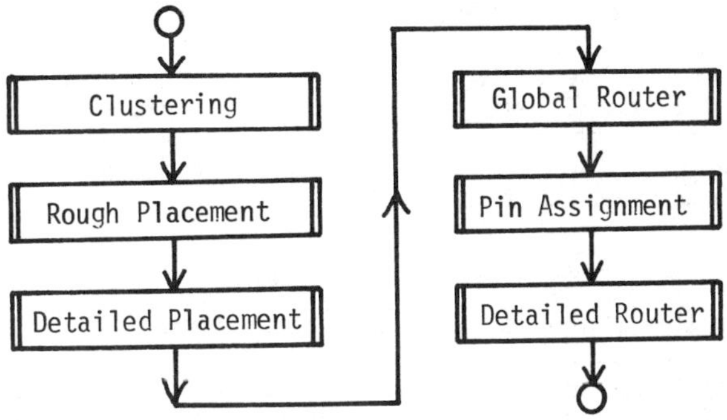

Fig. 2 A general flow.

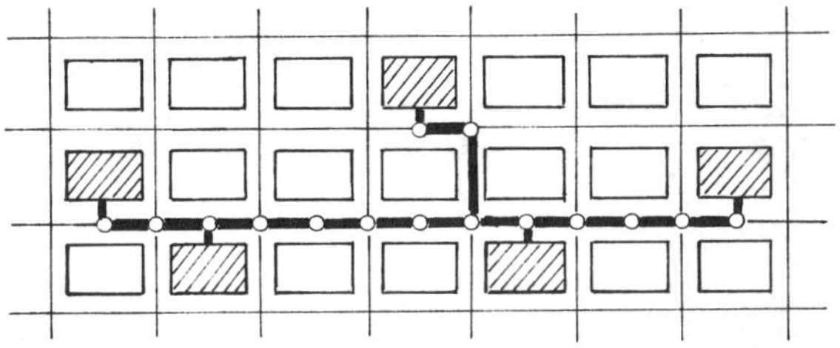

Fig. 4 An interconnection pattern of net N.

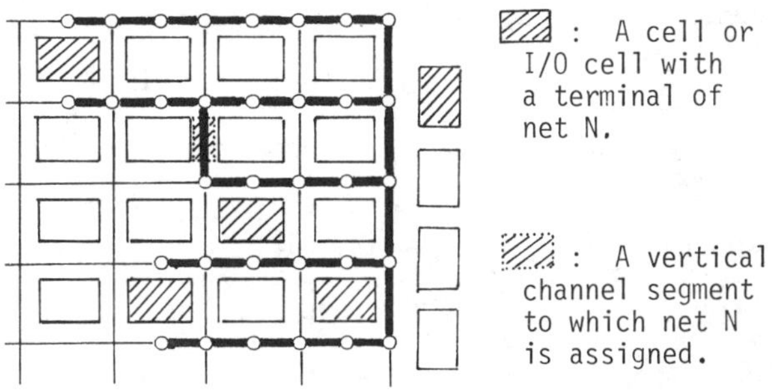

Fig. 5 Rough sketch S(N) of net N.

LCAD/GA : LSI CAD SYSTEM FOR VERY LARGE GATE ARRAYS

Y. Hatano, K. Kawauchi, T. Aikyo and J. Ishii

IC Development Division, FUJITSU Ltd. , Kamikodanaka,
Nakahara-KU, Kawasaki, Kanagawa, 211, Japan

ABSTRUCT: LCAD/GA is an integrated computer aided design system for gate arrays and has been put to practical application at Fujitsu to design 8000-gate CMOS gate array,2000-gate STTL gate array and other LSI gate arrays. A key feature of the system is performance prediction and autmatic layout capability taking hierarchical design approach and enables an LSI layout controllable on the initial logic design stage. This paper describes the configuration and the basic concept of the LCAD/GA system, and gives the outline of feature and function of four major subsystems: logic data entry, logic simulation,automatic layout, and final validation.

INTRODUCTION

With the increasing needs of large scale LSI,the demand for gate arrays has been remarkable in various fields. Because they can realize the desired LSI with shorter design cycle time and smaller chip development cost.

On the other hand, the evolution of LSI process thechnology has enabled the performance and integration density of large scale gate arrays to be enhanced markedly. Fujitsu has already produced 8000-gate CMOS gate array with 2.5 ns per gate, 2000-gate STTL gate array with 1.5 ns per gate and other twenty types of gate arrays. [1-2] Furthermore, the scale of several tens of thousands of gates will be feasible in near feature.

In VLSI design the effect of chip layout on the performance becomes significant. Therefore VLSI design has become impossible without chip layout examination on the initial logic design stage. Final validation which performs ordinary delay simulation or DC drop check with an actual wire length cannot be a solution for VLSI design. A design methodology and CAD system that unite logic design and layout design are required.

To realize this design integration,the LCAD/GA system described in this paper employs an approach which uses logic hierarchy structure for performance prediction and automatic layout control. This approach has achieved the desired goal in the application to the 8000-gate CMOS gate array and others. It is expected to be applicable to future gate arrays with several tens of thousands of gates.

This paper describes the basic concept and configuration of LCAD/GA which has made long record to develop numerous options of gate arreys. Following this,feature and function of four major subsystems, logic data entry, logic simulation, automatic layout and final validation are outlined. Finally, the application results to 8000-gate CMOS gate array and 2000-gate STTL gate array are summarized.

BASIC CONCEPTS

The LCAD/GA system has been developed based on the following concepts to cope with rapidly evolving gate array LSI technology and to establish design automation system in short development time.

(1) Generalization for device technology and masterslice structure

To be applicable to new device technology and increasing scale of gate arrays, CAD tools must be independent of device technology and masterslice structure to reduce system development efforts. For the purpose of embodying this generalization, physical shape and electrical characteristics of transistor determined by device technology, such as CMOS, STTL or ECL , are handled as cell library in which each group of transistors is standardized as a unit cell. In addition , basic cell arrangement, channel configuration, wiring tracks and wiring conditions are prepared as masterslice structure library.

(2) Integration of CAD tools by a centralized data base

A system in which CAD tools are integrated from logic design stage to mask data generation stage is required to minimize overall design process turnaround time and development costs of gate arrays. To establish this integrated CAD system, automated and interactive CAD tools are organized with a common database. This common database management enables a new CAD tool to be added with no effect on other CAD tools and makes it easy to perform phased addition of functions.

(3) Implementation of high performance LSI by integrating logic design and layout design

Because the performance of VLSI depends

significantly on the layout design, a system in which logic design and layout design are united is required to design high performance LSI. Control condition for layout design has to be set on the initial logic design stage , and performance prediction and automatic layout functions should be provided according to the control condition. As a means for this the LCAD/GA system uses hierarchical design information and large scale cells such as register, counter, etc. This mean is also effective for improving logic design efficiency.

(4) Turnaround time reduction by minimizing man-machine interface

In the system, the automatic technique, design procedure, and masterslice structure are set up so that fully automated layout design can be performed successfully. Because numerous gate array options have to be designed in a short period of time. Logic data can be entered using either the FLDL design language or logic diagram. The FTDL test language, which is capable of systematically describing Function, DC and delay tests, is prepared for test specification description. The language permits CAD to CAD interface and shoten the turnaround time.

SYSTEM OVERVIEW

Figure 1 shows the overall configuration of the LCAD/GA system. Each subsystem is integrated with a centralized common data base, which is composed of control section,library section and design data section. Contents of each section is given in Table-1.

Table-1 Data base contents

DATA BASE CONTROL	MANAGEMENT FILE Technology management file
LIBRARY DATA	CELL LOGICAL LIBRARY Logic function Electric characteristics
	CELL PHYSICAL LIBRARY Physical pattern Shape and prohibited area I/O pin position
	MASTERSLICE STRUCTURE LIBRARY Basic cell arrangement Channel configuration Wiring tracks per channel Wiring condition
DESIGN DATA	MODULE LOGICAL FILE Hierarchically structured logic design data
	EXPANDED MODULE FILE Expanded logic design file
	MODULE PHYSICAL FILE Layout design data
	TEST DATA FILE Function/DC/Delay test data

LOGIC DATA ENTRY SUBSYSTEM

Logic data entry is the first stage in design process and classified to logic input, design

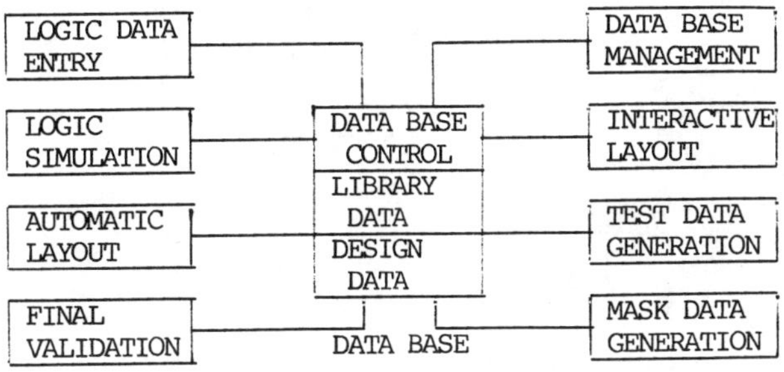

Figure-1 LCAD/GA system configuration

rule check, and wire length estimation. Using the estimated length the performance can be predicted by the initial stage simulation. Therefore the estimation method becomes an important factor for a high performance VLSI design.

Logic input

Logic design is represented using the hierarchical structure specified with cells as the lowest level. The cells are prepared in their respective families from simple gate to large scale gates such as register, counter, etc. so that design can be carried out efficiently.

Hierarchically represented logic data are written in the FLDL language and registered on the design database. In addition, an input tool is prepared to input logic diagrams directly through an interactive editor.

Design rule check

Hierarchical structure of the logic data is expanded and checked initially through a design rule check program. This check is performed for design specifications determined by each device technology and masterslice structure.

Wire length estimation

A wire length can be estimated successfully as a function of the number of the cells contained in the highest hierarchy to which the wire belongs. This approach is based on a hierarchical layout method (described later), by which cells contained in the same hierarchy are closely laid out on chip.

In reference to the logic hierarchical representation shown in Figure-2, an example is given to estimate wiring length. If Lij is the estimated wiring length between the cell i and j, then Lab, Lac and Lad are given by the followings.

Lab = F(No.of BCs contained in hierarchy "D")
Lac = F(No.of BCs contained in hierarchy "B")
Lad = F(No.of BCs contained in hierarchy "A")

LOGIC SIMULATION SUBSYSTEM

Logic simulation is indispensable for performance prediction on the initial logic design stage and detailed timing verification after the chip layout .

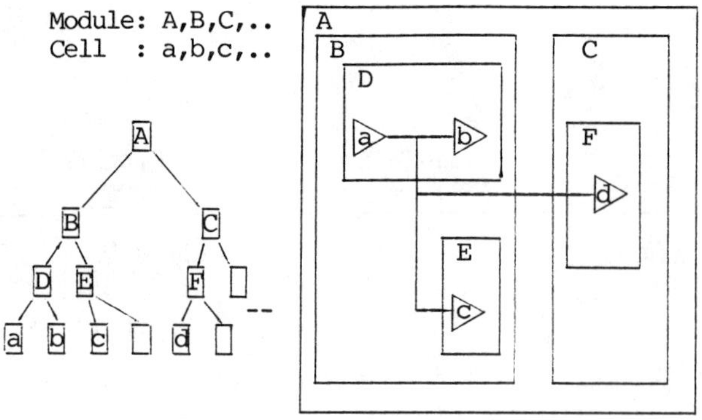

Figure-2 Logic hierarchical structure

As the layout design has been fully automated in gate arrays,logic simulation process becomes predominant parts of design cycle. Consequently requirements imposed to the logic simulator increase more and more. [4] The requirements are:more accurate modeling of timing,higher speed of simulation and automated facility of error detection. This section describes features of the simulator that has been developed to satisfy above requirements.

Logic network model

The logic network model has two kinds of properties to be used by the simulator. These are interconnection of logic elements and propagation delay values. Accuracy of the model depends entirely upon the delay value which is computed from the device parameter and the net length ,estimated or actual,using specified delay equation. The delay value is specified by the six types,that are Tpd0-NOM,Tpd0-MIN,Tpd0-MAX, Tpd1-NOM,Tpd1-MIN and Tpd1-MAX. The delay types to be used during simulation can be selected by the simulation mode. The minimum resolution of the delay value can be set to 0.01 ns so that high accuracy of modeling is realized.

Dotted functions are automatically generated from signal interconnection and design technology. Bus modeling is also done automatically as a subset of the dotted function process. This automated modeling process contributes to flexible hierachical design .

Up to eleven different signal values are handled, depending on the selected simulation mode and the design technology. These are: 0,1 , X,Z,E,U,D,Z0,Z1,ZX and ZE. Signal values U and D are used when worst case timing analysis is performed. Z0,Z1,ZX and ZE are used only for MOS transistor element.

Simulation technique

To enhance the delay resolution ,extension of the time wheel is required. Simple extention causes the scanning of empty time slot and increases the simulation run time.

As a solution to this problem this simulator takes a multi-wheel technique [5] to skip time segmuents during which no events occure. This technique provides detailed timing simulation with minimum increse of run time.

Automated error detection

This simulator has the following automated error detection capabilities that require little efforts from the logic designer and allow him VLSI verification efficiently.

1) To compare expected signal value and simulated signal value
2) To check permissible time interval of Bus confliction
3) To detect Setup,Hold and pulse width violation
4) To detect Oscillation,Hazard and spike pulse occurence

AUTOMATIC LAYOUT SUBSYSTEM

Automatic layout subsystem consists of automated placement and routing process.

Placement

Sophisticated placement method has been developed to conform to hierachical structure to be set on the initial logic design stage. This method carries out placement in three steps and applies the min-cut technique [6-7] to first and second step so that hierachical top down placement and high density placement for variable size cells are achieved. These steps are illustrated in Figure-3.

In first step,the top down block placement is carried out using the block oriented min-cut technique. At this step the relative locations of the blocks are assigned by repeated hierachical logic partitioning process and the size of each block is computed as a function of the number of basic cells,terminals and nets contained in the block. Following this,rectangle area to each block is determind after size ajustment on a chip.

In second step,the concrete initial position of cells are determind by partitioning of a chip using the cut oriented min-cut technique. During partitioning process,cell packing check which assures complete placement of variable size cells is performed. As the result the boundary line among the block areas becomes intricate in some cases.

In third step,the initial placement is improved to distribute wiring density and to reduce wiring length. This improvement technique includes pairwise interchange,external cell movement,and internal cell rotation.

Routing

In contrast to above mentioned hierachical placement approach the routing process is carried out over the whole chip simultaneously. The reason why adopting this approach is to use channel area efficiently as a whole,since the tracks per channel is predetermind in gate array LSI. To improve wirability,it is important to make the wiring density on channels uniform. Thus the routing method which evaluates wiring density and equalizes it in actual processing is developed. This realizes high wirability and accomplishes fully automated layout result together with the placement method.

FINAL VALIDATION SUBSYSTEM

Final validation subsystem has been developed to achieve complete error detection prior to manufacturing. The functions of this stage are classified as the following four categories.

1) To check the consistency between the layout design results and the logic design data base

2) To check various physical design restriction such as minimum space between layout objects,occupation of the prohibited area by line segments,etc.

3) To check various electrical design restriction such as the limit of wire and/or net length,the maximum value of DC voltage drop,etc.

4) To verify performance by the worst case timing simulation based on the layout design results and the variation of manufacturing condition.

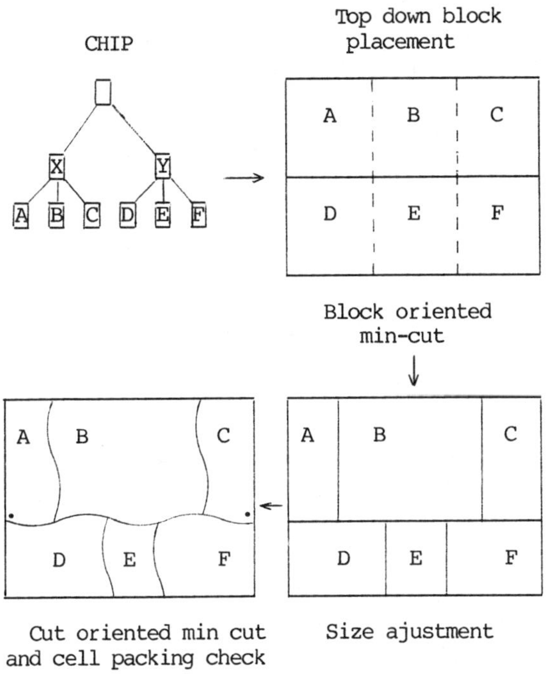

Figure-3 Placement process

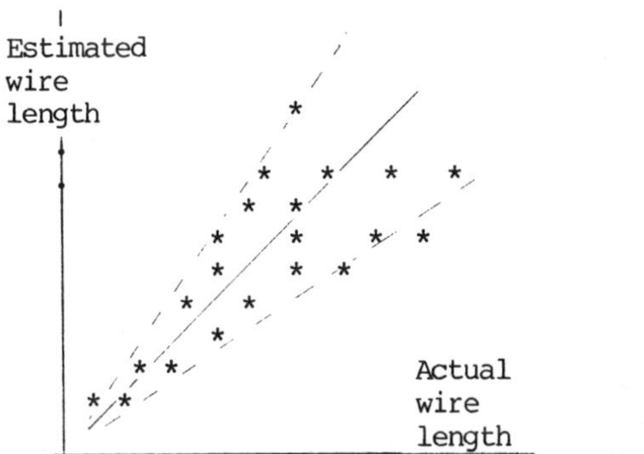

Figure-4 Relationship between estimated and actual wire length

SUMMARY

An overview of the LCAD/GA system based on the hierarchical design approach has been presented.

This system employs the method to use logic hierarchical structure for performance prediction and automatic layout control on the initial logic design stage.

Figure 4 shows the relationship between the estimated wire length on the logic design stage and the actual wire length after layout. Although the difference increases according to the magnitude of the value, this approach satisfies the requirements in actual application.

Futhermore this approach will become a driving force to design a high performance very large gate arrays in short design cycle time.

Table 2 shows the average CPU time (on 4 MIPS machine) for 8000-gate CMOS gate array and 2000-gate STTL gate array design. The automatic layout method has accomplished 100% wirability and satisfied electrical restrictions in the case of 90% cell usage.

Table-2 Application result

	CMOS 8000-gate	STTL 2000-gate
Logic simulation	1.5 MIN / K clock	0.4 MIN / K clock
Automatic layout	60 MIN	15 MIN

REFERENCES

[1] R.Hoshikawa,N.Matsumura and K.Kawauchi: 'Introduction of an ultra fast 8000-gate CMOS gate array',The 2nd International Conference of Semi-Custom ICs,London,1982.

[2] M.Ashida,R.Hoshikawa,J.Ishii, H.Ichikawa, Y.Sugiura and K.Inayoshi:'A 3000-gate CMOS Masterslice LSI',Proc.11th Conf.on Solid State Devices,Tokyo,1979.

[3] B.W.Kernighan and S.Lin:'An efficient Heuristic Procedure for Partitioning Graphs',Bell System Technical Journal',Vol.49,1970,pp291-307.

[4] M.A.Breuer and A.C.Parker:'Digital System simulation:Current Status and Future Trend', Proc.18th Design Automation Conference ,1981 ,pp269-275.

[5] E.G.Ulrich:'Exclusive Simulation of Activity in Digital Networks',Communications of the ACM,Vol.12,No.2,February,1969,pp102-110

[6] D.G.Schweikert and B.W.Kernigahan:'A Proper Model for Partitioning of Electrical CirCuits', Proc 9th Design Automation Workshop,1972,pp57-62.

[7] M.A.Breuer:'A Class of Min-Cut Placement Algorithms',Proc 14th Design automation Conf,1977

[8] Korenjak,A.J,and Teger,A.H:'An integrated CAD data base system',Proc.12th Design Automation Workshop,June 1975,Boston,Massachusetts.

[9] L.I.Corrigan:'A Placement Capability Based on Partitioning',Proc.16th Design Automation Conf,1979,pp406-413

AN LSI MASK ARTWORK VERIFICATION SYSTEM

Takashi Mitsuhashi*, Hisashi Yamada* and Kenji Yoshida**

* : Research and Development Center
TOSHIBA Corporation
Kawasaki, 210 Japan

** : Integrated Circuit Division
TOSHIBA Corporation
Kawasaki, 210 Japan

Abstract

This paper describes a layout verification system used at TOSHIBA for LSI layout design. A new resistance calculation algorithm based on finite element method and a generalized resistor model used in this system are also discussed. Contact resistance effects, pinch resistors, and other kinds of resistance appearing in LSIs are considered in the proposed resistor model.

The resistance value for an arbitrarily shaped, multi-regional resistivity and multi-port resistor can be calculated efficiently.

1. Introduction

Along with increase in LSI complexity, debugging activities after the fabrication become unacceptably long and almost impossible for LSI designers. This fact makes design verification in the pre-fabrication stage indispensable for LSI designers.

LSI layout design verification consists of the following three functions: (1) Functional verification, which checks LSI functions with the intended function. (2) Performance verification, by which electrical characteristics, including power dissipation, timing constraints and other electrical performances, are checked against LSI specifications, (3) Design rule verification, every shape in the layout database is checked against the dimensions and tolerances specified by the technology design rule. In many cases, the designed LSI function and performance are finally verified by checking the circuit data extracted from the layout database [1], [12].

Figure 1 shows an LSI layout verification system example. Connectivity checking or functional behavior checking are required to verify LSI functions. In this system, a connectivity rule checking program [2] is used which detects open nodes, short circuits of signals, unusual use of transistors and other violations in the formal connectivity rules. A logic simulator [2], [3] based on extracted circuit description is also used for precise functional behavior verification. Through the authors' several years experience, these two methods have been proved highly effective. For performance verification, the extracted circuit is checked by a macro model timing simulator [4] and a circuit simulator [5].

Accurate device parameter estimation from mask artwork is extremely important for performance verification. Parasitic device parameters, including wiring resistors and capacitors, as well as designed device parameters must be estimated accurately. Resistance calculation for an arbitrary artwork shape is considered to be one of the most difficult problems in circuit parameter extracton. Several algorithms pertinent to this goal have been reported [6]-[9]. Few resistance calculation programs are provided with both complete resistor model and efficient fully automated algorithm. In practical use, careful consideration for contact resistance, pinch resistor and other kinds of resistance which appear in LSIs must be given.

This paper gives a practical resistor model which represents a variety of resistances in LSIs and an efficient algorithm based on FEM (Finite Element Method).

2. Resistor model for LSI

In order to calculate both intended and parasitic resistance accurately, the following requirements for a resistor model must be taken into consideration; (1) arbitrary shape resistor, (2) contact resistance with sufficient accuracy, (3) multi-port resistors and (4) multi-regional resistors which consist of mutually distinct sheet resistivity regions.

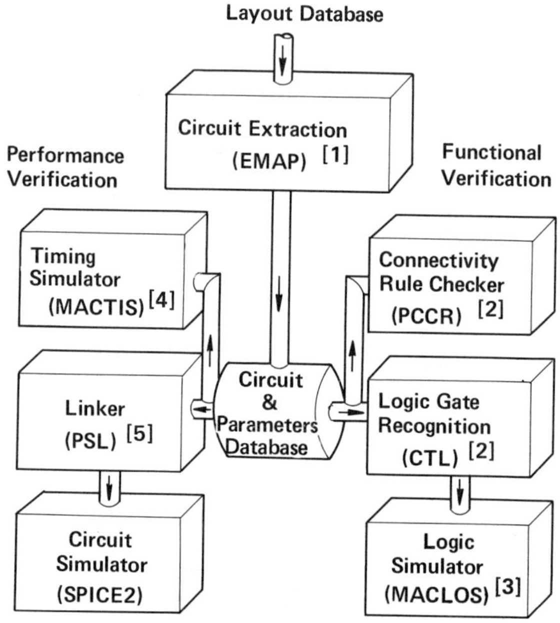

Fig. 1 LSI Mask Artwork Verification System

CH1935-6/83/0000/0604$01.00 © 1983 IEEE

Contact resistance

Contact resistance cannot be a negligible factor in resistance calculation, but no accurate and efficient mathematical model has been proposed. One author adopted a model, in which contact resistance is assumed to be contact material sheet resistance divided by contact area [7]. In this model, current crowding effects in a contact area on resistance value are not evaluated [10]. Murrmann et al. proposed a transmission line model for metal-semiconductor contact resistance [10]. Their model showed a good agreement with actual contact resistance. However, it is disadvantageous to use this model because three dimensional treatment is required to calculate the resistance for arbitrarily shaped resistors on the semiconductor. A contact resistance model, proposed in the present approach, is shown in Fig. 2. In this model, inherent three dimensional structure is embedded into a two dimensional shape. Characteristic contact resistivity ρ_c and characteristic contact length ℓ_c are defined to characterize a metal contact. In this model, a metal-semiconductor contact is assumed to take place within ℓ_c (See hatched area in Fig. 2). An external current is assumed to sink only through an inner boundary of the hatched area and flow out through a resistor with characteristic resistivity ρ_c and characteristic length ℓ_c. The value of ρ_c and ℓ_c are determined to approximate resistance values measured in advance.

Resistor Complex

Pinch resistance is modeled as a two dimensional resistor, in which a pinched region is characterized as a different resistivity area. A resistor model used in the present program is called a resistor complex, and schematically shown in Fig. 3. The resistor complex consists of an arbitrary number of contact terminals and non-overlapping sub-regions. A specific sheet resistivity is assigned to each subregion, and external current flow occurs only through the inner boundary of the assumed contact part. Hereafter this boundary is referred to as a contact terminal boundary.

Behaviors of current **i** and electric field **E** in the resistor complex are described by

$$\mathrm{div}\,\mathbf{i} = 0, \tag{1a}$$

$$\mathbf{i} = \frac{1}{\rho_s}\,\mathbf{E}, \tag{1b}$$

where ρ_s is the sheet resistivity. A potential ϕ exists, such that E = -grad ϕ, which satisfies Laplace's equation. The diffusion current is neglected in (1b) for the steady current state.

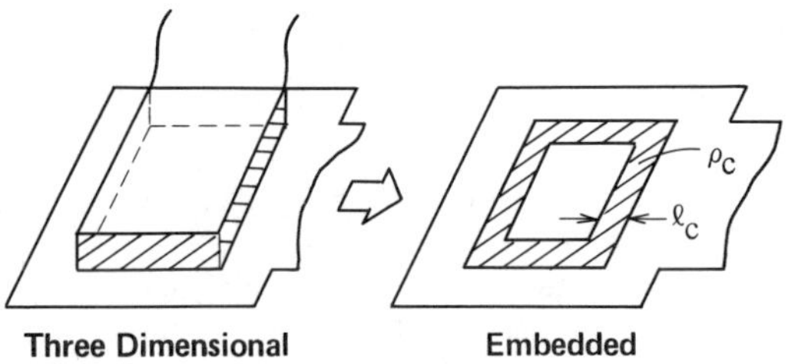

Fig. 2 Contact Resistor Model

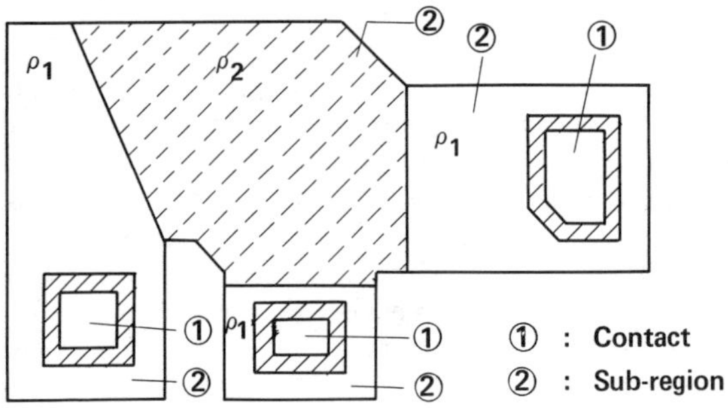

Fig. 3 Resistor Complex

3. Resistance Calculation

This section describes the algorithm and program scheme for resistance calculation. An advantages in the proposed algorithm are as follows: (1) Resistance value for an arbitrarily shaped, multi-port and multi-regional resistor can be calculated fully automatically. (2) No complicated treatment of boundary condition is required [8] to set up an equivalent lump circuit. In other words, an admittance matrix for the resistor complex is derived in a straight-forward manner. (3) Equations to be solved are described by reduced, well-conditioned positive definite band matrices, so efficient Cholesky decomposition can be adopted.

Algorithm [13]

As described in the previous section, the current and potential distributions in the resistor complex are governed by Laplace's equation. The present algorithm is based on FEM [11]. A shape function

$$\phi(x,y) = a_0 + a_1 x + a_2 y, \tag{2}$$

is applied to approximating the potential ϕ in each finite element. The stationary condition for the functional corresponding to Laplace's equation and boundary conditions in each finite element yields

$$\mathbf{i}_e = \mathbf{Y}_e\,\mathbf{v}_e\,, \tag{3}$$

where $\mathbf{i}_e$, $\mathbf{v}_e$ are vectors of external current flow and potential at each node for a triangular element and $\mathbf{Y}_e$ is an element matrix. An analogy between finite element and electric circuit, shown in Fig. 4, is established [11].

Assembling the element matrix Eq. (3), a system matrix equation for resistor complex is derived;

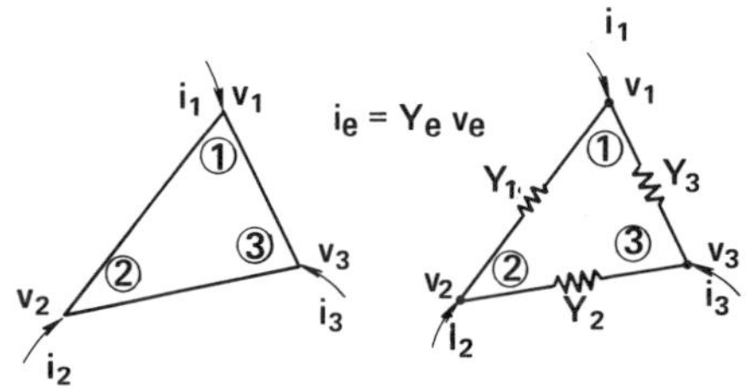

Fig. 4 Finite Element and Circuit Interpretation

$$\mathbf{i} = \mathbf{Y}\mathbf{v}\,, \qquad (4)$$

where $\mathbf{i}$ is a current vector, each component corresponding to the external current at a finite element node, and $\mathbf{v}$ is a potential vector corresponding to the potential of nodes. It is assumed that the contact terminals have the following properties: (i) External current flows only through contact terminal boundary. (ii) Electric potential at a contact terminal boundary is constant. Nodes on contact terminal boundaries are called contact nodes. Other nodes are called non-contact nodes. Arranging the nodes, Eq. (4) leads to

$$\begin{bmatrix} \mathbf{i}_c \\ \mathbf{i}_n \end{bmatrix} = \begin{bmatrix} \mathbf{Y}_{11} & \mathbf{Y}_{12} \\ \mathbf{Y}_{21} & \mathbf{Y}_{22} \end{bmatrix} \begin{bmatrix} \mathbf{v}_c \\ \mathbf{v}_n \end{bmatrix} \qquad (5)$$

where $\mathbf{i}_c$, $\mathbf{v}_c$ are current and potential vectors for contact nodes, and $\mathbf{i}_n$, $\mathbf{v}_n$ are current and potential vectors for non-contact nodes, respectively. Assuming that a resistor complex consists of m contacts, contact nodes can be divided into m groups. The same arrangement of vectors $\mathbf{i}_c$ and $\mathbf{v}_c$ yields;

$$\mathbf{i}_c^T = (i_{c_1 1}, i_{c_1 2}, \cdots i_{c_1 m_1}\; i_{c_2 1}, i_{c_2 2}, \cdots, i_{c_2 m_2}, \cdots i_{cm1}, i_{cm2}, \cdots, i_{cm\, mm})$$

$$\mathbf{v}_c^T = (v_{c_1 1}, v_{c_1 2}, \cdots, v_{c_1 m_1}, v_{c_2 1}, v_{c_2 2}, \cdots, v_{c_2 m_2} \cdots, v_{cm1}, v_{cm2}, \cdots, v_{cm\, mm})$$

where T denotes the transposition of vector. From assumptions (i) and (ii), current and potential at contact nodes satisfy

$$v_{c_i i_1} = v_{c_i i_2} = \cdots = v_{c_i m_i} \qquad \text{for } i = 1, n\,, \quad (6)$$

$$i_{c_i} = \sum_{j=1}^{m_j} i_{c_i j} \qquad \text{for } i = 1, n\,, \quad (7)$$

$$\mathbf{i}_n = 0. \qquad (8)$$

Using Eqs. (6), (7) and (8), Eq. (5) is reduced to

$$\begin{bmatrix} \mathbf{i}_c' \\ 0 \end{bmatrix} = \begin{bmatrix} \mathbf{Y}_{11}' & \mathbf{Y}_{12}' \\ \mathbf{Y}_{21}' & \mathbf{Y}_{22} \end{bmatrix} \begin{bmatrix} \mathbf{v}_c' \\ \mathbf{v}_n \end{bmatrix}, \qquad (9)$$

where $\mathbf{i}_c'$ and $\mathbf{v}_c'$ are current and potential vectors for a resistor complex with dimension m, $\mathbf{Y}'_{11}$, $\mathbf{Y}'_{12}$ and $\mathbf{Y}'_{21}$ come from $\mathbf{Y}_{11}$, $\mathbf{Y}_{12}$ and $\mathbf{Y}_{21}$ by summing the row and column in each contact node group. $\mathbf{Y}'_{11}$ and $\mathbf{Y}'_{12} = (\mathbf{Y}'_{21})^T$ are m x m and m x $(t - \sum_{\ell=1}^{m} k_\ell)$ submatrices, respectively, where t is the number of mesh nodes, and k_ℓ is the number of contact nodes on contact ℓ. By deleting $\mathbf{v}_n$, the current and potential equation can be obtained from Eq. (9)

$$\mathbf{i}_c' = (\mathbf{Y}_{11}' - \mathbf{Y}_{12}'\mathbf{Y}_{22}^{-1}\mathbf{Y}_{21}')\,\mathbf{v}_c'. \qquad (10)$$

From Eq. (10), admittance matrix $\mathbf{Y}_A$ for a resistor complex is obtained,

$$\mathbf{Y}_A = \mathbf{Y}_{11}' - \mathbf{Y}_{12}'\mathbf{Y}_{22}^{-1}\mathbf{Y}_{21}'. \qquad (11)$$

Matrix $\mathbf{Y}_{22}^{-1}\mathbf{Y}'_{21}$ can be obtained more efficiently by solving

$$\mathbf{Y}_{22}\,x = \mathbf{Y}_{21}', \qquad (12)$$

where $\mathbf{Y}_{22}$ is a $(t - \sum_{\ell=1}^{m} k_\ell) \times (t - \sum_{\ell=1}^{m} k_\ell)$ matrix. $\mathbf{Y}_{22}$ is a positive definite matrix*, so Cholesky decomposition can be adopted to solve Eq. (12).

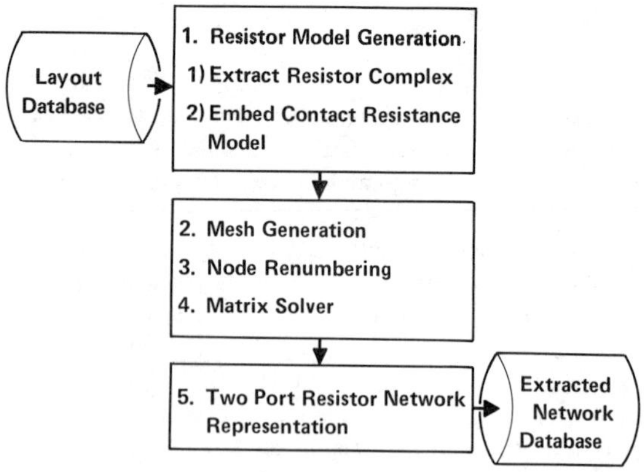

Fig. 5 Automatic Resistance Calculation Scheme

Program Scheme

A resistance calculation program, based on the above mentioned algorithm, has been developed. This program extracts resistor complexes from the layout database, derives the admittance matrix, represents the matrix as a network of two port resistors, and stores the network in the network database, automatically.

Figure 5 shows a program scheme which consists of the following five steps: (1) Resistor model generation step - a group of polygonal shapes which define contact holes, diffusion area and other parts of a resistor complex is extracted. Finally, the contact resistance is embedded in the defined resistor complex. (2) Mesh generation step - triangular elements which cover a resistor complex without overlapping is generated to adopt FEM. (3) Renumbering step - reduction in the bandwidth for a system matrix equation will result in efficient solution. In order to reduce band-width, the node numbers assigned by step (2) are renumbered. The strategy is renumbering the nodes along the longest current flow path. Node renumbering is performed by breadth-first search. To select the starting node, the complete graph is produced, whose edges connect the contact terminals for the resistor complex and are weighted by the spatial distance between the terminals at the edges. The starting node is selected from terminals with maximum weighted edge on the graph. (4) Matrix solver - admittance matrix $\mathbf{Y}_A$ is derived by the described algorithm. The band form of the matrix allows the use of external storage to solve the large system equation. (5) Two port resistor network representation - the derived admittance matrix $\mathbf{Y}_A$ is represented as a circuit consisting of two port resistors, and the circuit is stored in the network database. This is because the resistor circuit representation is suitable for both designers' understanding and interfacing ciruit simulator - like SPICE2.

*: Physical consideration of $\mathbf{Y}$ shows that electric dissipation $\mathbf{v}^T\mathbf{i}$ for a resistor complex is always non-negative. So, $\mathbf{v}^T\mathbf{i} = \mathbf{v}^T\mathbf{Y}\mathbf{v} \geq 0$ for all $\mathbf{v}$, or $\mathbf{Y}$ is positive semi-definite matrix with rank t-1, where t is the number of mesh nodes. This yields that $\mathbf{Y}_{22}$ is positive definite, which is derived from $\mathbf{Y}$ by deleting row and column.

4. Program Evaluation

The present resistance calculation program has been installed in an LSI mask artwork analysis system called EMAP [1], [12]. Three years experience gained in its use shows that this program is really practical and efficient. An example of required CPU time to extract resistors and to calculate resistance values is 390 sec. (using a 1.1 MIPS computer) for a block containing 36 resistors. In Fig. 6, a circuit schematic plotted with calculated resistance values is shown.

In a practical program, average time complexity is a critical issue. A principal term of the time complexity is that of matrix decomposition in the present program. The computer experiments show that the average time complexity for matrix decomposition is $O(N^{1.4 \sim 1.8})$, where N is size of the matrix. Compared with the time complexity $O(N^3)$ for the naive decomposition algorithm, the observed result is fairly good. This is because the matrix appearing in the algorithm is a positive definite band matrix with small bandwidth, and an efficient Cholesky decomposition algorithm is implemented.

The calculation accuracy was evaluated by using a square coaxial transmission line example [6], in which an exact resistance value is known. A 0.8 % accuracy result was obtained by 125 mesh nodes, and 12.8 second calculation time, including mesh generation, node renumbering, matrix solving and other setup times.

5. Conclusion

An efficient resistance calculation algorithm in which no boundary condition treatment is required, and a resistor model which represents arbitrarily shaped, multi-terminal and multi sheet resistivity resistors have been proposed. In this model, the accurate contact resistance model is embedded into a two dimensional shape. The algorithm has been implemented in the artwork analysis system-EMAP. Resistor extraction and resistance calculation are performed fully automatically. The resistance calculation has become a key program in the layout verification system shown in Fig. 1.

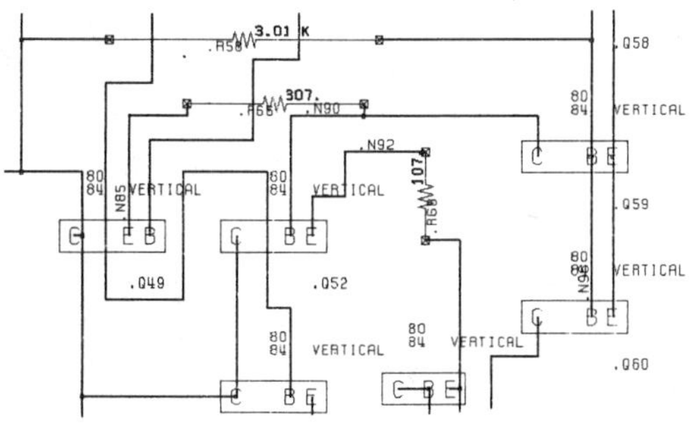

Fig. 6 A Circuit Schematic Example

Three years application shows that the resistance calculation program is practically useful and a cost effective tool for LSI performance verification.

6. References

[1] T. Mitsuhashi, et al., "An Integrated Mask Artwork Analysis System", 17th DAC Proc., pp. 277 - 284, 1980.

[2] M. Takashima, et al., "Programs for Verifying Circuit Connectivity of MOS/LSI Mask Artwork", 19th DAC Proc., pp. 544-550, 1982.

[3] K. Hirabayashi, et al., "MACLOS - Mask Checking Logic Simulator", IEEE J. Solid-State Circuits Vol. SC-15, June 1980.

[4] M. Kawamura, et al., "MACTIS - Mask Checking Timing Simulator", IEEE Trans. on CAS, Vol. CAS-27, No. 12, p. 1276, 1980.

[5] T. Watanabe, et al., "PSL-Pattern Analyzer to Circuit Simulator Linker -", 1982 ICCC Proc. pp. 316 - 319, 1982.

[6] B. R. Chawla, et al., "A Boundary Technique for Calculation of Distributed Resistance", IEEE Trans. on ED, Vol. ED-17, No. 10, pp. 915-925, Oct. 1970.

[7] H. Yoshimura, et al., "An Algorithm for Resistance Calculation from IC Mask Pattern Information", 1979 ISCAS Proc., pp. 478-481, 1979.

[8] C. M. Sakkas, "Potential Distribution and Multi-Terminal DC Resistance Computation for LSI Technology", IBM J. Res. Develop. Vol. 23, No. 6., pp. 640-651, Nov. 1979.

[9] Y. Okamura, et al., "LAS : Layout Pattern Analysis System with New Approach", 1982 ICCC Proc., pp. 308-311, 1982.

[10] H. Murrmann, et al., "Current Crowding on metal Contacts to Planar Devices", IEEE Trans. on ED, Vol. ED-16, No. 12, pp. 1022-1024, Dec. 1969.

[11] Y. Kagawa, "Introduction to Finite Element Mehtod for Electronics and Electric Engineering", Ohm Publishing, Tokyo, 1977, (in Japanese).

[12] T. Chiba, et al., "A Mask Artwork Analysis System for Bipolar Integrated Circuits", 1981 COMPCON FALL, pp. 175-185, 1981.

[13] T. Mitsuhashi, et al., "A Computational Mehtod for Resistance Calculation from IC Mask Artwork Data", 21st National Convention Record of IPSJ, pp. 1109 - 1110, 1980, (in Japanese).

Session 3-2d
Tradeoffs in Design

Chairman

F. Colon Osorio
Digital Equipment Corporation

DIRECTLY EXECUTED LANGUAGE ARCHITECTURES FOR VLSI PROCESSOR DESIGN

by Donald Alpert, Michael J. Flynn, and Scott Wakefield

Computer Systems Laboratory
Stanford University, Stanford, California 94305

Abstract

Directly Executed Language (DEL) architectures are derived as idealized representations of high-level languages. They have advantages in both simplifying compilation and improving execution of high-level language source programs. DEL architectures for FORTRAN and Pascal have been designed, and their implementations have been compared to conventional stack and general-purpose register architectures with highly favorable results. In this paper we review the concepts of DELs and report some recent results. Trends in VLSI processor design argue that DEL based architectures are a good match to VLSI technology.

Introduction

The discipline of computer architecture is by analogy similar to that of building architecture. In both cases the architect must know the function for which the design is intended and the technology with which the design will be constructed. The architect must match the function to the technology, recognizing cost constraints and employing an aesthetic sense of form and symmetry.

During the past decade many studies have identified the inefficiencies of conventional computer architectures for supporting applications programmed in high-level languages. The objects and operators specified by high-level language programs have no direct representation in the machine resources of registers and ALU that define conventional architectures. Execution of even a simple high-level statement requires a proliferation of instructions to position the data in registers and store results back to memory. For IBM/370 and DEC PDP-11 architectures about one in four instructions perform a function required by the high-level source program; the remaining instructions are overhead introduced by the architecture[1].

Researchers in our group at Stanford University have developed a new class of computer architectures called Directly Executed Languages (DELs)[2], which we believe to be an effective match of VLSI technology with high-level programming languages. DEL architectures directly represent the objects and operators specified in a high-level language program in a form that is theoretically minimal, but still permits ready execution.

DEL Principles

In a DEL architecture the operations exactly correspond to the operators specified in the programming language, e. g., integer and floating point arithmetic, array address calculation, complex arithmetic for FORTRAN, and set operations for Pascal. The operands in a DEL architecture are objects (i.e., variables, constants, and labels) named in the source program. For each subroutine or procedure in the program a table, known as the *contour*, is used to access the DEL operands. Every object referenced in the source subroutine has an entry in the contour to hold the value or address of the corresponding DEL operand. (The contour entry holds the value for simple local variables, constants, and labels. The address is used for non-local variables, arrays, and records, all of which are indirectly accessed.) The name contour is derived from Johnston's[3] use of the term to model the effects of lexical scoping in block-structured languages.

The close correspondence between the source program and its DEL representation makes compilation straightforward. Every operator in the source program generates a single DEL instruction, which consists of several bit-packed fields specifying the operands, operation, and format. The format specifies the number, order, and use of operands in the instruction. The DEL instruction formats have two properties, called *transformational completeness*, that minimize the number of operand fields in an

The work described herein was supported by the Army Research Office-Durham under contract #DAAG-29-82-K-0109.

instruction. First, an operand is specified only once in an instruction even if it serves more than once as source or destination. Second, an expression evaluation stack is implicitly used to avoid introducing temporary operands that are not mentioned in the source program. For example, Table 1 lists the transformationally complete set of formats for dyadic operators. The most general case

a+b->c

is represented in stack architectures with four instructions.

```
push a
push b
add
pop c
```

A simple, register-oriented architecture replaces the pushes with register loads and the pop with a register store. Three-address architectures capture the case above concisely; but, the evaluation of complex expressions or statements with repeated identifiers, such as a+a->b, introduces significant redundancy. The DEL formats have the advantages of stack architectures for evaluating expressions and the advantages of one, two, and three-address architectures for reducing instruction count.

The number of bits used to encode a field in a DEL instruction is $\lceil \log_2(N) \rceil$, where N is the number of distinct field values. This is the minimum number of bits required without resorting to frequency encoding. Thus, the DEL representation for the source program is minimal because only explicit operators and objects from the source program appear in DEL instructions and the instruction fields are fully packed.

Format	Transformation	No. Explicit Operands
F1	A op B -> C	3
F2	A op B -> B	2
F3	A op B -> A	2
F4	A op A -> B	2
F5	A op A -> A	1
F6	A op B -> T	2
F7	A op T -> B	2
F8	T op A -> B	2
F9	T op A -> A	1
F10	A op T -> A	1
F11	A op A -> T	1
F12	A op T -> T	1
F13	T op A -> T	1
F14	T op T -> A	1
F15	T op U -> A	1
F16	T op U -> U	0

A, B, and C are explicit operands.
T is the top of stack; U is below T in the stack.

Table 1: Transformationally Complete Dyadic Formats

For a concrete example of DEL representation, consider the following procedure in a Pascal program.

```
procedure p(a:integer; var b:integer);
    begin
    b:=(b+a)*3;
    end;
```

The DEL contour for the procedure (see Figure 1) has three entries for the parameters *a* and *b* and the constant 3. The single statement in the procedure body generates two DEL instructions.

```
F6 a b +        (a+b -> T)
F8 3 b *        (T*2 -> b)
```

Compare this to IBM/370, which uses six instructions.

```
L   8,b         address of b -> register 8
L   7,0(8)      value of b -> register 7
A   7,a         a+b -> register 7
LA  5,3(0)      3 -> register 5
MR  6,5         (a+b)*3 -> register 7
ST  7,0(8)      register 7 -> b
```

Experimental Results

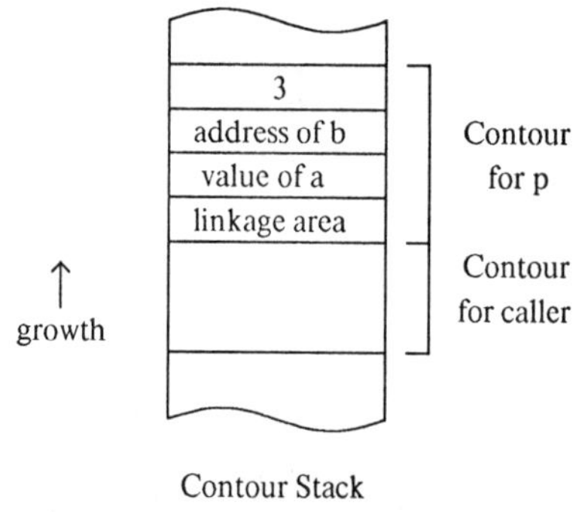

Figure 1: DEL Contour

DEL architectures have been developed for FORTRAN (DELtran[2]) and Pascal (Adept[4]). The Adept architecture was compared against IBM/370, HP1000, and P-Code on a sample of five Pascal programs. The programs were FFT, Kalman filter, puzzle (an unofficial benchmark developed by Forest Baskett), quicksort, and a program that compiles and interprets a subset of Pascal. Table 2 shows average results for several measures of the programs.

By eliminating the overhead instructions of the conventional architectures, the DEL architecture has reduced the static program size by a factor of 2 and the number of instructions

Measure	IBM/370	HP1000	P-Code
Code size	2.40	2.22	3.60
Instructions Executed	3.42	3.55	3.57
Inst. bytes fetched	3.29	2.57	4.12
Data bytes fetched	5.82	4.80	5.81
Data bytes stored	20.15	12.56	9.19

Table 2: Architectural Measures Relative to Adept

executed by a factor of 3, while lowering the bandwidth for fetching instructions by a factor of 3. In addition, the DEL contour has reduced main memory bandwidth for data traffic by a factor of 5. These results encourage the belief that DEL processors can be highly cost-effetive.

VLSI Processor Design

As VLSI processing technology has improved in recent years, a trend has developed in microprocessor design toward increased use of design resources for integrated memory functions. For example, the 32-bit microprocessor from Hewlett-Packard is the most highly integrated CPU to date. This circuit contains 450,000 devices, of which 350,000 are in the microcode ROM[5]. Another example is the U. C. Berkeley RISC I microprocessor, which uses 45% of chip area for an integrated register stack buffer[6].

This trend, originally identified by Patterson and Sequin[7], has two major causes. First, designing a VSLI processor requires a high degree of regularity to manage the complexity, and memory design is very regular. Second, as devices become smaller, the time to access on-chip memory decreases faster than off-chip access time. This is because the off-chip access time is determined by capacitance from packaging and interconnections that do not scale with device technology.

The main types of memory used in microprocessor design are ROM or PLA for control storage, RAM for data storage, and CAM (Content Addressable Memory) for address tags in cache and address translation buffers. In microprocessor designs we have examined there is a startling difference in the cost (measured in area, power, or device count) per bit for the different types of memory. Figure 2 shows typical memory circuits used in nMOS microprocessor designs. The ROM cell uses a single transistor and dissipates no static power. The single-ported static RAM cell uses six transistors. The CAM cell includes the cross-coupled inverters of the RAM plus a built-in comparator for a total of nine transistors. In practical designs the area of the RAM cell is typically larger than the ROM cell by a factor of 10 to 25. The area of the CAM is larger than the RAM by a factor of 1.5 to 2.5.

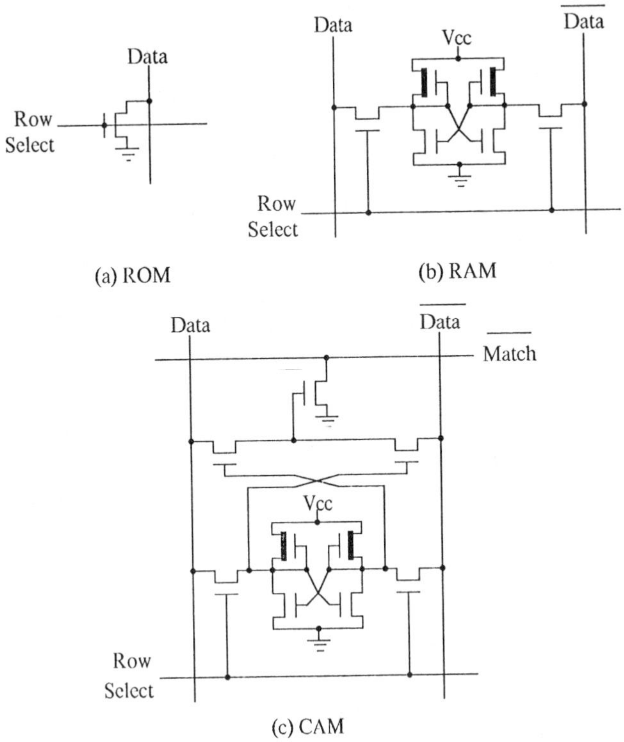

Figure 2: Memory Circuit Diagrams

This disparity in memory costs is favorable to DEL architectures compared to conventional and reduced instruction set architectures. The compact code of DELs makes effective use of integrated cache memory, which is built from relatively expensive RAM and CAM circuits. The high level of functionality in DELs, which include all the operators and run-time support of the programming language, makes effective use of dense, ROM-based control storage. In addition, the DEL contour can be stored on-chip in a buffer of variable size, thus permitting the quantity of data storage integrated with the processor to change with technology and cost/performance tradeoffs. In contrast, general-purpose register architectures require a fixed amount of data storage.

A DEL Processor

We have begun the design of a 32-bit DEL microprocessor using a 4 μm nMOS technology. Figure 3 shows a block diagram for the design. The design includes four major function units (each controlled by a PLA state machine), four major buses, and a contour memory buffer. The design is synchronous, using standard two-phase clocking, with an estimated cycle time of 250 ns.

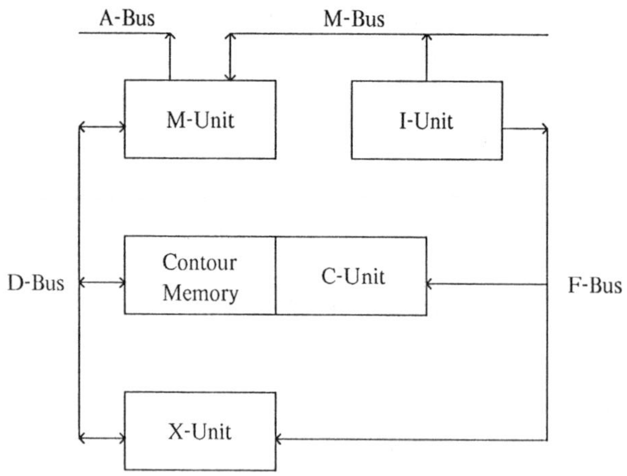

Figure 3: DEL Processor Block Diagram

The I-Unit extracts bit fields from the instruction stream. The I-Unit holds the current and next instruction words. A barrel shifter can extract an instruction field from any position in the instruction word. The I-Unit can extract a field on each cycle and drive it on the F-Bus to the C-Unit or X-Unit.

The C-Unit generates references to the contour memory. On each cycle a single contour entry, specified explicitly by the operand field on the F-Bus or implicitly using the evaluation stack pointer, can be fetched or stored on the D-Bus.

The X-Unit includes an ALU and temporary holding registers for source and destination operands. The X-Unit performs the operation specified by the field passed on the F-Bus from the I-Unit. On each cycle the ALU can operate on source operands in the temporary registers and return the result to a temproary register or on the D-Bus to the contour memory.

The M-Unit controls the memory interface. The M-Unit connects to the A-Bus, which is used to drive the memory address to the pins, and the M-Bus, which is used to drive or receive memory data from the pins. The M-Unit holds the program counter for instruction fetches. For data references the address is transferred on the D-Bus to the memory address register in the M-Unit. The data is transferred between the D-Bus and M-Bus through the memory data register, also in the M-Unit.

Although the design is still in its initial stages, preliminary evaluation indicates that a DEL can be implemented with cost and instruction execution time comparable to an IBM/370 and similar conventional architectures. Of course, compared to conventional architectures, the DEL processor executes fewer instructions and makes fewer memory references, resulting in potentially higher performance.

Conclusion

DEL architectures present a promising alternative to conventional architectures for VLSI technology. The feasibility of designing a single-chip DEL microprocessor is under study.

References

1. Michael J. Flynn, "Directions and Issues in Architecture and Language," *Computer,* Vol. 13, No. 10, October 1980, pp. 5-22.

2. Michael J. Flynn and Lee W. Hoevel, "Execution Architecture: The DELtran Experiment," *IEEE Transactions on Computers,* Vol. C-32, No. 2, February 1983, pp. 156-175.

3. John B. Johnston, "The Contour Model of Block Structured Processes," *Sigplan Notices,* Vol. 6, February 1971, pp. 55-82.

4. Scott Wakefield, "Studies in Execution Architectures," Technical Report No. 237, Computer Systems Laboratory, Stanford University, January 1983.

5. Joseph W. Beyers et al., "A 32-Bit VLSI CPU Chip," *IEEE Journal of Solid-State Circuits,* Vol. SC-16, No. 5, October 1981, pp. 537-542.

6. Daniel T. Fitzpatrick et al., "A RISCy Approach to VLSI," *Computer Architecture News,* Vol. 10, No. 1, March 1982, pp. 28-32.

7. David A. Patterson and Carlo H. Sequin, "Design Considerations for Single-Chip Computers of the Future," *IEEE Trans. Computers,* Vol. C-29, No. 2, February 1980, pp. 108-116.

A Perspective on the Processor Complexity Controversy*

Robert P. Colwell, Charles Y. Hitchcock III, E. Douglas Jensen

Computer Science Department
Department of Electrical and Computer Engineering
Carnegie-Mellon University
Pittsburgh, PA 15213

Abstract

The size and complexity of processor instruction sets has steadily increased over the last twenty years, with the intent of improving performance and reducing software costs. Recently there has been wide publicity for an alternative approach called the "Reduced Instruction Set Computer (RISC)", which contends that simple machines are not only less expensive to design and build but also offer greater performance. We feel that both viewpoints have merit but that their proponents have not produced much convincing scientific evidence to support their claims. The principles and practices of defining and implementing complex instructions appropriately are ad hoc, and questionable in outcome. RISC designs recognize that the current predominance of high level language programming argues for abandoning the distinction between instruction set architecture and implementation as an artifact of the assembly language software era. But focusing on simplicity in a narrow context, such as number of devices on a chip, overlooks the more nebulous yet equally real costs of pushing complexity elsewhere in the system. Furthermore, performance is often over-emphasized because it is more quantifiable than other metrics which are frequently as or more important, including modularity, fault tolerance, and cost of use. We argue that this controversy is in need of more research and a more balanced perspective.

1. Introduction

The increasing size and complexity of processor instruction sets has encompassed additional data types (e.g., floating point, decimal, character strings, arrays, priority queues, linked lists), operating system support (e.g., process management, synchronization, interprocess communication), and compensations for other disabilities (e.g., addressing modes to deal with insufficient instruction address field length). This trend is typified by such popular complex instruction set computers (CISC) as the VAX and the Motorola 68000, and currently culminates in the Intel 432. The intent has been to improve performance, and especially in the case of the 432, to reduce software costs.

Recently an alternative design approach has been widely publicized as the "Reduced Instruction Set Computer (RISC)". Three research machines in particular, the IBM 801 [Radin 83], the Stanford MIPS [Hennessy 82], and the Berkeley RISC I [Patterson 82a], are based on the belief that computers with simpler instruction sets are not only less expensive to design and build, but also offer greater performance than the more traditional complex computers.

We feel that both approaches have merit, but that neither is sufficiently scientific, and we do not find much credible evidence for the claims of either RISC or CISC proponents. Ad hoc designs and implementations have been done but not evaluated, the effects of orthogonal issues have not been separated out, systems which differ in kind have been "compared", important attributes which are difficult to quantify have been presumed not relevant, and complexities have been "removed" by moving them to different places in the system. Unfortunately, many important ramifications of this controversy have remained unappreciated due to unquestioning acceptance of either point of view.

* This research is sponsored in part by the Department of the Army under contract DAA B07-82-C-J164.

In this paper we briefly discuss some of the topics of contention, note where additional research is needed to attain better understanding, and generally argue for a view of the matter which is broader than just instruction set size and complexity.

We will use the term *RISC* to imply all research efforts concerning reduced instruction set computers. *RISC I* will be specifically used to refer to the research being pursued at Berkeley.

2. Notions of Simplicity

Perhaps the most fundamental RISC tenet is that the most primitive instructions dominate a computer's activity, and that their performance will be adversely affected by the inclusion of anything more complex in the instruction set. The tenet carries with it an implicit assumption, which is not necessarily true, that any loss in performance is inherently bad. We will discuss this further in Section 3. This section explores some of the implications and problems that follow from this tenet.

2.1. Perceiving Distinct Qualities

Unfortunately, the terms *reduced* and *complex* have been contraposed in the context of this first tenet of the RISC philosophy, as Clark has pointed out [Clark 80]. In fact, two orthogonal instruction set dimensions are at issue here: size (reduced vs. massive) and complexity (simple vs. complex). The first dimension concerns the number of instructions (addressing modes, number of possible values in instruction fields in general) that characterize an architecture. The other concerns the functional complexity of the instructions as might be represented by the number of "primitive" operations that would be needed to synthesize them. This dimension is much harder to quantify since mixtures of simple and complex instructions can exist within the same architecture.

It is true that *reduced* and *simple* take on a mutually reinforcing relationship in the context of RISC design, as *massive* and *complex* normally do in the CISC domain. This does not have to be the case. Simplicity means different things to chip designers, computer architects, and all other people involved in the design process. The VAX has often been singled out as being a complex architecture. Yet, from the designer's point of view, the VAX was to be a simple yet massive instruction set. The definition of simplicity used in this context was:

> those attributes (other than price) that make minicomputer systems attractive. These include approachability, understandability, and ease of use [Strecker 78].

It is questionable whether or not this goal was achieved, but it will be argued later that, in some ways, the issue of simplicity may not be of prime importance.

2.2. The Utility of Complex Instructions

RISC proponents warn of detrimental effects due to the use of complex instructions. Nevertheless, the popularity of installing support for specialized functions such as interprocess communication (IPC) seems to be undiminished. The designers of the ELXSI 6400 report [Olson 83], for instance,

> A key architectural feature which allows the operating system to cope with the tremendous variability in its hardware environment is the microcode and hardware implemented message system. The use of messages allowed us to make choices in the CPU and operating system architecture which greatly enhance the effectiveness of additional processors.

But we concur with one of the RISC criticisms of the published accounts of these machines: it is not enough to show that a complex instruction executes faster than an equivalent sequence of primitive instructions. It must also be shown that the net effect is to improve system performance. We believe that this aspect of the problem must be part of the design effort.

Even the premise that primitive instructions always dominate a computer's activities is not universally true. The instruction set interface of machines designed to run operating systems may be solely at the "system call" level, for example.

There are many other computing environments, such as real-time or signal processing systems, where it would be hard to argue against supporting complex functions directly in the computer architecture and implementation. More generally, Radin has written [Radin 83]:

> It is often true that implementing a complex function in random logic will result in its execution being significantly faster than if the function were programmed as a sequence of primitive instructions. Examples are floating point arithmetic and fixed point multiply. We have no objections to this strategy, provided the frequency of use justifies the cost and, more important, provided these complex instructions in no way slow down the primitive instructions [Radin 83].

We subscribe to this statement, but we assert that "frequency of use" is an insufficient criterion for justifying a given instruction. As Clark and Levy have pointed out [Clark 82]:

> Aggregate statistics alone cannot guide the design of an instruction set intended for different languages and applications. In particular, instructions that are infrequently used overall can be critical for some intended users.

The notion that complex functions slow down the simple actions of a computer seems to be the real problem that prevents us from having the best of all worlds. We believe that serious research efforts in the areas of functional partitioning, instruction interpretation, and distributed decoding will produce computer structures which reduce or eliminate this effect. Until research is directly aimed at this problem, a greater understanding of the scientific truths and principles involved, as opposed to the folklore currently being disseminated, is not possible.

2.3. Designing Simple Machines

The RISC simplicity tenet has a related side-effect in that simpler computers are thought to be easier and faster to design than complex ones. Unfortunately, the comparisons that have been published to substantiate this tenet are based on design times for a student project simple microprocessor versus the design times for some current complex commercial microprocessor products [Patterson 82a]. While these comparisons seem interesting, we do not find them relevant, since the objectives, constraints, and design tasks are significantly different between the academic and industrial environments. Design considerations such as yield, testability, and fault tolerance are not handled in the same way for both contexts. Logistical and administrative factors necessarily imposed by a large organization (e.g., synchronizing simultaneous development of support chips, software development systems, and fabrication facilities) cannot be disregarded. It strikes us as improper to make any such comparisons without first attempting to calibrate the units of measurement.

To make matters more confusing, comparing the hardware design times of processors of different scale is misleading since complexity shed by the processor design team could well be encountered by the system software designers or even the applications programmers. The tables of comparisons don't even hint at the tradeoffs. As Justin Rattner has said [Barney 82], "They say that the RISC (I) chip was developed in 6% of the time it took for the 432 ... My response is, 'Yes, and they only did 6% of the job.'" The hardware/software partitioning of a design begs for a more detailed analysis. In particular, an economic analysis of the hardware/software design cycle tradeoffs would be of strong practical interest.

2.4. Complexity Migration

One specific form of transferring complexity away from the hardware is by migrating functions to compile time which previously were considered run time activities that were supported by hardware. One of the three criteria for instruction set design in the 801 was that the operation could not be moved to compile time. The 801 approach also utilized a very sophisticated compiler to make some of these tradeoffs (for example, by precomputing functions wherever possible).

This concept of complexity migration is the basis for MIPS [Hennessy 82], which is based on a pipeline implementation having no hardware interlocks. The only means of ensuring proper sequencing of events in this machine's instruction stream is via a *pipeline reorganizer* program. Although a straightforward compiler can be used to generate valid code for this machine, it is only by using the reorganizer that the machine's pipeline can be fully utilized.

Of course, not all complex functions can be moved to compile time. Dynamic program activities, such as garbage collection and bounds checking, must of necessity be done at run time. But as further work is done to evaluate the merits of complexity migration to compile time, computer system designers will be able to make decisions based on evidence rather than educated guesswork.

3. Importance of the Performance Aspects of Computer Design

Throughout the RISC literature there is a largely unstated but pervasive bias towards those aspects of a computer system dealing with performance. Clearly, if all other attributes are equal, higher performance must be considered an improvement to a machine. However, we believe that it is possible to ascribe too much importance to the performance dimension of a computer system. Since performance is the most quantifiable measure of a machine, it is the most frequently discussed and measured -- not because performance is always inherently so much more valuable than other system parameters, but because benchmarking is the easiest way of comparing system alternatives. It is a mistake to pursue performance blindly without explicitly acknowledging what is being traded for it.

3.1. What is meant by performance?

System performance can be measured in different ways, and these differences can be significant, because they reflect the fundamental goals of the system. For example, performance can be measured in terms of peak instruction execution rate, a number which may be of interest in deciding the suitability of various supercomputers to some proposed task. Another performance measure is response time, which is of particular interest in real-time control systems. Yet a third performance measure is average system throughput. We assume that it is this measure which is being discussed in the RISC/CISC literature.

3.2. Performance vs. Other System Aspects

It is our view that a very wide range of performance is currently available in the marketplace, and that, except for the most demanding applications, a user with sufficient money can buy whatever performance is desired. We agree that the dramatic declines in the price/performance ratio have been largely responsible for the enormous economic growth in the field.

But it strikes us that significant performance gains will be given to us almost free by the semiconductor technologists, first by the constantly improving fabrication process (driving down gate delays), and second by the increasing integration densities that will allow more computational activity to occur without the need to go off-chip. Device technologists can not address other aspects of a computer system, however. For example, even if the military had arbitrarily large amounts of money, they could not buy systems with the level of modularity and expandability they desire, because we architects do not yet know how to provide it at any cost. Conversely, the Japanese Fifth Generation Project requires such large increases in performance that it is commonly assumed that no von Neumann architecture will ever be able to provide it (regardless of the complexity or simplicity of the instruction set). Hence, research is being pursued on multiple processor architectures, where the bulk of the performance results from combining large numbers of processors. Research to deal with these problems should be directed at the interconnect and usage problems as much as the processors themselves.

4. Ambiguous Performance Claims

Much of the interest generated by the RISC efforts comes from the reported performance improvements. One study [Patterson 82b], for example, lists the execution times of the simulated Berkeley RISC I chip vs. the 68000, the Z8002, the VAX 11/780, the PDP11/70, and the BBN C/70. For every benchmark measured (benchmarks included Ackerman's function, quicksort, and the puzzle program) the simulated RISC I chip promised faster execution times.

We do not find that the performance claims that have been published are conclusive evidence that a breakthrough in the price/performance ratio has been achieved. We will state some of our reservations in the next few sections.

4.1. All or Nothing

RISC machines have not only a reduced instruction set, but also many other items which affect performance. For example, as we have pointed out earlier [Colwell 83], we believe that the overlapping register window scheme used in RISC I accounts for a substantial amount of the performance expected from that machine, and can be of value to CISC machines as well. Likewise, we would like to know exactly what performance improvement to expect from the compiler and pipeline management techniques used in MIPS. We feel that it would be much more meaningful to compare reduced instruction set machines to CISCs on those aspects which are unique to each, factoring out those which can be utilized in either style.

4.2. Fair Comparisons

The Intel 432 would seem to be a very promising candidate for close scrutiny in this RISC/CISC controversy. It can be considered an archtypical CISC: it has a complex instruction set (including such instructions as **BROADCAST TO PROCESSORS** and **LOCK OBJECT**); it is programmable only in Ada, not assembler; and it is a complete computer system, including an operating system kernel. Although the 432 performance study [Hansen 82] did not mention RISC I, the same benchmarks were used and it is a simple matter to correlate the two reports to arrive at the conclusion that the 432 runs the benchmarks about two orders of magnitude slower, in general.

But as we pointed out in [Colwell 83], it is important not to overlook other aspects of the 432 that have affected these results. For example, the 432 is an object-oriented machine. This object orientation was provided to support the intended software programming environment, and is an attempt to minimize life cycle cost, not performance per se. However, the other machines measured in [Hansen 82] were not object oriented, so we are left unsure as to what part of the reported performance loss in the 432 is due to its object orientation. The object facilities cannot be removed from the 432 for comparison purposes, but they can be added in software to other machines to make the results useful.

All benchmarks reported in [Patterson 82b] were coded in the C programming language. We do not object to C as the HLL of choice. Given that machine-dependent aspects of the C language are avoided, this eliminates one source of uncertainty. However, not all comparisons are set up in this way. Our concern is that the 432 has only an Ada compiler. Thus the quality of the Ada compiler, as well as the efficiency of the Ada language itself, are in question here. We feel that these two variables alone render the 432 results inconclusive.

Another interesting aspect of the 432 that may have a negative effect on its single-processor performance is its innate multiprocessing support. This feature was designed at the system level so that processors can be added and automatically utilized without software participation. As far as we know, this is one of the few system architectures that have ever been produced with this capacity. Intel has estimated that support for multiprocessing takes approximately 13% of the available microcode space on chip, which we take as more evidence that this is not a trivial function to implement. Since the machines to which the 432 is being compared do not have this kind of support, we can only wonder to what extent this feature skews the conclusions that one might attempt to draw.

The 432 also provides some hooks for enhancing system reliability, such as fault handling and functional redundancy microcode. In a naive comparison of instruction execution rates this hidden functionality will appear as unseen baggage, dragging down the machine's performance on the benchmark. Benchmarking is a very interesting exercise, but unless the machines being compared differ only along one major dimension, it is difficult to make a fair comparison. An unfair comparison is inconclusive.

4.3. Justification and Analysis

The important question is whether these extra features in CISCs such as the 432 contribute enough to have made the apparent performance loss and extra design complexity worthwhile. One study [Cox 83] reports that the 432's support for its interprocess communication primitives do indeed speed up those operations by large amounts over the software approach used in comparable machines. This proves that a **SEND** can be executed faster if we are willing to devote system resources to it, but it leaves unanswered several other questions of equal importance. How were the complex functions like **SEND** chosen in the first place? What was gained on a system-wide basis by including these functions? What was the cost, both in resources and in low-level instruction performance? We suggest that one of the tasks facing computer architecture research is to find out how to assign better life-cycle cost models to the systems we build, so that the performance aspects don't receive improper weighting, either positive or negative. Especially in an system architecture as radically different as the 432, it is incumbent upon the designers to carefully justify the design tradeoffs they have made. It seems to us that they have done so at the system level, making the case that object orientation is a goal worth pursuing. However, they do not attempt this same justification at the architecture or implementation levels, nor do they analyze the resulting machine. A good understanding of existing CISCs is not possible without examining the tradeoffs at those levels.

5. Architectures and Implementations

Since the early sixties, computer designers have found it beneficial to conceptually separate a machine's architecture from its implementation. This dichotomy was useful when trying to decompose the design problem. However, its economic strength came then, as it does today, from software compatibility.

After 20 years of using this concept, a sense of good and bad computer architecture has developed around the notions of purity. To quote from Blaauw and Brooks [Blaauw 82]:

> The architecture must be comprehensible and consistent, so it will be easy to learn and use. The user beholds the whole system. It will be easy for him to master and use it only to the degree that it shows an integrity of concept, a consistency of viewpoint, tying together all the design decisions.

5.1. The Rules Have Changed

The recent research in RISC concepts has stretched the fabric of some purist computer architecture notions. To begin with, software development costs are still of major concern to most installations, but not at the assembly level. While there are probably more assembly programmers hacking today than we care to know, the world is being dominated by high-level language code. The importance of high-level language programming is reflected by the fact that almost all new general purpose computing machines, RISCs and CISCs alike, are founded on optimizing the execution of compiled code. (This is certainly no new idea in light of the long history of Burroughs high-level language machines.)

5.2. Departures From Purity

Purist notions notwithstanding, it seems indisputable that blurring aspects of architecture and implementation can often lead to better machines. Again, from the notes of Blaauw and Brooks:

> ... some of the genius of Seymour Cray's work ... lies precisely in his total personal control of architecture, implementation, and realization, and his consequent freedom in making trades across the boundaries.

While the separation of these factors produces conceptually cleaner architectures, and might aid in the partitioning of the design task, RISC research trades these advantages for possible performance improvements. For example, a cache is normally invisible to the software, yet the 801 has explicit instructions for cache control so that the computer does not perform unnecessary cache line loads and stores. Instruction ordering constraints, as imposed by a machine's implementation, are present in the 801, the RISC I, and especially MIPS with its non-interlocked pipeline. These, too, are attempts to optimize a machine's performance by trading across classical boundaries.

Traditionally, microcoding has been a powerful implementation technique for instruction interpretation which has made designing massive/complex machines like the 432 and the VAX tractable. In an interesting twist of concepts, many RISC researchers view their machine architectures as exposing what might otherwise be a hidden vertically coded microengine. While RISC instruction bits drive control lines every cycle via minimal decoding, which is reminiscent of traditional microcoded instructions, such a view ignores the conceptual and cultural differences between macrocode and microcode. Manufacturers have generally not supported machines with such exposed microengines (implementations) since they require individual compilers (and hence, cannot share object code with other machines) and are severely limited in the types of changes that can be made to them after their release. We do not believe that these problems will remain for long, as will soon be explained.

5.3. Moving to Higher Ground

Since programmers never see the machine-level interface, and since performance gains are possible by mixing architecture and implementation, do "computer families", in the traditional sense, have a place in the futures of computer companies? The designers of computers as diverse as the VAX, the 801, the 432, and RISC I, all wanted their machines to be good targets for compiled code. The natural question to ask becomes: "Why don't computer companies market 'families' of machines that are compatible at the system software level?" The HLL programs would be compatible on "families" of machines that could span a spectrum of price/performance ranges. Each member of the family would be free to trade architecture and implementation features to optimize performance. True, unique compilers would be needed for each family member, since instruction sets would differ, but this issue may decline in importance with the entrance of automated compiler-compilers [Wulf 80].

A possible problem arises with the definition of the common system software interface. Just compare a UNIX manual with almost any machine definition you can find. It seems a hard enough task just to define such a massive interface without having to ensure compatibility of that interface across many machines and their system software. Validation becomes a much larger issue than simply running a suite of test programs. This challenge is not necessarily insurmountable, but it is not well understood at present. The idea is not a new one, just one that has yet to succeed on a grand scale. Probably the largest hurdle will be in overcoming the electro-political status quo that has dictated how computer systems should be structured for the last 20 years.

6. Conclusion

The RISC advocates have put forth a perspective on processor architecture and implementation which is more coherent and concrete than those which seem to guide most CISC proponents. We feel that this perspective is both interesting and insightful in some ways, yet oversimplified and thus justly controversial.

7. Acknowledgements

We would like to acknowledge many of our colleagues, especially Frederick Brooks, George Cox, Kevin Kahn, David Patterson, and Justin Rattner, for helpful discussions on some of these topics.

References

[Barney 82] Clifford Barney. Fewer instructions speed up VLSI. *Electronics*:101-102, November 17, 1982.

[Blaauw 82] Gerrit A. Blaauw and Frederick P. Brooks Jr. Computer Architecture. Unpublished draft of a book; Fall, 1982.

[Clark 80] Douglas W. Clark and William D. Strecker. Comments on 'The Case for the Reduced Instruction Set Computer,' by Patterson and Ditzel. *Computer Architecture News* 6(6):34-38, October, 1980.

[Clark 82] Douglas W. Clark and Henry M. Levy. Measurement and Analysis of Instruction Use in the Vax-11/780. In *Proceedings of the Ninth Annual Symposium on Computer Architecture*, pages 9-17. IEEE and ACM, April, 1982.

[Colwell 83] R.P. Colwell, C.Y. Hitchcock III, E.D. Jensen. Peering Through the RISC/CISC Fog: an Outline of Research. *Computer Architecture News* 1(11):44-50, March, 1983.

[Cox 83] G.W. Cox, W.M. Corwin, K.K. Lai, F.J. Pollack. Interprocess Communication and Processor Dispatching on the Intel 432. *ACM Transactions on Computer Systems* 1(1), February, 1983.

[Hansen 82] P.M. Hansen, M.A. Linton, R.N. Mayo, M. Murphy, D.A. Patterson. A Performance Evaluation of the Intel iAPX 432. *Computer Architecture News* 10(4), June, 1982.

[Hennessy 82] John Hennessy, Norman Jouppi, John Gill, Forest Baskett, Alex Strong, Thomas Gross, Chris Rowen, and Judson Leonard. The MIPS Machine. In *Proceedings of the Spring CompCon*, pages 2-7. IEEE, February, 1982.

[Olson 83] Robert A. Olson, B. Kumar, Leonard E. Shar. Messages and Multiprocessing in the ELXSI System 6400. In *Proceedings of the Spring 1983 CompCon*, IEEE, March, 1983.

[Patterson 82a] David A. Patterson and Carlo H. Sequin. A VLSI RISC. *Computer* 15(9):8-21, September, 1982.

[Patterson 82b] D. Patterson, R.S. Piepho. RISC Assessment: A High Level Language Experiment. In *9th Annual Symposium on Computer Architectures*, April, 1982.

[Radin 83] George Radin. The 801 Minicomputer. *IBM Journal of Research and Development* 27(3):237-246, May, 1983.

[Strecker 78] William D. Strecker. VAX-11/780: a Virtual Address Extension to the DEC PDP-11 Family, *Computer Engineering: A DEC View of Hardware Systems Design*, pages 409-428, . Chapter 17. Digital Press, Bedford, MA, 1978.

[Wulf 80] Wm. A. Wulf. *PQCC: A Machine-Relative Compiler Technology.* Technical Report CMU-CS-80-144, Carnegie-Mellon University, 1980.

COMPILING HIGH LEVEL FUNCTION ON LOW LEVEL MACHINES

M. E. Hopkins

IBM T.J. Watson Research Center
P.O. Box 218
Yorktown Heights, New York 10598

Introduction

Low level computer architectures have been advocated by Radin (6), Paterson (5), Hennesey (3) and others. Paterson has characterized these machines as reduced instruction set computers or RISC machines. Like the advocates of high level machines, all of these authors assume that most programming should be done in a high level language. This paper provides some insights into a variety of compilation techniques for low level machines gained from the PL.8 compiler (1). It also suggests some desirable features for low level machines. In order to focus the discussion we will consider only one high level function, integer multiplication. However the techniques discussed can be used to implement a wide variety of function in many languages.

Low Level Machines Characterized

The RISC nomenclature for low level machines is somewhat unfortunate as reducing the number of instructions is not the primary issue. The advantage of low level machines derives from the fact that complex machines are often implemented in microcode, while low level machines can be implemented directly without paying the penalty for interpreting microcode. As complex machines spend much of their time executing simple instructions like load, store, add, shift and branch, for which direct counterparts exist in the micro engine, avoidance of the interpretive overhead provides substantially better performance for those instructions. However complex function such as multiply and storage to storage move require many instructions. One important issue with simple machines is how to gain acceptable performance for such function.

Low level machines may utilize a great deal of hardware in their implementation. However instead of using the added hardware for the implementation of complex ops the tendency is to make the memory hierarchy more efficient. One example is that data paths are wider. (It is surprising how many high level machines, which were implemented with a "hardware is free" philosophy have have ended up with a 16 bit data flow.) In addition, more registers, cache and translation look aside buffers have been seen as attractive ways to enhance performance.

Compiling Complex Function

Multiply is a good example of complex function that may be missing from a low level machine. The first decision is whether or not the function should be incorporated into the hardware. This decision can only be made with knowledge of the expected frequency of execution of multiply instructions, the hardware used and the expected speed of the instruction. Another important consideration is basic hardware regularity. If all instructions take exactly one machine cycle the computer is basically simpler and will run faster. Suppose one examines the frequency of integer multiply and finds it to be less the .1%. (This is a high number in many environments.) If the hardware to implement a fast multiply then seems too costly, perhaps the instruction can be deleted. However one then must still have a reasonable alternative. After all a machine with no multiply instruction could end up spending 20% of its time in the multiply function if it takes 200 instructions to develop the product of a 32 x 32 bit multiplication. One alternative is a multiply step instruction that performs a 2 x 32 bit multiply, making it possible to do a full 32 x 32 bit multiply in 16 instructions. Such an instruction is similar to the underlying micro code instructions used to implement multiply on some high level architectures. The difference lies in the fact that the primitive instruction is now directly available to the compiler writer, who now has the privilege and responsibility of making a number of trade offs which are the province of CPU architects on high level machines. The first choice is whether or not the function is to be compiled in line or turned into a subroutine. A subroutine has many advantages. First, it is clear we need not produce an in line loop consisting of multiply step, increment, compare and branch; instead there will be a subroutine with 16 multiply step instructions in a row. This trade off of space for time is is usually referred to as loop unrolling. To the extent that this loop is executed frequently the instructions will be in the cache. In this instance the cache is analogous to a pagable micro store. It is desirable to share these microcode type routines among all the independent programs in the system just as micro code is shared. Thus the micro code will be in shared virtual memory,

CH1935-6/83/0000/0617$01.00 © 1983 IEEE

accessible to all programs. It is also desirable to reduce the cost of reaching the routine to a single fast branch and link instruction. No load of an address constant should be required. This may require a special branch and link instruction that can get to the shared "micro code address space".

Inevitably there is some overhead associated with the use of a subroutine to perform a function. In line code will have to be introduced to copy the operands to the registers assumed by the subroutine if the compiler can not arrange for them to be in the appropriate spot. There will also have to be a branch and link, a branch register to return and possibly another register copy to move the result to an appropriate register. The idea is to pass the operands themselves in registers and return the result in a register. All work done by the multiply subroutine is done in registers so the standard linkage conventions must permit a subroutine to destroy the contents of a few registers; the multiply subroutine can not afford to save and restore any registers. It should make no storage references. The branch and link instruction must leave the return address in a register.

The compiler will have to develop strategies to treat the multiply subroutine like an operator. As multiplies occur in many subscript computations it is well worthwhile to apply the standard optimization techniques. The major problem is that the subroutine expects its inputs in, and produces its output in a particular register. Also some registers may be destroyed. A method for copying with this is described in (2, 1), the basic ideas are:

- Initially generate code as if there were a multiply operator with inputs and outputs in any register.

- After optimization lower the text to expose the copying of operands to the routine specific input and output registers.

- Modify this code to reduce the actual number of copies required. One method is discussed in (1), but others are possible.

Not all multiplies need to be done by the standard subroutine; other alternatives are possible. In some cases a full 32 x 32 bit multiply may not be required. This is often the case with subscript computations and it may be worthwhile for the compiler writer to design several multiply subroutines where the size of the multiplier is known to be constrained. Many other possibilities exist. If the multipliers are highly skewed in the low range, a multiply procedure might include a test to direct 90% of the multiplies to a routine that only had a few steps. One nice aspect of this approach is that it is possible to introduce instrumentation into the multiply subroutine, study the behavior of actual programs and then modify the compiler and the standard subroutines to exploit the properties of actual programs. The particular techniques chosen may vary depending on the language. The ability to tailor the function to exactly fit the rules of the particular high level language is particularly important with respect to detection of errors. Ada and Fortran need not use, but are not excluded from using, the same library. It may well be that two languages treat overflow in multiplication differently. The language specific multiply routines will reflect this.

Another strategy is to compile in line code to implement multiply. This is particularly appropriate for constant multipliers. Multiplication in that case becomes a series of adds, subtracts, and shifts. If this process is to be efficient it is particularly important that these instructions be non destructive. If an operand is destroyed in executing these ops as is the case on the IBM 370, MC 68000 and similar machines it will often be necessary to introduce another instruction to preserve an operand. Thus doing constant multiplication by adds, subtracts and shifts is less efficient on those machines.

In an optimizing compiler the common subexpression and code motion processes will operate on the individual instructions that make up a multiply by a constant. For example a multiply of n by 48 reduces to:

(n shift 5) + (n shift 4)

The shifts can thus eliminate multiplies of n by 32 or 16 which occur elsewhere.

It also is possible to capitalize on data flow that must be in the CPU. In order to do multiply step, a one bit shifter must be in front of the ALU. Thus the existence of multiply step implies the possibility of an instruction that shifts the contents of a register by one bit and then adds this value to another register. It is usually easy to assign an operation code and make such fortuitous op codes generally available. Let us call this particular instruction shift left one and add (sloa). Thus the computation (n*48) becomes:

(n sloa n) shift 4

The use of the dual function, but still single cycle sloa op, obviously saves an instruction. However there is a somewhat discouraging message for those looking for increased performance via complex instructions. Note that this higher function instruction is no longer able to eliminate a subsequent (n*32) or (n*16). The moral of this is that multifunction instructions inhibit optimization. In this case the instruction may be worth the code point but the resulting performance improvement will probably be somewhat less than was anticipated. This example illustrates a basic weakness of high level architectures; they force the reexecution of that which should be available.

Space VS Time

It is a long habit of programmers to be concerned about code space. Some of the techniques proposed above will result in more bulky code than than alternative strategies on high level machines. How can they be justified? There is an argument that code space will not be as an important issue in the future as it has been in the past. The validity of the argument that follows depends on actual execution frequencies, which will vary from design to design. Observation generally confirms that there there is more repetition and clustering of code than of data references. A consequence of the resulting high reuse of pages and cache lines that contain code is that there are relatively few misses for code compared with data. There is another problem associated with virtual memory. In many systems the address space that is available to a program is limited by the requirement to have resident page tables. Thus available, but seldom used, virtual address space always requires real memory for resident page tables. Most modern virtual memory computers have this problem, but it is not necessary. Houdek and Mitchell(4) describe an efficient method of implementing virtual memory, where the page table is proportional to the size of real, rather than, virtual memory. The above argument suggests that modest increase in code size, of say 30%, will not noticeably affect performance on hardware with the proper characteristics. In fact the ability of the compiler writers to develop many widely used subroutines that are shared over separate programs may well result in even better performance. The performance of the storage hierarchy may be more important than any details of the computational instruction set. This suggests an approach for those who want to exploit VLSI design. Expend your effort on the memory hierarchy, not on exotic instructions. This is appealing as memory function tends to be more regular and is thus easier to design.

Conclusion

We have shown that there are a number of ways in which high level function may be efficiently implemented on low level machines. In the process we have indicated that the low level machine design can assist in the implementation by having an appropriate computational instruction set. In addition the importance of the storage hierarchy in achieving satisfactory performance has been indicated.

References

1. Auslander, M. and Hopkins, M.E. "An Overview of the PL.8 Compiler," Proc. of the Sigplan '82 Symposium on Compiler Writing, Boston, Mass. June 23-25 1982

2. Chaitin, G. J., et al, "Register Allocation via Coloring," Computer Languages, Vol. 6, pp. 45-57, 1981, Great Britain.

3. Hennessy, J., Jouppi, N., Basket, F., Gross, T. and Gill, J. "Hardware/Software Tradeoffs for Increased Performance," Proc. of Symposium on Architechtural Support for Programming Languages and Operating Systems Palo Alto, California, March 1-3, 1982

4. Houdek, M.E. and Mitchell, G.R. "Translating a Large Virtual Address," Electronics, March 15,1979

5. Paterson, D.A. and Ditztel, D.R. "The Case for the Reduced Instruction Set Computer," Computer Architecture News, October 1980

6. Radin, G., "The 801 Minicomputer," Proc. of Symposium on Architectural Support for Programming Languages and Operating Systems Palo Alto, California, March 1-3, 1982

Tradeoffs In The Design of A System For High Level Language Interpretation

Fernando C. Colón Osorio +
and
Yale N. Patt ++

+ Large Vax Engineering Group,
Digital Equipment Corporation, Marlboro, Massachusetts 01752

++ University of California at Berkeley, and
Digital Equipment Corporation, Marlboro, Massachusetts 01752

ABSTRACT

This paper treats the problem of designing a system for High Level Language Interpretation (HLLI). First, a model of the design process is presented where several styles of design, e.g., Turing Machine Interpretation, RISC Architecture Interpretation, and CISC Architecture Interpretation are treated uniformly. Second, the most significant characteristics of HLLI are presented and analyzed vis-a-vis the different design styles for HLLI. We conclude by presenting a hypothesis regarding the applicability of each design style to the solution of high level language problems.

1.0 INTRODUCTION

The problem of **HIGH LEVEL LANGUAGE INTERPRETATION (HLLI)** has attracted the attention of Computer Architects, Logicians, Compiler Writers, and Programmers in general, since the introduction of the first **RISC** machine in 1948, the **Manchester Mark-I**, see Kilburn and Williams [1]. Simply stated, the problem is that of:

> Matching Application Environments (i.e., Real World Problems) to an Interpretative Environment (i.e., a combination of a compiler, a target architecture for the compiler, and an implementation of that target architecture); or
>
> Closing the semantic GAP between the high level language, and the environment where it executes.

In general, several design styles have emerged over the last thirty five (35) years to solve this problem. The most well known of these are:

1. **Classical Compiler Technology:** this design style is characterized by a combination of state of the art compiler technology running on top of a general purpose computing environment. Examples of such a design style include the VAX-11 with C(UNIX) and the IBM s/370 with PL/1.

2. **Direct Execution of High Level Languages (DEHLL):** this design style is characterized by the implementation of high level language semantics directly in the architecture of the target machine. The semantic gap is forced to zero. Early examples of this design style were the **EULER** Classical Compiler Technology approach. Here, the target architecture for the compiler is restricted such that its instruction set is limited to a rather primitive collection having the property that each instruction executes in a single, straightforward machine cycle. This design style is exemplified by the IBM TJ Watson experimental design, the IBM 801, see Radin [5], and by the **RISC** architecture from the University of California, Berkeley [6].

3. **CISC** Architectures: this design style is another variation of the Classical Compiler technology approach. However, in this case the target architecture is one whose design attempts to close the semantic gap between the HLL and the execution environment in that part of the design which the architect feels is most important. "Importance" in this context refers to that set of semantic functions whose frequency of execution is high. There are many examples of this design style in the computer industry today, and they reflect the general tendency from computer manufacturers to provide the users

CH1935-6/83/0000/0620$01.00

of their equipment with that set of semantic functions within their machine architectures. This design style is best exemplified by the VAX-11 -- COBOL(VMS) compiler pair, Intel's iAPX432 -- ADA pair, and the CDC CYBER 205 -- FORTRAN pair.

In Figure 1 below, the different design styles are modeled in terms of two functions, $f_1(.)$ and $f_2(.)$. In Figure 1, $f_1(.)$ represents the design of the target architecture, while $f_2(.)$ represents the design of the compiler and its associated computing environment for that target architecture. Note that at one end of the spectrum, as depicted in Figure 1, we have a **Turing Machine**, in which case the design of a target architecture is relatively simple if not trivial, while the design of its associated compiler that generates efficient code (the design of $f_2(.)$) is a difficult task. At the other extreme we have a **DEHLL** design, where the semantic gap has been closed (i.e., $f_2(.)$ is the identity operator), and the design of the target architecture is difficult. In terms of this model, the problem of "MATCHING HLL PROBLEMS TO ELECTRONIC SOLUTIONS" becomes that of selecting two functions, $f_1(.)$, and $f_2(.)$ in such a way that the resulting computing

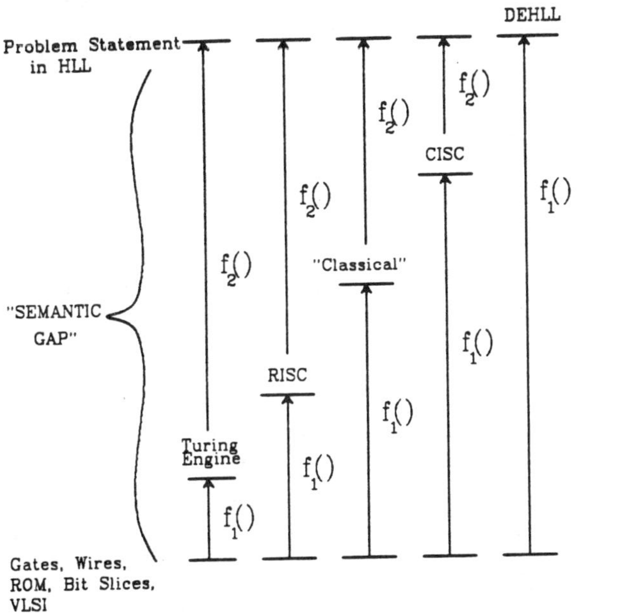

Figure 1: "Matching HLL Problems To Electronic Solutions" $f_1(.)$ is the realm of the computer architect, while, $f_2(.)$ is the realm of the Compiler designer.

environment is optimized. Although we recognize its limitations, our optimality criterion (at least for the present paper) will be cost/mip. Indeed all optimizations are made based on the performance X cost power product.

Recently, a large number of articles have appeared in the literature, see Patterson, et. al. [7], [8], Levy and Clark [9], and Larus [10], proposing that within the realm of design styles, i.e., solutions to the above problem, one approach is better than the other, universally. We suggest that these conclusions are premature. Evaluation of an "optimal design style" in terms of cost/mip is very complex, partly due to our uneasiness with the metric cost/mip itself, and partly due (given an acceptance of the metric) to the lack of understanding of how much of this figure of merit (cost/mip) is caused by the design function $f_1(.)$ and how much by $f_2(.)$. The purpose of this paper is to examine this question.

This paper is organized as follows. In section 2, we discuss our contention that "THE BEST DESIGN STYLE, i.e., RISC vs. CISC is dependent on the application environment on the one hand, and on the **Technology/ Performance/ Cost** tradeoffs on the other. In section 3, we attempt to characterize HLLI processing. In section 4, we present our preliminary conclusions and indicate areas in which we are continuing our research.

2.0 THE HIGH LEVEL LANGUAGE INTERPRETATION PROBLEM

The problem of **HLLI** involves the design of two functions $f_1(.)$ and $f_2(.)$ such that the cost/mip is minimized. Clearly the design of $f_1(.)$ involves much more than the selection of an instruction set. For example, it involves the design of system support functions, such as the memory management subsystem. Although it is an attractive concept to "GO BACK TO THE BASICS" and to apply such axioms as "SIMPLE IS BETTER" to the design of a computer architecture, it would be somewhat naive on the part of an architect to seriously propose an architecture where all high level constructs are removed from architectural direct support (hardware), and instead are implemented as a set of procedure calls, unless the application environment is correspondingly restrictive.

Consider, for example, the memory management hierarchy. Since the concept of one level storage was first introduced in the ATLAS computer, see [11], support for the memory management function has continued to increase, until today it is strongly supported by most architectures, almost universally (e.g., VAX-11 family, iAPX432, etc.). We cannot say, however, that all serious recent architectures support this function, the CRAY I being an obvious counterexample. Consequently, a proposal today for an architecture which does not include memory management support must be accompanied by performance requirements which overwhelmingly demand that this be the case.

On the other hand, it is also true that in the past, the tendency on the part of Computer Architects to move the support of high level constructs into the architecture (i.e., a combination of random logic and microcode) has sometimes had an overall negative effect on the performance of a given implementation of that architecture while executing a specific workload or benchmark. This performance degradation is usually due to a longer machine cycle for the CISC implementation needed to accommodate one or more complex operations (e.g., the VAX-11/750 450 nsec cycle time required by the rotate primitive). One must be careful, however, to correctly assign the fault for this performance degradation. It is often the case that the performance degradation is not at all due to the architecture, that rather it should be attributed to one of two causes, namely:

- **A SPECIFIC BENCHMARK or WORKLOAD**
- **IT IS IMPLEMENTATION DEPENDENT, NOT ARCHITECTURE DEPENDENT**

A specific benchmark can always be contrived to show performance improvement due to the inclusion of high level constructs in the architecture. Also, another implementation of the same architecture may or may not show a pernicious effect in its overall performance due to the inclusion of the complex construct, as example 2.1 shows.

Clearly, one cannot attribute performance improvement/degradation (at least in the above examples) to the fact that a high level construct has been included. Still the question remains:

> Given a HLLI environment, what design style, RISC or CISC, will result in an optimization of the cost/mip metric?

It is very important that in the design of a new architecture, or in an extension to an old architecture, the decision of whether or not to add a high level construct, **F**, must be evaluated in terms of its effect on the overall performance of the machine, see Radin [5]. That is,

> The implementation of **F** must result in a faster version than if **F** were implemented as a succession of simple instructions already present in the architecture; and
>
> Inclusion of **F** must result in an overall performance improvement which overcomes any performance degradation in all other instructions. (This is the so called "N+1 instruction set enhancement problem", see Hennessy [12]).

EXAMPLE 2.1 : PDP-11 SOB instruction

TABLE 1: Comparison of Two Implementations Of the PDP-11 SOB construct (time in usecs)

PDP-11 Construct	PDP-11/60	PDP-11/70
SOB Rn,TGT	2.00	0.60
DEC Rn BNE TGT	1.19	0.90

The object of this paper (and our subsequent work) is to attempt to characterize processing within the realm of Von Neumman processors, and to propose a mechanism for deciding when RISC makes sense and when CISC makes sense? We note, incidentally, that different **HLLI** environments will result in different RISC vs CISC cost/mip solutions. That is, in certain process control environments (e.g., a CMOS-8 processor controller for a TU-58 tape drive), a RISC architecture like the PDP-8 is a better choice than a CISC architecture like the VAX-11. Similarly, if the **HLLI** environment is that of program development support in a general purpose sixty four (64) user environment, the CISC architecture (VAX-11) is the better choice.

3.0 A CHARACTERIZATION OF HLLI PROCESSING

Consider the problem posed by Figure 2 below: **For which HLLI environment does a**

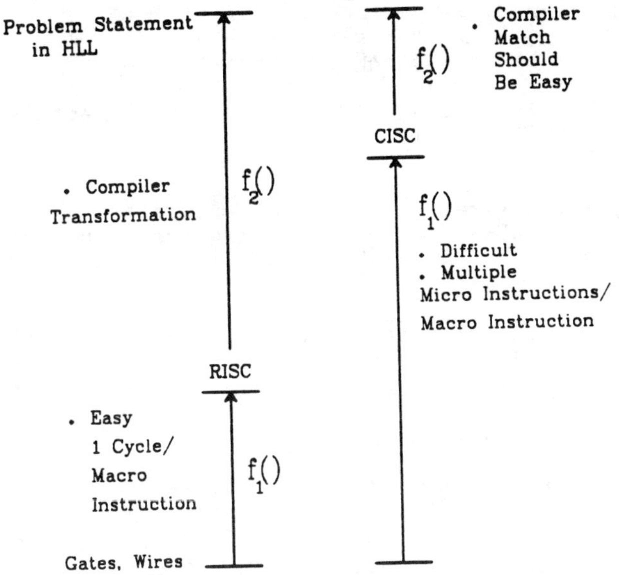

Figure 2: RISC vs. CISC Architectures

- **o What are the real tradeoffs?**
- **o Where does RISC makes sense?**
- **o Where does CISC makes sense?**
- **o What are the processing Characteristics of each?**

RISC architecture makes sense, and for which HLLI environment does a CISC architecture makes sense? In order to answer these questions, a clear understanding of the **intrinsic** characteristics associated with both RISC and CISC processing is needed. First consider the **FETCH/DECODE/EXECUTE** processing cycles in a RISC architecture, as shown in Figure 3 below.

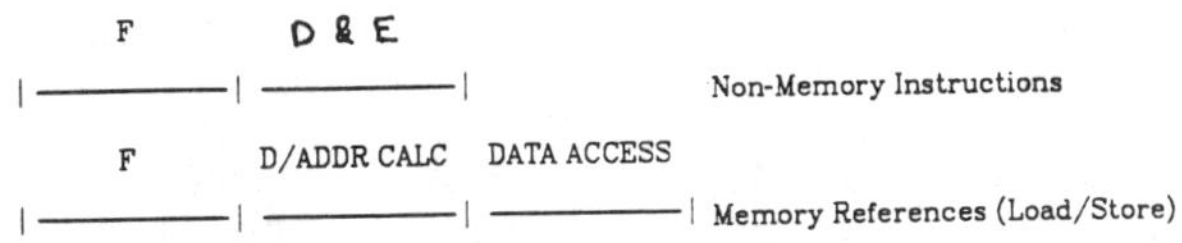

Figure 3: Model for The Execution of RISC instructions, without ISTREAM or DSTREAM fetch overlap.

Note from Figure 3, that several parameters determine the performance of the RISC architecture. These parameters are:

1. RISC cycle time := t_{e_r}

   t_{e_r} = max (cache access, RISC DP);

2. ISTREAM/DSTREAM Access Time := t_a

   $$t_a = p_{hit} * t_{e_r} + p_{miss} * t_M$$

   where t_M is the main memory access time

3. Instruction mix := p(Non-Memory) + p(load) + p(store) = 1

Hence, the time to execute a RISC (T_R) program if no pipelining or fetch overlap is considered is given by equation 1 below:

$$T_R = size(RISC) * \Big\{ p(non\text{-}memory) * \left[t_{e_r} + p_{hit} * t_{e_r} + p_{miss} * t_M\right] + p(load/store) * \left[t_{e_r} + 2 * \left[p_{hit} * t_{e_r} + p_{miss} * t_M\right]\right] \Big\} \qquad (1)$$

Furthermore, if all memory references are considered to be cache hits, equation 1 above becomes:

$$T_R = size(RISC) * \Big\{ 2 * p(non\text{-}memory) * t_{e_r} + 3 * p(load/store) * t_{e_r} \Big\} \qquad (2)$$

In a similar fashion, if we consider the model for the execution of CISC instructions

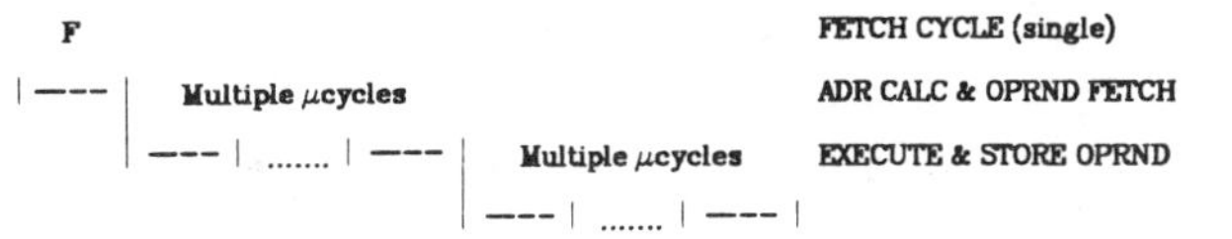

Figure 4: Model for The Execution of CISC instructions, without ISTREAM or DSTREAM fetch overlap.

shown in Figure 4, we note that several parameters determine the performance of the CISC architecture. These parameters are:

1. CISC cycle time := t_{e_c}

   t_{e_c} = max (cache access, CISC DP);

2. ISTREAM/DSTREAM Access Time := t_a

   $$t_a = p_{hit} * t_{e_r} + p_{miss} * t_M$$

where is the main memory access time

3. Average # of operands per CISC macro instruction := A_0

4. Average # of ucycles to perform the operand specifier calculation, and single operand fetch (μ_a), where

$$\mu_a = \sum_{i=1}^{i=N} g_i * \mu_i$$

where

g_i is the frequency of occurence of the ith operand specifier,and

μ_i is the number of cycles required to process the ith specifier

5. Average # of ucycles to execute a CISC macro instruction (μ_E),

where

$$\mu_E = \sum_{k=1}^{k=M} f_k * \mu_k$$

and,

f_k := frequency of occurrence of the kth instruction;

μ_k := number of cycles required to execute the kth instruction

Hence, the time to execute a CISC (T_C) program if no pipelining or fetch overlap is considered, and if we furthermore consider all memory references to have cache hits, is given by equation 3 below:

$$T_C = size(CISC) * \left\{ t_{e_c} + \left[A_0 * \mu_a \right] * t_{e_c} + \mu_E * t_{e_c} \right\} \tag{3}$$

Therefore, in order to determine the set of application environments where one architecture will do "better" than the other, the ratio T_R to T_C must be computed, as shown in equation 4.

$$RATIO \ \frac{RISC}{CISC} = \frac{size(RISC)}{size(CISC)} * \frac{t_{e_r}}{t_{e_c}} * \left\{ \frac{2p(non\text{-}mem) + 3p(Load/Store)}{1 + A_0 * \mu_a + \mu_E} \right\} \tag{4}$$

or,

$$RATIO \ \frac{RISC}{CISC} = \alpha * \beta * \left[\frac{2p(non\text{-}mem) + 3p(Load/Store)}{1 + A_0 * \mu_a + \mu_E} \right] \tag{5}$$

where

α := is the compaction ratio, and

β := is the RISC to CISC basic machine cycle time ratio

Equation 5 represents a characterization of the design tradeoffs required in the selection of an architecture. That is,

USE A RISC architecture when the right hand side of equation 5 $<$ 1;

otherwise,

USE A CISC architecture when the right hand side of equation 5 $>$ 1.

4.0 CONCLUSIONS AND RECOMMENDATIONS

In this paper, we have tried to describe the problem of high level language interpretation, and to identify a spectrum of design styles, each of which could accommodate a solution to the problem, depending on the particular application environment and the technology/performance/ cost criteria which must be satisfied. We have shown situations where a reduced instruction set computer is appropriate and situations where a complex instruction set computer is appropriate.

We suggest that this analysis only begins to scratch the surface of this question. If the RISC design style were in fact the panacea of the eighties, we would offer (with tongue appropriately in cheek) our own candidate VHRSC architecture, see Appendix A. Unfortunately, we suspect it is not. Several next steps are in order, including studying the pipeline case for both RISC and CISC architectures postulating more realistic application environments, measuring the performance of competing designs in equivalent application environments, and measuring the effects of competing enhancement features relative to their inclusion or omission from a particular design. We are continuing to work on these issues.

5.0 REFERENCES

[1] Williams, F.C., and T. Kilburn, "Electronic Digital Computers" letter, Nature, vol. 162, September 1948, p. 487.

[2] Wirth, N., and H. Weber, "EULER: A Generalization of Algol, and its Formal Definition, Parts I and II, Comm. ACM, vol. 9, no(s). 1 and 2, January and February 1966.

[3] Wilner, Wayne T., "Design of the Burroughs B1700", Proc. AFIPS Fall Joint Computer Conference, Vol. 41, pp. 489-497.

[4] Sussman, Gerald J., Jack Holloway, Guy Lewis Steele, Jr., and Alan Bell, "Scheme-79--LISP on A Chip," Computer, vol. 14, no. 7, pp. 10-21.

[5] Radin, George, "The 801 Minicomputer", Proc. of the Symposium on Architectural Support for Programming Languages and Operating Systems, ACM, Palo Alto, California, March 1-3, 1982, pp. 39-47.

[6] Patterson, David, and Carlo H. Sequin, "A VLSI RISC", Computer, vol. 15, 9 (September 1982), pp. 8-21.

[7] Patterson, David A., and David R. Ditzel, "The Case for the Reduced Instruction Set Computer", Computer Architecture News 8, 6 (October 1980), pp. 25-33.

[8] Patterson, David A., and Richard S. Piepho, "Assesing RISCs in High Level Language Support", IEEE Micro 2, 4 (November 1982), 9-19.

[9] Levy, Henry M., and Douglas W. Clark, "On the Use of Benchmarks for Measuring System Performance", Computer Architecture News, December 1982.

[10] Larus, James R., "A Comparison of Microcode, Assembly Code, and High-Level Languages on the VAX-11 and RISC I," Computer Architecture News, September 1982.

[11] Kilburn, T., D.G.B. Edwards, M.J. Lanigan, and F.H. Summer, "One Level Storage System", IRE Trans., EC-11, vol. 2, April 1962, pp. 223-235.

[12] Hennessy, John, Norman Jouppi, John Gill, Forest Baskett, Alex Strong, Thomas Gross, Chris Rowen, and Jud Leonard, "The MIPS Machine,", Proceedings of the Spring CompCon, IEEE, February 1982, pp. 2-7.

Appendix A
ISP Description for VHRSC

Pc State:

```
AC<0:31>                    ! Accumulator
L                           ! Link bit/AC extension for
                            ! overflow and carry
PC<0:31>                    ! Program counter
Run                         ! 1 when Pc is interpreting
                            ! instructions
Interrupt_State             ! 1 when Pc can be interrupted
                            ! under program control
Io_pulse_1;                 ! IO Pulses to IO Devices
IO_pulse_2;
IO_pulse_4;
```

Instruction Format

```
instruction/i<0:31>         := i<0:2>      ! opcode
indirect_bit/ib             := i<3> ! 0, direct; 1, indirect memory ref.
system_space_bit            := i<4> ! 0 selects system space
                                           ! 1 selects user program space
page_address<0:26>  := i<5:31>
user_page<0:4>              := PC<0:4>
IO_select<0:5>              := i<3:8>      ! selects a T or Ms device
sma                         := i<5> ! micro bit for skip on minus AC
                                           ! operate 2 group
sza                         := i<6> ! micro bit for skip on zero AC
sni                         := i<7> ! micro bit for skip on non-zero
                                           ! link
```

Effective Address Calculation Process

```
z<0:31>       := (
   ¬iB → z'';
  iB AND (A₁₆ ≤ z'' ≤ 10₁₆ → M[z''] ← M[z''] + 1; next)
  iB → M[z''])
z'<0:31>      := ( ¬iB → z''; iB → M[z''])
z''<0:31>     := (system_space_bit → user_page • page_address;
                  ¬system_space_bit → 0 • page_address)
```

Instruction Interpretation Proccess

```
Run AND ¬(interrupt_request AND interrupt_state) → (
   instruction ← M[PC]; PC ← PC + 4; next
   instruction_execution);

Run AND interrupt_request AND interrupt_state → (
   M[0] ← PC; interrupt_state ← 0; PC ← 4)
```

Instruction Set and Instruction Execution Process

```
          instruction_execution := (
    and (:= op = 0) →    (AC ← AC AND M[z]);
    tad (:= op = 1) →    (L • AC ← L • AC + M[z]);
    isz (:= op = 2) →    (M[z'] ← M[z] + 1; next
                          (M[z'] = 0) → PC ← PC + 4);
    dca (:= op = 3) →    (M[z] ← AC; AC ← 0);
    jms (:= op = 4) →    (M[z] ← PC; next PC ← z + 1);
    jmp (:= op = 5) →    (PC ← z);
    iot (:= op = 6) →    (
         io_p1_bit → IO_pulse_1 ← 1; next
         io_p2_bit → IO_pulse_2 ← 1; next
         io_p4_bit → IO_pulse_4 ← 1);
    opr (:= op = 7) →    Operate_execution)
```

Operate Instruction Set

```
          Operate_execution := (
    cla (:= i<4> = 1) →                  (AC ← 0);
    opr_1 (:= i<3> = 0) →                  (
    cll (:= i<4> = 1) →                  (L ← 0); next
    cma (:= i<5> = 1) →                  (AC ← ¬AC);
    cml (:= i<6> = 1) →                  (L ← ¬L); next
    cml (:= i<7> = 1) →                  (L ← ¬L); next
    iac (:= i<11> = 1) →                 (L • AC ← L • AC + 1); next
    ral (:= i<8:10> = 2) →   (L • AC ← L • AC × 2 {rotate}); next
    rtl (:= i<8:10> = 3) →   (L • AC ← L • AC × 2^2 {rotate}); next
    rar (:= i<8:10> = 4) →   (L • AC ← L • AC / 2 {rotate}); next
    rtr (:= i<8:10> = 5) →   (L • AC ← L • AC / 2^2 {rotate}); next
          opr_2 (:= i<3, 11> = 10) →   (
    skip condition XOR (i<8> = 1) → (PC ← PC + 1); next
          skip condition := ((sma AND (AC < 0))
                             OR (sza AND (AC = 0))
                             OR (snl AND L))
    osr (:= i<9> = 1) →                  (AC ← AC OR Data switches);
    hlt (:= i<10> = 1) →                 (Run ← 0));

FAE (:= i<3, 11> = 11) →                 EAF_instruction_execution))
```

Session 3-3a
Novel Design Issues I

Chairman

A. L. Rosenberg
Duke University

Systolic and Semisystolic Design

(Extended Abstract)

Charles E. Leiserson

Laboratory for Computer Science
Massachusetts Institute of Technology
Cambridge, Massachusetts 02139

Abstract—This paper identifies many architectural transformation techniques which can be used to help produce efficient systolic and semisystolic architectures for signal processing, image processing, numerical computation, and raster graphics. The transformations presented include retiming, slowdown, broadcast and census elimination, coalescing, and interlacing, as well as a brief discussion of code motion, resetting, register elimination, and parallel/serial compromises. All the transformations are at the architectural or algorithmic level, as distinct from the electrical circuit level, and provide high-level tools for a designer. In addition, they provide a step toward automatic compilation of systolic and semisystolic systems.

1. Introduction

This paper addresses the problem of hardware system design in the context of *systolic* and *semisystolic* architectures [3], [4], [7]. It offers a methodology of transformations which can be used at the architectural or *algorithmic* level of system design in order to produce an efficient implementation from a high-level algorithmic description. The transformations can be applied directly by the knowledgeable system builder, but they also offer a foundation for a "systolic" compilation.

Systolic system were introduced in [3] and are defined formally in [7]. Kung gives a nice overview of systolic algorithms in [4], and he argues persuasively in [5] that systolic architectures are suitable for numerical and signal processing applications.

Informally, a systolic system is a network of processors that possesses three primary characteristics:

1. *The processors are small.*
2. *There is an external interface (host).*
3. *All communication between processors is clocked through a register.*

Characteristic 3 gives the system "blood pressure" because a flow of data is kept up in the system. There is no combinational rippling, and so the clock period of a systolic system is guaranteed to be short. A *semisystolic system* is the same as a systolic system, but without Characteristic 3. Thus, in a general semisystolic system, there may be combinational rippling.

To illustrate these two kinds of systems, consider the problem of computing a *finite-impulse response* (FIR) filter. Given a set of fixed weights $a_0, a_1, \ldots, a_4$ and a stream of inputs $x_1, x_2, \ldots$, we wish to compute $y_1, y_2, \ldots$, where

$$y_i = \sum_{j=0}^{4} a_j x_{i-j} .$$

(We assume $x_i = 0$ for $i \leq 0$.)

Figure 1 shows a systolic array which computes an FIR filter. The functional elements of the array are inner-product-step processors which perform a multiplication and an addition. The small black rectangles represent globally clocked registers. The design is systolic because all communication between the processors is clocked through a register.

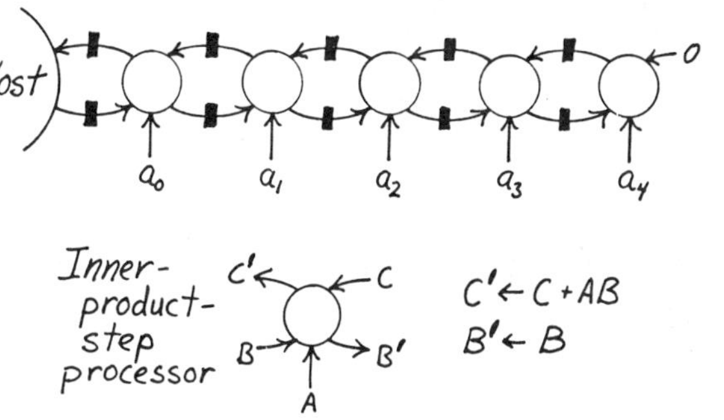

Figure 1. A systolic array implementation of an FIR filter assembled from inner-product-step processors.

Figure 2 shows how the systolic array works. The y_i are initialized at zero on the right and move to the left, while the x_i are input from the host on the left and move to the right. Processor j multiplies its stored weight a_j by x_{i-j} and adds the product to y_i. The timing is synchronized so that each y_i is able to accumulate all of its terms by the time it is output on the left, which is accomplished by inputting each x_i on the $2i$th clock tick. A side-effect of this timing strategy is that processors are doing useful work only half the time. With a little study, the reader can verify that the systolic array indeed computes the FIR filter.

This research was supported in part by the Defense Advanced Research Projects Agency under Contract N00014-80-C-0622.

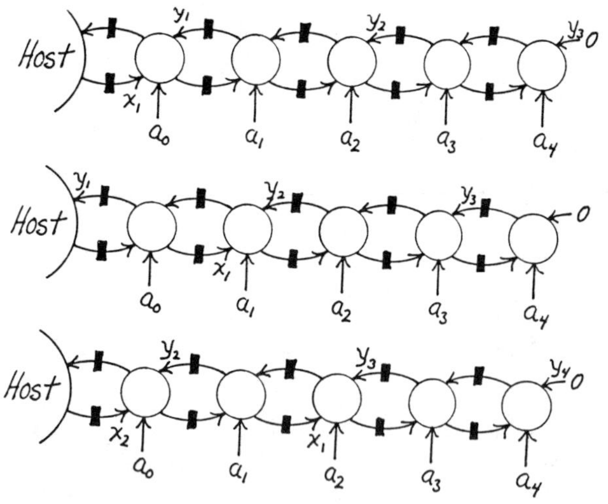

Figure 2. Several steps of the systolic array from **Figure 1.**

A simpler *semi*systolic design for the FIR filter is shown in Figure 3. Using the same inner-product-step processors as before, this design makes it obvious that the proper y_i value is output. The registers along the bottom act like a shift register. With each tick of the system clock, the x_i are shifted to the right and the $a_j x_{i-j}$ terms are accumulated leftward toward the host. Unlike the systolic design, an x_i is input and a y_i is output on every, not just every other, clock tick.

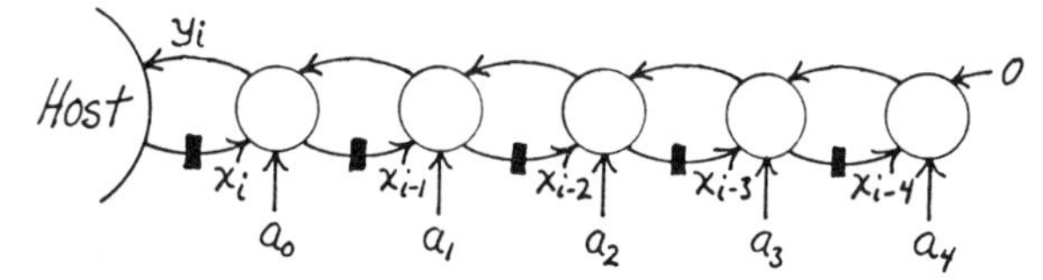

Figure 3. A semisystolic array implementation of an FIR filter assembled from inner-product-step processors.

The semisystolic design is somewhat easier to understand than the systolic design because the addition of terms for a single y_i occurs during a single clock period. On the other hand, because this accumulation toward the host ripples the entire length of the array, the clock period of the semisystolic design grows with the number of processors. For the example shown in the figure, the clock period is 5 functional elements. The clock period of the systolic design, however, is guaranteed to be short because every interconnection passes through a globally clocked register. Every systolic design has clock period 1. In summary, the semisystolic design is easier to understand, but the systolic design is more efficient.

Remarkably, the systolic design for the FIR filter in the last section can be produced automatically from the semisystolic design using two techniques, *retiming* and *slowdown,* which are presented in Section 2. Section 3 shows how these two techniques can be combined to implement *broadcasting* and *census* efficiently in a systolic system. Section 4 then presents *coalescing* and *interlacing* as two means of achieving hardware efficiency by making more effective use of processing elements. Section 5 discusses other high-level transformations for systolic and semisystolic architectures, and some conclusions are offered in Section **6.**

2. Retiming and slowdown

This section discusses retiming and slowdown [8], [9] which are two powerful transformation techniques that can be used to ease the design of systolic and semisystolic systems. Retiming is a means of relocating the registers in a circuit while preserving the exact I/O behavior of the circuit. Slowdown multiplies the number of registers everywhere in a circuit by a constant.

To illustrate retiming, consider the structure of the semisystolic FIR algorithm which is shown again in Figure 4(a). Observe that the circled portion of the circuit communicates with the rest of the circuit through one input and one output. Suppose the register on the input to the circled portion is removed as has been done in Figure 4(b). Then every input to the circled portion arrives one clock tick earlier, which means that every computation in that portion occurs one clock tick earlier, and thus each output from that portion will be produced one clock tick earlier. Therefore, if we add a register to the output of the circled portion, as has been done in Figure 4(c), we can compensate for the timing change. The host will notice no difference in the behavior of the semisystolic array in terms of how the circuit is clocked, but the clock period has been reduced. The clock period of the original circuit needed to be long enough to allow rippling through five functional elements. The new circuit has a clock period of only three functional elements.

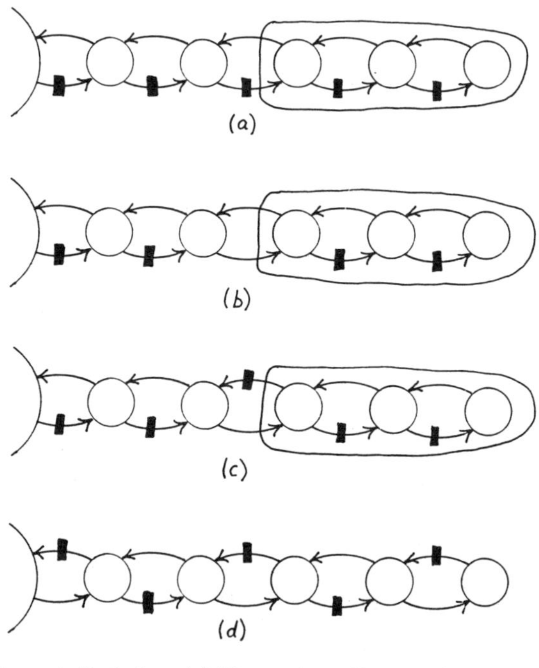

Figure 4. Retiming. (a) The semisystolic array from Figure 3. (b) Removal of a register causes all inputs to the circled portion to arrive one clock tick earlier. (c) The output is from the circled portion is delayed one clock tick to compensate. (d) A retiming that minimizes the clock period of the circuit.

It turns out that no matter how long the semisystolic array is made, properly retiming it yields a clock period of exactly two functional elements. Figure 4(d) shows the optimal retiming of the semisystolic array. Such an optimal retiming can be determined automatically by the algorithms in [8] and [9].

Using another kind of transformation, retiming can be used to produce the systolic FIR filter from the semisystolic design. Figure 5(a) shows once again the semisystolic design. In Figure 5(b) each register has been replaced by a pair of registers. This transformation, which is called *slowdown* [9], has the effect that two clock ticks are required in the new 2-slow system to accomplish the same thing as one clock tick in the original. Now, retiming the 2-slow system gives the systolic system in Figure 5(c), which is the same as the system in Figure 1.

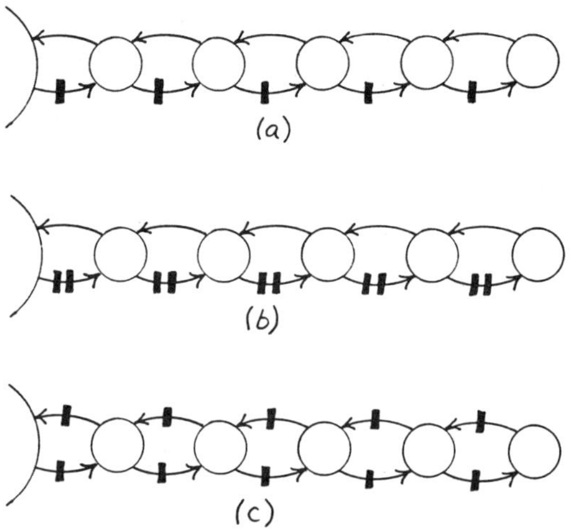

Figure 5. Slowdown. (a) The semisystolic array from Figure 3. (b) An equivalent 2-slow system. (c) Retiming the 2-slow system yields the systolic system from Figure 1.

To study retiming more precisely, we can represent a semisystolic system as a graph. Each functional element is a vertex, and each interconnection is an edge between two vertices. We label each edge with a numerical weight indicating the number of registers on that interconnection.

The graph representation allows us an economy of notation. For example, if G is the graph of the semisystolic system in Figure 5(a), then the 2-slow system in Figure 5(b) can be characterized as $2G$. The following conversion theorem, adapted from [8], [9], and [12] and based on this graph representation, shows when a semisystolic circuit can be efficiently retimed.

Theorem 1. *Let k be a given positive integer, and let G be the graph of a semisystolic system. Then the following are equivalent:*

1. *The constraint graph $G - 1/k$ has no cycle of edges whose total weight is negative.*
2. *There is a retiming of G whose clock period is less than or equal to k.*
3. *There is a retiming of the k-slow system kG which is systolic.*

3. Broadcast and census elimination

The most important ramifications of Theorem 1 involve broadcasting and census in a systolic system. Broadcasting is the communication of information from the host to every processor in a system within a single clock period. Census [10] is the reporting to the host a value which is an associative and commutative function of the information in all the processors.

The advantage of using broadcast and census when designing a system is that it is easy to think about the host directing the processors much as a mæstro conducts an orchestra. There are no timing considerations to worry about—data that is broadcast arrives at the destination processors immediately.

Retiming can be used to eliminate broadcasting and replace it with local communication. To illustrate, Figure 6(a) shows the graph G of a systolic system. Each interconnection has at least one register on it, and the processors are numbered breadth-first from the host. In Figure 6(b), a broadcast has been implemented according to a breadth-first spanning tree from the host. Any cycle of edges in the graph of the system so augmented must include at least as many nonbroadcast edges as broadcast edges. Furthermore, in the constraint graph $G - 1/2$ from Theorem 1, each broadcast edge has weight $-1/2$ and each nonbroadcast edge has have weight at least $1/2$, and thus any cycle of edges has nonnegative weight. According to Theorem 1, therefore, the system G can be retimed to have clock period 2, as is shown in Figure 6(c), or alternatively, the 2-slow system $2G$ can be retimed to be systolic.

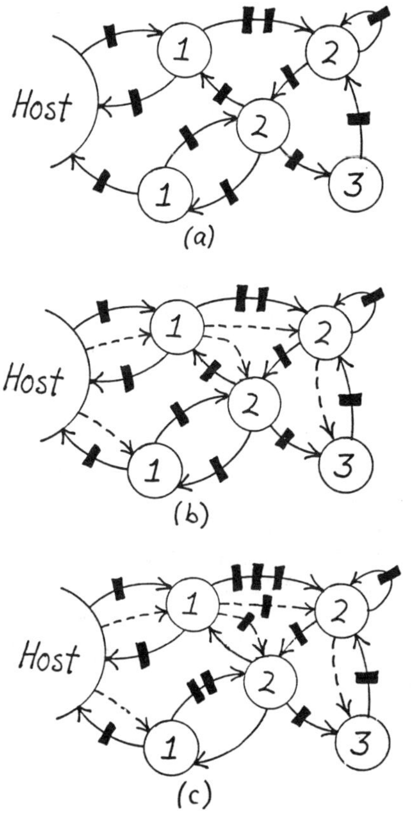

Figure 6. Broadcasting. (a) A systolic system with processors numbered breadth-first from the host. (b) Implementing a broadcast according to a breadth-first spanning tree. (c) Retiming yields a system with clock period 2.

If the edges are reversed in the breadth-first spanning tree and the processors compute some associative and commutative function, census can be implemented. As with broadcasting, the system can be retimed efficiently to eliminate the global census and replace it with local communication only.

The power of these techniques can be illustrated by designing a simple read-time priority queue. A priority is a device which maintains a set of numbers. There are two basic operations on the device—inserting a number into the queue and removing the smallest number in the queue.

With the capability of broadcasting, the design is easy. The numbers are stored in a linear array in sorted order with the smallest number next to the host. To insert a number, the host broadcasts the number. All numbers that are larger than the given number shift away from the host, and the broadcast number falls into place in the sorted order. To remove the smallest number, the host directs all processors to shift their numbers toward the host. The host gets the smallest number. The broadcast can now be eliminated to yield a system whose clock period is 2, or alternatively, the 2-slow system can be made systolic.

Other designs based on this methodology can be found in [7] and [9].

4. Coalescing and interlacing

Many systolic designs have the attribute that some fraction of the processors are idle at each clock tick. For example, in the FIR filter from Figure 1, half the processors are idle on each clock tick. The band-matrix multiplication algorithm from [3] only uses each processor once every three clock ticks. Also, if the broadcast/census methodology from Section 3 is used to design a 2-slow systolic system, processor utilization will be only fifty percent. This section describes two techniques, *coalescing* and *interlacing,* which can be used to improve processor utilization. It also shows how these techniques can be used to tune an algorithm to meet various constraints imposed by the operating environment of the implementation.

A 2-slow linear systolic array such as the one from Figure 1 supports two independent "equivalence classes" of computation, and it looks like two separate devices from the point of view of the host. Communication between the host and the device on even clock ticks is entirely independent from the communication on odd clock ticks. The host may be able take advantage of both equivalence classes of computation. (For example, some two-phase clocking methodologies require two equivalence classes, one of which is unusable.) Sometimes, however, only one equivalence class is actually needed.

Coalescing can save hardware at the expense of one equivalence class of comptutation. The idea is to implement a group of processors as a single processor plus a multiplexor and some memory. In a 2-slow systolic array, for example, the $2i$th and $2i+1$st processors can be paired up, and each two such "virtual" processors implemented as one "real" processor. If the processors are at all complicated, there will be a net savings in hardware size after coalescing. The performance of the coalesced system will be the same as that of the original, except for the minor overhead of the multiplexors. Thus coalesing will almost halve the amount of hardware at the expense of one equivalence class of computation.

Interlacing is another way of achieving similar goals. Suppose that the number of processors in the linear array is odd. Then the end of the linear array can be piped back to the beginning to take advantage of the processors on the offbeats. An array of length n can implement a virtual array of length $2n$. The ith processor in the real array implements the ith and $n+i$th processors in the virtual array which operate on different clock ticks if n is odd.

The advantage of interlacing the computation in this way—as opposed to coalescing—is that only one multiplexor is needed for the whole array, instead of one multiplexor for each pair of processors. The multiplexor causes the first processor in the array to communicate with the host on even clock ticks and with the last processor on odd clock ticks. Because of the lower hardware overhead, interlacing is often more desirable than coalescing.

There are situations when coalescing is better, however, because coalescing preserves locality of information. For example, if a pair of virtual processors is coalesced as one real processor on a chip, the processor requires only half the pin bandwidth that a processor in an interlaced array requires. This is because the communication between the two adjacent virtual processors can be carried out entirely on chip. A compromise can be achieved when several processors fit on a chip—partition the array into groups of contiguous processors and interlace each group independently.

Coalesing has other uses beyond saving hardware expense. Since systolic designs are characterized by enormous throughputs, it is easy to overdesign a system by making the processor computation rates faster than can be supported by pin, memory, or bus bandwidths. Suppose, for example, that in a linear systolic array design, the computation rate of a processor is 100 bits per unit time, but the I/O rate across pins is only 20 bits per unit time. Since the pins are the bottleneck, we can slow the computation rate of a processor without harming the overall computation rate of the device. By implementing five virtual processors in the linear array as a single real processor, the I/O and computation rates will be balanced. The real processor performs the actions of each of the five processors in turn. Whenever communication occurs between two of the five virtual processors, no off-chip communication is needed. Off-chip communication occurs only when the virtual processors on the ends need to communicate with their off-chip neighbors. Of course, if there is more room on the chip, $5k$ virtual processors might be implemented as k real processors.

Also, balancing I/O by coalescing can be particularly effective for two-dimensional array algorithms. As an example, the columns of a two-dimensional array can be coalesced to make a one-dimensional array implementation. By grouping two-dimensional subarrays, on the other hand, the original topology can often be maintained, and the amount of I/O goes up only as the square root of the number of coalesced processors. (Similar coalescing techniques are presented for other networks in [2].)

Coalescing and interleaving can also be used in conjunction with retiming to take advantage of pipelined processors. Figure 7(a) shows the graph of a typical systolic array. Suppose it were almost as efficient to manufacture the basic processor of the array pipelined to depth three as to manufacture the basic processor itself. The processor would then have three registers embedded in it, and any computation started on one clock tick would not settle until three ticks later. With a little overhead, we can make all of the equivalence classes of computation in the pipelined processor cooperate as part of the same array. The result can be a substantial hardware savings for implementing an array of a given length.

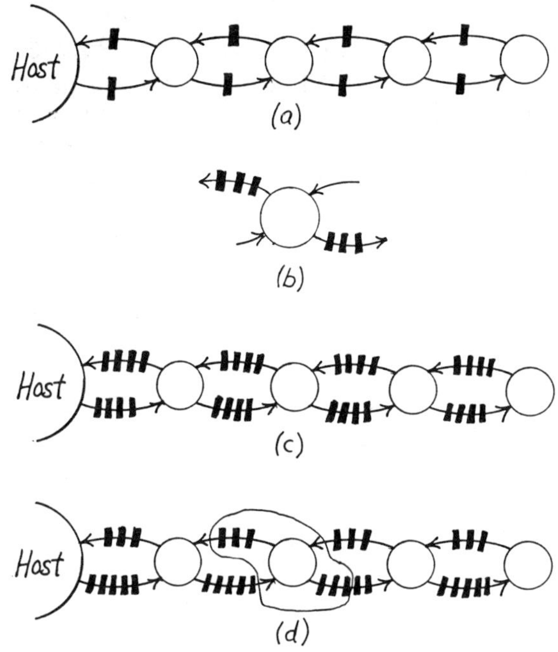

Figure 7. Taking advantage of pipelined processors. (a) A typical systolic array. (b) A pseudopipelined processor which is not really pipelined, but has a latency of 3 clock ticks. (c) The systolic array slowed down by a factor of 4. (d) Retiming yields an array made up of the pseudopipelined processors and some registers. Replacing these by the real pipelined processors, and then interlacing or coalescing will produce an efficient implementation.

In order to study the timing issues more easily, we use retiming to move the three registers out of the pipelined processor logic. This transformation does not affect the function of the pipelined processor, and results in the "pseudopipelined" processor shown in Figure 7(b). The function of the pseudopipelined processor is the same as the original processor, but its performance is diminished (because it is not really pipelined—its outputs are just delayed). For our purposes, the pseudopipelined processor will help us to keep track of the equivalence classes of computation more easily. When we are done, we shall be able to replace each pseudopipelined processor with a real one.

The first step in making the systolic array in Figure 7(a) take advantage of a processor pipelined to depth three is to slowdown the array by a factor of four, as shown in Figure 7(c). In this system, four clock ticks are required to perform the same computation as in the original system, but there are four equivalence classes of computation. Retiming the system from Figure 7(c) yields the one in Figure 7(d) which maintains the four equivalence classes of computation, but now the timing of each processor is skewed by one clock tick from its neighbor. Notice that the new system is composed of several copies of the pseudopipelined processor from Figure 7(b) plus a few registers. Now, by replacing the pseudopipelined processor with the real pipelined processor and using interlacing or coalescing, the system can be made to operate with a single equivalence class of computation.

5. Other transformations

This paper has presented a sample of design techniques for systolic and semisystolic architectures which have included retiming, slowdown, broadcast and census elimination, coalescing, and interlacing. To conclude, we mention some others: register elimination, code motion, resetting, and parallel/serial compromises.

Register elimination [8] is like the inverse of slowdown and retiming, and can be used instead of coalescing or interlacing to improve the cost-effectiveness of a design with several equivalence classes of computation. Register elimination, which can be performed automatically [8], reduces both the number of registers and the number of clock ticks required to perform the same function. It is particularly economical when the processors are small compared to registers.

Code motion [1, p. 924] is the transformation used in optimizing compilers for programming languages to move loop-independent code outside the loop. The same technique can be used in a systolic array, for example, to move computations performed by all processors to newly created processors on the boundary of the array. Occasionally, however, it may be worthwhile to use code motion to accomplish the opposite goal—moving computation to all processors–because it may be cheaper to recompute locally than to distribute globally.

Another useful technique is resetting—setting all the registers in a circuit to predefined values. Surprisingly, it is possible to perform the functional equivalent of a reset in time which is essentially independent of the size of the system. The reset action may take some time, but to the host, it appears to have occurred instantly. There is not space here to describe the technique fully, but the idea is to propagate a wavefront of reset through the system. The wavefront acts as a barrier between those processors that have been reset and those that are still performing the old computation. The interested reader is referred to [7] for a more complete description and examples of how it can be used in design.

In most numerical applications, there is always the choice between using serial arithmetic or parallel arithmetic. Parallel arithmetic, though more costly in terms of hardware, is usually more appropriate for high performance applications. For lower performance applications, the serial arithmetic should be considered together with techniques like coalescing in order to balance I/O and save hardware. Often, however, there is a substantial overhead in control complexity for serial arithmetic compared with the savings in datapath width. It is therefore worthwhile considering parallel/serial compromises where, for example, all computations are performed in k-bit bytes.

6. Conclusion

To date, most VLSI design aids have focused on the integrated circuit or printed circuit level. This paper advocates a thrust toward design aids at the system level. A system builder should be given the ability to design a custom integrated circuit component such that its cost is amortized by a single system which uses hundreds of that same component. In order to embark on large scale hardware development projects, however, the system builder needs high-level design support to help him organize his design effectively.

This paper has described several high-level transformations that address the design problem for systolic and semisystolic architectures. Systolic design is an error-prone activity because the timing and coordination of processors requires substantial bookkeeping. The transformations presented here to ease the design task can be used by a knowledgeable system builder, but ideally, the transformations should be implemented in a systolic compiler.

References

[1] Alfred V. Aho and Jeffrey D. Ullman, *The Theory of Parsing, Translation, and Compiling,* Volume II: *Compiling,* Prentice-Hall, Inc., Englewood Cliffs, New Jersey, 1973.

[2] John P. Fishburn and Raphael A. Finkel, "Quotient networks," *IEEE Transactions on Computing,* April 1982, pp. 288–295.

[3] H. T. Kung and Charles E. Leiserson, "Systolic arrays (for VLSI)," *Sparse Matrix Proceedings 1978,* I. S. Duff and G. W. Stewart, ed., Society for Industrial and Applied Mathematics, 1979, pp. 256–282. (An earlier version appears in Chapter 8 of [11] under the title, "Algorithms for VLSI processor arrays.")

[4] H. T. Kung, "Let's design algorithms for VLSI systems," *Proceedings of the Caltech Conference on Very Large Scale Integration,* Charles L. Seitz, ed., Pasadena, California, January 1979, pp. 55–90.

[5] H. T. Kung, "Why systolic architectures," *Computer Magazine,* IEEE, January 1982, pp. 37–46.

[6] Monica Lam and Jack Mostow, "A transformational model of VLSI systolic design," *Proceedings of the Sixth International Symposium on Computer Hardware Description Languages and Their Applications,* IFIP, May 1983.

[7] Charles E. Leiserson, *Area-Efficient VLSI Computation,* Ph.D. dissertation, Department of Computer Science, Carnegie-Mellon University, October 1981. Published in book form as part of the ACM Doctoral Dissertation Award Series by the MIT Press, Cambridge, Massachusetts, 1983.

[8] Charles E. Leiserson, Flavio M. Rose, and James B. Saxe, "Optimization of synchronous circuitry by retiming," *Third Caltech Conference on Very Large Scale Integration,* Randal Bryant, ed., pp. 87–116.

[9] Charles E. Leiserson and James B. Saxe, "Optimizing synchronous systems," *Twenty-Second Annual Symposium on Foundations of Computer Science,* IEEE, October 1981, pp. 23–36.

[10] Richard J. Lipton and Jacobo Valdes, "Census functions: an approach to VLSI upper bounds," *Twenty-Second Annual Symposium on Foundations of Computer Science,* IEEE, October 1981, pp. 13–22.

[11] Carver A. Mead and Lynn A. Conway, *Introduction to VLSI Systems,* Addison-Wesley, Reading, Massachusetts, 1980.

[12] Flavio M. Rose, *Models for VLSI Circuits,* Masters Thesis, Department of Electrical Engineering and Computer Science, Massachusetts Institute of Technology, March 1982. Also available as MIT VLSI Memo No. 82–114.

Automatic Generation of Three-Dimensional Circuit Layouts *

Extended Abstract

Frank Thomson Leighton

Mathematics Department and
Laboratory for Computer Science
Massachusetts Institute of Technology
Cambridge, Massachusetts 02139

Arnold L. Rosenberg

VLSI Research Laboratory
Department of Computer Science
Duke University
Durham, North Carolina 27706

Abstract: In this paper, we examine the savings in material volume and wire length afforded by the development of three-dimensional VLSI fabrication technology. We prove upper and lower bounds on the volume and maximum wire length of the best three-dimensional layout of a circuit in terms of the area and maximum wire length of its best two-dimensional layout. In some cases, the upper and lower bounds differ by only a constant factor. In the worst case, the bounds differ by a factor of $O(\log \frac{N^2}{A})$.

1. Introduction

1.1 Background

Recent advances in fabrication technology render imminent the fabrication of three-dimensional chips, wafers and circuit boards. Chip-like devices with up to thirty layers of wiring have been fabricated, and researchers have recently succeeded in fabricating chips with two layers of transistors. The rapid rate of progress in VLSI technology suggests that multilayer (three-dimensional) chips and PC boards will be commonplace in the not-distant future. Indeed, the president of Texas Instruments (quoted in [3]) predicts the production of such chips by the end of the decade.

One expects at least three benefits to accrue from the availability of the third dimension in circuit layouts. First, wire routing should become easier and more systematic. Next, by using the third dimension to avoid obstacles, one should be able to have shorter runs of wire, at least in the worst case. Finally, since avoiding obstacles in the plane requires area-consuming circuitous routing of the wires, one would expect savings in material: the volume of a compact three-dimensional realization of a circuit should be less than the area of any two-dimensional realization of the circuit.

The benefits of three-dimensional VLSI can be realized only if we develop effective techniques for devising and analyzing three-dimensional circuit layouts. It is crucial that these techniques not be materially more complicated than two-dimensional layout techniques, for even the latter are stretching our computing and analytical capabilities. The main result of the present work is that, at least in the idealized world of VLSI layout theory, the problem of three-dimensional layout is no harder than the problem of two-dimensional layout: any layout of the latter kind can be algorithmically be transformed into a layout of the former kind with close to optimal efficiency.

1.2 Related Work

One finds rather little work in the literature on the three-dimensional layout problem. Only a handful of studies can be construed as studying three-dimensional chip layout. Of these: [9] investigates channel routing in the presence of three layers; [2] concentrates on very constrained aspects of the problem of routing in multilayer media, specifically the question of how to assign wire to layers given a fixed placement of the nodes of the graph; [8, 10, 11, 14] study layouts of a few special networks or (in the case of [11]) of the class of "hardest-to-realize" networks. Only [6] (which is the completion of the work begun here) attempts to develop a general theory of three-dimensional layouts, equivalent in scope to the two-dimensional theory developed by Bhatt [1], Leighton [1, 4, 5], Leiserson [7], Thompson [12] and Valiant [13] (among others).

1.3 The Model

VLSI theory views circuits as undirected graphs whose nodes correspond to active devices (transistors, gates, etc.) and whose edges correspond to wires connecting these devices. The media in which the circuits are laid out are viewed as (two- or three-dimensional) rectangular grids. A *circuit layout* is an embedding of the circuit-graph in the grid. In the two-dimensional case, the nodes of the circuit are located at *intersection points* of horizontal and vertical tracks of the grid. The grid consists of two *layers* of interconnect. The top layer is used to route vertical segments of wire and the bottom layer is used to route horizontal segments of wire. Vertical and horizontal segments of the same

* The research of the first author (FTL) was supported in part by a Bantrell Fellowship, in part by DARPA Contract N00014-80-C-0622, and in part by Air Force Contract OSR-82-0326. The research of the second author (ALR) was supported in part by NSF Grant MCS-8116522.

wire are connected with *contact cuts.* Wires cannot overlap nodes to which they are not adjacent, and at most one contact cut can be made at any intersection point. (The routing of wires in this fashion is also known as *Manhattan routing* and *layer per direction routing.*) The *area* of a layout is the product of the number of vertical and horizontal tracks that contain a node or wire segment of the circuit. The *maximum wire length* is simply the length of the longest wire in the layout. For example, the layout in Figure 1 has area 15 and maximum edge length 8.

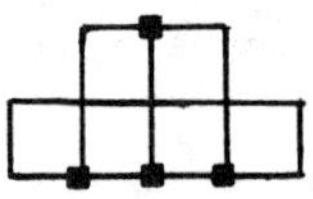

Figure 1: *A two-dimensional layout of a 4-node graph.*

Two models have been proposed for three-dimensional circuit layouts. In the *one-active-layer model,* the nodes of the circuit are placed on the bottom layer only. In the *unrestricted* or *many-active-layer model,* nodes can be located on any layer. In either case, the three-dimensional grid used for embedding the graph consists of H *levels* with two layers of interconnect on each level. Each level corresponds to a two-dimensional grid in which horizontal wire segments are routed in the bottom layer and vertical wire segments are routed in the top layer. Wire segments in different levels can be connected by a wire segment which passes between the intervening levels through intersection points. No other wire can pass through those intersection points on these levels, and no node can be located at such points. However, it is possible for contact cuts on different levels to overlap, and it is possible for a wire not on the bottom level to overlap a node to which it is not adjacent. The *base area* of a layout is the product of the number of vertical planes and the number of horizontal planes that contain a node or wire segment of the circuit. The *volume* of a layout is the product of the base area and the number of levels in the layout. The *maximum wire length* is the length of the longest wire in the layout. (A wire segment passing between adjacent levels is defined to have length one, the same as a wire passing between adjacent tracks.)

As described, the model imposes several limitations on the circuits that can be considered. In particular, we require a circuit to satisfy the following constraints:

1) every node is incident to at most three wires,
2) every wire connects at most two nodes, and
3) every node has unit area.

Like others, we have made these restrictions for simplicity. Results similar to those proved in the paper can also be proved for circuits that do not satisfy the constraints, but the details are more complicated.

1.4 Results

In this paper, we are primarily concerned with one-active-layer layouts. Specifically, we study the problem of transforming a two-dimensional layout (i.e., a 1-level layout) with area A and maximum wire length L_2 into an H-level one-active-layer layout with volume V and maximum wire length L_3. In Section 2, we prove lower bounds for V and L_3 by showing that

$$V \geq \max(\sqrt{AN}, \frac{A}{H})$$

and that

$$L_3 \geq \frac{L_2}{H}$$

where N is the number of nodes in the circuit. In Section 3, we describe two simple algorithms that, in certain cases, come within a constant factor of achieving the lower bounds. In Section 4, we briefly describe a general procedure that always comes within an $O(\log \frac{N^2}{A})$ factor of achieving the lower bounds. We conclude in Sections 5 and 6 with some remarks and references.

2. Proof of Lower Bounds

In what follows, we prove the lower bounds for volume and maximum edge length stated in Section 1.4. The key step in the proof is to transform an H-level layout with volume $V = HB$ into a 1-level layout with area H^2B where B denotes the base area of the layout. As a result, we can conclude that the minimum two-dimensional layout area A satisfies

$$A \leq H^2B = HV$$

and thus that

$$V \geq \frac{A}{H}.$$

Since all the nodes in a one-active-layer layout are on the bottom level, we know that $B \geq N$ and thus that

$$A \leq H^2B = \frac{V^2}{B} \leq \frac{V^2}{N}$$

and that

$$V \geq \sqrt{AN}$$

The transformation is produced by projecting the H-level grid onto a two-dimensional surface. We have illustrated how this is done for a 3-level, 4-by-4 grid in Figure 2. Except for the diagonal edges, an H-by-W-by-L grid is converted into an HW-by-HL grid by the projection.

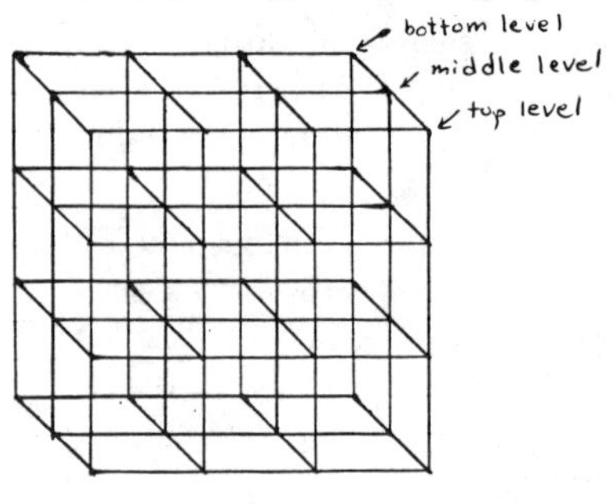

Figure 2: *Projection of a 3-level, 4-by-4 grid on the plane.*

In order to remove the diagonal edges, we must reroute the wire segments that they contain. These are precisely the segments that run between adjacent levels of the original three-dimensional layout. By the rules of the wiring model, at most one wire can pass through an intersection point that contains a wire that changes levels. Hence, a wire contained in a diagonal can be simply rerouted in the neighboring "right angle". (If a wire is already contained in a segment of the right angle, then it must be electrically equivalent to the wire being rerouted.) The area of the resulting layout is $H^2WL = H^2B$, as claimed.

The maximum edge length lower bound is proved with the same transformation. Unit-length vertical and horizontal segments are transformed into H-length segments by the projection. Unit-length segments running between levels are transformed into segments of length $2 \leq H$ when rerouted. Thus a wire of length L_3 in the H-level layout is transformed into a wire of length at most $L_2 \leq HL_3$ in the two-dimensional layout. Hence $L_3 \geq \frac{L_2}{H}$, as claimed.

3. Special-Case Algorithms

In this section, we describe two algorithms that transform two-dimensional layouts into one-active-layer layouts. The algorithms essentially reverse the projection transformation described in Section 2, and come within a constant factor of achieving the lower bounds for volume and wire length when the two-dimensional layout has a special form. Each method preserves the structure of the two-dimensional layout as it is transformed to three dimensions.

3.1 Algorithm I

The input to the algorithm is a 1-level, W-by-L layout of a circuit. The layout is partitioned into k-by-k squares where $k = H - 3$. For the algorithm to succeed, the following conditions on the layout must hold:

1) every k-by-k region contains at most one node, and
2) every k-by-k region contains at most one wire that crosses the boundary of the region both vertically and horizontally (i.e., at most one wire "bends" in the region).

The output of the algorithm is an H-level layout with base dimensions $\frac{3W}{k}$-by-$\frac{3L}{k}$ where $H = k + 3$.

The algorithm proceeds in four steps, as follows.

Step 1: Partition each wire into maximum-length segments so that each wire segment is contained in a single row or column of k-by-k regions.

Step 2: Collapse each k-by-k region into a single point of the base. The wire segments created in Step 1 should now appear as straight line segments in each row or column. (Segments that overlap themselves are replaced with equivalent segments that do not.)

Step 3: In each row and column, assign wire segments to levels by the following procedure: moving from left to right in each row (and from top to bottom in each column) assign each segment encountered to the lowest possible level. It is well known that this procedure requires d levels to route all the segments in a given row (or column) where d, the *density*, is the maximum number of segments that overlap at any point of the row (or column). (Segments that begin and end at the same point are considered to overlap.) Since segments in different rows and columns are routed independently (even those that cross), this means that the entire routing takes d levels. Since at most $k + 2$ horizontal segments can enter or leave a k-by-k region ($k - 2$ segments might pass straight through the region, three segments might be incident to the node in the region, and one segment could be part of a wire that bends in the region), the density d in any row is at most $k + 2$. The same is true for vertical segments, and thus $k + 2$ levels are sufficient to complete the routing.

Step 4: Expand the base grid by a factor of three in each dimension and add a new level at the bottom of the resulting three-dimensional grid. In each 3-by-3 region, there is now a totally empty 2-by-2 subregion on every level. Position the node (if one exists) for each region on the bottom level as shown in Figure 3. Also route wires on the bottom level from the node to the other three intersection points of the empty 2-by-2 grid as shown in Figure 3. Next make interlevel connections between these points and the corresponding points on the levels that contain the segments to which the node should be connected. In addition, run an interlevel wire between the two levels that contain segments of the bending wire (if it exists). This wire should be routed through the intersection points lying directly above the node. It remains only to make the appropriate connections between the wire segments routed in Step 3 and the interlevel segments just inserted. Since at most one vertical and one horizontal segment could have to be connected on any level, it is not difficult to show that the connections can all be made within the allotted space. The resulting layout has $H = k + 3$ levels and base dimensions $\frac{3W}{k}$-by-$\frac{3L}{k}$, as claimed.

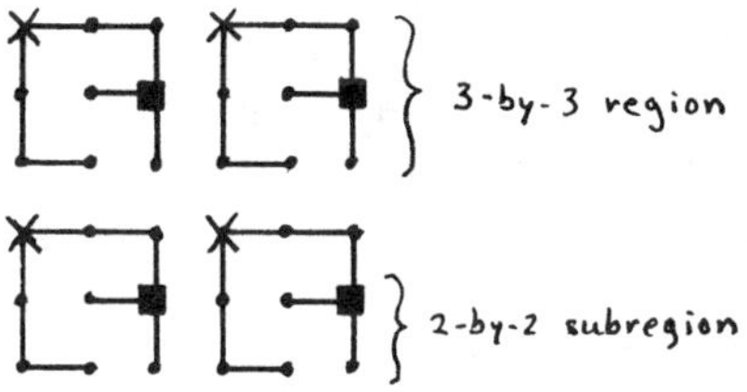

Figure 3: *Placement of the nodes and wires on the bottom level: X's denote intersection points after Step 2; dots denote intersection points after three-fold expansion; squares denote location of nodes; lines denote wires.*

3.2 Algorithm II

The input to the algorithm is a 1-level, W-by-L layout of a circuit. The layout is partitioned into k-by-k squares where $k = \frac{H-3}{5}$. For the algorithm to work, the following constraints on the layout must hold:

1) every k-by-k region contains at most one node, and
2) macroscopically speaking (i.e., ignoring features inside each k-by-k region), every wire has one of the two forms shown in Figure 4.

The output of the algorithm is an H-level layout with base area $\frac{3W}{k}$-by-$\frac{3L}{k}$ where $H = 5k + 3$.

Figure 4: *Wire shapes handled by Algorithm II.*

The algorithm proceeds in three steps, as follows.

Step 1: Collapse each k-by-k region into a single point of the base. As before, wires that overlap themselves are replaced with wires that do not. Every wire will then have the shape shown in Figure 4.

Step 2: First consider the type 1 wires. Route the wires as they are encountered by moving from left to right in the first row, from left to right in the second row, and so forth. For each wire, route it in the lowest available level so that the whole wire is routed on the same level. By using arguments similar to those in Section 3.1, it can be shown that $\frac{5}{2}k+1$ levels are sufficient for this portion of the routing to be completed. By a symmetric argument, it can be shown that an additional $\frac{5}{2}k+1$ levels are sufficient to route the type 2 wires. Hence, a total of $5k+2$ levels suffice to route all of the wires.

Step 3: Expand the base grid by a factor of three in each dimension and add a new level at the bottom of the resulting grid. The remaining connections between nodes and wires are then made as in Step 4 of Algorithm I. The resulting layout has $H = 5k+3$ levels and base dimensions $\frac{3W}{k}$-by-$\frac{3L}{k}$, as claimed.

4. The General Algorithm

Although the procedures described in Section 3 are quite simple and come close to achieving the lower bounds for volume and maximum edge length, they do not work for all two-dimensional circuit layouts. In the full paper [6], we describe a substantially more sophisticated procedure that works for all two-dimensional layouts, and still comes close to achieving the lower bounds.

The general procedure proceeds in three phases. In the first phase, the two-dimensional layout is ripped apart in a divide-and-conquer-like fashion to form a *decomposition tree* [1, 4] for the circuit. In the second phase, the decomposition tree is *fully balanced* [1, 4]. In the third phase, a three-dimensional layout is constructed for the circuit by recursively merging larger and larger pieces of the circuit. The resulting H-level layout has volume at most

$$V = O\left(\frac{A\log^2 \frac{N^2}{A}}{H}\right)$$

and maximum edge length at most

$$L_3 = O\left(\frac{\sqrt{A}\log \frac{N^2}{A}}{H}\right)$$

for any $H = O(\sqrt{\frac{A}{N}}\log\frac{N^2}{A})$. In general, these bounds are within a logarithmic factor of optimal. It is an interesting open question whether or not the logarithmic factors are really necessary in the upper bounds.

5. Conclusions

The algorithms for producing one-active-layer layouts described in this paper come close to achieving the lower bounds for volume and maximum wire length. When the number of levels used is small (in particular, when $H = O(\sqrt{\frac{A}{N}})$), the same lower bounds also hold for unrestricted three-dimensional layouts. Hence, we could not hope to do much better, even if we were allowed to place nodes on arbitrary levels! In fact, our findings indicate that (at least in this idealized model) the capability to locate transistors on several levels is only useful when the total number of levels is large.

In addition, our results provide an estimate of the number of levels needed to construct an efficient one-active-layer layout for a circuit. This number is (very roughly) $\sqrt{\frac{A}{N}}$. Since the volume of a one-active-layer layout is always at least $\Omega(\sqrt{AN})$, additional levels are useless. It is not surprising that this value is related to the number levels needed to produce an efficient layout since the value $\frac{A}{N}$ is a natural measure of the complexity of a two-dimensional layout. (When non-unit-area nodes are considered, the corresponding measures are $\sqrt{\frac{A}{N^*}}$ and $\frac{A}{N^*}$, respectively, where N^* denotes the total area of the nodes.)

6. Abbreviated List of References

[1] S. N. Bhatt and F. T. Leighton, "A framework for solving VLSI graph layout problems," *JCSS*, to appear.

[2] S. Even, A. Pnueli and A. Lempel, "Permutation graphs and transitive graphs," *JACM* 19 (1972), pp. 400-410.

[3] J. F. Gibbons, "SOI - a candidate for VLSI?," *VLSI Design III* (1982), pp. 54-55.

[4] F. T. Leighton, "A layout strategy for VLSI that is provably good," *Proc. 14th ACM Symp. on Th. of Comp.* (1982), pp. 85-98.

[5] F. T. Leighton, *Complexity Issues in VLSI: Optimal Layouts for the Shuffle-Exchange Graph and Other Networks*, MIT Press, to appear.

[6] F. T. Leighton and A. L. Rosenberg, "Three-dimensional circuit layouts," in preparation.

[7] C. E. Leiserson, *Area-Efficient VLSI Computation*, MIT Press (1983).

[8] F. P. Preparata, "Optimal three-dimensional VLSI layouts," *Math Systems Theory*, to appear.

[9] F. P. Preparata and W. Lipski, "Optimal three-layer channel routing," *Proc. 23rd IEEE Conf. on Found. of Comp. Sci.* (1982).

[10] A. L. Rosenberg, "Three-dimensional integrated circuitry," *VLSI Systems and Computations*, Computer Science Press (1981), pp. 69-80.

[11] A. L. Rosenberg, "Three-dimensional VLSI: a case study," *JACM*, to appear.

[12] C. D. Thompson, "A complexity theory for VLSI," *Proc. 11th ACM Symp. on Th. of Comp.* (1979), pp. 81-88.

[13] L. G. Valiant, "Universality considerations in VLSI circuits," *IEEE Trans. on Comp.* C-30 (1981), pp. 135-140.

[14] D. Wise, "Efficient layouts for FFT and Banyon networks," *VLSI Systems and Computations*, Computer Science Press (1981), pp. 186-195.

Heuristics in the SLAP Layout System

John E. Savage

Department of Computer Science
Brown University
Providence, RI 02912

ABSTRACT The development of compilers for the translation of functional chip specifications to chip geometry requires an investigation of good heuristics for this purpose. In this paper, we report on the use of level graphs for chip layout, a method used in the system *SLAP*. A linear time algorithm for chip layout from Boolean equations based on level graphs is given, a time efficient channel router is presented, and analysis is done of two problems that have to be addressed when level graphs are used, namely, building good row orders and placing given row orders to minimize the channel area between rows.

1. Introduction

Research in silicon compilation has as its primary goal to develop automatic means of translating high-level descriptions of chips into chip geometry. Functional, behavioral and architectural methods have been proposed for this purpose. In this paper we examine and analyze the underlying techniques used in a functional procedure for silicon layout, *SLAP* [8], that the author has developed jointly with S. P. Reiss. This procedure accepts Boolean equations, parses them, builds a level graph, a graph with the property that all edges pass between vertices whose levels differ by one, and produces layouts from these graphs. In this paper we analyze methods for translating level graphs to layouts, examine the heuristics that might be used for this purpose and those used in *SLAP*, and demonstrate that level graphs offer the potential of extremely fast layout algorithms.

Programmable Logic Arrays (PLAs) [7], one-dimensional Weinberger Arrays [12], and *SLAP* are methods for producing layouts from functional descriptions, in particular, Boolean equations. PLAs are well known, and Weinberger arrays place gates on a line and use the second dimension for wiring. *SLAP* puts all wires and gates on a virtual two dimensional grid. Each imposes some restrictions on layouts, with those imposed by *SLAP* being weakest [9]. The MacPitts compiler [10], and the bristle blocks compiler [3], compile specifications to some particular architectural form. One set of authors propose a behavioral approach that lets inputs be specified as regular expressions [2] from which PLAs are generated.

SLAP is a highly interactive system that provides an environment in which to experiment with layout heuristics at a time when the interaction between functional specifications and geometry in silicon compilers is not well understood. The heuristics it provides alter the order of elements on individuals levels, which are placed in rows on the chip, and affect the relative placement of elements in adjacent rows. Statistics are collected and displayed whenever requested. *SLAP* also provides: 1) automatic translation of functional specifications into chip specifications, 2) efficient heuristics for placement, routing and control of a layout, 3) satisfaction of design rules by synthesis, and 4) eliminates the need for node extraction for verification of the design.

In the next section we demonstrate that a variant of the *SLAP* method can be used to construct layouts in a time linear in the number of Boolean equations provided, and with an area that is at most quadratic in this number. In Section 3 we describe a channel routing method very similar to that used in *SLAP* and demonstrate that it can be made to route in time linear in the number of terminals on the channel boundaries. In Sections 4 and 5 we examine the row ordering and placement problems and present optimal or near optimal methods for dealing them. These results support the intuition that led to the selection of the heuristics that are used in *SLAP*. Section 5 states conclusions.

2. A Linear Time Layout Method

Given a bounded degree directed acyclic graph G = (V,E) with vertices V and edges E, the objective of a layout algorithm is to place the vertices V and route the edges E. We give a linear time algorithm for this purpose. It uses depth-first search (DFS) and assigns a level to a vertex which is the length of the longest path to a source vertex. Edges that connect vertices whose levels differ by one are called *short edges* and others are called *long edges*. We assume that each vertex has the same height and width, and that input and output connections are made using single vertical connections. We ignore power and ground lines since they are easily accommodated.

This research was supported in part by NSF under Grant ECS-80-24637.

CH1935-6/83/0000/0637$01.00

LVL - An O(E)-Time Layout Algorithm

(1) Assign levels to vertices with DFS.

(2) Build columns consisting of the *ith* elements in each row, beginning with the first row.

(3) Place elements in columns so that no conflicts occur across channels.

(4) Route nets of "short" edges with the J-router described below, adding horizontal tracks where necessary to connect "long" edges.

(5) Route each "long" edge with horizontal tracks in channels containing its endpoints and with a single vertical track on the right side of the layout.

In Figure 1 we illustrate the type of layout that will result from this algorithm.

This description is not complete. Simple DFS has to be modified for multiple output dags, and the building of columns without column conflicts must be done. Each is easily accommodated. The running time of this algorithm is linear, as stated below, if we assume that the elements in rows are in linked lists, as well as the first elements in each row. Such linked lists can be built during DFS. We state the following result without proof.

Theorem 1: The layout algorithm LVL will produce a layout for a bounded degree dag (V,E) in which each vertex has unit height and width in time O(E) and in area of $O(V^2)$.

SLAP uses a conflict resolution algorithm that is a bit more sophisticated than that given here. It also treats long edges differently, for which reason it cannot be guaranteed to use only O(E) time. In *SLAP*, "delay" elements are inserted in long edges so that all edges become short.

A number of bounds on the area required by layouts have been derived assuming that a separator theorem or something akin to one applies to a family of graphs [6,5,4,11]. They imply a worst case upper bound on area of $O(V^2)$. This is comparable to the worst case bound for LVL. There are examples of graphs for which the area given by LVL is much larger than that required using separator theorems. Open is the question of how little area is required by graphs that can be laid out with linear time algorithms.

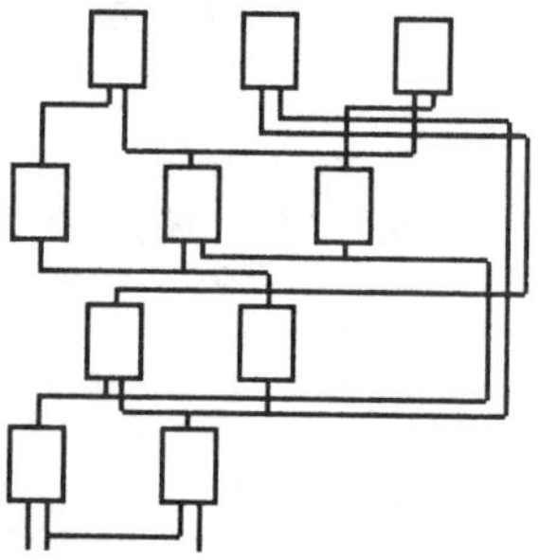

Fig. 1 A Layout Produced by LVL

3. The J-router

The *J-router* is a channel routing algorithm similar to the router used in *SLAP*, except for the use of the concept of parent described below. It assumes that there are no column conflicts and it connects a *net* (or set of connected terminals) using a single horizontal track. It assumes that the terminals on both sides of the channel are maintained in linked lists ordered by their distance from the left edge of the channel. We let *net*(v) be the net associated with a terminal v, *parent*(v) be the single element in *net(v)* that is parent to all elements in it, and *track*(v) be the single horizontal track connecting v with the other terminals in *net*(v).

The J-Router

Until the two lists of terminals are exhausted, select as the next vertex v to be visited that which is leftmost of those at the heads of the two lists.

If v has not been routed

a) Assign a free track to *net*(v).

b) Declare all terminals in *net*(v) routed.

c) Record at *parent*(v) the dimensions of *track*(v).

d) At terminals in *net*(v) record the vertical connection to *track*(v).

If v has been routed

a) If v is the last terminal in a net, add its horizontal track to the free list.

b) If v is not the last terminal in a net, continue.

To determine the length of a horizontal track, it is sufficient to visit each of the vertices in a net once. A "routed" flag can also be set at that time. It is easy to see that the running time of this procedure is linear in the number of terminals.

The set of terminals on channel boundaries and the edges between them define a bipartite graph. *Channel density* is defined as the maximum number of nets cut by any line perpendicular to the sides of a channel. Every channel of channel density d requires at least d horizontal tracks. The following result is stated without proof.

Theorem 2: The J-router achieves channel density for bipartite graphs (V,E) with no column conflicts in O(V) time.

We note that the area used by this channel router is minimal, given the positions of the terminals and the fact that no column conflicts occur.

4. Row Placement

The *row placement* problem is the following: given an order for elements on both sides of a channel, give them a placement that that minimizes channel density without creating column conflicts. Elements on both sides of the channel may be shifted left or right to minimize channel density. Each element has a number of inputs attached to different nets.

The order between elements on the sides of a channel and the connections between elements are characterized by an (n,m) bipartite *connection graph* which has n vertices on the top and m on the bottom. The intervals between vertices on the top and bottom sides of the channel are denoted by $\{x_i\}$, $\{y_j\}$, respectively. The order of the vertices on each side of the channel is fixed but their placement is not. Placement amounts to creating a total order for the vertices in the two sets.

Consider the *net-cost matrix* $[c(i,j)]$ indexed by the intervals x_i and y_j and containing as value $c(i,j)$ the number of nets cut by a straight line from interval x_i to interval y_j in the order graph. If the placement puts the *ith* and *jth* intervals above one another, $c(i,j)$ is the number of nets that will be cut by a line perpendicular to the sides of the channel. The following lemma establishes a connection between the placement graph and a path through the matrix $[c(i,j)]$ from which an algorithm can be derived for computing the best placement graph.

Lemma 1: A placement graph for the row placement problem on an (n,m) bipartite connection graph corresponds to a path from entry (1,1) to entry (n,m) in the net-cost matrix $[c(i,j)]$.

An efficient algorithm based on dynamic programming to construct an optimal placement graph can be generated based on this information, thus leading to the following result.

Theorem 3: There exists an O(nm) time algorithm to construct a conflict-free layout of minimal channel density for an arbitrary (n,m) bipartite connection graph.

A channel route produced by this algorithm will occupy area proportional to $d_{min} max(n,m)$, where d_{min} is the minimal channel density.

In *SLAP* the EXPAND heuristic is a crude approximation to the method proposed above for placement, but in our experience it almost always reduces the area required of a layout. EXPAND tries to make straight-wire connections between connected elements and it has a running time that is linear in the number of elements in a graph.

5. Row Orders

The *row ordering problem* concerns the assignment of an order to the vertices on the top side of a channel so as to minimize channel density when the method of the preceding section is used to place elements. We assume that the order of elements on the bottom of the channel is fixed and that each element on the top side of the channel has indegree of 1 or 2. We also assume that each net connects one element on the bottom side of the channel with elements on the top side. This exactly models the problem treated by several heuristics in *SLAP*.

Consider a set of distinct ordered elements that reside on the bottom of the channel, labeled a, b, ..., z, and where a < b < ... < z. Since each net has one element on the bottom side, these can also be taken as the names of nets. The problem consists of ordering a set of singlets (a) or pairs (a,b) denoting elements on the top side of the channel with one or two inputs that are to be connected to the named nets for the purpose of minimizing channel density.

To simplify the analysis without much loss in generality, we ignore singlets and repeated pairs, and assume that in every pair (a,b) a < b. If the inputs to elements do not satisfy this order, we may need to add two tracks to properly connect elements. All elements needing such a permutation can use the same two tracks. We now define the *pair ordering relation*.

Definition 1: Two pairs (x,y) and (u,v) over the set {a,b,...,z} where x < y and u < v are in the order (x,y) ≤ (u,v) if and only if x < u or x = u and y < v.

Definition 2: A pair (x,y) *contains* a pair (u,v) if x < u < v < y.

The following result demonstrates that if the pairs are sorted using this ordering of pairs, the resulting ordering has a channel density that is close to the minimal.

Theorem 4: A row ordering obtained by sorting pairs (x,y) using the pair ordering relation provides a channel density d satisfying

$$d_{min} \le d \le d_{min} + c + 2$$

where d_{min} is the minimal achievable channel density, and c is the number of pairs that contain other pairs.

The number of pairs that typically satisfy the containment property and for which more than the minimum number of tracks is used is likely to be small. Thus, this algorithm is a very good heuristic for the row ordering problem.

A row ordering heuristic that is most effective in *SLAP* is the INSERT heuristic. It takes an initial row order and builds a new order by inserting elements into a list so as to try minimize the number of edges that cross other gates in the list at that time. Shown in Figures 2 and 3 are the *SLAP* layouts of a 16-bit carry lookahead adder proposed by Brent and Kung [1] with the *SLAP* default row orders and placements, and after both EXPAND and INSERT heuristics have been applied.

6. Conclusions

A provably fast algorithm, LVL, for producing chip layouts from Boolean equations using level graphs has been presented that uses no more area in the worst case than algorithms based on divide and conquer. The J-router, a linear-time channel routing algorithm and one

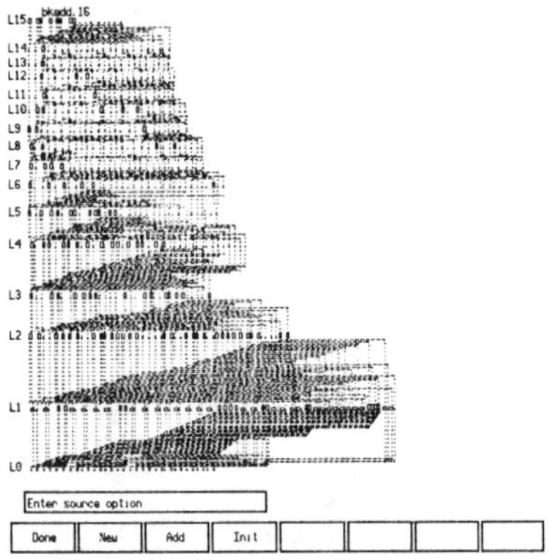

Fig. 2 Initial *SLAP* Layout of 16-bit Adder

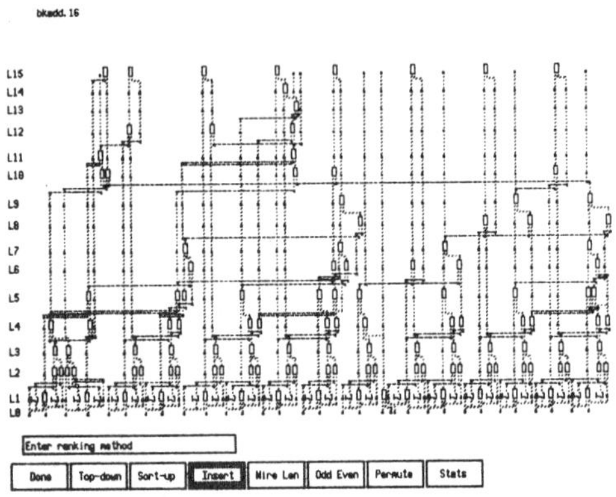

Fig. 3 Layout of Adder After EXPAND and INSERT

very close to that used in *SLAP*, has also been presented.

We have studied the the row ordering and placement problems and have found an area-optimal, quadratic-time algorithm for the row placement problem and have given a provably good algorithm for the row ordering problem.

7. Acknowledgements

The system *SLAP* is the result of joint work between the the author and S. P. Reiss. The *SLAP* heuristics discussed here are the product of that joint effort. The author also gratefully acknowledges helpful conversations with Gerard Baudet on the row ordering problem.

8. References

1. R. P. Brent and H. T. Kung, "A Regular Layout for Parallel Adders," *IEEE Trans. on Computers* C-31(3) pp. 260-264 (March 1982).
2. R. W. Floyd and J. D. Ullman, "The Compilation of Regular Expressions into Integrated Circuits," *JACM* **29** (3) pp. 606-622 (July 1982).
3. D. Johannsen, "Bristle Blocks: A Silicon Compiler," *Proc. 16th Design Automation Conf.*, pp. 310-313 (June 1979).
4. Frank Thomson Leighton, "A Layout Strategy for VLSI which is Provably Good," *14th Ann. ACM Symp. on Theory of Computing*, pp. 85-98 (May 1982).
5. C. E. Leiserson, "Area-Efficient Graph Layouts (for VLSI)," *Procs. 21st Ann. Symp. on Fnds. of Comp. Sci.*, pp. 270-281 (Oct. 13-15, 1980).
6. R. J. Lipton and R. E. Tarjan, "Applications of a Planar Separator Theorem," *SIAM J. Comput.* **9** (3) pp. 615-627 (Aug. 1980).
7. C. Mead and L. Conway, *Introduction to VLSI Systems*, Addison-Wesley (1980).
8. S. P. Reiss and J. E. Savage, "SLAP - A Methodology for Silicon Layout," *Procs. of the Int. Conf. on Circuits and Computers*, pp. 281-285 IEEE, (Sept. 28 - Oct. 1, 1982).
9. J. E. Savage, "Three VLSI Compilation Techniques: PLA's, Weinberger Arrays, and SLAP, A New Silicon Layout Program," in *Algorithmically-Specialized Computers*, ed. L. Snyder, L. J. Seigel, H. J. Seigel, D. Gannon, Academic Press (1983). to appear
10. J. M. Siskind, J. R. Southard, and K. W. Crouch, "Generating Custom High Performance VLSI Designs from Succinct Algorithmic Descriptions," *Procs. Conf. on Advance Research in VLSI*, pp. 28-40 MIT, (Jan. 1982).
11. L. G. Valiant, "Universality Considerations in VLSI Circuits," *IEEE Trans. Comps.* C-30(2) pp. 135-140 (Feb. 1981).
12. A. Weinberger, "Large Scale Integration of MOS Complex Logic: A Layout Method," *IEEE Jnl. of Solid State Circuits* SC-2(4) pp. 182-190 (Dec. 1967).

Session 3-3b
High Level Aids

Chairman

P. A. Subrahmanyam
University of Utah

Mathematical Approach to Computational Networks[1]

Danny Cohen
Information Sciences Institute
University of Southern California
Marina Del Rey, California 90291

Lennart Johnsson
Computer Science
California Institute of Technology
Pasadena, California 91125

Abstract

Fast Fourier Transform, **FFT**, algorithms are interesting for direct hardware implementation in VLSI. Typically, the description of FFT algorithms is made either in terms of graphs illustrating the dependency between different data elements or in terms of mathematical expressions without any notion of how the computations are implemented in space or time. Expressions in the notation used in this paper can be given an interpretation in the implementation domain. The notation in this paper is used to derive a description of array implementations of **DIF** (Decimation-In-Frequency) FFT algorithms. The correctness of the implementations is guaranteed by the way of derivation.

Notation

Define the operator Z by: $Zx(j) = x(j-1)$.

If $\{x\}$ is a sequence of data elements such that $x(j)$ precedes $x(j+1)$, then j can be interpreted as time, Z as a delay, and its inverse as a prediction.

If Z is a clocked storage cell, such as a flip-flop, then Z^N is a shift register of length N.

A function is "time-independent" if its value does not depend on the time evaluation. If F(X) is such a function then $F(ZX)=ZF(X)$. Hence, F and Z commmute.

Capital letters are used to denote functions, operators, and sequences. Small letters are used for the individual elements of sequences. The operator Z acts on the ordering index of the elements.

[1]Some of the material in this paper has been published in "A Formal Derivation of Array Implementation of FFT Algorithms". *Proceedings of the USC Workshop on VLSI & Modern Signal Processing*, 1982.

The Fast Fourier Transform, FFT

The definition of the Discrete Fourier Transform, **DFT**, is for $m = 0{:}(N-1) = \{0,1,2...N-1\}$

$$y(m,k) = \sum_{j=0}^{N-1} w^{mj} x(k+j) \qquad (1)$$

The reduction in arithmetic operations in the FFT algorithm, compared with a direct DFT evaluation, is accomplished by using the special properties of the coefficients.

$$w = w_N = e^{-2\pi i/N} \text{ yields } w^{N/2} = -1.$$

Therefore,

$$w^{mj} = \begin{cases} w^{m(j-N/2)} & \text{if } m \text{ is even} \\ -w^{m(j-N/2)} & \text{if } m \text{ is odd} \end{cases} \qquad (2)$$

In all the networks developed here the data arrives sequentially--one sample per cycle. The data is treated as a sequence of non-overlapping contiguous windows of size N. In each cycle the network computes one frequency component. Hence, a complete N-component spectrum is computed for an N-sample window every N cycles.

In our derivations of FFT algorithms both m and j are mapped to the time index k. j is mapped onto the time index in increasing order such that the input is in normal order. As a result of the design decision to always handle "even before odd", the output is produced in bit-reversed order.

The window size is $N=2^n$. FFTs are computed for all windows starting at values of k which are multiples of N. Hence, the running (with time) value of k can be expressed as $k=[k]+\langle k\rangle$ where $[k]$ is the reminder of k divided by N, and $\langle k\rangle$ is a multiple of N.

CH1935-6/83/0000/0642$01.00 © 1983 IEEE

With these definitions equation (1) is rewritten as:

$$y(m,\langle k\rangle) = \sum_{[k]=0}^{N-1} w^{m[k]}\, x(\langle k\rangle+[k]) = \sum_{[k]=0}^{N-1} w^{m[k]} x(k) \qquad (3)$$

There are N different sums in (3), one for each m. In every cycle a new value of X is available and the corresponding terms in all sums can be computed. The relative order in which computations for different frequency components are carried out is not specified in (3). In the space-time descriptions of FFT algorithms that follows, this is used to determine a regular data flow that reduces the arithmetic operations, in the manner common for FFT algorithms.

Using (2) in (3) yields for m = 0:(N/2−1):

$$y(2m,\langle k\rangle) = \sum_{[k]=N/2}^{N-1} w^{2m[k]}(x(\langle k\rangle+[k]) + x(\langle k\rangle+[k-N/2])) =$$

$$= \sum_{[k]=N/2}^{N-1} w^{2m[k]}(I + Z^{N/2})x(\langle k\rangle+[k]) =$$

$$= \sum_{[k]=N/2}^{N-1} w^{2m[k]} v0(k) = \sum_{j=0}^{N/2-1} (w^2)^{mj} v0(\langle k\rangle+N/2+j)$$

Hence,

$$Z^{N/2}y(2m,\langle k\rangle) = \sum_{j=0}^{N/2-1} w_{N/2}^{mj}\, v0(\langle k\rangle+j) \qquad (4)$$

and

$$y(2m+1,\langle k\rangle) =$$

$$= \sum_{[k]=N/2}^{N-1} w^{2m[k]}w^{[k]}(x(\langle k\rangle+[k]) - x(\langle k\rangle+[k-N/2])) =$$

$$= \sum_{[k]=N/2}^{N-1} w^{2m[k]}w^{[k]}(I - Z^{N/2})x(\langle k\rangle+[k]) =$$

$$= \sum_{[k]=N/2}^{N-1} w^{2m[k]} v1(k) = \sum_{j=0}^{N/2-1} (w^2)^{mj} v1(\langle k\rangle+N/2+j)$$

Hence,

$$Z^{N/2}y(2m+1,\langle k\rangle) = \sum_{j=0}^{N/2-1} w_{N/2}^{mj}\, v1(\langle k\rangle+j) \qquad (5)$$

Expressions (4) and (5) define two summations: the set of even and the set of odd frequency components. Each summation is performed over the time interval $N/2 \le [k] < N$ which is the second half of the N-cycle window. The even components are computed from one intermediate signal, V0, the odd components from another, V1, where for $N/2 \le [k] < N$:

$$\begin{aligned} v0(k) &= (I + Z^{N/2})\, x(k) \\ v1(k) &= w^{[k]}(I - Z^{N/2})\, x(k) \end{aligned} \qquad (6)$$

The expressions (6) define the familiar DIF butterfly computation. It is implemented by the network in Figure–1.

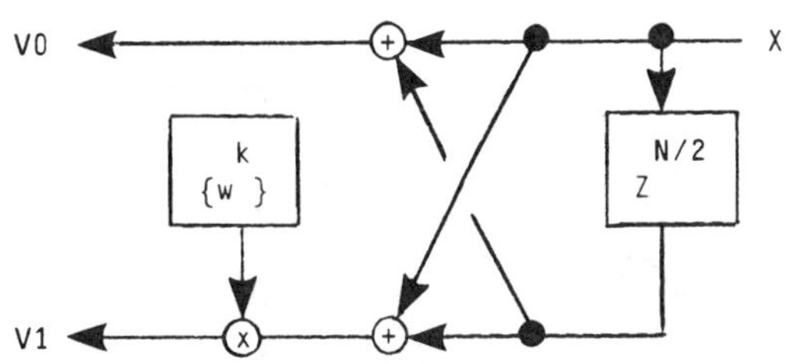

Figure 1: A DIF butterfly, as defined in (6)

(1) defines Y as a single transform of order N on windows of X. (4) and (5) redefine Y as 2 transforms of order N/2 each on windows of V's. These V's are computed in the interval $N/2 \le [k] < N$.

If (1) defines Y = T(N) X , then (4) and (5) define Y = T(N/2) **DIF**(N) X, where **DIF**(N) is the computation of the **DIF**-butterflies on a window of length N. By continuing the iteration logN times the entire computation is transformed into Y = T(1) **DIF**(2) **DIF**(4)...**DIF**(N) X. Since T(1)= I the entire FFT computation is reduced to a sequence of butterfly computations only.

Each **DIF**(L) processes L input values to produce two V sequences of length L/2 each. (4) and (5) suggest that V0 and V1 are computed concurrently in the second half of the particular window of each **DIF** operator. In order to stream the operation the **DIF** operations we delay V1 (the odd components) and output it after V0. Therefore, each **DIF**(L) introduces a delay of L/2 cycles.

In the following, a subscript on a signal name indicates that the signal is treated as a sequence of "windows" each of that length.

Define the signal, $X_{N/2}$, for the merged signals V0 and V1 by:

$$x_{N/2}(k) = (I + Z^{N/2})x_N(k) \qquad N/2 \leq [k] < N$$

and

$$x_{N/2}(k) = Z^{N/2}\{w^{[k]}(I - Z^{N/2})\}x_N(k) \quad N/2 \leq [k-N/2] < N$$

This selection mechanism can be implemented by a binary multiplexor controlled by a signal U_N defined by:

$$u_N(k) = \begin{cases} 0 & \text{when } N/2 \leq [k] < N \\ 1 & \text{when } N/2 \leq [k-N/2] < N \end{cases} \qquad (7)$$

The intermediate signal $X_{N/2}$ can now be expressed as:

$$x_{N/2}(k) = \{\, u_N(k)\ Z^{N/2}w^{[k]}[I - Z^{N/2}] + \qquad (8)$$
$$+\ u'_N(k)\ [I + Z^{N/2}] \,\}\ x(k)$$

The network for implementing (8) is depicted in Figure–2.

The DFT components, $y(m,\langle k\rangle)$, $m=0:(N-1)$, are obtained in bit-reversed order in the stream X_1.

Each DIF(N) needs a control signal, U_N. The control signals for the different units are in phase with each other in the sense that when the signal with the lowest frequency has a 0/1 transition, so have all the others. Since it is easier to half the frequency of square waves than to double it, U_N is derived from $U_{N/2}$ unlike $X_{N/2}$ which is derived from X_N. Therefore, the control signal propagates in the opposite direction to the data, X.

A simplified drawing of Figure–2 is shown in Figure–3, and an array for N = 8 is shown in Figure–4.

Table–1 shows the intermediate results in the computation of an 8–point transform.

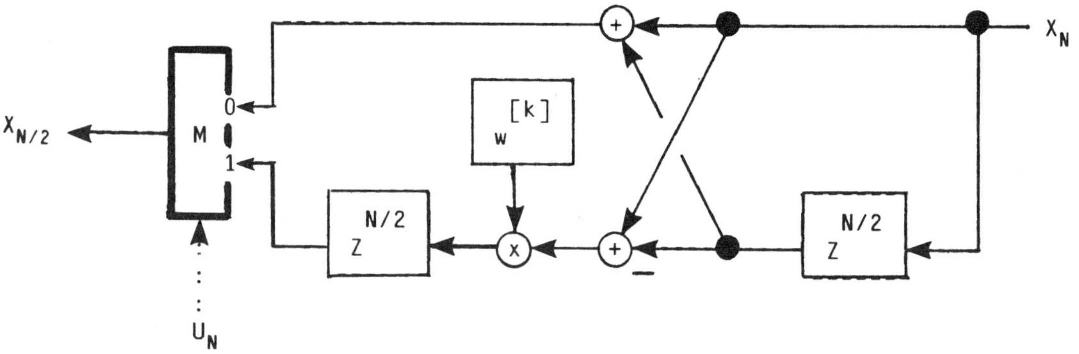

Figure 2: Computing $X_{N/2}$ from X_N, as defined in (8)

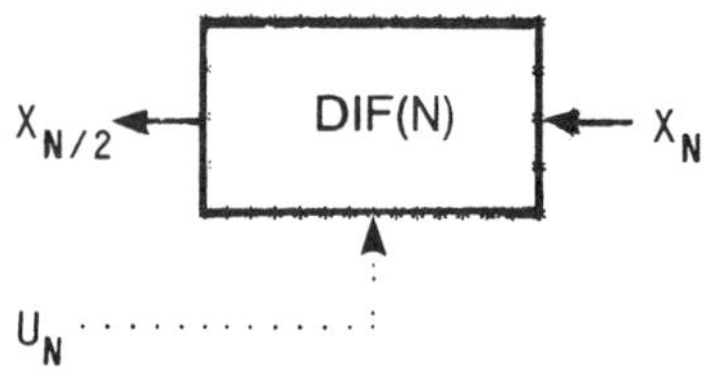

Figure 3: Computing $X_{N/2}$ from X_N

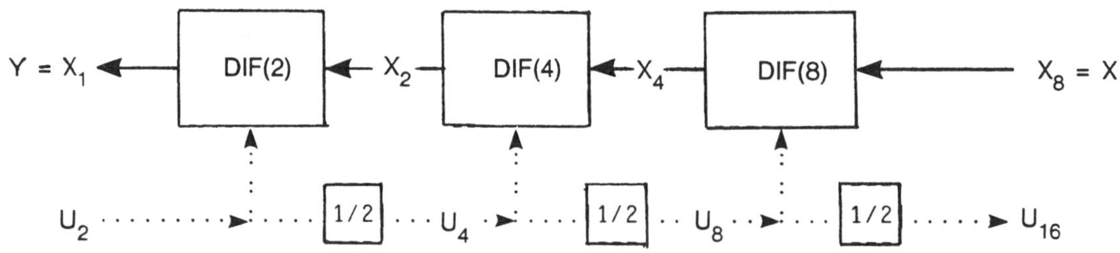

Figure 4: A decimation-in-frequency array for N = 8

k	U_8	X_4	U_4	X_2	U_2	X_1
0	1		1		1	
1	1		1		0	
2	1		0		1	
3	1		0		0	
4	0	0(4+0)	1		1	
5	0	0(5+1)	1		0	
6	0	0(6+2)	0	0[0(6+2)+0(4+0)]	1	
7	0	0(7+3)	0	0[0(7+3)+0(5+1)]	0	0[0[0(7+3)+0(5+1)]+0[0(6+2)+0(4+0)]]
8	1	4(4-0)	1	4[0(6+2)-0(4+0)]	1	4[0[0(7+3)+0(5+1)]-0[0(6+2)+0(4+0)]]
9	1	5(5-1)	1	6[0(7+3)-0(5+1)]	0	0[6[0(7+3)-0(5+1)]+4[0(6+2)-0(4+0)]]
10	1	6(6-2)	0	0[6(6-2)+4(4-0)]	1	4[6[0(7+3)-0(5+1)]-4[0(6+2)-0(4+0)]]
11	1	7(7-3)	0	0[7(7-3)+5(5-1)]	0	0[0[7(7-3)+5(5-1)]+0[6(6-2)+4(4-0)]]
12	0		1	4[6(6-2)-4(4-0)]	1	4[0[7(7-3)+5(5-1)]-0[6(6-2)+4(4-0)]]
13	0		1	6[7(7-3)-5(5-1)]	0	0[6[7(7-3)-5(5-1)]+4[6(6-2)-4(4-0)]]
14	0		0		1	4[6[7(7-3)-5(5-1)]-4[6(6-2)-4(4-0)]]
15	0		0			

	0	4	2	6	1	5	3	7	0	1	2	3	4	5	6	7
0[0[0(7+3)+0(5+1)]+0[0(6+2)+0(4+0)]]	0	0	0	0	0	0	0	0	0	0	0	0	0	0	0	0
4[0[0(7+3)+0(5+1)]-0[0(6+2)+0(4+0)]]	0	0	0	0	4	4	4	4	0	4	0	4	0	4	0	4
0[6[0(7+3)-0(5+1)]+4[0(6+2)-0(4+0)]]	0	0	4	4	2	2	6	6	0	2	4	6	0	2	4	6
4[6[0(7+3)-0(5+1)]-4[0(6+2)-0(4+0)]]	0	0	4	4	6	6	2	2	0	6	4	2	0	6	4	2
0[0[7(7-3)+5(5-1)]+0[6(6-2)+4(4-0)]]	0	4	2	6	1	5	3	7	0	1	2	3	4	5	6	7
4[0[7(7-3)+5(5-1)]-0[6(6-2)+4(4-0)]]	0	4	2	6	5	1	7	3	0	5	2	7	4	1	6	3
0[6[7(7-3)-5(5-1)]+4[6(6-2)-4(4-0)]]	0	4	6	2	3	7	1	5	0	3	6	1	4	7	2	5
4[6[7(7-3)-5(5-1)]-4[6(6-2)-4(4-0)]]	0	4	6	2	7	3	5	1	0	7	6	5	4	3	2	1

The notation m(i+j) is a short hand for $w^m[X(i)+X(j)]$.

Table 1: Exponents and arguments for DIF(8), DIF(4), and DIF(2) in the computation of 8-point transform

Improving the Storage Utilization

The computation of each DIF(L) requires L cycles in which L/2 additions, L/2 substractions, and L/2 multiplications are performed. Since the network contains two adders and one multiplier, the arithmetic units are only half-utilized. By employing some multiplexing the L/2 additions and the L/2 subtractions could be performed by a single adder rather than by two, as shown in all the figures. Since this trade-off highly depends on the implementation at hand, we do not pursue it in this paper.

Since only L/2 multiplications are needed the half-utilization of the multiplier is expected and accepted.

On the other hand, in addition to the above arithmetic operations, storage is also required. In each window period L/2 input values are delayed and so are L/2 output values (the V1). In total L values have to be delayed, i.e., written and read. Since storage can be both read and written in each cycle one storage bank should be sufficient for both delays. The need to delay by L/2 cycles dictates that the size of this storage bank is L/2 complex-words.

This requirement for a single storage bank of size L/2 is only half of the storage used in Figure-2 which uses two storage-banks, each of size L/2. The storage-bank on the right is written into during the first half of the window, and read from during the second half. The storage-bank on the left is read from during the first half and written into during the second half.

Hence, both banks may be easily merged into one bank which performs both read and write operations in every cycle. The additional data steering required for this merge is achieved by a single multiplexor as shown in Figure-5.

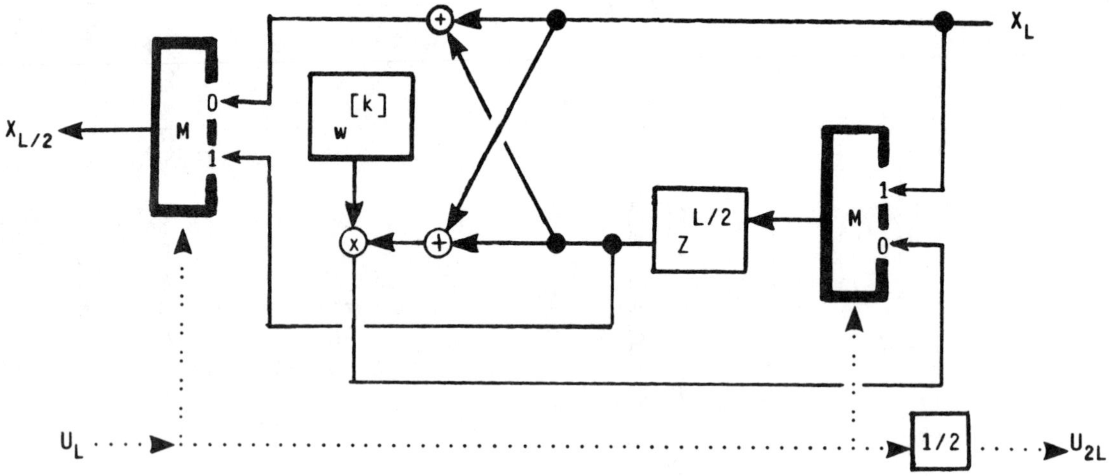

Figure 5: Storage Efficient Computation of $X_{L/2}$ from X_L

Conclusions

From the intrinsic symmetry of the FFT algorithm it is easy to derive similar computation networks also for Decimation-In-Time, DIT, algorithms.

The paper proposes networks for real-time computation of FFT which is well suited for VLSI implementation.

The formal derivation of these networks from the mathematical definition of the Fourier Transforms guarantees the correctness of these networks.

Acknowledgement

This work was sponsored by the Defense Advanced Project Agency (DARPA) under Contract No. MDA903-81-C-0335 with USC/Information Sciences Institute. Views and conclusions contained in this paper are the authors' and should not be interpreted as representing the official opinion or policy of DARPA, the U.S. Government, or any person or agency connected with them.

Overview of a Conceptual and Formal Basis for An Automatable High Level Design Paradigm for Integrated Systems

P.A.Subrahmanyam [1]
Department of Computer Science, University of Utah, Salt Lake City, Utah 84112

1. Introduction

It is important that design methodologies for special purpose VLSI circuits enable a smooth embedding of the resulting circuits into larger systems that consist of both software and specialized hardware components e.g., on–board control systems. Furthermore, we believe that it is advantageous if automated design aids that support such methodologies are based on a uniform set of principles — ideally, on a unifying theoretical basis. In this paper we delineate a conceptual and theoretical basis intended to aid both in the analysis and in the mechanical synthesis and verification of special purpose VLSI systems, proceeding from high level behavioral specifications. [2]

1.1. Motivation

Before discussing the formalism itself, it is useful to explore further the motivations for having such a formalism. In order to do so, it is pertinent to examine the major steps involved in the design process. These include: acquisition of the system specifications from the end user, "debugging" of such specifications (to ensure that they indeed embody the user's requirements), decomposing the specifications into a detailed system level architecture, and finally, mapping the system components into either software that is executable on a general purpose host computer, or special purpose hardware.

There already exist an increasing number of design aids that primarily address the low level tasks in this spectrum. A random sampling of tools available in the hardware domain includes: editors at various levels e.g., stick diagram editors and composite editors, channel routers, silicon "assemblers" that transform state machine descriptions into PLAs (and thence to mask descriptions), mixed–mode simulators of various flavors, database support facilities etc., to name only a few. A corresponding complement of tools exists for supporting software design, although these are typically much more mature in several areas. Examples include editors for various languages (both textual, and "syntax–directed"), compilers for high level languages, debugging aids, version control systems, data base support facilities, etc., again to name only a few.

Unfortunately, the various hardware design aids in existence usually have very little in common. This is because they are often designed in an *ad hoc* manner, based on almost unrelated sets of concepts, and therefore unrelated associated terminology. The result is that it is very difficult, if not impossible, to master the complicated (and distinct) user interfaces, design concepts, and tool idiosyncrasies of more than a few tools. Not surprisingly, there exist few people who even attempt to understand the overall spectrum of the design process, and who can therefore provide the "conceptual integrity" that has been shown to be key to the successful design of large systems [1]. In order to improve the reliability of systems developed and to reduce the cost of development it is important to remedy this situation.

2. An Outline of the Framework. Applications

In this context, we have been investigating a unifying theoretical basis to support the analytic and synthetic tasks of integrated system design, and the salient features of which are as follows:

- The formalism allows various levels of abstraction involved in the software/hardware design process to be modelled. For example, three levels of abstraction —functional (behavioral), architectural (system and chip level), and electrical (switch–level)— are explicitly modelled as being typical of the levels of abstraction that human "expert designers" work with.
- The formalism allows for explicit reasoning about performance criteria. A "duality principle" in guiding the design of intermediate models is explicated in this connection. An example of an application of this duality principle is the explicit incorporation of space–time performance metrics into a model at the chip architecture level: this enables reasoning about the topological properties of VLSI circuits and the progress of loci of computations (or "computational wavefronts").

2.1. Applications

There are several economic, technical, and managerial benefits that we believe will accrue in the long run from having a unifying theoretical basis that supports a wide spectrum of design activities. We list a few of these below.

- A common body of concepts can be used to understand different facets of the overall system, from high level (axiomatic) behavioral specifications, down to the electrical (transistor) level.
- A unifying theoretical basis provides for the development of a family of "parameterizable" design aids that can potentially be used at different levels of abstraction (provided, of course, appropriate care is given in designing their interface). A familiar analogy is the recognition that "editing" activities on diverse kinds of objects e.g., texts, programs, circuit schematics, stick diagrams, data flow programs etc. — all have certain commonalities that, once explicated, can be exploited in designing general purpose editors that have "modes" catering to the idiosyncrasies of the more specialized activities. Examples of other parameterizable design aids addressed by the formalism will include debugging aids, rapid prototyping systems, mixed mode simulators, testers, silicon assemblers and compilers, etc.
- A common basis for manipulating descriptions of a problem becomes available. This can be applied, for example, in proving the correctness of source–to–source program transformations, in proving the correctness of optimizations at a symbolic level of IC design [7], in proving the equivalence of descriptions of a problem at two different levels of abstraction e.g., in proving that a specific layout implements a 4–bit counter, or that a special purpose systolic architecture implements matrix multiplication, etc.

---

[1] This is a summary of an invited talk presented at the 1983 International Conference for Computer Design and VLSI, Westchester. This research was sponsored in part by the Office of Naval Research under contract N00014–83–K–0317 and in part by Defense Advanced Research Projects Agency, US Department of Defense under contract MDA903–81–C–0414

[2] While the modelling of software systems is also addressed by the formalism, this aspect will not be emphasized here.

- A common basis for the analysis of properties of specifications at different levels enables discussion of various flavors of completeness and inconsistencies e.g., logical inconsistencies in high level specifications, design rule verification, performance violations, etc.
- We have found that the development of the theory affords deeper insights into the nature of a problem and its solution techniques. For example, the notion of the progress of loci of computations arises naturally out of the formal development at the chip architecture level, and is useful both in the analysis and synthesis of designs. Such insights can sometimes be translated into "idioms" and can guide the design of corresponding linguistic constructs that enable succinct expression of these idioms e.g., "wavefront" notations to express the progress of the loci of computations.

2.2. An Example Application in Synthesis

We have been exploring the use of high level languages as a medium for specifying the desired behavior of special purpose systems, as well as paradigms for mapping such specifications into VLSI architectures [11]. We have also been developing a set of automated tools for transforming axiomatic and/or high level language program specifications into integrated software–hardware systems [4].

The initial specifications are in a high level language that has been augmented to allow axiomatic (behavioral) specifications (treated as "annotations" [2, 8]). This enables us to state the following:

1. Abstract axiomatic specifications of the behavior of a system, including statement of temporal characteristics.
2. Performance requirements to be met by an acceptable implementation along various dimensions of interest e.g., area, time, response time, throughput, reliability etc.
3. Relevant characteristics of the external environment a system is designed to operate in e.g., external timing constraints, relative function application frequencies, etc.

2.2.1. An Example in Outline

To crystallize a few of the ideas discussed herein, consider a "Queue Chip". Its skeletal specification is given below:

```
System Interface of QueueChip is
        q: Queue;
    — description of global I/O ports
        port INITIALIZE;
        port INSERT(x: in Item);
        port DELETE(x: out Item);
end QueueChip Interface;

— Semantics of behavior exhibited at the I/O ports

Behavior of QueueChip is
  loop
     select
        accept INITIALIZE do q := NEWQ(Size);
        or
        accept INSERT(x: in Item) do q := ADDQ(q,x);
        or
        accept DELETE(x: out Item := FRONTQ(q))
                          do q := DELETEQ(q);
        end select;
  end loop;
end QueueChip Behavior;

Performance Constraints on QueueChip are
— O(1) denotes constant time complexity
        Time(INITIALIZE) = O(1);
        Time(ADDQ(q,x)) = 20 milliseconds;
        ...
        Time(FRONTQ(q)) = O(1);
end QueueChip Performance Constraints;

Application Patterns on QueueChip are
    Frequency(INITIALIZE) = 1;
    Frequency(INSERT) > Frequency(INITIALIZE);
        ...
end Application Patterns;

Syntax of type Queue[Item] is
        ...
end Queue Syntax;

Semantics of type Queue[Item] is
        ...
end Queue Semantics;
```

Skeletal Specification of a QueueChip

Given the above definition, it is possible to follow one of several implementation paths. Ignoring the performance criteria for the moment, we summarize the outcome of some prototypical paths.

1. The specification above can be transformed into a program in a high level language that uses, for example, an array to represent the queue, two "pointers" that point to the head and tail of the queue etc.. This program may be compiled into machine code executable on a general purpose computer.
2. The specification may directly be transformed to yield an architecture suited for hardware implementation, e.g., a distributed "systolic" architecture. An appropriate description of this architecture can then be implemented in hardware as sketched below.
3. A high level language program e.g., the output produced in (1) above (or, alternatively, a program that is used as an initial "specification"), can be mapped directly into a combination of hardware structures that represent (1) the objects used in the program e.g., a RAM or register bank to represent the queue elements, counters that represent the up–down pointers to the head and tail of the queue, etc.; and (2) a state machine that controls various data transfers e.g., between the input data lines and the RAM. Note that this hardware embodiment can be designed to operate either synchronously or asynchronously at a low level.

Which of the above routes should be adopted in practice? The answer obviously depends upon the application context and the desired performance characteristics. Obviously, the second and third alternatives will result in faster implementations, while a choice between the latter two must rest on a more detailed analysis of requirements (e.g., whether extensibility is critical, whether minimizing area used is important etc.), as well as the costs of the target primitives chosen.

2.2.2. Transformation Strategy

In essence, the annotated specifications may be transformed into any desired mixture of software programs and special purpose hardware. If the initial specifications are axiomatic behavioral specifications, then a system architecture is first obtained [10]. The symbolic description of the hardware implementation is couched in our specification language syntax, by using derived language constructs (or "macros") for describing specialized hardware structures and algorithms tailored to them. The model dependent (structural) descriptions at this level are transformed into state machine ensembles that execute in an environment that is defined by the declarative part of the program; these state machines control data transfers between the hardware representations of the data structures used by the program. The state machine descriptions are then automatically translated into symbolic circuit layouts using path programmable logic arrays (alternative low level target structures, e.g., PLAs, may also be used). The resulting layouts are integrated with data paths using an interactive layout editor/design rule checker. Such layouts are finally automatically translated into fabrication masks in the desired format. Currently, the system can provide output in Caltech Intermediate Form (CIF 2.0 and CIF 2.7), ComputerVision EXDB and a few other formats.

Our attempts at applying the methodology and tools in conjunction to generate several kinds of hardware implementations are discussed elsewhere [4]. These include the implementation in NMOS of an Ada program fragment that describes a part of the Internet Protocol, as well as various classes of systolic designs.

2.2.3. Software/Hardware System Integration

In order to enable actual experimentation with this development methodology, it is necessary to have a test–bed that allows a program module that is a component of larger system to be replaced by an equivalent hardware module i.e., a "transparent" interface between software and hardware is needed. The development of such a test bed is part of our ongoing research effort [5]. We expect that the resulting prototyping ability will enable a closer study of issues related to the evaluation of performance and functionality of modules in such combined environments.

3. Model Features

In this section, we summarize some of the key features of an algebraic framework for VLSI design (see [11] for mathematical details).

A VLSI circuit/distributed software system is viewed as an interconnected network of processing elements (communicating processes). Each processing element has a set of I/O ports; it responds to external stimuli (electrical signals/messages) at its input ports, and provides a set of results at its output ports.

General Syntax and Semantics. The behaviors of networks of processing elements are described using a generalized form of "BehaviorExpressions" [3] that enable the expression of various flavors of parallelism and also allow processes to be passed as parameters. These BehaviorExpressions may be viewed as being instances of a specific parameterized algebraic data type. The syntax and semantics associated with these modified BehaviorExpressions provide the "general" syntactic and semantic basis for our modelling.

Detailed Models. Specialized Syntax and Semantics. Using the underlying algebraic framework, detailed models for various points in the design spectrum have been developed e.g., architectural descriptions [11, 9], symbolic layouts (e.g., PPLs, SLAs and PLAs [7]), self timed system protocols [6], transistor networks [11]. These intermediate models are characterized by (1) specialized ("derived") syntactic constructs, and (2) specialized semantics i.e., specialized reasoning methods e.g. specialized axiom systems and theorems in this system, fixed–point semantics, etc.

Derived Constructs in Models. Linguistic Primitives. Derived syntactic constructs that are useful in describing systems in intermediate models often offer insights in designing language primitives that are useful for succinctly describing constructs at these levels. Examples at the architectural level include data streams, wavefronts, progress of loci of computation, etc. The formulation of the model at the PPL level has led to a proposal of derived constructs for expressing special kinds of parallelism [7].

Levels of Abstraction. Abstraction in modelling is provided at 2 levels:

1. The behavior expressions that result from combining the syntactic primitives available can be shown to have denotation models. In particular, fixed point semantics can be developed for the recursive definitions of behaviors of processes.

   However, we believe that it is not necessary to reason at this ("low") semantic level when doing "day–to–day" reasoning and analysis. Rather, a useful body of theorems can be developed in a model–specific calculus. It is these theorems that can then be used more advantageously for manipulations at various levels.

2. As we progress from geometric layouts to transistors networks, to schematics, to symbolic layouts, to chip architectures, to high level language descriptions, to axiomatic behavior specifications, more and more detail is abstracted away. When combined with appropriate modularization strategies, such abstraction provides a way to significantly alleviate the complexity of (top–down) design and/or (bottom–up) analysis. The presence of specialized semantics further re–enforces this encapsulation at the theoretical level, while the presence of a uniform underlying set of principles aids both in the comprehension of concepts and in the incremental augmentation of model–specific knowledge.

3.1. A Specialized Model for Chip Architectures

As an example we outline a model at the chip architecture level that incorporates the potential for explicitly embodying spatial and temporal characteristics of the architecture.

Recall that the function accomplished by a system can be viewed in general as providing a set of functions operating on a set of objects i.e., as implementing a parameterized abstract data type. In a simple case, a chip may implement a *single* function on some domain e.g. a matrix multiplication chip. A common example of a chip that provides more than one function on a single object is the Queue chip mentioned earlier: it provides operations for inserting an element at the head of the queue, removing an element from the tail of the queue, etc. Another common example is an ALU.

We will concentrate here on the discussion of the case wherein a system achieves a map from a domain of values D_1 to a domain of values D_2. An implementation of this map at the chip architecture level can, at the first level, be dissected into the following components:

- A map from the (abstract) input domain values into data streams that are fed into the global input ports of a chip, i.e., an appropriate partitioning of the input.
- A transformation on the set of input data values (stream components) performed by the primitive processing elements comprising the system.
- A "reassembly" of the data values made available at the output ports i.e., a map from the data streams at the output ports onto the output domain values.

The 3 components may be expressed as

1. The input transformation I: D_1 –> S
2. The computational transformation C: S –> S.C will be further decomposed as elaborated upon below.
3. The output transformation O: S –> D_2.

A description of architecture involves (1) describing the behavior of the component processing elements and (2) detailing the interconnection structure of the processing elements.

The component processing elements, in order to achieve the transformation required in C, must (1) represent some (possibly none) abstract objects concretely inside the chip (perhaps in a distributed manner) e.g. the set of queue elements; and (2) perform appropriate computations on the input data (streams) to yield the desired output (streams).

The concrete representation of objects inside the chip may be viewed as comprising the "state" of the system (and, in turn, of the processing elements that are the finer grain components of a network). If no such objects are represented internally, the individual processing elements do not have any state component. *This is often the case for several systolic algorithms.*

Processor Interconnections. A succinct description of processing element interconnections is possible if there is regularity in the nature of the interconnections and the processors. This is particularly relevant for several existing systolic implementations.[3] If the interconnections are irregular, then no convenient abbreviations are possible, and it becomes necessary to detail the nature of each processing element and its interconnections.

In case of regular interconnections, the BehaviorExpressions can be parameterized (indexed) by co–ordinates describing the location of processing elements in space. The progress of computations can explicitly be associated with a temporal sequence. This automatically leads to a mathematical characterization of *wavefronts* that serve to describe the progress of *loci of computations*; different loci can be characterized by invariants that thought of as the phase of a wave.

The *semantics* of a given processing element network can then be viewed as the "solution" of the set of BehaviorExpressions describing the network. (Since there is a denotational semantics underlying the BehaviorExpressions, this can be obtained by computing the output streams that are least fixed points of the associated set of equations.)

3.2. Example Applications of the Theory

Instances wherein the formalism has found use include:

- Synthesis of circuit level architectures, proceeding from axiomatic behavioral specifications.
- Proving the *a posteriori* correctness of systolic designs.
- Proving the correctness of optimizations of circuits transformations at the PPL level.
- Proving the consistency of a PPL circuit with a higher level circuit description.
- Proving the correctness of transformation from simple logical expression into layouts.
- Deriving solutions of switch level networks that serve as the basis for simulation algorithms.

etc.

3.3. Performance Metrics. A Duality Principle in Modelling

One of the distinctions between models of distributed systems in software and a network of processing elements in hardware is that the spatial interrelationships of the processing elements play a more central role in hardware. There is therefore an advantage in bringing this aspect to the "foreground" by explicitly including it in a model for this level of design. Thus, while the analog of interconnection topology is not unimportant in distributed systems, this aspect has traditionally been de–emphasized in typical formal models.

This offers an example of what we will here refer to as the *metric–model duality principle.*[4] This "duality" reflects the fact that while on the one hand a performance metric can be viewed as a function of the other parameters of the model, it may also be embodied as an explicit parameter of the model. In practice, the perspective adopted depends upon the emphasis that is needed in a context: if the performance metric plays a major role, then it should be incorporated explicitly in the model.

If a performance metric is not a parameter of the model, then "reasoning" about the metric must be handled by the "general" algebraic rules governing BehaviorExpressions and parameterized types. Explicit incorporation of the performance metric into the model permits "specialized" rules to be formulated, which typically offer more power in manipulations. The drawback of explicit incorporation of metrics into a model is that reasoning in the model must then always carry along this "extra baggage" (which may result in undue overhead).

Examples of the use of this duality principle in modelling various points of the design spectrum include:

- Explicit incorporation of topological and temporal aspects in the model at the chip architecture level. This enables statements about spatial adjacency of processing elements and the progress of computations with time can be explicitly made in such models.
- Incorporation of "relative" geometric metrics (the number of cells in a rectangular grid) and topology

[3] Studies relating to symmetry groups in various dimensions are of relevance here [12].

[4] This could also be thought of as a *function–parameter* duality [11].

(adjacency of cells in the grid) in a model for PPL/SLA designs (a symbolic IC design technique [7]).

- Explicit incorporation of absolute geometric metrics at the layout level (where design rule checking is relevant).

4. Summary and Conclusions

We have outlined some features of a conceptual and formal basis for modelling various aspects of the task of designing integrated systems. We believe that in the long run such models will have important analytic and synthetic applications in the overall design process. In particular, we (optimistically) envision "second generation" software/hardware design aids to be based on variants of such a theory. While encouraging progress has been made, there is a lot that remains to be done, both in theory and in practice.

References

1. F.P.Brooks. *The Mythical Man–Month: Essays on Software Engineering*. Addison–Wesley, Reading, Mass., 1975.

2. Krieg–Bruckner, B., Luckham, D.C., von Henke, F.W., Owe, O. (Draft) Reference Manual for Anna, A language for Annotating Ada Programs. Unpublished, Reviewer's Copy, October 1982.

3. Milner, Robin. *A Calculus of Communicating Systems*. Springer–Verlag, Berlin, Heidelberg, 1980.

4. Organick, E.I., Carter, T., Hayes, A.B., Lindstrom, G., Nelson, B.E., Smith, K.F., Subrahmanyam, P.A. Transformation of Ada Programs into Silicon. Second SemiAnnual Technical Report. Tech. Rept. UTEC–82–103, University of Utah, November, 1982.

5. Organick, E.I., Keller, R.M., Lindstrom, G., Smith, K.F., Subrahmanyam, P.A., Carter, T., Klass, D., Maloney, M.P., Nelson, B.E., Purushothaman, S., Rajopadhye, S. Transformation of Ada Programs into Silicon. Third SemiAnnual Technical Report. Tech. Rept. UTEC–83–026, University of Utah, April, 1983.

6. Subrahmanyam, P.A. and S.Purushothaman. An Algebraic Basis for Specifying and Reasoning about Protocols For Designing Self Timed Circuits. In F.Anceau, Ed., *VLSI 83*, North Holland, Amsterdam, 1983.

7. Rajopadhye, S.and P.A. Subrahmanyam. Formal Semantics for a Symbolic IC Design Technique: Examples and Applications. Tech. Rept. UTEC # 83–041, University of Utah, May, 1983.

8. Subrahmanyam, P.A. From Anna + to Ada: Automating the Synthesis of Ada Package and Task Bodies. Tech. Rept. Internal Report, University of Utah, March, 1982.

9. Subrahmanyam, P.A. and Rajopadhye, S. Automated Design of VLSI Architectures: Some Preliminary Explorations. Tech. Rept. UTEC # 82–067, University of Utah, October (Revised), 1982.

10. Subrahmanyam, P.A. Synthesizing VLSI Circuits from Behavioral Specifications: A Very High Level Silicon Compiler and its Theoretical Basis. In F.Anceau, Ed., *VLSI 83*, North Holland, Amsterdam, 1983.

11. Subrahmanyam, P.A. An Algebraic Basis for VLSI Design. Draft of a Research Monograph, April 1982. Available from the Department of Computer Science, University of Utah.

12. Wigner, Eugene P.. *Group Theory and its Applications to the Quantum Mechanics of Atomic Spectra*. Academic Press, New York, 1959.

The Circuit Design Language Xi (Ξ)

Stuart I. Feldman

Bell Laboratories
Murray Hill, New Jersey 07974

ABSTRACT

This paper describes a programming approach to circuit design. The language Xi is an extension of the algorithmic language C. A Xi program can describe the connections among many components in a readable yet compact fashion. Though the user's program is independent of technology, the Xi system is capable of creating forms that can be simulated, and realizations that can be fabricated in different technologies. It enables a person who understands computer programs but does not have any great knowledge of digital electronics to design a significant silicon device.

Introduction

This paper describes a programming language approach to the design of VLSI devices. The language is called Xi (the Greek letter Ξ). Xi is a fairly conventional language — syntactically it is an extension of C.

A prototype system has been built and shown to be useful: complicated circuits designed by neophytes have been manufactured and tested successfully. The original form of the language was defined by A. G. Fraser. The current compiler, a version of the Portable C compiler for the VAX, was written by S. C. Johnson. The author of the paper extended the language and implemented the run-time system that supports the language.

The Xi approach takes advantage of standard concepts of programming languages and the environments in which they run that seem to be directly applicable to the circuit design process. A Xi program is a readable description or specification of a circuit. The ideas of hierarchy come naturally in this world, since procedures are a well understood part of most languages. Data structures are a convenient vehicle for packaging related objects and for hiding implementation details. Libraries and assembly language windows have analogs in the design world, since it is useful to re-use parts of other designs and sometimes necessary to use technology-dependent designs that cannot be described easily in the higher level language.

The use of a programming language encourages structured thinking and design. It is easy to build and use libraries of procedures that generate circuits with particular functions, and to combine the results of these procedures. If a sub-circuit is defined by an extant procedure, that component can be thought of as a building block. The designer should think at a high level in terms of a simple circuit model and an idea of an executing computer program. The current research system has access to library procedures that generate specialized subcircuits such as RAMs and ALUs optimized to a particular technology; a Xi program can connect these pieces without concern for their internal structure.

Computer programmers should find this language a natural way to specify electronic devices. Logical design is separated from the graphic, geometric (layout) and physical (processing) aspects of VLSI. Of course there may be a penalty for working this way. The language we describe will meet most needs, so that escapes to other techniques will be common only in library routines. But the Xi approach is sufficiently flexible to permit use of sub-components created using special purpose circuit generators that achieve maximum performance from a limited resource.

A program written in Xi is first compiled and loaded to form an ordinary computer program, which is then executed. Most parts of the program have the same effect as in a C program, and can perform loops, tests, calls, and input/output. But those statements that involve values of a new type, *signal*, have special meanings. A statement containing an operation on a signal conveys information about electronic components and connections between them in a device to be constructed. The program describes in piecemeal fashion the graph of the circuit, and the Xi system builds data structures to capture these relations. After the user's program terminates, the resulting design is written out in a form suitable for further processing and perhaps fabrication as a chip. At present, Xi can drive a logic simulator, a gate array and polycell layout system, and an nMOS geometric layout program.

People often misunderstand a programming language by confusing what happens at 'compile' time and what happens at 'run' time. This confusion is compounded in Xi because there are three epochs to consider: some decisions are made at the time the Xi program is compiled, others when a particular statement is executed, and other actions take place as values are computed in a physical circuit. A little thought usually suffices to clear up any such difficulties.

Use of Xi to Design Real Circuits

It is easy to write, test, and modify Xi programs, so it is feasible to try out several approaches to a circuit before concentrating on a single tight design. It is possible to test all of the prototype designs using the same simulation drivers and to get some feel for the area each requires. Xi programs are independent of technology, and form quite readable documentation.

The Xi language assumes a simple circuit model: normally all clocked elements are synchronized, and all feedback paths (cycles) must include a a clocked element. The language is best at expressing irregular (random) logic, especially cases of related but not quite identical paths in which the special cases can be easily handled by programming. The limited set of builtin operators is not really suited for describing such regular and well understood constructs as RAMs and ALU's, but these are precisely the parts of designs that can be generated by separate, technology-specific, programs. In a significant circuit, most of the area will probably be occupied by such regular structures, as is commonly the case in microprocessors today; but the Xi system is very good at describing the glue that normally occupies so much of the designer's time.

Combining sections created using different methodologies is feasible only if the different tools speak the same language. Otherwise different programs would be needed to store or simulate each piece, and it might not be possible to have the parts fabricated onto a single chip. Our research software uses the same language for describing layouts, so we can share many of our tools.

CH1935-6/83/0000/0652$01.00 © 1983 IEEE

Informal Description of the Xi (Ξ) Language

The Xi language is an extension of C. In addition to the features of C it has

- an additional data type, *signal*
- some old operators that have new meanings when applied to the new type
- some new operators that apply only to the new type
- new functions in the standard library

A statement that uses none of the new features will execute as if it were part of a C program.

A signal represents a connection between components. Most of the components are represented in a technology-independent manner by operators in the Xi language. When two signals are attached by the = operator, the nets represented by the two signals are inextricably linked. Thus assignment of two signals is somewhat different from assignment of other data in C: after assignment the two *values* have identical meaning. The implications of this rule will soon be made clearer.

The standard Boolean operations of *and, or, exclusive-or,* and *not* (symbolized by the C operators &, | , ^, and !) apply to signal values. The value of the expression a & b is a signal whose electrical value will be equal to the *and* of the signals a and b. In the digital device being described by the Xi program, an *and* gate will be created with inputs attached to those signals, and output connected to the signal attached to the expression a&b. The following Xi program declares three variables of type signal, and attaches the conjunction (&) of two of them to the third.

```
main()
{
signal a,b,c;
c = a & b;
}
```

One difficulty with this program is that none of the signals is attached to anything in the outside world. This can be remedied, for example, by associating them with input/output connectors (pins) on the device. The Xi library contains some functions that return a signal that is attached to the pin with the given external name or number. The following program describes a chip that computes the popular logical functions of its inputs. (A function call that returns a signal may appear on the left side of an assignment in Xi.)

```
main()
{
signal a,b;
a = insig("pina");
b = insig("pinb");
outsig("and") = a&b;
outsig("or") = a | b;
outsig("xor") = a^b;
outsig("nota") = !a;
outsig("notb") = !b;
}
```

Now we describe a chip that accepts 4 pairs of inputs and outputs the *and* and *or* of each pair. One obvious way to do this is by using some arrays of signals. Signals, like other data types, may be used in arrays and structures and may be returned as the value of a function.

```
main()
{
int i;
for(i=0; i<4; ++i)
  ops(inpin(i),inpin(4+i),outpin(8+i),outpin(12+i));
}

ops(a,b,c,d)
signal a,b,c,d;
{
c = a&b;
d = a | b;
}
```

(The functions inpin and outpin permit I/O pins to be referenced by number rather than by name). Signal variables are passed by value as are all other arguments in C and Xi. But assignment (attachment) of signals is a special case: since the *values* are associated, it is possible to have a side effect on an argument to a procedure: though the cell containing an actual argument is not affected by the operation, the significance of its contents is changed by the call.

A neater way to describe this chip uses the Xi concept of a 'bundle'. By default, a signal value represents a single connection. But a signal value can represent a group of simple connections. The easiest way to establish such a value is to use the bundle function which takes as argument the width of the bundle and returns a signal of that width. A single connection (sub-signal) of a bundle may be extracted using a pun on C's subscript notation: the sub-signals are numbered 0 through $width-1$, and the ith element of a bundle b is written b[i]. A sub-bundle formed from the elements $b_i, b_{i+1}, \ldots, b_j$ is denoted b[i:j]. The width of a signal can be determined by using the built-in function width.

```
main()
{
signal a,b,c,d;
a = bundle(4);
b = bundle(4);
c = a & b;
d = a | b;
}
```

This program is incomplete because the signals have not been connected to pins. This can be done using the inbundle and outbundle functions which take as arguments a name and a width and return a bundle of that width connected to pins whose names begin with the given name. (If the name argument is zork, the pins will be named zork_0, zork_1, ...). The program may be written without declaring any variables at all if you prefer:

```
main()
{
outbundle("c",4) = inbundle("a",4) & inbundle("b",4);
outbundle("d",4) = inbundle("a",4) | inbundle("b",4);
}
```

Note that in the above two examples the *and* (&) operator is applied to bundles and returns a bundle as value. The Boolean operators work on bundles so long as the operands have the same widths.

Another C operation that has a new interpretation when applied to signals in Xi is the conditional expression. The terse notation a?b:c is the way C and Xi express the value **if** *a* **then** *b* **else** *c*. The meaning of the statement

```
d = a ? b : c;
```

depends on the type of a. If none of the operands in this statement is of type signal, this is an ordinary C statement. If any of b, c, or d is a signal, all three must be signals with the same width. If a is not a signal, then the selection is made when the Xi program executes this statement, and d is identified either with b or with c at that time. If a is a signal, it must be of width 1, and then this statement describes a switch (multiplexor): in the electronic device, signal d is attached to signal b at those times that a is true, and is attached to signal c when a is false.

The is-equal (==) and is-not-equal (!=) operations also apply to signals, and have a signal of width 1 as value. If a and b are bundles, then a==b is true (equal to 1) iff each sub-signal of a equals the corresponding sub-signal of b. The bundleval function creates a bundle whose width is specified by the second argument and whose value is the low-order bits of the first argument at the time the function is called. This function provides coercions from integers to signals. Two constant signals are defined, FALSE and TRUE each of width 1, with values 0 and 1 respectively. The following fragment raises an output signal every time the input eight-bit byte is an exclamation point. Note that a character constant acts as an integer in Xi.

```
. . .
outsig("bang") =
        inbundle("inbyte", 8) == bundleval('!', 8);
```

So far we have described 'combinatorial' logic. Xi also has some operators with memory. An implicit clock underlies these operators, so all act in synchrony. All memory devices are set to zero by an asynchronous reset. (The controlling signals can be changed by library procedures.) The unary operator prev provides a simple delay. If

```
b = prev a;
```

then b has the value that a did at the previous clock time. The binary operator when is a memory element. If

```
c = a when b;
```

then c has the value that a did the last clock time when b was true. The following chip delays its input by two clock times:

```
main()
{
outsig("delayedx") = prev prev insig("x");
}
```

To illustrate the use of some of these language features, consider a simple encoder that puts out the exclusive-or of the input byte and a 'key' byte. If a question mark ('?') appears in the input, the following character is used as a new key.

```
main( )
{
signal in, key;
in = inbundle("input", 8);
key = in when prev ( in == bundleval('?',8) );
outbundle("output", 8) = prev in ^ key;
}
```

The statement with key on the left side causes the key to change whenever the previous input was a question mark. The last statement encodes the input with a one-character delay so that the new key always masks itself.

A shift register can be built using a chain of prev operators. To illustrate various Xi programming techniques we will describe four ways of coding the same circuit. Each is a procedure that builds a shift register that returns its input signal delayed by *n* clock times:

```
signal shift(n, insig)
int n;
signal insig;
{
signal reg = bundle(n);
if(n == 0)
     return insig;
reg[0] = prev insig;
for(i=1; i<n; ++i)
     reg[i] = prev reg[i-1];
return reg[n-1];
}
```

Note that shift is a function that returns a value of type signal. Also note that the signal variable reg is initialized in the declaration statement, and that its bundle width is determined at the time the shift procedure is invoked. The loop attaches each element of the bundle to the delayed value of its neighbor. The first element (reg[0]) is identified with the delayed input signal, and the last element (reg[n-1]) is returned as the value of the procedure. After the procedure returns, the variable reg disappears as does any local variable in C, but like the Cheshire cat the smile remains: the devices (delay elements) required by the loop will appear in the final circuit.

Alternately, we can use a recursive procedure which adds one level of delay per recursive level of invocation:

```
signal shift(n, in)
int n;
signal in;
{
if(n>0)
     return shift(n-1, prev in);
else
     return in;
}
```

Instead of using a bundle to hold the intermediate connections, we could use an array of signals. The Xi (and C) languages do not allow the declaration of arrays of varying size, but it is possible to allocate such an array using the standard C allocation function calloc. The rest of the code would appear identical to the first shift example, since the same notations are used in Xi for indicating an element of a bundle and an element of an array.

When a signal variable is created (at load time for static and external variables, at procedure invocation time for automatic variables) its width is undefined and it is not yet associated with any signal. The first use of the variable that assumes it has a value causes it to be initialized to a new signal of width 1. But if a signal is assigned to an uninitialized one, the empty one gets a copy of the other. This is the reason a statement like

```
signal a = bundle(4);
```

can work. (If a were not empty and contained a bundle of width other than 4, the assignment would be illegal). There are times when it is convenient to redefine a variable of type signal. The NEW operator does this. (The operator is spelled in capital letters to avoid confusion with the **new** operator of Stroustroup's Class preprocessor.) Then the following is a concise way to express our shifter:

```
signal shift(n, in)
int n;
signal in;
{
while ( --n >= 0 )  /* loop n times */
     NEW in = prev in;
return in;
}
```

In the special form NEW `x = ... x ...` the previous value of the variable `x` is used on the right side before the NEW operator clears it.

It is sometimes desirable to break a device into parts. For example, at present the LGEN layout program generates a single circuit each time it is invoked. If complicated, the result is long but skinny, and will not fit into a square frame. But several skinny parts can be created by separate LGEN runs, and these can be placed above one another in a square frame. The output of a Xi program may be broken into blocks. At any instant in the execution of the program, there is a current block. A component (gate, memory element, etc.) is put into the block that is current when the statement that produces the component is executed. Blocks have character string names. The block that is current can be changed by a call to the function `useblock` which accepts a character string as argument and returns a pointer to a character string containing the previously current block as value. A procedure that puts its components in a particular block every time it is called might then look like

```
piece()
{
char *old = useblock("piece");
 . . . /* statements that generate components */
useblock(old);
}
```

The following program produces 3 blocks:

```
main()
{
useblock("b1");
 . . . /* parts of the circuit in block b1 */
useblock("b2");
 . . . /* parts of the circuit in block b2 */
useblock("b3");
 . . . /* parts of the circuit in block b3 */
useblock("b1");
 . . . /* more parts of the circuit in block b1 */
}
```

A component specified by a Xi program or some other circuit definition technique can be included in a circuit produced by a Xi program. The subroutine call

```
extblock (blockname,fileprefix,nin,nout,in,out);
```

asserts that there is a block with a given name, the information is in files whose names begin with `fileprefix`, that it has `nin` input signals in the array `in` and produces `nout` output signals in the array `out`.

For example, the Xi library has a procedure called `sram` that invokes one of these generators to produce a static RAM. Such a cell can be incorporated in a Xi design by a statement like

```
outdata = sram (readcycle, address, indata);
```

If the bundle argument `address` is of width n and the bundle argument `indata` is of width w, a RAM containing 2^n words of width w will be created. The RAM's address lines will be attached to the signals in `address`, data going into the RAM are taken from the bundle `indata`, and data read from the RAM are put onto the bundle that is returned by the `sram` function. The signal `readcycle` specifies whether this particular clock cycle is a read or a write for the RAM. This is the simplest RAM call in the library; interfaces with more arguments permit control over clocks and memory sizes that are not powers of two. In the implementation, the geometry produced by the RAM generator is associated with the signals by means of a call to `extblock` in the interface routine.

EXPERIENCE WITH A REGULAR EXPRESSION COMPILER†

Anna R. Karlin
Howard W. Trickey
Jeffrey D. Ullman
Stanford Univ., Stanford CA

The language of regular expressions is a useful one for specifying certain sequential processes at a very high level. They allow easy modification of designs for circuits, like controllers, that are described by patterns of events they must recognize and the responses they must make to those patterns. This paper discusses the compilation of such expressions into reasonably compact layouts. The translation of regular expressions into nondeterministic automata by two different methods is discussed, along with the advantages of each method. A major part of the compilation problem is selection of good state codes for the nondeterministic automata; one successful strategy is explained in the paper.

I. The Regular Expression Language

We shall give a brief introduction to the language of regular expressions; for more information on this language and on nondeterministic finite automata, the reader is referred to Hopcroft and Ullman [1979]. Regular expressions consists of operators and operands. The operands are abstract symbols that represent events in terms of combinations of wires. Events are assumed to occur at discrete times, so regular expressions define synchronous systems.

Operators

The operators of our language are:

1. Juxtaposition (no operator) standing for sequencing of events. That is, if regular expressions R and S each represent a set of events, then RS represents the set of events consisting of an event of R followed by an event of S.
2. The + operator standing for union. Thus, $R+S$ stands for the set of events that are either events of R or events of S.
3. The unary postfix operator * standing for the closure, or "any number of" operator. Thus, R^* stands for any sequence (including the null sequence, denoted ϵ) of events in R.

In our language we also use the shorthand operators:

4. R^{++} stands for one or more occurrences of events in R, that is, RR^*.
5. $R?$ stands for zero or one occurrence of an event in R, that is, $R+\epsilon$.

Example 1: Let a and b be events, i.e., abstract symbols of our language. Then ab stands for an a followed immediately by a b; ab^* stands for an a followed by any number of b's, i.e., $\{a, ab, abb, \ldots\}$; ab^{++} stands for $\{ab, abb, \ldots\}$. Also, $a+b$ stands for either an a or a b, and $(a+b)^*$ stands for any sequence of a's and b's, in any order. □

† Research supported by DARPA contract MDA-80-C-0107 and NSF grant MCS-82-03405.

Special Operands

In addition to abstract symbols as operands, we allow two other operands.

1. A dot (.) is an operand that matches any combination of input wires, i.e., it is always seen.
2. The symbol # is never seen, no matter what input wires are on. Although seemingly without purpose, this symbol is essential when we use state names in our expressions, as described below.

Line Declarations

Our expression language has declarations of five types: lines, symbols, outputs, states, and subexpressions. The lines are wire names, from which the symbols are constructed. For example,

line x, y[8]

declares x to be the name of a wire, and y to be the name of a group of eight wires, which may be referred to individually in symbol definitions as $y[1], \ldots, y[8]$.

Symbol Declarations

Symbols are the operands of regular expressions mentioned above. Each is defined by a set of wires that must be on and a set that must be off. Thus,

symbol zap(x, y[1], -y[3])

declares that abstract symbol *zap* is seen whenever wires x and $y[1]$ are on, and wire $y[3]$ is off. Any other wires are ignored when deciding whether event *zap* is seen. Note that, unlike the usual conventions in automata theory, we allow more than one symbol to be seen at the same time. for example, we could define

symbol zip(y[1], y[2])

and see both *zip* and *zap* at the same time, if x, $y[1]$, and $y[2]$ are on, while $y[3]$ is off.

Output Declarations

Output symbols are embedded in the regular expression and represent output wires. The exact rules determining when an output wire is raised are complicated, and the details appear in Ullman [1983]. However, the general idea is that if R is a regular expression, U is an output symbol and RU a subexpression of the complete regular expression, then we raise output U immediately after seeing a sequence of inputs that forms an event of R. An example will help make the

CH1935-6/83/0000/0656$01.00 © 1983 IEEE

ideas known. If we declare

output U, V

and write the regular expression

zip (zip U + zap U V)*

then after the first event, which must be *zip* or no output is ever made, we look for any sequence of events *zip* and *zap*. Each time we see *zip*, we raise output U, and each time we see *zap* we raise both U and V. If at any time we see neither *zip* nor *zap*, we not only raise no output, but recognition of the regular expression has "derailed," and we never make any more output.

State Declarations

State names are used like "goto's" in the regular expression. While the regular expression language is most appropriately used in situations where there is little need for explicit state transitions, we have found that the occasional use of such transitions is almost essential. With this feature, our language has all the power of deterministic finite automaton languages, like SLIM (Hennessy [1981]), while also offering the expressive power of regular expressions where that is more appropriate. We can declare s to be a state by the declaration

state s

Then, in the regular expression, there will be one occurrence of s followed by a colon; thus s : marks the "label" position of s, where control transfers whenever it is determined that state s is entered. Often, we find s : preceded by #, the symbol that is never matched, so that control does not accidentally reach state s without an explicit transfer to that state.

In the expression there can be any number of occurrences of s not followed by a colon; these are the "goto's" to state s. As with output symbols, a state symbol is activated when a match for the preceding regular expression is recognized. The reader should be reminded that because of the inherent nondeterminism in the input recognition process defined by regular expressions, the use of states can be more general than in a deterministic finite automaton language. For example, two or more goto's to different states could be activated at the same time, causing us to be in both states at once.

Program Structure

The fifth kind of declaration is a subexpression, where an identifier is declared to stand for a regular expression. Thus

subexp string = (zip + zap)*

declares *string* to stand for any sequence of *zip*'s and *zap*'s, so

string zip zip string

stands for any sequence of *zip*'s and *zap*'s with at least two *zip*'s in a row.

After all declarations for a program, there is a single semicolon, followed by a single regular expression. The limitation to one expression is not significant, since there can be any number of output symbol occurrences in the expression. We can even use the # operand to simulate a multistate automaton by using an expression of the form

```
line x
symbol
      zero(-x)
      one(x)
output OUT
;
.* one one (one zero? zero?)++ OUT
```

Fig. 1. Bounce filter regular expression.

state1: expression1
+ # state2: expression2
...
+ # state*n*: expression*n*

Presumably, each of the expressions has within it one or more state symbols, which cause transitions to other states.

Example 2: A *bounce filter* is a device with a single input and single output; the output will generally agree with the input, but we wish to ignore small "bounces," where for a small number of cycles the input changes and then returns to its original value. For our example, we shall ignore one or two consecutive 0 or 1 inputs that do not match their surroundings. The regular expression that defines this output as a function of the input is shown in Fig. 1.

The first line of text declares x to be a wire; this wire is the only input wire in this example. The next two lines declare *zero* and *one* to be abstract symbols, seen, respectively, when the input wire is off and on. The fourth line declares OUT to be an output signal, the only output in this example. Then comes the obligatory semicolon, and finally the expression itself. The expression says that the output OUT is to be raised whenever we see an input pattern that requires the output to be 1. That is, we may see anything at all (indicated by the .*), then two *one*'s and one or more groups represented by the expression

one zero? zero?

that is to say, each group consists of a 1 followed by up to two optional 0's. The 1 from the first group forms the third consecutive 1, along with the 1's that match the two earlier symbols *one* in the expression. After these three 1's, there cannot be three 0's in a row, no matter how many groups are present, so the output in response to any input that matches the pattern of the expression given in Fig. 1 should be 1. □

Example 3: Now, let us see how to write as a regular expression the famous traffic light controller from Mead and Conway [1980]. This problem involves a light at the crossing of a highway and farm road. A sensor detects cars waiting on the farm road to cross the highway; its output is the line *cars* in Fig. 2. Two timing signals are also used to control the light. The short time out signal, *tos* in Fig. 2, indicates that enough time has elapsed from the last time the $RESET$ output signal was raised that a yellow light may be turned to red. The long time out, *tol* in Fig. 2, is used to measure the minimum time, from the last $RESET$, that we shall allow the highway to be green, even when cars are waiting on the farm road, and also to measure the maximum time that we shall allow the farm road to be green, even if there is a steady stream of cars on that road.

Let us examine the parts of the expression in detail. Starting from the *highway* state, the subexpression

```
line cars, tol, tos
output RESET, HWYGREEN, HWYYEL,
    FARMGRN, FARMYEL
state highway, farm
symbol
    carstol(cars, tol)
    carsntol(cars, -tol)
    nocars(-cars)
    timeup(tol)
    notol(-tol)
    switch(tos)
    wait(-tos)
;
highway: (nocars+notol)* HWYGREEN
    carstol RESET
    wait* HWYYEL switch farm RESET
+ # farm: carsntol* FARMGRN
    (nocars+timeup) RESET
    wait* FARMYEL switch highway RESET
```

Fig. 2. Traffic light controller.

(nocars + notol)*

matches any sequence of events in which either there are no cars waiting, or the long timeout interval has not elapsed. As long as that is the case, the highway stays green, as reflected by the fact that the HWYGREEN output follows this subexpression. Then if both the *cars* and *tol* signal are on, the input no longer matches

(nocars + notol)*

but it matches the longer expression

(nocars + notol)* carstol

because the *carstol* symbol is seen whenever both the wires *cars* and *tol* are on. Note that the output and state symbols, such as HWYGREEN, are ignored when considering subexpressions of the complete expression that might match the input.

In response to a match of the above expression we emit signal RESET, which starts the counter for the purpose of measuring the time of the yellow light. Any input that matches the above expression also matches

(nocars +notol)* carstol wait*

since *wait* can occur zero times in a match of *wait**. Thus, at the same time RESET is signaled, HWYYEL is also signaled, and the highway light turns yellow, while the farm road remains red.

As long as the short timeout period has not elapsed, *wait* will continue to be seen, so the above expression will be matched, and HWYYEL, but not RESET, will be emitted continuously. Then, when *switch*, the abstract symbol that represents the *tos* wire going on, is seen, we can no longer emit HWYYEL, because the input seen since we entered the *highway* state no longer matches the above expression. However,

(nocars + notol)* carstol wait* switch

is matched. Thus, we emit the two following signals, *farm* and RESET. The first takes us to the farm state, and since that state is followed by subexpression *carsntol**, which is matched by the empty string, we immediately signal FARMGRN as well. That causes the farm road to become green and the highway red. The events following the farm state are similar to those just discussed for the highway state, and we shall omit a detailed description.

We might note that although the traffic light is inherently a four-state device, we used only two states, and in fact, we did so only for convenience; we could do without states altogether. There is really no need for the farm state, because whenever we enter it, we would "fall through" to it anyway. We can do without the *highway* if we put the closure operator around the whole expression, thus causing the cycle to repeat indefinitely. The state-free expression for the traffic light is

```
((nocars + notol)* HWYGREEN carstol RESET
    wait* HWYYEL switch RESET
    carsntol* FARMGRN
    (nocars + timeup) RESET
    wait* FARMYEL switch RESET)*
```

□

II. The Compilation Strategy

Figure 3 outlines the way regular expressions are compiled into PLA's. The language of nondeterministic finite automata (NFA's) is used as an intermediate language. We shall not detail the language here, as it is fairly conventional. The important thing to remember is that the nondeterministic states each correspond to a single operand of the expression. There are two reasons we prefer to work from the NFA, rather than converting regular expressions into deterministic automata and using standard state-coding heuristics.

1. Sometimes the regular expression is short, yet the number of states of the deterministic automaton is enormous. We worked with one example of an expression, that described pattern matching with don't care's, where the regular expression has 72 operands, yet the deterministic automaton has over eight million states. By coding the NFA directly, we were able to get a PLA with 24 feedback wires, which is only one more than the minimum possible for the implementation of an 8,000,000 state machine.
2. The regular expression gives us important clues to a good state coding. In particular, we shall see below that we can always find a PLA implementation with one term per operand, i.e., one term per NFA state. If we converted to a deterministic automaton, we might lose some of the useful information and wind up with a PLA with more terms, unless we spent a great deal of effort optimizing the coding.

The "before" and "after" type compilers are really implemented by a switch on a single compiler; we shall discuss the difference below. The details of the compiling algorithms involved in translating regular expressions to NFA's by either method are found in Trickey [1982] and Ullman [1983].

We have experimented with several strategies for the state coder. They all depend on knowing the conflict matrix of the NFA, i.e, which pairs of states can be on at the same time. We shall have more to say about these strategies later.

The output of the state coder is a PLA personality. This personality has the number of its terms reduced by a

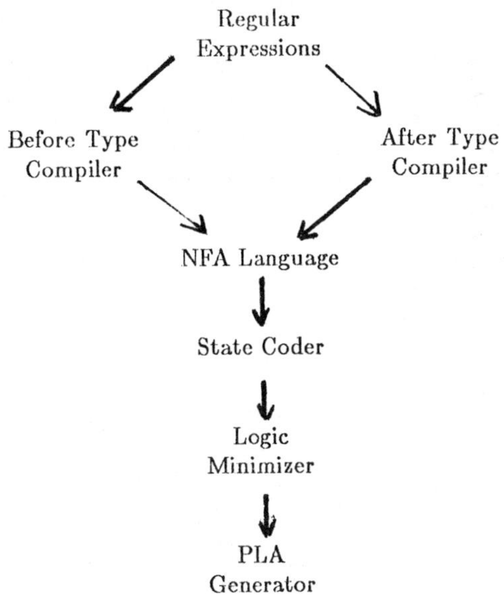

Fig. 3. Outline of RE Compiler.

program called GRY, written by Hemachandra [1982] and based on algorithms described in Hachtel et al. [1982]. The output of the minimizer is fed to a PLA generator written by Kevin Karplus.

III. The Partition of Regular Expressions

The first thing the regular expression compiler does is break up the given expression into manageable pieces; we try to have each piece represent about fifty operands. One of the important features of the regular expression approach to design is that expressions can be broken up into subexpressions that have very little interaction; in essence the outer expression "calls" the subexpression at exactly one place, and the call can be represented by a pair of wires carrying a startup signal to the subexpression recognizer and a completion signal back to the caller. For example, the bounce filter of Example 1 could have its expression broken into

$$main = .^{*}\ one\ one\ sub^{++}\ OUT$$
$$sub = one\ zero?\ zero?$$

It is important to realize that the circuit recognizing the subexpression can receive start signals more than once, and may even be working on more than one "call" at a time, but this activity is a correct implementation of regular expressions. We should also be aware that if there are state "goto's" connecting a subexpression to its environment, then more than one pair of wires will be necessary for the interconnection.

Before translating the subexpressions into NFA's, the compiler does a certain amount of algebraic manipulation of the subexpression to reduce the number of NFA states needed, if possible. For example, we left- and right-factor expressions, so

$$abc + adc$$

becomes

$$a(b + d)c$$

The motivation for splitting the expression into small pieces is that the PLA implementation degrades in both speed and in area used per regular expression operand, as the number of operands grows. That is, the area of a PLA for an n-operand expression could be proportional to n^2. The reason we do not therefore break the expression into PLA's of size 1 is that the PLA cost also has an overhead term. The result is that to get the lowest ratio of operands to PLA area we should use subexpressions of about 15–25 operands for each PLA. However, because of the wasted area involved in putting many PLA's together, we prefer somewhat more than the optimal number of operands per PLA to reduce the overhead due to PLA's that don't quite fit together.

A previous incarnation of the compiler attempted to translate the expression directly to a layout using an algorithm of Floyd and Ullman [1982] that requires area that grows only proportionally to the number of operands. However, the network-of-PLA's implementation was found superior in practice. The reader should also be aware of another approach to laying out regular expression recognizers in linear area due to Kung and Foster [1982], which we have not tried.

Another style of implementation is described in Ullman [1982], where regular expressions were translated into the Igen logic language (Johnson [1983]) and thus implemented as Weinberger arrays. The area of such implementations was found to be comparable to the PLA implementation. Theoretically we might expect the Weinberger array approach to use less area than PLA's, but to form circuits of very large aspect ratio as the size of the expressions compiled grows.

The reader is also referred to Trickey [1981] for a description of some experiments with the systematic exploration of the different ways a regular expression could be partitioned into subexpressions, and the subexpressions converted to PLA's that would fit together with little wasted area. It was found that significantly improved layouts could be obtained, but the computation time grew exponentially with expression size. That makes it doubtful the method could be applied to expressions with more than a few hundred operands, unless some way of focussing the search for partitions were found.

IV. Before and After NFA Constructions

Now, let us return to the two methods whereby NFA's are constructed from regular expressions. We begin either process by identifying each operand with a state.

Example 4: Consider the bounce filter of Example 2. We may number the operands of the expression from left to right as follows.

$$._1{}^{*}\ one_2\ one_3\ (one_4\ zero_5?\ zero_6?)^{++}\ OUT_7$$

We may then associate with operand i the state N_i.

In Fig. 4 we see what looks like a transition diagram for a finite automaton. It actually represents the successor relationship among the states or operands, i.e., which operands can follow which in the regular expression. For example, there is an arc from N_5 to N_4 because after seeing a 0 corresponding to $zero_5$, we could begin another group consisting of a 1 and up to two 0's, and such a group must begin with a 1 that matches one_4. We also have an arc from N_5 to N_6, because after matching $zero_5$ we could see another 0 that matched $zero_6$. Finally, there is an arc from

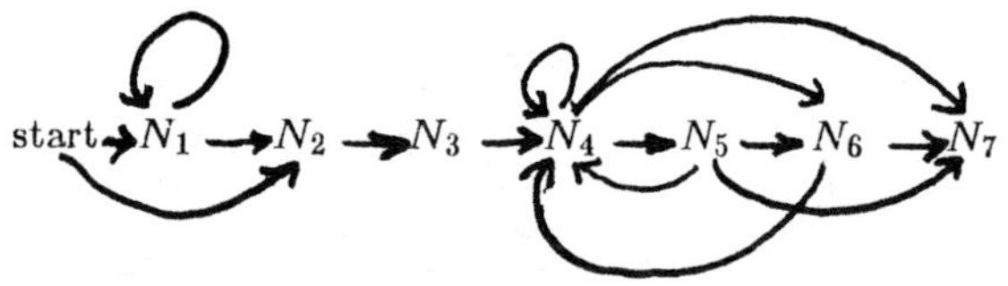

Fig. 4. Successor relation for bounce filter.

N_5 to N_7 because after seeing a match for $zero_5$ we have seen an input that matches the subexpression prior to the OUT output and therefore must make the OUT signal. See Ullman [1983] for the details on the algorithm used to compute the successor function. □

Whether we use the before or the after interpretation of states, we can see transition diagrams like Fig. 4 as representing places that can be "active," which we might represent by putting a marker on a subset of the nodes. When we use the *before* interpretation, a marker at state N_i tells us we are ready to recognize the operand corresponding to that state. Thus, if state N_5 of Fig. 4 is active at a given time unit, it will activate for the next time unit the states N_4, N_6, and N_7, provided an input 0 is seen. If the input is not 0, those states will not be activated by N_5. N_5 will not be active at the next time unit, unless it is activated by a transition from N_4, its only predecessor.

Figure 5(a) shows the before interpretation of the NFA for the bounce filter. Each transition in Fig. 4 is made on the input that corresponds to the state at the tail of the transition. Those states that correspond to the operands that could match the first input seen, namely N_1 and N_2, are designated initial. A transition into N_7 causes the associated output, OUT, to be raised.

In the *after* interpretation of NFA's, each state represents a situation where we have already seen the corresponding operand of the regular expression and are ready to recognize the symbol corresponding to any of its successor states. Figure 5(b) shows the after interpretation of the bounce filter. In general, each transition is labeled by the symbol corresponding to the state entered by the transition, while in the before interpretation the same transition is executed when the input matches the operand of the state that the transition leaves. In the after interpretation, the start is a state itself, with transitions to its successors on the appropriate inputs, as shown in Fig. 5(b). Finally, states associated with output symbols are no longer states in any useful sense. Rather, transitions into such states, shown in Fig. 5(b) as associated with ϵ, mean that the states from which such a transition is made are to raise that output signal as soon as they themselves are entered.

Comparison of Methods

Neither method is uniformly superior to the other. The advantage of the before method is that when we convert the NFA to a PLA, we need only one term per state (plus extra terms corresponding to transitions from the initial states when the start signal is raised). To see why, we have to understand that each NFA state is coded by turning on a subset of the feedback wires of the PLA; we shall discuss the method of selecting the representation of each state shortly. In the before interpretation, we need for each state N a term that checks

1. The code bits representing N were turned on at the previous time unit, and
2. The input corresponding to N is seen.

This term must turn on all the wires in the or-plane that are needed to represent any of the successors of state N. It may be unclear at the moment how one represents NFA states (which may be on simultaneously in various combinations), unambiguously by turning on sets of bits; we shall cover the method in the next section.

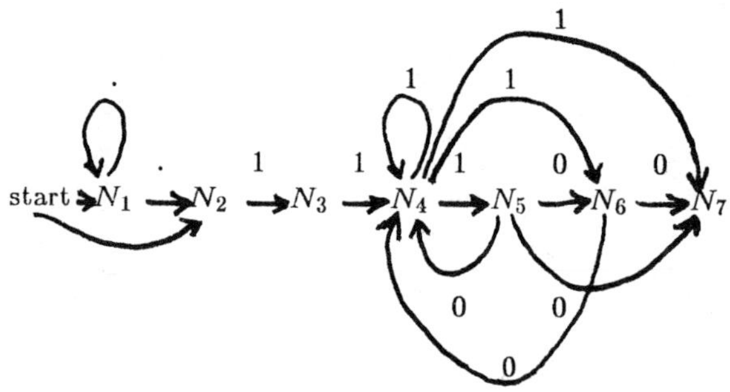

(a) Before method interpretation

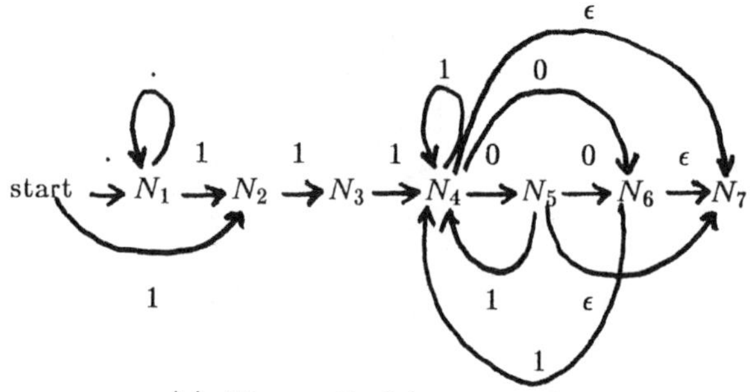

(b) After method interpretation

Fig. 5. Interpretations of bounce filter NFA.

Example 5: Figure 6(a) shows the PLA constructed from Fig. 5(a) if we code states N_1 through N_7 with a single wire each. (This turns out to be as good as we can do for the bounce filter NFA.) The left and right groups are the and- and or-planes. A 0 or 1 in the and-plane means that the wire represented by the column must be off or on, respectively, for the term corresponding to the row to be seen. A 2 in the and-plane means "don't care." In the or-plane, 1's represent taps, so each column is the logical "or" of the terms with 1's in that column. Note that N_7, the output, need not be fed back.

The first six rows are the terms for states N_1 through N_6. For example, row one says that if we are in N_1, and the input "dot" is seen (i.e., the input may be 0 or 1, represented by the 2, or "don't care, in the input column, X), we turn on N_1 and N_2 for the next time unit. Row two says that if state N_2 is on, and the input is 1, turn on N_3 for the next time unit.

The last two rows duplicate rows one and two, but with the start signal S replacing the wires for N_1 and N_2, respectively, in the term's conditions. Thus, these last two wires express the fact that N_1 and N_2 are on initially. □

S	X	N_1	N_2	N_3	N_4	N_5	N_6		N_1	N_2	N_3	N_4	N_5	N_6	N_7
2	2	1	2	2	2	2	2		1	1	0	0	0	0	0
2	1	2	1	2	2	2	2		0	0	1	0	0	0	0
2	1	2	2	1	2	2	2		0	0	0	1	0	0	0
2	1	2	2	2	1	2	2		0	0	0	1	1	1	1
2	0	2	2	2	2	1	2		0	0	0	1	0	1	1
2	0	2	2	2	2	2	1		0	0	0	1	0	0	1
1	2	2	2	2	2	2	2		1	1	0	0	0	0	0
1	1	2	2	2	2	2	2		0	0	1	0	0	0	0

(a) Before method PLA.

S	X	N_1	N_2	N_3	N_4	N_5	N_6		N_1	N_2	N_3	N_4	N_5	N_6	N_7
1	2	2	2	2	2	2	2		1	0	0	0	0	0	0
1	1	2	2	2	2	2	2		0	1	0	0	0	0	0
2	2	1	2	2	2	2	2		1	0	0	0	0	0	0
2	1	1	2	2	2	2	2		0	1	0	0	0	0	0
2	1	2	1	2	2	2	2		0	0	1	0	0	0	0
2	1	2	2	1	2	2	2		0	0	0	1	0	0	1
2	1	2	2	2	1	2	2		0	0	0	1	0	0	1
2	0	2	2	2	1	2	2		0	0	0	0	1	1	1
2	1	2	2	2	2	1	2		0	0	0	1	0	0	1
2	0	2	2	2	2	1	2		0	0	0	0	0	1	1
2	1	2	2	2	2	2	1		0	0	0	1	0	0	1

(b) After method PLA

Fig. 6. Before and after PLA's for bounce filter.

If we use the after interpretation of NFA's then we must create for each state N, and each symbol a labeling a transition out of that state, a term to check that the state is on and that the input is seen; if so, the successors of N on input a are turned on in the or-plane for the next time unit. If successor M has an ϵ-transition to an output signal, then those terms that turn on M in the or-plane also turn on that output wire.

Example 6: Figure 6(b) shows the after method PLA for the bounce filter. For example, rows three and four represent the transitions from N_1 on "dot" and 1 to states N_1 and N_2, respectively. Note that since N_4, N_5, and N_6 have ϵ-transitions to N_7 in Fig. 5(b), the last five rows of Fig. 6(b), which represent transitions into those states, also turn on the output wire, which is N_7. □

If we compare Examples 5 and 6 we might get the impression that the before method is superior to the after; each uses the same number of columns, and the after method uses more rows. While it is typical that the before method saves rows, it is often true that the after method saves columns because it allows better NFA state codes. It just happens that for the bounce filter, no better state code is possible with the after method.

V. Selecting NFA State Codes

We shall now take up the matter of how the compiler selects codes for states of an NFA. We first discuss the notion of conflicting states, that is, pairs of states that can be on at the same time. We show how the conflict information determines the permissible state codes and we discuss a particular method for finding legal codes.

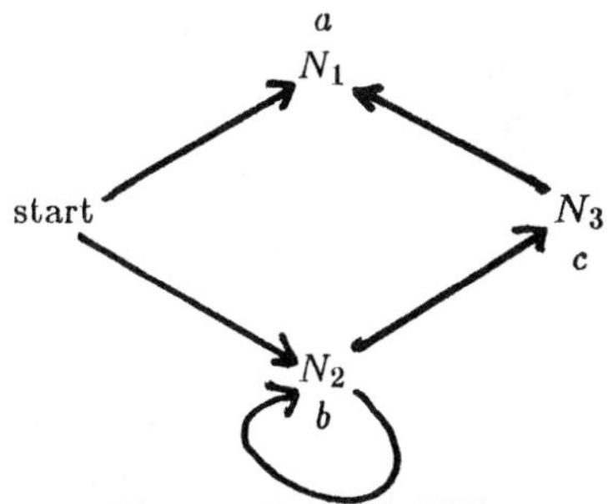

Fig. 7. Example NFA.

Conflicting Symbols and States

Before discussing conflicting states, we need to define *conflicting* input symbols. Symbols a and b conflict if both can be on at the same time, i.e., there is no wire x that is on in the definition of a and off in the definition of b, or vice versa.

If we are using the before interpretation of an NFA, then we use the following two rules to compute pairs of states that conflict. Rule (1) initializes the set of conflicts; we then add conflicting pairs by rule (2) until no more can be added.

1. Each state conflicts with itself. All initial states conflict with one another.
2. Suppose N and M are states that conflict, and they are associated with conflicting symbols a and b. (Note $N = M$ is allowed.) Then for each successor P of N and each successor Q of M, P and Q conflict.

There are similar rules that can be applied if we use the after interpretation of NFA's; they are:

1. Each state conflicts with itself. If N and M are initial states that are associated with conflicting symbols, then N and M conflict.
2. Suppose N and M conflict, P is a successor of N and Q is a successor of M. Also suppose that P is associated with symbol a and Q with b, and a and b conflict. Then P and Q conflict.

Note that the set of conflicts under the after interpretation is always a subset of those found under the before interpretation. It is this effect that explains why we often get better state codes with the after method.

Example 7: Suppose we declare symbols by

```
line x, y
symbol a(-x, -y), b(x), c(y)
```

Then b and c conflict. However, a and b do not conflict, because of wire x, and a and c do not conflict because of wire y.

Consider the NFA shown in Fig. 7. In the before interpretation, (N_1, N_2) is a conflicting pair because both are initial. Next, by rule (2) we find that (N_2, N_3) is conflicting because they are both successors of the conflicting "pair" (N_2, N_2). Then, we find (N_1, N_3) is a conflicting pair because they are, respectively, successors of the states N_3 and N_2, which are conflicting and associated with conflicting symbols.

In the after interpretation, rule (1) yields no nontrivial conflicting pairs, because the start states N_1 and N_2 are associated with nonconflicting symbols. However, the successors N_2 and N_3 of the trivial conflict between N_2 and itself are associated with conflicting symbols, so (N_2, N_3) is a nontrivial conflicting pair. There are no other nontrivial conflicts in the after interpretation. □

Legal Codes for NFA States

It is useful to think of the conflict information as a *conflict graph*, with states for nodes and edges between pairs of states that conflict. The compiler makes the simplifying assumption that any clique† in the conflict graph represents a set of states that can all be on at the same time. Surely if some set of states can all be on at once, then each pair of states in the set conflicts, but the converse is not true; there could be three different input sequences that lead, respectively, to states M and N, to M and P, and to N and P, yet no one input sequence turns on M, N, and P together.

Our decision to consider only conflicts between pairs, rather than all subsets, was so that the amount of information handled by the compiler would grow only quadratically with the regular expression size, not exponentially as it would if we considered conflicts among arbitrary sets of states. The assumption that all cliques represent conflicting sets is conservative, in the sense that it may prevent us from taking advantage of some good codes for states but will not lead us into an error where we design a malfunctioning PLA.

When choosing codes for states, we make the following hypothesis, which is oriented toward the PLA implementation of NFA's. We suppose that associated with each state is a vector of k 0's, 1's, and 2's, with 2 standing for "don't care." Let $C(N,i)$ be the ith position in the vector for state N. If state N is to be on, then we turn on the ith feedback bit whenever $C(N,i) = 1$. If $C(N,i)$ is 0 or 2, we do not turn on the ith bit because of N, although it could be turned on because of some other state.

In the and-plane, when we must recognize that we are in state N, perhaps among others, we examine the feedback bits. If $C(N,i) = 1$, we check that the ith feedback bit is 1; if $C(N,i) = 0$, we check that it is 0, and if $C(N,i) = 2$, we do not check the ith feedback bit. We must consider under what conditions the code C allows us to interpret all possible combinations of feedback bits correctly. There are two conditions that together ensure that we shall make the proper inferences from the feedback bits.

1. *When state N is on, we detect N.* If $C(N,i) = 2$, we do not check bit i, so there are no constraints on i as far as N is concerned. If $C(N,i) = 1$, and N is on, we know bit i will be turned on, so the test for N will be met at bit i. Finally, if $C(N,i) = 0$, then we must be assured that no other state M that conflicts with N, and could therefore be on at the same time as N, has $C(M,i) = 1$. For if there were such an M, then we could find bit i equal to 1, and fail to detect N even though it is on.
2. *If N is detected, then N is on.* Here, we must check that there is no (not necessarily maximal) clique $\{M_1, \ldots, M_r\}$ that does not contain N but can forge the code for N. That is, for no such clique is it the case that for all i, $1 \le i \le k$:
   a) If $C(N,i) = 0$, then for all j, $C(M_j,i) \ne 1$.
   b) If $C(N,i) = 1$, then there is some j for which $C(M_j,i) = 1$.

   If no clique satisfies (a) and (b), then the code C satisfies condition (2).

† A *clique* is a set of nodes with an edge between any two nodes in the set. The clique is *maximal* if no node outside the clique has an edge to each member of the clique.

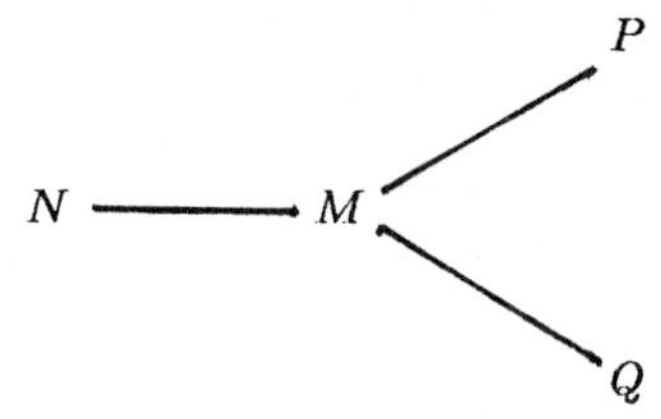

Fig. 8. Example conflict graph.

Example 8: Figure 8 shows a conflict graph. A possible 3-bit code for this set of states is:

	1	2	3
M	2	2	1
N	1	0	2
P	0	1	2
Q	1	1	2

To check condition (1) we have only to examine the 0's. For example, $C(N,2) = 0$, but there is no other state conflicting with N that has 1 in its second bit. That is, only M conflicts with N, but $C(M,2) = 2$.

We must also check condition (2). For example, it looks like N and P together could forge Q, but $\{N,P\}$ is not a clique, because N and P do not conflict. □

Simple Coding Methods

The first coding method implemented, which we call the *greedy* method, is to look for maximal *independent sets* in the conflict graph. An independent set is a set of nodes no two of which are connected by an edge. We may partition the nodes into maximal independent sets by starting with any node and adding nodes that do not conflict with any of the nodes previously added, until no more can be added. The result is one maximal independent set. We then remove the nodes of this set from the graph and start with another node to grow another independent set, and so on. This method has been used for similar purposes in several other works, such as Haskin [1980] and Nagle, Cloutier, and Parker [1982].

Having obtained a partition into independent sets, we may binary code the states in each set, omitting the all-zero code, so a set of m states can be coded with $\lceil \log_2(m+1) \rceil$ bits. Each of the independent sets uses bits of its own in the state code, and the code for each state has don't care's in the bits belonging to the independent sets other than its own. This coding method works because the only possible combinations of states have at most one from each independent set. The bits for each set tell us which, if any, state from that set is on. By not using the all-zero code for any state, we can detect the case where no member of an independent set is on.

Example 9: Consider the conflict graph of Fig. 9. We might start growing an independent set with state N. We may add P and Q, because neither conflicts with N or with each other. However, we cannot add M because it conflicts with N. Thus, we start a second independent set with M, and the partition of Fig. 8 is $\{\{N,P,Q\},\{M\}\}$.

The first of these sets requires two bits and the second one bit. The resulting state code is that given in Example

8. □

The Clique Compatability Class Method

A second coding method tried is described in Ullman [1982]. It gave better codes than the greedy method in some cases. We shall omit a description of that method and instead describe the most recent and most successful coding method. This approach partitions the states into *maximal clique compatibility classes* (MCCC's). An MCCC consists of a collection of cliques, such that no node of one clique is connected to a node of another clique by an edge of the conflict graph.

We grow MCCC's by starting with maximal independent sets and growing them by adjoining nodes whenever possible. We can adjoin node N to clique Q if N is adjacent to every node in Q but N is adjacent to no node in any of the other cliques in the MCCC. After partitioning the conflict graph into MCCC's, we code each MCCC in a manner to be described. We then find the overall state code by using a separate set of bits for each MCCC, just as we did for independent sets in the greedy algorithm. Each state has its code in the bits of its own MCCC and don't care's elsewhere.

Coding MCCC's

The basic idea is that we try to use the same code bits for as many cliques in the MCCC as we can. We start off coding each clique individually, and then try to combine the codes for different cliques. As the states of one clique can be on or off in any combination, there is nothing better than to use a one-hot code, i.e., use as many bits as there are states in the clique, with each state given a code consisting of a 1 in a unique position and 2's (don't care) elsewhere.

Then, we combine cliques, in pairs, until we have combined all pairs. The priorities for which pair to combine are as follows.

1. If there are two cliques or sets of cliques that have codes with the same number of bits, combine them. However, that number of bits must be at least two, i.e., we do not apply step (1) to a pair of cliques consisting of one state each. If there are two or more pairs that may be combined, pick a pair that have the shortest codes.
2. If no sets of cliques may be combined by rule (1), find the two cliques or sets of cliques with the shortest codes (including codes of one bit), and combine them. However, do not choose a singleton clique unless the number of singleton cliques is at least as great as the number of other sets of cliques remaining.

The combination of two sets of cliques may have the side effect of combining with them a singleton clique, as we shall see. That is the reason we do not wish to pick singleton cliques for combination unless there are so many that they cannot all be consumed in the combination of other cliques. In general, the rules for combining the codes for two sets of cliques are the following.

1. If one code is shorter than the other, pad the shorter out with leading 0's until they are of equal length.
2. Place a 0 in front of all the codes in one set and a 1 in front of all the codes of the other set.
3. If there are remaining cliques consisting of a single state, choose one and give it the code 10···0. It is easy to prove by induction on the number of cliques combined that we never produce an all-zero's code. Thus, 10···0 will not be the same as the code of any other state in the set.

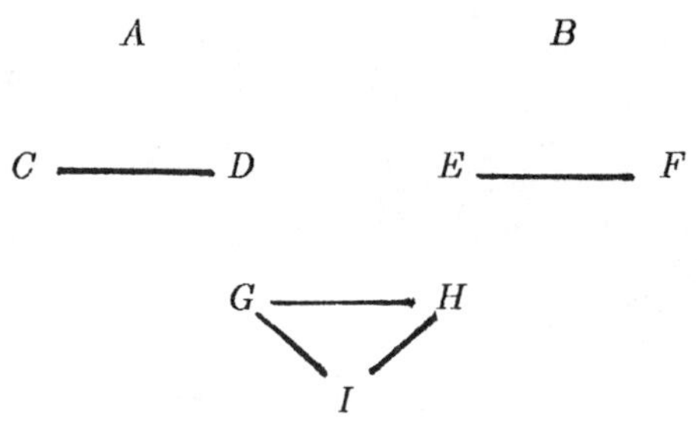

Fig. 9. An MCCC.

Example 10: Consider the MCCC shown in Fig. 9, where there are two singleton cliques, two doubletons, and one clique of size three. The initial codes for the cliques are the following.

A	1	E	21
		F	12
B	1		
		G	221
C	21	H	212
D	12	I	122

Note that the bits for different cliques do not yet bear any relation to each other, so there is nothing wrong with assigning the code 1 to both A and B, for example.

According to rule (1), our first task is to combine cliques CD and EF, since they are not singletons, but have the same code length. Let us say we put 0 in front of the codes for C and D and 1 in front of the codes for E and F. We then add a singleton, say B, giving it the code 100. The result is a set of five states with the following code:

B	100
C	021
D	012
E	121
F	112

Now, the set $BCDEF$ and the clique GHI have the same length code, so we combine them, consuming the singleton A in the process. The resulting code is:

A	1000	F	0112
B	0100	G	1221
C	0021	H	1212
D	0012	I	1122
E	0121		

□

VI. Evaluation of the Compiler

We believe that the principal reason to express designs in the regular expression language is its ability to accept descriptions of the patterns it must recognize and the responses it must make, in a flexible manner. For example, additional patterns may be added to the description of a controller, and the compiler will produce the necessary modifications without the user having to worry about the possibility of interactions between the new patterns and the old ones. Design systems based on deterministic automata do not have this robustness.

```
line input[2]
symbol
      zero(-input[1])
      symbol one(-input[2])
output MISMATCH
;
.*(
(. . . . (
. . (. (zero .* one + one .* zero) +
      (zero .* one + one .* zero) .)
+
(. (zero .* one + one .* zero) +
      (zero .* one + one .* zero) .) . .
)
+
(. . (. (zero .* one + one .* zero) +
      (zero .* one + one .* zero) .)
+
(. (zero .* one + one .* zero) +
      (zero .* one + one .* zero) .) . .
) . . . .
)
) MISMATCH
```

Fig. 10. Pattern matching regular expression.

However, it is also important that the design produced by the compiler be of good quality. We have run several test cases, and these indicate that the compiler performs well, in some cases better than obvious hand designs of PLA's. We shall mention some of these trials here.

The Bounce Filter

Here we do not do well; a PLA with about half the area of that of Fig. 6(a) can be designed.

The Traffic Light Controller

Using either the before or after method, the compiler comes up with essentially the same PLA as appears in Mead and Conway [1980]. The only difference is that the compiler introduces an initialization signal, which is not really needed for the perpetually running traffic light.

The Pattern Matcher

We alluded above to a regular expression with 72 operands and an 8,000,000 state deterministic automaton. This expression is shown in Fig. 10. The problem is to signal mismatches between the first eight symbols read and the last eight read. Input 1 is represented by turning *input*[1] on and *input*[2] off; input 0 is represented by the opposite, and don't care, which matches anything, is represented by turning both input wires on. The expression appears complicated, but the idea is that a mismatch between the first and last eight symbols can be expressed recursively as either a mismatch between the first four and the next-to-last four, or a mismatch between the second four and the last four; these mismatches can be expressed as mismatches of two pairs, and so on.

An obvious hand implementation of a PLA was attempted, using the straightforward idea that each of the symbols to be remembered, the first eight and the last seven, would be coded by two bits, for 0, 1, and don't care. This

```
line x[3]
symbol
      in0(x[1] -x[2])
      in1(x[2] -x[1])
      badin(x[1] x[2])
      ack(x[3])
      noin(-x[1] -x[2])
      noack(-x[3])
output OUTA, OUTB, OUTC, ERROR
state statea, stateb, statec
subexp somein = in0 + in1 + badin
subexp waitin = noin + badin
subexp allbut01 = ack + badin
;
waitin* (
      allbut01 ERROR +
      in0 statea +
      in1 stateb
      )
+ # statea: noack* OUTA (
            somein ERROR +
            ack waitin* (
                  allbut01 ERROR +
                  in0 stateb +
                  in1 statec
                  )
            )
+ # stateb: noack* OUTB (
            somein ERROR +
            ack waitin* (
                  allbut01 ERROR +
                  in0 statec +
                  in1 statea
                  )
            )
+ # statec: noack* OUTC (
            somein ERROR +
            ack waitin* (
                  allbut01 ERROR +
                  in0 statea +
                  in1 stateb
                  )
            )
```

Fig. 11. Regular expression for transmitter.

approach requires 16 feedback wires for the first eight inputs, 14 more to remember the most recent seven inputs, and four feedback wires to represent a counter that counts up to eight, to tell the PLA whether to remember its current input as one of the first eight symbols. As for terms of the PLA, we need 16 to feed back the first eight inputs, 14 more to feed back and shift the seven most recent inputs, 16 to load the first eight symbols originally, 32 to detect mismatches, and eight to implement the counter. Thus, the hand design uses 34 feedback wires and 86 terms.

In comparison, using the before method and the greedy state coder, we require 28 feedback wires and 62 terms. When we implemented the MCCC method of state coding, this number was reduced to 24 feedback wires, which is only one more than the theoretical minimum. This problem is an example where the before method yields significantly better

	And Cols.	Or Cols.	Terms	Area
Hand-Unoptimized	13	7	29	580
Hand-Optimized	13	7	23	460
Before Compiler	17	11	23	644
After Compiler	13	7	25	500

Fig. 12. PLA's for communication protocol.

results than the after method, as well as significantly better results than the obvious hand design.

A Communication Protocol

We designed the transmitter portion of the protocol for handling lost bits discussed in Aho, Ullman, and Yannakakis [1979]. Briefly, that transmitter has two inputs, $x[1]$ and $x[2]$ telling it to send a 0 or 1 down the channel, while $x[3]$ is another input wire, used as an acknowledgement signal. The protocol works because the transmitter sends one of three signals, *a*, *b*, and *c*, which we may view as arranged in a circle. To send a 0, the transmitter steps one around the circle and to send 1, it steps twice. States ***statea***, ***stateb***, and *statec* are the states in which the transmitter is trying to send *a*, *b*, and *c*, respectively. We assume that signals are not mutated, so any signal acknowledged must be the correct one. The regular expression program is shown in Fig. 11.

In Fig. 12 we see the results of hand and mechanical generation of PLA's for the transmitter. A straightforward hand design was optimized by GRY, to reduce the number of its terms. The results of the before and after methods are shown after optimization by GRY. We should be aware that when we count columns in the and-plane, we count one column if a signal is needed either true or complemented, but not both; if needed both ways we count two columns. This method of counting is realistic, provided the PLA generator used does not force all wires to become true and complemented pairs in the and-plane.

The "area" of the PLA in Fig. 12 is the product of rows and total columns, which is not precisely accurate, but serves to measure approximately the area actually used by the PLA.

The reader should note that the results shown in Fig. 12 for the compiler are the result of the greedy algorithm, not the MCCC algorithm. In both the before and after interpretations, the greedy method achieves the smallest possible number of feedback wires. The MCCC algorithm uses the same number of bits to code the states as the greedy algorithm does, i.e., the number of feedback wires is the same, but because of differences in the coding used, the number of terms after optimization was slightly larger when the MCCC coder was used.

References

Aho, A. V., J. D. Ullman, and M. Yannakakis [1979]. "Modeling communications protocols by automata," *Proc. Twentieth Annual ACM Symposium on the Theory of Computing*, pp. 267–273.

Floyd, R. W. and J. D. Ullman [1982]. "The compilation of regular expressions into integrated circuits," *J. ACM* **29**:2, pp. 603–622.

Brayton, R. K., G. D. Hachtel, L. A. Hemachandra, A. R. Newton, and A. L. M. Sangiovanni-Vincentelli [1982]. "A comparison of logic minimization strategies using EXPRESSO: an APL program package for partitioned logic minimizalization," *Proc. IEEE Intl. Conf. on Circuits and Computers.*

Haskin, R. L. [1980]. "Hardware for searching very large text databases," Ph. D. Thesis, Dept. of Computer Science, Univ. of Illinois, Urbana, Ill.

Hemachandra, L. A. [1982]. "GRY: a PLA minimizer," unpublished memorandum, Dept. of Computer Science, Stanford Univ., Stanford, CA.

Hennessy, J. L. [1981]. "SLIM: a simulation and implementation language for VLSI microcode," *LAMBDA*, April, 1981, pp. 20–28.

Hopcroft, J. E. and J. D. Ullman [1979]. *Introduction to automata theory, languages, and computation*, Addison-Wesley, Reading Mass.

Johnson, S. C. [1983]. "Code generation for silicon," *Proc. Tenth ACM Symposium on Principles of Programming Languages.*

Foster, M. J. and H.-T. Kung [1981]. "Recognize regular languages with programmable building blocks," in *VLSI-81* (J. P. Gray, ed.), Academic Press, New York, pp. 75–84.

Mead, C. A. and L. A. Conway [1980]. *Introduction to VLSI Systems*, Addison-Wesley, Reading Mass.

Nagle, A. W., R. Cloutier, and A. C. Parker [1982]. "Synthesis of hardware for the control of digital systems," *IEEE Trans. on Computer-Aided Design of Integrated Circuits and Systems* **CAD-1**:4, pp. 201–212.

Trickey, H. W. [1981]. "Good layouts for pattern recognizers," *IEEE Trans. on Computers* **C-31**:6, pp. 514–520.

Trickey, H. W. [1982]. "Using NFA's for hardware design," unpublished memorandum, Stanford Univ., Dept. of C. S.

Ullman, J. D. [1982]. "Combining state machines with regular expressions for automatic synthesis of VLSI circuits," STAN-CS-82-927, Computer Science Dept., Stanford Univ., Stanford, CA.

Ullman, J. D. [1983]. *Algorithmic Aspects of VLSI*, Computer Science Press, Rockville, Md.

Session 3-3c
Interactive Design

Chairman

H. Fleisher
IBM

IDL (INTERACTIVE DESIGN LANGUAGE) FEATURES AND PHILOSOPHY

Leon Maissel and Raymond L. Phoenix

IBM Corporation
D/C14, B/704, P. O. Box 390
Poughkeepsie, New York 12602

ABSTRACT

IDL is a hardware design and description language suitable for use in the VLSI environment. It allows hardware to be designed, documented, modified and synthesized. A brief overview of the language is presented sufficient to facilitate an understanding of the design examples that follow.

Hardware Description Languages

Many hardware description languages have been reported in the literature (Shiva 1979). Of these, a fair percentage are still in use either as aids for the documentation of designs or as teaching tools. Usually, such languages are also capable of driving simulators and this is an important aspect of the assistance they provide towards creating a design. A hardware description language cannot, however, be used for the efficient synthesis of the design from its own description alone. One approach that has been used successfully is to compare (in a formal manner) the language description with a conventional design done by hand, thus verifying that the latter is correct (Monachino 1982).

The reason that hardware description languages are not suitable for creating hardware is that designers tend to minimize portions of their designs as they go along. Furthermore, designers constantly make decisions that affect the final hardware but which have nothing to do with its function. As a result, any hardware that could be generated from the description would be too non-optimal to make the exercise worth it.

Hardware Design Languages

Recently, a new generation of hardware DESIGN languages has begun to appear. By definition such languages not only allow the hardware to be described but they also allow it to be synthesized directly from the description. As already mentioned, the state of the design art is such that many decisions still require human intervention. A case in point is the general question of space-time trade-offs. For example, should a particular logical operation be performed in one machine cycle or should it be spread out over several machine cycles in order to save hardware? Although both forms can be simulated a designer who is limited to a hardware description language has no way of knowing what penalty in hardware he is paying by opting to have something occur in only one cycle. Even more difficult for a description language is the closely related question of the number of logical levels in which a given hardware function should be implemented. This is one aspect of the general problem of serial decomposition (particularly going from two levels of logic to multiple levels).

A related problem is the question of encoding. Very often it is known that certain outputs are mutually exclusive (never come up at the same time), so it may be possible to represent them by using fewer real outputs and encoding the information that they carry. The cost is an extra level of delay since a decoder is needed to recreate the original outputs.

Another general problem is that of partitioning logic from one large piece into several smaller pieces. Machines can deal with particular cases extremely well but a general program without any associated human guidance is not yet available.

CH1935-6/83/0000/0667$01.00

Hardware Design Systems

The above comments imply that it is not sufficient if a hardware design language is limited to a language and a synthesizer. The hardware design language must be part of a hardware design SYSTEM so that after describing, simulating, and synthesizing the hardware, the user has available to him facilities for manipulating the result. This includes minimizing and partitioning (both in serial and in parallel), and performing such important tasks as state encoding (with optimum state assignment wherever possible), etc.

The hardware design system is thus one which performs various difficult tasks for the user but only under the user's guidance. In general the system offers the user a number of alternative paths which he can follow. The system will also give him expert assistance along these paths, but it is the user's decision as to where he wants to go with a particular design. Systems of this sort have come to be called expert systems and, for the foreseeable future, hardware design languages will work best in association with some type of expert design system rather than as stand-alone entities.

Interactive Design Language (IDL)

An example of the type of system we have been describing is IDL. Hardware described in IDL may be simulated under the control of one of three simulators. Which one is used depends on how close the description is to the hardware. Once the user is satisfied with the simulation, the system will synthesize the logic into hardware (as described below) and, finally, the system offers an assortment of facilities to allow that logic to be manipulated by the user for purposes of optimization.

The standard hardware format which IDL produces is two level logic (sum of products). Two level logic is immediately useful for defining the contents of a programmable logic array (PLA) but is non-optimum for many other types of logic. To this end IDL allows the user to represent his design as a collection of two level 'logic boxes' that drive one another. Programs are available which allow such descriptions to be translated into 'true' random logic or they could be used to personalize a linked PLA (a PLA which includes zero delay feedbacks).

In principle it would be useful if the entire design description could initially be represented as one piece of two level logic (Fleisher 1979). The problem associated with this is that many times such a representation is so large that it cannot be stored, even in very large systems. Even if it could be temporarily stored, its conversion to an optimum piece of hardware would require suitable serial decomposition algorithms which are not yet available. It is therefore very important for the user to be able to indicate (as part of the description) which parts of the design are to be implemented in more than two levels of logic.

IDL source code may be entered either as text or through a graphical assist format. This is done under the control of one of several available editors. A distinction is made between 'logic boxes' (the pieces of two level logic described above) and what IDL calls the external environment. The latter consists of items whose design is already well known and thus need not be redesigned by the user. Many of them are not well suited for representation as two level logic. Examples are registers, buses and random access memories. These are functionally referenced in the design and are therefore available during simulation of the logic boxes. They are not synthesized (by the IDL system), being inserted as hardware macros that have been predesigned.

IDL Constructs

The IDL user has a large number of language constructs available to him (Maissel 1982). The basic statement around which a design is built is the IF THEN ELSE statement. To improve readability it is possible to define groups of variables which can be referred to as single entities. This is analogous to records in software. In a number of cases, names of variables are created automatically for the user if such names are meaningful only to the system.

An assortment of functions which will automatically be synthesized as hardware are built into the language. These include relational operators, control signals, and counters. Truth tables (several types) are a standard part of the language, representing in many cases a shorthand format for the user. Truth functions (i.e. functions that describe truth tables) allow a user to delay the creation of a truth table while still making its function available to him. The actual truth table can be inserted in its place at a later time. Users are encouraged to indicate the presence of output don't-cares wherever possible. These help the system to perform effective minimization when the time comes.

Considerable attention has been given in IDL to dealing with state control and parallel processes. IDL allows a user to employ parallelism to any extent that he desires. In particular, the language facilitates the representation and simulation of parallel sequences (sometimes referred to as automata). These can give birth to other sequences or they can merge with existing sequences. One sequence can be influenced from inside another sequence, if desired. If the hardware can do it the language will allow it.

A related hardware feature which IDL handles is timing. While the basic notion of timing in IDL is built around individual delay times and (in certain cases) repetition rates for each piece of two level logic, other times not directly related to these can be included in the design and then simulated. The user is given many opportunities to decide such issues as whether to implement the logic in more than two levels, more than one machine cycle, etc. Whenever the user feels that he can do a better job of encoding a set of outputs than can the machine, he is encouraged to do so.

Once the user is satisfied with the simulation of his design he has available to him a significant number of tools for manipulating his logic. These include facilities for minimizing the number of product terms (IF THEN statements), factoring inputs to minimize wiring, minimizing the number of feedbacks used to represent a finite state machine, partitioning and/or merging logic boxes to create different sets of logic boxes.

Certain technology transformations are also available to the user. For example, a particular type of latch may be cheaper to implement in the technology of his choice, than a different latch-type originally called for in his or somebody else's high level design. In certain cases, transformation of logic implemented with one latch type to logic implemented using a different type can be performed automatically. Folding (of PLAs) (Hachtel 1982) is also available to the user as a built-in feature.

In general this manipulation of the logic is done under user control and he needs to have in mind a clear picture as to what the final implementation technology will be. The IDL output prior to the manipulation stage is technology independent. This is a very important aspect of the language since it allows a single IDL design to be implemented in many different technologies. Once the manipulation stage is over, there exist a number of programs which will take the result and convert it to the actual physical design needed to build the logic. These programs are outside the scope of IDL but interface with it to produce the desired result.

IDL has been used for the design of several pieces of hardware. Some of these are discussed in succeeding papers. Users report that IDL provides an improvement in design time by factors of 7 to 10.

References

H. Fleisher, L. I. Maissel, and D. L. Ostapko, "4th International Conference on CHDL," 1979, pp. 89-95.

G. D. Hachtel, A. R. Newton, and A. L. Sangiovanni-Vincentelli, "19th D. A. Conference," Las Vegas, 1982, pp. 147-155.

L. I. Maissel and D. L. Ostapko, "19th D. A. Conference," Las Vegas, 1982, pp. 193-201.

M. Monachino, "19th D. A. Conference," Las Vegas, 1982, pp. 83-90.

S. G. Shiva, "Proceedings IEEE," Vol. 67, pp. 1605-1615, 1979.

Acknowledgements

The authors would like to thank Dr. H. Fleisher for his encouragement and support through all phases of this work.

Top-Down Design Using IDL

K.R. Woodruff and P.S. Balasubramanian*

IBM Information Products Division, Boulder
Colorado 80302

* IBM Thomas J. Watson Research Center
Yorktown Heights, New York 10598

ABSTRACT

Interactive Design Language IDL[1] is a valuable tool for the top-down design approach. By using two design experiences as examples, we discuss some practical considerations and benefits that were learned while using an IDL based top-down approach. The two design problems discussed are an implementation of the Modified READ Image Data COMpression/DECompression (COMDEC) Algorithm[2] and design of a VLSI macro which is functionally equivalent to the Intel 8041 Universal Peripheral Interface.[3] Both examples are included because each experience covers different practical considerations and benefits.

COMDEC Design Experience

Problem Description

The Modified READ data compression algorithm is the CCITT two-dimensional standard for facimile transmission. It is a line-by-line coding method in which the position of the "changing pel" on the current line is coded with respect to the position of a corresponding reference pel located on either the current line or the reference line which lies immediately above the current line. A "changing pel" is defined as a pel whose "color" (i.e. black or white) is different from that of the previous pel along the same scan line.

The job of the compression system is to locate the position of the changing pels in the input image data, and to assign codewords to each of the "image features." An "image feature" is defined as the particular relationship between the changing pel and the reference pel. A few example image features defined by the algorithm are shown in Figure 1. The codewords are Modified Huffman codes which vary in length from 1 bit to as many as 13 bits. The image features having a higher probability of occurence are assigned shorter length codes while the lower probability features have the longer length codes.

The job of the decompression system is to decode the input codewords and then reconstruct the original image.

The compressor and decompressor implementations are similar; only the compressor will be discussed here. The compressor is implemented as a large finite state machine as illustrated in Figure 2. The ROS (Read-Only-Storage) contains next-state information and the codewords for the image features. The PLA serves two purposes. The first is to generate the address for the ROS from the ROS feedback and the input data/control lines. The second is to assemble the variable length codewords into bytes.

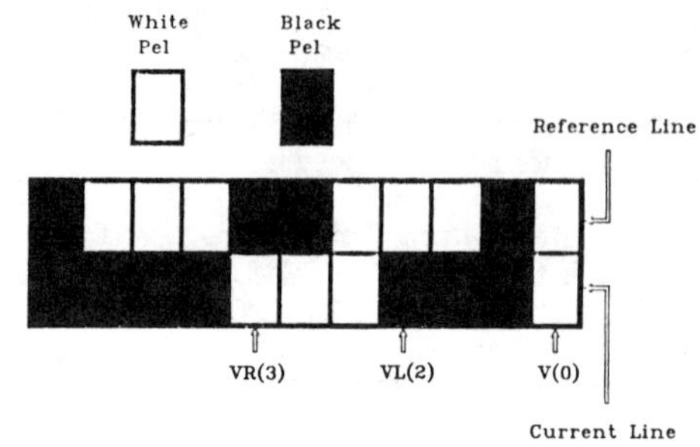

Vertical reference features: vertical right by 3 VR(3), vertical left by 2 VL(2), and vertical reference with 0 skew V(0)

Figure 1: Examples of Image Features

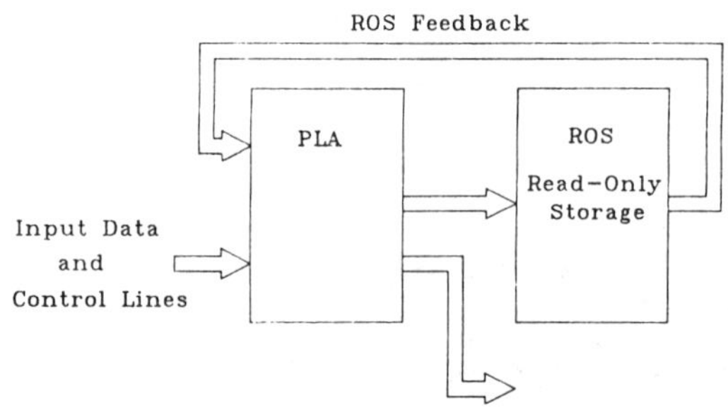

Figure 2: COMDEC Implementation Approach

COMDEC Design Approach

The first step in the design is to define the finite state machine description using a symbolic state table. Next we use special purpose APL programs (not part of IDL) to convert the symbolic description into the bit-level ROS patterns. Conversion from the symbolic state table to the bit-level description is analogous to generating machine language from an assembly language program. At this point, given the ROS description we describe the PLA function using IDL statements of the form

If (input conditions) THEN (output results)

CH1935-6/83/0000/0670$01.00 © 1983 IEEE

Design Verification

Initial simulation is accomplished in IDL, but due to the nature of the algorithm, large amounts of test image data are required to verify the design. Since it is impractical to continue simulation within IDL, we continue simulation using a high-speed software model of the system. The high-speed model is easily written since the PLA tables from IDL are readily converted into sum-of-products logic equations. The design methodology is summarized in Figure 3.

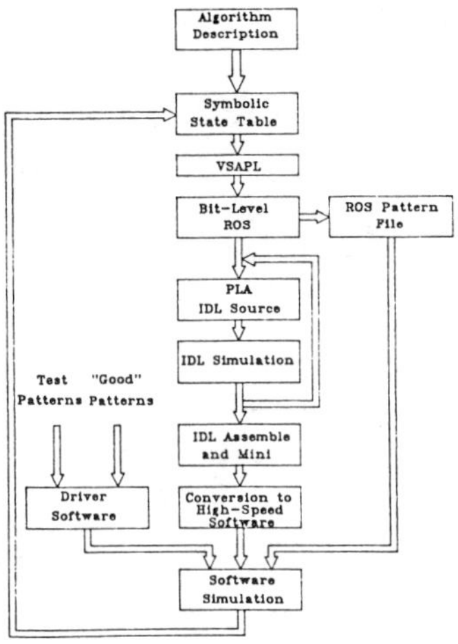

Figure 3: COMDEC Design Methodology

Physical Design

At this point in the design cycle we have a logic description that is verified to be correct. Using the minimization and partitioning tools in IDL, mapping the design into usable PLA personalities is automatic.

The COMDEC design was implemented two ways:

1. 2 PLA masterslice chips and 1 ROS chip
2. 1 custom chip and 1 ROS chip

First, the IDL-minimized PLA is partitioned to suit the technology. For the masterslice technology this yielded two PLAs and for the custom chip this yielded four PLAs. The PLA description files were input directly into the DA systems. For the masterslice PLAs the description files simply defined the personality of the existing chips. For the custom chip the PLA files defined macros which were placed and wired using the DA tools.

COMDEC Design: Conclusions

The design approach used to implement COMDEC proved to be very effective yielding an error free first pass design. Actually, two design passes were done but the first implementation was a different algorithm. Both design passes yielded error free hardware. The design was shown to be technology independent as it was easily mapped into two different technologies.

8041 Microcomputer Design Experience

Problem Description

The Intel 8041 is an intelligent peripheral controller that has the following main features: 8-bit CPU and data flow, 8-bit bus interface, 1K x 8 ROS, 64 x 8 RAM, Interval timer/event counter, and Two 8-bit I/O ports.

See Figure 4.

The objective is to implement a logical macro, functionally equivalent to the 8041, to be used as a VLSI "building block". The design must be mappable in a straightforward manner into different technologies.

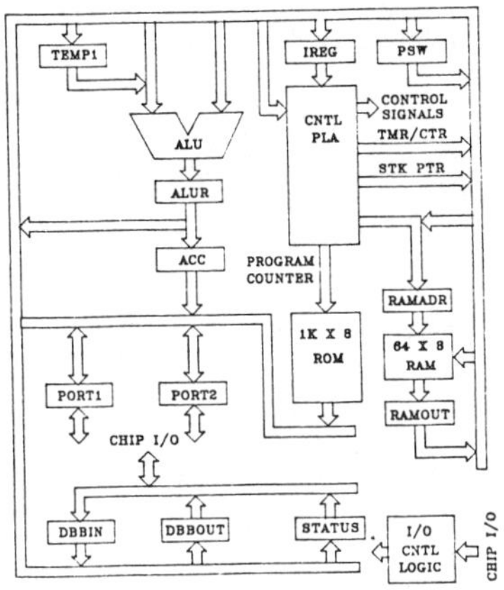

Figure 4: 8041 Data Flow

8041 Design and IDL Simulation

We started by using existing IDL source code for a microcomputer functionally similiar to the Intel 8048.[4] The 8048 has essentially the same CPU, ROS, RAM, Timer and I/O ports as the 8041; the only difference being the 8-bit peripheral interface. We chose to use the existing code to see if we could understand a design done by other designers well enough to save time. Typically, picking up someone elses design can take more time than starting over since the high-level design documentation often does not match the actual logic closely enough to be useful. We found that the IDL code for the 8048 was very easy to understand and after only a few days we were able to understand how to make the necessary changes.

Initially, we divided the design into the following parts: (1) Instruction Decode and Control, (2) Program Counter, (3) Stack Pointer, (4) Arithmetic-Logic-Unit (ALU), (5) Timer/Counter. (6) Data-flow elements: RAM, ROS, registers and busses.

After running the compile and assemble programs in IDL the resulting logic boxes were simulated using the multibox simulator in IDL. The multibox simulator applies stimulus from a test pattern file and records all signal changes in an all-events-trace (AET). After the simulation run, the AET is analyzed. This mode of operation proved to be tedious and time consuming so after some thought we saw that we could make the simulation interactive and higher-level by defining a logic box that monitored the system registers and busses and displayed them on the terminal CRT as the simulation was running. We also modified the control ROS model to take the instructions op-codes from the terminal keyboard instead of a file. In this way we were able to simulate each instruction at a higher level and to use the AET only when detailed analysis of a problem was required.

This "tailoring" of the IDL simulation is extremely easy since the logic boxes can be linked to APL. Any signal, register or bus in the system can be passed to an APL "black box" and once the information has been passed to the APL program it can be processed in any way, e.g., display it ,print it ,write to a file, test for conditions, etc. Also, any register or bus in the system can be forced with data from an APL "black box". Thus the simulation can be made to be as sophisticated as necessary to verify the design.

Mapping to the Hardware Level

After the design was verified, we began considering the problem of mapping it into vendor HMOS technology. This is when we learned many practical considerations that should be considered from the beginning.

The first problem is explicitly defining the data flow elements. IDL leaves the designer great freedom in defining the data flow. In fact, the data flow does not have to be "hard-wired" in the IDL code, rather it can be implied by the control description. When beginning a new design in IDL, if the data flow architecture is not initially specified, one must pay close attention to the data flow that is implied by the code. For example, data transfers from multiple sources to a single register implies a multiplexer or bus structure which take hardware to implement. For new problems, it may be necessary to iterate through the design in order to arrive at a data flow that is practical.

Using DA tools we coded the data flow elements (registers, bus drivers, ROS, RAM, and non-PLA control logic) as logic macros. Although this was a manual process, it was very straightforward. The PLAs were partitioned and minimized using IDL and input to the DA system as macros.

At this point we faced the problem of how to meet the peculiar requirements of the register latches, the RAM and ROS and how to handle the interface to asynchronous signals. The main problems encountered were: (1) The available latch macros were level-sensitive master slave latches requiring a multi-phase system clock. (2) The RAM and ROS required various clocks and control signals (3) System clocks needed to be defined and derived from the oscillator. (4) The asynchronous peripheral interface port had to be synchronized to the system clocks.

All of these problems come under the problem of system timing. Like the data flow problem it is better to have some definition of system clocking to start. But, since this is not always possible using a top-down design approach, the definition of system clocking may have to be delayed until later in the design cycle. The following example illustrates how the same IDL code could be implemented differently based on the clocks required by the particular data flow elements.

Consider the IDL expression:

LABEL: If (conditions) THEN BUS<-PORT1|ACC<-BUS

where PORT1 and ACC are registers
and BUS is a data bus

In this example we have a register to register transfer that takes place through a data bus. The subcycle bar "|" is used to indicate that the BUS must be given time to settle before the data is strobed into the ACC register. The subcycle bar "|" does not necessarily imply two separate clocks on the PLA output latches.

Given edge-triggered latches, one might implement the example above as shown in Figure 5. In this case, the PLA outputs are latched control signals that are used directly to gate the data to the bus and to clock the ACC register. The setup time for the ACC register latches is provided by the skew introduced by the system clocks PLA CLK 1 and PLA CLK 2.

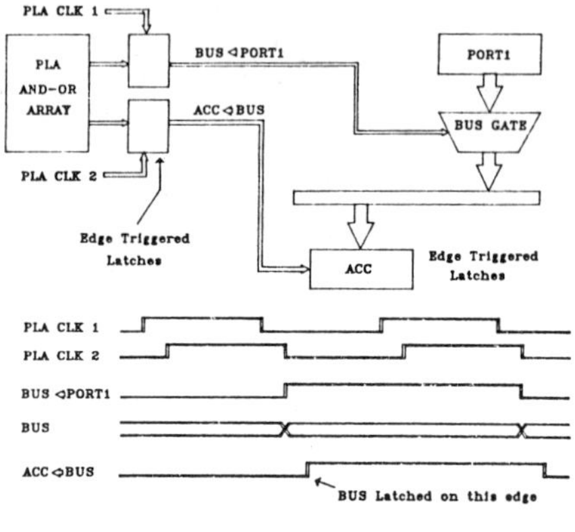

Figure 5: Edge-Triggered Implementation

Now, given registers and PLA latches that must be implemented using LSSD SRLs how do we clock the system? Given the level-sensitive latches, one possible implementation is shown in Figure 6. We arrive at Figure 6 by first assuming a PLA/SRL network with a single pair of C1/C2 clocks. Next we see that this assumption results in two control signals with no skew. We solve the bus setup time problem by using the control signal "ACC<-BUS" as a gate instead of a clock.

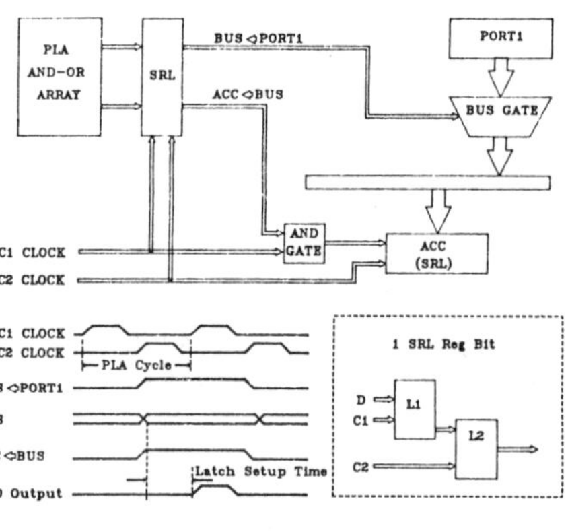

Figure 6: Level-Sensitive Implementation

It is clear from this example that considerable differences in the actual implementation could exist as the design is mapped into different technologies. While this is true, however, the IDL description that has been well simulated provides a point from which the actual implementation flows in a straightforward manner. Furthermore, given a specific technology with its unique timing requirements or limitations, the bridge from the IDL description to the complete gate-level implementation could be automated.

After the bridge from IDL to the hardware-level implementation has been crossed, more simulation must be performed. For the 8041 logic macro example, we simulated the system at the gate-level using estimated delays. Very few logic errors were found in this simulation. (The logic errors found could have been found easily in the IDL simulation had we spent more time.) Most of the problems found by the gate-level simulation and other checking programs were due to electrical problems caused by such things as excessive fanout or loading. These problems could have been avoided if we had taken fanout and loading rules into account when we mapped the data flow logic. Instead we chose to map blindly and fix the problems indicated by the checking programs.

A summary of the design methodology is illustrated in Figure 7.

8041 Design: Conclusions

IDL appears to be an excellent way to design logic macros that are to be mapped into a variety of technologies. Even though a considerable amount of work exists in going from the IDL description to a specific technology implementation, the work is straightforward and has the potential of being automated. Since we arrived at our 8041 design by modifying existing IDL source code for a similiar microcomputer that was generated years earlier at another IBM location, we observed the "self-documenting" nature of IDL system descriptions.

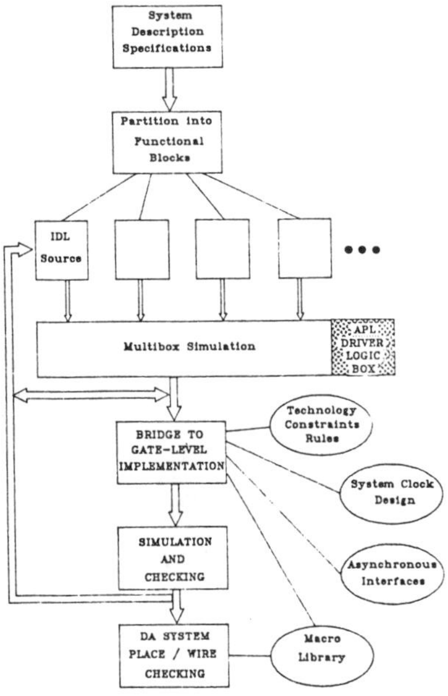

Figure 7: 8041 Macro Design Methodology

General Conclusions

As we continue to have design experiences like these it is evident that top-down design using IDL is a very effective approach, especially when it is applied to the idea of logical macros for VLSI design. As shown above, the designers do not have to select the technology of the DA system early in the design cycle. In addition, mapping into different technologies is straightforward and in some cases automatic. In the case of COMDEC, we see that the high-level IDL design makes it possible to do an error free first pass design. In both cases, the IDL source code provides design documentation that makes understanding the design very easy. We also see that by using the PLA macro approach, physical design is simplified and accurate physical sizings early in the design cycle are possible for selecting the best technology. Overall, we see the IDL design approach as one that reduces development time and greatly increases productivity.

Acknowledgments

The COMDEC state machines and APL programs were the work of Randall Alexander (IBM Charlotte Lab) ,Gerald Goertzel and Joan Mitchell (IBM T. J. Watson Research Center). The COMDEC high-speed software simulation work was done by Art Roberts (IBM Boulder Lab). The IDL source code for the 8048 processor was provided by Tak Ng (IBM Kingston Lab). Many thanks also to all the authors of IDL especially to Leon Maissel and Ray Phoenix for their great support of both designs.

References

1. L.I.Maissel and D.L. Ostapko, "Interactive Design Language: A Unified Approach to Hardware Simulation, Synthesis and Documentation", Proc. ACM IEEE 19th Design Automation Conference, pp. 193-201 (1982).

2. CCITT Recommendation "Standardization of Group 3 Facsimile Apparatus for Document Transmission", Vol. VII, Fascicle VII.2, Rec T.4.

3. UPI-41A User's Manual, Santa Clara, Calif.: Intel Corp., 1980.

4. MCS-48 Family of Single Chip Microcomputers User's Manual, Santa Clara, Calif.: Intel Corp., 1978.

MULTIPLE IMPLEMENTATIONS OF A MICROPROCESSOR FROM A SINGLE HIGH-LEVEL DESIGN

Mark W. Brown* and M. Jay Kimmel**

*IBM System Products Division
Highway 52 and 37th Street Northwest
Rochester, Minnesota 55901

**IBM T. J. Watson Research Laboratory
P.O. Box 218
Yorktown, New York 10598

Abstract

This paper discusses the use of the Interactive Design Language (IDL) for the design of three distinctly different physical implementations of a control microprocessor. The high-level, implementation independent design and the simulation were done once in IDL. The paper discusses the various ways that this one high-level design was transformed (more or less routinely) into an IBM NMOS custom version, into an IBM NMOS masterslice version, and into a TTL lab version driven by IBM masterslice PLAs.

Goals for High-Level Language Modeling

Several characteristics are required of useful high-level hardware design languages such as IDL. First is a measure of technology independence. It is very useful to be able to change the technology of a design implementation without having to redefine the high-level language description. On the other hand, the loss of a reasonable solution due to automatic but inefficient compilation is also to be avoided. IDL solves this problem in two major ways. First the compilation and assembly of the high-level source language description produces an intermediate "universal" form, the sum of products form of a PLA. Subsequent programs, not part of IDL, either directly personalize an existing PLA or map a PLA into a given technology. However, the control portion of the design constitutes only a small part of the total gate count of a typical design. The portions not covered by this type of logic, such as ALUs, RAMs, multiplexors, registers, etc., are best designed as macros. These parts are modeled for simulation in IDL through the facility of the "black box," a functionally described portion of logic for which there is no compilation or assembly facility. The black box approach is IDL's second major solution to the problem of technology independence, essentially by allowing technology-dependent solutions only through the prescribed black box interfaces. The functional description is in APL, the language in which the IDL system is written. These APL black boxes allow total simulation without dictating or implying the underlying implementation. This means that the design of each implementation-dependent black box portion is reduced to a separate design exercise and can be produced in parallel. Hence, the black box facility of IDL encourages an hierarchical design approach. (See the paper by Maissel and Ostapko[1] and papers being presented in this session for more details on IDL.)

Some Aspects of IDL

The global model assumed by an IDL design, written in IDL source code, is that of a finite state machine (FSM), called the logic box by IDL. Independent inputs come from the simulation driver. Multiple logic boxes are possible in one version of the IDL simulator. The outputs are combinatorially generated. They consist of the independent outputs, which go to the simulator, and the next state variables, which are fed back as dependent inputs through a single clock time delay. This model is also exactly that of a PLA that has latched internal feedback paths. It is possible that an IDL logic box consists only of logic which will be totally implemented in a PLA with output latches. In this case, the IDL compiler and assembler can synthesize the PLA personalization for a variety of physical PLAs. On the other hand, the black boxes with associated registers and memory elements can also be used to functionally describe some portion of the logic box, most often the data flow portions. Hence, the IDL programmer controls which portions of the IDL design will be handled by IDL automatically with compilation to some target PLA technology and which portion will be treated functionally with the implementation to be defined in another way and merged with the whole-hardware solution later.

The primary IDL source language statement syntax is:

Label: IF alpha THEN beta ELSE gamma

In each clock or simulation cycle, all lines whose labels are active are executed in parallel. Alpha designates a combinatorial expression of PLA inputs. For each active line whose alpha expression is satisfied, the beta expression is enabled, otherwise the gamma expression is enabled. Alpha, beta, and gamma expressions in any combination may be null. For all inactive lines, neither the beta nor gamma expressions will be enabled. The form of both the beta and gamma expressions is a time sequence of one or more subexpressions. Their syntax allows both parallel and sequential activation of subexpressions. Each subexpression can be one of several forms. It may be activation of a black box, specification of an output variable, or specification of the next label processing. For example, the subexpression OUT1 specifies output OUT1 is turned on. The subexpression ZREG←AREG indicates a register transfer, this function being an IDL supplied black box. BREG←BBOX ZREG specifies the activation of the black box named BBOX whose functional description is supplied by the IDL programmer. Finally, the subexpression →NEWLABEL is one possible next-state specification for the label activation and deactivation logic.

The simulation order is as follows: All output variables are set by the PLA first. Then black box activations occur in the designated sequential subcycles. Finally, the parallel execution of all next-label subexpressions from all selected beta and gamma expressions determine the next state of the finite state machine. The remaining IDL program constructs generally supply additional information about the structure of the hardware environment in which these label expressions are executed. Examples are DIMension statements which supply the widths of registers and the sizes of memories, and UDL statements which supply the application equations for user defined lataches of the PLA in tabular form, etc. The facility also exists in IDL to include as the alpha statement a truth table form which itself may have been generated as an IDL PLA.

All these facilities of IDL work together to provide a clear way to define a multitude of operations, which may or may not occur in parallel. At the same time, these facilities provide an environment for defining the function of the operations via the black box notation without binding the system to any predetermined implementation for such black boxes. Hence, it is easy to see how several hardware versions of the same IDL design can be conceptually described in one program. Each technology will have its own data file for the description of a macro function. IDL will have only one APL functional description of the same macro as a black box. The internal structure of the various technology data files are unknown and unnecessary to the functioning of the IDL simulation or to the functioning of the IDL compilation of the PLA control logic which provides the state sequencing logic and the control inputs to the black box.

The Practical Application of IDL

It should now be apparent how one IDL specification for the operation of a microprocessor could be detailed enough to provide automatic generation of all control logic, but still general enough to allow diverse physical implementations. As a matter of history, the processor was first described in IDL for implementation on a NMOS custom chip. Some time after the custom version was completely designed and simulated, a separate project implemented a nearly equivalent processor, first in PHILO (an NMOS masterslice technology), and then in TTL with separate personalized PLAs for easy and fast build in the laboratory. A few changes were required in the instruction set, but this only required small modifications to the IDL description. The function of each of the principal macros (black boxes) was unchanged. Each were implemented in the desired technology.

The Control Microprocessor

Brown and Dix[2] extensively discussed the data flow, instruction set, etc., of this processor. The data path widths are 16 bits in the instruction interpretation portion and 8 bits in the ALU path. The approximate complexity is 5000 equivalent three-way NOR gates. Typical black boxes include the ALU, the local store registers, and the incrementer/adder. The approximate size of the control PLA is 120 product terms, 50 inputs, and 60 outputs with slight variations between implementations.

Masterslice and Custom Design Methodologies

The design methodologies of the masterslice and custom implementations are very similar. The differences between the two will be discussed when they occur. The masterslice used is known within IBM as PHILO[3]. The basic design system used for both technologies is IBM's Engineering Design System (EDS). To obtain a high-level design system, IDL output is coupled to EDS. The translation to EDS is accomplished in two ways (Figure 1). First, the two-level logic array output from IDL is logically minimized and converted into a personality data format that EDS programs can interpret. As part of the macro generation capability of EDS, the array is first folded to physically minimize it. The physical layout and logical description then are created and stored in the main design-data libraries. Second, the black box portions of the IDL design that contain the data flow are designed in EDS's hardware description language (HDL) with the aid of a graphic data-entry program. It is at this point that the implementation of the black boxes actually occurs. In actual practice, this portion of the design is done in parallel with the IDL design, and thus any changes due to the implementation must be reflected into the black box functional descriptions. The HDL points to the proper technology-dependent circuits and macros in the design-data libraries. Here the masterslice design system has a fixed set of circuits, whereas the custom design system must have the circuits and macros designed. The PLA macro HDL and the HDL of the data flow are then linked and stored in the specific design-data file for a particular part.

The next step of the design is to perform preliminary checks to ensure that the logic described by the HDL conforms to the technology rules. After the checking has provided a technology correct HDL, low-level logic simulation can take place to verify the translation of IDL black boxes into the HDL. If any errors occur, the HDL can be edited, rechecked, and simulated again. Once the total design has been verified, physical design and test pattern generation can take place in parallel.

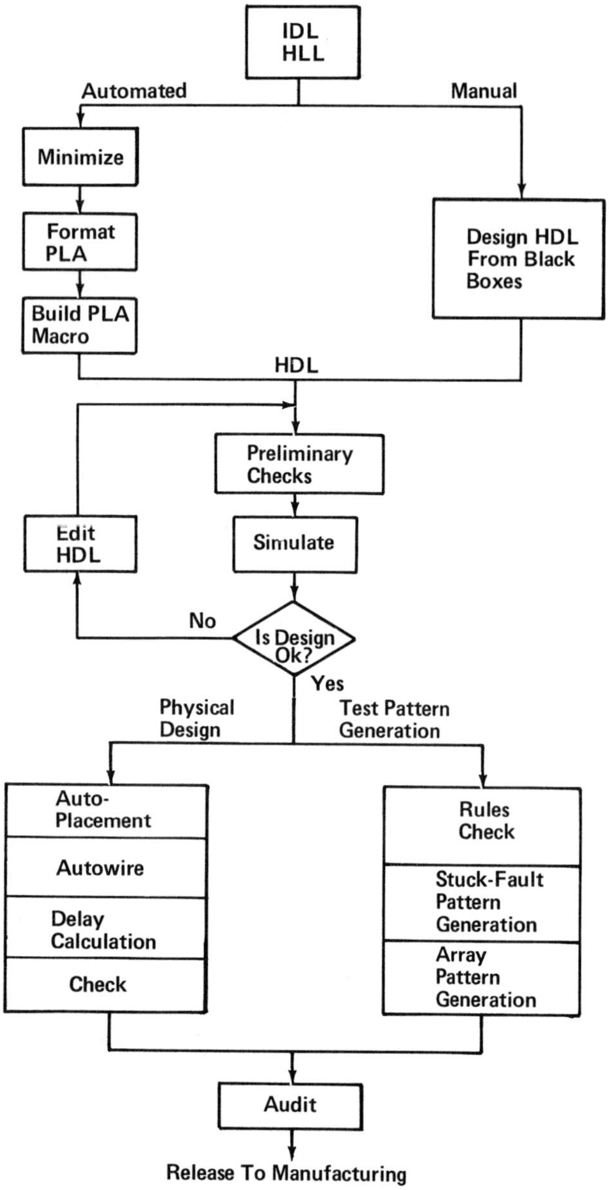

Figure 1. Masterslice/Custom Design Methodology

Physical design includes placement of the circuits, wiring, and delay calculation/ optimization. The predesigned circuits are automatically placed on the chip while the macros are manually placed. All wiring is automatic with overflows manually embedded. The delay calculator calculates the delays of each logic block and reduces the delay of a block by increasing its power if the user-requested performance is not met. This is only done for the masterslice technology since the custom circuits have fixed power levels. Checking functions at this point in the design cycle include physical-to-logical checks and topology or ground rule checks for the particular technology. Since most of the circuits of the masterslice design system are predesigned, the checking procedures have a high probability of success while the custom design can easily run into problems and may require thorough checking and possible redesign.

The test pattern generation includes design rules checking, automatic stuck fault pattern generation, and automatic RAM array pattern generation. The design rules checker ensures that the logic and its clocking are designed so that stuck fault testing can be used. The automatic stuck fault pattern generator and the automatic RAM array pattern generator create tester-independent instructions which are then converted to specific tester instructions at the manufacturing site. After the physical and test information is created, an audit is run. This program determines that the right release of the programs and technology definitions are used, and that all the programs have run successfully and in the right order. Success of this procedure proves that the part is acceptable for manufacturing.

Short-Turnaround-Time Methodology

The IDL to EDS conversion of the short-turnaround-time methodology is similar to the masterslice/custom methodology (Figure 2). First, the two-level logic array of IDL is logically minimized. Then a special partitioning and folding program is run to create the personality data for each of the special PLAs. This personality data is then used as input into specialized programs that create the physical and logical descriptions that are stored in the design data libraries. Manufacturing uses this information to build and test the short-turnaround-time PLAs. The black boxes that contain the data flow are mapped into standard off-the-shelf TTL modules via the HDL. The HDL of the PLAs and the HDL of the data flow are linked and stored in the specific design-data file for that particular part. The physical design of this methodology includes automatic placement of all modules on printed circuit (PC) cards and the creation of a wirewrap data file to be used to wirewrap the cards in manufacturing.

Conclusions

If one makes the analogy between the process of program compilation and the process of conceptual logic design followed by specific hardware implementation, it should be easy to see that the black box facility of IDL provides the function of delayed binding, only requiring specific technology details to be worked out after the high-level logic design, data flow, and macro specification have been specified and simulated. Add to this the rich nature of IDL in specifying parallel processes, and we see how IDL is a very useful high level, technology independent, hardware design language.

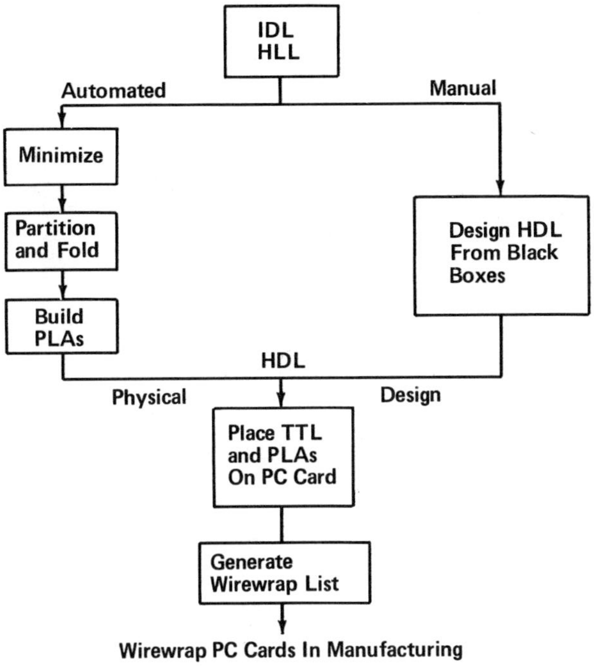

Figure 2. Short-Turnaround-Time Methodology

Acknowledgements

The authors thank Dr. Leon Maissel and Ray Phoenix for their help and encouragement in using IDL. They also thank Bob Dubke and Max Brown for their close cooperation on the projects described in this paper.

References

1. L. I. Maissel and D. L. Ostapko, "Interactive Design Language: A Unified Approach to Hardware Simulation, Synthesis and Documentation," 19th Design Automation Conference, Las Vegas, June 14-16, 1982, pp. 193-201.
2. G. L. Dix and M. D. Brown, "Common Chip for Use in Disk and Diskette Controllers," IBM Journal of Research and Development, Vol. 26, Number 4, July 1982, pp. 440-445.
3. R. L. Donze, J. J. Sanders, J. Jenkens, G. Sporzynski, "PHILO--A VLSI Design System," proceedings of the 19th Design Automation Conference, June 14-16, 1982, pp. 163-169.

Design of a Memory Controller for the MIT Tagged Token Dataflow Machine

Steven K. Heller Arvind

Laboratory for Computer Science Massachusetts Institute of Technology

ABSTRACT

I-structure storage is a new type of random access storage for multi-processor systems. By employing extra *status bits*, data slots can be tagged as *data-present* or *data-absent*, allowing reads which precede writes to be deferred until the data arrives. In order to implement an I-structure Storage in hardware, a sophisticated controller is required. A functional overview of the I-Structure Storage Controller (ISC) is presented. Based on our experience developing the ISC in IDL (Interactive Design Language), we discuss certain hardware design language issues which will challenge the hardware design languages of the future.

Keywords: Hardware design languages, IDL, Multiprocessors, I-Structure Storage, Storage Controller, Dataflow.

1 Introduction

The IDL design system [5], [6] allows a designer to design, simulate, and modify a VLSI design. The system includes facilities for specifying, testing, and manipulating a design as well as a path directly to silicon through PLA personality. The IDL system is particularly well suited for designing Finite State Machines (FSMs). The IDL system was used to design a controller for a new type of storage for multiprocessor architectures. This storage is known as the I-structure Storage or simply I-store and constitutes an integral part of the Tagged-token Data Flow Machine [1], [2], being developed in the MIT Laboratory for Computer Science by the Functional Languages and Architecture group. The machine is currently being simulated in Pascal, and the work on emulating a 64 processor version of the machine on 64 Lisp machines (Symbolics 3600's) has already begun. Depending upon the success of the emulated machine, a custom VLSI version of the machine may be built after 2 to 3 years.

In this paper we will discuss our experience in developing the I-structure Storage Controller (ISC) in IDL. It had become clear early on that ISC was too complex to be designed at the logic level directly. Since many important hardware parameters (*e.g.*, RAM speed, controller technology) could not be specified when the design was undertaken, we decided to separate the control logic of ISC from the rest of the design with the understanding that the control logic (PLA) may have to be suitably modified as other design decisions became firm. Because of our inexperience in hardware design, we underestimated the complexity of the control logic program. The fact that original specification of the I-store operations were incomplete and in some cases inconsistent also contributed to enlarging the design. We think that one person (Steve Heller in this case) could not have completed this design without using IDL. We are convinced that IDL (or similar design aids) are indispensable if ISC or the rest of the machine design is to be committed to silicon.

Specification of ISC in IDL took approximately 1500 lines of code. The ISC design did however, stress the IDL system to the limit, and in order to facilitate even larger designs, another generation of behavioral design systems is needed. We will describe several design issues which provide a difficult challenge for the hardware design languages of the future. On one hand the level of abstraction in digital design has to be raised if design of ISC magnitude have to be done routinely by 1 or 2 designers, and on the other hand a greater control over the hardware generated must be provided if the designs are to be of practical value.

2 I-Structure Storage and Associated Operations

A data structure known as I-structure has been proposed by Arvind and Thomas to efficiently manipulate arrays in functional languages [3]. I-store and associated operations form an implementation model for I-structures (see [4] for more details).

2.1 I-Structures

An I-structure can be viewed as an array of slots, where slots can be filled in any order, and each slot is filled at most once. The array name can be treated as the address of the first slot (*i.e.*, a descriptor), not unlike the name of an array in Fortran. In fact an I-structure descriptor (ISD) also contains information about the length (number of elements), width (size of an element), and the type of the structure. However, unlike a Fortran array, the semantics of I-structures permit passing the descriptor of an array to other parts of the code even before all the slots have been filled. (This feature increases parallelism in a program, especially when the program is decomposed into a producer and a consumer of a data structure.) A read request (*i.e.*, x[i]) to a slot is processed whenever the slot is filled. If the slot is never filled, then the read request never gets satisfied.

There are two kinds of I-structures: *uniform* and *mixed*. Each element of a *uniform* structure has the same type and hence length. Since the type can be easily stored in the ISD, type information does not need to be stored with each entry. Uniform I-structures are stored in a format called *u-fix* (untyped fixed-length); they are written using a **store-u-fix* operation and read using a **fetch-notype-stored* operation. Since no type information is stored with each entry, a type must be provided when reading a structure of uniform type.

In a *mixed* I-structure, each element can have a different type and length. We have developed two formats for storing mixed I-structures. The *t-fix* (typed fixed-length) format allocates to each element as much storage as the largest element requires. The *t-var* (typed variable-length) format allocates a minimum amount of space to each element, and if a value doesn't fit, an internal pointer points to a larger slot where the value has actually been stored. This use of invisible pointers was inspired by LISP machines. The type is stored along with each element in the t-fix and t-var formats. The **store-t-fix* and **store-t-var* operations are used to write into t-fix and t-var formats respectively, and both formats can be read using the **fetch-type-stored* operation.

2.2 I-Store Implementation

The implementation of an I-store is fundamentally different from that of a conventional storage in two ways. Locations are tagged as *data-present* or *data-not-present*; and if data is not present and a read occurs, the read is *deferred* until the data arrives. The information required in order to monitor the "status" of a location is stored in a hidden *status* field.

In our model, a destination (48 bits) is associated with each **fetch* instruction, and indicates the computation to which the fetched data is to be sent. If a **fetch* precedes the **store* operation for a piece of data, the ISC remembers the destination carried by the **fetch*. The destination is used by the token switching network to route tokens to their "destinations." Other multiprocessor systems that wish to use an I-store might simply combine a processor number with a locally unique computation identifier to form a destination. A computation identifier is needed since processors should expect to send many requests for data before the first response is received. (The use of 48 bits to name a destination computation may seem excessive, but it is needed in our machine because each instruction constitutes a separate computation and many copies of the same instruction may execute concurrently.) The ISC will assist storage managers by providing additional status information and accommodating variable length data. While the presence bit and the deferred bit are the only status bits required for a minimal implementation of I-structures, additional status information is used to assure consistency against possible compiler bugs and physical hardware boundaries.

The current ISC design assumes 288K bytes of Random Access Memory (RAM), registers, a Finite State Machine (FSM), and some random logic. The I-structure Memory (ISM) is divided between two physical memories: a 256K byte data storage, and a 32K byte status storage. The data storage has 16 address bits and has a 32 bit wide word. The status storage also has 16 address bits, but has a 4 bit wide word. Together they store 64K words of 32 bits each, with 4 bits of hidden status associated with each word. The 64K words are divided into two sections: the *I-store* and the *free list*. The location at which the I-store ends and the free list begins is programmable and is set when the system is initialized. Using the **initialize* operation, the free list section is initialized as a linked list of double words (free-links). The status field of the second word is used to indicate a cell at the end of the list, and the second half of the second word is used to point to the next free-link. The **initialize* operation also sets the status of each word in the I-store section to *not-allocated*.

2.3 Normal I-Store Operations

Normal I-store operations can be understood with the help of the status transition diagram for storage cells given in Figure 1. For example, suppose an array of n elements where each element is 2 words long is allocated starting with location 0. This operation will cause the status of words 0,2,4,...,2(n-1) to be changed to *empty-nowait* (see Figures 1 and 2a). The status of words 1,3,5,...,2n-1 will be changed to *middle*. If the machine and the compiler are working correctly there is no possibility of generating the address of a word whose status is marked as *middle*. When a **store* operation is performed, say on the first element (*i.e.* words 0 and 1) of the array, the status of word 0 is changed to *untyped-data* if the operation is **store-u-fix*, and to *typed-data* if the operation is **store-t-fixed* (see Figure 2b). The *invisible* status is used only when a **store-t-var* operation attempts to store data which is longer than 1 word into a slot that is one word long. In this case, data is stored indirectly on the free list.

If a **fetch* operation follows the corresponding **store*, the data structure remains intact, and the data is sent to the destination provided by the **fetch*. If the **fetch* precedes the **store* however, the destination is remembered in a cell from the free list, a pointer to the cell is left behind, and the status bits are set to *deferred* (see Figure 2c). When the data arrives on the **store* operation, the data is sent to all deferred destinations, the storage occupied by the deferred destinations is put back on the free list, and the status of the data word is changed to either *untyped-data*, *typed-data*, or *invisible* as shown in Figure 2b.

Thus normal operations on I-store can be summarized as follows. The **initialize* operation reinitializes the ISM using two parameters: the boundary between the data section and the free list section, and the size at which the free list should spawn a warning message; the **allocate* operation is used to prepare an array of locations for stores and fetches; the **clear* operation is used to recover a sequence of locations, and return them to the unallocated state; the **store-t-fix*, **store-t-var*, and **store-u-fix* operations are used to store data of types t-fix, t-var, and u-fix respectively; the **fetch-type-stored* operation retrieves data of type t-fix or t-var (data stored along with its type); and the **fetch-notype-stored* operation retrieves data of type u-fix (data stored without its type); the **reset* operation is used to recover a word from an error state; and the **allocate-free-space* operation increases the size of the free list by allocating a space in the I-store section for free cells. This space can be returned only by reinitializing the entire ISM.

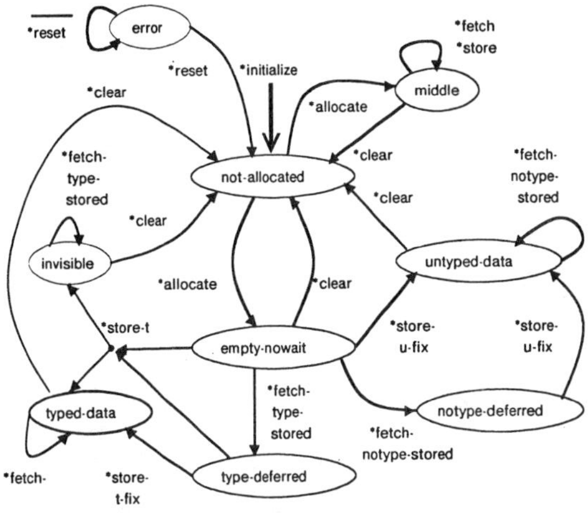

All omitted arrows are branches to error

Figure 1: Status Transition Diagram

2.3.1 Error Handling in the I-Store Controller

Since the ISC operates at a fairly high level, it is possible to detect certain errors. There are three types of errors that can occur in the ISC: *status errors, data type errors,* and *free list errors.* A *status error* indicates that the operation being performed is inconsistent with the status of the addressed word; an attempt to overwrite causes a status error. *Data type* errors indicate an inconsistency in the structure of the data being operated on; data being forced into a slot which is too small to contain it is a data type error. *Free list* errors indicate that the free list is low or out of free cells. In our analysis, we assumed the existence of an *error manager* which can be thought of as a piece of software that deals with errors and inconsistencies as they occur. If an error occurs at any location in the I-store, the high order bit of a status field is set to one, the lower order bits are left unchanged, and information is sent to the error manager. Information is thereby preserved about the structure of a data element that must be untangled by the error manager.

The operations needed by the error and storage managers are summarized below. The **read-absolute* and **write-absolute* operations fetch and store a word directly; the **load-error-destination* operation loads the contents of a six byte register, called the error destination register which contains the address of the manager to which all errors are sent; and the **get-free-size* operation sends the contents of the *free-size* register to the specified destination. The true size of the free list can be determined given the value stored in *free-size* and a constant (provided the last time the ISM was initialized).

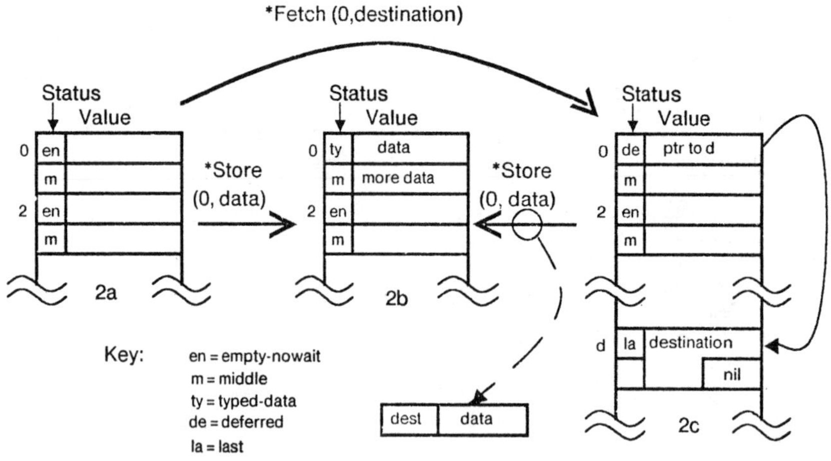

Figure 2

3 ISC Implementation

In designing the ISC, several stages of the design were developed in succession. An informal paper design describing the function and structure of the ISC was the first step. Then a formal high level design in the form of non-executable Pascal-like code was drawn up. Detailed descriptions of the algorithms were specified, and all of the data structures were then presented. The final stage of the design was the IDL code. The IDL description can be implemented in hardware, and both algorithmic and bookkeeping details are present. Since this code is executable, it can be simulated and verified.

The ISC design required about 1500 lines of IDL code. IDL compiled and assembled the ISC to PLA personality (two level logic), and the following statistics were gathered:

Controller:
- Product terms: 887
- Inputs (no feedbacks): 151
- Outputs (no feedbacks): 316
- Feedbacks: 91

Registers: 718 bits
Comparators: 40 bits

The logic derived in this exercise is too big to implement. Also, all of the logic has not yet been compressed to two level form. The size of the design is a result of our approach to the problem; we were more concerned with developing the ISC design and studying IDL, and less concerned with the size of the PLA generated.

3.1 Useful Features of IDL

Several IDL constructs were helpful in the ISC design. In IDL, several cooperating processes can be designed to run in parallel with varying degrees of communication. There are basically two kinds of parallelism that we want to deal with when designing hardware, low level and high level parallelism. Low level parallelism involves actions occurring simultaneously at the lowest level (*e.g.*, "Load register 1 and register 2 in parallel"). By high level parallelism we mean communication between concurrent processes which may be considered sequential as a first approximation. Low level parallelism is handled naturally by IDL statements which allow for the expression of parallel actions. Furthermore, primitives to handle high level parallelism can be easily constructed using available IDL mechanisms. The communication required may come from shared feedbacks, testing and setting the state of another process, or through external connections. The ISC design takes advantage of this facility as the token receiving, the instruction execution, and the token dispatch sections all operate in parallel. When a **fetch* is taking place there is no reason that the next token cannot be received and the last token sent off, all in parallel. Shared feedbacks were used to synchronize the sections. We were very successful in expressing the high level parallelism of ISC in IDL. Essentially serial processes were written (more or less) independently and later merged to get the total design. In contrast, the Pascal code does not have any sense of time (other than sequences). As a direct result of this problem there was no way to express the idea of a hard reset in Pascal.

IDL Truth tables were used to concisely describe several boolean functions that would have otherwise unnecessarily cluttered the code. IDL allows truth tables for testing inputs as well as for setting outputs. In the ISC design, the input field carying the size of the datum (the second half of the type field) was tested in many ways. For example, the precise length of the data has a great bearing on how many words must be referenced while processing a request; truth tables allowed the design to hide the boolean details from the main source, making the code more readable.

IDL strings also were a useful tool in the ISC code. By using strings exclusively in output assignment and input testing, we were able to control the ways in which variables were manipulated. This technique is analogous to the renaming of constants in a program. As a result of this organization, a change in the ISC address space from 16 to 24 bits (planned for the next ISC) will have only small changes in the IDL design. In a similar fashion, IDL groups allowed data to be reorganized from the bit level into more useable objects. IDL strings, however, are semantically unconstrained entities and allow for the possibility of drastic misuse. Strings can be improved by using some of the ideas from modern programming theory.

The most useful capabilities of the IDL System are its ability to generate and manipulate two level logic and its ability to simulate and verify the logic designed. The simulation comes in three flavors, allowing the logic designer to verify the design at different levels of abstraction. For several reasons, the IDL code for the ISC was not simulated. The functionality of the design was being verified by Pascal programs derived from the High Level Code. Also, the magnitude of the ISC design severely stressed the IDL system. Simulation using the implementation of IDL as it existed at that time, while possible, would have been difficult. IDL did not, however, have any fundamental problems with the ISC design. If the design were "real", we would have simulated the ISC, perhaps breaking the design into several pieces.

3.2 IDL as a High Level Language

Designing in IDL is a quantum leap above designing at the gate level. Writing IDL is in some ways similar to writing microcode with the aid of a powerful micro-assembler. Yet we were unable to design the ISC directly in IDL without first specifying it in some detail in the Pascal-like pseudo code. The fact that the high level pseudo code was written at all indicates that IDL has not gone far enough in raising the level of abstraction for the hardware designer. The ways in which the IDL specification differs from the Pascal specification indicate some of the difficulties the that designers of hardware design languages will face in raising the level of abstraction of a design language. We will explain this point with a few examples.

The IDL code tended to be non-reentrant. When copying code to a similar section of the design, extreme caution had to be used in changing all references to state (labels). Handling errors provides a good illustration of this difficulty. Often similar actions had to be taken in each case that an error was detected, and in these situations, the code had to be copied from place to place. The Pascal did not suffer from this general problem, and it is not an intrinsic property of many hardware designs.

Pascal data structures could not be used directly to model the hardware, that is, without loosing the advantages of having the data structure in the first place. When a **variant-record** is used in Pascal, there is implied storage (to discriminate the cases). The very bits which are beyond our control are the most critical ones. Those are the bits by which decisions are to be made. We must have close control over them or know that the compiler will do a particularly good job in dealing with them.

Many important details (from the hardware point of view) were glossed over in Pascal. One of these issues involve passing tokens to the system, and transferring values within it. When we say x:=y in software, no one asks how the value travels. The details of value transferral cannot be ignored in hardware.

4 Language Design Issues

The dual exercise of writing the ISC in IDL and Pascal has convinced us of two facts: (1) Behavioral design languages such as IDL are essential for hardware design, and (2) Sequential languages such as Pascal are inadequate for expressing parallelism, which is present in almost all hardware designs. However, IDL (and we suspect all other hardware design languages in use) lacks many features of modern programming languages, features which are essential to manage the complexity of any large design. IDL was designed more than 10 years ago when many concepts of modern programming languages were not well understood. It has been successful in providing a design tool which allows synthesis of good hardware. Several of the ideas from modern programming languages, especially in the area of data structures, can be incorporated into IDL fruitfully and without much difficulty. However, we discuss two issues related to the quality of hardware generated from a behavioral language, that pose some difficulty for all hardware design languages.

4.1 The Enumeration Assignment Problem

The problem of state assignment (discussed extensively in the literature, see for example [7], [9], [8]) is well known and difficult. A similar problem arose in our study of the ISC. In the design of the I-store section, eight distinct structures arose for the data word: *not-allocated, empty-nowait, typed-data, untyped-data, middle, invisible, type-deferred,* and *notype-deferred.* The actual assignment of binary values to the various enumerations of the status bits have no significance for the functionality I-store operations. The assignment will, however, have a significant impact on the logic required to operate on these bits.

This Enumeration Assignment Problem is common to the designer of digital systems. The designer must collect all decisions that will be asked of these bits and make an assignment that will optiminally implement the decisions. In practice, it is not particularly difficult to make some reasonably efficient assignment. In a design language such as IDL, however, we would like to defer the assignment as much as possible, and allow the assignment to change (automatically perhaps) as we modify the decisions made.

In the design, we wish to operate conceptually in a multiple valued logic. All decisions represent two valued predicates in the larger logic. When the time comes to implement our design in (boolean) hardware, we wish to use the list of predicates to make an assignment that yields the simplest realization of these predicates. The problem is slightly more complex than presented so far. Various variables may be related to each other in such a way that the best enumeration assignment cannot be found independently.

4.2 The Condition Don't Care Assignment Problem

When certain combinations of inputs arise, we may not care which branch of a decision to follow. There is no way of indicating in IDL that a don't care should be associated with a decision rather than an output. Even if we express this fact using some construct, the minimization problem generated by this kind of a don't care for conditions is different from the problem of minimizing several multi-variable functions (with *output* don't cares).

Consider the following Karnaugh Map, which should be considered a boolean function on two variables, and the code fragment that uses it.

(decision)
Karnaugh Map 1

b \ a	0	1
0	0	?
1	1	1

Code Fragment 1

```
if (Map 1) (a b)
    then x <- 1 / y <- 0
    else x <- 0 / y <- 1
```

The don't care at the ab=10 position means that both the **then** and the **else** branch provide an acceptable assignment of values to x and y. If we expand the statement into maps for x and y, propagating the don't care, the following Karnaugh maps result:

(action)
Karnaugh Map 2
x

b \ a	0	1
0	0	?
1	1	1

(action)
Karnaugh Map 3
y

b \ a	0	1
0	1	?
1	0	0

The don't cares at the ab=10 positions have been decoupled incorrectly in the following sense. There are four ways of assigning the two don't cares, but only two of those assignments preserve the semantics of the decision structure. If the don't cares at ab=10 on Map 1 and Map 2 are both assigned the same value (either 1 or 0), we encounter a peculiar situation. When ab=10, the values of outputs x and y will be equal to each other. These values are inconsistent with both the **then** and **else** branches of code fragment 1.

If we want to move the *condition* don't cares to the output side (of the **if-then-else**), we must also carry along some constraints on their assignments. In the example above we must assert that the assignments of the don't cares for x and y must be 01 or 10, hence of opposite values.

We can compress logic to two levels and preserve the semantics of our decision structures as follows. Uniquely label all input don't cares. When we multiply out the multiple level logic, the don't cares are linked by their labels (some will be in inverted form). Whenever we assign a don't care, all instances of the don't care must be assigned the same value. This is by no means a minimization algorithm. This problem is clearly harder than the "classical minimization" problem.

This problem would never have arisen if we could not express don't cares as a part of conditions. In fact, code fragment 1 cannot be expressed using only output don't cares. Another syntax for expressing condition don't cares is the **if-either-or** statement:

```
if <condition> then
   either <actions>
   or     <actions>
```

Even though we don't know how to solve the input don't care assignment problem precisely, we can apply the heuristic of locally minimizing the boolean map associated with a decision. We might then consider the sizes of the **then** and **else** branches. When a series of decisions are compressed into 2 level logic, it becomes more important that each decision node is as simple as possible since the size of each term has a multiplicative effect. (*i.e.* the terms associated with the *on set* ((true condition)) have to be "anded" with the **then**-*subtree*, and *off set* ((false condition)) terms have to be "anded" with the **else**-*subtree*. A subtree itself may contain a decision to be resolved. . . .). Hence a don't care involving a *decision* in a decision tree provides a possibility for immense savings. Consider the following Karnaugh maps.

(decision)
Karnaugh Map 4

C \ AB	00	01	11	10
0	0	0	?	1
1	?	0	1	1

assign 00 1 = 0

and 11 0 = 1

so that ON SET = A

OFF SET = $\bar{A}$

(decision)
Karnaugh Map 5

FG \ DE	00	01	11	10
00	0	0	?	?
01	0	?	1	?
11	?	1	1	?
10	?	?	?	?

In the case of Karnaugh map 4, the best assignment seems to be assigning the don't cares in such a way that the *on set* = A, and the *off set* = $\bar{A}$.

The most promising assignment for Karnaugh Map 5 is assigning the 0101 spot to one boolean value, and assigning the rest of the spots to the other value. If 0101 is assigned 0, then the *off set* requires the single term F·D and the *on set* requires the terms D and F. If 0101 is assigned 1, then the *on set* requires the single term E·G and the *off set* requires the terms E and G. Depending on the complexity of the **then** and **else** branches, one of the two choices can be made. If the **then** branch is "heavier," we assign the 0101 spot to 1. If the **else** branch is "heavier," we assign the 0101 spot to 0. In any case, assigning all zeros or all ones is probably non-optimal.

Acknowledgements

This work was done during 1982 when Steve Heller was working in Dr. Harold Fleisher's group at IBM Poughkeepsie. We gratefully acknowledge Dr. Fleisher's support and encouragement, and Dr. Leon Maissel's valuable insights into hardware design languages. However, this work could not have been done without Mr. Ray Phoenix, who besides being an IDL expert, knows its implementation best.

This research was also supported by the Advanced Research Projects Agency of the Department of Defense under Office of Naval Research contract N00014-75-C-0661.

References

1. Arvind and Members of the Functional Languages and Architecture Group. The Tagged Token Dataflow Architecture. Tech. Rep. TR (unpublished), Laboratory for Computer Science, MIT, Cambridge, Mass., August, 1982.

2. Arvind, and Iannucci, R. A. Instruction Set Definition for a Tagged Token Data Flow Architecture. Tech. Rep. TM-212, Laboratory for Computer Science, MIT, Cambridge, Mass., December, 1981. updated May 1982

3. Arvind, and R. E. Thomas. I-Structures: An Efficient Data Type for Functional Languages. Tech. Rep. TM-178, Laboratory for Computer Science, MIT, Cambridge, Mass., October, 1981.

4. Heller S. K. An I-Structure Memory Controller (ISMC). Master Th., Dept. of Electrical Engineering and Computer Science, MIT, Cambridge, Mass., May, 1983.

5. Maissel, L. I. and Ostapko, D. L. Interactive Design Language: A Unified Approach to Hardware Simulation, Synthesis, and Documentation. Proceedings of the ACM IEEE 19th Design Automation Conference, June, 1982, pp. 193-201.

6. Maissel, L. I. and Phoenix, R. L. Interactive Design Language. IEEE ICCD, 1983. These Proceedings.

7. Marcus, Mitchell P. *Switching Circuits for Engineers.* Prentice-Hall, 1975.

8. Miller, Raymond E. *Switching Theory, Volume II: Sequential Circuits and Machines.* John-Wley and Sons, Inc., 1965.

9. Wood, Jr., Paul E. *Switching Theory.* Lincold Laboratory Publications, 1968.

Session 3-3d
Computer Arithmetic

Chairman

M. Ercegovac
University of California

AN AREA-TIME EFFICIENT VLSI DESIGN OF A RADIX-4 MULTIPLIER

M.D. Ercegovac +

UCLA Computer Science Department
University of California, Los Angeles

J.G. Nash

Hughes Research Laboratories,
Malibu, California

Abstract

A scheme and a VLSI (NMOS) implementation of an area-time efficient 2-bit-at-a-time (radix-4) 2's complement multiplier are described. The scheme has a highly modular bit-slice organization and it is suitable for bus-oriented chip designs. The logic specification and the circuit design details are discussed and analyzed in terms of area-time complexity.

I. INTRODUCTION

A large class of applications, such as digital signal processing and robotics control, require extensive arithmetic capabilities. By far the most important arithmetic operation is multiplication. There are two basic approaches to multiplication algorithms: recursive (sequential) and parallel (combinational). Parallel multipliers have higher speed and larger area requirements than the recursive multipliers.

Recursive multiplication schemes are attractive with respect to the circuit area requirements but often unacceptably slow due to their sequential mode of operation. The number of steps in the recursive algorithm is linearly proportional to the precision and the step time depends on the partial product representation and the adder type. The speed can be improved by a) recoding the multiplier into a higher radix $r=2^k$, and b) by reducing the step time using a carry-save adder. The recursion can be implemented in an obvious manner by an iterative network of carry-save adders in order to eliminate clocking overhead [3]. However, such an iterative (combinational) multiplication scheme requires approximately an n-fold area increase compared to a sequential multiplier.

One of the primary motivations for the design discussed here was the need for a fast, area-efficient multiplier that can fit well within a bus-oriented chip organization. A common criticism of the bus-oriented approach is that it is much slower than "hardwired" versions of arithmetic processors, which offer much higher speeds at reduced flexibility and programmability. We have pursued an alternative approach that combines the advantages of each. Our chip design integrates a high-speed carry-save multiplier with a conventional, slower bus structure. The design is also highly modular so that its use in other custom chips is straightforward. The bit-slice carry-save approach provides, in addition, flexibility in increasing the word lengths without large speed penalties and costly redesign.

Supported in part by the ONR Contract No. N00014-79-C-0866 (Research in Distributed Processing)

II. THE SCHEME

The multiplicand X and the multiplier Y are n-bit fractions in 2's complement system:

$$X = (x_0, x_1, \ldots, x_{n-1})$$

$$Y = (y_0, y_1, \ldots, y_{n-1})$$

The multiplier is recoded by a triplet scanning method into the radix-4 multiplier z=y using the modified Booth's algorithm [1,2,6]:

$$z = (z_0, z_1, \ldots, z_{m-1}) \quad z_i \epsilon \{-2, -1, 0, 1, 2\}$$

where

$$m = n/2 \quad (n \ even)$$

$$z_i = y_{2i+1} + y_{2i+2} - 2y_{2i} \text{ for } i=0,1,\ldots,m-1$$

and $y_n = 0$

The corresponding switching expressions can be obtained from the recoding table in terms of multiplier bits y_{n-2}, y_{n-1}, and y_n :

$$M2 = y_{n-1} \oplus y_n = \begin{cases} 0 & Select\ 2X \\ 1 & Select\ X \end{cases}$$

$$M1 = y_{n-2} = \begin{cases} 0 & Select\ direct \\ 1 & Select\ complement \end{cases}$$

$$M0 = y_{n-2}y_{n-1}y_n + \bar{y}_{n-2}\bar{y}_{n-1}\bar{y}_n = \begin{cases} 0 & No\ clear \\ 1 & Clear \end{cases}$$

The recoder and the generator of Xz_i are organized as shown in Fig.1.

The product is obtained by the following recursion

$$P(k+1) = \frac{1}{4}(P(k) + X \cdot z_{m-1-k}) \quad k = 0,1,\ldots,m-1$$

where the initial partial product is $P(0)=0$. The addition operation is carried out using a carry-save adder so that the partial product $P(k)$, for $k=0,\ldots,m-1$, is represented as a pair of bit-vectors (C(k), S(k)) where C is the partial carry and S is the partial sum bit-vector. The product $P=XY$ requires assimilation of carries using a carry-propagate adder (CPA). Since the signed operands are implicitly handled by the recoding no correction is required in the case of a negative multiplier.

CH1935-6/83/0000/0684$01.00 © 1983 IEEE

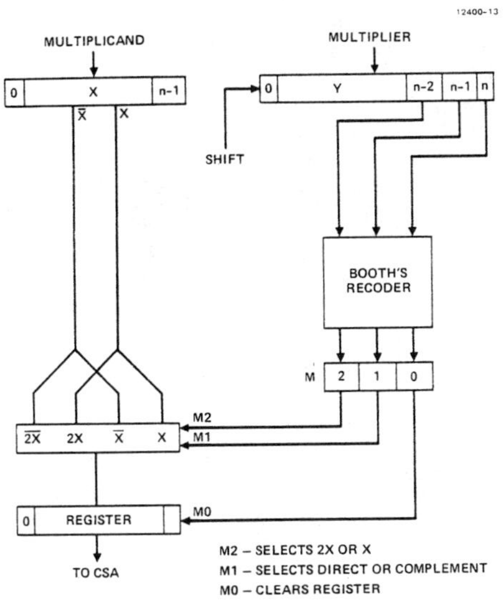

Fig. 1. Recoder and Multiple Generator

The multiplication recursion is implemented as a pipeline consisting of three stages: stage S1 performs recoding, S2 generates required multiple of the multiplicand X, and S3 performs the carry-save addition. The timing of the pipeline is shown in Fig.2.

	0	1	2	3	4	. . .	m+1	m+2	(CPA)
	Load X								
	Load Y								
S1	0	z_{m-1}	z_{m-2}	z_{m-3}	z_{m-4}	. . .	-	-	
S2	0	0	A_{m-1}	A_{m-2}	A_{m-3}	. . .	A_0	-	
S3	0	0	0	P(1)	P(2)	. . .	P(m-1)	P(m)	

Fig. 2. Timing Diagram of the Multiplier Pipeline

When $z_i < 0$, a negative multiple of X is formed in a standard manner by complementing the shifted/nonshifted multiplicand X and adding one in the least significant position of the adder. Since the carry-save addition operation is associative and it never causes a carry into the least significant position, the 1 required in the negation can be inserted into the LSB position *after* the step in which the negation was required. This can be conveniently implemented by inserting a delayed value of the M1 output of the recoder into the least significant bit of the partial carry register.

III. THE DESIGN

The multiplier was designed under assumption of a bus-oriented, bit-slice chip organization. As a result, custom VLSI chips requiring multiplier can be rapidly assembled by attaching required modules to the chip bus. The use of a carry-save, bit-slice multiplier scheme also provides important flexibility in increasing the operand length.

The multiplier organization and its interface to the carry-propagate adder (CPA) is shown in Fig.3. It consists of a linear array of bit-slice sections, all of which are controlled by a set of circuits outside the array. In order to accommodate shifted multiplicand $2X$ it was necessary to append an extra full adder cell to the most significant position. In addition to the basic full adder logic, each bit-slice contains storage registers for both X and Y operands and partial sum S and carry C. The design produces only a single precision product but it can be easily extended to accommodate double precision outputs.

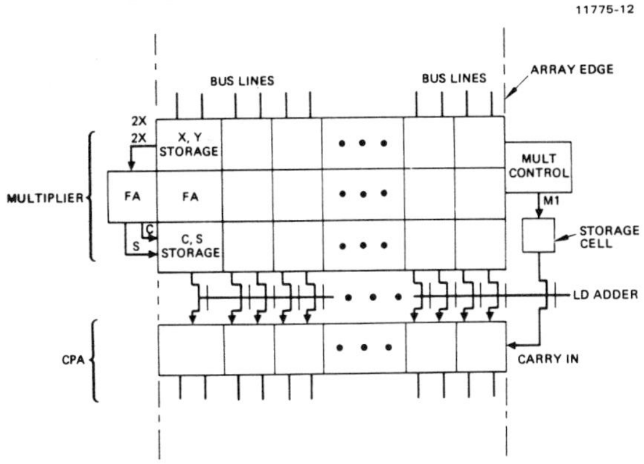

Fig. 3. Bit-Slice Organization and Interface

The relation of the bit-slice sections to the controller circuit is shown in Fig.4. As mentioned above, the circuit pipeline has three stages so that three two phase clock cycles are required to fill the pipeline. During the first clock cycle, the low order multiplier bits are shifted into the storage cells, y_{n-2}, y_{n-1}, and y_n. On the ϕ_1 phase of the next clock cycle, these multiplier bits are recoded, producing M0, M1, and M2. On the ϕ_2 phase of the second clock cycle, these inputs are used to generate the corresponding multiple of the multiplicand X as discussed in Section 2. The 1-of-4 decoder performs the appropriate selection and also functions as a control line driver. The M0 or clear signal overrides the select operation. The output of the select/clear function box is latched at the end of the second clock cycle. The third clock cycle consists of addition during the ϕ_1 phase followed by storage of the partial sum and carry during the ϕ_2 phase. These storage registers are initialized to zero when the multiplier is loaded with its operands.

The dual shift register, shown in Fig.5, is arranged in a fashion that allows two multiplier bits to be examined by the recoder each clock cycle. It consists of two identical shift registers, each of which holds n/2 bits. One shift register holds the odd digits, and the other holds the even digits. Each shift register is spread over two bit-slices, so that every clock cycle the data in a shift register cel advances two positions. As a result, two bits of the original radix-2 multiplier are scanned each clock.

The movement of data between the partial sum and carry storage cells and the full adder is illustrated in Fig.6. The actual addition is done in two parts, one for the carry and one for the sum. Since two multiplier bits are examined each clock cycle, it is necessary for the sum outputs of each bit-slice to be shifted to the right two bit-slices as well. It can also be seen that the carry bit is shifted to the right one-bit slice.

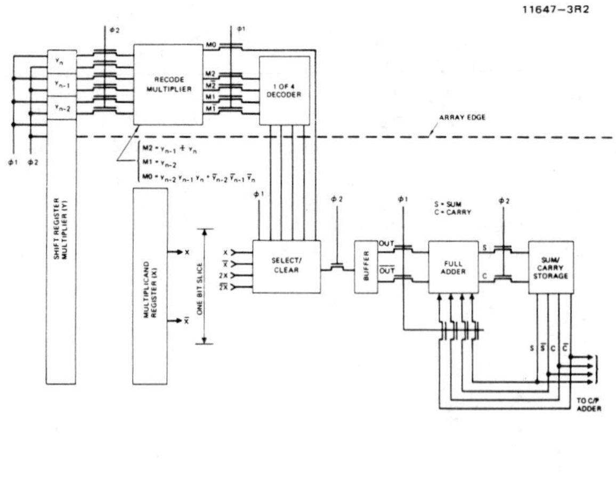

Fig. 4. Schematic of Multiplier Pipeline

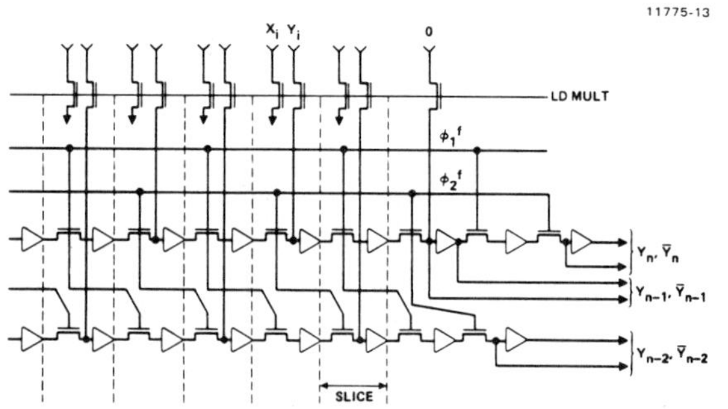

Fig. 5. Dual Shift-Register

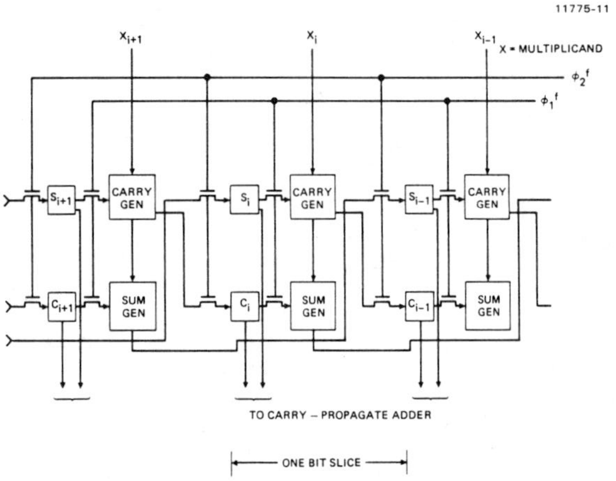

Fig. 6. Carry-Save/Shift Section

A bit-slice layout of the multiplier is shown in Fig.7 with the pipeline stages indicated. This circuit was designed based on a NMOS depletion mode, five mask level process. All controll and I/O lines running across the slice are in metal. Note that two bus lines per slice, to support a two-port memory, are available for data transfer. Circuit area, not including the controller is approximately $12\lambda \; x \; 2\lambda$ *mil*² *where* λ is the Mead-Conway scaling parameter [4]. For example, with $\lambda = 2$, the bit-slice area is 24x4 *mils*². In Fig.8 we show an example of a chip design that uses this multiplier [5]. As can be seen the multiplier efficiently uses available space in that it takes up approximately the same amount of area as the CPA. The control circuitry is not shown.

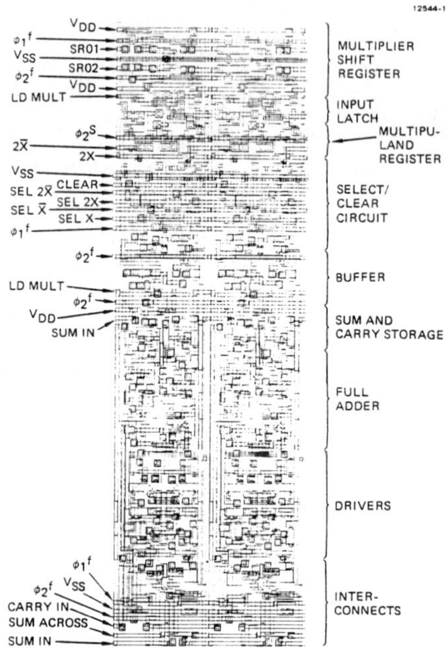

Fig. 7. Layout of Bit-Slice Section
(NMOS, depletion load, five mask level process)

Fig. 8. (a) Chip Organization, (b) Chip Microphotograph

IV. DISCUSSION

In this section we estimate the area-time efficiency of this multiplier. The maximum operating speed of the multiplier is determined by the slowest stage in the pipeline. There are six phases of activity in the three-stage pipeline as listed in the following table.

Clock Phase	Activity
1	Shift
2	Shift
1	Recode
2	1 of 4 Decode
1	Precharge Select/Clear
2	Drive Select/Clear
1	Partial Product Addition
2	Partial Product Storage

The delays in all sections of the pipeline are determined by approximately two gating levels per half clock cycle. From Fig.4 it is clear that the largest drive requirements are placed upon the 1-of-4 decoder, which must charge the select/clear lines across all bits. However, because the controller circuits are outside the bit-slice array, there is space enough to make them as large as desired. As a result, the recoder section can also be used as an intermediate driver stage for the 1-of-4 decoder/driver. This arrangement provides a capability to activate the select/clear blocks with delays comparable to the delay through the full adder and the shift register. Simulations of the multiplier stages, including layout parasitics, indicated that for 6μ gate lengths, the pipeline slowest stage of 33ns was in the full adder bit-slice section. Thus, the clock speed for the multiplier is expected to be about 16MHz.

The area-time complexity of the multiplier (apart from the CPA) can then be expressed as:

$$AT_{(r=4)} = 3nA_{FA}\left(\frac{n}{2}+3\right)t_{FA}$$

where A_{FA} is the full adder area and t_{FA} is the propagation time through the full adder. Here, we have used as an estimate for the bit-slice area, three times the full adder area in the bit-slice. Assuming 6μ gate lengths, a 32-bit fixed-point multiplier would require about 500ns. Using a 3μ technology we estimate that about 200-250ns would be required to perform this multiplication.

The corresponding area-time for a radix-2 (no recoding) iterative or combinational (array) multiplier, also excluding the CPA, is approximately

$$AT_{(r=2)} = n(n-1)A_{FA}\ nt_{FA}$$

For large n, the radix-4 recursive multiplier approach provides a factor of n/3 improvement in the area-time product with respect to the combinational array multiplier.

Acknowledgement

The authors are indebted to G.R. Nudd of HRL for his comments and support.

References

[1] Anderson, S.F. et al., "The IBM System 360/ Model 91: Floating-Point Execution Unit", IBM J.Res.Develop., January 1967, pp.35-53.

[2] Booth,A.D., "A Signed Binary Multiplication Technique", Quart. J. Mech. Appl. Math., Vol.4, part 2, 1951.

[3] Habibi,A. and Wintz, P.A., "Fast Multipliers", IEEE Trans. on Computers, Vol.C-19, February 1970, pp.153-157.

[4] Mead, C. and L. Conway, Introduction to VLSI Systems, Addison-Wesley, 1980.

[5] Nash, J.G., Nudd, G.R., and Hansen, S. "Concurrent Architectures for Toeplitz Linear System Solution", Proc. Govt. Microcircuit Appl. Conf., Orlando, Fla., November 2, 1982.

[6] Rubinfield, L.P., "A Proof of the Modified Booth's Algorithm for Multiplication", IEEE Trans. on Computers, Vol.C-24, October 1975, pp.1014-1015.

THE IMPLEMENTATION OF A RADIX–16 DIGIT–SLICE USING A CELLULAR VLSI TECHNIQUE

Tony M. Carter and Lee A. Hollaar
Department of Computer Science
University of Utah
Salt Lake City, Utah 84112

Abstract

This paper presents the design of a radix–16 digit–slice arithmetic processor using a structured tile approach to integrated circuit design. It treats the logic design of a set of arithmetic operators which conform to a theory of decomposition of structures for arithmetic, their NMOS implementation as tiles and the implementation of the digit–slice using these tiles.

Introduction

Robertson has proposed a theory of decomposition of structures for binary addition and subtraction[1]. This theory leads to a method of designing arithmetic units which can be relatively easily implemented using a cellular VLSI technique. Based on the notion of digit–sets, this theory involves the use of binary (indicated by a lower case a) and ternary (indicated by a lower case b) digit sets. The implementation of ternary digit sets involves the use of two binary digits, necessarily introducing redundancy into the representation. There are a variety of possible *formats* for the binary implementation of the ternary digit set. Chow's dissertation[2] and King's thesis[3] present a design for a variable precision processor module (digit–slice). In a paper entitled "Design of the Combinational Logic for a Radix 16 Digit Slice for a Variable Precision Processor Module", Robertson postulates that there are three format combinations worthy of consideration in the implementation of the Radix–16 digit–slice,

1. format 4 (sign and magnitude) for the digit set $\mathbf{b}^1$ = {–1,0,1}, and and format 2 for the digit sets $\mathbf{b}^0$ = { 0,1,2} and $\mathbf{b}^2$ = {–2,–1,0}
2. format 2 for all three digit sets
3. format 1 for all three digit sets.

In this paper, we discuss only the implementation of the Radix–16 Digit–Slice using the first option (the use of formats 4 and 2). First we deal with the logic design of the various operators implied by the use of the redundant ternary digit sets. Next, the cellular NMOS implementations of these operators are discussed. Following that, the tile–level design of the digit–slice is presented and finally the the use of the NMOS arithmetic cells is demonstrated in the implementation of the digit–slice. In the remainder of this paper, the following symbols will be used:

- $\mathbf{a}^0$ — the digit set { 0,1}
- $\mathbf{a}^1$ — the digit set {–1,0}
- $\mathbf{b}^0$ — the digit set { 0,1,2}
- $\mathbf{b}^1$ — the digit set {–1,0,1}
- $\mathbf{b}^2$ — the digit set {–2,–1,0}.

Logic Designs of Operators

In the binary implementation of the ternary digit sets, Robertson has introduced the notation of combinations of paired greek letters and their english equivalents. Thus, the digit set $\mathbf{b}^2$ could be represented by the symbols τ and **t**. In this paper, we will use **bg** and **be** in their place. Superscripts will indicate the digit–set offset and subscripts will differentiate between ambiguous **bg** or $\mathbf{be}_i$. Thus, $\mathbf{bg}^1_1$ would represent the greek symbol in the digit set $\mathbf{b}^1$, and the subscripted 1 indicates that it is the first such symbol defined in a given operator.

Table 1 describes the various operators and their symbolic name in the integrated circuit tiling scheme. In this table, CN refers to converters, HA refers to generalized half–adders, CG refers to carry generators, RCG refers to redundant carry generators, IRCG refers to the inverse redundant carry generator, MPY refers to multipliers, and CC refers to conditional complementation.

Table 1: Arithmetic Operators

Name	Set Equation
CN0	$\mathbf{a}^0 + \mathbf{a}^0 <- \mathbf{b}^0$
CN1	$\mathbf{a}^0 + \mathbf{a}^1 <- \mathbf{b}^1$
CN2	$\mathbf{a}^1 + \mathbf{a}^1 <- \mathbf{b}^2$
HA0	$\mathbf{b}^0 <- \mathbf{a}^0 + \mathbf{a}^0$
HA1	$\mathbf{b}^1 <- \mathbf{a}^0 + \mathbf{a}^1$
HA2	$\mathbf{b}^2 <- \mathbf{a}^1 + \mathbf{a}^1$
CG0	$2\mathbf{a}^0 + \mathbf{a}^0 <- \mathbf{b}^0 + \mathbf{a}^0$
CG1a	$2\mathbf{a}^0 + \mathbf{a}^1 <- \mathbf{b}^0 + \mathbf{a}^1$
CG1b	$2\mathbf{a}^0 + \mathbf{a}^1 <- \mathbf{b}^1 + \mathbf{a}^0$
CG2a	$2\mathbf{a}^1 + \mathbf{a}^0 <- \mathbf{b}^1 + \mathbf{a}^1$
CG2b	$2\mathbf{a}^1 + \mathbf{a}^0 <- \mathbf{b}^2 + \mathbf{a}^0$
CG3	$2\mathbf{a}^1 + \mathbf{a}^1 <- \mathbf{b}^2 + \mathbf{a}^1$

CH1935-6/83/0000/0688$01.00

RCG0	$\mathbf{2a}^0 + \mathbf{b}^0 \leftarrow \mathbf{b}^0 + \mathbf{b}^0$
RCG1	$\mathbf{2a}^0 + \mathbf{b}^1 \leftarrow \mathbf{b}^0 + \mathbf{b}^1$
RCG2a	$\mathbf{2a}^1 + \mathbf{b}^0 \leftarrow \mathbf{b}^1 + \mathbf{b}^1$
RCG2b	$\mathbf{2a}^0 + \mathbf{b}^2 \leftarrow \mathbf{b}^1 + \mathbf{b}^1$
RCG2c	$\mathbf{2a}^1 + \mathbf{b}^0 \leftarrow \mathbf{b}^0 + \mathbf{b}^2$
RCG2d	$\mathbf{2a}^0 + \mathbf{b}^2 \leftarrow \mathbf{b}^0 + \mathbf{b}^2$
RCG3	$\mathbf{2a}^1 + \mathbf{b}^1 \leftarrow \mathbf{b}^2 + \mathbf{b}^1$
RCG4	$\mathbf{2a}^1 + \mathbf{b}^2 \leftarrow \mathbf{b}^2 + \mathbf{b}^2$
IRCG	$\mathbf{b}^1 + \mathbf{b}^1 \leftarrow \mathbf{2a}^1 + \mathbf{b}^0$
MPY1	$\mathbf{b}^1 \leftarrow \mathbf{b}^1 * \mathbf{b}^1$
MPY2	$\mathbf{2a}^1 + \mathbf{b}^0 \leftarrow \mathbf{b}^1 * (\mathbf{2a}^1 + \mathbf{b}^0)$
CC1	$\mathbf{b}^1 \leftarrow \pm\, \mathbf{b}^1$, conditional on κ
CC2	$\mathbf{2a}^1 + \mathbf{b}^0 \leftarrow \pm\, (\mathbf{2a}^1 + \mathbf{b}^0)$, conditional on κ

In designing the logic implementations of these operators, the following format mappings were used:

bg	be	$\mathbf{b}^0$	$\mathbf{b}^1$	$\mathbf{b}^2$
0	0	0	0	0
0	1	1	1	–1
1	0	2	0	–2
1	1	X	–1	0

In the logic design of these operators, the notion of *coupled don't–cares* is introduced. Consider the design of CN0. Here we are taking a single ternary digit and converting it into two binary digits. For the input value 1 ($\mathbf{bg}^0$ = 0 and $\mathbf{be}^0$ = 1) there are two possible outputs — either $\mathbf{a}^0$ output can be 1 (but not both). Thus, the entry in the logic design table for input 01 is 01 OR 10. In the resulting karnaugh map, this is indicated by a *coupled don't–care* where the entry in the first $\mathbf{a}^0$ map is the complement of the entry in the second $\mathbf{a}^0$ map. The notion of *coupled don't–cares* introduces additional difficulty in the logic design since each individual output variable cannot be independently minimized.

It should be noted that the final logic equations cannot be picked until an implementation strategy is defined. In this particular case, implementations using as few inverters as possible were preferred in order to reduce transistor count. The design of MPY2 used 16 *coupled don't–cares* and 24 *don't–cares.*

NMOS Implementations of Operators

Arithmetic operations inherently require the extensive use of exclusive–or gates. In the NMOS technology, there is a five transistor implementation of the exclusive–or gate which requires no inverted inputs. Each operator was designed so as to select the logic implementation which required the least number of transistors (with a couple of exceptions where layout and fanout requirements dictated otherwise). In several cases, the logic equations used to implement the desired function were selected primarily for *layout* considerations, not for minimality of transistors.

The layout of these operators conforms to a structured tiling approach to integrated circuit layout. In this approach, tiles must cover integral multiples of *unit–tile* locations but there is no restriction on shape. Power and ground routing is included in each tile and the placement of tiles next to each other automatically completes power and ground connections between adjacent tiles. In these arithmetic tiles, data flows through the tile with the inputs entering on one side and the outputs emerging on the other. Figure 1 shows the schematic and final shape of the CG0 tile. CG0 has three inputs ($\mathbf{bg}^0$, $\mathbf{be}^0$ and $\mathbf{a}_1^0$) and two outputs ($\mathbf{2a}^0$ and $\mathbf{a}_1^0$). The corresponding logic equations are:

$$\mathbf{2a}^0 = \mathbf{bg}^0 + \mathbf{be}^0 \cdot \mathbf{a}_1^0$$

$$\mathbf{a}_2^0 = \mathbf{be}^0 \oplus \mathbf{a}_1^0$$

This tile is three *unit–tiles* wide on the input side and two on the output side, making it L–shaped. The *carried* output (the $\mathbf{2a}^0$) emerges from two different *unit–tile* locations, facilitating the placement and interconnection of other tiles.

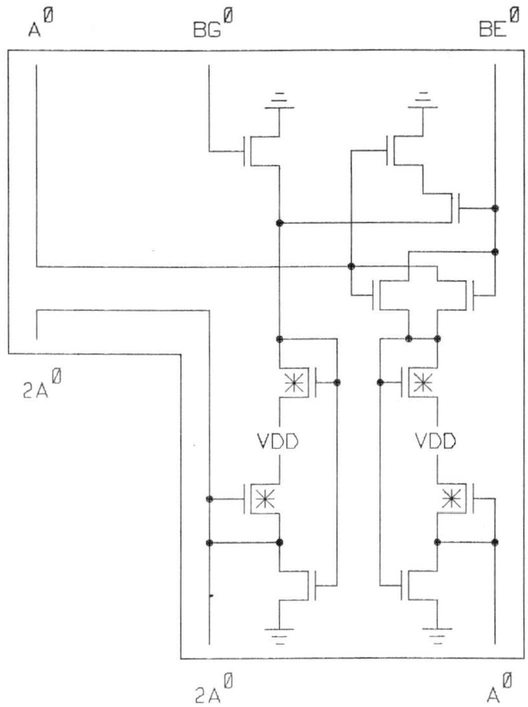

Figure 1: The CG0 Arithmetic Tile

An important feature of the layouts is the availability of inputs and outputs on wires running perpendicular to the data propagation direction. This facilitates the easy routing of the sometimes very random tile interconnect patterns encountered in the design of the Radix–16 Digit–Slice.

Tiled NMOS Implementation of the Digit–Slice

The original design of the digit–slice is presented as a block diagram (figure 2) and a set of associated information loss charts. The use and design of information loss charts is discussed in Robertson's paper[1]. The translation of information loss charts to block–level interconnections of arithmetic tiles is mechanical.

The reduction of the block–level arithmetic tile diagrams to actual circuit tile placements and interconnections proceeds relatively quickly as a result of the structured nature of the tiles. The layout of the digit–slice circuit required approximately two man–days using the arithmetic tiles. While no significant effort has been made to minimize the area either of the individual arithmetic tiles or the layout of the digit–slice, the circuit occupies an area of 2647 square mils using 4 micron design rules. SPICE simulations indicate that the combinational logic portion of the circuit would consume a maximum of about 30 milli–watts power. Preliminary transient analyses indicate that the on–chip resolution time for the multiply would be around 400 nano–seconds. Figure 3 shows the organization of the final layout of the digit–slice.

Conclusions

The implementation of the combinational logic portion of the radix–16 digit–slice using a structured tile approach to integrated circuit design points out the usefulness of Robertson's decomposition techniques in designing arithmetic units. It should be noted that the structured tile approach taken for arithmetic can also directly incorporate the automatically generated control–units produced by ASSASSIN[4, 5], and other functions provided by Path–Programmable Logic[6, 7]. The use of this tiling approach allows large reductions in design time while resulting in circuits which are not unacceptably larger or slower than full–custom designs.

Given the structure of the radix–16 digit–slice and the power and sizes indicated above, it should be possible to implement a 32–bit–equivalent arithmetic processor on a single chip which does add, subtract or multiply in less than 500 nano–seconds and which can be easily cascaded or pipelined to operate on much larger word sizes.

Acknowledgments

The authors wish to thank James E. Robertson for his help and encouragement in this effort. This work was sponsored in part by the Defense Advanced Research Projects Agency (DARPA) under contract number MDA903–81–C–0414 and by the National Science Foundation (NSF) under contract number ECS–81–16742.

References

1. J. E. Robertson, "A Theory of Decomposition of Structures for Binary Addition and Subtraction", This is a document in preparation for publication from the University of Illinois at Urbana–Champaign.

2. C. Y. F. Chow, *A Variable Precision Processor Module,* PhD dissertation, University of Illinois at Urbana–Champaign, July 1980.

3. S. B. King, "Use of a Systematic Approach in the Design of a Variable Precision Processor Module," Master's thesis, University of Illinois at Urbana–Champaign, June 1982.

4. T. M. Carter, "ASSASSIN: An Assembly, Specification and Analysis System for Speed–Independent Control–Unit Design in Integrated Circuits Using PPL," Master's thesis, Department of Computer Science, University of Utah, June 1982.

5. T. M. Carter, "ASSASSIN: A CAD System for Self–Timed Control–Unit Design," Technical Report UTEC–82–020, Department of Computer Science, University of Utah, October 1982, A version of this paper has been accepted for publication in IEEE Transactions on Computer–Aided Design of Circuits and Systems.

6. T. M. Carter; K. F. Smith; B. E. Nelson; A. B. Hayes; and D. A. Fisher, "Path–Programmable Logic and the Use of CADDS2/VLSI," *Proceedings of the Fourth Annual International Computervision User Conference,* Computervision Corp., Sep. 1982, pp. 523–528.

7. K. F. Smith; T. M. Carter; and C. E. Hunt, "Structured Logic Design of Integrated Circuits Using the Stored Logic Array," *IEEE Transactions on Electron Devices,* Vol. ED–29, No. 4, April 1982, pp. 765–776.

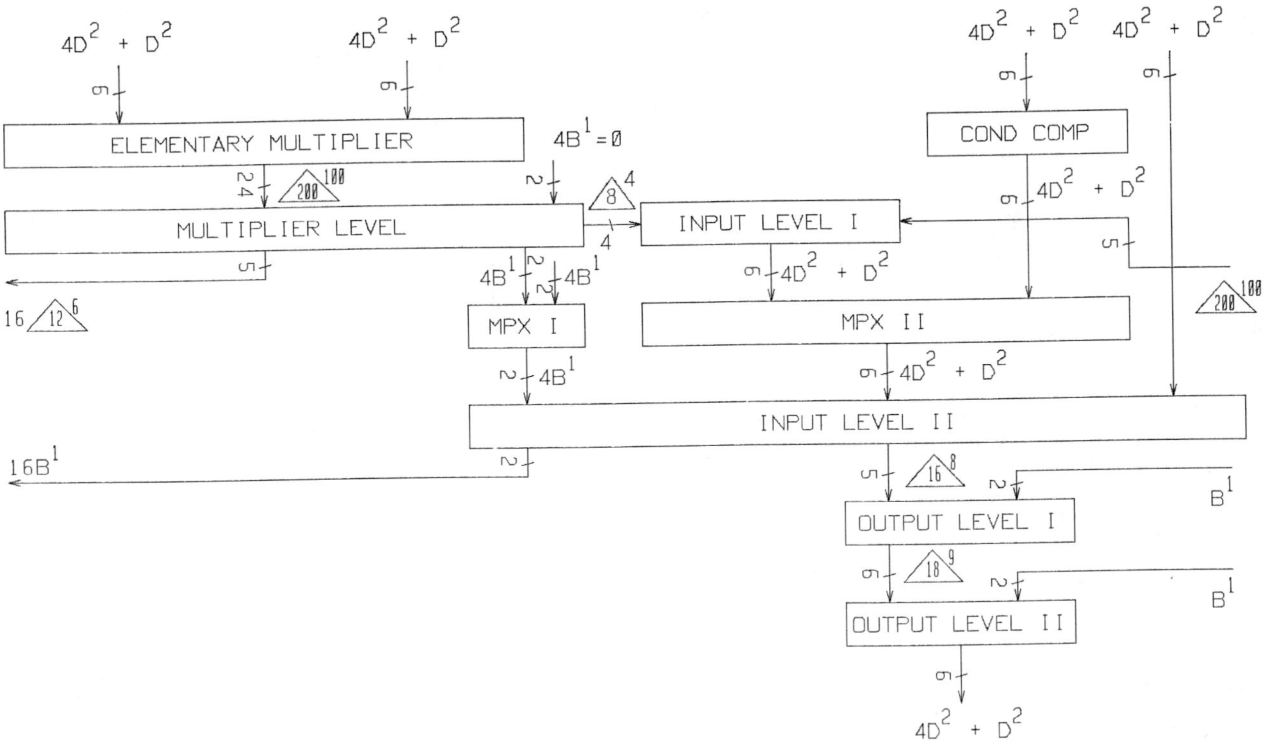

Figure 2: Block Diagram of the Radix–16 Digit–Slice

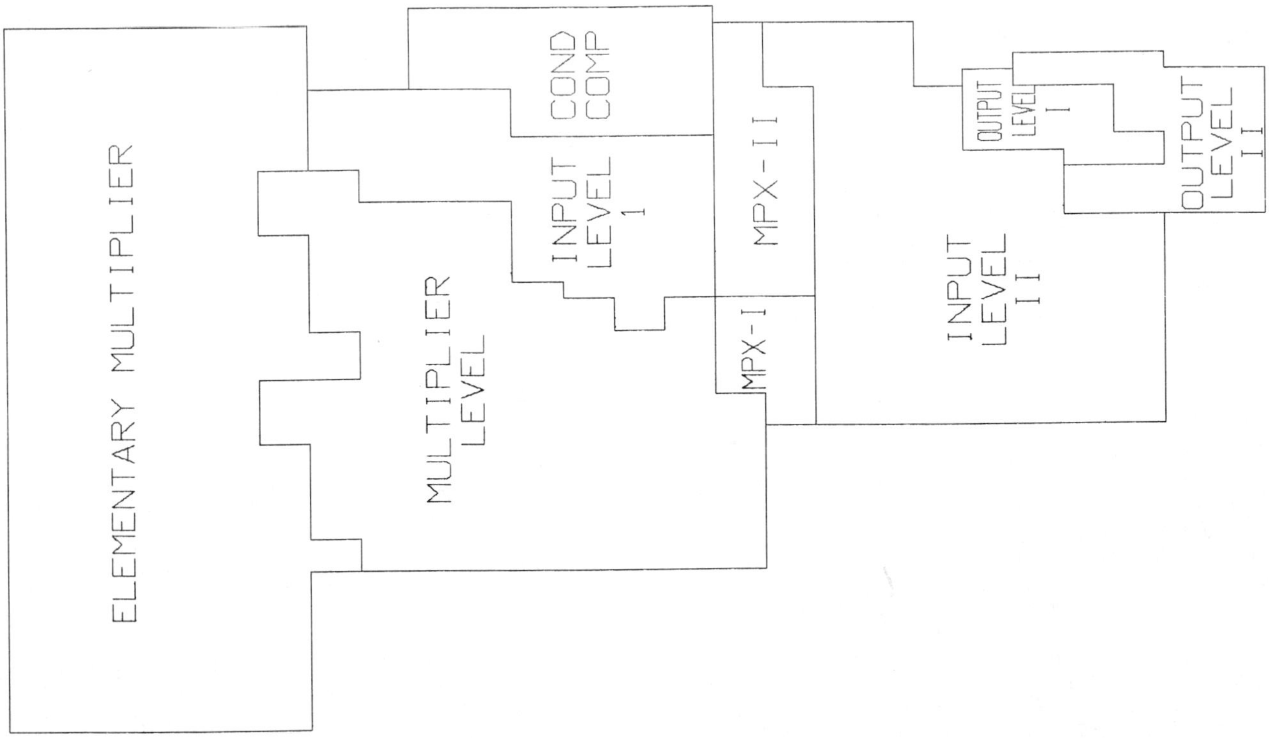

Figure 3: Floor Plan of the Radix–16 Digit–Slice

A VARIABLE PRECISION PROCESSOR MODULE

Catherine Y. Chow

International Business Machines Corporation
General Products Division
Tucson, Arizona 85744

ABSTRACT

This paper presents one method of designing a variable precision processor module for the four basic floating-point operations. The module can be used serially by itself or in parallel with identical modules. Both the precision of operands and the number of modules can be varied without requiring additional programming effort or affecting the control of the modules.

The signed-digit number representation is used. The general direction of mantissa processing is from the most to least significant digits. For most of the allowed digit sets, a result digit depends on a maximum of three adjacent digit positions. The digit set affects the cost of the quotient-digit selector as well as the complexity of the algorithms. The module is proposed for VLSI implementation because of its wide domain of application and modularity.

1. INTRODUCTION

This paper describes the major features of a family of variable precision (VP) processor modules that can satisfy the following requirements:

(1) an individual module can process floating-point numbers of any specified precision;
(2) two or more modules can be connected to run in parallel for any specified precision;
(3) the modules perform the four basic floating-point arithmetic operations efficiently.

A subset of the symmetric redundant signed-digit (SD) representations is used for the VP module [Aviz61]. These representations allow addition to be totally parallel. The number of modules and the precision of operands do not affect the time to perform one addition step. Each result digit is dependent only on a fixed number of adjacent digit positions. The detection and correction of hardware errors can be independently implemented for a digit position. The "round-off" error from truncation is without bias. The shift input of any arithmetic shift is always zero. The leading digit indicating the sign can be processing like any other digits. The overflow condition can be detected as soon as the leading result digits are available. Hence, the processing direction used for serial operations is from the most to least significant digits.

2. ARITHMETIC CONSIDERATIONS

The signed-digit representation for the module uses a symmetric digit set D_ρ containing the digits from $-\rho$ to ρ. For a radix r, the maximum digit ρ must satisfy the constraint:

$$\lceil (r-1)/2 \rceil + 2 \leq \rho \leq r - 2.$$

This, in turn, limits the radix r to be at least 7. The digit set is redundant since it contains more than r values. For Radix 16, D_{10} with digits from -10 to 10 is one of the feasible digit sets conforming to the above definition. The maximum digit ρ is more restricted here than in other signed-digit designs [TuAv70, Pist70, Goya76] to avoid a complex normalization procedure and a costly quotient-digit selector. Furthermore, the multiplication algorithm can use a single precision adder even when the partial product digits are generated most significant first.

The module is designed to perform the four basic operations, addition, subtraction, multiplication and division. Three supporting operations, assimilation, conversion and normalization, are also included to assist data conversion. Except for conversion and normalization, all operands are assumed to be normalized floating-point numbers in the form

$$X = r^{(E_x)} \cdot \sum_{i=0}^{m} X_i r^{-i},$$

where E_x is an integer with an e-bit two's complement representation, X_0 is in D_1 and X_i with i at least 1 is in D_ρ. X is normalized if all X_i's are zero or if

$$|X_0 + X_1 r^{-1}| \leq 1 \text{ and } r^{-1} \leq |X_0 + X_1 r^{-1} + X_2 r^{-2}|.$$

From the requirements of the arithmetic algorithms [Chow80], the allowed digits sets are classified into two groups. The first group uses a three-level digit slice to handle all arithmetic manipulations related to one digit position and the second group uses a four-level digit slice to do the same (Table 1). Only the three-level digit slice is considered here. However, the four-level digit slice is also useful because it applies to Radix 10.

CH1935-6/83/0000/0692$01.00 © 1983 IEEE

Radix r	Maximum digit ρ	
	Group 1 (3-level)	Group 2 (4-level)
7		5
8		6
9		6,7
10		7,8
11	7,9	8
12	10	8,9
13	8,10,11	9
14	11,12	9,10
15	9,11,12,13	10
16	10,12,13,14	11
17	11,12,13,14,15	10
18	11,13,14,15,16	12
>18	$\lceil (r-1)/2 \rceil +2$ to $(r-2)$	

Table 1 The allowed maximum digits

For all operations, a numeric result and four indicator bits for exponent-underflow, exponent-overflow, zero-divide and internal hardware error are available. In the assimilation of signed digits to a non-redundant representation A, the result A is expressed as $r^{(E_A)} \cdot \sum_{i=0}^{m} A_i r^{-i}$ with the integer A_0 ranging from -2 to 1 and the digits A_i ranging from 0 to r-1 for i at least 1. For conversion , the non-redundant number A is converted to a normalized signed-digit number X defined previously.

The division scheme used in the module is basically the non-restoring division. The divisor and the quotient have single precision mantissas with m digits and the dividend has a double precision mantissa with 2m digits. A one-digit scaling constant is used on both the divisor and the dividend before the quotient digits are generated. The scaling procedure reduces the complexity of the quotient-digit selector by a factor of about $\log_2 r$.

All other operations have single precision operands. Multiplication yields a double precision product; the rest of the operations gives a single precision result.

Since the networks for the normalization test, the overflow test and the quotient-digit selector communicate with the first three mantissa digits, the module has included four digit slices to eliminate any initial sequencing of operand digits. More digit slices can be used if the implementation environment permits. The computational requirements dictate that the operand precision m be expressible as S(zn) - 2, where S is any positive integer, z is the number of digit slices in the module and n is the number of modules used together.

3. THE DIGIT SLICES

The digit slice handles all arithemetic manipulations required for one digit position [Chow80]. The three-level digit slice described here is applicable for the digit sets listed in Group 1 of Table 1. Four digit slices are included in the module. A digit slice contains three arithmetic levels: a digit multiplier M, a multiple-input summation level S1 and an adder level S2 (Figure 1).

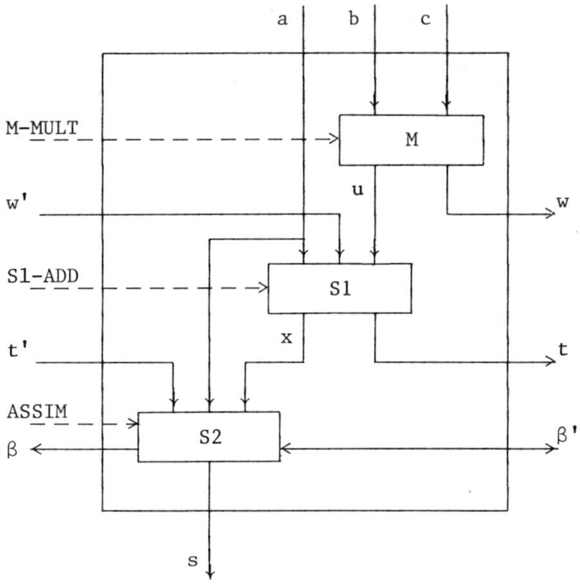

Figure 1 The three-level digit slice

Three control signals M-MULT, S1-ADD and ASSIM direct the levels to perform the appropriate operations. The first level M acts as a digit multiplier only when M-MULT is active. The summation level S1 does an addition or subtraction, if the signal S1-Add is active or inactive, respectively. The control ASSIM directs the level S2 to function either as an assimilator or as an adder. The data inputs consist of three digits a, b and c from operands or intermediate results and three transfers w', t' and β' from adjacent digit positions. The data outputs consist of one result digit s and three transfers w, t and β. Table 2 further defines the operations of the digit slice. Normalization does not involve the digit slice directly. The defined maximum values for the digits and transfers represent one feasible combination intended to make the digit slice as widely applicable as possible. For a given radix r and a given maximum digit ρ, the optimum physical implementation may require an adjustment of the maximum values for the intermediate transfers.

A Radix-16 digit slice with 10 as the maximum digit ρ can be constructed with 16 cascaded sum-of-product networks. The number of gates and the total number of gate inputs are less than 400 and 1000, respectively. The values used for w_{max}, t_{max}, u_{max} and x_{max} defined in Table 2 are 8, 8, 6 and 1, respectively.

(a) Definitions

$w_{max} = \lceil (r-1)/2 \rceil$.

$t_{max} = \lceil (r-1)/2 \rceil$ if $\rho - \lceil (r-1)/2 \rceil$ is even,
$= \lceil (r-1)/2 \rceil + 1$, otherwise.

$u_{max} = \lceil (\rho^2 - w_{max})/r \rceil$.

$x_{max} = \lceil (\rho + u_{max} + w_{max} - t_{max})/r \rceil$.

D_h contains digits from -h to h for any h.

$D_{(r-1)^+}$ contains digits from 0 to r-1.

$D_\rho \cup D_{(r-1)^+}$ contains digits from ρ to r-1.

(b) Input digits

The input digits a, b, and c belong to the digit sets $D_\rho \cup D_{(r-1)^+}$, D_ρ and D_ρ, respectively.
The transfer digits t' and w' belong to the digit sets $D_{(\rho - x_{max})}$ and $D_{(w_{max})}$, respectively.
The borrow β' is either 0 or 1.

(c) Functions of the three levels

[M] If M-MULT = 1
then $ru + w \leftarrow bc$ with u in $D_{(u_{max})}$ and w in $D_{(w_{max})}$,
else $u \leftarrow b$ and $w \leftarrow 0$.

[S1] If S1-ADD = 1
then $rx + t \leftarrow a + w' + u$,
else $rx + t \leftarrow a - w' - u$.
In either case, x is in $D_{(x_{max})}$ and t is in $D_{(t_{max})}$.

[S2] If ASSIM = 1
then $-r\beta + s \leftarrow a - \beta'$ with s in $D_{(r-1)^+}$ and β is 0 or 1,
else $s \leftarrow t' + x$.

Table 2 Operations of the digit slice

4. THE MAJOR LOGIC COMPONENTS

OPERAND REGISTERS AND INITIAL CONSTANTS

The module contains a set of internal operand registers for operands with an e-bit exponent and a 15-digit mantissa. Since the most significant mantissa digit is the sign digit, the actual precision supported by the internal operand registers (m) is 14. No external operand registers are necessary if the entire operands can be stored within the provided registers. If the precision m exceeds 14 or if more than one module is used, external operand registers must be used. They must be organized in a specific manner with three (m+2)-digit mantissa registers and two e-bit exponent registers and must execute the shift, rotate and load commands issued by the leading VP module.

Five constants INTERNAL, LEADER, m, n and S have to be initialized before the module can be used for computation. INTERNAL should only be 1 if the internal operand registers are used. LEADER is 1 only if the module is the leading module or the only module. The constants m, n and S are e bits long, and should be initialized with the precision of the operands, the number of modules and the number of passes (m+2)/4n. The value S must be an integer.

SUPPORT LOGIC FOR COMPUTATION AND CONTROL

After the module is initialized, its Control Sequencer executes a given opcode independently. When two or more modules are used, the one with LEADER set to 1 has its Control Sequencer active and the non-leading modules carry out the controls issued by this leading Control Sequencer. The Sequencer issues the same control signals regardless of the location of the operand registers and the number of modules. The indicator INTERNAL is crucial in directing the signals to the appropriate registers.

The controls to the four digit slices and the supporting memory elements are selected by a Control Selector. Depending on the value of LEADER, these controls come either from the Control Sequencer or from the module inputs.

In order for the Control Sequencer to process multi-precision operands, counting and branching facilities are provided around the Exponent Adder which also processes the exponents of the operands. Mantissa manipulation is supported by a normalization test network, a quotient-digit selector and a scale-constant selector. The rest of the module mainly consists of digit storages, bit storages and control connections.

INTERFACE SIGNALS

The module communicates with the external logic through three buses, four transfer digits, two borrow bits, two input controls and one output indicator (Figure 2). The module interface contains a total of 29 signals and 8 digits. For Radix 16, a digit requires 5 bits. The number of interface signals is 69.

The three buses, Bus 1, Bus 2 and Bus 3, are all bi-directional. Bus 1 is a 4-digit wide data bus. Its width depends on the number of digit slices included in the module. Binary data can pass through Bus 1 by ignoring the digit boundaries. Although Buses 2 and 3 are bi-directional, their directions are fixed by the initialization step.

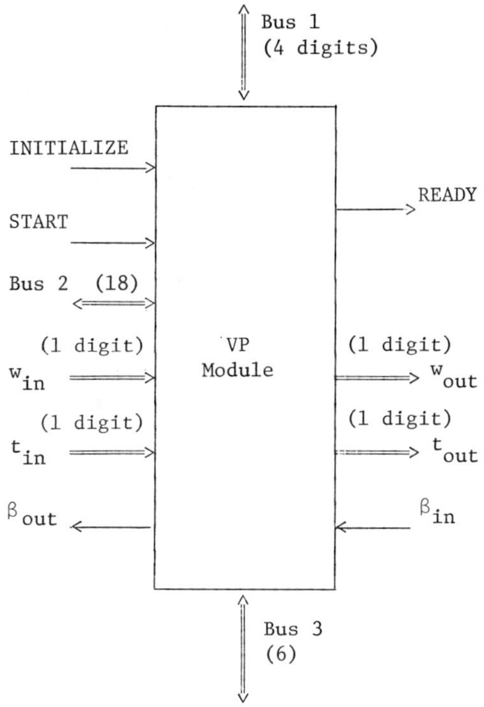

Figure 2 Interface signals of the module

When the input INITIALIZE is active, Bus 2 and Bus 1 are set for inputs. Control signals for loading the constants m, n, S, INTERNAL and LEADER can be successively applied to the module through Bus 2 while the corresponding values are supplied to the module through Bus 1.

After the initialization, Bus 2 is set for inputs if INTERNAL is 1. In this case, the loading of the internal operand registers can be controlled externally through Bus 2. If INTERNAL is 0, Bus 2 allows the leading module to control the shifting and loading of the external operand registers and to select the appropriate data onto Bus 1 of the n modules.

Bus 3 is for passing controls from the leading to the non-leading modules. LEADER equal to 1 sets Bus 3 for outputs and LEADER equal to 0 sets the bus for inputs. Bus 3 signals mainly co-ordinates the gating of data from Bus 1 into the internal buffers feeding the digit slices.

The four transfer digits, t_{in}, w_{in}, t_{out} and w_{out}, and the two borrow bits, β_{in} and β_{out}, are used for inter-module connections. The input START tells the controlling module that the opcodes are at Bus 1. The module executes the commands and indicates their completion with the output READY.

5. SUMMARY

One organization of the variable precision processor module has been discussed. The module supports the four basic floating-point operations. Both the number of modules and the precision of operands can be varied without additional programming effort because the internal control sequence is based on parameters initialized by the environment.

The complexity of the module can be adjusted over the radix, the digit set, the number of digit slices and the complexity of the communication arrangements. A module can be tailored to fit the limits of the implementation technology. The module has a small number of input and output signals and can be implemented as a single unit or a few building blocks. Further study can be in the direction of expanding the operation set and improving the communication scheme.

REFERENCES

[Atki68] Atkins, D.E.,"Higher Radix Division Using Estimates of Divisor and Partial Remainders," IEEE Transactions on Computers, Vol. C-17, No. 10, pp. 925-934, Oct. 1968.

[Aviz61] Avizienis, A., "Signed Digit Number Representation for Fast Parallel Arithmetic," IRE Transactions on Electronic Computers, Vol. EC-10, No. 3, pp. 389-400, Sept. 1961.

[Chow80] Chow, C.Y., "A Variable Precision Processor Module," Ph.D. Thesis, Department of Computer Science, University of Illinois, DCS Report 1032, Oct. 1980.

[Goya76] Goyal, L.N., "A Study in the Design of an Arithmetic Element for Serial Processing in a Linear Iterative Structure," Ph.D. Thesis, Department of Computer Science, University of Illinois, DCS Report 797, May 1976.

[Pist70] Pisterzi, M.J., "A Limited Connection Arithmetic Unit," Ph.D. Thesis, Department of Computer Science, University of Illinois, DCS Report 398, June 1970.

[TuAv70] Tung, C., and A. Avizienis, "Combinational Arithmetic Systems for the Approximation of Functions," AFIPS Conference Proceedings, Spring Joint Computer Conference, pp. 95-107, 1970.

ACKNOWLEDGMENT

This paper is based on the author's thesis research in 1980, at the University of Illinois, under the direction of Professor James E. Robertson.

Design of the Combinational Logic for a Radix 16 Digit Slice for a Variable Precision Processor Module

James E. Robertson

University of Illinois
Department of Computer Science
Urbana, IL

Abstract

This paper is concerned with the details of design of the arithmetic unit necessary for a variable precision processor module. A recently developed design procedure is described, and the results for a radix 16 digit slice are given.

1. Introduction

The purpose of this report is to describe the status of investigations into the possibility of a VLSI realization of the variable precision processor module of Catherine Chow (ref. 9.2). Given a module of η bits precision, the design is such that a precision of $\kappa\eta$ bits (κ an integer) can be achieved by serial use of a single module κ times or by parallel use of κ modules. The operations include addition, subtraction, multiplication and division. The motivation for the research is two fold:

1) A VLSI realization of the module is worth-while in its own right.
2) The design provides a complex practical test of the recently developed procedures for decomposition of structures for addition and subtraction. (ref. 8.4)

2. Representation of Numbers for Signed Digit Arithmetic

In order to achieve the design objectives, it is necessary to use signed-digit arithmetic. The notation employed is as follows:

a. A digit set $\triangle_{\delta}^{w}$ is a sequence of consecutive integers and is specified by two parameters, the diminished cardinality δ and the offset w. δ is the number of non-zero values in the digit set and w is the magnitude of the most negative value.

b. If $\delta = 2w$, the digit set is symmetric, and has the useful property that if x is an element of the digit set, then -x is also an element.

c. A simpler notation is used for digit sets of low cardinality. The letter a is used for two valued digit sets, with a superscript indicating the offset. Thus $a^0 = \{0,1\}$ and $a^1 = \{\bar{1},0\}$. Similarly, b is used for three valued digit sets, with $b^0 = \{0,1,2\}$, $b^1 = \{\bar{1},0,1\}$, and $b^2 = \{\bar{2},\bar{1},0\}$.

d. If $\delta > r-1$, where r is the radix, the digit set is redundant. In general, $r-1 \leq \delta \leq 2(r-1)$, and $0 \leq w \leq \delta$.

For the signed-digit arithmetic of the radix 16 digit slice, the following symmetric digit set is used:

$$\triangle_{20}^{10} = \{\overline{10},\bar{9},\ldots,\bar{1},0,1,\ldots,9,10\}$$

To simplify normalization and quotient digit selection, the specific representation is

$$\triangle_{20}^{10} = 4d^2 + d^2, \text{ where } d^2 = \triangle_{4}^{2} = 2a^1 + b^0 = \{\bar{2},\bar{1},0,1,2\}$$

Each radix 16 digit requires 6 bits, or three bits for each radix 4 digit, with a^1 requiring 1 bit and b^0 requiring 2 bits.

3. The Radix 16 Digit Slice

The basic operation for addition and subtraction transforms $\triangle_{20}^{10} \pm \triangle_{20}^{10}$ into $\triangle_{20}^{10}$. This requires two levels and requires a transfer of the form b^1. For multiplication and division the basic operation is $\triangle_{20}^{10} \times \triangle_{20}^{10} + \triangle_{20}^{10}$. In other words, the product of a multiplier digit and a multiplicand digit is formed and added to a digit of the partial product. Three levels are required, with a second transfer of the form $\triangle_{12}^{6}$.

The portion of the digit slice required exclusively for multiplication and division is shown in Figure 3.1. The product is formed by the elementary multiplier, and transformed by the multiplier level into three outputs, a transfer 16 $\triangle_{12}^{6}$, and outputs $4b^1$ and $\triangle_{8}^{4}$. For compatibility with addition and subtraction, input level I then combines the output $\triangle_{8}^{4}$ with the transfer $\triangle_{12}^{6}$ to form the compatible digit set $\triangle_{20}^{10}$.

The remainder of the digit slice which is shared by all four elementary operations of arithmetic is shown in Figure 3.2. For multiplication and division, the multiplexers select $4b^1$ from the multiplier level and $\triangle_{20}^{10}$ from input level I as inputs to input level II. For addition and subtraction, 0 and the conditionally negated digit are selected. In either case, another digit of the form $\triangle_{20}^{10}$ is added by input level II to form a transfer $16b^1$ and an output 16^{8}. Output level I then combines the transfer b^1 with the sum from the input level to to form 18^{9}. Output level II is necessary to provide for overflow during the steps of multiplication, which is performed most significant digit first, and generates the sum in the consistent form 20^{10}.

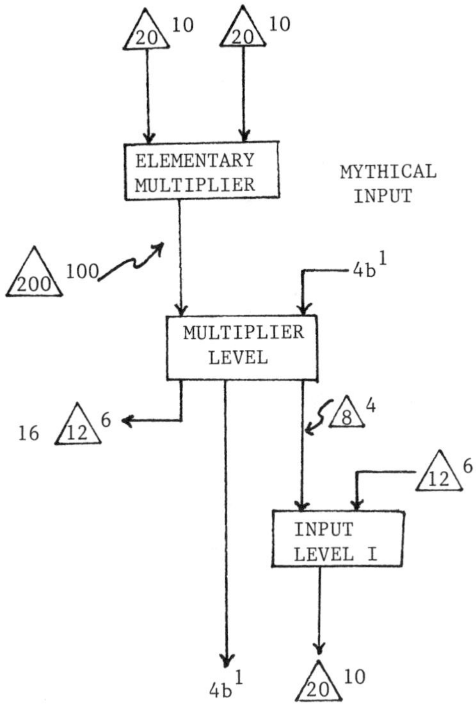

Figure 3.1

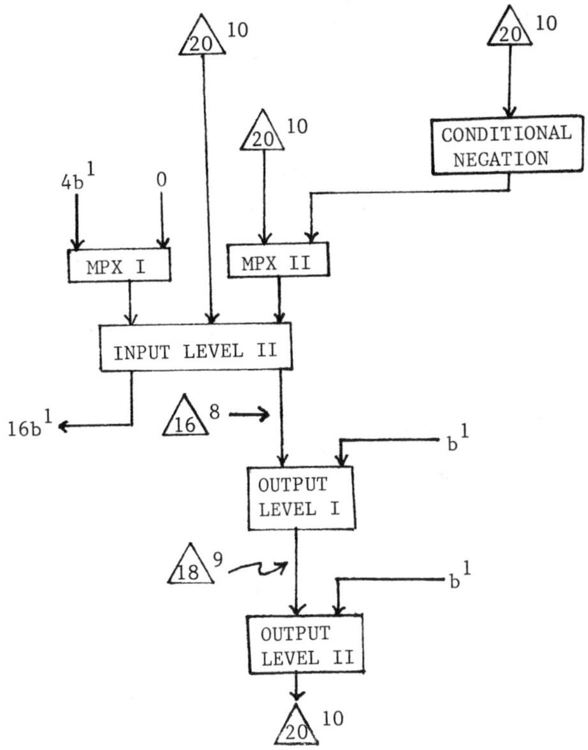

Figure 3.2

4. Basic Structures for Decomposition

With the exception of conditional negation, the elementary multiplier, and the multiplexers, the function of all levels of Figures 3.1 and 3.2 is addition or subtraction. They can therefore be designed by a recently developed decomposition procedure.

An important observation is that structures for addition and subtraction preserve diminished cardinality and offset. For example, a binary subtracter is characterized by the expression

$$2a^1+a^0 \leftarrow a^1+a^1+a^0, \text{ where } a^0=\{0,1\} \text{ and } a^1=\{\bar{1},0\}.$$

Since each type a digit set has a diminished cardinality of 1, the sum of the input diminished cardinalities is 3 and the weighted sum of the output diminished cardinalities is also 3. Similarly, the offsets of the inputs and outputs are both 2.

A second observation is that a complex structure for addition and subtraction can be decomposed into a limited number of basic structures. The basic structures employed for the radix 16 digit slice are:

$b \leftarrow a+a$ GENERALIZED HALF ADDER

$2a+a \leftarrow b+a$ CARRY GENERATOR

$2a+b \leftarrow b+b$ REDUNDANT CARRY GENERATOR

These, with their offset variants, are sufficient for addition and subtraction. For the elementary multiplier, the basic multiplication is performed by $d^2 \leftarrow d^2 \times b^1$, which requires another structure:

$b^1+b^1 \leftarrow 2a^1+b^0=d^2$ INVERSE REDUNDANT CARRY GENERATOR

Conditional negation can be achieved by using the inverse redundant carry generator to generate $b^1+b^1 \leftarrow d^2$, and then forming $b^1 \leftarrow \underline{+}b^1$, conditional on κ, for each b^1.

The remaining structure needed is the two-way multiplexer, for which two bits are needed for MPX I and 6 bits are needed for MPX II (Fig. 3.2).

The costs of these basic structures is very much a function of the correspondence chosen between the three values of b^0, b^1, or b^2, and the two binary digits employed to represent them for logical design. The particular correspondences

b^0	η	y
0	0	0
1	0	1
2	1	0
D.C.	1	1

b^1	ξ	x
0	0	0
1	0	1
0	1	0
$\bar{1}$	1	1

b^2	η	y
0	0	0
$\bar{1}$	0	1
$\bar{2}$	1	0
D.C.	1	1

Table 4. Binary Representations of Three-Valued Digit Sets

selected are given in Table 4. The choice for b^1 is a sign and magnitude representation, which simplifies conditional negation and elementary multiplication, at the expense of an extra gate for the carry generators.

The third important observation is that each digit set has a unique minimal representation involving a weighted sum of type a and type b digit sets. The representations of the digit sets of Figures 3.1 and 3.2 are:

$$\triangle_{8}^{4} = 4a^1+2a^0+b^0$$

$$\triangle_{12}^{6} = 4b^1+2a^1+b^0$$

$$\triangle_{16}^{8} = 8a^1+4a^0+2a^0+b^0$$

$$\triangle_{18}^{9} = 8a^1+4a^0+2b^0+b^1$$

$$\triangle_{20}^{10} = 8a^1+4b^0+2a^1+b^0=4d^2+d^2$$

With these conventions for the representation of digit sets, the design parameters of the levels of Figures 3.1 and 3.2 are completely specified.

5. The Decomposition Procedure

The process of decomposition of complex structures into basic structures is fundamentally a matter of accounting for the number of type a and type b digit sets of each binary weight. Each basic structure employed in the decomposition process has associated with it input values in terms of the number of digit sets, and the corresponding number of output values. These are shown, for the basic structures of section 4, in Table 5.1.

The decomposition procedure thus becomes a procedure for recognizing the change each type of basic structure causes. The changes are obvious, and are indicated in Table 5.1.

	$2^{\kappa+1}$ b	$2^{\kappa+1}$ a	2^{κ} b	2^{κ} a
b ← a+a			0	2
			1	0
2a+a ← b+a	0	0	1	1
	0	2	0	1
2a+b ← b+b	0	0	2	0
	0	1	1	0

Table 5.1

It is possible to predict in advance, subject to some limitations, the cost of a structure when the design parameters specifying the nature of the inputs and outputs of a structure are known. Consider a complex structure, consisting of y generalized half-adders, z carry generators, and w redundant carry generators. The values of w, y, and z are initially unknown, but it is possible to determine some information about their values from the specifications of the structure.

Table 5.2 indicates in detail the changes resulting from the use of each of the basic structures. The three parameters employed are α, the number of type a digit sets, β, the number of type b digit sets, and λ, the information content, defined as $\lambda = 2\beta+\alpha$. For a complex structure, the specifications make it possible to determine the values

	N_{01}	λ_{in}	λ_{out}	α_{in}	α_{out}	β_{in}	β_{out}	$\Delta\lambda$	$\Delta\alpha$	$\Delta\beta$
b ← a+a	y	2	2	2	0	0	1	0	2	-1
2a+a ← b+a	z	3	2	1	2	1	0	1	-1	1
2a+b ← b+b	w	4	3	0	1	2	1	1	-1	1

Table 5.2

of $\Delta\lambda$, $\Delta\alpha$, and $\Delta\beta$ for the structure. With reference to table 5.2, it can be determined that

$$\Delta\lambda = w+z$$

$$\Delta\alpha = zy-(w+z)$$

$$\Delta\beta = (w+z)-y$$

From these equations, it is possible to solve for $w+z = \Delta\lambda$ and for $y = \Delta\lambda-\Delta\beta$. The solution is incomplete, in that it is possible to determine only the sum w+z.

6. Example: Design of Output Level II

The function of output level II is

$\triangle_{20}^{10} \leftarrow \triangle_{18}^{9}+b^1$, or $8a^1+4b^0+2a^1+b^0 \leftarrow 8a^1+4a^0+2b^0+b^1+b^1$

The design parameters are $\lambda_{in}=8$, $\lambda_{out}=6$, $\alpha_{in}=2$, $\alpha_{out}=2$, $\beta_{in}=3$, $\beta_{out}=2$, so that $\Delta\lambda=2$, $\Delta\alpha=0$ and $\Delta\beta=1$. It is thus known in advance that $w+z=2$ and $y=1$.

The design is summarized by Table 6.1. The input is translated into the numerical entries

	8		4		2		1					STEP
	b	a	b	a	b	a	b	a	λ	β	α	NO.
INPUT		1^1		1^0	1^0		2^2		8	3	2	0
$2a^1+b^0 \leftarrow b^1+b^1$		1^1		1^0	1^0	1^1	1^0		7	2	3	1
$2a^0+a^1 \leftarrow b^0+a^1$		1^1		2^0		1^1	1^0		6	1	4	2
$b^0 \leftarrow a^0+a^0$		1^1	1^0			1^1	1^0		6	2	2	3

Table 6.1

of step 0. For step 1, use of a redundant carry generator transforms two type b digit sets of weight 1 into one type b digit set of weight 1 and and one type a digit set of weight 2. For step 2, one type b digit set and one type a digit set of weight 2 are transformed by a carry generator into one type a digit set of weight 2 and one type a digit set of weight 4. For step 3, a generalized half adder converts 2 type a digit sets into one type b digit set, all of weight 4. The result of step 3 represents the desired output. The design requires $w=1$, $z=1$, and $y=1$, with $w+z=2$ and $y=1$ as initially predicted.

The other structures for addition and subtraction are as easily designed, although more complex. The design process is reduced to a straightforward series of numerical transformations, and the end result is the equivalent of a connectivity diagram involving the three types of basic structures.

7. The Elementary Multiplier and Negation

The elementary multiplier performs the function $(4d^2+d^2)\times(4d^2+d^2)$. The design is reduced to a manageable level by reducing each radix 4 digit of the multiplier by use of an inverse redundant carry generator:

$$b^1+b^1 \leftarrow 2a^1+b^0 = d^2$$

Thus the elementary multiplier becomes

$$(4d^2+d^2)\times[4(b^1+b^1)+(b^1+b^1)]=$$

$$16(b^1\times d^2)+16(b^1\times d^2)+4(b^1\times d^2)+4(b^1\times d^2).$$

$$+4(b^1\times d^2)+4(b^1\times d^2)+(b^1\times d^2)+(b^1\times d^2).$$

Each element of the multiplier $d^2 \leftarrow b^1\times d^2$ is a logical design, somewhat more complex than a binary full adder.

Conditional negation may be simplified in a similar manner. Each radix 4 digit is reduced to the form $b^1+b^1 \leftarrow d^2$, and each b^1 is then conditionally negated. Use of the sign and magnitude format for b^1 (Table 4) simplifies this process, since complementing the sign digit 5 changes the sign.

8. Summary and Conclusions

The basic purpose of this paper is to indicate how a complex design can be reduced to a combination of relatively few basic structures. The summary of costs is shown in Table 8.

	ELEM. MULT.	MULT. LEVEL	INPUT I	INPUT II	OUTPUT	MPX	COMP. CKT.	TOTAL
2. $b \leftarrow a+a$		11	2	4	2			19
3. $2a+a \leftarrow b+a$		12	2	5	1			20
4. $2a+b \leftarrow b+b$		4	1	2	2		2	11
5. $b+b \leftarrow 2a+b$	2						2	4
6. $b^1 \leftarrow \pm b^1$, conditional on κ							4	4
8. $2a^1+b^0 \leftarrow b^1\times(2a^1+b^0)$	8							8
2 INPUT MULTIPLEXER						8		8

Table 8 Summary of Costs of Radix 16 Digit Slice

9. Bibliography

9.1 Avizienis, A. "Signed-Digit Number Representations for Fast Parallel Arithmetic", IRE Transactions on Electronic Computers, Vol. EC-10, No. 3, pp. 389-400, Sept., 1961.

9.2 Chow, Catherine, "A Variable Precision Processor Module", Ph.D. Thesis, July, 1980, Report Number UIUCDCS-R-80-1032.

9.3 King, Stanley B., "Use of a Systematic Approach in the Design of a Variable Precision Processor Module", M.S. Thesis, May, 1981.

9.4 Robertson, James E., "A Systematic Approach to the Design of Structures for Arithmetic", Proc. 5th Symposium on Computer Arithmetic, Ann Arbor, Mich., May, 1981.

9.5 Robertson, James E., "Conditions for the Distributivity of Multiplication with Respect to Set Addition and their Affect on the Design of Array Multipliers", Proc. 6th Symposium on Computer Arithmetic, Aarhus, Denmark, June, 1983.

Session 3-4a
Novel Design Issues II

Chairman

A. L. Rosenberg
Duke University

Optimal Resource Allocation in VLSI

Zvi M. Kedem and Alessandro Zorat

Department of Computer Science
State University of New York at Stony Brook
Stony Brook, NY 11794

1. Introduction

Thompson [Tho79] developed a mathematical model of a silicon chip implementation of a boolean circuit. The model was sufficiently powerful to allow meaningful analysis, while at the same time it was not overburdened by the specific details of the actual circuits. The basic VLSI model of Thompson has been used, with some variations, by many researchers to analyze the VLSI complexity of circuits designed to execute (in silicon) certain specific algorithms, or classes of algorithms to solve a given problem.

This survey paper will present some of the recent formal results dealing with the amount of resources required by integrated circuits using standard planar technology for the computation of certain (special purpose) functions. We will concentrate on two measures of interest mentioned above: the time required for the computation and the area of the chip. Some research has been started on two additional measures: the length of the period that can be achieved [ChM81, Vui83] and the required power [Kis82]. However, the research on these two additional measures is much less extensive, and therefore we will be not discussing them.

2. Detailed Analysis of Circular Shifts Complexity

For the purpose of our discussion we assume that some specific (easily analyzable formally!) function is to be computed. It seems reasonable to assume (and we will show this later) that in order to compute this function in a relatively short time, a larger chip is needed than if we are willing to accept lower time performance. Indeed, the bigger the chip, the more computational devices it may be possible to put on the chip, thus increasing the potential parallelism, resulting in a shorter overall computation time.

Ideally, we would like to solve the following problem: *Given a function to be computed and a chip of area A, generate a provably optimal design for a chip computing this function*, namely give a complete design of the chip that minimizes the time T required to compute the function. We are not able to solve such a problem now, but we are able to provide some related formal results which, one hopes, may enhance the intuition of the designers of chips. For the purpose of our discussion, we assume, oversimplifying the technology, that the chip is made of a constant number of layers. (Rosenberg in [Ros81] discussed the possibility of building a 3-dimensional chip and showed that the complexity measures for the 2-dimensional model can be extended to three dimensions.) We assume that the area is devoted to the following types of "devices":

- I/O (Input/Output) pads of some constant area (independent of the chip size),
- memory cells of some constant area, each of them capable of storing a single bit,
- wires of some constant width, each capable of "moving" one bit during one "clock time" unit. (Other choices on wire delay are possible. [BPP81, MeC80])
- computational elements are gates of constant area that compute binary or unary boolean functions.

We will ignore the constants in our discussion, and therefore our results will be of asymptotic nature. Specifically, we assume that the clock time unit is one; the area of a memory cell, a gate, and an I/O pad is one, and the width of a wire is one also. We are making these assumptions in order not to carry a large number of constants in our derivations; of course, an I/O pad is actually much larger than a memory cell. Within this context we will attempt to answer a question of the following form: For a specific chip area value A, which of the possible devices enumerated above are *critical* for minimizing the computation time T? In other words: If we could increase the area of the chip by some small amount, which of the devices it would be most useful to add in order to minimize the computation time?

This research was supported in part by NSF, under grants ECS 80-25376, MCS 81-04882, and MCS 81-10097.

Perhaps surprisingly, the problem of optimal allocation of resources in the design of a single chip device seems to be easier than that of designing "optimal" devices of other types, such as boolean circuits that do not have to satisfy any geometric constraints. The reason for that seems to be the fact that we are dealing with "homogeneous stuff," namely the silicon on which we can "draw" any device we like, while having to satisfy stringent geometric constraints. As our standard example we will consider the problem of designing an optimal circular shifter. This is an interesting computation on its own right and not just a toy example, as it is very intimately related to important computations, such as the multiplication of standard binary numbers. To see that, note that every chip multiplying binary numbers has to be able, as a special case, to multiply a number by a power of two, which is really a boolean (regular) shift; on the other hand, such shifts can be used to perform a binary multiplication. Boolean shifts are themselves very closely related to circular shifts.

Therefore we assume that we wish to compute a function of the form: $f(x_0,x_1,...,x_{n-1};\ s_0,s_1,s_{p-1}) = (y_0,y_1,...,y_{n-1})$, where $(x_0,...,x_{n-1})$ is the boolean vector to be shifted, $(s_0,...,s_{p-1})$ are some boolean parameters specifying the amount of the shift, and $(y_0,...,y_{n-1})$ is the resulting circularly shifted vector. Furthermore, we assume that the chip is so called *where and when oblivious*, [LiS81] that is, for each input (output) pad and each time instance during the computation it is determined as part of the chip specification which input (output) bit variable is read (written) into (from) the pad. This is indeed the case for the standard design methodology of chips and systems.

We now concentrate on the special case in which it is assumed that each input variable is read only once during the computation; this also is satisfied by the vast majority of designs, although there are notable exceptions [FPP82] as we will indicate later, when we remove this assumption. Using this assumption we briefly present results originally developed by, among others, [BrK81, Tho79, Vui83].

Assume that the chip is a square of area A. We will determine what is (asymptotically) the minimum amount of time required for the computation of circular shifts of n-bit vectors, as stated above. Consider two cases:

- *There exists an I/O pad from which at least $n/3$ of the elements of $(y_0,...,y_{n-1})$ are written during the computation.* As to read a bit in takes an I/O pad one time unit, we have that $T \geq n/3$, and as $A \geq 1$, $AT^2 \geq n^2/9$.
- *Every I/O pad writes out fewer than $n/3$ of the elements of $(y_0,...,y_{n-1})$.* Under this assumption it is possible to divide the chip by a "perpendicular" cut with at most one "knee" (see Fig. 1) so that each of the two parts into which the square chip is divided writing at least $n/3$ elements of $(y_0,...,y_{n-1})$. Define $D(i,s)=0$ if the I/O pads reading in x_i and write out y_{i+s} are on the same side of the cut, and $D(i,s)=1$, otherwise, for $i,s=0,...,n-1$. (The + in the subscripts will be modulo n.) Consider some specific x_i, read into some output pad P. For at least $n/3$ values of s, x_i has to be output from some output pad P' (this is the output pad from which y_{i+s} is written) and that P' is on the other side of the cut from P. Thus, $\sum_{s=0}^{n-1} D(i,s) \geq n/3$. From here summing over all values of i, $\sum_{i=0}^{n-1}\sum_{s=0}^{n-1} D(i,s) \geq n^2/3$, and thus $\sum_{s=0}^{n-1}\sum_{i=0}^{n-1} D(i,s) \geq n^2/3$. It follows that for some "difficult" value of the shift s, $\sum_{i=0}^{n-1} D(i,s) \geq n/3$. Therefore, at least $n/3$ bits have to cross the cut of length $\sqrt{A}+1$ during the computation. As at most $\nu(\sqrt{A}+1)$ (ν is the number of layers of the chip) bits can cross the cut during one time unit, it follows (ignoring constants) that $T \geq n/3\sqrt{A}$. From here again we have that $AT^2 \geq n^2$. We have thus proved a result of [Vui83]:

Theorem 1: Under the assumptions stated above (note especially that each input variable can be read only once during the computa-

CH1935-6/83/0000/0701$01.00 © 1983 IEEE

tion), the area of the chip A and the time of the computation T for any chip computing all circular shifts of a boolean vector of length n are related by $AT^2=\Omega(n^2)$. (We remind the reader that $f(n)=\Omega(g(n))$ means that for some positive constants c and n_0, $f(n)\geq cg(n)$, for all $n\geq n_0$.)

It is interesting to note that it is possible to design chips actually satisfying this bound exactly, so we have a strong formal result. Moreover, we can learn something even more interesting from the *proof* given above. Reviewing the "non-degenerate" second case above, we notice that the only argument used in the proof was based on the fact that a large area needs to be devoted to the wires. *In effect, the proof would hold even if we could store unlimited amount of information in one unit of area.* This essentially shows that, in this case, wires are a critical resource, whereas memory is not.

Let us now see what can be proved using the assumptions on memory cells. As the chip must be able to compute all circular shifts it has to store all of $(x_0,...,x_{n-1})$ before any element of $(y_0,...,y_{n-1})$ can be output. Thus the number of memory cells on the chip must be at least n, showing that $A=\Omega(n)$. [BrK81]

Summarizing the situation, under the above assumptions the chip has to be of area n at least and wires but not memory are a critical resource; we do not need more memory cells than n. However, if we increase the area devoted to the wires to above n the preformance of the chip may be increased (T will be smaller).

This leads us to the conclusion that we have to consider a more general framework if we wish to study the consequences of having to work with a smaller chip area, as there it may not be possible to store the whole problem's definition in the chip. This may not be realistic in the context of the problem we study here - circular shifts - but consider the problem of sorting a sequence using a chip. Generalizing the results proved so far, it is easily to see that in this case too we must have $A=\Omega(n)$. However, the maximal area of the chip is limited by technology, and a straightforward application of the results above would imply that it is impossible to sort large sequences using the computational power of a single chip! What, of course, should be done in this case is to assume the existence of additional memory "outside the chip" in which information may be stored *but not manipulated*. Indeed, by studying the classical results on the complexity of sorting, see e.g., [Knu73 Vol. 3], we note that if the size of the RAM is sufficient to store the whole sequence to be sorted, then we are concerned with minimizing the movement of the information in the computer, tests of the form $x_j<x_k$, etc. This is reminiscent of the results we have derived so far, as wires, in effect are used for internal transfer of data. On the other hand, if the RAM available is small with respect to the size of the problem, then the main cost seems to be the transfer of information between RAM and the auxiliary storage, the amount of which depends on the size of the available RAM. We wish to analyze formally what happens under such circumstances.

In order to derive formal results we will assume a weaker capability than may be actually implemented. Instead of assuming that some "intermediate results" during the computation of a circular shift of a "long" vector on a "small" chip are stored "outside," we will assume instead that the inputs may be read by the chip more than once. We still assume that the chip is where and when oblivious, thus the locations and times in which the inputs are provided are independent of the values of the inputs. Let us start with some illustrative toy design.

Consider the drawing in Fig. 2. There we have a shifter which consists of two I/O pads, two memory cells, and some wires. (For simplicity, we assume that the memory cell is "inside" the I/O pad.) This chip is essentially of area 2 and can perform circular shifts of vectors of length $n=6$ as follows. It first reads the value of s, the length of the sift; it is stored in some small finite automaton, which will manage the computation. We will ignore the area of this automaton, as it is not important in the context of this example, and it grows only slowly with n. The computation can be schematically described by the following sequence of operations where each row describes a single time instance:

read x_0 into P_0; read x_1 into P_1
read x_2 into P_0; read x_3 into P_1
read x_4 into P_0; read x_5 into P_1
circularly shift, if necessary, the values in the pads
write y_0 from P_0; write y_1 from P_1
read x_0 into P_0; read x_1 into P_1
read x_2 into P_0; read x_3 into P_1
read x_4 into P_0; read x_5 into P_1
circularly shift, if necessary, the values in the pads
write y_2 from P_0; write y_3 from P_1
read x_0 into P_0; read x_1 into P_1
read x_2 into P_0; read x_3 into P_1
read x_4 into P_0; read x_5 into P_1
circularly shift, if necessary, the values in the pads
write y_4 from P_0; write y_5 from P_1

Instead of defining the computation formally, let us examine what happens when the shifter executes a shift by 5. In this case $(y_0, y_1, y_2, y_3, y_4, y_5)=(x_1, x_2, x_3, x_4, x_5, x_0)$. The computation is divided into 3 *phases*; in each phase all the inputs are read and two of them are "captured" and output. In the first phase the inputs x_1 and x_2 will be stored. They will be output at the end of the phase as y_0 and y_1; however, they are stored in the "wrong" memory cells. (x_1 is stored in P_1 but has to be output from P_0.) Thus, an additional time unit is used if the s is odd to place the outputs in the right output pads. In the second phase x_3 and x_4 are captured, shifted, and output as y_2 and y_3, etc.

What is T in this case? It is easily seen that for a (more general shifter) containing k I/O pads and k memory cells, and shifting a sequence of length n:

T = *(number of phases)*(length of one phase) = (number of phases)*(time required to read all the inputs once + time required to perform internal shift between the memory cells + time to output the partial result computed in a single phase) = (n divided by (number of memory cells)) * (n divided by (number of I/O pads) + number of memory cells - 1 + 1*, or $T=\frac{n}{k}(\frac{n}{k}+k)$.

Now, if k is small (the chip is small) then k is smaller than n/k, and therefore the bulk of computation time is spent on I/O *because of limited memory size and limited number of I/O pads.* We have not considered the case where the number of I/O pads is different from the number of memory cells, but if both of them are small compared to the size of the problem (n), then generalizing the arguments we have that approximately $T=\frac{n^2}{\mu\pi}$ where μ is the number of the memory cells and π is the number of I/O pads. Thus, in this case *both memory and I/O are critical resources.* On the other hand, wires are not a critical resource. It is possible to decrease, *but not asymptotically*, the computation time by providing additional wires as it is possible to use them to decrease the amount of time required for the internal shifting. Note, however that if $k<\sqrt{n}$ (formally $k=o(\sqrt{n})$), then in our original design (in which wires were not used in order to make local shifting very efficient) we already had $k=o(n/k)$ and thus the local shifting took less time than the rest of the computation. We conclude that $T=n^2/k^2$ asymptotically *no matter how much additional area is devoted to wires.* (For detailed constructions of circular shifters, see [KeZ81].)

Formally, we have [Ked82]:

Theorem 2: Under the assumptions stated above (note that an input variable can be read more than once during the computation), the area of the chip A and the time of the computation T for any chip computing all circular shifts of a boolean vector of length n are related as follows:

$$\begin{array}{ll}
\text{if } A \leq n^{2/3}, & \text{then } A^2T = \Omega(n^2), \\
\text{if } n^{2/3} \leq A \leq n, & \text{then } A^3T^3 = \Omega(n^4), \\
\text{if } n \leq A \leq (\frac{n}{\log n})^{3/2}, & \text{then } A^2T^3 = \Omega(n^3), \\
\text{if } (\frac{n}{\log n})^{3/2} \leq A, & \text{then } T = \Omega(\log n).
\end{array}$$

(Note, that at least from theoretical point of view, it is useful to replicate inputs even if the chip area is greater than n, and thus the lower bound of $AT^2=\Omega(n^2)$ does not hold in this case.)

The difference between the example and the theorem is due to the fact that our toy design was not optimal, and it would have been better, though more difficult to describe, to arrange the cells in a mesh and not on a line.

It is worthwhile to consider our example again. We have stated that if the chip is small, then both memory and I/O pads are the critical resources, and thus it is worthwhile to balance the area devoted to I/O pads with the area devoted to memory. One can object and say that whereas the chip designer can allocate the chip area between (internal) wires and memory, he cannot assume that the system designer (who is concerned with building a complete system from chips, boards, etc.) will supply to the shifter chip a large number of (external) wires to take advantage of the many I/O pads. We do not consider system level complexity in this presentation, but it may be worthwhile to mention that if the area devoted to to the pads is constant, which is a reasonable assumption because of the packaging standards, then only memory is the critical resource and the lower bound of $AT=\Omega(n^2)$instead of $A^2T=\Omega(n^2)$ holds for this case.

In [FPP82] the authors present a special purpose "graphics engine" in which internal memory is saved by providing the same inputs twice during the computation. Essentially half the memory is saved; and as the number of I/O pads is kept the same, this is within the spirit of the result above.

3. Survey of Results on Complexity of Certain Functions

We devote this section to an anthology of results dealing with relation between the size of the chip and its efficiency for computation of various functions. We will not discuss which resources are critical, although in the vast majority of cases these are the wires.

3.1. Discrete Fourier Transform and Sorting

In [Tho79, Tho80] it was shown that the complexity of computing the discrete Fourier transform (DFT) of n elements from a finite ring or sorting n elements taken from the set of integers $\{0, 1, ..., n\text{-}1\}$, must have a complexity such that $AT^2 = \Omega((n \log n)^2)$. The proof technique was based on arguments similar to those used in the previous section. An important assumption was that the inputs to the function to be computed were given only once and only in one location of the chip during the course of the computation. This implies that $A = \Omega(n)$ as before, and hence one can show that $AT^{2\alpha} = \Omega(n^{1+\alpha} \log^{2\alpha} n)$, for any $0 < \alpha < 1$.

Thompson then showed constructions of circuits to solve the DFT and the sorting problems that exhibited near optimal performance in the AT^2 measure of complexity. One construction was based on the *perfect-shuffle* graph [Sto71] and required $A = O(n^2/\sqrt{\log n})$ to compute the in $T = O(\log^2 n)$ or to sort n elements in time $T = O(\log^3 n)$. The other construction was based on a mesh-connected network and required an area $A = O(n \log^2 n)$ and a time $T = O(\sqrt{n} \log\log n)$ for both problems. The gap between upper and lower bounds that was left open by Thompson was closed later in [KLL81], where a construction for the perfect shuffle with $A = O(n^2/\log^2 n)$ was presented.

3.2. Integer Product and Matrix Multiplication

Abelson and Andreae in [AbA80] showed, by applying arguments similar to those of [Tho80], that the chip area A and execution time T of any chip computing the product of two n-bit integers must be such that $AT^2 = \Omega(n^2)$. They left the question open on whether a circuit with this complexity could indeed be designed.

Brent and Kung [BrK81], generalized the lower bound of Abelson and Andreae by showing that $AT^{2\alpha} = \Omega(n^{1+\alpha})$, for $\alpha \in [0,1]$, by considering that the $2n$ inputs must be read into the chip and hence $A = \Omega(n)$, as well as the information transfer requirements of the chip implementing the integer multiplier. In the rest of that paper the authors presented an abstract design of a circuit based on carry lookahead adders that performed integer multiplication with $A = O(n \log n)$ and $T = O(\sqrt{n} \log n)$. While their circuit, with a combined complexity of $AT^{2\alpha} = O(n^{1+\alpha} \log^{1+2\alpha} n)$, for $\alpha \in [0,1]$, fails to match the theoretical lower bound of $AT^{2\alpha} = \Omega(n^{1+\alpha})$, it does much better, in the asymptotic sense, than the serial pipeline multipliers used in digital signal processing [JKM68], for which $A = O(n)$ and $T = O(n)$, and hence $AT^{2\alpha} = O(n^{1+2\alpha})$.

The gap between upper and lower bounds was closed later by Preparata and Vuillemin [PrV81b], who presented a circuit based on the Cube Connected Cycles [PrV81a] that met the lower bound $AT^2 = O(n^2)$, for all values of T in the interval $\Omega(\log^2 n) \leq T \leq O(\sqrt{n})$, but required connections of greatly differing lengths. Preparata in [Pre81, Pre83] presented a circuit based on a mesh connected network of processors that emulated the cube-connected-cycles interconnection network and was optimal, but only for $T = O(\sqrt{n})$. Both these two solutions were based on applying the DFT to the vectors to be multiplied, performing the product componentwise, performing the inverse DFT, and finally transforming the resulting convolution product into an ordinary product.

Luk [Luk81] argued that these methods were not of practical use and presented a circuit based on the well known "divide-and-conquer" method of computing the product: The n-bit operands x and y are divided in two halves: $x = 2^{n/2}a + b$ and $y = 2^{n/2}c + d$, and hence $x \times y$ can be rewritten as $x \times y = 2^n a \times c + (a \times d + b \times c) + b \times d$. Now the problem is reduced to the computation of the four $n/2$-bit products, two shifts, and three additions. The computations of the products can be carried out by recursion until the size of the operands is sufficiently small that the product can be computed by table look up. While this circuit is not optimal, it can be easily implemented in practice. It has $A = O(n^2 \log^2 n)$, $T = O(\log^2 n)$; the area can be further reduced to $A = O(n^2)$.

It is worth mentioning that Yao [Yao81] pointed out that the relative complexity of integer addition versus multiplication depends on the representation chosen for the presentation, and for example in the logarithmic number system addition is more difficult than multiplication. However, they cannot be both "easy" in any system.

The results on integer multiplications have been extended to the multiplication of n by n matrices by Savage [Sav79], who showed that the any chip performing the matrix product of two such matrices, whose entries are elements of a finite ring, must satisfy the $AT^2 = \Omega(n^4)$ tradeoff. This lower bound is met by the complexity of the circuit presented by Kung and Leiserson in [MeC80].

In [RFS82] a design of a chip implementing matrix multiplication was proposed, the salient feature of the design was the fact that linear array was used for implementation thus keeping the number of I/O pads was kept constant.

4. References

[AbA80] H. Abelson and P. Andreae, "Information Transfer and Area-Time Tradeoffs for VLSI Matrix Multiplication", *Comm. ACM*, **23**, 1 (Jan. 1980), 20-23.

[BPP81] G. Bilardi, M. Pracchi and F. P. Preparata, "A Critique and an Apprisal of VLSI Models of Computation", in *VLSI Systems and Computations*, H.T. Kung, B. Sproull and G. Steele, (eds.), Oct. 1981, 81-88.

[BrK81] R. P. Brent and H. T. Kung, "The Area-Time Complexity of Binary Multiplication", *J. ACM*, **28**, 3 (July 1981), 521-534.

[ChM81] B. Chazelle and L. Monier, "A Model of Computation for VLSI with Related Complexity Results", *Proc. 13-th Annual ACM Symp. on Theory of Computing*, May 1981, 209-213.

[FPP82] H. Fuchs, J. Poulton, A. Paeth and A. Bell, "Developing Pixel-Planes, a Smart Memory-Based Raster Graphics System", *Proc. Conf. on Advanced Research in VLSI*, Jan. 1982, 137-146.

[JKM68] L. B. Jackson, S. F. Kaiser and H. S. McDonald, "An Approach to the Implementation of Digital Filters", *IEEE Trans. on Audio and Electro-acoustic*, **AU-16**, (Sep. 1968), 413-421.

[KeZ81] Z. M. Kedem and A. Zorat, "Replication of Inputs May Save Computational Resources in VLSI", in *VLSI Systems and Computations*, H.T. Kung, B. Sproull and G. Steele, (eds.), Oct. 1981, 52-60.

[Ked82] Z. M. Kedem, "Optimal Allocation of Computational Resources in VLSI", *Proc. 23-rd Annual Symp. on Foundations of Computer Science*, Nov. 1982, 379-385.

[Kis82] G. Kissin, "Measuring Energy Consumption in VLSI Circuits: a Foundation", *Proc. 14-th Annual ACM Symp. on Theory of Computing*, May 1982, 99-104.

[KLL81] D. Kleitman, F. T. Leighton, M. Lepley and G. L. Miller, "New Layouts for the Shuffle-Exchange Graph", *Proc. 13-th Annual ACM Symp. on Theory of Computing*, May 1981, 278-292.

[Knu73] D. Knuth, *The Art of Computer Programming*, Addison Wesley, Reading, MA, 1973.

[LiS81] R. J. Lipton and R. Sedgewick, "Lower Bounds for VLSI", *Proc. 13-th Annual ACM Symp. on Theory of Computing*, May 1981, 300-307.

[Luk81] W. K. Luk, "A Regular Layout for Parallel Multiplier of $O(\log^2 N)$ Time", in *VLSI Systems and Computations*, H.T. Kung, B. Sproull and G. Steele, (eds.), Oct. 1981, 317-326.

[MeC80] C. Mead and L. Conway, *Introduction to VLSI Systems*, Addison-Wesley, Reading, Mass., 1980.

[Pre81] F. P. Preparata, "A Mesh-Connected Area-Time Optimal VLSI Integer Multiplier", in *VLSI Systems and Computations*, H.T. Kung, B. Sproull and G. Steele, (eds.), Oct. 1981, 311-316.

[PrV81a] F. P. Preparata and J. Vuillemin, "The Cube-Connected-Cycles: a Versatile Network for Parallel Computation", *Comm. ACM*, **24**, 5 (May 1981), 300-309.

[PrV81b] F. P. Preparata and J. Vuillemin, "Area-Time Optimal VLSI Networks for Computing Integer Multiplication and Discrete Fourier Transforms", *Proc. of I.C.A.L.P. Symposium*, Haifa, Israel, July 1981.

[Pre83] F. P. Preparata, "A Mesh-Connected Area-Time Optimal VLSI Multiplier for Large Integers", *IEEE Trans. on Computers,* **C-32**, 2 (Feb. 1983), 194-198.

[RFS82] I. V. Ramakrishnan, D. S. Fussell and A. Silberschatz, "Systolic Matrix Multiplication on Linear Array", *Proc. 20-th Annual Allerton Conf. on Computing, Control and Communication,* Oct. 1982.

[Ros81] A. L. Rosenberg, "Three-Dimensional Integrated Circuitry", in *VLSI Systems and Computations,* H.T. Kung, B. Sproull and G. Steele, (eds.), Oct. 1981, 69-80.

[Sav79] J. E. Savage, "Area-Time Tradeoffs for Matrix Multiplication and Related problems in VLSI", Tech. Rept. CS-50, Dept. of Computer Science, Brown Univ. Providence, RI, 02912, Aug. 1979.

[Sto71] H. S. Stone, "Parallel Processing with the Perfect Shuffle", *IEEE Trans. on Computers,* **C-20**, 2 (Feb. 1971), 153-161.

[Tho79] C. D. Thompson, "Area-Time Complexity for VLSI", *Proc. 11-th Annual ACM Symp. on Theory of Computing,* Apr. 1979, 81-88.

[Tho80] C. D. Thompson, "A Complexity Theory for VLSI", Ph.D. Diss., Dept. of Computer Science, Carnegie-Mellon University, Aug. 1980.

[Vui83] J. Vuillemin, "A Combinatorial Limit to the Computing Power of VLSI Circuits", *IEEE Trans. on Computers,* **C-32**, 3 (Mar. 1983), 294-300.

[Yao81] A. C. Yao, "The Entropic Limits of VLSI Computations", *Proc. 13-th Annual ACM Symp. on Theory of Computing,* May 1981, 308-311.

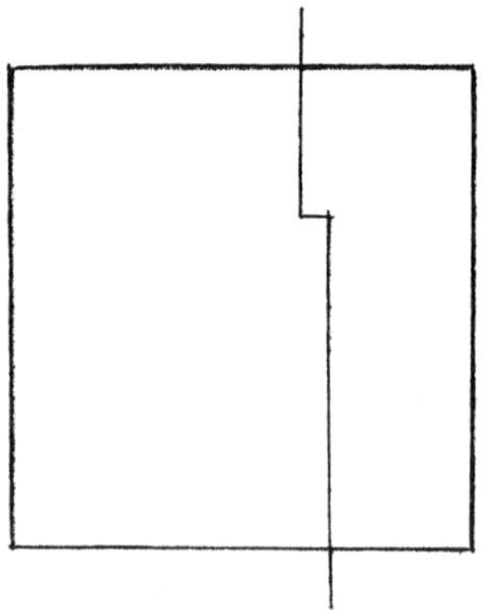

Figure 1.

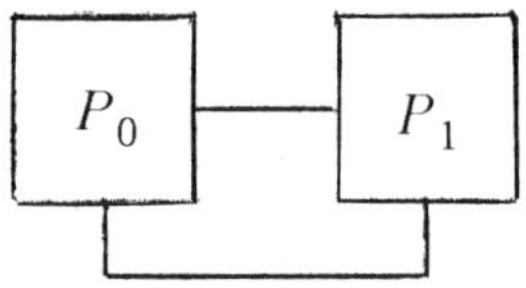

Figure 2.

LAYOUT STRATEGIES FOR NMOS-CMOS VLSI

F. ANCEAU

IMAG Computer Architecture Group
BP 68 - 38402 ST MARTIN D'HERES - France

ABSTRACT

The design of the layout of complex ICs at the level of quality needed for mass production needs to follow some topological rules about the organization of their floor plan. The shape of the blocks must be adjusted to their environment by using their deformability function. A general deformability function which may be parametrized to each kind of functional block will be presented. The wiring strategies of the blocks come from their organization by the crossing of two straight data and control streams. The allocation of these streams to conductive layers depends strongly on the features of the technologies used. The case of NMOS and CMOS simple or double poly/metal layers will be considered . The organization of the layout by strips of overlapped logic cells and connections will be presented as a good engineering solution. These results allow the prediction of the organization of the floor plan of future VLSI from their functional specifications.

1.INTRODUCTION

The optimization of the layout of VLSI designed for mass production follows some simple rules:

- The shape of blocks which constitute a VSLI must be adjusted to their destination place. The modification of the shape of blocks is controlled by their deformability function.
- Blocks interconnections must be optimised by:
  - designing blocks which may be directly "plugged" together.
  - passing-through a block by a connection instead of turning around it. The availability of a block to be crossed by a connection without any dimensional change is given by its "transparancy" functions per direction and per conductive layer.
  - using the space under the connections to put the transistors. Statistical results [1] show that in VLSI circuits the area occupied by the connections is greater than, or equal, to the area occupied by the functional cells.

Obviously, following these rules leads to designing larger layout for the cells than their minimum possible area (by aproximatly a factor of 1.5). But the total area for the IC is reduced (by aproximately a factor of 0.25).

This layout strategy means that it is better to consider global optimization than local (cells) optimization. It is strongly linked to the structured approach for VLSI design by looking for the topological relationship between blocks at every level of the top-down approach.

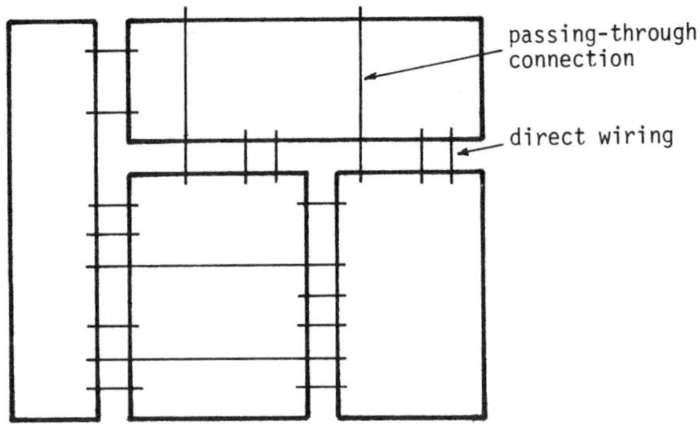

fig. 1 Typical organization of blocks and connections in a VLSI circuit

The drawbacks of this layout strategy are:

- a very low rate for the reusing of the layout of previously designed cells. The adaptation of the layout of a block to its environment makes it specific to a given application. The use of program-generated layout for blocks increases their generality and their reusability.
- the speed of a block may be decreased by the use of a non-optimal internal organization which is used to give it a requested shape. This is a trade-off situation where the designer uses his Art to balance a circuit between speed and silicon area. From a practical point of wiew, all the block of a VLSI do not need to be optimized in the same way (speed or silicon area).

Applying these rules implies the use of a floor plan to manage and control the design of a whole VLSI circuit. The floor plan is a data structure which allows the solving of topological and connection problems before the layout of the VLSI circuit will be completed. Drawing a floor plan needs to predict the area and the shape of the main blocks and their connections with a raisonable accuracy [2].

2.CELL DEFORMABILITY

The prediction of the area, shape and connections of a cell need the selection of an internal organization which satisfies these requirements. Experiments show that the general deformability function of the layout of a block is given by:

$$y = \frac{K}{x - delt} - delt$$

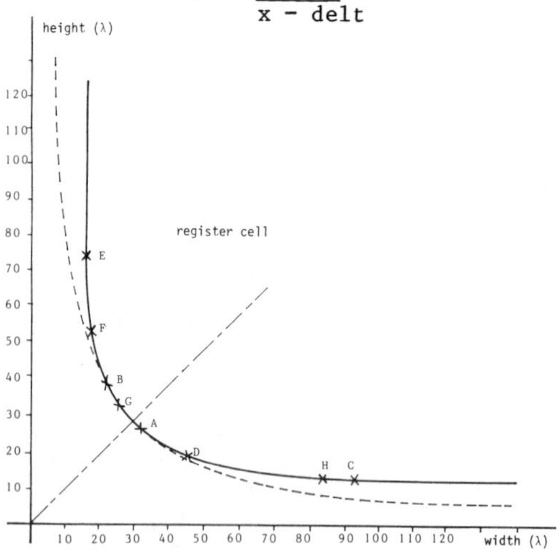

fig. 2 deformability curve of a layout block

Where "delt" is the minimum size of a block and K a coeficient linked to its minimum area. The value of delt is linked to the minimum size of elementary devices (transistors, contacts) like a layout granulometry and to the maximum levels of wiring in each direction. The internal organisation and the area of a block varies when its shape varies.

3.WIRING STRATEGY

The wiring strategy of a VLSI is the technique for using as efficiently as possible the conductive layers that technology provide. These conductive layers may be considered in the decreasing order of their availability to transmit electrical signals (metal, silicide, polysilicon, diffusion). The main problem to be solved by the wiring strategy is the use of these layers to make the connections according to their length.

Unfortunatly the practical number of conductive layers a technology provides which may be used to realize the connections is small because the first polysilicon layer and the diffusion layer are already used to make transistors.

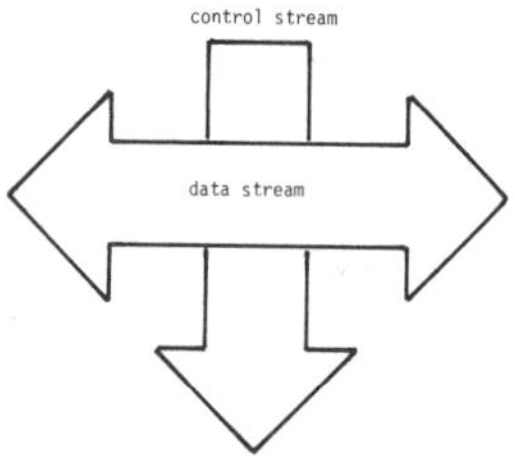

fig. 3 data and control streams

The layout of the main blocks of a VLSI is designed either by using well-known techniques elaborated for some specific blocks (ROM, RAM, PLA) or by using a general strategy (random logic) which consists in organizing the block as the crossing area of two data and control streams.

These streams are implemented by using the two best conductive layers that technology provides.

This strategy leads to the organization of the block as an array of a cells located at the crossing of each data wire and each control wire.

The transparancy of the cells to these conductive layers determines their wiring organization. An high transparancy (e.g. for the metal layer) allows the implementation of the corresponding stream as a set of straight wires running above the cells. A low transparancy (e.g. for the polysilicon layer used both for making transistors and wiring) requests the implementing of the corresponding stream into routing chanels between cells.

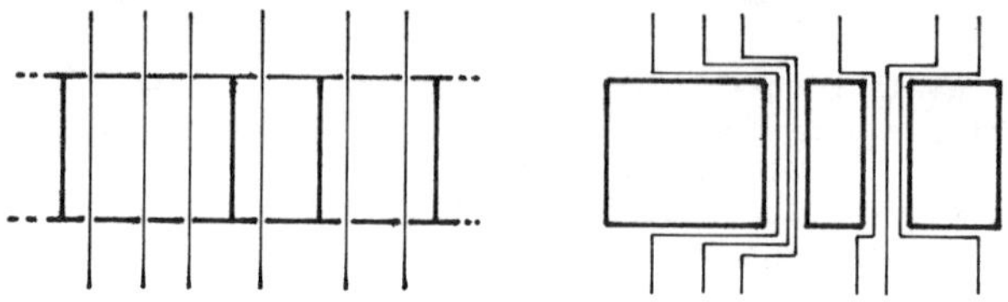

high transparancy low transparancy

fig. 4 wiring strategy depending upon cell transparancy

4.NMOS/HMOS TECHNOLOGIES

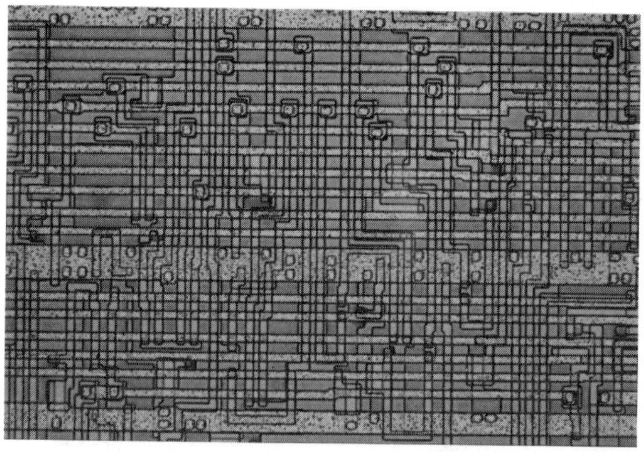

fig. 5 strip layout (from MC 6800 LSI processor)

The dissemblance of cell transparancy between the metal and polysilicon layers of the NMOS/HMOS technologies leads us to organize the layout for random logic blocks as strips of cells located between the teeth of the metalic power supply combs. The straight metalic wires, which implement a stream, run parallel with the power supply lines, above the cells which are transparent for this layer. The polysilicon wires which implement the other stream are located into routing chanels between cells which are not transparent for this layer.

The resolution of the topological problems in cells layout (e.g. feedback connections) and some technological features (e.g. butting contacts) introduces perturbations in the organization of the upper conductive layers which should be fully allocated to the realization of a stream. These perturbations are small for NMOS/HMOS technologies but higher for CMOS technologies.

When the technology which is used provides two conductives layers which may be exclusively used to implement connections (double poly or double metal), the contacts between the upper layer and the cells which are beneath make perturbations into the intermediate layer. In order to reduce these perturbations, it is necessary to allocate to the upper conductive layer the stream which is less connected to the cells. The distribution of power supply also must be allocated to the upper layer which is made of the thicker metal layer. In the case of the use of two metal layers, it is not possible to connect directly the upper metal layer to the cells. The intermediate metal layer must be used as a relay. The layout organization (e.g. HP 9000) look like those used for single poly and single metal technology. Strips of cells are developed in the direction of the intermediate metal layer which contains straps which make the relays between cells and the upper metal layer.

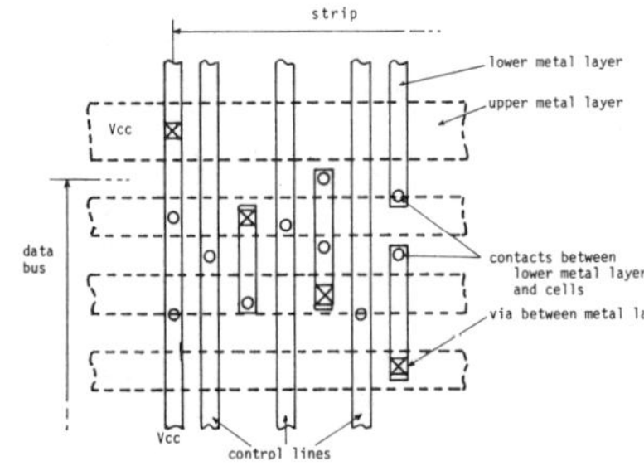

fig. 6 double metal NMOS/HMOS layout

5.TRANSPARANCY FUNCTION

The transparancy "T" of a set of cells in a strip, for the metal wires, in NMOS technology is given by a decreasing exponent function which indicates how many wires may manage to pass through this set of cells.

$$T = 0.7\ (Xp\ /\ pm)\ e^{-\ Ym\ /\ 20\ pp}$$

Where Xp and Ym indicates the block sizes in the poly and metal directions and pp and pm are the polysilicon and metal pitches (including alternate contacts). Statistical study [2] has shown that the mean number of logical metal wires in a strip is between 6 and 8. This value is used to estimate the strip pitch during the floor plan prediction step.

6.LAYOUT DENSITY FOR TRANSISTORS

The same statistical study has shown that the product of polysilicon and metal pitches is a good unit to measure the layout density for transistors for several NMOS/HMOS technologies.

The mean layout area per transistor is close to 4 pp.pm for hand optimized cell layout and 13 pp.pm for a large layout area. Large and regular structures which obey the optimization rules (e.g. data path) have a mean layout area of 6 pp.pm per transistor.

7.CMOS TECHNOLOGIES

The use of CMOS technology introduces new constraints which are:

- static circuitry requests that the gate of each n-transistor must be connected to a gate of a p-transistor. This feature increases the topological problems to layout gates.
- the polysilicon layer (n-doped) cannot be directly connected to p^+ diffusion without making a parasitic diode.

7.1.CMOS single poly and single metal technologies

Using these technologies brings us to the point of organizing each CMOS logical gate as two facing diffusion strips (n^+ and p^+) crossed by common polysilicon strips making the gates of the p and n type of transistors.

Every static gate (except inverter) contains transistors in parallel which are obtained by making strap(s) on one of the diffusion strips. The alignment of each type of diffusion strip requests that the teeth of the power suply combs be parallel to these diffusion strips to feed each type of transistor. wells are also organized as large strips in the same direction.

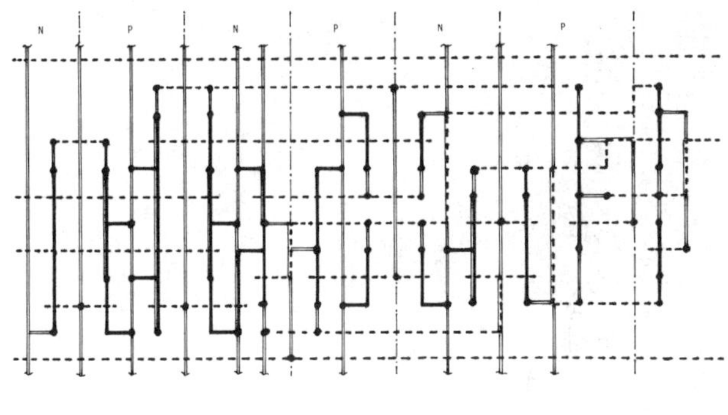

— dif n^+,p^+ --- poly ═ metal —·— well

fig. 7 CMOS single poly single metal layout

The layout is also organised in the form of strips. Metal lines run parallel to diffusion strips between the teeth of metal power supply combs. The straps used to put transistors in parallel are also made of metal because they run also in the same direction. The n and p transistor networks of each gate are connected by a poly strip connected to the n^+ diffusion area by a buried contact and to the p^+ diffusion area by a butting contact. This dissymetrical feature suggests the use of NAND gates for logic because their p-transistors are put in parallel by a metal wire directly connected to the butting contact using the same layout track. Electrical considerations also suggest to use NAND gates.

7.2.CMOS double poly or double metal technologies

The use of a second polysilicon layer allows to

use the first polysilicon layer only to make transistor gates (which may be independent), which eliminates the topological problems of CMOS single poly. That allows the use of diffusion straps to put transistors in parallel. The two pure conductive layers are used to implement the data and control streams. The output strap of a gate connecting n^+ and p^+ diffusion area is made with the metal layer which may be directly connected to these booth diffusions. Then this wiring layer becomes parallel to the gates of the transistors and is used to connect n and p transistor networks. The other wiring layer and local power supply must be implemented by the second polysilicon layer. The low power consumption of static gates allows them to be fed locally by poly wires. The main distribution of power supply is made with the upper metal layer. In the case of the use of two metal layers it seems better to organize the layout as in the case of single poly and single metal technology where intermediate metal layer plays the role of the single metal layer. As for NMOS/HMOS, straps in this layer are used as relays to connect cells to the upper metal layer.

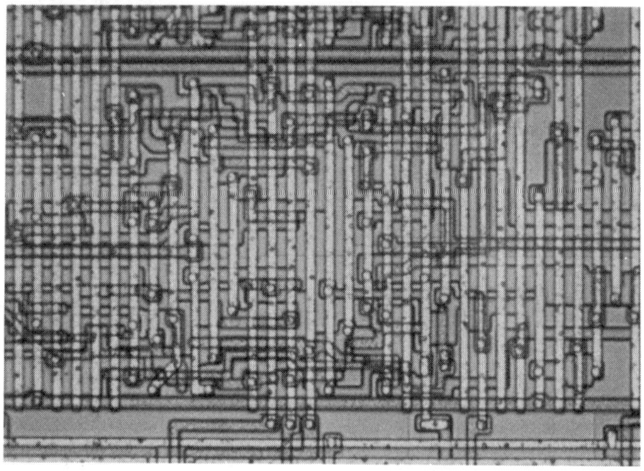

fig. 8 CMOS double poly layout (from NSC 800 LSI processor)

8.A TOOL FOR FLOOR PLANING

The design of the floor plan of a VLSI must be done with an accuracy level depending upon the advance of the design process for each block. These levels are:

- a first level of estimation of the size and the shape of the main blocks is useful during the preliminary stage of the floor planning in order to create the main topological structure of the chip. This first estimation level is based on the estimation of the block complexity. The shape of the block may be adjusted to its destination place by using the standard deformability function.

At this level the connections are given as a set of equipotential names connected to the ports of the blocks.

- a more accurate second level of estimation is based on the choice of internal organizations for the blocks. Each block organization gives a possible location of its ports. This second level is used to improve the floor plan accuracy and extend the topological structure inside the blocks.

At this level a pre-routing of the connections consists in the allocation of the equipotential to the routing chanels, pass-through wires, etc.. These informations allows the estimation of the size of the routing chanels and the change in the size of the blocks due to the connections passing through them.

- the third level consists of the real layout of the blocks and connections.

These three levels are used to progressively create and extend the chip structure. The tool we are developing is called FLOPE-TESS [2]. It consist of three main parts:

- a data structure representing:
  - the topological relationship between the blocks
  - the hierachical structure of the imbrication of the blocks
  - the functional structure of the connections
- a graphical editor used to dialogue with the designer for:
  - moving, turning, mirroring the blocks
  - getting the topological constraints the designer applies to the blocks
  - getting the information about the pre-routing of the connections
- a set of estimation routines consisting of:
  - the standard routines for size and deformability prediction
  - the specific evaluation routines which decide the suitable internal organization of each type of block

9.CONCLUSION

The layout strategy which is presented here is an extension of the engineering approach used to layout commercial LSI. Its application gives a signifiant silicon area reduction and a decreasing of layout design cost by decreasing the amount of connection to route the layout blocks, and by an increasing of the layout regularity. The knowledge of statistical results allows the prediction of the size and shape of the main blocks of a VLSI circuit in order to build progressively its floor plan.

10.REFERENCES (in English)

[1] W.R. HELLER, W.F. MIKHAIL, W.E. DONATH
Prediction of wiring space requirement for VLSI
Journal of Design Automation and Fault-tolerant Computing, pp.117/144, 1978

[2] F. ANCEAU, R. REIS
Complex integrated circuit design strategy
IEEE Journal of Solid State Circuits, vol.SC 17, n.3, pp.458/464, june 1982

[3] A. GUYOT, I. SUPRIANA
Top-down design of masks layout
submited to ESSIRC 83 conf.

Optimization Techniques for IC Layout and Compaction

Gershon Kedem and Hiroyuki Watanabe

Computer Science Department
University of Rochester
Rochester, NY 14627

Abstract

This paper describes a new approach for IC layout and compaction. The compaction problem is translated into a mixed integer-linear programming problem of a very special form. A graph based optimization algorithm is used to solve the resulting problem. An experimental program that uses the above techniques is described. The program could be used either as an aid to hand layout or as the bottom part of an automatic layout generation program.

1. Introduction

One of the most time consuming and error prone parts of custom integrated circuit design is low level cell layout. The logic designer can design a topological layout of a digital circuit in the form of a "stick diagram" [Mead & Conway 1979; Williams 1977] quite rapidly. However, the translation of this design into an actual layout is tedious and error prone. Moreover, the same stick diagram can be mapped to a mask level layout in many different ways. On different occasions it is desirable to layout the same functional block differently. For example, the aspect ratio of the block might need to be changed so it could fit other parts of the design, or transistor sizes might need to be changed depending on the "fanout" load they have to drive.

Having a library of cells in stick diagram form together with a set of procedures to translate them into layout is much more flexible and therefore more useful than a standard cell library. The cells could be scaled automatically according to fan in and fan out considerations, power buses could be scaled according to the current requirements, their shapes could be altered automatically to fit other parts, and so on. Moreover changes in design rules will not render the cells obsolete.

Several algorithms that translate stick diagrams into layout have appeared in the literature [Williams 1978; Dunlop 1978; Hsueh 1979a]. However, these algorithms are heuristic and *ad hoc*. Our approach is to translate the layout compaction problem into an optimization problem.

The compaction problem is stated as a minimization problem; that is,

find the minimum of the area function subject to linear and nonlinear constraints.

The optimization problem resulting is a mixed integer linear programming problem of a very special kind. The integer variables are {0,1} decision variables. For every fixed set of decision variables, one needs to solve two independent longest-path problems, one for the horizontal direction and the other for the vertical one. The interaction between the horizontal and vertical compaction is via the integer decision variables. More details on the problem formulation are given in Section 2.

The linear constraints come from geometrical design rules, from connectivity information and possibly from user specified constraints.

The ability to specify additional constraints to the problem is very useful. For example, when one designs a cell that is to be replicated many times (a popular example is the shift register cell in [Mead & Conway]), it is important to be able to specify that the center of an input node should be at the same height as the center of a given output node. Using our technique, it is easy to specify relations between connection points. It is also easy to compact cells in one direction while keeping the other dimension fixed. This feature is useful when one makes cells fit one another. Moreover, by manipulating the objective function it is possible to influence the final shape of the compacted cells.

We would like to stress that our compaction scheme is truly a two-dimensional compaction. Previous methods mentioned in the literature [Williams 1978; Dunlop 1978; Hsueh 1979a] use two one-dimensional compaction schemes.

The circuit compaction problem is made of three parts:

1) Translation of stick diagrams into an optimization problem.

2) The solution of the optimization problem

3) Making modifications to the stick diagrams and mathematical optimization to try and improve the layout; for example, inserting jog points or symbol rotations.

In Section 2, we discuss the problem formulation and the translation from stick diagrams to the mathematical optimization problem. In Section 3, we describe the branch and bound technique we use to solve the optimization problem. In Section 4, we give examples and some data about the experimental program we have developed.

2. Problem Formulation and Constraint Generation

Stick diagrams are a well known symbolic form for describing integrated circuits (see [Williams 1977; Mead & Conway 1979]). Stick diagrams provide a topological description of the circuit specifying transistor size and type, the connecting lines, the layers they are made of and their connections to other lines or transistors. Loosely speaking, the layout is made of hard fixed-shape cells. Cells are pull-up and pull-down transistors, pass transistors, capacitors, and contacts. Interconnecting these cells are rubber-band like lines that could be stretched and contracted along the horizontal or vertical axis, but have a fixed width. The width of lines is either the default minimal width or larger.

In our system, the input is a CIF file. Each transistor or contact is an instance of a predefined symbol from a library. Interconnecting these cells are line segments. Each line segment is terminated either by touching a symbol or another line segment or by touching the outside boundary of the global cell. By convention, lines that are drawn with width less than or equal to the minimal width are assumed to be of minimal width. Lines that are wider are assumed to be constrained to be the width at which they are drawn. All line segments are either horizontal or vertical. All vertical or horizontal line segments stay vertical or horizontal respectively, under the compaction process. Symbol orientations stay the same, too. Figure 1 shows an example of the input to the system.

Circuit Compaction as Mixed Integer Linear Programming

We are now ready to describe the translation of the compaction problem into an optimization problem:

Each symbol in the diagram contributes two variables to the problem: the x coordinate and the y coordinate of its center. Each vertical line contributes one variable: the x coordinate of its center. The length of the line is determined by the position of the two other objects to which it is connected. Similarly, each horizontal line contributes one variable, namely the y coordinate of its center. By assigning values to these variables, one can generate a layout.

The design rules and the connectivity requirements are encoded as a set of inequalities.

CH1935-6/83/0000/0709$01.00

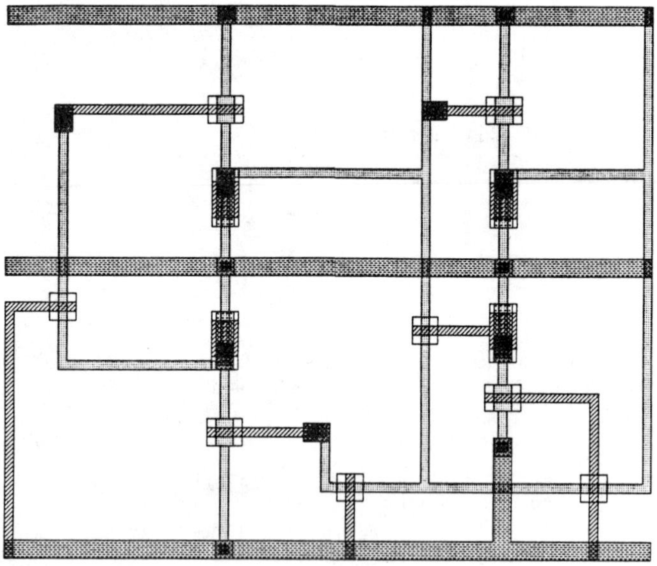

Figure 1: Stick diagram of T-flipflop

Geometrical design rules come in three varieties, separation rules, line-width rules, and overlap rules. Since all transistors and contacts are predesigned, overlap rules are automatically satisfied. Line-width rules are built-in; that is, the width of each line is computed and is assumed to be fixed. Line-width enters into the calculations as constants in the constraint generation part.

Separation rules give rise to two kinds of inequalities: simple linear inequalities or a pair of inequalities that are mutually exclusive. In Figure 2, we illustrate a simple constraint. The distance requirement between a polysilicon line and a poly-metal contact is encoded as

$$x_1 - x_2 \geq 5$$

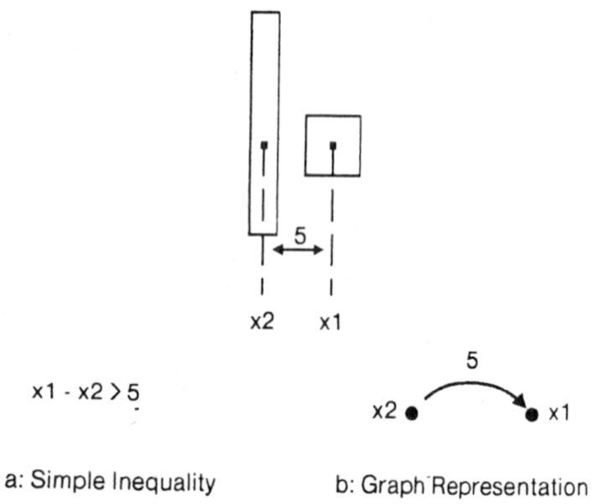

Figure 2: Simple Constraint

A more complicated case arises when two pieces interact but one could be to the right of the other *or* below it. Figure 3 shows such a case. Symbol 2 could be either d_1 units to the right of symbol 1 or d_2 units below it.

$$x_2 - x_1 \geq d_1$$

or

$$y_1 - y_2 \geq d_2$$

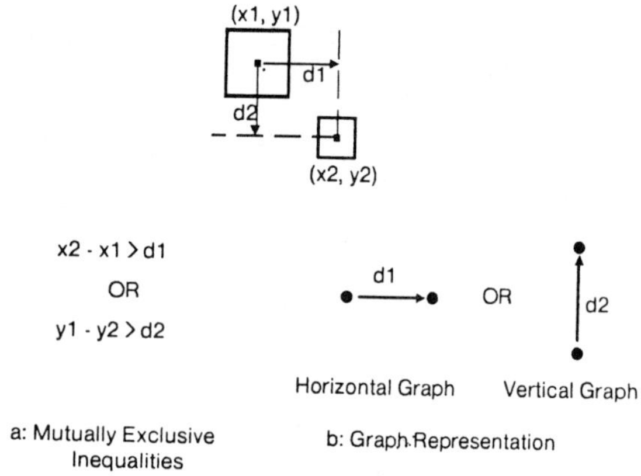

Figure 3: Mutually Exclusive Constraints

This pair of inequalities gives rise to a non-convex feasible region which cannot be coded by linear inequalities alone. It is possible however to encode these requirements as mixed linear-integer inequalities.

$$x_2 - x_1 \geq d_1 - \delta K$$

$$y_1 - y_2 \geq d_2 - (1-\delta)K$$

δ is 0 or 1 and K is a sufficiently large number.

Another type of constraints comes from connectivity requirements. In the course of compaction, one has to maintain connections between the lines and the corresponding symbols. A connection point between a line and a symbol does not have to be fixed. It can slide along the edge of the connecting layer, as long as it does not violate line-width or distance rules. The connectivity requirement is coded as a pair of inequalities. For example, Figure 4 shows a vertical line connected to a symbol. The vertical line could move along the bottom edge of the symbol. It could be connected at the left corner (Figure 4.a), the right corner (Figure 4.b), or in between. These connection requirements are coded as the pair of inequalities: $x_1 - x_2 \leq d_1$; $x_2 - x_1 \leq d_2$. As long as these inequalities are satisfied, the line will be properly connected to the symbol.

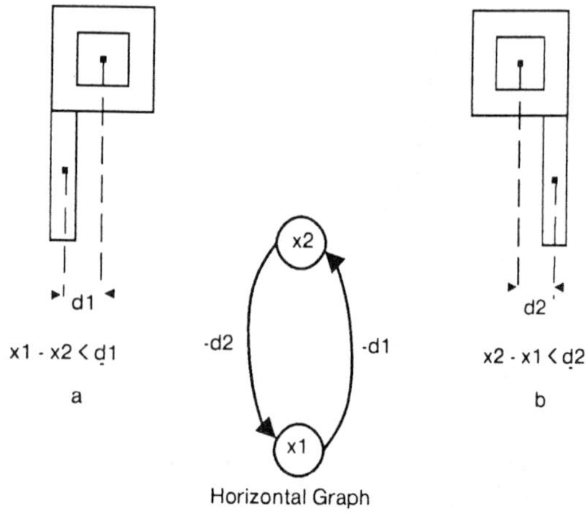

Figure 4: Connectivity Constraints

In addition to the horizontal and vertical variables, there are four special variables, S_x, T_x, S_y and T_y. S_x is the x coordinate of the left boundary of the bounding box and T_x is the coordinate of the right boundary of the box. S_y is the bottom and T_y is the top. Simple inequalities are generated between each bounding box boundary and all the objects that it faces. The objective function that one is to minimize is the area of the cell $(T_x - S_x) \cdot (T_y - S_y)$. This function could be approximated by the linear function $(T_x - S_x) + (T_y - S_y)$.

Circuit Compaction as a Graph Optimization Problem

Although the compaction problem could be solved as a general mixed integer-linear programming problem, doing so would be very inefficient and expensive. Since all the linear (non-integer) inequalities are of the form $x_i - x_j \geq d$, the optimization can be translated into a simpler graph optimization problem.

There are two graphs G_x and G_y that encode the set of constraints, one for horizontal constraints, G_x, and one for vertical constraints, G_y.

Each horizontal variable x_i is a node in the graph G_x, and each y_i is a node in G_y. The graphs are directed graphs. Each arc in the graph corresponds to an inequality. Each arc (x_i, x_j) has a weight w_{ij} attached to it. The arc corresponds to the inquality

$$x_i - x_j \geq w_{ij}.$$

In addition to the two graphs, there is a set of decision variables I. Each decision variable d_i, corresponds to a pair of arcs (e^x_i, e^y_i). Each decision variable can take the values {0,1}. Each pair of weighted arcs (e^x_i, e^y_i) and the corresponding decision variable $d \in I$ encode the dual pair of inequalities

$$x_{i1} - x_{i2} \geq w^x_{i1,j2} \text{ or } y_{i1} - y_{i2} \geq w^y_{i1,i2}$$

If the variable d is equal to 1, the first inequality applies; if d is equal to 0, the second one applies.

The optimization problem

For each node $x_i \in G_x$ and $y_i \in G_y$, we assign values x_i and y_i respectively. (We use the node name and its value interchangeably, hopefully without confusing the reader).

The vector of values $z = \langle x, y, d \rangle$ is said to be feasible if the following conditions are satisfied:

For each simple or connection arc, $e = \langle x_i, x_j \rangle$

$$x_i - x_j \geq w^x_{ij}$$

Similarly, for each arc, $e = \langle y_i, y_j \rangle$

$$y_i - y_j \geq w^y_{ij}$$

and for each $d_i \in I$

$$x_i - x_j \geq w^x_{ij} \text{ if } d_i = 1 \qquad y_i - y_j \geq w^y_{ij} \text{ if } d_i = 0$$

where d_i corresponds to the pair of dual arcs (e^x_{ij}, e^y_{ij}).

Let $\Omega = \{z \mid z = \langle x, y, d \rangle ; z \text{ is feasible}\}$

be the set of feasible points. Our problem is

Problem 1:

$$\min_{z \in \Omega} f(x, y) .$$

f(x, y) is the objective function. Ordinarily the objective function is the area occupied by the cell. In that case,

$$f(x, y) = (Tx - Sx)(Ty - Sy).$$

It is possible, however, to manipulate the objective function in order to get certain effects. For example, if compacting in the horizontal direction is more desirable than compacting in the vertical one, $f = (Tx - Sx)(Ty - Sy) + C(Tx - Sx)$ with large C could be used.

3. Methods for Solving the Mathematical Problem

In the previous section a detailed description of the graph optimization problem to be solved was given. A full description and analysis of the solution methods and algorithms one could use for solving the above problem cannot be given here; it will be reported elsewhere [Watanabe]. We are, however, going to outline the major ideas and algorithms that are used to solve the problem.

Problem 1 is made of two parts - an integer part and a graph part. Let **d** be a vector of decision variables $\mathbf{d} = \langle d_1, ..., d_{|I|} \rangle$ $d_i \in \{0,1\}$. Clearly **d** can take $2^{|I|}$ different values. A few observations are in order:

1) For each fixed value of **d** one has two separable graph problems, one for the x-variables and one for the y-variables.

2) Since the objective function is monotone in $(T_x - S_x)$ and $(T_y - S_y)$, each one of the graph optimization problems is simply the longest path algorithm [Lawler 1976].

3) Although the graphs G_x and G_y are not acyclic, they are "almost" acyclic. Only a simple modification of the longest path algorithm is needed.

4) If $\mathbf{d}_x^1$ is different from $\mathbf{d}_x^2$ only in one component, say $d_i^1 = 1$ and $d_i^2 = 0$, then $g_x(\mathbf{d}^1)$ is different from $g_x(\mathbf{d}^2)$ only by one arc. Namely, $g_x(\mathbf{d}^2)$ is obtained from $g_x(\mathbf{d}^1)$ by removing the arc that corresponds to the decision variable d_i. Similarly $g_y(\mathbf{d}^2)$ can be obtained from $g_y(\mathbf{d}^1)$ by adding the arc that corresponds to d_i. Therefore, if one has solutions for the longest path problems for $g_x(\mathbf{d}^1)$ and $g_y(\mathbf{d}^1)$, finding the longest path solution for $g_x(\mathbf{d}^2)$ and $g_y(\mathbf{d}^2)$ could be done by updating the longest path solution at hand with fewer steps than otherwise would be required.

The algorithm used to find the best decision variables combination is "branch and bound" (see [Garfinkel and Nemhauser 1972]). The "branch and bound" technique is basically an exhaustive search by enumeration. The set of all possible values of the decision variables can be thought of as a complete binary decision tree. The leaves of the decision tree correspond to a fixed set of values of the decision variables. Branch and bound technique is a depth first search for the best answer. At each stage of going down the decision tree, one more 0-1 variable is being fixed. The main idea behind branch and bound is to compute at each stage a global upper bound and a local lower bound for the subtree that is searched. Each upper bound is a feasible solution. The lower bound is not. The moment a lower bound exceeds the best upper bound that was found so far, the corresponding subtree can be ignored since no feasible solution for that subtree will improve the global result.

Each lower bound is computed by dropping the constraints that correspond to decision variables that have not been fixed yet. An upper bound is computed by picking a specific value for the decision variables that agrees with the set of decision variables that have been fixed.

The algorithm starts by computing an initial guess. The Branch and Bound (BB) is used as a *correction* to this initial guess. At the beginning of the BB algorithm, all the decision variables, d_i, are assigned values but are not fixed. At each step as one descends down the decision tree, one more decision variable is being fixed. At each branching point, upper and lower bounds are computed using subgraphs of G_x and G_y; an upper subgraph pair and a lower subgraph pair. The upper subgraphs consist of all simple and connection arcs and one arc from each dual pair of arcs.The lower subgraphs contain simple arcs, dual arcs, and only the dual arcs whose corresponding decision variables are fixed. Their value will not change by going deeper into the search tree. In order to decide which decision variable to fix next, the critical paths of the current upper graphs are searched. The variable that is not fixed and is estimated to give the best improvement is selected and its value changed from 0 to 1 or vice versa.

The fathoming conditions are:

a) The lower bound value is greater or equal to the best solution found so far.

b) The lower bound is equal to the current upper bound.

In either case no better global solution could be found in the current subtree and therefore this tree is pruned.

A few observations are in order:

1) The algorithm starts by computing an initial guess. The better the initial guess, the better branch and bound performs. We use a heuristic method for finding the initial guess. It has been observed that the initial guess yields a solution that is about 10% worse than the optimal one.

2) At each step of the branch and bound, an arc from the critical path of G_x and G_y is removed. Therefore, if some decision variable corresponds to arcs that do not become part of a critical path, that decision variable never becomes part of the branch and bound tree.

3) The choice of where to branch next is arbitrary. We use a heuristic to decide where to branch. In all cases tested so far, the optimal solution was found after few branchings (less than 200). However, verifying that a solution is optimal took a long time. How exhaustive a search to make, however, could be set by the user.

4) By modifying the pruning procedure of the branch and bound algorithm, it is possible to speed up the search tremendously but at the expense of perhaps finding only a suboptimal solution. In our experience, even the initial solution is a good suboptimal solution.

Adding User Defined Constraints

One of the advantages of formulating the computation as an optimization problem is that additional constraints imposed by the user or other programs could easily be incorporated into the formulation.

One useful feature is the ability to stretch cells to match the height of other cells while compacting them in the other direction. In Figure 5 we show two cells A and B that abut. Compacting Cell B without regard to the height of Cell A will not yield a good layout. By stretching B to match the height of Cell A, a smaller global area will result. To force Cell B to be at least the same height as A, a constraint could be put between the source and the sink of B with weight equal to the height of A. This constraint will automatically force the compaction to concentrate on reducing the width of B. To force B to be exactly at the same height as A, a penalty function could be used.

Another constraint that is easy to add is the requirement that two different points (say input and output) will be at the same height. This requirement is natural if one wants to design cells that can be replicated many times. This constraint can be encoded by putting connection arcs between the two points (see Figure 6). No modifications to the algorithm are needed.

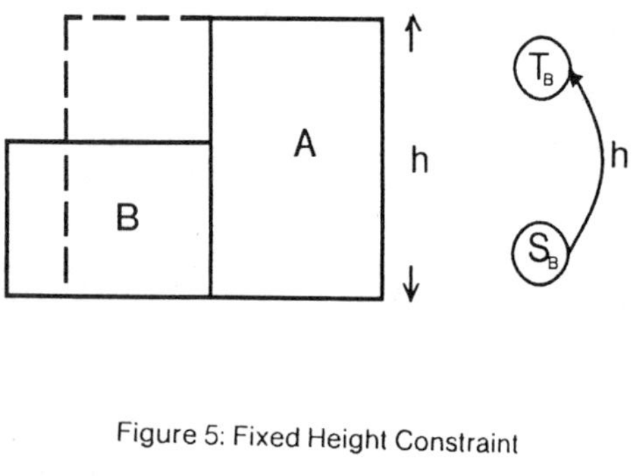

Figure 5: Fixed Height Constraint

4. Implementation, Examples, Conclusions

SQUASH, an experimental program to test our ideas was constructed. The program is written in PASCAL and running on a VAX 11/780 under the UNIX operating system. The input to the program is written in CIF and the output from the program is a CIF file. The program compacts NMOS stick diagrams using Mead & Conway design rules. A decision was made not to provide a fancy user interface. However, using CIF as the input/output language allowed us to use existing tools. Although the program works only with Mead & Conway design rules for NMOS, it can easily be adapted to other technologies and different design rules. Most of the technology and process-dependent rules are in table form. Also, there are no restrictions on the predefined library of cells. That set can have any user defined cells.

Our experience with the program has been very good. First, the method used for constructing an initial guess yields very good results. In most cases tested, the initial guess was within 15% of the optimal solution. Second, the heuristics used in the branch and bound search worked very well. Even though the number of decision variables can be large (100-4000 or more), an optimal solution is found quite fast, usually after less than 100 steps. Verifying that the solution is optimal, however, can be very expensive. Thousands of branch and bound steps may be necessary. To date, a good stopping strategy, except limiting the number of branch and bound steps, has not been found. Automatic insertion of jog points has not been implemented as yet.

In Figure 8 we show 3 super buffers that were generated from the same stick diagram. Figure 8.b shows a super buffer with driving stage double the size of the one in 8.a. 8.c shows the same super buffer as in 8.a but stretched to match the height of 8.b.

In Table 1 we give a summary of some representative test cases we used. The table gives the number of elements (pieces) in the stick diagram, the number of nodes and arcs in G_x and G_y, the number of decision variables, the area (in λ^2) that correspond to the initial guess, the best area found, the number of iterations in BB at the time BB stopped, the cumulative time it took to preprocess the cell, find the initial guess, find the best answer and final time. The final time includes the translation from optimization solution to an actual layout.

The compaction algorithms we presented here proved to be of significant practical value. The algorithms, as presented here, are suitable for layout of "leaf-cells," that is, cells of medium size. In our experience, the layout achieved is competitive with "hand-crafted" layout and the amount of computing required is small.

The compaction problem, in general, is Np complete [Lengauer 1982]. This however does not imply that good layouts for large circuits cannot be found. Moreover, our method could be "bootstraped" into an hierarchical compaction scheme. Preliminary experiments show that the same formulation can be very effectively used for packing blocks together (see Figure 9). Therefore the method could be used as part of a placement and routing procedure [Ciesielski 1983].

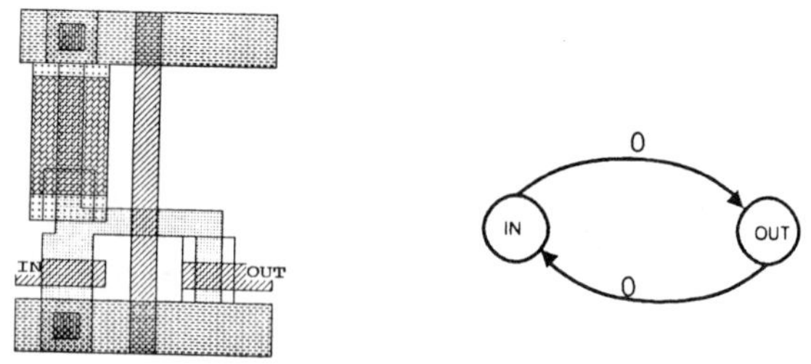

Figure 6: Periodicity Constraints

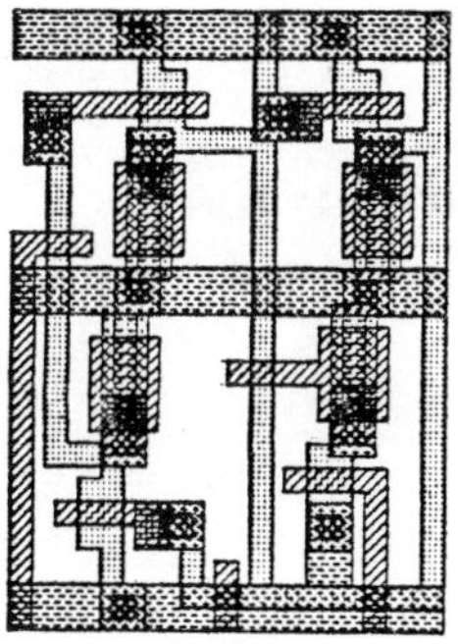

Figure 7: T-flipflop compacted

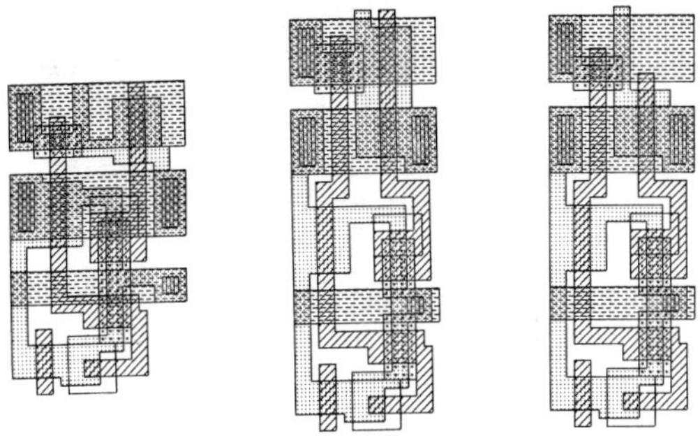

a b c

Figure 8

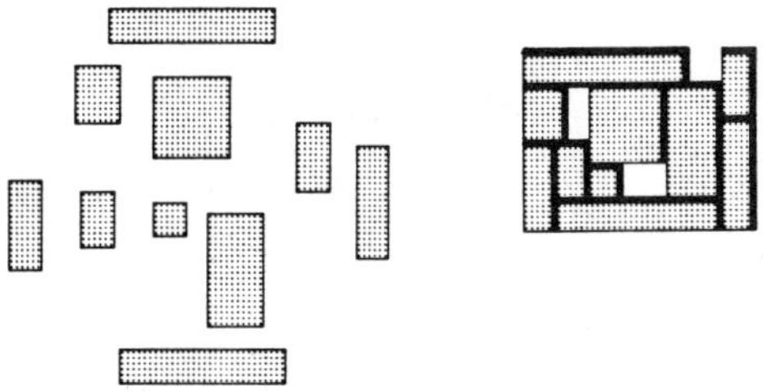

Figure 9: Block Packing

	sbuf	tff	prcell	alu	minmax	nemo
# of elements	57	69	210	221	269	394
x-nodes	31	47	125	137	182	253
x-arcs	93	156	492	544	838	1050
y-nodes	37	43	133	144	159	265
y-arcs	138	113	476	656	599	1122
0-1 variables	187	39	1052	1896	3005	2445
initial area	988	2067	8307	8320	9948	32512
best area	836	2014	7381	7272	7467	30988
best BB	14	1	89	48	84	19
final BB	300	12	300	300	300	19
cumulative time in seconds:						
preprocessing	15.97	12.93	60.46	82.32	141.27	158.95
initial guess	18.17	13.85	78.53	105.63	187.90	190.17
best BB	19.00	13.93	91.91	119.65	222.45	195.15
final BB	29.43	16.86	114.22	164.13	275.80	198.63

Table 1

Acknowledgements

The research reported in this article was supported in part by the Defense Advanced Research Projects Agency Grant No. N00014-82-K-0193, in part by ONR Grant No. N00014-80-C-0197, in part by Defense Advanced Research Projects Agency Grant No. N00014-78-C-0164 and in part by Semiconductor Research Corporation Grant No. 83-01-028.

References

Agrawala, A.K. and Kulkarni, A.U., "A Sequential Approach to the Extraction of Shapes Features," *Computer Graphics and Image Processing, 6*, 6, Dec. 1977, 538-557.

Ciesielski, M.J. and Kinnen, E., "Digraph Relaxation for Placement Modifications," *Proceedings, IEEE ISCAS*, 1983.

Dunlop, Alfred E., "SLIP: Symbolic Layout of Integrated Circuits with Compaction", *Computer Aided Design*, Vol. 10, No. 6, November 1978, 387-391.

Garfinkel, R. S. and Nemhauser, G. L., *Integer Programming*, John Wiley & Sons, Inc., 1972.

Hsueh, Min-Yu, "Symbolic Layout and Compaction of Integrated Circuit", Memorandum No. UCB/ERL M79/80, Electronics Research Lab., University of California Berkeley, December 1979a.

Hsueh, Min-Yu and Pederson, Donald O., "Computer-Aided Layout of LSI Circuit Building-Blocks", *Proceedings*, 1979 International Symposium on Circuits and Systems, Tokyo, Japan, 1979, 747-477.

Kedem, Gershon and Watanabe, H., "Graph-Optimization Techniques for IC-Layout and Compaction," *Proceedings*, 20th Design Automation Conf., Miami, FL., June 1983, 113-120.

Lawler, Eugene, *Combinatorial Optimization: Networks and Matroids*, Holt, Rinehart and Winston, 1976.

Lengauer, Thomas, "The Complexity of Compacting Hierarchically Specified Layouts of Integrated Circuits," *Proceedings*, 23rd Annual Symp. on Foundations of Computer Science, Nov, 1982, 358-368.

Mead, C. and Conway, L., *Introduction to VLSI Systems*, Addison-Wesley, 1979.

Mosteller, R. C., "REST A Leaf Cell Design System," *VLSI81*, J.P. Gray (Ed.), Academic Press, 1981.

Otten, R. H. J. M., and van Lier, M. C., "Automatic IC-Layout: The Geometry of The Islands", *Proceedings*, 1975 IEEE International Symposium on Circuits and Systems, Newton, Mass., April 1975, 231-234.

Williams, John D., "STICKS--A New Approach To LSI Design", Master's Thesis, Massachusetts Institute of Technology, June 1977.

Williams, John D., "STICKS--A Graphical Compiler for High Level LSI Design", *AFIPS Conference Proceedings*, Vol. 47, June 1978, 289-295.

Watanabe, Hiroyuki, "IC Layout Compaction Using Mathematical Optimization", Ph.D.Thesis, Department of Computer Science, University of Rochester, In preparation, 1983.

AN INTEGRATED STRUCTURE FOR DESIGN AUTOMATION

Hobart Cable, Alan Ross*, and Harold Carter**

Aeronautical Systems Division,
Wright-Patterson Air Force Base, Ohio

ABSTRACT

This paper proposes a construct for integrating Computer-Aided Design tools into a complete design system. The project described here provides an organization and framework for that integration so the tools can communicate with each other and be part of a system that is easy for the user to operate. The structure is organized around a partitioning of the system design process into three stages: Specification, Implementation, and Realization. Design problems at each stage, and the integration of the tools into a complete design environment, are also discussed.

INTRODUCTION

Very Large Scale Integrated (VLSI) circuit technology holds the promise of hundreds of thousands of transistors on a single Integrated Circuit (IC) chip, making system design an order of magnitude more difficult. The field of Computer-Aided Design (CAD) or Design automation (DA) has evolved to assist digital designers in manipulating the large amounts of data associated with complex designs. Many design tools have been developed, but significant effort is often required to reformat the output of one tool before it can be used as input to another. To deal with the complex and sophisticated IC's of the VLSI era, completely integrated systems of design aids still need to be developed. The research described in this paper has proposed such a system, designated Super-CAD (1).

GENERAL APPROACH

The Super-CAD design system has been characterized but not yet implemented. The intent of the project is to provide an integrated system of design tools such that a designer can input a set of requirements and interact with the system to produce a design, including a mask set for hardware fabrication and the software and/or firmware to run it. The goal is a system that is very easy to use, no matter which parts are selected. To achieve this goal, the system must exhibit certain characteristics.

First, a large system must be developed hierarchically, step-by-step, and be adaptable to changing technologies. Many current CAD tools are drawn together and new ones are added as they are developed. A master program, or Executive, controls these tools and the translations between them, and provides a continuous interface to the user. Even with the use of artificial intelligence concepts, interactive user inputs and decisions are required throughout the design process. An effective database system provides data capturing at all levels of this process and maintains libraries of design building blocks, from individual cells in an IC to entire subsystems comprised of many such cells. Finally, a crucial element of the system is the definition of the model, which emphasizes partitioning of the design process or modularity, and which must be highly flexible to permit easy modifications or additions.

THE MODEL

Design Stages

The design of a digital system goes through three distinct stages as the statement of a problem is transformed into a completed design: Specification, Implementation, and Realization [2]. (See Figure 1) First, designers analyze the problem to define the requirements and arrive at a set of specifications. Then they attempt to implement those specifications with currently available circuits, where possible, or newly designed ones. Finally, circuit layouts are accomplished to produce mask sets for fabrication of the IC's and/or circuit boards. The Super-CAD model addresses these three stages, with the major emphasis on the Implementation stage.

Specification. The Specification stage is most difficult to automate. It must transform a relatively simple, unstructured statement of the designer's requirements into a well-structured behavioral specification that can be used by the Implementation stage. Top-down and structured specification procedures are mandatory, to insure later stages in the design process can be completed. Modularity of the specification enhances technology-independence and

*Naval Postgraduate School, Monterey, California

**Air Force Institute of Technology, Wright-Patterson AFB, Ohio

CH1935-6/83/0000/0714$01.00

reduces the number of required design cycles. Developments in artificial intelligence will have a significant impact on this stage of the process.

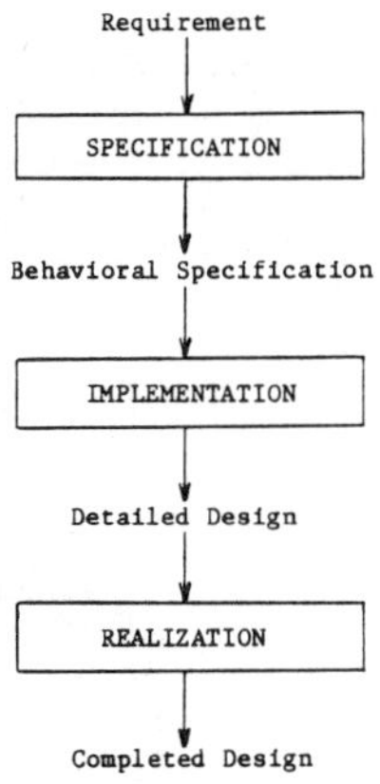

FIGURE 1. Design Stages

The goal is to produce a specification in a form that can be mapped to implementations at different levels in the Implementation stage. To facilitate this, the design specification is partitioned into subsets. Several approaches might be used for the partitioning, from requiring the designer to input his requirements in a hardware-description or register-transfer language, to having the system accomplish the partitioning with automated algorithms. When partitioning is complete, the results are sent to the Implementation stage.

Realization. In the Realization stage--where most design aids are currently available--the design produced during Implementation is turned into an actual physical design, with associated software, ready for fabrication. The detailed design is divided according to individual IC's and the hardware and software portions are separated. Then the layout of the IC's and/or circuit boards is completed and the software is coded. Finally, the complete hardware and software designs are combined to provide the total system package for the user.

Implementation.

The major emphasis of the Super-CAD project is on the Implementation stage. The input for this stage is in the form of a specification, structured so that it can be separated into subsets. A hierarchical approach to implementing the specification is based on different levels of abstraction: the Behavioral, Functional, and Logical levels [3] (See Figure 2). A design solution is sought at the highest, most general level of abstraction possible, leaving management of the lower levels of detail to the computer.

The highest Behavioral level provides a conceptual definition of the digital design, specifying the overall behavior of the system. The Functional level begins to deal with some of the specifics of the design, generally at the register-transfer level. Details of the actual implementation, however, are not considered until the Logical level, which specifies interconnections of primitive logic elements such as gates or flip-flops.

BEHAVIORAL LEVEL
Conceptual definition.

FUNCTIONAL LEVEL
Operational definition; register transfers.

LOGICAL LEVEL
Primitive logic elements.

FIGURE 2. Levels of Abstraction

Figure 3 demonstrates the hierarchical approach for completing a design at as high a level as possible. (The number labels are part of the hierarchical notation from the original report.) If all or part of the problem cannot be solved first with existing implementations at the Behavioral level, what remains is sent to the Functional level. If any part of the problem still cannot be implemented, it goes to the Logical level for final implementation.

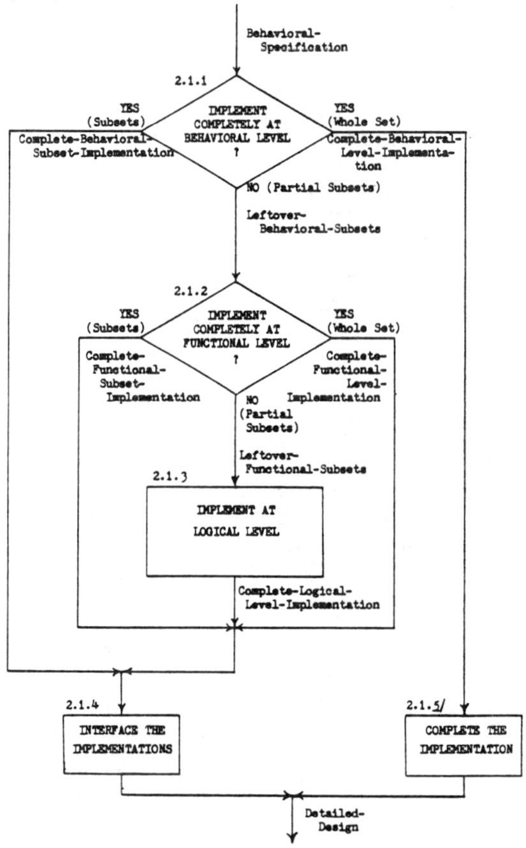

FIGURE 3. Design Development

At any step, whenever all subsets of the specification have been satisfied, the design is finished. All that remains is to interface the various implementations to complete the design.

The result of this approach is a top-down design which maps as many specification subsets as possible to existing implementations in the database libraries. It filters down to the level where enough details are present to complete the design, including--in some cases--synthesis of new circuits at the logic level. The process is supported by libraries of components at all levels, with new families being added as they become available.

At each level of abstraction, the Super-CAD system attempts to implement the complete behavioral specification by mapping it to an available implementation or set of implementations. This is demonstrated in Figure 4, using the Logical level as an example. If no suitable circuits can be found, the designer may choose to modify the specification to fit a "close" solution, as shown in Figure 5. This figure, incidentally, is an example of some of the lower levels of detail that are a part of the hierarchical Super-CAD model. If no implementation can satisfy the whole specification, the system attempts to implement subsets (Figure 4). This process is highly detailed, involving many separate steps. The system examines each specification subset individually and attempts to map it to an implementation. If successful, it saves the implementation in the database and goes on to the next subset. If unsuccessful, it searches for a "close" implementation and interacts with the designer to determine if it is worth modifying the specification of that subset. When previously implemented subsets are affected, their implementations may have to be modified or redone.

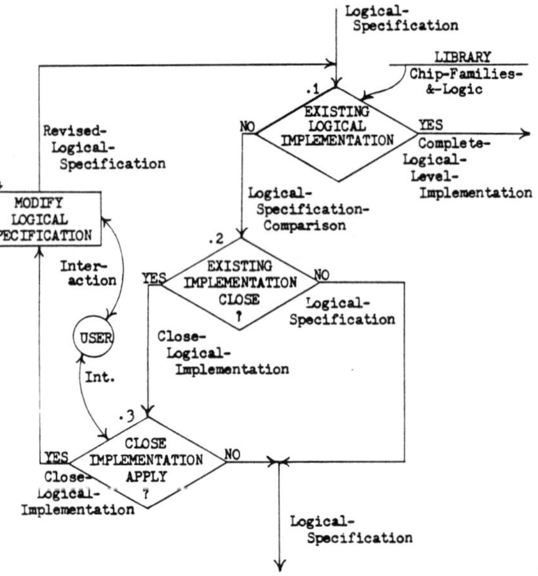

FIGURE 5.

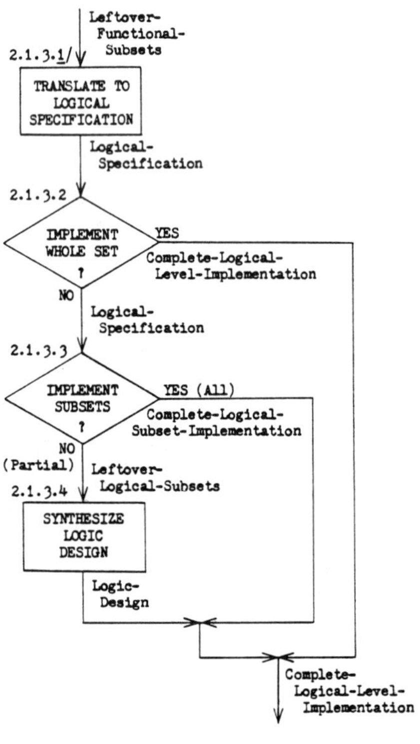

FIGURE 4. Logical Level

When all of the subsets have been examined at one level, the system checks for any that were not implemented and sends them down to the next level. If any subsets still remain after all the implementation steps have been completed at the Logical level, the system transforms them into logic design for new IC's (bottom block in Figure 4).

After the design has gone through all three levels of abstraction, the various implementations from the different levels are gathered together. Note in Figure 3 that any one design problem follows only a single path to arrive at a complete implementation. Simulations are then run to see if the design does indeed satisfy the specification requirements. While this can be a very complex process for VLSI circuits, designing primarily with existing families families of IC's means that those blocks have already been tested. The only simulations required are for integration into the whole design and any new components or IC's that have been added. Also, the system performs multi-level simulations to test all three levels of abstraction at once. When the design if finalized, it is sent to Realization.

The Implementation stage, then, is the heart of the Super-CAD model, where a partitioned behavioral specification is molded into a detailed design output. This is a complex operation with the design effort following but one of many possible paths. For the system to be effective, many design tools must be developed to support each step in the operation.

Summary. The Super-CAD model proposes a method to produce a top-down, structured design in three stages. First, the designer inputs his requirements and assists the system in developing a behavioral specification. The system next traverses the three levels of abstraction in search of existing implementations, calling on the designer to rule on "close" ones. When all possibilities have been exhausted, any remaining portions of the specification are made into new circuit designs. Finally, the hardware and software components of the design are realized separately and integrated into a complete system design.

FUTURE WORK

The result of the project described here is a high-level definition of the system [1]. A significant amount of additional work is required to advance it toward realization. Many aspects of the definition are general in nature and not necessarily the only possible approaches to solving the problems. Recommend directions for future work include the following: further develop Specification; enhance Implementation, for example, to create new IC's at Functional and Behavioral levels; refine Realization for integrating current CAD tools; add user interaction to subset partitioning; investigate the use of artificial intelligence. The Air Force Institute of Technology is committed to strong research efforts in design automation in the coming years, including a number of these areas.

CONCLUSIONS

The approach to design automation represented by this project has grown out of the need to integrate many available computer-aided design tools and develop new ones for the era of highly complex integrated circuits. Digital designers can no longer efficiently create designs for such circuits through manual methods. To be effective, future design processes must rely on integrated sets of tools to assist the designer in all stages and permit complete and clear interaction at all levels of abstraction. The system must allow continuing evolvement as more of it is defined and implemented and to adapt to changing technologies and changing requirements.

REFERENCES

1. Cable, Hobart S. II. Super-CAD: An Integrated Structure for Design Automation. MS thesis. Wright-Patterson AFB, Ohio: Air Force Institute of Technology, June 1982. (AD--A118 039) (Available from National Technical Information Service, 5285 Port Royal road, Springfield, Virginia 22121 (703) 487-4650. Use reference AD--A118 039. Registered Users of DTIC can obtain it from Defense Technical Information Center, Attn: DORA, Cameron Station--Building 5, Alexandria, Virginia 22314 (202) 274-7633 or Autovon 284-7633. Use same reference number.)

2. Siewiorek, Daniel P. and Larry Kwok-woon Lai. "Testing of Digital Systems," Proceedings of the IEEE, 69(10): 1321-1333 (October 1981).

3. Thomas, Donald E. "The Automatic Synthesis of Digital Systems," Proceedings of the IEEE, 69(10): 1200-1211 (October 1981).

Session 3-4b
VLSI Architecture II

Chairman

G. X. Ritter
University of Florida

ALGORITHMS FOR CELLULAR IMAGE PROCESSING

Azriel Rosenfeld

Center for Automation Research
University of Maryland
College Park, MD 20742

Abstract

Algorithms for performing basic image processing tasks on cellular machines are surveyed. Emphasis is on pixel-level algorithms for performing local operations, transforms, geometric operations, and statistics computation. Operations involving non-array structures (region representations, region graphs) are also discussed.

1. Introduction

It was proposed about 25 years ago [1] that many algorithms for image processing and analysis (IPA) could be implemented in parallel using a "cellular array" machine - i.e., a two-dimensional array of processors ("cells"), operating synchronously, each of which can communicate with its neighbors in the array. Several machines of this type, with array sizes of up to 128×128, have actually been constructed. Numerous IPA algorithms suitable for implementation on a cellular array have been developed; for general discussions of this subject see [2,3]. Recently, there has been some interest in using "pyramids" of cellular arrays, of sizes $2^n \times 2^n$, $2^{n-1} \times 2^{n-1}$, ..., 2×2, 1×1, where each cell can communicate not only with its neighbors ("brothers") on its own level, but also with its four "sons" on the level below and with its "father" on the level above [2,4].

Cellular arrays or pyramids are suitable for many types of IPA operations at the pixel level. On the other hand, some image analysis operations, involving regions in an image, do not make use of pixel arrays, but rather use other types of data structures to represent regions and their relationships. For such operations, a more general class of graph-structured cellular machines would be appropriate, in which the cells correspond to the nodes of a graph, and can communicate with their neighbors as defined by the arcs of the graph. On such "cellular graph" machines see [5,6]; on architectures corresponding to more specific types of data structures see [7-9].

The support of the U.S. Air Force Office of Scientific Research under Contract F49620-83-C-0082 is gratefully acknowledged, as is the help of Janet Salzman in preparing this paper. An expanded version of this paper will appear in S. Y. Kung, ed., "VLSI and Modern Signal Processing," to be published by Prentice-Hall, Inc.

2. Pixel-level operations

2.1 Point and local operations

Most of the operations commonly performed in IPA take images into images, where the value of a pixel in the output image depends only on the value(s) of the corresponding pixel, and possibly of some of its neighbors, in the input image (or images). Some examples of such operations are: contrast enhancement by grayscale transformation; sharpening by Laplacian filtering; smoothing by local averaging or median filtering; segmentation by thresholding or pixel classification; edge (or other local feature) detection; expanding or shrinking.

Operations of these types can be performed very efficiently on a cellular array machine in which (ideally) there is a processor associated with each pixel. Each processor collects the values of its neighbors, if necessary, and then computes the required function of these values. The time required to do this depends on the neighborhood size and the complexity of the operation, but not on the size of the image. If there are not enough processors, we can process the image blockwise, using enough overlap between blocks to avoid border effects.

2.2 Transforms

Various types of integral transforms (or their discrete versions) are often performed on images; the Fourier transform is the most common example. Here again the output image is an array of the same size as the input image, but there is no longer a correspondence between their pixels. The transforms are usually separable, so that they can be performed first row-wise, then column-wise; the value of a pixel in the transform is then a linear combination of the values of the pixels in that row (or column) of the image.

When transforms are done on a conventional computer, one can use efficient algorithms (e.g., the "fast Fourier transform") which require O(n log n) operations, rather than $O(n^2)$, on each

CH1935-6/83/0000/0719$01.00

row (or column); the total computational cost for an n×n image is thus $O(n^2 \log n)$ (We ignore here the problem of accessing the image from peripheral storage, and the possible need to transpose the image in order to access it efficiently column-wise as well as row-wise.) On a cellular array machine, the rows (or columns) can be transformed in parallel, and the time required for each row is $O(n)$ (each pixel must be multiplied by n coefficients, and the results must be grouped and summed); thus the overall time is also $O(n)$.

2.3 Geometric operations

Another class of image-to-image operations involves geometric transformations of an image - e.g., rescaling, rotation, or arbitrary "warping" (to correct geometric distortions, or to achieve registration with another image). Here the output pixels do correspond to the input pixels, but not in a simple one-to-one fashion (even a transformation such as rotation, when performed digitally, is not one-to-one). To perform such a transformation on an image, one must compute, for each pixel in the output array, the corresponding position in the input array (which will not, in general, coincide with the position of an input pixel). One must then assign a value to that output position by interpolation on the nearby input values.

The basic method of performing a geometric transformation on a cellular array machine is to assign an output pixel to each processor, and scan the input image over the array in such a way that each processor can collect the input values that it needs to compute its output value. To illustrate how this can be done in $O(n)$ steps, let S_{ij} denote the set of input pixels needed to compute the output value at pixel (i,j); let R_i denote the ith row, and C_j the jth column. Let us assume that, for all i and j, the set $|(\bigcup_i S_{ij}) \cap R_i|$ is of bounded size - in other words, the number of input pixels in a given row R_i whose values are needed by all the output pixels in a given column C_j is bounded. (This is not true for all possible transformations; for example, it is false for the operation of transposing. However, for any of the commonly encountered geometric transformations, there exist a pair of directions - not necessarily the "row" and "column" directions - for which the analogous property holds.) Then we proceed as follows: Cyclically shift the input data along the rows, and let each cell (i,j) collect all the values in $|(\bigcup_i S_{ij}) \cap R_i|$ as they pass it. Then cyclically shift the collected data along the columns, and let each cell (i,j) select the values it needs, i.e., those in S_{ij}.

2.4 Property measurement

We now consider operations that map an input image into a (set of) property value(s), rather than into an output image. Examples of such operations include: determining the presence or absence of a particular pixel value in the image, or computing statistics of the values; counting the number of occurrences of particular values (histogramming, etc.); counting the number of connected components of pixels having a particular value (this is the standard method of counting objects in a segmented image). On a cellular array machine, such operations require $O(n)$ time, since the pixel values must be brought together in one place in order to count them or compute their statistics, and this requires a number of communication steps proportional to the array diameter. Counting connected components requires a preliminary step in which each component is reduced to a single pixel, but this too can be done in time $O(n)$ using a special type of shrinking process.

Statistics computation and counting can be done in time $O(\log n)$ on a cellular pyramid machine (see Section 1). Each pixel passes its value to its "father" on the level above it, which counts or consolidates the values received from its sons and passes on the results to its own father; thus after log n steps (the number of levels), the cell at the apex of the pyramid has the final desired value. Note that this process makes use only of the vertical connections (between levels) in the pyramid, but not of the connections within a given level; thus it requires only a cellular tree machine having the pixels at its leaves.

3. Region-level operations

3.1 Region representations

A region in an image, or in fact any subset of an image, can be represented by a two-valued "overlay" image in which the pixels belonging to the subset have value 1, and all other pixels have value 0. This representation has the advantage of being in registration with the original image, but it has the disadvantage of requiring n^2 bits of memory no matter how simple the given region may be. Regions can be represented in other ways which require less memory for simple regions. Moreover, we can compute properties of regions, and derive new regions from given ones, by operating directly on the representations. Such operations too can be implemented in parallel, but an array of processors is no longer the appropriate architecture, since the representation is no longer array-like. In this section we discuss some standard region representations and the possibility of operating on them in parallel using appropriate multiprocessor architectures.

A region can be defined by specifying its borders (there is more than one border if the region has holes); for each border, this requires the coordinates of a starting point, together with a sequence of codes defining the succession of moves from pixel to pixel around the border (3 bits per move, since successive pixels are neighbors, and a pixel has only 8 immediate neighbors). A natural architecture for parallel processing of border codes [7] consists of processors connected in ring structures, with each ring representing a border (one code per processor). Computing properties of the regions represented in this way takes time $O(m)$, where m is the border length, as it would on a sequential machine; but certain tasks that take time $O(m^2)$ when done sequentially can be done in $O(m)$ on a ring machine - e.g., computing the border codes of the union or intersection of two regions, given the codes of the regions.

Another way to represent a region is to regard each row of the image that meets the region as a succession of runs (=maximal sequences) of 0's alternating with runs of 1's. Each row is determined by specifying the starting value (1 or 0) and the sequence of run lengths. Region properties and run length codes of derived regions can be computed directly from the code(s) of the given region(s). A simple architecture for processing run length codes in parallel might consist of strings of processors, where each string contains the run lengths for a given row. Greater efficiency could be achieved by allowing direct connections between strings representing adjacent rows, with the processor representing a given run connected directly to the processors representing runs on the adjacent rows that overlap the given run. This approach to parallel region processing does not seem to have been systematically investigated (but see [8]).

Runs are maximal horizontal "strips" of constant-value pixels; a more compact way of representing a region is to use maximal two-dimensional blocks of constant-value pixels. Each such block is defined by specifying its center and radius, and the region is then the union of the blocks. A representation of this type, known as the medial axis transformation, was introduced about 20 years ago; but it has not been used extensively for region processing, because it is difficult to compute region properties or to derive new regions from it directly, due to the fact that the blocks overlap one another and are not organized in a systematic way. In some cases, it may be possible to represent a region as a union of "generalized ribbons," where each ribbon is a union of maximal blocks whose centers all lie on a curve. Such a representation would be much more manageable, and could be processed in parallel by assigning the code of each curve (i.e., the sequence of moves and the corresponding radii) to a string of processors; this possibility has not been investigated.

Another type of maximal-block representation can be constructed by recursively subdividing the given two-valued image into quadrants, subquadrants, ... until blocks of constant value are reached. The resulting block structure can be represented by a tree of degree 4 (a quadtree) in which the root corresponds to the entire image, and the sons of a node correspond to its quadrants. For an $n \times n$ image, where n is a power of 2, the height of the tree is at most $\log_2 n$, and each leaf of the tree represents a block of the image consisting entirely of 0's or 1's. Region properties, and quadtree representations of derived regions, can be computed from the quadtree(s) of the given region(s) sequentially by traversing the tree(s). Quadtree-connected sets of processors can be used to perform many of these operations in parallel very efficiently [9]. A generalization of this approach can be used to represent a multivalued image as a union of homogeneous blocks (e.g., blocks of constant value, or blocks in which the standard deviation of pixel values is low), where we divide a block into quadrants iff it is nonhomogeneous.

3.2 Region properties and relations

The region representations described above are especially useful in manipulating data bases of regions, e.g., in digital cartography. In image analysis, such representations are used for measuring region properties and for deriving new regions from given ones. In this section we consider a more abstract level of processing in which regions are not completely specified, but are represented by lists of their properties. An image segmentation can be represented, at this level, by a graph structure in which the nodes correspond to regions, labeled with lists of property values; and the arcs correspond to related pairs of regions (e.g., adjacent), labeled with relation values (e.g., length of common border).

A segmentation can be modified by merging pairs of regions based on information provided by the graph representation, without any need to refer to the original image; and the graph of the new segmentation can be constructed directly from that of the given segmentation. For example, suppose the graph contains information about the area, perimeter, and average pixel value of each region, and the length of common border of each adjacent pair of regions. The following are some possible criteria for merging a pair of adjacent regions: their average values are very similar; their areas are very different; their length of common border is a large fraction of (one of) their perimeters. These criteria can be checked directly from the graph. Moreover, if we decide to merge two regions, we can construct the new graph directly from the old one, by replacing the two old nodes with a single node connected to all of the old nodes' other neighbors. The properties of the new node can be computed as follows: its area is the sum of the old nodes' areas; its average pixel value is the weighted average of the old nodes' averages, weighted by their areas; its perimeter is the sum of the old nodes' perimeters minus the length of their common border. Finally, the lengths of common border between the new node and its neighbors can be computed immediately from these lengths for the old nodes; they remain the same except in the case of a neighbor common to both of the old nodes, where the two lengths must be added.

Region merging processes such as that just described can be carried out in parallel using a network of processors in which a processor is assigned to each region, and the processors corresponding to adjacent regions can communicate directly - in other words, the processor network is isomorphic to the region adjacency graph. Thus each processor can examine the information stored at its neighbors (and at their arcs) and decide whether merging is possible. It should be pointed out that when merging is done in parallel, the decision to merge a pair of regions must be agreed to by both of them; if a region were allowed to make such a decision on its own, we might find that region A merges with region B and at the same time B merges with C, leading to an inconsistency (there are new nodes representing A+B and B+C, but no node for A+B+C, which in any case may not be an appropriate merge). To avoid this, only disjoint pairs should be allowed to merge. For further discussion of parallel region-level processing see [5].

4. Concluding remarks

This paper has reviewed some of the basic types of operations used in image processing and analysis, at both the pixel and region levels, and has described idealized cellular machines suitable for carrying out such operations in parallel.

We have seen that for pixel-level operations taking images into images, a cellular array architecture, with processors connected in a regular grid, is very natural. For image property measurement, on the other hand, greater efficiency can be achieved by using tree-structured connections, with the processors at the leaves of the tree. For parallel processing of regions defined by border codes, ring-connected processors are appropriate. Other connection schemes are suitable if the regions are defined by maximal blocks, e.g., by run length codes or by quadtrees. At a more abstract level, when regions are represented by lists of properties, region merging can be carried out in parallel using a network of processors connected in the same way as the region adjacency graph.

Parallel region-level processing generally requires a much smaller number of processors than parallel processing at the pixel level. A cellular array machine for parallel processing of a 512×512 pixel image, one processor per pixel, requires $\frac{1}{4}$ million processors, which is not yet practical; but a region-level processor might require only a few hundred processors per region (depending on their complexity), or even fewer processors to handle a region adajcency graph (depending on the complexity of the segmentation). These numbers of processors are quite manageable, but their interconnections pose a problem. For pixel-level processing, the images to be processed will all be of the same size, and the neighbor interconnections are the same for every image, so that a cellular array machine can be hard-wired once and for all. For processing at the region level, on the other hand, the interconnections vary from image to image, since the shapes of the regions cannot be predicted in advance. Worse yet, we may even want the interconnections to vary in the course of a computation, as new regions are defined or old regions merged. This calls for some type of reconfigurable multiprocessor architecture [6], where ideally the reconfiguration itself should take place in parallel. For some types of representations (e.g., border codes, for which linked rings of processors can be used), such reconfiguration may be relatively easy; but for other representations, requiring tree or graph interconnections, parallel reconfiguration may not be easy to realize in such a way as to avoid serious interprocessor communication bottlenecks. As advances in hardware technology make it possible to build large multiprocessor networks, the problems involved in designing efficient systems for parallel image processing and analysis, both at the pixel and region levels, will have to be addressed.

References

1. S. H. Unger, A computer oriented toward spatial problems, Proc. IRE 46, 1958, 1744-1750.

2. A. Rosenfeld, Picture Languages, Academic Press, NY, 1979.

3. C. R. Dyer and A. Rosenfeld, Image processing by memory-augmented cellular automata, IEEE Trans. PAMI-3, 1981, 29-41.

4. C. R. Dyer and A. Rosenfeld, Triangle cellular automata, Info. Control 48, 1981, 54-69.

5. A. Rosenfeld and A. Wu, Parallel computers for region-level image analysis, Patter Recognition 15, 1982, 41-50.

6. A. Rosenfeld and A. Wu, Reconfigurable cellular computers, Info. Control 50, 1982, 64-84.

7. T. Dubitzki, A. Wu, and A. Rosenfeld, Parallel computation of contour properties, IEEE Trans. PAMI-3, 1981, 331-337.

8. J. S. Todhunter and C. C. Li, A new model for parallel processing of serial images, Proc. 5th Intl. Conf. on Pattern Recognition, 1980, 493-496.

9. T. Dubitzki, A. Wu, and A. Rosenfeld, Region property computation by active quadtree networks, IEEE Trans. PAMI-3, 1981, 626-633.

IMAGE PROCESSING ARCHITECTURES AND LANGUAGES

S. Chen and G. X. Ritter

Department of Computer and Information Sciences,
Department of Mathematics, University of Florida
Gainesville, Florida 32611

ABSTRACT

VLSI technology has great impact on I.P. (image processing) architectures. An unified approach toward I.P. architecture design is proposed. A standard I.P. language based on common grounds of various existing ones is suggested. Its influence on architectures is explored and leads to the design of data flow I.P. architectures.

1. INTRODUCTION

VLSI technology has influenced both the computer hardware and software systems. On one hand, computation of small processes may be assigned to microcomputers to achieve cost-effectiveness. On the other, complex modular systems are built from VLSI chips to carry out extremely large parallel processes. I.P. (image processing) requires a great deal of computing and memory. Both the processor and memory will need the parallelism offered by the modularity and extendability of VLSI systems. So far, I.P. architecture design is motivated by a particular set of image processing operations together with the specific tasks that these operations have to perform. Once built, the I.P. language is expressed in terms of the architecture's features. For instance, cellular logic computers are used for computation of neighborhood functions in digital I.P.. Due to increase in I.P. activities around the globe, many different architectures have been developed (9), (10), (17). This diversity of architectures has given rise to a great variety of I.P. languages many of which are used to solve similar tasks. In order to concentrate on the impact of VLSI on I.P. architectures, we suggest an unified and systematic approach which is "language-based" and can be considered as "image data flow " systems. In this paper, we examine common grounds of existing architectures and languages and propose a roadmap for the development of a standard I.P. language based on these common features. We then turn to architectures and in particular data flow architectures. The influence of the standard I.P. language on these architectures is explored.

2. THE NEED FOR A STANDARD I.P. LANGUAGE.

For simplicity and limited space, we focus our attention on cellular logic computers. These computers are better suited for solving spatially arranged data problems than general purpose computers (26). Because of advances in VLSI technology, they can now be implemented with identical processing modules arranged in two dimensional arrays (8), (10), (17). Although many types of cellular computers have already existed, their full potential has not yet been realized. This is partially due to the non-existence of an unified, rigorous and mathematical structure for cellular I.P.. If such a structure existed-not only in terms of an uniform I.P. language, but also as a mathematical environment for the testing, evaluation of performance and extrapolation of new algorithms - then it would be feasible to find the optimal capabilities of existing as well as hypothetical cellular I.P. systems. Instead, due to the vast increases in I.P. and pattern recognition activities throughout the military, industrial and academic communities, we are being deluged with an ever increasing proliferation of different operations and architectures that often perform similar tasks. There are probably as many I.P. languages as there are architectures, and all of them differing in capabilities.

In view of this ever growing diversity of I.P. architectures and languages, the development of a standard I.P. language for aiding the design, development, optimization and testing of I.P. algorithms and hardware configurations is highly desirable. A brief roadmap for the development of a standard cellular I.P. language is outlined in the next two sections.

3. THE CRITERIA FOR A STANDARD I.P. LANGUAGE

It follows partially from the rational of the necessity for a standard I.P. language that such a language should satisfy the following criteria:

1. transportable

2. machine independent

3. uniform in definitions, procedures and functions

4. capable of interactive mode

5. algebraically based and structured with images as basic operands

Some of these criteria are self-evident. However, actual implementation is a very difficult task. We have some solutions and ideas as to how these requirements can be met. Uniformity and transportability, for instance, can be achieved by extracting common factors from existing languages and architectures. Due to limited space here, we shall not address each individual criterion, but only focus our attention on the last one.

In our scheme, the solution to the overall problem will yield a standard level algebraic I.P. language at the userend, while at the machine end, virtual devices accept the I.P. instructions and insure their execution on the appropriate physical device. Thus, we propose a three-level approach. On the first level, we envision a high level algebraic I.P. language (the focus of our study). The second level will be a virtual device definition language, while the third level provides the interface between the second level and the physical or assembler level.

4. PROPOSAL FOR AN ALGEBRAIC I.P. LANGUAGE

An algebraic structure is a set together with a finite number of closed operations. Well known examples of algebraic structures are semigroups, groups, rings, modules, fields and vector spaces. An I.P. language is an algebraic I.P. language if it implements such algebraic structures in order to express I.P. operations. The relational formalism imposed by an algebraic structure provides many convincing reasons for an algebraic language for I.P.. We only list a few: (1) an extensive mathematical basis exists, (2) simplification and optimization of algorithms due to relational formalism, (3) I.P. operations are expressible in concise algebraic notation, resulting in clarity and transparent coding, (4) increased productivity due to simplification of programmers' and users' tasks. We also note that methods using precise algebraic and mathematical operations such as Fourier analysis and Kalman filtering have made outstanding contributions in global I.P.. Last but not the least is the fact that a common set of algebraic I.P. operations for cellular computers already exists.

There are numerous I.P. languages capable of Boolean and other cellular logic operations. We have made an extensive survey of these languages, many (but not all) of which can be found in (6). Most languages, such as PIXAL (12), SLIP (16), PAX (16), PAL(6), PICASSO (11) and SUPRIC (16), are not algebraically based and, in addition, failed to meet some of the other criteria mentioned earlier.

GLOL (16) and languages developed by Sternberg's group in Michigan (23), (24), Miller at Perkin-Elmer (14) and Serra's group in France (20), (21) treat images as basic operands and use algebraic operations. However, all are machine dependent. Furthermore, GLOL restricts itself to Boolean images and rotationally invariant templates. While GLOL is based on the Golay transform, the latter three approaches all have one common feature. Even though their basic I.P. operations are differently defined, they are all mathematically equivalent and can be based on the Minkowski algebra of sets (7), (15). In fact, many of the theorems independently developed by Sternberg, Miller and Serrs's groups, which deal with the effects of their basic operations on images, can be found in (7). Some of the results date back to the turn of the century (3), (15).

In view of this common basis with its well established mathematical roots, we propose a set consisting of images (both gray level and binary) and neighborhoods (surrounds, templates, etc) as operands for an algebraic structure. The operations on this set should include the Minkowski operations of addition and subtraction (with their various combinations) and Boolean operations. Obviously other operations (such as arithmetic operations) need to be investigated and added. However, in order to preserve the overall coherence of the algebraic structure, great care needs to be exercised when adding new operations. A good guideline is to add only those neighborhood operations that do not result in elaborate notation and loss of clarity while extending the structure.

The host language of the computer system plays a role in the design of our algebraic I.P. language. These algebraic operations and their combinations must be embeddable in the host language. For the usual computer systems, PASCAL and ADA support structured programming and may be good choices for the host language. For multiprocessing systems, new languages beyond vectorization may be needed. We intend to discuss these considerations and further details in a future report.

5. DATA FLOW SYSTEMS FOR I.P.

There are three kinds of I.P. architectures: pipeline vector processors, SIMD array processors and MIMD multiprocessor systems. Pipeline vector processors include cytocomputer, GOP and ESL/systolic processor. SIMD (single instruction and multiple data stream) array processors are ILLIAC IV, Staran, the CLIP series and MPP. MIMD (multiple instruction and multiple data stream) multiprocessor systems are collections of processors with interconnection networks. Some developed systems are PICAP II, TOSPICS, ZMOB and PUMPS. We refer to (8), (9), (10), (17) for details and references of these systems. Data flow systems deal with only data and operations. An operation is executable if and only if all the required input data are available. Output data are sent to other operations or users. The programming languages based on data flow concepts are functional. In fact, data flow computers are language-based architectures. Data flow program graphs serve as an interface between system architecture and user programming language. The standard I.P. language will be the basis of our data flow machines.

A data flow program graph consists of operators which are connected by arcs. Input and output arcs carry tokens containing images as operands. Tokens are placed on and removed from the arcs according to some firing rules. For the operator to be enabled, each input arc must have one token and each output arc must not have any token. Any enabled operator may be fired. These tokens contain images which may be global ones or local neighbors. Thus sizes may vary from 1024 by 1024 to 9 by 9. Since operations may be unary, binary or have more arguments, the number of input arcs as well as out arcs may be different for different operators. One can also represent data flow programs by tables with pointers. Each table has several fields: one operation code, some receiving fields and destination fields. Conditionals and iteration are implemented also. Data flow programs can be executed in parallel.

The computer architecture should implement the behavior of program graphs. An one to one mapping from a program graph into the computer architecture has to be performed for each run. Reconfigurability of an architecture is often needed for a given set of programs. Also, source language programs are translated into program graphs. The technique is similar to path analysis of data dependency in compiler optimization. By connecting several processors through an interconnection network, a higher level of parallelism is achieved. The interconnection network carries out the routing task, while each processing component serves one operator. This data flow multiprocessing system can be implemented on any interconnection network (22), (25). The crossbar meets the requirements for many purposes but it has a high complexity. The ring structure has been used by Texas Instruments to interconnect four processors for their data flow machine. The Banyan and tree structures are also useful. We are investigating the switch lattice in (22) for its flexibility and reconfigurability.

6. CONCLUSIONS

An unified and system approach is suggested for the design and development of I.P. architectures and languages. A standard algebraic I.P. language and data flow architectures based on this language are proposed.

ACKNOWLEDGEMENTS. This research is in part supported by AFSOR grant 83-00 65.

7. REFERENCES

1. W. B. Ackerman, Data flow languages, Computer, 15 (2), Feb. 1982, pp.15-25.

2. T. Agerwala and Arvind, Data flow systems, Computer, 15 (2), Feb. 1982, pp.10-14.

3. W. Blaschke, Kreis und Kugel, Veit Co., Leipzig, 1911.

4. S. Chen and G. X. Ritter, Data flow systems for image processing, in preparation.

5. J. B. Dennis, Data flow supercomputers, Computer, 13 (11), Nov. 1980, pp. 48-56.

6. M. J. Duff and S. Levialdi, eds., Languages and Architectures for Image Processing, Academic Press, New York, N.Y., 1981.

7. H. Hadwiger, Vorlesungen ueber Inhalt, Oberflaeche und Isoperimetrie, Springer-Verlag, Berlin, 1957.

8. K. Hwang, Computer architectures for image processing, Computer, 16 (1), Jan. 1983, pp. 10-12.

9. K. Hwang and K.S. Fu, Integrated computer architectures for image processing and database management, Computer, 16 (1), Jan. 1983, pp. 51-60.

10. M. Kidode, Image processing machines in Japan, Computer, 16 (1), Jan. 1983 , pp. 68-80.

11. Z. Kulpa, PICASSO, PICASSO-SHOW and PAL: A development of a high level software system for image processing, Languages and Architectures for Image Processing, Academic Press, New York, N.Y., 1981.

12. S. Levialdi et al., PIXAL: A high level language for image processing, Real Time Parallel Computing: Image Analysis, Plenum Press, New York, N. Y., 1981.

13. P. E. Miller, An investigation of Boolean image neighborhood transformations, Ph.D. thesis, Ohio State University, 1978.

14. P. E. Miller, Image transformation algebra, Perkin-Elmer Tech. Report, USAF Eglin AFB, 1983.

15. H. Minkowski, Gesammelte Abhandlungen, Teubner Verlag, Leipzig, 1911.

16. K. Preston, Image manipulation languages, Pattern Recognition in Practice, North Holland, New York, N.Y., 1980.

17. K. Preston, Cellular logic computers for pattern recognition, Computer, 16 (1), Jan. 1983, pp. 36-47.

18. G. X. Ritter, Development of a mathematical foundation for cellular image processing, Tech. Report USAF, Eglin AFB, AFOSR F49620-82-C-0035.

19. A. Rosenfeld and A. Kak, Digital Picture Processing, Academic Press, New York, N. Y., 1982.

20. J. Serra, Image Analysis and Mathematical Morphology, Academic Press, London, 1982.

21. J. Serra and J. C. Klein, The texture analyzer, J. Microscopy, 95, 1972, pp. 349-356.

22. L. Snyder, Introduction to the configurable highly parallel computer, Computer, 15 (1), Jan. 1982, 47-56.

23. S. Sternberg, Languages and architectures for parallel image processing, Pattern Recognition in Practice, North Holland, New York, N.Y., 1980.

24. S. Sternberg, Cellular Computer and Biomedical Image Processing.

25. E. E. Swartzlander, VLSI networks for image processing, Proc. Workshop Computer Architecture for PAIDM, 1981, pp. 161-167.

26. S. H. Unger, A computer oriented toward spatial problems, Proc. IRE., 46 (10), 1958, pp. 1744-1750.

CELLULAR ARCHITECTURES FOR IMAGE PROCESSING

Kendall Preston, Jr.

Department of Electrical Engineering
Carnegie-Mellon University, Pittsburgh, PA 15213
Department of Radiation Health
University of Pittsburgh, Pittsburgh, PA 15213

Abstract

This paper reviews basic architectures now in use for cellular logic computers as applied to the field of image processing. The simplest architecture, which was initiated by the author in the CELLSCAN system in the early 1960s is to use a single array of processing elements (1-bit computers) to operate in parallel upon the value of a picture element and the values of the directly adjacent picture elements. Such a configuration is now called a "subarray processor" in that it causes the array of processing elements to scan across the image data thus computing a cellular logic transform. Most modern processors built on this architecture have several processing elements. Examples are the diff3 of Coulter and the PHP of Carnegie-Mellon. An alternative architecture is the "full array" wherein the entire image and several result images are stored in a large array of processing elements. Examples are CLIP (University College London) and MPP (Goodyear Aerospace and NASA Goddard).

Introduction

Cellular computers were inspired by the work of von Neumann in the 1950s.[1] Von Neumann envisaged arrays of thousands of computers connected so that each computer (called a "processing element") could communicate with its directly adjacent neighboring computers in either a square or hexagonal tessellation. It was not until the late 1970s that VLSI semiconductor fabrication techniques permitted the realization of such arrays. In the meantime cellular computer architectures were simulated by means of "sub-array" computers consisting of 9 (3x3) processors and shift registers with which a full-array cellular architecture could be emulated. The first of these cellular architecture emulators (cellular logic machines) was CELLSCAN designed by the author in 1961.[2] CELLSCAN had a single sub-array in the square tessellation which was capable of carrying out a neighborhood transform in 250 microseconds. CELLSCAN was followed in 1968 by another machine designed by the author called GLOPR (Golay Logic Processor) which, again with a single sub-array system, performed the same operation in 3 microseconds and then by BIP (Binary Image Processor) designed and built by Information International which performed the same operation in 100 nanoseconds.[3,4]

A typical cellular logic operation is performed digitally on an array of numbers P(I,J) so as to transform this array of numbers into a new array having the values P'(I,J) in such a way that the new value of each element is determined by the former value of this element in the original array and the original values of its nearest neighbors. The neighborhood configuration is called the "cell" from which follows the term "cellular computer." When the array P is binary-valued, then the operation is called "cellular logic." CELLSCAN, GLOPR, and BIP were all cellular logic machines using 1-bit computational elements. They represented the major developments of the 1960s and witnessed a speed increase over that interval of time in cellular logic computational rates by a factor of 2500. Further speed increases occurred in the 1970s with the commercialization of the cellular logic machine in the form of the Coulter diff3. Also, at this time, the first full-array cellular computers were constructed in the University College of London series of CLIPs (Cellular Logic Image Processors). Then, in the 1980s, another, still faster, full-array processor was built called the MPP (Massively Parallel Processor). These machines are described below.

diff3

The diff3 analyzer is a robotized microscope which is being used in the automation of visual hematology.[8] Over the interval 1977-1983 two hundred of these robot microscopes have been deployed world-wide. Collectively they perform the analysis of approximately one thousand images per second examining the texture, shape, color, etc. of human blood cells from approximately 25,000 patients per day. Each of the diff3 analyzers contains an advanced version of the original GLOPR cellular logic machine which employs eight processing elements in parallel. By means of this architecture it is possible to conduct a single picture element neighborhood transform in 25 nanoseconds. The GLOPR of the diff3 contains four dedicated binary image memories two of which may be used as sources (and a third as destination) in conducting the Golay hexagonal parallel pattern transform.[5]

CH1935-6/83/0000/0727$01.00

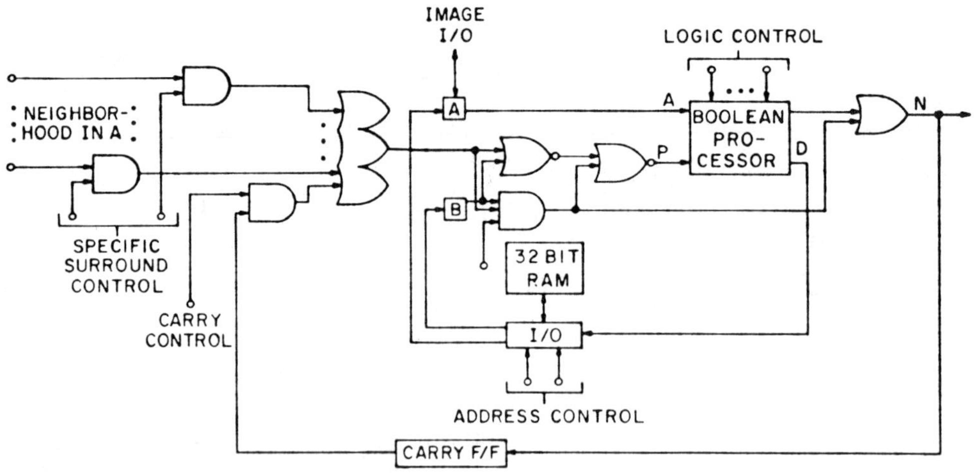

Fig. 1 Logic diagram of one processing element of CLIP

CLIP

The diff3 analyzer was designed in the early 1970s and did not benefit from custom LSI/VLSI whereas this is not true for either the CLIP or the MPP. The current CLIP system has a processing element as diagrammed in Figure 1. This is the processing element for the fourth in the series of CLIP computers (CLIP4).[6] A design study was commenced in the mid-1970s to determine whether eight of these processing elements could be integrated on a single chip using LSI/VLSI technology. This was finally accomplished in the late 1970s with the assemblage of an array of 9216 processors to accomplish the completion of the CLIP4 system. This required the incorporation of several thousand transistors on a single substrate and to obtain yield sufficient to produce over one thousand working chips.

As can be seen from Figure 1 the CLIP4 processing element is capable of storing 32 images (in a dynamic RAM). By addressing this RAM the control circuitry in the processing element is capable of storing one image in the latch A with another in the latch B. The contents of A is directly input into a "Boolean Processor" while the other (B) is combined with either the neighborhood inputs (selected by means of the surround control word) or the "carry" input (selected by carry control). This combination is carried out by an AND gate whose output feeds an OR gate producing the signal P. Once the A and P signals are generated, the Boolean processor produces one of sixteen possible outputs according to "logic control" word. This output in the form of a signal D is returned to the dynamic RAM for storage and also maybe broadcast to all neighbors as the signal N which may alternatively be stored in the "carry" latch.

By means of this processing element architecture, the CLIP4 system may carry out any Boolean neighborhood operation or may be used as a 1-bit central processing unit for the purpose of executing 1-bit arithmetic. By suitably organizing data within the 9216 RAM's of the system, integer arithmetic may be performed on graylevel images or cellular logic transforms may be conducted on binary (black and white) images. By means of this technique it is possible to perform edge detection, gradient analysis, to cause boundary propagation for the purpose of forming either the convex hull or convex kernel, the skeleton or residues. The effective picture point operation time for most of these operations is one nanosecond. Thus the CLIP4 is 25 times faster than the diff3 analyzer (250,000 times faster than the original CELLSCAN).

MPP

The Massively Parallel Processor (MPP) was originally conceived as a super-high-speed image processing system for the purpose of automating the analysis of images generated by satellite cameras by the staff of the National Aeronautics and Space Administration (Goddard Space Flight Center). A multi-million dollar procurement was placed with Goodyear Aerospace Corporation (Akron, Ohio) in the 1970s and delivery was accomplished in mid-1983.[7] In its present form the MPP is an array of 128x132 elements which are arranged in two-row-four-column subarrays on a custom VLSI chip. Eight processing elements are integrated per chip but each processor chip requires two additional 1Kx4 memory chips so as to supply each processor with 1K bits of storage. The array normally operates as 128x128 with a block of processing elements by-passed (unless a fault has developed) with all connections to and from by-pass processing elements automatically routed to subsequent neighbors so that the user is not aware of this re-arrangement.

The processing element is capable of executing arithmetic operations performed in a full adder which has a three 1-bit inputs buffered by three latches. The 1-bit sum may then be stored in another latch as with the output carry. In order to avoid the extra instruction cycles required to return the partial results of integer sums (or multiplication) to local storage, a shift register of variable length (2,6,10,14,18,22,26,30) is provided which may be used in conjunction with the full adder as desired.

Logical operations are performed in a logical unit using one of the input latches as the storage for input data. A data bus interconnects all latches in the processing element and interfaces to the RAM's. In distinction to CLIP, the MPP processing element has connections only to four immediate neighbors. Therefore eight-neighbor logic must be performed by means of shifting the data over the array. Finally, there is a "activity bit" upon whose state the entire operation of the processing element is dependent, i.e., sets of processing elements may be disabled at any time by having their activity bits set to zero. In the MPP image data transfer rates and arithmetic capability surpass in speed those available in CLIP4 but not as much as one would surmise from a simple comparison of instruction execution times. This is because of the fact that MPP pays the penalty of not providing parallel inputs from all neighborhood processing elements in the square or hexagonal tessellations as does CLIP4. Since the cycle time of the MPP is 200 nanoseconds, and the array consists of 16384 processing elements, the effective single-picture-point processing time is about 10 picoseconds.

Acknowledgements

The author wishes to acknowledge information received from University College London (from Dr. M.J.B. Duff) and from the Goodyear Aerospace Corporation (Drs. K. Batcher, and J.Potter) in preparing this manuscript. Furthermore, many helpful discussions were held with Dr. M.D. Graham (Coulter Biomedical Research). The illustration was prepared by Mr. M. Westfall of the Mellon Institute and Ms. Linda Daigle, Executive Suite, Tucson, Arizona, was responsible for transcribing this dictation and preparing the manuscript.

References

(1) Von Neumann, J. "The general logical theory of automata," in Cerebral Mechanisms In Behavior - The Hixon Symposium (Jeffress L.A. ed.) Wiley, New York (1951).

(2) Preston, K., Jr., "The CELLSCAN System - A Leukocyte Pattern Analyzer," Proc. Western Joint Comput. Conf., 173-176 (1961).

(3) Preston, K., Jr., and Norgren, P.E., "Interactive image processors speeds pattern recognition," Electronics 45, 89-99 (1972).

(4) Gray, S.B., "Local properties of binary images in two dimensions," IEEE Trans. Comput. C-20, 551-562 (1971).

(5) Golay, M.J.E., "Hexagonal Parallel Pattern Transformations," IEEE Trans. Comput. C-18, 733-740 (1969).

(6) Fountain, T.J., "CLIP4: A progress report," in Languages And Architectures For Image Processing (Duff, M.J.B. and Levialdi, S., eds.) Academic Press, London (1981) pp. 283-293.

(7) Potter, J.L., "MPP architecture and programming," in Multicomputers And Image Processing (Preston, K., Jr. and Uhr, L., eds.) Academic Press, New York (1982) pp. 275-290.

(8) Graham, M.D., and Norgren, P.E.,"The diff3 Analyzer: A parallel/serial Golay image processor" in Real-Time Medical Image Processing (Onoe, M., Preston, K. Jr., and Rosenfeld, eds.) Plenum, New York (1981) pp 163-182.

Session 3-4c
VLSI Logic Design

Chairman

T. Mangir
University of Southern California

I/O OVERLAPPED SORTING SCHEMES FOR VLSI

H. Shin, A. J. Welch and M. Malek

Department of Electrical Engineering
University of Texas at Austin
Austin, Texas 78712

Abstract

A class of sorting schemes for VLSI are surveyed in which the sorting time completely overlaps input/output time. These I/O overlapped sorting schemes considered the most attractive for on-line, real-time sorting include the ripple sorter, rebound sorter and parallel enumeration sorter. Comparison analysis demonstrates that the ripple sorter based on a linear systolic array is the most efficient for sorting and the most suitable for VLSI implementation. A hardware algorithm is presented for the ripple sorter.

1. Introduction

Intensive research has been conducted on the sorting problem because of its importance in compiler and assembler design, operating systems and database systems. Recently, a class of sorting schemes for VLSI has been developed in which the sorting time completely overlaps input/output time in a real-time environment with a single I/O port. The systolic priority queue [1], rebound sorter [2] and parallel enumeration sorter [3] represent this class of sorters. They sort N items in O(N) time rather than O(NlogN) time required by a uniprocessor [5].

Each processing element in the sorter is a functional and physical building block, typically consisting of a comparator, registers and control circuits. The sorting of an input sequence starts when the first item is entered. Upon completion of the input operation, the first sorted item is ready for output, and thereafter all the items are returned in proper order. Hence, the minimum sorting time is 2N in a single I/O on-line, real-time system.

The linear array for systolic priority queue envisioned by Leiserson and Kung [1] is the most attractive scheme for VLSI implementation because of its structural and operational simplicity. In this scheme, the sorting can be accomplished by performing N "insert" operations followed by N "extract-minimum" operations [6]. That is, as N items to be sorted are entered through the first processor one at a time, even and odd numbered processors perform comparison-exchange and shift operations alternately so that the larger values "ripple" through the array of N processors until they are properly ordered. Thus, in this paper, it will be called "ripple sort".

The purpose of this paper is two-fold: to survey I/O overlapped sorting schemes for VLSI, and to develop a hardware algorithm for the ripple sort. In Sections 2 to 4, three different I/O overlapped sorters are discussed. In Section 5, comparison analysis is performed. The realization of the ripple sorter is illustrated in Section 6. Our discussions are concluded in Section 7. Throughout this paper the following convention is adopted:

-- An input sequence is sorted in ascending order.
-- A cycle is defined as the time required to compare two items and push one or both items to neighboring processing elements (PEs).
-- N denotes the number of input items to be sorted. MAX(or MIN) denotes the largest (or smallest) value a sorter can have.

2. Rebound Sorter

The rebound sorter [2], the first known I/O overlapped sorter, consists of N-1 PEs plus the bottom cell and the top cell as shown in Fig. 1. Each PE which is called a steering unit has two registers, P[i] and Q[i], 1<=i<=N-1, a comparator, and control and routing circuits. Let P[0] and Q[N] denote the top cell and bottom cell, respectively. The sorter is configured in such a way that P[i] is connected to P[i-1] and Q[i+1] and that Q[i] to P[i+1] and Q[i+1]. The sorting procedure is as follows:

```
PROCEDURE ReboundSort

  BEGIN
    FOR EACH i: 0<=i<=N PARDO
      Initialize P[i] and Q[i] to MIN;
    FOR i:=1 TO 2N DO
      FOR j:=1 TO 2 DO
        FOR EACH k: 1<=k<=N-1 PARDO
          BEGIN
            Enter an input item into Q[1]
              if j=1 and i<=N;
            IF Q[k]<P[k]
```


CH1935-6/83/0000/0731$01.00 © 1983 IEEE

```
            THEN P[k-1]:=Q[k], Q[k+1]:=P[k]
            ELSE Q[k+1]:=Q[k], P[k-1]:=P[k];
          P[N-1] := Q[N];
          Set the emptied registers to MAX;
          Output a sorted item in P[0]
            if j=1 and i>=N;
        END;
  END.
```

The language construct FOR EACH m: <condition> PARDO <statements> indicates that all statements are executed simultaneously for the values of m which satisfy the given condition. The total I/O time is 4N cycles. It is claimed to be efficient for those cases where the presentation of items at the input is inherently slower than the comparison operations. In particular, it is amenable to shift register technology, such as bubble memory and CCD. Chen et al. [2] have shown implementation of the rebound sorter on a uniform ladder which is a storage structure based on shift register technology. However, its relatively large number of interconnections require complex control and routing circuits, reducing the area efficiency of this sorter for VLSI implementation.

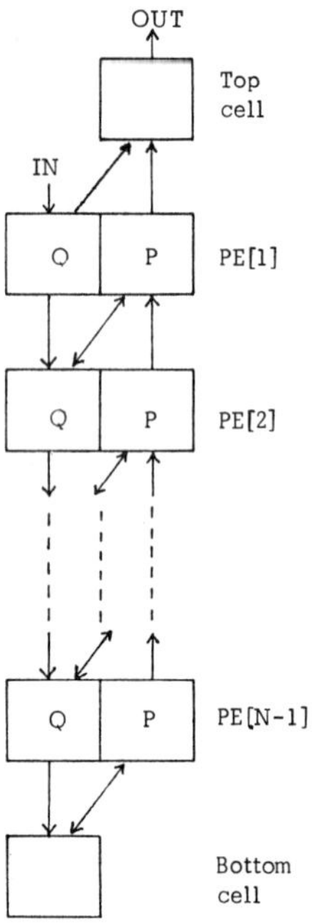

Figure 1.

Rebound Sorter

3. Ripple Sorter

The ripple sorter consists of N/2 PEs configured as a linear array as shown in Fig. 2. Each PE has two registers, P[i] and Q[i], a comparator and control circuits. Thus P[i] (or Q[i]) is connected to P[i-1] (or Q[i]), 1<=i<=N/2. The sorting algorithm is shown below. Assume that a register P[N/2+1] is always set to MAX, and let Q[0] and P[0] denote the input and output buffer registers, respectively.

```
PROCEDURE RippleSort

  BEGIN
    FOR EACH i: 1<=i<=N/2 PARDO
      Initialize P[i] and Q[i] to MAX;
    FOR i:=1 TO N DO
      FOR EACH k: 1<=k<=N/2 PARDO
        BEGIN
          Enter an input item into Q[0];
          Q[k] := Q[k-1];
          IF Q[k]<P[k]
            THEN Swap P[k] and Q[k];
        END;
    FOR i:=1 TO N DO
      FOR EACH k: 1<=k<=N/2 PARDO
        BEGIN
          IF Q[k]<P[k]
            THEN Swap P[k] and Q[k];
          P[k] := P[k+1];
          Output a sorted item in P[0];
        END;
  END.
```

The total I/O time required is 2N cycles. Data structures, especially a priority queue [1], are easily implemented on this sorter. Slight modification of its structure and algorithm allows an input sequence of N>m items to be sorted in O(N*N/m) time where m is the number of PEs. Simplicity and ease of VLSI implementation of this systolic architecture makes it very attractive and powerful proposition [4,6]. A fault tolerant version of the ripple sorter has been proposed [7] which ensures more robust implementations.

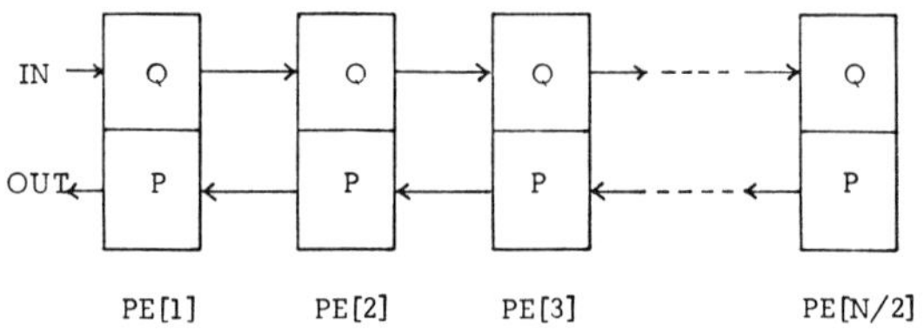

Figure 2. Ripple Sorter

4. Parallel Enumeration Sorter

Parallel enumeration sorting [3] is accomplished in two stages: ordering stage and rearranging stage. The ordering stage is the enumeration process to determine the order of each item in the given input sequence. Afterwards, in the rearranging stage, the host rearranges the sequence in proper order. A bus connected linear array is used to implement this scheme (Fig. 3). N PEs are required to process N items. Each PE consists of two registers, P[i] and Q[i], a comparator, a counter C[i] and control circuits. Q[i] or a shift register is connected to Q[i+1]; all P[i] which holds the i-th item throughout the enumeration process is connected to the input bus; all counters to the output bus. The enumeration process is accomplished using the following procedure. Let Q[0] denote the input buffer register.

```
PROCEDURE ParallelEnumerationSort

  BEGIN
    FOR EACH i: 1<=i<=N PARDO
      Initialize C[i] to 0;
    FOR i:=1 TO 2N DO
      FOR EACH k: 1<=k<=N PARDO
        BEGIN
          Enter an input item into
            Q[0] and P[i] if i<=N;
          Q[k] := Q[k-1];
          IF i<2k-1
            THEN IF P[k]>=Q[k]
                   THEN C[k]:=C[k]+1
            ELSE IF P[k]>Q[k]
                   THEN C[k]:=C[k]+1;
          Output C[i-N] if i>N;
        END;
  END.
```

The total I/O time is 2N cycles. Although a linear array is most suitable for VLSI implementation [6], a counter, buses and relatively complex control are the overhead caused by parallel enumeration algorithm itself. This sorter can be easily modified to be flexible so that a sequence of N>m items is sorted in O(N*N/m) time. It may also be applied to the merging and multikey sorting for the database operations.

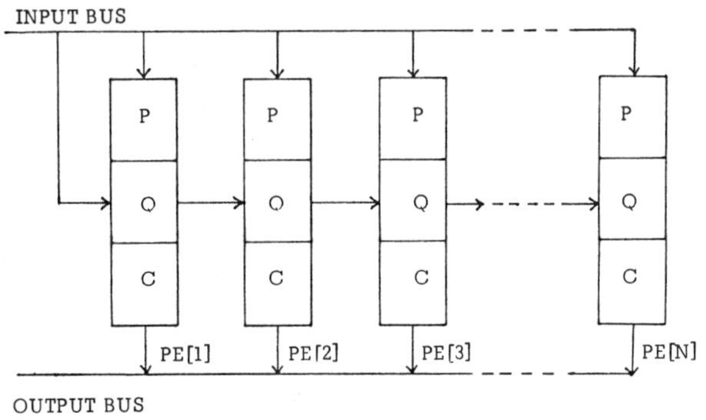

Figure 3. Parallel Enumeration Sorter

5. Comparison Analysis

The I/O overlapped sorting schemes share a number of properties to achieve real-time sorting during the I/O period in the single I/O, on-line environment. First, computational concurrency is achieved by the use of pipelining. The sorted sequence is produced as input items travel through a linear path. Second, all input items are exhaustively compared with each other exactly once. Third, they all need O(N) identical processing elements which have a comparator and no more than two registers to sort a sequence of numbers. Last, they require O(N) area for planar mapping on the VLSI chip.

The three sorters differ widely in structure and operational mode from one another. The most conspicuous dissimilarity originates from the different sorting algorithms, particularly, between the parallel enumeration and the other two sorting schemes. Since the parallel enumeration sorter only generates the order of each item in the sequence, it requires not only additional rearranging process, but more hardware components such as the input and output buses and counter. Although it does not need routing of data based on the comparison, there must be appropriate control signals to properly synchronize the actions of receiving and emitting data via I/O buses.

The rebound sorter has a simple configuration except for a relatively large number of interconnections for each PE. Nevertheless, it sorts N items in 4N cycles with N-1 PEs. Based on area-time product cost, it is approximately four times as costly as the ripple sorter whose sorting time is 2N cycles with N/2 PEs.

Consequently, the ripple sorter shows the best performance in many aspects. Its structural and operational simplicity makes a significant improvement for VLSI implementation [6] in comparison to other I/O overlapped sorting schemes. For this reason, a hardware algorithm for the ripple sorter is considered in the next section. Table 1 compares the various properties of the sorting schemes.

6. Realization of Ripple Sorter

Pipelining is indispensable for all I/O overlapped sorting schemes. Without pipelining it is impossible to compare every item with all others in a parallel fashion. The ripple sorter uses a linear systolic array which is a natural means of pipelining.

The elementary building block is a PE which consists of a comparator, two shift registers P[i] and Q[i], 1<=i<=N/2, 2x2 crosspoint switch and control logic and signals X and RESET, as shown Fig. 4. A control signal RESET is used to set the registers to MAX; X indicates whether the sorter is in the input or output state. A clock with two phases synchronizes proper comparison-exchange operations by controlling the shift registers. The 2x2 switch consists of four tristate drivers or pass transistors in NMOS, controlled by an internal control signal C.

The system is configured in a linear array with PEs cascaded as shown in Fig. 5. The array is preceded by an input/output data shift register (RIO), and followed by a register (RMAX) permanently set to MAX. The sorting operation is performed in the following way. RESET signal is asserted until the first input item is latched in so that all P[i] and Q[i] are initially set to MAX. During the input period, X is asserted, enabling Q[i] to receive the data from the output of PE[i-1]. When the data are ready in both registers, P[i] and Q[i] are compared. Based on the comparison result, the 2x2 switch routes the data properly: the larger value is routed to Q[i-1] and the smaller one back to P[i]. These operations are repeated until the N-th item enters Q[1]. This is immediately followed by negation of

X so that the data may flow in the opposite direction and that the complementary comparison may occur during the output period. In other words, Q[i] receives the data from PE[i+1] and outputs the smaller of P[i] and Q[i]. Q[N/2] always accepts MAX from the register attached to PE[N/2]. In this way, all sorted items are output through PE[1].

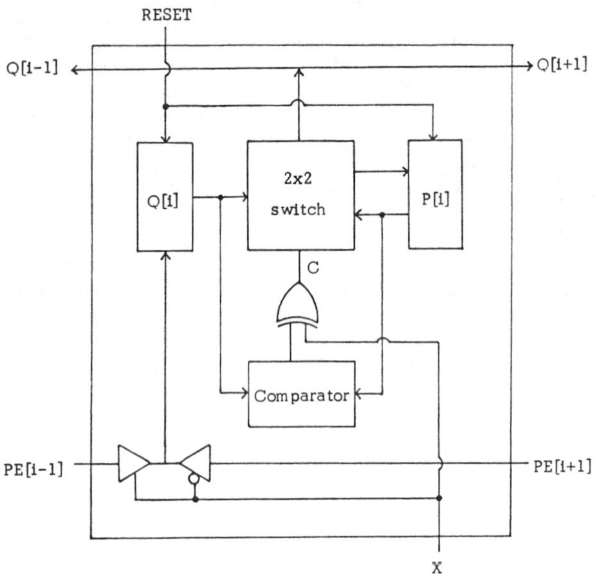

Figure 4. Organization of PE

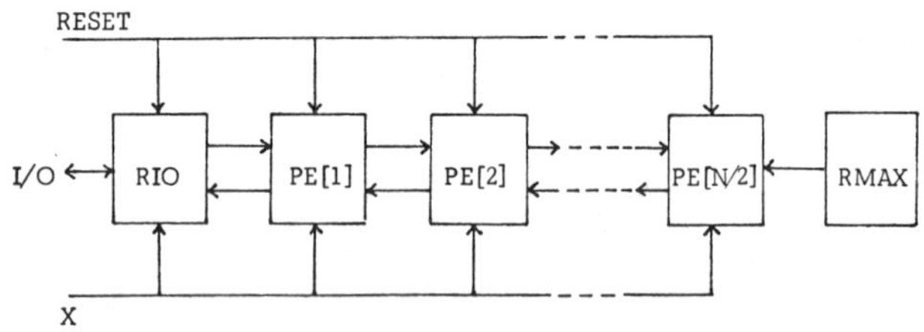

Figure 5. Ripple Sorter Circuit

7. Conclusion

I/O overlapped schemes for VLSI, such as the ripple sorter, rebound sorter and the parallel enumeration sorter have been surveyed and compared. Whereas all the sorters of this class are suitable for VLSI implementation, the ripple sorter on a linear systolic array makes the most of the VLSI technology because of its intrinsic structural simplicity and regularity. It makes a significant improvement in performance in comparison to other schemes. The circuit constructed for the ripple sorter is compact and amenable to straightforward VLSI implementation. Therefore, we claim that the ripple sorter is the most efficient of all I/O overlapped sorting schemes and the most suitable one for the VLSI implementation.

References

1. C. E. Leiserson, "Systolic priority queues," Proc. Caltech Conference on VLSI, Jan. 1979, pp. 199-214.

2. T. C. Chen, V. Y. Lum and C. Tung, "The rebound sorter: An efficient sort engine for large files," Proc. Fourth Int'l Conference on Very Large Data Bases, 1978, pp. 312-318.

3. H. Yasuura, N. Takagi and S. Yajima, "The parallel enumeration sorting scheme for VLSI," IEEE Trans. Comput., Vol. C-31, Dec. 1982, pp. 1192-1201.

4. C. A. Mead and L. A. Conway, Introduction to VLSI Systems, Addison-Wesley, Reading, MA, 1980.

5. D. E. Knuth, The Art of Computer Programming, Vol. 3, Sorting and Searching, Addison-Wesley, Reading, MA, 1973.

6. H. T. Kung, "Let's design algorithms for VLSI systems," Proc. Caltech Conference on VLSI, Jan. 1979, pp. 65-90.

7. D. Fussell and P. Varman, "Fault-tolerant wafer-scale architectures for VLSI," Proc. 9th Annual Symposium on Computer Architecture, April 1982, pp. 190-198.

Table 1. Comparison of the Sorting Schemes

	Ripple sorter	Rebound sorter	Parallel enumeration sorter
No. of PEs	N/2	N-1	N
Degree of connectivity of PE	4	6	4
Global bus	None	None	2
Topology	Linear array	Ladder	Linear array
Data routing necessary?	Yes	Yes	No
Sorting time	2N	4N	2N
Rearranging necessary?	No	No	Yes

IMPACT AND LIMITATIONS OF INTERCONNECT TECHNOLOGY ON VLSI AND RESTRUCTURABLE VLSI DESIGN

Tulin Erdim Mangir

School of Engineering and Applied Science
Computer Science Dept., UCLA
LA, CA., 90024

ABSTRACT

In this presentation, impact of interconnect technologies on VLSI will be presented. Specifically, interconnect technologies for VLSI will be reviewed, interconnect techniques for restructurable and programmable VLSI will be presented.

With the scaled technologies for VLSI, there is considerable emphasis on the gate delays being reduced as well, for higher processing speeds. However, for VLSI, gate delays are not the limiting factors for speed of computations.Interconnect RC time delays do not scale down with scaled technology and become the limiting factor for VLSI designs.

For VLSI, proposed signal processing architectures (such as systolic arrays, wavefront processors, array processors), suggest that operands arrive from different processing elements, to be operated on at a specific processing site. The delays introduced by the interconnections between stages imply that they will set the the physical limits to communicationation between processors, and to the overall performance of VLSI .

INTRODUCTION:

Very large scale integration has been achieved by scaling the lithographic dimensions of circuits and corresponding vertical dimensions. This has been achieved by technological advances in several areas such as lithography, device development, materials technology. Currently there are further challenges : - to reduce vertical dimensions further, low temperatuer processing techniques are required. - to reduce lateral dimensions further better isolation techniques needs to be developed. - diffusion and polysilicon are too high resistivity for high speed interconnects for VLSI, therefore new interconnection layer technologies must be developed.

The emphasis of this paper is on the last of these challenges. In the following sections, candidate technologies for VLSI will be reviewed, technologies for programmable interconnections for restructurable VLSI will be presented.

Interconnect Technologies for VLSI:

For VLSI, the interconnect should provide two desirable features:

1. High transfer capability (both bandwidth and speed)[1,2,19];
2. Reconfigurability (for testing, for restructurable architectures)[13-16,21].

With increasing chip sizes and complexities,propagation delays and clock skews in the signal lines has become the limiting factors to the use of VLSI technology. Exact modelling of RC delays in the interconnect and its effects on overall design and performance limitations has been studied in [1,2,5,19,20]. Since interconnect time delays do not scale with scaled technologies, they become the limiting factor for performance of the VLSI designs. In Figure 1, relative gate and interconnect technologies are plotted for various materials [20].

Polysilicon ,more than any other material has enabled the increase in the density of the circuits to LSI and VLSI complexity. This was because it is suitable for both gate and interconnect material. However, its high resistivity is a definite drawback to the further increase in the performance of the VLSI circuits. This has caused increased efforts to develop various metal silicide technologies with the MOS processes [9,24].

High electrical conductivity allows these materials to be used for narrower and thinner conductors contributing to higher packing densities and higher resolution geometries without becoming the source of series resistance (limiting performance). This is especially significant for single metal systems, where the second layer of interconnect has been the polysilicon up to now. Processing incompatibilities are the reason for the MOS chips not having silicides for their gates and as a second level interconnect. The silicide material has to be put on the gate over a very thin oxide and this deposition process needs to be followed by etching and high temperature oxidation.

The factors that go into choosing a silicide include resistivity and formation sequence as well as compatibility with the particular manufacturers process sequence such as etchants and temperatures used. In Table 1, some of the candidates for the next interconnect level are listed with their resistivities and deposition techniques. Most of the research activity has centered around the refractory metal silicides.Titanium ($TiSi_2$) and Tantalum ($TaSi_2$) are two of the prime silicides for low resistivities and lower temperature annealing

CH1935-6/83/0000/0735$01.00

required. $TaSi_2$ is also a good process for compatibilty with the currently existing wet etch processes. Another very popular material is the PlatinumSilicide (PtSi). In addition to being metal rich, it also provides Schottky barrier contacts for the source and drain junctions without requiring implants and diffusions at all [10].

In addition to the silicides, metals such Titanium, Tungsten, and Molybdenum are the metal interconnect materials which are being favored due to their high conductivity and lower resistivity. The trouble with metal technologies is that they can melt , evaporate, or form silicides(unintentional) when heated. So for these materials to be commonly used for VLSI low temperature chemical deposition techniques need to be developed.

Interconnection Technologies For Restructurable VLSI:

In this section, we consider the technologies applicable for restructuring a VLSI circuit. We also discuss some alternatives for VLSI architectures amenable to the method suggested in [13] , namely bypassing a faulty module on chip, or a chip on a wafer for WSI (Wafer Scale Integration), or RVLSI (Restructurable VLSI) [21].

The ability to reconfigure the logic of an integrated circuit may be used for several purposes:

a) Fault tolerance to improve the yield for higher levels of integration;
b) Fault tolerance to improve the reliability;
c) Customization;
d) Dynamic reconfiguration.

In the first case, reconfiguration is done at the probe level, and is required for the levels of integration for VLSI and above. In the second case, restructuring is done in the field, for maintenance, and requires that restructuring information be available outside the packaged chip. The third case permits customization of designs using standard modules for specific products, thereby providing quick turnaround for custom and semicustom [8,11,13,22] designs. Dynamic reconfiguration allows the change of the system configuration (dynamic routing of information) to give better utilization of resources, or to improve speed of processing in order to meet changing requirements of applications.

As it can be seen, each of these applications impose different requirements for implementation. However, these requirements can be grouped basically into five categories as follows:

—localization or diagnostic techniques to identify working modules;
—interface requirements for isolating the modules;
—permanence of reconfiguration;
—frequency of reconfiguration;
—overhead (area, power dissipation, and performance).

In the following sections, we will present the techniques used for restructuring for yield improvement, and discuss these requirements.

Programmable Interconnections:

In previous work, redundant elements to improve the processing yield have been first proposed and implemented for high density memories [3,4,6,13]. The approach is to substitute spare word or bit lines (address lines) for defective ones during testing of the chips. In Table 2, classes of repairable defects and methods for a metal bit line dynamic or static RAM [23]. Feasibility of extending redundant designs to VLSI logic has been recently studied by several researchers [8,13-16,21]. The information which identifies good and bad locations can be stored on the chip by one of several alternatives [13]:

1) Latches;
2) Laser personalization ;
3) Electrically programmable storage elements ;
4) Electrically programmable links ;
5) e- beam programmable fuses.

In Tables 3 and 4, state changes for various interconnect techniques are shown In Tables 5 and 6, these technologies are compared .

Latches and other electrically alterable memory elements can be reprogrammed in the field, if necessary. Therefore, the redundancy included on the chip for yield improvement can also be used for field maintenance/reliability improvement.

Laser personalization involves blowing the fuses to spare elements [4] or disconnecting the contacts to faulty elements [3], or joining interconnect layers together [12,26,21]. Thus it is permanent, and once done can not be changed. In [4], laser pulses are used to burn fuses to isolate the faulty elements from the power supply. This technique is particularly suited to cases where interconnections between modules are done in serial or pipeline fashion.

Electrically programmable approaches, e-beam, and laser personalization are favored because they are non-volatile. Otherwise, the chip interconnect pattern has to be stored and redone after a power failure. Latches are volatile. More detailed comparisons of these techniques are provided in [13].

Testing, Diagnosis, Restructuring:

Depending on the degree of restructurability desired, the hardware for testing, diagnosis, routing, and restructuring will vary considerably. When only fault tolerance for yield improvement is required, all of these functions can be located off the chip. This of course simplifies the problem of verifying the correctness of the tester. However, accessing, testing, and re-routing around failed areas will require sophisticated computer design aids.

To connect the modules in a final configuration, modules which are tested and determined to be good are connected to the horizontal and vertical interconnect lines through programmable links.

Another approach that can be used to access and test internal modules is implementing a module select mechanism on the chip and then using self testing designs where the module design is modified to generate "good" or "faulty" signals for individual modules. This arrangement then can be used for multiple input/multiple output testing of individual modules. The use of self testing modules or modules with built-in test would be preferred in this case to help the testing problem [16].

In the last case, the select lines can be used to tristate (disable the outputs of the module by putting in a high impedance state) the failed module interconnections to the bus. Another way to bypass the failed modules is to disconnect the power connections to the module.

One of the most important design parameters affecting restructurability is the degree of connectivity available around failed modules. As was shown in [13,-15], there are basic tradeoffs between module complexity and connectivity. As the intermodule interconnection requirements become more complicated, the probability of success of bypassing failed regions and restructuring diminishes not only because of the very large interconnection area which affects yield, but also because of the large number of substitutions necessary. As the number of connections to be programmed increases, the probability of success of restructuring diminishes. As an example, consider the probability of one programmed link working to be given by p, and numebr of links required to substitute a working module for a defective one to be given by m. Then the probability that the substitution is successful will be p^m, since all links must work for the restructuring to be successful. Note that p, probability of successful programming of a link is directly a function of the technology used for the programmable interconnections.

When we consider the interconnections of modules on a chip, there are a number of possibilities:

a) modules are not connected prior to testing;
b) a primary set of modules are connected and set of extra (redundant) modules are unconnected prior to testing;
c) all modules are connected prior to testing.

In the first case, after the testing good modules can be connected to achieve a required function capability. Location of the good and bad modules can be stored in a computer "map" and the final interconnection pattern can be developed automatically, and the links to be programmed are identified this way. This approach is especially suited for customization as well as the cases where interconnections between the modules are strictly random (unstructured, irregular) in nature. Therefore, connecting tested "working" modules is preferable. The disadvantage of this case is the large number of interconnects to be programmed.

In the second case, faulty modules are disconnected and good redundant modules are substituted for them. And in the last case, faulty modules are disconnected along with the extra redundant modules. Note that in both these cases some decision as to how the modules are replaced must be made prior to the testing. This is also dependent on how the signals are communicated between modules. For interconnection patterns which are serial or pipeline in nature, or parallel the substitution is easier.

To do the substitutions, modules can be disconnected (connected) from (to) the power lines on the chip which may be designed on a second level metal, with the intermodule interconnections running on first level metal and other interconnect layers. A different alternative is to design the interconnections between modules to be done at the second level metal. The modules are connected together by using vias (interlayer connections), for example, between the different metal layers [13].

Another possibility to provide a working intermodule interconnections, and to bypass the defective modules is, to provide each module with a set of input and output lines, and then enable or disable them [13]. This is particularly attractive for parallel designs, bitslice architectures.

CONCLUSION:

The RC time constants of VLSI chips implemented in scaled technologies remain constant or increase, while the circuit speeds are increasing due to scaled gate delays. A major problem in chip design is that interconnect quickly dominates the overall speed. These problems are especially significant when polysilicon or diffused lines are used for power or signal distribution. This has resulted in the development of multilayer metal and silicide interconnect technologies. To preserve intrinsic speed capability, long distance interconnect time must not dominate gate delay. This suggests limiting the length of the interconnections by using modular architectures with localized communications. To facilitate easy testing, design, and customizing these designs, as well as improving the IC processing yield of VLSI, restructurable interconnections have become the focus of recent research.

Acknowledgement:

The author would like to thank R.J.Smith of Intel Corporation, for providing copies of his slides. We would also like to acknowledge the help of Ms.Vicki Graham and Ms. Cheryle Childress in preparation of the manuscript.

Material	Deposition	Resistivity $\mu\Omega cm$
$TiSi_2$	*/†	15/25
$CoSi_2$	*/†	19/25
PtSi	*	31
$ZrSi_2$	*	37
$NbSi_2$	*	50
VSi_2	*	50-55
$NiSi_2$	*/†	50-55
$TaSi_2$	*	35-70
WSi_2	*	70
$MoSi_2$	†	100
FeSi	*	200
$CrSi_2$	*	600
$FeSi_2$	*	1000
PolySi	pyrolysis	1000

*=Metal on polysilicon

†=Co-sputtered alloy

TABLE 1

TABLE 2

DEFECT TYPE AND REPAIR ELEMENTS
(METAL BIT LINE SRAM OR DRAM)

TYPE OF DEFECT	BIT FAILURE PATTERN	PREFERRED REPAIR ELEMENT
DIFFUSION SHORT OR OPEN	SINGLE, MULTIPLE BIT	ROW OR COLUMN
POLYSILICON WORD LINE SHORT OR BREAK	FULL OR PARTIAL ROWS	ROW
OPEN OR CLOSED CONTACT	SINGLE, MULTIPLE BIT	ROW OR COLUMN
METAL LINE SHORT OR BREAK	FULL OR PARTIAL COLUMN	COLUMN
INTERLEVEL SHORT	ROW	ROW
GATE OXIDE SHORT	SINGLE BIT OR ROW	ROW OR COLUMN
CELL LEAKAGE	SINGLE BIT	ROW OR COLUMN

TABLE 3

STATE CHANGES FOR REDUNDANT PROGRAMMING ELEMENT
(CONNECT TECHNIQUES)

	PRE-PROGRAMMED	POST-PROGRAMMED
LASER PROGRAMMED POLY RESISTOR	$R \cong 10^9\ \Omega$	$R<10^3\ \Omega$
ELECTRICALLY PROGRAMMED POLY DIODES		
-DOPANT REDISTRIBUTE	$R \geq 10^7\ \Omega$	$R \cong 10^2 - 10^3\ \Omega$
-AI MIGRATION	$R \geq 10^7\ \Omega$	$R \cong 10-30\ \Omega$
LASER RESTRUCTURED VIAS		
-INTERLEVEL METALS	DIELECTRIC ISOLATION	$R \cong 0.3\ \Omega$
-METAL-TO-DIFFUSION	DIELECTRIC ISOLATION	$R \cong 10-15\ \Omega$
LASER CONNECTED METAL GAP	$R>10^9\ \Omega$	$R \cong 1-10\ \Omega$
ELECTRICALLY PROGRAMMED OXIDES	DIELECTRIC ISOLATION	$R \cong 500-2000\ \Omega$
LASER FORMED LINKS	$R \sim 10^9\ \Omega$	$R<10\ \Omega$
E^-BEAM PROGRAMMED AL FUSES	$R>10^9\Omega$	$R \sim 1-10\Omega$

TABLE 4

STATE CHANGES FOR REDUNDANT PROGRAMMING ELEMENT
(DISCONNECT TECHNIQUES)

	PRE-PROGRAMMED	POST-PROGRAMMED
ELECTRICAL ON-CHIP TRANSISTOR/POLY FUSES	$R \cong 200\Omega$	$R > 10^7\ \Omega$
LASER PROGRAMMED POLY FUSES	$R \cong 200\Omega$	$R\ 10^9\ \Omega$
EPROM CELL	$V_T<2V$	$V_T \cong 5V$
LASER LINKS	$<10\Omega$	$R \sim 10^9\ \Omega$
e^--BEAM PROGRAMMED METAL FUSE	$<10\Omega$	$R \sim 10^9\Omega$
e^--BEAM PROGRAMMED POLY FUSE	$R \sim 200\Omega$	$R \sim 10^9\Omega$

TABLE 5

RELATIVE BENEFITS OF REDUNDANCY TECHNOLOGIES
(CONNECT TECHNIQUES)

TECHNIQUES	PROCESS REQUIREMENTS	TEST EQUIPMENT	HIGH VOLTAGE NODE ISOLATION REQUIRED
LASER PROGRAMMED POLY RESISTOR	POLY RESISTOR	TESTER/ LASER	NO
ELECTRICALLY PROGRAMMED POLY DIODES	POLY DIODES	TESTER	YES
LASER RESTRUCTED VIAS			
-INTERLEVEL METAL	POLY, α -SI, QUARTZ	TESTER/ LASER	NO
-METAL-TO-DIFFUSION	POLY INTERPLAYER	TESTER/LASER	NO
LASER CONNECTED METAL GAP	SIO_2 ON POLY/ SIO_2	TESTER/LASER	NO
ELECTRICALLY PROGRAMMED OXIDES	THIN GATES	TESTER	YES
LASER LINKS	POLY, AlCuSi	TESTER/LASER	NO
e^-BEAM CONNECTED METAL GAP		TESTER/e^-BEAM SYSTEM	NO

TABLE 6

RELATIVE BENEFITS OF REDUNDANCY TECHNOLOGIES
(DISCONNECT TECHNIQUES)

TECHNIQUES	PROCESS REQUIREMENTS	TEST EQUIPMENT	HIGH VOLTAGE NODE ISOLATION REQUIRED
ELECTRICAL ON-CHIP TRANSISTORS WITH POLY FUSE	POLYSILICON	TESTER	YES
LASER PROGRAMMED POLY FUSES	POLYSILICON	TESTER/ LASER	NO
EPROM CELL	FAMOS TECHNOLOGY	TESTER	YES
ELECTRICAL OFF-CHIP SIGNAL WITH METAL FUSES	DOUBLE METAL	TESTER	YES
LASER FORMED LINKS	POLYSILICON & AlCuSi	TESTER/ LASER	NO
e^--BEAM PROGRAMMED POLY FUSES	POLYSILICON	TESTER/ e^-BEAM SYSTEM	NO
e^--BEAM PROGRAMMED AL FUSES		TESTER/e^-BEAM SYSTEM	NO

REFERENCES

1. Antinone, R., Brown, J., "The Modelling of Resistive Interconnects for Integrated Circuits", IEEE JSSC, V SC-18, No. 12, pp.200-203, April 1983.
2. Bilardi, G., et.al., "A Critique of Network Speed in VLSI Models of Computation", IEEE JSSC V SC-17, No. 4, pp. 696-703.
3. Cenker, R. P., et.al., "A Fault Tolerant 64K Dynamic RAM", IEEE ISSCC 1979.
4. Elmer, B. R., "Fault Tolerant 92160 Bit Multiphase CCD Memory", ISSCC 1977, pp. 116-117, February 1977.
5. Elmasry, M. I., "Interconnect Delays in MOSFET VLSI", IEEE JSSC V SC-16, pp. 585-591, Oct. 1981.
6. Fitzgerald, B. F., and Thoma, E. P., "Circuit Implementation of Fusible Redundant Addresses on RAMs for Productivity Enhancement", IBM, J. Res. Develop, V24, No. 3, pp. 291-298, May 1980.
7. Frohman-Bentchkowsky, D., "A Fully Decoded Electrically Programmable MOS-RAM", ISSCC 1971, pp. 80-81, February 1971.
8. Garvick, S., Pierce E., "A Single Wafer 1G Point 16 M Hz. FFT Processor", CICC Digest of Papers, pp. 244-248, May 1983.
9. IEDM '82 Digest of Papers, Session 24, "Advanced Interconnect Technologies", San Francisco, December 1982.

10. Koeneke, C. J., Lynch W. T., "Lightly Doped Schottky MOSFGT", IEDM '82, Digest of Papers, pp. 466-469, Dec. '82.

11. Koide, Kazuo, et.al, "Building Block Approach and Variable Size Memory for CMOS VLSIs", IEEE, ISSCC '83, February 24, 1983, pp. 148-149.
12. Kuhn, L., "Experimental Study of Laser Formed Connections for LSI Wafer Personalization", IEEE JSSC, SC-10, pp. 219-228, August 1975.
13. Mangir, T. E., "Use of On Chip Redundancy for Fault-Tolerant VLSI Design", Ph.D. Thesis, June 1981, UCLA, Also published as a technical report, UCLA Computer Science Department, No. CSD-82-02-01.
14. Mangir, T. E., and Avizienis, A., "Effect of Interconnect Requirements on Yield Improvement of VLSI Design", IEEE Trans. Comp., V.C-31, No. 7, pp. 609-615, July 1982.
15. Mangir, T. E., and Avizienis, A., "Effect of Interconnect Requirements on VLSI Circuit Yield Improvement by Means of Redundancy", Proc. IEEE COMPCON '81, pp. 322-326, San Francisco, February 1981.
16. Mangir,T.E.,"Design for Testability-An Integrated Approach to VLSI Testing",to appear in Proceedings of ICCAD 1983,September 1983.
17. Mano, T., et.al., "A Fault Tolerant 256K RAM Fabricated with Molybdenum-Polysilicon Technology", JSSC, Vol. SC-15, No. 10 pp. 865-872, October 1980.
18. Mano, T., et.al., "256K DRAMs Submicron VLSI Memory Circuits", ISSCC '83, February 25, 1983, pp. 234-235.
19. Mead, C., Rem, M., "Minimum Propagation Delays in VLSI", IEEEJSSC, VSC-17, No. 4, 773-774, August '82.
20. Meindl, J. et.al., "Circuit Scaling Limits for Ultra Large Scale Integration", ISSCC 1981, pp. 36-37, February 1981.
21. Raffel, J., "Restructurable Logic", Proc. Caltech VLSI Conf., 1979.
22. Sano, Toshi, et.al., "A 20ns CMOS Functional Gate Array with a Configurable Memory", ISSCC '83, February 24, 1983, pp. 146-147.
23. Smith, R. J. "Redundancy-The New Device Technology", IEDM '82, Digest of Papers, pp. 608-611., December 1982.
24. Special Issue on VLSI, IEEE Trans. on Electron Devices, V ED-26, No. 4, April 1979.
25. Schinoda, Takashi, et.al., "A 1Mb ROM with On-Chip ECC for Yield Enhancement", ISSCC '83, February 24, 1983, pp. 158-159.
26. Yasaitis, J. A. et al., "Low Resistance Laser Formed Links", IEEE EDL, VEDL-3, No. 7, 184-187, July 1982.

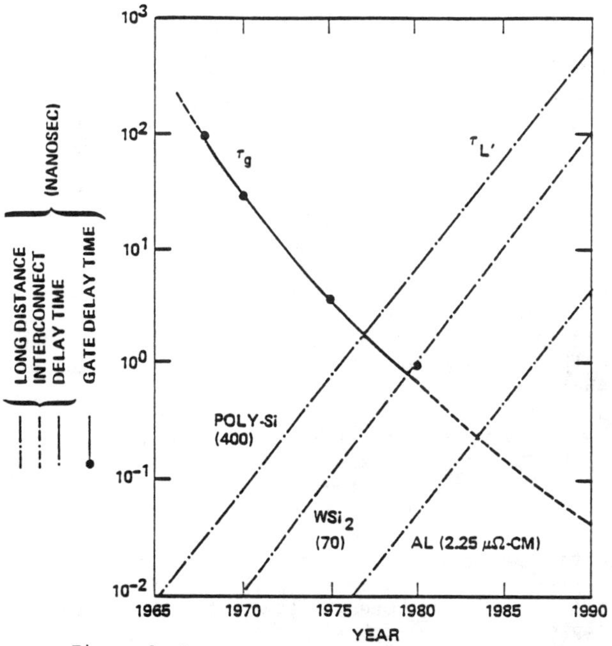

Figure 1. Interconnect and gate delays for various materials

COMPARISON OF VARIOUS CHIP-I/O SCHEMES FOR INTERCONNECTING VLSI SYSTOLIC ARRAY PROCESSOR CHIPS*

Ming-Yang Chern and Tadao Murata

Department of Electrical Engineering and Computer Science
University of Illinois at Chicago, Chicago, Illinois 60680

Abstract -- This paper proposes three chip-I/O schemes and two I/O timing modes for VLSI systolic array processor chips. The required number of I/O pins, chip-I/O delay, difficulty of interchip connection and the resultant delay on the entire system operation are evaluated for each scheme. The chip wafer area is estimated under certain assumptions and the average required wafer area for a successful chip is calculated based on the Seeds model. With a few more assumptions, the combined cost of the required wafer area and the system delay caused by chip-I/O for different I/O schemes are further expressed in an equivalent unit and compared.

I. Introduction

It has been five years since the systolic array was first proposed [1, 2]. During this period, many algorithms based on this architecture have been developed. The systolic array was considered to be suitable for VLSI implementation from the regularity and interprocessor communication points of view [3]. There is, however, an upper limit on the number of processing elements (PEs) that can be implemented in one chip. To construct a large array from smaller unit chips, we must carefully design a chip-I/O scheme for chip interconnection. Design of the I/O scheme greatly influences the ease of interchip connection, data transfer efficiency, and the number of I/O pins in one chip (and hence the cost of the chip). It also affects the system performance, when used for host system interfacing.

The feasibility of VLSI implementation of the systolic array has been accepted by many researchers. However, the chip I/O schemes of VLSI systolic arrays were not well investigated. The scheme of serial communication to the systolic array by connecting its virtual rows and columns [4] is not appropriate for VLSI chip interconnection. Hwang and Cheng [5] showed some I/O demultiplexing and multiplexing designs for their processor arrays. Their I/O scheme is good only for the one-chip array while many aspects of this scheme were not evaluated. Hsu and Shanblatt [6] evaluated three chip-I/O circuit design strategies for the required wafer area and I/O delay. Nevertheless, the investigation of different I/O schemes is not emphasized. The chip interconnection problem can not be solved without an appropriate I/O design. And the system performance can not be correctly evaluated without knowing the array chip I/O scheme and system-array interfacing. Therefore, the I/O schemes for VLSI systolic arrays must be examined before the applicability of systolic arrays can be justified.

* This work was supported by the National Science Foundation under Grant ECS 81-05649.

II. Proposed Chip-I/O Schemes

This paper presents three chip-I/O schemes for hex-connected systolic array processors. The completely parallel I/O scheme is not considered, because it requires too many I/O pins. Assume that a VLSI chip contains an n x n systolic PE array. Let the word length of all PEs be b bits (b is assumed to be a multiple of 8 for the convenience of analysis in this paper). A total of 4n-1 input channels and 4n-1 output channels on the border of the array will be required. In case of a small array in which n = 4 and b = 32, the completely parallel I/O scheme would require 960 I/O pins in one chip. This number is unacceptable to the current or near-future VLSI technology. Obviously we have to reduce the number of I/O pins using an appropriate I/O scheme. The following three I/O schemes are proposed:

(1) Bit-serial I/O scheme

Fig.1 illustrates this scheme. Each input or output channel has one serial-to-parallel or parallel-to-serial shift register. The resultant pin number for each I/O channel is only one. This scheme is convenient for chip interconnection. However, the interfacing between this chip and the RAM memory needs extra chips to do the multiplexing/demultiplexing (mpx/dmpx) and shifting. And the number of shifts, which is the word length b, will cause certain fixed time delay. On the other hand, the number of I/O pins, which is 8n-2, increases linearly as the array size n increases.

(2) Byte-serial I/O scheme

CH1935-6/83/0000/0740$01.00 © 1983 IEEE

The block diagram of this scheme is also illustrated in Fig.1, where each I/O pin in the previous scheme now becomes 8 pins. This scheme is faster since only one eighth of shifts in the previous case are needed for data transfer. And it does not require external shifting circuits to interface with system RAM memory. Nevertheless, external mpx/dmpx chips may still be needed, because the system may not have enough memory modules to interface with all the I/O channels in a one-to-one corresponding way. The number of I/O pins in this scheme is 8(8n-2) which is 8 times that required in the previous scheme. The large number of I/O pins (for n > 3) not only causes the high cost of pin connecting, but also prohibits the use of this scheme in the current VLSI design.

(3) Byte-serial grouped I/O scheme

This scheme for n = 3 is shown in Fig.2. For any array size n, only 35 pins for I/O are required. In this diagram, the I/O lines in the diagonal direction are designed to transfer data in a bit-serial way. These two lines are used to connect the array chips in the diagonal direction. In case the chip interfaces with the system memory, the diagonal input data can be obtained from the bottom demultiplexer by controlling the "SELECT". The diagonal output of the PE at the upper-left corner is connected to both the output multiplexer on the top portion and the parallel-to-serial converter. The timing sequence of the data I/O is arranged such that the neighboring array chips can be interconnected directly. Under this scheme, the problems of memory interfacing and chip interconnection can be solved easily. No matter how large the array size n is, only 35 I/O pins for each chip are to be connected. And the number of the memory modules required for direct interfacing is fewer than the last scheme. This scheme is more attractive in all the above aspects. However, its main drawback is the loss of interconnection configurability. The individual I/O channels on the same border (side) have been preconfigured in a fixed way. In addition, the number of shifts for one burst of data I/O is the maximum of {b, (n*b)/4}. The time required for the data I/O is proportional to the array size for n > 4.

Some systolic arrays have diagonally oriented data flow in the direction opposite to that shown in Fig. 2. The I/O scheme for this case can be made similar to the above design except that the input and output channels on the same side must be grouped together. Here, the number of I/O pins and the time delay would be still the same.

III. Required Wafer Area

To estimate the wafer area required for a single array chip, we assume that each PE (processing element) in the array occupies a square of certain fixed size. We also assume that the total number of pin connections is ten plus the number of I/O pins. According to the recent practical designs [7, 8], the size of an I/O pad is about 230 μm square. As shown in [6], there are several ways to design the I/O circuits while the circuit area and layout crossover are highly dependent on the I/O circuits. Thus, in this paper, we estimate the wafer area required for I/O circuits in a much simplified way for the convenience of analysis. The area required for the I/O buffer registers at the border of the array chip is neglected.

For the bit-serial I/O scheme, the serial input or output of each buffer register is connected directly to an I/O pad. No extra area for line connections is required. For the other two I/O schemes, the I/O pads are shared by I/O lines from different buffer registers or different bytes of the same buffer register. An acceptable way of layout connection in this case is to have 8 metal bus lines along each edge corresponding to the 8 bits in an I/O byte. If we use $\lambda = 2$ μm and the design rule of 3λ for metal lines and their spacing [3], then an area of 100 μm width along each edge of the chip will be occupied by the metal bus lines.

Based on the above assumptions, we can estimate the wafer area of a single chip if the PE size and the array size n are given. Let Lp be the edge length of the PE. Then the edge length of the chip Lc can be calculated by the following formula:

For the bit-serial I/O scheme,

$$Lc = \max\{n*Lp + 2*0.23\text{mm},\ (\lceil \#pin/4 \rceil + 1)*0.23\text{mm}\}$$

For the other two I/O schemes,

$$Lc = \max\{n*Lp + 2*(0.23\text{mm} + 0.10\text{mm}),\ (\lceil \#pin/4 \rceil + 1)*0.23\text{mm}\}$$

where #pin is the total pin number of a chip and the symbol ⌈ ⌉ represents the ceiling function. Note that when there are too many I/O pins, the wafer area of the given chip is mainly determined by the wafer edge length required to accommodate all I/O pads along the four edges.

The average wafer area required to produce a successful chip is determined by the fabrication yield and the utilization of the wafer disk. According to the current VLSI technology [9], it is reasonable that we let the wafer defect density $D = 1/\text{cm}^2$ and the diameter of the wafer disk Dw = 150mm. With the wafer defect density given, the yield can be estimated based on the Seeds model [9, 10]. From the above given size of the wafer disk, we can estimate the percentage of useful disk wafer area by the formula (approximated from a plot in [9]):

useful percentage = 100 - 2.3 * Lc

where Lc is expressed in units of mm. Based on the above model and the given parameters, the average required wafer area for a successful array chip can be calculated for different array size n and PE size Lp.

IV. I/O Timing Arrangement

To evaluate the performance of the computational array, it is not enough to consider only the array structure and its chip-I/O scheme. The timing arragement of the chip I/O must be specified. There are two possible scheduling modes:

Mode P: I/O in parallel with computation

Under this scheduling, the I/O time delay is not critical as long as it is shorter than one unit of computation time. However, it causes one unit of computation time delay for data to flow from one chip to its neighboring array chip. And this time delay may invoke a data mismatching problem in the systolic array operations. To solve this problem, the timing control scheme would be quite complex. Therefore, this mode is good only for the operation of a single-chip systolic array.

Mode S: I/O in series with computation

In this case, the chip I/O delay has a direct effect on the system speed by adding itself to the unit computation time. This mode does not have the array data mismatching problem.

To show the effect of the chip-I/O delay on the system performance, we may express the system turnaround time for a computational task by a formula. Let Tc be the time for one slot of computation and Td be the time delay due to the data I/O of the chip for one slot of computation. The chip I/O delay Td can be calculated by multiplying the unit shift time by the number of shifts required for one slot of I/O. Assume that a given task requires m slots of computation. Operating in Mode P, the turnaround time would be (m+2)*Tc, if Tc $\geq$ Td; it would be (m+2)*Td, if Td > Tc. Operating in Mode S, the turnaround time is then m*(Tc+Td) + Td. Substracting m*Tc from these expressions, we can evaluate the effect of the chip-I/O delay. Note that Mode P is not always faster than Mode S. For 2*Tc > (m+1)*Td, the turnaround time for Mode P is longer.

V. Cost Comparisons

It rarely happens that an I/O scheme is better than another in all aspects. And it is hard to compare the effect of different factors. In this paper, we try to combine several factors into one measure for comparison. We assume that the total cost of IC packaging and chip interconnection is proportional to the total number of pins. We also assume that the cost of 1 cm^2 disk wafer area is equivalent to the total cost due to 50 connection pins. Given the PE size, wafer disk diameter and the wafer defect density, we can evaluate the resultant cost as a function of the array size n based on the above two assumptions.

To estimate the cost of the chip I/O time delay, the effect of the I/O delay on the resultant array performance must be translated into an equivalent unit of cost. It is assumed that when the total array computation time is doubled due to the chip I/O delay, the cost of this I/O delay is equivalent to the cost of the hardware array. Based on this, we can estimate the cost of the I/O time delay. For example, if the chip I/O causes 20% extra time to complete a computation, the cost due to this chip-I/O delay is considered to be equivalent to 20% of the hardware cost. Combining the above cost estimations, we can evaluate and compare the relative costs of VLSI systolic arrays for different I/O schemes.

As an example, let the word length b = 32 bits, PE size = 1 mm square, wafer defect density = 1/cm^2, disk wafer diameter = 150 mm, one I/O shift time = 1% of Tc (unit computation time). Also assume that the given computational task has m = 20. The resultant costs for the I/O schemes (1) and (3) in the two I/O timing modes can be calculated and are plotted as shown in Fig.3. The byte-serial I/O scheme is not evaluated here, because it obviously has too many I/O pins for n $\geq$ 3. Fig.3 shows: for mode S, the bit-serial I/O scheme is always better than the byte-serial grouped I/O scheme. For mode P, the bit-serial I/O scheme is the best for n < 5 or n > 13, while the byte-serial grouped I/O scheme is for 5 $\leq$ n $\leq$ 13; the costs of these two schemes are about the same at n = 5.

VI. Conclusions

The design of chip-I/O schemes is important in designing a VLSI array processor. An acceptable chip-I/O scheme is one which is good for chip interconnection. In this sense, the three chip-I/O schemes proposed in this paper for hex-connected systolic arrays are acceptable and possibly there are other acceptable schemes. To compare these I/O schemes, the number of I/O pins, ease of chip interconnection, the required wafer area and the chip I/O delay are evaluated. However, it is difficult to tradeoff between the factors in several different aspects. In Section V, we have tried to combine the cost due to the required wafer area, the number of connection pins and the chip-I/O delay into one measure. This measure indicates, to some extent, which I/O scheme is better and in what sizes of arrays. Nevertheless, the effect of the chip-I/O delay on the entire system turnaround time is dependent on the given computational tasks. And the VLSI chip fabrication cost is technology-dependent and time-varying. Consideration of all these factors is beyond the scope of this paper.

References

[1] H.T. Kung, "Let´s design algorithms for VLSI systems", in Proc. Caltech Conf. VlSI, Jan. 1979, pp.65-90.

[2] H.T. Kung and C.E. Leiserson, "Systolic Arrays (for VLSI)", Sparse Matrix Proc., 1978,

Society for Industrial and Appl. Math., 1979, pp.256-282.

[3] C. Mead and L. Conway, Introduction to VLSI Systems, Reading, MA: Addison-Wesley, 1980.

[4] K. Bromley, et al., "Systolic array processor developments", in VLSI System and Computations, ed. by H.T. Kung, B. Sproull, and G. Steelle; Rockville, MD: Computer Science Press, 1981, pp.273-284.

[5] K. Hwang and Y.H. Cheng, "VLSI computing structures for solving large-scale linear system of equations", Proc. 1980 Int. Conf. on Parallel Processing, pp.217-227.

[6] W.C. Hsu and M.A. Shanblatt, "Evaluation of a single VLSI chip algorithm for triangulating large band form matrices", Research report MSU-ENGR 82-015, Michigan State Univ., Aug. 1982.

[7] J. Iwamura, et al., "A 16-bit CMOS/SOS multiplier-accumulator", Proc. 1982 Int. Conf. on Circuit and Computers, pp.151-154.

[8] T. Tokuda, et al., "16b CPU design by a hierarchical polycell approach", Proc. 1982 Int. Conf. on Circuit and Computers, pp.102-105.

[9] D.J. McGreivy, "VLSI chip trend -- size, complexity, cost", in VLSI Technologies: through the 80s and beyond, ed. by D.J. McGreivy and K.A. Pickar, IEEE Computer Society Press, 1982, pp.31-39.

[10] S. Muroga, VLSI System Design, John Wiley & Sons, 1982.

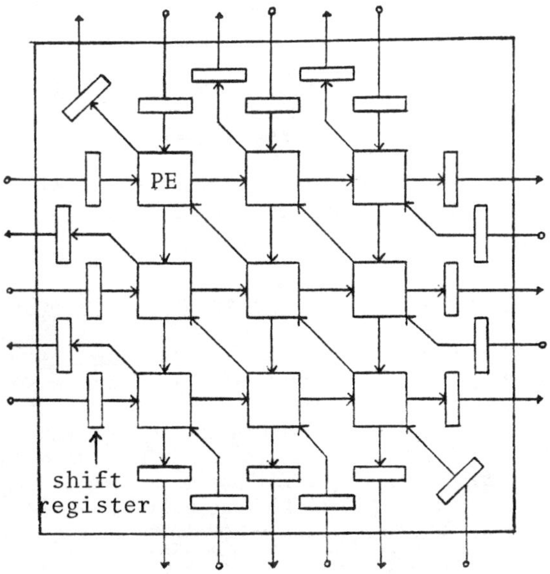

Fig. 1 Bit-serial I/O scheme for a hex-connected systolic array chip

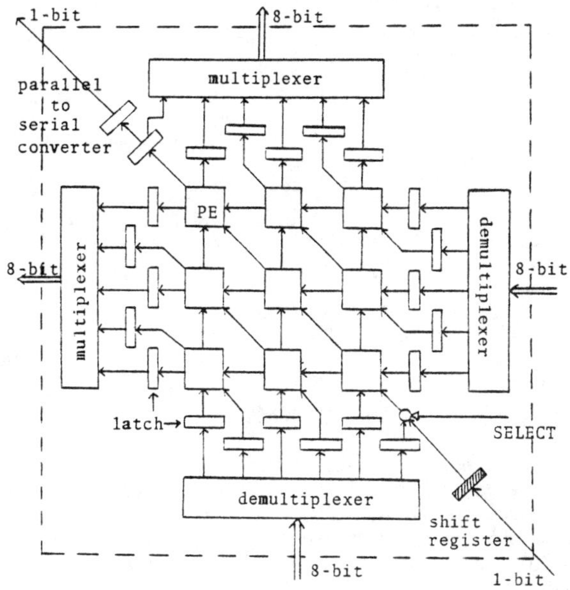

Fig. 2 Byte-serial grouped I/O scheme for a hex-connected systolic array chip

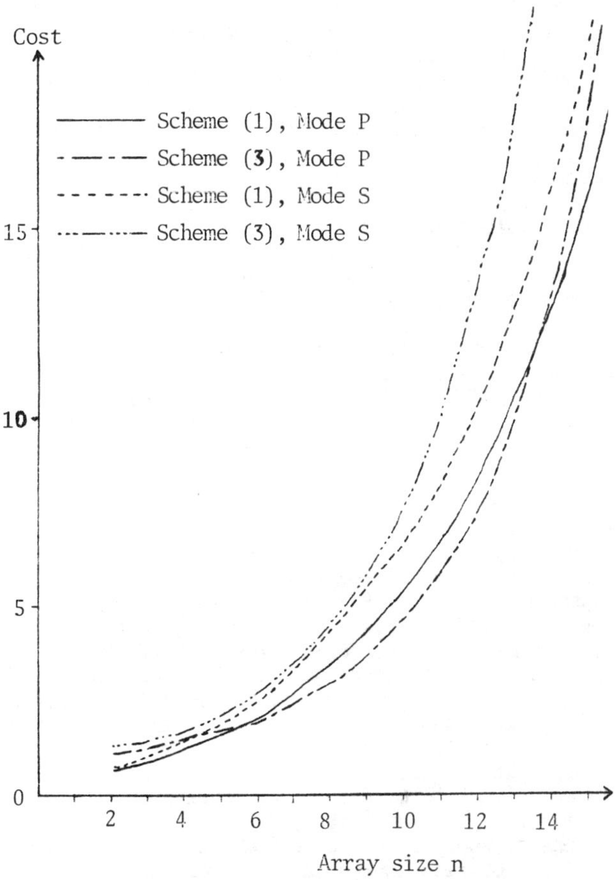

Fig. 3 A plot of the combined cost v.s. array size for Chip-I/O schemes (1) and (3) under the two timing modes P and S.

ON THE VLSI IMPLEMENTATION OF FAULT-TOLERANT ARCHITECTURES

C. S. Raghavendra

Electrical Engineering-Systems Dept.
University of Southern California
Los Angeles, CA 90089

T. E. Mangir

UCLA Computer Science Dept.
University of California
Los Angeles, CA 90024

ABSTRACT

The next challenge for VLSI is the alternative architectures and layouts which reduce the communication delays and other limitations interconnect technology imposes on the designers. Recent studies indicate that to be able to cope with the exploding interconnect complexity and delay problems, VLSI chips and systems should be composed of hierarchical connections of modules. Tree architectures and array architectures are two of the current alternatives which keep the propagation delays to a minimum by using localized communications. In addition, they are also easily partitioned for reducing design complexity and verification. These architectures are also suitable to be designed and implemented redundantly for both reliability and yield improvement. In this paper, implementation of fault-tolerant binary tree and arrays are attempted with consideration to interconnect requirements between processing modules. We show that it is possible to implement highly fault-tolerant structures in VLSI with only two layers of interconnections. We discuss the issues of yield, reliability, and area overhead in the implementation of systems with redundancy. We also perform the yield analysis to find out the improvement obtainable with the redundancy in hardware.

1. INTRODUCTION

To manage the associated complexity problems of VLSI design, verification and low yields, implementation of architectures designed using large number of identical or similar processing modules onto VLSI has attracted many researchers [SEQU 83, MANG 82]. It has been shown that binary tree, systolic arrays etc. map quite well to silicon because of their planar, and hierarchical interconnect requirements [MEAD 82, REM 80]. Hierarchical connections between modules are essential to limit the propagation delays and to reduce the interconnect complexity between modules. In addition this kind of architectures can be implemented with semicustom, quick turnaround methodologies, where we can extend and exploit the capabilities of restructurable and repairable interconnect technologies for VLSI and WSI [MANG 83]. When we want to design fault-tolerant binary tree or arrays several redundant links will be required and the resulting structure may not be planar. However, it is interesting to study whether highly redundant systems can be mapped to VLSI with very small layers of interconnections (planar structures requires only one). In this paper, we show that it is possible to implement fault-tolerant binary tree and arrays with two layers of interconnections. Also, the interconnections are quite uniform and therefore it is very easy to implement larger systems.

Fault-tolerance issues in tree architectures have been studied by several researchers [DESP 78, HAYE 76, KWAN 80, RAGH 83]. In [HAYE 76] design procedures for k-fault tolerant tree structures are presented; in particular, design of optimal 1-fault-tolerant trees are discussed. The optimal design can sustain any single failure with the minimum number of redundant links. In [RAGH 83] fault-tolerance schemes are proposed for tolerating any single failure in each level of the tree as well as trading performance for reliability. These schemes can be extended to arrays in a straightforward manner.

The implementation of VLSI circuits with redundancy has been suggested to improve the processing yield [MANG 82].

In the next section we present feasible designs of fault-tolerant binary tree and arrays with discussion of interconnect requirements, failure modes, and test procedures. Then we perform reliability and yield improvement analysis and estimate area overhead for providing redundancy.

2. IMPLEMENTATION OF FAULT-TOLERANT STRUCTURES

We consider the VLSI implementation of a fault-tolerant binary tree which is shown in Figure 1. There is a spare node for each level in the tree (shown by circle within circle) and there are redundant links shown by dashed lines. With this arrangement any single failure in a tree level can be tolerated. Any link failure will be equivalent to a node failure. In an n-level tree (numbered 0, 1, 2,...n-1) three are $2 \cdot 2^i$ non-redundant links and $3 \cdot 2^i + 1$ redundant links at the ith level. Therefore the amount of redundancy in this fault-tolerant structure is 1 node per level and about 150% extra links.

CH1935-6/83/0000/0744$01.00

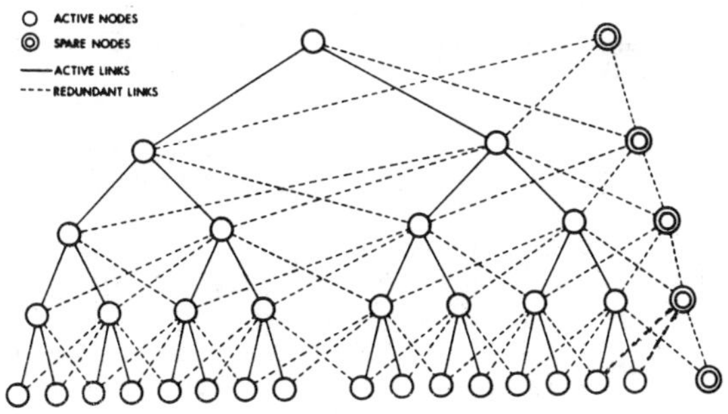

Figure 1. Fault-Tolerant Binary Tree

Implementation of fault-tolerant binary tree in VLSI is a special case of implementing any system with arbitrary interconnects among modules. Here, we precisely know the interconnect requirements. Our aim would be to optimize on the number of horizontal and vertical channels needed for inter-processor communication. There is also some degree of freedom as to the placement of modules on the chip. We are interested in compacting the modules as much as possible and at the same time consider failure modes of chip. In particular, we would like to place the modules so that more common faults will not destroy more than one module per level of tree. It is also desirable to use interconnect channels such that a fault does not destroy communication between several modules.

Figure 2 shows a possible implementation of a 3-level fault-tolerant binary tree. Thick lines represent vertical connections and thin lines represent horizontal connections. Horizontal and vertical lines are channels in different layers. The circuit pattern is quite regular and can be easily extended to larger trees. There is some extra hardware for testing modules and for switching right channels for inter-module communication. However, the redundancy pays off in terms of improved reliability and higher chip yield. This implementation is a direct hierarchical mapping, and therefore always possible if there are sufficient modules on the chip.

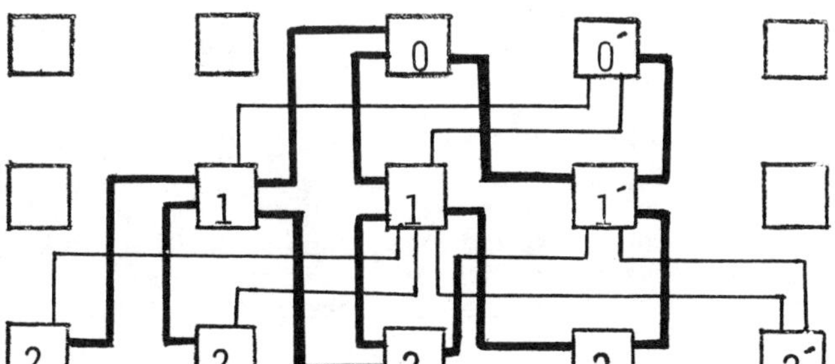

Figure 2. VLSI Implementation of Fault-Tolerant Binary Tree

We try to estimate the chip area for this design which would give us an upper bound as we are not minimizing area in this layout. The number of rows of modules is same as the number of levels in the tree, i.e., n-1. The number of modules per row is equal to the number of leaf nodes in the tree, which is 2^{n-1}. Let k_h and k_v be the average number of horizontal channels between two consecutive rows and the average number of channels between two consecutive columns of modules respectively. Let A_m^2 be the area required for a module and A_c be the width of a channel (a link may be several wires). Now the estimated width of chip for an n level tree is:

$$W_U = 2^{n-1} * A_m + (2^{n-1} - 1) * A_c * k_v$$

The estimated height of chip is:

$$H_U = n * A_m + (n-1) * A_c * k_h$$

An upper bound on the required area is:

$$A_U = W_U * H_U$$

A lower bound on the area requirement can be estimated by considering the most compact layout which is possible by mapping the fault-tolerant tree to nearest square array of modules. The number of modules in row and column of such a square array will be the nearest integer greater than

$$\sqrt{(2^n + n - 1)}$$

where n is the number of levels in the tree. Let a be the number of modules per row and column of such an array. Then width and height respectively are:

$$W_L = a * A_m + (a-1) * A_c * k_v$$

$$H_L = a * A_m + (a-1) * A_c * k_h$$

A lower bound on the chip area is:

$$A_L = W_L * H_L$$

In a fault-tolerant binary tree we are primarily interested in protection against module and link failures. In VLSI we cannot make the independent module failure assumption. It is likely that certain faults may destroy more than one neighboring modules. Since the fault-tolerant binary tree can sustain more than one failure, but in different levels, we would like to implement the system such that modules of same level are not neighbors on the chip. Also, we may skip over some modules in a square array of modules for added protection against multiple nearest neighbor failures. One such implementation is depicted in Figure 3.

In the fault-tolerant binary tree of Figure 1, a link failure is treated as equivalent to a node failure. In a VLSI implementation link failures occur due to defects on the communicating channels between modules. One way of protecting against complete channel failures is to draw two different wires between every connected modules by

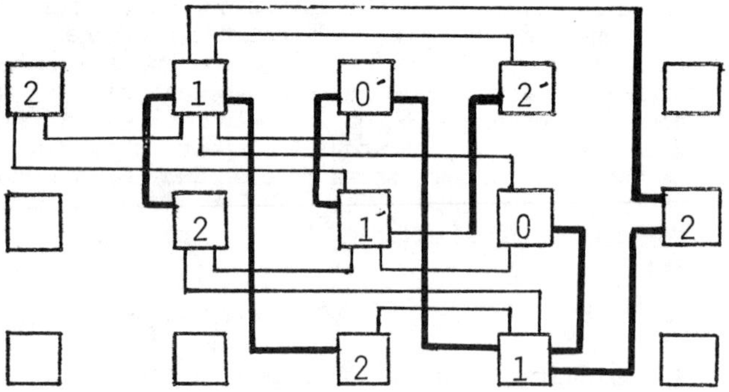

Figure 3. Alternate Layout of Fault-Tolerant Binary Tree

using separate channels. If the channels are used judiciously the penalty of area for providing this redundancy is not large. Further, the channels can be built using segments with restructurable capability by switching. Such a restructurable channels are possible with some technologies and have been used for wafer scale arrays [GREE 82]. With this type of channels we not only require fewer numbers but also have higher fault-tolerance.

Figure 4 shows a compact implementation of a redundant square array. There is a spare processing module for each row of array. Redundant links are provided to bypass failed modules. This system tolerates any single failure in each row of array. If it is an nxn array then the system can tolerate n failures all in different rows. The amount of redundancy is n spare modules (i.e. 100%) and 3n(n-1) redundant links (i.e. about 300%).

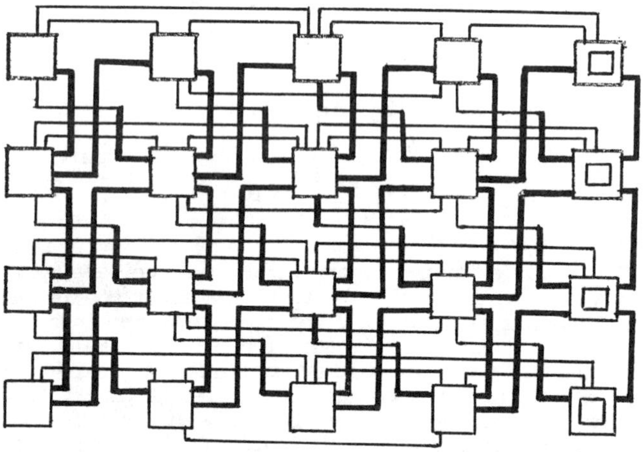

Figure 4. Implementation of Fault-Tolerant Array

This mapping of fault-tolerant array to VLSI is quite compact and we do not have any unused modules. The interconnections are quite regular and there are at most nine interconnects from any module. If a complete column of modules are destroyed still the chip will have an active array configuration. However, the system can tolerate only one module failure per row. If we would like to provide more protection in each row against adjacent module failures, there should be more columns of modules with the rows skewed similar to what is shown in Figure 3.

3. YIELD, RELIABILITY, AND AREA OVERHEAD

The implementation of fault-tolerant architectures in VLSI is aimed at achieving both higher reliability and improved yield in the fabrication process. A fabricated chip may have defects in modules or channels. In Figure 5, relative yield improvement is plotted against percent redundancy. In this case the array is assumed not to have any redundant interconnect. FM, Figure of Merit is defined as $\left(\frac{A_0}{A_R}\right)\left(\frac{Y_R}{Y_0}\right)$, where A_0 and A_R are nonredundant and redundant areas of the chip, and Y_R and Y_0 are the redundant and nonredundant yields, respectively.

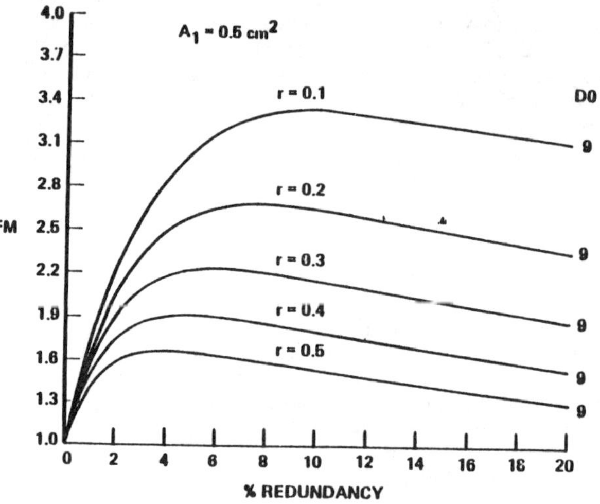

Figure 5. FM Figure of Merit vs. % redundancy with defect density $D_0 = 9cm^{-2}$. for a square array, of 100 modules, module size is 1000gates/module

To connect the modules in a final configuration, modules need to be tested and connected to the horizontal and vertical interconnect lines with the programmable links. In [MANG 83], possible technologies to facilitate this restructuring are described.

We have studied the effects of redundant interconnections and module sizes on the yields of redundant and nonredundant binary trees which are mapped onto a square array. The method for estimating the yield is similar to the one used in [MANG 82], except for the redundant links. We have also assumed that the connection requirements of the tree will be proportional to the levels in the tree. The results indicate that, for binary trees with very large module sizes (10^5 gates) the nonredundant yields are low enough to benefit from the use of the redundancy to improve the yield of processing. As module sizes decrease, (we have used 10^2 to 10^5 gates per module), the nonredundant yields become high enough that redundancy does not contribute to the yield significantly. It is concluded that, for smaller module sizes the redundancy can be used for reliability improvement, and for large modules it will actually be beneficial to include redundancy for yield improvement. Another interesting side to this result is that for high complexity modules, it predicts the advantages of wafer scale integration of binary tree architectures. In Figure 6, yield of a three level redundant tree is plotted against the width of the wiring channel between modules. As is seen on the figure as the communication path widens between the modules, which dictates the width of the channels, yield decreases sharply. Also note that the 'yield ' in this figure is the relative yield after the gross yield losses.

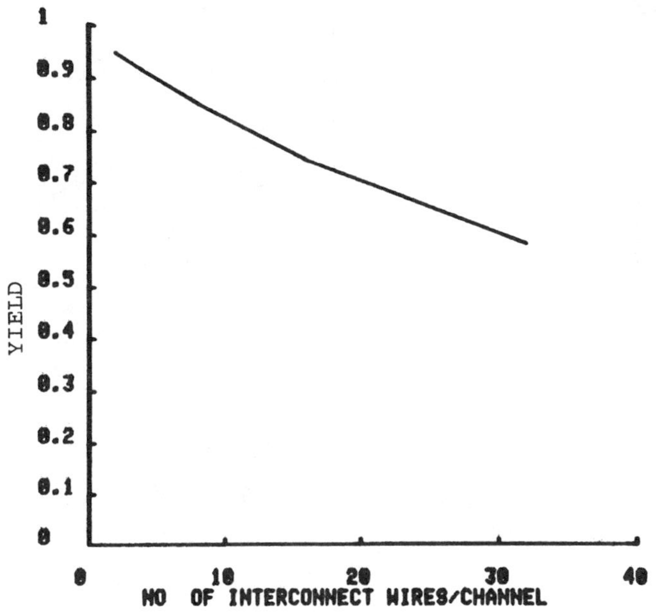

Figure 6. Yield vs. wiring channel width (in bits) for a three level redundant binary tree, with module size 10K gates/module

Implementing a fault-tolerant binary tree assures a greater chance of obtaining a working binary tree. It is possible to analytically model and estimate the yield improvement obtainable with this scheme. Certain defects may destroy only one module or several adjacent modules. Some other defects may destroy complete interconnect channels. When that happens the chip is not usable. When performing yield improvement calculations one has to carefully consider the different failures and their effects. However, it is possible to implement trees such that multiple adjacent module failures do not destroy the chip.

When fault-tolerant architectures are implemented on VLSI the fabricated chips will have higher reliability. A good percentage of chips will have the active elements and some or all of the spare elements working. Then the chip can tolerate one or more module failures in the field test which implies improved reliability. It is possible to perform reliability analysis of fault-tolerant binary trees and arrays [RAGH 83] with independent module failure assumption. However, for VLSI a more general failure model is required. A model suitable for VLSI can be mixed exponential failure rates although this problem is still under research.

4. CONCLUSIONS AND FUTURE WORK

Implementation of fault-tolerant binary tree and array are presented. Issues such as yield improvement due to redundancy, reliability, and area overhead are discussed. We find that yield improvement for fault-tolerant binary tree is not significant when modules are not very complex. However, reliability of system can be improved. Further work involves in studying accurate reliability models for VLSI fault-tolerant structures and in the use of restructurable channels.

REFERENCES

[DESP 78] A. Despain, D. Patterson, "X-Tree: A Structured Multiprocessor Computer Architectures", Proceedings of the 5th Symposium on Computer Architecture, April 1978, pp 144-151.

[GREE 83] J. W. Greene, A. E. Gamal "Area and Delay Penalties in Restructurable Wafer-Scale Arrays", Proceedings 1983 Caltech VLSI Conference.

[HAYE 76] J. P. Hayes, "A Graph Model for Fault-Tolerant Computing Systems", IEEE Transactions on Computers, Vol. C-25, No. 9, September 1976, pp 875-884.

[KWAN 80] C. L. Kwan, S. Toida, "Optimal Fault-tolerant Realizations of Hierarchical Tree Systems", Proceedings 1980 Int. Symp. Fault-Tolerant Computing, October 1980, ppl 176-178.

[MANG82] T. E. Mangir, "Use of On-Chip Redundancy for Fault-Tolerant VLSI Design", Technical Report No. CSD-820201, UCLA Computer Science Department, February 1982.

[MANG83] T.E.Mangir, "Impact and Limitations of Interconnect Technology and Restructurable Connections on VLSI Design,"This Proceedings

[MEAD 82] C. Mead, M. Rem, "Minimum Propogation Delays in VLSI", IEEE Journal of Solid State Circuits, Vol. Sc-17, No. 4, August 1982, pp 772-775.

[RAGH 83] C. S. Raghavendra, A. Avizienis, M. Ercegovac, "Fault-Tolerance in Binary Tree Architectures", Proceedings 1983 Int. Symp. Fault-tolerant Computing, June 1983.

[REM 80] M. Rem, "Mathematical Aspects of VLSI", Proceedings of the Caltech Conference on VLSI, January 1980, pp 55-64.

[SEQU 83] C. Sequin, "Managing VLSI Complexity", Proceedings IEEE, Vol. 71, No. 1, January 1983, pp 149-166.

Session 3-4d
Signal Processing

Chairman

G. C. Brown
CEC

DIGITAL FILTERING USING DISTRIBUTED VLSI DIGITAL SIGNAL PROCESSORS

K.Y. Liu

Department of Electrical Engineering
National Taiwan University
Taipei, Taiwan, Republic of China

ABSTRACT In this paper, two techniques for decomposing a one-dimensional as well as a multi-dimensional cyclic convolution into many smaller and identical cyclic convolutions are presented. Distributed VLSI digital signal processor architectures utilizing these two decomposition techniques for digital filtering are then presented.

1. Introduction

Digital filtering have found many applications in radar, sonar, image processing, speech processing, etc.[1-2] Currently, the most commonly used method for achieving high-speed digital filtering is to implement the algorithm in software on general purpose minicomputers which use peripheral array processors for number-crunching. One way to reduce size, power dissipation, and to increase processing speed is to used a network of VLSI digital signal processor chips to improve the processing speed [3]. The purpose of this paper is to present the distributed VLSI digital signal processor network architectures, which are based on FFT decomposition and polynomial factorization, for accomplishing one dimensional (1-D) and multi-dimensional (M-D) digital filtering.

2. Distributed Network Architecture Based on FFT Decomposition for Achieving Digital Filtering

Digital filtering is quite often accomplished using cyclic convolutions [4]. A 1-D cyclic convolution of two N-point sequences $\tilde{a} = \{a_0, a_1, \cdots, a_{N-1}\}$ and $\tilde{b} = \{b_0, b_1, \cdots, b_{N-1}\}$ is defined as

$$c_k = \sum_{j=0}^{N-1} a_j\, b_{(k-j)} \text{ for } k = 0, 1, 2, \cdots, N-1 \qquad (1)$$

Eq (1) can be computed using the FFT algorithm. The procedure is the well known one, where Fourier transforms are applied to sequences $\tilde{a}$ and $\tilde{b}$ and an inverse Fourier transform is applied to the product of these two transforms.

The transform of sequence $\tilde{a}$ can be expressed as

$$A(f) = \sum_{t=0}^{N-1} a(t)\, W^{ft} \text{ for } 0 \le f \le N-1 \qquad (2)$$

where $a(t)$ are data points of sequence $\tilde{a}$ and W is a Nth root of unity. Suppose $N=r^n$, where r is the radix used in the FFT. Then integers f and t in (2) can be expressed as

$$f = f_{n-1} r^{n-1} + f_{n-2} r^{n-2} + \cdots + f_1 r + f_0 \qquad (3)$$

$$t = t_{n-1} r^{n-1} + t_{n-2} r^{n-2} + \cdots + t_1 r + t_0 \qquad (4)$$

where $f_{j-1} = 0, 1, \cdots, r-1$ for $1 \le j \le n$, and $t_j = 0, 1, \cdots, r-1$ for $0 \le j \le n-1$. Then by the same procedure given in Ref. 5 for deriving a radix-r, decimation-in-frequency (DIF), FFT algorithm, one can express (2) as

$$A(f_{n-1}, f_{n-2}, \cdots, f_1, f_0) = \sum_{t_0}\sum_{t_1}\cdots\sum_{t_{n-i-1}} a^{(i)}(f_0, f_1, \cdots, f_{i-1}, t_{n-i-1}, \cdots, t_0) \times W^{[f_{n-1} r^{n-1} + f_{n-2} r^{n-2} + \cdots + f_i r^i]\, t} \qquad (5)$$

In (5)

$$a^{(i)}(f_0, f_1, \cdots, f_{i-1}, t_{n-i-1}, \cdots, t_0) = \left\{ \sum_{t_{n-i}=0}^{r-1} a^{(i-1)}(f_0, f_1, \cdots, f_{i-2}, t_{n-i}, \cdots, t_0)\, W^{f_{i-1} t_{n-i} (N/r)} \right\} \times W^{f_{i-1}[t_{n-i-1} r^{n-i-1} + \cdots + t_1 r + t_0]\, r^{i-1}}$$

where $i = 1, 2, \cdots, n$ is the FFT stage number. The initial input data $a^{(0)} = a(t)$ is in normal order and the resulting data $a^{(n)} = A(f)$ is in digital-reversed order.

Note that for a given $f_0, f_1, \cdots, f_{i-1}$, (5) is a N/r^i - point FFT. Moreover, these r^i FFT's can be computed in parallel. Similar procedure can be applied to an inverse FFT (IFFT) with digital-reversed input. In this case, the first n-i stage consists of r^i, N/r^i - point independent FFT's. Thus this FFT-IFFT combination for computing a 1-D cyclic convolution lends itself to a natural partition of independent computational channels. As an example, an 8-point 1-D cyclic convolution using a radix-2, DIF, FFT-IFFT combination with two-channel and four-channel decomposition is shown in Fig. 1.

As an example, a distributed VLSI digital signal processor network for computing a 256-point 1-D cyclic convolution with two-level of decomposition is shown in Fig. 2. A detailed description of this system is given as follows. In Fig. 2, each digital signal processor module (SPM) consists of an input processor chip (IPC), a signal processor chip (SPC), an output processor chip (OPC) and three memory units-one serves as the input buffer for IPC, one serves as the data memory for SPC, and the third one serves as the instruction memory for SPC.

The loading of input data is controlled by IPC of SPM No.1 . The input data is stored in the input buffer first and then loaded into the data memory when SPC and OPC are not busy. The SPC then performs the first two stages of a normal-order input, 256-point, radix-2, DIF or DIT FFT, using the data stored in the data memory, to complete the convolution decomposition.

After the decomposition is completed the data memory contents are transferred to the input buffers of SPM's No.2 to No.5. This data transfer is controlled by the OPC of SPM No.1 and IPC's of SPM's No.2 to No.5. At the same time of output data transfer, new data is loaded into the input buffer of SPM No.1.SPM's No.2 to No.5 then compute four 64-point, FFT's in parallel. The resulting data are transferred to SPM No.6 which performs the windowing and weighting function multiplications. The weighting function is down-loaded prior to the processing from the control processor via the control bus.

The results of windowing and weighting function multiplications are transferred from SPM No.6 to SPM's No.7 to No.10. Each of these SPM's then performs a 64-point IFFT. The results of these four IFFT's are transferred to the combine processor, SPM No.11, where the last two stages of the bit-reversed input, 256-point, radix-2, DIF or DIT IFFT are performed. The output is then taken from the data memory of SPM No.11 while new data is loading into the input buffer of the same processor.

The above decomposition, FFT, windowing, reference multiplications, IFFT, and combine operations can be pipelined by proper load balancing to yield high system throughput. "No Op" can be put into SPM instructions to maintain a uniform pipeline speed. To hande different convolution size using to same architecture, one needs to provide enough data memory in each SPM, load in proper processing parameters via the control bus to each SPM, and balance the processing load among processors.

The above architecture can also be extended to compute a multi-dimensional cyclic convolution.

3. Distributed Network Architecture Based on Polynomial Factorization

Both 1-D and 2-D cyclic convolutions can be accomplished using architectures based on polynomial factorization. Let us consider the 2-D cyclic convolution case.

Let F be the field of complex numbers. Also let $a_{t1,\,t2}$ and $b_{t1,\,t2}$ be two $d_1 \times d_2$ arrays, where $0 \le t_i \le d_i - 1$ for $i = 1, 2$. Then the two-dimensional cyclic convolution of $a_{t1,\,t2}$ and $b_{t1,\,t2}$, where $a_{t1,t2}$ and $b_{t1,\,t2} \in F$, is defined by

$$C_{n1,\,n2} = \sum_{t_1=0}^{d_1-1} \sum_{t_2=0}^{d_2-1} a_{t1,\,t2}\, b_{(n_1-t_1),\,(n_2-t_2)} \quad (6)$$

$0 \le n_i \le d_i - 1$ for $i = 1, 2$

where $(n_i - t_i)$ denotes the residue of $(n_i - t_i)$ modulo d_i for $i = 1, 2$.
Define

$$A_{t1}(Z) = \sum_{t_2=0}^{d_2-1} a_{t1,\,t2}\, Z^{t_2}\,;\; B_{t_1}(Z) = \sum_{t_2=0}^{d_2-1} b_{t_1,\,t_2}\, Z^{t2}$$

$$C_{n_1}(Z) = \sum_{n=0}^{d_2-1} C_{n1,\,n2}\, Z^{n2} \quad (7)$$

where $0 \le n_i,\ t_i \le d_i - 1$ for $i = 1,2$. It was shown[6] that the 2-D convolution in (6) can be expressed as a one-dimensional polynomial convolution, i.e.,

$$C_{n1}(Z) = \sum_{t_1=0}^{d_1-1} A_{t1}(Z)\, B_{(n_1-t_1)}(Z) \bmod (Z^{d_2} - 1) \quad (8)$$

for $0 \le n_1 \le d_1 - 1$, where $(n_1 - t_1)$ denotes the residue of $(n_1 - t_1)$ modulo d_1.

If $d_2 = 2^m$, with $m > 1$, then one can factor $Z^{d_2} - 1$ into

$$Z^{d2} - 1 = (Z^{d2/2} + 1)(Z^{d2/2} - 1)$$

Since $(Z^{d2/2} + 1)$ and $(Z^{d_2/2} - 1)$ are relatively prime, by the Chinese Remainder Theorem (CRT) for polynomials[7], the polynomial congruences

$$C_1^{(1)}(Z) \equiv C_{n_1}(Z) \bmod (Z^{d_2/2} + 1) \quad (9a)$$

$$C_2^{(1)}(Z) \equiv C_{n_1}(Z) \bmod (Z^{d_2/2} - 1) \quad (9b)$$

have a unique solution

$$C_{n1}(Z) = C_1^{(1)}(Z)\left(-\tfrac{1}{2}\right)\left(Z^{d_2/2}-1\right) + C_2^{(1)}(Z)\left(\tfrac{1}{2}\right)\left(Z^{d_2/2}+1\right) \bmod \left(Z^{d_2}-1\right) \quad (10)$$

Thus we have decomposed a d_2-point 1-D polynomial convolution into two $d_2/2$ - point 1-D polynomial convolutions. Note that in (10) the arithmetic required to compute $C_{n1}(Z)$ from $C_1^{(1)}(Z)$ and $C_2^{(1)}(Z)$ requires only cyclic shifts and additions.

If one uses the transformation $Z = w_1 u_1$ given in Ref. 8, where w_1 is a $d_2/2$th root of -1, on $C_2^{(1)}(Z)$, then (9b) can be expressed as

$$C_2^{(1)}(w_1 u_1) = \widetilde{C}_2^{(1)}(u_1) = C_{n1}(w_1 u_1) \bmod \left(u_1^{d_2/2}+1\right) \quad (11)$$

Thus we have decomposed a d_2-point 1-D polynomial convolution into 2, $d_2/2$-point 1-D polynomial convolutions given in (9a) and (11), respectively. In the computation of a 2-D convolution the input polynomial $A_{t_1}(Z)$ is decomposed into 2 polynomials, $A_{t_1}^{(1)}(Z)$ and $A_{t_1}^{(2)}(Z)$ through the modulo operation by the polynomials $(Z^{d_2/2}+1)$ and $(Z^{d_2/2}-1)$, respectively. $A_{t2}^{(2)}(Z)$ is then changed to $A_{t1}^{(2)}(u_1)$ through the transform $Z=w_1 u_1$. $A_{t1}^{(1)}(Z)$ and $A_{t1}^{(2)}(u_1)$ are then convolved with the corresponding weighting function polynomial obtained likewise from $B_{t1}(Z)$. The results of these two polynomial convolutions are combined using the CRT for polynomials to form the final result $C_{n1}(Z)$.

If $d_1 = d_2 = 2^n$, then one can use the fast polynomial transform (FPT) technique [9], which uses only 1/2 of the multiplications required by the 2-D FFT approach, to compute the small 1-D polynomial convolutions given in (9a) and (11). This approach uses a FFT-like algorithm to perform the polynomial transforms, and the FFT algorithm to transform the coefficients of a polynomial. Hence the FFT decomposition technique discussed previously can by applied to the polynomial transform and the FFT computations. A decription of the polynomial decomposition, FPT, and CRT combine circuits can be found in Refs. 10.

4. Conclusions

We have presented two distributed processing techniques and their associated VLSI digital signal processor network architectures for computing the cyclic convolution. It has been shown that using these network architectures, one can achieve high throughput digital filtering.

REFERENCES

1 A.V. Oppenheim, Ed., Applications of Digital signal Processing, Prentice-Hall, Inc., 1978.

2 C. Wu, K.Y. Liu, and M. Jin, "On Modeling and Processing of Spaceborne SAR Signals," IEEE Trans. on AES, Sept. 1982.

3 F. Mintzer, "Parallel and Cascade Microprocessor Implementations for Digital Signal Processing," IEEE Trans, on ASSP, Oct. 1981.

4 A.V. Oppenheim and R.W. Schafer, Digital Signal Processing, Prentice-Hall Inc., 1975.

5 G.D. Bergland, "The Fast Fourier Transform Recursive Equations for Arbitrary Length Records," Math. of Comput., Vol. 21, 1967.

6 S. Winograd, "Some Bilinear Forms Whose Multiplicative Complexity Depends on the Field of Constants," IBM Watson Research Center Report RC5669, Oct. 10, 1975.

7 E.R. Berlekamp, Algebraic Coding Theory, McGraw-Hill Book Company, New York, 1968.

8 B. Arambepola and P.J W.Rayner, "Efficient Transform for Multidimensional Convolutions," Electronic Letters, 15 March 1979, Vol. 15, No.6.

9 H.J. Nussbaumer and P Quandalle, "Computation of Convolutions and Discrete Fourier Transforms by Polynomial Transforms," IBM J. Res. Develop., Vol. 22, No.2, Mar. 1978.

10 K.Y. Liu, "A Pipelined Tree Machine Architecture for Computing a Multi-Dimensional Convolution," IEEE Trans. on Circuits and Systems, April, 1982.

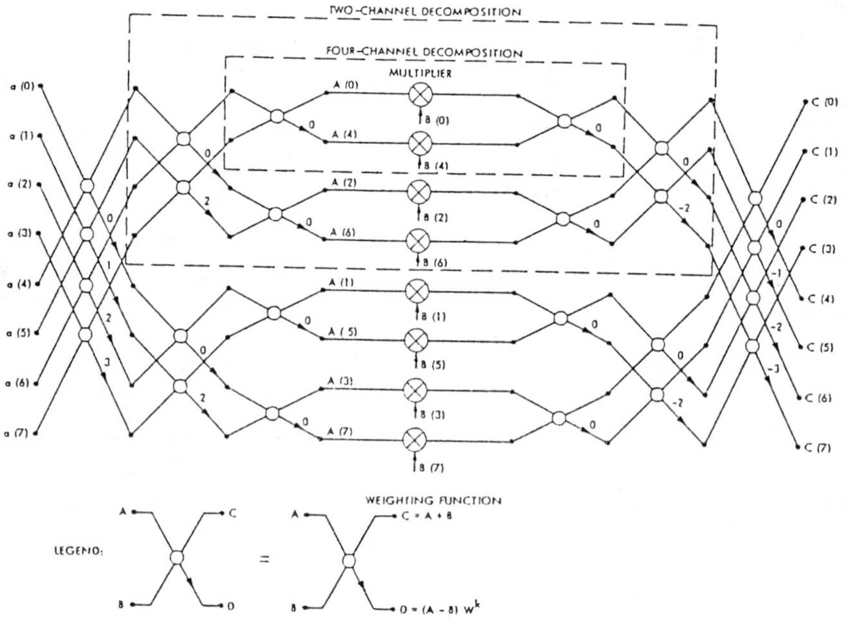

Figure 1. 1-D Cyclic Convolution Decomposition Based on FFT Decomposition.

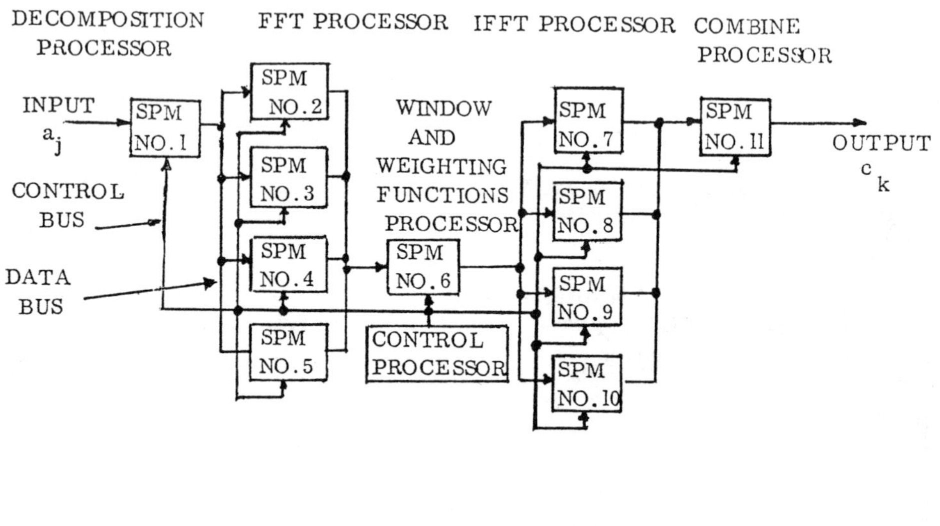

Figure 2. A Distributed VLSI Digital Signal Processor Network for Computing a 1-D Cyclic Convolution with Two-Level of Decomposition Using the FFT Decomposition Technique.

VLSI COMPUTER ARCHITECTURE FOR SCENE MATCHING

Yee-Hong Yang*
Mellon Institute
Computer Engineering Center
4616 Henry Street
Pittsburgh, PA 15213

Tsung-Wei Sze
Department of Electrical Engineering
University of Pittsburgh
Pittsburgh, PA 15261

ABSTRACT

A simple algorithm for scene matching, called the Edge Guided Scene Matching (EGSM), was developed in this paper. Six computer architectures, the cluster-connected and the mesh-connected, were evaluated for possible implementation of EGSM. Results indicate that the mesh-connected computer architecture is the best candidate. A one-dimensional model was developed to determine a strategy for implementing the EGSM on a mesh-connected architecture. It was found that by utilizing idle Processor Elements (PEs), considerable speed up could be obtained. Results for the two-dimensional case was obtained using simulation. Similar trends were found in the speed up as in the one dimensional case.

1. INTRODUCTION

Scene matching techniques have been developed mostly because of their potential applications in areas such as target identification, earth resource change detection, and industrial automation. In general, the performance (speed) of an algorithm is affected by (1) the configuration of the algorithm and (2) the supporting hardware. In the past, an algorithm is usually tailored for a specific piece of hardware. As an example, the Fast Fourier Transform (FFT) is tailored for conventional sequential and pipelined machines by reducing the number of multiplications and additions. Recently, because of the advancement of VLSI/VHSIC technology, more attention is being drawn to designing and developing special purpose hardware for specific applications. In this paper, we will present a simple variation of a well-developed scene matching algorithm and its possible implementation on a parallel computer architecture. A tradeoffs study between six parallel computer architectures was performed to determine the best candidate for implementation.

Scene matching can be defined as a procedure to locate a subimage within a sensed image using a template. There are two commonly used figures of merit of best matched. One of the measures is the correlation coefficient. The main disadvantage of this measure is that it requires a large amount of computations. FFT techniques can be used to reduce the amount of computations at each window. However, the number of windows required in the computation is still $(N-M+1)^2$.

Another technique developed to overcome this problem is known as the sequential decision scheme by Barnea, et al. [Barnea,72] The main idea of this method is to use a simple-to-compute measure, the absolute difference, instead of the correlation coefficient.

Another very attractive scheme, developed by Wong [Wong,78], employs a hierarchical scheme to reduce the amount of computations. This scheme is similar in principle to the divide-and-conquer scheme by solving the scene matching problem at reduced spatial resolution. After locating all the probable locations of best match, the matching is performed at a higher resolution. This process is repeated until the best match is obtained at full resolution. The main advantage of this scheme is the reduced amount of computations as compared with computing in a brute force manner using the correlation coefficient. The method reduces the amount of computations logarithmically. A limitation of this method is that it may not be able to locate small reference images at low resolution because at low resolution the features will be lost. The method to be presented is to complement the hierarchical scheme in cases when full resolution has to be employed.

2. SCENE MATCHING USING SEGMENTATION

The algorithm of the new scheme will now be described as follows:

1. First the sensed image is segmented using an available segmentation scheme. There are in general two main segmentation approaches, namely, region-based and edge-based, which were discussed in detail in most standard text books in image processing. The region-based approach requires more computations than the edge-based approach. For time critical applications, it is expected that an edge-based approach segmentation scheme is to be used. In the following analysis, the edge-based approach is used.

2. It is assumed that the center of the template is within a region that constitutes the object. The template is processed initially by an edge operator. After the sensed image is processed using the same edge operator, the boundaries of regions in both edge maps of the template and the sensed image are expanded until the center of the template is reached. Experimentally, it was found that a simpler scheme can be used. The same effect can be obtained by a suitable choice of threshold. The threshold to be used can be chosen such that the gradient at the center of the template is above the threshold, i.e. the center is considered as an edge point in the edge map. In general, this threshold can be predetermined because the template is assumed to be known *a priori*.

3. Then the adaptive sequential scheme is applied to the expanded boundaries. This means that each point in the expanded boundary of an edge map will be a center of the matching window. Suppose the number of edge points after expansion is n in a sensed image of size NxN. The saving that can be obtained is then $n(L-M+1)^2$. Further savings in computations can be obtained by updating the threshold. The updating rule can be expressed as follows: Let T be the current threshold, and (r,s) be the center of the matching window, then the new threshold T' is:

$$T'=\min\{T,E(r,s)\}$$

The total amount of saving will then be of the order of the ratio of the area of the expanded boundary region to the area of the sensed image. The overhead involved in preprocessing the image using segmentation is comparatively small as compared with the actual scene matching operation. Furthermore, the segmentation operation can be easily adapted for implementation on a parallel machine.

3. PARALLEL COMPUTER ARCHITECTURES

During the past twenty years, a significant number of high speed computer architectures has been proposed. [COMPUTER,83] Some of these architectures that were thought to be impractical to build are now considered not only viable but also necessary for speed enhancement. This is made possible due to the advancement of solid state technology. Signal and image processing require a large number of computations. In particular, scene matching algorithms are among those algorithms. An advantage that all scene matching algorithms share is the opportunities of concurrent operations exist in the algorithms and hence favors the use of innovative parallel computer architectures for further speed enhancement.

Six topologies of architectures were analysed, namely, cluster-connected, mesh-connected, cluster-cluster-connected, cluster-mesh-connected, mesh-cluster-connected and mesh-mesh-connected. The various configurations can be

* Current Address: Department of Computational Science, University of Saskatchewan, Saskatoon, Saskatchewan, Canada, S7N OWO.

This work was sponsored by the Department of Defense, contract number N00039-80-C-0640, and monitored by the Naval Air Development Center.

CH1935-6/83/0000/0753$01.00

seen in Figure 1. A detailed tradeoffs study can be found in [Yang, 83]. Results indicate that the mesh-connected architecture is the best candidate for implementation of the scene matching algorithm.

3.1. Theoretical Analysis of EGSM in a MCA Environment

In a sequential processing environment, the advantage of EGSM is the reduction of search space. In a MCA environment, the reduction of search space appears to be irrelevant to the performance of the system. This is so only when the search space for each PE is of one pixel in size. As was discussed earlier, this is still difficult to achieve technologically. In general, it is expected to have one PE to search a number of pixels. Without using the EGSM, all the PEs will be engaged in searching the same number of pixels. However, using the EGSM, a number of PE may be idling. The following strategy takes advantage of the idle PEs.

In order to simplify discussion, the analysis is first restricted to one-dimensional cases. Analysis of two-dimensional cases will be obtained using simulation which will be discussed in the next section.

The process of a PE searching over its search space is similar to the process of emptying a bucket of water where the search points are equivalent to the water in the bucket. Without using EGSM, all PEs will be like a mosaic of buckets having the same amount of water. The water will be emptied at the same rate. Using EGSM, a number of these buckets will be empty, i.e. the PEs will be idle.

Consider a one-dimensional case as shown in Figure 2. Suppose, initially, there are m buckets. Each of the buckets can hold up to q liters of water. All the buckets beyond the mth one are empty. All but the boundary bucket are leaking at a constant rate r liter/sec. The boundary bucket is leaking at a faster rate of kr liter/sec. The increase in leaking rate is due to the presence of empty buckets beyond the mth bucket. It is assumed that water can be transferred to the empty buckets from the boundary bucket and thus effectively increase the leaking rate. Furthermore the time for the transfer between buckets is assumed to be negligible as compared with the time required to empty a bucket.

The time required for the mth bucket to empty is:

$$t_m = \frac{L - (m-1)q}{kr}$$

where L is the total amount of water in the system. After the mth bucket is empty, the (m-1)th bucket is then a boundary bucket and thus is capable of leaking at a rate of kr liters/sec. The time required for the (m-1)th bucket to empty its content is:

$$t_{m-1} = \frac{q - t_m r}{kr}$$

The same argument can be repeated for other buckets, and hence:

$$t_{m-2} = \frac{q-(t_m+t_{m-1})r}{kr}$$

....

$$t_1 = \frac{q-r(\sum_{i=2}^{i=m} t_i)}{kr}$$

The solution to the above recurrence equations gives:

$$t_{m-1} = \frac{1}{k}\left(\frac{q}{r}-t_m\right)$$

$$t_{m-2} = \frac{k-1}{k} t_{m-1}$$

Figure 2. A one-dimensional model for analyzing a parallel scene matching strategy.

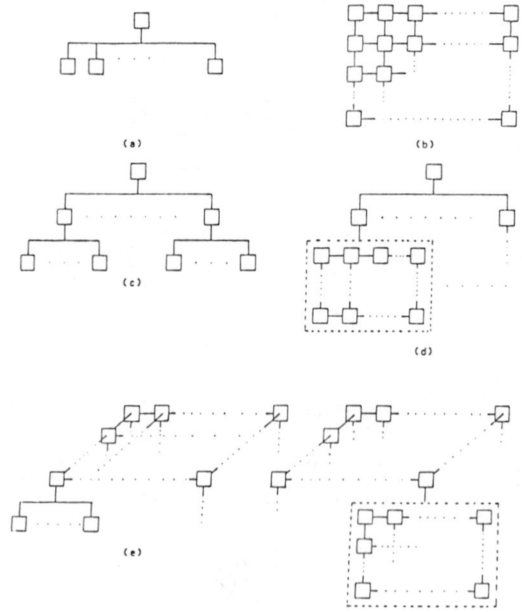

Figure 1 Vaious architectures for analysis.
(a) Cluster-Connected Architecture,
(b) Mesh-Connected Architecture,
(c) Cluster-Cluster-Connected Architecture,
(d) Cluster-Mesh-Connected Architecture,
(e) Mesh-Cluster-Connected Architecture and
(f) Mesh-Mesh-Connected Architecture.

...

$$t_1 = \left(\frac{k-1}{k}\right)^{m-2} t_{m-1}$$

The total time required to empty all the water in the system is then:

$$T = \sum_{i=1}^{i=m} t_i$$

$$= \frac{q}{r} - \left(1-\frac{1}{k}\right)^{m-1} \frac{(m+k-1)q-L}{kr}$$

Define R_s as:

$$R_s = \frac{\frac{L}{r}-T}{\frac{L}{r}}$$

The ratio R_s can be interpreted as the speed-up obtained. Hence the speed-up is then:

$$R_s = 1-\frac{q}{L} + \left(1-\frac{1}{k}\right)^{m-1} \frac{(m+k-1)q-L}{kL}$$

By defining p=q/L, then

$$R_s = 1-p+\left(1-\frac{1}{k}\right)^{m-1} \frac{(m+k-1)p-1}{p}$$

The variation of R_s with respect to k for various values of m and p can be seen in Figure 3. It is apparent from the figures that EGSM is not only beneficial for conventional sequential machines, it is also useful for a MCA machine. Analytical generalization of the above to two-dimension is tedious. Instead, a different approach, using simulation, will be used to evaluate the implementation of EGSM on a MCA machine.

3.2. Simulation of EGSM in MCA

Simulation of a computer system can be performed at various levels of detail. The goal of the present study was to investigate tradeoffs between data transfer and computation of adopting EGSM in a MCA environment.

It is assumed that each PE works in an autonomous

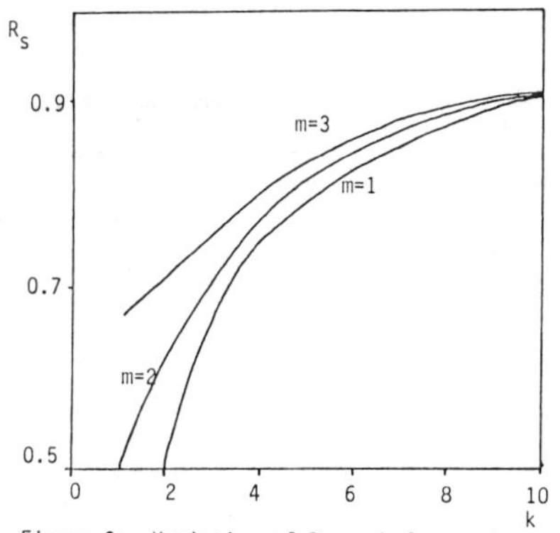

Figure 3. Variation of R_s vs k for various values of m and p.

fashion and there is a central controller to keep track of all the activities of each PE.

Each PE has all the data it requires. Depending on whether there are any search points in the search space, the PE can be either busy or idle. If a PE is idle, it can look at its four nearest neighbors to see if their search spaces are empty or not. If a neighbor is found to have a non-empty search space, a request to that neighbor will be made for a transfer. The purpose of the transfer is to partition search points among available PEs. Upon receiving a request, a PE will acknowledge the requesting PE by setting an "acknowledge" flag. A transfer will then be made. The search space of the transferring PE will partition its search points evenly among requesting PEs. If there is only one request, then the requesting PE will only receive half of the search points. This strategy is to avoid unnecessary thrashing. There are simple handshaking rules implemented in the simulation program. The experimenter can specify the locations of the search points and the number of them, the number of cycles required for each transfer, the number of cycles required to process one search point and the size of the array of PEs. All PEs will be initialized to their initial states at the beginning of each simulation.

Results of two simulations are shown in Figure 4. In this simulation, it was assumed that there was only one non-empty search space with 321 search points locating at the center of the array. The size of the array is k by k. The goal was to investigate a general trend in which the system will behave by varying the size of the array. It can be seen from the figure that implementing EGSM on MCA environment can provide significant percentage of speedup when k increases from 1 to 3. This indicates that speed up is possible if the idle PEs are utilized. It is also seen from the figure that by increasing k from 3 to 5 provides comparatively small amount of speed up. This is due to the fact that the overhead in the transfer is predominant.

Another simulation was performed. In this simulation, it was assumed that nine PEs have non-empty search points. The distribution of the search points can be seen in Figure 5. The total number of search points in the system was the same as in the previous simulation. Results of the simulation can be seen in Figure 6. The relative speedup with respect to k=3 can be seen in Figure 7. It can be seen that considerable speedup can be obtained by increasing k from 3 to 5. From the experiments, one can also observe that the general trend in performance of a two-dimensional case agrees with the general trend in performance of a one-dimensional case deduced using a hypothetical model. It is evident that if idle PEs are utilized as was shown in the above experiments, considerable speed up can be obtained without significant increase in the hardware complexity.

4. SUMMARY

Scene matching algorithms have long been known to be computationally intensive. The edge-guided scene matching offers possible savings in computations. The results, which were not described in this paper, were confirmed by implementing the algorithm on a sequential machine. (PDP 11/55)

Due to the advancement of VLSI/VHSIC technology, it is now practical to design, develop and build special purpose parallel machines for specific applications. Six computer architectures were analyzed for suitability of implementing scene matching algorithms. Theoretical analysis indicates that the mesh-connected architecture is more favorable.

An analytic model was developed to analyze the advantage of adopting EGSM on a mesh-connected computer architecture. The model was developed for the one-dimensional case and extrapolation to two dimensions was confirmed using simulation.

5. ACKNOWLEDGEMENT

The authors would like to thank Dr. R.L. Krutz and Mrs. Lorna L. Cheng of the Computer Engineering Center, and Mr. John Lawler of NADC for their suggestion and encouragement.

BIBLIOGRAPHY

Antonsson, D., et al., "PICAP - A System Approach to Image Processing," IEEE Trans. C-31, 1982, pp. 997-1000.

Backus, J., "Can Programming be Liberated from the von Neumann Style: A Functional Style and its Algebra of Pragrams," CACM, Vol. 21, 1978, pp. 613-641.

Barnea, D.I. and H.E. Silverman, "A Class of Algorithms for Fast Digital Registration," IEEE Trans. C-21, 1972, pp. 179-186.

Barnes, G.H., el. al., "The ILLIAC IV Computer," IEEE C-17, 1978, pp.746-757.

Batcher, K., "Bit-Serial Parallel Processing Systems," IEEE Trans. C-31, 1982, pp. 377-384.

COMPUTER, IEEE Computer Society monthly magazine, Vol. 16, January, 1983. This is an issue on computer architectures for image processing.

Hunt, D.J., "Application Techniques for Parallel Hardware," Infotech State of the Art Report on Supercomputer, Vol. 2, 1979.

Svedlow, M., C.D. McGillem and P.E. Anuta, "Image Registration: Similarity Measure and Preprocessing Method Comparisons," IEEE Trans. AES-14, 1978, pp. 141-149.

Wong, R.Y. and E.L. Hall, "Sequential Hierarchical Scene Matching," IEEE Trans. C-27, 1978, pp. 359-366.

Wong, R.Y., "Sequential Scene Matching Using Edge Features," IEEE Trans. AES-14, 1978, pp. 128-140.

Unger, S.H., "A Computer Oriented Toward Spatial Problems," Proc. of the IRE, October 1958, pp. 1744-1750.

Yang, Y.H. and T.W. Sze, "An Evaluation Study of Six Topologies of Parallel Computer Architectures for Scene Matching," Presented at the 1983 International Conference on Parallel Processing in Bellaire, Michigan.

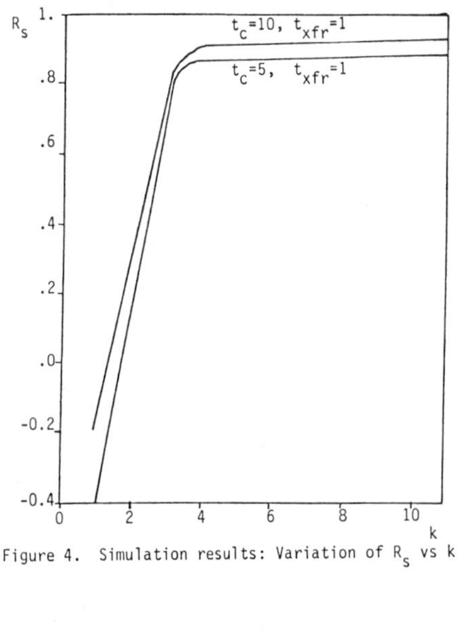

Figure 4. Simulation results: Variation of R_s vs k.

35	35	35
35	41	35
35	35	35

Figure 5 Distribution of search points among 9 PEs (Total number of search points is equal to 321.)

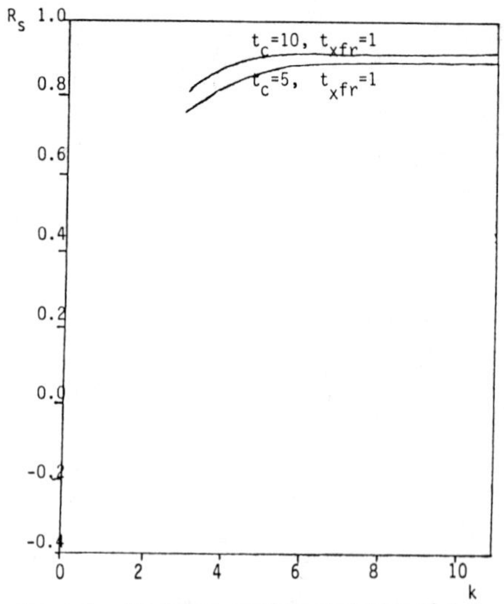

Figure 6. Simulation results: Variation of R_s vs k using the distribution of search points as is shown in Figure 5.

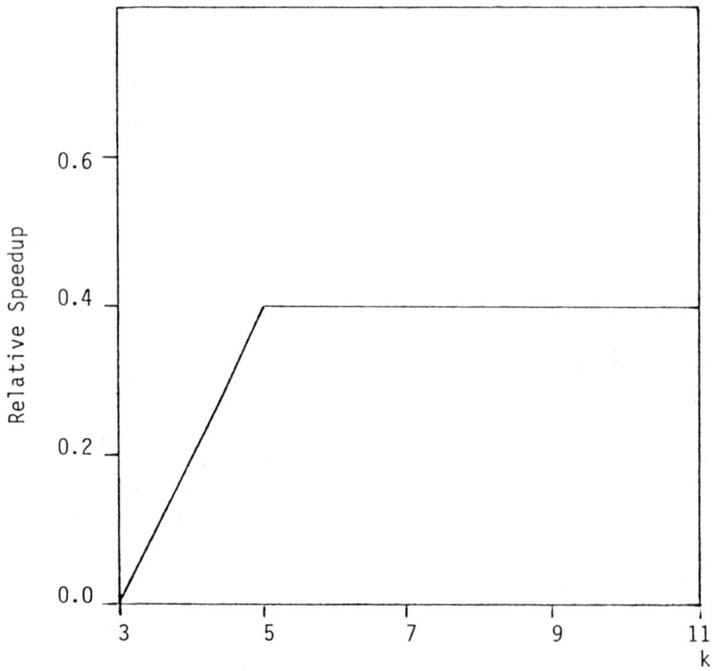

Figure 7. Relative speedup with respect to k=3 vs k.

CUSP: A CUSTOM VLSI PROCESSOR FOR DIGITAL SIGNAL PROCESSING BASED ON THE FAST FOURIER TRANSFORM

Peter P. Reusens*, Richard W. Linderman and Walter H. Ku

School of Electrical Engineering and
National Research and Resource Facility for Submicron Structures
Cornell University
Phillips Hall
Ithaca, NY 14850

ABSTRACT

The Cornell University Signal Processor (CUSP) is a high performance, CMOS processor, custom designed to efficiently compute digital signal processing algorithms based on the Fast Fourier Transform. One CUSP chip can be used as a stand-alone peripheral in a micro-processor system, or CUSP units can be combined into arrays in order to process signals with sampling rates of several MHz. It can attain a high numerical accuracy while maintaining a throughput superior to most available systems. We describe the CUSP architecture, the advantages of its bit-serial approach, its simple interface mechanism, its easy programmability and its error detecting features.

INTRODUCTION

The Cornell University Signal Processing chip is designed for a broad spectrum of applications from digital audio to radar signal processing. Its basic algorithm is the Fast Fourier Transform (FFT). Most monolithic systems of this type have been designed around a single 16bx16b parallel multiplier [1,2]. In this paper we describe the CUSP architecture and the many advantages of its bit-serial approach over the bit-parallel approach. We show how the CUSP architecture employs 16 20bit bit-serial multipliers and 24 serial adders to compute similar results much faster than the comparable monolithic bit-parallel processors and with extra bits of precision. The architecture overlaps high throughput computations with high bandwidth bit-parallel I/O to obtain over 90% hardware utilization.

CUSP maintains a very simple asynchronous control interface with its host computer allowing the user to choose the best system architecture for the particular application. This can range from a CUSP peripheral in a microprocessor system which performs filtering on audio inputs, to a small fault-tolerant array of CUSP chips computing FFTs for radar applications. In designing CUSP we have tried to maximize speed while maintaining a level of numerical accuracy (signal-to-noise ratio) sufficient for most foreseen applications and superior to what is generally available at present [3].

*Presently with Bell Telephone Manufacturing Company, Antwerp, Belgium.

FUNCTIONALITY

CUSP can operate on blocks of data containing up to 16k complex points in a programmed sequence of up to five possible modes called an "assignment". The five modes of operation are windowing, FFT, multiplication, Inverse FFT and scaling. These modes allow the supervisor to assign CUSP fundamental signal processing tasks including FFTs, Finite Impulse Response (FIR) filtering, convolution, and correlation. After completing an assignment, CUSP can be reprogrammed by the supervisor by loading a new state vector. This allows the user to quickly change between various transform sizes and algorithms which CUSP can perform.

The control interface between the host computer and CUSP consists of only the following lines: OPERATE, DONE, and LOAD_ASSIGNMENT. After the host has loaded the input data into the memory, it asynchronously activates CUSP via the OPERATE line. Upon completion of the assignment CUSP will interrupt the host by raising DONE. The host, in turn, lowers OPERATE which causes the CUSP to reset and the DONE signal to fall. During any period when OPERATE is low the host may reprogram the CUSP by placing a new assignment on the address bus and raising LOAD_ASSIGNMENT.

FFT ALGORITHM

The primary goal of the CUSP chip is the efficient computation of the Fast Fourier Transform (FFT). The chip is designed to execute a Cooley-Tukey power of 2, decimation-in-time FFT on at least 4 and up to 16384 complex data points.

The basic algorithmic step executed by the chip is a radix-4 butterfly. The radix-4 operation was chosen for the following reasons:

1. The execution speed of an FFT built with high speed arithmetic hardware is limited by the bandwidth of the channel between storage elements and arithmetic units [4]. In the radix-4 version the necessary communication between memory and arithmetic unit is reduced by 50%.

2. One radix-4 butterfly replaces four radix-2 butterflies. Therefore, in a bit-serial architecture a radix-4 system achieves a fourfold increase in parallelism.

3. It is also possible to improve the numerical precision considerably by using the radix-4 butterfly. This reduces the errors caused by the

CH1935-6/83/0000/0757$01.00

rounding and scaling of the data, and by the use of finite precision coefficients.

The radix-4 butterfly does not restrict the possible FFT sizes. The chip can calculate FFTs with sizes which are not a power of four (e.g. 512).

ARCHITECTURE OF THE CHIP

Figure 1 shows the simple system architecture required by the CUSP chip. External RAM and ROM are used for the storage of data, coefficients and the time and frequency domain windows. Thus, the memory does not consume on-chip area and the system can be configured for a wide range of problem sizes between 4 and 16384 points.

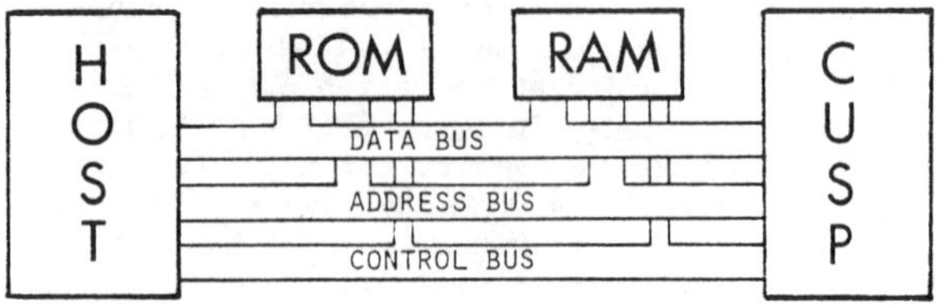

Figure 1: Simple System Configuration

Figure 2 shows the schematics of the internal data flow: The CUSP chip has two parallel data ports to increase the I/O bandwidth. This allows simultaneous access of the real and imaginary components of data words and coefficients. The combined use of dual ports and the radix-4 algorithm cuts the I/O time by a factor of four when compared with a standard radix-2 implementation.

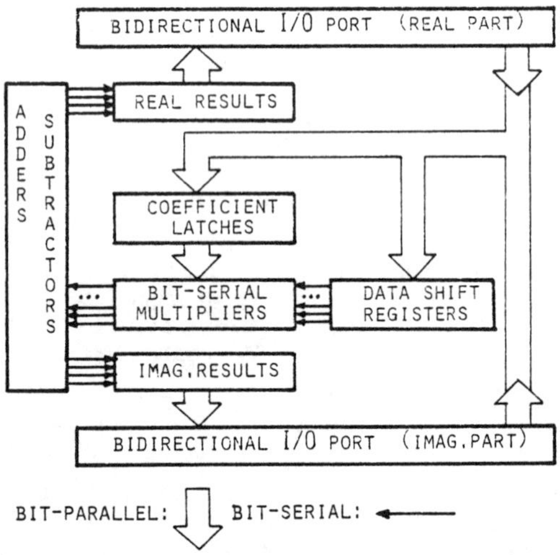

Figure 2: Internal Organization of the Data Flow

Bit parallel I/O was chosen over bit-serial I/O because it can achieve high bandwidth with a simple scheme for address generation and memory partitioning.

Four data words and four coefficients (all complex) are fetched from memory. Four complex multiplications, requiring 16 real multiplications are executed using 16 serial multipliers, all operating simultaneously. The multiplication results are combined by bit-serial adders and subtractors and the results are temporarily stored in the result shift registers before transfer to the external data RAMs. The pipelining of I/O and arithmetic operations is shown in figure 3.

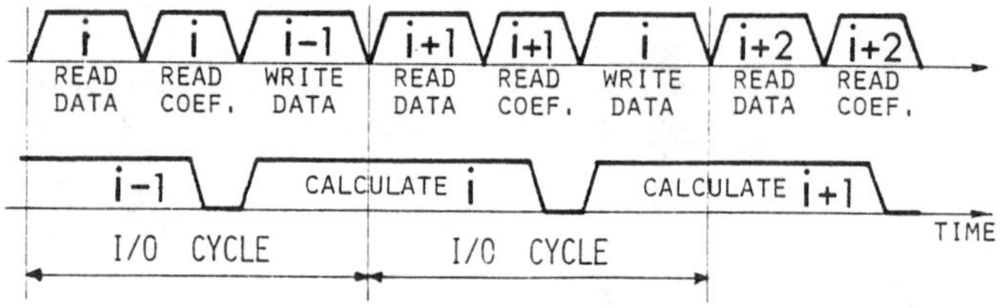

Figure 3: Pipelining of I/O and Calculations

CUSP is a completely self-contained FFT processor except for an external clock generator and RAM and ROM, which are function of the transform size. All address generation for data and coefficients is done on-chip, as well as all mode control necessary for carrying out the assignment specified by the user.

THE BIT-SERIAL HARDWARE AND ITS ADVANTAGES FOR DEDICATED, HIGH THROUGHPUT SYSTEMS

By using bit-serial arithmetic hardware it was possible to layout a 50000 transistor high throughput VLSI chip with a minimal amount of design time.

The advantages of serial hardware are:

1. The smaller size of the bit-serial multipliers allows us to include 16 multipliers which have a throughput equivalent to one 70ns 20x20 bit-parallel multiplier in continuous operation.

2. In a bit-parallel architecture several scratch-pad registers would be required to pipeline the arithmetic section effectively. One needs parallel busses to bring results to and from registers and arithmetic units. The control of the communications is elaborate, the energy and time to distribute operands is also considerable.

In the CUSP design all data transfers between cells are along dedicated paths. These communications are local, regular and simple. Parallel busses are absent, except for one bus to bring operands to their respective input registers. This results in a better use of the chip area. The local communications allow much higher clock rates and less power dissipation.

3. Special hardware for rounding and 2's-complementing, parity checking and generation is expensive in bit-parallel systems.

With bit-serial hardware all this is much simpler. The CUSP chip can do all arithmetic in double-precision with only one rounding for every output word.

4. Because of the simple control of the arithmetic section we can generate all control signals on-chip in a fast Programmable Logic Array.

5. The bit-serial architecture allows us to construct the arithmetic section by merging a small number of predesigned cells. Interconnections

between cells are established by placement of the cells; there is no need for extra routing of wires.

To summerize, on the CUSP chip 16 multiplications, 24 additions or subtractions, 16 parity checks, 8 roundings and 8 parity generations are pipelined between input and output registers with only four dynamic control signals, allowing simple, fast control. In a bit-parallel machine a microprogram or complicated hardwired pipelines would be required.

NUMERICAL ISSUES

Number Representation

The CUSP chip is designed to operate on 2's-complement, 16 bit fixed-point input data. During the FFT execution the intermediate results can overflow the fixed-point representation. Therefore, a scaling operation is necessary and the intermediate results are effectively stored as a set of block-floating-point numbers [5].

In order to provide at least 16 bits of precision for the largest results even in the case where no overflow occurs, 4 extra guard bits are stored at the least significant side of the results. The intermediate results are therefore stored with at most 20 bits of precision. The scaling operation will guarantee that at least one result will be stored with (on average) 18 bits of precision.

The chip uses coefficients for the FFT, IFT and windows in 2's-complement format with up to 20 bits of precision.

Internal Precision

The bit-serial arithmetic operations multiply the 20 bit 2's-complement data and coefficients together to form a 39 bit double-precision result. These serial results are summed in a serial adder tree to form the result of the butterfly with up to 42 bits. The necessary scaling is done by stopping the bit-serial calculation early.

Bit-serially these oversized results are rounded to the 20 bit results which are stored. Note that only one rounding operation is performed on every butterfly result, compared to more than twice as many rounding or truncation operations in similar systems (such as discussed in [5]).

Numerical performance

The noise caused by the coefficient rounding is negligible compared to the rounding noise. If 16 bit coefficients are used, a noise with an average power about 90dB below the power of the input signal is introduced for an FFT of length 16384 points followed by an inverse FFT. Using 20 bits for the coefficients the noise will be at -114dB.

Much more important is the noise caused by the adjustment of the multiplication results from double-precision to single-precision. This noise depends largely on the adjustment method used, on when the adjustment is done during the algorithm, and on the actual data input. Our benchmark input is a white noise with no DC component, flat distribution and 16 bits of precision at the input.

If the CUSP chip performs a 1024-point FFT on this benchmark input followed by an inverse FFT, then the original sequence is reconstructed with an added noise. This noise has a power 88dB below the power of the original input.

If the CUSP chip would use truncation instead of rounding, the signal to noise ratio would degrade by 8dB. If the CUSP chip would adjust the double-precision results right after the multiplications instead of delaying the adjustment, then the signal to noise ratio would degrade an additional 6 to 8 dB.

The extra 4 guard bits help to improve the numerical performance an extra 12 dB (on average) over the best achievable performance by a 16 bit block-floating-point system.

PERFORMANCE AND FEATURES

Because the FFT computation is a bottleneck in many signal processing systems, most notably radar processing, it was desirable to obtain the highest possible performance per processor. Due to the bit-serial architecture and high speed control logic of the CUSP, a 1024 point FFT on complex data is expected to need 1.4ms. This performance rivals that of other multichip architectures and other monolithic systems [1,2,6]. The expected CUSP performance for typical applications is given in table 1.

	EXECUTION TIME	MAXIMUM DATA RATE
RADIX-4 BUTTERFLY (COMPLEX DATA)	1.1μs	3.6MHz
1024 POINT FFT (COMPLEX DATA)	1.4ms	710kHz
512 TAP FIR FILTER (1024 REAL DATA INPUTS)	3.4ms	290kHz
EFFECTIVE 20BITx20BIT MULTIPLY TIME	70ns	14.5MHz
256 POINT HAMMING WINDOW (REAL DATA)	35μs	7.2MHz

TABLE 1: CUSP PERFORMANCE

The CUSP chip is designed in a p-well CMOS technology with 2.5μm gate lengths. The die contains about 50000 devices in a 7.5mmx7.5mm area. There are 128 pads on the chip of which 84 are bounded and 40 are used solely for testing purposes. Table 2 gives an overview of the CUSP features. The regularity of the bit-serial design allowed the layout to be completed by 7 student designers within 8 months after an initial architecture study.

CUSP incorporates parity checking and generating hardware on-chip for both data and coefficients. It can also function as a watchdog processor for another CUSP chip to detect errors in either the data or address output of the active CUSP chip. These two features enable the construction of a very high performance fault

NUMBER OF DEVICES	50,000
CHIP SIZE	7.5x7.5mm^2
NUMBER OF PINS	84
TECHNOLOGY	2.5μm BULK CMOS
DATA REPRESENTATION	16bit fixed-point input 20 bit block-fl.-point internal 16 bit block-fl.-point output

TABLE 2: CUSP FEATURES

detecting and fault tolerant array element with 2 CUSP chips, as shown in figure 4.

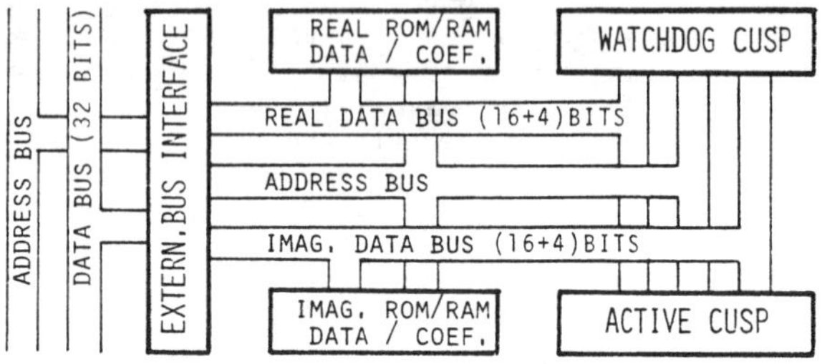

Figure 4: Array Element with 2 CUSP chips

If a watchdog error occurs it is possible to run an extensive test on the processor duo that signaled the error. Faulty processor duos can be disconnected from the external bus.

Even small arrays of CUSP modules can give dramatic performance improvements over current state-of-the-art. For example, in figure 5 we show a 5 by 5 array of CUSP elements with one spare element per row, and one spare row of CUSP elements. This arrangement is expected to perform a 1024 point FFT at a sampling rate in excess of 10MHz.

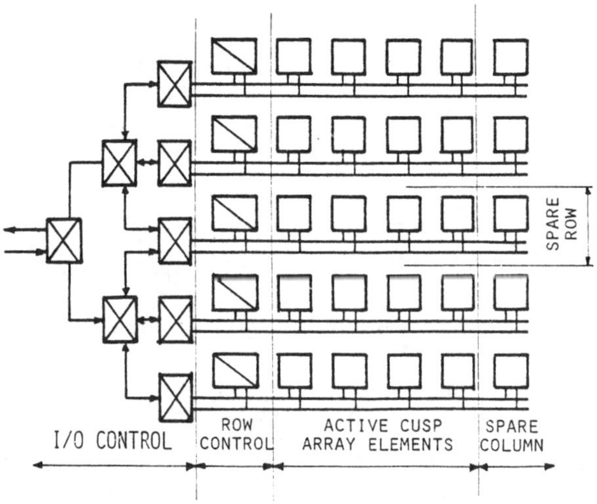

Figure 5: Array of CUSP Elements

CONCLUSIONS

Existing monolithic systems generally use the radix-2 butterfly as the basic building block for FFT computations and rely on parallel multipliers to attain the desired performance.

Our novel approach is to execute a radix-4 butterfly on a single chip with bit-serial computations being performed with maximal parallelism. The choice of both the radix-4 and the serial hardware gives us improved throughput and numerical precision. Moreover, the reduced control complexity for the bit-serial arithmetic allowed us to include extra control logic to permit the chip to be completely autonomous.

We have developed an architecture with a simple control interface, which allows CUSP to meet a broad range of applications from peripherals for micro-processors to ultra-high performance fault-tolerant arrays.

ACKNOWLEDGEMENTS

The authors wish to acknowledge the assistance obtained from the National Research and Resource Facility for Submicron Structures, supported by Grant ECS-8200312 of the National Science Foundation.

The following Master of Engineering students took part in the design: Victor Peng and Steve Geisler (architecture study), Randy Fahcy, Anand Jhaveri, and Marita Smith. We also received irreplaceable help from the Ph.D. students Paul Chau and James Lyons.

REFERENCES

[1] "A 2μm CMOS/LSI 32-point Fast Fourier Transform Processor", B. L. Troutmans et al., Proc. 1982 IEEE ISCC, pp. 26-27 and 282-283.

[2] "Digital Signal/Array Processing Products", New Product Information, Advanced Micro Devices, 901 Thompson Place, Sunnyvale, Ca 94086.

[3] "VLSI Signal Processors Based on the Fast Fourier Transform", Peter P. Reusens, Ph.D. Thesis, Cornell University, Electrical Engineering, (1983).

[4] "Fourier Transforms in VLSI", C. D. Thompson, 1980 IEEE Circuits and Computers Conf., pp. 1046-1051.

[5] Theory and Applications of Digital Signal Processing, L. Rabiner and B. Gold, Prentice Hall, 1975.

[6] "Frequency Domain Filtering with VLSI", E. E. Swartzlander, Jr and G.H. Hallnor, Proc. of the Workshop on VLSI and Modern Signal Processing, University of Southern California, Nov 1-3, 1983, pp 25-30.

Author Index